Linear Circuit Analysis

Linear Circuit Analysis

Artice M. Davis

San Jose State University

PWS Publishing Company

I(T)P   An International Thomson Publishing Company

Boston • Albany • Bonn • Cincinnati • London • Melbourne • Mexico City • New York
Paris • San Francisco • Singapore • Tokyo • Toronto • Washington

PWS PUBLISHING COMPANY
20 Park Plaza, Boston, MA 02116-4324

I(T)P   International Thomson Publishing
         The trademark ITP is used under license.

Windows is a registered trademark of Microsoft, Inc.
Electronics Workbench is a registered trademark of Interactive Image
    Technologies, Ltd.
PSPICE is a trademark of MicroSim Corporation
AT&T Worldnet is a trademark of AT&T

Sponsoring Editor: *Bill Barter*
Technology Editor: *Leslie Bondaryk*
Marketing Manager: *Nathan Wilbur*
Production Editor/Interior Designer: *Andrea Goldman*
Manufacturing Buyer: *Andrew Christensen*
Production: *Helen Walden*
Compositor: *Progressive Information Technologies*
Interior Artist: *Academy Artworks*
Cover Designer: *Diana Coe*
Cover Image: *Waterfall by M.C. Escher © 1997 Cordon Art, Baarn,
Holland. All rights reserved.*
Cover Printer: *Phoenix Color Corp.*
Text Printer/Binder: *R. R. Donnelley & Sons Company/Crawfordsville*

Library of Congress Cataloging-in-Publication Data

Davis, Artice M.
    Linear circuit analysis / Artice M. Davis
        p.   cm.
    ISBN 0-534-95095-7
    1. Electric circuit analysis   2. Electric circuits. Linear.
I. Title
TK454.D414   1998   97-30067
621.319'2--dc21   CIP

Printed in the United States of America.

  This book is printed on recycled, acid-free paper.

For more information, contact:

PWS Publishing Company
20 Park Plaza
Boston, MA 02116-4324

Nelson Canada
1120 Birchmount Road
Scarborough, Ontario
Canada M1K 5G4

International Thomson Publishing Asia
221 Henderson Road
#05-10 Henderson Building
Singapore 0315

International Thomson Publishing Europe
Berkshire House
168–173 High Holborn
London SC1V 7AA

International Thomson Editores
Campos Eliseos 385, Piso 7
Col. Polanco
11560 Mexico D.F., Mexico

International Thomson Publishing Japan
Hirakawacho Kyowa Building, 31
2-2-1 Hirakawacho
Chiyoda-ku, Tokyo 102
Japan

Thomas Nelson Australia
102 Dodds Street
South Melbourne, 3205
Australia

International Thomson Publishing GmbH
Königswinterer Strasse 418
53227 Bonn, Germany

98 99 00 01 02 — 10 9 8 7 6 5 4 3 2 1

This book is dedicated to the memory of *Dr. Oliver Heaviside,* F.R.S. (1850–1925), who developed the theory of differential operators used here. The theory was not understood by others of his day, yet it was later cited by the respected British mathematician Dr. Edmund T. Whittaker as "one of the three most important discoveries of the late nineteenth century." Dr. Heaviside's last home was a cottage high on a hill overlooking the bay in Torquay, Devonshire, England. On the gatepost there is a plaque dedicated to "Dr. Oliver Heaviside, mathematician and engineer" that cites his many accomplishments. When the author visited the site a few years ago, there was a crack in the plaque, which his fancy recalls as being between the words "mathematician" and "engineer." It is the author's sincere hope that this text will help mend the crack.

Contents

7 Time Response of First-Order Circuits 313

8 Complex Signals and Systems 395

9 Time Response of Higher-Order Circuits 423

10 Stability and Forced Response 491

14 The Fourier Transform 807

PART IV   SELECTED TOPICS   879

15 Two-Port Subcircuits 881

16 The Transformer 971

Appendices 1027

Preface

This book is intended for the undergraduate course in circuit analysis, customarily the first course taken by aspiring electrical engineers. For many students entering the field of electrical engineering, circuit analysis can be a rude awakening. It tests the depth of their background knowledge in mathematics and physics and carries a full load of serious analytical content. Typical texts for this course devote only a few pages to fundamental concepts (such as circuit topology, basic laws, and the why's and how's of network analysis) before plunging into the mechanics of solving complicated circuits. Students unable to make this transition quickly and easily are challenged to make sense of such basic ideas as equivalent resistance and current and voltage division, even as they struggle to master more advanced topics, such as Thévenin/Norton equivalents, time domain response methods, and the phasor analysis of circuits. As a result, many students who would benefit from a more gradual learning curve at the outset of the course are doomed to drop out of it, and out of the field of electrical engineering as well. *Linear Circuit Analysis* is designed to address this problem by enabling students to retain more concepts and skills in circuit analysis as they work through the early chapters so that they can complete the course with confidence and with a better awareness of how circuit analysis relates to succeeding courses in electronics, signals and systems, and control. The remainder of this preface will outline how that goal has been accomplished.

Part I (Dc Analysis)

In Part I, Chapters 1–3, basic concepts of dc circuit analysis are presented more deliberately and in greater detail than in traditional texts, while a solid mathematical foundation for the analysis and solution of circuit problems is established. Here is a small, yet significant example—one borrowed from the authoritative text *Classical Circuit Analysis* by Vitold Belevitch: *simply erase the body of each element in the circuit and the connected "islands" of conductor that remain are the circuit nodes.* Students are encouraged to do this exercise at the outset of their attack on a given circuit problem until node identification becomes second nature. Here's another example: the use of multiple ground reference symbols is introduced early, and circuits are occasionally drawn using this symbolism to prepare students for later courses in electronics. Still another example is the stress placed in Chapter 3 upon the idea of a subcircuit and the closely linked idea of an equivalent subcircuit. This forces students to come to understand these important ideas before they encounter the "big machinery" of nodal and mesh analysis in Chapter 4. I have often seen students fail, not only in this course, but in succeeding electronic circuits courses as well, because they have not grasped these basic ideas. Such a lack of understanding causes the student difficulty in dealing with simple series and parallel resistors, to say nothing of an active subcircuit exhibiting negative resistance. The topics often referred to as "circuit theorems," superposition and the Thévenin-Norton equivalents for instance, are presented before nodal and mesh analysis so that the student is not sidetracked from these important ideas by the lure of more general and algorithmic methods.

Linear Circuit Analysis bases its approach on three fundamental physical quantities: time, voltage, and current. These variables are defined operationally, that is, by specification of the measuring instruments and the procedures by which they are measured. All other variables are derived quantities; their definitions are based on the three fundamental quantities and mathematical formulas. For instance, power is defined as the product of voltage and current, and energy as the integral of power (the product of power and time). Therefore, the concept of mechanical work is not required. This approach has two very important advantages: it eliminates the physics prerequisite and it lends a practical flavor to the theory being developed because of its close connection with the way measurements are actually made in the laboratory.

Chapter 4 covers mesh and nodal analysis. There are some novel features here. For instance, each example is illustrated with several figures showing a circuit at different stages of analysis. Students are encouraged to make such drawings as a "preamble" to the writing of nodal and mesh equations. A carefully constructed node (and mesh) cataloguing scheme is introduced to assist students in making these drawings and in deciding how many (and where) equations should be written. In a later section of the development, an original technique of nodal and mesh analysis by inspection is developed that enables students to write the matrix form of these equations directly from the circuit diagram. Appendix A offers a "minicourse" on linear algebra, matrices, and determinants to complement the methods developed here. Selected portions should be either covered in lecture or assigned for home study. Though students have typically been exposed on several occasions to this material, they have not yet seen an organized development of it.

Chapter 5 is devoted to the analysis of active circuits. The topic has been postponed until this point to allow students enough time to become better acquainted with the basic ideas of circuit analysis before they encounter the complication of dependent sources. A novel method called *taping* is presented for analyzing circuits with dependent sources. The idea is to visualize a dependent source as having a label on it marked with the dependency relationship. One imagines a small piece of masking tape being placed over this label and a literal unknown value written on it. Because the *v-i* characteristic of the device is exactly the same as that of an independent source of the same type, one can then proceed to analyze the circuit just as if it only contained independent sources and resistors. After the circuit equations are written, one "untapes" the source and invokes the dependency relationship. This turns the analysis of active circuits into a procedure whose main subprocedure is the one already developed for passive circuits. Sections 5.1 and 5.2 were written so that an instructor can cover them before Chapter 4 if so desired. Section 5.3 requires a mastery of nodal and mesh analysis. Sections 5.4 through 5.6 deal with the op amp, and Section 5.5 is devoted to a careful explanation of the need for negative feedback and the conditions under which the ideal op amp model is valid. Section 5.6 discusses the derivation of models for electronic devices: diodes and bipolar junction transistors. Later in the text, all exercises, examples, and problems using op amps or transistors are marked appropriately with (O) and (T), respectively.

Part II (Time Domain Analysis)

Part II, Chapters 6–10, presents a method of time domain analysis that unifies the three major divisions of circuit analysis (dc, ac, and transient) based on the use of differential operators. This technique was developed in response to the question often posed by students when they are applying the classical theory of differential equations to determine the transient response of an RLC circuit: "Isn't there a general method for determining the arbitrary constants in the solution from circuit initial conditions?" A careful analysis reveals that there is—state variable theory—but, unfortunately, it is too complicated for the introductory circuits course. This seems to be the main reason that most introductory texts do not deal with circuit responses of higher order than second in the first course material.

The differential operator approach does not have this deficiency, and its development here from first principles means that a differential equations course is not a prerequisite (the only prerequisite is freshman calculus). The approach relies upon the fact that the inductor and the capacitor obey a generalized form of Ohm's law in the time domain and, therefore, that all of the methods previously derived for dc analysis continue to hold, just as they do for the computation of ac forced response using phasors. This has obvious pedagogical advantages, a major one being that the student uses the same techniques twice more after being exposed to them in the dc material. Though it is unusual for a course syllabus to permit the repetition of techniques while significant new material is being introduced, this benefit does accrue when one adopts the operator method of transient analysis. This method is straightforward, algorithmic, and easy to teach. It has been class tested for some time by the author and his colleagues and has been received very favorably by the students. The reason for this is probably due to the important fact that the operator approach is precisely the same as the Laplace transform method, except that no transform is required (with its attendant difficulties of dealing with improper integral convergence and the necessity of translating each problem into a completely different domain). The operator method proceeds just like the Laplace transform technique—including the partial fraction expansion—in a manner that is strictly algebraic for the waveforms normally encountered in circuit analysis. Students like this algorithmic and constructive approach, and it is not intrinsically limited to circuits of low order.

The material on transient analysis was written, however, with flexibility in mind. At the option of the instructor, transient analysis can be covered only through second-order circuits, with the remaining time domain material being postponed to the second course. In either case, the instructor should cover the basic ideas of impulse and impulse response in Chapters 6 and 7 because these computations replace the evaluation of arbitrary constants (in the classical approach) from circuit initial conditions. I have taught this material with both course organizations, and both work quite well.

Part III (Frequency Domain Analysis)

Part III, Chapters 11–14, develops the frequency domain viewpoint and methods of analysis. Chapter 11 deals with the phasor analysis of ac circuits. It was written so that an instructor can teach the material at any desired point after the material in Part I on dc analysis and thus does not rely upon the material in Part II on time domain analysis. There are several new approaches in this material, a notable example being the interpretation of complex power. Complex power is introduced with more physical motivation than usual, and a slightly different definition than the conventional one is given: it is defined as $S = \overline{V} * \overline{I}$, rather than as the more conventional $S = \overline{VI}*$, because the former has a number of notational advantages and creates no difficulties in application. Appendix B states and proves Tellegen's theorem, which is necessary for the discussion of conservation of complex power in this chapter (and for making the proof of solvability using the methods of Chapter 4 logically complete).

Chapter 12 starts with an overview of the frequency domain point of view motivated by the design problem of removing a signal from noise, followed by a detailed analysis of tuned circuits and a practical discussion of filter design. A new, rapid method for sketching Bode plots is presented here.

Chapter 13 develops circuit analysis techniques using the Laplace transform. Because the circuits and systems literature of the past half-century has used this notation, it is important that students today be aware of this different language. For those who have already covered the operator techniques presented in Part I, Section 13.0 is presented as a "quick fix" in the basic ideas and vocabulary. This section proves that the operator method is merely the time domain equivalent of the Laplace method and that one can interpret operator equations as Laplace-transformed equations by merely replacing p at each occurrence by s and simply eliminating all delta functions. The rest of the chapter develops the entire theory of circuit analysis using the Laplace transform in a more leisurely manner. It is written in such a way that an instructor can elect to cover it immediately after Part I on dc analysis if he or she wishes to use a Laplace orientation for transient analysis. Needless to say, there is quite a bit of duplication of material between this chapter and the methods already developed in Part II. A somewhat more sophisticated, though still largely intuitive, treatment of generalized functions (than was given earlier in Part I) is presented in the body of this chapter. Appendix C is a more rigorous mathematical complement that has been included for the instructor and the more analytically inclined student. It has previously been published only in the form of a paper that I delivered at a conference, and I felt that some would find it of use in a more accessible place.

The subject of Chapter 14 is Fourier analysis. The Fourier transform is defined to be the Laplace transform evaluated (either directly or as a limit) on the imaginary axis of the s-plane. The Fourier transform of periodic functions is derived, and this leads to the Fourier series representation. The important topic of spectrum analysis is covered, and the concept of negative frequency is discussed in a novel way. This discussion then leads naturally into a treatment of various types of amplitude modulation schemes, developed briefly as practical applications of transform theory.

Part IV (Selected Topics)

Part IV contains what I consider to be special topics, though other instructors might wish to cover them sooner than their placement in the book might indicate. This part contains only two chapters, the first on two ports and the second on transformers. Two ports are treated in somewhat more depth than usual in Chapter 15, in that discussions are given as to why some parameter sets fail to exist for a specific subcircuit. A rather extensive development of the application of two-port methods to the analysis of feedback systems is given. I feel that most texts fail to provide motivation for the study of two ports, and the analysis of feedback systems is a major application. This material is, of course, optional. If it were chosen for coverage, one could "download" a significant amount of material from succeeding electronics courses. The material on transformers in Chapter 16, too, is a bit different from the usual fare. It deals with the transient response of circuits having mutual inductance as well as with their more commonly treated ac steady-state behavior. A discussion of the reciprocity of transformers is given that is more sound than the usual one and is based upon a novel idea that only recently appeared in print in a professional journal.

In summary, this book uses differential operators to unify the treatment of the three major divisions of introductory circuit analysis. This student-friendly approach results in focused discussions of dc, ac, and transient analysis in a logical and parallel fashion. It presents methods for solving specific circuits in a concrete manner and then shows their abstract mathematical nature in a more general setting that makes the transition to linear systems easier for the student to understand. Discussion of the systems-level generality of the methods being developed is aided by the use of simulation diagrams. Finally, it seeks to help students retain concepts and problem-solving skills, and help instructors retain more students in the course, through a more gradual, careful, and systematically integrated discussion of the fundamental ideas underlying circuit analysis.

Instruction Support Materials

In addition to the Answers to Selected Problems at the back of the book, a complete solutions manual is available from the publisher.

Web-Based Interactive Study Guide

The CD-ROM bound into the back of this book contains a number of tools that are designed to help the student learn about circuits more effectively and that may aid students with different learning styles.

A CD-ROM icon (shown at left) is distributed throughout the book to indicate which examples are given treatment in the electronic files on the CD-ROM in the back of the book. The letters in the four quadrants on the icon indicate the kind of file(s): ML for MATLAB, MC for MATHCAD, EW for Electronics Workbench, and PS for PSPICE. Some examples will have all four types of files to choose from, some fewer than four.

The CD includes:

- Web-based self-quiz software that tests understanding of each chapter and provides explanations of right and wrong answers, plus browser software and a Web connection package from AT&T Worldnet™, if Internet access is not already available.

- The evaluation version of MicroSim PSPICE® for Windows®-based computers, and electronic copies of all the SPICE netlists printed in this book.

- The evaluation version of the Student Edition of Electronics Workbench® for Windows-based computers, which will load a set of files keyed to the book and allow students to work their own problems.

- MATLAB and MATHCAD files that work and extend specific examples in the book.

The contents of the CD provide a broad range of electronic ancillary support from which the instructor and the student can pick and choose to best support the course and the book.

Acknowledgments

At PWS Publishing, I would like to single out William Barter, who showed interest, faith, and commitment to the approach to circuit analysis I have presented here. It is not easy for a person trained in one field to understand the technical intricacies of another specialization. Bill did, and his continued faith and moral support were invaluable in the completion of this text. As everyone who has written a text knows, it is a long and arduous task and nerves begin to become a bit frayed toward the end of the process. Bill always provided a cool-headed, sane approach to each problem as it arose, and his solutions were always rational and considered. Nathan Wilbur provided quite a bit of assistance in "putting it all together": articulating what the book is all about. He has aided me in several ways with a very perceptive and analytical evaluation of "how to say things" at just the right moment.

Leslie Bondaryk is the "web guru" who took on the onerous task of putting some of the text material on line. Yet she did even more. She has been fiendishly clever at thinking up ways of presenting somewhat abstract concepts in a design-oriented manner. She also made several valuable suggestions about the text presentation itself.

Andrea Goldman has provided a lot of behind the scenes support, taking care of details in such a manner that they were invisible to me (one of the most meaningful kinds of support, but one that too often goes unacknowledged). Helen Walden provided very competent support with the editing through many vagaries in the process, such as the sudden UPS workers strike that impeded the flow of manuscript for a time. She, too, was a cool head to whom I turned several times for that type of assistance that is so valuable, yet is "not in the job description." Torrey Lee Adams did a superb job of line editing, and I learned several things about tight, clean writing from her. She helped by clarifying my writing in a number of ways.

At the top of the list of those who provided technical assistance go my students over the years, who have been the best critics of this work. Their questions, suggestions, error corrections, and enthusiastic response to the methods contained in this text were invaluable. Without them, the book would simply not have been written. There is one student, however, whom I would like to single out for a special thank you: George Wong. George studied circuit analysis with me while acquiring the background to make the transition from his undergraduate field of chemical engineering to the graduate program at San Jose State University. Not only was George an exemplary student, he was (and is) a very hard-working and precise person. His support has been invaluable in pointing out errors in the manuscript and in writing the solution manual that accompanies the text.

Among my San Jose State University colleagues, several have provided support in one manner or another. Gene Moriarty taught from the manuscript and provided valuable suggestions while it was being developed, as did Patricia Johnson and Richard Duda. Michael O'Flynn showed me an interesting derivation of the Fourier transform of the unit step function that led to the entire approach for defining the Fourier transform in Chapter 14. Jack Kurzweil pointed out the fact that the Fourier series can be treated as a special topic in Fourier transform theory. I learned the "quick method" of doing linearized Bode gain plots from Evan Moustakas, though the extension of the technique to phase plots is my own. I also would like to say thanks to Moustakas for sharing several of his pedagogical devices with me when I was a neophyte instructor. Ray Chen, the department chair, provided support by allowing me to use the text in class as it was being developed and by providing a somewhat lightened teaching load in the last stages of the writing.

As for colleagues not affiliated with San Jose State, I would like to select several who contributed in significant ways, in one manner or another, to the development of *Linear Circuit Analysis*. Armen Zemanian at New York State University at Stony Brook has provided a mentoring type of support over the years. He is a person whom I admire and respect for his truly significant accomplishments in the field of circuit theory and for his willingness to support younger colleagues. Gary Ford, at the University of California at Davis, a fine teacher and circuit theorist, helped greatly in several ways. As a personal favor, he reviewed an early version of a manuscript for a paper upon which the Heaviside operator technique is based (as did Zemanian), and made valuable comments that led to significant improvements, directly in the paper and more indirectly in this book. Of special note is his contribution to my understanding of the relationship between mathematical and physical operations, as exemplified in the treatment of complex variables in Chapter 8. This treatment was also motivated to a large extent by fairly recent research results in filter design published by Sanjit Mitra at U.C. Santa Barbara, P. P. Vaidyanathan at Cal Tech, Adel Sedra at the University of Toronto, and others. Over the years, I have considered the work of these people to be inspiring in molding my own thinking on the subject.

P. M. Lin of Purdue also reviewed the operator paper mentioned above and offered substantive comments. Furthermore, the method used in this text for presenting the reciprocity of mutual inductance is due to Dr. Lin. I learned it as an associate editor for the *IEEE Transactions on Circuits and Systems*, when I was charged with the responsibility of seeking out good tutorial papers for that journal's Circuits and Systems Expositions (CASE) section. Dr. Lin contributed a paper on the reciprocity of mutual inductance during that time, and I benefited greatly by corresponding with him on the subject. Much the same can be said of Wai-Kai Chen, whose texts and papers have influenced me over the years and who also offered valuable suggestions leading to the improvement of my earlier paper on circuit analysis using Heaviside operators.

Ron Rohrer is responsible for stimulating me to think about the relation between two-sided and one-sided waveforms and the ideas of forced responses and steady state through a couple of what seemed at the time to be passing comments about his classroom approach to these subjects. These comments were made to me at a professional meeting a number of years ago and proved to be quite fundamental, contributing greatly to the frequency domain analysis used in *Linear Circuit Analysis*. I have also been greatly influenced by Dr. Rohrer's more advanced circuits texts, as well as the aforementioned one of Dr. Vitold Belevitch.

In a broader manner, I have benefited from discussions with many other colleagues. Ken Jenkins at the University of Illinois provided several opportunities for professional growth in the discipline of circuit theory through the IEEE Circuits and Systems Society, and his oral presentations of technical papers influenced by own style of speaking and writing. Terry Cotter at the University of New Bruswick at Saint John inspired me to think in a fundamental way about the physical significance of complex power. Mohammed Ismael at Ohio State, Igor Filanovsky at the University of Edmonton in Alberta, Peter Aronheim at the University of Louisville, and William Stephenson at Virginia Tech influenced my thinking about active networks that is presented in these pages.

Several reviewers also provided invaluable feedback as the manuscript for this book was being developed. I would like to thank:

- Peter Aronheim, University of Louisville
- Wai-Ken Chen, University of Illinois at Chicago
- Douglas Draper, Portland Community College
- Gary Ford, University of California-Davis
- John Nyenhuis, Purdue University
- S. N. Prasad, Bradley University
- Theresa Tuthill, University of Dayton
- Ravi Warrier, GMI Engineering and Management Institute

Finally, I would like to express appreciation to my wife, Lalah, the "practical engineer" of the family (she is an electrical safety engineer, with a P.E. license). She offered quite a few valuable technical suggestions at several critical points, thereby keeping me from straying too far off into the ethereal realm of theory, and listened with at least half an ear to my fumbling attempts to explain (more to myself than to her, perhaps) why I was doing each thing in a certain way. In fact, her critique was responsible for my decision to completely rewrite a major portion of the first half of the text. She also accompanied me in an interesting search for the worldly traces of Oliver Heaviside in Devonshire, England, and she was the one who found his last home (with the plaque at the gate) and his final resting place. Heaviside operators form the central pivot around which this text turns, so perhaps its title should really be "Searching for Ollie, with Lalah."

Part I

DC ANALYSIS

1 Basic Concepts

This chapter introduces the basic ideas of circuit analysis: element and circuit, as well as current and voltage and their integrals: charge and flux linkage. Time, voltage, and current are taken as fundamental. They are defined operationally; that is, by a description of how to measure them. Their units, the second, the volt, and the ampere, are therefore fundamental units. All other quantities and units are derived. For instance, power is the product of voltage and current. Its unit is the volt-ampere, or watt. The energy absorbed by a two-terminal subcircuit is defined as the time integral of the absorbed power. Its unit, the joule, is the watt-second. A passive element or subcircuit is one that only absorbs energy; an active element or subcircuit is one that is capable of generating energy. The two most fundamental active elements, the independent voltage source and the independent current source, are defined and discussed.

This chapter is basic to all of the others in the text. It is therefore prerequisite reading for each of the others. The background required is at least two semesters of calculus.

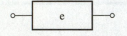

1.1 | Elements and Circuits

Some Basic Concepts

Figure 1.1 A circuit element or branch

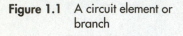

The basic objects of circuit analysis are called *elements,* or sometimes *branches*. A circuit element, or simply *element,* is an object to which two or more wires called *leads* (pronounced "leeds") are connected. We show this in a stylized way in the sketch in Figure 1.1. The leads are assumed to be ideal conductors. The small "e" inside the rectangle is just a label for the given element. Other symbols, such as e_1, a, R_1, or L_5 might also be used. The small circles at the ends of the leads are called *terminals*. They are conceptual in nature and indicate where leads from other elements can be attached to form a circuit.

A *circuit* is simply a collection of elements in which each lead of a given element is connected to a lead of at least one other element. An example of a circuit consisting entirely of two-terminal elements is shown in Figure 1.2. The term "two-terminal element" is synonymous with the phrase "two-lead element"—though the latter is never used. We will discuss circuits having elements with more than two terminals later in the text; for now, when we use the term element we will mean a two-terminal element. Figure 1.3 shows a collection of elements that is not a circuit. Why is it not a circuit? Because the element labeled "a" has one free lead that is not connected to a lead of any other element. We will assume that the shape of the elements and their leads have no effect on the behavior of the circuit.[1] We will furthermore assume that we can stretch the leads to any length and bend them in any way without affecting the circuit behavior. All that will count are the characteristics of the elements themselves and whether or not a given lead of a given element is connected to a specified lead of another element. This last characteristic of a circuit is called its *topology*.

Figure 1.2
A circuit consisting of two-terminal elements

Figure 1.3
A collection of elements that is not a circuit

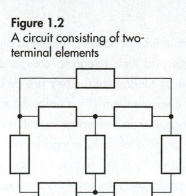

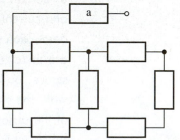

Notice the black dots in Figure 1.2. When two conductors meet each other, there is often an ambiguity as to whether they actually make contact or merely "cross over" each other. A black dot (called a *joint*) means that a connection is made. Figure 1.4 shows some of the commonly encountered conventions. Figure 1.4(a) shows two conductors that connect. So does Figure 1.4(b). Figure 1.4(c), on the other hand, shows two that do not connect. The small semicircular part of the conductor is called a *jumper.* Figure 1.4(d) shows two crossing conductors that are assumed not to connect; this, however, is somewhat poor notation—the one in Figure 1.4(c) is preferred.

---

[1] Our theory will thus not work at very high frequencies (or, equivalently, at short wavelengths.)

Figure 1.4
Some connection conventions

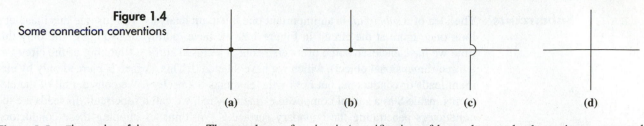

(a)   (b)   (c)   (d)

Figure 1.5   The nodes of the circuit in Figure 1.2

The topology of a circuit (specification of how element leads are interconnected, as previously discussed) is shown by its *nodes*. These are the connected segments of conductor that remain when the element bodies are simply erased. In Figure 1.5 we show the nodes for the circuit drawn in Figure 1.2; as per the directions just given, we have erased the rectangles representing the element bodies, leaving the leads and other interconnecting conductor segments. We will use the symbol N for the number of nodes and B for the number of branches (elements). For the example here, $N = 6$ and $B = 8$.

Circuits are quite often constructed on what are called *pc boards* (pc is short for "printed circuit"). The printed circuit process is very similar to the one used to make this printed page. Figure 1.6(a) shows how the circuit in Figure 1.2 might look when constructed on a printed circuit board. The small dots represent holes that have been drilled in the board (an insulating material such as fiberglass or phenolic). The element leads have been bent at right angles and inserted in the holes. Figure 1.6(b) shows the bottom of the board. We have flipped the circuit board from right to left—the elements on the left of the board in the top view are on the right in the bottom view. The black dots in the bottom view are the holes in the board through which the conductors protrude. The darkly shaded areas represent a conductive film (often copper) through which the holes have been drilled. The holes are filled with solder, thus bonding the leads to the film; then the leads are snipped off with wire cutters. Thus, the connected areas of metal film form the circuit nodes. You can now see the advantage of a circuit diagram. It is a two-dimensional sketch of the circuit that shows all the connection information necessary to construct the physical three-dimensional circuit; however, it is easier and more compact to draw.

Figure 1.6
Printed circuit board fabrication of the circuit in Figure 1.2

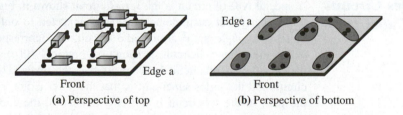

(a) Perspective of top   (b) Perspective of bottom

Let's look once more at the collection of elements in Figure 1.3. We have outlined the nodes and shaded them for emphasis in Figure 1.7. Element a has one lead that is not connected to any other element lead, so the collection of elements is not a circuit. The node formed by the unconnected lead of element a is called a "floating" node.

Figure 1.7
The concept of a floating node

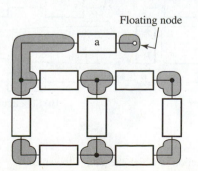

Subcircuits   The idea of a *subcircuit* is an important one in circuit analysis. To illustrate this idea, let's
                    look once more at the circuit in Figure 1.2. We have redrawn it in Figure 1.8, only this
                    time we have enclosed some of its elements in a closed surface (thinking of the circuit as
                    a three-dimensional object), which we have labeled S. This surface is pierced only by ele-
                    ment leads or conductors, but not by the elements themselves. We consider all of the ele-
                    ments inside S as a sort of composite element, which we call a subcircuit. Its leads are the
                    conductors penetrating the boundary surface S. We think of clipping these conductors,
                    thus forming terminals x and y, as shown in Figure 1.9, to which the other elements are
                    connected. As we will soon see, one can very often replace the subcircuit by a much sim-
                    pler subcircuit—often consisting of only one element—as far as the rest of the circuit
                    (the part outside S) is concerned. The subcircuit we have shown in our example has two
                    terminals. In general, a subcircuit can have any number of terminals, but we will find that
                    the two-terminal subcircuits are by far the most important we will encounter. Notice, by
                    the way, that a two-terminal subcircuit has exactly two nodes (the ones to which the ter-
                    minals are connected) that are shared with the rest of the circuit.

Figure 1.8
Forming a subcircuit

Figure 1.9
The subcircuit itself

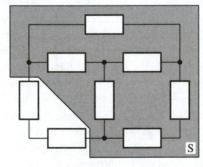

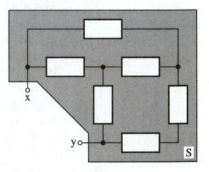

Series Circuits   A special type of circuit is the *series* circuit shown in Figure 1.10(a). A series circuit has
and Subcircuits   the property that each element lead is connected to only one other element lead; thus,
                    there is a single path[2] around the circuit. The corresponding two-terminal subcircuit,
                    called the series subcircuit, is shown in Figure 1.10(b). As we see, it has only a single
                    path inside the subcircuit between its two terminals. The difference between the series
                    circuit and the series subcircuit is that the latter is not a circuit, but the former is. Thus,
                    the path in the subcircuit is not closed; that is, the endpoints of the path are different
                    nodes. The path in the circuit, on the other hand, is closed. In other words, the endpoints
                    of the path are the same node. This special type of path is called a *loop*. The paths in both
                    the subcircuit and the circuit are shown by lines with arrows to indicate the direction as-
                    sociated with each path. We will study these ideas about path and loop in more detail later.

Figure 1.10
Series circuits and
subcircuits

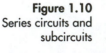

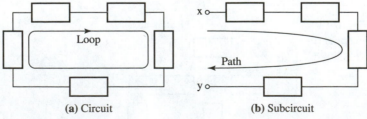

                  (a) Circuit   **(b)** Subcircuit

---

[2] Not counting the direction in which the path is traversed.

Parallel Circuits and Subcircuits   Another important concept is that of a *parallel* circuit or subcircuit. Figure 1.11(a) shows a parallel circuit. This is a circuit that has precisely two nodes, no more and no fewer. The same is true of the parallel subcircuit, shown in Figure 1.11(b). Series and parallel circuits and subcircuits are the simplest type to study; therefore, we will investigate them in the next chapter before embarking on a study of more complex configurations.

Figure 1.11
Parallel circuits and subcircuits

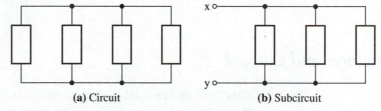

(a) Circuit   (b) Subcircuit

Section 1.1 Quiz

Q1.1-1. What is the small circle labeled a in Figure Q1.1-1 called?

Figure Q1.1-1

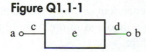

Q1.1-2. What is the straight line labeled c in Figure Q1.1-1 called?

Figure Q1.1-2

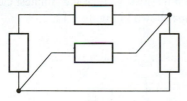

(a)

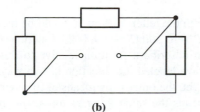

(b)

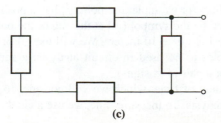

(c)

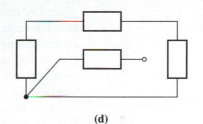

(d)

Q1.1-3. What is the entire object (labeled e) in Figure Q1.1-1 called?

Q1.1-4. For each of the collections of elements in Figure Q1.1-2, state whether or not it is a circuit and identify the nodes and floating nodes (if any).

Q1.1-5. For each of the subcircuits inside the shaded surface in Figure Q1.1-3, draw the subcircuit and state how many terminals it has and whether it is series, parallel, or neither.

Figure Q1.1-3

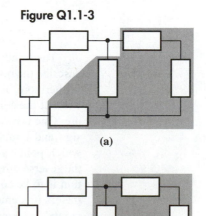

(a)

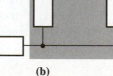

(b)   *(cont.)*

Figure Q1.1-3 *(cont.)*

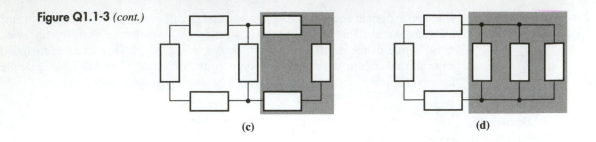

(c)   (d)

1.2 | Voltage and Current

We will define the variables used in circuit analysis in an *operational* manner. An operational definition consists of two things: a specification of a measuring instrument and a procedure for using it. To illustrate, let's think about what is required to measure the length of a classroom. We specify the instrument—a meter stick—and a procedure—laying the meter stick off a number of times end-to-end from one end of the room to the other. The result is a number and a unit, say 15 meters. The number is the outcome of our measurement, and the unit is defined by the measuring instrument. We will not have need for measurements of distance in circuit analysis because we are assuming that any circuit with which we will deal has a behavior that is independent of size. It does, however, serve very well to illustrate all of the important concepts involved in the definition of a fundamental quantity in terms of a measurement procedure.

Length itself is never negative; however, we often need to consider position—a concept related to length—which requires the additional idea of a *signed variable.* It would not be very precise to say that San Francisco is at a position of about 50 miles from San Jose because a lot of other points on the earth also fit this description. To indicate position, one needs the idea of an origin and a reference. To treat the simplest case, let's consider the measurement of position on the line shown in Figure 1.12.

Figure 1.12
A signed position
variable x

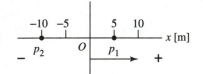

The origin in this figure is labeled with the letter O. Now it would be meaningless to describe the position of a point without a "zero point," which is the function of the origin; furthermore, it would be meaningless to describe a position relative to this zero point without the concept of a direction. We have labeled the direction in two ways in the figure. One is by an arrow pointing to the right, the other is by means of two symbols: a plus sign and a minus sign. The plus sign means the same thing as the arrow; it designates which points are to be considered as positive. The negative sign, which is equivalent to the reverse direction relative to the arrow, means that the position of a point in this direction is to be considered as negative. Thus, for example, the point p_1 has a position of $+5$ meters (the unit "meter" being indicated by the symbol [m] at the end of the positive horizontal axis), and the point p_2 has a position of -10 meters. We will use all of these elements in defining the electrical variables to be used in circuit analysis: a zero point, a unit, and a reference (either an arrow or a pair of $\pm$ signs).

Another example is the measurement of time, which we will consider to be one of the fundamental quantities of circuit analysis. To measure time, we use a clock, which we start at some instant in time and stop at another. The unit of time is considered to be the second, s. In many cases, such as when we graph two quantities relative to each other, we will place brackets around the unit as we did above for position. For example, we might label the horizontal axis of a graph with the expression t[s], meaning "time in seconds."

Sometimes we feel the need to use units of a more appropriate size. As an illustration, the distance from New York to Paris is not conveniently measured in meters, but in terms of kilometers. The distance from the nucleus of an atom to one of its orbital electrons is much more conveniently measured in nanometers. For these measurements, we introduce prefixes, which act as multipliers on the basic unit. Table 1.1 shows the commonly used engineering prefixes and their names and abbreviations. Thus, we would abbreviate the time 1.5×10^{-9} second as 1.5 nanosecond (ns).

Table 1.1
Standard
engineering prefixes

Name	Value	Abbreviation
tera	10^{12}	T
giga	10^{9}	G
mega	10^{6}	M
kilo	10^{3}	k
milli	10^{-3}	m
micro	10^{-6}	μ
nano	10^{-9}	n
pico	10^{-12}	p
femto	10^{-15}	f

Units are named and abbreviated, with a few exceptions, according to a simple rule. All of the units are considered to be common nouns and so are written in lowercase letters. If the unit is the name of a person, usually a person who figured greatly in scientific or engineering discoveries related to the unit, the abbreviation starts with an uppercase letter. Thus the unit of time, the second, is abbreviated s; the unit of voltage, which we will discuss below, is called the *volt* and abbreviated V after the Italian, Alessandro Volta, the inventor of the electric battery.

Voltage   In addition to time, we will use only two fundamental quantities in our study of circuit analysis: *voltage* and *current*. All other quantities are defined in terms of these three, and so are referred to as *derived* quantities. The instrument with which we measure voltage is called a *voltmeter*. In fact, a modern voltmeter is merely one function of a general test instrument called a *multimeter*. We show a sketch of a multimeter in Figure 1.13(a). Rather than sketch such a complicated drawing each time we wish to discuss a measurement, we will use the symbol shown in Figure 1.13(b). Most modern multimeters are *autoranging;* that is, the scale changes automatically to an appropriate value for the voltage being measured. A digital number displays the result of the measurement. We have shown both plus and minus signs in our drawing to emphasize that the reading can be negative, though only one of these appears for a given measurement. There are two *probes,* which are touched to two points in the circuit in which the voltage is to be measured. One of the probes is usually red and the other black. In our abbreviated voltmeter symbol in Figure 1.13(b), we show a plus sign near the voltmeter lead with the red probe; by default, the other lead is attached to the black probe. The unit of voltage is the *volt* and is named after the Italian Volta, as we have previously discussed. Therefore, it is abbreviated V. Figure 1.13(a) shows a voltage with a prefix: 24.732 mV.

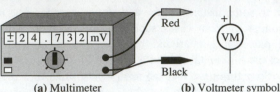

(a) Multimeter   (b) Voltmeter symbol

Figure 1.14(a) shows a circuit with a voltage being measured between two nodes. We have numbered the nodes for ease of reference. Thus, we are measuring the voltage between nodes 2 and 1. Sketching the voltmeter symbol each time we wish to refer to a voltage would lead to a messy circuit diagram, so we choose to abbreviate it using the notation shown in Figure 1.14(b). The + sign means that the positive side of the voltmeter (the + on the voltmeter symbol or the red probe on the actual meter) would be connected to the node carrying that sign. This completes our operational definition of voltage, for we have specified the measuring instrument (the voltmeter) and the way a measurement is taken. The result of such a measurement consists of three pieces of information: a real number magnitude (such as 24.732), a unit (such as mV), and a sign (either plus or minus). Thus, a given voltage might be $+24.5$ V or $-35.6\ \mu$V. The sign depends upon the manner of connection. For instance, suppose that the voltage in Figure 1.14 is $v = +2.5$ V. If we reverse our meter leads, we define the *new* voltage v' shown in Figure 1.15. Its value will be -2.5 V. You should be sure you thoroughly understand this fact before continuing.

Figure 1.14
A voltage measurement
and its symbol

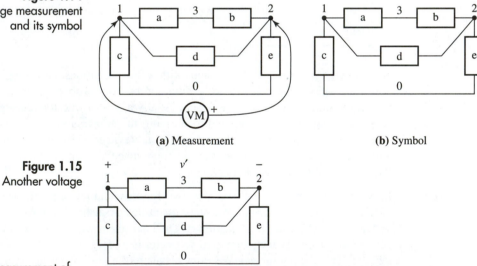

(a) Measurement   (b) Symbol

Figure 1.15
Another voltage

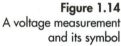

Figure 1.16   Measurement of
the voltage v with
different probe
locations

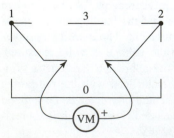

The two voltages we have just defined are called *element voltages* because they are voltages *across* an element—element d, directly connected between nodes 1 and 2. Sometimes a voltage between two points is given a double subscript notation. The voltage v we have just discussed would be written v_{21}. In this notation, the first subscript carries the plus sign and the second, the negative. Thus v' would be written v_{12} and, in general, it is true that $v_{12} = -v_{21}$.

Figure 1.16 shows another important concept relating to our definition of voltage. There, we show a voltmeter being used to measure the voltage v as discussed in connection with Figure 1.14. The only difference is that we have moved the probes to another part of each node. (To focus attention on the nodes themselves, we have not drawn the element bodies. They are there, but for clarity are simply not shown.) The main point is this: there is only one voltage associated with each pair of nodes, and moving a probe from one point on a given node to another point on the same node has no effect on the voltage being measured.

On many occasions it is convenient to measure all voltages relative to a single arbitrarily chosen node. To facilitate this, many voltmeters have a clip attached to the black probe, as shown in Figure 1.17 (using our shorthand voltmeter symbol). The red probe is moved from one node to another, and the value of voltage at each is recorded. We show the voltage at node 2, denoted v_2, being measured in the figure. In general, the symbols for node voltages carry the same subscript as the node at which they are defined. By convention, we do not show a plus sign at each node. The black probe is clipped to the fixed reference node for all measurements, and each of the other node voltages is tacitly assumed to carry a positive sign at the node for which it is defined. A perfectly legitimate node voltage *value* at node 2, however, is -5 V. Notice, by the way, that the value of the voltage at the reference node is 0 V. Thus, all of the other node voltage values are relative to this reference; if we change our choice of reference node, all of the other node voltages will in general change in value.

Figure 1.17
Definition of node voltages

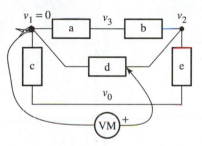
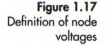

Once we have defined the node voltages with symbols, the meter connection is irrelevant. We will therefore not draw the meter symbol explicitly (as we also decided for element—or *node pair*—voltages); we will, however, indicate our choice of reference node with the symbol shown in Figure 1.18. We often call this the "ground reference." We hasten to point out, however, that it has nothing to do with grounding a circuit to an earth ground for safety from electrical shock hazard. That is quite another matter.[3] Two other symbols that you might encounter in reading other circuit analysis literature are shown in Figure 1.19. On occasion, different symbols are given different meanings; one might mean "chassis ground," another "power line ground," and still another "signal ground." You should be careful to determine the precise meaning of the various symbols in any circuit schematic diagram.

Figure 1.18
The "ground reference" symbol

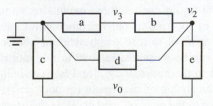

Figure 1.19
Two alternate ground reference symbols

---

[3] The author is stressing this fact because his wife, Lalah, a *safety engineer,* has insisted on it.

A valuable use of the ground reference symbol is to simplify the drawing of circuit diagrams. We often split this symbol into multiple copies, all of which signify the same physical node. Figure 1.20 shows the situation. Each of the three ground reference symbols refers to a connection at the same node—node 1 in the circuit we have been working with. To see that this circuit is the same as our original one in Figure 1.18, just recall that one of our basic assumptions about circuits is that we can stretch and pull and bend their interconnecting leads and conductors without affecting the electrical behavior. Thus, we can go even further and move the circuit around so that it has the form shown in Figure 1.21. We have not, however, altered the topology, i.e., the specification of which leads of which elements are connected to which others. It might be worth your time at this point to verify this statement by reconnecting the ground reference symbols with a conductor and convincing yourself that it is the same circuit as the one with which we started. It will be vital in the future for you to be able to visualize circuit transformations such as this because our circuit analysis procedures often make it necessary. Furthermore, practical schematic diagrams of such appliances as televisions use this notation to reduce the number of conductors that must be drawn.

Figure 1.20
The use of multiple ground reference symbols to refer to a single node

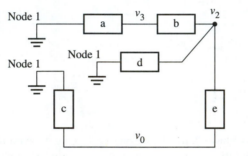

Figure 1.21
The same circuit, redrawn

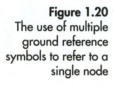

An Analogy for Voltage

An extremely useful analogy for voltage is height. In Figure 1.22 we pursue this idea just a bit. If you make it a habit to think of a voltage as a height, you will find circuit analysis a lot easier. In the figure, we see three mountains standing near a lake. If we arbitrarily pick the surface of the lake as our reference point (we have somewhat whimsically attached a ground symbol to it to emphasize our choice), we see that the heights of the mountain peaks can be measured relative to this altitude. We have called them, reading from left to right in the figure, h_3, h_2, and h_0. The heights of the peaks are analogous to node voltages. The analog of an element (or node pair) voltage is the height of one of the mountains relative to the peak of another. Thus, h_{02} in our diagram is the equivalent of the voltage v_{02} in the circuit we have been discussing. Observe that in making this measurement, we have effectively shifted our altitude reference to the peak of mountain number 2 and therefore h_{02} is negative.

Figure 1.22
The altitude-voltage analogy

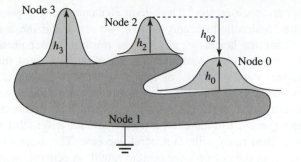

Figure 1.22
The altitude-voltage analogy

Current

Figure 1.23 The ammeter symbol

Current is measured with the same multimeter as that used for voltage measurement; for current, however, the instrument must be adjusted to serve as an *ammeter*—which has the symbol shown in Figure 1.23. The arrow marks the lead corresponding to the red probe on the actual meter, and the unmarked side to the one carrying the black probe. Although voltage is measured *across* an element or *between* two nodes, current is measured *through* (or *along*) a conductor. We specify this measurement in the following way. We cut the conductor in which the current is to be measured and insert the ammeter—as shown in Figure 1.24(a). The red probe is attached to one of the free ends, and the black one to the other. The reading, of course, can then be either positive or negative. In Figure 1.24(b) we show the symbolic shorthand way of representing this measurement. The arrow near the conductor points in the same direction as the ammeter reference arrow—hence, the tail of the arrow is associated with the red probe on the actual meter, and the point of the arrow with the black probe. When we use this shorthand description of a current on a circuit diagram, we will mean that i can be measured by the setup shown in Figure 1.24(a)—merely this and nothing more. The unit of current is called the *ampere,* abbreviated A, after the French electrical researcher André Marie Ampère. The definition of a current carries three pieces of information: a magnitude (such as 6.2), a sign (plus or minus), and a unit with possibly a prefix (such as mA). Figure 1.25 illustrates this in a bit more detail. The two different measurements shown result in values that are negatives of each other.

Figure 1.24
The operational definition of current

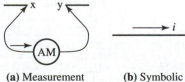

(a) Measurement

(b) Symbolic representation

Figure 1.25
Two different (but associated) current measurements

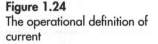

$$\longrightarrow i = 2 \text{ mA}$$
$$i' = -2 \text{ mA} \longleftarrow$$

An Analogy for Current

Just as we presented an analogy for voltage, we can also give one for current. Figure 1.26 shows a pipe carrying water. The flow rate, which we have labeled f, is analogous to current. Faster water flow corresponds to larger current. In fact, an even more extensive analogy is worthwhile. In Figure 1.27(a) we show a simple circuit; in Figure 1.27(b) we illustrate the hydraulic analogy. The currents correspond to water flow rates, as we have said, and the reference node to the altitude reference—which we have taken to be the surface of a small pond of water. The nodes in the circuit are much like tanks of water

Figure 1.26 Water flow rate
analogy for
current

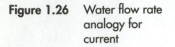

that are open on top.[4] The circuit elements are like pieces of hydraulic machinery: tur-
bines, hydraulic lifts, and the like. We see that if the tanks are lifted higher, thereby increasing the heights h_1 and h_2, the resulting water pressure (the "head") will generally
increase—with a consequent increase in flow rates through the pieces of equipment
marked a and b (unless, of course, one is a pump that is pumping water upward). In the
original circuit, increasing the node voltages v_1 and v_2 generally results in an increase in
the currents i_a and i_b. In the water system, if we increase the heights h_1 and h_2, the flow
rates f_a and f_b will increase; if we increase the height difference *between* the two tanks, the
water flow rate f_c will, in general, increase. The same thing holds for increasing the voltage v_{12} in the circuit—the current i_c will, in general, increase. Further, if we insert a flow
rate meter to measure f_c' in Figure 1.27(b) (as shown, with $h_1 > h_2$), we will measure a
negative flow rate. The analogous statement would be true about the current i_c' (if $v_1 > v_2$).
At any rate, whatever the values of f_c and i_c, f_c' and i_c' will have values that are their
negatives.

Figure 1.27
Circuit interpretation of
flow rate analogy

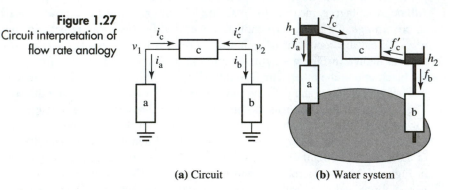 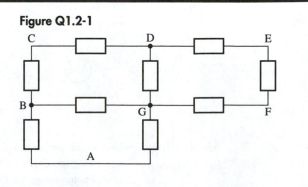

(a) Circuit   (b) Water system

Summary of the
Fundamental Circuit
Variables and
Their Units

In this section we have defined all three of the fundamental variables of circuit analysis:
time, voltage, and current. All others, as we will see, are derived quantities because they
are defined in terms of these three with mathematical formulas. The basic units, therefore,
are the second [s], the volt [V], and the ampere [A].

Section 1.2 Quiz

Q1.2-1. Express 4.54 volts in mV, kV, MV, and μV.

Q1.2-2. Express -2.4 nanoamperes in A, MA, kA, mA, pA,
and fA.

Q1.2-3. Show the voltmeter connection for measuring v_{DB} in
Figure Q1.2-1. Is v_{DB} an element voltage, a node voltage, or neither?

Figure Q1.2-1

---

[4] This analogy should not be pursued *too* far. Charge, to be discussed in the next section, cannot accumulate at a node.

Q1.2-4. If G is chosen as the reference node in Figure Q1.2-1, show the voltmeter connections for measuring v_F, v_B, and v_C.

Q1.2-5. Identify the nodes in Figure Q1.2-1 by erasing the element bodies.

Q1.2-6. What is the value of v_4 in Figure Q1.2-2.

Figure Q1.2-2

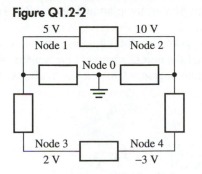

Q1.2-7. If the reference node in Figure Q1.2-2 is changed to node 4, what is the value of v_0, the former reference node voltage?

Figure Q1.2-3

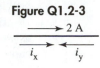

Q1.2-8. What is the value of i_x in Figure Q1.2-3? i_y?

Q1.2-9. Name the three fundamental circuit quantities (without peeking!).

1.3 | Charge and Flux Linkage

Charge Let's continue just a bit with the hydraulic analogy that we discussed at the end of Section 1.2 by looking at Figure 1.28, in which a pipe is filling a tank with water. Current is analogous to water flow rate, so we would expect there to be an electrical quantity analogous to the amount of water that accumulates in the tank. There is. It is called *charge*. Figure 1.29 shows a conductor—the electrical analog of the water pipe—carrying a current $i(t)$. We define the electric charge to be the accumulation, or integral of current,

Figure 1.28 Hydraulic analogy for electric charge

$$q(t) = \int_{-\infty}^{t} i(\alpha)\, d\alpha. \tag{1.3-1}$$

In this formulation[5] we allow for the fact that the current might have been nonzero for negative values of time; an alternate formulation is to write

Figure 1.29 Direction of flow of charge

$i(t), q(t)$

$$q(t) = \int_{-\infty}^{t} i(\alpha)\, d\alpha = \int_{-\infty}^{0} i(\alpha)\, d\alpha + \int_{0}^{t} i(\alpha)\, d\alpha, \tag{1.3-2}$$

or

$$q(t) = q_0 + \int_{0}^{t} i(\alpha)\, d\alpha, \tag{1.3-3}$$

where

$$q_0 = \int_{-\infty}^{0} i(\alpha)\, d\alpha \tag{1.3-4}$$

is the *initial charge*. This second form is useful for circuit problems in which we know

---

[5] Notice that α, although a time variable, must be different from the upper limit t. Because we consider t to be a parameter that can change, we call this integral a *running integral*.

the current for positive values of time and the only information we have about the past history is the initial charge q_0. This type of problem is referred to as an *initial condition problem*. The constant q_0 is sometimes referred to as the *initial condition*.

Because charge is defined in terms of current, we can associate a direction with it just as we did for the current. Figure 1.29, discussed previously, shows a conductor carrying current and therefore charge. The arrow denotes the direction of the current, and this can be interpreted—just as we think of water flow in the pipe of Figure 1.28—as a flow of charge in the same direction.

Recalling that the integral in (1.3-1) is merely the limit of a sum of current values multiplied by time increments, we see that the unit of charge is the ampere·second, A·s, which we call the *coulomb* after the French engineer Charles Coulomb. Its abbreviation, therefore, is C.

Example 1.1   Suppose the current in a conductor is given by

$$i(t) = u(t)\,\mathrm{A} = \begin{cases} 1; & t > 0 \\ 0; & t \le 0 \end{cases} \mathrm{A}. \tag{1.3-5}$$

Find the charge in coulombs.

Solution   First, let us discuss the notation a bit. The symbol $u(t)$ means the *unit step function,* which has the definition given by the second expression in (1.3-5). That is, it is unity for positive values of time and zero for negative values. We plot this function in Figure 1.30(a). The meanings of the solid dot and open circle are illustrated in Figure 1.30(b). If we observe such a waveform in the laboratory, there is always a finite "rise time" from zero to one. Thus, the solid dot means that the idealized waveform is continuous from the left, and the open dot means that it is unity immediately to the right of the origin, but is zero at $t = 0$. The unit step function is the ideal model of anything that "turns on" suddenly.

Figure 1.30
The unit step function
and its practical
realization

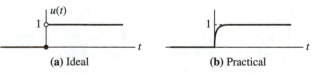

(a) Ideal   (b) Practical

The charge is the area under the current waveform from $-\infty$ to the running parameter t. If $t \le 0$, we see that this area is zero. If $t > 0$, it is the area of a rectangle of unit height and base t, as we see in Figure 1.31. There, we have used the variable of integration α as our independent variable and t as the upper limit. We see that this area is simply t. Thus, we can write

$$q(t) = \begin{cases} t; & t > 0 \\ 0; & t \le 0 \end{cases} \mathrm{C} = t u(t)\,\mathrm{C}. \tag{1.3-6}$$

We have written this function in abbreviated form after the second equality sign by using the unit step function. Because $u(t) = 0$ for $t \le 0$, when we multiply the function t by $u(t)$ it forces the resulting product to be zero for $t \le 0$—as we desire. For $t > 0$, this multiplier is unity, so we get t—also as we desire. This is a useful property of the unit step function: forcing waveforms described by a formula to be zero for negative values of time. The charge waveform is plotted in Figure 1.32. The function $tu(t)$ is called the *unit ramp function.*

Figure 1.31
The running integral for
positive time

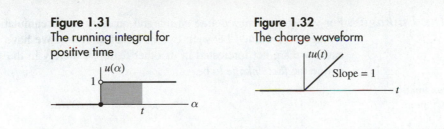

Figure 1.32
The charge waveform

Charge is the integral of the current, so the current can be expressed as the time derivative of the charge:

$$i(t) = \frac{dq(t)}{dt}. \tag{1.3-7}$$

Example 1.2 illustrates the computation of current from a known charge waveform.

Example 1.2   Assuming that the charge in a conductor is given by

$$q(t) = \sin(2t)u(t) \text{ C}, \tag{1.3-8}$$

find the current $i(t)$.

Solution   It is usually a good idea when working with waveforms to plot them. Thus, we plot our charge waveform in Figure 1.33. Computation of the current is actually quite simple. We have

$$i(t) = \frac{dq(t)}{dt} = \frac{d}{dt}[\sin(2t)u(t)] = 2\cos(2t)u(t) \text{ A}. \tag{1.3-9}$$

Checking the units, we see that taking the derivative is equivalent to dividing the unit of charge by the unit of time, so we have C/s = A. The current $i(t)$ is plotted in Figure 1.34. The derivative of $q(t)$ at $t = 0$ is not defined mathematically because the slope of $q(t)$ is zero to the left of $t = 0$ and unity immediately to the right. However, if we observe such a current waveform in the laboratory, we will see that it is zero at $t = 0$ and has a finite rise time to unity at a small value of positive time—before the sinusoidal nature becomes apparent for increasing time. This is the reason for the solid and open dots in the figure.

Figure 1.33
The charge waveform

Figure 1.34
The current waveform

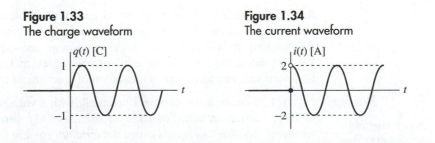

We caution you here that had we presented the waveform in Figure 1.34 as the charge waveform and computed *its* derivative we would have run into difficulty. Why? Because a function does not have a derivative at a discontinuity. (This is substantially different from the fact that our derivative in Example 1.2 was, itself, discontinuous.) We address this problem in another chapter.

Flux Linkage For voltage, we also define an integral quantity (an accumulation). Figure 1.35 shows a voltage $v(t)$ defined between two nodes of a circuit. We have not drawn the circuit because we are not interested in its other features, merely in the voltage measurement. We define the *flux linkage* to be

Figure 1.35 Flux linkage definition

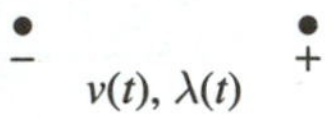

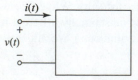

$$\lambda(t) = \int_{-\infty}^{t} v(\alpha)\, d\alpha. \tag{1.3-10}$$

It can be considered to have a set of plus and minus reference polarities, just as for the voltage. We have shown this in the figure. Also, just as for charge, there is an equivalent initial condition formulation:

$$\lambda(t) = \lambda_0 + \int_{0}^{t} v(\alpha)\, d\alpha, \tag{1.3-11}$$

where

$$\lambda_0 = \int_{-\infty}^{0} v(\alpha)\, d\alpha \tag{1.3-12}$$

is the initial condition, or initial flux linkage. The unit of flux linkage is the volt·second, which we call the weber after Ernst Weber, a German electrical pioneer. Thus, its abbreviation is Wb. It should be apparent that if we know the flux linkage as a function of time, we can obtain the voltage from

$$v(t) = \frac{d\lambda(t)}{dt}. \tag{1.3-13}$$

Section 1.3 Quiz

Q1.3-1. If $i(t) = e^{-2t}u(t)$ A, find and sketch $q(t)$ in C.

Q1.3-2. If $q(t) = (1/2)(t^2 u(t))$ C, find and sketch $i(t)$ in A.

Q1.3-3. If $v(t) = 4\cos(2t)u(t)$ V, find and sketch $\lambda(t)$ in Wb.

Q1.3-4. If $\lambda(t) = (1/6)(t^3 u(t))$ Wb, find and sketch $v(t)$ in V.

1.4 | Power and Energy

Section 1.2 dealt with the fundamental quantities of circuit analysis: time, voltage, and current. In Section 1.3, we discussed two derived quantities, charge and flux linkage. In this section, we will define and discuss two more derived quantities: power and energy. We will define them strictly in terms of voltage, current, and time; thus, we will not be burdened with any technical consideration of mechanical power and energy.

Power Figure 1.36 shows a two-terminal subcircuit with a voltage v defined across its terminals (and for this reason called its *terminal voltage*) and a current i defined into one of these terminals (and for this reason called the *terminal current*). As a consequence of a general circuit law to be discussed in Chapter 2, the current into one terminal is always the same as the current leaving the other. Notice that we have defined the positive current reference going *into* the terminal with the positive voltage reference. This choice is called the *passive sign convention*. We will discuss reasons for our choice shortly.

Figure 1.36 A two-terminal subcircuit—The passive sign convention

Relative to the subcircuit—which, we point out, could consist of a single element—we define the *power absorbed by the subcircuit* to be

$$P(t) = v(t)i(t). \tag{1.4-1}$$

The unit of power is, therefore, the product of the unit of voltage and the unit of current. We call it the *watt* in honor of James Watt, a Scottish engineer. Thus, we abbreviate it W, and have the unit equation W = V·A. Power is normally defined in physics courses as the product of force and distance, but we will have no need of this formulation here. All of the quantities we use will be based strictly on time, voltage, and current.

Example 1.3   Find the power absorbed by each of the subcircuits in Figure 1.37.

Solution   The first subcircuit, the one in Figure 1.37(a), is a special case of the general situation depicted in Figure 1.36 with $v(t) = 10$ V (a constant) and $i(t) = 2$ A (also a constant). Thus, the power absorbed by the subcircuit is constant at $P(t) = 10 \times 2 = 20$ W. In Figure 1.37(b), we still have $v(t) = 10$ V; now, however, the current is *coming out* of the terminal carrying the positive reference for the voltage. But this is equivalent to a negative value for the current going *into* this same terminal, as we show in Figure 1.38. Thus, the power absorbed by the subcircuit is $P(t) = v(t)i(t) = 10 \times (-2) = -20$ W. When a subcircuit has a negative absorbed power, it is said to be *delivering* or *generating* power.

Figure 1.37
Example subcircuits

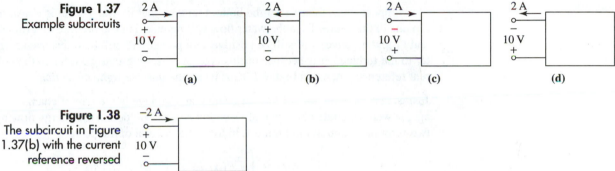

(a)   (b)   (c)   (d)

Figure 1.38
The subcircuit in Figure 1.37(b) with the current reference reversed

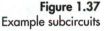

The basic idea we used in conjunction with Figure 1.38 can be used in general. For the subcircuit in Figure 1.37(c) we have $P(t) = v(t)i(t) = (-10) \times 2 = -20$ W, so it is delivering power. In this case, we have merely taken the positive voltage reference at the top node and used a negative value. We observe here that a negative *value* is not the same as a negative *reference polarity*. The latter tells us where to connect the black probe on our voltmeter, whereas the former is the result of the measurement; that is, the sign that appears on the meter display. In Figure 1.37(d) we have $P(t) = (-10)(-2) = 20$ W, an absorbed power.

A Physical Analogy for the Passive Sign Convention   Our choice of the positive current reference as being *into* the terminal carrying the positive voltage reference is best explained with an analogy. In Figure 1.39 we show a hydraulic turbine connected between two reservoirs of water. The turbine is *passive;* that is, it only transforms power being supplied to it by the water flow into another form of energy, say as rotational energy of a mechanical shaft protruding from one end (not shown). It does not produce any net energy on its own. The height h of the water level in the top tank relative to that of the bottom one is analogous to the terminal voltage of an electrical element or subcircuit (a positive value of h is like a positive value of v), the water flow rate is analogous to the terminal current, and the turbine is analogous to the two-terminal element or subcircuit itself. If h is positive, then water will flow out of the top tank, through the turbine, and into the bottom tank. Thus, the flow rate f will be positive and the product of h and f will also be positive. Furthermore, we can see that hydraulic power is being transferred from the water flow into the turbine—the turbine is absorbing hydraulic power.

Figure 1.39
Interpretation of the
passive sign convention

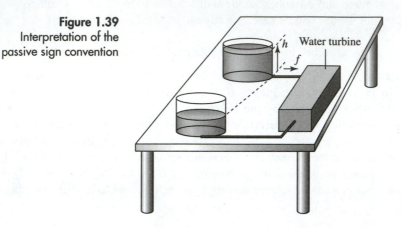

Suppose, now, that we fill the bottom tank so that its water level is above that of the top one in the figure. Then the water flow will reverse, but the product of h (now negative) and f (now negative) will still be positive, and power will still be transferred from the water to the turbine.[6] It is because of this type of reasoning that we refer to the voltage/current reference scheme in Figure 1.36 as being the *passive sign convention.*

Energy   Just as current is the rate of flow of charge, and voltage is the rate of change of flux linkage, power is the rate of change of electrical *energy.* We define the energy flowing into the two-terminal subcircuit in Figure 1.36 to be the integral of the power:

$$w(t) = \int_{-\infty}^{t} P(\alpha)\, d\alpha = \int_{-\infty}^{t} v(\alpha)i(\alpha)\, d\alpha. \qquad (1.4\text{-}2)$$

The integral is the limit of a sum of power values multiplied by incremental time intervals. Thus the unit of energy is the watt-second (W·s), which we call the *joule (J)* after the British physicist James Prescott Joule. The unit equation is J = W·s. An alternative formulation of (1.4-2) is

$$w(t) = w_0 + \int_{0}^{t} P(\alpha)\, d\alpha = w_0 + \int_{0}^{t} v(\alpha)i(\alpha)\, d\alpha, \qquad (1.4\text{-}3)$$

where

$$w_0 = \int_{-\infty}^{0} P(\alpha)\, d\alpha = \int_{-\infty}^{0} v(\alpha)i(\alpha)\, d\alpha. \qquad (1.4\text{-}4)$$

This form of expression, like those for charge and flux linkage, is called an initial condition formulation.

Example 1.4   If the terminal voltage and current of a two-terminal subcircuit are given by $v(t) = 10u(t)$ V and $i(t) = 2u(t)$ A (passive sign convention), respectively, find and sketch the power and energy absorbed as a function of time.

---
[6] We assume that the turbine will run "backwards."

Solution   The power computation is simple. We have

$$P(t) = v(t)i(t) = 10u(t) \text{ V} \times 2u(t) \text{ A} = 20u(t) \text{ W}. \qquad (1.4\text{-}5)$$

The energy absorbed is clearly zero for $t \leq 0$ because $P(t)$ is zero for all negative time. Hence, for $t > 0$,

$$w(t) = 0 + \int_0^t P(\alpha)\, d\alpha = \int_0^t 20\, d\alpha = 20t \text{ J}. \qquad (1.4\text{-}6)$$

Using the unit step function to "turn off" the energy expression for $t \leq 0$ (after all $20t$ is a valid expression for negative t also), we get

$$w(t) = 20tu(t) \text{ J}. \qquad (1.4\text{-}7)$$

The waveforms are shown in Figure 1.40.
    There is a small item of notation about which you should be careful. As you can see in Figure 1.40, the symbol for energy, w, is the lowercase version of the unit of power, W. As energy and power are so closely related, you should be on guard not to confuse these two items.

Figure 1.40
Power and energy
waveforms

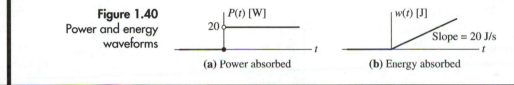

(a) Power absorbed   (b) Energy absorbed

Section 1.4 Quiz

Q1.4-1. Is the single element subcircuit in Figure Q1.4-1 absorbing or generating power for $t > 0$? Assume that $v(t) = 40u(t)$ and $i(t) = 10u(t)$.

Figure Q1.4-1

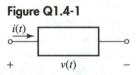

+ $v(t)$ −

Q1.4-2. Find the energy absorbed by the element in Figure

Q1.4-1 at $t = 1$ s if $v(t)$ and $i(t)$ are the waveforms specified in Question Q1.4-1.

Q1.4-3. Assume that the terminal variables $v(t)$ and $i(t)$ of the element in Figure Q1.4-1 are connected by the relationship $v(t) = L\, di(t)/dt$. Find $w(t)$ for $t > 0$ if $w(0) = 0$ and $i(t) = 0$ for all $t \leq 0$.

Q1.4-4. Assume that $v(t)$ and $i(t)$ in Figure Q1.4-1 are connected by the relationship $i(t) = C\, dv(t)/dt$. Find $w(t)$ for $t > 0$ if $w(0) = 0$ and $i(t) = 0$ for all $t \leq 0$.

1.5 | Independent Sources

Active and   Figure 1.41 shows, once more, a typical two-terminal element. The question we now ad-
Passive Elements dress is this: When does it act as a *load,* merely absorbing energy, and when does it act as
                    a *source,* producing energy? Evidently, we need only to compute the absorbed energy
                    from

$$w(t) = \int_{-\infty}^{t} P(\alpha)\, d\alpha = \int_{-\infty}^{t} v(\alpha)i(\alpha)\, d\alpha. \qquad (1.5\text{-}1)$$

Figure 1.41 A general two-
terminal element

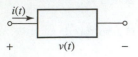

If $w(t) \geq 0$ for each and every value of t, we say that the element is *passive*. This simply means that it can only absorb energy. The voltage and current waveforms, of course, must be related by an element constraint that dictates—for each $v(t)$ waveform—which $i(t)$ waveform is possible, or the converse. If we can find *any one instant of time t_0*, at which $w(t_0) < 0$, we say that the element is *active*. This term just means that we can extract net energy from the element. At other instants, the element perhaps has absorbed a total energy that is positive, but if we merely stop things at t_0, we have extracted a net energy from the element. This means that it serves as a *source* of energy.

We will refer to such sources of energy simply as *sources*. In many cases, such sources produce voltage or current waveforms spontaneously without being influenced to any degree by the circuit in which they are connected. In this case, we call them *independent sources*. If they produce waveforms that are functions of other voltages or currents in the circuit, we refer to them as *dependent* (or *controlled*) *sources*. Sources of the second type are discussed in Chapter 5. Here, we will only treat independent sources.

The Independent Voltage Source

In Figure 1.42(a) we show an independent voltage source. The symbol for any independent source is a round circle, meant to evoke an image of a dynamo or generator such as one sees at a hydroelectric plant. Such a unit is round because it contains a massive rotor that spins in a magnetic field to generate electricity. The plus and minus signs inside the circle indicate that it is a voltage source.

Figure 1.42
The independent
voltage source

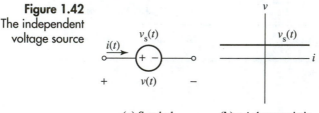

(a) Symbol   **(b)** *v-i* characteristic

The independent voltage source, which we will simply call a voltage source until we study those of the dependent variety, is the first circuit element to be defined. For this reason, it is perhaps well to discuss element definitions a bit. We will make a basic assumption: *the v-i characteristic of an element completely describes its behavior when it is connected together with other elements to form a circuit.* Furthermore, we assume that this *v-i* characteristic can be determined by experiment on the element when it is isolated from other elements. In plain terms, this means that we can test it on a lab bench to determine its characteristics and those characteristics continue to be valid when we wire up our circuit.

An element such as our voltage source has a *v-i* characteristic that can be defined by a graph: a plot of v versus i in a two-dimensional plane. The specific way in which $i(t)$ or $v(t)$ changes with time does not destroy this relationship. For the voltage source, this graph is a horizontal line in the v-i plane, as we show in Figure 1.42(b). The value of $v_s(t)$ can perhaps change with time, but the shape of the graph does not. If $v_s(t)$ becomes larger, the horizontal line moves up on the voltage axis; if it becomes smaller, the line moves down. The main idea is this: no matter what the value of the current $i(t)$, the voltage $v(t)$ is always specified to be equal to $v_s(t)$. Thus, the voltage source *does not constrain the current at all*. As we will see, the current depends upon the circuit into which the voltage source is connected.

Here is an item of notation that you should remember. If we wish to emphasize that a

Figure 1.43 The battery—A dc source

Figure 1.43 The battery—A dc source

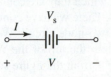

voltage $v(t)$ does not change with time, we write V (that is, we use an uppercase letter).[7] If we simply write $v(t)$, we mean that the voltage can change with time or can be constant; this notation is more general than the use of the uppercase V. Often, we will simply write v without the (t). If the voltage actually changes with time, the (t) functional dependence is just understood to hold. We call constant voltages and currents *dc* quantities. This is short for *direct current,* and is used for both currents and voltages. Sometimes one sees the symbol shown in Figure 1.43 for a dc voltage source. This symbol designates a *battery*. The long line represents the positive terminal, and the short line the negative terminal. We will not have much occasion to use it in this text.

The Independent Current Source

An independent current source, or simply current source, is shown in Figure 1.44. Figure 1.44(a) shows the symbol; Figure 1.44(b) shows the v-i characteristic. Again, the symbol is round; now, however, the plus and minus signs have been replaced by an arrow to denote the direction of the current. As you can see from Figure 1.44(b), the current is independent of the voltage. The voltage is determined by the circuit into which the current source is connected, but the current source itself does not constrain this variable. It only specifies that the current has the value $i_s(t)$.

Figure 1.44 The independent current source

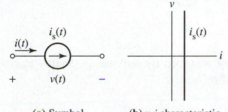

(a) Symbol   **(b)** v-i characteristic

Analogies for the Independent Sources

Let's use our hydraulic analogy once again, this time to gain a physical feel for the voltage source and the current source. In Figure 1.45 we see a sketch of a water mill, which lifts buckets of water from a stream up to the flume above. The water (the electric charge) is lifted through a height h—independent of the size of the buckets and thus independent of how fast the water is flowing along the flume. The mill is analogous to a voltage source. Figure 1.46, on the other hand, shows a constant volume flow rate electrical water pump being used to pump the water up through a pipe. This provides a flow rate of say 10 gallons per minute regardless of the height h through which it is pumped. The water pump is analogous to a current source.

Figure 1.45 Hydraulic analogy for a voltage source

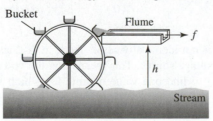

Figure 1.46 Hydraulic analogy for a current source

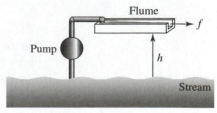

---

[7] We will use a similar notation for current and other variables.

Independent Sources as Active Elements

Voltage and current sources are active; that is, they are able (given the right circumstances) to provide energy for delivery to the rest of a circuit in which they are connected. To see this, let's look at the simple circuit shown in Figure 1.47. There we have connected together a voltage source and a current source, thus forming a two-node circuit. We have defined a voltage v between these two nodes and a current i into the top of the voltage source. Now let us keep the current source in mind, but remove it from the figure so as to focus our attention on the voltage source alone, as we have done in Figure 1.48. We assume the current source is still connected, but we have simply chosen not to draw it. We will do this quite often in the future, sometimes without spelling things out in quite such detail, so you should always keep this possible interpretation of a figure in mind.

Figure 1.47
A simple two-node circuit

Figure 1.48
Focusing on the voltage source

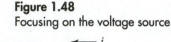
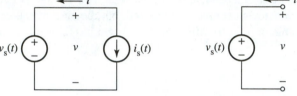

The energy absorbed by our voltage source is given by equation (1.5-1), repeated here for ease of reference:

$$w(t) = \int_{-\infty}^{t} P(\alpha)\, d\alpha = \int_{-\infty}^{t} v(\alpha) i(\alpha)\, d\alpha. \tag{1.5-1}$$

The voltage source dictates the voltage to be $v(t) = v_s(t)$. A glance back at Figure 1.47 and perhaps also at Figure 1.44 (which pertained to the definition of the current source) shows that $i(t) = -i_s(t)$. Be sure that you understand the negative sign! Thus,

$$w(t) = \int_{-\infty}^{t} v_s(\alpha)[-i_s(\alpha)]\, d\alpha = -\int_{-\infty}^{t} v_s(\alpha) i_s(\alpha)\, d\alpha. \tag{1.5-2}$$

Now, whatever the value or waveform of the voltage source, we can adjust our independent current source to be the same numerical value (the units, of course, will be different). This gives

$$w(t) = -\int_{-\infty}^{t} v_s^2(\alpha)\, d\alpha < 0, \tag{1.5-3}$$

so long as $v_s(t)$ is not identically zero.[8] Thus, in this case, we are constantly extracting energy from the voltage source. This means that the voltage source is an active element, for we only needed to find one value of time for which $w(t) < 0$ for this conclusion to be valid.

The situation we have just described is typical. A car battery normally delivers both power and energy to the engine, the headlights, and the accessories (such as radios and fan motors in the heating and cooling systems). When the battery runs down, however, we

---

[8] Technically, it must be nonzero at a sufficiently "large" set of time values for the integral to be nonzero.

connect it to a battery charger. This piece of electrical equipment is equivalent, electrically, to a current source such as the one in Figure 1.47; now, however, the current source is reversed in direction so that the current $i(t)$ in that figure is positive. In this case the integral in (1.5-1) is positive and the voltage source is absorbing energy; that is, it is *recharging*. Notice that it is not the particular situation that defines the voltage source to be an active element, but its *capability* of delivering net energy under some set of circumstances. We will leave it to you to show that the current source, like the voltage source, is also an active element.

The Implied Source Convention

We will finish this section with a small, but highly important, item of notation. Figure 1.49(a) shows a grounded element with a terminal labeled with the value 12 V. Figure 1.49(b) shows the meaning. This is called the *implied source* convention because the voltage source is simply understood to be present but is not drawn. The reason for this, like the splitting of the reference node symbol, is merely to simplify circuit drawings.

Figure 1.49
The implied source convention

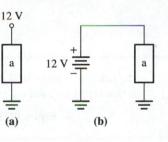

(a)   (b)

Section 1.5 Quiz

Q1.5-1. Write down the indicated voltage and/or current for each of the parts of Figure Q1.5-1.

Figure Q1.5-1

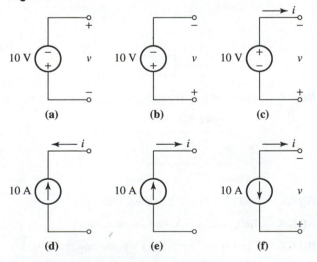

(a)   (b)   (c)

(d)   (e)   (f)

Q1.5-2. For the voltage source in part(c) and the current source in part (f) of Figure Q1.5-1, determine if the source is delivering power to the rest of the circuit or absorbing it from the rest of the circuit.

Q1.5-3. Prove that the independent current source is an active element.

Chapter 1 Summary

In this chapter we have covered the most fundamental concepts upon which all of circuit analysis is based. We introduced time, current, and voltage by means of operational definitions: specification of the instrument by which each is measured and a description of the procedure for measuring it. The other circuit variables—*charge, flux linkage, power,* and *energy*—were then defined in terms of these fundamental quantities by equations. For this reason, these other variables are referred to as *derived quantities*.

We discussed the basic ideas of an *element* and a *subcircuit* composed of elements. The important idea of *passive sign convention* was discussed and related to the power absorbed by an element or subcircuit.

The ideas of circuit and subcircuit were related to the concept of element by means of the related concept of *topology;* namely, the way elements are *interconnected* to form circuits and subcircuits. The ideas of *node, loop,* and *series* and *parallel connections* were then introduced as topological concepts.

Finally, we defined an *active element* as one that is capable of giving up a net energy to its surrounding environment and a *passive element* as one that cannot. The prototypes of all active elements, the *independent current source* and the *independent voltage source,* were defined in terms of their *v-i characteristic,* an equation or plot relating their two terminal variables: v and i.

Chapter 1 Review Quiz

RQ1.1.   Find and sketch the nodes in Figure RQ1.1.

Figure RQ1.1

RQ1.2.   One of the circuits in Figure RQ1.2 is series; the other is parallel. Identify them.

Figure RQ1.2

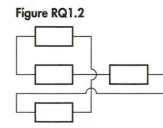

(a)

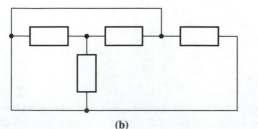

(b)

RQ1.3.   In Figure RQ1.3 show the voltmeter connection for measuring v_{24}, v_1, and v_4.

Figure RQ1.3

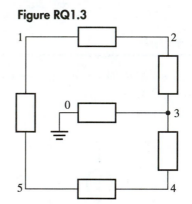

RQ1.4.   Find i_x and i_y in Figure RQ1.4.

Figure RQ1.4

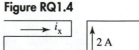

RQ1.5.   If $q(t) = 6t^2u(t)$ C, find and sketch $i(t)$.

RQ1.6.   If $i(t) = 6t^2u(t)$ A, find and sketch $q(t)$.

RQ1.7.   If $\lambda(t) = e^{0.2t}\sin(3t)u(t)$ Wb, find and sketch $v(t)$.

RQ1.8.   If $v(t) = te^{-2t}u(t)$ V, find and sketch $\lambda(t)$.

RQ1.9.   Express each of the waveforms in Figure RQ1.5 as a weighted sum of step functions.

Figure RQ1.5

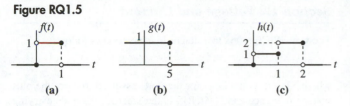

(a)   (b)   (c)

RQ1.10. Find the power absorbed by subcircuits A and B in Figure RQ1.6.

Figure RQ1.6

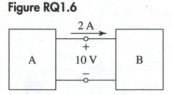

RQ1.11. Find the energy delivered to subcircuit A in Figure RQ1.6 at $t = 2$ s. (Assume that both the voltage and current are zero for $t \leq 0$.)

RQ1.12. Assuming that the terminal voltage and current for a subcircuit are—relative to the passive sign convention—related by

$$v(t) = 2 \int_{-\infty}^{t} i(\alpha) \, d\alpha,$$

find the energy absorbed, $w(t)$. (Assume that $v(-\infty) = 0$.)

RQ1.13. Is the subcircuit in Question RQ1.12 passive or active?

RQ1.14. Answer Question RQ1.12 under the assumption that the subcircuit has the v-i relationship $v(t) = 2i(t)$.

RQ1.15. Is the subcircuit in Question RQ1.14 passive or active?

RQ1.16. For the simple circuit in Figure RQ1.7, find the power absorbed by the voltage source, the power absorbed by the current source, and the total power absorbed by the circuit.

Figure RQ1.7

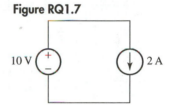

Chapter 1 Problems

Section 1.1 Elements and Circuits

A popular computer simulation program called SPICE, an acronym for *Simulation Program with Integrated Circuit Emphasis,* uses the following format to describe the topology of a circuit:

`EXX NPLUS NMINUS.`

The period merely ends our sentence—it is not part of the format. The label EXX is the name of the particular element being described to SPICE. It can be somewhat arbitrarily assigned and consists of any sequence of letters and or numbers without spaces. The very first letter designates the particular type of element, which we will not discuss now. For the present, we will simply use E as the first letter of a generic element. The labels NPLUS and NMINUS refer to the nodes between which the element is connected. We will discuss the significance of the ordering of these labels in the problems of Section 1.2. Any set of letters and/or numbers can be used to identify the nodes, but there must be a unique correspondence between the nodes in the circuit and these labels. Furthermore, a reference node must be chosen and assigned the numeral 0 (zero). The collection of all element descriptions is called a netlist, which is short for network listing.

1.1-1. Draw a circuit diagram that has the following netlist:

NETLIST

e1	1	0
e2	0	1
e3	1	2
e4	1	0
e5	0	2
e6	2	0

1.1-2. Draw a circuit diagram that has the following netlist (note, by the way, that SPICE is not sensitive to case; therefore, E is interpreted to be the same as e):

NETLIST

E1	1	3
E2	1	2
E3	1	2
E4	2	0
E5	0	2
E6	3	0

1.1-3. Write down the netlists for the circuits shown in Figure P1.1-1(a) and (b). They should be identical except for possible differences in the ordering of the element descriptions and in the order of the nodes describing each element connection. Remember that the ground symbols both refer to the single node 0.

Figure P1.1-1

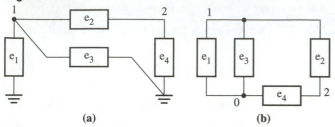

(a)   (b)

1.1-4. Write down the netlist for the circuit in Figure P1.1-2.

Figure P1.1-2

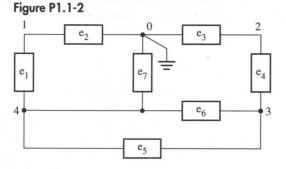

1.1-5. Write down the netlist for the circuit in Figure P1.1-3. Compare it with the netlist of the circuit in Figure P1.1-2.

Figure P1.1-3

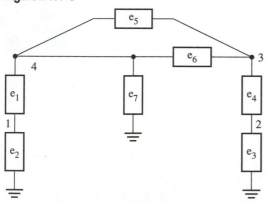

1.1-6. Write down the netlist for the collection of elements in Figure P1.1-4. *Note:* Node 2 is a *floating node.* The SPICE compiler generates an error message and will not attempt to simulate a circuit with such a floating node.

Figure P1.1-4

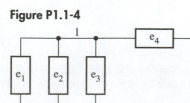

Section 1.2 Voltage and Current

The way SPICE refers to voltages in a circuit is as follows:

`V(NPLUS, NMINUS)`

where NPLUS is the label for the node assumed to carry the plus voltage reference and NMINUS is the label for the node carrying the minus reference. We show this in Figure P1.2-1. As you can easily see, this is equivalent to the *double subscript* convention for node pair voltages. Likewise, $v(k)$ refers to the node voltage at the node with the label k relative to ground reference. Like all node voltages, $v(k)$ is assumed to be positive at node k relative to the reference.

Figure P1.2-1

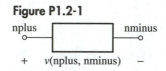

$+ \quad v(\text{nplus, nminus}) \quad -$

1.2-1. Draw the circuit corresponding to the following netlist and label the plus and minus polarity references for the following voltages: $v(1, 2)$, $v(3, 2)$, and $v(2)$.

NETLIST

e1	1	0
e2	1	2
e3	2	1
e4	2	0
e5	3	2
e6	3	0

The way SPICE refers to a current is as follows:

`I(EK)`

where EK is an element label. Thus, only element currents can be referred to. This excludes currents in conductors unless they carry the same current as that in an element. This is elaborated on at greater length in SPICE material in other chapters. Here is the convention for interpretation of i(ek): it is the current whose reference points through element ek from the node labeled nplus to the one labeled nminus. We show this in Figure P1.2-2. As you can see, this means that SPICE uses the passive sign convention for its voltage and current references.

Figure P1.2-2

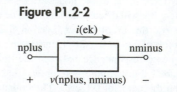

$+ \quad v(\text{nplus, nminus}) \quad -$

1.2-2. Draw the circuit described in the netlist given in Problem 1.2-1 (if you have not already done so) and mark the current references for i(e1), i(e2), i(e3), i(e4), i(e5), and i(e6).

1.2-3. Suppose $v = 3 \text{ nV}$. Express v in MV, kV, mV, μV, and pV.

1.2-4. Suppose $i = 3.4 \times 10^9 \ \mu A$. Express i in MA, kA, A, mA, and nA.

1.2-5. Compute the number of picoamperes in a current of 10 amperes.

1.2-6. A voltmeter reads 100 μV. Express this in mV, V, and nV.

1.2-7. The speed of light is 3×10^8 m/s. How long does it take for a light beam to travel 1 inch (2.53 cm)? What is the distance it travels in 1 μs?

1.2-8. A distance is measured to be $4536.454 \times 10^9 \ \mu m$. Round this distance to a precision of three decimal digits and three decimal places.

1.2-9. Show the ammeter connection for measuring the current i in Figure P1.2-3.

Figure P1.2-3

1.2-10. Show the voltmeter connections for measuring v_1 and v_2 in Figure P1.2-4.

Figure P1.2-4

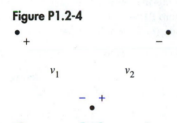

Section 1.3 Charge and Flux Linkage

1.3-1. Suppose that the current $i(t)$ in a conductor has the waveform shown in Figure P1.3-1. Find and graph the charge $q(t)$.

Figure P1.3-1

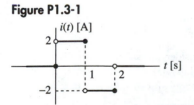

1.3-2. Suppose that the current $i(t)$ in a conductor has the waveform shown in Figure P1.3-2. Find and graph the charge $q(t)$.

Figure P1.3-2

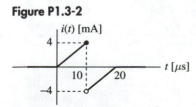

1.3-3. Suppose that the charge $q(t)$ in a conductor has the waveform shown in Figure P1.3-3. Find and graph the current $i(t)$.

Figure P1.3-3

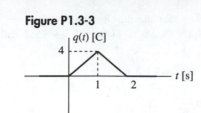

1.3-4. Suppose that the charge $q(t)$ in a conductor has the waveform shown in Figure P1.3-4. Find and graph the current $i(t)$.

Figure P1.3-4

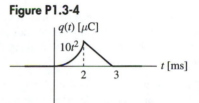

1.3-5. Suppose that the voltage $v(t)$ between two points has the waveform shown in Figure P1.3-5. Find and graph the flux linkage $\lambda(t)$.

Figure P1.3-5

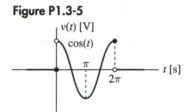

1.3-6. Suppose that the voltage $v(t)$ between two points has the waveform shown in Figure P1.3-6. Find and graph the flux linkage $\lambda(t)$.

Figure P1.3-6

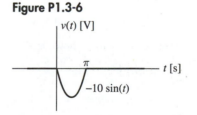

1.3-7. Suppose that the flux linkage $\lambda(t)$ between two points has the waveform shown in Figure P1.3-7. Find and graph the voltage $v(t)$.

Figure P1.3-7

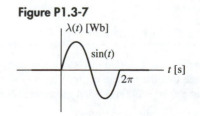

1.3-8. Suppose that the flux linkage $\lambda(t)$ between two points has the waveform shown in Figure P1.3-8. Find and graph the voltage $v(t)$.

Figure P1.3-8

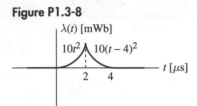

1.3-9. A charge of 2 nC has flowed along a conductor in a given direction prior to $t = 0$. If $i(t) = 4$ mA (dc) in the same direction, find the total charge that has flowed along the conductor in that direction at $t = 2$ μs.

1.3-10. The charge along a conductor is given by $q(t) = 10(t - 2)^2 u(t)$ μC with t in ms. At what time does the current reach a value of 100 mA?

1.3-11. A current $i(t)$ has the waveform shown in Figure P1.3-9. Find and sketch $q(t)$.

Figure P1.3-9

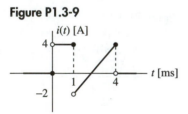

1.3-12. A voltage $v(t)$ has the waveform $v(t) = 10 \cos(5t)u(t)$ nV. Find $\lambda(t)$ in μWb.

1.3-13. If $v(t) = 4e^{-2t}$ mV for $t \geq 0$ and $\lambda(0) = 2$ mWb, find the time t at which $\lambda(t) = 3$ mWb.

1.3-14. If $\lambda(t) = 10 \sin(5t)$ μWb, find $v(t)$.

1.3-15. If $\lambda(t)$ has the waveform shown in Figure P1.3.10, find and sketch $v(t)$.

Figure P1.3-10

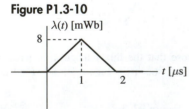

Section 1.4 Power and Energy

1.4-1. Consider the two-terminal subcircuit shown in Figure P1.4-1. Assume that the voltage is given by $v(t) = 10u(t)$ V and the current by $i(t) = -2u(t)$ A. Find the power absorbed by the subcircuit.

Figure P1.4-1

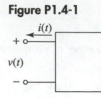

1.4-2. Show voltmeter and ammeter connections for measuring the power absorbed by the electrical load in Figure P1.4-2.

Figure P1.4-2

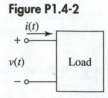

1.4-3. For the load in Figure P1.4-2, $v(t) = u(t)$ V and $i(t) = 10e^{-2t}u(t)$ mA. Find the power absorbed by the load at $t = 0.5$ s. Find and sketch $w(t)$ in mJ. What is the total energy generated by the load?

1.4-4. Assuming the passive sign convention for $v(t)$ and $i(t)$ as shown in Figure P1.4-2, find the energy delivered to the load in terms of $v(t)$ and $q(t)$. Repeat for $i(t)$ and $\lambda(t)$.

1.4-5. Find $\lambda(t)$ and $q(t)$ for Problem 1.4-3.

1.4-6. Express a voltage of 4 V in terms of joules and coulombs. Express it in terms of μJ and mC.

1.4-7. Consider the two-terminal subcircuit shown in Figure P1.4-1. Assume that the voltage is given by $v(t) = -10u(t)$ V and the current by $i(t) = -2u(t)$ mA. Find the power absorbed by the subcircuit.

1.4-8. Consider the two-terminal subcircuit shown in Figure P1.4-1. Assume that the voltage is given by $v(t) = 100 \sin(2t)u(t)$ V and the current by $i(t) = 2u(t)$ μA. Find the power absorbed by the subcircuit.

1.4-9. Consider the subcircuit in Figure P1.4-1. If it satisfies the relationship $v(t) = -2i(t)$ and if $i(t) = 3tu(t)$ A, find the power and energy absorbed for all time.

1.4-10. Find $P(t)$ and $w(t)$ for each of the subcircuits in Figure P1.4-3. For parts (a) and (b), assume that both voltage and current are zero for $t \leq 0$.

Figure P1.4-3

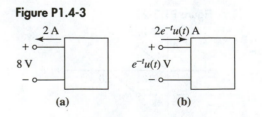

(a)   (b)

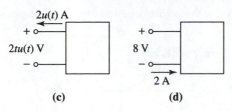

(c)

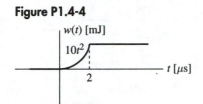

(d)

1.4-11. If the energy absorbed by the subcircuit in Figure P1.4-2 has the time waveform shown in Figure P1.4-4, find the absorbed power $P(t)$. If $v(t) = 10u(t)$ KV, find and sketch $i(t)$.

Figure P1.4-4

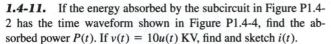

1.4-12. Suppose the load of Figure P1.4-2 is described by $v(t) = \int_{-\infty}^{t} i(\alpha)\, d\alpha$. Show that the load is passive. Assume that all waveforms are one sided, that is, zero for $t \leq 0$.

1.4-13. Suppose that the load of Figure P1.4-2 is described by $v(t) = 2u(t)$ V, independently of $i(t)$. Show that the load is active.

1.4-14. Let the load of Figure P1.4-2 be described by $v(t) = 2di/dt$. Let $i(t)$ have the waveform shown in Figure P1.4-5. Find and sketch $v(t)$, $P(t)$, and $w(t)$. What is $w(\infty)$, the total energy delivered to the load?

Figure P1.4-5

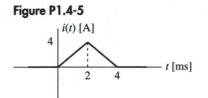

Section 1.5 Independent Sources

1.5-1. Two of the simple circuits in Figure P1.5-1 are disallowed. Which ones and why?

Figure P1.5-1

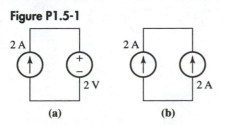

(a)   (b)

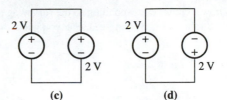

(c)   (d)

The way one describes a voltage source to SPICE is to specialize the general element description that we discussed in Sections 1.1 and 1.2 in these problems. An ideal voltage source must begin with the letter V; furthermore, after the specification of the two nodes between which it is connected, one must specify the value. Thus, one writes

VXXX NPLUS NMINUS VALUE

If the source has a dc (constant) value, one merely writes in this value. If it is a waveform, such as an exponential, a step function, a ramp, or a sinusoid, a more complicated description is required. Here, we will only work with dc values. The convention is that the positive voltage reference of the voltage source is connected to the node named NPLUS, and the negative reference to the one named NMINUS. A sketch of this convention is shown in Figure P1.5-2.

Figure P1.5-2

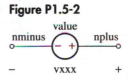

1.5-2. Draw a circuit corresponding to the following netlist and determine the values of v(1), v(3), and v(4, 1).

NETLIST

vs1	1	0	8
vs2	0	2	4
vs3	0	3	-6
vs4	1	4	16
e1	1	2	
e2	3	2	
e3	4	3	

A current source element in SPICE has a name that begins with the letter I. It is specified by

IXXX NPLUS NMINUS VALUE

The only difference between this nomenclature and that of the voltage source is the fact that the reference arrow for the current is assumed to go through the element from the node labeled NPLUS to the one labeled NMINUS. We show this in Figure P1.5-3.

Figure P1.5-3

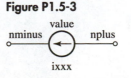

1.5-3. Draw a circuit corresponding to the following netlist and determine the values of i(e3) and i(e2). Can you guess the value of i(e1)?

NETLIST

is1	1	2	10
is2	3	2	10
is3	2	3	2
is4	0	4	8
e1	1	2	
e2	0	3	
e3	1	4	

1.5-4. Write the netlist corresponding to the circuit in Figure P1.5-4. What are the values of v(1, 2) and v(0, 4)?

Figure P1.5-4

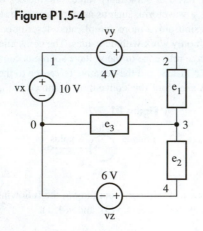

1.5-5. Write the netlist corresponding to the circuit in Figure P1.5-5.

Figure P1.5-5

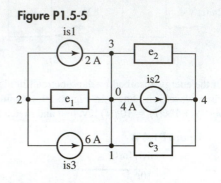

1.5-6. Find the current *I* in Figure P1.5-6.

Figure P1.5-6

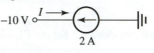

2 Simple Resistive Circuits and Subcircuits

This chapter introduces the resistor circuit element, along with its resistance and conductance. It discusses two basic circuit laws: Kirchhoff's current law (KCL) and Kirchhoff's voltage law (KVL) and their application, along with Ohm's law, in solving simple circuits. Two special types of simple subcircuit are investigated: the series subcircuit and the parallel subcircuit. An important tool for working with the former, the voltage divider rule, is discussed; similarly, the corresponding tool for working with the latter, the current divider rule, is presented. A number of examples are worked in detail to illustrate the application of the basic concepts. Dc analysis using SPICE is covered. Finally, we summarize the basic axioms of circuit analysis.

The prerequisite for this chapter is an understanding of the material in Chapter 1.

2.1 | Ohm's Law

Review of Independent Sources

In Chapter 1 we introduced two basic circuit elements: the independent voltage source and the independent current source. Their symbols and defining *v-i* characteristics are shown in Figure 2.1. Our basic assumption is that these graphs of voltage versus current completely describe these two elements when they are connected into a circuit. The voltage source constrains the voltage to be the given value v_s regardless of the current flowing through it; likewise, the current source constrains the current to be i_s independent of the voltage across it.

Figure 2.1
The independent source elements

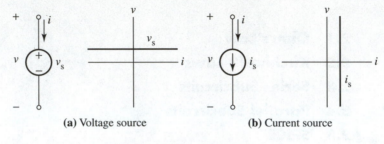

(a) Voltage source   **(b)** Current source

The Resistor Element

The circuit element that we call a *resistor* is similarly specified by a graph of its *v-i* constraint. The *v-i* characteristic and the symbol for a resistor are shown in Figure 2.2. The *v-i* characteristic is a straight line through the origin with a constant slope, which we call the *resistance R*. The equation describing the resistor element is

$$v = R\,i, \tag{2.1-1}$$

which we call *Ohm's law*. The unit for voltage is the volt, V, and that for the current is the ampere, A; therefore, we see that the unit of resistance is V/A, which we call the *ohm*. For this unit, we depart from our usual convention of abbreviating units named after persons with the capitalized first letter of that person's name. For reasons unknown to the author, the ohm is referred to by the (uppercase) Greek letter *omega*, Ω (although this *is* the Greek equivalent for the English 0).

Figure 2.2
The resistor

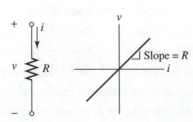

It is important to observe that we have assumed the passive sign convention in defining the *v-i* characteristic of the resistor—a glance back at Figure 2.2 confirms this. Suppose, however, that we define the current in the opposite direction, as shown in Figure 2.3(a). What is the equation for the *v-i* constraint? We have not only shown the new current reference, we have also indicated in Figure 2.3(b) that by reversing the current direction, and thus forcing the passive sign convention to hold, we change its sign. Thus, for either of our two figures, we have

$$v = R(-i) = -Ri. \tag{2.1-2}$$

Figure 2.3   Ohm's law with an alternate current reference

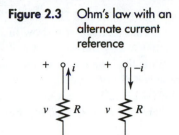

(a)   (b)

Figure 2.4   A nonlinear resistor

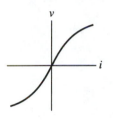

We emphasize, however, that Ohm's law is equation (2.1-1), the one for the passive sign convention on voltage and current. To determine the sign for alternate references, we merely force them into obeying the passive sign convention by affixing negative signs as necessary.

The circuit element that we have called the resistor is an idealization of a physical resistor. This is often a small cylinder made either of wire or a carbon material or a thin film of metal or semiconductor. For such elements the resistance parameter is always positive, $R > 0$. For higher values of voltage and current, a physical resistor exhibits *nonlinear* behavior; that is, its *v-i* characteristic is not a straight line. We show this in Figure 2.4. Such an element is often called a *nonlinear resistor,* though we will insist that our resistor element be linear. Unlike the characteristic for the (linear) resistor shown in Figure 2.2, the one for the nonlinear resistor does not have a constant slope and thus the ratio of *v* to *i* is different at different points on the graph.

It was shown in Chapter 1 that the independent voltage and current sources are active; that is, they can supply energy to an external circuit. The resistor, on the other hand, is passive because it cannot do this. In fact, the absorbed power is

$$P(t) = v(t)i(t) = Ri(t) \times i(t) = Ri^2(t). \tag{2.1-3}$$

The resistance parameter is always positive, so we see that the absorbed power is always positive. This, in turn, means that the energy absorbed is also positive:

$$w(t) = \int_{-\infty}^{t} P(\alpha)\, d\alpha = \int_{-\infty}^{t} Ri^2(\alpha)\, d\alpha > 0. \tag{2.1-4}$$

Thus, the resistor is a passive element. (Note that we cannot integrate $i^2(t)$ unless we know explicitly what the current waveform is—but we do know that the integral is positive.)

The voltage source and the current source are both *noninvertible* elements. Knowledge of the voltage for the voltage source does not permit us to find the current because the voltage is independent of the current and the converse is true for the current source. The resistor, however, is *invertible.* We can solve (2.1-1) for the current, obtaining

$$i = \frac{1}{R} v = G v, \tag{2.1-5}$$

where $G = 1/R$ is called the *conductance.* The unit of conductance, the inverse ohm, is called the *siemens,* named after two German engineering brothers, Werner and William Siemens. Thus, we abbreviate the unit of conductance S. Do not confuse it with lowercase s, the abbreviation for the second.

Equivalent Resistance

The idea of equivalent resistance is closely related to the resistor element. In Figure 2.5 we show a two-terminal subcircuit and one possible *v-i* characteristic. In the case shown the graph is a straight line passing through the origin with a slope R_{eq}. We consider the two-terminal subcircuit, therefore, to be *equivalent* to a resistor, as far as any external elements are concerned, having an *equivalent resistance* R_{eq}. An equivalent resistance, unlike that of a resistor element, can be negative. Indeed, because of the negative slope, this is the case shown in Figure 2.5.

Figure 2.5
A two-terminal subcircuit and equivalent resistance

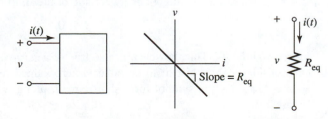

Testing a Subcircuit   How does one determine the *v-i* characteristic of an element or a subcircuit? The answer is simple: one applies a *test source* of either the voltage or current variety, as we show in Figure 2.6. The idea is to vary the applied source to all possible values and record it and the other (response) variable at each point. The resulting graph is then the *v-i* characteristic. If the subcircuit is a 2-Ω resistance element, as we show for example in Figure 2.7, we get a straight line graph going through the origin of the *v-i* plane having a slope of 2 Ω. This is true regardless of whether a voltage source or a current source is used for the test.

Figure 2.6
Testing an element or
subcircuit

Figure 2.7
Testing a resistor
element

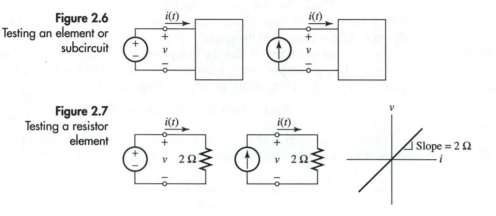

If we try to test an independent voltage source with a test voltage source or an independent current source with a test current source—as we show in Figure 2.8—we will, however, run into problems. Why? Simply because there can only be one voltage between two nodes of a circuit and only one current in a conductor. We can, therefore, not adjust either the voltage source or the current source to all possible values. We must be able to do this in order to determine the complete *v-i* characteristic. Thus, we must pick a test source consistent with subcircuit constraints.

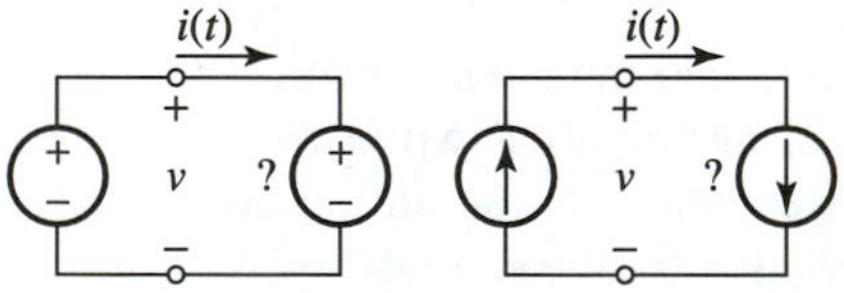

Figure 2.8
Testing sources

The Open Circuit and   There are two *degenerate* circuit elements: the short circuit and the open circuit. They are
the Short Circuit;   special cases of the resistor element. The short circuit is a resistor having zero resistance,
Degenerate Elements   and the open circuit is a resistor having a resistance of infinity. Suppose we look at Ohm's law for the short circuit. We have

$$v = Ri = 0 \tag{2.1-6}$$

if $R = 0$. Thus, the defining *v-i* characteristic of the short circuit is

$$v = 0, \tag{2.1-7}$$

independently of i. This, however, is also true of a *zero-valued voltage source*. Hence, we can consider a short circuit to be either a zero-valued resistor or a zero-valued voltage source. We show the symbol for a short circuit, along with these two interpretations, in Figure 2.9.

Figure 2.9
The short circuit

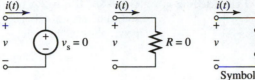

Ohm's law for infinite resistance assumes the form

$$i = \frac{1}{R} v \longrightarrow 0 \qquad (2.1\text{-}8)$$

as $R \to \infty$. Thus, the defining v-i relationship for an open circuit is

$$i = 0, \qquad (2.1\text{-}9)$$

independently of v. Thus, we can also describe the open circuit as a zero-valued current source as well as an infinite-valued resistor (or, equivalently, as a zero-valued conductance). These two interpretations for an open circuit are shown in Figure 2.10.

Figure 2.10
The open circuit

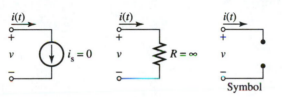

As we have mentioned, the open circuit and the short circuit are considered to be degenerate elements; that is, they are not elements in their own right. They are useful in certain circuit manipulations in which one wishes, for instance, to deactivate (reduce to zero) a voltage source or a current source and compute some variable in the resulting circuit.

Element Summary   We now have a total of three circuit elements: the independent voltage source, the independent current source, and the resistor. The two independent sources are active; the resistor is passive. We will work with only these three elements for some time to come, though we will later add to our catalog of elements. With only these three, however, we can develop essentially all of the methods of circuit analysis, which we will then use to analyze circuits containing other types of elements as well.

Section 2.1 Quiz

Q2.1-1. The element or subcircuit shown in Figure Q2.1-1(a) has the v-i characteristic sketched in Figure Q2.1-1(b). Draw the single circuit element to which it is equivalent.

Q2.1-2. Answer Question Q2.1-1 for the v-i characteristic in Figure Q2.1-2.

Figure Q2.1-2

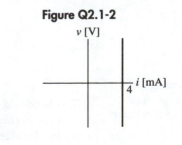

Figure Q2.1-1

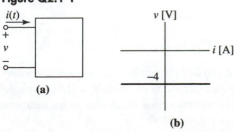

(a)

(b)

Q2.1-3. Answer Question Q2.1-1 for the v-i characteristic in Figure Q2.1-3.

Figure Q2.1-3

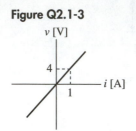

Q2.1-4. Answer Question Q2.1-1 for the v-i characteristic in Figure Q2.1-4.

Figure Q2.1-4

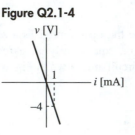

Q2.1-5. A resistor has a resistance of 1 kΩ. Find its conductance in S, mS, μS, and nS.

Q2.1-6. Suppose the subcircuit in Figure Q2.1-1(a) contains a short circuit. Should it be tested with a voltage source or a current source? Explain your answer.

Q2.1-7. Suppose the subcircuit in Figure Q2.1-1(a) contains an open circuit. Should it be tested with a voltage source or a current source? Explain your answer.

Q2.1-8. If $i_s = 2$ A (dc) in Figure Q2.1-5, find the power absorbed by the resistor and by the current source. What is the total power absorbed by both?

Figure Q2.1-5

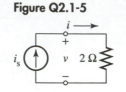

Q2.1-9. Answer Question Q2.1-8 if $i_s = -2$ A.

Q2.1-10. If $i_s(t) = -2u(t)$ A in Figure Q2.1-5, find and sketch $v(t)$, and $P(t)$ and $w(t)$ (absorbed).

Q2.1-11. Find the unspecified variable in each of the circuits in Figure Q2.1-6.

Figure Q2.1-6

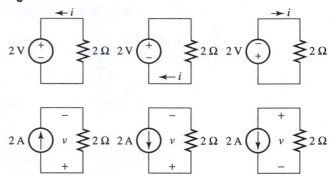

2.2 | Kirchhoff's Laws

In this section we will develop two basic circuit laws that, together with the element relations, form a sufficient set of tools for solving any circuit containing sources and resistors.

Kirchhoff's Current Law (KCL) The first law we will present concerns charge and current. Figure 2.11 shows an ideal conductor in which we have defined a current (and consequently a charge flow) entering one end and another current (and charge) leaving at the other end. We will assume that charge does not "stack up" on the conductor. This means that

$$q_1 = q_2 \qquad (2.2\text{-}1)$$

and

$$i_1 = i_2. \qquad (2.2\text{-}2)$$

Figure 2.11 Current and charge in an ideal conductor

i_1, q_1 ⟶ i_2, q_2 ⟶

This is one manifestation of the general physics principle of *charge conservation*. In conforming to (2.2-1) and (2.2-2) we note that our study of circuits will not include *electrostatics,* in which charge can be placed in isolation on a conductor or an insulator. We are studying the theory of charges in motion, not at rest.

Figure 2.12 Current and charge in a circuit element

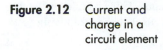
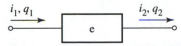

Figure 2.12 shows a similar situation, but one in which the conductor has been replaced by a general circuit element. Here again we assume that equations 2.2-1 and 2.2-2 hold—the charge "stored" inside the element e is zero. We will now say a word or two about signed charges because you have probably studied them in physics. Our assumption is that current is a flow of *positive* charge. In metals the actual particles flowing are electrons, which carry a negative charge. In some types of semiconductor compounds, on the other hand, the moving particles are positively charged. We, however, have based our definition of charge on the fundamental quantity current. Thus, positive charge is defined to be consistent with the direction of current. We will adhere to that definition throughout this text. In another chapter, we will find that a certain element, the *capacitor*, seems to store charge; what it does, however (in terms of the positive and negative charges learned in physics), is to maintain in equilibrium two balanced quantities of equal and oppositely charged particles. This is entirely consistent with our demand that the (positive) charge stored in it be zero.

If the current and charge waveforms are made to vary very rapidly, our assumptions about zero stored charge will no longer be true. Thus, we would not expect the techniques we are developing in this text to be valid at very high frequencies—and this is true. At high frequencies (or for very rapid pulsed waveforms) one must turn to another scientific theory: the theory of electromagnetic waves. This highly mathematical study can be considered as a generalization of circuit theory.

Figure 2.13 shows a closed surface S in three dimensions, which we will refer to as a Kirchhoff surface.[1] Here are our basic assumptions: it is closed in three dimensions and it is penetrated only by conductors, possibly including element leads. More specifically, we do not allow the surface to slice through an element. Though allowing this to occur would not give us problems now, it would create them later on when we study the capacitor. We have shown only four conductors in Figure 2.13, though any number can penetrate S in general.

Figure 2.13
The concept of a Kirchhoff surface

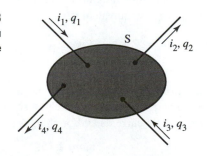

Now look at the current and charge reference arrows associated with each of these conductors. Some of them (i_1, q_1 and i_3, q_3 to be specific) point inward toward the surface S. We will simply use the adjective "in" to represent such a reference pointed in this direction. We will use the term "out" to represent a current-charge reference pointed outward, away from S. The variables i_2, q_2 and i_4, q_4 have such reference directions. Note that these reference arrows are simply specifications for connecting the ammeter to measure the current—the *values* can be either positive or negative. Our assumption is that

---

[1] In electromagnetic theory, it is called a Gaussian surface. For our purposes, however, we will only use such a surface to enclose circuit elements and conductors (which do not store charge). For this reason, we will refer to it as a Kirchhoff surface.

current and charge only flow through conductors and elements. This, coupled with the fact that charge is not stored on a conductor or in an element, leads to the following fact:

$$\sum_S q_{in} = \sum_S q_{out},$$ (2.2-3)

where the subscript "in" refers to a charge reference pointed toward S and "out" refers to an oppositely directed reference. If we differentiate (2.2-3) with respect to time, we have what is called *Kirchhoff's current law,* which we will abbreviate KCL:

$$\sum_S i_{in} = \sum_S i_{out}.$$ (2.2-4)

For the specific set of currents shown in Figure 2.13, we have

$$i_1 + i_3 = i_2 + i_4.$$ (2.2-5)

We remind you that these currents are *variables;* one possible set of *values* is $i_1 = -2$ A, $i_2 = 4$ A, $i_3 = 8$ A, and $i_4 = 2$ A. Alternately, we could have $i_1(t) = \cos(3t)$ A, $i_2(t) = e^{-t}$ A, $i_3(t) = -\cos(3t)$ A, and $i_4(t) = -e^{-t}$ A. Thus, *KCL is assumed to hold for each and every instant of time* for time-varying waveforms.

Figure 2.14 shows the result when we single out the inward reference arrows and turn them around, thus changing the signs of the associated variables. A moment's reflection will convince you that equation (2.2-4) can be written as

$$\sum_S i_{out} = 0.$$ (2.2-6)

We could, of course, just as well have turned around the references pointed outward and changed their signs. The resulting configuration is shown in Figure 2.15. We see that both equations (2.2-4) and (2.2-6) are equivalent to

$$\sum_S i_{in} = 0.$$ (2.2-7)

The corresponding equations in terms of charge are also clearly valid. We will refer to any of them in the future as KCL. For practice, you should write down the explicit form of each of these for the specific currents and/or charges shown in the figures.

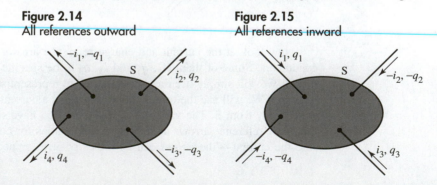

Figure 2.14
All references outward

Figure 2.15
All references inward

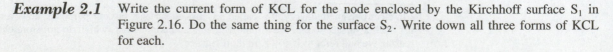

Example 2.1   Write the current form of KCL for the node enclosed by the Kirchhoff surface S_1 in Figure 2.16. Do the same thing for the surface S_2. Write down all three forms of KCL for each.

Figure 2.16
An example circuit

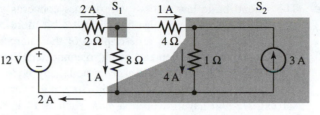

Solution   For S_1, using the "sum of currents in equals sum of currents out" form, we have

$$2 = 1 + 1. \qquad (2.2\text{-}8)$$

Using the "sum of currents out equals zero" form, we have

$$-2 + 1 + 1 = 0. \qquad (2.2\text{-}9)$$

Finally, the "sum of currents in equals zero" form gives

$$2 - 1 - 1 = 0. \qquad (2.2\text{-}10)$$

Clearly, these three forms are equivalent to each other. For S_2, the "sum of currents in equals the sum of the currents out" form is

$$1 + 1 = 2. \qquad (2.2\text{-}11)$$

The "sum of the currents out equals zero" gives

$$-1 - 1 + 2 = 0. \qquad (2.2\text{-}12)$$

Finally, the "sum of the currents in equals zero" yields

$$1 + 1 - 2 = 0. \qquad (2.2\text{-}13)$$

Notice, by the way, that we have actually drawn several of the current reference arrows by the side of elements, rather than by their leads or interconnecting conductors. Now that we know that the current into one lead is identical with the current leaving the other, the element current is unambiguously defined to be that same current. Notice also that currents either totally inside or totally outside the given surface do not appear in the KCL equations; only those that penetrate it appear.

Perhaps a word or two about our hydraulic analogy for current would be appropriate here. Considering the conductors to be water pipes and the elements to be pieces of water operated machinery, we see that our analogy breaks down just a trifle in the charge form of KCL because such physical devices *can* store water. If we assume, however, that the entire system of pipes and machines is completely filled to capacity, then we see that any added water in one or more pipes going through a closed surface must be balanced by water flowing out in other pipes penetrating that same surface. Water is an incompressible fluid. Thus, the current (derivative) form of KCL does have an exact analog: the rate of flow of water into a closed surface (the current) must be balanced by the same rate of flow out of the same surface.

KCL for a general surface, by the way, can be derived from KCL applied only to nodes. If you enclose the two nodes inside the surface S_2 of Example 2.1 and apply KCL to them, you will see that the sum of the two KCL equations is the same as the KCL equation(s) we presented for the surface S_2 itself. In general, KCL for any closed surface is the sum of the KCL equations applied at the enclosed nodes.

Kirchhoff's Voltage Law (KVL)

The second basic law that we will discuss concerns voltages (and, therefore, flux linkages). To develop it, we must first discuss a bit more circuit topology. The topology of a circuit, you will recall, is a description of the interconnections without concern for the specific nature of the elements being so connected. Figure 2.17 shows an example circuit. If we think of starting out at a given node, say node 1, and walking around the circuit along conductors and element leads (stepping over the bodies of the elements themselves), we will have traversed a *path,* such as P_1 in the figure. Technically, we can describe a path as *an ordered sequence of elements, each two successive elements of which share a common node.* Thus, $P_1 = \{e_6, e_1, e_2\}$ is a path. $P_2 = \{e_5, e_4, e_3\}$ is also a path. The first element in a path listing is connected to a node not shared with its successor called the *initial node.* Similarly, the last element is connected to a node not shared with its predecessor called the *final node.*[2] Thus, we see that a path has a direction from the initial node to the final node; hence, we can designate the path by an arrow on the circuit diagram passing close to each element traversed. We see that path P_1 has node 1 as its initial node and node 4 as its final node. Path 2 also has node 1 as its initial node and node 4 as its final node—but the two paths are different.

Figure 2.17
An example circuit

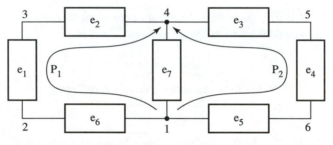

Figure 2.18 shows the same circuit with two element voltages v_x and v_y arbitrarily defined as shown. We will say that v_x is a *voltage rise* and that v_y is a *voltage drop* relative to the path P_1. Neither is a rise nor a drop for P_2 because P_2 does not pass through the corresponding elements. If we use our height analogy for voltage, we see that we climb from a minus upward to a plus when we go along the path through a voltage rise, and go down from the plus to the minus when we traverse a drop. Now look at Figure 2.19, where we have indicated the same voltages v_x and v_y by reversing the plus and minus signs and changing the sign of the voltage variables themselves. This indicates that a negative voltage rise is a voltage drop, and vice versa.

Figure 2.18
Voltage rise and drop

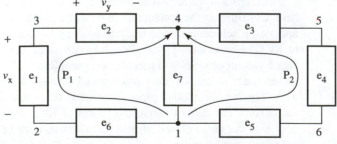

---

[2] Our definitions of initial and final nodes break down for paths consisting of only two elements connected between the same two nodes. We do not wish to complicate our definition to take care of such a simple technicality, so we will use the definitions we have given here.

Figure 2.19
Negative rises and
drops

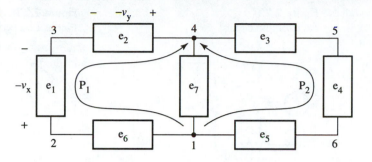

We can now state *the path form of Kirchhoff's voltage law,* which we abbreviate for convenience as KVL: *if two paths P_1 and P_2 have the same initial node and the same final node, then the sums of the voltage rises along the two paths are the same.* Thus, in Figure 2.20, we can write

$$v_x - v_y = v_r + v_s. \tag{2.2-14}$$

Figure 2.20
KVL example

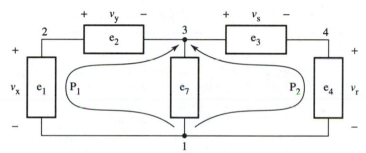

The voltage rise along the two paths must be the same. Think of the old adage about the Roman Empire: "all roads lead to Rome." No matter what the path, we must scale the same total height (see Figure 2.21). If we merely integrate equation (2.2-14) with respect to time, we obtain the *flux linkage form* of KVL:

$$\lambda_x - \lambda_y = \lambda_r + \lambda_s. \tag{2.2-15}$$

Figure 2.21
Height analogy for KVL

Example 2.2   Find the voltage v in Figure 2.22 by using KVL.

Solution   We have redrawn the circuit in Figure 2.23 and numbered the nodes for convenience. We have also defined two paths: one starting from node 1 and going directly across the current source to node 2, and the other across the three resistors immediately to the right of the current source. We picked the first path because the only element voltage along it was the unknown v; we picked the other for two reasons: first, because it has the same initial and final nodes and, second, because all the element voltages along this path are known. All three of these voltages are rises relative to P_2. Thus, we can immediately write KVL in the form

$$v = 4 + 2 + 8 = 14 \text{ V}. \tag{2.2-16}$$

Figure 2.22
An example circuit

Figure 2.23
Nodes and paths defined

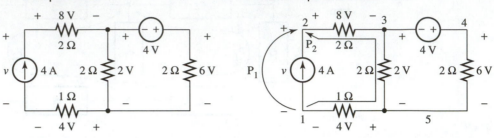

We could also have picked our path P_2 to go through the same 1-Ω resistor at the bottom, but this time to go through the 2-Ω resistor on the far right and the voltage source, then back through the top 2-Ω resistor. Noting that the voltage source is a drop, we would then have gotten

$$v = 4 + 6 - 4 + 8 = 14 \text{ V}. \tag{2.2-17}$$

Again, we see that "all roads lead to Rome!"

Just as there are several alternate forms of KCL, there are several alternate forms of KVL as well. To discuss them, we need the concept of a loop. A *loop* is a specialized type of path whose initial and final nodes are identical. Thus, the path P shown in Figure 2.24 is a loop. We can write its listing of elements in a number of ways; one way to do this is $P = \{e_4, e_3, e_7, e_5\}$. In this case node 6 is both the initial node and the final node. We could have started with any of the other elements in the listing, and the node it shares with the last in the list would become both the initial and the final node.

Figure 2.24
A loop

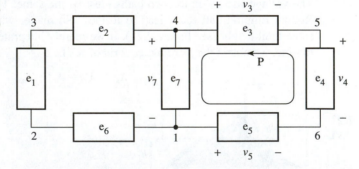

Relative to the loop P, voltages v_4 and v_3 are rises and v_7 and v_5 are drops. Here is the way we will phrase the *loop forms* of KVL:

a. The sum of the voltage rises around any loop is zero.

b. The sum of the voltage drops around any loop is zero.

These statements are assumed to hold for any instant of time if the voltages change with time. We also remind you that a negative rise is a drop, and vice versa. Thus, for our example, we can write (form a)

$$v_4 + v_3 - v_7 - v_5 = 0, \tag{2.2-18}$$

and (form b)

$$-v_4 - v_3 + v_7 + v_5 = 0. \tag{2.2-19}$$

Let's once more use our height analogy for voltage. Figure 2.25 shows the same mountain climb with which we illustrated our first form of KVL, which we termed its *path form*. If we begin at the base of the mountain, climb to the peak along one route, then descend to our starting point via another, the sum of the gains in height must be zero because we return to our starting point; alternatively, the sum of the losses in height must be zero for the same reason. In the first form, we would have $h_1 + h_2 - h_3 = 0$. In the second form, $-h_1 - h_2 + h_3 = 0$.

Figure 2.25
Height analogy for the loop form of KVL

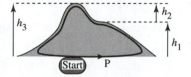

It will be well for you to keep these analogies in mind as we progress.

Section 2.2 Quiz

Q2.2-1. Find i_x in Figure Q2.2-1.

Figure Q2.2-1

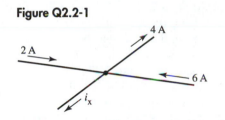

Q2.2-2. Find i_x in Figure Q2.2-2.

Figure Q2.2-2

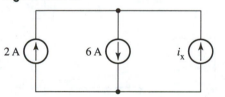

Q2.2-3. Find i_x and i_y in Figure Q2.2-3. (Note that the complete circuit is not shown.)

Figure Q2.2-3

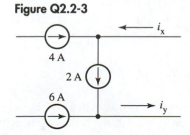

Q2.2-4. Find i_x and i_y in Figure Q2.2-4.

Figure Q2.2-4

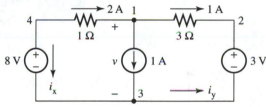

Q2.2-5. Write the KCL equation at node 1 in Figure Q2.2-4 in the form $\sum_{\text{node 1}} i_{\text{in}} = 0$.

Q2.2-6. Repeat Question Q2.2-5 for the form $\sum_{\text{node 1}} i_{\text{out}} = 0$.

Q2.2-7. Repeat Question Q2.2-5 for the form $\sum_{\text{node 1}} i_{\text{in}} = \sum_{\text{node 1}} i_{\text{out}}$.

Q2.2-8. Find v_x in Figure Q2.2-5. (Remember that the ground symbols refer to a single node!)

Q2.2-9. Find v in Figure 2.2-4. Use the path form of KVL.

Figure Q2.2-5

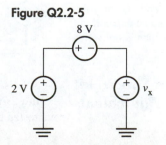

Q2.2-10. Write KVL for the indicated loop in Figure 2.2-6 in the form $\sum\limits_{\text{loop 1}} v_{\text{rises}} = 0$. Note that you must also apply Ohm's law to determine the resistor voltages.

Figure Q2.2-6

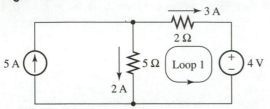

Q2.2-11. Write KVL for the indicated loop in Figure 2.2-6 in the form $\sum\limits_{\text{loop 1}} v_{\text{drops}} = 0$. Note that you must also apply Ohm's law to determine the resistor voltages.

Q2.2-12. Write KVL for the indicated loop in Figure 2.2-6 in the form $\sum\limits_{\text{loop 1}} v_{\text{rises}} = \sum\limits_{\text{loop 1}} v_{\text{drops}}$. Note that you must also apply Ohm's law to determine the resistor voltages. (This is a form that we did not present in the body of the lesson—but you can figure it out, can't you?)

2.3 | Series Subcircuits

Review of Series Subcircuits

A *series subcircuit,* we recall, is one with two terminals and only one path within the subcircuit between these two terminals. Such a subcircuit is shown in Figure 2.26(a). A series subcircuit can have any number of elements, though we have only shown five in our figure. In Figure 2.26(b) we show how we might test the subcircuit with a current source. As there is only one path within the subcircuit, *the addition of the current source creates a series circuit: a circuit having exactly one loop.* By KCL, the current in each element is the same as that in any other and is equal to i in the direction shown by the arrow. Some texts define a series circuit (or subcircuit) as one in which all elements carry the same current. This, however, is somewhat ambiguous, as Figure 2.27 shows. The two currents are the same. If we change one of the voltage source values (say the 4-V one), however, the associated current becomes different from the other. Clearly the collection of all the elements does not form a series circuit. Therefore, we will use our preceding "single path" definition.

Figure 2.26
A series subcircuit

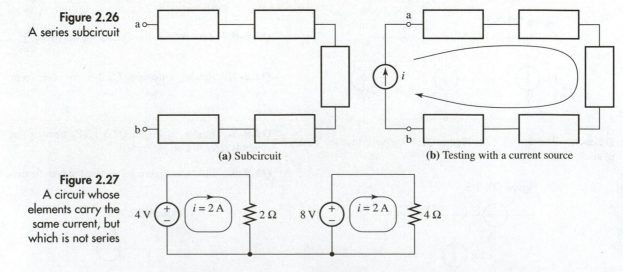

(a) Subcircuit   (b) Testing with a current source

Figure 2.27
A circuit whose elements carry the same current, but which is not series

Series Subcircuit Equivalents

Now let's have a look at some specific types of series subcircuits. The one we show in Figure 2.28(a) consists of voltage sources only. We ask the following question. Suppose we tested the v-i characteristic of this subcircuit relative to its two terminals a and b. What would be the result? Using KVL (the easiest version is perhaps the path form), we find that

$$v = v_{s1} + v_{s2} \tag{2.3-1}$$

independently of i! This, however, is simply the *v-i* characteristic of a single voltage source with value

$$v_{eq} = v_{s1} + v_{s2}. \tag{2.3-2}$$

We therefore say that our original subcircuit is *equivalent* to the simpler single voltage source subcircuit shown in Figure 2.28(b).

Figure 2.28
Series voltage sources

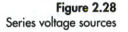
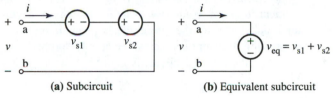

(a) Subcircuit   (b) Equivalent subcircuit

Figure 2.29 A subcircuit with a grounded internal node

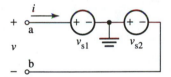

This idea of equivalence plays a major role in circuit analysis and its applications. One of the basic assumptions of circuit analysis is that *an element or a subcircuit can affect other elements in the circuit in which it is located only by means of its terminal v-i characteristic.* Thus, two elements or subcircuits are equivalent if they have the same *v-i* characteristic because they result in exactly the same voltages and currents *in the rest of the circuit.* Now these italics are important. To see why, just notice that our two voltage sources in Figure 2.28(a) have disappeared and have been replaced by only one. This means that the node between them is lost. Suppose that this node were our ground reference in the original circuit in which the two sources are located. We show this by the ground symbol in Figure 2.29. When we combine the two sources, we lose our ground reference. In fact, however, a moment of reflection will convince you that this choice of ground reference actually converts the subcircuit into a *three-terminal* one because it now shares three nodes with the external circuit, and, according to our definition, a series subcircuit must be a two-terminal one. In the future, by the way, we will occasionally use the shorthand expression "v-source" for a voltage source.

A series subcircuit consisting of only current sources (which we will often term "i-sources") is shown in Figure 2.30(a). To what, if anything, is it equivalent? Let's first have a look at the "if anything" part of our question. Suppose we apply KCL at the common node between the two current sources. The result is

$$i_{s1} = i_{s2}. \tag{2.3-3}$$

Therefore—if the two source values are not equal—the circuit is disallowed. In practice, our models of the two elements as ideal current sources would not be valid; in fact, one would never wittingly construct a circuit in this fashion, though it could occur in practice as an equivalent circuit under some set of conditions. Assuming that the two sources have equal value, we know that this is the same as the terminal current *i*, for there is only one current in a series subcircuit. Thus, we obtain the equivalent subcircuit shown in Figure 2.30(b).

Figure 2.30
Series current sources

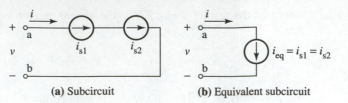

(a) Subcircuit   (b) Equivalent subcircuit

The series subcircuit in Figure 2.31(a) contains a voltage source and a current source. Is it, like the previous two, equivalent to a single element? It is, because the current i is given by

$$i = i_{eq} = i_s \qquad (2.3\text{-}4)$$

independently of the terminal voltage v. Thus, our subcircuit is equivalent to the single element (current source) subcircuit shown in Figure 2.31(b). In fact, we can extend our logic to the more general subcircuit in Figure 2.32(a). *Any* element[3] in series with a current source can be simply "pruned away" and replaced by a short circuit as far as the external circuit is concerned, thus giving the current source equivalent in Figure 2.32(b).

Figure 2.31
Series current and
voltage sources

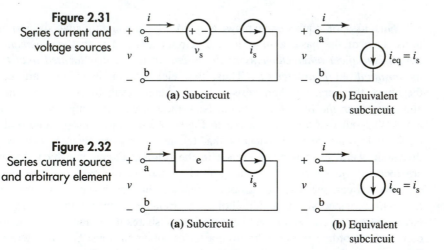

(a) Subcircuit   (b) Equivalent
                                     subcircuit

Figure 2.32
Series current source
and arbitrary element

(a) Subcircuit   (b) Equivalent
                                     subcircuit

Although we have only looked at subcircuits with two elements, the extension to any number of elements is easy. If there is a single current source, it "wins" and the equivalent is that of a single current source having the same value. If there are two or more current sources, they must have the same values so that KCL is not violated. If the subcircuit contains only voltage sources, the equivalent is a single voltage source whose value is the algebraic sum of the values of the individual voltage sources.

The Series Resistor
Subcircuit and
Voltage Division

Now let's investigate a very important special case: that in which all of the elements are resistors. We will first assume that there are two, as we show in Figure 2.33(a). An application of KVL shows at once that

$$v = v_1 + v_2. \qquad (2.3\text{-}5)$$

---

[3] To be more exact, any element that does not violate KCL at the common node.

Figure 2.33
Series resistors

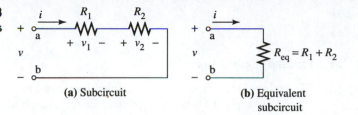

(a) Subcircuit   (b) Equivalent
                                      subcircuit

According to KCL, though, we know that both resistors carry the same current i (which is, by the way, consistent with the voltages and the passive sign convention). Thus, by Ohm's law, we have

$$v = R_1 i + R_2 i = (R_1 + R_2)i = R_{eq}i, \tag{2.3-6}$$

where

$$R_{eq} = R_1 + R_2. \tag{2.3-7}$$

The result is that the v-i characteristic is the same as that for a single resistor having a value that is the sum of the two individual resistances. We show the equivalent resistor in Figure 2.33(b).

Ignoring our equivalent resistance for the moment, let's go back to Figure 2.33(a) and concentrate on the original subcircuit. There is a rule for finding the voltage across each resistor that is important in solving circuit problems. We write

$$v_1 = R_1 i. \tag{2.3-8}$$

If we solve (2.3-6) for i, however, and use it in (2.3-8), we obtain

$$v_1 = \frac{R_1}{R_1 + R_2} v. \tag{2.3-9}$$

A similar relation holds for v_2:

$$v_2 = \frac{R_2}{R_1 + R_2} v. \tag{2.3-10}$$

In words, the voltage across a given resistor is the ratio of that resistance value to the sum of the two multiplied by the terminal voltage. Equations (2.3-9) and (2.3-10) are called *voltage division,* or the *voltage divider rule.*

The preceding analysis can be easily extended to the case of N resistors in a series subcircuit. We leave the analysis to you, but show the resulting equivalence in Figure 2.34. The voltage divider rule takes the form

$$v_k = \frac{R_k}{R_1 + \cdots + R_k + \cdots + R_N} v. \tag{2.3-11}$$

Figure 2.34
N series resistors

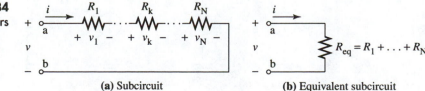

(a) Subcircuit   (b) Equivalent subcircuit

Series Resistor Examples

Example 2.3   Find the voltage v in Figure 2.35.

Solution   We can apply the theory of series subcircuits we have learned, but to which subcircuit should we apply it? If we include the resistor in such a subcircuit, then apply an equivalence, we will lose the physical identity of the voltage across it—and this is what we are looking for. Therefore, let's apply a series equivalence to the subcircuit consisting of the v-source and the i-source, as we show in Figure 2.36. Noting that the voltage source can be replaced by a short circuit (a conductor), we see that we have merely an equivalent current source connected in series with a resistor. This is shown in Figure 2.37. If we are to apply Ohm's law with a positive sign for the resistance, we must assure ourselves that the resistor current is pointed into the resistor terminal with the plus sign for the voltage reference. Applying KCL at the top node in our circuit, we get

$$i = -2\,\text{A}. \tag{2.3-12}$$

Thus,

$$v = 3\,\Omega \times (-2\,\text{A}) = -6\,\text{V}. \tag{2.3-13}$$

Figure 2.35   An example series circuit

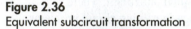

Figure 2.36
Equivalent subcircuit transformation

Figure 2.37
Defining the resistor current

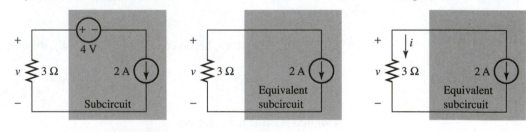
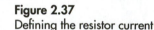

We were somewhat "tricky" in the last example. We purposely made the current source have a direction opposite to that of the passive sign convention for the resistor to illustrate the thought process involved. We will take a facility with this type of sign manipulation for granted in what follows, so you should make sure you understand it very thoroughly here. Furthermore, we will not always draw nice boxes around our subcircuits as we did in the last example. Thus, you should practice doing this until it becomes second nature. Here's another example to help you practice these manipulations.

Example 2.4   Find the voltage v in each of the simple series circuits in Figure 2.38.

Figure 2.38
Two example circuits

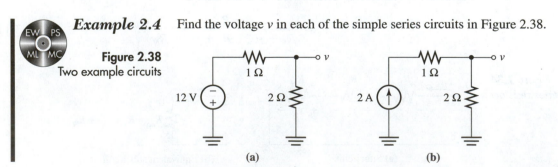

(a)   (b)

Solution   Remember that the two ground reference symbols refer to a single node. To emphasize the topology of the circuits, we have redrawn both of them in Figure 2.39. The two terminals, by the way, have no effect on the circuit. They serve only to identify where the voltage is to be defined. This is a common convention for circuit analysts and designers, so you should become used to it. Nothing is connected between these two terminals external to the subcircuit, so there is no current in their associated conductors; hence, both circuits are series circuits. For the circuit in Figure 2.39(a), we apply the voltage divider rule to get

$$v = \frac{2}{2 + 1} \times (-12) = -8 \text{ V} \qquad (2.3\text{-}14)$$

For the circuit in part (b), we note that the subcircuit consisting of the 1-Ω resistor and the current source can be replaced by an equivalent subcircuit consisting of the current source alone. Thus, we have

$$v = 2 \text{ A} \times 2 \, \Omega = 4 \text{ V}. \qquad (2.3\text{-}15)$$

Figure 2.39
The same two
circuits redrawn

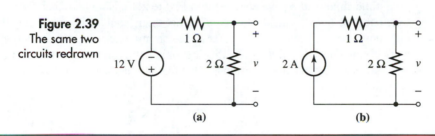

(a)   (b)

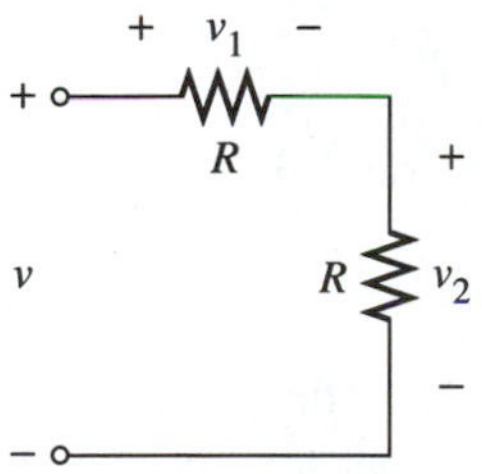

It often happens that a series resistor subcircuit consists of two resistors having equal values. In this case, shown in Figure 2.40, one has

Figure 2.40   Series subcircuit
with equal-valued
resistors

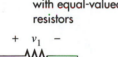

$$v_1 = \frac{R}{R + R} v = \frac{v}{2}, \qquad (2.3\text{-}16)$$

and likewise for v_2.

Another common occurrence is that in which the conductances of the resistors are given rather than the resistances. In this case, one can divide the top and bottom of the voltage divider expression in equation (2.3-9) by $R_1 R_2$ to obtain

$$v_1 = \frac{G_2}{G_1 + G_2} v. \qquad (2.3\text{-}17)$$

Likewise, equation (2.3-10) becomes

$$v_2 = \frac{G_1}{G_1 + G_2} v. \qquad (2.3\text{-}18)$$

Figure 2.41   An example
subcircuit

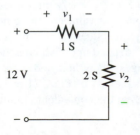

Thus, using conductances, one can express the voltage divider rule in the form "the conductance of the *other* resistor divided by the sum of the two conductances times the terminal voltage." Figure 2.41 shows an example. Notice that there must be additional elements not shown that produce the 12 V across the terminals. This is another convention to which you should become accustomed—showing only a subcircuit with a voltage across its terminals and/or a current into one of them. In this case, we can apply the voltage divider rule in the conductance form to get

$$v_1 = \frac{2}{2 + 1} \times (-12) = 8 \text{ V} \qquad (2.3\text{-}19)$$

and

$$v_2 = \frac{1}{2+1} \times 12 = 4 \text{ V} \qquad (2.3\text{-}20)$$

Example 2.3   Find the values of i and v in Figure 2.42.

Figure 2.42
An example subcircuit

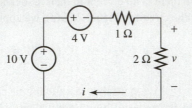

Solution   To find the current, we can simply note that the two v-sources form a series subcircuit, as do the two resistors. Replacing them by their single element equivalent subcircuits, we obtain the simplified equivalent circuit in Figure 2.43. Note here that we have for the first time used the term "equivalent circuit," rather than "equivalent subcircuit." It is, however, only equivalent to the original insofar as the circuit external to the two equivalent subcircuits is concerned. But this means that it is only valid for finding the current i (or the voltage between the two pairs of terminals, but that was not asked for.) We easily compute the current to be

$$i = \frac{6 \text{ V}}{3 \text{ } \Omega} = 2 \text{ A}. \qquad (2.3\text{-}21)$$

Figure 2.43
The two subcircuits

v-source subcircuit

Resistive subcircuit

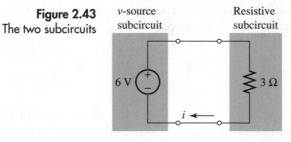

In the process of forming the two equivalent subcircuits, we lost the physical identity of the voltage v for which we were asked. We know, however, that the current i in the original circuit of Figure 2.42 is the same value as that which we just found. Thus, we can return to the original circuit and use this value to compute v:

$$v = 2i = 4 \text{ V}. \qquad (2.3\text{-}22)$$

Let's play around a little with the circuit in this last example. Suppose we choose a ground reference at the bottom of the 10-V v-source in Figure 2.42. The result is shown in Figure 2.44. We will leave it to you to use Ohm's law and KVL to verify the values of the node voltages shown. If we choose another reference node, the values of these node voltages will change. Again, we will leave it to you to verify the values shown in Figure 2.45. Not only do we see that the values of the node voltages depend upon which node is chosen as the reference, but we lose any of these node voltages when their nodes are included inside an equivalent subcircuit. Thus, the voltage at the top of the 2-Ω resistor and at the top of the 10-V v-source disappeared when we drew the equivalent circuit of Figure 2.43.

Figure 2.44
The example circuit with one choice for the reference node

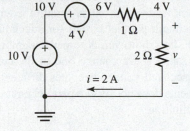

Figure 2.45
The example circuit with another choice for the reference node

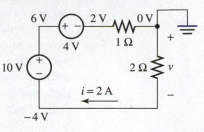

Section 2.3 Quiz

Q2.3-1. Find v_x in Figure Q2.3-1.

Figure Q2.3-1

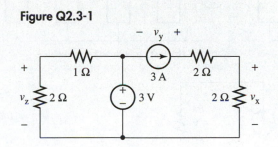

Figure Q2.3-3

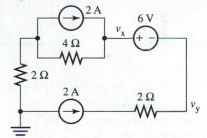

Q2.3-2. Find v_y in Figure Q2.3-1.

Q2.3-3. Find v_z in Figure Q2.3-1.

Q2.3-6. Find v_x in the circuit shown in Figure Q2.3-4.

Figure Q2.3-4

Q2.3-4. Find an equivalent subcircuit for the subcircuit shown in Figure Q2.3-2.

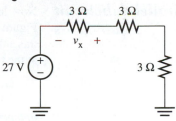

Figure Q2.3-2

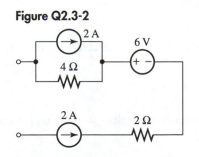

Q2.3-7. Find v_x in the circuit shown in Figure Q2.3-5.

Figure Q2.3-5

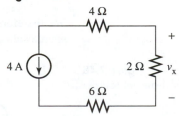

Q2.3-5. Find the node voltage v_x in the circuit shown in Figure Q2.3-3. (*Hint:* Use equivalent subcircuits, Ohm's law, KCL, and KVL. This is not a simple series circuit!)

2.4 | Parallel Subcircuits

Review of Parallel Subcircuits A *parallel subcircuit* is one with precisely two nodes (those attached to its two terminals); there can be many paths between its two terminals. Such a subcircuit is shown in Figure 2.46(a). A parallel subcircuit can have any number of elements, though we have only shown four in our figure. In Figure 2.46(b) we show how we might test the subcircuit with a voltage source; as there are still only two nodes, the addition of the voltage source creates *a parallel circuit: a circuit having exactly two nodes.*[4] By KVL, the voltage across each element is the same and is equal to *v*, with the polarities defined by the v-source. Some texts define a parallel circuit (or subcircuit) as one in which all elements have the same voltage. This, however, is somewhat ambiguous, as Figure 2.47 shows. The two voltages are the same. If we change one of the current source values (say the 2-A one), however, the associated voltage becomes different from the other. Clearly, the collection of all the elements does not form a parallel circuit. Therefore, we will use our preceding "two-node" definition.

Figure 2.46
A parallel subcircuit

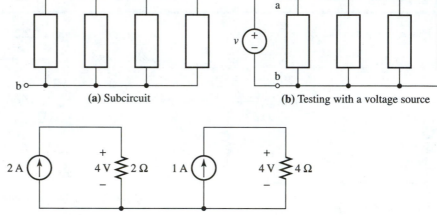

(a) Subcircuit   (b) Testing with a voltage source

Figure 2.47
A circuit whose
elements have the same
voltage, but which is
not parallel

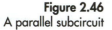

Parallel Subcircuit Equivalents

Now let's have a look at some specific types of parallel subcircuits. The one we show in Figure 2.48(a) consists of current sources only. Suppose we tested the *v-i* characteristic of this subcircuit relative to its two terminals, a and b. What would be the result? Using KCL at the top node, we find that

$$i = i_{s1} + i_{s2} \tag{2.4-1}$$

independently of v! This, however, is simply the *v-i* characteristic of a single current source with value

$$i_{eq} = i_{s1} + i_{s2}. \tag{2.4-2}$$

We therefore see that our original subcircuit is equivalent to the simpler single current source subcircuit shown in Figure 2.48(b).

Figure 2.48
Parallel current sources

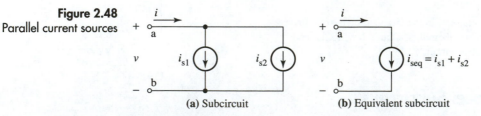

(a) Subcircuit   (b) Equivalent subcircuit

---

[4] Notice that a parallel subcircuit, unlike the series subcircuit, is *already* a circuit before application of the external source.

A parallel subcircuit consisting of only voltage sources is shown in Figure 2.49(a). To what, if anything, is it equivalent? Let's first have a look at the "if anything" part of our question. Suppose we apply KVL in path form between the two nodes. The result is

$$v_{s1} = v_{s2}. \tag{2.4-3}$$

Therefore—if the two source values are not equal—the circuit is disallowed. In practice, our models of the two elements as ideal voltage sources would not be valid; in fact, one would never wittingly construct a circuit in this fashion—though it could occur in practice as an equivalent circuit under some set of conditions. Assuming that the two sources have equal value, we know that it is the same as the terminal voltage v, for there is only one voltage (or, of course, its negative) in a parallel subcircuit. Thus, we obtain the equivalent subcircuit shown in Figure 2.49(b).

Figure 2.49
Parallel voltage sources

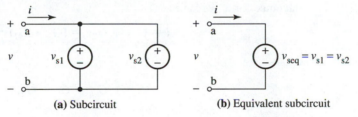

(a) Subcircuit   (b) Equivalent subcircuit

The parallel subcircuit in Figure 2.50(a) contains a voltage source and a current source. Is it, like the previous two, equivalent to a single element? It is, because the voltage v is given by

$$v = v_s \tag{2.4-4}$$

independently of the terminal current i. Thus, our subcircuit is equivalent to the single element (voltage source) subcircuit shown in Figure 2.50(b). In fact, we can extend our logic to the more general subcircuit in Figure 2.51(a). *Any* element[5] in parallel with a voltage source can be simply "pruned away" and replaced by an open circuit as far as the external circuit is concerned, thus giving the voltage source equivalent in Figure 2.51(b).

Figure 2.50
Parallel voltage and
current sources

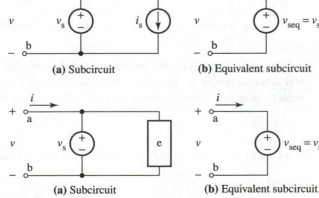

(a) Subcircuit   (b) Equivalent subcircuit

Figure 2.51
Parallel voltage source
and arbitrary element

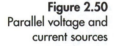

(a) Subcircuit   (b) Equivalent subcircuit

---

[5] To be more exact, any element that does not violate KVL (between the two nodes in, say, path form).

Although we have only looked at subcircuits with two elements, the extension to any number of elements is easy. If there is a single voltage source, it "wins" and the equivalent is that of a single voltage source having the same value. If there are two or more voltage sources, they must have the same values so that KVL is not violated. If the subcircuit contains only current sources, the equivalent is a single current source whose value is the algebraic sum of the values of the individual current sources.

The Parallel Resistor Subcircuit and Current Division

Now let's investigate a very important special case: that in which all of the elements are resistors. We will first assume that there are two, as we show in Figure 2.52(a). An application of KCL shows at once that

$$i = i_1 + i_2. \tag{2.4-5}$$

According to KVL, though, we know that both resistors have the same voltage v (which is consistent with the currents and the passive sign convention). Thus, by Ohm's law in conductance form, we have

$$i = G_1 v + G_2 v = (G_1 + G_2) v = G_{eq} v, \tag{2.4-6}$$

where

$$G_{eq} = G_1 + G_2. \tag{2.4-7}$$

The net result is that the v-i characteristic is the same as that for a single resistor having a conductance that is the sum of the two individual conductances. We show the equivalent resistor in Figure 2.52(b). We have labeled it with the equivalent resistance, which we obtain from (2.4-7) by inversion:

$$R_{eq} = \frac{1}{G_{eq}} = \frac{1}{G_1 + G_2} = \frac{1}{\dfrac{1}{R_1} + \dfrac{1}{R_2}} = \frac{R_1 R_2}{R_1 + R_2}. \tag{2.4-8}$$

Figure 2.52
Parallel resistors

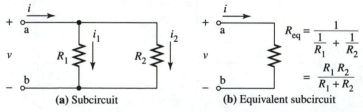

(a) Subcircuit   (b) Equivalent subcircuit

Ignoring our equivalent resistance for the moment, let's go back to Figure 2.52(a) and concentrate on the original subcircuit. There is a rule for finding the current through each resistor that is very important in solving circuit problems. We simply write

$$i_1 = G_1 v. \tag{2.4-9}$$

If we solve (2.4-6) for v and use the result in (2.4-9), however, we can write this as

$$i_1 = \frac{G_1}{G_1 + G_2} i. \tag{2.4-10}$$

A similar relation holds for i_2:

$$i_2 = \frac{G_2}{G_1 + G_2}i. \tag{2.4-11}$$

In words, the current through a given resistor is the ratio of its conductance value to the sum of the two conductances. Equations (2.4-10) and (2.4-11) are called *current division*, or the *current divider rule*. If we divide the top and bottom of these equations by $G_1 G_2$, we obtain the resistance form, which is often preferred:

$$i_1 = \frac{R_2}{R_1 + R_2}i \tag{2.4-12}$$

and

$$i_2 = \frac{R_1}{R_1 + R_2}i. \tag{2.4-13}$$

In this case, the *opposite* resistance value goes in the numerator.

The preceding analysis can be easily extended to the case of N parallel resistors. We leave the analysis to you, but show the resulting equivalence in Figure 2.53. The equivalent conductance is the sum of the individual conductances. The current divider rule takes the form

$$i_k = \frac{G_k}{G_1 + \cdots + G_k + \cdots + G_N}i. \tag{2.4-14}$$

Figure 2.53
N parallel resistors

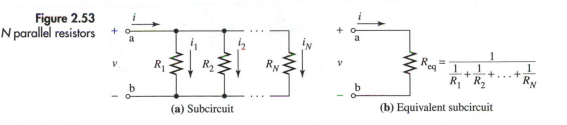

(a) Subcircuit   (b) Equivalent subcircuit

Parallel Resistor Examples

Example 2.6   Find the current i in Figure 2.54.

Figure 2.54
An example parallel circuit

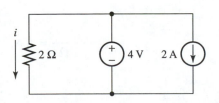

Solution   We can apply the theory of parallel subcircuits we have learned, but to which subcircuit should we apply it? If we include the resistor in such a subcircuit, then apply an equivalence, we will lose the physical identity of the current through it—and this is what we are looking for. Therefore, let's apply a parallel equivalence to the subcircuit consisting of the v-source and the i-source, as we show in Figure 2.55.

Figure 2.55
Equivalent subcircuit
transformation

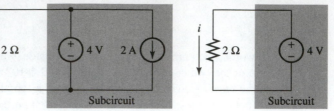

Subcircuit   Subcircuit

Noting that the current source can be removed (replaced by an open circuit, an element that does not violate KVL) we see that we have only an equivalent voltage source connected in parallel with a resistor. A single application of Ohm's law gives

$$i = \frac{4\text{ V}}{2\ \Omega} = 2\text{ A}. \qquad (2.4\text{-}15)$$

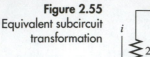

Example 2.7   Find the current i in each of the simple parallel circuits in Figure 2.56.

Figure 2.56
Two example circuits

12 V   1 Ω   2 Ω   6 A   1 Ω   2 Ω

(a)   (b)

Solution   Remember that the three ground reference symbols refer to a single node. To emphasize the topology of the circuits, we have redrawn both of them in Figure 2.57. Both circuits are parallel ones. In the circuit in Figure 2.57(a), we notice that the voltage source and 1-Ω resistor can be replaced by the voltage source alone—as far as the 2-Ω resistor is concerned. We can therefore draw the equivalent circuit shown in Figure 2.58. This circuit is equivalent to the original *as far as the 2-Ω resistor is concerned*. We have redefined the current direction to be consistent with the passive sign convention. Ohm's law now gives

$$-i = \frac{12\text{ V}}{2\ \Omega} = 6\text{ A}, \qquad (2.4\text{-}16)$$

so

$$i = -6\text{ A}. \qquad (2.4\text{-}17)$$

For the circuit in part (b), we merely use the current divider rule:

$$i = \frac{1\ \Omega}{1\ \Omega + 2\ \Omega} \times 6\text{ A} = 2\text{ A}. \qquad (2.4\text{-}18)$$

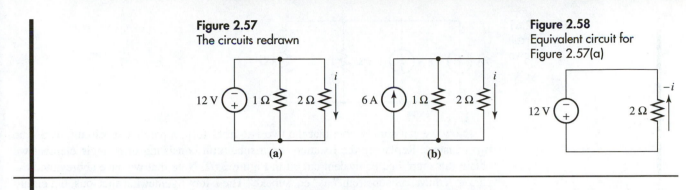

Figure 2.57
The circuits redrawn

(a)   (b)

Figure 2.58
Equivalent circuit for
Figure 2.57(a)

When applying the current divider rule, you should always keep in mind our water flow analogy: water flowing into a "tee" junction through one pipe results in water flow out through the other two. This is a useful image to use in order to keep the signs straight. If the 6-A current source had been reversed in Figure 2.56(b), the result would have been $i = -2$ A.

It often happens that a parallel resistor subcircuit consists of two resistors having equal values. In this case, shown in Figure 2.59, one has

Figure 2.59   Parallel subcircuit
with equal-valued
resistors

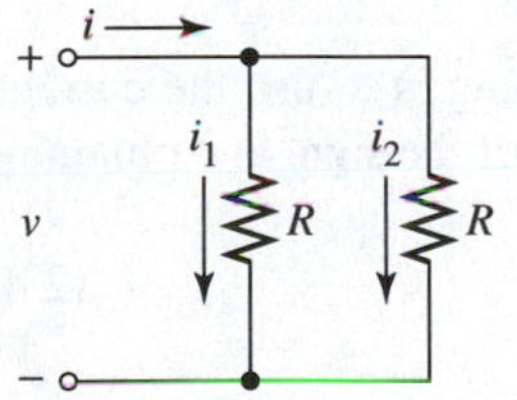

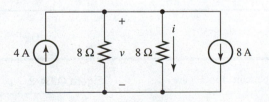

$$R_{eq} = \frac{RR}{R + R} = \frac{R}{2},$$ (2.4-19a)

$$i_1 = \frac{G}{G + G} i = \frac{i}{2},$$ (2.4-19b)

and likewise for i_2. In general, for N parallel equal-valued resistors, one has

$$R_{eq} = \frac{1}{\dfrac{1}{R} + \dfrac{1}{R} + \cdots + \dfrac{1}{R}} = \frac{R}{N}$$ (2.4-20a)

and

$$i_k = \frac{G}{G + G + \cdots + G_i} = \frac{i}{N}.$$ (2.4-20b)

Example 2.8   Find the values of i and v in Figure 2.60.

Figure 2.60
An example circuit

4 A   8 Ω v 8 Ω i 8 A

Solution   This example is deceptively simple. The solution itself is simple, but in carrying it out, we must use some of the basic circuit concepts we have learned. We recall first that neither the positions nor the sizes of the elements and their interconnecting conductors affect the voltages and the currents in the circuit. This allows us to redraw our circuit as shown in Figure 2.61. It would be worth your while to verify for yourself that each element is connected between the same nodes and that we have not altered the definition of any of our circuit variables.

Figure 2.61
The example circuit
redrawn

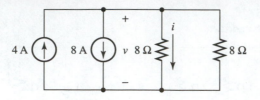

Now we can simply note that the two i-sources form a parallel subcircuit, as do the two resistors. Replacing the i-sources by a subcircuit consisting of a single element we obtain the simplified equivalent circuit in Figure 2.62. Note that we have abbreviated the phrase "equivalent subcircuit" as "eq. subckt." The k looks somewhat specious, but circuit analysts do use "ckt" as the abbreviation for "circuit." It is one of those ineffable things that we simply cannot explain, but that we use.

Figure 2.62
Equivalent subcircuit
for the i-sources

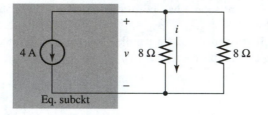

Eq. subckt

We now have a number of options. One way of proceeding is to use the current divider rule for the equal-valued resistors, taking careful notice of the sign, and obtaining

$$i = \frac{4\,\text{A}}{2} = -2\,\text{A}. \tag{2.4-21}$$

We can then apply Ohm's law to obtain

$$v = 8i = -16\,\text{V}. \tag{2.4-22}$$

An alternative would have been to combine the two resistors using an equivalent subcircuit, thus obtaining a single resistor of value 4 Ω, then to use Ohm's law to get the voltage ($v = 4\,\Omega \times (-4\,\text{A}) = -16\,\text{V}$). Then, we would have been forced to return to the original circuit to compute the current, again using Ohm's law ($i = -16\,\text{V}/8\,\Omega = -2\,\text{A}$).

Section 2.4 Quiz

Q2.4-1. Find the simplest equivalent subcircuit for Figure Q2.4-1.

Figure Q2.4-2

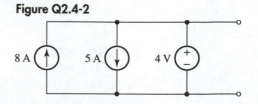

Figure Q2.4-1

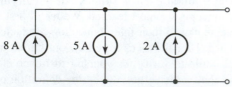

Q2.4-2. Find the simplest equivalent subcircuit for Figure Q2.4-2.

Q2.4-3. Find the simplest equivalent subcircuit for Figure Q2.4-3.

Figure Q2.4-3

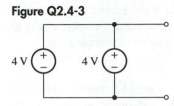

Q2.4-4. Find the simplest equivalent subcircuit for Figure Q2.4-4.

Figure Q2.4-4

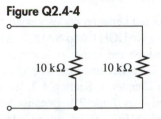

Q2.4-5. Find the simplest equivalent subcircuit for Figure Q2.4-5.

Figure Q2.4-5

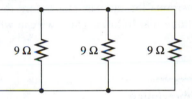

Q2.4-6. Find i in the subcircuit shown in Figure Q2.4-6.

Figure Q2.4-6

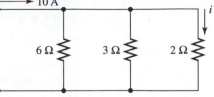

Q2.4-7. Find v in the circuit shown in Figure Q2.4-7.

Figure Q2.4-7

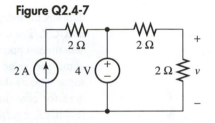

Q2.4-8. Find i in the circuit shown in Figure Q2.4-8.

Figure Q2.4-8

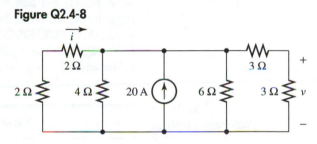

Q2.4-9. Find v in the circuit shown in Figure Q2.4-8.

2.5 | SPICE

The Netlist The idea of a SPICE netlist was discussed in the problems at the end of Chapter 1. SPICE, as we pointed out, is a computer simulation program whose acronym stands for Simulation Program with Integrated Circuit Emphasis. We also discussed the way elements are described to SPICE. Specifically, the line

```
EXXX...X    NPLUS    NMINUS    VALUE
```

tells SPICE that an element whose name is EXXX...X (the X's stand for any number of alphanumeric characters) is connected between two nodes named NPLUS and NMINUS and that the element has a value of "VALUE" (that is, the symbol "VALUE" in the statement is to be replaced with the numerical value of the element). Furthermore, the voltage across the element is assumed to have its plus sign on the node NPLUS and its negative sign on node NMINUS. The current reference arrow for the element goes from node NPLUS, through the element, to NMINUS. The leading E in the element name indicates the type of element.[6]

---

[6] We should perhaps warn you that the E actually means a particular type of voltage-dependent controlled source. This topic is discussed in Chapter 5. Until then, we will simply use e as a generic element symbol.

As discussed in the Chapter 1 problems, if the letter E is replaced with the letter V the element will be interpreted as an independent voltage source; if it is replaced with the letter I, it will be interpreted as an independent current source. Thus,

```
V1      5       19      25
```

represents an independent voltage source connected between nodes 5 and 19, with its positive reference on node 5 and its negative reference on node 19. Its value is 25 V. Similarly,

```
ISOURCE    NAUGHTY    SAUCY    −3.4UAMPERES
```

denotes an independent current source connected between nodes named NAUGHTY and SAUCY. The current is assumed to flow from NAUGHTY to SAUCY. SPICE is case insensitive; uppercase letters are interpreted to be the same as lowercase ones. A word about the value is perhaps in order. It is -3.4 microamperes. The U is a prefix, of the same type as those described in Table 1.1 in Chapter 1, Section 1.2. We repeat this table here for convenience—with a few alterations—as Table 2.1. Because SPICE is case insensitive, a different prefix for mega is used: MEG (or, of course, meg). Furthermore, SPICE does not support a character set having the Greek mu (μ) so the letter u is used instead. Finally, the femto prefix is not used because the letter f is reserved for the unit of capacitance, the farad. Notice, too, that in specifying ISOURCE we also wrote out the word AMPERES. SPICE correctly interprets the number just before this word and the following prefix, then just ignores anything that follows. Thus, we can write out units for values and the like for clarity.

Table 2.1
Engineering prefixes for SPICE

Name	Value	Abbreviation
tera	10^{12}	T
giga	10^{9}	G
mega	10^{6}	MEG
kilo	10^{3}	k
milli	10^{-3}	m
micro	10^{-6}	u
nano	10^{-9}	n
pico	10^{-12}	p
femto	10^{-15}	f (not used)

In Section 2.1 we discussed another circuit element, the resistor. A resistor element is described in the following way:

```
R53     23      38      3.6E3
```

We have given the value in standard scientific exponential notation, which is understood by SPICE. The E3 means $\times 10^3$, so our resistor has a value of 3.6 kΩ. Because a resistor has no voltage or current polarities "built in" we might anticipate that the order of the nodes is completely irrelevant. This is not entirely true, however. As is discussed in the chapter 1 problems, when we refer to an element current as i(R53) we mean the current flowing from its first-named node to its second. Thus, i(R53) means the current from node 23 to node 38.

Solution Control Statements

So much for the SPICE netlist, the description of the circuit topology. Now we must discuss two other items. SPICE expects you to use the first line in its input file as a program name. This line can consist of anything. It is simply ignored by the SPICE compiler. Good practice, though, is to use this line to identify the circuit for later reference. A typical example from this text might be

End of Chapter Problem: Section 2 Number 4--A Series Circuit

Figure 2.63 Structure of a SPICE input file

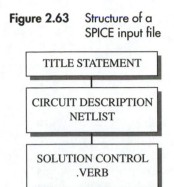

TITLE STATEMENT

CIRCUIT DESCRIPTION NETLIST

SOLUTION CONTROL .VERB

Any line beginning with an asterisk, *, is a comment line and is also disregarded. If you follow any valid statement with a semicolon (;) the remaining text on that line is ignored.

The second item expected by SPICE is an indication of what to do with the circuit. For example, you might want a dc analysis performed; in this case, you must tell SPICE to do this type of simulation. Such statements are like verbs in an English sentence — they tell SPICE what to do. We will describe these "verbs" shortly. Now we pause to show you the typical structure of a SPICE input file in Figure 2.63. We have grouped these sets of statements in logical order; SPICE, generally, is order insensitive. Conceivably, though it would not be advisable to do so, one could mix the element description statements with the "verb" statements. The title statement, however, must be the very first statement in the file.

Now let's discuss the verb statements. Each starts with a period (.), but after that the similarity ends. We will not discuss them all at this time; rather, we will only describe those pertaining to the circuit analysis techniques we have discussed at a given point. Thus, the way we tell SPICE to perform a dc analysis is as follows:

```
.DC     VS2     -10V     +10V     0.1V
```

This statement tells SPICE to leave all the other independent sources set at the values specified in the netlist, but to readjust the voltage source named VS2 to -10 V (note that SPICE ignores the V after the -10). SPICE then runs a simulation at that value. After it has finished, it increments VS2 by 0.1 volt and checks the new value against the $+10V$ in the next-to-last entry. If this new value of VS2 is less than that value, it performs another simulation for the new value of VS2. This is repeated until the test fails, at which time the simulations are stopped.

Finally, an "end the simulation" statement is required. Like the title statement, its order is important; it must be the last one in the file. It has the form

```
.END
```

At this point we have covered enough SPICE material to write a simulation except for the statements specifying output. We will cover such a statement in Example 2.9.

Examples

Example 2.9 Write a SPICE file that will perform a simulation of the circuit shown in Figure 2.64 for the determination of the current i and the voltage v.

Figure 2.64 An example circuit

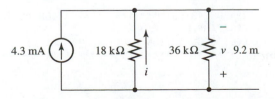

4.3 mA   18 kΩ   36 kΩ v 9.2 m

Solution Our first step is to name the nodes and the elements, with an arbitrarily chosen reference node assigned as node 0. We have done this in Figure 2.65. Did you notice that our circuit is a parallel one? For this reason, we can check our simulation with pencil and paper analysis. One possible SPICE file is as follows:[7]

```
FIRST   SPICE    EXAMPLE
IS1     0        1       4.3MA
IS2     1        0       9.2MA
R1      0        1       18KOHMS
R2      1        0       36KOHMS
.DC     IS1      4.3MA       4.3MA      1MA
.PRINT           DC      I(R1)V(0, 1)
.END
```

Figure 2.65
The example circuit prepared for SPICE

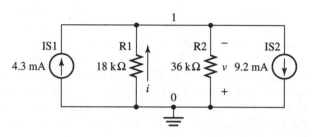

A few comments are perhaps in order. First, notice the ordering of the nodes in our current source statements. If we had used the reverse ordering, we would have been forced to specify negative values. We picked the ordering shown on R1 because we were looking ahead toward having its current as an output, and the I(R1) in the print statement gives the current relative to this ordering. The specification of the voltage V(0, 1) results in printing of the voltage with node 0 carrying the positive reference and node 1 carrying the negative reference. In general,

```
V(N1, N2)
```

specifies the voltage between node N1(+) and N2(−). A number of different analyses can be run with one SPICE file, so we had to specify that the dc results were to be printed in the .PRINT statement. In general, the format of this output statement is

```
.PRINT     DC     OUTPUT LIST
```

where OUTPUT LIST is simply a listing of the output variables desired, in the format previously specified for currents and voltages, separated by spaces.

The netlist statements specifying the independent sources are more generally of the form

```
SOURCENAME     NPLUS     NMINUS     TYPE     VALUE
```

Thus, for dc analysis, one would put the letters DC in the place of TYPE. In case the VALUE is a constant, the TYPE is DC by default.

The output from a SPICE run for our example (using the evaluation version of PSPICE, version 5.1) is shown here (with spacing reduced for compactness):

```
****06/22/94 08:43:53***********EVALUATION PSPICE (JAN 1992)*****
FIRST SPICE EXAMPLE
****  CIRCUIT DESCRIPTION
****************************************************************
******
IS1     0     1     4.3MA
IS2     1     0     9.2MA
R1      0     1     18KOHMS
```

----

[7] Ordering of nodes and ordering of statements are not unique.

```
R2      1    0    36KOHMS
.DC     IS1    4.3MA     4.3MA     1MA
.PRINT DC      I(R1)   V(0, 1)
.END
****06/22/94 08:43:53**********EVALUATION PSPICE (JAN 1992)
**************
FIRST SPICE EXAMPLE
****   DC TRANSFER CURVES          TEMPERATURE= 27.000 DEG C
**********************************************************************
******
IS1          I(R1)       V(0,1)
4.300E-03  3.267E-03  5.880E+01
        JOB CONCLUDED
        TOTAL JOB TIME            .42
```

There is one inconvenient aspect of SPICE that we should mention. Currents can only be defined in elements. Thus, one cannot directly ask SPICE to provide a current such as i in Figure 2.66 as an output. Therefore, we must use a trick. We must insert a "dummy" element, such as the zero-valued v-source in Figure 2.67 in order to "sense" the current. The resulting netlist is

```
ANOTHER EXAMPLE CIRCUIT
IS     0    1       DC      62M
VDUMMY      1    2       0
R1     1    0       2K
R2     2    0       5K
R3     2    0       3K
.DC    IS    62M    62M    1M
.PRINT DC      I(VDUMMY)
.END
```

Figure 2.66
An example circuit

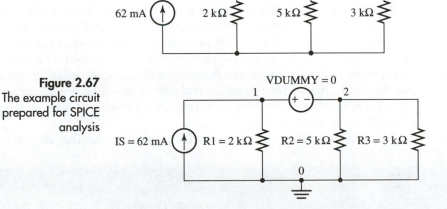

Figure 2.67
The example circuit
prepared for SPICE
analysis

Notice that VDUMMY has node 1 as its positive node and node 2 as its negative node. This means that I(VDUMMY) will be the current flowing from node 1 to node 2 through the zero-valued v-source, which, we remind you, is equivalent to a short circuit. A zero-valued resistor is also equivalent to a short circuit. It cannot be used, however, because for technical reasons SPICE does not permit resistors to have a value of zero.

    The actual SPICE output from the preceding file is as follows:

```
****02/15/94 13:57:13******* EVALUATION PSPICE (JAN 1992)********
ANOTHER EXAMPLE CIRCUIT
```

```
****CIRCUIT DESCRIPTION
*********************************************************************
******
IS      0    1    DC    62M
VDUMMY 1     2    0
R1      1    0    2K
R2      2    0    5K
R3      2    0    3K
.DC  IS  62M  62M  1
.PRINT DC    I(VDUMMY)
.END
****02/15/94 13:57:13******EVALUATION PSPICE (JAN 1992)******
ANOTHER EXAMPLE CIRCUIT
****   DC TRANSFER CURVES          TEMPERATURE= 27.000 DEG C
*********************************************************************
******
        IS          I(VDUMMY)
        6.200E-02   3.200E-02

        JOB CONCLUDED
        TOTAL JOB TIME      .37
```

As an exercise, you might wish to check this result with a hand calculation.

Section 2.5 Quiz

Q2.5-1. Compute (by hand) the values generated as output by the following SPICE file.

```
SPICE QUIZ PROBLEM QUESTION 1
VS1   1    0    DC    20
VS2   2    3    DC    40
R1    1    2    2K
R2    3    4    3K
R3    4    0    5K
.DC   VS1  20   20   1
.PRINT    DC    V(3, 4)    I(R3)
.END
```

2.6 | Basic Axioms of Circuit Theory

In this short section we will summarize what we have accomplished thus far. We have introduced the basic concepts of circuit theory: the conductor, the circuit element, the clock, the ammeter, and the voltmeter. These cannot be defined—their existence is presumed. Based upon the conductor and the circuit element, we defined a number of other concepts: node, path, loop, circuit, etc. Based upon the clock, the ammeter, and the voltmeter, we defined the fundamental quantities that they measure: the second, the ampere, and the volt. From these basic quantities, we then defined a number of derived quantities: charge, flux linkage, power, and energy.

The basic concepts of circuit theory are assumed to obey four axioms:

Definition   *Axiom 1* The *v-i* characteristic completely describes the behavior of a circuit element. It can be determined by tests on the element in isolation from others, and it continues to be valid when the element is connected with others to form a circuit.

Axiom 2 The electrical behavior of a circuit is unaffected by either the shape or the size of its nodes, conductors, and elements.

Axiom 3 (Kirchhoff's Current Law) The sum of all the currents (and charges) flowing into any Kirchhoff surface is zero at any instant of time.

Axiom 4 (Kirchhoff's Voltage Law) The sum of the voltage (and flux linkage) rises around any loop is zero at any instant of time.

These four axioms allow us to analyze any circuit we wish, so long as we know the circuit topology and the element characteristics. Much of the remainder of this book will develop the techniques for doing this using the particular elements we have discussed thus far: the independent voltage source, the independent current source, and the resistor. We will introduce others in later chapters; however, for the next several chapters we will concentrate on those just mentioned—extending the methods we have already described for series and parallel circuits.

Our reason for presenting the basic concepts and axioms here is simply this: we are developing circuit analysis as a branch of mathematics. There are two very good reasons for doing this. First, we limit the number of concepts with which we must deal. The usual circuit analysis text presents in an early chapter quite a long list of basic concepts from physics, such as lumens, calories, and oersteds.[8] Unfortunately, these concepts are never actually used later in the text, and the reader is left with a vague feeling that something important is sure to be overlooked at a later stage. Second, it allows us to concentrate on the essential circuit manipulations, which, after all, are inherently mathematical in nature. Of course, other concepts (such as, say mechanical power, temperature, or light intensity) are quite useful in the applications of circuit theory; the process of solving a circuit, however, is a mathematical one that does not rely upon them.

Axiom 1 basically means that energy is not radiated through free space from one circuit element to another and that there is no way, other than through the influence of terminal voltages and currents, of influencing other elements.[9]

Axiom 2 says that the size and orientation of elements are not important. This axiom fails when the independent sources vary too rapidly (high frequencies) and when the dimensions of a circuit become too large (such as, for example, in the ac power distribution system whose conductors travel from coast to coast).

Axiom 3 (KCL) is merely charge conservation, but in a particular form. We are assuming that charge does not accumulate on a conductor surface or inside an element. This means that the science of electrostatics is not covered in our theory. Though important (MOS integrated circuits are easily damaged by static charge, for example) you should be prepared to go back to your physics text to deal with that subject.

Axiom 4 (KVL) is one way of stating that we do not have to concern ourselves with magnetic fields. It is equivalent to Faraday's and Lenz's laws, which together state that the sum of the voltage rises around a closed loop is equal to the time derivative of the magnetic flux (a concept we have no need of in this text). If this derivative is zero, we have KVL; if the magnetic flux itself is identically zero, then the integral of the voltage rises—which we have called flux linkage—is zero.

Chapter 2 Summary

This chapter has been primarily devoted to the basic laws upon which the theory of circuit analysis will be constructed in chapters to follow. We presented *Kirchhoff's current law (KCL)* and *Kirchhoff's voltage law (KVL)*. These have to do with the constraints that

---

[8] We are not belittling these texts. Many are worthwhile for you to consult for clarification of points you find, for some reason, vague in this one. In reading them, however, you should be aware of this difference.

[9] Dependent sources and transformers, elements to be considered in later chapters, must be thought of as *multiterminal* elements; in this light, they will be seen to obey this axiom.

the topology (the way the circuit is wired together) places upon the currents (or charges) and voltages (or flux linkages). We also introduced the *resistor element,* which obeys *Ohm's law.* These three laws, along with the definitions of the basic circuit variables, completely suffice for our future development of circuit analysis methods.

We also developed the idea of an *equivalent subcircuit,* that is, a subcircuit that has the same *v-i* characteristic as some known subcircuit—but which is usually simpler. (We concentrated mainly on two-terminal subcircuits.) One type of equivalent subcircuit is the *equivalent resistor.* Like the resistor element, this type of subcircuit has a *v-i* characteristic that is a straight line passing through the origin of the *v-i* plane; unlike the resistor element, its *resistance* can be negative. We discussed a variety of other types of subcircuit. Among them, we considered the *series subcircuit* (one having a single path inside the subcircuit between the two terminals) and the *parallel subcircuit* (one having exactly two nodes—no more and no fewer); we noted as well that there are corresponding concepts for complete circuits. Thus, either a circuit or a subcircuit can be series or parallel (or, of course, neither).

We developed simple equivalents for circuits consisting of a single type of element (resistor, v-source, or i-source) and, for series and parallel resistor subcircuits, we developed two important circuit analysis tools: *voltage division* and *current division* (the former for the series resistor subcircuit and the latter for the parallel subcircuit).

Finally, we began our formal development of a very common computer tool used by circuit designers and analysts: SPICE. The acronym stands for Simulation Program with Integrated Circuit Emphasis.

With these tools in hand, we can now begin to develop the mathematical analysis techniques we need to analyze actual circuits.

Chapter 2 Review Quiz

RQ2.1. Consider the circuit in Figure RQ2.1.
 a. What is v when $R = 1\ \Omega$?
 b. What is v when $R = 0\ \Omega$ (short circuit)?
 c. What is v when $R = \infty\ \Omega$ (open circuit)?

Figure RQ2.1

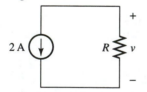

RQ2.2. Consider the circuit in Figure RQ2.2.
 a. What is i when $R = 1\ \Omega$?
 b. What is i when $R = 0\ \Omega$ (short circuit)?
 c. What is i when $R = \infty\ \Omega$ (open circuit)?

Figure RQ2.2

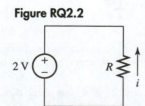

RQ2.3. Find v in Figure RQ2.3.

Figure RQ2.3

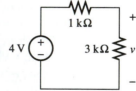

RQ2.4. Find v_1 and v_2 in Figure RQ2.4.

Figure RQ2.4

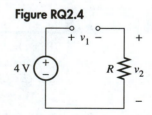

RQ2.5. Find i in Figure RQ2.5.

Figure RQ2.5

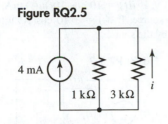

RQ2.6. Find i in Figure RQ2.6.

Figure RQ2.6

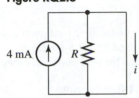

RQ2.7. Find R_{eq} for the subcircuit in Figure RQ2.7.

Figure RQ2.7

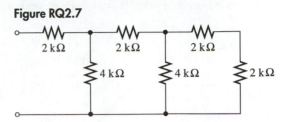

RQ2.8. Write a SPICE file for solving Question RQ2.5.

Chapter 2 Problems

Note: A simple request to "solve the circuit" means you are to find all the element voltages and currents.

Section 2.1 Ohm's Law

2.1-1. Tests on a two-terminal element result in the *v-i* characteristic of Figure 2.1-1. What is the value of the resistance in ohms, kilohms, and megohms?

Figure P2.1-1

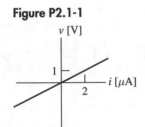

2.1-2. Find the conductances (in S, mS, and μS) of resistors having the following values of resistance:
   a. 0.25 Ω   *b.* 5000 Ω   *c.* 10 MΩ

2.1-3. A piece of wire is 1000 meters long. A 10-V battery is applied between its two ends and a current of 100 mA results. What is the resistance of the wire in Ω/m? In Ω/ft? (There are 39.37 inches in one meter.)

2.1-4. For the simple circuit in Figure P2.1-2, find the current i if:
   a. R = 1 kΩ   *b.* R = 1 Ω   *c.* R = 0.001 Ω

What is the value of i if $R \rightarrow 0$? Is the current in a short circuit *always* infinite? What is i if $R \rightarrow \infty$?

Figure P2.1-2

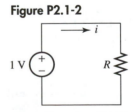

2.1-5. For the simple circuit in Figure P2.1-3, find the voltage v if:
   a. R = 1 kΩ   *b.* R = 1 MΩ   *c.* R = 10^{15} Ω

What is the value of v if $R \rightarrow \infty$? Is the voltage across an open circuit *always* infinite? What is v if $R \rightarrow 0$?

Figure P2.1-3

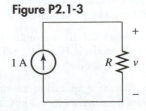

2.1-6. Find the power absorbed by the resistor in Problems 2.1-4 and 2.1-5, (a), (b), and (c).

2.1-7. The subcircuit shown in Figure 2.1-4(a) has the v-i characteristic shown in Figure 2.1-4(b). To what single element, if any, is it equivalent?

Figure P2.1-4

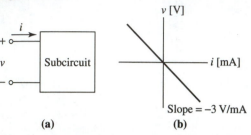

(a)   (b)

2.1-8. The subcircuit in Figure P2.1-4(a) has the v-i characteristic shown in Figure 2.1-5. To what single element, if any, is it equivalent?

Figure P2.1-5

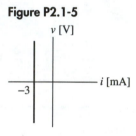

2.1-9. The subcircuit in Figure P2.1-4(a) has the v-i characteristic shown in Figure 2.1-6. To what single element, if any, is it equivalent?

Figure P2.1-6

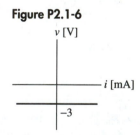

2.1-10. Suppose a source (either current or voltage) is applied to the subcircuit in Figure P2.1-4(a) so as to adjust v to a value of $+6$ V. If the v-i characteristic is as shown in Figure P2.1-4(b), what is the power absorbed by the subcircuit?

2.1-11. Repeat Problem 2.1-10 assuming that the subcircuit has the v-i characteristic shown in Figure P2.1-5. Must the applied source be a v-source or an i-source?

2.1-12. Suppose a source is applied to the subcircuit shown in Figure P2.1-4(a) to set the current i to $+4$ mA. Assuming that the v-i characteristic is that shown in Figure P2.1-6, what is the power absorbed by the subcircuit? Must the applied source be a v-source or an i-source?

2.1-13. The voltage source shown in Figure P2.1-7 has $v_s(t) = 100 \cos(10t)u(t)$ V. If $i(t) = 10 \cos(10t)u(t)$ mA, find the power absorbed by the source. If $i(t) = -10 \cos(10t)u(t)$ mA, find the power absorbed by the source.

Figure P2.1-7

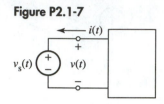

2.1-14. For each of the four configurations in Figure P2.1-8, find the power absorbed by the v-source P_v, the power absorbed by the i-source P_i, and the total power absorbed P_{tot}.

Figure P2.1-8

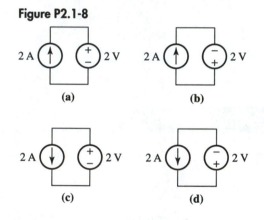

(a)   (b)

(c)   (d)

2.1-15. Identify the nodes for the circuit shown in Figure P2.1-9.

Figure P2.1-9

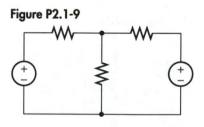

2.1-16. Repeat Problem 2.1-15 for the circuit shown in Figure 2.1-10. How is this circuit related to the one in Problem 2.1-15?

Figure P2.1-10

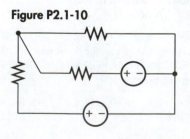

2.1-17. Identify the nodes in Figure P2.1-11. What specific type of circuit is this?

Figure P2.1-11

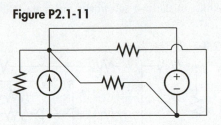

Section 2.2 Kirchhoff's Laws

2.2-1. Find the value of i_x in Figure P2.2-1.

Figure P2.2-1

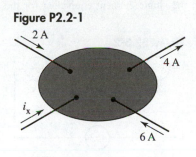

2.2-2. Find the value of i_x in Figure P2.2-2.

Figure P2.2-2

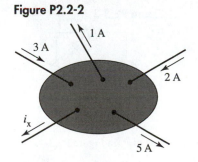

2.2-3. Find i_x and v_x in Figure P2.2-3.

Figure P2.2-3

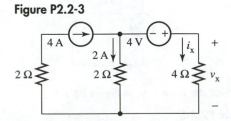

2.2-4. Figure P2.2-4 shows the same circuit as that in Problem 2.2-3 redrawn with ground reference symbols for the bottom node. Write KCL for this ground reference node in the form

$$\sum_{\text{gnd node}} i_{\text{in}} = 0$$ using the value for i_x found in Problem 2.2-3.

Figure P2.2-4

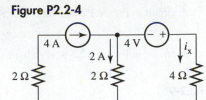

2.2-5. Find i_x in Figure 2.2-5.

Figure P2.2-5

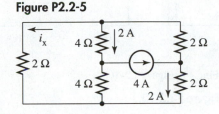

2.2-6. Find the value of v_x in Figure 2.2-6.

Figure P2.2-6

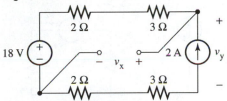

2.2-7. Find the value of v_y in Figure 2.2-6.

2.2-8. Find i_x in Figure 2.2-7.

Figure P2.2-7

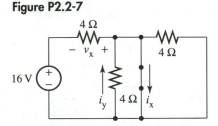

2.2-9. Find the value of i_y in Figure P2.2-7.

2.2-10. Find the value of v_x in Figure P2.2-7.

2.2-11. For each circuit in Figure P2.2-8, find the current I for $R = 1\ \Omega, 0.01\ \Omega$, and $0.001\ \Omega$.

Figure P2.2-8

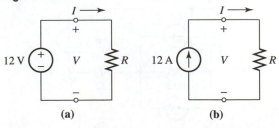

2.2-12. For each circuit in Figure P2.2-8, find V for $R = 10\ \Omega$, $1\ k\Omega$, and $1\ m\Omega$.

2.2-13. For each circuit in Figure P2.2-9, sketch the equivalent *deactivated* circuit.

Figure P2.2-9

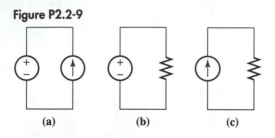

(a)   (b)   (c)

2.2-14. Find I in Figure 2.2-10.

Figure P2.2-10

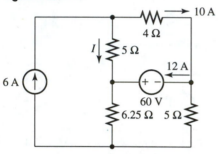

2.2-15. Find the power absorbed by each element in Figure 2.2-10. What is the total power absorbed by the circuit?

2.2-16. For the circuit in Figure 2.2-11, find I by selecting an appropriate Kirchhoff surface.

Figure P2.2-11

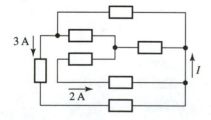

2.2-17. Use KCL to find I and V in Figure 2.2-12.

Figure P2.2-12

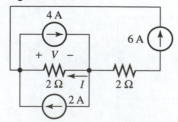

Section 2.3 Series Subcircuits

2.3-1. Which of the two subcircuits in Figure P2.3-1 is (are) series?

Figure P2.3-1

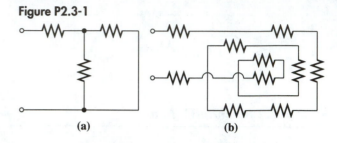

(a)   (b)

2.3-2. Find the single element equivalent for the subcircuit in Figure 2.3-2.

Figure P2.3-2

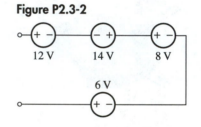

2.3-3. Find the value of v_x in Figure P2.3-3.

Figure P2.3-3

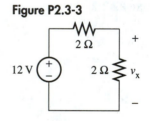

2.3-4. Find the value of v_x in Figure P2.3-4.

Figure P2.3-4

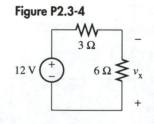

2.3-5. Find the value of v_x in Figure P2.3-5.

Figure P2.3-5

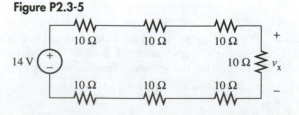

2.3-6. Find the value of v_x in Figure P2.3-6.

Figure P2.3-6

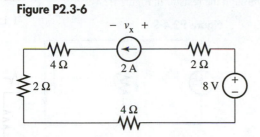

2.3-7. Find the value of the node voltages v_x and v_y in Figure P2.3-7. Then find the value of the "node-to-node" voltage v_{xy}.

Figure P2.3-7

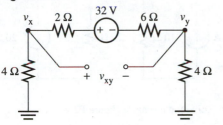

2.3-8. Find the voltage across the open circuit, v_x, in Figure P2.3-8 and the voltage across the resistor, v_y.

Figure P2.3-8

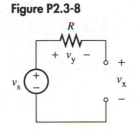

2.3-9. Solve the circuit in Figure P2.3-9.

Figure P2.3-9

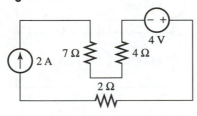

2.3-10. Solve the circuit in Figure P2.3-10.

Figure P2.3-10

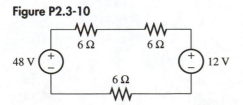

Section 2.4 Parallel Subcircuits

2.4-1. Find a single element equivalent for the parallel subcircuit in Figure P2.4-1. Note that it *is* a parallel subcircuit in spite of a slightly different drawing arrangement than we have previously used.

Figure P2.4-1

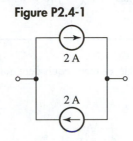

2.4-2. Find a single element equivalent for the subcircuit shown in Figure P2.4-2.

Figure P2.4-2

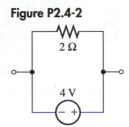

2.4-3. Find a single element equivalent for the subcircuit in Figure P2.4-3. What is i_x?

Figure P2.4-3

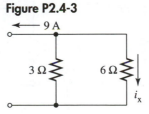

2.4-4. Find a single element equivalent for the subcircuit in Figure P2.4-4 and find i_x.

Figure P2.4-4

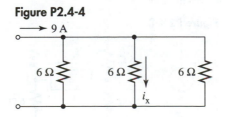

2.4-5.   Find v_x and v_y in Figure P2.4-5.

Figure P2.4-5

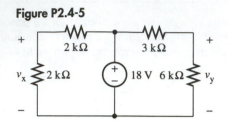

2.4-6.   Find a single element equivalent for the subcircuit in Figure P2.4-6.

Figure P2.4-6

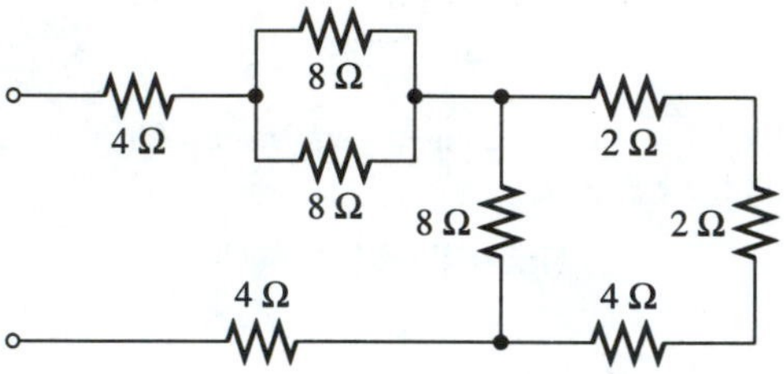

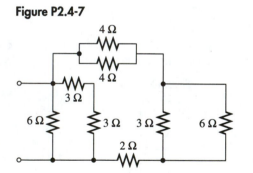

2.4-7.   Find a single element equivalent for the subcircuit in Figure P2.4-7.

Figure P2.4-7

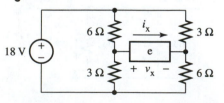

2.4-8.   Find i_x in Figure P2.4-8 assuming that the element e is a short circuit. Find v_x if e is an open circuit. Then find an equivalent circuit for everything except the element e if the 18-V source is reduced to zero in value.

Figure P2.4-8

2.4-9.   Find the current i_x through the short circuit and the current i_y through the resistor in Figure P2.4-9.

Figure P2.4-9

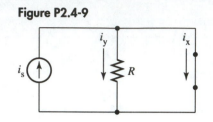

2.4-10.   Solve the circuit in Figure P2.4-10.

Figure P2.4-10

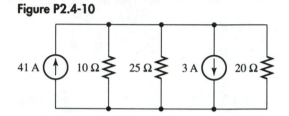

2.4-11.   Solve the circuit in Figure P2.4-11.

Figure P2.4-11

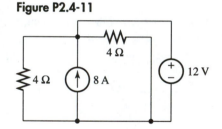

2.4-12.   Solve the circuit in Figure P2.4-12. Use units of mA, kΩ, and V.

Figure P2.4-12

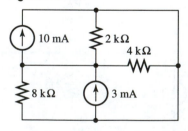

Section 2.5 SPICE

2.5-1.   Write a SPICE file for the circuit shown in Figure 2.5-1 using the labels shown and run it to find the voltage across RA (plus sign on the right).

Figure P2.5-1

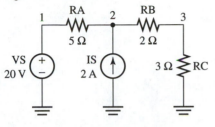

2.5-2.   Solve Problem 2.2-8 using SPICE.

2.5-3.   Solve Problem 2.4-8 using SPICE. Assume that e is a short circuit.

3

Circuit Analysis Using Subcircuits

This chapter develops analysis techniques for circuits that are not simple: those having more than two nodes or more than one loop. Analysis is based on network properties, most of which follow from the basic linearity of the resistor relationship. We develop two extremely important equivalent subcircuits: the Thévenin equivalent and the Norton equivalent. These permit a complicated two-terminal subcircuit to be replaced by a simpler one consisting of one independent source and one resistor. This is useful for the simplification of complicated circuits. In addition, we develop several other equivalences that are useful in both computation and future theoretical development of circuit analysis techniques.

It is assumed that the reader understands the material in Chapters 1 and 2.

3.1 | Source Transformation

Several types of equivalent subcircuit were discussed in Chapter 2: series sources, parallel sources, series resistors, parallel resistors, and so on. For the most part, these were "single element type" subcircuits, all the elements were voltage sources, all were current sources, or all were resistors. The only type of mixed element subcircuit considered in Chapter 2 was the current source in series with any other element, which is equivalent to the current source alone, and the voltage source in parallel with anything, which is equivalent to the voltage source acting alone. When we use the term "equivalence," we mean that the *v-i* characteristic of the original subcircuit is identical with that of the named type of subcircuit. In this section we will consider the equivalence of two-terminal subcircuits consisting of an independent source and a resistor.

Figure 3.1(a) shows a voltage source connected in series with a resistor. We can test its *v-i* characteristic by attaching an independent source to its terminals, varying that source to all possible values, and recording the resulting value of the response variable. We have somewhat arbitrarily chosen a current source for our test in Figure 3.1(b), but a voltage source would have worked just as well. We can compute *v* very simply using KCL and Ohm's law:

$$v = v_\text{s} + R_\text{s}i. \tag{3.1-1}$$

Let's invert this relation and express *i* in terms of *v*. The resulting expression is

$$i = -\frac{v_\text{s}}{R_\text{s}} + \frac{v}{R_\text{s}}. \tag{3.1-2}$$

This can be interpreted as a KCL equation for the subcircuit shown in Figure 3.2(a). Suppose that we apply a current source as in Figure 3.2(b) and vary it to all possible values. Then (3.1-2) is precisely the KCL equation at the top node. Thus, just as equations (3.1-1) and (3.1-2) are equivalent mathematical expressions, the subcircuits in Figures 3.1(a) and 3.2(a) are equivalent. No test at the terminals can distinguish one from the other. Notice that we have arbitrarily chosen the terminal voltage and the test current source direction to be consistent with the passive sign convention for the subcircuit. You should remember this when applying the equivalence. One can go either way; that is, one can replace the subcircuit in Figure 3.1(a) with the one in Figure 3.2(a), or vice versa.

Figure 3.1
Testing the *v -i*
characteristic of a
series v-source and
resistor subcircuit

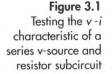

(a) Subcircuit   **(b)** *v-i* test

Figure 3.2
An equivalent circuit
and its *v -i*
characteristic test

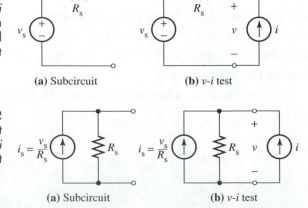

(a) Subcircuit   **(b)** *v-i* test

The source equivalence we have just derived is a powerful tool in analyzing circuits. We will demonstrate this in the following example.

Example 3.1 Find the voltage v and the current i in the circuit of Figure 3.3 using an equivalent subcircuit.

Figure 3.3
An example circuit

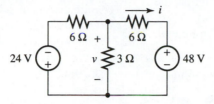

Solution We see that there are two subcircuits of the form shown in Figure 3.1(a). We show them in boxes with terminals in Figure 3.4. We now envision the operation of "unplugging" these two boxes and replacing them with another pair of boxes containing the equivalent parallel current source-resistor combination of Figure 3.2(a). We then obtain the circuit shown in Figure 3.5. Here is an important observation: notice that when we replace the two subcircuits by their equivalent ones, we *lose the physical identity of the current i* that we are trying to find—*the subcircuit is only equivalent as far as the external part of the circuit is concerned!* The voltage $v(t)$, however, retains its physical identity because it is the actual voltage across the 3-Ω resistor external to the two subcircuits. We can find the voltage v by applying one KCL equation at the top node (noting that the voltage across each element is v):

$$\frac{v}{6} + \frac{v}{6} + \frac{v}{3} = 8 - 4. \tag{3.1-3}$$

This gives $v = 6$ V. We must now return to the original circuit in Figure 3.3 and use this result to find the current we are looking for. Figure 3.6 shows that circuit again with the known voltage shown explicitly. We have also added a ground reference at the bottom node to expedite our thought process. We can now write KCL at the top node, say in the form $\sum\limits_{\text{node}} i_{\text{out}} = 0$, to obtain

$$\frac{6}{3} + \frac{6 - (-24)}{6} + i = 0. \tag{3.1-4}$$

This results in $i = -7\,A$.

Figure 3.4
Identifying two subcircuits

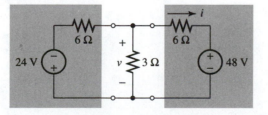

Figure 3.5
After source transformation

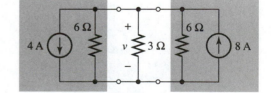

Figure 3.6
The original circuit

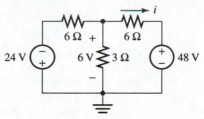

As you can see, the source transformation is a powerful tool. With it we can analyze circuits that do not consist merely of a single loop or a single pair of nodes—circuits that are neither series nor parallel.

Example 3.2   Find the current i in Figure 3.7.

Figure 3.7
An example circuit

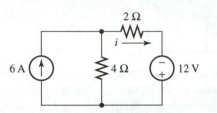

Solution   We will apply our source transformation this time to the two-terminal subcircuit on the left consisting of the 6-A i-source and the 4-Ω resistor. This gives the equivalent circuit shown in Figure 3.8—as far as the 2-Ω resistor and 12-V v-source are concerned. Notice that we have applied our source transformation in the "reverse" direction relative to that of the last example; that is, we are going from Figure 3.2 to Figure 3.1 in our original derivation. Analysis of this simple series circuit gives

$$i = \frac{24 - (-12)}{4 + 2} = 6 \, A. \qquad (3.1\text{-}5)$$

Figure 3.8
After source
transformation

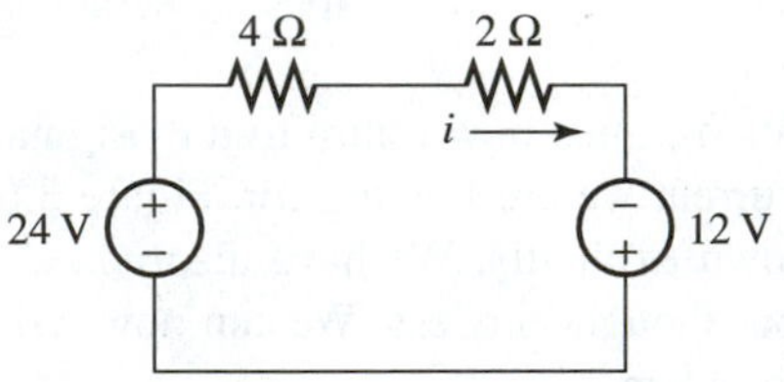

Section 3.1 Quiz

Q3.1-1.   Perform source transformation on the subcircuit in Figure Q3.1-1 to obtain the equivalent parallel subcircuit.

Figure Q3.1-1

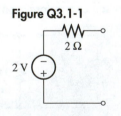

Q3.1-2.   Repeat Question Q3.1-1 for the subcircuit in Figure Q3.1-2.

Figure Q3.1-2

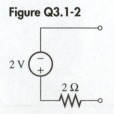

Q3.1-3.   Perform the "reverse" source transformation on the subcircuit in Figure Q3.1-3.

Figure Q3.1-3

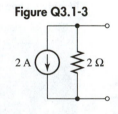

Q3.1-4.   Figure Q3.1-4 shows two equivalent subcircuits being tested with a current source. Compute v and v'. Then compute the power absorbed by each element. Show that the powers absorbed by the subcircuits are identical, but that there is no correspondence between the powers absorbed by the equivalent elements in the two subcircuits.

Figure Q3.1-4

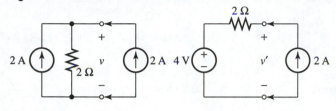

Q3.1-5. Perform source transformations on the subcircuit in

Figure Q3.1-5 to get an equivalent circuit consisting of a single v-source and a single resistor.

Figure Q3.1-5

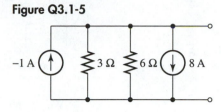

3.2 | Linearity and Superposition

Figure 3.9 shows a simple circuit whose independent sources have known, but literal, values. What is the influence of each source on the indicated response variable *i*? In fact, this has a very specific form that we will now derive. Using source transformation on the parallel subcircuit consisting of the current source and its nearest resistor, we derive the equivalent circuit shown in Figure 3.10—as far as all other elements are concerned, and the desired response is the current in one of those elements. We easily analyze this single loop (series) circuit to obtain

$$i = \frac{1}{3}i_\mathrm{s} + \frac{1}{6}v_\mathrm{s}. \tag{3.2-1}$$

Thus, i is a linear combination of the two source values. This is a property that holds in general for all voltage and current responses in circuits with only independent sources and resistors. This statement will be proved in Chapter 4 in a manner that does not use the results of this section; thus, the reasoning will not be circular.

Figure 3.9
An example circuit

Figure 3.10
After source transformation

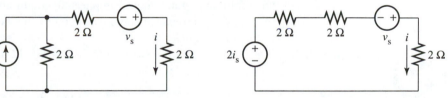

Now notice the following small, but important, fact. We can write (3.2-1) in the form

$$i = i_1 + i_2, \tag{3.2-2}$$

where

$$i_1 = i]_{v_\mathrm{s}=0} \tag{3.2-3}$$

and

$$i_2 = i]_{i_\mathrm{s}=0}. \tag{3.2-4}$$

We call i_1 and i_2 the *partial responses*. In words, i_1 is the response with the voltage source *deactivated* (that is, reduced to zero) and i_2 the response with the current source deactivated. Recalling that a deactivated voltage source is a short circuit and a deactivated current source is an open circuit, we obtain the two *partial circuits* shown in Figure 3.11. We see quickly that $i_1 = 1/3(i_\mathrm{s})$ and $i_2 = 1/6(v_\mathrm{s})$ as in equation (3.2-1).

Figure 3.11
The partial circuits

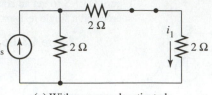
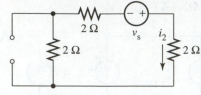

    (a) With *v*-source deactivated   (b) With *i*-source deactivated

We can now generalize. Any voltage or current response y in a circuit having only re-sistors and n independent sources with values $x_1, \ldots, x_n$ has the form

$$y = a_1x_1 + a_2x_2 + \cdots + a_nx_n, \tag{3.2-5}$$

where the a_i's are constants determined by the resistive portion of the circuit. Thus, if we define the n partial responses by

$$y_i = a_ix_i, \tag{3.2-6}$$

we can write

$$y = y_1 + y_2 + \cdots + y_n, \tag{3.2-7}$$

where

$$y_i = y]_{x_k=0}, k \neq i. \tag{3.2-8}$$

Thus, we can compute each partial response y_i from the circuit that results when all the independent sources other than the ith are deactivated.

    Why is this procedure useful? There are two reasons. One is theoretical. It is often nice to know the form of response of a circuit for the purpose of deriving other useful re-sults. As a case in point, Section 3.3 discusses two important equivalent subcircuits that are based on this result. Furthermore, it is often a useful calculational tool. Notice that we have split an original, more complicated, circuit analysis problem into a number of sim-pler ones. The process of solving a circuit by this method is called *superposition*.

Example 3.3   Find the indicated response current in Figure 3.12 using superposition.

Figure 3.12
An example circuit

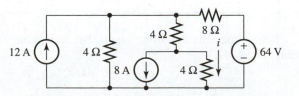

Solution   Deactivation of the 12-A i-source and the 8-A i-source gives the partial circuit shown in Figure 3.13. Using parallel and series equivalents, Ohm's law, and voltage and current di-vision, one easily computes the partial response to be $i_1 = 2\,A$.

Figure 3.13
The partial circuit with the v-source active

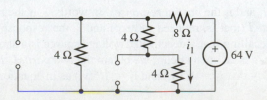

Deactivation of the 12-A i-source and the v-source gives the second partial circuit shown in Figure 3.14. Again, the simple techniques covered in Chapter 2 lead to $i_2 = -5\,A$. We then allow the 12-A source to remain active and deactivate the 8-A i-source and the v-source. The resulting third partial circuit is shown in Figure 3.15. This single source circuit can easily be analyzed to determine that $i_3 = 3\,A$.

Figure 3.14
Another partial circuit

Figure 3.15
A third partial circuit

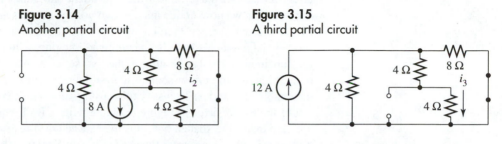

Finally, we add each of the three partial response currents to obtain the actual current in the original circuit: $i = i_1 + i_2 + i_3 = 2\,A - 5\,A + 3\,A = 0\,A$.

Many beginning circuit analysts seem to be wary of an answer of zero for voltage or current; it is, however, a legitimate value like any other. You might verify for yourself that, with $i = 0$ in the original circuit of Example 3.3, KCL and KVL hold for all loops and nodes.

Section 3.2 Quiz

Q3.2-1. Use source transformation on the circuit in Figure Q3.2-1 to show that $i = a_1 v_s + a_2 i_s$ and find the values of a_1 and a_2.

Figure Q3.2-1

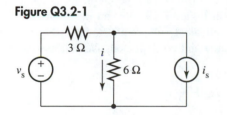

Q3.2-2. If $v_s = 18\,V$ and $i_s = 27\,A$ in Figure Q3.2-1, use superposition to find i.

Q3.2-3. Find v_x in Figure Q3.2-2 by superposition.

Figure Q3.2-2

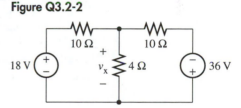

3.3 | Thévenin and Norton Equivalent Subcircuits

The Thévenin Equivalent

The fact that all variables in a circuit (or a subcircuit, as we will shortly see) are linear combinations of the independent source values leads to a striking equivalence. We will derive this equivalence using the subcircuit shown in Figure 3.16. We assume that this

Figure 3.16
An example subcircuit

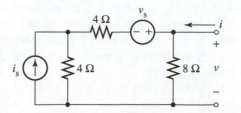

subcircuit is connected into a larger circuit; that is, there are elements external to this sub-circuit connected to it through the two terminals shown. Therefore, the terminal voltage v and the terminal current i have nonzero values in general.

If we apply a source transformation to the current source and its parallel 4-Ω resistor, we obtain the equivalent shown in Figure 3.17—as far as the remainder of the subcircuit is concerned. We have shown the equivalent subcircuit in a shaded box for emphasis. Suppose that we now notice that the two voltage sources and the two 4-Ω resistors are se-ries connected. We can then enlarge our box slightly, as shown in Figure 3.18. The two series voltage sources have been combined into one, with no resulting effect on the rest of the subcircuit variables. Similarly, the two series resistors have also been combined into one. (Although the two v-sources are on opposite ends of the two resistors, they can still be combined in series. To justify this, just assume a current i_x in the series elements to the left of the 8-Ω resistor in Figure 3.17 and write a KVL equation in path form for the sin-gle path inside the shaded box plus the external 4-Ω resistor and v-source and compare it with the similar equation for the shaded box in Figure 3.18.)

Figure 3.17
After one source transformation

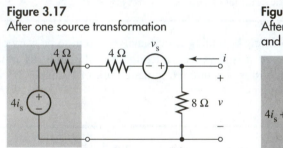

Figure 3.18
After combination of the v-sources and resistors

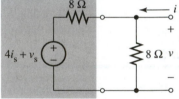

Now we apply one last source transformation to the elements in the shaded box in Figure 3.18, obtaining the equivalent in Figure 3.19. We can now combine the two 8-Ω parallel resistors to obtain the final equivalent subcircuit shown in Figure 3.20. Notice that all of our transformations have left i and v unaffected. We can now write one KCL equation at the top node to get

$$v = 2i_s + \frac{1}{2}v_s + 4i. \qquad (3.3\text{-}1)$$

Figure 3.19
After second source transformation

Figure 3.20
The final equivalent subcircuit

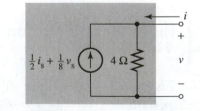

We could have also obtained this result by attaching a test i-source to the terminals of the subcircuit and applying superposition. In any event, we see that the voltage is a linear combination of the internal independent sources *and the terminal current i.* Notice that the first two terms on the right side of (3.3-1) are independent of i and that the third is di-rectly proportional to the terminal current i. A moment of reflection should convince you that the superposition property implies that any two-terminal subcircuit composed of re-sistors and independent sources has a v-i relationship of the form in equation (3.3-1).

Therefore, we can write this terminal relationship in general as

$$v = v_{oc} + R_{eq}i, \qquad (3.3\text{-}2)$$

where v_{oc} is a voltage independent of the terminal current i and R_{eq} is a resistance. The reason for our choice of subscripts will be discussed shortly.

We see that equation (3.3-2) describes not only the original subcircuit but also the series v-source and resistor combination shown in Figure 3.21. This subcircuit is, therefore, *equivalent* to the original because it has the same *v-i* characteristic. It is called the *Thévenin equivalent* (subcircuit) after the French telegraph engineer Charles Thévenin.[1] Its importance cannot be overestimated. A quite complex subcircuit can be "boiled down" to one consisting of only a v-source and a resistor.

To gain an understanding of the two parameters v_{oc} and R_{eq}, let's return to the subcircuit we have previously discussed. As we see from equation (3.3-1),

$$v_{oc} = 2i_s + \frac{1}{2}v_s. \qquad (3.3\text{-}3)$$

Although only a part of the total terminal voltage *v*, it becomes the *actual* terminal voltage under the condition that $i = 0$. This, however, is equivalent to simply removing the subcircuit from the elements to which it is connected. We show this in Figure 3.22. It is important to notice that this is precisely the same as the original subcircuit with only two small notational changes: we have explicitly specified that the terminal current is zero, and we have labeled the terminal voltage with the symbol v_{oc} to denote the *o*pen *c*ircuit voltage whose value is given in (3.3-3). This voltage comes from the independent sources. This is to be contrasted with Figure 3.16, in which we show only the subcircuit, but assume that other external components are present and that the current *i* is not necessarily zero. Hence, the terminal voltage in that case is *not* v_{oc}; rather, it has the term $R_{eq}i$ added to it.

Figure 3.21 The Thévenin equivalent subcircuit

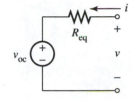

Figure 3.22
The open circuit voltage

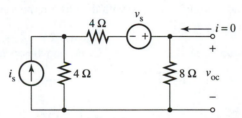

What is the significance of the term $R_{eq}i$? Going back to equation (3.3-1) once more, we see that if we deactivate the internal sources (i_s and v_s in our example) the v_{oc} term will be forced to zero, leaving only the term under consideration. The resulting subcircuit, called the *deactivated subcircuit*,[2] is shown for our example in Figure 3.23. In general, all independent v-sources are replaced by short circuits and all i-sources by open circuits—the equivalents of the corresponding deactivated sources. Because v_{oc} comes from the independent sources, we see that deactivation forces it to zero; hence, the Thévenin equivalent becomes the subcircuit in Figure 3.24. We hasten to note that the terminal variables *v* and *i* in Figures 3.23 and 3.24 are different from those in the *activated* subcircuit (that is, the subcircuit with its independent sources restored) with the original external components in place; however, the *v-i* relationship for the deactivated subcircuit shows that it is equivalent to a resistor of value R_{eq} (the reason for our choice of subscript). Thus, one can deactivate all the independent sources and determine the *v-i* characteristic to de-

---

[1] The French pronunciation is something like "Tay-vee-naah," spoken nasally.

[2] Many texts refer to it as the *dead network,* and the process as *killing* the *circuit.* We prefer our less cruel term and note that one should not confuse a *circuit* (a synonym is network) with a *subcircuit.*

termine this resistance. For instance, in our example one can easily see (using series and parallel equivalent resistances) that $R_{eq} = 4\ \Omega$. (To compute v_{oc} we would have to know the values of the independent current and voltage sources.)

Figure 3.23
The deactivated subcircuit

Figure 3.24
The equivalent resistance

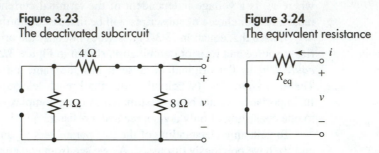

Thévenin Examples

Example 3.4   Find the Thévenin equivalent subcircuit for the subcircuit shown in Figure 3.25.

Solution   We have several options as to how we proceed. We can either compute the v-i characteristic directly, as we did in the example circuit in Figure 3.16, or we can use superposition, as we did when we discussed the significance of v_{oc} and R_{eq}. The latter technique shows the physical nature of these parameters, so let's use it. We first "clip the leads" connecting the subcircuit to the external elements (not shown in Figure 3.25), thus forcing i to zero and v to v_{oc}. This is shown in Figure 3.26. There, we show the result of applying KCL at the top left node. The current up through the 9-Ω resistor is 3 A, as shown, resulting in a resistor voltage of 27 V (positive on the bottom). Now we can apply KVL from the bottom node to the top left node, then to the top right node, to deduce that the open circuit voltage is $v_{oc} = -27 + 2 = -25$ V. Notice carefully that this sign is relative to our choice for the terminal voltage v with the plus on the top terminal.

Figure 3.25
An example subcircuit

Figure 3.26
Computing the open circuit voltage

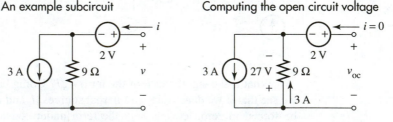

To compute R_{eq} we deactivate the subcircuit, thus replacing the 2-V source with an equivalent short circuit and the 3-A source with an equivalent open circuit. The resulting deactivated subcircuit is shown in Figure 3.27. Clearly, $R_{eq} = 9\ \Omega$. The Thévenin equivalent subcircuit is shown in Figure 3.28.

Figure 3.27
Equivalent resistance

Figure 3.28
The Thévenin equivalent subcircuit

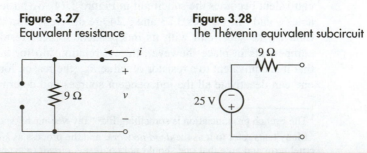

Example 3.5   Use the "direct method" to find the Thévenin equivalent subcircuit for the original subcircuit in Example 3.4 (Figure 3.25).

Solution   In Figure 3.29 we show the original subcircuit with a test current source attached (a voltage source would have worked as well). By showing this source we are explicitly signifying that we are testing the subcircuit for its v-i characteristic.[3] Furthermore, simply for variety, we have chosen to place the plus sign for the terminal voltage on the bottom terminal and the terminal current reference into that terminal. Our choice is quite arbitrary — *except for the fact that our derivation requires that the passive sign convention be observed relative to the subcircuit.* We now use KCL at the bottom node and KVL from the top right node, across the 2-V source, and down through the 9-Ω resistor to derive the v-i terminal relationship. It is

$$v = 9(i + 3) - 2 = 25 + 9i. \tag{3.3-4}$$

We now need only identify terms: the constant, 25 V, is v_{oc} and the constant multiplier of i is R_{eq}, 9Ω. Being careful to keep to our choice for the terminal variables in our interpretation, we see that Figure 3.28 once again is the Thévenin equivalent subcircuit — as it should be.

Figure 3.29
Testing the subcircuit
with an i-source

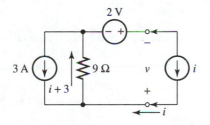

Which of the two preceding methods should you use? As is often the case in circuit analysis, the answer is not clear-cut. The superposition technique clearly shows the physical significance of the two parameters in the Thévenin equivalent; often, however, the direct technique is less work — particularly in more complicated circuits. You should be warned, however, that the terminal variables v and i in the direct approach must be *literal* (they must be letters). One cannot simply assign a value to i or v and compute the other because the v_{oc} and R_{eq} parameters could not then be separated — the voltage v (or current i) resulting from an analysis would just be a number.

The Norton Equivalent   We can obtain another equivalent subcircuit quite quickly and effortlessly from the Thévenin equivalent, though it was originally derived independently. It is called the Norton equivalent[4] and results from merely applying a source transformation to the Thévenin equivalent. Thus, by using a source transformation on the subcircuit in Figure 3.21, we have the Norton equivalent of Figure 3.30. We note that, in order for this subcircuit to be equivalent to the original circuit (to which we have already shown the Thévenin version to be equivalent), we must have

$$i_{sc} = \frac{v_{oc}}{R_{eq}}. \tag{3.3-5}$$

---

[3] It is conventional, particularly in more advanced courses, to simply show the terminal voltage and current v and i as we did in our derivation of the Thévenin equivalent and not explicitly show the test source.

[4] E. L. Norton was an American engineer and designer of circuits for telephony.

Figure 3.30   The Norton equivalent subcircuit

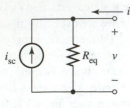

Notice that the terminal v-i characteristic for the Norton equivalent is

$$i = \frac{v}{R_{eq}} - i_{sc}.$$
(3.3-6)

Using (3.3-5), we see that this is merely a rearranged form of (3.3-2), the equation describing the Thévenin terminal v-i relationship.

Once more, we have selected our notation with some forethought. To see the significance of the "sc" subscript, suppose the subcircuit is shorted—that is, a short circuit is placed across its terminals. This is the same as shorting the Norton equivalent, as we have done in Figure 3.31. (We have shown arrows on the shorting conductor to denote that it is merely a temporary, or test, connection.) The terminal voltage and therefore the resistor voltage are both zero, so the resistor *current* is zero also. Thus, all of the source current flows out of the top terminal and down through the short circuit.

Figure 3.31
The short circuit current

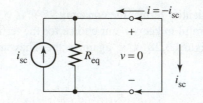

We now see that the Thévenin and Norton equivalents are organically connected and that there are three parameters: v_{oc}, R_{eq}, and i_{sc}. These three parameters are connected through the source transformation equation (3.3-5); thus, they are not independent. One need only compute two from the circuit, then use (3.3-5) to find the third. We can see the connection graphically if we return for a moment to the terminal characteristic given in equation (3.3-2), repeated here for convenience of reference:

$$v = v_{oc} + R_{eq}i.$$
(3.3-2)

If we plot a graph of this equation—or, equivalently, of (3.3-6)—Figure 3.32 results. We have sketched this graph under the assumption that v_{oc} (and hence i_{sc}) is positive. You might be puzzled as to why we have shown the straight line characteristic going into quadrants II and III. The reason, however, is quite simple and practical. Depending upon the nature of the external elements to which the subcircuit is attached, either of the terminal variables might be forced to go negative. The value of v that occurs when i is zero, however, is by definition v_{oc}; similarly, the value of i resulting from forcing v to be zero is $-i_{sc} = -v_{oc}/R_{eq}$. Thus, R_{eq} is the slope of the line.

Figure 3.32
Graph of subcircuit terminal v-i relationship

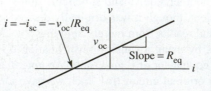

We will conclude this section with an example exploring all possible parameters of the two equivalent subcircuits we have derived.

A Norton Example

Example 3.6   Find v_{oc}, i_{sc}, and R_{eq} by direct tests on the subcircuit in Figure 3.33; then derive the Norton and Thévenin equivalents by testing it with a general voltage source.

Figure 3.33
An example subcircuit

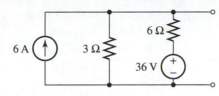

Solution Notice that the terminal variables have not been labeled. This is strictly up to us as the investigators; however, once we choose a set of references (obeying the passive sign convention, to be sure) we must stick with them. Let us, therefore, choose the positive terminal voltage at the top and the terminal current reference consistently. Then the open circuit voltage can be computed by analyzing Figure 3.34. Among the many techniques available to us is superposition. Let's use it by first deactivating the 36-V v-source and then finding the corresponding partial response. Figure 3.35 shows the partial subcircuit that results. We note that the two resistors are connected in parallel, giving an equivalent resistance of 2 Ω. Thus, we have $v_{oc1} = 6 \text{ A} \times 2 \text{ Ω} = 12 \text{ V}$. Next, we deactivate the 6-A i-source, resulting in the partial subcircuit of Figure 3.36. Here, we see that the two resistors and the v-source form a series circuit and that the voltage across the 3-Ω resistor is the partial terminal voltage. Thus, we have $v_{oc2} = (3/(3 + 6)) \times 36 = 12 \text{ V}$. Adding, we obtain the total (actual) open circuit voltage: $v_{oc} = v_{oc1} + v_{oc2} = 12 + 12 = 24 \text{ V}$.

Figure 3.34
The open circuit voltage

Figure 3.35
Partial subcircuit with v-source deactivated

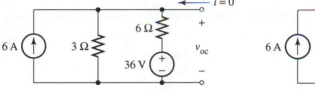

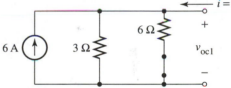

Now let's compute the value of R_{eq}. To do so, we must first deactivate the entire subcircuit. This gives the equivalent shown in Figure 3.37. The 3-Ω and 6-Ω resistors are connected in parallel, giving $R_{eq} = 2 \text{ Ω}$.

Figure 3.36
Partial subcircuit with i-source deactivated

Figure 3.37
Equivalent resistance—The deactivated subcircuit

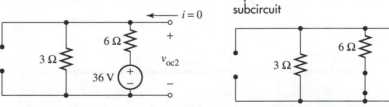

We could now, of course, compute the short circuit current from the first two parameters; we were, however, asked to compute it directly by testing the subcircuit. Therefore, we place a short circuit (an ideal conductor) across the terminals and identify the short circuit current as shown in Figure 3.38.[5] Now this might not be immediately clear to you—but work at it until you are sure it is so: the short circuit current is merely the sum of the current source current and the current upward through the 6-Ω resistor (apply KCL at the top terminal out of which the reference arrow for i_{sc} is directed). (Notice that the current in the 3-Ω resistor in Figure 3.38 is zero because the short circuit constrains its

_____
[5] Be sure you understand the reason we have chosen the reference direction shown by comparing with Figure 3.31.

voltage to be zero.) Thus, we have $i_{sc} = 6 + (36/6) = 12$ A. A quick check shows that, indeed, $v_{oc}/R_{eq} = 24\text{ V}/2\text{ }\Omega = 12$ A, as expected.

Figure 3.38
Computing the short
circuit current

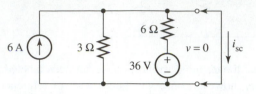

Finally, as per our instructions, we will attach a test voltage source having a general literal value and find the responding terminal current, then compare with equation (3.3-6) to identify i_{sc} and R_{eq}. The circuit being thus tested is shown in Figure 3.39. A bit of thought should convince you that each of the resistor voltages is known;[6] therefore, we need only to use Ohm's law and KCL to find the terminal current i. It is

$$i = \frac{v - 36}{6} + \frac{v}{3} - 6 = \frac{v}{2} - 12. \tag{3.3-7}$$

Comparison with (3.3-6) gives $i_{sc} = 12$ A and $R_{eq} = 2\text{ }\Omega$ as expected.

Figure 3.39
Testing the subcircuit

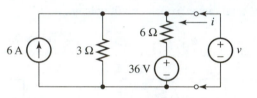

Regardless of which computation scheme we use, the Norton equivalent is the subcircuit shown in Figure 3.40(a) and the Thévenin that of Figure 3.40(b).

Figure 3.40
The two equivalent
subcircuits

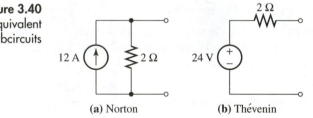

(a) Norton   (b) Thévenin

Section 3.3 Review Quiz

Q3.3-1. Find the Thévenin and Norton equivalents for the subcircuit in Figure Q3.3-1. Compute v_{oc}, R_{eq}, and i_{sc} by individual tests on the subcircuit.

Figure Q3.3-1

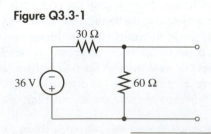

Q3.3-2. Find the Thévenin and Norton equivalents for the subcircuit in Figure Q3.3-2 using a test current source with a general value i.

Figure Q3.3-2

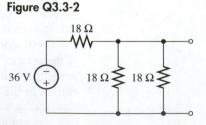

---

[6] It is for this reason, computational convenience, that we chose a v-source rather than an i-source—to emphasize the fact that these voltages were all known values.

Q3.3-3. Find the Norton equivalent subcircuit for all elements to the left of terminals ab, then use this subcircuit to find the voltage v across the 3-Ω resistor in Figure Q3.3-3.

Figure Q3.3-3

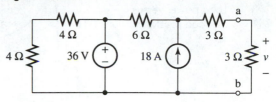

3.4 | Practical Sources and Matching

Resistance Matching a Subcircuit

This section, though relatively brief, considers some issues that have great practical importance. We will start by considering the following problem. A crystal microphone, for our present purposes, can be assumed to possess the Thévenin equivalent shown in Figure 3.41. Furthermore, a loudspeaker can be modeled to a good degree of approximation by the simple resistor R_L shown there. The current in the speaker is

Figure 3.41 An illustration of the matching problem

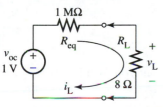

Microphone   Loudspeaker

$$i_L = \frac{1\text{ V}}{1\text{ M}\Omega + 8\text{ }\Omega} \cong 1\text{ }\mu\text{A}. \tag{3.4-1}$$

This is quite a problem, because such a current is insufficient to produce a reasonable sound level. A loudspeaker consists of a permanent magnet with a small coil of wire wrapped around the small end of a paper cone placed in front of the magnet. When current is passed through the coil, the cone is thereby attracted to, or repelled from, the permanent magnet. As the current varies, the position of the paper cone varies in step, thus producing pressure variations in front of it: sound waves. The production of sound waves of sufficient intensity requires at least several milliamperes of current. As a matter of fact, the auditory power is proportional to the power delivered to the speaker. In the case shown in Figure 3.41, the power absorbed by the load is

$$P_L = v_L i_L = i_L^2 R_L = 8\text{ pW!} \tag{3.4-2}$$

(Remember that 1 pW = 10^{-12} W!) Clearly, this is not a very effective circuit.

Let's explore the situation with more convenient values. Figure 3.42 shows the same circuit as Figure 3.41, but with more general values. We will address two different problems. In the first, we assume that the Thévenin equivalent parameters v_{oc} and R_{eq} are specified and we are asked to determine the value of R_L that results in the power absorbed by R_L being the maximum possible.

Figure 3.42 A more general circuit

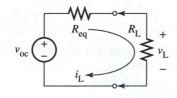

Before solving the problem analytically, let's pick some convenient numbers and do a numerical study. Suppose that $v_{oc} = 12$ V and $R_{eq} = 4\text{ }\Omega$. The power absorbed by R_L is

$$P_L = v_L i_L = i_L^2 R_L = \frac{v_{oc}^2 R_L}{(R_{eq} + R_L)^2} = \frac{144 R_L}{(4 + R_L)^2}. \tag{3.4-3}$$

If we evaluate this expression at $R_L = 0\text{ }\Omega$ we get 0 W; at $R_L = \infty$, it is again 0 W. (An open circuit dissipates zero power, as does a short circuit.) In fact, if we evaluate (3.4-3) at several values of R_L between zero and infinity, we obtain the results shown by the plot in Figure 3.43. We see that the power absorbed by the load starts at zero, increases, reaches a relative maximum for some finite value of R_L, and then decreases toward zero at $R_L = \infty$, that maximum value being around 9 W.

Figure 3.43
Load power versus
load resistance

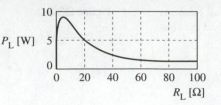

Now let's solve the problem analytically. We take the general expression in (3.4-3) and manipulate it slightly into the form

$$P_L = \frac{v_{oc}^2 R_L}{(R_{eq} + R_L)^2} = \frac{v_{oc}^2}{\left(\dfrac{R_{eq}}{\sqrt{R_L}} + \sqrt{R_L}\right)^2}. \tag{3.4-4}$$

We note that maximizing P_L is equivalent to minimizing the denominator of (3.4-4). Furthermore, we can minimize the quantity inside the denominator parentheses in order to minimize the entire denominator. For convenience, we can change variables to the more compact $x = \sqrt{R_L}$ and minimize

$$f(x) = \frac{R_{eq}}{x} + x. \tag{3.4-5}$$

We will leave it to you to take the derivative and show that $f(x)$ is minimized when $x = \sqrt{R_{eq}}$. (Inserting this value into the second derivative gives a positive value, and so the result is, indeed, a relative minimum.) This value of x gives

$$R_L = R_{eq}. \tag{3.4-6}$$

Therefore, we see that the load resistance must be numerically the same as the Thévenin equivalent resistance.

Now let's look at our second problem. Here, we assume that the load resistance R_L and the Thévenin equivalent voltage v_{oc} are fixed, known values, and the problem is to pick R_{eq} such that the power absorbed by the load is maximized. This one is easy. We simply look at the first expression in (3.4-4). Because R_{eq} occurs only in the denominator as a term that is added to R_L, we immediately see, assuming both resistances are nonnegative, that

$$R_{eq} = 0 \tag{3.4-7}$$

is necessary for maximum power into the load. This, put simply, means that the subcircuit represented by the Thévenin equivalent must be an ideal voltage source.

The problem we have just investigated is generally known as the *matching problem*. In Chapter 5, we will treat circuits that include other elements in addition to resistors and independent sources—and the problem becomes somewhat more complicated.

Practical Sources   We now propose another, equally practical, issue. We will initiate discussion by making the somewhat provocative statement, "Some subcircuits are more Thévenin than Norton, and vice versa!" Here's what we mean. As long as R_{eq} is nonzero and noninfinite, the Thévenin and Norton equivalents are equally valid. But consider the case of, let us say, an electronic device known as a photodiode. This is a solid-state device that has the Norton equivalent shown in Figure 3.44(a); these values are in the correct general range. If we transform to the Thévenin equivalent, we obtain the values shown in Figure 3.44(b). Now

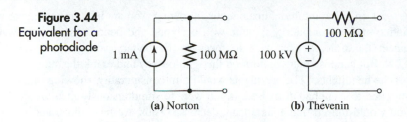

Figure 3.44
Equivalent for a
photodiode

(a) Norton   (b) Thévenin

100 kV is a very large voltage, indeed! The point is simply this: such a device is much more efficiently portrayed as a current source than as a voltage source. This fact is reflected in the two equivalents. Ideally, one could let the equivalent resistance in the Norton equivalent approach infinity without destroying the effectiveness of the model; in the Thévenin equivalent, on the other hand, the resulting open circuit would appear in series with an infinite voltage. Thus the photodiode is "more Norton than Thévenin." Similarly, one can easily come up with subcircuits that are "more Thévenin that Norton." The battery in an automobile is an example. The internal resistance is very small, and the voltage is around 12 V. The equivalent Norton current source would have a very large value.

Let's keep the car battery in mind while we discuss the behavior of nonideal sources. In the Thévenin equivalent the nonideal nature appears because the series equivalent resistance is not zero; in the Norton equivalent, it is the noninfinite nature of the parallel resistance that is nonideal. Let's investigate the situation using the Thévenin equivalent shown in Figure 3.45.

Figure 3.45   A nonideal source

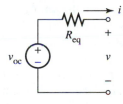

Before we proceed, let's discuss the origin of the word "source." The Latin root means "spring," as in a "source of water flowing from the ground." Therefore, when current is coming out of the terminal with the positive voltage reference (that is, the subcircuit is delivering power to external elements), we say that it is "sourcing current." When the current is reversed—with the current going into the positive reference so that the subcircuit is absorbing power—we say that is "sinking current." In this sense, the image is that of a kitchen sink with water disappearing down the drain. Thus, we have drawn the subcircuit in Figure 3.45 to be sourcing current. We are tacitly assuming that there are external elements sinking this current.

Another term that is often used is "load." A large load is synonymous with a large current and vice versa. We say that the external elements are loading the subcircuit shown and that they form the load for it. Thus, we think of the following experiment. We increase the load on the subcircuit by applying an external resistor, thereby causing the current to increase from zero (when the load resistance is infinite—an open circuit) up to the maximum value it can have (which occurs when the load resistance is zero—a short circuit). Recognizing that[7]

$$v = v_{oc} - R_{eq}i, \tag{3.4-8}$$

we have the situation depicted graphically in Figure 3.46.

Figure 3.46   The loading effect

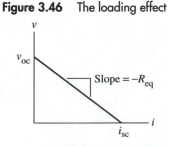

For light loads (small i) the voltage is approximately equal to the v-source value, v_{oc}; for heavier loads, the voltage falls appreciably. Finally, it becomes zero when the current has increased sufficiently. At this point the external load is equivalent to a short circuit, and in fact simply placing a short circuit across the terminals results in exactly this situation.

---

[7] Notice that the positive reference for i in Figure 3.45 is opposite to that for the current i in Figure 3.21; hence, equation (3.4-8) has a negative sign on the second term as opposed to the positive sign in equation (3.3-2), which resulted in our general Thévenin equivalent.

As an aid to intuition, imagine the following. You are in your parked car with the headlights on. The car battery is quite well designed and hence has a very low internal resistance; therefore, if you were to measure the battery voltage you would obtain about 12 V. But suppose that you turn on the ignition key and crank the starter—without turning off the headlights.[8] The headlights would dim considerably, showing that the voltage being supplied to them had decreased. If you were to simultaneously measure the battery voltage, you would indeed find a significant decrease. Thus, we have illustrated in a very down-to-earth manner an important conceptual application of the Thévenin equivalent subcircuit.

Section 3.4 Quiz

Q3.4-1.   Find the value of R in Figure Q3.4-1 that draws maximum power and find that power.

Figure Q3.4-1

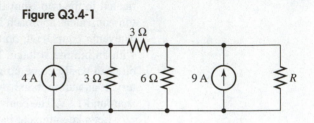

3.5 | Source Transportation

Figure 3.47   A three-terminal subcircuit

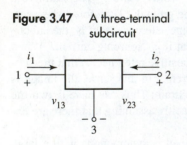

Most of the discussion in this text regards two-terminal elements and subcircuits and their equivalent elements or subcircuits. In this section, however, we would like to focus on a rather powerful pair of equivalences involving three-terminal subcircuits. Such a subcircuit is shown in Figure 3.47. We have used the double subscript notation for the voltages between terminals 1 and 3 and between 2 and 3. By KVL, we see that these two voltages suffice to determine any other voltage relative to the three terminals. Similarly, KCL shows that the terminal currents i_1 and i_2 suffice to determine any other terminal current (say that coming out of terminal 3). Recalling that one of the basic axioms of circuit theory is that the relationship among v_{13}, v_{23}, i_1, and i_2 suffices to completely describe the influence of the subcircuit on the rest of any circuit in which it is imbedded, we will set out to determine this characteristic for two different independent source subcircuits.

Voltage Source Transportation

Figure 3.48(a) shows a three-terminal v-source subcircuit. We note that

$$v_{13} = v_{23} = v_s \tag{3.5-1}$$

independently of the currents i_1 and i_2. We also quickly see that the same statement is true of the subcircuit shown in Figure 3.48(b). Thus, the two are equivalent.

Figure 3.48
V-source transportation

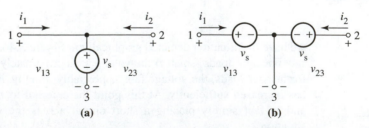

(a)   (b)

The equivalence we have just derived is called *voltage source transportation* because we are "transporting" a v-source through a joint. The following example shows just how useful it is in solving circuits.

---

[8] Not a recommended practice.

Example 3.5-1   Find the current *i* in the circuit shown in Figure 3.49 using v-source transportation.

Figure 3.49
An example circuit

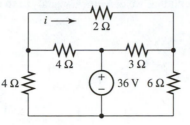

Solution   This circuit is not a simple one; that is, it is neither a parallel circuit nor a series circuit. We see, however, that the v-source with the conductor and lead at its top form a subcircuit like the one shown in Figure 3.48(a). We can therefore replace it with the circuit shown in Figure 3.48(b) without affecting our desired current *i*. We show the resulting circuit in Figure 3.50.

Figure 3.50
After v-source
transportation

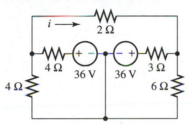

We now see that, insofar as the current *i* is concerned, this circuit is equivalent to the one shown in Figure 3.51.

Figure 3.51
After a slight
rearrangement

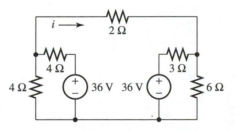

Still another equivalent rearrangement is shown in Figure 3.52. These are valid operations because we have not altered the topology of the circuit. As you will recall from Chapter 2, this is one of the basic axioms of circuit theory: that such shape alterations do not affect the operation of the circuit.

Figure 3.52
After another
rearrangement

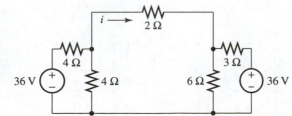

At this point we can apply the Thévenin equivalent transformation to the v-sources and their closely associated resistors to obtain the equivalent in Figure 3.53. This, at last, is a simple series circuit that we can quickly analyze to determine i: $i = -1$ A.

Figure 3.53
After yet another
rearrangement

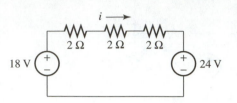

Current Source Transportation

There is an i-source version of the source transportation equivalence. In this case, we deal with a current source attached to two nodes, as shown in Figure 3.54(a), and a third terminal not connected to the source at all (though in a circuit it will be connected to external elements). It is easy to see that the terminal currents are identical in the two subcircuits 3.54(a) and 3.54(b) and *are independent of the voltages between the terminals*. Because the v-i characteristics are the same, the two three-terminal current source subcircuits are equivalent as far as any external elements connected to the terminals are concerned.

Figure 3.54
I-source transportation

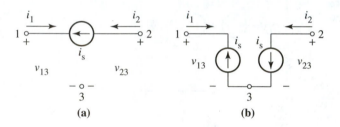

(a)   (b)

Example 3.8   Find the current i in Figure 3.55.

Figure 3.55
An example circuit

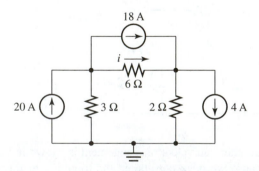

Solution   One might say that this circuit is "i-source intensive." Thus, we can perhaps expect to simplify it by i-source transportation. Transporting the 18-A source to the ground node yields the circuit shown in Figure 3.56. This circuit, of course, is entirely equivalent to the original insofar as the current i is concerned.

Figure 3.56
After i-source
transportation

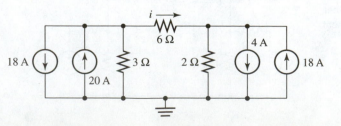

At this point you might object that we have not simplified the circuit because there is one more i-source than in the original. Notice, however, that there are two sets of parallel i-sources. We can simplify the circuit by replacing each with its parallel single source equivalent as we show in Figure 3.57.

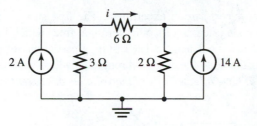

Figure 3.57
After applying parallel
i-source equivalence

Finally, let us apply source transformations (or, in more general terms, Norton-to-Thévenin equivalences) to the two i-sources and their associated parallel resistors. This generates the equivalent circuit—as far as i is concerned—of Figure 3.58. We easily find that $i = -2$ A.

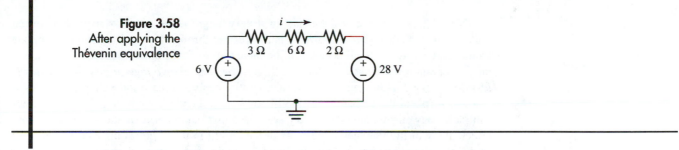

Figure 3.58
After applying the
Thévenin equivalence

Section 3.5 Quiz

Q3.5-1. Apply v-source transportation through node a to find the voltage v in Figure Q3.5-1.

Q3.5-2. Apply i-source transportation to the i-source on top in Figure Q3.5-2 to find i.

Figure Q3.5-1

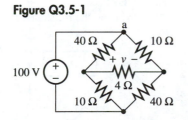

Figure Q3.5-2

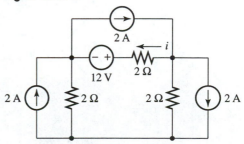

3.6 | Source Substitution

**Voltage Source
Substitution**

We will now discuss an equivalence that is very handy in developing other theoretical tools—as well as for the rigorous justification of a number of operations that one does intuitively. We begin with a portion of a circuit as shown in Figure 3.59. There we see a two-terminal element or subcircuit e and two nodes, i and j, to which other elements (not shown) are connected. We indicate this by means of the conductors connected to these nodes. We assume that the voltage v across e is a known quantity.

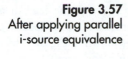

Figure 3.59
A partial circuit

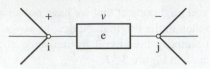

We now perform an experiment. The first step is to connect one lead of a voltage source to node j, leaving its other terminal free (which forms a floating node, i′), and adjust its value to be exactly *v*. The resulting partial circuit is shown in Figure 3.60. Because of the floating node the current through the added source is clearly zero—*hence its addition cannot have affected any of the voltages or currents in the circuit.*

Figure 3.60
After the v-source has
been connected

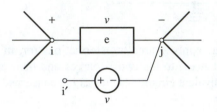

Now let us attach a resistor having a very large value, say $R = 10^{15} \, \Omega$, between nodes i and i′ as in Figure 3.61. What is the current through this resistor? Of course, one would expect it to be small because of the large value of resistance. But here is an important observation: *the current is identically zero because the voltage across the resistor is identically zero!* This means that we can continuously decrease the value of this resistance to zero and the resistor current will continue to be zero. We have neither added nor subtracted any current at nodes i and j, so we see that our experimentation thus far has not affected any of the voltages or currents in the circuit in the slightest manner. Figure 3.62 shows the situation after we have reduced the resistance to zero, thus producing a short circuit. At this point, we recall that—again insofar as the external voltage and current variables are concerned—the two-terminal subcircuit consisting of the voltage source and the element e can be replaced by the voltage source alone. This is shown in Figure 3.63.

Figure 3.61
After the resistor has been connected

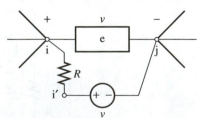

Figure 3.62
After reducing the resistance to zero

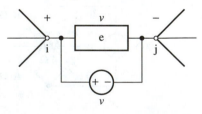

Figure 3.63
After removing the element e

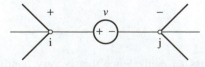

Here is our conclusion: *if the voltage across any two-terminal element or subcircuit is known, it can be removed and replaced by a voltage source having that known value without affecting any of the other voltages or currents in the circuit.* This statement is often called the voltage form of the *substitution theorem.*

Current Source Substitution

There is a current source form of the preceding result that we will now derive with the aid of Figure 3.64. Again, we show the partial circuit; this time, however, we assume that it is the current *i* that is known. (There should be no confusion between the current *i* and the node labeled i.) Suppose that we add an independent i-source, as shown in Figure 3.65, and adjust its value to be *i*. What is the effect of this on the rest of the circuit? The answer is: there is no effect! Why? If we place a closed surface around the added source and the segment of conductor to which it is attached, we see that the current *i* is unaffected on either side. But notice one interesting fact: the current in the segment of conductor between the i-source connections is zero! Because this current is identically zero, the conductor segment can be clipped out and replaced by an open circuit without affecting any of the other voltages or currents in the rest of the circuit. The resulting partial circuit is shown in Figure 3.66.

Figure 3.64
The partial circuit again

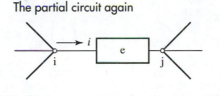

Figure 3.65
After the i-source has been connected

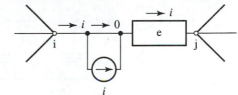

Figure 3.66
After opening the conductor segment

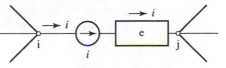

At this point we recall the basic fact that any subcircuit consisting of an i-source connected in series with any other element is equivalent—as far as the external voltage and current variables are concerned—to the current source by itself. This means that the partial circuit in Figure 3.67 is equivalent to the original one.

Figure 3.67
The final equivalent

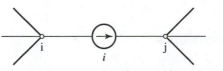

We express our conclusion in the following way, which is known as the current source form of the substitution theorem: *if the current in any two-terminal element or subcircuit is known, that element or subcircuit can be replaced by a current source having the known value as far as the external voltages and currents are concerned.*

Example 3.9   Find the value of v_s required to adjust the current *i* to zero in Figure 3.68.

Figure 3.68
An example circuit

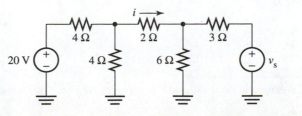

Solution   There are many ways to solve this problem, but it is very effective to apply substitution. Just suppose that v_s has been adjusted to the value required to make $i = 0$. Then, because $i = 0$, the 2-Ω resistor can be replaced by an open circuit (an i-source whose value is zero), as shown in Figure 3.69. Now we can find the node voltages v_a and v_b by application of the voltage divider rule. This gives $v_a = 10$ V and $v_b = (2/3)v_s$. But now recall that the element we removed was a resistor. Because its current is zero, we can also say—by applying Ohm's law—that its voltage is zero as well. However, this means that the resistor can be replaced by a short circuit (a voltage source having zero value)! This results in the equivalent circuit shown in Figure 3.70, which shows that $v_a = v_b$. Together with our preceding results, this gives $v_s = 15$ V.

Figure 3.69
After application of i-source substitution

Figure 3.70
After application of v-source substitution

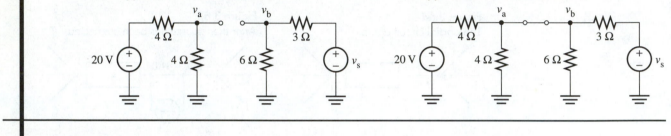

Section 3.6 Quiz

Q3.6-1. By applying substitution to the circuit in Figure Q3.6-1, find the value of i for which $v = 6$ V; then find v_x.

Q3.6-2. By applying substitution to the circuit in Figure Q3.6-1, find the value of i for which $v_x = 6$ V, then find v.

Figure Q3.6-1

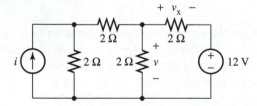

3.7 | SPICE: Thévenin Equivalents

The SPICE .TF Command   This section will be very brief—in fact, we will present only one basic idea: the determination of Thévenin and Norton equivalents with SPICE. We will discuss the main points of this procedure with the example circuit sketched in Figure 3.71. Now SPICE will not accept "dangling" nodes such as the top right-hand terminal of our circuit. Our solution, therefore, is to add a very large-valued resistor across the terminals. We show this, as well as the resulting circuit prepared for SPICE analysis, in Figure 3.72.[9] The SPICE listing is:

```
THEVENIN EXAMPLE 1
VS   1   0   DC   12
IS   2   0   DC   12
```

---

[9] Addition of the resistor RHI converts the subcircuit into a true circuit.

Figure 3.71
An example subcircuit

Figure 3.72
The example subcircuit prepared for SPICE

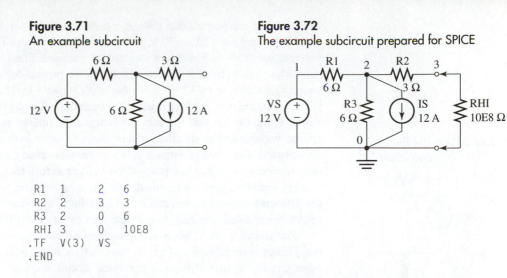

```
R1  1    2    6
R2  2    3    3
R3  2    0    6
RHI 3    0    10E8
.TF  V(3)   VS
.END
```

All statements are familiar except the .TF, or *transfer function,* statement in the second from last line.

Here is what the .TF statement tells SPICE to do. The routine will compute and print out all node voltages relative to the reference. In addition, it will print out the ratio of the voltage specified in the first field after .TF (V(3) in our example) to the independent source variable specified in the second field (in our example, this is VS). It will also print out the alternate variable for the specified independent source. By this we mean that it will be the current for a v-source (as is our case) or the voltage for a current source.

The SPICE run for our example is:

```
****03/01/94 15:53:35***********EVALUATION PSPICE (JAN 1992)*****
THEVENIN EXAMPLE 1
****CIRCUIT DESCRIPTION
***************************************************************
******
 VS  1    0    DC    12
 IS  2    0    DC    12
 R1  1    2    6
 R2  2    3    3
 R3  2    0    6
 RHI 3    0    10E8
.TF  V(3)   VS
.END
****03/01/94 15:53:35***********EVALUATION PSPICE (JAN 1992)*****
THEVENIN EXAMPLE 1
****SMALL SIGNAL BIAS SOLUTION        TEMPERATURE = 27.000 DEG C
***************************************************************
*****
NODE VOLTAGE   NODE VOLTAGE    NODE VOLTAGE    NODE VOLTAGE
( 1) 12.0000   ( 2) -30.0000   ( 3) -30.0000
   VOLTAGE SOURCE CURRENTS
   NAME  CURRENT
   VS   -7.000E+00
 TOTAL POWER DISSIPATION 8.40E+01 WATTS
 ****SMALL-SIGNAL CHARACTERISTICS
  V(3)/VS = 5.000E-01
  INPUT RESISTANCE AT VS = 1.200E+01
  OUTPUT RESISTANCE AT V(3) = 6.000E+00
   JOB CONCLUDED
   TOTAL JOB TIME  .35
```

Let's look at the output, one item at a time. First, we see the node voltages V(1), V(2), and V(3). They are 12 V, −30 V, and −30 V, respectively. The current in the independent voltage source VS is 7 A out of its positive terminal. The total power dissipation result is somewhat misleading: it is actually the total power delivered by the independent v-sources (in this case by VS).[10] The ratio of V(3) to VS is 0.5. Note that 0.5 × 12 V = 6 V is the voltage across the 6-Ω resistor between node 2 and ground *with the i-source deactivated*. Thus, we see that the transfer function computed is the ratio of the specified response variable to the specified independent variable *with all other independent sources deactivated*. The input resistance at VS is the resistance "seen by" the specified independent source; that is, it is the ratio of the voltage across that source divided by the current coming out of its positive terminal. For our example, this value is 12 Ω. Finally, one has the Thévenin equivalent resistance "looking into" the specified pair of nodes at which the response variable is defined. In our case, this resistance is 6 Ω.

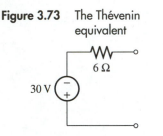

Figure 3.73   The Thévenin equivalent

For drawing the Thévenin equivalent subcircuit, we only need the last item—the Thévenin resistance and the voltage across RHI (which is our approximation to an open circuit, so that voltage is the open circuit voltage). This equivalent is shown in Figure 3.73.

Section 3.7 Quiz

Q3.7-1.   Find the Thévenin equivalent for the subcircuit shown in Figure Q3.7-1 using the SPICE .TF command.

Figure Q3.7-1

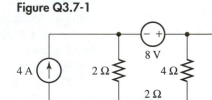

Chapter 3 Summary

This chapter has presented a number of different techniques for analyzing circuits, several using the idea of a subcircuit and its equivalent as an important tool. These techniques should not be looked upon simply as efficient computational methods (though this is frequently the case for specific circuits); rather, they form the theoretical foundation for a number of more advanced results to be developed in this text and particularly in later applications-oriented courses such as electronics.

The first method we looked at was the *source transformation*. This consists of replacing a v-source and a series resistor with an equivalent composed of an i-source and a resistor connected in parallel with it—or vice versa. One can often apply this idea effectively to transform a given circuit to a simpler one, such as a series or parallel connected circuit. Furthermore, one often looks upon practical sources (discussed in Section 3.4) as being of one or the other variety of subcircuit.

Our next major topic was that of *linearity* and *superposition;* these are related concepts. The basic idea is this: the resistive part of a circuit acts as a *combiner,* forming *linear combinations* of all the independent source values to produce the other voltages and currents in the circuit. This means that one can find the *partial response* for any voltage or current due to a given independent source acting alone (with the others deactivated) and then sum up these responses to produce the actual value. This process is called *superposition.*

---

[10] Observe that 12 V × 7 A = 84 W, and 7 A is the magnitude of the current in VS.

Next, we considered a very far-reaching pair of equivalents for an arbitrary two-terminal subcircuit: the *Thévenin* and *Norton equivalents*. The Thévenin equivalent for any two-terminal subcircuit is another two-terminal subcircuit consisting of a voltage source connected in series with a resistor; the Norton equivalent consists of a current source in parallel with a resistor. The Thévenin equivalent can be obtained by a source transformation on the Norton equivalent—and vice versa. These two equivalences can truthfully be said to be one of the most important ideas in circuit analysis, for they allow one to replace a perhaps very, very complicated subcircuit with another that is much simpler.

A discussion of practical sources was given in Section 3.4 and applied to an important problem: that of *matching* the load to a given source for *maximum power transfer*.

Two useful circuit transformations were then derived: *voltage source transportation* and *current source transportation*. We also discussed *source substitution*.

Finally, our discussion of SPICE was continued, considering the computer derivation of Thévenin and Norton equivalents.

Chapter 3 Review Quiz

RQ3.1. Draw the source-transformed equivalent for the subcircuit in Figure RQ3.1.

Figure RQ3.1

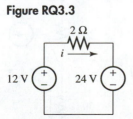

RQ3.2. Draw the source-transformed equivalent for the subcircuit in Figure RQ3.2.

Figure RQ3.2

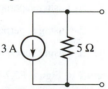

RQ3.3. Find i in Figure RQ3.3 by means of superposition.

Figure RQ3.3

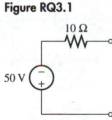

RQ3.4. Find the Thévenin equivalent of the subcircuit drawn in Figure RQ3.4 by attaching a test i-source.

Figure RQ3.4

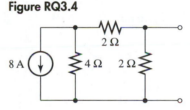

RQ3.5. Find the Thévenin equivalent for the subcircuit in Figure RQ3.4 by separately computing v_{oc} and R_{eq}.

RQ3.6. Find the Norton equivalent for the subcircuit in Figure RQ3.4 by attaching a general v-source to its terminals.

RQ3.7. Find the Norton equivalent for the subcircuit in Figure RQ3.4 by separately computing i_{sc} and R_{eq}.

RQ3.8. What value of R will absorb a maximum amount of power when attached to the terminals of the subcircuit in Figure RQ3.4? What is this maximum power?

RQ3.9. Use source substitution to replace the subcircuit in Figure RQ3.5 with an equivalent current source and compute v.

Figure RQ3.5

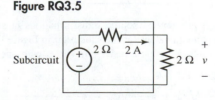

RQ3.10. Write a SPICE file to find the Thévenin equivalent for the subcircuit in Figure RQ3.4.

Chapter 3 Problems

Section 3.1 Source Transformation

3.1-1. Find i_x in Figure P3.1-1 by performing a source transformation of the 6-A source and its parallel 30-Ω resistor.

Figure P3.1-1

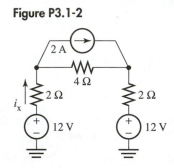

3.1-2. Find i_y in Figure P3.1-1 by first performing a source transformation of the 90-V source and its associated 60-Ω resistor.

3.1-3. By first applying a source transformation on the 2-A i-source and its associated 4-Ω resistor, find i_x in Figure P3.1-2.

Figure P3.1-2

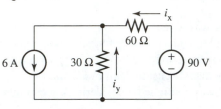

3.1-4. Using two source transformations in succession, find an equivalent subcircuit consisting of a voltage source in series with a resistor for the subcircuit in Figure P3.1-3.

Figure P3.1-3

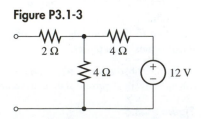

3.1-5. By using several source transformations in succession, find an equivalent for the subcircuit in Figure P3.1-4 that consists of a single resistor in series with one v-source.

Figure P3.1-4

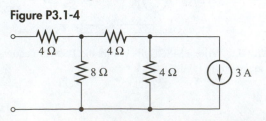

3.1-6. Find i_x in Figure P3.1-5 by performing source transformations on the subcircuit produced by (temporarily) removing the 4-V v-source.

Figure P3.1-5

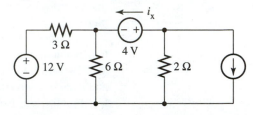

3.1-7. Find the voltage v in Figure P3.1-6 using source transformations.

Figure P3.1-6

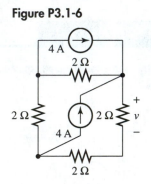

Section 3.2 Linearity and Superposition

3.2-1. Solve Problem 3.1-1 using superposition.

3.2-2. Solve Problem 3.1-3 using superposition.

3.2-3. Solve Problem 3.1-6 using superposition.

3.2-4. Solve Problem 3.1-7 using superposition.

3.2-5. Find the current i in Figure P3.2-1 using superposition.

Figure P3.2-1

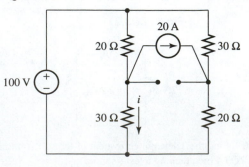

3.2-6. Find the voltage v in Figure P3.2-2 using superposition.

Figure P3.2-2

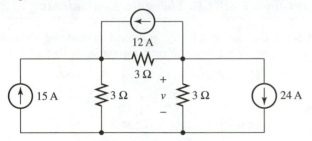

Section 3.3 Thévenin and Norton Equivalents

3.3-1. Find the Thévenin and Norton equivalents for the subcircuit in Problem 3.1-4.

3.3-2. Find the Thévenin and Norton equivalents for the subcircuit in Problem 3.1-5.

3.3-3. Solve Problem 3.1-6 by replacing all elements except the 4-V source by a Thévenin equivalent subcircuit.

3.3-4. Solve Problem 3.1-7 by replacing all elements except the central 4-A source by a Norton equivalent subcircuit.

3.3-5. Solve Problem 3.2-5 by replacing all elements except the bottom left 30-Ω resistor by a Norton equivalent subcircuit.

3.3-6. Find the Thévenin equivalent subcircuit for the one in Figure P3.3-1 by applying a test current source having a general, literal value i at terminals a and b.

Figure P3.3-1

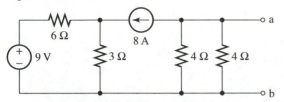

3.3-7. Compute the open circuit voltage of the subcircuit in Figure P3.3-1 (plus on terminal a).

3.3-8. Compute the short circuit current of the subcircuit in Figure P3.3-1 (arrow out of terminal a).

3.3-9. Compute the equivalent resistance of the subcircuit in Figure P3.3-1 "looking into" terminals a and b.

Section 3.4 Practical Sources and Matching

3.4-1. Compute and plot the power absorbed by R versus the value of R in Figure P3.4-1 for $R = 0, 5, 10, 15,$ and 20 Ω.

Figure P3.4-1

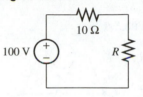

3.4-2. In Figure P3.4-1, what value of R maximizes the power delivered to the 10 Ω resistor?

3.4-3. Find the value of R for maximum power delivery to R in Figure P3.4-2.

Figure P3.4-2

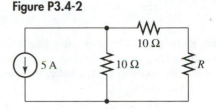

3.4-4. Find the value of R for maximum power delivery to the vertical 10-Ω resistor in Figure P3.4-3. (Be careful here and work from first principles!)

Figure P3.4-3

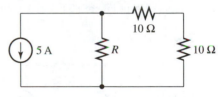

Section 3.5 Source Transportation

3.5-1. Solve Problem 3.1-1 using current source transportation.

3.5-2. Find v in Figure P3.5-1 using voltage source transportation.

Figure P3.5-1

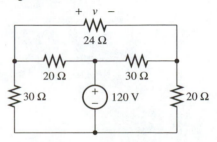

3.5-3. Find v in Figure P3.5-2 using current source transportation.

Figure P3.5-2

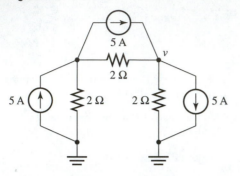

3.5-4. Solve Problem 3.1-3 by first using source transformations on the 12-V sources and their series connected 2-Ω resistors, then applying current source transportation followed by superposition.

3.5-5. Solve Problem 3.1-7 by first using current source transportation of the top 4-A source and then additional source transformation(s).

Section 3.6 Source Substitution

3.6-1. Find the value of v_s in Figure P3.6-1 required to adjust i to zero.

Figure P3.6-1

3.6-2. Find the value of v_s in Figure P3.6-1 required to adjust i to 2 A.

Section 3.7 SPICE: Thévenin Equivalents

3.7-1. Use the SPICE .TF command to find the Thévenin equivalent for the subcircuit shown in Figure P3.7-1.

Figure P3.7-1

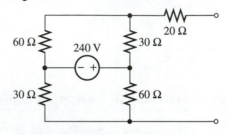

3.7-2. Use the SPICE .TF command to find the Norton equivalent for the subcircuit shown in Figure P3.7-1.

3.7-3. Use the SPICE .TF command to find the Thévenin equivalent for the subcircuit consisting of all elements except the 2-A source in the circuit of Problem 3.1-3.

3.7-4. Use the SPICE .TF command to find the Thévenin equivalent for the subcircuit consisting of all elements except the 20-A source in the circuit of Problem 3.2-5.

4 Nodal and Mesh Analysis

This chapter develops general analysis techniques for resistive circuits—those having only resistors and independent sources. These techniques are quite general in that they are guaranteed to always be successful under only very mild restrictions that are easy to test. For simple circuits or for circuits having special features in their topology, the methods described in Chapter 3 often give the answer faster and more efficiently. They depend, however, upon the pattern recognition capability of the human eye and brain. Nodal and mesh analysis, on the other hand, proceed algorithmically. It is comforting to know methods that are guaranteed to work should the simpler methods fail; furthermore, the nodal method in particular is absolutely essential if a circuit analysis algorithm is to be programmed on a digital computer. For these reasons, we explore nodal and mesh analysis in some detail in this chapter, which can be considered to be basic for most of the succeeding chapters in the text.

It is assumed that the reader understands the material in Chapters 1 and 2 and Appendix A (covering matrices and determinants).

4.1 | Nodal Analysis of Circuits Without Voltage Sources

Given a specific circuit, it is often true that we can analyze it quickly and efficiently by recognizing certain features possessed by that circuit that others do not exhibit. For instance, because of its topology, a given circuit might require only one equation using a given technique but perhaps five or more using another. Recognition of which method to use, however, requires skill and the pattern recognition capability of the human brain. On the other hand, we also need a method that we know will always work regardless of the specific nature of the circuit. Then, if our more intuitive methods fail, we can fall back on the general algorithm; furthermore, if our job is to write a computer program, we need a method that is algorithmic. Nodal analysis, which we will develop in this section, is such a method.

Figure 4.1 shows a circuit having a reasonable level of complexity. It is complicated enough that we might prefer an algorithmic method of analysis. We will use it as motivation for our explanation of the nodal method and develop its solution gradually as we progress. Though it has only resistors and current sources, the algorithm we develop for solving it will be applicable to circuits with voltage sources as well, as we will see in Section 4.2. We have labeled all the elements of our circuit in Figure 4.1 with both literal and numeric values because we want to show how each element enters into the analysis.

Figure 4.1
An example circuit

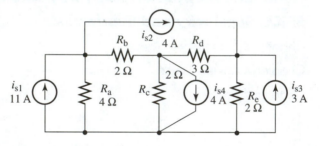

Review of Nodes and Node Voltages

As you might expect from the name, nodal analysis concentrates on the circuit nodes as being of fundamental importance. Remember that the nodes are the connected islands of conductor (including element leads) that remain when we erase the bodies of all the circuit elements. The nodes of our example circuit are shown in Figure 4.2. Identifying and counting the nodes will be an important part of nodal analysis. There are $N = 4$ nodes in our circuit, as you can verify. We have labeled them 0 through 3 in the figure (rather than 1 through 4) because a reference node is often assigned the index zero in computer analysis algorithms. (This will not, however, be necessary in our "pencil and paper" analysis.)

Figure 4.2
The nodes of the example circuit

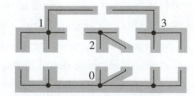

Now look at Figure 4.3. Like Figure 4.2 it shows the circuit nodes, but we have also added a voltmeter measuring the voltage at node 3 relative to an assumed reference at node 0. We call this node voltage v_3. Its positive reference (the plus sign) is assumed to be at node 3; its negative reference (the minus sign) is assumed to be at the reference node, node 0. We will not show these reference signs explicitly in the future, but will assume tacitly that each of the nonreference nodes carries a plus reference sign. We show the complete set of node voltages relative to reference node 0 in Figure 4.4. We assume that the original elements are present; we are simply not showing them so that we can better focus on the nodes and the node voltages.

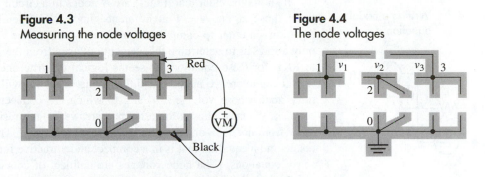

Figure 4.3
Measuring the node voltages

Figure 4.4
The node voltages

The Ground Reference: Grounded and Floating Elements

In Figure 4.4 we have introduced the *ground reference,* which is merely a symbol on the circuit diagram indicating that the node to which it is connected is the one chosen to be the reference node for all the other node voltages. To see why the node voltages are so important, look at Figure 4.5. There, we have shown only two of the resistor elements, but have still not displayed the other elements. The resistor labeled R_c is a *grounded* element because one of its leads is connected to the ground reference node. The other resistor, R_b, does not have this property. Both of its leads are connected to nonreference nodes; hence, it is called a *floating* element. These terms should conjure up a picture of two swimmers in a pool — one with his or her feet on the bottom of the pool and the other floating on the surface.

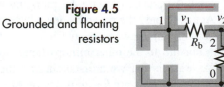

Figure 4.5
Grounded and floating resistors

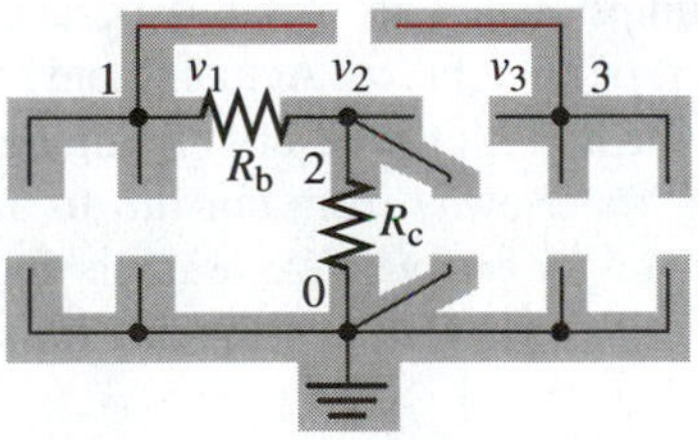

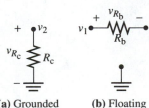

Figure 4.6   Element voltages

(a) Grounded   **(b)** Floating

We focus on our two resistors in a bit more detail in Figure 4.6. Figure 4.6(a) shows the grounded resistor. Its element voltage, v_{R_c}, is the same as the node voltage v_2 at the nonreference node to which it is connected. The only other way we could define this element voltage would be to reverse its polarity — in which case it would be the negative of the nonreference node voltage: $-v_2$. Figure 4.6(b) shows the floating resistor and its element voltage v_{R_b}. This element voltage is the *difference* of two node voltages: $v_1 - v_2$. (To see this, just write a KVL equation in path form with one path going from node 2 across the resistor to node 1 and the other from node 2 down to ground, then up to node 1.) The only alternative would be to reverse the polarities defining the element voltage,[1] in which case it would be equal to $v_2 - v_1$.

From this discussion, we can draw the following important conclusion: *each and every element voltage in the circuit is completely determined by the node voltages; in*

---

[1] Here is an excellent memory device: just call the voltage of the node connected to the + side of R_b v_+ and the voltage of the node connected to its negative side v_-. Then write $v_{R_b} = v_+ - v_-$. In the particular case shown in Figure 4.6, $v_+ = v_1$ and $v_- = v_2$. Even if we reverse our definition of v_{R_b}, however, we can still use the same equation — now v_+ will be v_2 and v_- will be v_1.

fact, each is either a node voltage, the negative of a node voltage, or the difference of two node voltages. Thus, if we are able to find the node voltages, we will then be able to compute all element voltages (and hence the element currents as well).

It should be clear that if there are N nodes in a circuit there will be $N - 1$ node voltages: those at the $N - 1$ nonreference nodes. We therefore need $N - 1$ independent equations in order to compute them. From where do they come? Well, we see that KVL only allows us to determine the element voltages from the node voltages, so we must turn to KCL for these equations. It seems logical to write one KCL equation at each of the $N - 1$ nonreference nodes—and that is just what we will do. Figure 4.7 shows an arbitrary node whose voltage is v_i and to which are connected some resistors and current sources. We have chosen the reference arrows for the resistor currents to all be pointed away from node i—or, as we will say, "leaving node i." This is an arbitrary choice, but it results in an equation that is in a compact and attractive form for solution of the resulting set of equations. (The node voltages are defined, of course, relative to a reference node not shown explicitly in the figure.)

Expressing these resistor currents in terms of element voltages using Ohm's law and at the same time expressing the element voltages in terms of the node voltages,[2] we can write a KCL equation at node i in the form

$$\frac{v_i - v_j}{R_a} + \frac{v_i - v_m}{R_b} = i_{s1} - i_{s2}. \tag{4.1-1}$$

If we do this for each of the $N - 1$ nonreference nodes, we will obtain $N - 1$ equations whose unknown variables are the $N - 1$ node voltages. Notice that the node voltages on the "far sides" of the i-sources have no effect; they do not appear in the equation at all.

The hydraulic analogy in Figure 4.8 is a good memory aid.[3] There we see two constant flow rate water pumps analogous to the i-sources in Figure 4.7 pumping water to a hill (analogous to node i) of "height" v_i. Actually, only the pump having a "value" of i_{s1} is pumping water *toward* the hilltop; the other is pumping water *away* from it. Furthermore, there are pipes carrying water away from our hill to the other hilltops (analogous to the nodes j and m) of "height" v_j and v_m. We assume that we are standing on top of hill i when we write KCL at node i and that all the other hills are *lower*. If one of the other hills

Figure 4.7   Writing a nodal equation

Figure 4.8
Hydraulic analogy

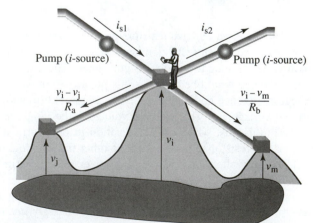

---

[2] Refer once more to footnote 1.

[3] In fact, current was once thought to be a "fluid" passing through wires much like water through a pipe—so our analogy has some basis in history.

happens to be higher than ours, the *value* of the water flow rate in the corresponding pipe will possibly have a negative sign. Our nodal equation in words, then, is analogous to the observation that the total flow rate of water away from our hill through the pipes connected to the other hilltops is equal to the total flow rate toward our hill from water pumps. In purely circuit terminology, the total current away from node i through resistors is equal to the total current toward node i from current sources.

We now have enough mathematical background to solve the circuit we have been investigating, the one we drew in Figure 4.1. We arbitrarily choose the ground reference as before to be at the bottom node and assign unknown node voltages to nodes 1, 2, and 3. We show the circuit "prepared for nodal analysis" in Figure 4.9. We remind you that the six ground symbols all refer to the single node chosen as reference: node 0.

Figure 4.9
The example circuit prepared for nodal analysis

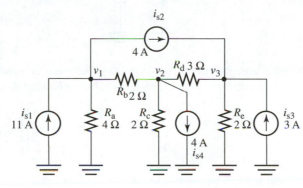

We will now write three KCL equations. The portion of the circuit pertaining to the equation at node 1 is shown in Figure 4.10. Once more, we note that the voltage v_3, being on the "other side" of the 4-A i-source, has no effect on the KCL equation, which is

$$\frac{v_1}{R_a} + \frac{v_1 - v_2}{R_b} = i_{s1} - i_{s2}. \tag{4.1-2}$$

Figure 4.10
The partial circuit concerning node 1

Figure 4.11
The partial circuit concerning node 2

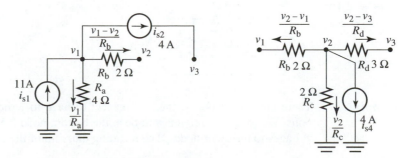

The partial circuit concerning node 2 is shown in Figure 4.11. Only one current source is attached directly to node 2; thus, the others do not appear in the KCL equation for that node. This equation is

$$\frac{v_2}{R_c} + \frac{v_2 - v_1}{R_b} + \frac{v_2 - v_3}{R_d} = -i_{s4}. \tag{4.1-3}$$

The last partial circuit, that concerned with node 3, is shown in Figure 4.12. Once more, the node voltage on the "far side" of the current source at the top (v_1) has no effect:

$$\frac{v_3}{R_e} + \frac{v_3 - v_2}{R_d} = i_{s2} + i_{s3}. \tag{4.1-4}$$

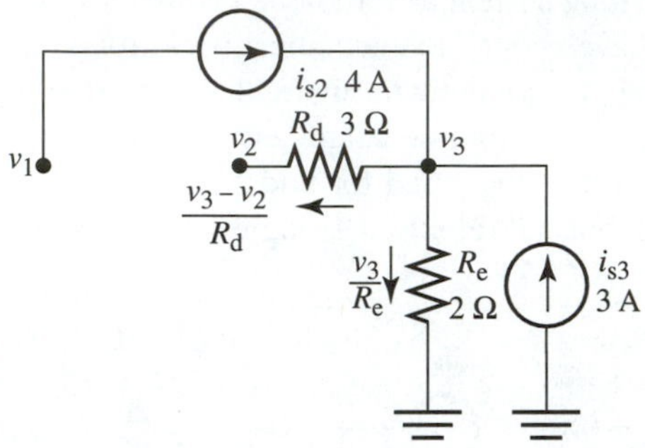

Figure 4.12
The node 3 partial circuit

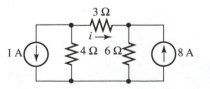

Matrix Form for the Nodal Equations

We can write our three nodal equations in more compact form as a single matrix equation. This equation, with the element values replacing their symbols, is

$$\begin{bmatrix} \frac{1}{4} + \frac{1}{2} & -\frac{1}{2} & 0 \\ -\frac{1}{2} & \frac{1}{2} + \frac{1}{2} + \frac{1}{3} & -\frac{1}{3} \\ 0 & -\frac{1}{3} & \frac{1}{3} + \frac{1}{2} \end{bmatrix} \begin{bmatrix} v_1 \\ v_2 \\ v_3 \end{bmatrix} = \begin{bmatrix} 7 \\ -4 \\ 7 \end{bmatrix}. \tag{4.1-5}$$

We will leave it to you to use the matrix techniques[4] covered in Appendix A to show that the solution is $v_1 = 12$ V, $v_2 = 4$ V, and $v_3 = 10$ V. Notice that each diagonal term in position (i, i) is the sum of the conductances connected to node i and that each off-diagonal term in position (i, j) is the negative of the conductance connecting nodes i and j. The conductance connecting nodes j and i is precisely the same, which means that the coefficient matrix is symmetric. We will discuss this more fully in Section 4.5.

A Nodal Analysis Example

Example 4.1   Find the current i for the circuit in Figure 4.13 using nodal analysis.

Figure 4.13
An example circuit

Solution   There are $N = 3$ nodes; thus, we expect to have to write $N - 1 = 2$ KCL nodal equations. We select the reference to be at the bottom node (somewhat arbitrarily)[5] and assign voltages to the other nodes. The resulting diagram of the circuit prepared for nodal analy-

---

[4] Perhaps the easiest method is to first multiply each row (both left and right sides of the equation) by the least common denominator, then to use row reduction.

[5] There is some slight savings in complexity by choosing the reference node as the one to which the most elements are connected (the bottom node in our example); however, if one uses the height/voltage and water flow rate/current analogy, there is an intuitive advantage in choosing the bottom node as the reference.

sis is shown in Figure 4.14. We remind you that the four ground symbols all refer to a single node: the reference node.

Figure 4.14
The example circuit prepared for nodal analysis

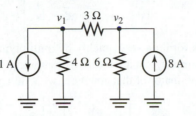

Now let's write one KCL equation at each nonreference node. Figure 4.15 shows the part of the circuit pertaining to the equation at node 1. You should sketch such a partial circuit until you become comfortable with the technique, after which you will probably want to simply write down the equation. The equation (everything that goes in has to come out) is

$$\frac{v_1}{4} + \frac{v_1 - v_2}{3} = -1. \tag{4.1-6}$$

Figure 4.15
The KCL equation at node 1

Figure 4.16
The KCL equation at node 2

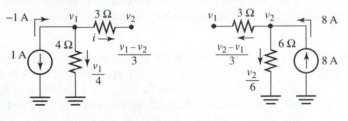

Moving along to node 2, we have the partial circuit shown in Figure 4.16. The KCL equation at this node is

$$\frac{v_2}{6} + \frac{v_2 - v_1}{3} = 8. \tag{4.1-7}$$

One thing that confuses many beginners is the current in the floating 3-Ω resistor: we assumed it to be leaving node 1 when we wrote the KCL equation at that node, and now we are assuming the opposite! Is this a mistake? No, it is not—because we have also reversed the order of the node voltages v_1 and v_2; this means that our two current definitions actually represent the same physical current.

Now let's discuss the solution a bit. We could simply rationalize the two scalar equations in (4.1-6) and (4.1-7) and solve them, perhaps, by substitution. However, this can become quite difficult for more complicated circuits. In fact, if we use a calculator with a matrix equation solver or a personal computer with such a software package, we *must* be able to write our equations in matrix form. The matrix form for our example circuit is

$$\begin{bmatrix} \frac{1}{4} + \frac{1}{3} & -\frac{1}{3} \\ -\frac{1}{3} & \frac{1}{6} + \frac{1}{3} \end{bmatrix} \begin{bmatrix} v_1 \\ v_2 \end{bmatrix} = \begin{bmatrix} -1 \\ 8 \end{bmatrix}. \tag{4.1-8}$$

For clarity of explanation of the basic concepts, many of our examples will involve only one or two equations. The single equation case is easy, and you can easily handle two equations in two unknowns if you commit the formula for the inverse of a two-by-two matrix derived in Appendix A to memory. It is

$$\begin{bmatrix} a & b \\ c & d \end{bmatrix}^{-1} = \frac{1}{ad - bc} \begin{bmatrix} d & -b \\ -c & a \end{bmatrix}. \tag{4.1-9}$$

You can verify this formula very quickly by multiplying the right-hand side by the original matrix. If we simplify equation (4.1-8) by multiplying each row on both sides by the least common denominator of that equation, we will have

$$\begin{bmatrix} 7 & -4 \\ -2 & 3 \end{bmatrix} \begin{bmatrix} v_1 \\ v_2 \end{bmatrix} = \begin{bmatrix} -12 \\ 48 \end{bmatrix}. \tag{4.1-10}$$

Then, after applying the inverse formula in (4.1-9), we get

$$\begin{bmatrix} v_1 \\ v_2 \end{bmatrix} = \frac{1}{13} \begin{bmatrix} 3 & 4 \\ 2 & 7 \end{bmatrix} \begin{bmatrix} -12 \\ 48 \end{bmatrix} \begin{bmatrix} 12 \\ 24 \end{bmatrix}. \tag{4.1-11}$$

Thus, $v_1 = 12$ V and $v_2 = 24$ V. Going back to our original circuit (Figure 4.13), we have

$$i = \frac{v_1 - v_2}{3} = \frac{12 - 24}{3} = -4 \text{ A.} \tag{4.1-12}$$

Notice that we did not use i in our solution process until the last step. *Nodal analysis always treats the node voltages—and only the node voltages—as the unknowns.*

Independence and Solvability

At this point we have developed the nodal analysis procedure completely for circuits with resistors and i-sources. However, we asserted earlier that we required $N - 1$ *independent* equations in order to solve for the $N - 1$ unknown node voltages. How do we know that our procedure always results in equations that are independent? It turns out that in fact they always will be—provided that the circuit meets a very general condition that is satisfied by all practical realizations. We will now discuss this condition.

In order to make our thinking concrete, let's go back to our original example, the one shown in Figure 4.1. The same circuit, you will recall, was also drawn in Figure 4.9 with the ground reference node selected at the bottom and the nonreference node voltages marked. We repeat the latter figure here as Figure 4.17 for ease of reference. We recall that the nodal equations could be written in the matrix form of equation (4.1-5), which we now write more compactly and more generally as

$$G\bar{v}_N = \bar{i}_s. \tag{4.1-13}$$

G will be called the *nodal conductance matrix*. It is of dimensions $(N - 1) \times (N - 1)$ and has entries that involve only the conductances of the resistors. The matrix $\bar{v}_N$ is the $(N - 1) \times 1$ column matrix (or vector) of node voltages, and $\bar{i}_s$ is an $(N - 1) \times 1$ col-

Figure 4.17
The example circuit prepared for nodal analysis

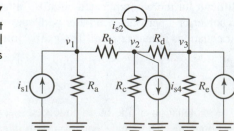

umn matrix (or vector) whose *i*th entry is the algebraic sum of all the current source values connected to node i, with a plus sign attached to any directed toward that node. As we show in Appendix A, equation (4.1-13) will have a unique solution if (and only if) *G* is nonsingular—that is, if its determinant is nonzero.

Now suppose we *deactivate* the circuit by reducing all the i-sources to zero. This produces the *deactivated network* shown in Figure 4.18.[6] The right-hand side of our matrix nodal equation (4.1-13) is now zero (a zero column matrix); therefore, that equation now has the form

$$G \bar{v}_N = \bar{0}. \tag{4.1-14}$$

G is, of course, unchanged. *G* will be nonsingular and our original nodal equations will therefore have a unique solution if and only if the only solution of (4.1-14) is $\bar{v}_N = \bar{0}$; that is, if all the node voltages in the deactivated network are zero (see Appendix A). We will show that in fact all the node voltages are zero provided that the circuit satisfies one simple condition.

Figure 4.18
The deactivated network

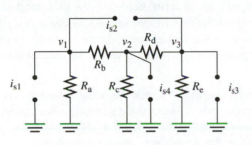

This condition is best stated in terms related to the topology of the circuit. We say that a circuit is *connected* if there is a path of elements from any node to any other node. In our deactivated network in Figure 4.18, for example, there are several paths between the reference node and node 3. We show one in Figure 4.19. A simpler one, of course, is the path consisting of the single element R_e. If we check paths for each of the four nodes, we will see that the deactivated network of our example circuit is, in fact, connected.

Figure 4.19 One path between the reference node and node 3

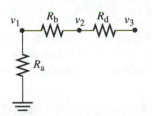

Here's how connectedness and the positivity of the resistors assure us that the nodal equations are always solvable. We already know that the node voltages determine all the element voltages. The reverse statement is also true—if the network is connected. To see this, consider a general circuit and look at any path from the reference node to an arbitrary node k. We sketch the situation in Figure 4.20. The reference node, by definition,

Figure 4.20
A resistive path in a connected deactivated network

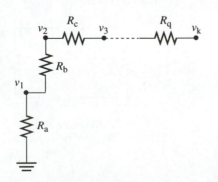

---

[6] You will often find it referred to in the literature as the "dead network;" we, however, prefer the less violent term "deactivated." This should convey the impression that it is possible to "reactivate" it.

has zero voltage. We notice that v_1 is the same as the element voltage for R_a. To find v_2 we add the element voltages of R_a and R_b, and so on. Last, we express the node voltage v_k as the sum of element voltages along the path from the reference node to node k. This shows that each node voltage can be written as the algebraic sum (with signs considered) of the element voltages because there is a path of resistors between the reference node and each nonreference node in a connected circuit.

Now we recall that the power absorbed by a resistor of value R is $v^2/R = Gv^2$, where v is the branch voltage and $G = 1/R$ is the conductance. What is the total power absorbed by the deactivated network of Figure 4.18? Why, zero, of course![7] But this total absorbed power in our example circuit is also given by the formula (conservation of energy)

$$P_{\text{total absorbed}} = g_a v_a^2 + g_b v_b^2 + g_c v_c^2 + g_d v_d^2 + g_e v_e^2 = 0, \qquad (4.1\text{-}15)$$

where we have used the element voltages; that is, v_a is the voltage of element a, v_b the voltage of element b, and so on. (A similar equation holds in general for any resistive circuit.) But because each term is nonnegative, this means that each branch voltage in the deactivated network is zero;[8] hence, by the preceding argument, so are all the node voltages—provided the deactivated network is connected. Because zero is the only solution of (4.1-14), the nodal conductance matrix has to be nonsingular. We are therefore assured that the "reactivated" network, in which the current sources have their original values, is solvable.

Example: A Circuit Not Solvable by Nodal Analysis

To illustrate a circuit for which the nodal equations do *not* possess a solution, we need only look at one whose deactivated network is not connected. A simple circuit of this type is shown in Figure 4.21(a). Figure 4.21(b) shows the deactivated network. The matrix nodal equation for the circuit can be easily shown (meaning we leave the work to you) to be

$$\begin{bmatrix} g_a & 0 & 0 \\ 0 & g_b & -g_b \\ 0 & -g_b & g_b \end{bmatrix} \begin{bmatrix} v_1 \\ v_2 \\ v_3 \end{bmatrix} = \begin{bmatrix} -i_s \\ i_s \\ -i_s \end{bmatrix}. \qquad (4.1\text{-}16)$$

As you can see, the last two rows of the nodal conductance matrix are negatives of each other; this means that the matrix is singular and consequently there is no solution for the node voltages. In terms of the node voltages, the total power absorbed by the deactivated network is

$$P = g_a v_1^2 + g_b(v_2 - v_3)^2. \qquad (4.1\text{-}17)$$

Figure 4.21
A circuit whose nodal equations are unsolvable

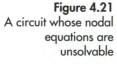
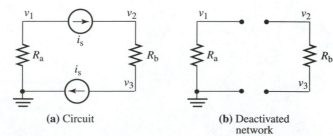

(a) Circuit   (b) Deactivated network

---

[7] This follows quite rigorously from Tellegen's theorem, which is proved in Appendix B and does not depend upon the existence of the solution to the nodal equations (so our argument is not circular).

[8] The only way a number of nonnegative terms can sum to zero is for each, individually, to be zero!

Both sources have been deactivated, so P is zero; however, although this implies that $v_1 = 0$, it only leads to $v_2 = v_3$. Such a circuit seems to require two references, and it actually *can* be treated as two independent circuits for the purpose of analysis. In fact, it should be clear that any circuit whose deactivated network is not connected consists of "islands" (or partial circuits) of connected resistors between which only i-sources (or nothing) are connected. We can choose a reference node in each part—and nodal analysis will then work.

The practically oriented reader might question the usefulness of such a circuit as the preceding one. After all, he or she might argue, one would not knowingly design such a circuit. This is perhaps true, but we point out that more practical circuits are often modeled with ideal elements, and such circuits can occur—particularly in computer simulation algorithms. One must be aware of these circuits and know how to test for the nonexistence of solutions.

Section 4.1 Quiz

Q4.1-1. Draw the deactivated network for the circuit in Example 4.1, Figure 4.13.

Q4.1-2. Write the nodal equation at node 3 in Figure Q4.1-1 after choosing the reference to be at node 0.

Figure Q4.1-1

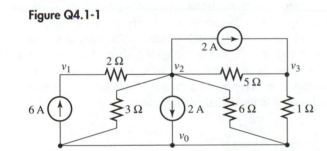

Q4.1-3. Choosing the reference node to be at node 1, write the nodal equation at node 0 for the circuit shown in Figure Q4.1-1.

Q4.1-4. Choosing the reference node to be at node 2, write the complete set of nodal equations for the circuit shown in Figure Q4.1-1.

Q4.1-5. Find the voltage v in the circuit shown in Figure Q4.1-2 using nodal analysis relative to the ground reference shown.

Figure Q4.1-2

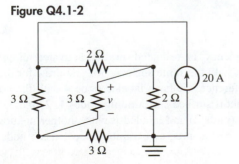

Q4.1-6. Do the nodal equations for the circuit in Figure Q4.1-3 have a solution? Why, or why not?

Figure Q4.1-3

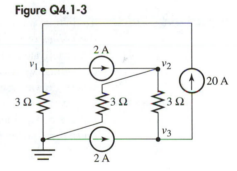

Q4.1-7. If possible, solve for the node voltages v_1, v_2, and v_3 in the circuit shown in Figure Q4.1-3.

Q4.1-8. Solve for the element voltages in Figure 4.21a using KCL, KVL, and Ohm's law. Note that this shows that some circuits can be analyzed for their *element* variables even when one cannot solve the nodal equations.

4.2 | Nodal Analysis of Circuits with Voltage Sources

We will now discuss the analysis of circuits having voltage sources. The one sketched in Figure 4.22 is an example. We have labeled the four nodes, which you can identify by erasing the body of each of the elements as we discussed in Section 4.1. If the two v-sources were replaced by i-sources, we would have to write $4 - 1 = 3$ nodal equations in order to solve for three independent nonreference node voltages. With the v-sources in the circuit, however, the node voltages are not independent. Let's see how many *are*.

Figure 4.22
An example circuit

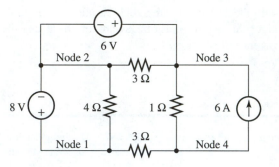

Generalized Nodes   To do this, we make the following observation: our analysis method should not depend on the *values* of the independent sources. Therefore, let us deactivate all of them to obtain the deactivated network shown in Figure 4.23. We see that there are two nodes in this circuit; thus, we would anticipate only $2 - 1 = 1$ nodal equation. The only question is, where do we write it? Notice that nodes 1, 2, and 3 merge together when the voltage sources are deactivated. This reflects the fact that their node voltages are not independent, but are constrained by the v-sources. We have drawn a closed Kirchhoff surface around these three nodes and one around node 4 in the deactivated network.

Figure 4.23
The deactivated circuit

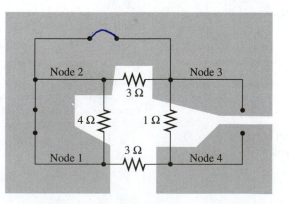

Now let's see how this dependency is reflected in the activated circuit itself. Figure 4.24 shows the original circuit with the same surfaces we just drew for the deactivated network. Without regard for reference, we have labeled node 4 with the node voltage v_y; furthermore, within the Kirchhoff surface surrounding nodes 1, 2, and 3, we have chosen one of these nodes arbitrarily (node 2) and labeled it with another unknown node voltage v_x. We now see the effect of the v-source constraints: node 1 has a node voltage of $v_x + 8$ and node 3 a node voltage of $v_x + 6$.[9] We call any *set* of nodes connected by a path of v-sources a *generalized node*. We see that *there is only one independent node voltage for each generalized node*. We will refer, temporarily, to nodes to which no v-source is connected, such as node 4, simply as nodes (or "ordinary" nodes).

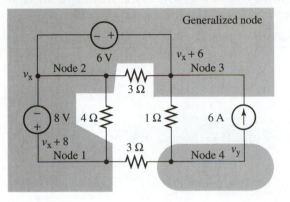

Figure 4.24
An example circuit

Node Classification

We have already observed that only one nodal equation is required to solve our example circuit because it is essentially the same as a two-node circuit. It therefore seems logical to anticipate writing one KCL equation for either the node at the lower right or the generalized node. Therefore, let us choose a ground reference node. We will first consider making our choice within the generalized node at one of its three constituent nodes (remember that we have defined a generalized node to be a *set* of nodes, not a single node). Figure 4.25 shows the result of choosing the reference at node 2. This forces $v_x = 0$; at the same time, we see that node 1 now carries the known value 8 V and node 3 the known value 6 V. We can now sort the nodes within the generalized node further than we have already done, into the reference node and the others, which we *now* term *nonessential nodes*. Their voltages are known values. We now call the ordinary nonreference node an *essential node* to distinguish it from the others. The essential node carries an unknown voltage.

Figure 4.25
Essential and
nonessential nodes

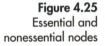
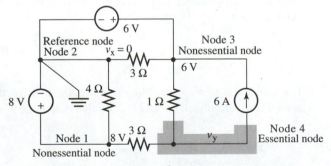

We now write one KCL equation at the essential node (shown shaded) in the form

$$\frac{v_y - 8}{3} + \frac{v_y - 6}{1} = -6. \tag{4.2-1}$$

---

[9] To verify these values, apply KVL in path form from node 2 to nodes 1 and 3, noting that they are held at voltages 8 and 6 volts, respectively, *above* the voltage of node 2.

The solution is easily seen to be $v_y = 2$ V. At this point, we know all of the node voltages and can compute any circuit variable of interest.

Now let's explore what happens if we choose the ground reference to be at node 4. Figure 4.26 shows the result. We now refer to the generalized node as a *supernode*. *A supernode, then, is a generalized node not containing the reference node.* KCL for the entire surface surrounding the supernode is

$$\frac{v_x + 8}{3} + \underset{(+)}{\frac{v_x - (v_x + 8)}{4}} + \underset{(-)}{\frac{(v_x + 8) - v_x}{4}} + \frac{v_x + 6}{1} + \underset{(+)}{\frac{v_x - (v_x + 6)}{3}} + \underset{(-)}{\frac{(v_x + 6) - v_x}{3}} = 6. \qquad (4.2\text{-}2)$$

Figure 4.26
The concept of a
supernode

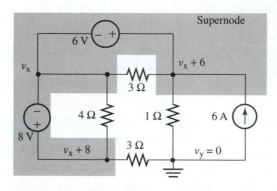

Before we solve this equation, let's make a simple observation that will save quite a bit of computational labor in our future work: there are two pairs of terms that are equal but opposite in sign (the second and third and the fifth and sixth). This is no accident. They correspond to the currents through the two resistors connected directly between two of the nodes making up the supernode. Each resistor current appears once leaving the Kirchhoff surface around the supernode and once—with opposite sign—reentering. We can make our work simpler at the outset by merely including any such element within the Kirchhoff surface itself, as shown in Figure 4.27, thereby eliminating the offending terms completely. Application of KCL to this surface results in the much simpler (but equivalent) equation

$$\frac{v_x + 6}{1} + \frac{v_x + 8}{3} = 6. \qquad (4.2\text{-}3)$$

The solution is $v_x = -2$ V. Notice that this is consistent with our earlier result in which node 4 was determined to be at $+2$ V relative to node 2.

Figure 4.27
Enclosing "redundant"
elements in the
supernode surface

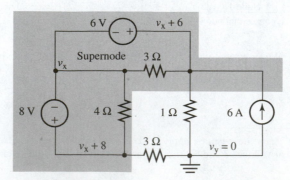

The Nodal
Analysis Algorithm

The following algorithm summarizes the nodal method as a procedure:

Nodal Analysis

1. Deactivate the circuit and determine all of the ordinary nodes and generalized nodes. The latter are those that merge together when the v-sources are deactivated.

2. Choose the ground reference. If it is chosen within a generalized node, all of the other nodes within that generalized node are called nonessential nodes, for they carry known node voltages. If it is chosen at an ordinary node, each generalized node is referred to as a supernode. Ordinary nonreference nodes are now referred to as essential nodes.

3. Assign node voltages—one unknown node voltage to each essential node and one unknown node voltage to an arbitrarily chosen node in each supernode. Each of the other node voltages within each supernode is expressed in terms of the arbitrarily chosen one and the v-source values. Any nonessential nodes are labeled with their known values of node voltage.

4. Write one KCL equation for each essential node and one for each supernode in terms of the unknown node voltages assigned in step 3.

5. Solve the resulting equations for the unknown node voltages.

6. Solve for any desired circuit variable.

Required Number of
Nodal Equations

Before continuing, we should clear up one question. In general, how many KCL nodal equations are necessary? Refer to the deactivated version of the circuit we have been working with, Figure 4.23, which is repeated here for ease of reference as Figure 4.28. The dark dots represent the terminals of the original sources, where each v-source has been replaced by a short circuit and the current source by an open circuit. Notice that when each v-source was deactivated it merged two nodes, thus reducing the number of nodes by one; hence, if there are N_v v-sources, the number of nodes is reduced by N_v. Therefore, the number of unknown node voltages (and also the number of equations) is

$$N_{ne} = N - 1 - N_v. \qquad (4.2\text{-}4)$$

Each v-source reduces the complexity of nodal analysis by one equation. Notice, however, that deactivation of the i-source did not affect the number of nodes.

Figure 4.28
The deactivated circuit

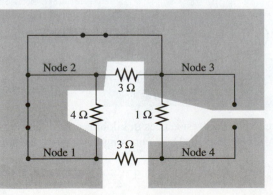

A Nodal Analysis Example with Voltage Sources

Example 4.2   Find the current i in Figure 4.29.

Figure 4.29
An example circuit

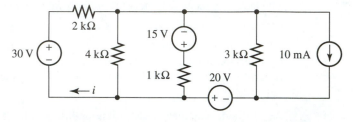

Solution   First, we identify nodes and generalized nodes. The result is given in Figure 4.30. Notice that there are no ordinary nodes, only two generalized nodes. Next, we choose our ground reference node at the bottom left node within generalized node 1. This is an arbitrary choice. The nodes can now be further categorized. We see that there are two nonessential nodes and one supernode. We next assign node voltages as shown in Figure 4.31. Known voltages (30 V and −20 V) are assigned to the two nonessential nodes, and an unknown voltage is assigned to one arbitrary node inside the supernode. The voltage at the other node in the supernode is expressed in terms of that unknown voltage. We next write one KCL equation at the supernode. Before we do, however, let's pause for an important observation. Ohm's law can be written in any convenient set of units desired. For instance, we have

$$V = mA \times k\Omega. \tag{4.2-5}$$

Figure 4.30
Identifying the node structure

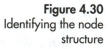

Figure 4.31
The example circuit prepared for nodal analysis

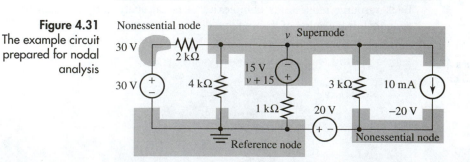

If we express all resistances in kilohms and all currents in milliamperes, then the voltages will still be in volts. Thus, choosing these units, the supernode equation is

$$\frac{v - 30}{2} + \frac{v}{4} + \frac{v - (-20)}{3} + \frac{v + 15}{1} = -10. \tag{4.2-6}$$

The solution is $v = -8$ V.

It is at this point—and only at this point—that we reintroduce the current i that we were asked to find. We have done this and labeled all nodes with their now-known voltages in Figure 4.32. Furthermore, we have used the differences of the node voltages (that is, KVL) and Ohm's law to find the resistor currents and KCL to find currents at joints, resulting in the values shown. By applying KCL at the ground reference node, we see that $i = 14 + 7 - 2 = 19$ mA. You might wish to check this by applying KCL at, say, the joint at the top of the 4-kΩ resistor.

Figure 4.32
Solving for the desired
quantity

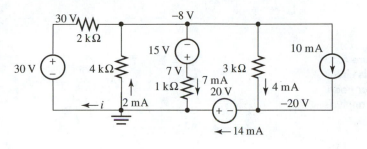

General Form
for the Matrix
Nodal Equation

Now that we have developed the technique of nodal analysis in a "shirtsleeves" way, that is, by writing a scalar equation at each appropriate node and supernode, we are in a position to investigate the general nature of the resulting equations in matrix form. We will analyze the circuit in Figure 4.33 as a concrete example. It is complex enough to illustrate the general method. We want to draw some conclusions about the mathematical form of the nodal equations, so we will use the literal values shown. Notice that each resistor has been labeled for convenience with its conductance rather than its resistance.

Figure 4.33
An example circuit

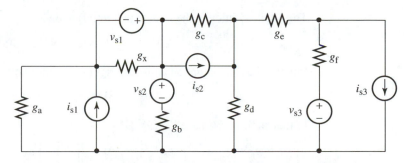

First, let's get some idea about the complexity to be expected. We count $N = 7$ nodes (don't overlook the one between v_{s2} and g_b and the one between v_{s3} and g_f) and $N_v = 3$ v-sources. Thus, we would expect to have to write $N - 1 - N_v = 3$ nodal equations. The only question is where? *To answer this, we identify the ordinary and generalized nodes by erasing all element bodies except those of the v-sources as shown in Figure 4.34.* We see that there are two ordinary nodes and two generalized nodes. Next, we select the ground reference. We pick it to be the bottommost node merely to keep our height-voltage analogy running smoothly. Otherwise, this choice is arbitrary. We then redraw the circuit prepared for nodal analysis, with the nodes further classified into essential, nonessential, and supernodes as shown in Figure 4.35. We have split the bottom node into several pieces using the ground reference symbol for convenience. Although we have shown a surface around the nonessential node, we will not write a KCL equation for it, only one for each of the three other surfaces shown.

Figure 4.34
Testing for node
structure

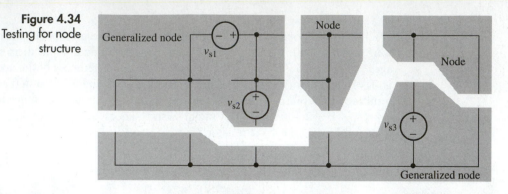

Figure 4.35
The example circuit
prepared for nodal
analysis

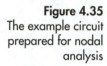
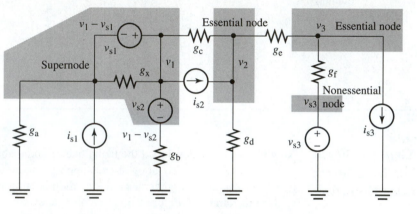

The KCL equation for the supernode is[10]

$$g_a(v_1 - v_{s1}) + g_b(v_1 - v_{s2}) + g_c(v_1 - v_2) = i_{s1} - i_{s2}. \qquad (4.2\text{-}7)$$

The KCL equation at the middle (essential) node is

$$g_c(v_2 - v_1) + g_d v_2 + g_e(v_2 - v_3) = i_{s2}. \qquad (4.2\text{-}8)$$

Finally, the KCL equation at the top right (essential) node is

$$g_e(v_3 - v_2) + g_f(v_3 - v_{s3}) = -i_{s3}. \qquad (4.2\text{-}9)$$

In matrix form, we have

$$\begin{bmatrix} g_a + g_b + g_c & -g_c & 0 \\ -g_c & g_c + g_d + g_e & -g_e \\ 0 & -g_e & g_e + g_f \end{bmatrix} \begin{bmatrix} v_1 \\ v_2 \\ v_3 \end{bmatrix} = \begin{bmatrix} i_{s1} - i_{s2} + g_a v_{s1} + g_b v_{s2} \\ i_{s2} \\ -i_{s3} + g_f v_{s3} \end{bmatrix}. \qquad (4.2\text{-}10)$$

(We have moved the terms involving the independent v-sources to the right-hand side of
this equation.)

Notice that the coefficient matrix on the left of equation (4.2-10)—the *generalized
nodal conductance matrix*—consists only of conductances (as in Section 4.1, wherein we
discussed circuits without v-sources). The right-hand side of (4.2-10) contains the currents entering each node due to the independent sources (both i-sources and v-sources).
Thus, we can rewrite (4.2-10) in the compact symbolic form

$$G \bar{v}_N = \bar{u}_s, \qquad (4.2\text{-}11)$$

---

[10] Notice that we have included the resistor g_x inside the supernode surface as per our earlier discussion.

where G is the generalized nodal conductance matrix, $\bar{v}_N$ is the column matrix of independent node voltages, and $\bar{u}_s$ is a column matrix, the ith entry of which is a linear combination of the independent sources affecting that node. Thus, if we deactivate all the independent sources (v-sources as well as i-sources), we obtain the deactivated network shown in Figure 4.36. The matrix form of the nodal equations for this circuit is

$$G\,\bar{v}_N = \bar{0}, \tag{4.2-12}$$

where G is the same matrix as the one in equation (4.2-11).

Figure 4.36
The deactivated network

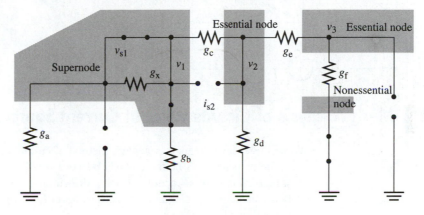

Solvability Condition for the General Nodal Equations

We can now use the same argument that we used in Section 4.1. If the deactivated network is connected (and for our example it is), the only solution to (4.2-12) is $\bar{v}_N = \bar{0}$. This implies that G is nonsingular and the solution to the original matrix nodal equation (4.2-11) is

$$\bar{v}_N = G^{-1}\bar{u}_s. \tag{4.2-13}$$

This means that each node voltage is a linear combination of the source values; furthermore, because we can compute any circuit voltage or current as a linear combination of node voltages using KVL, KCL, and/or Ohm's law, we see that all circuit variables are linear combinations of the independent source values. Superposition, the Thévenin and Norton equivalents, and many other circuit analysis procedures depend upon this fact. We assumed its truth in Chapter 3, where these concepts were developed. Our present development does not depend on the results in that chapter, so our assumption there was, in fact, quite valid.

Section 4.2 Quiz

Q4.2-1. For each of the partial circuits (generalized nodes) in Figure Q4.2-1, compute the voltages at the nodes not labeled with the unknown voltage. All voltage source values are 6 V.

Figure Q4.2-1

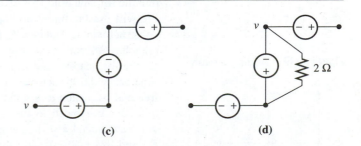

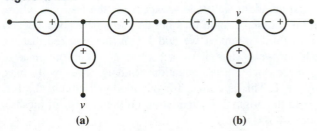

(a)   (b)   (c)   (d)

Q4.2-2. Consider the circuit in Figure Q4.2-2. By erasing the bodies of all elements except the v-sources, identify and classify the nodes into regular nodes and generalized nodes.

Figure Q4.2-2

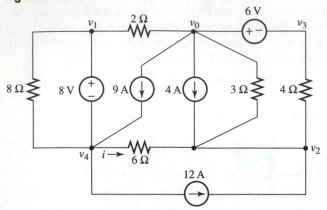

Q4.2-3. For the circuit in Figure Q4.2-2, draw the appropriate closed surfaces for writing KCL equations for a nodal analysis—one around each generalized node and one around each ordinary node.

Q4.2-4. Choosing the ground reference at node 0 in Figure Q4.2-2, write the nodal equations. (The node numbers are the subscripts on the node voltages shown in the figure.)

Q4.2-5. Find the current i in Figure Q4.2-2.

4.3 | Mesh Analysis of Circuits Without Current Sources

We will now develop another general method of circuit analysis called *mesh analysis*. It is not quite as general as nodal analysis because it requires that the circuit to be analyzed must be *planar*, that is, capable of being drawn on a sheet of paper in such a fashion that no two conductors or elements cross. It is nevertheless general enough for most applications involving hand analysis.

Meshes and Mesh Currents

Recall that a *loop* is a path of elements whose initial and final nodes are the same. We define a *mesh* to be a loop that does not contain any elements in its interior. Such a loop is often called a *window pane* loop. Figure 4.37 shows a "circuit skeleton," that is, it shows the nodes and merely uses lines in the place of the individual elements. Do not confuse these lines with short circuits. You can easily recognize the three meshes for they do, indeed, look like window panes. It is a fact that the number of meshes in a planar circuit is given by

$$N_m = B - N + 1, \qquad (4.3-1)$$

Figure 4.37   A circuit skeleton

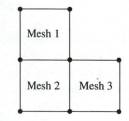

where B is the number of branches (elements) and N is the number of nodes. For our circuit skeleton, we see that there are $N = 8$ nodes and $B = 10$ branches. The number of meshes is $10 - 8 + 1 = 3$, as we expected from our inspection of the circuit.

Figure 4.38   A circuit with $B = N = 2$

We will go through the proof of equation (4.3-1), for it is not difficult. We first notice that the simplest circuit is one with two elements and two nodes, as in Figure 4.38. We can verify our formula at once: $N_m = B - N + 1 = 2 - 2 + 1 = 1$ mesh. Now assume that the formula holds for all circuits with any number of meshes less than or equal to N_m. We will use the three mesh circuit of Figure 4.37 to illustrate. From such a circuit, we construct another having $N_m + 1$ meshes, as shown in Figure 4.39, by adding nodes and branches. We start by adding a branch to one of the existing nodes of the circuit, then add another branch by connecting one of its ends to the free node of the first, and so on. Each time we add a branch we also add an additional node—except for the last one, whose free end we attach to a node in the original circuit, thus not adding a new node. Clearly this adds one mesh to the circuit. Furthermore, if we add k branches, we see that we will have added $k - 1$ new nodes as well. Thus, if we substitute $B + k$ in the place of B and $N + k - 1$ in the place of N in equation (4.3-1), we will have $B + k - (N + k - 1) + 1 = N_m + 1$. This is correct for our modified circuit; the formula will give one more mesh than the original. By induction on the number of meshes, therefore, we have proved equation (4.3-1).

Figure 4.39   Adding a mesh

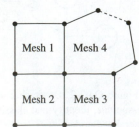

Figure 4.40   An example circuit skeleton

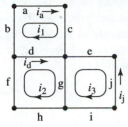

Figure 4.41   A "fictitious" mesh current

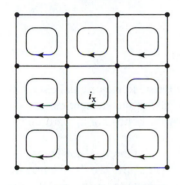

Figure 4.42   KCL and meshes

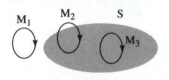

Mesh Equations

Figure 4.43   An example circuit

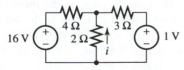

Now let's think about analyzing a circuit using its meshes. We modify Figure 4.37 as Figure 4.40 for reference here, by labeling the elements with letters. We also assume the existence of $B - N + 1 = 3$ fictitious *mesh currents* i_1, i_2, and i_3 to be circulating around the meshes as shown in the figure. We will discuss why we used the adjective "fictitious" in a moment. For now, picture the circuit in Figure 4.40 as an aerial view of a downtown city neighborhood, with each mesh being a city block. Suppose there is a flow of traffic around each block in a clockwise direction as shown by the arrows. That is, assume that cars continuously drive in a clockwise direction around the same block. The flow rate of circulating traffic around each block is analogous to the mesh currents. We now want to find the net flow rate of traffic in each street, which is the analog of an element in the actual circuit. We see that $i_a = i_1$ because i_1 is the only mesh current in the top branch, the one for which i_a is defined. Also, we see that $i_j = -i_3$ (note the sign!), for we are taking the traffic on that street (that element current) to be positive upward. Finally, we see that the *net* flow of traffic in the street marked with the i_d arrow (that is, the actual branch current in the analogous element) is given by $i_d = i_2 - i_1$. To summarize, we note that *each and every element current can be expressed as a mesh current, the negative of a mesh current, or the sum or difference of two mesh currents.*[11]

Figure 4.41 shows why we used the term "fictitious" mesh currents. The mesh current i_x does not appear alone in any element; hence, it cannot be directly measured. However, with our traffic flow analogy, it should be plausible that such a mesh current can be used. Notice, by the way, that the circuit skeleton shown in Figure 4.41 has $B = 24$ and $N = 16$, thus giving $B - N + 1 = 9$ meshes.

The mesh currents automatically satisfy KCL because they are assumed to flow along closed paths. To see this, just look at Figure 4.42. There, we have drawn an arbitrary three-dimensional closed surface S and three stylized mesh currents in the form of simple closed loops. We have shown three of them, labeled M_1 (completely outside S), M_2 (penetrating S), and M_3 (totally inside S). We see that any such mesh current must pierce the surface S an even number of times (if at all), half of them in a direction from inside to outside and half in the opposite direction. Thus, the mesh current will completely cancel in any KCL equation written for that surface.

Because KCL tells us nothing about a circuit in terms of its mesh currents, we turn to the only obvious alternative: KVL. In fact, we will write one KVL equation around each mesh. Though any form will suffice, it turns out that the most convenient form—the one leading to the solution in the smallest number of steps—is the following:

$$\sum_{\text{mesh } i} v_{\text{drops}}(R\text{'s}) = \sum_{\text{mesh } i} v_{\text{rises}}(v\text{-sources}), \qquad (4.3\text{-}2)$$

that is, we set the sum of the resistor voltage drops around mesh i in the direction of its mesh current equal to the sum of the voltage rises encountered in that same direction due to the v-sources. As an analogy, think of a ski lift. The tow ropes are analogous to voltage sources and the ski runs to resistors. The total altitude decrease one experiences in skiing to the bottom of the mountain must be equal to the total altitude gain imparted by the tow ropes. We are considering circuits that have no i-sources; therefore, (4.3-2) is a perfectly valid KVL equation.

Figure 4.43 shows an example circuit. We can see at once that it has two meshes, but we can quickly verify the formula for this number by noting that $B = 5$ and $N = 4$. Thus,

---

[11] Sum for two mesh currents in the same direction (cw or ccw) and difference for oppositely directed ones.

there are $B - N + 1 = 5 - 4 + 1 = 2$ meshes. We assume that we are to find the current i.

Our first step is to define mesh currents for these two meshes as we have done in Figure 4.44, which we will call the circuit "prepared for mesh analysis." We write one KVL equation for each mesh. In doing so, we treat the mesh currents as unknowns by expressing each resistor voltage—which we assume to be a drop in the direction of the mesh current—in terms of them. Thus, the KVL equation for mesh 1 is

$$4i_1 + 2(i_1 - i_2) = 16 \tag{4.3-3}$$

and the one for mesh 2 is

$$2(i_2 - i_1) + 3i_2 = -1. \tag{4.3-4}$$

In matrix form, we have

$$\begin{bmatrix} 6 & -2 \\ -2 & 5 \end{bmatrix} \begin{bmatrix} i_1 \\ i_2 \end{bmatrix} = \begin{bmatrix} 16 \\ -1 \end{bmatrix}, \tag{4.3-5}$$

which has the solution $i_1 = 3$ A and $i_2 = 1$ A. Finally, we return to the original circuit in Figure 4.43 and note that the quantity of interest, the current i, is the element current for the 2-Ω resistor, with its reference direction upward. Now i_2 is flowing upward and i_1 downward; hence, $i = i_2 - i_1 = 1 - 3 = -2$ A.

Let's discuss the signs of the terms in (4.3-3) and (4.3-4) a bit more. The 4-Ω resistor carries only the mesh current i_1, and its flow is from left to right, so we see (by the passive sign convention and Ohm's law) that the positive sign for voltage is on the left and the negative sign on the right. Hence, that voltage is a drop in the direction of the mesh current i_1. We have agreed to treat all drops across resistors as positive in our KVL equation, so it gets a plus sign in our equation. But what about the 2-Ω resistor that is shared by the two meshes? It, too, carries a positive sign because—as we are assuming its voltage is a drop in the direction of *mesh current i_1*—we must take the plus voltage reference on the top and the negative on the bottom. The passive sign convention then tells us that the reference direction for the current in that resistor must be downward. Furthermore, i_1 flows downward and i_2 upward. Thus, the voltage drop (with its associated positive sign in the KVL equation) is given by $2(i_1 - i_2)$.

"But wait!" you might object. "You have taken the plus sign on the bottom when you wrote the KVL equation for mesh 2!" We have indeed reversed our voltage reference for this element so that it is a drop in the second KVL equation; *however, we have also reversed the assumed direction of that element current so that it now points upward.* Therefore, we have also exchanged the positions of i_1 and i_2. The resulting voltage drop in the second mesh equation is $2(i_2 - i_1)$. This, of course, is the negative of the term in the equation for mesh 1, as it should be. Figure 4.45 shows a partial circuit to illustrate. The voltage v_x is a drop for mesh 1 and v_y a drop for mesh 2. We see that $v_x = 2(i_1 - i_2)$ and $v_y = 2(i_2 - i_1) = -v_x$. Thus, we are consistent in our approach.

Now let's look at a more complex circuit and derive the general form of the mesh equations. Such a circuit is shown in Figure 4.46. We have labeled all elements with literal values for generality. We have already assigned the mesh currents, all in the clockwise direction. We note that $B = 7$ and $N = 5$ (don't forget the node between v_{s3} and R_d!). Thus, the number of meshes is, as we expect, $B - N + 1 = 7 - 5 + 1 = 3$.

In Figure 4.47 we show the partial circuit of interest in writing the KVL equation for mesh 1. Taking both resistor voltages to be drops in the direction of mesh current i_1, we have

$$R_a(i_1 - i_3) + R_b(i_1 - i_2) = v_{s1}, \tag{4.3-6}$$

Figure 4.44 The example circuit prepared for mesh analysis

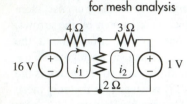

Figure 4.45 Concerning the shared resistor voltage

Figure 4.46 An example circuit

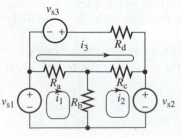

or
$$(R_a + R_b)i_1 - R_b i_2 - R_a i_3 = v_{s1}. \qquad (4.3\text{-}7)$$

Figure 4.48 shows that portion of the circuit involved in writing the KVL equation for mesh 2. That equation is

$$R_b(i_2 - i_1) + R_c(i_2 - i_3) = -v_{s2}, \qquad (4.3\text{-}8)$$

or
$$-R_b i_1 + (R_b + R_c)i_2 - R_c i_3 = -v_{s2}. \qquad (4.3\text{-}9)$$

Finally, in Figure 4.49, we show a sketch of the circuit elements entering into the KVL equation for mesh 3. That KVL equation is

$$R_d i_3 + R_c(i_3 - i_2) + R_a(i_3 - i_1) = v_{s3}, \qquad (4.3\text{-}10)$$

or
$$-R_a i_1 - R_c i_2 + (R_a + R_c + R_d)i_3 = v_{s3}. \qquad (4.3\text{-}11)$$

Figure 4.47
Pertaining to the mesh 1
KVL equation

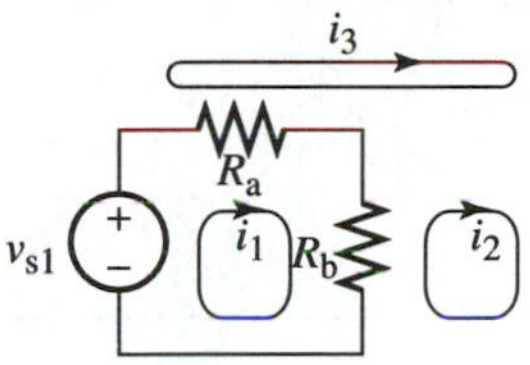

Figure 4.48
Pertaining to the mesh 2
KVL equation

Figure 4.49
Pertaining to the mesh 3
KVL equation

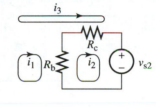
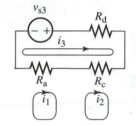

If we collect our three equations in matrix form, we see that

$$\begin{bmatrix} R_a + R_b & -R_b & -R_a \\ -R_b & R_b + R_c & -R_c \\ -R_a & -R_c & R_a + R_c + R_d \end{bmatrix} \begin{bmatrix} i_1 \\ i_2 \\ i_3 \end{bmatrix} = \begin{bmatrix} v_{s1} \\ -v_{s2} \\ v_{s3} \end{bmatrix}. \qquad (4.3\text{-}12)$$

In more compact form, we can write the mesh equations as

$$R\bar{i}_m = \bar{v}_s, \qquad (4.3\text{-}13)$$

where R is the $(B - N + 1) \times (B - N + 1)$ *mesh resistance matrix*, $\bar{i}_m$ is the $(B - N + 1) \times 1$ column matrix (vector) of mesh currents, and $\bar{v}_s$ is the $(B - N + 1) \times 1$ column matrix (vector) of source voltages.

A Mesh Analysis Example

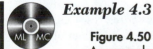

Example 4.3 Find the voltage v in the circuit shown in Figure 4.50.

Figure 4.50
An example
circuit

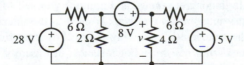

Solution We prepare the circuit for mesh analysis by assigning a mesh current to each of the $B - N + 1 = 7 - 5 + 1 = 3$ meshes in the same direction. This is shown in Figure 4.51. The KVL equation for the leftmost mesh is

$$6i_1 + 2(i_1 - i_2) = 28, \qquad (4.3\text{-}14)$$

the one for the middle mesh is

$$2(i_2 - i_1) + 4(i_2 - i_3) = 8, \qquad (4.3\text{-}15)$$

and the one for the rightmost mesh is

$$4(i_3 - i_2) + 6i_3 = -5. \qquad (4.3\text{-}16)$$

We now write down the matrix mesh equation in the form

$$\begin{bmatrix} 6+2 & -2 & 0 \\ -2 & 2+4 & -4 \\ 0 & -4 & 4+6 \end{bmatrix} \begin{bmatrix} i_1 \\ i_2 \\ i_3 \end{bmatrix} = \begin{bmatrix} 28 \\ 8 \\ -5 \end{bmatrix}. \qquad (4.3\text{-}17)$$

The solution is $i_1 = 13/3$ A, $i_2 = 10/3$ A, and $i_3 = 5/6$ A. Thus, $v = 4(i_2 - i_3) = 10$ V.

Figure 4.51
The example circuit
prepared for mesh
analysis

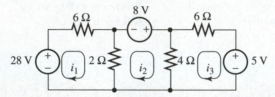

Solvability of the Mesh Equations

We now ask whether our mesh analysis procedure always works. To answer this question, we observe that the mesh resistance matrix does not involve the independent sources at all. Thus, let us deactivate the network of Figure 4.46, which we analyzed earlier. We show the result in Figure 4.52. The mesh equation for the deactivated network, in symbolic matrix form, is

$$R\bar{i}_m = \bar{0}. \qquad (4.3\text{-}18)$$

We know that the determinant of R is nonzero, and therefore (4.3-13) has a unique solution—provided that $\bar{i}_m = \bar{0}$ is the only possible solution to (4.3-18). (See Appendix A.)

To show that the only possible mesh currents in the deactivated network are those having zero value, we must look just a bit at the topology of the network. We define the *outer mesh* to be the set of conductors and/or elements making up the periphery (outside rim) of the network. In Figure 4.52, R_d is in the outer mesh, as are the short circuits representing the deactivated v-sources and the lengths of conductor connecting these elements. Note carefully, however, that the outer mesh is not a mesh. It is a loop, but it contains other elements. In this discussion only, we will use the term mesh to refer to either a true mesh or the outer mesh.

Figure 4.52 The deactivated network

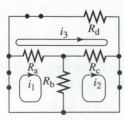

We will call resistors that are common to two meshes *shared* resistors. Now look at Figure 4.52 for a concrete image and imagine yourself to start in the interior of any mesh (or outside the outer mesh) and walk from it to the interior of any other mesh (or to the exterior of the outer mesh), allowing yourself to cross from one mesh to another *only by stepping over shared resistors*. For instance, in Figure 4.52, we can go from outside the outer mesh to the interior of mesh number 1 by stepping over R_d into mesh number 3, then across R_a into mesh number 1. We will say that a circuit is *coupled* if it is possible to thus travel from any arbitrary mesh to any other. It is easy to verify that the network in Figure 4.52 (and hence the original one in Figure 4.50 with the sources reactivated, as well) is coupled.

The concept of a coupled network[12] is important because it is for such networks that the mesh equations are guaranteed to be solvable. To show this, we first observe that the power absorbed by the deactivated network must be zero. In Figure 4.52, for instance, we can write the total power absorbed as

$$P_{\text{abs}} = R_a i_a^2 + R_b i_b^2 + R_c i_c^2 + R_d i_d^2 = 0, \tag{4.3-19}$$

where i_a, i_b, i_c, and i_d are the resistor element currents. Assuming that each of the resistors is positive, we see that each of these currents must be zero. Now, let's select an arbitrary mesh such as mesh 1 in the deactivated network of Figure 4.52 and consider ourselves to move from the outside mesh to the given mesh by stepping over shared resistors in the sequence just described. The first shared resistor we step over (R_d in Figure 4.52) carries only one mesh current (i_3), so that mesh current must be zero. This implies that the current in the next mesh we step into must be zero also ($i_1 = 0$), because the element current in the resistor we step across (R_a) carries only two mesh currents (i_1 and i_3) and one of them (i_3) is zero. Thus, we have shown that the only mesh currents possible in a deactivated coupled circuit have zero values; hence, the mesh resistance matrix is non-singular and the mesh equations are guaranteed to be solvable.

Example: A Circuit Not Solvable by Mesh Analysis

You might wonder what type of circuit cannot be solved by mesh analysis. According to our preceding argument, such a circuit cannot be coupled. Figure 4.53 shows a deactivated circuit that is not coupled. In this case, there is no resistor in the outer mesh. Furthermore, we see that any set of equal-valued mesh currents is possible. Each of the resistor currents is zero, even though the mesh currents are not. This is consistent with zero power being dissipated in the circuit. The mesh equations in matrix form are

Figure 4.53 A circuit that is not coupled

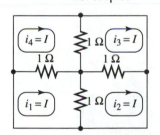

$$\begin{bmatrix} 2 & -1 & 0 & -1 \\ -1 & 2 & -1 & 0 \\ 0 & -1 & 2 & -1 \\ -1 & 0 & -1 & 2 \end{bmatrix} \begin{bmatrix} i_1 \\ i_2 \\ i_3 \\ i_4 \end{bmatrix} = \begin{bmatrix} 0 \\ 0 \\ 0 \\ 0 \end{bmatrix}. \tag{4.3-20}$$

A direct calculation shows that the determinant of the coefficient matrix is zero; more completely, if one uses Gaussian elimination, one can show that there are only three linearly independent equations among the set of four.

Section 4.3 Quiz

Q4.3-1. Write the mesh equation for mesh number 2 in the circuit in Figure Q4.3-1 in the form

$$\sum_{\text{mesh}} v_{\text{drops}}(R\text{'s}) = \sum_{\text{mesh}} v_{\text{rises}}(v\text{-sources}).$$

Figure Q4.3-1

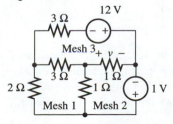

Define all mesh current references in the clockwise direction.

Q4.3-2. Write the matrix mesh equation for the circuit in Figure Q4.3-1.

Q4.3-3. Draw the deactivated network for the circuit in Figure Q4.3-1 and list the resistors shared by each pair of meshes, including the outer mesh.

Q4.3-4. Solve the circuit in Figure Q4.3-1 using mesh analysis and find the voltage v.

Q4.3-5. Select the outside mesh conductor in Figure 4.53 to be the reference node and use nodal analysis to find all the resistor voltages and currents. (This shows that a circuit that is not coupled can have a solution for its element voltages and currents even though the mesh equations are not solvable.)

---

[12] This term is also used later in this text *in a different context* to mean quite a different thing: "magnetically coupled."

4.4 | Mesh Analysis of Circuits with Current Sources

Mesh Classification

We will now extend the mesh analysis method to include circuits with current sources, such as the one in Figure 4.54. The problem is to find the current i. Counting branches and nodes, we see that $B = 10$ and $N = 6$ (don't forget the two nodes between the v-sources and their series resistors!). If the circuit had no i-sources, we would have to write $B - N + 1 = 10 - 6 + 1 = 5$ KVL equations, for there should be one for each of the five meshes; as it happens, however, the presence of i-sources reduces this number. To see just how, let's reason in the following way: our *method* of analysis (that is, where we write KVL equations and how many of them) should not depend on the precise *value* of the i-sources—only on how they are connected. Therefore, let's imagine the sources to have controls by which we can adjust their values. We will then imagine ourselves turning these controls to decrease the value of each to zero. Recalling that a zero-valued i-source is an open circuit and a zero-valued v-source a short circuit, we obtain the deactivated network in Figure 4.55. This deactivated circuit has only two meshes. Two have disappeared and two have merged together; only one was unaffected. We will call any mesh in the original circuit that disappears under the preceding test a *nonessential mesh,* any set of meshes that merge together a *supermesh,* and any mesh that remains unaffected an *essential mesh.* As we have already mentioned, we would analyze the original circuit in exactly the same manner as our test circuit, which has only two meshes. Thus, only two KVL equations are required in our example, and these equations should be written for the essential mesh and for the supermesh.

Figure 4.54
An example circuit

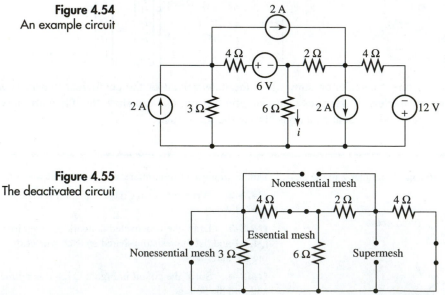

Figure 4.55
The deactivated circuit

The Required Number of Mesh Equations

In general, as a moment's reflection will show, deactivation of a current source results in the disappearance of one mesh (either it disappears completely or merges with another mesh), and deactivation of a v-source has no effect on the mesh structure; thus, the number of meshes in the test circuit, and therefore the number of required KVL equations, is

$$N_{me} = B - N + 1 - N_i, \tag{4.4-1}$$

where B is the number of branches, N is the number of nodes, and N_i is the number of i-sources in the circuit. There are three i-sources in our example. So $10 - 6 + 1 - 3 = 2$ KVL equations will be required.

Writing the Mesh Equations

To write these equations, we reactivate the sources to obtain the original network, then we assign a mesh current to *each* mesh. We label the known mesh currents in the nonessential meshes with their values. Furthermore, *we assign one unknown mesh current arbitrarily to each mesh making up the supermesh and express the other mesh currents in the supermesh in terms of that unknown.* We show the resulting circuit, which we will call the "circuit prepared for mesh analysis," in Figure 4.56. Note that we have arbitrarily chosen the leftmost of the two meshes making up the supermesh to carry the unknown value and have expressed the other in terms of it using the constraint imposed by the current source shared by the two: the two mesh currents must add up so as to produce a current source value of 2 A in the downward direction. Observe that $i_2 - (i_2 - 2) = 2$ in the downward sense.

Figure 4.56
The example circuit prepared for mesh analysis

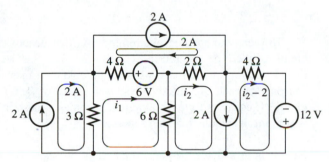

Now we write one KVL equation for the essential mesh, the one whose mesh current we have labeled i_1. The portion of the circuit influencing this equation is shown in Figure 4.57. We have only shown the elements in the essential mesh and have only indicated the mesh currents for the other meshes along with their values. The KVL equation for this mesh is

$$3(i_1 - 2) + 4(i_1 - 2) + 6(i_1 - i_2) = -6, \tag{4.4-2}$$

or

$$13i_1 - 6i_2 = 8. \tag{4.4-3}$$

Figure 4.57
The essential mesh KVL equation

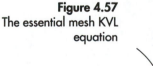

The portion of the circuit influencing the supermesh is shown in Figure 4.58. (Note, once again, that the two mesh currents are not independent, but add up to give the current source value.) The KVL equation for the supermesh is

$$6(i_2 - i_1) + 2(i_2 - 2) + 4(i_2 - 2) = 12, \tag{4.4-4}$$

or

$$-6i_1 + 12i_2 = 24. \tag{4.4-5}$$

In matrix form, these equations become

$$\begin{bmatrix} 13 & -6 \\ -6 & 12 \end{bmatrix} \begin{bmatrix} i_1 \\ i_2 \end{bmatrix} = \begin{bmatrix} 8 \\ 24 \end{bmatrix}. \qquad (4.4\text{-}6)$$

The solution is $i_1 = 2$ A and $i_2 = 3$ A. Returning to the original circuit, we see that the quantity of interest is given by $i = i_1 - i_2 = 2 - 3 = -1$ A.

Figure 4.58
The KVL equation for
the supermesh

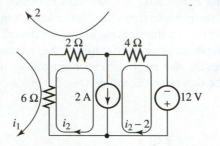

To recap our method, we remind you that deactivation of the i-sources produces a circuit of the type covered in Section 4.3—one that we can analyze without complication. In the original circuit, we use one unknown mesh current for each essential mesh and one for each supermesh. This gives a total of $B - N + 1 - N_i$ unknowns. We write one KVL equation for each loop that becomes a mesh in the test circuit: one for each essential mesh and one for each supermesh.

*The General Form for
the Mesh Equations*

Now let's investigate the general nature of our mesh equations. The circuit in Figure 4.59 is general enough for our purposes. (It is probably as complicated as any that one would elect to analyze by hand.) First, we test the circuit to determine its mesh structure by deactivating the sources. This produces the test circuit shown in Figure 4.60. *The closed loops do not represent loop currents.* They merely indicate the loops around which we will write KVL equations in the sense indicated by the arrowheads. We see that there are one nonessential mesh, one essential mesh, and two supermeshes.

Figure 4.59
An example circuit

Figure 4.60
Testing the circuit

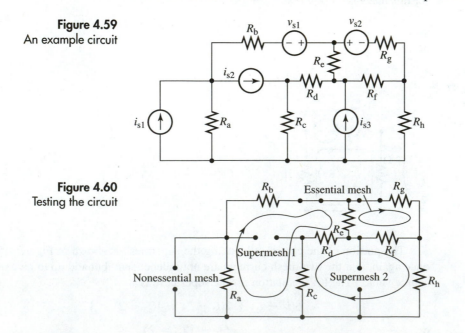

Let's return to the original circuit and arbitrarily assign one unknown mesh current to a mesh in each supermesh and one to the essential mesh. We next express the other mesh currents in each supermesh in terms of the one already chosen and the current source value(s). Finally, we label the nonessential mesh with its known mesh current. The circuit thus prepared for mesh analysis is shown in Figure 4.61. Note how we have expressed the "other" mesh current in each supermesh in terms of the chosen unknown mesh current and the current source value. In each case, the difference of the two mesh currents is the current source value. Now we write one KVL equation for each supermesh and one for the essential mesh.

Figure 4.61
The example circuit prepared for mesh analysis

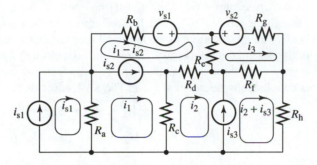

That portion of the circuit that influences the equation for supermesh 1 is shown in Figure 4.62. Notice that—now that we have already used the current source constraint—we no longer actually need the current source in the figure because its effect is taken care of by the upper mesh current dependence relation. The KVL equation is

$$R_a(i_1 - i_{s1}) + R_b(i_1 - i_{s2}) + R_e(i_1 - i_{s2} - i_3) + R_d(i_1 - i_{s2} - i_2) + R_c(i_1 - i_2) = v_{s1}, \quad (4.4\text{-}7)$$

or

$$(R_a + R_b + R_c + R_d + R_e)i_1 - (R_c + R_d)i_2 - R_e i_3 = v_{s1} + R_a i_{s1} + (R_b + R_d + R_e)i_{s2}. \quad (4.4\text{-}8)$$

Carefully observe that we are summing the voltage drops for resistors on the left-hand side of our equation and setting them equal to the sum of the voltage rises from the v-sources on the right-hand side; furthermore, *we never take a path including a current source.*

Figure 4.62
The KVL equation for supermesh 1

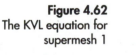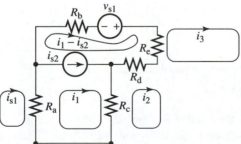

The partial circuit influencing the KVL equation for supermesh 2 is shown in Figure 4.63. Recall that we have decided to write KVL clockwise around this loop (just refer back to Figure 4.60 to check this). The KVL equation itself is

$$R_c(i_2 - i_1) + R_d(i_2 - [i_1 - i_{s2}]) + R_f(i_2 + i_{s3} - i_3) + R_h(i_2 + i_{s3}) = 0, \quad (4.4\text{-}9)$$

or $-(R_c + R_d)i_1 + (R_c + R_d + R_f + R_h)i_2 - R_f i_3 = -R_d i_{s2} - (R_f + R_h)i_{s3}. \quad (4.4\text{-}10)$

Figure 4.63
KVL for supermesh 2

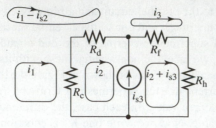

Finally, we tackle the easy one: the essential mesh. It and its associated currents are shown in Figure 4.64. Notice that the mesh current i_2 only influences the equation through the term $i_2 + i_{s3}$, which is the current in the right-hand mesh of supermesh number 2. The mesh current in the left-hand mesh of that supermesh does not induce a voltage in the essential mesh because it does not flow through a resistor shared with that mesh— it only flows through a small segment of conductor shared with that mesh, but no voltage is coupled into the essential mesh. The KVL equation is

$$R_g i_3 + R_f(i_3 - [i_2 + i_{s3}]) + R_e(i_3 - [i_1 - i_{s2}]) = -v_{s2}, \qquad (4.4\text{-}11)$$

or

$$-R_e i_1 - R_f i_2 + (R_e + R_f + R_g)i_3 = -R_e i_{s2} + R_f i_{s3} - v_{s2}. \qquad (4.4\text{-}12)$$

Figure 4.64
KVL around the
essential mesh

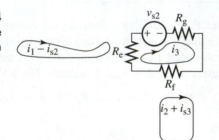

Collecting these equations into one matrix relation, we have

$$\begin{bmatrix} R_a + R_b + R_c + R_d + R_e & -(R_c + R_d) & -R_e \\ -(R_c + R_d) & R_c + R_d + R_f + R_h & -R_f \\ -R_e & -R_f & R_e + R_f + R_g \end{bmatrix} \begin{bmatrix} i_1 \\ i_2 \\ i_3 \end{bmatrix}$$

$$\begin{bmatrix} v_{s1} + R_a i_{s1} + (R_b + R_d + R_e)i_{s2} \\ -R_d i_{s2} - (R_f + R_h)i_{s3} \\ -R_e i_{s2} + R_f i_{s3} - v_{s2} \end{bmatrix}. \qquad (4.4\text{-}13)$$

This can be written more compactly as

$$R\bar{i}_m = \bar{u}_s. \qquad (4.4\text{-}14)$$

Here R is the symbol for the $(B - N + 1 - N_i) \times (B - N + 1 - N_i)$ mesh resistance matrix (the matrix of coefficients in (4.4-13). The vector, or $(B - N + 1 - N_i)$ column matrix, $\bar{i}_m$, contains the unknown mesh currents; similarly, $\bar{u}_s$ is a $(B - N + 1 - N_i)$ column matrix whose ith entry is a linear combination of the independent source values.

Now we deactivate the sources to obtain the deactivated network in Figure 4.65. The matrix form of the mesh equations for this circuit is

$$R\bar{i}_m = \bar{0}, \qquad (4.4\text{-}15)$$

where R is the same matrix as in (4.4-14). As we showed in Section 4.3, this matrix is nonsingular and (4.4-13) (and (4.4-14)) will have a solution if the deactivated (resistive) network is coupled. Our example circuit is clearly coupled because we can travel from the interior of any mesh (or the exterior of the outer mesh) to the interior of any other (or the exterior of the outer mesh) in the deactivated circuit by "stepping over" shared resistors. Furthermore, all resistors are positive, so our mesh equations in (4.4-14) are solvable.

Figure 4.65
The deactivated network

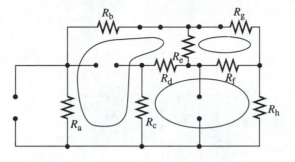

A General Mesh Analysis Example

Example 4.4 Write the mesh equations for the circuit shown in Figure 4.66 and solve them.

Figure 4.66
An example circuit

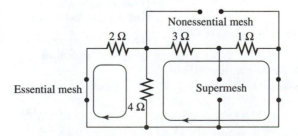

Solution Our first step will be to determine the mesh structure by deactivating all of the sources as shown in Figure 4.67. We see that there is one mesh of each type: essential, nonessential, and super. *The arrows in the figure are not mesh currents.* We have sketched them in to indicate the direction for which we will write our KVL equations. We see that there will be two such equations, one for the essential mesh and one for the supermesh.

Figure 4.67
The deactivated circuit

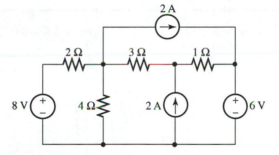

We now return to the original circuit and define the mesh currents: one unknown for the essential mesh and one for the supermesh, as in Figure 4.68. This defines the unknowns in our solution. We next write the KVL equations:

$$2i_1 + 4(i_1 - i_2) = 8, \tag{4.4-16}$$

and
$$4(i_2 - i_1) + 3(i_2 - 2) + 1(i_2 + 2 - 2) = -6, \tag{4.4-17}$$

In matrix form, our nodal equations become

$$\begin{bmatrix} 6 & -4 \\ -4 & 8 \end{bmatrix} \begin{bmatrix} i_1 \\ i_2 \end{bmatrix} = \begin{bmatrix} 8 \\ 0 \end{bmatrix}. \tag{4.4-18}$$

The solution is $i_1 = 2$ A and $i_2 = 1$ A.
This concludes our example.

Figure 4.68
The example circuit prepared for mesh analysis

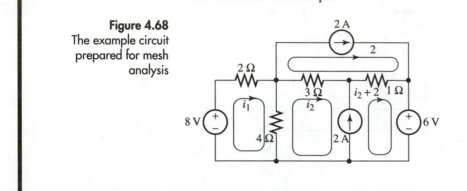

Section 4.4 Quiz

Q4.4-1. For the circuit in Figure Q4.4-1, express the mesh current i_2 in terms of the mesh current i_1 and the 2-A current source value.

Figure Q4.4-1

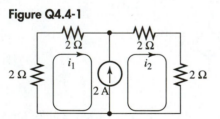

Q4.4-2. For the circuit in Figure Q4.4-1, express the mesh current i_1 in terms of the mesh current i_2 and the 2-A current source value.

Q4.4-3. Write KVL around the supermesh in Figure Q4.4-1 in terms of the unknown mesh current i_1.

Q4.4-4. Write KVL around the supermesh in Figure Q4.4-1 in terms of the unknown mesh current i_2.

Q4.4-5. Draw the deactivated network for the circuit in Figure Q4.4-2 and determine whether or not it is coupled.

Figure Q4.4-2

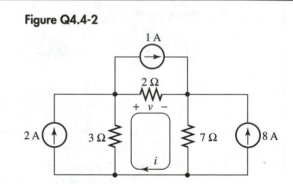

Q4.4-6. How many nodal equations are required for solution of the circuit in Figure Q4.4-2? How many mesh equations?

Q4.4-7. Solve the circuit in Figure Q4.4-2 for the mesh current i and use it to find the voltage v.

Q4.4-8. Write the mesh equations for the circuit in Figure Q4.4-3 and solve them.

Figure Q4.4-3

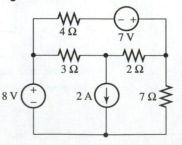

4.5 | Nodal and Mesh Analysis by Inspection

Now that we have developed the basic ideas about nodal and mesh analysis, let's streamline it! It is possible to write the nodal and mesh equations *by inspection,* providing one is willing to draw two simple circuits to assist the process. Let's have a look at the circuit in Figure 4.69 to develop the method. We will look at nodal analysis first. We have therefore shown the circuit already prepared for nodal analysis. Let's use KCL at the essential node and the supernode in the form

$$\sum_{\text{node}} i_{\text{out}} = 0. \tag{4.5-1}$$

The results for our example are

$$[g_a + g_b + g_c + g_f]v_1 - [g_c + g_f]v_2 - g_a v_{s1} - g_f v_{s3} - i_s = 0 \tag{4.5-2}$$

and

$$-[g_c + g_f]v_1 + [g_c + g_d + g_e + g_f]v_2 - g_e v_{s2} + g_f v_{s3} + i_s = 0. \tag{4.5-3}$$

There is, of course, a small amount of work involved in deriving these equations; we have left this to you. Moving the terms due to the independent sources to the right side, we obtain the matrix form:

$$\begin{bmatrix} g_a + g_b + g_c + g_f & -(g_c + g_f) \\ -(g_c + g_f) & g_c + g_d + g_e + g_f \end{bmatrix} \begin{bmatrix} v_1 \\ v_2 \end{bmatrix} = \begin{bmatrix} i_s + g_a v_{s1} + g_f v_{s3} \\ -i_s + g_e v_{s2} - g_f v_{s3} \end{bmatrix}. \tag{4.5-4}$$

Keeping this explicit example in mind, we can write it more compactly (and generally) as

$$G\,\bar{v}_N = \bar{u}_s, \tag{4.5-5}$$

where G is the generalized nodal conductance matrix, $\bar{v}_N$ is the column matrix of independent node voltages, and $\bar{u}_s$ is the column matrix of linear combinations of the independent source values.

Figure 4.69
An example circuit

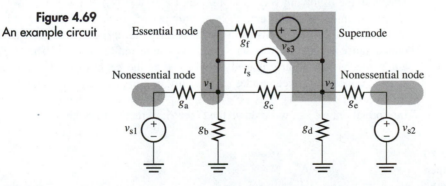

The Inspection Method of Nodal Analysis

The *nodal conductance matrix G* has some special characteristics, as you can see by inspection of equations (4.5-2) and (4.5-3) or the matrix form in (4.5-4) and looking at Figure 4.69:

1. The diagonal entry G_{ii} is the sum of the conductances connected to the nonreference node (or supernode) i.
2. The off-diagonal entry G_{ij} is the *negative* of the sum of the conductances connected between the nonreference node (or supernode) i and the nonreference node (or supernode) j.
3. The nodal conductance matrix G is symmetric, that is $G_{ji} = G_{ij}$.

The sum of the conductances connected to supernode 2, for instance, is $g_c + g_d + g_e + g_f$, and that is the entry in the (2, 2) position of the G matrix in (4.5-4). The sum of the conductances between supernode 2 and essential node 1 is $g_c + g_f$, and the negative of this quantity appears in both the (1, 2) and (2, 1) positions. *These special characteristics allow us to write the nodal conductance matrix by inspection. In what follows we will also develop the capability of writing down the right-hand side matrix $\bar{u}_s$ by inspection as well.*

Let's look at our original KCL equations (4.5-2) and (4.5-3) more closely. We see that they have the form

$$f(v_1, v_2, v_{s1}, v_{s2}, v_{s3}, i_s) = 0, \tag{4.5-6}$$

and

$$g(v_1, v_2, v_{s1}, v_{s2}, v_{s3}, i_s) = 0, \tag{4.5-7}$$

where f and g are *linear combinations* of the (unconstrained, or independent) node voltages and the independent source values. We can thus apply superposition to (4.5-6) and (4.5-7) to obtain

$$f(v_1, v_2, v_{s1}, v_{s2}, v_{s3}, i_s) = f(v_1, v_2, 0, 0, 0, 0) + f(0, 0, v_{s1}, v_{s2}, v_{s3}, i_s) \tag{4.5-8}$$
$$= f_1(v_1, v_2) + f_2(v_{s1}, v_{s2}, v_{s3}, i_s) = i_{out}(R\text{'s}) + i_{out}(\text{sources})$$

and

$$g(v_1, v_2, v_{s1}, v_{s2}, v_{s3}, i_s) = g(v_1, v_2, 0, 0, 0, 0) + g(0, 0, v_{s1}, v_{s2}, v_{s3}, i_s) \tag{4.5-9}$$
$$= g_1(v_1, v_2) + g_2(v_{s1}, v_{s2}, v_{s3}, i_s) = i_{out}(R\text{'s}) + i_{out}(\text{sources})$$

The first term in each equation is the sum of the currents out of the node (or supernode) through the resistors *due to the node voltages themselves—with the sources deactivated.* The second term is the sum of the currents out *due to the sources themselves—with the node voltages forced to zero.* Both f_1 and g_1 can be visualized by referring to the deactivated network in Figure 4.70. Noting that deactivation of the sources does not affect the coefficients of v_1 and v_2 in equations (4.5-2) and (4.5-3)—and, therefore, does not affect the coefficient matrix in (4.5-4)—we can interpret the generalized nodal conductance matrix in the following way. G is a square $(N - 1 - N_v) \times (N - 1 - N_v)$ matrix, where N is the number of nodes and N_v is the number of v-sources. Looking at it carefully, we see that deactivation permits us to easily identify the (i, i) diagonal entry as the sum of the conductances of all resistors with one lead connected to node i.[13] The (i, j) off-diagonal entry, too, is now clearly placed in evidence as the *negative* of the sum of the conductances connected between nodes i and j. Furthermore, as we have pointed out, G is a symmetric matrix; that is, the (i, j) entry is the same as the (j, i) entry. Using this interpretation, we can write down the G matrix directly from the deactivated network.

Figure 4.70
The deactivated network

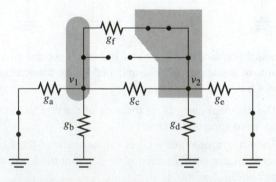

[13] In the deactivated network, there is no distinction between a node and a supernode. This makes identification of the terms in G easier.

Now let's look at the source matrix $\bar{u}_s$. If we move the f_2 and g_2 functions in (4.5-8) and (4.5-9) to the right-hand side, we obtain the form

$$i_{out}(R\text{'s}) = i_{in}(\text{sources}) \tag{4.5-10}$$

for each of these equations. The right-hand side, therefore, corresponds to $\bar{u}_s$. To compute it, we merely *force the independent node voltages to zero and compute the current entering the node or supernode to the sources acting alone*. We show the associated network, which we will call the *forced network*, in Figure 4.71. We note here that *the special node inside the supernode carrying the independent node voltage must have been previously selected*. We have already done this for our example and show the resulting zero values in shaded boxes in Figure 4.71. By inspection of the forced network in that figure, we have[14]

$$i_{in}(\text{sources at node 1}) = g_a v_{s1} + g_f v_{s3} + i_s, \tag{4.5-11}$$

which is the first entry (the one for the essential node) in $\bar{u}_s$ given in equation (4.5-4). Similarly, we find that the second entry (the one for the supernode) is

$$i_{in}(\text{sources at node 2}) = g_e v_{s2} - g_f v_{s3} - i_s. \tag{4.5-12}$$

We will now illustrate the "by inspection" analysis procedure we have just developed by reconsidering the same circuit we have been working with. To make our analysis procedure more concrete this time, however, we will use numerical element values.

Figure 4.71
The forced network

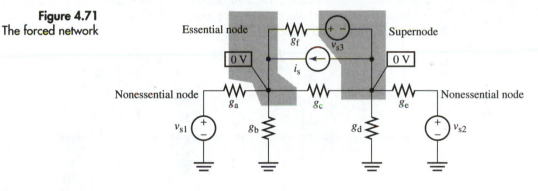

Examples of Nodal Analysis by Inspection

Example 4.5 Write the nodal equations for the circuit in Figure 4.72 by inspection and solve for the node voltages v_1 and v_2.

Figure 4.72
An example circuit

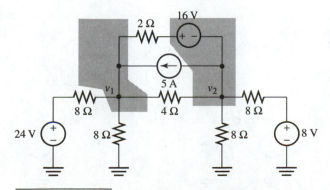

---

[14] Notice that the resistor currents due to the independent *node voltages* are zero; for instance, the current in g_c is given by $g_c(0 - 0) = 0$.

Solution   The deactivated network is shown in Figure 4.73. The corresponding nodal conductance matrix, by inspection is

$$
G\bar{v}_N = \begin{bmatrix} \dfrac{1}{8} + \dfrac{1}{8} + \dfrac{1}{4} + \dfrac{1}{2} & -\left(\dfrac{1}{4} + \dfrac{1}{2}\right) \\[2ex] -\left(\dfrac{1}{4} + \dfrac{1}{2}\right) & \dfrac{1}{8} + \dfrac{1}{8} + \dfrac{1}{4} + \dfrac{1}{2} \end{bmatrix} \begin{bmatrix} v_1 \\ v_2 \end{bmatrix} = \begin{bmatrix} 1 & -\dfrac{3}{4} \\[2ex] -\dfrac{3}{4} & 1 \end{bmatrix} \begin{bmatrix} v_1 \\ v_2 \end{bmatrix}. \quad (4.5\text{-}13)
$$

Notice that we have also written in the node voltage vector $\bar{v}_N$. This is not essential, but it helps to keep the ordering of the elements in the G matrix straight.

Figure 4.73
The deactivated
network

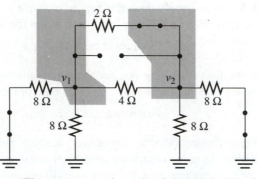

The source matrix—the right-hand side of our matrix nodal equation—is computed from the forced network shown in Figure 4.74. Computation of the currents entering the essential node and the supernode gives

$$
\bar{u}_s = \begin{bmatrix} \dfrac{24}{8} + \dfrac{16}{2} + 5 \\[2ex] \dfrac{8}{8} - \dfrac{16}{2} - 5 \end{bmatrix} = \begin{bmatrix} 16 \\ -12 \end{bmatrix}. \quad (4.5\text{-}14)
$$

Assembling these two parts gives the complete matrix nodal equation:

$$
\begin{bmatrix} 1 & -\dfrac{3}{4} \\[2ex] -\dfrac{3}{4} & 1 \end{bmatrix} \begin{bmatrix} v_1 \\ v_2 \end{bmatrix} = \begin{bmatrix} 16 \\ -12 \end{bmatrix} \quad (4.5\text{-}15)
$$

The solution is easily computed to be

$$
\begin{bmatrix} v_1 \\ v_2 \end{bmatrix} = \dfrac{16}{7} \begin{bmatrix} 1 & \dfrac{3}{4} \\[2ex] \dfrac{3}{4} & 1 \end{bmatrix} \begin{bmatrix} 16 \\ -12 \end{bmatrix} = \begin{bmatrix} 16 \\ 0 \end{bmatrix}. \quad (4.5\text{-}16)
$$

Figure 4.74
The forced network

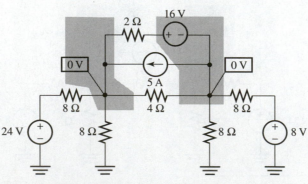

The following practice example proceeds in precisely the same fashion.

Example 4.6   Write the nodal equations in Figure 4.75 by inspection and solve them.

Figure 4.75
An example circuit

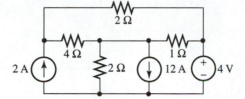

Solution   This is somewhat easier than the preceding example because we can choose our ground reference at the bottom and avoid supernodes. We show the circuit prepared for nodal analysis in Figure 4.76. This step is essential, for it defines the node voltages for which the equations are to be derived. The next step is to draw the deactivated network, shown in Figure 4.77. The nodal conductance matrix, by inspection, is

$$G\bar{v}_N = \begin{bmatrix} \dfrac{1}{4} + \dfrac{1}{2} & -\dfrac{1}{4} \\ -\dfrac{1}{4} & 1 + \dfrac{1}{2} + \dfrac{1}{4} \end{bmatrix} \begin{bmatrix} v_1 \\ v_2 \end{bmatrix} \begin{bmatrix} \dfrac{3}{4} & -\dfrac{1}{4} \\ -\dfrac{1}{4} & \dfrac{7}{4} \end{bmatrix} \begin{bmatrix} v_1 \\ v_2 \end{bmatrix}. \tag{4.5-17}$$

The forced network, on the other hand, is shown in Figure 4.78 (on p. 144). Summing the currents directed toward the two essential nodes gives

$$\bar{u}_s = \begin{bmatrix} 2 + \dfrac{4}{2} \\ -12 + \dfrac{4}{1} \end{bmatrix} = \begin{bmatrix} 4 \\ -8 \end{bmatrix}. \tag{4.5-18}$$

Assembling the component parts gives the complete matrix nodal equation:

$$\begin{bmatrix} \dfrac{3}{4} & -\dfrac{1}{4} \\ -\dfrac{1}{4} & \dfrac{7}{4} \end{bmatrix} \begin{bmatrix} v_1 \\ v_2 \end{bmatrix} = \begin{bmatrix} 4 \\ -8 \end{bmatrix}. \tag{4.5-19}$$

The solution is

$$\begin{bmatrix} v_1 \\ v_2 \end{bmatrix} = \dfrac{4}{5} \begin{bmatrix} \dfrac{7}{4} & \dfrac{1}{4} \\ \dfrac{1}{4} & \dfrac{3}{4} \end{bmatrix} \begin{bmatrix} 4 \\ -8 \end{bmatrix} = \begin{bmatrix} 4 \\ -4 \end{bmatrix}. \tag{4.5-20}$$

Figure 4.76
The circuit prepared for nodal analysis

Figure 4.77
The deactivated network

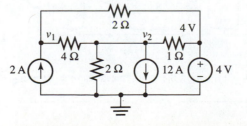

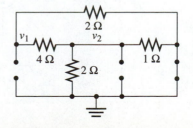

Figure 4.78
The forced network

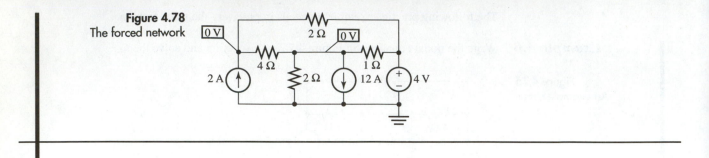

The Inspection Method of Mesh Analysis

We now turn our attention to mesh analysis. Our method of analysis by inspection is quite analogous to the one we have just developed for nodal analysis. We will work with the circuit shown in Figure 4.79. Because of the similarity just referred to, we will use numerical values. We will leave it to you to show that the mesh equations can be written as

$$16i_1 - 8i_2 = 24,\qquad\qquad(4.5\text{-}21)$$

$$-8i_1 + 20i_2 - 8i_3 = -20,\qquad\qquad(4.5\text{-}22)$$

and
$$-8i_2 + 16i_3 = -8.\qquad\qquad(4.5\text{-}23)$$

In matrix form, we have

$$\begin{bmatrix} 16 & -8 & 0 \\ -8 & 20 & -8 \\ 0 & -8 & 16 \end{bmatrix}\begin{bmatrix} i_1 \\ i_2 \\ i_3 \end{bmatrix} = \begin{bmatrix} 24 \\ -20 \\ -8 \end{bmatrix}.\qquad(4.5\text{-}24)$$

We can write this in more symbolic and compact form as

$$R\bar{i}_M = \bar{u}_s,\qquad\qquad(4.5\text{-}25)$$

where R is the square $(B - N + 1 - N_I) \times (B - N + 1 - N_I)$ *mesh resistance matrix.* The (i, i) diagonal entry is the sum of the resistances in mesh i,[15] and the (i, j) off-diagonal entry is the negative of the sum of the resistances shared by meshes i and j. Notice that R is a symmetric matrix: that is, the (i, j) entry is the same as the (j, i) entry. The $(B - N + 1 - N_I) \times 1$ column matrix $\bar{i}_M$ has the independent mesh current i_i for its ith entry and the $(B - N + 1 - N_I) \times 1$ column matrix $\bar{u}_s$ has for its ith entry the sum of the voltage *rises* around mesh i due to the sources acting alone (with the independent mesh

Figure 4.79
An example circuit

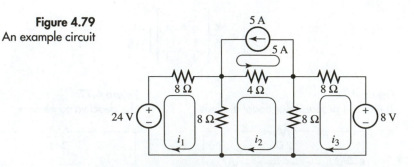

---

[15] You should interpret the word "mesh" here as denoting either a supermesh or an essential mesh.

currents forced to zero). Note that the resistor voltages are considered to be *drops* on the left side of the equation; thus, the matrix form of the mesh equations can be expressed as

$$\sum_{\text{mesh } i} v_{\text{drops}}(R\text{'s}) = \sum_{\text{mesh } i} v_{\text{rises}}(\text{sources}). \tag{4.5-26}$$

The derivation of our inspection method proceeds in exactly the same fashion as that used for nodal analysis, but with the equations being KVL rather than KCL. Because the KVL equations are linear combinations of the mesh currents and the source values, we see that we can find the mesh resistance matrix R from the deactivated network and $\bar{u}_s$ from the network with the mesh currents forced to zero (the "forced network"). Thus, let's first investigate the deactivated network shown in Figure 4.80. The mesh resistance matrix is, by inspection, [16]

$$R\bar{i}_M = \begin{bmatrix} 8+8 & -8 & 0 \\ -8 & 8+4+8 & -8 \\ 0 & -8 & 8+8 \end{bmatrix} \begin{bmatrix} i_1 \\ i_2 \\ i_3 \end{bmatrix} = \begin{bmatrix} 16 & -8 & 0 \\ -8 & 20 & -8 \\ 0 & -8 & 16 \end{bmatrix} \begin{bmatrix} i_1 \\ i_2 \\ i_3 \end{bmatrix}. \tag{4.5-27}$$

Next, we reactivate the sources and force the nonconstrained (independent) mesh currents to zero. This results in the forced network sketched in Figure 4.81. We sum the voltage *rises* around each mesh due to the sources, thus obtaining the source matrix. It is

$$\bar{u}_s = \begin{bmatrix} 24 \\ -5 \times 4 \\ -8 \end{bmatrix} = \begin{bmatrix} 24 \\ -20 \\ -8 \end{bmatrix}. \tag{4.5-28}$$

Putting these two results together gives

$$\begin{bmatrix} 16 & -8 & 0 \\ -8 & 20 & -8 \\ 0 & -8 & 16 \end{bmatrix} \begin{bmatrix} i_1 \\ i_2 \\ i_3 \end{bmatrix} = \begin{bmatrix} 24 \\ -20 \\ -8 \end{bmatrix}. \tag{4.5-29}$$

The solution (Gaussian reduction, Appendix A is convenient) is

$$\begin{bmatrix} i_1 \\ i_2 \\ i_3 \end{bmatrix} = \begin{bmatrix} 1 \\ -1 \\ -1 \end{bmatrix}. \tag{4.5-30}$$

Figure 4.80
The deactivated network

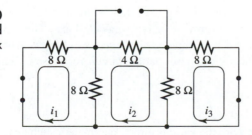

---

[16] As for nodal analysis, we carry along the mesh current column matrix simply to help in keeping the equations straight.

Figure 4.81
The forced network

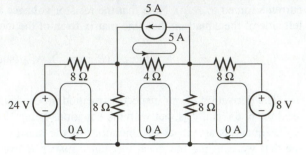

Section 4.5 Quiz

Q4.5-1. Write the nodal equations for the circuit in Figure Q4.5-1 by inspection. Use the reference node and node voltages shown.

Q4.5-2. Write the mesh equations for the circuit in Figure Q4.5-1 by inspection. Use the mesh currents shown.

Figure Q4.5-1

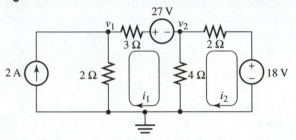

4.6 | SPICE: Sensitivity and Subcircuits

The fundamentals of SPICE are covered in Chapter 2, Section 2.5, and Chapter 3, Section 3.7. SPICE "doesn't care" how complex (within the limits imposed by the specific software and computer) a circuit is; thus, the analysis of a complicated circuit proceeds no differently than for the simpler ones covered in Chapters 2 and 3. However, SPICE uses nodal analysis as its basic algorithm.

Sensitivity In this section we will explore a few additional features of SPICE. One such feature is its ability to perform sensitivity calculations. The sensitivity of a circuit response is a measure of how rapidly that response changes when a specific element is varied. More specifically, suppose that the circuit response in which we are interested is written as

$$y = f(x_1, x_2, \ldots, x_n) \qquad (4.6\text{-}1)$$

where the x_i's are the various circuit parameters such as its resistor values. These parameter values are subject to variation: resistors have a finite precision (that is, they vary in a statistical way relative to a mean, or average, value), they change with temperature, and so on. You will recall from calculus that the total differential is

$$dy = \frac{\partial f}{\partial x_1} dx_1 + \frac{\partial f}{\partial x_2} dx_2 + \cdots + \frac{\partial f}{\partial x_n} dx_n. \qquad (4.6\text{-}2)$$

The differentials dx_i are the variations in the x_i, and, if these changes are small, dy is a good approximation to the actual change in y. Using the delta notation, we have

$$\Delta y \cong \frac{\partial f}{\partial x_1}\Delta x_1 + \frac{\partial f}{\partial x_2}\Delta x_2 + \cdots + \frac{\partial f}{\partial x_n}\Delta x_n. \tag{4.6-3}$$

We are now in a position to define the sensitivity of the network response y to a specified parameter. Calling it $S^y_{x_i}$, which we read as "the sensitivity of y with respect to x_i," we write

$$S^y_{x_i} = \frac{\partial y}{\partial x_i} = \frac{\partial f(x_1 \ldots, x_n)}{\partial x_i}. \tag{4.6-4}$$

The derivative is evaluated at the *nominal* (or ideal design value) of each of the parameters x_i.

SPICE will automatically compute such sensitivities if we merely include the verb .SENS in the file. Here is a simple example. Suppose we are designing a precision voltage divider, such as the one shown in Figure 4.82. Of course, we can compute the sensitivities rather quickly for such a simple circuit. In fact, by the voltage divider rule, we have

$$v_o = \frac{R_2}{R_1 + R_2}v_s. \tag{4.6-5}$$

Figure 4.82 An example circuit

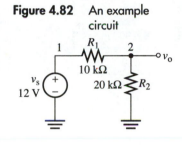

Thus v_o takes the place of y and R_1, R_2, and v_s take the place of the parameters x_i. Taking the partial derivative of v_o with respect to each of the parameters, we get

$$S^{v_o}_{R_1} = \frac{-R_2}{(R_1 + R_2)^2}v_s = \frac{-20\text{ k}\Omega}{(30\text{ k}\Omega)^2} \times 12\text{ V} = -0.267\frac{\text{V}}{\text{k}\Omega}, \tag{4.6-6}$$

$$S^{v_o}_{R_2} = \frac{R_1}{(R_1 + R_2)^2}v_s = \frac{10\text{ k}\Omega}{(30\text{ k}\Omega)^2} \times 12\text{ V} = 0.133\frac{\text{V}}{\text{k}\Omega} \tag{4.6-7}$$

and

$$S^{v_o}_{v_s} = \frac{R_2}{R_1 + R_2} = \frac{20\text{ k}\Omega}{30\text{ k}\Omega} = 0.667\frac{\text{V}}{\text{V}}. \tag{4.6-8}$$

Thus, if we keep R_2 and v_s fixed while we increase R_1, the response v_o will *decrease* (the meaning of the negative sign in (4.6-6)) by 0.267 volt for each kΩ increase in R_1. Increasing R_2, on the other hand, results in an *increase* in v_o by 0.133 volt for each kΩ increase in R_2. Keeping R_1 and R_2 fixed and increasing v_s, however, results in an increase of v_o by 0.667 volt for each 1-V increase in v_s. One might argue that a kΩ increase in one of the resistors or a 1-V increase in the input voltage is not a small change—which we assumed when we approximated the change in response by its differential. This might well be true; the sensitivity, however, is defined regardless. It is only the physical approximation in going from (4.6-2) to (4.6-3) that becomes dubious for large changes. In practice, the changes are actually often small. For instance, the change in resistance of a resistor would be only a fraction of a percent for a one °C rise in temperature. The SPICE file describing the circuit in Figure 4.82 is:

```
SENSITIVITY EXAMPLE  ;TITLE STATEMENT
VS   1    0    12    ;NOMINAL VALUE = 12V
R1   1    2    10K   ;NOMINAL VALUE = 10K
R2   2    0    20K   ;NOMINAL VALUE = 20K
.SENS     V(2)       ;RESPONSE IS V(2)
.END
```

The following is the result of the actual SPICE run:

```
****03/29/94 09:29:17**********EVALUATION PSPICE (JAN 1992)*****
SENSITIVITY EXAMPLE  ;TITLE STATEMENT
****CIRCUIT DESCRIPTION
***********************************************************************
VS  1   0   12    ;NOMINAL VALUE = 12V
R1  1   2   10K   ;NOMINAL VALUE = 10K
R2  2   0   20K   ;NOMINAL VALUE = 20K
.SENS   V(2)      ;RESPONSE IS V(2)
.END
****03/29/94 09:29:17**********EVALUATION PSPICE (JAN 1992)*****
SENSITIVITY EXAMPLE  ;TITLE STATEMENT
**** SMALL SIGNAL BIAS SOLUTION  TEMPERATURE = 27.000 DEG C
***********************************************************************
NODE VOLTAGE   NODE VOLTAGE   NODE VOLTAGE   NODE VOLTAGE
( 1) 12.0000   ( 2) 8.0000
 VOLTAGE SOURCE CURRENTS
 NAME    CURRENT
 VS   -4.000E-04
 TOTAL POWER DISSIPATION 4.80E-03 WATTS
****03/29/94 09:29:17**********EVALUATION PSPICE (JAN 1992)*****
SENSITIVITY EXAMPLE  ;TITLE STATEMENT
****DC SENSITIVITY ANALYSIS    TEMPERATURE = 27.000 DEG C
***********************************************************************
DC SENSITIVITIES OF OUTPUT V(2)
    ELEMENT   ELEMENT   ELEMENT      NORMALIZED
    NAME      VALUE     SENSITIVITY  SENSITIVITY
                        (VOLTS/UNIT) (VOLTS/PERCENT)
    R1       1.000E+04  -2.667E-04   -2.667E-02
    R2       2.000E+04  1.333E-04    2.667E-02
    VS       1.200E+01  6.667E-01    8.000E-02
    JOB CONCLUDED
    TOTAL JOB TIME     .35
```

There are several things about this output that we should discuss. One of these is the comment statement or part of a statement. Each semicolon in the preceding file listing is interpreted by the SPICE compiler to be the end of the statement on the same line. Any symbols that follow this marker are disregarded during execution. Furthermore, if we begin a line with an asterisk (*) the *entire* line is disregarded.

Another important item is the *operating point.* Even had we not specified a type of analysis, all the node voltages would have been computed and placed in the output file. The term "operating point" is used in the analysis of electronic circuits in which there is usually a constant, or dc, component plus a "small signal" time-varying signal component. At any rate, one gets all the node voltages "for free." Additional pieces of information supplied by SPICE are the current in the input voltage source and the total power delivered by the voltage source(s). This is perhaps not evident from the output, but the term "total power dissipation" means the total power delivered by the independent sources to the remainder of the circuit.

The SPICE verb that causes the computation of the sensitivity of a given response variable to each and every circuit parameter is of the following general form:

```
.SENS   OUTPUTVARIABLE
```

In our example, we have chosen the voltage at node 2 relative to ground, for it corresponds to v_o in the original circuit. As you can see, the output file consists of a listing of each and every circuit parameter and its nominal value, followed by a column giving the sensitivities as defined previously. In addition, the last column gives the sensitivity on a normalized basis, that is, normalized to the nominal value of the parameter. Thus, these numbers represent the response variation divided by the percentage change in the associated element. In our example the response voltage changes by -26.7 mV for each 1% change in R_1.

We note that $100\Delta a/a$ is the percentage variation in a parameter a; thus, we can write the limiting ratio of the change in y to a 1% change in the value of a as

$$\lim_{\Delta a \to 0} \frac{\Delta y}{100\Delta a/a} = \frac{a}{100} \frac{\partial y}{\partial a} = \frac{a}{100} S_a^y . \tag{4.6-9}$$

We thus see that we can get the numbers in the last column by multiplying those in the preceding column by the nominal parameter value divided by 100. Note that if the "output variable" is a current, it must be the current in a voltage source. Thus, one might be obliged to add a zero-valued dummy voltage source in series with an element to generate the required sensitivity. This was explained in Chapter 2, Section 2.5.

You might feel that it is just as easy to compute the sensitivities by hand, rather than by doing a computer simulation. In this you would be wrong, for the response function y is often a very, very complicated function of the network parameters. The next example shows this. You are invited to do the computations by hand!

Example 4.7   Find the sensitivities of the circuit in Figure 4.83 using SPICE. Assume that all resistors are nominally 10 kΩ in value, that i_s is nominally 10 mA, and that v_s is nominally 100 mV.

Figure 4.83
An example circuit

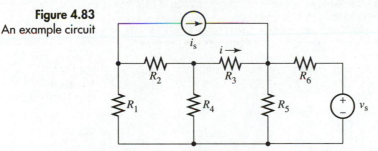

Solution   The circuit prepared for SPICE analysis is shown in Figure 4.84. We have numbered the nodes and given values to each element as specified previously; furthermore, we have inserted a zero-valued dummy voltage source in series with R_3 to sense the current i.

Figure 4.84
Circuit prepared for
SPICE analysis

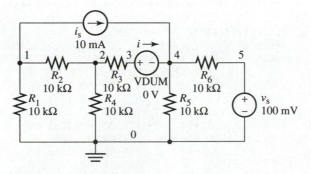

The SPICE file is:

```
ANOTHER SENSITIVITY EXAMPLE
VS    5    0      100E-3
VDUM       3    4     0
IS   1     4     10E-3
```

```
R1  1    0    10K
R2  1    2    10K
R3  2    3    10K
R4  2    0    10K
R5  4    0    10K
R6  4    5    10K
.SENS I(VDUM)
.END
```

The SPICE output file is:

```
****03/29/94 14:04:48***********EVALUATION PSPICE (JAN 1992)*****
ANOTHER SENSITIVITY EXAMPLE
****CIRCUIT DESCRIPTION
*********************************************************************
******
VS  5   0   100E-3
VDUM  3  4   0
IS  1   4   10E-3
R1  1   0   10K
R2  1   2   10K
R3  2   3   10K
R4  2   0   10K
R5  4   0   10K
R6  4   5   10K
.SENS I(VDUM)
.END
****03/29/94 14:04:48***********EVALUATION PSPICE (JAN 1992)*****
ANOTHER SENSITIVITY EXAMPLE
****   SMALL SIGNAL BIAS SOLUTION  TEMPERATURE = 27.000 DEG C
*********************************************************************
******
NODE VOLTAGE   NODE VOLTAGE   NODE VOLTAGE   NODE VOLTAGE
( 1) −53.8380 ( 2) −7.6769  ( 3) 30.8080  ( 4) 30.8080
( 5).1000
  VOLTAGE SOURCE CURRENTS
  NAME   CURRENT
  VS   3.071E-03
  VDUM   −3.848E-03
  TOTAL POWER DISSIPATION −3.07E-04 WATTS
****03/29/94 14:04:48***********EVALUATION PSPICE (JAN 1992)*****
ANOTHER SENSITIVITY EXAMPLE
**** DC SENSITIVITY ANALYSIS   TEMPERATURE = 27.000 DEG C
*********************************************************************
******
DC SENSITIVITIES OF OUTPUT I(VDUM)
    ELEMENT      ELEMENT      ELEMENT    NORMALIZED
    NAME        VALUE   SENSITIVITY  SENSITIVITY
                      (AMPS/UNIT)(AMPS/PERCENT)
    R1    1.000E+04  −8.283E-08  −8.283E-06
    R2    1.000E+04   7.102E-08   7.102E-06
    R3    1.000E+04   1.776E-07   1.776E-05
    R4    1.000E+04  −2.362E-08  −2.362E-06
    R5    1.000E+04  −7.109E-08  −7.109E-06
    R6    1.000E+04  −7.086E-08  −7.086E-06
    VS    1.000E−01  −2.308E-05  −2.308E-08
    VDUM    0.000E+00    −4.615E-05  0.000E+00
    IS    1.000E−02  −3.846E-01  −3.846E-05
    JOB CONCLUDED
    TOTAL JOB TIME       .40
```

Subcircuits

Figure 4.85 shows a situation that frequently occurs. There are several subcircuits that are identical in both topology and value—the three series combinations of 1-Ω and 3-Ω resistors. Notice, however, that the middle voltage divider is "upside down." We have drawn boxes around these three subcircuits to emphasize the fact that, in the overall circuit, we will consider them simply as elements. Notice that we have labeled the boxes starting with the letter X, which is the SPICE syntax for a subcircuit. Here is the SPICE code:

```
A SUBCIRCUIT EXAMPLE
VS    1    0    8
R1    1    2    4
X1    2    0       SERIES
X2    0    2       SERIES
X3    2    0       SERIES
.SUBCKT    SERIES      TOP   BOT
RTOP   TOP   INSIDENODE         1
RBOT   INSIDENODE   BOT         3
.ENDS
.OP
.END
```

The top part seems straightforward enough. We simply treat the subcircuits as elements, giving the nodes between which they are connected. In the place of a value we have put a name—one that is used later in the definition of the subcircuit. Because it is a series subcircuit, we have given it the name SERIES. Figure 4.86 shows the subcircuit topology.

Figure 4.85
An example circuit

Figure 4.86
Subcircuit definition

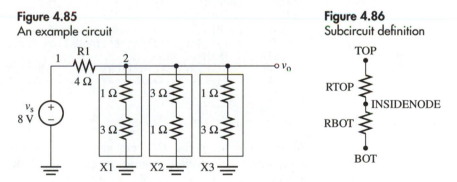

The portion of the code between the .SUBCKT statement and the .ENDS statement defines the subcircuit. In Figure 4.86 we have sketched our subcircuit with the resistors and nodes labeled. The .SUBCKT statement signals the SPICE compiler that the code that follows (and precedes the .ENDS statement) is to be considered as a subcircuit. The format of this statement is

```
.SUBCKT    SUBCKTNAME    NODELIST
```

where the nodelist is a listing of the nodes in the subcircuit through which the external circuit is to be attached. These nodes are much like the formal parameters in a subroutine or function call in a programming language like FORTRAN or C. The actual parameters are the nodes assigned to them in the X statements in the main body of the code. Thus, X1 specifies that node TOP is to be replaced by node 2 and node BOT by node 0 when the program is actually compiled—and similarly for the other X statements. Note that X2 has this assignment reversed relative to the two others. We should perhaps mention that node 0 is a global variable. It always means the ground reference, whether in a subcircuit or the main part of the circuit. Thus, if the subcircuit has an internal ground (which ours does not), it does not have to be "passed" as a parameter.

The SPICE output file corresponding to our preceding code is:

```
****03/29/94 14:54:12**********EVALUATION PSPICE (JAN 1992)*****
A SUBCIRCUIT EXAMPLE
**** CIRCUIT DESCRIPTION
*****************************************************************
******
VS   1   0   8
R1   1   2   4
X1   2   0   SERIES
X2   0   2   SERIES
X3   2   0   SERIES
.SUBCKT SERIES TOP BOT
RTOP  TOP     INSIDENODE  1
RBOT  INSIDENODE     BOT 3
.ENDS
.OP
.END
****03/29/94 14:54:12**********EVALUATION PSPICE (JAN 1992)*****
A SUBCIRCUIT EXAMPLE
**** SMALL SIGNAL BIAS SOLUTION        TEMPERATURE = 27.000 DEG C
*****************************************************************
******
NODE VOLTAGE    NODE VOLTAGE    NODE VOLTAGE    NODE VOLTAGE
( 1) 8.0000    ( 2) 2.0000 (X1.INSIDENODE) 1.5000
(X2.INSIDENODE)  5000    (X3.INSIDENODE) 1.5000
   VOLTAGE  SOURCE CURRENTS
   NAME     CURRENT
   VS      -1.500E+00
   TOTAL POWER DISSIPATION  1.20E+01 WATTS
****03/29/94 14:54:12**********EVALUATION PSPICE (JAN 1992)*****
A SUBCIRCUIT EXAMPLE
**** OPERATING POINT INFORMATION        TEMPERATURE =27.000 DEG C
*****************************************************************
******
     JOB CONCLUDED
     TOTAL JOB TIME    .40
```

We have introduced the .OP verb here. This command, which stands for operating point, directs the SPICE routine to compute all the node voltages, among other things.

Notice that the external node voltages have been given. In addition, look at the nodes labeled X1.INSIDENODE, X2.INSIDENODE, and X3.INSIDENODE. This notation means that the node called INSIDENODE inside the subcircuit whose name precedes the dot has the given value. Thus, X1.INSIDENODE is the node inside the X1 subcircuit and has the value 1.5 volts.

When the SPICE compiler compiles the input file, it looks at the subcircuit definition and adds that code to the code for the external circuit, after replacing the formal parameters with the actual node values. Thus, X1.TOP is given the value 2, X2.TOP the value 0, and X3.TOP the value 2.

We were fortunate in the preceding example that the three subcircuits all consisted of the same values; it was only a coincidence that the values of X2 were the reverse of the others. Hence, we had only to reverse the assignment of the nodes to be assigned to that subcircuit. We are not always so lucky. The circuit of Figure 4.87 has three subcircuits also, but the internal resistor values are all different. What do we do in this case? Fortunately, SPICE gives us a way out. We merely pass "parameters" to each subcircuit indicating the resistor values.

Here is the SPICE code for our circuit:

Figure 4.87
An example circuit

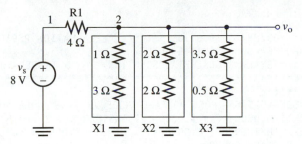

```
ANOTHER SUBCIRCUIT EXAMPLE
VS    1    0    8
R1    1    2    4
X1    2    0    SERIES  (PARAMS: TOPVAL = 1 BOTVAL = 3)
X2    0    2    SERIES  (PARAMS: TOPVAL = 2 BOTVAL = 2)
X3    2    0    SERIES  (PARAMS: TOPVAL = 3.5 BOTVAL = 0.5)
.SUBCKT SERIES TOP BOT  (PARAMS: TOPVAL = 1 BOTVAL = 3)
RTOP  TOP    INSIDENODE           {TOPVAL}
RBOT  INSIDENODE    BOT           {BOTVAL}
.ENDS
.OP
.END
```

This code is essentially the same as that for our last example; however, there are some added twists. In the .SUBCKT statement we have added the following information:

```
PARAMS:   TOPVAL = 1 BOTVAL = 3
```

The key word is PARAMS: (note the colon). This tells the compiler that we are going to pass parameters with the name TOPVAL and BOTVAL to the subcircuit; furthermore, if a specific subcircuit call (starting with X) does not actually specify the values of these parameters, they will default to TOPVAL = 1 and BOTVAL = 3. In our example, however, we have specified these parameters at each call. Look, for instance, at the line beginning X1. We have passed the parameters TOPVAL = 1 and BOTVAL = 3 (which we could just as easily have left to default); we have passed other values in the lines beginning X2 and X3. The parentheses are ignored by the compiler. They were used merely to make the code more readable. Commas can be similarly used.

Now look at the subcircuit definition part of the code. In the place of values for RTOP and RBOT, we have placed TOPVAL and BOTVAL, respectively, inside curly brackets (braces). These brackets tell SPICE to compute the expression inside and use it for the corresponding value. Thus, we could actually have computed more complex expressions for these parameters.

The SPICE output for our example is:

```
****03/29/94 15:37:03***********EVALUATION PSPICE (JAN 1992)*****
ANOTHER SUBCIRCUIT EXAMPLE
****  CIRCUIT DESCRIPTION
****************************************************************
******
VS 1  0  8
R1 1  2  4
X1 2  0  SERIES  (PARAMS: TOPVAL = 1 BOTVAL = 3)
X2 0  2  SERIES  (PARAMS: TOPVAL = 2 BOTVAL = 2)
X3 2  0  SERIES  (PARAMS: TOPVAL = 3.5 BOTVAL = 0.5)
.SUBCKT SERIES TOP BOT  (PARAMS: TOPVAL = 1 BOTVAL = 3)
RTOP  TOP  INSIDENODE    {TOPVAL}
RBOT  INSIDENODE  BOT    {BOTVAL}
.ENDS
```

```
.OP
.END
****03/29/94 15:37:03***********EVALUATION PSPICE (JAN 1992)*****
ANOTHER SUBCIRCUIT EXAMPLE
**** SMALL SIGNAL BIAS SOLUTION          TEMPERATURE = 27.000 DEG C
******************************************************************
******
NODE VOLTAGE  NODE VOLTAGE  NODE VOLTAGE  NODE VOLTAGE
( 1) 8.0000 ( 2) 2.0000 (X1.INSIDENODE) 1.5000
(X2.INSIDENODE) 1.0000        (X3.INSIDENODE) .2500
   VOLTAGE SOURCE CURRENTS
   NAME    CURRENT
   VS   -1.500E+00
   TOTAL POWER DISSIPATION  1.20E+01  WATTS
****03/29/94 15:37:03***********EVALUATION PSPICE (JAN 1992)*****
 ANOTHER SUBCIRCUIT EXAMPLE
****    OPERATING POINT INFORMATION     TEMPERATURE = 27.000 DEG C
******************************************************************
******
     JOB CONCLUDED
     TOTAL JOB TIME    .52
```

Section 4.6 Quiz

Q4.6-1. Write and execute a SPICE program using subcircuits for all elements except the three horizontal resistors at the top that will compute the sensitivities of the node voltage between the two 50-kΩ resistors in the circuit shown in Figure Q4.6-1.

Figure Q4.6-1

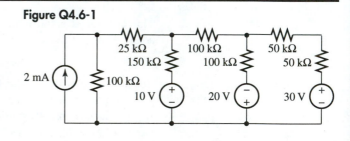

Chapter 4 Summary

This chapter has been, in many respects, the heart of Part I on dc analysis techniques because the methods it has developed are general—they work for essentially all circuits you will encounter.

Nodal analysis of an N-node circuit with only resistors and current sources consists of writing $N - 1$ KCL equations at each of the $N - 1$ *nonreference* nodes. The unknowns are the voltages at these nodes relative to an arbitrarily chosen *reference* node. If the circuit contains voltage sources, the number of required equations is reduced by one for each voltage source; thus, the number of equations required is $N - 1 - N_v$, where N_v is the number of voltage sources. After a reference node is chosen, the nonreference nodes are categorized as follows. A node to which no voltage source is connected is called an *essential* node; a node connected to the reference node through a path of voltage sources is called a *nonessential* node; and any set of nodes connected together by voltage sources, but not so connected to the reference, is called a *supernode*. (Note that a supernode, unlike the other types of node, is not one node but a collection of nodes.) The nonessential nodes carry known node voltages, and no KCL equation is written for them; the essential nodes carry a single unknown node voltage, and one KCL equation is written for each such node; and there is a single unknown node voltage for each supernode, and thus a single KCL equation is written for each supernode. If the *deactivated network* (the circuit resulting after all independent sources are deactivated) is *connected*—that is, there is a path of resistors between any two arbitrarily chosen nodes—then the nodal equations have a solution that is unique. If the deactivated network is not connected, one can choose a different reference node in each connected part, and the resulting system of nodal equations will have a unique solution.

Mesh analysis is very similar to nodal analysis, with *mesh currents* taking the place of node voltages, although it only works for *planar* circuits (those that can be drawn on a sheet of paper with no conductors crossing over any others). A *mesh* is a loop that contains no elements within it, and the corresponding mesh current is a fictitious current assumed to be circulating around that mesh. A *mesh equation,* then, is a KVL equation written around a mesh. A simple derivation is given in this chapter showing that the number of meshes in a circuit is $B - N + 1$, where B is the number of *branches* (synonymous with the word element) and N is the number of nodes. For circuits having only resistors and voltage sources, one writes one KVL equation for each mesh, expressing the voltage across each resistor in terms of the mesh currents using Ohm's law. For circuits with current sources, one catalogs the various types of mesh in a manner that is exactly analogous to the classification scheme used in nodal analysis: if the circuit is deactivated, any meshes that vanish are called *nonessential;* any that are unaffected are called *essential;* and any set of meshes that merge together but do not vanish is called a *supermesh.* The nonessential meshes carry known mesh currents in the original circuit, the essential meshes each carry one unknown mesh current, and there is only one unknown mesh current for each supermesh; the others are expressed in terms of that unknown by means of the current source constraints. The mesh current method is shown to always work for planar circuits that are *coupled;* that is, for those in which one can go from any mesh to any other mesh (including the *outside mesh,* the "outside rim" of the circuit) by crossing over only resistors.

After thoroughly developing the methods of nodal and mesh analysis from first principles, this chapter also presented a method by which one can write down the final matrix form for either the nodal or mesh equations by *inspection.* This method relies upon the fact that the nodal equations and the mesh equations are linear combinations of the unknowns (node voltages and mesh currents, respectively) and the independent source values. Thus, the method consists of first deactivating the circuit and writing the left-hand side of the matrix equation by merely looking at the resistor topology and then forcing the unknowns (node voltages or mesh currents) to zero and applying KCL (or KVL, as appropriate) to find the right-hand side. This method, then, is as firmly based in basic circuit properties as is the slower and more tedious method of writing a KCL (or KVL) equation at each node (or around each mesh).

Finally, we presented two new ideas about SPICE simulation: the *sensitivity* of various circuit voltages and currents to variation in the circuit parameters (resistance and independent source values) and the idea of a *subcircuit.* A SPICE subcircuit is much like a subroutine or function call in a programming language. If a given subcircuit occurs more than once in a given circuit, there is a programming (though not execution time) advantage in using a subcircuit.

Chapter 4 Review Quiz

Questions RQ4.1 through RQ4.7 pertain to the circuit in Figure RQ4.1.

RQ4.1. Identify the nodes by erasing the bodies of all elements. By counting nodes, branches, voltage sources, and current sources, determine the number of equations required for both nodal and mesh analysis.

RQ4.2. Identify the generalized and ordinary nodes by erasing the bodies of all elements except the voltage sources.

RQ4.3. Select the voltage reference at the bottom node and identify the resulting essential, nonessential, and supernodes.

Figure RQ4.1

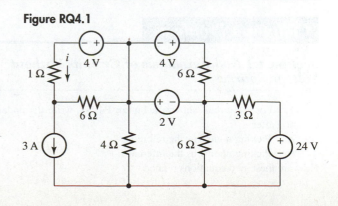

RQ4.4. With the ground reference at the bottom node, assign node voltages—one unknown for each essential node and one for each supernode. Express each of the node voltages in each supernode in terms of the single unknown and label the nonessential nodes with their known values. Then write one KCL equation at each essential node and one at each supernode.

RQ4.5. Draw the deactivated network and write down the nodal conductance matrix by inspection.

RQ4.6. Force the independent node voltages you assigned in Question RQ4.4 to zero and draw the forced network. Write down the right-hand side of the matrix nodal equation by inspection. Then write the complete matrix nodal equation.

RQ4.7. Solve the matrix nodal equation you derived in Question RQ4.6 and use the resulting node voltages and Ohm's law to find the value of i.

Questions RQ4.8 through RQ4.13 pertain to the circuit in Figure RQ4.2.

Figure RQ4.2

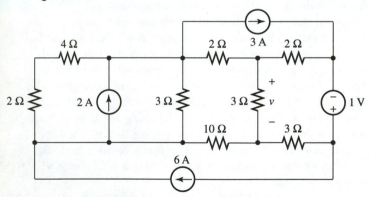

RQ4.8. Deactivate the current sources and classify the meshes: essential, nonessential, and super.

RQ4.9. Reactivate the circuit and assign mesh currents—one unknown for each essential mesh and one for each supermesh. Label the nonessential meshes with their known mesh currents and label each of the previously unlabeled meshes in each supermesh with mesh currents that have been expressed in terms of the single unknown mesh current using the appropriate current source constraint(s).

RQ4.10. Write the mesh equations: one KVL equation around each essential mesh and one around each supermesh.

RQ4.11. Draw the deactivated network and write down the mesh resistance matrix by inspection.

RQ4.12. Draw the auxiliary circuit that results when all independent mesh currents are forced to zero and write down the right-hand side of the matrix mesh equation by inspection. Then write the complete matrix mesh equation.

RQ4.13. Solve the matrix mesh equation derived in Question RQ4.12, and use the resulting mesh current values to find v.

RQ4.14. Draw the deactivated network for the circuit shown in Figure RQ4.3 and determine whether or not a solution exists for its nodal equations.

Figure RQ4.3

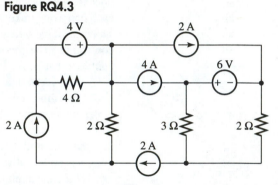

RQ4.15. Draw the deactivated circuit for the circuit shown in Figure RQ4.4 and determine whether or not a solution exists for the mesh equations.

Figure RQ4.4

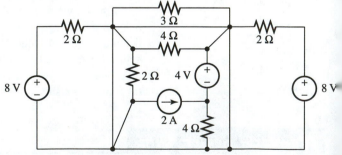

Chapter 4 Problems

Section 4.1 Nodal Analysis of Circuits Without Voltage Sources

4.1-1. For the circuit shown in Figure P4.1-1, write the nodal equation after:
   a. Picking node 0 as the reference
   b. Picking node 1 as the reference.
How are these two equations related?

Figure P4.1-1

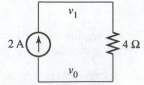

4.1-2. For the circuit shown in Figure P4.1-2, write the nodal equations after:
  a. Picking node 0 as the reference
  b. Picking node 1 as the reference.
How are these two equations related?

Figure P4.1-2

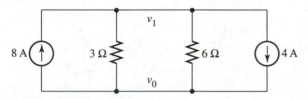

4.1-3. For the circuit shown in Figure P4.1-3, write the nodal equations after:
  a. Picking node 0 as the reference
  b. Picking node 1 as the reference
  c. Picking node 2 as the reference.

Figure P4.1-3

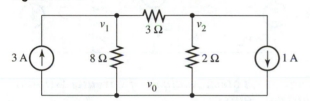

4.1-4. Using node 0 as the ground reference, write the matrix nodal equation for the circuit in Figure P4.1-3.

4.1-5. For the circuit shown in Figure P4.1-4, write the nodal equations after:
  a. Picking node 0 as the reference
  b. Picking node 1 as the reference
  c. Picking node 2 as the reference
  d. Picking node 3 as the reference.
How are these four equations related?

Figure P4.1-4

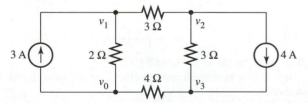

4.1-6. Using node 0 as the ground reference, write the matrix nodal equation for the circuit in Figure P4.1-4 and solve it for the node voltages.

Figure P4.1-5

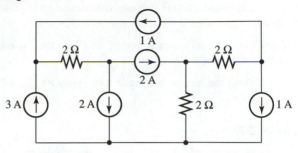

4.1-7. Pick the reference at the bottom node for the circuit of Figure P4.1-5 and write the matrix nodal equation. Does a unique solution of this equation exist? Why, or why not? Draw the deactivated network to explain your answer.

4.1-8. Noting that the resistors in Figure 4.1-6 are all marked with their *conductance* values, select the reference node at the bottom and write the matrix nodal equation.

Figure P4.1-6

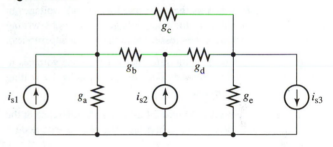

Verify the fact that the (i, i)th entry in the nodal conductance matrix is the sum of all the conductances attached to node i and that the (i, j)th entry is the negative of the sum of all the conductances connected between nodes i and j. Also verify that the ith entry of the right-hand side is the sum of the values of all the current sources connected to node i with references pointed toward node i.

Section 4.2 Nodal Analysis of Circuits with Voltage Sources

4.2-1. Consider the circuit drawn in Figure P4.2-1.
  a. By erasing the body of each element except for the voltage source, identify the ordinary nodes and the generalized nodes.
  b. Pick node 0 as the reference and further classify the nodes of the generalized nodes found in step a as essential nodes, nonessential nodes, and/or supernodes.

Figure P4.2-1

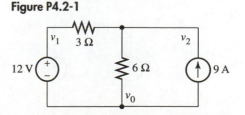

c. Pick node 2 as the reference and classify the nodes of the generalized nodes found in step a as nonessential nodes and/or supernodes.

4.2-2. Picking the reference at node 0 in Figure P4.2-1, solve for the node voltage v_2.

4.2-3. Picking the reference at node 2 in Figure P4.2-1, solve for the voltage v_0.

Figure P4.2-2

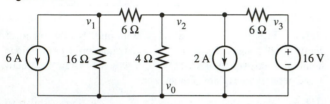

4.2-4. Choose node 0 as the reference for the node voltages in Figure P4.2-2 and solve for the node voltages v_1 and v_2 by writing KCL at the appropriate nodes.

4.2-5. Choose node 1 as the reference for the node voltages in Figure P4.2-2 and solve for the node voltages v_0 and v_2 by writing KCL at the appropriate nodes (essential nodes and/or supernodes).

4.2-6. Choose node 2 as the reference for the node voltages in Figure P4.2-2 and solve for the node voltages v_0 and v_1 by writing KCL at the appropriate nodes.

4.2-7. For the circuit in Figure P4.2-3, pick the reference at the bottom node and solve for v_1. Then pick the reference at node 1 and solve for v_0.

Figure P4.2-3

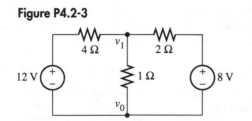

4.2-8. Assuming the reference to be at the bottom node in Figure P4.2-4, write the nodal equations, solve them, and use the resulting known node voltage values to find v.

Figure P4.2-4

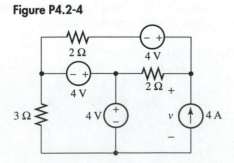

4.2-9. Using nodal analysis on the circuit in Figure P4.2-5, let $R = 2\ \text{k}\Omega$ and find v. What value of R results in $v = 0$? (*Note:* This circuit is called a *bridge* or one section of a *lattice* network.)

Figure P4.2-5

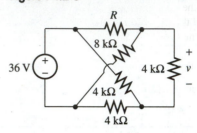

4.2-10. Using nodal analysis on the circuit in Figure P4.2-6, find i.

Figure P4.2-6

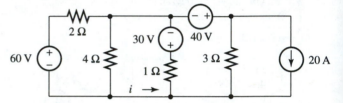

Section 4.3 Mesh Analysis of Circuits Without Current Sources

4.3-1. Consider the circuit in Figure P4.3-1.
   a. Assume a mesh current clockwise, write the mesh equation, and solve for that mesh current.
   b. Assume a mesh current counterclockwise, write the mesh equation, and solve for that mesh current.

Figure P4.3-1

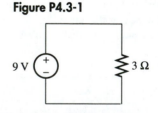

4.3-2. For the circuit in Figure P4.3-2:
   a. Define both mesh currents clockwise and write the mesh equations.
   b. Define both mesh currents counterclockwise and write the mesh equations.

Figure P4.3-2

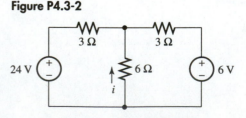

c. Define the left-hand mesh current clockwise and the right-hand mesh current counterclockwise and write the mesh equations.

d. Find the value of *i*.

4.3-3. For the circuit in Figure P4.3-3:

   a. Find the required number of nodal and mesh equations.

   b. Define clockwise mesh currents in the meshes and solve them.

   c. Use the solution found in step b to find *v*.

Figure P4.3-3

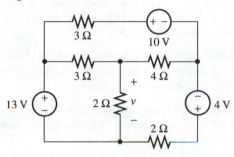

4.3-4. Use mesh analysis (cw mesh currents) to find *v* and *i* in Figure P4.3-4. (*Hint:* Notice that Ohm's law can be expressed in the form mA × kΩ = Ω, so it is consistent to use units of mA, kΩ, and V. This makes the "numbers" easier.)

Figure P4.3-4

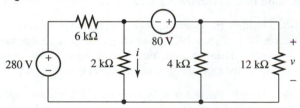

4.3-5. Use mesh analysis to find *i* and *v* in Figure P4.3-5.

Figure P4.3-5

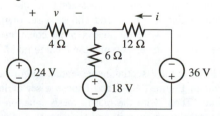

4.3-6. Solve Problem 4.2-7 using mesh analysis to find the current into the plus terminal of the 8-V source.

4.3-7. Solve Problem 4.2-9 using mesh analysis.

Section 4.4 Mesh Analysis of Circuits with Current Sources

4.4-1. Consider the circuit in Figure P4.4-1.

   a. How many nodal equations are required?

b. How many mesh equations are required?

c. Deactivate the current sources and identify the essential, nonessential, and supermeshes.

d. Write one KVL equation for each essential mesh and one for each supermesh.

e. Solve for the unknown mesh current(s) and find the value of *v*.

Figure P4.4-1

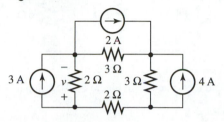

4.4-2. For the circuit in Figure P4.4-2:

   a. Deactivate all the sources and use the resulting deactivated network to identify the essential, nonessential, and supermeshes.

   b. In the original circuit, define a set of independent mesh currents in the clockwise direction—one unknown for each essential mesh and one for each supermesh. Use the current source constraints to express each other mesh current in each supermesh in terms of the single unknown mesh current, and label the nonessential mesh currents with their known values.

   c. Write one KVL equation for each essential mesh and one for each supermesh.

   d. Solve the mesh equations for the mesh currents.

Figure P4.4-2

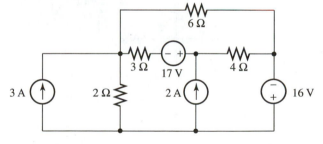

4.4-3. Find the voltage *v* in Figure P4.4-3 using mesh analysis.

Figure P4.4-3

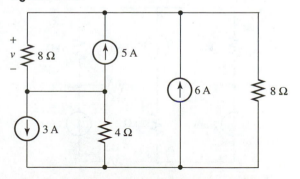

4.4-4. Find i in Figure P4.4-4 using mesh analysis.

Figure P4.4-4

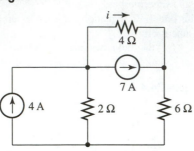

4.4-5. Find the value of v in Figure P4.4-5 using mesh analysis.

Figure P4.4-5

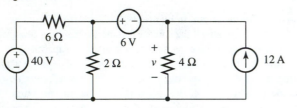

4.4-6. Find the value of v in Figure P4.4-6 using mesh analysis.

Figure P4.4-6

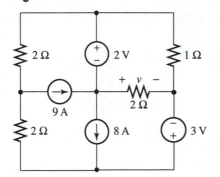

4.4-7. Find the value of v in Figure P4.4-7 using mesh analysis.

Figure P4.4-7

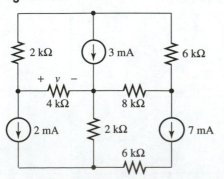

Section 4.5 Nodal and Mesh Analysis by Inspection

4.5-1. Select the voltage reference at node 0 in the circuit in Figure P4.1-3 and write the matrix nodal equation by inspection. Solve for the nonreference node voltages.

4.5-2. Select the voltage reference at node 0 in the circuit in Figure P4.1-4 and write the matrix nodal equation by inspection. Solve for the nonreference node voltages.

4.5-3. Select the voltage reference at node 0 in the circuit in Figure P4.2-1 and write the nodal equation by inspection. Solve for the nonreference node voltage.

4.5-4. Select the voltage reference at node 0 in the circuit in Figure P4.2-2 and write the matrix nodal equation by inspection. Solve for the nonreference node voltages.

4.5-5. Select the voltage reference at node 0 in the circuit in Figure P4.2-3 and write the nodal equation by inspection. Solve for the nonreference node voltage v_1.

4.5-6. Select the voltage reference at the bottom node in the circuit in Figure P4.2-4 and write the nodal equation by inspection. Solve for the nonreference node voltage.

4.5-7. Select the voltage reference at the bottom node in the circuit in Figure P4.2-6 and write the nodal equation by inspection. Solve for the nonreference node voltages.

4.5-8. Identify the essential, nonessential, and/or supermeshes in the circuit in Figure P4.3-2 and choose a set of independent mesh currents. Then write the matrix mesh equation by inspection. Solve for the mesh currents and use them to find v and i.

4.5-9. Identify the essential, nonessential, and/or supermeshes in the circuit in Figure P4.3-3 and choose a set of independent mesh currents. Then write the matrix mesh equation by inspection. Solve for the mesh currents and use them to find v.

4.5-10. Identify the essential, nonessential, and/or supermeshes in the circuit in Figure P4.3-4 and choose a set of independent mesh currents. Then write the matrix mesh equation by inspection. Solve for the mesh currents and use them to find i.

4.5-11. Identify the essential, nonessential, and/or supermeshes in the circuit in Figure P4.3-5 and choose a set of independent mesh currents. Then write the matrix mesh equation by inspection. Solve for the mesh currents and use them to find v and i.

4.5-12. Identify the essential, nonessential, and/or supermeshes in the circuit in Figure P4.4-1 and choose an independent mesh current. Then write the mesh equation by inspection. Solve for the mesh current and use it to find v.

4.5-13. Identify the essential, nonessential, and/or supermeshes in the circuit in Figure P4.4-2 and choose a set of independent mesh currents. Then write the matrix mesh equation by inspection. Solve for the mesh currents.

4.5-14. Identify the essential, nonessential, and/or supermeshes in the circuit in Figure P4.4-3 and choose an independent mesh current. Then write the mesh equation by inspection. Solve for the mesh current and use it to find v.

4.5-15. Identify the essential, nonessential, and/or supermeshes in the circuit in Figure P4.4-4 and choose an independent mesh current. Then write the mesh equation by inspection. Solve for the mesh current and use it to find i.

4.5-16. Identify the essential, nonessential, and/or supermeshes in the circuit in Figure P4.4-5 and choose an independent mesh current. Then write the mesh equation by inspection. Solve for the mesh current and use it to find v.

4.5-17. Identify the essential, nonessential, and/or supermeshes in the circuit in Figure P4.4-6 and choose a set of independent mesh currents. Then write the matrix mesh equation by inspection. Solve for the mesh currents and use them to find v.

4.5-18. Identify the essential, nonessential, and/or supermeshes in the circuit in Figure P4.4-7 and choose an independent mesh current. Then write the mesh equation by inspection. Solve for the mesh current and use it to find v.

4.5-19. Solve Problem 4.5-17 using nodal analysis "by inspection."

4.5-20. Solve Problem 4.5-18 using nodal analysis "by inspection."

Section 4.6 SPICE: Sensitivity and Subcircuits

4.6-1. Write a SPICE program to compute the sensitivity of the voltage v in Problem 4.4-7 to each of the circuit parameters.

4.6-2. Write a SPICE program using subcircuits to simulate the circuit in Figure P4.6-1 for the purpose of finding the value of the current i.

Figure P4.6-1

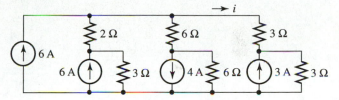

5

Active Circuits

In this chapter we introduce new types of active elements: dependent sources and the operational amplifier. There are four types of dependent source: the voltage-controlled voltage source (VCVS), the current-controlled voltage source (CCVS), the current-controlled current source (CCCS), and the voltage-controlled current source (VCCS). We start by considering the basic dependent source characteristics. The VCVS and CCVS are both voltage sources, so their v-i characteristics are the same as that for an independent voltage source. The only difference is that there is an additional algebraic constraint that makes the voltage source value a constant gain factor times a controlling voltage or current elsewhere in the circuit. Similarly, the dependent current sources abbreviated CCCS and VCCS have the same v-i characteristic as that of the independent current source. These two elements, however, have values controlled by a voltage or current somewhere else in the circuit. We first discuss the method of analysis of circuits containing dependent sources and suggest a procedure called "taping," which refers to the image of placing a piece of masking tape over an assumed label that gives the algebraic relationship of the dependent source. This turns it, temporarily, into an independent source of the appropriate type, albeit one having an unknown value. This is not a trivial step. By using it, we turn the analysis into one of more conventional type: one with only resistors and independent sources. We perform this analysis as usual, then "untape" the dependent source(s) and express its (their) value(s) in terms of the unknowns we have defined for the circuit. Thus, the taping and untaping steps are the only new elements in our circuit analysis procedure.

In this chapter we also consider circuits having one or more operational amplifiers (op amps). We define an op amp to be a VCVS with two special characteristics: one terminal of the VCVS is connected to ground reference (all op amp circuits therefore are required to possess a previously defined reference), and its voltage gain μ is infinite. This latter characteristic leads to the requirement for a more specialized analysis procedure because we cannot express the value of the VCVS directly as a multiple of another voltage in the circuit. We show that the analysis is expedited by the introduction of two new ideal elements: the nullator and the norator. Together these devices model the ideal op amp and they provide an easy visualization for the behavior of this element. Following the operational amplifier material, we develop the basic ideas of modeling within the context of semiconductor diodes and BJTs.

Sections 5.1, 5.2, 5.4, 5.5, and 5.7 depend only upon the material covered by the first three chapters of this text. Section 5.3 also requires the material in Chapter 4. Section 5.6 requires a knowledge of the material in Chapter 4 plus that in Sections 5.4 and 5.5 of this chapter. Finally, Section 5.8 requires knowledge of the SPICE material in all previous SPICE sections.

5.1 | Dependent Sources

Chapters 1 through 4 discuss circuits having only three types of elements: resistors, independent voltage sources, and independent current sources. The resistor is passive; that is, one cannot connect it to any circuit that will "draw" net energy from it. Independent voltage and current sources, on the other hand, are active. It is possible to extract a net amount of energy from them by connecting them to an appropriate load. We refer to circuits having only these three types of elements as passive circuits because sometimes the independent sources are considered to be external inputs to the circuit, and the "circuit" itself is considered to only consist of the resistive part. In this section we will introduce four additional two-terminal active elements called *dependent (or controlled) sources.* Unlike independent sources, they must receive a stimulus from somewhere else in the circuit in order to provide a voltage or a current.

The Dependent Voltage Source   Figure 5.1 shows the symbol and *v-i* characteristic for a *dependent voltage source.* Its outline is a diamond shape to distinguish it from the round-shaped independent voltage source. The terminal voltage of the dependent voltage source, when plotted as a function of its terminal current, is exactly the same as that of an independent voltage source. The difference is that the terminal voltage of an independent voltage source is independent[1] of all the other voltage and current variables in the circuit in which it is located. The dependent voltage source terminal voltage v_c (the subscript c standing for "controlled"), on the other hand, depends[2] upon either a voltage v_x or a current i_x somewhere else in the circuit. These circuit variables are called the *controlling voltage* or the *controlling current*, respectively. The voltage v_c is referred to as the *controlled voltage* (or *dependent voltage*). When the controlling variable is a voltage, the dependent voltage source is called a *voltage-controlled voltage source*, or *VCVS*, and a *dependency relationship* of the form

$$v_c = \mu v_x. \tag{5.1-1}$$

specifies the controlled voltage as a linear function of the controlling voltage. For this reason, a more proper term would be *linear* dependent voltage source. The parameter μ is a unitless real constant called the *voltage gain*.

Figure 5.1
The dependent voltage
source (VCVS and
CCVS)

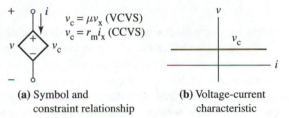

(a) Symbol and
constraint relationship

(b) Voltage-current
characteristic

When the controlling variable is a current, the dependent voltage source is called a *current-controlled voltage source* (or *CCVS*), and the linear dependency relationship takes the form

$$v_c = r_m i_x. \tag{5.1-2}$$

The real constant r_m has units of ohms and is called the *transresistance*. The prefix "trans" is used because it "transfers" the effect of a current somewhere else in the circuit to the dependent voltage source; the suffix "resistance" is used because it multiplies a current by a constant having the unit Ω to turn it into a voltage. (The subscript "m" denotes a term used in earlier times, "mutual," which is a synonym for "trans.") The CCVS, however, is quite different from the resistor element because the current in a resistor element must be defined in the *same* element as that across which the voltage is defined.

*The Dependent
Current Source*

Figure 5.2 shows the symbol for a *dependent current source*. It has the same diamond shape as that of the dependent voltage source; however, it has an arrow denoting the positive current reference, whereas the dependent voltage source contains a pair of voltage polarity indicators: plus and minus signs. A glance at the *v-i* characteristic in Figure 5.2 shows that, just like the dependent v-source, the dependent i-source has exactly the same terminal behavior as that of an independent one—the only difference is that the *value* of its current variable depends upon either a voltage v_x or a current i_x somewhere else in the circuit. In the first case, we refer to the dependent i-source as a VCCS, *for voltage-controlled current source*, and the dependency relationship is

$$i_c = g_m v_x. \tag{5.1-3}$$

---

[1] This is the reason for the term "independent" voltage source.

[2] This is the reason for the term "dependent" voltage source, and the alternative term "controlled" voltage source has a similar explanation.

The real constant g_m has units of siemens and is therefore called the *transconductance*. In the second case, we refer to the dependent i-source as a CCCS, *for current-controlled current source,* and the dependency relationship is

$$i_c = \beta i_x. \tag{5.1-4}$$

The real unitless constant β is called the *current gain.*

Figure 5.2
The dependent voltage source (VCVS and CCVS)

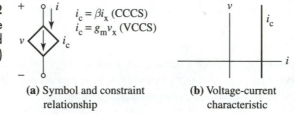

$i_c = \beta i_x$ (CCCS)
$i_c = g_m v_x$ (VCCS)

(a) Symbol and constraint relationship

(b) Voltage-current characteristic

An Intuitive Analogy for Dependent Sources

Dependent (or controlled) sources should be thought of intuitively as *amplifiers,* with the VCVS (for example) being the idealized model of a preamplifier for a home stereo system. The basic idea is nicely illustrated in Figure 5.3(a), in which we show an analogy for the operation of a VCVS. We imagine ourselves to connect a voltmeter between the two nodes for which the controlling voltage v_x is defined.[3] Such a connection does not directly influence the voltage between these two nodes because the electrical equivalent of an ideal voltmeter is an open circuit, as we show in Figure 5.3(b).

Figure 5.3
Operation of the VCVS

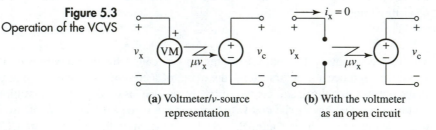

(a) Voltmeter/*v*-source representation

(b) With the voltmeter as an open circuit

The jagged arrow between the voltmeter and the "independent" v-source is meant to suggest that we can adjust an imagined control knob on the v-source so that it produces a voltage v_c that is exactly μ times the voltage we have measured for v_x. If v_x decreases for some reason, we simply adjust the control knob down to lower the value of v_c; if v_x increases, we adjust the knob so that v_c increases proportionally. Operation would be similar for a VCCS, but we would control the value of a *current* source with our knob.

For either of the current-controlled devices (CCVS or CCCS), we would monitor the current in a conductor with an ammeter[4] and use the value of its reading to adjust the value of the source on the right as we show for the CCCS in Figure 5.4. The crucial thing is this: simply think of the dependent source as an independent one—as far as the rest of the circuit is concerned—with one important difference. There is an *added algebraic constraint between its value and that of another circuit variable.*

---

[3] Remember that any voltage is defined relative to two nodes in a circuit.

[4] Remember that an ammeter must be inserted by cutting a conductor and inserting it. Because the ammeter is equivalent to a short circuit electrically, this does not directly have any effect on the currents or voltages in the circuit.

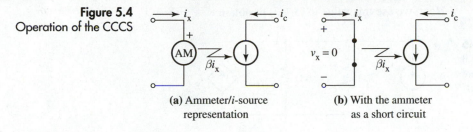

Figure 5.4
Operation of the CCCS

(a) Ammeter/i-source
representation

(b) With the ammeter
as a short circuit

Dependent Source Examples

Example 5.1 Find the value of current i in the circuit shown in Figure 5.5.

Figure 5.5
An example active
circuit

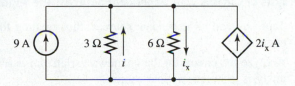

Solution Before tackling the solution, let's discuss an important fact: the controlling variable i_x is an important part of the definition of the dependent source (a CCCS here). Therefore, one cannot change its direction at will. We must always do our computation with i_x measured in the direction of its defining arrow; similarly, the upward arrow in the source symbol itself cannot be changed.[5]

Our circuit is a simple parallel one, so we have many options as to how to analyze it; however, we will use it to illustrate a general technique. We note that the dependent current source has a v-i characteristic identical to that for an independent current source. We will treat the additional algebraic constraint relationship $2i_x$ as a *label* on the source. We will imagine that we temporarily cover this label with a small piece of masking tape on which we write the symbol i_c, standing for the (unknown) value of an equivalent independent source. We will call this procedure *taping* the dependent source. We show the result in Figure 5.6 on p. 168. We retain the diamond-shape symbol for convenience to remind ourselves that there is an additional constraint that must be used sooner or later. For our simple parallel circuit, we can use the current divider rule:

$$i = -\frac{6\,\Omega}{3\,\Omega + 6\Omega}[9 + i_c] = -6 - \frac{2}{3}i_c. \qquad (5.1\text{-}5)$$

We would be through if i_c were a known quantity; however, it is not. Thus, we next *untape* the dependent source and *express its controlled variable in terms of the unknown variable* (in this case, i). When we untape the CCCS, we see using KCL that

$$i_c = 2i_x = 2[9 + i + i_c]. \qquad (5.1\text{-}6)$$

We solve this to obtain

$$i_c = -18 - 2i. \qquad (5.1\text{-}7)$$

---

[5] If the directions of *both* controlling and controlled references are changed, the resulting dependent source will produce the same effect; however, it is recommended that you not alter either.

Next, we use this value in (5.1-5) to obtain $i = -18$ A.

Figure 5.6
The example circuit
with the dependent
source "taped"

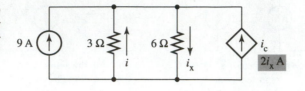

As we mentioned, there are perhaps faster ways of solving the circuit in the last example, but we were developing a general method for the analysis of circuits containing dependent sources. This method is stated as follows.

Analysis of Circuits with Dependent Sources: The Algorithm

1. Tape any and all dependent sources, thus treating them *temporarily* as *independent* sources.

2. Analyze the circuit for the unknown variable or variables you have chosen using the analysis technique of your choice.

3. Untape the dependent sources and express their values in terms of the unknowns you have chosen to use in step 2.

4. Solve the resulting equations.

5. Solve for the desired quantity.

Example 5.2 Solve the circuit in Figure 5.7 for the voltage v.

Figure 5.7
An example circuit

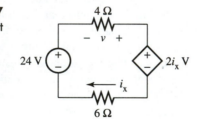

Solution This is a simple series circuit. We show it with the dependent source taped in Figure 5.8. We can use voltage division to compute v.[6] Its value is

$$v = \frac{4\ \Omega}{4\ \Omega + 6\ \Omega}[v_c - 24] = \frac{2}{5}v_c - \frac{48}{5}. \tag{5.1-8}$$

If v_c were a known quantity, we would be through. As it is not, we untape the dependent source and express its value in terms of the unknown voltage v. This gives

$$v_c = 2i_x = 2\left(\frac{24 - v_c}{10}\right). \tag{5.1-9}$$

We solve this to obtain $v_c = 4$ V. Then we use it in equation (5.1-8) to obtain $v = -8$ V.

---

[6] The two-terminal network "looking away from" the 4-Ω resistor is equivalent to one consisting of *any* reordering of its three constituent elements—as you can show by writing its v-i characteristic.

Figure 5.8
The example circuit
with the CCVS taped

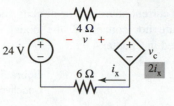

Nonexistence and
Nonuniqueness of
Solutions for
Active Circuits:
Singular Cases

We now ask whether each and every circuit having one or more dependent sources has a solution. As it happens, the answer is no. Such circuits do not always have solutions for certain values of the parameter in their constraint equations (that is, for certain values of μ, r_m, β, or g_m). In this, they are unlike their passive relatives, which were shown in Chapter 4 to always have solutions provided that the topology meets some very mild conditions. To review the situation for passive circuits, see Figure 5.9. Each of these simple circuits is disallowed because it violates either KVL or KCL. Now look at Figure 5.10(a), a generalization of the leftmost two in Figure 5.9. Unless v_s is exactly equal to 4 V, it too violates KVL. The problem is that there is a loop consisting of only independent voltage sources. In fact, one of the three v-sources is mislabeled because it cannot be adjusted independently to any value; thus, we should rather think of it as a dependent v-source. If we deactivate this circuit, thus turning the v-sources into short circuits, we obtain the deactivated network shown in Figure 5.10(b). As we showed in Chapter 4, this circuit does not have a unique solution in terms of mesh currents[7] because it is not coupled: one cannot go from the exterior of the entire circuit to the internal mesh by "stepping over" shared resistors.

Figure 5.9
Some disallowed
circuits

Figure 5.10
A disallowed circuit

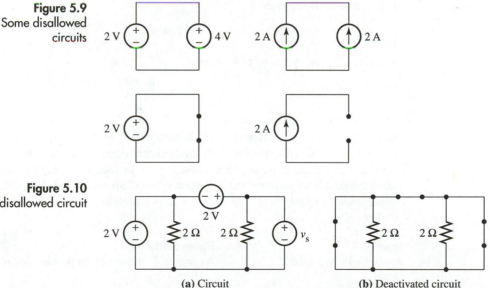

(a) Circuit   **(b)** Deactivated circuit

A similar situation, one that generalizes the two on the right-hand side of Figure 5.9, is shown in Figure 5.11. There we see that the circuit is split into two parts when the circuit is deactivated (see the deactivated network in Figure 5.12). We call such a set of current sources that separate a connected circuit into exactly two parts when they are

---

[7] It is apparent that the voltage—and hence the current—is known for each of the resistors; however, it is impossible to find the current in the v-sources.

deactivated a *cut set* of current sources. For circuits such as this, that is for those whose deactivated networks are not connected, the nodal equations do not have a unique solution[8]—as we showed in Chapter 4.

Figure 5.11
A disallowed circuit

Figure 5.12
Deactivated network

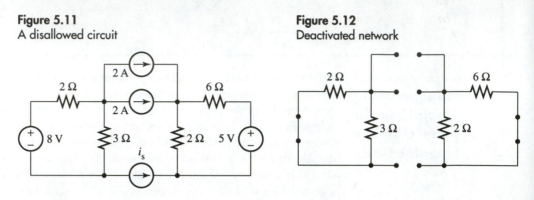

It is unfortunately true that circuits with dependent sources can fail to have solutions even if they are coupled and connected. A simple circuit that can exhibit such "maverick" behavior is shown in Figure 5.13. There we see a single loop (series) circuit containing one independent v-source, one dependent v-source, and one resistor. Because this circuit is so simple, we have already shown the dependent source in taped form. We can use KVL and untape to get

$$v_x = 8 - (-v_c) = 8 + \mu v_x \tag{5.1-10}$$

This can be rearranged to

$$(1 - \mu)v_x = 8. \tag{5.1-11}$$

Therefore—*provided $\mu \neq 1$*—we have

$$v_x = \frac{8}{1 - \mu}\text{V}. \tag{5.1-12}$$

If $\mu = 1$, however, equation (5.1-11) becomes $0 \cdot v_x = 8$. This is an impossible situation because any value of v_x results in the inconsistent demand that $0 = 8$. Thus, we can only conclude that there is no solution when $\mu = 1$. In fact, it is easy to show (and we will leave the details to you) that the 8-V source "sees" an equivalent short circuit in this case, as we show in Figure 5.14. This, of course, is a prohibited circuit combination because it violates KVL.

Figure 5.13
An example "maverick" circuit

Figure 5.14
A graphic illustration of the problem for $\mu = 1$

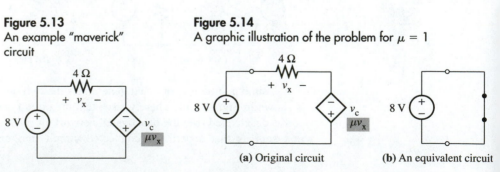

(a) Original circuit   (b) An equivalent circuit

---

[8] Here, the indeterminate quantity is the voltage across any of the current sources.

**Practical
Interpretation of
Nonexistence and
Nonuniqueness
of Solutions**

It might be well to pause for a bit of discussion about the nonexistence of solutions. After all, our physical intuition is that *every* circuit has a solution. We can certainly hook up the circuit in the lab and measure the voltages and currents. The voltage v_x in Figure 5.13 (or Figure 5.14), for instance, should be *something* even if $\mu = 1$, shouldn't it? The answer of course is yes. Physical circuits certainly have voltages and currents that can all be measured. The shortcoming is in our circuit *model* of the "real world." Elsewhere in the text, we point out that two v-sources with different values represent a poor model when connected in parallel as in Figure 5.9. In practice, an actual component such as a car battery is a complex device. Over a range of load currents it can be modeled fairly accurately by its Thévenin equivalent circuit. Over a still more restricted range of load currents, one can neglect the series resistance and approximate it as an ideal v-source. For heavier load currents, one cannot do this, and exactly this situation occurs when it is paralleled with a battery having a different terminal voltage.

In the case of the dependent source there is an equivalent range of conditions. Figure 5.15 shows a VCVS, along with a plot of how its (supposedly linear) algebraic constraint actually looks when measured in the laboratory. The "saturation values" $\pm V_s$ are the "power supply" voltages.[9] (In fact, the corners would probably be somewhat more rounded.) The basic idea is this: it is only over a finite range of values of v_x ($-\epsilon \leq v_x \leq \epsilon$) that the linear relationship $v_c = \mu v_x$ holds, and it is for this range of values that a solution can fail to exist mathematically. Suppose, for instance, that the linear range in Figure 5.15 is $-4\text{ V} \leq v_x \leq 4\text{ V}$ (that is, $\epsilon = 4$ V) and $\mu = 1$ (the "bad" value). When we see that our analysis with an ideal linear VCVS fails, we should suspect that operation is actually out of the linear range. Assuming that $v_x > 4$ V, therefore, we simply set $v_c = 4$ V ($\mu = 1$ and $\epsilon = 4$, so V_s has to be 4 V). Doing this, we see that $v_x = 12$ V is a perfectly legitimate solution for the circuit of Figure 5.13. KVL "works," and $v_x = 12$ V holds v_c at its saturation value of 4 V, as shown in Figure 5.16. (If we try $v_c = -4$ V, KVL will result in $v_x = 8 - 4 = 4$ V, which is *not* consistent with $v_c = -4$ V.)

Figure 5.15
A more realistic characteristic for a VCVS

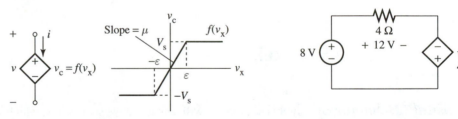

Figure 5.16
A "realistic" equivalent

In the rest of this text, we will assume that all dependent sources are linear—with the consequent possibility that solutions might not exist for particular parameter values.

---

[9] A dependent source is a model of an active device such as a transistor, which requires dc power supplies for proper operation. These power supplies, however, are not shown on circuit diagrams when the devices are being modeled as ideal elements.

Section 5.1 Quiz

Q5.1-1. Identify (without looking at the definitions in the text) the dependent sources shown in Figure Q5.1-1.

Figure Q5.1-1

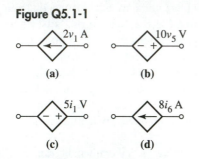

(a)   (b)

(c)   (d)

Q5.1-2. For the circuit in Figure Q5.1-2, find v_a if $g_m \neq 0.5$. Does a solution exist if $g_m = 0.5$? Is it a unique solution?

Figure Q5.1-2

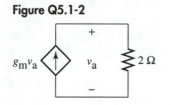

Q5.1-3. Find v_b and the power absorbed by the dependent source in Figure Q5.1-3.

Figure Q5.1-3

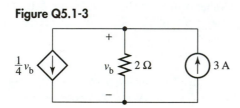

Q5.1-4. Find i and the power absorbed by the dependent source in Figure Q5.1-4.

Figure Q5.1-4

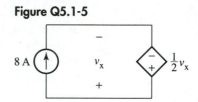

Q5.1-5. Find v_x in the circuit shown in Figure Q5.1-5. What is the equivalent circuit for everything except the 8-A current source?

Figure Q5.1-5

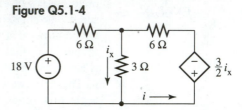

Q5.1-6. Prove the statement made in footnote 6 (relative to Example 5.2).

5.2 | Active Subcircuits

"Unusual" Behavior of Active Subcircuits

In this section we will discuss active subcircuits, which we will loosely define as any subcircuit containing one or more dependent sources.[10] Such subcircuits exhibit rather unusual and striking behavior relative to their passive cousins. We can see this fairly quickly by looking at the simple subcircuit shown in Figure 5.17(a). It is just a CCVS whose controlling variable i_x is one of the terminal currents. Because the terminal voltage is proportional to the terminal current, $v = -2i$, we see that this subcircuit is equivalent to a resistor—one, however, whose value is negative. Thus, unlike the resistor *element,* whose

---

[10] Our more precise definition of Chapter 1 requires that the subcircuit be capable of supplying net energy to the external environment through its terminals; however, our usage here is more common in the general circuit analysis literature when *loosely* categorizing circuits and subcircuits.

resistance parameter is always positive, a subcircuit can act like a resistor with a negative value. Notice that the power absorbed by the subcircuit is

$$P = vi = -2i^2, \tag{5.2-1}$$

which is strictly negative if the current is nonzero. Hence, the energy, being the time integral of power, is also negative and the negative resistor can deliver energy to the external circuit. The subcircuit is clearly active in our more rigorous sense.

Figure 5.17
A CCVS as a negative resistor

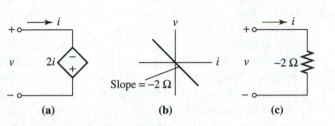

(a)   (b)   (c)

Figure 5.18(a) shows a subcircuit only slightly more complicated. We have just added an ordinary resistor element in series with the CCVS. We know from our preceding discussion (generalized slightly in that the transresistance r_m is a symbolic parameter rather than 2 Ω as in the preceding case) that the CCVS is equivalent to a negative resistor whose value is $-r_m$. This equivalent resistance is connected in series with the positive resistor of value R, giving the equivalent subcircuit shown in Figure 5.18(b). Finally, we know that two resistors connected in series are equivalent to a single resistor whose value is the sum of the two individual resistances. This gives the final equivalent subcircuit shown in Figure 5.18(c).

Figure 5.18
Another active subcircuit equivalence

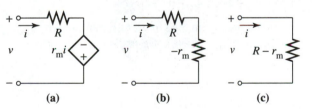

(a)   (b)   (c)

Figure 5.19(a) shows one of the interesting—and perhaps unexpected—things that can happen in a subcircuit with active elements. If r_m in our example subcircuit is set equal to R, the resistor element value, then the equivalent resistance is zero! This means that the subcircuit is equivalent to a short circuit. Let's look at the internal voltage variables, which we have labeled in Figure 5.19(b), a bit closer. By voltage division,

$$v = v_1 + v_2 = \frac{-r_m}{R - r_m}v + \frac{R}{R - r_m}v. \tag{5.2-2}$$

Thus, if we conceptually allow r_m to approach R from below we see that v_1 and v_2 approach $-\infty$ and $+\infty$, respectively, if $v \neq 0$. (We also see that a voltage divider can provide voltage *gain* when one resistor is negative.) Their algebraic sum, however, remains finite and equal to v. This type of behavior is, of course, not true of the physical short circuit (a length of conductor). The conclusion is that active networks can exhibit rather bizarre behavior internally, even though they might present an equivalent face to the world that is nicely behaved.

Figure 5.19
An unexpected
equivalence

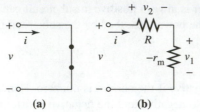

(a)   (b)

Practical Aspects of
Dependent Sources

We will digress just a bit to discuss some practical aspects of dependent sources. We are treating them as elements. In practice, however, they are either constructed from a number of other physical circuit elements or they occur as part of an equivalent circuit of a more complicated electronic device. Thus, one cannot go into a parts store and ask about the price of a VCVS.[11] In fact, this is true for a great many circuit elements. Although one can buy a battery at the local grocery store, a current source—even one of the independent variety—must be constructed from a number of electronic elements. In fact, even the voltage source is not always a physical battery; it is more often a symbolic representation of a *power supply,* a complicated unit that is equivalent in most contexts to a voltage source.

Another practical aspect of circuit elements is this: all of the active elements (independent and dependent sources) rely upon a power supply of some sort to provide the energy that they "generate." From this point of view, these elements do not create energy—they simply change its form from the energy supplied by the power supply to the energy supplied by the active element. When we draw an active element symbol, we are actually drawing a simple representation of this more complicated arrangement of a number of devices and a power supply.

Some Examples of
Unusual Behavior

Example 5.3   Find the voltage v in the circuit of Figure 5.20 by superposition. Evaluate v for $\beta = 2$ and $\beta = 3$. For what value(s) of β is there no solution?

Figure 5.20
An example circuit

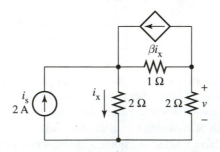

Solution   As there are two sources, we will find two partial responses: one for each source. Furthermore, as one of the sources is a dependent one, we will also find its controlling variable i_x by superposition. Figure 5.21 shows the partial circuit resulting from deactivating the independent current source. It is important to note that we have *taped* the dependent source, thus treating it temporarily as an independent source. We have also labeled the desired response *and the controlling variable* with the subscript 1 to denote the first partial response. Using current division and Ohm's law, we compute the values of i_{x1} and v_1 (the

---

[11] Or perhaps we should say that one *should not*—otherwise risk being considered somewhat strange!

two 2-Ω resistors are equivalent to a single 4-Ω resistor connected in parallel with the i-source and the 1-Ω resistor):

$$i_{x1} = \frac{1}{1 + 2 + 2}i_c = \frac{1}{5}i_c \qquad (5.2\text{-}3a)$$

$$v_1 = -2i_{x1} = -\frac{2}{5}i_c. \qquad (5.2\text{-}3b)$$

Figure 5.21
Partial response due to
the dependent source

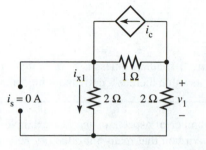

Figure 5.22 shows the partial circuit resulting when we deactivate the dependent source. We can once again use current division and Ohm's law to obtain

$$i_{x2} = \frac{1 + 2}{1 + 2 + 2}i_s = \frac{3}{5}i_s \qquad (5.2\text{-}4a)$$

and

$$v_2 = 2\frac{2}{2 + 1 + 2}i_s = \frac{4}{5}i_s. \qquad (5.2\text{-}4b)$$

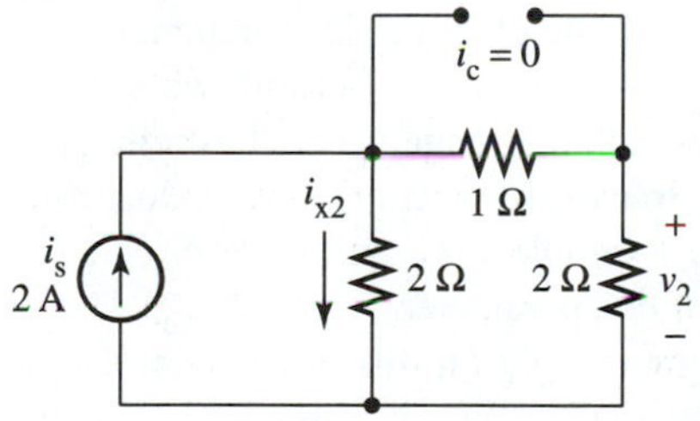

Figure 5.22
Partial response due to
the independent source

The principle of superposition now tells us to add the two partial responses we have just computed to obtain the total (actual) value of the associated variable. These two values are

$$i_x = i_{x1} + i_{x1} = \frac{1}{5}i_c + \frac{3}{5}i_s \qquad (5.2\text{-}5a)$$

and

$$v = v_1 + v_2 = -\frac{2}{5}i_c + \frac{4}{5}i_s. \qquad (5.2\text{-}5b)$$

We now *untape* the dependent source. This results in

$$i_c = \beta i_x = \beta\left[\frac{1}{5}i_c + \frac{3}{5}i_s\right]. \qquad (5.2\text{-}6)$$

This equation is implicit in i_c. Rearranging, we obtain

$$[5 - \beta]i_c = 3\beta i_s. \qquad (5.2\text{-}7)$$

Now if $\beta \neq 5$, we have

$$i_c = \frac{3\beta}{5 - \beta}i_s. \qquad (5.2\text{-}8)$$

We can use this value in equation (5.2-5b) to get

$$v = -\frac{2}{5}\left(\frac{3\beta}{5-\beta}\right)i_s + \frac{4}{5}i_s = \frac{2(2-\beta)}{5-\beta}i_s. \qquad (5.2\text{-}9)$$

For $\beta = 2$, we obtain $v = 0$ independently of the value of i_s. For $\beta = 3$, we have $v = -i_s = -2$ V. We also see at once that the special value $\beta = 5$ gives problems in equations (5.2-8) and (5.2-9). Therefore, we back up a step to equation (5.2-7). This equation becomes

$$0\,i_c = 15\,i_s, \qquad (5.2\text{-}10)$$

which *does not have a solution* unless $i_s = 0$. Is this case any value of i_c will work, and for any choice for its value there is a corresponding value of v from equation (5.2-5b).

There are a number of aspects of the last example that are worth exploring in more detail. First, we warn you that many introductory texts issue an injunction that deactivation of the dependent source is incorrect. In fact, it is a quite proper procedure. It is the linearity of the *resistive portion* of the circuit that allows superposition to be used. The concept of taping the dependent source is vital in our approach because it temporarily turns a dependent source into an independent one—for which superposition is valid. After applying superposition, we simply factor in the added dependent source constraint. You should use this technique with some care, however, for you must find the value of the controlling variable (i_x in the preceding example) as well as the specified variable desired. Furthermore, untaping must be done relative to the *total* (actual) value of the controlling variable—not one of its partial components.

Regardless of our method of solution, we see that a circuit with dependent sources can behave in some rather strange ways, depending on the value of its parameter (voltage gain, current gain, transresistance, or transconductance). As we pointed out in Section 5.1 (and illustrated in Example 5.3), an active circuit can fail to have a solution for some value (or values) of this parameter.

We can shed more light on this issue by considering the circuit of the last example for $\beta = 5$, the "singular" value. We show this situation in Figure 5.23. We have labeled the value of voltage across each of the resistors. We see that KVL around the resistor loop results in

$$0 = 2i_x - (4i_x + i_s) - 2(i_s - i_x) = -3i_s. \qquad (5.2\text{-}11)$$

Clearly, if i_s is nonzero, there is a violation of KVL. This is usually the case for a circuit having "singular" dependent source values, and it represents a conflict between the topological constraints (KVL, KCL) and the dependent source relationship. We will leave it as an exercise for you to show that the equivalent two-terminal subcircuit "seen" by the 2-A source is equivalent to an open circuit—a configuration prohibited by KCL.

Figure 5.23
The "singular" case

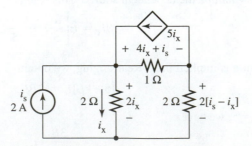

"Unusual" Aspects of Thévenin and Norton Equivalents for Active Subcircuits

The Thévenin and Norton equivalent subcircuits exhibit several differences when compared with those for passive subcircuits. We can illustrate these differences quite well by investigating the subcircuit in Figure 5.24. Although it has only one independent source (a v-source) and one dependent source (a VCVS), the general ideas work for two-terminal subcircuits with any number of independent and dependent sources of any type. As we show in the figure, we assume first that the *controlling* variable for the dependent source is located external to the subcircuit. We also show the dependent source as being taped.

Figure 5.24
An active subcircuit

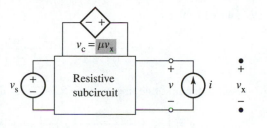

Because the resistive portion of the circuit is linear, we can apply superposition, as we have just illustrated. This results in

$$v = a_1 v_s + a_2 v_c + a_3 i$$
$$= a_1 v_s + \mu a_2 v_x + a_3 i \qquad (5.2\text{-}12)$$
$$= v_{oc} + R_{eq} i,$$

where the a's are constants whose values depend on the resistive subcircuit only. We see that the open circuit voltage is the sum of the values of an independent and a dependent v-source, as shown in Figure 5.25(a). If we solve (5.2-12) for i, we obtain

$$i = -\frac{a_1}{a_3} v_s - \frac{\mu a_2}{a_3} v_x + \frac{1}{a_3} v$$
$$\qquad (5.2\text{-}13)$$
$$= i_{sc} + \frac{1}{R_{eq}} v.$$

Equation (5.2-13) describes the Norton equivalent in Figure 5.25(b). We must again alert you to what you might encounter in other texts: many of them state *as an edict* (incorrectly) that one cannot determine a Thévenin or Norton equivalent if the controlling variable is exterior to the subcircuit under consideration.[12]

Figure 5.25
Thévenin and Norton equivalents for active subcircuits

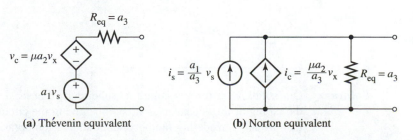

(a) Thévenin equivalent (b) Norton equivalent

---

[12] We have *proved* that you can.

Suppose, now, that the controlling variable v_x is *inside* our two-terminal subcircuit. Then we can apply superposition to express it in terms of v_s, v_c, and i:

$$v_x = b_1 v_s + b_2 v_c + b_3 i, \qquad (5.2\text{-}14)$$

where the b's are constants depending on the resistive part of the network only. Untaping the dependent source gives

$$v_c = \mu v_x = \mu b_1 v_s + \mu b_2 v_c + \mu b_3 i. \qquad (5.2\text{-}15)$$

If $\mu \neq 1/b_2$, we can solve for v_c:

$$v_c = \frac{\mu b_1}{1 - \mu b_2} v_s + \frac{\mu b_3}{1 - \mu b_2} i. \qquad (5.2\text{-}16)$$

Using this result in the first line in equation (5.2-12) results in

$$v = \left[a_1 + \frac{\mu a_2 b_1}{1 - \mu b_2} \right] v_s + \left[a_3 + \frac{\mu a_2 b_3}{1 - \mu b_2} \right] i \qquad (5.2\text{-}17)$$

$$v = v_{oc} + R_{eq}\, i.$$

The main result here is that the dependent source disappears from the Thévenin equivalent subcircuit when the controlling variable is in the same subcircuit as the dependent source; therefore, the Thévenin equivalent reverts to the conventional one we obtained for passive circuits in Chapter 3, Section 3.3. The only difference between the earlier one and the one we have just derived is that the dependent source parameter affects both the Thévenin voltage source value and the equivalent resistance. A similar conclusion, of course, holds for the Norton equivalent as well.

Thévenin and Norton Equivalent Examples

Example 5.4   For the two-terminal subcircuit shown in Figure 5.26:

a.   Find the Thévenin and Norton equivalents for general values of β.

b.   Evaluate the parameters of these equivalents for $\beta = 1$ and $\beta = 2$.

Figure 5.26
An example subcircuit

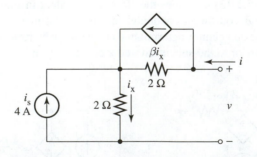

Solution   Our solution, as for any circuit or subcircuit of at least moderate complexity, starts with the taping of the dependent source, thus giving the circuit of Figure 5.27. We remind you that we are considering the dependent source temporarily to be an independent one, although we retain the diamond shape of the dependent source. In Figure 5.27 we have attached an independent current source to the terminals as a test instrument and have drawn an imaginary closed surface around the dependent source and its parallel resistor to aid you in visualizing the following KCL equation:

$$i_x = i_s + i. \tag{5.2-18}$$

Furthermore, KCL applied to the node at the junction of the applied test source and the dependent source gives a current of $i - i_c$ through the resistor inside the surface. Thus, by KVL around the loop consisting of the two resistors and the applied test source, we have

$$v = 2(i + i_s) + 2(i - i_c) = 4i - 2i_c + 2i_s. \tag{5.2-19}$$

Figure 5.27
Taping the dependent
source and testing the
subcircuit

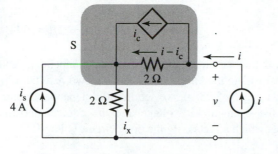

We next untape the dependent source, getting

$$i_c = \beta i_x = \beta(i + i_s). \tag{5.2-20}$$

Using this in equation (5.2-19) gives

$$v = 2(2 - \beta)i + 2(1 - \beta)i_s. \tag{5.2-21}$$

We now identify the Thévenin parameters:

$$R_{eq} = 4 - 2\beta \tag{5.2-22}$$

and

$$v_{oc} = 2(1 - \beta)i_s. \tag{5.2-23}$$

The Thévenin equivalent subcircuit is shown in Figure 5.28(a). The Norton equivalent is sketched in Figure 5.28(b). It is obtained by simply solving (5.2-21) for i:

$$i = \frac{v}{2(2 - \beta)} - \frac{1 - \beta}{2 - \beta}i_s. \tag{5.2-24}$$

We identify the constant term as the short circuit current i_{sc}:

$$i_{sc} = \frac{1 - \beta}{2 - \beta}i_s. \tag{5.2-25}$$

If $\beta = 1$, we see that the Thévenin parameters are

$$v_{oc} = 2(1 - \beta)i_s = 0 \tag{5.2-26}$$

Figure 5.28
Thévenin and Norton
equivalent subcircuits

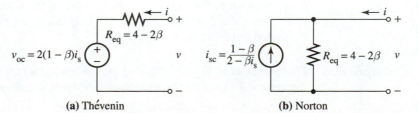

(a) Thévenin (b) Norton

and
$$R_{eq} = 4 - 2\beta = 2\ \Omega. \tag{5.2-27}$$

This is a rather interesting result, for it says that the entire subcircuit is equivalent to a pure 2-Ω resistance—even though there is an independent source!

When $\beta = 2$, we see that

$$v_{oc} = 2(1 - \beta)i_s = -2i_s = -8\ \text{V} \tag{5.2-28}$$

and

$$R_{eq} = 4 - 2\beta = 0\ \Omega. \tag{5.2-29}$$

Thus, the subcircuit is equivalent to an independent voltage source—even though there are resistors in the circuit. We show our two results in Figure 5.29. In general, if we look back at Figure 5.28(b), we see that the dependent source current gain parameter affects both the open circuit voltage and the equivalent resistance—unlike the independent source, which affects only the open circuit voltage.

Figure 5.29
Equivalents for special values of the current gain

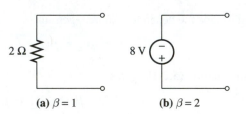

(a) $\beta = 1$ **(b)** $\beta = 2$

As usual, the Norton equivalent resistance parameter is the same as that of the Thévenin. We compute the short circuit current from equation (5.2-25) for $\beta = 1$. It is

$$i_{sc} = 0, \tag{5.2-30}$$

as we would have expected, for we already know that the subcircuit is purely resistive. If $\beta = 2$, we see that

$$i_{sc} = \infty\ . \tag{5.2-31}$$

This tells us that something is unusual. A check of (5.2-22) or (5.2-27) shows that the equivalent resistance is zero! But the case $\beta = 2$ consists of an ideal voltage source for the Thévenin equivalent, as we already know. Thus, the Norton equivalent does not exist in this special case for we cannot invert (5.2-21) to get (5.2-24).

For the subcircuit in the previous example, the controlling variable i_x was defined in an element within the subcircuit. Thus, as we see from equation (5.2-21), the dependent source affects both the open circuit voltage and the equivalent resistance. This is in accordance with our earlier general result. If the controlling variable is located outside the subcircuit, then the dependent source will only affect the open circuit voltage. This idea is illustrated further in the next example.

Example 5.5   For the circuit in Figure 5.30, find the current i_x by replacing the subcircuit to the left of the vertical 4-Ω resistor by its Norton equivalent subcircuit.

Figure 5.30
An example circuit

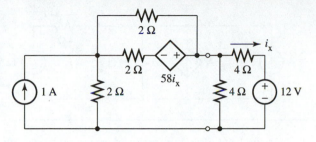

Solution   In Figure 5.31 we have removed the two 4-Ω resistors and the 12-V source, taped the
CCVS, and applied a test voltage source. We have also defined the ground reference node
and the node voltage at the single essential node in anticipation of nodal analysis. The
single nodal equation is

$$\frac{v}{2} + \frac{v - v_t}{2} + \frac{v - (v_t - v_c)}{2} = 1. \tag{5.2-32}$$

Solving for v_t and using it to compute i_t gives

$$i_t = -\left[\frac{2}{3} + \frac{v_c}{6}\right] + \frac{v_t}{3}. \tag{5.2-33}$$

The bracketed term represents the parallel combination of an independent current source
of value 2/3 A and a dependent one of value $v_c/6$ A. The coefficient of the remaining term
is the parallel conductance of the Norton equivalent. In Figure 5.32 (on p. 182) we show
this Norton equivalent and have replaced the elements attached to its terminals. We can
compute the desired current i_x in several ways. Choosing superposition, we have

$$i_x = \frac{-12}{4 + \dfrac{3 \times 4}{3 + 4}} + \frac{\dfrac{1}{4}}{\dfrac{1}{3} + \dfrac{1}{4} + \dfrac{1}{4}}\left[\frac{2}{3} + \frac{v_c}{6}\right] = \frac{v_c - 38}{20}. \tag{5.2-34}$$

(We used current division in conductance form to get the second term after the first equal
sign.) Untaping the dependent source, we have

$$v_c = 58i_x. \tag{5.2-35}$$

Using this in (5.2-34) gives $i_x = 1$ A.

Figure 5.31
Testing the subcircuit

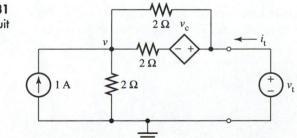

Figure 5.32
The Norton equivalent
subcircuit

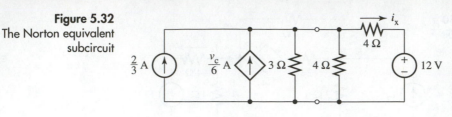

As we have already mentioned, many texts state that, for the Thévenin and Norton equivalents to exist, the controlling variable must be inside the subcircuit under consideration; they, however, do not include a dependent source in the equivalent subcircuit. (To be fair, we should point out that the Thévenin and Norton equivalents were first derived for passive circuits. But the restriction is simply not necessary.)

Example 5.6   Consider the circuit in Figure 5.33.

a.   If $\beta = 0.5$, find the value of R that results in maximum power absorbed by R.

b.   If $\beta = 1.5$, find the value of R that results in maximum power absorbed by R.

c.   If $R = 16\ \Omega$, sketch a graph of the power absorbed by R versus β.

Figure 5.33
An example circuit

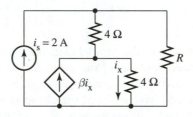

Solution   The first step, of course, is to tape the dependent source. Additionally, we attach a test source in the place of the resistor R. This is shown in Figure 5.34. We have shown the resistor currents obtained by applying KCL to the top and middle nodes. KVL now implies that

$$v = 4(i + i_s) + 4(i + i_s + i_c) = 8i_s + 4i_c + 8i. \tag{5.2-36}$$

Untaping the dependent source, we obtain

$$i_c = 0.5i_x = 0.5\,[i_s + i + i_c]. \tag{5.2-37}$$

Solving this last equation for i_c and using the resulting value in (5.2-36) gives

$$v = 24 + 12i, \tag{5.2-38}$$

where we have used the given value of $i_s = 2$ A. Thus, we know that $R_{eq} = 12\ \Omega$ and we must therefore make $R = 12\ \Omega$ for maximum power transfer.

Figure 5.34
Testing the circuit after
taping

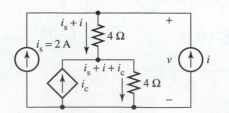

If we let $\beta = 1.5$ when we untape, the resulting equation (5.2-37) becomes

$$i_c = 1.5 i_x = 1.5 [i_s + i + i_c]. \tag{5.2-39}$$

Solving for i_c and using the resulting value in (5.2-36) now gives

$$v = -8 - 4i. \tag{5.2-40}$$

As the equivalent resistance is negative, we cannot blindly make the value of R equal to $-4\,\Omega$ because our original derivation of this maximum power transfer condition in Chapter 3, Section 3.4, was done under the assumption that both R and R_{eq} were nonnegative. Therefore, we will work directly with the circuit incorporating the Thévenin equivalent described in equation (5.2-40). We show this circuit in Figure 5.35. The power dissipated in R is given by

$$P_R = v_R i_R = \frac{R}{R-4}(-8) \times \frac{-8}{R-4} = \frac{64R}{(R-4)^2}. \tag{5.2-41}$$

The maximum value is infinity when $R = 4\,\Omega$!

Figure 5.35
Thévenin equivalent

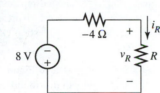

Now let's express the CCCS relationship of (5.2-37) in terms of a general value of β when we untape, getting

$$i_c = \beta i_x = \beta[i_s + i_c + i]. \tag{5.2-42}$$

Solving explicitly for i_c and using the resulting value in (5.2-36) with $i_s = 2$ A gives

$$v = \frac{8(2-\beta)}{1-\beta} + \frac{4(2-\beta)}{1-\beta}i. \tag{5.2-43}$$

Equation (5.2-43) now gives the Thévenin equivalent shown in Figure 5.36. The power transferred to the $R = 16$-Ω resistor is (after a small bit of work that we leave to you)

$$P_R = 64\left[\frac{2-\beta}{6-5\beta}\right]^2. \tag{5.2-44}$$

The graph of this function of β is shown in Figure 5.37. Notice that the dissipated power approaches infinity as β approaches 1.2. Note also that it is zero when $\beta = 2$. These facts show that the conditions for maximum power transfer in active circuits are quite different from those for passive circuits; also, the delivered power can have a quite complicated dependence upon the gain constant(s) of the dependent source(s).

Figure 5.36
General Thévenin
equivalent

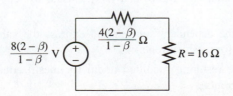

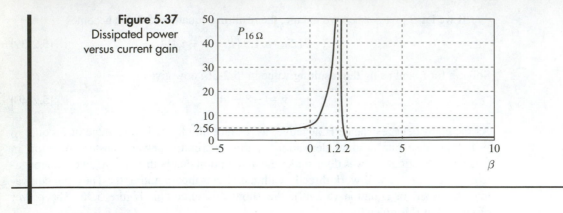

Figure 5.37
Dissipated power
versus current gain

Section 5.2 Quiz

Figure Q5.2-1 pertains to all questions.

Figure Q5.2-1

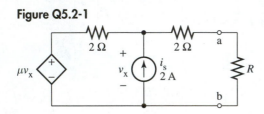

Q5.2-1. Use superposition to find v_x.

Q5.2-2. Remove the resistor labeled R, attach a test current source, and analyze the resulting circuit to determine the Thévenin equivalent subcircuit for all the elements to the left of the terminals labeled a and b.

Q5.2-3. For which value(s) of μ is the Thévenin equivalent a pure resistance? For which value(s) is it a pure voltage source? For which value(s) does a Thévenin equivalent not exist?

Q5.2-4. Answer Questions Q5.2-2 and Q5.2-3 by attaching a test *voltage* source and replacing the word "Thévenin" at each occurrence by the word "Norton." Also replace the phrase "pure voltage source" in Question Q5.2-3 with the phrase "pure current source."

Q5.2-5. If $\mu = 2$ in Figure Q5.2-1, find the nonnegative value of R that results in maximum power delivered to R.

Q5.2-6. If $\mu = 0.5$ in Figure Q5.2-1, find the nonnegative value of R that results in maximum power delivered to R.

Q5.2-7. If $R = 2 \, \Omega$ in Figure Q5.2-1, find and plot the power delivered to R as a function of μ.

5.3 | Nodal and Mesh Analysis of Circuits with Dependent Sources

In Section 5.1 we pointed out that dependent sources have exactly the same *v-i* characteristics as their independent cousins; the only difference is that there is an added algebraic constraint requiring that their values be proportional to some other voltage or current in the circuit. We will now use this observation to extend our techniques of nodal and mesh analysis to active circuits with the following step-by-step algorithm.

Nodal and Mesh Analysis of Active Circuits

1. *Tape* all the dependent sources, thus temporarily treating them as independent sources—albeit with unknown (symbolic) values.

2. Perform nodal (or mesh) analysis as usual to determine the nodal (or mesh) equations.

3. *Untape* the dependent sources and express the value of each in terms of the node voltages (or mesh currents).

4. Solve the resulting modified nodal (or mesh) equations for the node voltages (or mesh currents).

5. Solve for the desired circuit quantity in terms of the node voltages (or mesh currents).

As the procedures for nodal and mesh analysis for circuits with dependent sources are so similar to those for circuits without them, we will concentrate here on illustrating the method with examples. *We will assume in each case that the values of the dependent source parameters are such that a unique solution exists.* Though the determination of such values is an important issue in its own right, we have explored this situation sufficiently in Sections 5.1 and 5.2, so we will now focus on the solution procedures themselves.

Example 5.7   Find the value of i in the circuit in Figure 5.38 using nodal analysis.

Figure 5.38
An example circuit

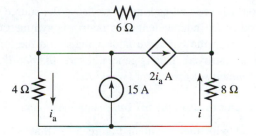

Solution   Our first step, as outlined in the algorithm, is to tape the single dependent source. We then prepare the circuit for nodal analysis, as usual, by assigning a reference node and labeling the nonreference nodes with node voltage symbols. The result is shown in Figure 5.39. We will leave it to you to apply KCL at the two labeled essential nodes or use the inspection method to show that the matrix form of the nodal equation is:

$$\begin{bmatrix} \frac{1}{4} + \frac{1}{6} & -\frac{1}{6} \\ -\frac{1}{6} & \frac{1}{8} + \frac{1}{6} \end{bmatrix} \begin{bmatrix} v_1 \\ v_2 \end{bmatrix} = \begin{bmatrix} 15 - i_c \\ i_c \end{bmatrix}. \tag{5.3-1}$$

Figure 5.39
The example circuit prepared for nodal analysis

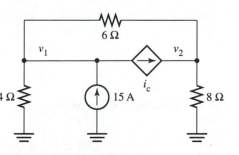

Next, we untape the dependent source and express its value in terms of the unknown node voltages v_1 and v_2:

$$i_c = 2 i_a = 2 \frac{v_1}{4} = \frac{v_1}{2}. \tag{5.3-2}$$

Inserting this in equation (5.3-1) gives

$$\begin{bmatrix} \frac{1}{4} + \frac{1}{6} & -\frac{1}{6} \\ -\frac{1}{6} & \frac{1}{8} + \frac{1}{6} \end{bmatrix} \begin{bmatrix} v_1 \\ v_2 \end{bmatrix} = \begin{bmatrix} 15 - \dfrac{v_1}{2} \\ \dfrac{v_1}{2} \end{bmatrix}. \tag{5.3-3}$$

If we move the terms involving v_1 in the right-hand side matrix back to the left-hand side of the equation, we see that it merely modifies the (1, 1) and (2, 1) entries:

$$\begin{bmatrix} \dfrac{1}{4} + \dfrac{1}{6} + \dfrac{1}{2} & -\dfrac{1}{6} \\[2ex] -\dfrac{1}{2} - \dfrac{1}{6} & \dfrac{1}{8} + \dfrac{1}{6} \end{bmatrix} \begin{bmatrix} v_1 \\ v_2 \end{bmatrix} = \begin{bmatrix} 15 \\ 0 \end{bmatrix}. \qquad (5.3\text{-}4)$$

The solution is $v_1 = 28$ V and $v_2 = 64$ V. Referring to Figure 5.38, $i = -v_2/8 = -8$ A.

The preceding example illustrates an important aspect of circuits with dependent sources: the nodal conductance matrix is no longer symmetric. If you look at equation (5.3-1), you will see that as long as we leave the source taped and therefore treat the dependent source as independent, this matrix *is* symmetric; however, once we untape the source and group the unknown node voltages on the left-hand side of the equation, the symmetry is destroyed. This is generally true: *the nodal conductance matrix for a circuit having dependent sources is not symmetric.*

Example 5.8   Solve the circuit in Example 5.7 for *i* using mesh analysis.

Solution   We will start with the circuit in Figure 5.39, that is with the dependent source taped. We test it by deactivating both current sources (including the dependent one, which we are temporarily treating as independent). This gives the deactivated network shown in Figure 5.40. We see that there is only one mesh in this circuit; thus, there is a single supermesh in the original circuit around which we will write KVL. *Note that the arrow in Figure 5.40 does not represent a mesh current,* merely a direction arbitrarily chosen around the supermesh for the purposes of defining voltage rises and drops in our KVL equation. Figure 5.41 shows the circuit with the *i*-sources reactivated and the mesh currents defined. Notice that we have chosen only one independent mesh current (the one in the lower left mesh) and have expressed the other two in terms of this one and the current source constraints. The KVL equation around the supermesh, now (using the inspection method), in the clockwise direction is

$$[4 + 6 + 8]i_1 = -6(15 - i_c) - 8(15), \qquad (5.3\text{-}5)$$

or

$$18i_1 = -210 + 6i_c. \qquad (5.3\text{-}6)$$

We next untape the dependent source and express its value in terms of the unknown mesh current i_1:

$$i_c = 2i_a = -2i_1. \qquad (5.3\text{-}7)$$

(Refer to Figure 5.38 to refresh yourself about the definition of i_a.) Using this in (5.3-6) gives the final equation

$$30i_1 = -210. \qquad (5.3\text{-}8)$$

Figure 5.40
The example circuit
with its i-sources
deactivated

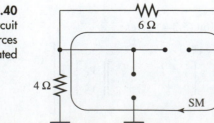

Figure 5.41
The example circuit
prepared for mesh
analysis

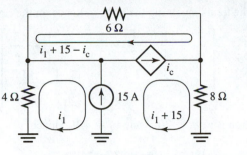

We solve this to obtain $i_1 = -7$ A. Referring once more to Figure 5.38 for the definition of i, we have $i = -(i_1 + 15) = -8$ A.

Example 5.9 Find the value of v in Figure 5.42 using nodal analysis.

Figure 5.42
An example circuit

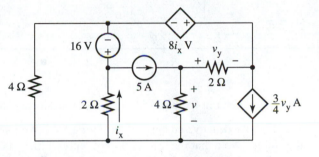

Solution We show the circuit prepared for nodal analysis in Figure 5.43. We have already identified a supernode (shown shaded) and an essential node (also shaded). A key step was for
us to tape both dependent sources. Now, using any method you like for deriving the nodal
equations, you can verify that the matrix form is

$$
\begin{bmatrix} \dfrac{1}{4} + \dfrac{1}{2} + \dfrac{1}{2} & -\dfrac{1}{2} \\[2mm] -\dfrac{1}{2} & \dfrac{1}{4} + \dfrac{1}{2} \end{bmatrix}
\begin{bmatrix} v_1 \\ v_2 \end{bmatrix} =
\begin{bmatrix} -5 - i_c - \dfrac{16}{2} - \dfrac{v_c}{2} \\[2mm] 5 + \dfrac{v_c}{2} \end{bmatrix}.
\tag{5.3-9}
$$

Untaping the CCVS, we have

$$
v_c = 8i_x = -8\,\frac{v_1 + 16}{2} = -4v_1 - 64.
\tag{5.3-10}
$$

For the VCCS, we have

$$
i_c = \frac{3}{4} v_y = \frac{3}{4} [v_2 - (v_1 + v_c)] = \frac{3}{4} v_2 + \frac{9}{4} v_1 + 48.
\tag{5.3-11}
$$

We have used (5.3-10) to obtain (5.3-11). The modified form of (5.3-9) is now

$$
\begin{bmatrix} \dfrac{1}{4} + \dfrac{1}{2} + \dfrac{1}{2} & -\dfrac{1}{2} \\[2mm] -\dfrac{1}{2} & \dfrac{1}{4} + \dfrac{1}{2} \end{bmatrix}
\begin{bmatrix} v_1 \\ v_2 \end{bmatrix} =
\begin{bmatrix} -29 - \dfrac{1}{4}v_1 - \dfrac{3}{4}v_2 \\[2mm] -27 - 2v_1 \end{bmatrix}.
\tag{5.3-12}
$$

Moving the terms involving the node voltages on the right-hand side to the left, we get

$$\begin{bmatrix} \dfrac{1}{4} + \dfrac{1}{4} + \dfrac{1}{2} + \dfrac{1}{2} & -\dfrac{1}{2} + \dfrac{3}{4} \\[3mm] 2 - \dfrac{1}{2} & \dfrac{1}{4} + \dfrac{1}{2} \end{bmatrix} \begin{bmatrix} v_1 \\[2mm] v_2 \end{bmatrix} = \begin{bmatrix} -29 \\[2mm] -27 \end{bmatrix}. \tag{5.3-13}$$

The solution is $v_1 = -20$ V and $v_2 = 4$ V. Hence, $v = v_2 = 4$ V.

Figure 5.43
The example circuit
prepared for nodal
analysis

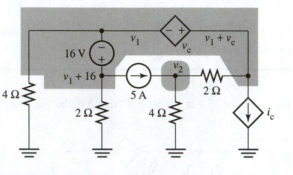

Example 5.10   Solve the circuit in Example 5.9 for v using mesh analysis.

       Solution   We show the deactivated network in Figure 5.44. We have identified the mesh structure in that figure. Note that the arrows do not represent mesh currents; to define these, we must return to the original circuit. We show the reactivated circuit in Figure 5.45 with one independent mesh current for each essential mesh and one for the supermesh. We have used the current source constraint to express the top right-hand mesh current in terms of the other in the supermesh. Using any method you like, you can convince yourself that the matrix mesh equation is

$$\begin{bmatrix} 6 & -2 \\ -2 & 8 \end{bmatrix} \begin{bmatrix} i_1 \\ i_2 \end{bmatrix} = \begin{bmatrix} 16 \\ v_c + 10 + 6i_c - 16 \end{bmatrix}. \tag{5.3-14}$$

Untaping, we find (referring to Figure 5.42) that

$$v_c = 8i_x = 8(i_2 - i_1) \tag{5.3-15}$$

and

$$i_c = \frac{3}{4} \times 2 \times [i_c - (i_2 - 5)]. \tag{5.3-16}$$

The latter equation is implicit in i_c. Solving, we have

Figure 5.44
The deactivated
network for the circuit
of Example 5.9

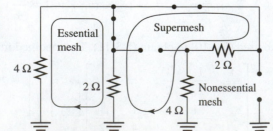

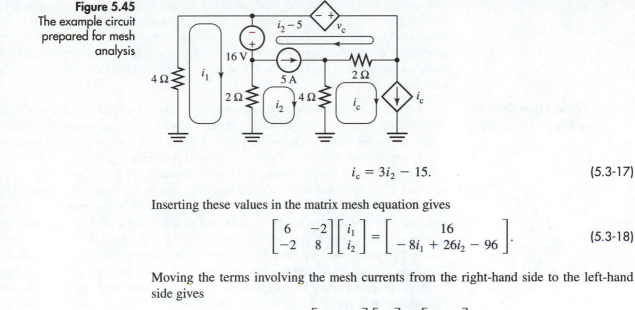

Figure 5.45
The example circuit prepared for mesh analysis

$$i_c = 3i_2 - 15. \tag{5.3-17}$$

Inserting these values in the matrix mesh equation gives

$$\begin{bmatrix} 6 & -2 \\ -2 & 8 \end{bmatrix} \begin{bmatrix} i_1 \\ i_2 \end{bmatrix} = \begin{bmatrix} 16 \\ -8i_1 + 26i_2 - 96 \end{bmatrix}. \tag{5.3-18}$$

Moving the terms involving the mesh currents from the right-hand side to the left-hand side gives

$$\begin{bmatrix} 6 & -2 \\ 6 & -18 \end{bmatrix} \begin{bmatrix} i_1 \\ i_2 \end{bmatrix} = \begin{bmatrix} 16 \\ -96 \end{bmatrix}. \tag{5.3-19}$$

The solution is $i_1 = 5\,\text{A}$ and $i_2 = 7\,\text{A}$. Thus, $i_c = 6\,\text{A}$ (from (5.3-17), and $v = 4 \times (i_2 - i_c)$, or $v = 4 \times (7 - 6) = 4\,\text{V}$.

Section 5.3 Quiz

Q5.3-1. Find v for the circuit shown in Figure Q5.3-1 using nodal analysis.

Q5.3-2. Find v for the circuit shown in Figure Q5.3-1 using mesh analysis.

Figure Q5.3-1

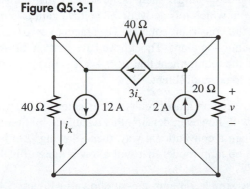

5.4 | The Operational Amplifier: Basic Concepts and Subcircuits

As we point out elsewhere in this chapter, the active circuit elements are actually equivalent representations of the ideal behavior of more complex things: an independent voltage source is often a representation of a "power supply" or a "signal generator." An independent current source is often designed and constructed with other elements; so too, are the dependent sources. As it happens, there is a complex electronic unit called the *operational amplifier* that can be used to construct each and every one of these two-terminal active elements. It is cheap and small and, for many situations, it actually possesses real characteristics that are close to being ideal. For these reasons, we will treat it as a circuit element in its own right. Furthermore, we will show how to construct many of the other elements we have discussed with operational amplifiers. It is common practice to shorten the term "operational amplifier" to "op amp" in informal situations, both written and spoken. We will follow this practice here.

The Operational Amplifier: Definition

We will define the *op amp* to be a *grounded* VCVS with a *voltage gain (μ) that is infinite*. We show the circuit symbol for the op amp in Figure 5.46(a) and an equivalent subcircuit, based on the definition just given, in Figure 5.46(b). The three terminal voltages v_+, v_-, and v_o are all *node voltages* relative to a ground reference that is *prespecified*. Thus, the ground reference definition is inherent in any op amp circuit as a given. The two terminals on the left are considered to be the input; the one on the right is the output. The basic idea is that a given signal is imposed on the input, and the output responds. If μ were finite, the output voltage would be proportional to the difference in the two input terminal voltages. Thus, the op amp is one manifestation of a *differential (or difference) amplifier*. The term "operational," by the way, arose because such amplifiers were used historically in analog computers to perform the operations of scalar multiplication, sign inversion, summation, integration, and differentiation in the solution of differential equations. Nowadays, they are considered to be circuit elements and are used in much different ways.

Figure 5.46
The (ideal) operational amplifier

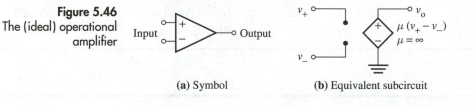

(a) Symbol   (b) Equivalent subcircuit

As the op amp is merely a VCVS, it would perhaps seem that there is no reason to devote an entire section to a discussion of op amps. There is, however, a major difference between the op amp and the VCVS from which it is derived: the voltage gain μ is infinite. Thus, we cannot merely write down the constraint equation (in practice) as $\mu(v_+ - v_-)$ when we are analyzing a circuit containing op amps. Though we have written this relation in Figure 5.46(b), we have also noted the requirement that $\mu = \infty$; it is this property that causes the analysis of op amp circuits to be so different.

The Op Amp Virtual Short Model

We will develop an analysis procedure for op amp circuits that is somewhat at variance with what we have done before. We will, however, develop this technique by first assuming that μ is finite, doing the analysis in a conventional way, then allowing μ to become infinite. From this, we will derive an op amp model that will allow a more effective approach to be taken.

We will start by investigating the simple op amp circuit shown in Figure 5.47. We will assume that our task is to compute the output voltage v_o. To do so, we will simply imagine ourselves to "unplug" the op amp symbol and replace it with its VCVS equivalent—but one having finite voltage gain. We show the resulting equivalent circuit in Figure 5.48. Notice that the op amp has its inverting input on top in Figures 5.48 and 5.49. There is a reason for this, which we will discuss in Section 5.5; in this section we will simply assume that a "feedback" resistor (or resistors) from the output to the input terminals is always connected to the input terminal labeled v_-. This "negative feedback" is necessary for practical reasons in order for the op amp to operate linearly.

Figure 5.47
An example circuit

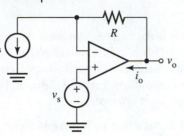

Figure 5.48
An equivalent circuit

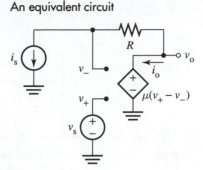

Because no current flows into the op amp input terminals (they are equivalent to an open circuit), we can apply KCL and KVL to see[13] at once that

$$v_o = v_- + Ri_s. \tag{5.4-1}$$

Now we introduce the VCVS constraint

$$v_o = \mu(v_s - v_-). \tag{5.4-2}$$

Solving these two equations simultaneously, we have

$$v_- = \frac{\mu v_s - Ri_s}{1 + \mu}. \tag{5.4-3}$$

Using this result in equation (5.4-2), we also have

$$v_o = \mu(v_+ - v_-) = \frac{\mu(v_s + Ri_s)}{1 + \mu}. \tag{5.4-4}$$

We now let μ become infinitely large. Doing so, we see that when $\mu = \infty$ we have

$$v_- = v_+ = v_s \tag{5.4-5}$$

and $$v_o = v_s + Ri_s. \tag{5.4-6}$$

This last equation is the solution we were after; however, it turns out that the preceding one ($v_+ = v_-$) is really the crucial one. The fact that the VCVS voltage gain is infinite implies that the voltage across the *input* terminals of the op amp is zero! In other words, the voltages (relative to ground reference) at the input terminals are identical. We also observe that the VCVS—*in itself*—says nothing about the output voltage v_o or the current i_o into the op amp output terminal. We note that the latter is actually $-i_s$ (assuming that there are no other elements between the output terminal and ground than the ones we have shown). It is the external circuit that determines v_o and i_o, not the op amp output terminal itself.

Figure 5.49   Op amp variables

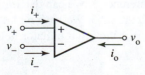

Let's look now at the op amp symbol in Figure 5.49 and pursue this thinking a bit further. We see that the input terminals are constrained in two ways. Because they actually form an open circuit, we have

$$i_+ = i_- = 0. \tag{5.4-7}$$

---

[13] We have not taped the VCVS because the circuit is so simple to analyze.

Because of the infinite value of voltage gain, we also have

$$v_+ = v_-. \tag{5.4-8}$$

This last equation is the same as that for a short circuit. (The voltage between the input terminals is zero.) This is quite astonishing! It means that these two terminals act simultaneously like both a short circuit and an open circuit. For this reason, it is often said that the two input terminals form a *virtual short,* and equations (5.4-7) and (5.4-8) are referred to as the *virtual short principle.*

The Nullor Model for the Op Amp

It is somewhat hard to visualize such a constraint mentally, so we will use the symbol shown in Figure 5.50(a) to represent a virtual short. This two-terminal element is called a *nullator* and can be thought of as a basic circuit element. The output terminal behavior of the op amp is equally strange: we simply must confess ignorance about the value of either the voltage or the current. For this reason we replace the infinite-gain VCVS at the output with the symbol shown in Figure 5.50(b), the *norator.* It, too, can be thought of as a circuit element. It offers a way of visualizing a two-terminal element that does not (by itself) have any effect whatsoever on the voltage across it or the current through it—though it does provide a path for current to flow. By drawing the nullator and the norator symbols on a circuit diagram in place of the op amp, we remind ourselves of the constraints imposed by the op amp.

Because the nullator and the norator always occur in pairs in an op amp circuit, we give the combination of these two elements (shown in Figure 5.51(a)) a name: the *nullor.* The infinite-gain VCVS at the output of an op amp is grounded, so the nullor model for the op amp has one terminal of the norator grounded as shown in Figure 5.51(b).

Figure 5.50   The nullator and the norator

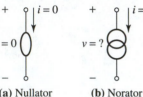

(a) Nullator   (b) Norator

Figure 5.51
The nullor: An op amp model

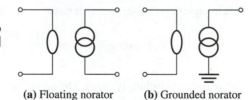

(a) Floating norator   (b) Grounded norator

Equivalents for Op Amp Subcircuits

Example 5.11   Find an equivalent[14] for the op amp subcircuit shown in Figure 5.52.

Figure 5.52
The inverting op amp subcircuit

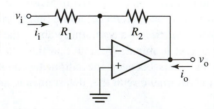

Solution   This is a three-terminal subcircuit (don't forget the ground reference!). We can determine its *v-i* characteristics by applying test sources as shown in Figure 5.53. We have replaced the op amp by its nullor equivalent and have noted the fact that the input nullator holds

---

[14] Note that we are dealing here with a three-terminal network as we were in Chapter 3, Section 3.5.

the node voltage at the top to zero. Furthermore, as the current into the nullator is zero, the current through R_2 is the same as that through R_1. We see at once that

$$v_i = R_1 i_i \tag{5.4-9}$$

and

$$v_o = -R_2 i_i = -\frac{R_2}{R_1} v_i. \tag{5.4-10}$$

The first equation is simply that of a resistance of value R_1 between the input terminal and ground, and we recognize the second to be that of a VCVS from the output terminal to ground (v_o is independent of the current i_o and proportional to v_i.). Thus, we have the resulting equivalent subcircuit shown in Figure 5.54. Note, by the way, that the norator current in Figure 5.53 is just the value necessary to make KCL at the output node "work."

Figure 5.53
Testing for the *v-i* characteristic

Figure 5.54
The equivalent subcircuit

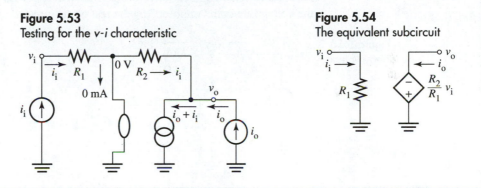

The practical importance of equivalent subcircuits such as the one derived in the last example is simply that they allow one to design and analyze more circuits quite rapidly by recognizing their topologies whenever they occur in a circuit. We illustrate this point with the next example.

Example 5.12   Compute the voltage v_o, and currents i_a and i_o in the circuit shown in Figure 5.55.

Figure 5.55
An example circuit

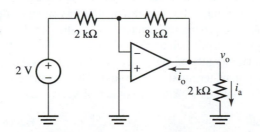

Solution   First, have a look at the resistor values. It is a fact that op amps limit the current flowing in their output terminals, usually to a value of a few tens of milliamperes. For this reason, a practical range of values for resistors in an op amp circuit is on the order of a few hundred ohms to a few megohms. Typical values are in the kilohm range, as we have shown. We can quickly analyze this circuit for v_o by simply noting that this quantity is the negative of the ratio of the "feedback" resistor (8 kΩ) to the "input" resistor (2 kΩ) multiplied by the value of voltage at the input terminal. Thus, it is −8 V. We can then use Ohm's law to find that $i_a = -8$ V/2 kΩ = −4 mA. We can find i_o by observing that the top middle node is held to zero volt by the op amp input terminals and that the current into the minus (or *inverting*) op amp terminal is zero. This gives a current of 1 mA from left to right through the input 2-Ω resistor and the same 1 mA from left to right through the feedback 8-kΩ resistor. Then KCL at the output node gives $i_o = 1$ mA $- (- 4$ mA$) = 5$ mA into the output of the op amp.

In case this was a bit fast for you, let's go back to fundamentals and use the nullor model for the op amp, as we have shown in Figure 5.56. We have done most of the analysis right on the circuit diagram. Just remember that the current into the nullator is zero and the voltage across it is also zero. Hence, the voltage across the input resistor is the entire 2 V supplied by the input source, and the current in that resistor is 1 mA in the direction shown. Because no current flows into the nullator, all of this 1 mA goes through the feedback resistor from left to right. Using KVL, we see that the output voltage is 0 V $- 1$ mA $\times 8$ kΩ $= - 8$ V and that the current through the "load" resistor (2 kΩ) is $- 4$ mA downward. We can then apply KCL at the output node to get $i_o = 4$ mA $+ 1$ mA $= 5$ mA. The norator at the output can either "source" or "sink" current of any value and bear any voltage whatsoever across its terminals—the current and the voltage are both "imposed" by the rest of the circuit.

Figure 5.56
The nullor equivalent

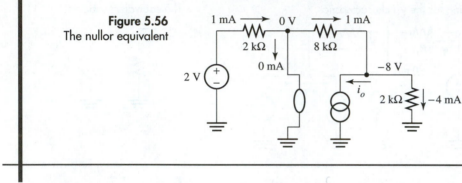

The Basic Inverting Amplifier Topology

Figure 5.57   An example circuit

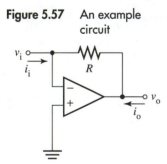

The subcircuit shown in Figure 5.57 appears in quite a wide range of op amp circuits encountered in practice; in fact, it is a component of the inverting op amp circuit that we just discussed. Let's see what its characteristics are by itself. To do this, we replace the op amp symbol by its nullor equivalent and attach two test current sources (a voltage source would not work at the input because of the nullator constraint of zero voltage). We show the resulting circuit in Figure 5.58. We see at once that $v_i = 0$ V, regardless of the value of the input current i_i. For this reason, there is an equivalent short circuit between the input terminal and ground reference. Because the current into the nullator is zero, the entire current i_i flows from left to right through the feedback resistor. Hence, we have $v_o = -Ri_i$—independently of the current i_o. This means that there is an equivalent CCVS between the output terminal and ground. The complete equivalent subcircuit is shown in Figure 5.59. Notice that R has become a *transresistance*.

Figure 5.58
Testing for the *v-i* characteristic

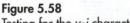
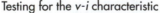

Figure 5.59
CCVS equivalent

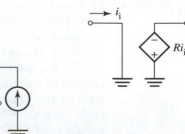

You will observe that all of the equivalent subcircuits we have derived thus far have one terminal grounded. There is an outstanding problem related to this in the design of active circuits: design an equivalent element (dependent source, etc.) using an op amp such that the equivalent element is "floating," that is, neither terminal is grounded. This is explored to an extent in the problems at the end of the chapter; however, for the most part we defer a discussion of this to more applied follow-on courses.

Example 5.13   Find the voltage v_o for the *summing amplifier* circuit in Figure 5.60.

Figure 5.60
The summing amplifier

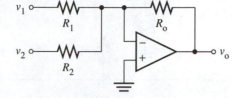

Solution   This circuit can easily be analyzed on the circuit diagram itself, as we show in Figure 5.61. From the computations shown there we deduce that the output voltage is given by

$$v_o = \left[\frac{v_1}{R_1} + \frac{v_2}{R_2} \right] R_o. \tag{5.4-11}$$

Thus, our circuit output is the weighted sum of the two inputs. Clearly, we could add additional resistors and sum additional inputs. The node to which the inverting input of the op amp itself is attached is called a *virtual ground* because it is held to zero volt. It is also called a *summing junction* because it is there that the two currents are summed together.

Figure 5.61
Analysis on the
diagram

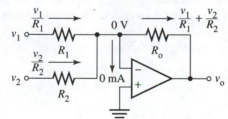

You should practice doing such an analysis as we have just done directly on the circuit diagram; however, in order to help you develop this capability we will rework the same example in two other ways. For the first, we merely recall the CCVS equivalent for the op amp and its feedback resistor shown in Figure 5.59. Thus, we can replace the op amp and feedback resistor in our example with the CCVS shown in Figure 5.62. We can immediately see from the circuit diagram that the currents in the two input resistors add together to produce i_i. We then obtain (5.4-11) quickly.

Figure 5.62
CCVS equivalent

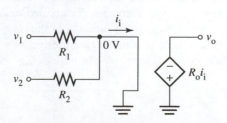

If all else fails, we can fall back to the original grounded nullor equivalent for the op amp. We show our summing amplifier with the nullor model substituted for the op amp in Figure 5.63. The inverting op amp input terminal is held to zero volts and the current into it to zero. We still see that the two currents through the input resistors sum and become the current in the feedback resistor R_o. Furthermore, the entire input voltage (v_1 or v_2, as appropriate) appears across the corresponding input resistor because of the voltage constraint of the nullator.

Figure 5.63
Using the nullor op
amp model

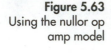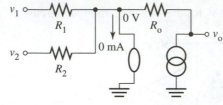

**An Important
Caveat for Circuits
with Op Amps**

Though we have pointed this out earlier, it is so important that we repeat it: you should never try to write an equation at the output terminal of the op amp in an attempt to represent the action of the op amp. The output voltage and current of the op amp are determined by the elements external to the op amp—not by the op amp itself. In fact, exactly this—nothing more and nothing less—is the meaning of the norator symbol.

**The Basic Noninverting
Amplifier Topology**

Thus far, each of the op amp circuits or subcircuits we have investigated centered more or less upon the equivalent CCVS shown in Figure 5.59; that is, the basic topology consisted of an op amp with its positive input terminal grounded and a feedback resistor connected between the output and the negative input terminal. The input signal, in some manner, was fed into the inverting input of the op amp. This is generally called the *inverting topology*. We will now investigate another topology: the *noninverting topology*. In this form of circuit, there is still a feedback resistor between the output and the inverting input, but the *input* signal is fed in some manner into the positive input terminal.

Example 5.14

Find an equivalent subcircuit for the subcircuit (called the *noninverting amplifier*) shown in Figure 5.64.

Figure 5.64
The noninverting
amplifier

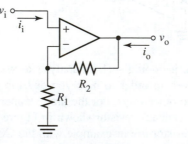

Solution

We attach a test source at the input and one at the output and use the nullor model for the op amp as shown in Figure 5.65. We have used a voltage source at the input for a reason that should be clear: the input current is constrained to be zero by the nullator. Similarly, the nullator forces the voltage at the junction of the two feedback resistors to be the same as the input voltage. These two resistors are effectively connected in series because no current leaves their junction. Thus, we can write

$$v_i = \frac{R_1}{R_1 + R_2} v_o. \tag{5.4-12}$$

Inverting, we obtain

$$v_o = \left[1 + \frac{R_2}{R_1}\right] v_i. \tag{5.4-13}$$

Figure 5.65
The nullor equivalent

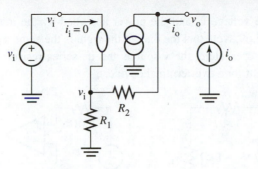

Figure 5.65
The nullor equivalent

Because the input current is zero independently of the input voltage, there is an equivalent open circuit between the input terminal and ground. The output voltage obeys (5.4-13) independently of the output current i_o, so we see that there is an equivalent VCVS connected between the output terminal and ground. Thus, the equivalent circuit is that shown in Figure 5.66. We see that it is noninverting (the plus sign is at the top and the voltage gain is positive). Unlike the inverting voltage amplifier, it presents an open circuit at the input; thus, it is more of an ideal dependent source. However, we see that the voltage gain cannot be less than unity for the present configuration,[15] whereas it *can* be less than unity for the inverting topology.

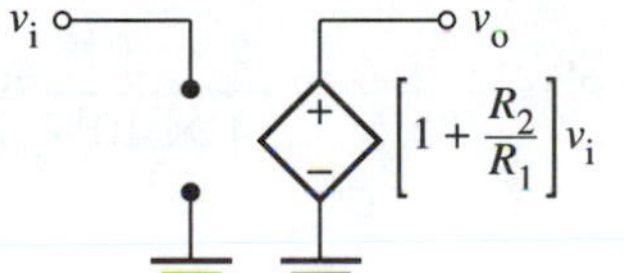

Figure 5.66
VCVS equivalent

$$v_i \circ \qquad \circ v_o$$

$$\left[1 + \frac{R_2}{R_1}\right] v_i$$

The Voltage Follower (or Unity Gain Buffer)

Let's look just a bit more closely at the noninverting amplifier subcircuit of the last example. Suppose we let R_1 become infinitely large. We see that the voltage gain approaches unity, independently of R_2. Thus, we simply let R_2 equal zero. This gives the subcircuit of Figure 5.67, which is called a *buffer* or *voltage follower*. It undoubtedly will look somewhat strange until you recall that there is a ground reference in the circuit that is not shown. The voltages at the input and output are defined relative to this ground reference. We note that because the buffer is derived from the noninverting amplifier circuit, the VCVS equivalent of Figure 5.66 is valid for it too—with a voltage gain of unity. We show the nullor equivalent for the buffer in Figure 5.68 for future reference. This equivalent makes the ground reference explicit.

Figure 5.67
Voltage follower or buffer

Figure 5.68
Nullor equivalent

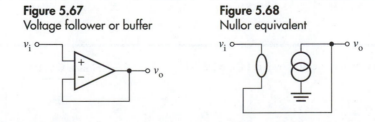

You might very well ask what the usefulness of a buffer is, as it does not provide any voltage gain. The following example answers this question.

---

[15] Assuming that R_1 and R_2 both have positive values.

Example 5.15   Find the voltage v_L and the power absorbed by the load resistor R_L for both circuits in Figure 5.69. Note that the one on the right is the same as the one on the left except for the added buffer. Here, the v-source and its series resistor are assumed to be the Thévenin equivalent for a two-terminal device.

Figure 5.69
Illustrating the use of a
buffer

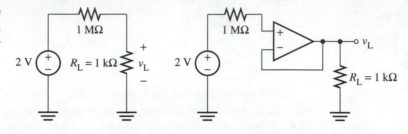

Solution   For the circuit on the left, we use the voltage divider rule to obtain the voltage v_L:

$$v_L = \frac{1 \text{ k}\Omega}{1 \text{ M}\Omega + 1 \text{ k}\Omega} \times 2 \text{ V} = \frac{1}{1001} \times 2 \text{ V} \cong 2 \text{ mV}. \qquad (5.4\text{-}14)$$

Not a very large voltage in comparison with that developed by the source! The power absorbed by the load resistor is

$$P_L = \frac{v_L^2}{1 \text{ k}\Omega} \cong \frac{4 \times 10^{-6}}{1 \times 10^3} = 4 \text{ nW}, \qquad (5.4\text{-}15)$$

a very small power indeed.

Now let's look at the circuit on the right side of the figure. It is a simple modification of the one on the left side in which we have just inserted a unity gain buffer between the Thévenin equivalent of the source and the load resistor R_L. The buffer presents an open circuit to the Thévenin equivalent of the source, so we see that the current through the 1-MΩ resistor is zero; thus, there is no voltage drop across it. Therefore, the voltage at the positive input terminal of the op amp is the v-source value, 2 V, and this same voltage is transferred to the load resistor. Thus,

$$v_L = 2 \text{ V} \qquad (5.4\text{-}16)$$

and

$$P_L = \frac{v_L^2}{1 \text{ k}\Omega} = \frac{4}{1 \times 10^3} = 4 \text{ mW}. \qquad (5.4\text{-}17)$$

This is an increase in delivered power of a factor of one million!

Section 5.4 Quiz

Q5.4-1.   Find the voltage v_o in the circuit shown in Figure Q5.4-1.

Figure Q5.4-1

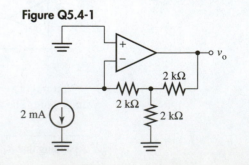

Q5.4-2. Find the voltage v_o in the circuit shown in Figure Q5.4-2.

Figure Q5.4-2

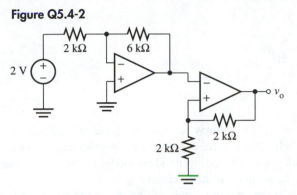

Q5.4-3. Find the equivalent "seen" between terminal a and ground in Figure Q5.4-3.

Figure Q5.4-3

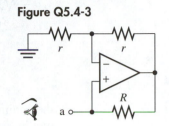

Q5.4-4. Find the voltage v_o in the circuit shown in Figure Q5.4-4.

Figure Q5.4-4

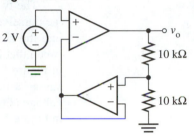

5.5 | Practical Aspects of Op Amp Circuits: Feedback and Stability

The fundamental theory of the operational amplifier was developed in Section 5.4. There, operation was assumed to be ideal: the input terminals were modeled with the nullator, a device that constrains the voltage across it and the current through it to be zero simultaneously. Another descriptive term often used for this element is the virtual short. The output of the op amp was modeled by the norator, a device that supports any current and any voltage imposed by the rest of the circuit. If you turn back through the pages of Section 5.4, you will see that, for each of the circuits studied, there was a "feedback resistor" connected between the output of the op amp and the negative input terminal. This is no coincidence, and we would now like to explore the consequences of *not* doing this.

Feedback and Loop Gain

We will start by investigating the two subcircuits shown in Figure 5.70. The only difference between the two is the feedback resistor R_2: it is returned to the negative op amp input in one and to the positive input in the other. Now suppose we assume that the op amp is ideal in both circuits. In that case, *both* are equivalent to the subcircuit shown in Figure 5.71, with the op amp replaced by its nullor equivalent. Thus, the response voltage at their outputs are the same: $v_o = (-R_2/R_1)v_i$. But surely this cannot be correct, can it? After all, the signal is fed into the inverting input of the op amp in one circuit and into

Figure 5.70
Two similar topologies

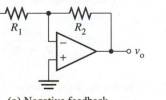

(a) Negative feedback

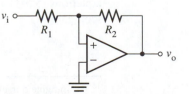

(b) Positive feedback

Figure 5.71
Nullor equivalent

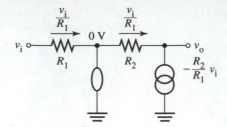

the noninverting input in the other. In fact, it turns out that the "positive feedback" topology in Figure 5.70(b) does not behave in a nice, expected way in practice—and our assumption that the op amp is well approximated by its ideal model is not a valid one at all. To see why, we must investigate things a bit more closely.

The two circuits in Figure 5.70 are, in reality, quite different. To see this, let's imagine ourselves to reduce the input voltage of each subcircuit to zero,[16] cut the lead feeding into the nongrounded input terminal of the op amp, and attach a test voltage source between that input and ground. This procedure is called "opening the loop," and is depicted in Figure 5.72(a) and (b). There, we represent the op amp (for the moment) by a VCVS having a finite voltage gain μ. Calling our inserted test voltage v_t (t for "test") and the voltage at the junction of the two resistors v_f (f for "feedback"), we see that for the circuit in Figure 5.72(a) we have

$$v_f = -\frac{\mu R_1}{R_1 + R_2} v_t = -\mu F v_t, \tag{5.5-1}$$

where F is the voltage divider ratio, called the *feedback factor*. For the circuit in Figure 5.72(b) we have

$$v_f = +\frac{\mu R_1}{R_1 + R_2} v_t = +\mu F v_t. \tag{5.5-2}$$

Figure 5.72
Measuring the loop
gain

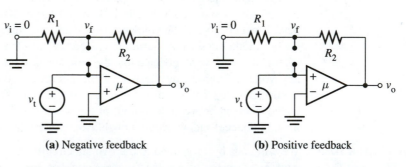

(a) Negative feedback   (b) Positive feedback

We define the *loop gain* by the equation

$$LG = \frac{v_f}{v_t}, \tag{5.5-3}$$

and say that the circuit has *positive feedback* if this ratio is positive and *negative feedback* if it is negative. (Of course we also say that "there is no feedback" if it is zero.) From equations (5.5-1) and (5.5-2), we see that the subcircuit in Figures 5.70(a) and 5.72(a) has loop gain

$$LG = -\frac{\mu R_1}{R_1 + R_2} = -\mu F. \tag{5.5-4}$$

---

[16] We imagine ourselves to be driving the subcircuit with a voltage source at the input.

But F is a resistor ratio, and hence is positive; furthermore, the voltage gain μ is positive. The subcircuit therefore has negative feedback. Conversely, the subcircuit in Figures 5.70(b) and 5.72(b) has a loop gain of

$$LG = +\frac{\mu R_1}{R_1 + R_2} = +\mu F, \tag{5.5-5}$$

thus exhibiting positive feedback.

Clearly, it would be highly desirable for the response voltage at the op amp output to be zero when the input voltage is zero. In fact, this is one of the prime characteristics of a linear system. If f is a linear function, then $f(0) = f(0 + 0) = f(0) + f(0) = 2f(0)$, and (if we subtract $f(0)$ from both sides of the equation) we see that we must have

$$f(0) = 0. \tag{5.5-6}$$

We will explore this a bit more and show that the response of the positive feedback topology is not zero in practice (because of "turn on" initial conditions) even if the input is zero. However, we must deal with the finite response time of the op amp to changes at its input.

A "Dynamic" Model for the Op Amp To do this, we must improve our op amp model just a bit. In Section 5.4 we defined an operational amplifier to be a grounded VCVS with infinite voltage gain. We will relax this condition temporarily and allow μ to be finite. We will also incorporate another nonideal factor, which is best explained by the model shown in Figure 5.73. There we have modified the basic scalar VCVS relationship by subtracting a "correction factor" that is proportional to the time rate of change of the voltage across the VCVS itself. If all voltages were constant (dc), we see that this correction factor would be zero; however, it adds a term to the VCVS v-i characteristic in case the voltages vary with time.

Figure 5.73
Dynamic op amp model

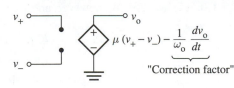

The parameter ω_0 clearly has the dimension of inverse time. It is an important measure of quality of an op amp called the "open loop bandwidth." For large values of ω_0, the op amp behaves in a more ideal manner (the correction factor is small); for smaller values the deviation from ideal is greater (the correction factor is larger). The most widely used op amp, called the 741 series, has a numerical value for ω_0 of approximately $10\pi\,\text{s}^{-1}$ and a value for μ of about 10^5. For higher-performance op amps, the value of ω_0 can be much higher.

We will use the op amp model in Figure 5.73 to determine the stability characteristics of an op amp circuit. We will analyze the topology in Figure 5.74, for by specializing its resistor values we can make it represent either of the topologies in Figures 5.70 and 5.72. For instance if we make $R_1 = 0$ (a short circuit) and $R_2 = \infty$ (an open circuit), we have the topology in Figure 5.70(a); if we make $R_3 = 0$ and $R_4 = \infty$, we have the topology in Figure 5.70(b). It would be worth your while to spend just a moment referring to Section 5.4 to convince yourself that it also represents both the inverting amplifier of Figure 5.52 (of Example 5.11) and the noninverting amplifier of Figure 5.64 (Example 5.14), provided that the input voltages are zero. Thus, our circuit is a fairly general one.

Figure 5.74 A general topology

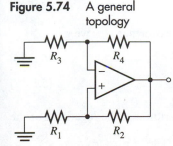

Figure 5.75 The dynamic
model

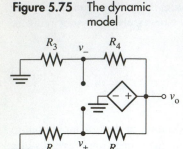

In Figure 5.75 we show our circuit with the dynamic op amp model of Figure 5.73 replacing the op amp symbol. For reasons of space, we have not shown the dependent source relationship in the figure; however, it is the one given in Figure 5.73. We define

$$F_+ = \frac{R_1}{R_1 + R_2} \tag{5.5-7}$$

and

$$F_- = \frac{R_3}{R_3 + R_4}. \tag{5.5-8}$$

Using these definitions, we see that

$$v_+ = F_+ v_o \tag{5.5-9}$$

and

$$v_- = F_- v_o. \tag{5.5-10}$$

Our dynamic op amp model specifies that the response voltage is

$$v_o = \mu(v_+ - v_-) - \frac{1}{\omega_o} \frac{dv_o}{dt}. \tag{5.5-11}$$

Using equations (5.5-9) and (5.5-10) in (5.5-11), we have

$$\frac{dv_o}{dt} + [1 - \mu(F_+ - F_-)]\omega_o v_o = 0. \tag{5.5-12}$$

Now, although all of Part I of this text is concerned primarily with the dc behavior of circuits, here we have encountered for the first time the need to analyze a circuit that has a more complex behavior: one that changes with time. We must solve the *differential equation* in (5.5-12); it is not difficult, however, and we can write it more compactly as

$$\frac{dv_o}{dt} + a v_o = 0, \tag{5.5-13a}$$

where

$$a = [1 - \mu(F_+ - F_-)]\omega_o. \tag{5.5-13b}$$

If we rearrange it slightly, we can write the equivalent equation

$$\frac{v_o'}{v_o} = -a, \tag{5.5-14}$$

where we have used the prime notation for the derivative. But the term on the left-hand side is just the derivative of the logarithm, so we can integrate both sides to obtain

$$\ln[v_o(t)] = -at + k, \tag{5.5-15}$$

where k is a constant of integration. If we evaluate this expression at $t = 0$ we get

$$\ln[v_o(0)] = k. \tag{5.5-16}$$

Thus,

$$\ln[v_o(t)] - \ln[v_o(0)] = -at, \tag{5.5-17}$$

which can be rearranged (noting that the difference of logarithms is the logarithm of the ratio) into

$$\ln\left[\frac{v_o(t)}{v_o(0)}\right] = -at. \tag{5.5-18}$$

Next, we recall that $x = y$ implies that $e^x = e^y$. Using this relationship with

$$x = \ln\left[\frac{v_o(t)}{v_o(0)}\right] \tag{5.5-19}$$

and

$$y = -at, \tag{5.5-20}$$

we get

$$e^{\ln[v_o(t)/v_o(0)]} = \left[\frac{v_o(t)}{v_o(0)}\right] = e^{-at}. \tag{5.5-21}$$

Finally, multiplying by $v_o(0)$, we obtain the time response we are looking for:

$$v_o(t) = v_o(0)\,e^{-at}. \tag{5.5-22}$$

The parameter $v_o(0)$ is the *initial condition* for the output voltage.[17]

We show positive time plots of the response voltage in Figure 5.76 for $v_o(0) > 0$ and for $a > 0$, $a < 0$, and $a = 0$. It is clear that if $a > 0$ the response voltage settles down to zero for large values of time, if $a < 0$ it becomes infinitely large, and if $a = 0$ it remains constant. In the first case ($a > 0$) we say the circuit is *stable;* in the second ($a < 0$) that it is *unstable;* and in the third ($a = 0$) that it is *marginally stable* or *conditionally stable.*

We are, however, interested in what happens as $\mu \to \infty$. In this case, we must look at the expression for a in (5.5-13b). We see that if $F_+ > F_-$ a will be negative for sufficiently large values of μ and the circuit will be unstable. If $F_+ < F_-$, a will be positive for sufficiently large values of μ and the circuit will be stable. We notice that if $F_+ = F_-$ exactly, the circuit will be stable for all values of μ; however, we also observe that, in practice, resistor ratios are never precisely equal. Thus, practically speaking, the "exactly equal" case never occurs. Therefore, we can phrase a general test for our ideal op amp models to hold with good precision: *if there is more negative feedback than positive feedback ($F_- > F_+$), an op amp circuit will be stable and we will be entirely justified in using the ideal op amp model.*

Figure 5.76   Natural response

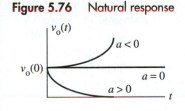

Saturation Effects

Now let's tie up one loose end that might be worrying the more practically oriented reader. Is it true that an unstable op amp circuit will have a response that "actually" approaches infinity? The answer, of course, is no—for op amps must be powered by dc power supplies in order to function properly, and they will limit the response to finite positive or negative values.[18] In fact, as we noted for dependent sources in another section in this chapter, even if voltages and currents are constant (or vary very slowly), their defining relations "saturate" for values of controlling variables outside a given range of values. Thus, neglecting for the moment our "correction factor," the VCVS has the more realistic characteristic shown in Figure 5.77. In the unstable case we see that the waveform at the output of the op amp would in practice look something like the one sketched

---

[17] As we have discussed previously, all active elements, including op amps, require dc power supplies to function properly. When the power supply is turned on (at $t = 0$), the "turn on" transient can cause $v_o(0)$ to be nonzero.

[18] If the initial voltage $v_o(0)$ is negative, the response will theoretically go to $-\infty$ if the circuit is unstable.

in Figure 5.78. Our mathematical analysis has, however, predicted the exact *circumstances* for which circuit behavior is undesirable.

Figure 5.77
"Saturation" characteristic

Figure 5.78
"Practical" unstable response

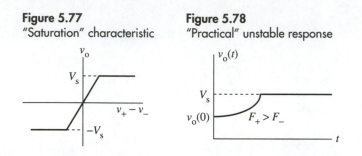

Stability Issues in the Testing of Active Subcircuits

There are a number of consequences of instability, many of which we must defer to another chapter. However, here is one that we think is important enough for discussion here. It has to do with the testing of an unknown subcircuit for its *v-i* characteristic. We have pointed out in the introductory sections of this chapter that one cannot use a v-source to test a v-source subcircuit (or a short circuit, a special case) or an i-source to test an i-source subcircuit (or an open circuit, again a special case), as indicated in Figure 5.79. Of course any active subcircuit that produces the *v-i* characteristic of a pure voltage or current source presents the same situation. (We demonstrated this several times in Section 5.2.) Now we must assert another condition that must be checked for op amp circuits: that of stability. Instability means that one obtains a response that is not defined at all by the test source.

Figure 5.79
Two disallowed test configurations

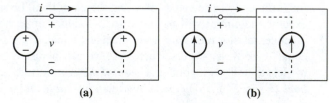

(a)   (b)

Figure 5.80 shows an example of such a subcircuit. In Figure 5.80(a) we show it being tested by a v-source, in Figure 5.80(b) by an i-source. (This particular subcircuit appears as a problem in the end-of-section quiz for Section 5.4. You did do that one, didn't you?) As it happens, these two circuits have different stability properties. You can easily

Figure 5.80
Testing a negative resistance subcircuit

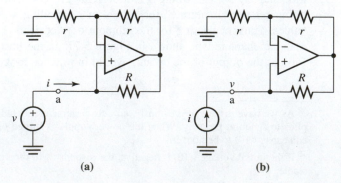

(a)   (b)

see this by glancing at the deactivated equivalents shown in Figure 5.81. In Figure 5.81(a) we show the one of Figure 5.80(a) with its test voltage source reduced to zero. We easily see that $F_+ = 0 < F_- = 0.5$; hence, the circuit is stable. Figure 5.81(b) shows the circuit of Figure 5.80(b) with its test current source reduced to zero. For it, $F_+ = 1$ (there is no voltage drop across R) and $F_- = 0.5$. Thus, *this circuit is inherently unstable and no useful test information would result.* Circuits of this type are said to be *short circuit stable* (or *open circuit unstable*). Those of the opposite type—*open circuit stable* and *short circuit unstable*—also exist.

Figure 5.81
Deactivated networks

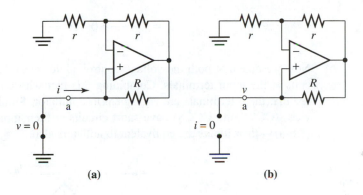

(a)   (b)

Stability Analysis for Multiple Op Amp Circuits

Although we have only analyzed a fairly simple op amp circuit for its stability properties, the technique is quite general and can be used for more complex active circuits and subcircuits. For circuits with more than one op amp or other active element, the analysis, however, becomes quite involved. It requires the solution of higher-order differential equations—which is rightly a topic for Part II of this text on time domain analysis techniques.

Section 5.5 Quiz

Q5.5-1. Determine whether or not the op amp buffer, or voltage follower, circuit with v-source input is stable by directly formulating and solving the differential equation for $v_o(t)$.

Q5.5-2. Replace the test v-source in Figure 5.80(a) with a Thévenin equivalent having finite equivalent resistance R_{eq} and determine the range of values of that equivalent resistance for which the resulting circuit is stable.

Q5.5-3. Consider the circuit shown in Figure 5.80(a), with the ideal v-source at the input replaced by a Thévenin equivalent having an equivalent resistance of R_{eq}. Assume that the resistor R has a value of $6 \text{ k}\Omega$. Use the 741 parameters $\mu = 10^5$ and $\omega_o = 20\pi$ rad/s to derive the time response $v_o(t)$ for $R_{eq} = 3 \text{ k}\Omega$. How long is required for $v_o(t)$ to drop to 10% of its initial value? Repeat this computation for $R_{eq} = 5 \text{ k}\Omega$.

5.6 | Nodal and Mesh Analysis of Circuits with Op Amps

Nullor Models for Infinite-Gain Dependent Sources

In Section 5.4 we derived the nullor model of the op amp from a grounded VCVS. Here, we would like to demonstrate that any of the four types of dependent source becomes a nullor in the limit as its parameter becomes infinite. For ease of reference, we show the dependent sources in Figure 5.82. We can argue in a quite intuitive way, as follows. We assume that the voltage across the dependent voltage source and the current through the dependent current source (that is, the dependent variable) always remain finite. Then we write

$$v_x = \frac{v_o}{\mu} \longrightarrow 0 \quad (\mu \longrightarrow \infty, \text{VCVS}), \tag{5.6-1}$$

$$v_x = \frac{i_o}{g_m} \longrightarrow 0 \quad (g_m \longrightarrow \infty, \text{VCCS}), \tag{5.6-2}$$

$$i_x = \frac{v_o}{r_m} \longrightarrow 0 \quad (r_m \longrightarrow \infty, \text{CCVS}), \tag{5.6-3}$$

$$i_x = \frac{i_o}{\beta} \longrightarrow 0 \quad (\beta \longrightarrow \infty, \text{CCCS}). \tag{5.6-4}$$

Now we note that both the voltage-controlled devices (VCVS and VCCS) have open circuits at the input terminals. Combining this fact with equations (5.6-1) and (5.6-2), we see that these terminals are equivalent to a nullator. Similarly, the current-controlled devices (CCVS and CCCS) have short circuits at their inputs, so—in light of (5.6-3) and (5.6-4)—their inputs are equivalent to nullators also.

Figure 5.82
The four dependent
sources

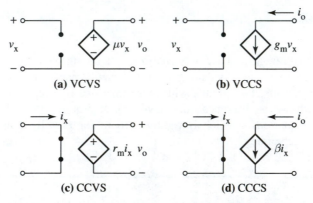

(a) VCVS   (b) VCCS

(c) CCVS   (d) CCCS

As for the output terminals, we see that

$$v_o = \mu v_x \longrightarrow \infty \times 0 = ? \quad (\mu \longrightarrow \infty, \text{VCVS}), \tag{5.6-5}$$

$$i_o = g_m v_x \longrightarrow \infty \times 0 = ? \quad (g_m \longrightarrow \infty, \text{VCCS}), \tag{5.6-6}$$

$$v_o = r_m i_x \longrightarrow \infty \times 0 = ? \quad (r_m \longrightarrow \infty, \text{CCVS}), \tag{5.6-7}$$

$$i_o = \beta i_x \longrightarrow \infty \times 0 = ? \quad (\beta \longrightarrow \infty, \text{CCCS}). \tag{5.6-8}$$

Figure 5.83   Nullor equivalent
of an infinite-gain
dependent source

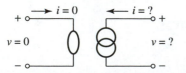

In the limit as the parameter becomes infinite, the dependent variable becomes indeterminate; that is, there is no constraint upon either the current or the voltage. But this is simply the characteristic of a norator. Hence, we have shown that each of the dependent sources is equivalent to the nullor shown in Figure 5.83. Notice that the norator is not necessarily grounded, although it would be if the nullor were the model of an op amp.

Nodal Analysis for
Circuits with Op Amps

The reason we have developed the foregoing equivalence is simply this: for the purposes of analysis, an op amp can be replaced with any of the four types of dependent sources, provided that the parameter is allowed to become infinite. Thus, when we are deciding upon the nodal structure for nodal analysis, we can choose any of the four dependent sources as a model. *Voltage constraints are easier to work with in nodal analysis, so we will choose a voltage-controlled model; because each voltage source reduces the number of nodal equations by one, we will choose it to be a voltage-controlled voltage source (VCVS).*

Figure 5.84 shows an example circuit containing an op amp. You are invited to deactivate both the 8-V and 2-V sources, compute the positive and negative feedback factors F_+ and F_-, and use the resulting values (1/3 and 2/3, respectively) to show that the circuit is stable using the results we developed in Section 5.5. Our first step in the analysis using the nodal method is to replace the op amp by its (grounded) nullor equivalent, shown in Figure 5.85. Although we could analyze this circuit as is, it is good practice to alter the circuit slightly by moving the elements around so that it is "squared off" as in Figure 5.86. (You should take the time to convince yourself that it *does* represent the same circuit.) Now, for the purpose of identifying the node structure, we temporarily replace the nullor with its VCVS equivalent. The topology of the resulting circuit is shown in Figure 5.87. We see there are two nonessential nodes, one supernode, and one essential node. Thus, we anticipate two KCL equations.

Figure 5.84 An example circuit

Figure 5.85 The nullor equivalent

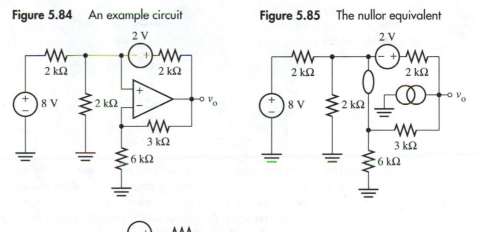

Figure 5.86
The nullor equivalent in
a more conventional
configuration

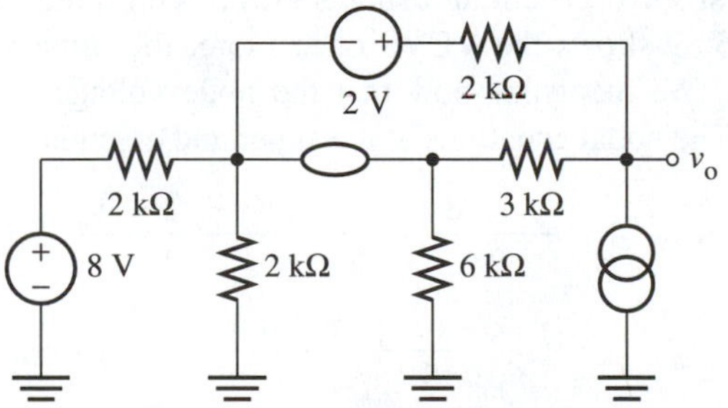

Figure 5.87
The VCVS equivalent

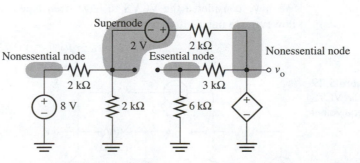

Returning to our nullor circuit of Figure 5.86, we prepare it for nodal analysis as shown in Figure 5.88. We see that we must write one equation at the supernode and one at the essential node. We use the nullator constraint, however, to recognize that the node voltages on either side of that element are the same. Thus, we have

$$\frac{v-8}{2} + \frac{v}{2} + \frac{v+2-v_o}{2} = 0 \quad \text{(supernode)} \tag{5.6-9}$$

and

$$\frac{v}{6} + \frac{v-v_o}{3} = 0 \quad \text{(essential node)}. \tag{5.6-10}$$

Note that we have used the self-consistent units of kΩ, mA, and V. We are treating the norator as a voltage source; thus, in setting up the equations we considered its value to be a known quantity. However, instead of untaping the dependent source, as we would for a VCVS, we now merely recognize the fact that its value is actually an unknown quantity. We can easily solve these equations, (5.6-9) and (5.6-10), to get $v = 4$ V and $v_o = 6$ V. Note that the nullator carries zero current, hence it does not influence the two nodal equations via a current component.

Figure 5.88
The nullor circuit
prepared for nodal
analysis

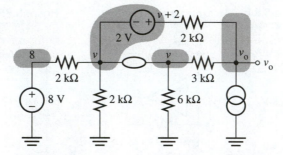

A More Careful Justification for the Nullor Model

The foregoing solution is an outline of the general method of analyzing any op amp circuit using nodal analysis. However, you might remain a bit concerned about our transition from the VCVS to the nullor equivalent. If so, perhaps your concern will be allayed if we first solve the circuit using a VCVS with finite gain, then let the gain go to infinity. Figure 5.89 shows the VCVS once more, this time with the node voltages labeled for analysis. We recognize now that the node voltages v_1 and v_2 in general have different values. The nodal equations at the super and essential nodes are

$$\frac{v_1-8}{2} + \frac{v_1}{2} + \frac{v_1+2-v_o}{2} = 0 \quad \text{(supernode)} \tag{5.6-11}$$

$$\frac{v_2}{6} + \frac{v_2-v_o}{3} = 0 \quad \text{(essential node)}. \tag{5.6-12}$$

We have considered the VCVS to have been taped, with the unknown value v_o. Now, however, we untape it, obtaining

$$v_o = \mu(v_1 - v_2). \tag{5.6-13}$$

Figure 5.89
The finite-gain VCVS
equivalent

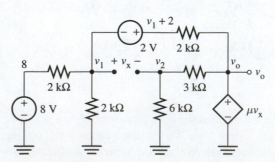

Together with equations (5.6-11) and (5.6-12), we have the matrix equation

$$\begin{bmatrix} 3 - \mu & +\mu \\ -2\mu & 3 + 2\mu \end{bmatrix} \begin{bmatrix} v_1 \\ v_2 \end{bmatrix} = \begin{bmatrix} 6 \\ 0 \end{bmatrix}. \tag{5.6-14}$$

Solving, we have

$$\begin{bmatrix} v_1 \\ v_2 \end{bmatrix} = \begin{bmatrix} \dfrac{2(3 + 2\mu)}{3 + \mu} \\ \dfrac{4\mu}{3 + \mu} \end{bmatrix}. \tag{5.6-15}$$

Consequently, we can write

$$v_o = \mu(v_1 - v_2) = \frac{6\mu}{3 + \mu}. \tag{5.6-16}$$

Computing the limit as $\mu \to \infty$, we see that $v_1 = v_2 = 4$ V and $v_o = 6$ V, as before. We will leave it as an exercise for you to use, for example, a VCCS and repeat the preceding analysis to show that it, too, leads to the same result.

Example 5.16   Solve the circuit in Figure 5.90 using nodal analysis.[19]

Figure 5.90
An example circuit

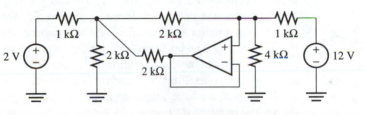

Solution   Note that there is a buffer subcircuit, so we could replace it by a VCVS having a unity voltage gain and solve the circuit using the techniques developed in Section 5.3 for dependent sources. However, our mission here is to practice nodal analysis using nullors. We show the circuit prepared for analysis in Figure 5.91. We have labeled the node at the top of the norator with the symbol v_c, for we are considering it (from a topological point of view) to be a VCVS. In order to emphasize that we are not to write a nodal equation at this nonessential node, we have not used the nullator voltage constraint immediately (the constraint enforces $v_c = v_2$). The KCL equations at the two essential nodes are, therefore,[20]

$$\frac{v_1 - 2}{1} + \frac{v_1}{2} + \frac{v_1 - v_2}{2} + \frac{v_1 - v_c}{2} = 0 \tag{5.6-17}$$

Figure 5.91
The nullor equivalent prepared for nodal analysis

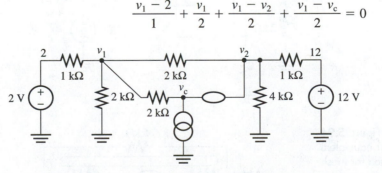

---

[19] The feedback factors are $F_- = 1$ and $F_+ = 1/14$; hence this op amp circuit is stable.

[20] Be careful to note that the current in the nullator is zero. We observe also that the method of inspection works for circuits with nullators and norators—just use it to directly verify equation (5.6-19) for confirmation of this fact. (Treat the norator as a v-source.)

and
$$\frac{v_2}{4} + \frac{v_2 - 12}{1} + \frac{v_2 - v_1}{2} = 0. \tag{5.6-18}$$

Now, instead of the untaping step, we apply the nullator constraint that $v_c = v_2$. This gives (in matrix form)

$$\begin{bmatrix} \frac{5}{2} & -1 \\ -\frac{1}{2} & \frac{7}{4} \end{bmatrix} \begin{bmatrix} v_1 \\ v_2 \end{bmatrix} = \begin{bmatrix} 2 \\ 12 \end{bmatrix}. \tag{5.6-19}$$

The solution is $v_1 = 4$ V and $v_2 = v_c = 8$ V.

We will now consider the mesh analysis of circuits with op amps, that is, with nullors. We can use any of the dependent sources in the place of the op amp when we are determining the mesh structure. Which one should we use? The unknowns are mesh currents, so we should probably choose a current-controlled type. This way, the input (nullator) constraint will directly constrain these variables. Furthermore, we observe that each current source in a circuit reduces the required number of mesh equations by one. For these reasons, we choose the current-controlled current source (CCCS).

Figure 5.92 shows an example circuit that we will solve using mesh analysis. You are urged to compute F_+ and F_- (they are 1/7 and 1/2, respectively) and show that the circuit is stable using the theory developed in Section 5.5. We show the circuit with the op amp replaced by a nullor in Figure 5.93. As we propose to do mesh analysis, we note that the circuit seems to be a bit "cramped" for definition of mesh currents, etc. Thus we "bend it around," to obtain the configuration of Figure 5.94. You should go through this carefully and convince yourself that it is correct. Because the ground reference is not necessary for mesh analysis, we have simply connected all of the ground symbols together with a conductor and eliminated the symbols. Additionally, we have assigned a mesh current to each mesh, using the nullator zero current constraint to express the top mesh current in terms of the middle one (the two are equal, a condition required for the nullator current to be

Figure 5.92
An example circuit

Figure 5.93
The nullor equivalent

Figure 5.94
The nullor equivalent
rearranged for mesh
analysis

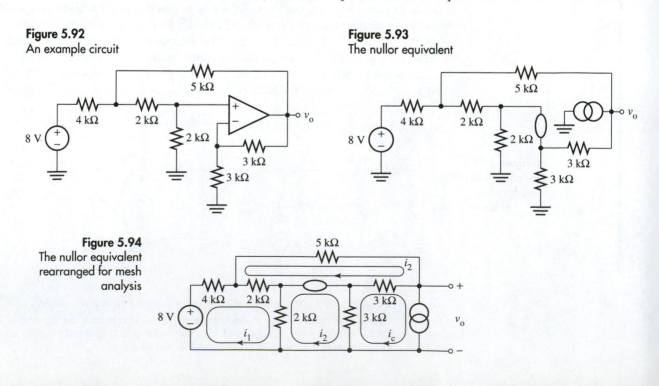

zero). We have labeled the current in the mesh containing the norator i_c. This mesh is a nonessential one because we have agreed to treat the norator as a current source. Thus, we will not write a KVL equation for that mesh. The central and top meshes are essential because we have agreed to treat the nullator as a short circuit. The resulting mesh equations are[21]

$$4i_1 + 2(i_1 - i_2) + 2(i_1 - i_2) = 8, \tag{5.6-20}$$

$$2(i_2 - i_1) + 3(i_2 - i_c) = 0, \tag{5.6-21}$$

and
$$5i_2 + 2(i_2 - i_1) + 3(i_2 - i_c) = 0. \tag{5.6-22}$$

The mesh current i_c would be considered to be a known, at least at first (after taping), if the nullor were replaced by a CCCS; however, with the nullor model it immediately becomes an unknown. Thus, there are three equations in the three unknowns i_1, i_2, and i_c. The matrix form of the KVL equations is

$$\begin{bmatrix} 8 & -4 & 0 \\ -2 & 5 & -3 \\ -2 & 10 & -3 \end{bmatrix} \begin{bmatrix} i_1 \\ i_2 \\ i_c \end{bmatrix} = \begin{bmatrix} 8 \\ 0 \\ 0 \end{bmatrix}. \tag{5.6-23}$$

We solve this (Gaussian reduction is convenient) to obtain $i_1 = 1$ mA, $i_2 = 0$, and $i_c = -2/3$ mA. Thus, using KVL around the rightmost (nonessential) mesh, we obtain

$$v_o = 3(i_2 - i_c) + 3(i_2 - i_c) = -6i_c = 4 \text{ V}. \tag{5.6-24}$$

Just as we did for nodal analysis, let's solve the same circuit by replacing the op amp by a CCCS, doing the mesh analysis, then allowing the current gain to become infinite. The appropriate circuit is shown in Figure 5.95, with the CCCS taped. We see that we must write three mesh equations: one for each of the three essential meshes. In doing so, we treat the mesh current i_c as a known because it is the mesh current in a nonessential mesh. The three equations are

$$4i_1 + 2(i_1 - i_3) + 2(i_1 - i_2) = 8, \tag{5.6-25}$$

$$2(i_2 - i_1) + 3(i_2 + i_c) = 0, \tag{5.6-26}$$

and
$$5i_3 + 3(i_3 + i_c) + 2(i_3 - i_1) = 0. \tag{5.6-27}$$

In matrix form,

$$\begin{bmatrix} 8 & -2 & -2 \\ -2 & 5 & 0 \\ -2 & 0 & 10 \end{bmatrix} \begin{bmatrix} i_1 \\ i_2 \\ i_3 \end{bmatrix} = \begin{bmatrix} 8 \\ -3i_c \\ -3i_c \end{bmatrix}. \tag{5.6-28}$$

Figure 5.95
The finite-gain CCCS equivalent

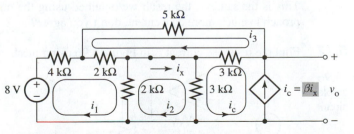

---

[21] Note that, as usual, we are using the convenient and consistent units of mA, kΩ, and V.

Our next step, as always in the analysis of circuits with dependent sources, is to untape them and express them in terms of the unknowns. Thus, we have

$$i_c = \beta i_x = \beta(i_2 - i_3) \tag{5.6-29}$$

Inserting this expression in (5.6-28), we obtain the matrix equation

$$\begin{bmatrix} 8 & -2 & -2 \\ -2 & 5 + 3\beta & -3\beta \\ -2 & 3\beta & 10 - 3\beta \end{bmatrix} \begin{bmatrix} i_1 \\ i_2 \\ i_3 \end{bmatrix} = \begin{bmatrix} 8 \\ 0 \\ 0 \end{bmatrix}. \tag{5.6-30}$$

Let's look at the solution in a bit more detail than usual. One step of Gaussian reduction gives

$$\begin{bmatrix} 8 & -2 & -2 \\ 0 & \dfrac{9 + 6\beta}{2} & -\dfrac{6\beta + 1}{2} \\ 0 & \dfrac{6\beta - 1}{2} & \dfrac{19 - 6\beta}{2} \end{bmatrix} \begin{bmatrix} i_1 \\ i_2 \\ i_3 \end{bmatrix} = \begin{bmatrix} 8 \\ 2 \\ 2 \end{bmatrix} \tag{5.6-31}$$

Thus, we see that the second two equations form a (2×2) system (the i_1 variable can be dropped):

$$\begin{bmatrix} \dfrac{9 + 6\beta}{2} & -\dfrac{6\beta + 1}{2} \\ \dfrac{6\beta - 1}{2} & \dfrac{19 - 6\beta}{2} \end{bmatrix} \begin{bmatrix} i_2 \\ i_3 \end{bmatrix} = \begin{bmatrix} 2 \\ 2 \end{bmatrix}. \tag{5.6-32}$$

Thus,
$$\begin{bmatrix} i_2 \\ i_3 \end{bmatrix} = \begin{bmatrix} \dfrac{80}{60\beta + 170} \\ \dfrac{40}{60\beta + 170} \end{bmatrix} = \begin{bmatrix} \dfrac{8}{6\beta + 17} \\ \dfrac{4}{6\beta + 17} \end{bmatrix}. \tag{5.6-33}$$

We see that both i_2 and i_3 approach zero as $\beta \rightarrow \infty$. Using the first equation in (5.6-28), (5.6-30), or (5.6-31)—the one we have until now discarded—we obtain $i_1 = 1$ mA regardless of the values of i_2, i_3, and β. Going back to the finite β case and using (5.6-33), we have

$$i_c = \beta(i_2 - i_3) = \frac{4\beta}{6\beta + 17}, \tag{5.6-34}$$

which approaches 2/3 mA as $\beta \rightarrow \infty$. (Notice that i_c here is the negative of the i_c we obtained earlier using the nullor model.) Finally, $v_o = 3(i_2 + i_c) + 3(i_3 + i_c) = 6i_c = 4$ V. This is the same as the result we obtained using the nullor equivalent, but the nullor approach is much more convenient, don't you agree?

Example 5.17   Find the response current i_o in Figure 5.96 using mesh analysis.

Figure 5.96
An example circuit

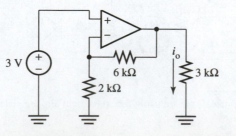

Solution   Figure 5.97 shows the nullor equivalent and Figure 5.98 the same circuit, "straightened out." Notice that we are considering the norator to be an i-source; thus, the mesh on the far right of the circuit, together with the middle one, forms a supermesh. We have used the nullator constraint in the essential mesh on the left side to label that mesh current with the value zero; additionally, we have recognized that the current we desire as the response is merely the right-hand mesh current, i_o. The mesh equations are[22]

$$2(0 - i_1) = 3 \quad \text{(essential mesh)}, \tag{5.6-35}$$

and

$$2(i_1 - 0) + 6i_1 + 3i_o = 0 \quad \text{(supermesh)}. \tag{5.6-36}$$

The solution is $i_1 = -3/2$ mA and $i_o = -8i_1/3 = 4$ mA.

Figure 5.97
The nullor equivalent

Figure 5.98
The nullor equivalent rearranged

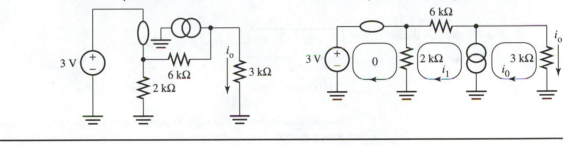

Section 5.6 Quiz

Q5.6-1.   For the circuit shown in Figure Q5.6-1, find v_o using nodal analysis. Before doing so, however, show that $F_+ = 6/13$ and $F_- = 1$ and hence that the circuit is stable.

Q5.6-2.   For the circuit shown in Figure Q5.6-1, find v_o using mesh analysis.

Figure Q5.6-1

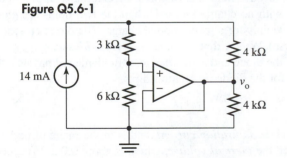

5.7 | Diodes, Transistors, and Device Models

This section will consider the modeling process: how one begins with the basic physical facts about a more complex device and derives the circuit elements that represent it. We will deal with devices that are inherently *nonlinear*. These devices—the diode and the bipolar transistor—are basic building blocks in many electronic circuits, so they are important. We will not explore the applications of these devices in great detail; that is the province of a course in electronic circuits.

---

[22] As usual, we are using the units of mA, kΩ, and V.

The Semiconductor Junction Diode

The *semiconductor junction diode* (or simply *diode*) is shown in Figure 5.99. Figure 5.99(a) shows a stylized conceptual picture of the construction of the diode, and Figure 5.99(b) shows the electrical symbol. It consists of two types of semiconductor ("semi" means "partial") joined to one another as shown. In the P-type material there are an abundance of free positive charge carriers, known as *holes,* that are free to move about under the influence of an electric field; in the N-type, there are an abundance of free negative charge carriers, known as *free electrons,* that are free to move about. We have symbolized these charge carriers by means of positive and negative signs enclosed in small circles. In both types of material, for each charge carrier that is free to move about there is a *fixed* or *bound* charge of the opposite type that cannot move. Thus, each infinitesimal region in each material is electrically neutral. The diode voltage is defined to be positive on the P material and negative on the N, and the diode current is chosen consistently with the passive sign convention.

Figure 5.99
The PN junction diode

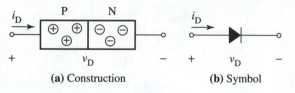

(a) Construction   (b) Symbol

Now suppose a voltage source is attached to the diode. If we make v_D positive, the holes move to the right under the influence of the resulting electric field, and the free electrons move to the left. These charges move across the boundary (called the *junction*) between the two materials. Once on the opposite side, they recombine with charges of the opposite variety and thus give rise to a relative large external current i_D. We say that the diode is *forward biased.*

Now suppose the voltage is reversed. The situation with $v_D < 0$ is shown in Figure 5.100, which shows the symbolic form on the right and a more detailed picture on the left. The voltage produces an electric field that tends to move the mobile charges away from the junction. This sets up an electric field that opposes the externally applied one. The diode will be in equilibrium, with no charge motion, when the two fields are equal and opposite to one another in the vicinity of a given mobile charge. The diode current i_D will be small (and negative of course). We say that the diode is *reverse biased.*

From this basic description of the diode and a knowledge of fundamental physics, the following equation can be derived for the diode *v-i* relationship:

$$i_D = I_s[e^{v_D/V_T} - 1]. \tag{5.7-1}$$

Here I_s is a constant, called the *reverse saturation current,* and is in the range of perhaps 10^{-14} A, and V_T is a constant called the *thermal voltage,* which is about 0.025 V at room temperature. This equation holds over a wide range of waveforms for the voltage and current. The constant I_s varies from one device to another. For simplicity, we will assume in our discussion that both are constant with the values just given.

Figure 5.100
The reverse biased PN junction diode

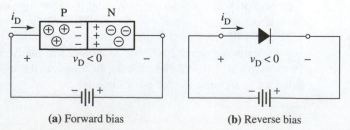

(a) Forward bias   (b) Reverse bias

Physics and mathematics have gotten us this far. We will now discuss the procedure by which we use the equation to derive a circuit model. First, let's do what we might do if we were in a laboratory. We might plot an experimental *v-i* characteristic. We will do this using the diode equation. Table 5.1 shows the result. Notice that the current ranges over a set of values whose extremes are fourteen powers of ten (or *decades*) apart. Figure 5.101 shows a plot of this *v-i* characteristic. We actually computed each point using a spreadsheet and graphed them with a graphics software package. It is important to notice how rapidly the current seems to change in the vicinity of 0.7 V and that it is almost zero for smaller values. For negative values of v_D, the current quickly approaches the asymptotic value $I_s = 10^{-14}$ A. Actually, the rapid change around 0.7 V for the forward biased diode is an artifact introduced by our eye: our vision tends to compress small and large things and to expand moderately sized things. However, if we are working with a scale of perhaps milliamperes, an approximation that the diode voltage does not change appreciably from 0.7 V will still be in order, as we will see.

Table 5.1
Values of v_D and i_D
from diode equation

v_D [volts]	i_D [amperes]
−0.2	0.9997×10^{-14}
−0.1	-0.98×10^{-14}
0.0	0.0
0.1	0.536 pA
0.2	29.8 pA
0.3	2.6 nA
0.4	89.0 nA
0.5	4.85 μA
0.6	0.265 mA
0.7	14.5 mA
0.8	0.79 A

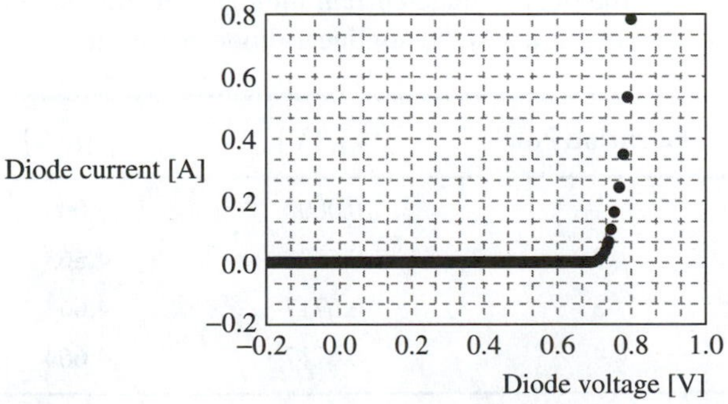

Figure 5.101
The diode *v-i*
characteristic

Diode current [A]

[plot with vertical axis from −0.2 to 0.8 and horizontal axis "Diode voltage [V]" from −0.2 to 1.0]

Example 5.18   Find the dc values V_D and I_D for the circuit in Figure 5.102.

Figure 5.102
A dc diode
circuit

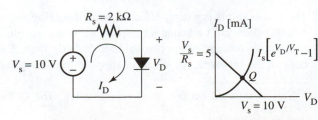

(a) Circuit   **(b)** Graphical solution

Solution   A single mesh equation (with currents in mA, resistances in KΩ, and voltages in V) gives

$$I_D = \frac{10 - V_D}{2} = 5 - 0.5\, V_D. \tag{5.7-2}$$

The resulting graph appears as a straight line in the figure with a vertical intercept of 5 mA (V_s/R_s in general) and a horizontal intercept of 10 V (V_s in general). This is one equation in two unknowns: V_D and I_D. We need another relation, one that is provided by the diode *v-i* characteristic of equation (5.7-1). That equation is highly nonlinear, so the solution is more difficult than it would be if it were linear. Our procedure will be as follows. We will solve equation (5.7-1) for the diode voltage V_D and use it in equation (5.7-2) in an *iterative* manner. Thus, the diode constraint is

$$V_D = V_T \ln\left[\frac{I_D + I_s}{I_s}\right]. \tag{5.7-3}$$

It is intuitively clear that the diode is forward biased, so that $I_D \gg I_s$. Thus, we can make the following approximation:

$$V_D = V_T \ln\left[\frac{I_D}{I_s}\right] = 0.025 \ln\left[I_D \times 10^{11}\right]. \tag{5.7-4}$$

We have used the value 10^{-14} A $= 10^{-11}$ mA for I_s and 0.025 V for V_T.

Now here is our method. We will guess a value for V_D, use (5.7-3) to compute the corresponding value of I_D, carry that value to (5.7-4), and use it to compute a corrected value of V_D. We will continue this process until we achieve the desired degree of accuracy. This we will decide to be the case when the last decimal digit in which we are interested ceases to change. What should be our initial guess for the diode voltage? Well, we know that appreciable current only occurs in the vicinity of 0.7 V (refer to the graph in Figure 5.101); thus, it would be logical to choose, say, 0.7 V to start. Just to illustrate the robustness of our procedure, however, we will assume an initial diode voltage of 0 V. Then our procedure generates Table 5.2. Notice that the values quickly converge to $V_D = 0.6717$ V and $I_D = 4.664$ mA. This is denoted by the Q label in Figure 5.102(b).[23] This is the only voltage-current number pair that simultaneously satisfies the linear KVL circuit equation and the nonlinear diode equation.

Table 5.2
Values of V_D and I_D

Iteration No.	V_D [V]	I_D [mA]
1	0.0000	5.000
2	0.6734	4.663
3	0.6717	4.664
4	0.6717	4.664

Piecewise Linearization   The last example is a fairly effective way of solving such a simple circuit, but we now ask whether there isn't a better way. After all, many circuits are more complex, and even for this simple one the computational load is fairly high. Even in this circuit, things can become quite involved; for instance, suppose the voltage source were time varying (that is,

[23] This is standard notation in electronics work. The Q stands for "quiescent," which means "stationary."

suppose the dc voltage source were replaced by a waveform source $v_s(t)$). Then, as both intercepts of the straight line in Figure 5.102(b) are proportional to the source voltage, the intersection point with the diode curve would move up and down, right and left, with time. We would be forced to choose a given finite number of time instants and find a solution for each using the method of the previous example. The work would be enormous!

Figure 5.103(a) shows an example of the type we have just been discussing. It is actually just a bit more specialized, for we will assume that the time-varying source amplitude is always much smaller than the dc source voltage. Figure 5.103(b) shows the nature of the total source voltage: the sum of the dc and the time-varying one. We have also introduced a bit of notation. An uppercase symbol for voltage or current denotes a dc quantity, a lowercase one a time-varying quantity. A lowercase subscript denotes the fact that the corresponding variable is ac, which means that its average value is zero. An uppercase subscript denotes the fact that the corresponding variable is a "total" one; that is, its average is not zero. Thus we refer to V_S as the dc value of the source, to v_s as its ac value, and to v_S as its "total instantaneous" value. We also clearly have

$$v_S(t) = V_S + v_s(t). \tag{5.7-5}$$

Figure 5.103
Definitions of dc, ac, and total instantaneous variables

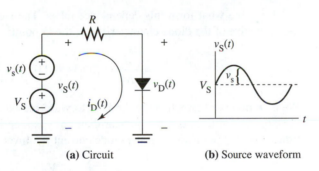

(a) Circuit   **(b)** Source waveform

Figure 5.104   Diode with time-varying waveforms

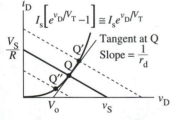

Let's develop a method for analyzing such a circuit. Figure 5.104 shows the diode v-i characteristic (shown with an exaggerated voltage scale for clarity in the following explanation). Suppose we deactivate the ac source, making $v_s(t) = 0$. Then we can solve the resulting dc circuit just as in our earlier example to obtain the Q point. As we pointed out in footnote 23, the term "Q point" means "quiescent" or dc operating point. Now if we reactivate the ac source, we see that the straight line will move back and forth around the Q point as shown in Figure 5.104.[24] When $v_s(t)$ achieves its maximum value, operation will be at the point Q'—the intersection of the topmost dashed line with the diode characteristic; when it achieves its minimum value, the dotted line at the bottom will result and the values of current and voltage will be those of point Q''. In general, we call the intersection of the straight line with the diode characteristic the *operating point*, whereas the special one that results when $v_s(t) = 0$ we call *the Q point* or the *dc operating point*.

Let us construct a straight line tangent to the diode characteristic at the Q point, as shown in the figure. Its slope clearly has the unit of reciprocal resistance. Calling this slope $1/r_d$ and the voltage axis intercept V_0, we have the approximation

$$i_D = \frac{v_D - V_0}{r_d}. \tag{5.7-6}$$

This actually gives the value of the ordinate on the tangent line and not on the diode curve; thus, perhaps we are being a bit loose with notation in using the same symbol

---

[24] The one in Figure 5.100 with negative slope and intercepts V_s and V_s/R is called the "load line."

for the two. However, consider the following fact. If the applied ac signal $v_s(t)$ is small, the points Q′ and Q″ are quite close to the Q point, and the tangent line is an excellent approximation for the diode characteristic itself. In this case, we are entirely justified in using the equation for the tangent line in place of the more complicated diode equation. The equation for this tangent line approximation can easily be shown, using KVL and Ohm's law, to also represent the v-i characteristic of the subcircuit shown in Figure 5.105(b). Thus, under the small signal condition, we can replace the diode symbol in Figure 5.105(a) by the equivalent subcircuit shown in Figure 5.105(b). This is called the *piecewise linear model* for the diode. It is also called a *quasilinear model* for, as we will see, the parameters r_d and V_0 depend on the Q point.

Figure 5.105
Quasilinear diode
model

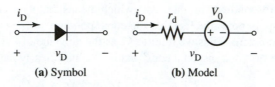

(a) Symbol   (b) Model

Let's see what form this dependence takes. The slope of the tangent line is equal to the derivative of the diode characteristic at the Q point:

$$\frac{1}{r_d} = \frac{d}{dv_D}[I_s[e^{v_D/V_T} - 1]]_Q = \frac{1}{V_T}I_s e^{v_D/V_T}. \qquad (5.7\text{-}7)$$

We now observe that when the diode is forward biased anywhere in the vicinity of 0.7 V, the exponential term is much greater than unity, so the last term in (5.7-7) is merely $1/V_T$ times the dc Q point value of the diode current I_D. Inverting (5.7-7), we have

$$r_d = \frac{V_T}{I_D}. \qquad (5.7\text{-}8)$$

We will discuss the value of V_0 shortly.

Figure 5.106
Using the quasilinearized diode
model

Figure 5.107
Application of superposition

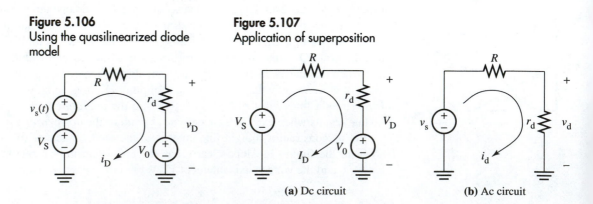

(a) Dc circuit   (b) Ac circuit

For now, let's go back and use our small signal equivalent subcircuit for the diode in our original circuit. The result is shown in Figure 5.106. We stress the fact that the diode voltage is the voltage across both the resistor r_d and the v-source V_0. There is one key characteristic of our approximate circuit: it is linear (for a given Q point)! Hence we can use superposition. We choose to first deactivate the ac source, resulting in the partial circuit shown in Figure 5.107(a), then the dc sources, resulting in the partial circuit of Figure 5.107(b). Once we solve the dc circuit in Figure 5.107(a), we can compute the value of r_d and find the ac values v_d and i_d in Figure 5.107(b). Now you may say, "Hold on a minute! We must use r_d in the dc circuit in order to find I_D, which determines r_d." This is a good point. We do know, of course, the functional dependence of r_d upon I_D, so we could actually solve the circuit (we would have to find the roots of a quadratic); however, practicality rescues us. Let us say that the typical value of I_D is 1 mA. Then a typical value of r_d is 25 mV/1 mA = 25 Ω,[25] and a typical value for the physical series resistor R in our circuit might be 1 kΩ. Thus, in practical circuits, r_d is entirely negligible in computing the dc current and voltage. To make this more palatable, refer to Figure 5.101, the graph of the diode current versus its terminal voltage. We see that *the characteristic is almost vertical and that V_0 is almost constant at 0.7 V for a wide range of diode currents.* Thus, we can not only neglect r_d in comparison with R, but we can also approximate V_0 as being the known value of 0.7 V. If $V_S >> 0.7$ V, this second approximation will certainly be valid. The result is the dc equivalent circuit shown in Figure 5.108. From it, we easily find the dc diode current to be $I_D = (V_S - 0.7)/R$. This is clearly much easier than the procedure of Example 5.18, wouldn't you agree? Once we have found I_D, the Q point current, we compute r_d and then analyze the "ac circuit" that we presented in Figure 5.107(b). This circuit is referred to, in general, as an *ac small signal equivalent circuit,* and the diode resistance r_d as the *small signal equivalent* for the diode.

Figure 5.108
The dc (approximate)
equivalent circuit

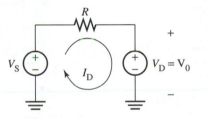

Example 5.19   Find the diode voltage $v_D(t)$ in Figure 5.109.

Figure 5.109
An example circuit

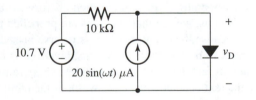

Solution   First, let's solve the dc circuit by deactivating the ac current source and replacing the diode by its voltage source approximation *without the r_d resistor.* This is shown in Figure 5.110. The dc current is readily seen to be $I_D = (10.7 - 0.7)/10$ kΩ = 1 mA. Thus, the diode small signal resistance is $r_d = 25$ mV/1 mA = 25 Ω. The resulting ac small signal equivalent circuit is shown in Figure 5.111. The diode small signal ac voltage is given by

$$v_d(t) = \frac{10 \text{ k}\Omega \times 25 \text{ }\Omega}{10 \text{ k}\Omega + 25 \text{ }\Omega} \times 20 \sin(\omega t) \text{ }\mu\text{A}$$

$$\cong 25 \text{ }\Omega \times 20 \sin(\omega t) \text{ }\mu\text{A} = 500 \sin(\omega t) \text{ }\mu\text{V}.$$

By superposition, because 0.7 V = 700 mV, we have $v_D(t) = 700 + 0.50 \sin(\omega t)$ mV. The ac value of the diode voltage is some 1400 times smaller than the dc voltage, so the small signal approximation is probably valid for any practical purpose.

---

[25] Note that $V_T = 0.025$ V = 25 mV.

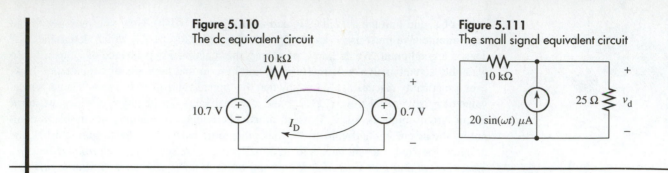

Figure 5.110
The dc equivalent circuit

10 kΩ

10.7 V I_D 0.7 V

Figure 5.111
The small signal equivalent circuit

10 kΩ

20 sin(ωt) μA   25 Ω v_d

The Bipolar Junction Transistor (BJT)

Another active device, the *bipolar junction transistor (BJT)* is depicted in Figure 5.112(a). It consists of three pieces of semiconductor material in the form of a sandwich. This particular type of BJT is called an NPN (a PNP would be the same with N and P interchanged). The circuit symbol for the BJT is shown in Figure 5.112(b). The three terminals are labeled E for *emitter,* B for *base,* and C for *collector.* As we can easily see, the BJT consists of two junction diodes, both pointed away from the base—toward the emitter and collector terminals. We will refer to these two diodes as the emitter diode and the collector diode, respectively.

Figure 5.112
The bipolar junction transistor (BJT)

i_E N   P   N $i_C = \alpha i_E$ i_E $i_C = \alpha i_E$

E α o C   E o   o C

$i_B = [1 - \alpha]i_E$ $i_B = [1 - \alpha]i_E$

B

(a) Physical operation   **(b)** Symbol

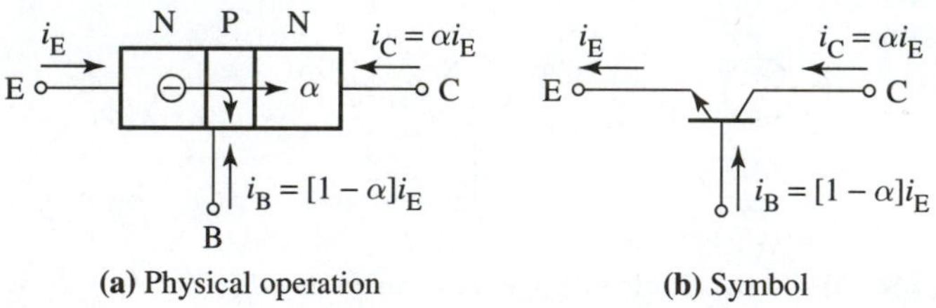

For normal operation[26] the emitter diode is forward biased and the collector diode is reverse biased. Thus, we see that electrons are propelled to the right and injected into the base region. Because the collector diode is reverse biased, we would normally expect the collector current to be essentially zero; this, however, is not the case because the base region is so thin. In fact, let's use a very rough analogy to indicate how thin. If the length of this page in the text is analogous to the width of the emitter and collector regions (the two N-type semiconductor regions), then the *thickness* is analogous to the base region width (the P-type region).

Now let's think on a statistical basis. For each 100 electrons that are injected from the emitter into the base, 99 of them might go entirely across the base region without meeting a hole (the type of positive mobile charge in the base)—because of the extreme thinness of the base. In this 100, however, a single one might run into a hole and recombine to produce base current. In general, we will call the fraction of injected electrons that survive α. Thus, the fraction of electrons that recombine with holes in the base will be $1 - \alpha$. Recalling that electron motion to the right constitutes a current to the left, we see that the three terminal currents have the directions and values shown in Figure 5.112.

BJT Dc Bias and Small Signal Models

We are now in a position to form a physical model for the BJT. As we show in Figure 5.113, it consists of two "back-to-back" diodes. But that is not all—if it were, the transistor would not have revolutionized modern society. Because of the extreme thinness of the base region—and the consequent transportation of electrons into the collector—we must

---

[26] "Normal operation" here means the *active mode,* suitable for use in linear applications.

Figure 5.113 The physical model for the BJT

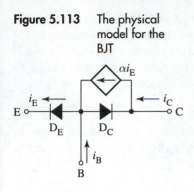

include another element: the current-controlled current source (CCCS) shown in the figure. It is this current source that makes the many interesting applications of the transistor possible. Much industrial effort and expense has been directed toward reducing the base width and consequently increasing the value of α, which is ideally unity (and practically on the order of 0.99). The importance of the BJT alone is sufficient to justify studying dependent sources thoroughly in a course on circuit analysis.

Now we know very well how to analyze circuits containing dependent sources, and we have just developed techniques for dealing with diodes; hence, we can immediately derive models for the BJT. If a BJT is contained in a circuit providing voltages that forward bias D_E, the emitter diode, and reverse bias D_C, the collector diode, we can replace these elements by a battery of $V_0 = 0.7$ V and an open circuit,[27] respectively, for the dc bias model. For the ac small signal model, we can similarly replace D_E by a resistor r_e that depends upon the dc value of emitter current,

$$r_e = \frac{V_T}{I_E}, \tag{5.7-9}$$

and D_C by an open circuit. We show the two models in Figure 5.114. The design of biasing circuits to adjust I_E to a given value and to stabilize it with respect to variations in circuit parameters, temperature, and the like is an art that we will leave to a course in electronics. Here, we will concentrate upon the small signal analysis of circuits with BJTs.

Figure 5.114 Circuit models for the BJT

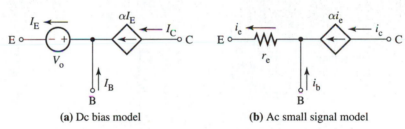

(a) Dc bias model **(b)** Ac small signal model

Example 5.20 Find the small signal "input resistance" r_{in} and the small signal voltage gain v_c/v_e of the BJT circuit in Figure 5.115. Note: This is a small signal circuit; any dc sources have already been deactivated for the analysis.

Figure 5.115 A common base BJT voltage amplifier circuit

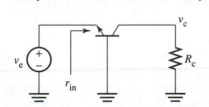

Solution As the title of the figure shows, we refer to this circuit as a *common base voltage amplifier circuit*. The term "common base" means that the base terminal is the common reference node for both the input and output voltages. The term "voltage amplifier" means that the output voltage is larger than the input voltage. By "input resistance" we mean the resistance we would measure if we were to clip the conductor at the corner of the arrow (where it makes a right angle turn) and measure the equivalent resistance. We call it the

---

[27] We showed earlier that even a very small reverse voltage causes the diode current to become constant at I_s, an exceedingly small value that we will approximate as zero—thus giving an open circuit equivalent.

"input" resistance, because we consider the source to be providing the input signal to the rest of the circuit. Finally, the small signal voltage gain is the ratio of the output (collector) voltage v_c to the input (emitter) voltage v_e. We are assuming that the input terminals yield a v-i characteristic equivalent to a resistor and that the ratio of the output voltage to the input voltage is independent of the values of these two variables. We will verify these two assumptions in the analysis process.

Analysis merely consists of mentally "unplugging" the transistor symbol and "plugging in" the small signal equivalent circuit. This results in the circuit shown in Figure 5.116. There is only one essential node—the output (the collector terminal of the BJT)—so only one KVL nodal equation is required. It is

$$\frac{v_c}{R_c} = -\alpha i_e. \qquad (5.7\text{-}10)$$

For this simple circuit, we do not take the trouble to tape the CCCS. Now we express the controlling variable of the CCCS in terms of the input source voltage:

$$i_e = -\frac{v_e}{r_e}. \qquad (5.7\text{-}11)$$

Using this in (5.7-10) results in

$$v_c = \frac{\alpha R_c}{r_e} v_e. \qquad (5.7\text{-}12)$$

Figure 5.116
The small signal
equivalent circuit

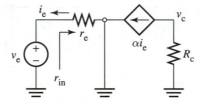

Thus, we see that the ratio of collector voltage to emitter voltage is, indeed, a constant and that this ratio—which we define to be the small signal voltage gain—is

$$A_v = \frac{v_c}{v_e} = \frac{\alpha R_c}{r_e}. \qquad (5.7\text{-}13)$$

The A stands for "amplification" and the v refers to the fact that it is voltage that is being amplified, or made larger. Notice that the current gain, the ratio of i_c to i_e, is equal to the constant $\alpha < 1$. Equation (5.7-11) is the v-i characteristic of the subcircuit formed by removing the input voltage source; clearly, it represents a resistor of value

$$r_{in} = r_e. \qquad (5.7\text{-}14)$$

We worked the last example in symbolic terms for generality, but let's get some idea of practical values. Let's assume that the dc value of emitter current I_E, which we assume to have been obtained from the original circuit (with the dc sources activated), is one milliampere. Then we can immediately compute $r_e = 25 \ \Omega$. Let's suppose that $R_c = 5 \ k\Omega$ (which is a practical value). Then, approximating α by unity, we have $r_{in} = 25 \ \Omega$ and $A_v = 200$. Thus, the collector voltage is much larger than the emitter voltage. Notice that the input source establishes a current i_e in the low-valued emitter diode resistance r_e, and the BJT reproduces that same current in the much larger resistance R_c through the action of the CCCS. It is from this principle that the name transistor comes: *trans*ferring current through a re*sistor*.

Though the voltage gain of the common base circuit is large, the input resistance is small; thus, if the Thevénin resistance of the input source is nonzero, a sizable voltage drop will occur at the input, and the *overall* voltage gain will be reduced. This is a disadvantage of the configuration. Another is the fact that the current gain is less than unity. Let's see how we can improve matters.

The Common Emitter Small Signal Model for the BJT

Suppose we twist the BJT around so that the base terminal is the input, as in Figure 5.117(a). Why do this? Well, remember that the base current is $(1 - \alpha)i_E$ and that $\alpha \cong 1$. Thus, i_B is typically very small. We would therefore expect that the current gain would be large and that the input resistance would be much higher than for the common base circuit. In order to quantify this, let's define a new BJT parameter. Because $i_C = \alpha i_E$ and $i_B = (1 - \alpha)i_E$, we can take their ratio and see that

$$\frac{i_C}{i_B} = \frac{\alpha}{1 - \alpha} = \beta. \tag{5.7-15}$$

Thus, $i_C = \beta i_B$. If a typical value for α is 0.99, then β will typically be perhaps 100 $(0.99/(1 - 0.99) = 99$. We have been working with total instantaneous variables, but the same relationship holds, of course, for small signal ones. Thus, expressing the CCCS value in terms of β, we have the small signal equivalent circuit shown in Figure 5.117(b).

Figure 5.117
The common emitter configuration

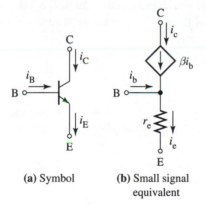

(a) Symbol

(b) Small signal equivalent

The Small Signal Hybrid π Model for the BJT

We can improve our equivalent circuit for the common emitter configuration. The one in Figure 5.117(b) has a "floating" (that is, ungrounded) current source. This makes hand analysis a bit more difficult. Thus, let's apply current source transportation[28] relative to the emitter terminal. This results in the circuit shown in Figure 5.118(a). Considering the

Figure 5.118
Development of the small signal hybrid π model

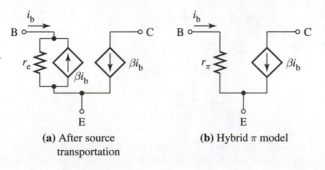

(a) After source transportation

(b) Hybrid π model

---

[28] See Chapter 3, Section 3.5.

r_e resistor and its parallel CCCS as a parallel subcircuit, we can easily use KCL, KVL, and Ohm's law to show that it is equivalent to the single resistor of value

$$r_\pi = (\beta + 1)r_e, \tag{5.7-16}$$

as in Figure 5.118(b). The resulting small signal circuit is called the *hybrid π model* because, with the addition of another element across the top, it looks like the Greek letter.

Example 5.21   Find the small signal voltage gain $A_v = v_c/v_b$ and the small signal input resistance of the common emitter voltage amplifier circuit shown in Figure 5.119. Note that this is a small signal equivalent circuit, with the dc sources deactivated. Thus these sources do not appear explicitly.

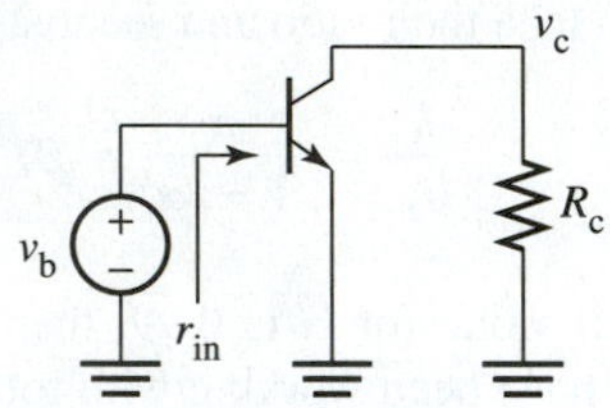

Figure 5.119
A common emitter
voltage amplifier

Solution   Just as for the common base circuit, we mentally unplug the BJT symbol and replace it with its small signal equivalent—this time the hybrid π model. The result is shown in Figure 5.120. Again, there is only one essential node: the collector terminal of the BJT. The nodal equation there is

$$\frac{v_c}{R_c} = -\beta i_b. \tag{5.7-17}$$

As for the common base circuit, we have not taped the CCCS because the analysis is so simple. We now express the controlling variable i_b in terms of the independent source variable v_b:

$$i_b = \frac{v_b}{r_\pi}. \tag{5.7-18}$$

These two equations together imply that the voltage gain is

$$A_v = -\frac{\beta R_c}{r_\pi}. \tag{5.7-19}$$

Figure 5.120
The small signal
equivalent

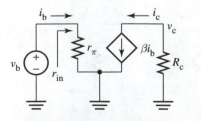

At first glance, it appears that this voltage gain is β times as large as that for the common base amplifier, but this is not true. The two voltage gains give exactly the same value! To see this, just remember that $r_\pi = (\beta + 1)r_e$ and convince yourself that

$\dfrac{\beta}{\beta + 1} = \alpha$. Actually, the two voltage gain expressions do not give exactly the same value, only the same magnitude. In fact, there is a negative sign in (5.7-19) that was not present for the common base circuit. What does this mean? Recalling that we are dealing with small signal quantities, that is, variables whose average value is zero, we see that they assume both positive and negative values. The negative sign means that when the input source has a positive value, the output voltage is negative and vice versa. Remember, though, that we are only dealing with the small signal components of total instantaneous signals, and it is these small signals that are inverted in sign.

It should be clear that the input resistance is $r_\pi = (\beta + 1)r_e$, a factor of $\beta + 1$ times as large as for the common base circuit. The current gain is clearly equal to β, also a factor of $\beta + 1$ times as large as for the common base circuit $(\beta/\alpha = 1/(1 - \alpha) = \beta + 1)$. Thus, the common emitter circuit has the same voltage gain magnitude, a higher current gain, and a higher input resistance. Thus, for many purposes it is preferred over the common base configuration.

The Ideal BJT Model ($\beta \rightarrow \infty$)

Based upon our study of the common emitter circuit in the last example, we see that it is desirable for the current gain β to be as high as possible.[29] Thus, an ideal BJT would have an infinite beta: $\beta \rightarrow \infty$. But how do we analyze a circuit with $\beta = \infty$? We would be dealing with a CCCS having an infinite current gain. This is reminiscent of the op amp, which is a VCVS with infinite voltage gain.

Let's get some idea of how to treat such a case by looking more closely at the hybrid π model, reproduced here for convenience in Figure 5.121(a). Let's use the r_π resistance relationship to convert the controlling variable from i_b to v_{be}. This gives, for the CCCS constraint equation,

$$\beta i_b = \beta \frac{v_{be}}{r_\pi} = \beta \frac{v_{be}}{(\beta + 1)r_e} = \frac{\alpha}{r_e} v_{be} = g_m v_{be}, \qquad (5.7\text{-}20)$$

where

$$g_m = \frac{\alpha}{r_e} \cong \frac{1}{r_e} \qquad (5.7\text{-}21)$$

is called the *transconductance*. The resulting hybrid π model with a VCCS is shown in Figure 5.121(b). Notice that, even for finite β, it is approximately independent of that pa-

Figure 5.121
Development of the transconductance hybrid π model

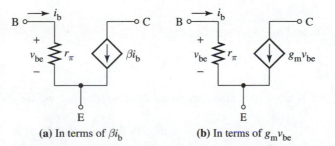

(a) In terms of βi_b (b) In terms of $g_m v_{be}$

[29] As $\beta = \alpha/(1 - \alpha)$, we see that this is equivalent to α being as close to unity as possible.

Figure 5.122   The ideal transistor model

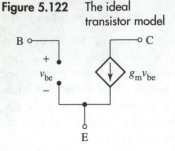

rameter ($\alpha = \beta/(1 + \beta)$). If we let $\beta \to \infty$, the preceding approximation becomes exact:

$$g_m = \frac{1}{r_e} \quad (\beta \longrightarrow \infty). \tag{5.7-22}$$

Furthermore, the input resistance $r_\pi = (\beta + 1)r_e \to \infty$. The resulting *ideal transistor model* is shown in Figure 5.122. Notice that the transconductance does have the unit siemens (inverse ohm); hence, we use the symbol g as for an ordinary conductance. The subscript m stands for "mutual," which means that the current in one pair of terminals has something "mutual" (in common) with the voltage between another pair. It is synonymous with the "trans" in transconductance.

The PNP BJT

Before we leave this fascinating topic of transistor models, we will introduce one more device, then work one last example. For the additional device, let's first look briefly at the PNP BJT, whose symbol and model are shown in Figure 5.123(a) and (b). Because the two diodes and the CCCS are merely reversed, we see that operation is identical to that of the NPN BJT—except that all voltages and all currents are reversed. A moment's thought should convince you that *this does not affect the small signal model at all!*[30]

Figure 5.123
The PNP BJT

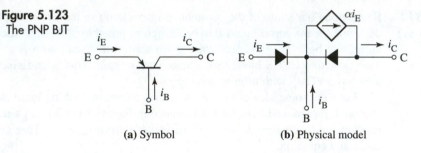

(a) Symbol   (b) Physical model

Voltage, Current, and Power Gain Relationships

Before closing this section on the interesting and important topic of device modeling, we will develop a generalization of the technique we have been applying to amplifier circuits. Figure 5.124 shows a general three-terminal device or subcircuit. The left side is assumed to be the input, and the right side the output. We assume that the subcircuit is linear. The voltage gain is defined by

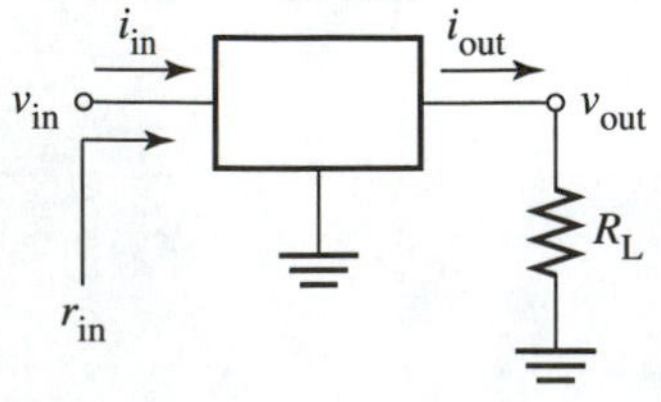

$$A_v = \frac{v_{out}}{v_{in}}, \tag{5.7-23}$$

the current gain by

$$A_i = \frac{i_{out}}{i_{in}}, \tag{5.7-24}$$

Figure 5.124   A general loaded three-terminal subcircuit or device

and the input resistance by

$$r_{in} = \frac{v_{in}}{i_{in}}. \tag{5.7-25}$$

If we apply Ohm's law to the *load resistor* R_L, we see that

$$v_{out} = R_L i_{out} = R_L A_i i_{in} = A_i \frac{R_L}{r_{in}} v_{in}. \tag{5.7-26}$$

---

[30] The simple reason for this is that reversal of both the controlled and the controlling variable in a dependent source has no effect on the external circuit. Just draw any of the small signal models we have discussed and go through this reversal and compute external quantities such as voltage gain.

Thus,

$$A_v = A_i \frac{R_L}{r_{in}}. \qquad (5.7\text{-}27)$$

This relates the voltage gain to the current gain. For the BJT, the current gain is either α (common base) or β (common emitter), independently of the load resistance. Thus, (5.7-27) is a generalization of our previous results for voltage gain.

Let us define one last term: the *power gain*. It is

$$A_P = \frac{P_{out}}{P_{in}} = \frac{v_{out} i_{out}}{v_{in} i_{in}} = A_v A_i. \qquad (5.7\text{-}28)$$

Thus, the power gain is the product of the voltage and current gains. We can use (5.7-27) to express the power gain in terms of the current gain and load resistance:

$$A_P = A_i^2 \frac{R_L}{r_{in}}. \qquad (5.7\text{-}29)$$

Section 5.7 Quiz

Q5.7-1. Find the current I and voltage V in the diode circuit in Figure Q5.7-1. $I_s = 10^{-14}$ A.

Figure Q5.7-1

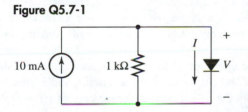

Q5.7-2. Find the small signal voltage gain of the *emitter follower* circuit in Figure Q5.7-2. Assume that $I_E = 0.5$ mA and note that the circuit shown is an ac small signal one; that is, all dc sources have been deactivated. Assume that $\beta = 49$.

Figure Q5.7-2

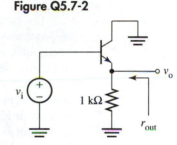

Q5.7-3. Find the small signal output resistance r_{out} (subcircuit Thevénin resistance) of the emitter follower circuit shown in Figure Q5.7-2.

Q5.7-4. Find the small signal power gain of the circuit in Figure Q5.7-2. (Assume that the 1-kΩ resistor shown is the load.)

5.8 | SPICE: Active Circuits

SPICE Models for Active Elements

SPICE uses the reserved letters E, F, G, and H as the first symbol in a name representing a dependent source. We will consider them one at a time; however, the order E, G followed by F, H is preferable because the first two represent voltage-controlled devices, and the latter two, current-controlled devices. A current-controlled element requires a bit more programming effort than do the voltage-controlled ones.

Figure 5.125 shows a voltage-controlled voltage source (VCVS). The SPICE syntax for this element is shown in the figure. The name of the device must start with an E (either upper or lowercase, because SPICE is not sensitive to case). After a space the *controlled* nodes (those between which the VCVS is connected) are listed, with the first being the one assumed positive. Then come the *controlling* voltage nodes, the first being the assumed positive reference for v_x and the next being the negative. Finally, the voltage gain μ is specified.

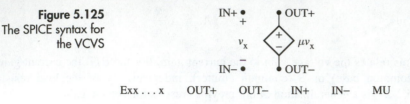

Figure 5.125
The SPICE syntax for the VCVS

Exx . . . x   OUT+   OUT−   IN+   IN−   MU

If we exchange the dependent voltage source for a dependent current source, we obtain the VCCS shown in Figure 5.126. The only differences are these: the initial letter must be a G, and the first node listed thereafter is the node from which the controlled current flows, with the next being that to which it flows. Finally, the transconductance g_m is listed rather than the voltage gain μ. The unit for g_m is the siemens, whereas μ is unitless.

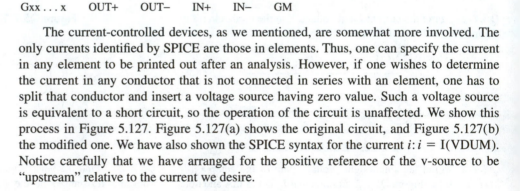

Figure 5.126
The SPICE syntax for the VCCS

Gxx . . . x   OUT+   OUT−   IN+   IN−   GM

The current-controlled devices, as we mentioned, are somewhat more involved. The only currents identified by SPICE are those in elements. Thus, one can specify the current in any element to be printed out after an analysis. However, if one wishes to determine the current in any conductor that is not connected in series with an element, one has to split that conductor and insert a voltage source having zero value. Such a voltage source is equivalent to a short circuit, so the operation of the circuit is unaffected. We show this process in Figure 5.127. Figure 5.127(a) shows the original circuit, and Figure 5.127(b) the modified one. We have also shown the SPICE syntax for the current i: $i = $ I(VDUM). Notice carefully that we have arranged for the positive reference of the v-source to be "upstream" relative to the current we desire.

Figure 5.127
Monitoring a nonelement current in SPICE

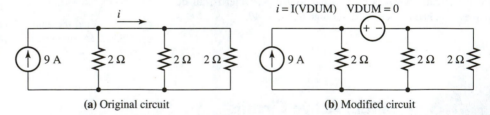

(a) Original circuit   (b) Modified circuit

Our conclusion is this: the controlling current for a dependent current source must be one in a v-source. If the controlling current is actually that flowing through a v-source, fine. Otherwise, we insert a dummy v-source. Figure 5.128 shows a CCVS. The syntax is almost the same as for the two we have already covered. The first letter of a CCVS, however, starts with an H, the two controlling nodes (the second pair listed) are replaced by the name of the voltage source through which the controlling current is monitored, and the last entry is the transresistance (whose unit is the ohm).

Figure 5.128
The SPICE syntax for the CCVS

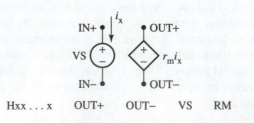

Hxx . . . x   OUT+   OUT−   VS   RM

Finally, we show the CCCS in Figure 5.129. Its first letter is an F, and the current gain beta is listed as the last item in the specification.

Figure 5.129
The SPICE syntax for the CCCS

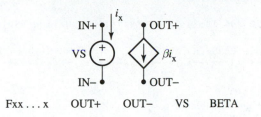

Fxx ... x   OUT+   OUT–   VS   BETA

Example 5.22   Write a SPICE file for the circuit in Figure 5.130 to find v_x.

Figure 5.130
An example circuit

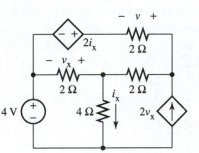

Solution   We show the circuit prepared for SPICE analysis in Figure 5.131, that is, with the elements named, the nodes numbered, and a ground reference assigned. Note that the current-controlled voltage source at the top carries the designation H1. Similarly, we recognize that the dependent source on the right is of the VCCS variety and hence carries the designation G1. Furthermore, as usual, we have labeled the ground reference with the number zero. Notice that node number 5 was created when we added the dummy v-source. A SPICE file that simulates this circuit is:

```
EXAMPLE 1
VS1      1     0      4
VDUM     5     0      0
H1   4   1     VDUM         2
G1   0   3     2        1   2
R1   1   2     2
R2   2   3     2
R3   4   3     2
R4   2   5     4
.DC VS1   4    4        1
.PRINT    DC   V(4,3)
.END
```

The corresponding output file resulting from a SPICE run is:

```
****05/05/94 13:39:17**********EVALUATION PSPICE (JAN 1992)*****
EXAMPLE 1
****   CIRCUIT DESCRIPTION
****************************************************************
******
VS1      1     0   4
VDUM     5     0   0
H1   4   1     VDUM    2
G1   0   3     2   1   2
R1   1   2     2
R2   2   3     2
R3   4   3     2
```

```
R4  2    5    4
.DC VS1    4    4   1
.PRINT  DC   V(4, 3)
.END
****05/05/94 13:39:17**********EVALUATION PSPICE (JAN 1992)*****
EXAMPLE 1
****   DC TRANSFER CURVES          TEMPERATURE= 27.000 DEG C
******************************************************************
*******
VS1        V(4, 3)
4.000E+00  -8.000E+00
     JOB CONCLUDED
     TOTAL JOB TIME     .42
```

As an exercise, you might wish to verify this result by performing either a nodal or mesh analysis (which requires the fewer equations?).

Figure 5.131
The example circuit prepared for SPICE analysis

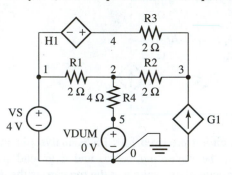

Ideal Op Amp Models for SPICE

Now we turn to the SPICE simulation of circuits having op amps. This is actually rather simple, for we know that an (ideal) op amp is merely a grounded VCVS with infinite voltage gain. SPICE, however, does not have such an element in its catalog of components. Thus, we must simulate it with a VCVS having a "very large" voltage gain. This is shown in Figure 5.132. We have chosen an arbitrary value of 10^9 for the voltage gain; you might find that your particular version of SPICE (and computer) will require a somewhat smaller or larger value. The only solution is to "simulate and see."

Figure 5.132
SPICE op amp model

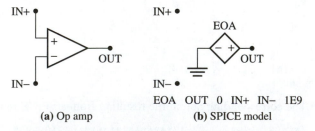

(a) Op amp   (b) SPICE model

Because a circuit often has multiple op amps, one can generate it as a subcircuit:

```
.SUBCKT IOA  IN+  IN-   OUT
EOA  OUT   0  IN+  IN-   1E9
.ENDS
```

Here, IOA stands for ideal op amp. Notice that we do not have to explicitly specify the fact that the VCVS has a ground connection in the node list we "pass" to the subcircuit code—the ground reference is a "global variable." The op amp would then be invoked with a statement such as

```
XOPAMP1  3  8  2  IOA
```

Example 5.23   Write a SPICE file to simulate the op amp circuit in Figure 5.133.

Figure 5.133
An example
circuit

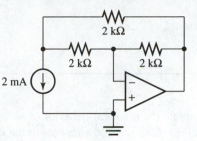

Solution   We treat the op amp as a subcircuit and show it prepared for SPICE coding in Figure 5.134. The SPICE file is as follows:

```
EXAMPLE 2
IS 1  0   2E-3
X1 0  2   3    IOA
R1 1  2   2K
R2 2  3   2K
R3 1  3   2K
.SUBCKT  IOA  IN+  IN-  OUT
EOA  OUT  0  IN+  IN-  1E9
.ENDS
.DC  IS  2E-3  2E-3  1E-3
.PRINT  DC   V(3)
.END
```

Figure 5.134
The example circuit
prepared for SPICE
analysis

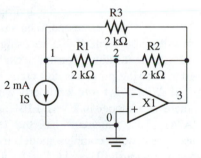

The SPICE output for this input file is:

```
****05/05/94 14:17:20***********EVALUATION PSPICE (JAN 1992)*****
EXAMPLE 2
****  CIRCUIT DESCRIPTION
****************************************************************
*******
IS    1 0 2E-3
X1    0 2 3    IOA
R1    1 2 2K
R2    2 3 2K
R3    1 3 2K
.SUBCKT  IOA  IN+  IN-  OUT
EOA  OUT  0  IN+  IN-  1E9
.ENDS
.DC  IS  2E-3  2E-3  1E-3
.PRINT
.END
****05/05/94 14:17:20***********EVALUATION PSPICE (JAN 1992)*****
EXAMPLE 2
****   DC TRANSFER CURVES          TEMPERATURE= 27.000 DEG C
```

```
*************************************************************
*******
IS          V(3)
2.000E-03 1.333E+00
    JOB CONCLUDED
    TOTAL JOB TIME   .42
```

Nonideal Op Amp Models for SPICE

If one wishes to simulate the nonideal behavior of an op amp in a particular circuit, there are a number of options. One can add elements to the defining subcircuit, or one can use one of the (much more complicated) models included with SPICE in the form of a library. One invokes this library in different ways for different versions of SPICE and different computer systems. For PSPICE on a Macintosh, one uses the following syntax:

```
.LIB    EVAL.LIB
```

(For DOS machines, use quotes: i.e., .LIB   "EVAL.LIB".) This tells the SPICE simulator to look in the file EVAL.LIB (assumed to be in the same directory or folder as the executable SPICE file) for any subcircuits not contained in the SPICE file itself. EVAL.LIB is a library of standard parts supplied by the company marketing PSPICE. There are other statements available if one wishes to store his or her own libraries in other directories or files. If you store subcircuits you have coded yourself, you must change the file name to something like

```
.LIB    MYLIB.LIB
```

To generate a library file, you simply create the appropriate code with a text editor, say for the subcircuit IOA we have just discussed, then save it as a file called MYLIB.LIB. Note that you must save it in text form. If you are using a word processing program such as Microsoft Word as your text editor, you must explicitly use the "save as text" (or ASCII) because word processing programs add other formatting control characters that must be "stripped off" if PSPICE is to work properly with the resulting file.

As it happens, there is a model in EVAL.LIB for a 741 type op amp. It has the subcircuit name UA741. This subcircuit has the terminal configuration shown in Figure 5.135. Because it is a quite complex model, it has more nodes than the simple (ideal) one we presented previously. The added nodes are labeled in the figure with VS+ and VS−, symbols for positive and negative "supply" voltages (dc "battery" voltages). In Figure 5.136 we show the way these power supplies, or batteries, are connected, using the circuit of Example 5.23. The power supplies are required to "power up" the op amp and make it work. They are not required by the circuit itself in a theoretical way. They are responsible, however, for nonideal behavior such as limitation of the output voltage "swing." Thus, if we wanted to use this op amp model, the SPICE code for the circuit in Example 5.23 would be:

Figure 5.135 The EVAL.LIB op amp subcircuit (macromodel)

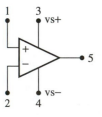

```
EXAMPLE 2.NONIDEAL
IS 1  0    2E-3
X1 0  2    5   4   3   UA741
R1 1  2    2K
R2 2  3    2K
R3 1  3    2K
.LIB EVAL.LIB
.DC IS  2E-3  2E-3   1E-3
.PRINT DC    V(3)
.END
```

The output from a SPICE run on this input file is:

```
****05/05/94 14:56:07**********EVALUATION PSPICE (JAN 1992)*****
EXAMPLE 2.NONIDEAL
****  CIRCUIT DESCRIPTION
******************************************************************
******
IS   1  0  2E-3
X1   0  2  5   4  3  UA741
R1   1  2  2K
R2   2  3  2K
R3   1  3  2K
.LIB  EVAL.LIB
.DC  IS  2E-3  2E-3  1E-3
.PRINT  DC  V(3)
.END
****05/05/94 14:56:07**********EVALUATION PSPICE (JAN 1992)*****
EXAMPLE 2.NONIDEAL
****  DIODE MODEL PARAMETERS
******************************************************************
******
     X1.DX
     IS 800.000000E-18
     RS   1
****05/05/94 14:56:07**********EVALUATION PSPICE (JAN 1992)*****
EXAMPLE 2.NONIDEAL
****  BJT MODEL PARAMETERS
******************************************************************
******
     X1.QX
     NPN
     IS 800.000000E-18
     BF 93.75
     NF 1
     BR 1
     NR 1
****05/05/94 14:56:07**********EVALUATION PSPICE (JAN 1992)*****
EXAMPLE 2.NONIDEAL
****  DC TRANSFER CURVES          TEMPERATURE= 27.000 DEG C
******************************************************************
*******
 IS         V(3)
 2.000E-03 1.335E+00
     JOB CONCLUDED
     TOTAL JOB TIME   1.32
```

Notice that these are almost the same as the results we obtained in Example 5.23 with an ideal op amp.

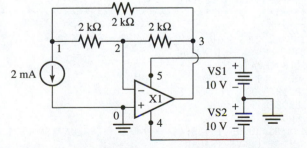

Figure 5.136
The circuit of Example
5.23 using the EVAL.LIB
macromodel

You might have noticed in the preceding SPICE output that when SPICE "expanded" the macromodel for the op amp UA741 contained in the library EVAL.LIB, it gave us some parameters of a diode model contained in the op amp "macromodel" (a fancy name for an equivalent subcircuit). You can open EVAL.LIB with a text editor and look at the statements defining each of the library subcircuits. Let's take a look at using the diode models in SPICE.

SPICE Diode Models

Perhaps a simple circuit would be the best way to get acquainted with diodes in SPICE, so we will write a SPICE simulation for the "curve tracer" shown in Figure 5.137 and use it to plot the exponential *v-i* characteristic of the diode D1. The SPICE file is

```
DIODE CURVE TRACER
IS     0    1    1MA
D1     1    0    D1N4148
.LIB    EVAL.LIB
.DC     IS   0    20MA      0.2MA
.PRINT DC    V(1), I(D1)
.PROBE
.END
```

Any element starting with the letter D (as in our D1) is interpreted by the SPICE compiler to be a diode. The "value" we have assigned to it—D1N4148—happens to be one whose model appears in EVAL.LIB. We have "swept" the diode current in steps of 0.2 mA from 0 to 20 mA and have asked for a printout of both the diode voltage and the diode current. In actuality, we would not have had to specifically request the diode current value for it is the same as that of the current source.

Figure 5.137
Diode "curve tracer"

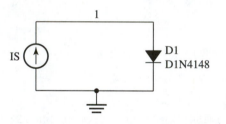

We have also inserted the SPICE verb .PROBE, which causes a graphics data file to be generated for the software routine Probe, a graphics postprocessor that comes packaged with SPICE. The simple .PROBE command we have used dumps all circuit variables into this data file for later retrieval with Probe. We will not discuss Probe here except to say that, when we run it, it looks for the graphics data file we have just generated with our .PROBE command. It is menu driven, and operation is straightforward.

The output of a SPICE run with the preceding file as its input is:

```
****01/13/96 15:18:48***********EVALUATION PSPICE (JULY 1993)****
DIODE CURVE TRACER
****   CIRCUIT DESCRIPTION
****************************************************************
******
IS   0   1   1MA
D1   1   0   D1N4148
.LIB  EVAL.LIB
.DC   IS  0   20MA      0.2MA
.PRINT  DC   V(1)   I(D1)
.PROBE
.END
****01/13/96 15:18:48***********EVALUATION PSPICE (JULY 1993)****
 DIODE CURVE TRACER
****   DIODE MODEL PARAMETERS
****************************************************************
******
```

```
      D1N4148
       IS 100.000000E-15
       BV  100
      IBV 100.000000E-15
       RS 16
       TT 12.000000E-09
      CJO 2.000000E-12
****01/13/96 15:18:48***********EVALUATION PSPICE (JULY 1993)****
DIODE CURVE TRACER
****   DC TRANSFER CURVES        TEMPERATURE= 27.000 DEG C
*****************************************************************
IS      V(1)     I(D1)
 0.000E+00 5.006E-18 3.236E-21
 2.000E-04 5.571E-01 2.000E-04
 4.000E-04 5.783E-01 4.001E-04
 6.000E-04 5.919E-01 6.004E-04
 8.000E-04 6.026E-01 8.000E-04
 1.000E-03 6.115E-01 1.000E-03
 1.200E-03 6.195E-01 1.201E-03
------------------- (DATA POINTS NOT PRINTED TO CONSERVE SPACE)
 1.980E-02 9.896E-01 1.980E-02
 2.000E-02 9.930E-01 1.999E-02
      JOB CONCLUDED
      TOTAL JOB TIME        2.33
```

The graphics output from Probe is shown in Figure 5.138.

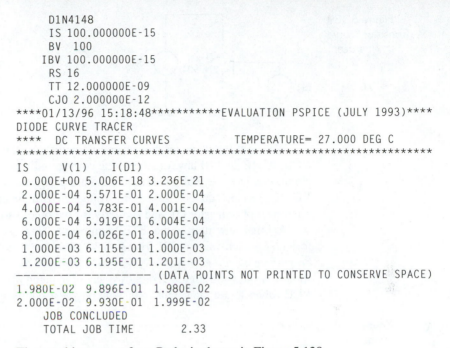

Figure 5.138
Probe output

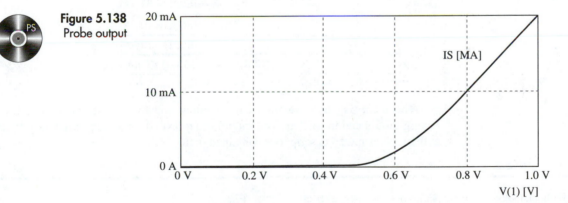

Transistors are handled in SPICE in a similar manner. EVAL.LIB has, for instance, an NPN transistor macromodel referred to by the label Q2N3904. (A name starting with Q always signifies a BJT, either NPN or PNP, depending on the specific model stored in the library.) We show a transistor curve tracer in Figure 5.139 that is similar to the one we discussed previously for the diode. The only difference here is that we must sweep two variables: the base current and the collector to emitter voltage. The SPICE file appears like this:

```
TRANSISTOR CURVE TRACER
IS1     0     1     1MA
VS1     2     0     1
Q1      2     1     0     Q2N3904
.LIB    EVAL.LIB
.DC     VS1   0     10    0.1
.STEP   IS1   0     30UA  10UA
.PRINT DC    IB(Q1)    VCE(Q1)    IC(Q1)
.PROBE
.END
```

Figure 5.139
Transistor "curve
tracer"

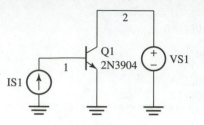

We should perhaps discuss several items in this file. First, the variables VCE, VBE, VCB, IC, IE, and IB always refer to the corresponding transistor variables. However, as a circuit can have several transistors, we must designate which one; thus, we have written VCE(Q1) as the "collector to emitter voltage" (with the collector terminal taking the plus reference) of transistor Q1, and similarly for the other variables.

Another new item is the .STEP command. This line causes a simulation to be performed for each value of IS1 from 0 to 30 microamperes, in steps of 10 microamperes, to be performed *for each value of VS1 called for in the .DC command.* The .PROBE response is shown in Figure 5.140. (We have truncated the right-hand side of the plot—for VCE values larger than eight volts—merely to make the plot a reasonable size.)

Figure 5.140
Probe output

The preceding two examples should allow you to simulate quite complex circuits having diodes and transistors. You should use a text editor to inspect EVAL.LIB and look at the various model descriptions contained there.

Section 5.8 Quiz

Q5.8-1. Without looking at their definitions, identify each of the following as the label of a VCVS, CCVS, VCCS, or CCCS:

a. E123

b. HOFF

c. F32

d. G3

Q5.8-2. Write a SPICE file to simulate the circuit shown in Section 5.6, Figure 5.84, and run it to determine v_o.

Chapter 5 Summary

Chapter 5 has introduced the idea of an *active element:* the various types of *dependent* (or *controlled*) *source,* the *operational amplifier,* and the *bipolar junction transistor (BJT).* Circuits containing such elements are called *active circuits.*

The dependent source is the circuit idealization of a common amplifier: a component that makes the amplitude of a signal larger. The type, of course, depends upon the nature of the signal. A dependent source that accepts a voltage as the *controlling variable,* multiplies it by a constant, and thus produces another voltage—the *controlled variable*—is called a *voltage-controlled voltage source* (or VCVS). The unitless constant of proportionality μ is called the *voltage gain.* If the controlling variable is a current, the constant of proportionality r_m has the unit ohm and is called the *transresistance*. This type of dependent source is called the *current-controlled voltage source* (or CCVS). If the controlled variable is current and the controlling variable is another current, the dependent source is called a *current-controlled current source* (or CCCS). The unitless gain constant β is called the *current gain*. Finally, if the controlled current depends upon a voltage, the constant g_m has the unit siemens and is called the *transconductance*, whereas the dependent source itself is called a *voltage-controlled current source* (or VCCS).

The *operational amplifier*, on the other hand, can be considered as any of the above four types of dependent source in which the constant of proportionality is allowed to become infinite. The operational amplifier (op amp) was shown in this chapter to have a mathematical characterization in which the input terminals are considered simultaneously to be a short circuit and an open circuit. This type of behavior is called the *virtual short principle*. If one of the input terminals is grounded, the other is referred to as a *virtual ground*.

The *nullator* was introduced as a graphical symbol to represent the rather unusual behavior of the virtual short at the input of an op amp. The *norator* was similarly introduced to symbolize the fact that the op amp output does not in any manner constrain either the output voltage or current. The combination of a nullator and a norator, called the *nullor*, was shown to effectively represent the op amp, provided that one terminal of the norator is grounded. This has two desirable consequences: the ground, which always exists (but is hidden by the op amp symbol), is thereby shown explicitly on the circuit diagram; furthermore, the nullator and the norator offer specific mnemonics for remembering the conditions at the op amp terminals. Without the explicit depiction of the ground reference, the op amp seems to violate KCL.

Circuits and subcircuits containing active elements were shown here to exhibit some rather bizarre types of behavior. For instance, a two-terminal circuit can exhibit *negative equivalent resistance,* and entire circuits can either fail to have a solution or can exhibit an infinite number of solutions. Furthermore, a voltage divider can have a voltage gain greater than unity. The meaning of these peculiarities was discussed at some length and related to the practical behavior of actual devices.

The concept of *taping* a dependent source was explained here, and it was shown that this operation (that of temporarily considering a dependent source to be an independent one) turns the analysis of active circuits into that of analyzing circuits having only resistors and independent sources—which, by now, we well understand—with the added step of *untaping* the dependent source(s) and expressing the controlled variable(s) in terms of the unknown(s) in the analysis procedure (e.g., node voltages for nodal analysis and mesh currents for mesh analysis). This means that all the usual analysis techniques are applicable to active circuits: superposition, Thévenin and Norton equivalents, etc.

It was pointed out that the application of superposition to active circuits required that the *partial responses* for the controlling variable(s) must be carried along in addition to those for the variable to be found. It was also shown that the Thévenin and Norton equivalents contain a dependent source (in addition to the usual independent one) if the controlling variable of any dependent source lies outside the subcircuit for which the equivalent is being determined. However, if all the controlling variables are inside this subcircuit, then the dependent source disappears.

The use of the nullor model of the op amp was shown to make the processes of nodal and mesh analysis methodical for circuits containing op amps. The concepts of feedback, loop gain, and stability were introduced and discussed for op amp circuits. Conditions were derived under which op amp circuits are stable, and hence for which the ideal op amp model is suitable.

An entire section was devoted to modeling procedures for nonlinear devices, in particular for the junction diode and the bipolar junction transistor, and *small signal models* were derived for them via a process called *quasilinearization,* or *piecewise linearization.* These models will be called upon in certain future optional sections and problems in the remainder of the text.

Finally, SPICE simulation of circuits with active elements was discussed.

Chapter 5 Review Quiz

RQ5.1. Solve both circuits in Figure RQ5.1 for v (and i_c for the circuit on the bottom), thus showing that the taped dependent current source is equivalent (for the particular value $g_m = 0.25$ S) to the independent one and verifying the taping process.

Figure RQ5.1

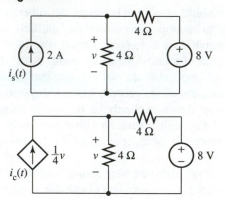

RQ5.2. Find the equivalent resistance R_{eq} shown in Figure RQ5.2 for $\beta = 1, 3,$ and 4.

Figure RQ5.2

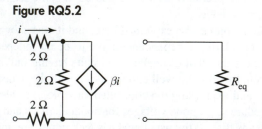

RQ5.3. Find the voltage v in Figure RQ5.3 using nodal analysis. (If you have not covered nodal analysis yet, use source transformations and Thévenin/Norton equivalents.)

Figure RQ5.3

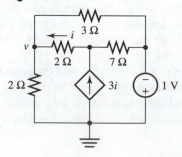

RQ5.4. Find the voltage v in Figure RQ5.4. (Skip this problem if you haven't covered Section 5.4.)

Figure RQ5.4

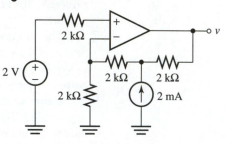

RQ5.5. Find the range of values of R in Figure RQ5.5 for stable operation. (*Hint:* Deactivate the input i-source and find F_+ and F_-. (Skip this problem if you haven't covered Section 5.5.)

Figure RQ5.5

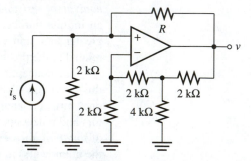

RQ5.6. Find the voltage v in Figure RQ5.6 using nodal analysis. *Note:* This is a fairly complex circuit, requiring four nodal equations (though they are simple ones). To simplify your work, use a convenient set of units for resistance, current, and voltage. (Skip this problem if you haven't yet covered nodal analysis.)

Figure RQ5.6

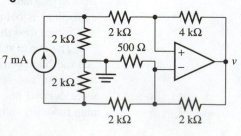

RQ5.7. Find v in Figure RQ5.7 using the method of iteration presented in Example 5.18 of Section 5.7. Assume that I_s for the diode is 10^{-14} A. (Skip this problem if you haven't covered Section 5.7.)

Figure RQ5.7

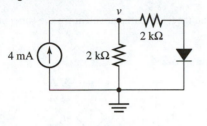

RQ5.8. Answer Question RQ5.6 using SPICE. (Skip this problem if you haven't covered Section 5.8.)

Chapter 5 Problems

Section 5.1 Dependent Sources

5.1-1. Find the current i in Figure P5.1-1 using elementary techniques (KVL, KCL, Ohm's law, etc.).

Figure P5.1-1

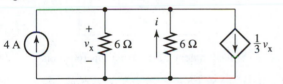

5.1-2. Find the voltage v in Figure P5.1-2 using elementary techniques (KVL, KCL, Ohm's law, etc.).

Figure P5.1-2

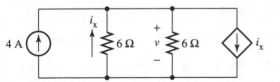

5.1-3. If $v_i(t) = 2 \sin(\omega t)$ V and $g_m = 2$ mS, find $v_o(t)$ for the circuit in Figure P5.1-3.

Figure P5.1-3

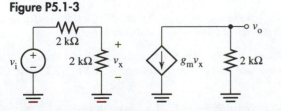

5.1-4. Find the voltage v in Figure P5.1-4 using elementary techniques (KVL, KCL, Ohm's law, etc.).

Figure P5.1-4

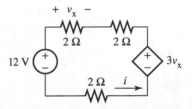

5.1-5. Find the current i in Figure P5.1-5 using elementary techniques (KVL, KCL, Ohm's law, etc.). If the VCVS relationship is replaced by μv_x, which (if any) values of μ cause the solution to not exist?

Figure P5.1-5

5.1-6. Find the current i in Figure P5.1-6 using elementary techniques (KVL, KCL, Ohm's law, etc.). If the CCCS relationship is replaced by βi_x, which (if any) values of β cause the solution to not exist?

Figure P5.1-6

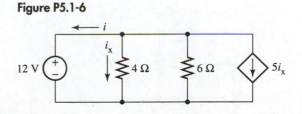

5.1-7. Find the voltage v_y in Figure P5.1-7 using elementary techniques (KVL, KCL, Ohm's law, etc.). If the VCVS relationship is replaced by μv_x, which value(s) of μ cause the circuit to be unsolvable?

Figure P5.1-7

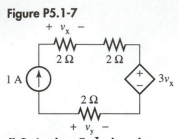

Section 5.2 Active Subcircuits

5.2-1. Find the voltage V in Figure P5.2-1 using equivalent subcircuits.

Figure P5.2-1

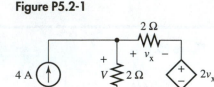

5.2-2. If the VCVS relationship in Figure P5.2-1 is replaced by μv_x, find the value(s) of μ for which the solution for V does not exist.

5.2-3. Find the current i_x in Figure P5.2-2 using equivalent subcircuits.

Figure P5.2-2

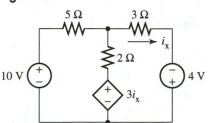

5.2-4. If the CCVS relationship in Figure P5.2-2 is replaced by $r_m i_x$, find the value(s) of r_m for which the solution for i_x does not exist.

5.2-5. Solve Problem 5.2-1 using superposition.

5.2-6. Solve Problem 5.2-3 using superposition.

Figure P5.2-3

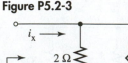

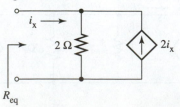

5.2-7. Find the equivalent resistance for the subcircuit shown in Figure P5.2-3.

5.2-8. If the CCCS value in Figure P5.2-3 is replaced by βi_x, find the value(s) of β for which the equivalent resistance is zero and those for which it is infinite.

5.2-9. Find the Thevénin equivalent for the subcircuit shown in Figure P5.2-4.

Figure P5.2-4

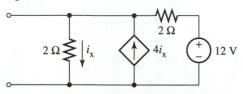

5.2-10. Solve the circuit in Figure P5.2-5 for V using equivalent subcircuits. (*Hint:* Use the consistent units of kΩ, mA, and V.)

Figure P5.2-5

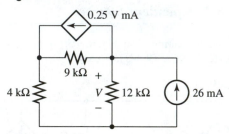

5.2-11. Solve Problem 5.2-10 using superposition.

5.2-12. If the VCCS in Figure P5.2-5 is replaced by one whose value is $g_m V$, find the value(s) of g_m for which a solution does not exist for V.

Section 5.3 Nodal and Mesh Analysis of Circuits with Dependent Sources

5.3-1. Solve Problem 5.2-1 using nodal analysis.

5.3-2. Solve Problem 5.2-3 using nodal analysis.

5.3-3. Solve Problem 5.2-10 using nodal analysis.

5.3-4. Solve Problem 5.2-7 by attaching a test i-source having a general value of i and using nodal analysis.

5.3-5. Solve Problem 5.2-9 by attaching a test i-source having a general value of i and using nodal analysis.

5.3-6. Solve Problem 5.3-4, replacing the test i-source by a test v-source and using simple techniques (KVL, KCL, and Ohm's law).

5.3-7. Solve Problem 5.2-1 using mesh analysis.

5.3-8. Solve Problem 5.2-3 using mesh analysis.

5.3-9. Solve Problem 5.2-10 using mesh analysis.

Section 5.4 The Operational Amplifier: Basic Concepts and Subcircuits

5.4-1. Find the voltage v_o in Figure P5.4-1. All resistors are 10 kΩ.

Figure P5.4-1

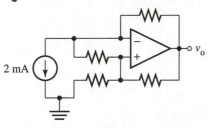

5.4-2. If $v_i(t) = 3 \sin(\omega t)$ V, find $v_o(t)$ in Figure P5.4-2.

Figure P5.4-2

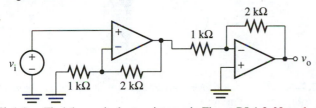

5.4-3. Find the equivalent resistance in Figure P5.4-3. Note that the two subcircuit terminals are the one shown and the ground reference.

Figure P5.4-3

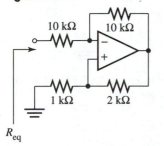

5.4-4. Find the output voltage v_o of the op amp circuit in Figure P5.4-4 in terms of the two input voltages v_1 and v_2.

Figure P5.4-4

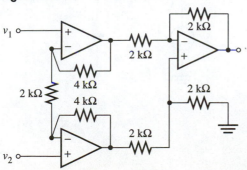

5.4-5. Derive the equivalent resistance "seen" between the terminal marked v_i and ground in Figure P5.4-5.

Figure P5.4-5

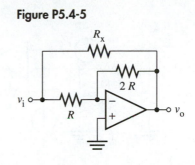

Section 5.5 Practical Aspects of Op Amp Circuits: Feedback and Stability

5.5-1. Deactivate the 2-mA source in the circuit of Figure P5.4-1 and determine whether or not the output voltage v_o is stable.

5.5-2. Deactivate the voltage source v_i in the circuit of Figure P5.4-2 and determine whether or not the output voltage v_o is stable. (*Hint:* If the subcircuit on the left is stable, then its output voltage is stable and can therefore be replaced by a deactivated v-source in determining stability of the one on the right.)

5.5-3. Investigate the short-circuit and open-circuit stability of the circuit in Figure P5.4-3 relative to the output voltage of the op amp.

5.5-4. Investigate the short-circuit and open-circuit stability of the circuit in Figure P5.4-5 relative to the output voltage of the op amp.

Section 5.6 Nodal and Mesh Analysis of Circuits with Op Amps

5.6-1. Solve the circuit in Figure P5.6-1 using nodal analysis and find v_o.

Figure P5.6-1

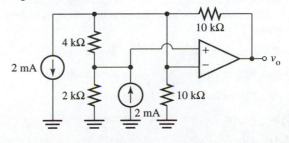

5.6-2. Solve the circuit in Figure P5.6-2 using nodal analysis and find v_o.

Figure P5.6-2

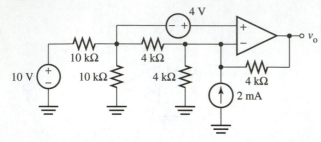

5.6-3. Solve the circuit in Figure P5.6-3 using nodal analysis and find v_o.

Figure P5.6-3

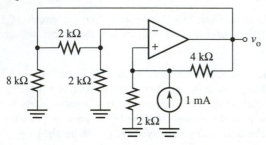

5.6-4. Solve the circuit in Figure P5.6-4 for v_o in terms of v_i. Use $k = 1 + R_2/R_1$.

Figure P5.6-4

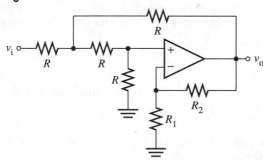

5.6-5. In the preceding problem, your solution undoubtedly assumed that the circuit was stable. Find the range of values of k for which this assumption is valid. Assume that the input is driven by a v-source.

5.6-6. Repeat Problem 5.6-1 using mesh analysis.

5.6-7. Repeat Problem 5.6-2 using mesh analysis.

5.6-8. Repeat Problem 5.6-3 using mesh analysis.

5.6-9. Repeat Problem 5.6-4 using mesh analysis.

5.6-10. Repeat Problem 5.6-5 assuming that the input is driven by an i-source.

Section 5.7 Diodes, Transistors, and Device Models

5.7-1. Use the method of iteration to find the diode current I_D in Figure P5.7-1. Assume that $I_s = 10^{-14}$ A.

Figure P5.7-1

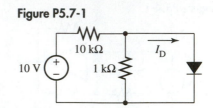

5.7-2. Assuming that diode D_1 has $I_s = 10^{-14}$ A and that diode D_2 has $I_s = 4 \times 10^{-14}$ A, find the voltage V in Figure P5.7-2.

Figure P5.7-2

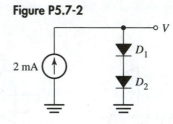

5.7-3. The diodes in Figure P5.7-3 are the same as those in Problem 5.7-2. Find V.

Figure P5.7-3

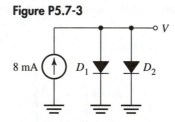

5.7-4. The diode in Figure P5.7-4 has $I_s = 10^{-14}$ A. Find $v(t)$ under the following conditions:

    a. $V_S = -1$ V
    b. $V_s = 1$ V
    c. $V_s = 10$ V

Assume that $v_s(t) = 0.1 \sin(\omega t)$ V.

Figure P5.7-4

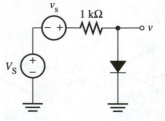

5.7-5. For the small signal circuit in Figure P5.7-5, find the voltage gain $A_v = v_c/v_s$. Assume that $r_e = 50\ \Omega$ and that $\beta = 99$.

Figure P5.7-5

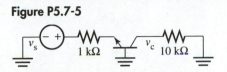

5.7-6. Assuming that the BJT has $\beta = 49$ and $r_e = 100\ \Omega$ in Figure P5.7-6, find the small signal voltage gain $A_v = v_o/v_i$ and the small signal input resistance R_{in}.

Figure P5.7-6

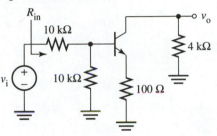

5.7-7. Assume that the BJT has a transconductance of $g_m = 40\ mS$ in Figure P5.7-7 and that, otherwise, it is ideal $(\beta = \infty)$. Find the small signal transresistance $R_m = v_o/i_i$ and the input resistance R_{in}.

Figure P5.7-7

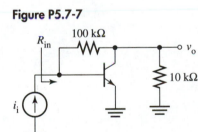

5.7-8. If the transconductance is the same for both transistors in Figure P5.7-8 with $g_m = 40\ mS$, find the small signal voltage gain $A_v = v_o/v_i$ and the input resistance R_{in}. Assume that $\beta = 99$ for each BJT.

Figure P5.7-8

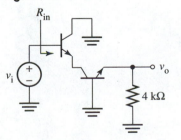

Section 5.8 SPICE: Active Circuits

5.8-1. Write and run a SPICE program to solve Problem 5.1-1.

5.8-2. Write and run a SPICE program to solve Problem 5.1-5.

5.8-3. Write and run a SPICE program to solve Problem 5.2-1.

5.8-4. Write and run a SPICE program to solve Problem 5.2-3.

5.8-5. Write and run a SPICE program to solve Problem 5.2-9.

5.8-6. Write and run a SPICE program to solve Problem 5.6-1. Use an ideal op amp subcircuit.

5.8-7. Write and run a SPICE program to solve Problem 5.6-3. Use the op amp model UA741 found in EVAL.LIB as a subcircuit. Assume power supply voltage of ± 10 V.

5.8-8. Write and run a SPICE program to solve Problem 5.7-1. For the diode, use the PSPICE model D1N4148 located in EVAL.LIB.

5.8-9. Write and run a SPICE program to find the voltage v_C as a function of the voltage v_i in Figure P5.8-1. Use the PSPICE model for an NPN transistor labeled Q2N3904 to be found in EVAL.LIB.

Figure P5.8-1

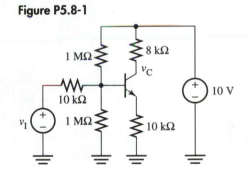

Part II

TIME DOMAIN ANALYSIS

6

Energy Storage Elements

Until now we have only discussed resistive circuits; that is, in addition to independent and dependent sources, the only other elements have been resistors. Furthermore, we have concentrated on dc independent sources: sources that produce values that are constant with time. In this chapter we begin to explore circuits in which the independent sources have waveforms that vary with time and elements that modify both their amplitude and shape. Section 6.1 shows that the resistive part of a circuit only modifies waveforms from its independent sources such that all voltages and currents are linear combinations of these waveforms. The capacitor and the inductor, treated in Sections 6.2 and 6.3, are quite different: they have memory. This is reflected in the fact that circuits having inductors or capacitors modify the independent source waveforms. In Section 6.4 we place circuits containing such energy storage elements into the earlier dc context we have studied so thoroughly with a discussion of the dc steady state, into which all stable circuits settle as time becomes large. In Section 6.5, we then develop important models for the capacitor and inductor that include initial conditions. These are important for solving circuits containing switches. In Section 6.6, we discuss impulse functions. The impulse function is a generalization of an ordinary waveform that allows us to find the response of circuits (such as the important switched capacitor network) in which instantaneous charge transfer takes place. Furthermore, this concept will allow us to develop in succeeding chapters an analysis procedure that is purely algebraic. Section 6.7 deals with circuits containing capacitors or inductors in addition to independent and dependent sources and resistors. Finally, Section 6.8 discusses the selection of compatible systems of units to make computations easier for circuits with practical element values. In this chapter (and all following) op amp and transistor examples are flagged with an O or a T, respectively.

The prerequisite for reading Sections 6.1 through 6.6 is the material in Chapters 1 through 4, with the exception of the sections on SPICE. In order to read Section 6.7, one also needs as a minimum the dependent source material in Sections 5.1 through 5.3 of Chapter 5; however, some of the examples require the op amp material in Sections 5.4 and 5.6 and some require the material on BJTs in Section 5.7.

6.1 | Waveform Properties of Resistive Circuits

In Part I, we rather thoroughly investigated the analysis of resistive circuits: those whose only passive elements were resistors. That is, the resistor was the only type of element in addition to dependent and independent voltage and current sources.[1] Furthermore, we concentrated primarily on circuits in which the independent sources were dc (constant in value). In the next two sections we will introduce two additional passive elements, the capacitor and the inductor, that behave quite differently from the resistor. To gain an understanding of this difference, it is essential that we investigate the way resistive circuits behave when the independent sources have values that vary with time—that is, that are *waveforms*. Let's start with a simple example as a refresher.

Example 6.1   Find the voltage $v(t)$ in Figure 6.1.

Figure 6.1
An example circuit

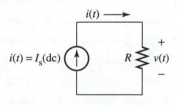

Solution   To derive the voltage, we simply apply Ohm's law, thus obtaining

$$v(t) = Ri(t) = RI_s. \tag{6.1-1}$$

The "waveforms" for $i(t)$ and $v(t)$ are shown in Figure 6.2. We have placed quote marks

---

[1] The operational amplifier is a VCVS, though, admittedly, one with special characteristics.

around the term "waveforms" because they are rather dull. Dc waveforms do not fluctuate with time, but remain constant.

Figure 6.2
The "waveforms"

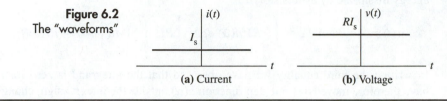

(a) Current   (b) Voltage

The type of behavior in the last example is typical of the resistive circuits that we discussed at length in Part I. But the concept of a dc circuit is actually quite idealized. Here is one reason. Suppose we ask for the energy dissipated in the resistor of the preceding example. It is given by

$$w(t) = \int_{-\infty}^{t} P(\alpha)\,d\alpha = \int_{-\infty}^{t} RI_S^2\,d\alpha = RI_s^2 \int_{-\infty}^{t} 1\,d\alpha = \infty! \qquad (6.1\text{-}2)$$

Clearly, this is a nonphysical result. Thus, it is much more realistic to insist that the circuit be "turned on" at some point in time—which, we will almost always agree, occurs at the time origin, $t = 0$—and it is for this reason that we become concerned with waveforms that vary with time.

Example 6.2   Find the voltage $v(t)$ in Figure 6.3.

Figure 6.3
An example circuit

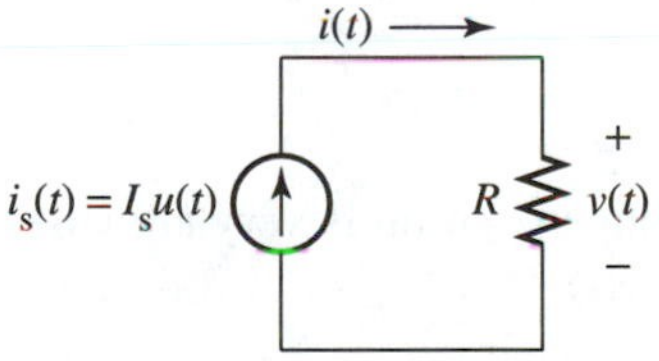

$i(t) \longrightarrow$

$i_s(t) = I_s u(t)$ R $v(t)$

Solution   First, we recall that the step function has the waveform shown in Figure 6.4(a). The meaning of the solid dot and open circle are motivated by the "practical" laboratory realization of the same waveform, shown in Figure 6.4(b), which requires a finite amount of time to change from zero to the final value. Thus, the solid dot in Figure 6.4(a) means that the associated value is attained, whereas the open circle means that it is not attained. To derive the voltage, we simply apply Ohm's law, thus obtaining

$$v(t) = Ri(t) = Ri_s(t) = RI_s u(t) = V_s u(t). \qquad (6.1\text{-}3)$$

The waveform for $v(t)$ is shown in Figure 6.5.

Figure 6.4
The step function waveform

Figure 6.5
The voltage response waveform

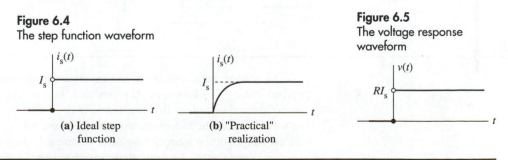

(a) Ideal step function   (b) "Practical" realization

Now Example 6.2 perhaps does not seem terribly exciting, but it does emphasize the fact that practical waveforms must turn on and off. We can now, for example, compute the energy dissipated by the resistor. It is

$$w(t) = \int_{-\infty}^{t} P(\alpha) \, d\alpha = \int_{-\infty}^{t} RI_s^2 u(\alpha) \, d\alpha = \left[RI_s^2 \int_0^t 1 \, d\alpha \right] u(t) = RI_s^2 t u(t). \qquad (6.1\text{-}4)$$

In writing the third equality we have observed that the integrand is zero for $t \leq 0$ and have therefore moved the unit step function $u(\alpha)$ outside the integral sign, changing its argument to t in the process. This is a practical use of the unit step function: "turning off" waveforms for $t \leq 0$. You should study this last equation carefully because it contains a technique we will be using relatively often in Part II.

Although Examples 6.1 and 6.2 involve only a very simple analysis procedure, they tell us something important about resistive circuits: they do not modify the waveforms of the independent sources to any great extent. The voltage waveforms were exactly the same in shape as that of the independent current source. We will now tackle a somewhat more complicated example.

Example 6.3   Find the voltage waveform $v(t)$ in the circuit shown in Figure 6.6 if the waveform of the v-source is $v_s(t) = 9u(t)$ V and that of the i-source is $i_s(t) = 4u(t - 1)$ A.

Figure 6.6
An example circuit

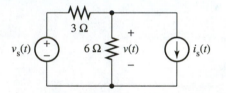

Solution   We will leave it to you (superposition is very attractive here) to analyze the circuit to obtain the response voltage $v(t)$. It is

$$v(t) = \frac{2}{3} v_s(t) - 2i_s(t) = 6u(t) - 8u(t - 1). \qquad (6.1\text{-}5)$$

The waveform for $v(t)$ is plotted in Figure 6.7. Again, we caution you to take the time to understand how we derived this sketch for we will rely upon such an ability in the future.

Figure 6.7
The voltage response
waveform

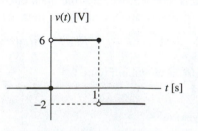

The voltage waveform in our last example was not exactly like either of the two independent source waveforms, but it resembled both to a remarkable extent. In fact, any response voltage or current in a resistive circuit is a linear combination (weighted sum) of the waveforms of the independent sources. To see this in general, look at Figure 6.8. We assume that all of the sources have been "extracted," thus leaving only resistors inside the box. For convenience of representation we have shown only one source of each type, but we will allow any number of sources. After we perform the first step of taping the dependent sources, we see that, by superposition, we can write any voltage or current response as

$$y(t) = a_1x_1(t) + a_2x_2(t) + \cdots + a_Nx_N(t), \tag{6.1-6}$$

where the $x_k(t)$'s are source waveforms and the a_k's are *constants*. Therefore, each voltage or current in a resistive circuit is a linear combination of the source values.

Figure 6.8
A general resistive
circuit

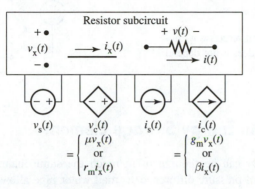

The right-hand side of (6.1-6) contains the dependent source values for they have temporarily been turned into independent sources (of unknown value) by the taping process. When we now untape these sources, however, we replace the associated $x(t)$ variable on the right side of (6.1-6) with the controlling relationship and express it in terms of the independent variables associated with the independent sources. This eliminates the dependent source waveforms from the right-hand side. Assuming, then, that a solution exists, we can interpret the right-hand side of (6.1-6) as consisting of a linear combination of the waveforms of the *independent* sources only. In fact, each "partial response,"

$$y_k(t) = a_kx_k(t), \tag{6.1-7}$$

is a linear function of the associated source value.[2] For this reason, we sometimes refer to (6.1-6) as a *multilinear* relationship.

The "bottom line" of our development here is simply this: all voltages and currents in a resistive circuit are linear combinations of the *independent* source values in which the coefficients of combination are real constants, that is, multilinear relationships. In Section 6.2 we will begin to study circuits having dynamic elements: those that alter waveforms in a more intricate manner.

Section 6.1 Quiz

Q6.1-1. Let a circuit response be given by $y(t) = 4 + 2x(t)$. Is this a linear relationship? Why, or why not? Is it a multilinear relationship? (Assume that $x(t)$ is the only independent source waveform in the circuit.)

Q6.1-2. Find $i(t)$ in terms of $v_s(t)$ for the circuit shown in Figure Q6.1-1. Is it a linear function of $v_s(t)$? Is it a multilinear relationship?

Figure Q6.1-1

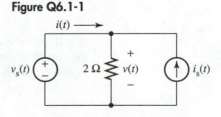

---

[2] Remember that a linear function $f(x)$ is one for which $f(ax) = af(x)$ and $f(x + y) = f(x) + f(y)$ for any constant a and any values—or waveforms—x and y (or $x(t)$ and $y(t)$).

Q6.1-3. For the circuit shown in Figure Q6.1-2:

a. Tape the dependent source and derive $v(t)$ in the form

$$v(t) = a_1 v_s(t) + a_2 i_c(t) + a_3 i_s(t).$$

b. Untape the dependent source and solve for $v(t)$ in the form

$$v(t) = b_1 v_s(t) + b_3 i_s(t).$$

For what value(s) of β does the circuit response $v_{(t)}$ not depend on v_s? For what values of β does it not depend on i_s?

c. Letting $\beta = 1$, $v_s(t) = 9u(t)$ V, and $i_s(t) = -9u(t)$ A, find $v(t)$.

Figure Q6.1-2

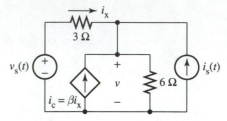

6.2 | The Capacitor: An Energy Storage Element

A resistor can be thought of in much the same manner as a water pipe. For a given amount of pressure difference, a small water pipe allows only a small flow rate of water, and a large one allows a larger flow rate. Likewise, for a given voltage difference, a resistor with a large value of resistance (small pipe) allows only a small current to flow, and one with a small value of resistance (large pipe) permits a large current. The circuit element we will introduce here, the capacitor, is very much like a water storage tank. It stores charge.

The Capacitor:
A Dynamic Element
with Memory

The symbol for a capacitor is shown in Figure 6.9(a). It is meant to convey the idea of a pair of parallel metal plates—the simplest type of capacitor construction. The "plates" are often a thin metal foil, separated by a thin and pliable dielectric insulator, rolled into the form of a tube. In times gone past, the curved plate meant that the capacitor was polarized; that is, the curved plate had to always be at a negative voltage relative to the straight one. Two straight lines meant that the capacitor was not polarized. This convention is no longer widely used, and the one shown in the figure represents either type. An older term for the capacitor is "condenser," but this is no longer used.

Figure 6.9
The capacitor

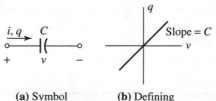

(a) Symbol

(b) Defining
characteristic

If we were to connect a voltage source, say, and measure the total charge flowing into the positive terminal of the capacitor, the result would be the graph shown in Figure 6.9(b). It is a linear relationship,

$$q = Cv. \tag{6.2-1}$$

If v is a time-varying waveform, then $q(t)$ is merely a scaled version of that waveform.

This equation serves to define the *capacitance* of the capacitor, C. The capacitance is the slope of the q-v characteristic. The unit of capacitance is the *farad*, named after the

British electrical experimenter Michael Faraday. Thus, it is abbreviated F. In terms of fundamental units, we have

$$F \text{ (faraday)} = \frac{C}{V} \left(\frac{\text{coulomb}}{\text{volt}} \right). \tag{6.2-2}$$

Our use of the uppercase C for capacitance is somewhat unfortunate, for this symbol is also the abbreviation for the unit of charge, the coulomb, as in equation (6.2-2); however, it is conventional to use this symbol and we will do so.

The circuit analysis techniques we developed in Part I relied upon the description of each element in terms of its v-i characteristic, not its q-v relationship. Therefore, we will convert the latter into the former. There are two ways we can do this. First, we note that current is the time derivative of charge, so if we differentiate both sides of equation (6.2-1) and assume that C is constant, we obtain

$$i = C \frac{dv}{dt}. \tag{6.2-3}$$

Alternatively, we can recognize that charge is the integral of current and divide both sides of (6.2-1) by the capacitance to obtain

$$v(t) = \frac{1}{C} \int_{-\infty}^{t} i(\alpha) \, d\alpha. \tag{6.2-4}$$

Both of these equations present certain difficulties that must be resolved before we can use them with confidence. We can only use (6.2-3) if the voltage is a differentiable waveform. Similarly, we can only use (6.2-4) if the improper integral exists, and this places certain limitations on the behavior of $i(t)$ as $t \rightarrow -\infty$. For now, we will simply assume that all voltages are differentiable and deal with those that are not in a later section of this chapter. Thus, we will apply (6.2-3) without further comment here. To make (6.2-4) work without difficulty, we will adopt the convention we mentioned in Section 6.1: that all waveforms "turn on" at $t = 0$. As we pointed out in that section, there are compelling reasons for doing this, anyway, so that our analysis techniques correspond to reality more closely. We will call such time functions, those that are zero for $t \leq 0$, *one sided*. This means that in our computations we can replace the lower limit in (6.2-4) with 0; however, we will continue to write it as $-\infty$ to stress that we are including all of the current waveform in the integration process.

Example 6.4   The current source in Figure 6.10(a) has the waveform shown in Figure 6.10(b). Find and sketch the waveform for the capacitor voltage $v(t)$.

Figure 6.10
An example circuit

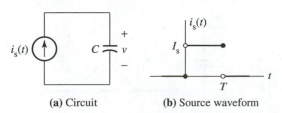

(a) Circuit   (b) Source waveform

Solution   Let's first discuss the current source waveform. It is called a "pulse," and can be written as

$$i_s(t) = I_s P_T(t), \tag{6.2-5}$$

where $\qquad\qquad\qquad\qquad P_T(t) = u(t) - u(t - T).$ $\qquad\qquad$ (6.2-6)

This can be verified by merely noting that $u(t)$ is zero for $t \le 0$ and that $u(t - T)$ is zero for $t \le T$; thus, the difference is unity for $0 < t \le T$ and zero everywhere else. We refer to $I_sP_T(t)$ as a pulse of "height" (or amplitude) I_s and "duration" T.

We now use the integral relationship in equation (6.2-4), repeated here for convenience, as equation (6.2-7):

Figure 6.11   Computing the running integral

$$v(t) = \frac{1}{C}\int_{-\infty}^{t} i(\alpha)\, d\alpha.$$ $\qquad$ (6.2-7)

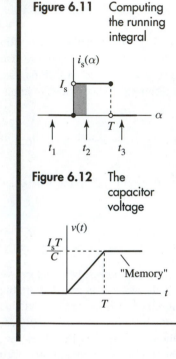

Figure 6.12   The capacitor voltage

We recall that the integral is the area under the graph of the integrand between the limits on the integral. Because we are regarding t as a variable parameter, we refer to our integral as a *running integral*. We show the computation of the area in Figure 6.11, where we have changed the argument of the current pulse function to α and have shown three values of t: one negative, one between 0 and T, and one larger than T. For the first, the area under the curve from $-\infty$ to t_1 is clearly zero; for the second, the area from $-\infty$ to t_2 is clearly I_st_2; and for the third (this is very important to understand), the area from $-\infty$ to t_3 is *constant* at I_sT—*independent of t_3!* Thus, letting t be a variable that "runs" from $-\infty$ to $+\infty$, we obtain the voltage waveform in Figure 6.12. (We have scaled the running integral just discussed by division by the capacitance as required by equation (6.2-7)).

Notice that we have labeled the flat portion of the waveform for $t > T$ as "memory," for this is the basic principle used in solid-state digital memory chips. $I_s = 0$ represents a logic zero, and a nonzero value of I_s represents a logic one. An electronic circuit inside the chip tests the voltage at some later time. If this voltage is nonzero, it decides that the stored bit is a one; otherwise, the stored bit is a zero.

The basic principles involved in the last example are ones you should master quite thoroughly. Not only are they fundamental to the analysis procedures we will develop in later chapters, they also are working tools of the electronic designer. The ability to work with and visualize waveforms is essential. It is tempting for the beginner to always try to write an analytical expression for the current waveform, then integrate it by using the Fundamental Theorem of Calculus to obtain an antiderivative and evaluate that function between the two limits on the integral. This is often more difficult than our graphical method, particularly when the integrand is piecewise constant, as in our preceding example. You should resist this temptation until you have determined that graphical integration is simply too difficult without analytical help.

Example 6.5   Find the current waveform for the circuit in Figure 6.13.

Figure 6.13
An example circuit

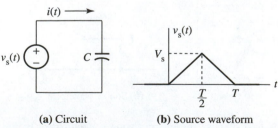

(a) Circuit $\qquad\qquad$ (b) Source waveform

Solution   Here, the capacitor voltage waveform is specified and we must determine the current waveform; thus, equation (6.2-3) for the current in terms of the derivative of the voltage is appropriate. Here, we recognize that the derivative is merely the slope of the waveform at each point, and our particular waveform consists of four regions of constant slope: $t \leq 0$, for which the slope is zero, $0 < t \leq T/2$, for which the slope is $2V_s/T$; $T/2 < t \leq T$, for which the slope is $-2V_s/T$; and $t > T$, for which the slope is zero once more. We multiply by the capacitance to obtain the waveform in Figure 6.14. We remind you that the open circles merely mean that the associated values are—in practice—not attained instantaneously, but require a small amount of time (which we are assuming to be negligible) to reach that value.

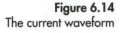

Figure 6.14
The current waveform

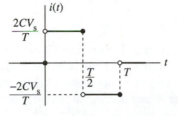

Once again, we emphasize that we did not solve this example by writing a mathematical expression for the voltage waveform; we simply used the fact that it consisted of several segments of constant slope. In some situations, of course, one *should* resort to analytical methods, but they should be avoided unless they are clearly easier to work with or until they indeed become necessary. There is another point we would like to make. We assumed a current source in Example 6.4 and a voltage source in Example 6.5. We did this to emphasize that the current was known in the first example and the voltage was known in the second. These choices, however, were not essential. What *is* essential is the recognition that the current is proportional to the time rate of change of the voltage and that the voltage is proportional to the integral of the current (that is, to the charge).

Operator Notation for Derivative and Integral

The symbols for derivative and, to an even greater extent, integral are cumbersome to write and work with. For this reason we will use the following notation:[3]

$$p = \frac{d}{dt} \tag{6.2-8}$$

and

$$\frac{1}{p} = \int_{-\infty}^{t} (\ ) \, d\alpha. \tag{6.2-9}$$

These symbols obviously have no meaning if they appear by themselves; rather, when applied *from the left side* to a time function $f(t)$, they mean

$$pf(t) = \frac{d}{dt} f(t) = f'(t) \tag{6.2-10}$$

and

$$\frac{1}{p} f(t) = \int_{-\infty}^{t} f(\alpha) \, d\alpha. \tag{6.2-11}$$

Our capacitor relationships of (6.2-3) and (6.2-4), in terms of this *operator notation*, become

$$i(t) = Cpv(t) \tag{6.2-12}$$

and

$$v(t) = \frac{1}{Cp} i(t), \tag{6.2-13}$$

---

[3] This notation was originally used by Oliver Heaviside, a British pioneer in circuit analysis.

respectively. *Cp*, of course, means that the derivative is taken and then the result multiplied by *C*; similarly, 1/*Cp* means that the integral is taken and the result then multiplied by 1/*C*. The symbols *p* and 1/*p* are referred to as *differential operators* or *Heaviside operators*. Notice how compact this notation is relative to the more cumbersome conventional signs of differentiation and integration.

An Example of the Use
of Operators

Example 6.6   Find *p* [sin (2*t*)*u*(*t*)] and 1/*p* [sin (2*t*)*u*(*t*)].

Solution   The waveform to be operated on is shown in Figure 6.15. The *u*(*t*) multiplier "turns off" the sinusoid for $t \leq 0$. We note that the derivative does not exist at $t = 0$ because the slope changes abruptly. This is not a severe problem, though, as we will see. For $t < 0$, we see that the derivative is zero because the slope is zero. For $t > 0$, we have

$$p[\sin(2t)] = 2\cos(2t). \tag{6.2-14}$$

Putting these two results together, we have

$$p[\sin(2t)u(t)] = \begin{cases} 2\cos(2t); & t > 0 \\ 0; & t \leq 0 \end{cases} = 2\cos(2t)u(t). \tag{6.2-15}$$

This waveform is plotted in Figure 6.16. Notice that we have *defined* the result to be zero for $t = 0$ even though the mathematical derivative is undefined there. Why? Remember that we have stressed several times that waveforms in the laboratory do not change instantaneously from one level to another. Thus, we have the left continuous waveform shown in the figure.

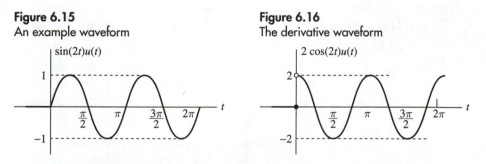

Figure 6.15
An example waveform

Figure 6.16
The derivative waveform

The integral computation can be done as follows. We observe that the integral from $-\infty$ up to any negative time is zero. Then, for positive values of *t*, the integral from $-\infty$ to *t* can be written as the sum of two integrals: the first from $-\infty$ to 0 and the second from 0 to *t*. Because the first integral is zero, we have

$$\frac{1}{p}[\sin(2t)u(t)] = \int_{-\infty}^{0} 0 \, d\alpha + \int_{0}^{t} \sin(2\alpha) \, d\alpha \tag{6.2-16}$$

$$= [-0.5\cos(2\alpha)]_0^t = -0.5\cos(2t) + 0.5.$$

Putting these results together, we have

$$\frac{1}{p}[\sin(2t)u(t)] = \left[\int_{0}^{t} \sin(2\alpha) \, d\alpha\right] u(t) = [-0.5\cos(2t) + 0.5] \, u(t). \tag{6.2-17}$$

Notice our use of the unit step function multiplier outside the integral to "turn off" the result for $t \leq 0$. (After all, the formula $-0.5 \cos(2t) + 0.5$ yields nonzero values for $t \leq 0$!) The waveform is shown in Figure 6.17. Notice that even though both the derivative and the integral waveforms involve the cosine function, there is an additive constant in the latter that makes the waveforms look quite different.

Figure 6.17
The integral waveform

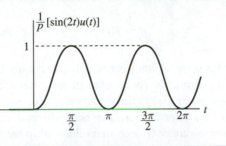

An Operator Model for the Capacitor

Differential operators, being simply alternative notation for the derivative and integral, have all of the properties of these operations. Thus, they are *linear* operators:

$$p[af(t) + bg(t)] = apf(t) + bpg(t) \qquad (6.2\text{-}18)$$

and

$$\frac{1}{p}[af(t) + bg(t)] = a\frac{1}{p}f(t) + b\frac{1}{p}g(t), \qquad (6.2\text{-}19)$$

where a and b are arbitrary constants. There is a great convenience symbolically in using differential operators, as we can see by referring once more to equations (6.2-12) and (6.2-13) for the capacitor. We can write (6.2-13) in the compact form

$$v(t) = Z(p)i(t), \qquad (6.2\text{-}20)$$

where

$$Z(p) = \frac{1}{Cp} \qquad (6.2\text{-}21)$$

is called the *impedance operator* or *operator impedance* of the capacitor. Further, (6.2-12) can be written as

$$i(t) = Y(p)v(t), \qquad (6.2\text{-}22)$$

where

$$Y(p) = Cp = \frac{1}{Z(p)} \qquad (6.2\text{-}23)$$

Figure 6.18 The capacitor impedance operator

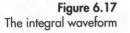

is called the *admittance operator* or *operator admittance* of the capacitor. Notice that equations (6.2-20) and (6.2-22) are generalizations of Ohm's law: $v = Ri$ and $i = Gv$, respectively. Thus, the impedance operator $Z(p)$ is a generalization of resistance, and the admittance operator $Y(p)$ is a generalization of conductance. *Just as we decided by convention to label resistors with their resistance, we will label capacitors with their impedance operator, as in Figure 6.18.*

Differentiation and Integration: Inverse Operations

We know that conductance and resistance are inverses of one another. Is the same true for operator impedances? Let's investigate this question by asking whether the differential operators p and $1/p$ are inverses of each other. That is, is it true that

$$p\frac{1}{p} = \frac{1}{p}p = 1? \qquad (6.2\text{-}24)$$

In other words, is it true that for any function $f(t)$ we have

$$p \frac{1}{p} f(t) = \frac{1}{p} pf(t) = f(t)? \tag{6.2-25}$$

If we investigate the leftmost expression first, we require that

$$p \frac{1}{p} f(t) = \frac{d}{dt} \int_{-\infty}^{t} f(\alpha) \, d\alpha = f(t). \tag{6.2-26}$$

In your calculus courses you learned that this is only true at time points for which the integrand $f(t)$ is continuous. As a matter of fact, this relationship can be extended so that it holds at points where $f(t)$ is discontinuous (with a jump discontinuity) also. We will investigate this in a later section. For now, we will simply assume that it is true for all waveforms we encounter. The second relationship we require to be valid is

$$\frac{1}{p} pf(t) = \int_{-\infty}^{t} f'(\alpha) \, d\alpha = f(t) - f(-\infty) = f(t). \tag{6.2-27}$$

This means that we require

$$f(-\infty) = \lim_{t \to -\infty} f(t) = 0. \tag{6.2-28}$$

In fact, we will go a bit further. We have already noted in several places that two-sided waveforms (those that are not identically zero for all sufficiently large and negative values of time) are not as realistic as one-sided ones: those that are identically zero for $t \leq 0$. For emphasis and ease of reference, let's phrase this as a formal definition.

Definition 6.1   **One-Sided Waveforms**   A waveform $f(t)$ is said to be *one sided* if it is identically zero for nonpositive values of time; that is, if $f(t) = 0$ for all $t \leq 0$.

We will assume (unless otherwise stated) that all waveforms are one sided. Furthermore, we will suppose that differentiation and integration are *causal* operations; that is, their responses must be zero for $t \leq 0$ if this property holds for their input waveforms. Again, we will state this formally as a definition.

Definition 6.2   **Causality**   Suppose that a circuit, a system, or a mathematical operation has the property of always producing a one-sided waveform response when the input (stimulus or argument) is one sided. Then we will say that the circuit, system, or operator is *causal*.

In plain language, this means that a response is not produced until a stimulus is introduced. *We will assume causality for all operators with which we will work.* Therefore, we can assert with confidence that (6.2-25) holds for all waveforms to be encountered.

Energy Storage and Losslessness

Now let's investigate the energy absorption properties of the capacitor. We recall that the power absorbed by a capacitor is the product of its terminal voltage and current, assuming the passive sign convention, and that the power is the rate of change of the energy absorbed. Thus, we have

$$P(t) = v(t)i(t) = v(t) \, Cpv(t) = p \left[\frac{1}{2} Cv^2(t) \right] = pw(t). \tag{6.2-29}$$

Let's look at the last two terms and apply the $1/p$ operator from the left (equivalent to integrating both sides). Then, assuming that $w(t)$ is zero for $t \leq 0$ along with the voltage and current, we have

$$w(t) = \frac{1}{2} Cv^2(t). \tag{6.2-30}$$

If you wish to use the conventional symbols to further strengthen your understanding of operators, follow the following computation:

$$w(t) = \int_{-\infty}^{t} P(\alpha)\, d\alpha = \int_{-\infty}^{t} \frac{d}{d\alpha}\left[\frac{1}{2}Cv^2(\alpha)\right] d\alpha = \left[\frac{1}{2}Cv^2(\alpha)\right]_{-\infty}^{t} = \frac{1}{2}Cv^2(t). \tag{6.2-31}$$

We have simply repeated our argument that integration and differentiation are inverses of each other—so one "cancels" the other in either order.

The interesting fact shown by (6.2-30) is this: if the voltage is zero, the total energy absorbed by the capacitor is zero. The next example illustrates this concretely.

Example 6.7   Compute and sketch the waveforms for the power and energy absorbed by the capacitor in Example 6.5.

Solution   Figure 6.19 shows the circuit of Example 6.5, as well as the voltage source and response current waveforms for ease of reference. We merely multiply the voltage and current point-by-point to obtain the absorbed power waveform in Figure 6.20(a). Its running integral gives the absorbed energy waveform shown in Figure 6.20(b). By the way, we did not have to use an analytical expression and evaluate the integral using a formula; rather, we noted that a linear function integrates to a quadratic (second order), then evaluated the area under the $P(t)$ curve at 0, $T/2$ (using the formula for the area of a triangle), and T. This gives us the waveshape in Figure 6.20(b), except for one piece of information: whether it opens upward (as shown) or downward. We see from Figure 6.20(a), however, that the slope of the $w(t)$ curve becomes larger as t gets closer to $T/2$ from below, so $w(t)$ must open upward rather than downward.

Figure 6.19
The circuit and the voltage and current waveforms for Example 6.5

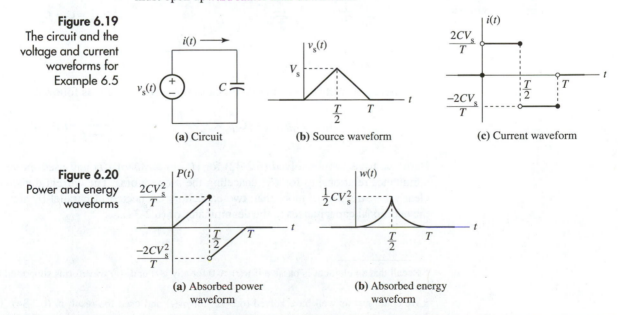

(a) Circuit   (b) Source waveform   (c) Current waveform

Figure 6.20
Power and energy waveforms

(a) Absorbed power waveform   (b) Absorbed energy waveform

The energy absorbed by a capacitor as given by (6.2-30) is nonnegative, so we see that this element is passive.[4] The result in Figure 6.20(b) shows that it has an additional interesting property. If the voltage returns to zero as t approaches infinity, the total energy absorbed is zero. We call this property *losslessness*.

If you look back at Example 6.4, Figure 6.12, you will see that even though the current waveform is zero for large values of time, the voltage waveform (and hence the absorbed energy waveform) is not. Thus, we must require that *both* the voltage and current waveforms approach zero at $t = \infty$ in order to test an element to see if it is lossless.

Parallel and Series Capacitor Subcircuits

Capacitors, like other types of elements, can be connected into either a series or parallel configuration. Figure 6.21(a) shows a parallel subcircuit consisting of capacitors only. We have labeled the capacitors with their impedance operators. We will, however, use the admittances in a KCL equation to write

$$i(t) = i_1(t) + i_2(t) = C_1 pv(t) + C_2 pv(t) \tag{6.2-32}$$
$$= (C_1 + C_2)pv(t) = C_{eq}pv(t).$$

Thus, the two parallel capacitors are equivalent to the single capacitor of Figure 6.21(b) having an equivalent capacitance equal to the sum of the two:

$$C_{eq} = C_1 + C_2. \tag{6.2-33}$$

If there are more than two capacitors, one simply adds their capacitances to C_{eq}.

Figure 6.21
A parallel capacitor subcircuit equivalent

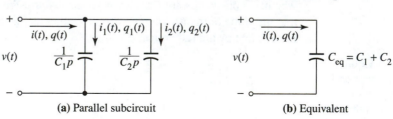

(a) Parallel subcircuit   **(b)** Equivalent

There is a current division rule for capacitors. We derive it as follows:

$$i_1 = C_1 pv(t) = C_1 p \times \frac{i(t)}{C_1 p + C_2 p} = \frac{C_1}{C_1 + C_2} i(t). \tag{6.2-34}$$

Here, we have simply solved (6.2-32) for $v(t)$ in terms of $i(t)$ and used the result in the admittance relationship for C_1, canceling the p operators.[5] The current division rule can clearly be extended to more than two capacitors connected in parallel by simply adding the additional capacitances to the denominator of (6.2-34).

---

[4] Recall that an element is passive if $w(t) \leq 0$ for any $v(t)$ and $i(t)$ waveforms supported by the element.

[5] We could just as well have solved (6.2-32) for $pv(t)$ and used the result in (6.2-34). This would have avoided the necessity of canceling the p operators—though we now know that this is a legitimate operation.

We can derive the preceding results in terms of charge, as follows. We simply use KCL in charge form and the q-v relationship for each capacitor to write

$$q(t) = q_1(t) + q_2(t) = C_1 v(t) + C_2 v(t) \tag{6.2-35}$$
$$= (C_1 + C_2)v(t) = C_{eq}v(t),$$

$$q_1(t) = C_1 v(t) = C_1 \frac{q(t)}{C_1 + C_2} = \frac{C_1}{C_1 + C_2} q(t). \tag{6.2-36}$$

We note that (6.2-36) is a "charge division rule." If we differentiate both sides of (6.2-36), equivalent to multiplying both sides from the left by the p operator, we get (6.2-34) once more.

A series subcircuit consisting only of capacitors is shown in Figure 6.22(a). We can write KVL, note that $i(t)$ is the same for both capacitors, and use the impedance operators of the two capacitances to get

$$v(t) = v_1(t) + v_2(t) = \frac{1}{C_1 p} i(t) + \frac{1}{C_2 p} i(t) \tag{6.2-37}$$
$$= \left[\frac{1}{C_1} + \frac{1}{C_2} \right] \frac{1}{p} i(t) = \frac{1}{C_{eq} p} i(t).$$

But this represents the equivalent single capacitor of Figure 6.22(b) with a capacitance of

$$C_{eq} = \frac{1}{\dfrac{1}{C_1} + \dfrac{1}{C_2}}. \tag{6.2-38}$$

Here, we see that the effect of placing more capacitors in series with the first two would be to add the *reciprocals* of their capacitances to the denominator of the right-hand side of (6.2-38). If, however, there are only two capacitors, we obtain

$$C_{eq} = \frac{C_1 C_2}{C_1 + C_2}. \tag{6.2-39}$$

Figure 6.22
A series capacitor equivalent subcircuit

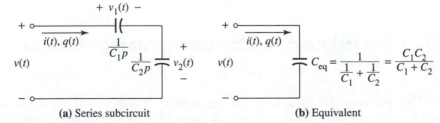

(a) Series subcircuit   (b) Equivalent

The voltage divider rule for capacitors can be derived by solving (6.2-37) for $i(t)$, then writing the voltage across C_1, for instance, as

$$v_1(t) = \frac{1}{C_1 p} i(t) = \frac{1}{C_1 p} C_{eq} p v(t) = \frac{\dfrac{1}{C_1}}{\dfrac{1}{C_1} + \dfrac{1}{C_2}} v(t). \tag{6.2-40}$$

This works for any number of capacitors, with the additional ones merely adding reciprocal terms to the denominator. For the case of two, we can rationalize to obtain

$$v_1(t) = \frac{C_2}{C_1 + C_2}\, v(t). \tag{6.2-41}$$

Once more, we can obtain these relationships strictly in terms of the q-v capacitor relationships. KCL as in the charge form of (6.2-35) gives the following version of KVL as in (6.2-37):

$$v(t) = \frac{q(t)}{C_1} + \frac{q(t)}{C_2} = \left[\frac{1}{C_1} + \frac{1}{C_2}\right] q(t) = \frac{1}{C_{\text{eq}}}\, q(t). \tag{6.2-42}$$

The voltage divider rule can be derived as

$$v_1(t) = \frac{q(t)}{C_1} = \frac{1}{C_1}\, C_{\text{eq}} v(t) = \frac{\dfrac{1}{C_1}}{\dfrac{1}{C_1} + \dfrac{1}{C_2}}\, v(t) = \frac{C_2}{C_1 + C_2}\, v(t). \tag{6.2-43}$$

We can summarize by saying that capacitances add like conductances; that is, one adds capacitance for parallel circuits and forms the reciprocal of the reciprocals (or, for two only, the product over the sum) for series circuits.

Section 6.2 Quiz

Q6.2-1. Find and sketch $v(t)$, $q(t)$, $P(t)$, and $w(t)$ for the circuit in Figure Q6.2-1.

Q6.2-2. Find and sketch $v(t)$ and $i(t)$ in Figure Q6.2-2. Assume that $v_s(t) = 9 \sin(\pi t/T)$ for $0 < t \le T$.

Figure Q6.2-1

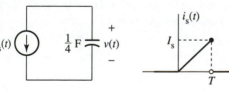

Figure Q6.2-2

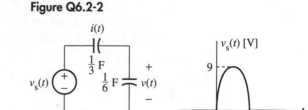

6.3 | The Inductor: Another Energy Storage Element

The Inductor: Another Dynamic Element with Memory

The second energy storage element we will discuss is the *inductor*. Just as the capacitor stores energy in the form of charge, the inductor stores energy in the form of *flux linkage*. Recall that the flux linkage is the running integral of the voltage: $\lambda(t) = (1/p)\, v(t)$. The inductor itself is the idealization of a coil of wire, with the symbol shown in Figure 6.23(a). This symbol is meant to be a stylized representation of the coil of wire. The defining relationship is shown in Figure 6.23(b), which can be taken as an experimental record of the flux linkage versus current relationship. We see that it is linear with a slope of L. Thus, we write

$$\lambda(t) = Li(t) \tag{6.3-1}$$

and call L the *inductance* of the inductor. The unit of inductance is the *henry*,

$$\text{H (henry)} = \frac{\text{Wb}}{\text{A}} \left(\frac{\text{weber}}{\text{ampere}} \right), \tag{6.3-2}$$

named after the American physicist, Joseph Henry.

Figure 6.23
The inductor

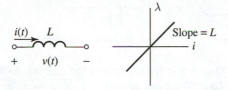

(a) Symbol   **(b)** λ-i relationship

As usual, we want a v-i relationship, so we use (6.3-1) in two ways. First, we recall that

$$v(t) = p\lambda(t) \tag{6.3-3}$$

and differentiate (6.3-1), assuming that L is constant, to obtain

$$v(t) = Lpi(t); \tag{6.3-4}$$

then we merely invert this equation to obtain

$$i(t) = \frac{1}{Lp} v(t). \tag{6.3-5}$$

Clearly, (6.3-4) and (6.3-5) are generalized forms of Ohm's law. Thus, we write

$$v(t) = Z(p)i(t) \tag{6.3-6}$$

and

$$i(t) = Y(p)v(t), \tag{6.3-7}$$

where

$$Z(p) = Lp \tag{6.3-8}$$

is the *impedance operator* (or *operator impedance*) of the inductor and

$$Y(p) = \frac{1}{Lp} \tag{6.3-9}$$

is its *admittance operator* (or *operator admittance*).

The Capacitor-
Inductor Duality

A bit of thought and reference to Section 6.2 should convince you of the following state- ment: the inductor relationships are precisely the same as those of the capacitor if one swaps voltage and current, capacitance and inductance, and charge and flux linkage. (This is called *duality*.) Thus, all of the mathematics used in analyzing one of these elements applies to the other with these changes. For this reason, we will only offer one example showing how to manipulate the basic relationships. As we did for the capacitor, we will label each inductance with either its inductance or its impedance operator.

Example 6.8   Given the voltage waveform shown in Figure 6.24, find and sketch the current waveform.

Figure 6.24
An example circuit

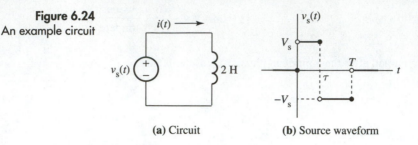

(a) Circuit   **(b)** Source waveform

Solution   The voltage is known and we are to find the current, so the admittance operator is the appropriate one to use. Thus, because $Y(p) = 1/Lp = 1/2p$, we write

$$i(t) = \frac{1}{2p} v_s(t) = \frac{1}{2} \int_{-\infty}^{t} v_s(\alpha) \, d\alpha; \tag{6.3-10}$$

that is, we must compute the running integral of the voltage. Recalling that this is the area under the voltage graph from $-\infty$ to any given time t, we see that $i(t)$ is zero for $t \leq 0$, climbs linearly for $0 < t \leq \tau$, then drops linearly for $\tau < t \leq T$. (The integral of a constant is a linear, or ramp, function.) This defines the shape of the graph in Figure 6.25. We now need only to compute the values at the "break points." The first one is the area under the positive part of the waveform, namely $V_s\tau$. The second is the area under that part plus the area under the negative part (which carries a negative sign). Thus, this value is given by $V_s\tau - V_s(T - \tau) = V_s(2\tau - T)$. These values are labeled in the figure. We note that if $\tau > T/2$ the final value of (t) will be positive and if $\tau < T/2$ it will be negative (as shown). Also, if $\tau \neq T/2$, the final current will remain nonzero—even though the voltage is zero for $t > T$. This is the reason we refer to the inductor (like the capacitor) as a "storage" device and say that it has "memory."

Figure 6.25
The current response
waveform

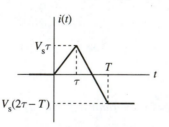

Energy Storage and Losslessness   The energy absorbed by an inductor can be computed as follows. First, we compute the power. Then we notice that it is an exact derivative. The power is the derivative of the energy absorbed, so we have

$$P(t) = v(t)i(t) = [Lpi(t)]i(t) = p\left[\frac{1}{2}Li^2(t)\right] = pw(t); \tag{6.3-11}$$

Multiplying both sides from the left by $1/p$ (equivalent to integrating) and assuming that all waveforms are one sided (identically zero for $t \leq 0$), we find that

$$w(t) = \frac{1}{2}Li^2(t). \tag{6.3-12}$$

Because $w(t)$ is always nonnegative, we see that the inductor is a passive element; furthermore, if $i(t)$ approaches zero for $t \to \infty$, we see that $w(\infty) = 0$. This means that the inductor is also a lossless element. By the way, we note here that the voltage waveform in Figure 6.24(b) would not suffice to check for losslessness unless $\tau = T$, for otherwise the current is not zero for large values of time. For any test of losslessness, we insist that both $v(\infty) = 0$ and $i(\infty) = 0$.

Series and Parallel Inductor Subcircuits   A series subcircuit consisting entirely of inductors is shown in Figure 6.26(a). We use KVL and our generalized Ohm's law to write

$$v(t) = v_1(t) + v_2(t) = L_1 p i(t) + L_2 p i(t) = (L_1 + L_2)\, p i(t). \tag{6.3-13}$$

But this simply means that the entire subcircuit is equivalent to the subcircuit in Figure 6.26(b) consisting of a single inductor of value

$$L_{eq} = L_1 + L_2. \tag{6.3-14}$$

That is, the equivalent inductance is the sum of the individual inductances. Clearly, if there are more than two inductors, we merely add the additional inductance values.

Figure 6.26
A series inductor subcircuit equivalence

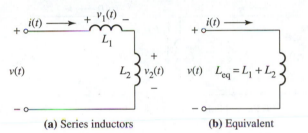

(a) Series inductors   (b) Equivalent

The voltage divider rule for inductors is derived as follows:

$$v_1(t) = L_1 p i(t) = L_1 p \, \frac{v(t)}{(L_1 + L_2)p} = \frac{L_1}{L_1 + L_2} v(t). \tag{6.3-15}$$

We have simply canceled the p and $1/p$ operators.[6] If you prefer to work with flux linkages, you can write (6.3-13) in the form

$$\lambda(t) = (L_1 + L_2)\, i(t) = L_{eq} i(t) \tag{6.3-16}$$

and

$$\lambda_1(t) = L_1 i(t) = L_1 \frac{\lambda(t)}{L_1 + L_2} = \frac{L_1}{L_1 + L_2} \lambda(t). \tag{6.3-17}$$

Although we might refer to this as the "flux linkage divider rule," that term is a bit more cumbersome; however, we can differentiate (6.3-17) to produce (6.3-15) once more.

A parallel inductor subcircuit is shown in Figure 6.27(a). The voltage is the same across both inductors, so—using our generalized version of Ohm's law, along with KCL this time—we find that

$$i(t) = i_1(t) + i_2(t) = \frac{v(t)}{L_1 p} + \frac{v(t)}{L_2 p} \tag{6.3-18}$$

$$= \left[\frac{1}{L_1} + \frac{1}{L_2} \right] \frac{1}{p} v(t) = \frac{1}{L_{eq} p} v(t).$$

---

[6] We could also solve (6.3-13) for $pi(t)$ and use the result in (6.3-15), thus avoiding operator cancellation. We hope that you are by now, however, comfortable with the use of differential operators.

We should perhaps stress one thing: we have written some of the operators directly beneath the voltage symbols. This is merely for compactness of notation; however, you should always be aware that all operators are assumed to be applied from the left to any time function. Thus, we are multiplying by an admittance operator. Here, we see that the equivalent inductance is given by the equation

$$L_{eq} = \frac{1}{\dfrac{1}{L_1} + \dfrac{1}{L_2}}. \tag{6.3-19}$$

If there are more than two inductors in our subcircuit, we merely add their reciprocals to the denominator of the fraction on the right-hand side. For the case of two inductors only, we can rearrange this result into the formula

$$L_{eq} = \frac{L_1 L_2}{L_1 + L_2}. \tag{6.3-20}$$

Figure 6.27
A parallel inductor
subcircuit equivalence

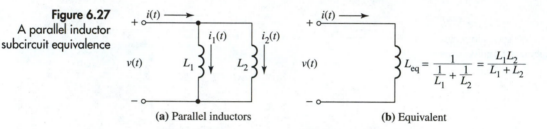

(a) Parallel inductors   **(b)** Equivalent

The current divider rule for inductors can be derived as follows:

$$i_1(t) = \frac{v(t)}{L_1 p} = \frac{1}{L_1 p} \frac{i(t)}{\left[\dfrac{1}{L_1 p} + \dfrac{1}{L_2 p}\right]} = \frac{\dfrac{1}{L_1}}{\left[\dfrac{1}{L_1} + \dfrac{1}{L_2}\right]} i(t). \tag{6.3-21}$$

This is a general result that can be immediately extended to N parallel inductors by adding the reciprocals of the additional inductance values to the denominator. For the case of two only, however, we can rearrange to get the simpler result

$$i_1(t) = \frac{L_2}{[L_1 + L_2]} i(t). \tag{6.3-22}$$

As we did for the series circuit, one can work with flux linkages directly to get both the equivalent inductance and the current divider rule; we will leave this for you to do as an exercise in the review quiz. We can summarize our results for series and parallel inductors by saying that "inductors add like resistors."

Operator Analysis of
the General Series
Subcircuit with a
Single Type of Element

Now that we have worked with equivalent subcircuits and current and voltage division rules for series and parallel subcircuits consisting of only resistors, only capacitors, or only inductors, we can see that there is a strong similarity among the three cases. In fact, a general series subcircuit is shown in Figure 6.28(a), wherein we have labeled each element with its impedance operator: it is the resistance R if the element is a resistor, the impedance operator $1/Cp$ if it is a capacitor, and the impedance operator Lp if it is an inductor. For any of these elements, we know that

$$v(t) = Z(p)i(t). \tag{6.3-23}$$

This is a generalized version of Ohm's law. We can therefore write KVL in the form

$$v(t) = Z_1(p)i(t) + Z_2(p)i(t) = [Z_1(p) + Z_2(p)]i(t) = Z_{eq}(p)i(t), \quad (6.3\text{-}24)$$

where

$$Z_{eq}(p) = Z_1(p) + Z_2(p). \quad (6.3\text{-}25)$$

For the resistor, inductor, and capacitor, respectively, we have

$$Z_{eq}(p) = R_1 + R_2, \quad (6.3\text{-}26a)$$

$$Z_{eq}(p) = [L_1 + L_2]p, \quad (6.3\text{-}26b)$$

$$Z_{eq}(p) = \cfrac{1}{\left(\cfrac{1}{\cfrac{1}{C_1} + \cfrac{1}{C_2}}\right)p} = \cfrac{1}{\left(\cfrac{C_1 C_2}{C_1 + C_2}\right)p} \quad (6.3\text{-}26c)$$

This is shown as the general two-terminal equivalent element in Figure 6.28(b).

Figure 6.28
A general series
subcircuit equivalence

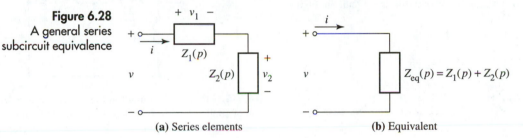

(a) Series elements   **(b)** Equivalent

The voltage divider rule for v_1, for example, can be obtained by solving (6.3-24) for i, then computing $v_2(t)$ as

$$v_1(t) = Z_1(p)i(t) = \frac{Z_1(p)}{Z_{eq}(p)}v(t) = \frac{Z_1(p)}{Z_1(p) + Z_2(p)}v(t). \quad (6.3\text{-}27)$$

This reduces to the usual forms when we substitute specific impedance operators.

Operator Analysis of the General Parallel Subcircuit with a Single Type of Element

A general parallel subcircuit is shown in Figure 6.29(a). We will leave it to you to apply KCL and the element relationships to derive the equivalent impedance:

$$Z_{eq}(p) = \cfrac{1}{\cfrac{1}{Z_1(p)} + \cfrac{1}{Z_2(p)}} = \frac{Z_1(p)Z_2(p)}{Z_1(p) + Z_2(p)}. \quad (6.3\text{-}28)$$

The current divider rule takes the form (again, we leave it to you to verify this)

$$i_1(t) = \frac{Z_2(p)}{Z_1(p) + Z_2(p)}i(t). \quad (6.3\text{-}29)$$

Figure 6.29
A general parallel
subcircuit equivalence

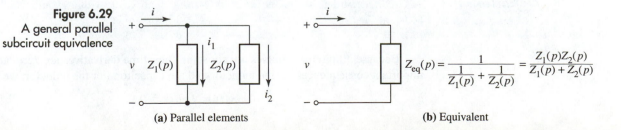

(a) Parallel elements   **(b)** Equivalent

Notice that we could have worked with admittance operators, showing that the equivalent admittance operators are the sum of the two individual admittance operators and that the current through $Z_1(p)$ is given by the ratio of $Y_1(p)$ to the sum of the two admittance operators multiplied by the current $i(t)$.

The Advantages of Operators

The nice feature of operators is evident: no matter what type of element is involved, we always form the equivalent impedance in the same manner. Furthermore, the voltage and current divider rules always take the same form. In fact, these manipulations are much more general than we have indicated—they continue to hold for general subcircuits not consisting of a single type of element. We will, however, not deal with such general subcircuits until we have developed much more sophisticated analysis techniques.

Section 6.3 Quiz

Q6.3-1. Derive the equivalent inductance and current divider rule for the parallel inductor subcircuit using flux linkage and current.

Q6.3-2. Find the voltage waveform for the circuit in Figure Q6.3-1.

Figure Q6.3-1

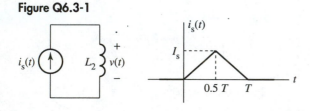

Q6.3-3. Compute and plot the instantaneous power and the energy absorbed by the inductor in Figure Q6.3-1. Is the $i_s(t)$ waveform shown appropriate for checking losslessness?

Q6.3-4. Find the equivalent inductance for the two-terminal subcircuit in Figure Q6.3-2.

Figure Q6.3-2

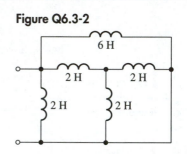

Q6.3-5. Reduce the circuit in Figure Q6.3-3 to a series connection of single-element-type, two-terminal subcircuits. Assume that each inductor has a value of 1 H and each capacitor has a value of 1 F.

Figure Q6.3-3

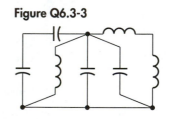

6.4 | The Dc Steady State

Circuits having energy storage elements behave in much more complex ways than those having only resistors as the passive elements. However, if all of the independent sources are dc, then all responses in many cases will approach constants. We define this as the dc steady state.

Definition 6.4 A circuit is said to be in the *dc steady state* if all of its voltages and currents are constant.

Because all circuit variables are constant, their time derivatives are zero, and this has important consequences for the inductor and the capacitor. For the inductor, we have

$$v(t) = Lpi(t) = 0. \qquad (6.4-1)$$

But $v(t) = 0$ defines a short circuit. Therefore—in the dc steady state—an inductor is equivalent to a *short circuit*. For the capacitor, we have a similar situation:

$$i(t) = Cpv(t) = 0. \tag{6.4-2}$$

Thus, the capacitor is equivalent to an *open circuit* in the dc steady state. We show these equivalents in Figure 6.30. Recall that the current in a short circuit and the voltage across an open circuit are both established by the external circuit—not by the short circuit and the open circuit themselves.

Figure 6.30
Dc steady-state models for *L* and *C*

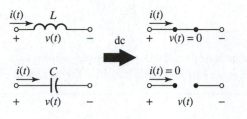

The following example shows a couple of simple situations in which the circuit is known to be operating in the dc steady state.

Example 6.9 Show that the two circuits in Figure 6.31 are operating in the dc steady state.

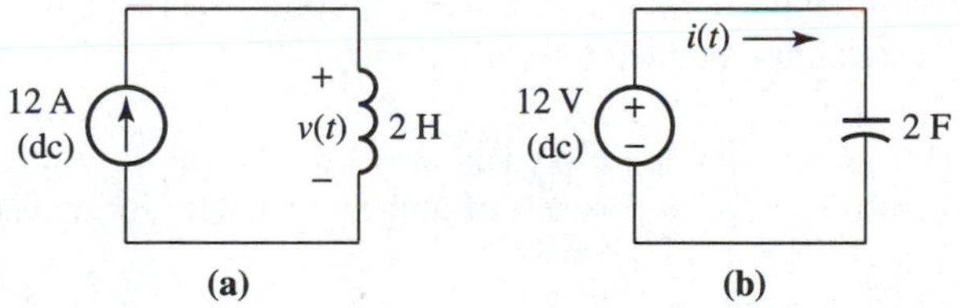

Figure 6.31
Two simple circuits in the dc steady state

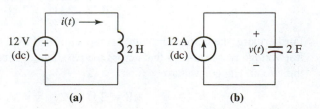

Solution For the circuit in Figure 6.31(a) we have $v(t) = 2p[12] = 0$ V. Notice carefully that it is the derivative of the *constant* 12 we are computing—not the derivative of the 12-V step function $12u(t)$. Thus, the voltage is constant (and not one sided, by the way) and so is the current. Because these (and their negatives, of course) are the only two circuit variables, we see that the circuit is, indeed, operating in the steady state. Similarly, for the circuit in part (b) of the figure, we have $i(t) = 2p[12] = 0$. Thus, both current and voltage are constant; hence, this circuit too is in the dc steady state.

At first thought, it might seem that all circuits having only dc independent sources operate in the dc steady state. The next example shows this to be untrue.

Example 6.10 Show that the two circuits in Figure 6.32 are not operating in the dc steady state.

Figure 6.32
Two simple circuits *not* in the dc steady state

Solution   Notice that the only difference between this example and the last one is that the v-source and the i-source have been swapped. We are now driving the inductor with a v-source and the capacitor with an i-source. This, however, has a remarkable effect. We see that for the circuit in part (a) we have $i(t) = 1/Lp$ [12] $= \infty$! Similarly, the circuit in part (b) has $v(t) = 1/Cp$ [12] $= \infty$. The point is that a constant does not integrate to a finite number, so the assumption of dc steady-state operation is incorrect. To pursue the point a bit further, let's suppose we "turn on" the voltage source in part (a) at $t = 0$. This is equivalent to replacing the 12-V dc source with a waveform source having the time variation $12u(t)$ V. Now we compute the integral to be a finite value; the current is

$$i(t) = \frac{1}{2p} [12u(t)] = 6tu(t) \text{ A.} \tag{6.4-3}$$

As the current is not constant—and never becomes so—the circuit never reaches the dc steady state. A similar argument holds for the circuit in Figure 6.32(b).

The question of the existence of a dc steady state is intimately related to the question of stability, which was discussed briefly in connection with op amp circuits and feedback in Section 5.5 of Chapter 5, and will be treated much more thoroughly in Chapter 10. The two circuits in Example 6.10 are not *stable. A circuit is stable if the fact that all the independent sources are bounded (do not become infinitely large) implies that each of the circuit response voltages and currents has this property also.* Clearly, the circuits in the last example are not stable because we found a bounded source waveform, $12u(t)$, that produced an unbounded response, $6tu(t)$.

Example 6.11   The circuit in Figure 6.33 is operating in the dc steady state. Find each of the labeled quantities. (Notice our use of uppercase letters for dc variables.)

Figure 6.33
An example circuit

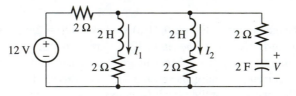

Solution   Though the circuit looks quite complicated, the solution is actually quite easy. The information that it is in the dc steady state means that we can replace each inductor by an equivalent short circuit and the capacitor by an open circuit, as we have done in Figure 6.34. It is now easy to see that the middle pair of resistors form a parallel subcircuit; its equivalent resistor is then series connected relative to the resistor in series with the source. Thus, we find that

$$I_1 = I_2 = \frac{12 \text{ V}}{2\,\Omega + 2\,\Omega \parallel 2\,\Omega} \times \frac{2\,\Omega}{2\,\Omega + 2\,\Omega} = 2 \text{ A.} \tag{6.4-4}$$

Furthermore, the current through the capacitor and its series resistor is zero, so the capacitor voltage is the same as that across the parallel-connected 2-Ω resistors in the middle of the circuit. Hence, we have

$$V = 2\,\Omega \times 2 \text{ A} = 4 \text{ V.} \tag{6.4-5}$$

Figure 6.34
The dc equivalent circuit

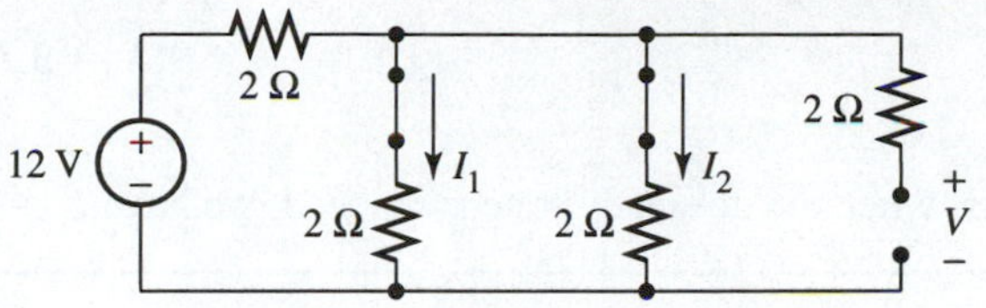

"Singular" Dc Steady-State Circuits

Sometimes a seemingly ambiguous situation can arise in circuits operating in the dc steady state that have parallel inductors or series capacitors. The next example illustrates this ambiguity and its solution.

Example 6.12 The circuit in Figure 6.35 is (for some reason) known to be operating in the dc steady state. Find each of the labeled currents.

Figure 6.35
An example circuit

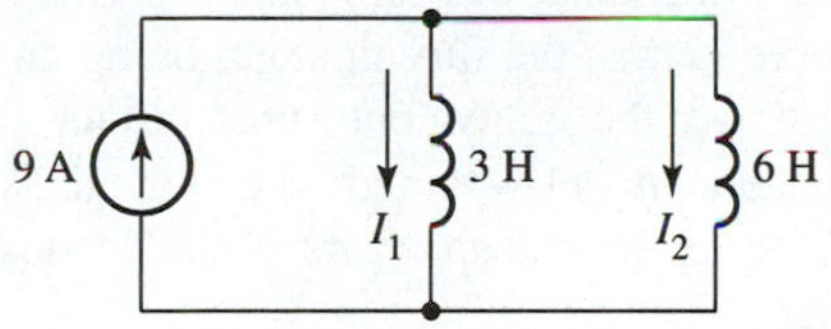

Solution Replacing both inductors with their dc equivalent short circuits, we obtain the circuit shown in Figure 6.36. The problem is clear to see: though we know by KCL that the sum of the two currents is 9 A, we do not know how that 9 A splits up! What do we do? The answer is that we must discuss how the circuit arrived at the dc steady state. Let's assume that the 9-A source is "turned on" at $t = 0$ and its value is gradually increased from zero to its final value of 9 A. We show the situation in the circuit of Figure 6.37, in which we have merely replaced the assumed constant currents with time-varying ones.

Figure 6.36
The dc equivalent circuit

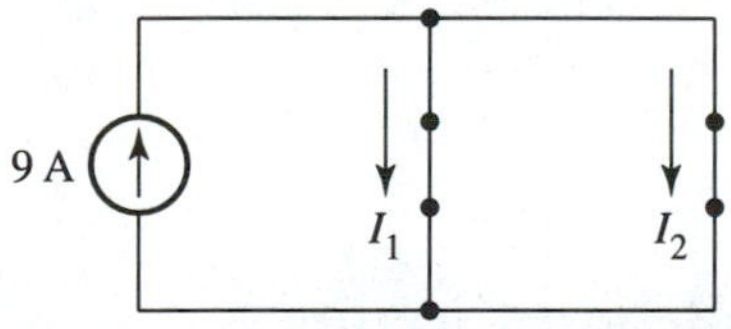

Figure 6.37
The example circuit with time-varying waveforms

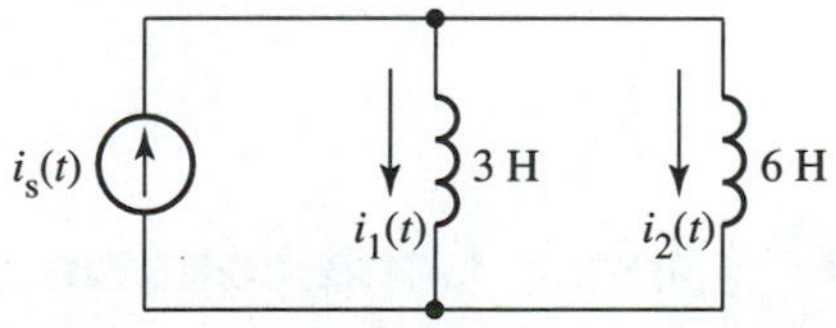

Now we merely apply the general current division rule to obtain

$$i_1(t) = \frac{6p}{6p + 3p} i_s(t) = \frac{2}{3} i_s(t) \qquad (6.4\text{-}6)$$

and

$$i_2(t) = \frac{3p}{6p + 3p} i_s(t) = \frac{1}{3} i_s(t). \qquad (6.4\text{-}7)$$

Now, if we let $i_s(t)$ "settle down" to its constant value of 9 A, we have

$$i_1(t) = \frac{2}{3} \times 9\text{ A} = 6\text{ A} \qquad (6.4\text{-}8)$$

and
$$i_2(t) = \frac{1}{3} \times 9 \text{ A} = 3 \text{ A}. \tag{6.4-9}$$

Thus, when the dc steady state is reached, we have $I_1 = 6$ A and $I_2 = 3$ A.

Let's summarize the method outlined in the last example. When a circuit has parallel inductors, we merely assume all variables to be time varying, find the current through each of the parallel inductors in terms of the independent source value(s), and then evaluate the results for the dc steady-state case of constant currents.

We can also use flux linkage instead of voltage to solve the problem. For instance, in the last example, we could write

$$\Lambda = L_1 I_1 = L_2 I_2 = \frac{L_1 L_2}{L_1 + L_2} I_s \tag{6.4-10}$$

and solve these algebraic equations for I_1 and I_2, where Λ is the dc value of flux linkage.[7] The thing to observe is this: the flux linkage, being an integral (an accumulation), is not necessarily zero across the equivalent short circuit that replaces an inductor in a dc steady-state equivalent model—though the voltage is. Finally, we note that a similar analysis can be done for series capacitors in the dc steady state, but we will leave this as an exercise.

Section 6.4 Quiz

Q6.4-1.   Find I and V for the circuit shown in Figure Q6.4-1, assuming the dc steady state.

Figure Q6.4-1

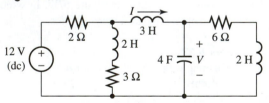

Q6.4-2.   Find the dc steady-state voltage V in Figure Q6.4-2. (*Hint:* Follow the procedure of Example 6.12 or the procedure outlined immediately after that example with flux linkage.)

Figure Q6.4-2

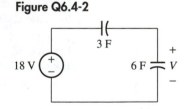

6.5 | Initial Conditions and Switched Circuits

Thus far in this chapter we have assumed that a circuit was constructed an infinite amount of time before we began observing it and that any signal we injected into it via an independent source was one sided; that is, it was zero for all $t \leq 0$. Now we will look at another situation of practical interest: a circuit that has been in operation in some manner for a long time, but that we begin to observe at $t = 0$. This is referred to as an *initial condition problem*. We discussed this to some extent in our earlier discussion of basic quantities such as charge, flux linkage, and energy in Chapter 1.

---

[7] Note that Λ is the uppercase lambda (λ), the proper notation for *constant* flux linkages.

One-Sided Truncation of Two-Sided Waveforms

Figure 6.38(a) shows a capacitor and its terminal variables. As usual, we have defined them consistently with the passive sign convention, for that is the assumption made in presenting the capacitor's v-i characteristic. As we show in Figure 6.38(b), however, we assume that we only observe the waveform for $t > 0$—the past is unknown. Can we find the voltage $v(t)$ for $t > 0$?

Figure 6.38
The initial condition problem for a capacitor

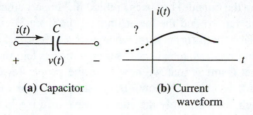

(a) Capacitor

(b) Current waveform

We know that the current is related to the voltage by the equation

$$i(t) = Cpv(t) = C\frac{dv(t)}{dt} = Cv'(t), \tag{6.5-1}$$

where we have shown three different symbols for the time derivative. If the voltage were specified and we were to find the current, there would be no difficulty: we would merely differentiate the voltage waveform for $t > 0$ and multiply by the capacitance. Our problem, however, is significantly different. The integral relationship for the capacitor is

$$v(t) = \frac{1}{Cp}\,i(t) = \frac{1}{C}\int_{-\infty}^{t} i(\alpha)\,d\alpha. \tag{6.5-2}$$

The problem? We cannot compute the integral because the current waveform is not specified for $t \leq 0$. Here's what we do: we just return to the derivative relationship in equation (6.5-1) and integrate[8] both sides—this time over the finite interval $(0, t]$. Thus,

$$\int_{0}^{t} Cv'(\alpha)\,d\alpha = \int_{0}^{t} i(\alpha)\,d\alpha. \tag{6.5-3}$$

Multiplying by $1/C$, integrating the derivative on the left side, and moving the *initial condition* $v(0)$ to the right gives

$$v(t) = \frac{1}{C}\int_{0}^{t} i(\alpha)\,d\alpha + v(0). \tag{6.5-4}$$

The result is this: we can solve for the voltage in terms of the positive-time portion of the current waveform, but we must know the initial value of the capacitor voltage $v(0)$.

   In the next few chapters we will find that if we use the Heaviside operators p and $1/p$ for differentiation and integration, respectively, and restrict all waveforms to be one sided, the solution of circuits having capacitors and inductors can be carried out in a manner that is strictly algebraic. But we still have a problem because the integral in equation (6.5-4)

---

[8] To be technically accurate, we must consider the integral to be improper at $t = 0$.

has a lower limit of zero. We can remedy this, however, in the following way. We simply multiply both sides of (6.5-4) by $u(t)$:

$$v(t)u(t) = \left[\frac{1}{C} \int_0^t i(\alpha)\, d\alpha \right] u(t) + v(0)u(t) = \frac{1}{C} \int_{-\infty}^t i(\alpha)u(\alpha)\, d\alpha + v(0)u(t). \quad (6.5\text{-}5)$$

A bit of attention to the last equality might be in order. The $u(t)$ multiplying the entire integral (from the outside) has been "pulled in," the argument of $u(t)$ changed to α (the variable of integration) and the lower limit of the integral extended to $-\infty$. If $t \leq 0$, we see that this makes both forms zero, whereas if $t > 0$, it makes both have the value of the integral taken from zero to t; hence, our operation is legitimate.

Figure 6.39 The one-sided equivalent waveform

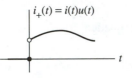

$i_+(t) = i(t)u(t)$

The net result is that we now have the proper lower limit on our integral, but the waveforms have changed. How? To see this, let's look once more at Figure 6.38(b). This time, though, let's multiply the current waveform by the unit step function $u(t)$. This results in the waveform shown in Figure 6.39. It is now known for all time and is identically zero for $t \leq 0$. A similar picture would, of course, hold for the one-sided equivalent of the voltage waveform. We have thrown away the unknown portion of the waveforms and have substituted their effect by the initial voltage $v(0)$; this single quantity carries all the information about the past history of the capacitor to allow us to compute the voltage for all positive values of t. Rewriting (6.5-5) in terms of our one-sided variables and the integration operator $1/p$, we have

$$v_+(t) = \frac{1}{Cp} i_+(t) + v(0)u(t). \quad (6.5\text{-}6)$$

Capacitor and Inductor Initial Condition Models

Now look more closely at equation (6.5-6). It is a v-i relationship for our capacitor, but it now consists of two terms: two voltages, in fact. Because the voltages add, we see that they represent two elements connected in series. The first is simply a capacitor with one-sided voltage and current waveforms, and the second is an independent voltage source, for its value does not depend on the current. We show this *initial condition model* for the capacitor in Figure 6.40. It is important to remember that the actual capacitor voltage is the sum of the two equivalent element voltages. This is easy to forget when using the model to solve circuits.

Figure 6.40 Initial condition model for the capacitor

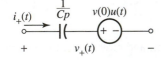

The inductor can be represented by a similar model. The v-i characteristic is the same as that of the capacitor (with the voltage and current and inductance and capacitance swapped), so the derivation proceeds in exactly the same mechanical way. Thus, we will do it more quickly; however, the very symmetry in characteristics makes it essential that we be careful in our interpretation.

Figure 6.41 shows the inductor with its terminal variables defined, again according to the passive sign convention. This time we use the relationship

$$v(t) = Lpi(t) \quad (6.5\text{-}7)$$

Figure 6.41 The inductor

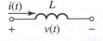

and ask for the current $i(t)$ when the voltage $v(t)$ is known only for positive values of t. In this case, using the prime notation for the derivative of the current, we integrate both sides of (6.5-7) to obtain

$$L \int_0^t i'(\alpha)\, d\alpha = \int_0^t v(\alpha)\, d\alpha \quad (6.5\text{-}8)$$

or

$$i(t) = \frac{1}{L} \int_0^t v(\alpha)\, d\alpha + i(0), \quad (6.5\text{-}9)$$

where $i(0)$ is the initial condition on the inductor current. Once more, we multiply both sides by the unit step function and bring the one on the right-hand side of the equation into the integral. This results in

$$i_+(t) = \frac{1}{L} \int_{-\infty}^{t} v_+(\alpha) \, d\alpha + i(0)u(t). \qquad (6.5\text{-}10)$$

In operator notation, this is

$$i_+(t) = \frac{1}{Lp} v_+(t) + i(0)u(t). \qquad (6.5\text{-}11)$$

Figure 6.42   Initial condition model for the inductor

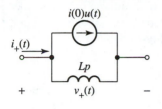

Thus far, our derivation has been exactly the same as for the capacitor; now, however, we see that (6.5-11) is the sum of two *currents* (rather than two voltages as for the capacitor). We can interpret it as the KCL equation for two elements connected in parallel. The first is merely an inductor with one-sided terminal variables; the other is an independent current source. This gives the equivalent model shown in Figure 6.42.

Switches   Before we use our two initial condition models for practical purposes, let us discuss where the initial conditions come from. There are several situations, but the one we will look at here involves *switches*. Although everyone is familiar with what a switch does, there are several types. The two shown in Figure 6.43 are called SPST, for *single-pole single-throw*, switches. (The movable part, represented in the figure as a straight line segment, must have looked to someone like a bamboo fishing pole, and one "throws" it from one side to the other.) There is one more type of switch that we will often use. It is a combination of the preceding two and is called SPDT, for *single-pole double-throw* (see Figure 6.44). It opens the connection from the left-hand contact to the bottom one and makes the one from the right-hand contact to the bottom when it is activated. One item should be stressed here: the switching action in each type of switch is assumed to occur only once—at $t = 0$. It is in the initial position for all $t \leq 0$ and in the alternate position for all $t > 0$. The assignment of the condition at $t = 0$ is an arbitrary idealization of actual switch operation, which takes some time to complete. Our idealization insists that it occur in zero time. We now give an example for solving a switched circuit.

Figure 6.43
Normally open and normally closed SPST switches

Figure 6.44
The SPDT switch

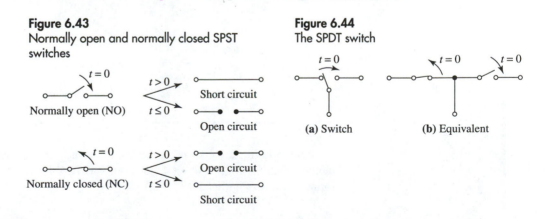

Example 6.13   Find the voltages $v_x(t)$ and $v_y(t)$ in Figure 6.45 for all values of t.

Figure 6.45
A switched circuit

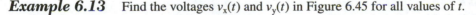

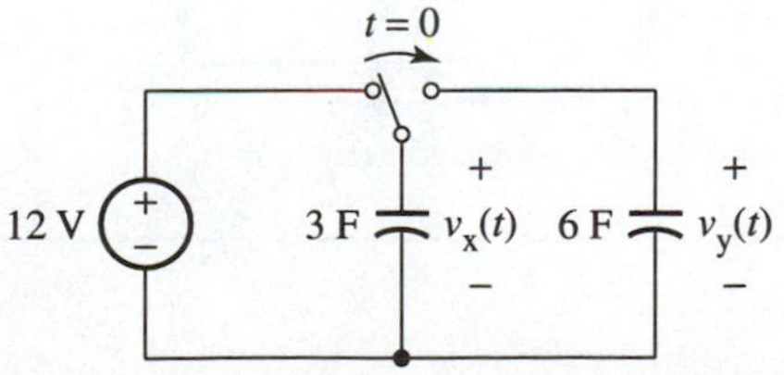

Solution   A circuit with switches is more complicated to solve than one without them for there are, in actuality, two circuits to solve: one before the switching action and one afterward. We will refer to them as the $t \leq 0$ circuit and the $t > 0$ circuit. The first is shown in Figure 6.46(a). Because the capacitor is directly across the v-source, its voltage is constant at 12 V. Thus, its current is zero. Furthermore, the capacitor on the right has had zero current since $t = -\infty$. We will take it as a given fact that no element can respond before it is stimulated to do so; that is, it is *causal* (as given by Definition 6.2). This means that this capacitor voltage is identically zero. For these reasons, we see that the circuit is in the dc steady state. Thus, it has the equivalent shown in Figure 6.46(b) with

$$v_x(t) = 12 \text{ V} \tag{6.5-12}$$

and

$$v_y(t) = 0 \text{ V} \tag{6.5-13}$$

for all values of $t \leq 0$.

Figure 6.46
The $t \leq 0$ circuit

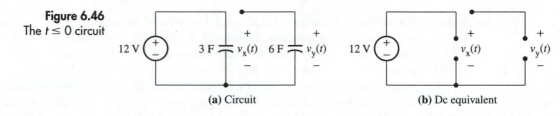

(a) Circuit   (b) Dc equivalent

Now we "flip the switches" and obtain the $t > 0$ circuit shown in Figure 6.47. The independent source value has been multiplied by $u(t)$ because we have agreed to make all voltages and currents one sided; however, it is uncoupled from the circuit and has no effect itself. We can now use the voltage divider rule to obtain

$$v_{x+}(t) = v_{y+}(t) = \cfrac{\cfrac{1}{6p}}{\cfrac{1}{3p} + \cfrac{1}{6p}} \, 12u(t) = 4u(t) \text{ V}. \tag{6.5-14}$$

Figure 6.47
The $t > 0$ circuit

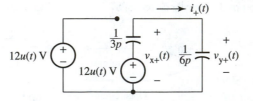

This solution for $v_y(t)$ is correct for *all* values of t because that variable was determined to be zero for $t \leq 0$; $v_x(t)$, however, was 12 V for $t \leq 0$. Thus, we obtain the waveforms shown in Figure 6.48.

Figure 6.48
The capacitor voltage
waveforms

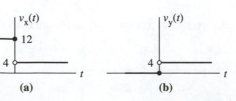

(a)   (b)

The preceding example illustrates a network that is of some practical importance: the *switched capacitor circuit*. In practical switched capacitor circuits the switches operate periodically rather than just once, so the analysis gets a bit more complicated; however, that analysis is based upon the technique we have just illustrated.

A First Look at Impulse Waveforms

Here, however, is a question we must resolve. What is the waveform for the current $i_+(t)$ that we have labeled in Figure 6.47—the current in the 6-F capacitor? (Notice that it is the same as the original two-sided current $i(t)$.) We find it from the basic capacitor relationship, as follows:

$$i_+(t) = 6pv_{x+}(t) = 6p[4u(t)] = 24pu(t) \text{ A.} \tag{6.5-15}$$

But what exactly is the derivative of the unit step function? It is not defined in classical mathematics. We can get a clue, however, by noting that the charge is proportional to the voltage:

$$q_+(t) = 6v_{x+}(t) = 24u(t) \text{ C.} \tag{6.5-16}$$

(The C here means the unit of charge, the coulomb.) Ideally, the charge stored on the capacitor increases from zero to 24 coulombs instantaneously. We know, however, that practical waveforms do not change instantaneously; it was for this reason that we chose our open circle and filled dot convention for discontinuous jumps. Thus, practically speaking, the current would start from zero amperes at $t = 0$, become very large, then drop back to zero in a very short interval of time. We call such a waveform an *impulse function* and the derivative of the unit step function the *unit impulse*. We give the latter the symbol $\delta(t)$. We will investigate impulses more thoroughly in Section 6.6. For now, we will simply symbolically write

$$\delta(t) = pu(t) \tag{6.5-17}$$

and

$$u(t) = \frac{1}{p}\delta(t). \tag{6.5-18}$$

Thus, we can say that

$$i_+(t) = 24\delta(t). \tag{6.5-19}$$

Stability and the Dc Steady State

Let's discuss another aspect of switched circuits. Figure 6.49 shows the same circuit we investigated in the last example, with one difference: we have added a 2-Ω resistor in series with the independent v-source. What effect does it have? The addition of the resistor means that it is no longer obvious that the $t \leq 0$ circuit is in the dc steady state because the 3-F capacitor is no longer directly in parallel with the v-source. In fact the circuit *is* in the dc steady state, but we cannot justify this statement rigorously until we develop analysis procedures for circuits with resistors in addition to the energy storage elements.

Figure 6.49
A switched circuit with a resistor

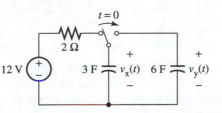

We can, however, justify our statement *intuitively*. Just assume that the 3-F capacitor is an open circuit (its dc equivalent) for $t \leq 0$. Then its current is zero, and so is the current through the 2-Ω resistor. But then the voltage drop across the resistor, by Ohm's law, is also zero, and therefore the voltage across the 3-F capacitor is the constant 12 V. This means that our assumption of dc steady-state operation is, at least, self-consistent. For some time, however, we will merely provide the fact of dc steady-state operation as an added bit of information in each problem statement and defer justification until we have studied the stability of circuits more thoroughly in Chapter 10. We now provide another example for solving a switched circuit.

Example 6.14 Find the current waveform $i(t)$ in Figure 6.50.

Figure 6.50
An example circuit

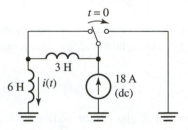

Solution The $t \leq 0$ circuit is shown in Figure 6.51. Here, as in the last example, we see that the circuit is operating in the dc steady state because the current is constant in each inductor (0 A in the 3-H inductor because it is paralleled with a short circuit,[9] and 18 A, the source current in the 6-H inductor). The $t > 0$ circuit is shown in Figure 6.52. We note that the switching operation shunts the i-source to ground on the right. Writing one nodal equation, we obtain

$$\frac{v_+(t)}{6p} + \frac{v_+(t)}{3p} = -18u(t). \tag{6.5-20}$$

Therefore, $$v_+(t) = -2p[18u(t)] = -36\delta(t). \tag{6.5-21}$$

Thus, $$i_+(t) = \frac{v_+(t)}{6p} + 18u(t) = \frac{1}{6p}[-2p(18u(t))] + 18u(t) = 12u(t) \text{ A}. \tag{6.5-22}$$

Figure 6.51
The $t \leq 0$ circuit

Figure 6.52
The $t > 0$ circuit

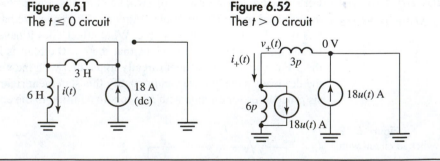

At the end of the last example, we could have merely used the fact that the integral of the impulse function is the step function, but we wanted to show that the p operators cancel. Thus, we have not actually needed the concept of the impulse function yet.

---

[9] More precisely, the inductor has been shorted since $t = -\infty$; thus, its voltage is identically zero for $t \leq 0$. Hence, by our causality assumption, the current must also be zero for $t \leq 0$.

Section 6.5 Quiz

Q6.5-1. Find $v_x(t)$ and $v_y(t)$ for the switched circuit in Figure Q6.5-1. Assume that the circuit is in the dc steady state for $t \leq 0$.

Figure Q6.5-1

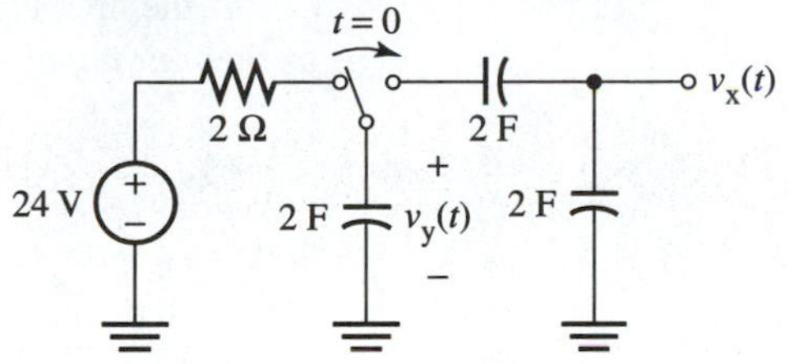

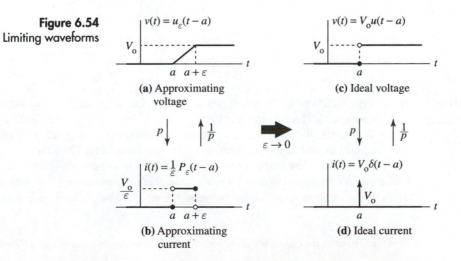

6.6 | Impulse Waveforms

Near the end of Section 6.5 we introduced the idea of an impulse function, the time derivative of a step function. We will now develop this idea in more detail. We first consider the problem of finding the current in the simple circuit of Figure 6.53, which consists of a v-source having a step function waveform occurring at $t = a$ and a 1-farad capacitor. If we formally compute the current $i(t)$, its waveform will be

$$i(t) = pv(t) = V_o\, pu(t - a) = V_o\delta(t - a), \tag{6.6-1}$$

where we have used the formal symbol $\delta(t - a)$ for the *unit impulse function*, or *delta function*, occurring at $t = a$. The charge stored on the capacitor is proportional to the voltage ($q = Cv$) and this quantity must change instantaneously; thus, as the current is the time derivative of the charge, the $\delta(t - a)$ symbol should represent an infinite value occurring at the leading edge of the step function.

Now let's make our argument more analytical by approximating the voltage step by the differentiable waveform $u_\epsilon(t - a)$ shown in Figure 6.54(a). As we show, it has the finite "rise time" ϵ. More specifically, assume that this waveform increases linearly from zero, starting at $t = a$, to V_o at $t = a + \epsilon$. Its derivative is shown in Figure 6.54(b); it consists of a pulse starting at $t = a$ and ending at $t = a + \epsilon$, with height CV_o/ϵ. Regardless of the value of ϵ, the area under the current pulse has the value $CV_o = V_o$: the total charge delivered to the capacitor. As $\epsilon \to 0$, the voltage waveform approaches the step of height V_o at $t = a$, and the current waveform tends toward a waveform that is zero everywhere, but has a finite, nonzero area of V_o! We call this waveform an *impulse function*, or

Figure 6.53   An example circuit

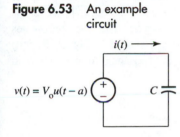

Figure 6.54
Limiting waveforms

delta function, at $t = a$. Its symbolic representation is the vertical arrow in Figure 6.54(d). The V_o by its side represents the area. We assume the derivative relationship between $v(t)$ and $i(t)$ remains the same in the limit. This is our meaning for equation (6.6-1).

If the area under the approximating pulse is unity, we refer to the limiting waveform as the *unit impulse,* or *unit delta function.* Thus, we have the defining relations

$$\delta(t - a) = pu(t - a) \tag{6.6-2}$$

and

$$u(t - a) = \frac{1}{p}\,\delta(t - a). \tag{6.6-3}$$

The linearly increasing approximation is not the only one leading to the unit impulse function. In Figure 6.55(a) we show another. It is called the *exponential function at* $t = a$ *with time constant* τ. The time constant[10] is an important parameter in the exponential time function: when $t = a + \tau$ the waveform has dropped by a factor of $1/e$ from its initial value (or by 63%). Thus, large values of the time constant mean that the waveform is changing very slowly, and very small values mean that it is changing very rapidly. The running integral of this waveform is shown in Figure 6.55(b), whose equation is shown in the figure and which we will leave for you to verify. As τ becomes smaller and smaller, the original waveform becomes taller and taller in the immediate vicinity of $t = a$ but (mathematically) approaches zero everywhere. The area under its graph, however, is unity, the justification of which we will once more leave to you. Thus, it approaches the unit impulse drawn in Figure 6.55(c). Its running integral shown in Figure 6.55(d) approaches the ideal unit step function at $t = a$. Thus, the exponential waveform in Figure 6.55(a) also represents the unit impulse function.

Figure 6.55
Another unit impulse approximation

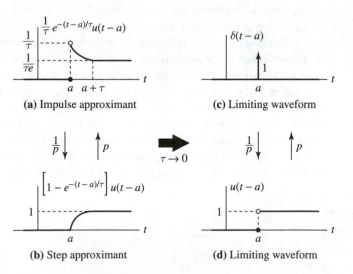

(a) Impulse approximant

(c) Limiting waveform

(b) Step approximant

(d) Limiting waveform

A Practical Interpretation of Impulse Functions

Here is the situation. Each circuit has an appropriate time scale. Thus, some circuits will exhibit noticeable changes in their voltage and current variables only after a few seconds, whereas others will exhibit the same amount of change over only a few microseconds. We will interpret the unit impulse function in the following way. The unit impulse function at $t = a$ is any waveform starting at $t = a$ and returning to zero sufficiently quickly that the entire pulse cannot be observed *on the time scale we are using,* and that has unit area. Thus, the waveform itself "drops between the cracks," but its area makes itself felt in the

---

[10] We will discuss the concept of an exponential time constant in more detail in Chapter 7.

response variables. These are intuitive ideas that can be made quite rigorous by a careful application of mathematics. We now present an example of a switched circuit with impulsive charge transfer.

Example 6.15   Find the current $i(t)$ and the voltage $v(t)$ in Figure 6.56. Assume that v_x has a value of 12 V just prior to the switch flip at $t = 0$, $C_1 = 100$ nF, and $C_2 = 50$ nF.

Figure 6.56
An example circuit

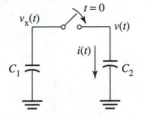

Solution   The switch has been open since $t = -\infty$, so we apply our causality assumption (that there is no response before an input) to see that $v(t)$ is zero just prior to the switch flip. We do not, therefore, need to do any further analysis of the $t \leq 0$ circuit because we are given the initial voltage on the capacitor C_1.[11] The $t > 0$ circuit is shown in Figure 6.57. A KCL equation at the top node gives

$$C_1 p[v_+(t) - 12u(t)] + C_2 p v_+(t) = 0. \qquad (6.6\text{-}4)$$

We can multiply both sides by $1/p$ (from the left—the same as integrating) to give

$$v_+(t) = \frac{C_1}{C_1 + C_2} 12u(t) = 8u(t) \text{ V}. \qquad (6.6\text{-}5)$$

This is the same as an application of the voltage divider rule to either the impedances or capacitances. The current $i_+(t)$ is now determined by using the admittance operator for the capacitor C_2:

$$i_+(t) = C_2 p[8u(t)] = (400 \text{ nC}) \, \delta(t). \qquad (6.6\text{-}6)$$

Note that the coefficient of the delta function has units of charge (coulombs) because that constant is the integral of the current—which is charge.

Figure 6.57
The $t > 0$ circuit

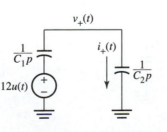

The unit impulse function is seen (by the last example) to be quite useful in switched energy storage circuits. If all independent sources are dc, or if there are no independent sources in the $t > 0$ circuit, all waveforms in such circuits are either step functions or impulses.

---

[11] In actuality, C_1 would have to have been connected to other elements not shown prior to $t = 0$ to "charge it up." We are just assuming that these elements are disconnected by opening other switches at $t = 0$. This additional circuitry is simply not shown in our figure.

The Sampling Property of the Unit Impulse Function

We will now look at an extremely useful result concerning the product of a continuous time function and a unit impulse. In Figure 6.58(a) we show a general time function $f(t)$ that is continuous at $t = a$ and the pulse approximation to the unit impulse. If we form the product for each time instant, we will have the waveform shown in Figure 6.58(b). It is a pulse type of waveform, although its top is not flat. Because $f(t)$ is continuous though, one can approximate the top as being constant, provided that ϵ is small enough. Then we can approximate the area beneath the pulse as that of a rectangle: $f(a)$. In the limit as $\epsilon \to 0$, we see that the product waveform approaches an impulse function at $t = a$ of area $f(a)$; that is,

$$\lim_{\epsilon \to 0} f(t)\delta_\epsilon(t - a) = f(a)\delta(t - a). \tag{6.6-7}$$

We will interpret this as the product of $f(t)$ and the ideal impulse function. Thus, we have shown that

$$f(t)\delta(t - a) = f(a)\delta(t - a). \tag{6.6-8}$$

This is called the *sampling property*. To relate it to something with which you might be familiar, we point out that in the recording process of a CD (compact disk), the music signal is first sampled,[12] then the coefficient of the impulse is converted to digital form with a finite number of bits precision (this requires "holding" the signal at a given value for a finite length of time) and then recorded on the disk. A famous result in communication, called the *sampling theorem,* states that under certain conditions the original information can be recovered from the sample. Example 6.16 to follow shows one way the sampling and holding operations can be done.

Figure 6.58
The sampling property

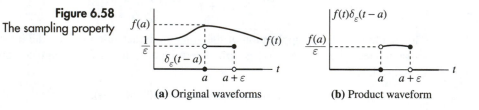

(a) Original waveforms   (b) Product waveform

Example 6.16   Find the voltage $v(t)$ in Figure 6.59.

Figure 6.59
A sample and hold circuit

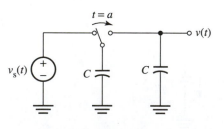

Solution   The capacitors have been specified (strictly to make the math simple) to be equal in value, though this is not necessary. Noting that the switch is closed to the left for $t \le a$ (note that a is not necessarily zero here!), the left-hand capacitor "tracks" the input signal in voltage until the time of the switch flip at $t = a$. At that time, the switch flips to the right. This uncouples the left-hand capacitor from the input signal. Observing that there has been no current through the right-hand capacitor for $t \le a$, we see that its voltage is zero

---

[12] That sampling process is repetitive. $f(t)$ is multiplied by a train of equally spaced impulses.

until that instant. For $t > a$, we have the circuit shown in Figure 6.60. Notice that we have moved our time origin to $t = a$. We can use the voltage divider rule to obtain

$$v(t) = 0.5v_s(a)u(t - a). \tag{6.6-9}$$

Figure 6.60
The $t > a$ circuit

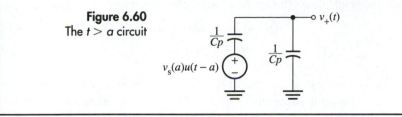

Generalized Differentiation of Discontinuous Waveforms

The following three examples illustrate the generalized differentiation of discontinuous waveforms.

Example 6.17   Compute $p[\cos(t)u(t)]$ and $p[\sin(t)u(t)]$.

Solution   We use the Leibniz rule:

$$p[\cos(t)u(t)] = -\sin(t)u(t) + \cos(t)pu(t) = -\sin(t)u(t) + \cos(t)\delta(t) \tag{6.6-10}$$
$$= -\sin(t)u(t) + \cos(0)\delta(t) = -\sin(t)u(t) + \delta(t).$$

For the second waveform, we compute

$$p[\sin(t)u(t)] = \cos(t)u(t) + \sin(t)pu(t) = \cos(t)u(t) + \sin(t)\delta(t) \tag{6.6-11}$$
$$= \cos(t)u(t) + \sin(0)\delta(t) = \cos(t)u(t).$$

The original waveforms and their derivatives are shown in Figure 6.61. There is a delta function in the derivative of the cosine because it has a jump discontinuity at $t = 0$, whereas there is none in the derivative of the sine for it does not have such a discontinuity.

Figure 6.61
Derivatives of sinusoids with and without discontinuities at the origin

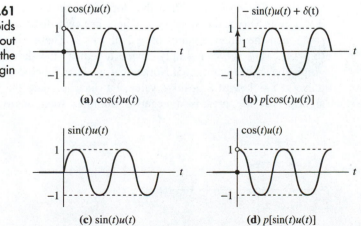

(a) $\cos(t)u(t)$

(b) $p[\cos(t)u(t)]$

(c) $\sin(t)u(t)$

(d) $p[\sin(t)u(t)]$

Figure 6.62   A discontinuous
function

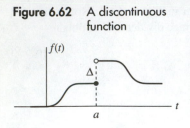

We can be a little more general with the results of the last example. Suppose that we have a waveform with a step discontinuity, such as $f(t)$ in Figure 6.62, to be differentiated. We see that it can be decomposed into the sum of a continuous function and a step function, as shown in Figure 6.63. The waveform labeled $f_c(t)$ is clearly continuous for we have removed the single discontinuity. Analytically, we can write

$$f(t) = f_c(t) + \Delta u(t - a), \tag{6.6-12}$$

where

$$f_c(t) = f(t) - \Delta u(t - a). \tag{6.6-13}$$

If the waveform $f(t)$ has a finite number of jumps in each interval of finite length, we can repeat the process for each discontinuity. In this case, the discontinuous function would be the sum of step functions occurring at different times. In any event, if we take the derivative of (6.6-12) we get

$$pf(t) = pf_c(t) + \Delta \delta(t - a). \tag{6.6-14}$$

In general, $pf(t)$ will have an impulse function of "strength" (area) Δ at each time point at which there is a jump discontinuity of size Δ.

Figure 6.63
The two constituent
functions

Example 6.18   Compute and sketch the derivative of the waveform in Figure 6.64.

Figure 6.64
An example waveform

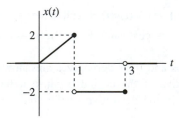

Solution   The solution is actually quite easy if we make use of graphical methods and the result in equation (6.6-14). We see that the slope (the derivative) is zero everywhere except for $0 < t \le 1$. There, the slope is 2. There are two discontinuities: one a jump of -4 at $t = 1$ and the other a jump of $+2$ at $t = 3$. These correspond to the constant Δ in (6.6-14). Thus we can simply "piece together" the derivative waveform to obtain the graph shown in Figure 6.65. Notice that we have labeled the strength of the negative impulse at $t = 1$ with a positive value, but have pointed the impulse arrow downward to signify that the area is a negative number. Analytically, we would write this im-

Figure 6.65
The derivative
waveform

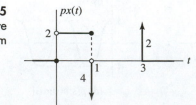

pulse waveform as $-4\delta(t - 1)$ and the one at $t = 3$ as $2\delta(t - 3)$. Notice, by the way, that the open circle at $t = 1$ has nothing to do with the impulse there; it represents the discontinuous waveform that drops from 2 to 0 at $t = 1$. This component can be written as $2u(t) - 2u(t - 1)$, or alternatively as $2P_1(t)$. Thus, we can write our derivative waveform as

$$px(t) = 2P_1(t) - 4\delta(t - 1) + 2\delta(t - 3). \tag{6.6-15}$$

Notice, by the way, that we did not compute the derivative analytically.

Example 6.18 illustrates an important fact that we did not mention explicitly. If the derivative does not exist because of a change of slope, rather than because of a discontinuity, there is no impulse ($\Delta = 0$). This is the case at $t = 0$ in the example.

Example 6.19 Using nodal analysis, find the node voltages $v_1(t)$ and $v_2(t)$ in Figure 6.66 assuming that $i_{s1}(t) = 6\delta(t)$ and $i_{s2}(t) = 12\delta(t) + 6\cos(t)u(t)$.

Figure 6.66
An example circuit

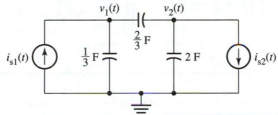

Solution We show the circuit again with the capacitors labeled with their impedance operators in Figure 6.67. The nodal equations are

$$\frac{p}{3} v_1(t) + \frac{2p}{3} [v_1(t) - v_2(t)] = i_{s1}(t) \tag{6.6-16}$$

and

$$2pv_2(t) + \frac{2p}{3} [v_2(t) - v_1(t)] = -i_{s2}(t). \tag{6.6-17}$$

If we multiply both sides of each equation by $1/p$ from the left (that is, we integrate both sides), we will have

$$v_1(t) - \frac{2}{3} v_2(t) = \frac{1}{p} i_{s1}(t) \tag{6.6-18}$$

and

$$-\frac{2}{3} v_1(t) + \frac{8}{3} v_2(t) = -\frac{1}{p} i_{s2}(t). \tag{6.6-19}$$

Now we integrate the right-hand sides to obtain

$$\frac{1}{p} i_{s1}(t) = \frac{1}{p} [6\delta(t)] = 6u(t) \tag{6.6-20}$$

and $\quad \dfrac{1}{p} [-i_{s2}(t)] = -\dfrac{1}{p} [12\delta(t) + 6\cos(t)u(t)] = -12u(t) - 6\sin(t)u(t).$ (6.6-21)

Notice, in conjunction with the last term in (6.6-21), that we have evaluated a definite integral, but the $\sin(t)$ term evaluated at $t = 0$ is zero. Using these results and placing

(6.6-18) and (6.6-19) in matrix form, we have

$$\begin{bmatrix} 1 & -\dfrac{2}{3} \\ -\dfrac{2}{3} & \dfrac{8}{3} \end{bmatrix} \begin{bmatrix} v_1(t) \\ v_2(t) \end{bmatrix} = \begin{bmatrix} 6u(t) \\ -12u(t) - 6\sin(t)u(t) \end{bmatrix}. \qquad (6.6\text{-}22)$$

Inverting, we obtain

$$\begin{bmatrix} v_1(t) \\ v_2(t) \end{bmatrix} = \dfrac{9}{20} \begin{bmatrix} \dfrac{8}{3} & \dfrac{2}{3} \\ \dfrac{2}{3} & 1 \end{bmatrix} \begin{bmatrix} 6u(t) \\ -12u(t) - 6\sin(t)u(t) \end{bmatrix} = \begin{bmatrix} 3.6u(t) - 1.8\sin(t)u(t) \\ -3.6u(t) - 2.7\sin(t)u(t) \end{bmatrix}. \qquad (6.6\text{-}23)$$

Figure 6.67
Impedance operator form

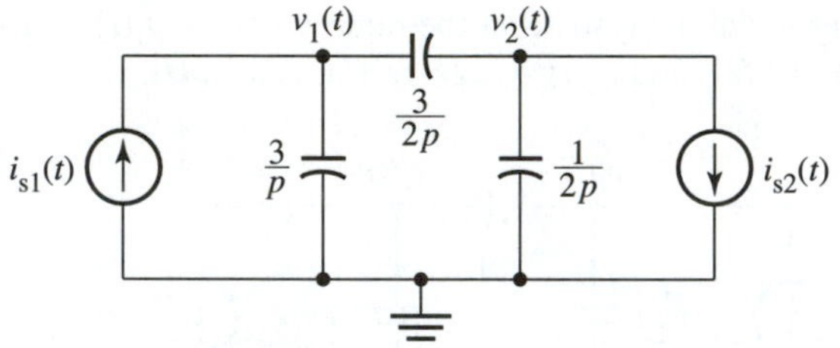

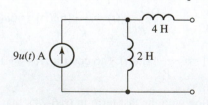

KCL in Charge Form for Switched Capacitor Circuits

If you inspect the nodal equations in the last example, you will see that—once they have been integrated—they become nothing more than KCL equations in *charge form.* The mathematics of solution, therefore, is exactly the same as that for resistive circuits, except that the current sources must be integrated to turn them into "charge sources." This is generally true of the analysis of any *single element kind circuit,* which we will define to be a circuit that has only one kind of element other than the sources. For inductive circuits, one integrates voltage to get flux linkage (rather than charge). However, it should be clear that there is not a great deal of difference between resistive circuits and capacitive or inductive circuits. One, of course, must deal with waveforms for any circuit (including resistive ones) that have waveform, rather than dc, sources. Each passive element we have discussed (resistor, inductor, capacitor) obeys Ohm's law, so all of the methods we have developed to date continue to work for any single element kind circuit. (Kirchhoff's laws do not depend upon the particular element relationships, but are only constraints imposed by the way they are interconnected.)

Thévenin/Norton Equivalents for Inductor and Capacitor Subcircuits

Example 6.20   Find the Thévenin and Norton equivalents for the inductive subcircuit in Figure 6.68.

Figure 6.68
An inductive subcircuit

Solution   If we define the voltage to be positive on the top and the current reference arrow into the top terminal as in Figure 6.69, we can compute the open circuit voltage $v_{oc}(t)$ using impedance operators:

$$v_{oc}(t) = 2p[9u(t)] = 18\delta(t). \qquad (6.6\text{-}24)$$

Notice, by the way, that the integral of the delta function is $18u(t)$ Wb (flux linkage). To find the equivalent of the deactivated network, we simply replace the i-source by an open circuit (as for resistive circuits), getting the configuration in Figure 6.70. We see that the subcircuit consists of two inductors in series, so it is equivalent to a single inductor whose impedance is the sum of the two individual impedances,

$$Z_{eq}(p) = 6p. \qquad (6.6\text{-}25)$$

Thus, the equivalent inductance is 6 H. We show the resulting Thévenin equivalent subcircuit in Figure 6.71. We can now simply do a source transformation to find the Norton equivalent subcircuit. This is equivalent to computing the short circuit current coming out of the top terminal, that is,

$$i_{sc}(t) = \frac{1}{6p}[18\delta(t)] = 3u(t), \qquad (6.6\text{-}26)$$

and assigning it to the Norton current source. The result is shown in Figure 6.72.

Figure 6.69
The open circuit voltage

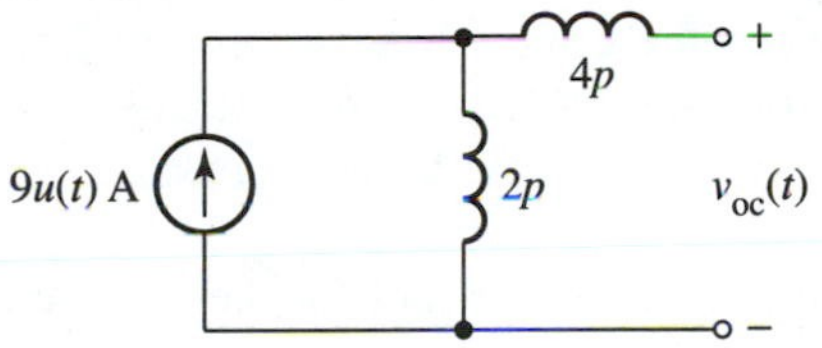

Figure 6.70
The deactivated network

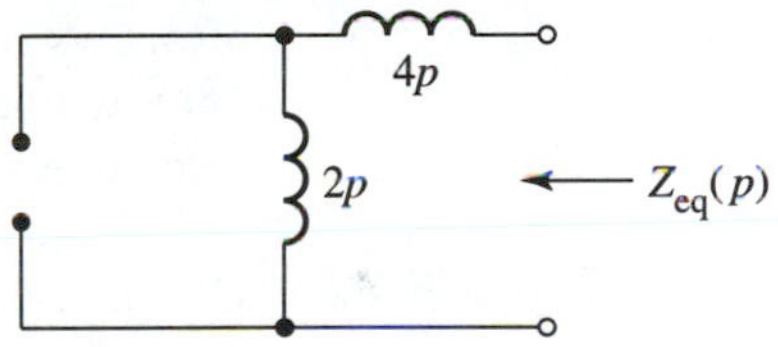

Figure 6.71
The Thévenin equivalent subcircuit

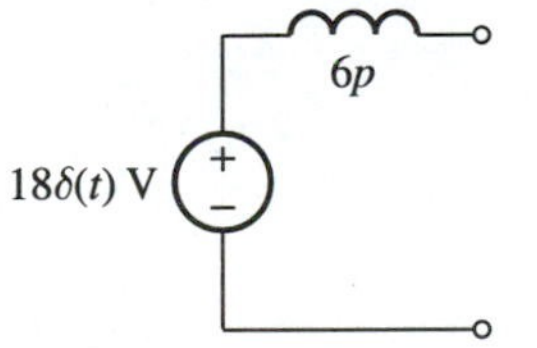

Figure 6.72
The Norton equivalent subcircuit

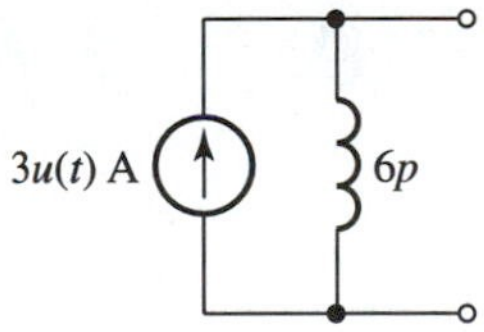

Impulsive Initial Condition Models

Our familiarity with the impulse function now makes it possible for us to find new initial condition models for the capacitor and inductor. In Section 6.5 we derived the one for the capacitor shown in Figure 6.73, in which the initial condition is established by a step voltage source. We will refer to it as the *series initial condition model* because its two elements are series connected. Suppose we compute the current, though, in terms of the voltage. We get

$$i_+(t) = Cp[v_+(t) - v(0)u(t)] = Cpv_+(t) - Cv(0)\delta(t), \qquad (6.6\text{-}27)$$

because, as we now know, the time derivative of the unit step function is the unit impulse function. This is an algebraic sum of currents, so we can interpret it as the KCL equation of the *parallel initial condition model* shown in Figure 6.74. Notice carefully the direction of the initial condition current source. It is in a direction that establishes the proper initial voltage across the capacitor. Furthermore, carefully observe that the current $i_+(t)$ is the total into the *combination* of elements—not into the capacitor symbol alone.

Figure 6.73 Series initial condition model for the capacitor

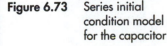
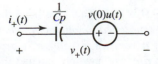

Figure 6.74   Parallel initial
condition model
for the capacitor

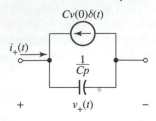

We can also derive an alternate model for the initial condition behavior of the inductor. Figure 6.75 shows the one we derived in Section 6.5. It is a parallel model. If we solve for the voltage $v_+(t)$, however, we find that

$$v_+(t) = Lp[i_+(t) - i(0)u(t)] = Lpi_+(t) - Li(0)\delta(t). \tag{6.6-28}$$

We can interpret this as the KVL equation of the series subcircuit shown in Figure 6.76.

Figure 6.75
Parallel initial condition
model for the inductor

Figure 6.76
Series initial condition model
for the inductor

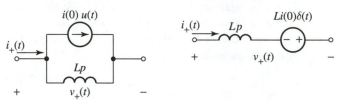

Choice of Initial Condition Model

A natural question is: "Which model do I use?" However, there is no clear-cut answer. For instance, both v-sources and i-sources can be easily accommodated in both nodal and mesh analysis. Because each v-source reduces the number of equations by one for nodal analysis, one might think that the series models would be more appropriate for that type of analysis; however, these models have three nodes, whereas the parallel models have only two. In fact, one should learn to manipulate all the initial condition models and decide upon which to use on a "case-by-case" basis.

Example 6.21   Use the series initial condition model for the inductors to determine the current $i(t)$ and the voltage $v(t)$ in the circuit shown in Figure 6.77. Assume that the circuit is operating in the dc steady state prior to $t = 0$.

Figure 6.77
An example circuit

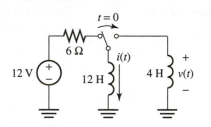

Solution   The $t \le 0$ circuit (dc steady state) is shown in Figure 6.78. It is clear that the values are as shown in the figure. Now we flip the switch and use the series model we have just derived for each inductor; this results in the $t > 0$ circuit of Figure 6.79. We see at once that

$$i_+(t) = \frac{1}{16p}[24\delta(t)] = 1.5u(t) \text{ A} \tag{6.6-29}$$

and

$$v_+(t) = -4pi_+(t) = -4p\frac{1}{16p}[24\delta(t)] = -6\delta(t) \text{ V}. \tag{6.6-30}$$

Actually, of course, we know that the value of $i(t)$ for $t \le 0$ is 2 A. Thus, we have the waveforms shown in Figure 6.80 for $i(t)$ and $v(t)$. *Notice the practical implication: when one attempts to switch a current suddenly in an inductor, a very large voltage is produced. In practice, this would cause the switch contacts to arc.* We have not labeled the unit on the vertical voltage axis in Figure 6.80(b) because the impulse function reaches to negative infinity in height; however, we note that its area ("strength") of 6 has units of flux linkage (Wb). We also remind you that when we turn the impulse arrow upside down, we do not use the negative sign on its area.

Figure 6.78
The $t \le 0$ circuit

Figure 6.79
The $t > 0$ circuit

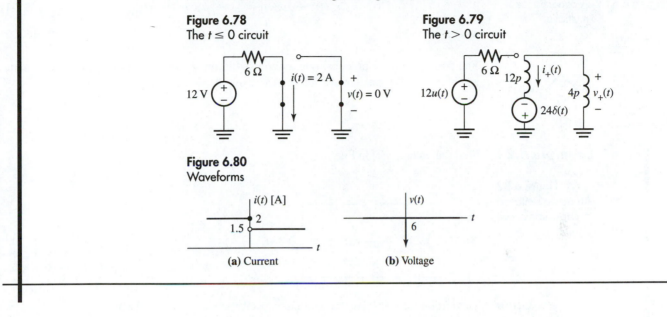

Figure 6.80
Waveforms

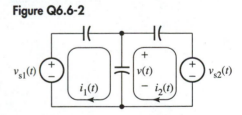

(a) Current

(b) Voltage

Section 6.6 Quiz

Q6.6-1. Find the Thévenin equivalent for the portion of the circuit in Figure Q6.6-1 to the left of the terminals marked a and b.

Figure Q6.6-1

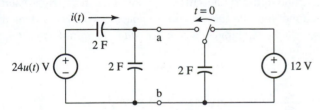

Q6.6-2. Derive the Norton equivalent to the left of terminals a and b in Figure Q6.6-1.

Q6.6-3. Find $i(t)$ in the circuit of Figure Q6.6-1.

Q6.6-4. Assume that $v_{s1}(t) = 12u(t)$, that $v_{s2}(t) = 24u(t)$, and that all three capacitors have a value of 1 F in the circuit shown in Figure Q6.6-2. Using mesh analysis, find the mesh currents $i_1(t)$ and $i_2(t)$. Use these mesh currents to find $v(t)$.

Figure Q6.6-2

6.7 | Active Energy Storage Circuits

In this section we will investigate capacitive and inductive circuits having active elements, both with and without switches. We will also develop a network property that actually applies to passive circuits as well, though it is most often used to analyze active networks. Although some circuits will contain resistors, their basic behavior will be that of a single-element-kind circuit; the resistors will only be present to set voltage gains for op amps, etc.

A Capacitance
Multiplier

Figure 6.81 shows a subcircuit consisting of a single capacitor and a VCVS. We can determine its *v-i* characteristic immediately by writing

Figure 6.81   A capacitance
multiplier

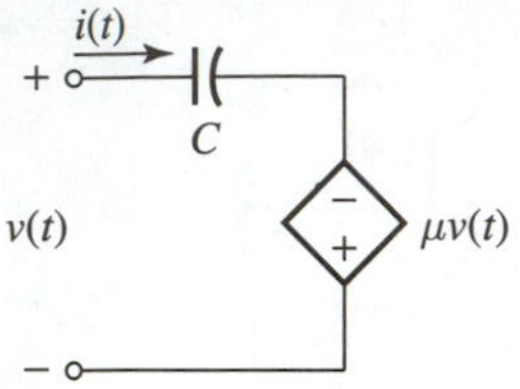

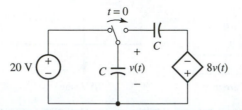

$$i(t) = Cp[v(t) + \mu v(t)] = (1 + \mu)Cpv(t) = C_{eq}pv(t), \qquad (6.7\text{-}1)$$

where

$$C_{eq} = (1 + \mu)C. \qquad (6.7\text{-}2)$$

The circuit is a very practical one, often used in integrated circuit designs to increase the value of a small capacitance without increasing the area exorbitantly.

Example 6.22   Find the voltage $v(t)$ in Figure 6.82.

Figure 6.82
An active switched
capacitor circuit

Solution   The circuit for $t \le 0$ has the vertical capacitor connected directly across the dc v-source, so the circuit is in the dc steady state. The horizontal capacitor is then in series with an open circuit, so its current is identically zero. By causality, therefore, its voltage must also be zero. The circuit for $t > 0$ is shown in Figure 6.83. There, we have replaced the capacitance multiplier subcircuit consisting of the horizontal capacitor and its series VCVS with the equivalent capacitance of $9C$. We have labeled each capacitor with its operator impedance, have used the initial condition models for the capacitors, and have made the independent v-source one sided (though it does not affect the variables in the right-hand part of the circuit at all). We remind you that the one-sided capacitor voltage $v_+(t)$ that we are looking for is that across the C farad capacitance *and its initial condition v-source*. The voltage divider rule gives

$$v_+(t) = \frac{\dfrac{1}{9Cp}}{\dfrac{1}{Cp} + \dfrac{1}{9Cp}} \, 20u(t) = 2u(t) \text{ V}. \qquad (6.7\text{-}3)$$

Actually, $v(t)$ is not zero for $t \le 0$—but 20 V, as we saw above. Thus, $v(t)$ has the waveform shown in Figure 6.84.

Figure 6.83
The $t > 0$ equivalent circuit

Figure 6.84
The response waveform

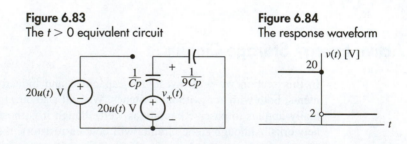

Miller Equivalents   We will now develop two forms of a network equivalence that is most often used in active circuit analysis, even though it is applicable to passive networks as well. In Figure 6.85 we show a three-terminal subcircuit having an element "bridging" its two nongrounded terminals. We have labeled that element $Z(p)$; we assume that the element is either a resistor ($Z(p) = R$), an inductor ($Z(p) = Lp$), or a capacitor ($Z(p) = 1/Cp$). When we have developed analysis procedures for circuits having all types of elements, we will see that our development here will work if this element is replaced by a general two-terminal subcircuit consisting of any number of resistors, capacitors, inductors, and/or dependent sources. For now, however, we will assume that it is a single element and *that the voltage ratio is a real constant μ*, that is

$$\frac{v_2(t)}{v_1(t)} = \mu. \tag{6.7-4}$$

The third terminal, by the way, does not have to be grounded, though it often is in applications.

Figure 6.85
Subcircuit for derivation of the Miller equivalent

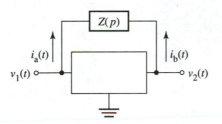

We compute the currents $i_a(t)$ and $i_b(t)$. They are

$$i_a(t) = \frac{1}{Z(p)}[v_1(t) - v_2(t)] = \frac{1}{\dfrac{Z(p)}{1 - \dfrac{v_2(t)}{v_1(t)}}} v_1(t) = \frac{1}{Z_{m1}(p)} v_1(t) \tag{6.7-5}$$

and

$$i_b(t) = \frac{1}{Z(p)}[v_2(t) - v_1(t)] = \frac{1}{\dfrac{Z(p)}{1 - \dfrac{v_1(t)}{v_2(t)}}} v_1(t) = \frac{1}{Z_{m2}(p)} v_2(t). \tag{6.7-6}$$

Using (6.7-4), we have

$$i_a(t) = \frac{1}{Z(p)/(1 - \mu)} v_1(t) = \frac{1}{Z_{m1}(p)} v_1(t) \tag{6.7-7}$$

and

$$i_b(t) = \frac{1}{Z(p)/\left(1 - \dfrac{1}{\mu}\right)} v_2(t) = \frac{1}{Z_{m2}(p)} v_2(t). \tag{6.7-8}$$

Because these two currents are proportional to the two node voltages $v_1(t)$ and $v_2(t)$, respectively, independently of the other voltage, we see that the equivalent circuit shown in Figure 6.86 is valid. The single floating element bridging the two terminals in the original subcircuit has become two *grounded* elements having the impedance operators

$$Z_{m1}(p) = \frac{Z(p)}{1 - \mu} \qquad (6.7\text{-}9)$$

and

$$Z_{m2}(p) = \frac{Z(p)}{1 - \dfrac{1}{\mu}}. \qquad (6.7\text{-}10)$$

Figure 6.86
The Miller equivalent
subcircuit

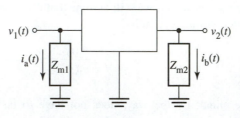

Example 6.23   Find the Miller equivalent of the subcircuit shown in Figure 6.87.

Figure 6.87
An example subcircuit

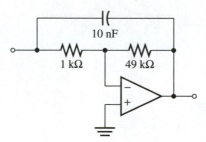

Solution   Here, although we have resistors in the circuit as well as a capacitor and an active element, we will see that our theory of Miller equivalents works. The only problem is identifying the three-terminal subcircuit. To do this, we simply lift out the capacitor and identify everything that remains as this subcircuit, which is shown in Figure 6.88. We see that

$$\mu = \frac{v_2(t)}{v_1(t)} = -49 \qquad (6.7\text{-}11)$$

independently of whether or not the capacitor is connected. Therefore,

$$Z_{m1}(p) = \frac{1/Cp}{1 - \mu} = \frac{1}{50Cp} \qquad (6.7\text{-}12)$$

and

$$Z_{m2}(p) = \frac{1/Cp}{1 - \dfrac{1}{\mu}} = \frac{1}{0.98Cp}. \qquad (6.7\text{-}13)$$

Thus, each of the Miller impedances is that of an equivalent capacitor, as shown in Figure 6.89, where we have merely swung the capacitors to ground and replaced them with their equivalent elements of $50 \times 10 \text{ nF} = 500 \text{ nF} = 0.5 \ \mu\text{F}$ and $0.98 \times 10 \text{ nF} = 9.8 \text{ nF}$. As you can see, this circuit (relative to the input terminals) is a capacitance multiplier—though a "lossy one" because of the 1-kΩ resistor at the input.

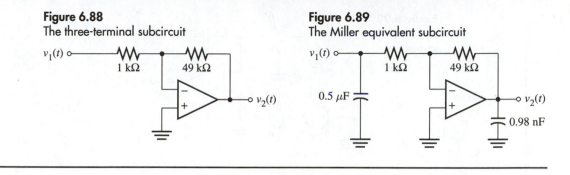

Figure 6.88
The three-terminal subcircuit

$v_1(t)$ ○——WW——•——WW—— 1 kΩ   49 kΩ   —○ $v_2(t)$

Figure 6.89
The Miller equivalent subcircuit

$v_1(t)$ ○——•——WW——•——WW—— 1 kΩ   49 kΩ   —○ $v_2(t)$

0.5 μF

0.98 nF

The preceding Miller equivalent was derived on a nodal analysis basis; there is a similar one that can be derived on a mesh basis. Figure 6.90 shows the situation: a three-terminal circuit with the third terminal connected in series with a two-terminal element. The KVL equations are

$$v_1(t) = v_a(t) + Z(p)[i_1(t) + i_2(t)] \tag{6.7-14}$$

and

$$v_2(t) = v_b(t) + Z(p)[i_2(t) + i_1(t)]. \tag{6.7-15}$$

We factor $i_1(t)$ from the first and $i_2(t)$ from the second. If

$$\frac{i_2(t)}{i_1(t)} = \beta, \tag{6.7-16}$$

a real constant, then we have

$$v_1(t) = v_a(t) + Z_{m1}(p)i_1(t) \tag{6.7-17}$$

and

$$v_2(t) = v_b(t) + Z_{m2}(p)i_2(t), \tag{6.7-18}$$

where

$$Z_{m1}(p) = (1 + \beta)Z(p) \tag{6.7-19}$$

and

$$Z_{m2}(p) = \left[1 + \frac{1}{\beta}\right]Z(p). \tag{6.7-20}$$

We can interpret equations (6.7-17) and 6.7-18) as shown in Figure 6.91. We will not work an example here, but leave it for a review quiz question. The next one is merely an exercise in solving circuits with op amps, switches, and energy storage elements.

Figure 6.90
Another configuration

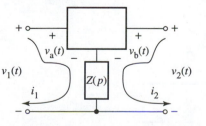

Figure 6.91
The mesh form of the Miller equivalent

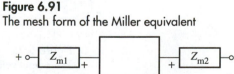
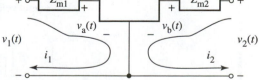

Analysis Examples

Example 6.24   Find the response voltage of the *noninverting sc* (sc stands for "switched capacitor") *integrator* circuit shown in Figure 6.92, assuming that $v_o(t) = 2$ V for $t \le 0$.

Figure 6.92
The noninverting
sc integrator

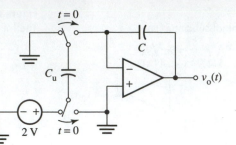

Solution   The circuit for $t \leq 0$ is shown in Figure 6.93. The virtual short constraint of the op amp dictates that the voltage across C is of the polarity and value shown in the figure. The $t > 0$ circuit is shown in Figure 6.94. We note that the current $i_+(t)$ is the current in the input capacitor and—by the virtual short principle of the op amp—it is the current in the feedback capacitor as well. We compute

$$i_+(t) = C_u p[-2u(t)] = -2C_u \delta(t). \tag{6.7-21}$$

Then we find the output voltage:

$$v_{o+}(t) = 2u(t) - \frac{1}{Cp} i_+(t) = 2u(t) - \frac{1}{Cp}[-2C_u\delta(t)] = 2\left[1 + \frac{Cu}{C}\right] u(t) \text{ V}. \tag{6.7-22}$$

Now $v_{o+}(t)$ is zero for $t \leq 0$, but we know this is not true of the original waveform; in fact, we have $v_o(t) = 2$ V for all $t \leq 0$. Thus, $v_o(t)$ has the waveform shown in Figure 6.95.

Figure 6.93
The $t \leq 0$ circuit

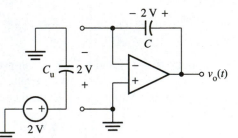

Figure 6.94
The $t > 0$ circuit

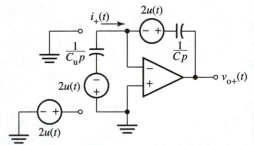

Figure 6.95
The output voltage

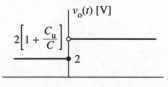

Example 6.25   Determine the equivalent subcircuit between terminal a and ground for the subcircuit shown in Figure 6.96.

Figure 6.96
An impedance converter

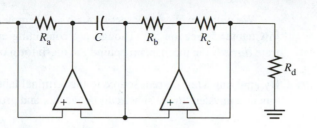

Solution We attach a test source (arbitrarily chosen to be a v-source) between terminal a and ground, as shown in Figure 6.97. The virtual shorts at the op amp inputs dictate that the three nodes labeled $v(t)$ do, indeed, carry that voltage. Now we can compute

$$i_1(t) = \frac{v(t)}{R_d}. \tag{6.7-23}$$

This, along with the zero current virtual short constraint on the right-hand op amp, gives

$$v_1(t) = v(t) + R_c i_1(t) = \left[1 + \frac{R_c}{R_d}\right]v(t). \tag{6.7-24}$$

Then
$$i_2(t) = \frac{v(t) - v_1(t)}{R_b} = -\frac{R_c}{R_b R_d}v. \tag{6.7-25}$$

Recognizing that the current into the lead attached to the junction of the two op amp input terminals is zero gives

$$v_2(t) = v(t) + \frac{1}{Cp}i_2(t) = \left[1 - \frac{R_c}{R_b R_d C p}\right]v(t). \tag{6.7-26}$$

Finally, we compute
$$i(t) = \frac{v(t) - v_2(t)}{R_a} = \frac{R_c}{R_a R_b R_d C p}v(t). \tag{6.7-27}$$

The coefficient of $v(t)$ is the equivalent admittance. Inverting, we find that

$$Z(p) = L_{eq}p, \tag{6.7-28}$$

$$L_{eq} = \frac{R_a R_b R_d}{R_c}C. \tag{6.7-29}$$

Thus, the capacitance has been "converted" into an equivalent inductance. This inductor, unfortunately for some applications, has one terminal grounded as in Figure 6.98.

Figure 6.97
Finding the equivalent impedance

Figure 6.98
Equivalent subcircuit

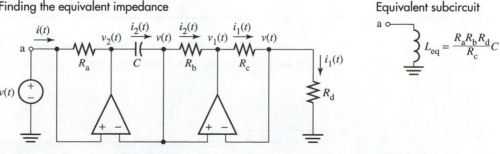

We will now close this section with an example involving a BJT, which we discussed in Chapter 5, Section 5.7. Though not containing energy storage elements, it does illustrate *directly* the mechanism behind the mesh form of the Miller equivalent.

Example 6.26   By applying a test current source to the terminal labeled x in Figure 6.99, find a single element equivalent to the subcircuit between x and ground.

Figure 6.99
An example BJT small signal subcircuit

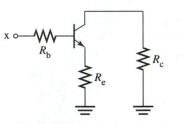

Solution   We remind you that "small signal" means that all the dc sources required to provide power to the BJT have been deactivated; thus, in the lab, dc "bias" voltages would be added to any voltages we might cause by applying a signal source to produce the actual total instantaneous values. In Figure 6.100 we show the circuit that results when we apply a test current source having the general literal value i and replace the BJT symbol by its equivalent hybrid π model, which we developed in Chapter 5, Section 5.7. We have identified mesh currents in anticipation of performing a mesh analysis.[13] In fact, we count $B - N + 1 - N_\mathrm{I} = 6 - 5 + 1 - 2 = 0$ mesh equations required! Thus, we must only apply KVL around the left-hand mesh to find the node voltage v_x:

$$v_\mathrm{x} = R_\mathrm{b}i + r_\pi i + R_\mathrm{e}(i + \beta i) = (R_\mathrm{b} + r_\pi + [1 + \beta]R_\mathrm{e})i. \qquad (6.7\text{-}30)$$

Notice that we have also identified i_b with i in writing this equation. Thus, as v_x is proportional to i, the constant of proportionality is an equivalent resistance with value

$$R_\mathrm{eq} = R_\mathrm{b} + r_\pi + [1 + \beta]R_\mathrm{e}. \qquad (6.7\text{-}31)$$

Figure 6.100
Using the hybrid π equivalent for the BJT

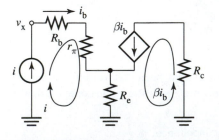

We will leave it to you to apply the mesh form of the Miller equivalent shown in Figure 6.91 to verify that it works by producing equation (6.7-31) with less computation.

---

[13] We remind you again that the three ground symbols represent only one node. Thus, each mesh current that flows into that node must flow out.

Section 6.7 Quiz

Q6.7-1.   Find the Miller equivalent for the subcircuit shown in Figure Q6.7-1.

Figure Q6.7-1

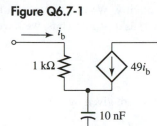

Q6.7-2.   If the capacitor C is initially uncharged, find the voltage $v(t)$ in Figure Q6.7-2.

Figure Q6.7-2

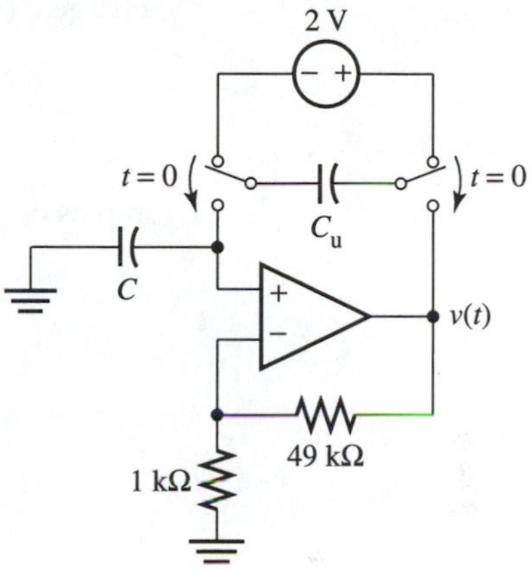

6.8 | Compatible Systems of Units

In many circuit analysis problems the notation can be simplified greatly by using systems of convenient units. Let's start the discussion of this idea by considering Ohm's law for the resistor, $v = Ri$, considered as an equation relating units. We can write it in the form

$$\frac{v}{Ri} = 1. \tag{6.8-1}$$

The numeral 1 is, of course, itself unitless. The "unit form" of equation (6.8-1) is

$$\frac{\text{V}}{\Omega\text{A}} = 1. \tag{6.8-2}$$

Now look at some of the things we can do—all without changing the ratio:

$$\frac{\text{V}}{\text{k}\Omega\text{ mA}} = \frac{\text{V}}{\text{M}\Omega\ \mu\text{A}} = \frac{\text{mV}}{\Omega\text{ mA}} = 1. \tag{6.8-3}$$

We call any set of units for voltage, current, and resistance that make the above ratio equal to unity a *system of compatible units*. We have already used this idea in a somewhat informal manner earlier in the text.

Let's look at the basic capacitor relation:

$$q = Cv. \tag{6.8-4}$$

We can rewrite it as

$$\frac{q}{Cv} = 1. \tag{6.8-5}$$

This means that the following systems of units are compatible:

$$\frac{C}{FV} = \frac{\mu C}{\mu F\ V} = \frac{nC}{\mu F\ mV} = \frac{pC}{nF\ mV} = 1. \tag{6.8-6}$$

So far, time has not entered the picture explicitly. We can factor it in by expressing the charge q (in a form only concerned with units) as $q = it$.[14] Thus, we have

$$\frac{q}{it} = 1. \tag{6.8-7}$$

Examples of consistent systems of units, therefore, are given by

$$\frac{C}{As} = \frac{\mu C}{mA\ ms} = \frac{nC}{mA\ \mu s} = 1. \tag{6.8-8}$$

We can, of course, combine this with our equation defining capacitance in the form

$$\frac{q}{Cv} = \frac{it}{Cv} = 1. \tag{6.8-9}$$

We can now find compatible systems of units for time, voltage, current, and capacitance:

$$\frac{As}{FV} = \frac{mA\ ms}{\mu F\ V} = \frac{mA\ \mu s}{nF\ V} = \frac{\mu A\ \mu s}{pF\ V} = 1. \tag{6.8-10}$$

Many other examples are possible. If it proves to be convenient, you could even use units of 10 μA, μs, 10 pF, and V. The ten's would cancel in the above ratio, and the units would therefore be compatible.

Examples of the Use of Compatible Units

Example 6.27   Find the node voltages $v_1(t)$, $v_2(t)$, and $v_3(t)$ in the circuit shown in Figure 6.101 assuming that $i_s(t) = 30\delta(t)$ nC.

Figure 6.101
An example circuit

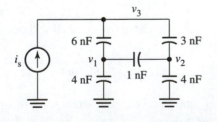

Solution   Let's use the charge form of KCL and integrate the current source waveform, thus producing

$$q_s(t) = \frac{1}{p} i_s(t) = 360u(t)\ nC. \tag{6.8-11}$$

---

[14] Of course charge is actually the time integral of current; however, an integral is basically a sum of products of current and time increments, so the "unit form" is $q = it$.

The charge form of KCL that we will use is

$$\sum_{\text{node}} q_{\text{out}}(C\text{'s}) = \sum_{\text{node}} q_{\text{in}}(\text{``charge'' sources}). \tag{6.8-12}$$

We have referred to the current source as a "charge" source because we have specified its waveform in terms of charge rather than current. In matrix form, using $q = Cv$, we have

$$\begin{bmatrix} 11 & -1 & -6 \\ -1 & 8 & -3 \\ -6 & -3 & 9 \end{bmatrix} \begin{bmatrix} v_1 \\ v_2 \\ v_3 \end{bmatrix} = \begin{bmatrix} 0 \\ 0 \\ 360 \end{bmatrix} u(t). \tag{6.8-13}$$

The "nodal capacitance matrix" on the left has the usual symmetry properties: the diagonal elements are the sum of the capacitances attached to the associated node, and the off-diagonal elements are the negatives of the capacitances connected between the respective node pair. We will leave it to you to solve (6.8-13) (row reduction, as covered in Appendix A, is perhaps the easiest) and thereby verify that

$$\begin{bmatrix} v_1 \\ v_2 \\ v_3 \end{bmatrix} \begin{bmatrix} 51 \\ 39 \\ 87 \end{bmatrix} u(t) \text{ V}. \tag{6.8-14}$$

The circuit of the preceding example is representative of an entire class of modern integrated circuits called "charge distribution" networks. The ideal nature of the step function node voltage waveforms is only an approximation. In practice, the charge will begin to "leak off" due to the parasitic resistance of the dielectric between the capacitor plates. This idea is illustrated in more detail in Chapter 7, Section 7.6, where we discuss realistic models for the passive elements (R, L, and C).

The next example is one concerned with a much simpler circuit, but with more complicated relationships between the units in a compatible system.

Example 6.28   Find the voltage $v(t)$ in the circuit shown in Figure 6.102.

Figure 6.102
An example circuit

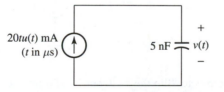

Solution   The greater complexity of units lies in the specification of the current source waveform. That specification means that (for instance) the current at $t = 2$ microseconds is 40 milliamperes. We look at the dimensionless ratio

$$\frac{\text{mA } \mu\text{s}}{\text{nF V}} = 1. \tag{6.8-15}$$

This means that we can use the compatible units of mA, μs, nF, and V. The basic capacitor equation in terms of this system is

$$v(t) = \frac{1}{5p} [20tu(t)] = 2t^2 u(t). \tag{6.8-16}$$

This result, for instance, means that $v(2\ \mu s) = 2(2)^2 = 8$ V. Notice that $u(t) = 1$ for $t > 0$ and $u(t) = 0$ for $t \le 0$ regardless of the units used for the time variable.

Example 6.29   Find the voltage $v(t)$ in the circuit shown in Figure 6.103.

Figure 6.103
A switched capacitor
example

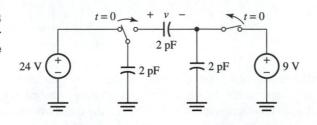

Solution   This is an example of another charge distribution circuit often referred to as a *switched capacitor* network. We can manipulate our dimensionless ratio into the form

$$\frac{pC}{pF\ V} = \frac{\mu A\ \mu s}{pF\ V} = 1. \tag{6.8-17}$$

Thus, we can elect to use the compatible system of units consisting of μA, μs, pC, pF, and V. Otherwise, we solve this switched circuit as usual. We begin by investigating the $t \le 0$ circuit shown in Figure 6.104. The voltage across the horizontal capacitor is zero, because the current through it has been zero since $t = -\infty$ and causality says that the voltage must therefore be zero (the charge, being the integral of the current, is zero, and therefore $v = q/c$ is also zero). The other two capacitors are connected directly in parallel with voltage sources; hence, their voltages are constrained to be the respective voltage source values.

Figure 6.104
The $t \le 0$ circuit

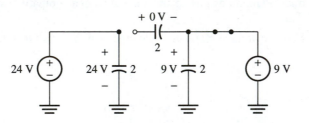

Our next step is to "flip the switches" and draw the "one-sided" equivalent circuit using the initial condition models for the capacitors. The result is shown in Figure 6.105. The voltage $v_+(t)$ is given by voltage division,

$$v_+(t) = \frac{\dfrac{1}{2p}}{\dfrac{1}{2p} + \dfrac{1}{2p} + \dfrac{1}{2p}}[24u(t) - 9u(t)] = 5u(t). \tag{6.8-18}$$

Checking our units, we see that the result is in volts. Because $v(t)$ was zero for all $t \le 0$, we see that $v(t) = v_+(t)$ as given in (6.8-18); that is, the equation is actually valid for all time. Notice once again that $u(t) = 1$ for $t > 0$ and $u(t) = 0$ for $t \le 0$ regardless of the units used for the time variable.

Figure 6.105
The $t \geq 0$ circuit

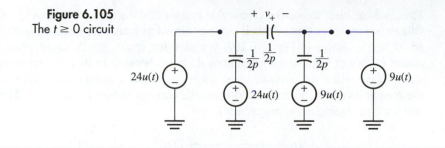

Although charge transfer and switched capacitor circuits are somewhat more important (at least in integrated circuit electronics) than those involving inductors, the latter are important as well. So let's have a look at the basic inductor unit-defining relationship, $\lambda = Li$. In dimensionless form, we have

$$\frac{\lambda}{Li} = 1. \tag{6.8-19}$$

We can therefore select a compatible set of units according to this equation in terms of

$$\frac{\text{Wb}}{\text{HA}} = \frac{\text{mWb}}{\text{H mA}} = \frac{\mu\text{Wb}}{\text{mH mA}} = 1, \tag{6.8-20}$$

as well as any other set that leaves this ratio invariant.

Factoring time in via the relationship $\lambda(t) = (1/p)v(t)$, we have[15]

$$\frac{vt}{Li} = \frac{vt^2}{Lq} = 1. \tag{6.8-21}$$

This allows us to use (for example) the following compatible units:

$$\frac{\text{Vs}}{\text{HA}} = \frac{\text{Vms}}{\text{H mA}} = \frac{\text{V}\mu\text{s}}{\text{mH mA}} = 1. \tag{6.8-22}$$

As of now, we do not need the alternate expression in (6.8-21) in terms of charge for we are only considering circuits with a single kind of element (R, L, or C). When we deal with circuits containing a mixture of all three types in Chapter 9 we will be able to use it to relate the units involved in the capacitor relationship with those involving inductors and resistors.

Example 6.30   Find the current $i(t)$ in the circuit shown in Figure 6.106.

Figure 6.106
A switched inductor circuit

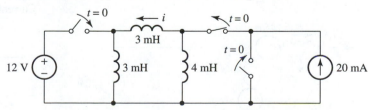

Solution   We can use our dimensionless ratio in (6.8-22) in the form

$$\frac{\text{V}\mu\text{s}}{\text{mH mA}} = 1. \tag{6.8-23}$$

---

[15] Note once again that the integral equation—with regard to units or dimensions—is the same as $\lambda = vt$.

Thus, we can elect to use the compatible system of units consisting of mA, μs, H, and V. Otherwise, we solve this switched inductor circuit as usual. We begin by investigating the $t \leq 0$ circuit shown in Figure 6.107. We have not drawn the dc steady-state equivalent circuit (although steady-state conditions do hold) because in this "inductor only" circuit we must use either the operator impedances or the flux-current relationships. Either way, the current division rule that results gives the currents shown in the figure. The computation for i, for example, is (in operator form)

$$i(t) = \frac{4p}{3p + 3p + 4p} i_s(t) = \frac{4}{10} \times 20 = 8. \tag{6.8-24}$$

Figure 6.107
The $t \leq 0$ circuit

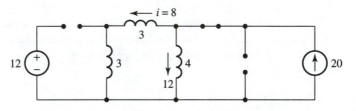

Our next step is to "flip the switches" and draw the "one-sided" equivalent circuit using the initial condition models for the inductors. The result is shown in Figure 6.108. Note the polarities of the initial condition v-sources and note that their values have been computed as the product of the initial current value and the inductance value. Using KVL around the loop having the top $24\delta(t)$ v-source, the $12u(t)$ v-source, and the $48\delta(t)$ v-source, we see that the current $i_+(t)$ is given by

$$-i_+(t) = \frac{12u(t) - 24\delta(t) + 48\delta(t)}{3p + 4p} = \frac{12}{7} tu(t) + \frac{24}{7} u(t) \tag{6.8-25}$$

$$= \frac{12}{7} (t + 2)u(t).$$

Checking our units, we see that the result is in mA. Because $i(t)$ was actually equal to 8 mA for $t \leq 0$, we see that

$$i(t) = \begin{cases} 8mA; & t \leq 0 \\ -\dfrac{12}{7} (t + 2)\text{mA}; & t > 0 \end{cases} \tag{6.8-26}$$

We have sketched this current in Figure 6.109.

Figure 6.108
The $t > 0$ circuit

Figure 6.109
Current waveform

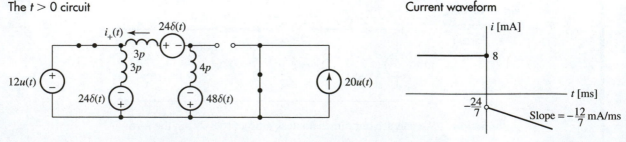

Notice that the response current $i(t)$ in the preceding example does not wind up in the dc steady state as t becomes large. This is because of the 12 V dc source.

Section 6.8 Quiz

Q6.8-1. Find a compatible set of units for the circuit shown in Figure Q6.8-1. Note that you are not asked to solve the circuit— merely to determine a convenient set of compatible units.

Figure Q6.8-1

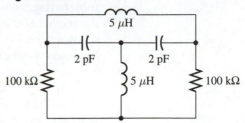

Chapter 6 Summary

Chapter 6 has presented the main ideas behind *time domain analysis:* finding a given response voltage or current variable in a circuit when the independent sources have prescribed *waveforms* (functions of time).

We started with the observation that resistive circuits do not drastically modify waveforms, but instead perform linear combinations on the waveforms of the independent sources. Another way of expressing this property is to say that they are *memoryless.*

Capacitors and *inductors,* on the other hand, have *memory.* The capacitor stores charge, the time integral of current; thus, it remembers (to a degree) how the current varies with time. The inductor performs a similar function for voltage, storing flux linkage. Both elements are *lossless,* and can therefore give back up to the surrounding circuit any energy they have stored. The energy stored in a capacitor is $0.5Cv^2$ joules and that stored in an inductor is $0.5Li^2$ joules.

We showed that capacitances in series and parallel subcircuits "add" like conductances and inductances "add" like resistances.

We defined the *dc steady-state* operation of a circuit with dc sources as the condition in which all voltages and currents are constants; thus, an inductor is equivalent to a short circuit and a capacitor to an open circuit in the dc steady state. We pointed out, however, that not all circuits with dc sources operate in the dc steady state. A requirement for such operation is that all circuit response variables be *stable.*

We discussed the *operator notation, p* for the time derivative and $1/p$ for the time integral. Its use makes the manipulation of time domain equations equivalent to those for the analysis of dc circuits. We defined the *impedance* and *admittance operators* for the inductor as $Z(p) = Lp$ and $Y(p) = 1/Lp$, respectively, and for the capacitor as $Z(p) = 1/Cp$ and $Y(p) = Cp$, respectively. In this form, both the inductor and the capacitor obey Ohm's law: $v(t) = Z(p)i(t)$, or $i(t) = Y(p)v(t)$.

We showed that the differentiation operator p and integration operator $1/p$ are *inverses* of each other, that is $p \times 1/p = 1/p \times p = 1$, provided that all waveforms upon which they operate (and produce) are *one sided,* that is, are identically zero for all $t \leq 0$. This means that one can manipulate these operators exactly like real or complex numbers, *provided they are kept on the left side of the time waveforms upon which they are to operate.*

We developed *initial condition models* for the inductor and the capacitor. These consist of a parallel step function current source for the inductor and a series step function voltage source for the capacitor. These sources provide all the information about the past history of the given element to enable the computation of a unique response for all $t \geq 0$, given the stimulus for all $t \geq 0$. (That is, knowing the voltage for all $t > 0$ and the initial current for an inductor, one can compute the future current waveform, and, similarly, knowing the current for all $t > 0$ and the initial voltage for a capacitor enables one to compute the future voltage waveform.)

We next discussed the concept of an *impulse function*, which occurs when a waveform having a discontinuity is differentiated, the *strength* (or area) of the impulse being the height of the discontinuity (the jump). This allowed us to develop alternate initial condition models for the inductor and capacitor in which the initial conditions are introduced by means of impulsive sources.

We analyzed a number of active energy storage circuits, in the process developing two forms of Miller equivalents, which are useful in electronic circuit design.

Finally, we discussed compatible systems of units, the use of which serves to make practical element values suitable for hand analysis.

Chapter 6 Review Quiz

RQ6.1   Solve the leftmost circuit in Figure RQ6.1 for v, thus showing that it is a linear combination of $v_s(t)$ and $i_s(t)$. Then tape the VCCS in the rightmost circuit and show that v is the same linear combination of source waveforms you found for the leftmost circuit (with i_c replacing i_s). Then untape the VCCS and find v in terms of $v_s(t)$. *Note:* This question illustrates the fact that a taped dependent source can be treated precisely the same as an independent one—but that a final untaping step must be performed. If both independent sources in the leftmost circuit are unit step functions, compute and sketch the waveform for $v(t)$. Then do the same with the rightmost circuit assuming a unit step waveform for the voltage source.

Figure RQ6.1

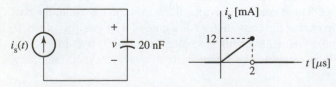

RQ6.2   Find the current waveform $i(t)$ for the circuit in Figure RQ6.2 and sketch it.

Figure RQ6.2

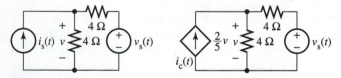

RQ6.3   Find the voltage waveform $v(t)$ for the circuit in Figure RQ6.3 and sketch it.

Figure RQ6.3

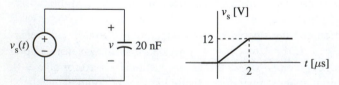

RQ6.4   Find the equivalent capacitance between terminals a and b in Figure RQ6.4. All capacitances are 1 μF.

Figure RQ6.4

RQ6.5   Find the voltage $v(t)$ in Figure RQ6.5.

Figure RQ6.5

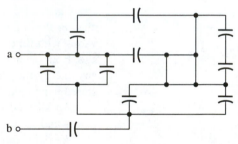

RQ6.6   Find the current $i(t)$ in Figure RQ6.6.

Figure RQ6.6

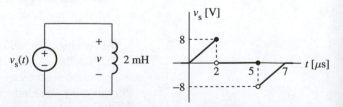

RQ6.7   Find the equivalent inductance in Figure RQ6.7. All inductances are 1 mH.

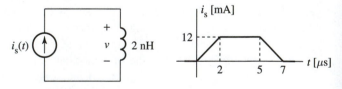

Figure RQ6.7

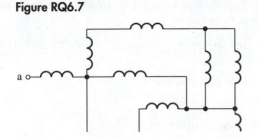

Figure RQ6.9

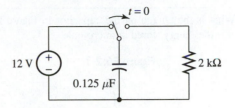

RQ6.8 Draw the dc steady-state equivalent circuit for the one shown in Figure RQ6.8 and compute v (under dc steady-state conditions).

RQ6.10 Compute the time derivative of the waveform $4e^{-3t}u(t-2)$. Note that there is an impulse in this derivative.

RQ6.11 Find a single element equivalent for the two-terminal subcircuit shown in Figure RQ6.10 for $\beta = 2, 1$, and -99.

Figure RQ6.8

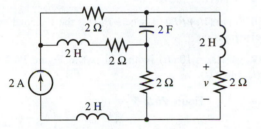

Figure RQ6.10

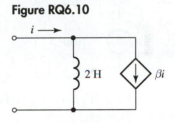

RQ6.9 Draw the equivalent circuit for $t > 0$ for the circuit shown in Figure RQ6.9.

Chapter 6 Problems

Note: **It might help to read Section 6.8 on compatible systems of units before tackling any of the problems with "practical values."**

Section 6.1 Waveform Properties of Resistive Circuits

6.1-1 Find i_x in Figure P6.1-1 in terms of i_s and v_s. Then find and sketch $i_x(t)$ if $v_s(t) = 90u(t)$ V and $i_s(t) = 6u(t)$ A.

Figure P6.1-1

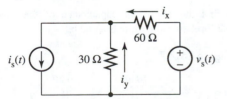

6.1-2 Find i_y in Figure P6.1-1 in terms of i_s and v_s. Then find and sketch $i_y(t)$ if $v_s(t) = 90u(t)$ V and $i_s(t) = 6u(t-1)$ A.

6.1-3 Find the voltage $v(t)$ in Figure P6.1-2 in terms of $i_s(t)$ and $v_c(t)$. Then untape the VCVS and find $v(t)$ in terms of $i_s(t)$. If $i_s(t) = 4 \sin(3t)$ A, find $v(t)$.

Figure P6.1-2

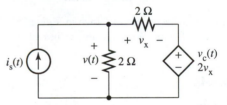

6.1-4 Find the Thévenin equivalent for the subcircuit shown in Figure P6.1-3 by attaching an i-source having the general value $i(t)$ to the terminals shown and solving for the terminal voltage. Assume that the CCCS is taped and has the value $i_c(t)$. Solve for the Thévenin voltage and resistance in terms of $v_s(t)$ and $v_c(t)$; then untape the dependent source and solve in terms of $v_s(t)$ only. From your result, find the value of Thévenin equivalent voltage and resistance. If $v_s(t) = 12e^{-2t}$ V, evaluate these parameters.

Figure P6.1-3

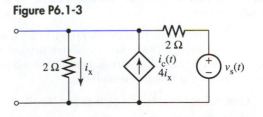

Section 6.2 The Capacitor: An Energy Storage Element

6.2-1   What is the charge on the capacitor in Figure P6.2-1 in μC? What is the energy stored in microjoules (μJ)?

Figure P6.2-1

$$2\ \mu\text{F}$$

$$+\ 10\ \text{V}\ -$$

6.2-2   If $i(t) = 10u(t)$ mA in Figure P6.2-2, find the voltage v and the energy stored w at $t = 2$ μs.

Figure P6.2-2

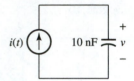

6.2-3   If $v(t) = 10tu(t)$ V with t in ns (e.g., $v(1$ ns$) = 10$ V) in Figure P6.2-3, find $i(t)$ and $w(t)$ at $t = 4$ ns.

Figure P6.2-3

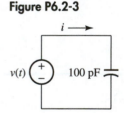

6.2-4   If $i(t)$ has the waveform shown in Figure P6.2-4, find and sketch $v(t)$ for all time. What are the values of v, q, and w at $t = 100$ μs?

Figure P6.2-4

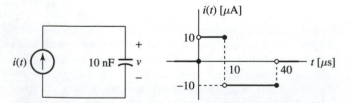

Problems 6.2-5 through 6.2-16 are exercises in differentiation and integration. For the latter you should review Appendix C and, where appropriate, use the "tabular method" of integration developed there.

6.2-5   If $x(t) = 10tu(t)$, find and sketch $px(t)$ for all t. Using the analytical expression you have just found for $px(t)$, compute $(1/p)[px(t)]$. Then compute $(1/p)x(t)$ analytically and use your result to find $p[(1/p)x(t)]$. Compare these results.

6.2-6   Compute $f(t) = p[4te^{-2t}u(t)]$, then find $(1/p)f(t)$.

6.2-7   Find the waveform $f(t) = (1/p)[8e^t \sin(2t)u(t)]$. Then use your result to find $p(1/p)f(t)$.

6.2-8   Let $f(t) = 4e^{-2t} \sin(2t)$ and compute $pf(t)$, $p^2f(t)$, and $p^3f(t)$.

6.2-9   Find $p[e^{2t} \sin(5t)]$.

6.2-10   Find $p^3[t^3e^{2t}]$.

6.2-11   Find $(1/p)[4te^tu(t)]$.

6.2-12   Find $(1/p)[(t^3 + 2t^2 + 3t + 4)e^tu(t)]$.

6.2-13   Find $(1/p)[16te^{2t}]$ if its value at $t = 0$ is 4.

6.2-14   Find $(1/p^2)[8 \sin(2t)u(t)]$.

6.2-15   Find $(1/p^n)[u(t)]$.

6.2-16   Find $(1/p)[t^2P_1(t)]$, where $P_1(t)$ is the 1-second-duration unit pulse function.

6.2-17   If $v(t) = 10tu(t)$ (with t in μs) in Figure P6.2-5, find $i(t)$, $q(t)$, and $v_x(t)$.

Figure P6.2-5

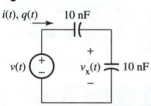

6.2-18   For the capacitive circuit in Figure P6.2-6 find $v_x(t)$, $i_x(t)$, and $q_x(t)$ if $i(t) = 10u(t)$ mA, with t in ms (e.g., $i(1$ ms$) = 10$ mA).

Figure P6.2-6

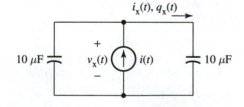

6.2-19   Find the equivalent capacitance of the subcircuit in Figure P6.2-7 relative to terminals a and b.

Figure P6.2-7

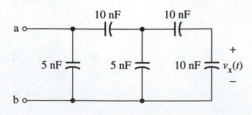

6.2-20   Attach a current source with its positive reference pointed toward terminal a of the subcircuit in Figure P6.2-7. If it has the waveform sketched in Figure P6.2-8, find the waveform for $v_x(t)$.

Figure P6.2-8

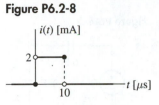

6.2-21 Find the equivalent capacitance "looking into" terminals a and b of the subcircuit shown in Figure P6.2-9. All capacitances are in farads.

Figure P6.2-9

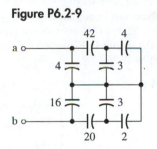

6.2-22 Let H be the operator $p + (1/p)$. Show that $H(p)$ is linear. If $x(t) = tu(t)$, find $y(t) = H(p)x(t)$.

6.2-23 A test run on the linear element e in Figure P6.2-10 with $v(t) = 10\cos(2t)u(t)$ V results in $i(t) = 5\sin(2t)u(t)$ mA. What is the admittance operator $Y(p)$? What is the impedance operator $Z(p)$?

Figure P6.2-10

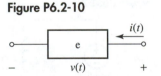

6.2-24 A test run on the linear element e in Figure P6.2-10 with $i(t) = 10u(t)$ mA results in $v(t) = 5tu(t)$ V. What is the impedance operator $Z(p)$? What is the admittance operator $Y(p)$?

6.2-25 Let $y(t) = (4/p)x(t)$. If $y(t) = 8te^{-2t}u(t)$, find $x(t)$.

6.2-26 Let $y(t) = 4px(t)$. If $y(t) = 8te^{-2t}u(t)$, find $x(t)$.

6.2-27 Let $y(t) = (4/p)x(t)$. If $y(t) = 8te^{-2t}$, find $x(t)$. How does this answer differ from the one you found in Problem 6.2-25?

6.2-28 If the element in Figure P6.2-10 has a v-i characteristic defined by $i(t) = 4pv(t)$, show that it is passive and lossless.

6.2-29 Show that an element with the impedance operator $Z(p) = p + (1/p)$ is passive.

6.2-30 Show that the element in Problem 6.2-29 is lossless.

Section 6.3 The Inductor: Another Energy Storage Element

6.3-1 If $L = 2$ H and $i(t) = 2$ A (dc) in Figure P6.3-1, find the flux linkage λ in webers.

Figure P6.3-1

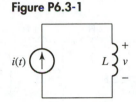

6.3-2 If $L = 2$ H and $v(t) = 4u(t)$ V in Figure P6.3-2, find the flux linkage $\lambda(t)$ in webers.

Figure P6.3-2

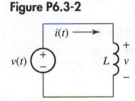

6.3-3 If $L = 2$ H in Figure P6.3-2 and $v(t)$ has the waveform shown in Figure P6.3-3, find $i(t)$ graphically.

Figure P6.3-3

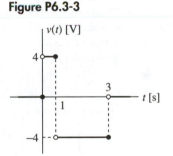

6.3-4 If $L = 2$ H and $v(t) = 4e^{-2t}u(t)$ in Problem 6.3-2, find $i(t)$ analytically. Find the energy stored at $t = 2$ s.

6.3-5 Find the equivalent inductance for the two-terminal subcircuit shown in Figure P6.3-4.

Figure P6.3-4

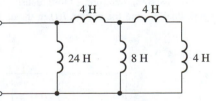

6.3-6 If $i(t) = 10tu(t)$ μA with t in ms (e.g., $i(1$ ms$) = 10$ μA) and $L = 2$ mH in Figure P6.3-1, find $v(t)$.

6.3-7 Let $L = 2$ H and $v(t) = 8\cos(2t)u(t)$ in Figure P6.3-2. Find and sketch $i(t)$, the instantaneous power $P(t)$ absorbed by the inductor, and the energy stored in the inductor, $w(t)$.

Section 6.4 The Dc Steady State

6.4-1 Find the values of v and i in the circuit shown in Figure P6.4-1 assuming that it is in the dc steady state.

Figure P6.4-1

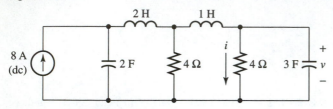

6.4-2 Find the value of i in the circuit shown in Figure P6.4-2 assuming dc steady-state conditions.

Figure P6.4-2

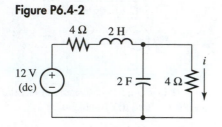

6.4-3 For the circuit in Figure P6.4-3, find the current i assuming dc steady-state conditions and $i_s(t) = 10$ A (dc). (*Hint:* Either use flux linkage or assume that $i_s(t)$ is time varying and "settles down" to the given dc value of 10 A.)

Figure P6.4-3

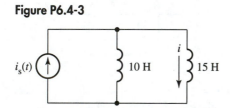

6.4-4 A more realistic model for an inductor is shown in Figure P6.4-4(a). At dc the resistance of the winding dominates, whereas for time-varying waveforms (that do not vary too slowly) its effects can be neglected in comparison with those due to the inductor. Figure P6.4-4(b) shows, therefore, a more realistic model for the circuit in Figure P6.4-3. Find the current $i(t)$ in the dc steady state.

Figure P6.4-4

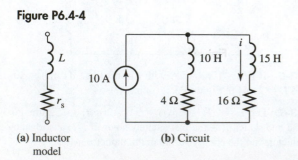

(a) Inductor   (b) Circuit
model

6.4-5 Find $v(t)$ for the circuit in Figure P6.4-5 assuming dc steady-state conditions and that $v_s(t) = 18$ V (dc). (*Hint:* Either use charge rather than current or assume that $v_s(t)$ varies from zero to a constant value of 18 V.)

Figure P6.4-5

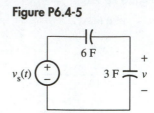

6.4-6 A more realistic model for a capacitor is shown in Figure P6.4-6(a). At dc the resistance of the dielectric between the plates dominates, whereas for time-varying waveforms (that do not vary too slowly) its effects can be neglected in comparison with those due to the capacitor. Figure P6.4-6(b) shows, therefore, a more realistic model for the circuit in Figure P6.4-5. Find $v(t)$ in the dc steady state if $v_s(t) = 18$ V.

Figure P6.4-6

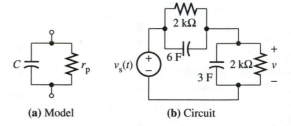

(a) Model   (b) Circuit

6.4-7 Determine whether or not dc steady-state conditions ever exist in the circuit shown in Figure P6.4-7.

Figure P6.4-7

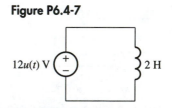

6.4-8 Determine whether or not dc steady-state conditions ever exist in the circuit shown in Figure P6.4-8.

Figure P6.4-8

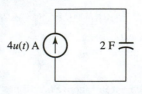

Section 6.5 Initial Conditions and Switched Circuits

6.5-1 Find and sketch $v(t)$ for all t in the circuit shown in Figure P6.5-1.

Figure P6.5-1

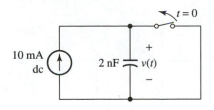

6.5-2 Find and sketch $v(t)$ in the circuit shown in Figure P6.5-2 for all t.

Figure P6.5-2

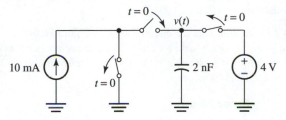

6.5-3 Find and sketch $v(t)$ for all t in the circuit shown in Figure P6.5-3. (*Hint:* First, assume dc steady-state conditions hold for all $t \le 0$ and find the voltage across *each* capacitor using the method developed in Question Q6.4-2 in the quiz for Section 6.4. Then "flip the switch" and use the appropriate initial condition models for the capacitors to find $v(t)$ for $t > 0$. The capacitors all have the same value (you don't need this value to solve the problem!).

Figure P6.5-3

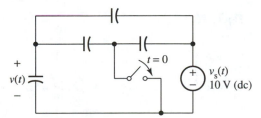

6.5-4 The circuit shown in Figure P6.5-4 is known to be operating in the dc steady state for $t \le 0$. Find $i(t)$.

Figure P6.5-4

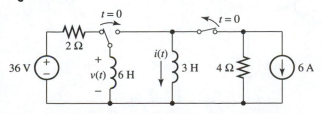

6.5-5 The circuit of Figure P6.5-5 is known to be operating in the dc steady state for $t \le 0$. Find and sketch $v(t)$ for all t. The capacitors have equal values.

Figure P6.5-5

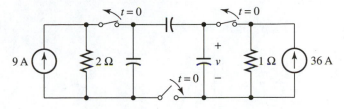

6.5-6 The circuit of Figure P6.5-6 is known to be operating in the dc steady state prior to $t = 0$. Find and sketch $i(t)$ for all t. All inductances have the same value.

Figure P6.5-6

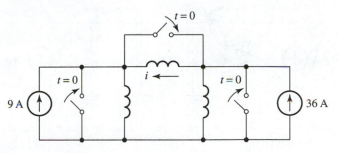

6.5-7 Assuming that $v_1(0-) = V_o$ and $v_2(0-) = 0$ in the switched capacitor circuit of Figure P6.5-7, find the total stored energy at $t = 0-$ (just before the switch flip), $w(0-)$, and that at $t = 0+$ (just after the switch flip), $w(0+)$. This is an "old favorite" of circuit theorists. We will ask you to explore an answer to the question, What happened to half the energy?, in a problem at the end of Chapter 7.

Figure P6.5-7

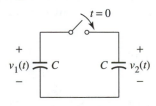

Section 6.6 Impulse Waveforms

6.6-1 For the waveform sketched in Figure P6.6-1, show that $(1/p)x(t) \to u(t)$ as $\epsilon \to 0$ and therefore that $x(t)$ defines the unit impulse function.

Figure P6.6-1

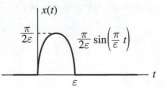

6.6-2 Repeat Problem 6.6-1 for the waveform shown in Figure P6.6-2.

Figure P6.6-2

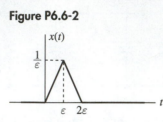

6.6-3 Assume that dc steady-state conditions hold in the circuit shown in Figure P6.6-3 for $t \leq 0$ and find and sketch $v(t)$.

Figure P6.6-3

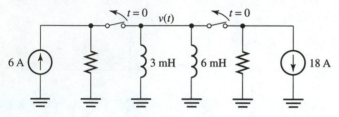

6.6-4 If $v_1(0-) = V_o$ and $v_2(0-) = 0$ in the circuit of Figure P6.6-4, find $i(t)$ for $t > 0$.

Figure P6.6-4

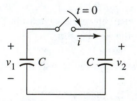

6.6-5 If $f(t)$ is specified by the waveform of Figure P6.6-5, find $f(t)\delta(t - 3)$.

Figure P6.6-5

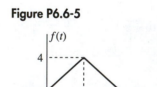

6.6-6 For the waveform in Figure P6.6-5, compute and sketch $pf(t)$ and $p^2f(t)$.

6.6-7 Find and sketch $(1/p)f(t)$ for the waveform shown in Figure P6.6-6. (Remember that our convention is to show impulses with negative area as pointing down with the area magnitude written by them; thus, the impulse at $t = 2$ has the formula $-4\delta(t - 2)$.)

Figure P6.6-6

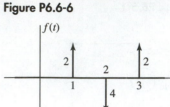

6.6-8 Compute and sketch the waveform $p\{4e^{-t}u(t)\}$.

6.6-9 Compute and sketch the waveform $p\{4e^{-(t-2)}u(t - 2)\}$.

6.6-10 Compute and sketch the following waveform: $p\{4 \sin(t)u(t - \pi/2)\}$.

6.6-11 Assume that the circuit shown in Figure P6.6-7 is operating in the dc steady state for $t \leq 0$. Draw the series initial condition model for the capacitor relative to the capacitor voltage $v_c(t)$ shown and label its values.

Figure P6.6-7

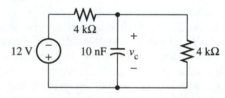

6.6-12 Assume that dc steady-state conditions exist in the circuit of Figure P6.6-8 for $t \leq 0$. Draw the series inductor initial condition model relative to the current $i_L(t)$ shown in the figure and label its parameter values.

Figure P6.6-8

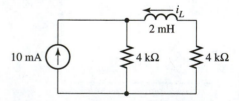

6.6-13 Determine the Thévenin and Norton equivalents for the subcircuit shown in Figure P6.6-9 relative to the terminal voltage polarity shown.

Figure P6.6-9

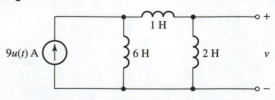

6.6-14 Determine the Thévenin and Norton equivalents for the circuit in Figure P6.6-10.

Figure P6.6-10

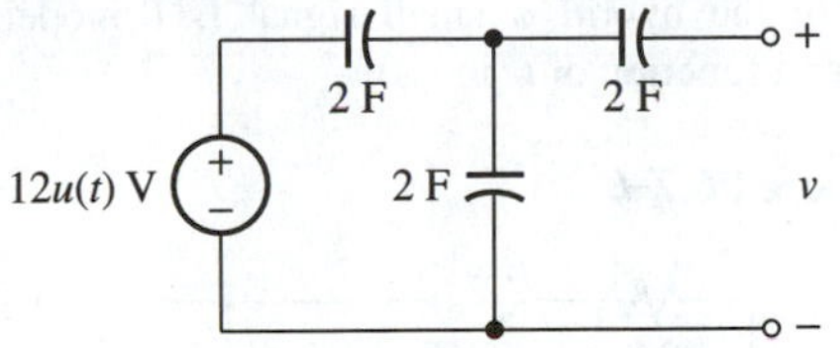

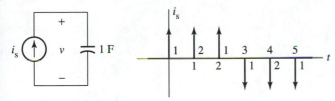

6.6-15 Find and sketch the waveform $v(t)$ in Figure P6.6-11. Remember that negative pointing arrows are impulses with negative areas: $-2\delta(t - 4)$ describes the one at $t = 4$.

Figure P6.6-11

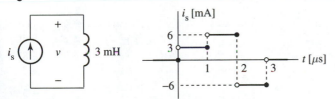

6.6-16 The current source waveform in the circuit of Figure P6.6-12 is specified by the graph shown. Find and sketch $v(t)$.

Figure P6.6-12

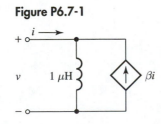

Section 6.7 Active Energy Storage Circuits

6.7-1 Determine the single element equivalent for the subcircuit shown in Figure P6.7-1. Evaluate this element value for $\beta = 10^6$, $\beta = -1$, and $\beta = -10^6$.

Figure P6.7-1

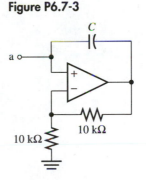

6.7-2 Find $i(t)$ for the circuit in Figure P6.7-2 assuming dc steady-state operation for $t \leq 0$.

Figure P6.7-2

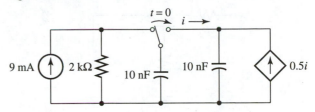

6.7-3(O) Use a Miller equivalent transformation to determine a single element that is equivalent to the subcircuit of Figure P6.7-3 between terminal a and ground.

Figure P6.7-3

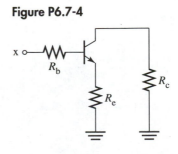

6.7-4(T) Using the mesh form of the Miller equivalent and the hybrid π equivalent (developed in Chapter 5, Section 5.7) for the BJT in Figure P6.7-4, find the single element that is equivalent to the subcircuit between terminal x and ground. We would like to note two items: this is an alternate approach to Example 6.26(T) in the body of the text, and we also emphasize that the circuit shown is a small signal one (with any dc biasing sources deactivated).

Figure P6.7-4

6.7-5(T)   Repeat Problem 6.7-4(T) for the circuit shown in Figure P6.7-5.

Figure P6.7-5

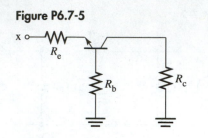

6.7-6(T)   Assuming that the small signal equivalent is valid for the BJT circuit shown in Figure P6.7-6 (with all dc power supply sources assumed to have been deactivated), that $v_i(t) = 0.12P_{10\mu s}(t) = 0.12\{u(t) - u(t - 10\ \mu s)]$ V,   and that $r_\pi = 2\ k\Omega$ for the hybrid π small signal BJT model, find and sketch $v_c(t)$ as a function of t (in μs).

Figure P6.7-6

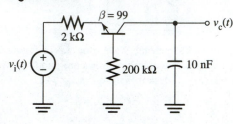

Time Response of First-Order Circuits

This chapter develops analysis procedures for any circuit response that can be described by a first-order differential equation. Though important in its own right, the first-order lowpass solution operator developed here turns out to be the basis for the analysis of higher-order circuits as well. In fact, a purely algebraic analysis procedure will be developed in Chapter 9 that depends upon the first-order lowpass solution operator as its building block. The idea of a system simulation diagram is presented and offered as a design procedure for first-order active circuits. The inherent nonidealities of the passive elements are discussed. Section 7.7 covers a method that is standard in introductory texts on circuit analysis. Though the operator method is more general and easier to apply, many practicing circuit analysts use the method covered in Section 7.7. Therefore, it is important to become familiar with this method in order to communicate with them. Finally, SPICE statements for the description of the capacitor and the inductor are covered, and the procedures for performing a transient analysis are described.

The prerequisite for this chapter is, generally, the core material in Part I on the dc analysis of resistive circuits—in particular, the basic material on the analysis of circuits that are purely passive (that is, which consist of only resistors and independent sources, but no active elements such as dependent sources, op amps, etc.). In order to read Section 7.5, one also needs the material in Chapter 5 on active circuits. The amount of background required from Chapter 5 for this section depends upon the extent of coverage. For instance, one can read only the material on time domain response of first-order circuits containing only dependent sources if one only covered these elements in Chapter 5. The same applies to the material on op amps and BJTs. Op amp and transistor examples are flagged with an (O) or a (T), respectively.

Section 7.8 on SPICE requires all the basic SPICE material in Part I.

7.1 | First-Order Lowpass Response

Operator Analysis of a First-Order Circuit

The circuit shown in Figure 7.1 consists of a single capacitor, a single resistor, and a single v-source. The voltage waveform $v(t)$ is called a first-order lowpass circuit response for reasons that we will ask you to develop in the quiz at the end of this section.

We would like to determine the voltage $v(t)$ for any waveform applied to the circuit by the voltage source. The desired variable is a voltage, so we will perform a nodal analysis. The circuit is shown prepared for analysis in Figure 7.2; that is, we have assigned the reference node and have labeled the essential node with the unknown voltage $v(t)$. Furthermore, we have tagged the capacitor with its impedance operator $1/Cp$. The KCL equation at the essential node is

$$i_C + i_R = 0, \tag{7.1-1}$$

or, in terms of the node voltage,

$$Cpv(t) + \frac{v(t) - v_s(t)}{R} = 0. \tag{7.1-2}$$

Rearranging slightly, we have

$$pv(t) + \frac{v(t)}{RC} = \frac{v_s(t)}{RC}. \tag{7.1-3}$$

In conventional notation (which we will not often use henceforth), it is

$$\frac{dv}{dt} + \frac{1}{RC}v = \frac{1}{RC}v_s(t). \tag{7.1-4}$$

We have succeeded at this point in deriving a differential equation whose solution is our desired response variable $v(t)$. Before looking at the solution, however, let's pause for a moment to present a "preview of things to come." As we know, the capacitor v-i charac-

Figure 7.1   A first-order lowpass circuit

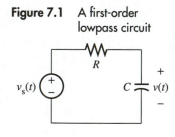

Figure 7.2   The circuit prepared for nodal analysis

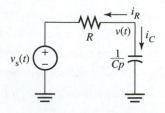

teristic obeys a generalized Ohm's law with the operator impedance $1/Cp$ taking the place of resistance; furthermore, it is connected in series with the resistor. It seems plausible that we should be able to use the voltage divider rule. If we did, we would get

$$v(t) = \frac{\dfrac{1}{Cp}}{R + \dfrac{1}{Cp}} v_s(t) = \frac{\dfrac{1}{RC}}{p + \dfrac{1}{RC}} v_s(t). \tag{7.1-5}$$

In writing the last term, we have multiplied both top and bottom by the operator p/R. If we now merely cross-multiply by the denominator, we obtain

$$\left[p + \frac{1}{RC} \right] v(t) = \frac{1}{RC} v_s(t), \tag{7.1-6}$$

or
$$pv + \frac{1}{RC} v = \frac{1}{RC} v_s(t). \tag{7.1-7}$$

This, however, is nothing more than the operator form of (7.1-4). We will justify these manipulations as we progress through this chapter. The advantage is obvious: if we can manipulate operators in p just as we manipulate resistances (as we have just done), we will save a lot of work in deriving our differential equation. In fact, we will show that the right-hand expression in (7.1-5) is actually the solution we are looking for; thus, we will not have to actually derive a differential equation. The solution will be generated (ultimately) by algebraic operations alone.

Operator Solution of the First-Order Lowpass Differential Equation

Now let's have a look at the solution of (7.1-4). In order to make our procedure more general, we will write it as follows:

$$\frac{dy}{dt} + ay = x(t). \tag{7.1-8}$$

We will call $x(t)$ the *forcing function* and $y(t)$ the *response variable*. We will call (7.1-8) the general *first-order lowpass differential equation*. In operator form, it is

$$py + ay = x(t). \tag{7.1-9}$$

It can be written in more condensed form as

$$[p + a]y = x(t). \tag{7.1-10}$$

The meaning of the operator $p + a$ is somewhat obvious. It is the sum of a differentiation operation and a scalar multiplication operation. It is defined as the left side of (7.1-9); that is, if we differentiate y and add to it the quantity ay, we obtain the result of operating on y with the operator $p + a$. Now what is the solution of (7.1-10)? It is

$$y(t) = \frac{1}{p + a} x(t), \tag{7.1-11}$$

where $1/[p + a]$ is the inverse operator for the operator $[p + a.]$ The problem is that we do not really know what this operator means. We must express it in terms of more basic operations if we are to compute with it — and that is what we will now do.

We start by looking at the result of differentiating the product of the exponential function e^{at} and the waveform $y(t)$. We have

$$p[e^{at}y(t)] = e^{at}py(t) + ae^{at}y(t) \tag{7.1-12}$$
$$= e^{at}[py(t) + ay(t)] = e^{at}[p + a]\,y(t).$$

We have merely used the "derivative of a product" rule (Leibniz's rule) and factored the exponential to the left and $y(t)$ to the right. Note that if we had factored both functions to the right, we would have gotten $[p + a]\,e^{at}y(t)$ or $[p + a]\,y(t)e^{at}$. These expressions are, however, ambiguous because it is not clear which time function the operator $p + a$ is to operate on. If you follow our derivation carefully, though, you will see that equation (7.1-12) is correct and unambiguous. Now we place upon $y(t)$ the requirement that it be the solution to our differential equation (7.1-10). We can then equate the very first term in (7.1-12) to the product of the exponential and $x(t)$, as follows:

$$p[e^{at}y(t)] = e^{at}x(t). \tag{7.1-13}$$

As the left side is nothing more than the derivative of a time function, we can integrate both sides (multiply both sides by the operator $1/p$) to get

$$e^{at}y(t) = \frac{1}{p}[e^{at}x(t)]. \tag{7.1-14}$$

Finally, we multiply both sides by e^{-at} to obtain

$$y(t) = e^{-at}\frac{1}{p}[e^{at}x(t)]. \tag{7.1-15}$$

This is the solution to our differential equation because we know how to multiply by an exponential, integrate the result, and then multiply by another exponential. Thus, we have also defined the operator $1/[p + a]$:

$$\frac{1}{p + a}x(t) = e^{-at}\frac{1}{p}[e^{at}x(t)]. \tag{7.1-16}$$

We call this operator the *first-order lowpass operator* or the *first-order solution operator.* We will now derive two very important lowpass responses: the impulse response and the unit step response.

Impulse Response

Example 7.1 (First-Order Lowpass Unit Impulse Response)

Compute the *unit impulse response*

$$h(t) = \frac{1}{p + a}\delta(t). \tag{7.1-17}$$

Solution　The symbol $h(t)$ is conventionally used to denote the unit impulse response. Before doing our computation, we observe that $h(t)$ is the solution to the differential equation (in conventional notation)

$$\frac{dh}{dt} + ah = \delta(t). \tag{7.1-18}$$

We do not need to rederive our solution, however. We need only to apply (7.1-16):

$$h(t) = \frac{1}{p+a}\delta(t) = e^{-at}\frac{1}{p}[e^{at}\delta(t)] \tag{7.1-19}$$

$$= e^{-at}\frac{1}{p}[e^{a0}\delta(t)] = e^{-at}\frac{1}{p}[\delta(t)] = e^{-at}u(t),$$

or

$$h(t) = e^{-at}u(t). \tag{7.1-20}$$

We performed two crucial steps. The first was our use of the "sampling property" of the unit impulse function, which evaluates e^{at} at $t = 0$. The second was our observation that the running integral of the unit impulse function is the unit step function.

Time Constant

Figure 7.3   The first-order lowpass unit impulse response

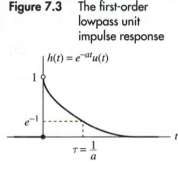

Example 7.1 is of more than passing interest. You should commit the result to memory because we will be using it many times throughout this entire part of the text. Because of its importance we plot it in Figure 7.3. It is zero for $t \le 0$, jumps to unity at $t = 0$, then "decays" to zero as time becomes large (assuming that $a > 0$). How fast does it decay? The answer is conveniently provided by the *time constant*, which we define by

$$\tau = 1/a. \tag{7.1-21}$$

If we evaluate $h(t)$ at $t = \tau$, we get

$$h(\tau) = e^{-a\tau}u(\tau) = e^{-a\cdot 1/a}u(1/a) = e^{-1}. \tag{7.1-22}$$

Figure 7.4   The physical interpretation of the time constant

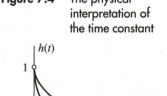

This means that it takes one time constant for the unit impulse response to fall from unity to the value of $e^{-1} = 0.368$. Therefore, if $\tau = 1/a$ is large, the unit impulse response drops slowly; if τ is small, it drops rapidly. We show this in Figure 7.4 for two such values. To clarify things a bit further, we note that τ does indeed have the dimension of time. Here is why: if we check our original differential equation in (7.1-9), repeated here for ease of reference as equation (7.1-23),

$$py + ay = x(t), \tag{7.1-23}$$

we see that each term must have the unit of $1/s$ because $p = p/dt$ has this unit. Then, because $\tau = 1/a$, it must have the unit s (second).

The Unit Step Response

Example 7.2 (Unit Step Response)

Compute the *unit step response*

$$s(t) = \frac{1}{p+a}u(t). \tag{7.1-24}$$

Solution   The symbol $s(t)$ is conventionally used for this response. Before doing this calculation, we note that $s(t)$ is the solution to the differential equation

$$\frac{ds}{dt} + as = u(t). \tag{7.1-25}$$

The computation goes as follows:

$$s(t) = \frac{1}{p+a}u(t) = e^{-at}\frac{1}{p}[e^{at}u(t)] = e^{-at}\int_{-\infty}^{t} e^{a\alpha}u(\alpha)\, d\alpha. \qquad (7.1\text{-}26)$$

If $t \leq 0$ the integrand is zero—hence, so is the integral itself. Notice that we must use another symbol than t for the variable of integration because we are reserving the latter for the upper limit of integration. For $t > 0$ we have

$$s(t) = e^{-at}\int_{0}^{t} e^{a\alpha}\, d\alpha = e^{-at}\frac{1}{a}[e^{at} - 1]. \qquad (7.1\text{-}27)$$

Multiplying through by the negative exponential gives

$$s(t) = \frac{1}{a}[1 - e^{-at}] \qquad (7.1\text{-}28)$$

for $t > 0$. Using the unit step function multiplier to "turn off" the waveform for $t \leq 0$ (after all, (7.1-28) gives an incorrect nonzero result for negative values of t), we have

$$s(t) = \frac{1}{a}[1 - e^{-at}]\, u(t). \qquad (7.1\text{-}29)$$

Forced and Natural Response

The unit step response is just as important as the unit impulse response, so we will now discuss its properties. Figure 7.5 shows this waveform with $a > 0$. Let's start our discussion by noticing that (7.1-29) can be written as the sum of two waveforms:

$$s(t) = \frac{1}{a}u(t) + [-\frac{1}{a}e^{-at}]\, u(t) = y_f(t) + y_n(t), \qquad (7.1\text{-}30)$$

with obvious definitions of $y_f(t)$ and $y_n(t)$. We call $y_f(t)$ the *forced response* and $y_n(t)$ the *natural response*. The forced response "looks like" the forcing function (the unit step), so it is what the response is being "forced to do." The natural response, on the other hand, involves the constant $a = 1/\tau$, which is a parameter of the differential equation, rather than of the forcing function. Thus, $y_n(t)$ is what the differential equation "wants to do" (naturally). We show these response components in Figure 7.6.

Figure 7.5
The first-order lowpass unit step response

Figure 7.6
Forced and natural responses

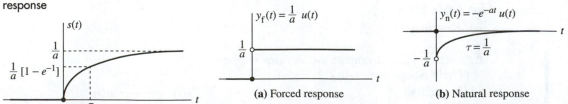

(a) Forced response   (b) Natural response

Stability

We see that the natural response has exactly the same form as the impulse response (the $-1/a$ constant multiplier does not change the fact that the waveform is an exponential). If $h(t)$ and $y_n(t)$ approach zero for large values of time, we say that the response and the differential equation are *stable*. This is the case if a (or τ) is positive. If a is negative, on the other hand, the response becomes infinite as t becomes infinitely large and we say that the response (differential equation) is *unstable*. If $a = 0$ (the time constant is infinite), $y_n(t)$ remains a constant for large values of time. The response and differential equation in this case are called *marginally stable*. We show these three cases in Figure 7.7. We have used $h(t)$, which has a positive constant multiplier of unity, for ease of interpretation. We again note that the constant multiplier has no bearing on stability.

Figure 7.7
Illustration of the impulse response for stable, unstable, and marginally stable DEs

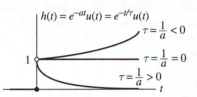

$$h(t) = e^{-at}u(t) = e^{-t/\tau}u(t)$$

$$\tau = \frac{1}{a} < 0$$

$$\tau = \frac{1}{a} = 0$$

$$\tau = \frac{1}{a} > 0$$

Dc Steady State   Looking once more at the unit step response (for $a > 0$) in Figure 7.5, we see that after a lapse of time equal to one time constant, the natural response component has decayed to e^{-1} times its initial value of $1/a$. When $a > 0$ (and the response is thus stable), the unit step response approaches a positive constant, shown in the figure as $t \to \infty$. This constant is called the dc steady-state value, and we say that the circuit response represented by the differential equation is in the dc steady state.[1]

Example 7.3   Find the unit impulse and step responses for $v(t)$ in Figure 7.8.

Figure 7.8
An example first-order lowpass circuit

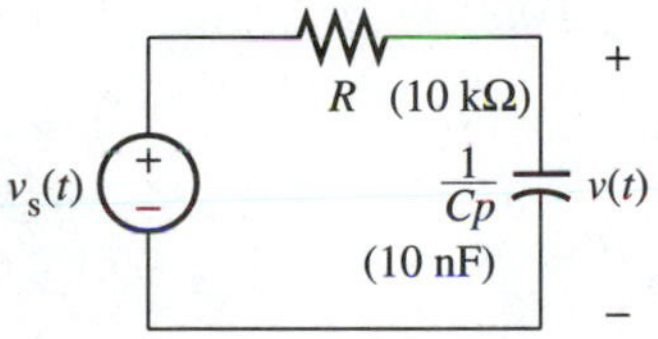

$$v_s(t) \quad R \ (10 \ \text{k}\Omega) \quad \frac{1}{Cp} \quad v(t) \quad (10 \ \text{nF})$$

Solution   Before beginning the solution, we comment that the resistor and capacitor values shown in parentheses are ones that might be practical if our circuit were used as part of an electronic device. This will give us a "feel" for the time constant of such circuits. This is the very circuit with which we started this section, so we will not repeat the derivation of the differential equation. It is given as equation (7.1-4), or in operator form, (7.1-6). The latter is repeated here for convenience of reference as (7.1-31):

$$\left[p + \frac{1}{RC} \right] v(t) = \frac{1}{RC} v_s(t). \tag{7.1-31}$$

This is a first-order lowpass differential equation with $a = 1/RC$, with $y(t) = v(t)$, and with $x(t) = v_s(t)/RC$. Thus, the unit impulse response is obtained by letting $v_s(t) = \delta(t)$:[2]

$$h(t) = e^{-t/RC} \frac{1}{p} \left[e^{t/RC} \frac{1}{RC} \delta(t) \right] \tag{7.1-32}$$

$$= \frac{1}{RC} e^{-t/RC} \frac{1}{p} [e^{t/RC} \delta(t)] = \frac{1}{RC} e^{-t/RC} u(t).$$

---

[1] You might wish to refer here back to Chapter 6, Section 6.4, for a discussion of the dc steady-state behavior of a circuit response.

[2] We adjust the independent source waveform, rather than $x(t)$, to be $\delta(t)$. This is conventional procedure for circuit analysis.

We see that there is a scaling factor of $1/RC$ when we consider $v_s(t)$ (rather than the waveform $x(t) = v_s(t)/RC$) as the input. When we speak of the unit impulse response of the circuit, this is what we mean. We observe that the time constant is given by

$$\tau = RC = 100 \ \mu s. \tag{7.1-33}$$

In general, large values of R and/or of C cause the circuit response to be slow, and small values result in a speedier response. Here, the speed of response has a definite value specified by $\tau = 100 \ \mu s$. The unit impulse response is shown in Figure 7.9. As is often the case with an impulse response, it is quite large at $t = 0+$.

Figure 7.9
The unit impulse
response

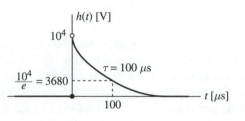

The unit step response is computed by adjusting $v_s(t)$ to be the unit step function. Notice how the scaling goes here:

$$s(t) = e^{-t/RC} \frac{1}{p} \left[e^{t/RC} \frac{1}{RC} u(t) \right] \tag{7.1-34}$$

$$= \frac{1}{RC} e^{-t/RC} \frac{1}{p} [e^{t/RC} u(t)] = [1 - e^{-t/RC}] \, u(t).$$

The scaling factor $1/RC$ on the right-hand side of (7.1-31) has been exactly canceled by the factor resulting from the integration so that the overall scale factor is unity. We show the step response in Figure 7.10. We see that it settles down into the dc steady state as $t \to \infty$ because the natural response approaches zero (and hence the circuit is stable). The capacitor is, therefore, equivalent to an open circuit as $t \to \infty$. We show this, that is the dc equivalent circuit, in Figure 7.11. Here is the reason we have shown this equivalent. We see that $\tau = RC > 0$, so we know in advance that the circuit is stable and will settle down into the dc steady state. Thus, we can replace the capacitor by an open circuit, draw Figure 7.11, and determine the final (dc steady state) value of $v(t)$ without solving the differential equation. We know that it starts from the origin at $t = 0$ and has a time constant of $100 \ \mu s$, so we can even sketch the step response in Figure 7.10 without solving the differential equation.

Figure 7.10
The unit step response

Figure 7.11
The dc steady-state equivalent

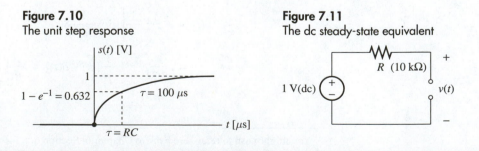

The last example illustrated several things. One is that the time constant of a simple *RC* circuit is given by the *RC* product. Another is that we can easily find and sketch the final asymptotic value by replacing the capacitor by an open circuit and analyzing a simple dc circuit. Finally, we see that the response scales: application of a general step function with amplitude V_o, say, results in a step response V_o times as large as the unit step response. The next example treats a first-order lowpass inductor circuit response.

Example 7.4   Find the unit impulse and step responses for $i(t)$ in Figure 7.12.

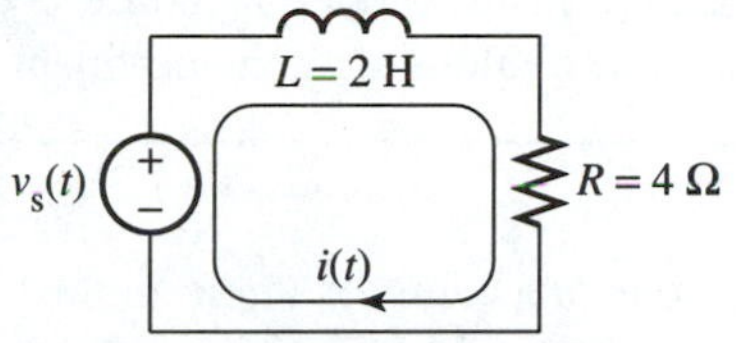

Figure 7.12
An example circuit

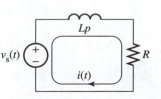

Solution   Because the mesh current is desired, we choose to do mesh analysis. The circuit is shown with the inductor impedance operator labeled in Figure 7.13. The mesh equation is

$$[Lp + R]i(t) = v_s(t). \tag{7.1-35}$$

The solution in terms of the first-order lowpass operator is

$$i(t) = \frac{\frac{1}{L}}{p + \frac{R}{L}} v_s(t). \tag{7.1-36}$$

Thus, we see that the time constant is

$$\tau = \frac{L}{R} = 0.5 \text{ s}. \tag{7.1-37}$$

This is a general result—you can check the dimension of *L/R* yourself to verify that it is time. Adjusting the v-source so that it has a unit impulse waveform (and using the given values for *L* and *R*), we compute the unit impulse response:

$$h(t) = \frac{\frac{1}{2}}{p + 2} \delta(t) = \frac{1}{2} e^{-2t} u(t). \tag{7.1-38}$$

The unit step response, on the other hand, is

$$s(t) = \frac{\frac{1}{2}}{p + 2} u(t) = \frac{1}{4}[1 - e^{-2t}] u(t). \tag{7.1-39}$$

Figure 7.13
Operator impedance form

Figure 7.14
Response waveforms
for Example 7.4

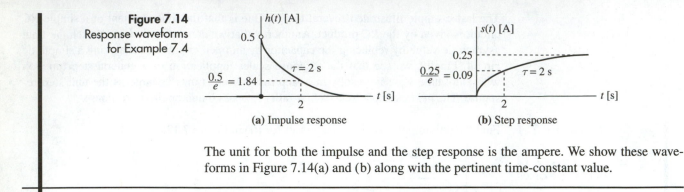

(a) Impulse response   (b) Step response

The unit for both the impulse and the step response is the ampere. We show these waveforms in Figure 7.14(a) and (b) along with the pertinent time-constant value.

Figure 7.15   The dc equivalent
for the last
example

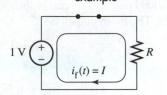

The main point in our preceding example is this: the time constant of a first-order lowpass circuit with an inductor is L/R, rather than RC as in the capacitive circuit. We note that when the inductive circuit is stable, as in the last example, the inductor current step response approaches a constant as $t \to \infty$; thus, the inductor becomes equivalent to a short circuit for large values of t. We show the resulting dc equivalent circuit in Figure 7.15. We see that the dc current can be immediately computed by inspection. With $R = 4\,\Omega$, the forced response is $i_f(t) = I = 0.25$ A. We caution you that not all circuits achieve the dc steady state. It is also a fact that if the independent source is not a step function, then one cannot compute and/or sketch the response without greater effort.

*First Order Circuits
with Nonconstant
Independent Sources*

We now give an example of a first-order lowpass response that requires more work to analyze.

Example 7.5

For the circuit shown in Figure 7.16, find the unit step and impulse responses; then find the response to the forcing function $i_s(t) = 4e^{-2t}u(t)$ A.

Figure 7.16
An example circuit

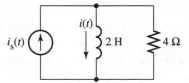

Solution

There are a number of ways of working this problem. Let's use mesh analysis on the operator equivalent shown prepared for mesh analysis in Figure 7.17. There is a nonessential mesh on the left and an essential one on the right. The KVL equation for the latter is

$$2p[i_x - i_s] + 4i_x = 0; \tag{7.1-40}$$

however, noting that $i = i_s - i_x$, or $i_x = i_s - i$, we have

$$2pi + 4i = 4i_s, \tag{7.1-41}$$

Figure 7.17
The operator form

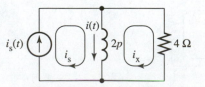

so

$$i(t) = \frac{2}{p + 2} i_s(t). \qquad (7.1\text{-}42)$$

The differential equation in (7.1-41) is a first-order lowpass one, and we know how to solve it using equation (7.1-42). We could also have simply used the current divider rule, treating the inductor impedance operator as a "generalized resistance" (it obeys Ohm's law), to obtain

$$i(t) = \frac{4}{2p + 4} i_s(t), \qquad (7.1\text{-}43)$$

which is equivalent to (7.1-42).

The unit impulse response is

$$h(t) = \frac{2}{p + 2} \delta(t) = 2e^{-2t}u(t), \qquad (7.1\text{-}44)$$

and the unit step response is

$$s(t) = \frac{2}{p + 2} u(t) = [1 - e^{-2t}]\, u(t). \qquad (7.1\text{-}45)$$

We have simply used the general result for the impulse and step responses of a first-order lowpass differential equation, having previously decided to commit them to memory.

The last part is somewhat more difficult because it is not one of these standard forms. We must actually compute an integral. We write

$$i(t) = \frac{2}{p + 2}[4e^{-2t}u(t)] = 8e^{-2t}\frac{1}{p}[e^{2t} \times e^{-2t}u(t)] \qquad (7.1\text{-}46)$$

$$= 8e^{-2t}\frac{1}{p}u(t) = 8te^{-2t}u(t).$$

We have used the fact that the integral of the unit step function is the *unit ramp function,* $r(t) = tu(t)$, shown in Figure 7.18. The unit for each of the above responses is the ampere.

Figure 7.18
The unit step and its
running integral

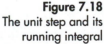

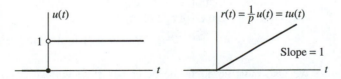

We will now pause for a couple of observations. The first has to do with *scaling.* We note that if we multiply the forcing function by a constant k, that is, we make the right-hand side of our original differential equation $kx(t)$ rather than $x(t)$, we get

$$\frac{1}{p + a}kx(t) = k\frac{1}{p + a}x(t) \qquad (7.1\text{-}47)$$

that is, the first-order lowpass operator and any constant commute. Thus, when we refer to the "impulse response" or the "step response" without the modifying word "unit," we mean that the forcing function is a constant times the unit impulse or the unit step.

The second observation has to do with *the running integral of time functions that are one sided*—those that are identically zero for $t \leq 0$. If $f(t)$ has this property, then

$$\frac{1}{p}f(t) = \int_{-\infty}^{t} f(\alpha) \, d\alpha = \left[\int_{0}^{t} f(\alpha) \, d\alpha\right] u(t). \tag{7.1-48}$$

We need only compute our integral for $t > 0$, then multiply by the unit step function to "turn off" the result for $t \leq 0$. Thus, in the case of our result in Example 7.5, equation (7.1-46), we need only note that

$$\frac{1}{p}u(t) = \int_{-\infty}^{t} u(\alpha) \, d\alpha = \left[\int_{0}^{t} 1 \, d\alpha\right] u(t) = tu(t). \tag{7.1-49}$$

We integrate the constant, then multiply by the unit step function.

Switched First-Order Circuits   We will look at the analysis of switched first-order circuits in the next example.

Example 7.6   Find the voltage $v(t)$ in Figure 7.19 for all t. What is the time constant τ for $t \leq 0$ and for $t > 0$?

Figure 7.19
A switched circuit example

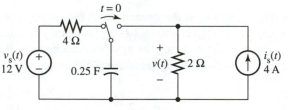

Solution   The dc steady-state equivalent for $t \leq 0$ is shown in Figure 7.20. The left-hand portion of the circuit is a simple RC circuit of the type with which we started this section; therefore, we see that the time constant is $\tau = RC = 4 \, \Omega \times 0.25 \, \text{F} = 1 \, \text{s}$. The time constant is positive, so $a = 1/\tau$ is positive also; hence, the circuit is stable for $t \leq 0$. Our assumption is that it was constructed (or "turned on") a long time before we began our observation of its behavior; thus we see that it is operating in the dc steady state for all finite values of $t \leq 0$. Our assumption that the capacitor is equivalent to an open circuit, then, is a good one. We compute the voltage of interest to be

$$v(t) = 2 \, \Omega \times 4 \, \text{A} = 8 \, \text{V}. \tag{7.1-50}$$

We also must compute the dc steady-state voltage across the capacitor v_c (it is clearly 12 V) because that value is needed in its initial condition model, which we will use in the $t > 0$ circuit (see Figure 7.21). We have used the series initial condition model for the capacitor and have multiplied all the independent source variables by $u(t)$. We have also chosen our reference node to be at the bottom in anticipation of nodal analysis.

Figure 7.20
The dc steady-state equivalent for $t \leq 0$

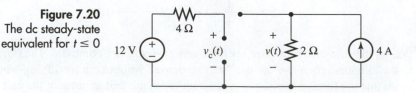

Figure 7.21
The $t > 0$ circuit

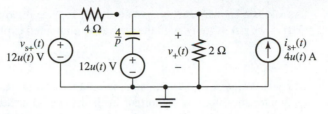

We can ignore the part of the circuit to the left side of the switch. There is only one essential node on the right-hand side, for which we write KCL as

$$\frac{p}{4}[v_+(t) - 12u(t)] + \frac{v_+(t)}{2} = 4u(t). \tag{7.1-51}$$

This can be rearranged into

$$(p + 2)v_+(t) = 16u(t) + 12\delta(t), \tag{7.1-52}$$

which can be considered as a first-order lowpass differential equation—with the forcing function being $16u(t) + 12\delta(t)$. The delta function comes from the derivative of the step function on the left-hand side of equation (7.1-51). We can now solve for $v_+(t)$:

$$v_+(t) = \frac{4}{p + 2}[4u(t) + 3\delta(t)] = 8[1 - e^{-2t}]u(t) + 12e^{-2t}u(t) \tag{7.1-53}$$

$$= [8 + 4e^{-2t}]u(t) \text{ V}.$$

We have used our known forms for the step and impulse responses in deriving the last term. Now this result shows that $v_+(t)$ is zero for $t \leq 0$; however, we know that this is not valid for the actual voltage $v(t)$. In fact, we saw previously that $v(t) = 8$ V for $t \leq 0$. This gives the graph in Figure 7.22. The time constant for $t > 0$, by the way, is different from that for $t \leq 0$. It is $\tau = 2 \ \Omega \times 0.25 \text{ F} = 0.5$ s, as shown in the graph. Notice that this is consistent—the circuit for $t > 0$ is stable because $a = 1/\tau > 0$ implies that the natural response goes to zero with increasing time, and we see from the graph that this is, indeed, what happens.

Figure 7.22
Time response

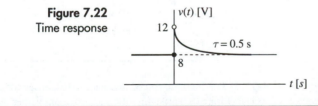

**Solution of
General First-Order
Differential Equations**

Look back at equation (7.1-52) in the preceding example. We see that the forcing function in that equation is

$$x(t) = 16u(t) + 12\delta(t). \tag{7.1-54}$$

In fact, this actually shows that any differential equation of first order can be solved by the techniques we have developed in this section. The most general first-order differential equation is

$$[p + a] y(t) = bpx(t) + cx(t). \tag{7.1-55}$$

As $x(t)$ is a known function, we need only replace the right-hand side by the new function $\bar{x}(t) = bpx(t) + cx(t)$, with the derivative computed explicitly, and solve

$$[p + a]\, y(t) = \bar{x}(t) \tag{7.1-56}$$

for $y(t)$. We will, however, examine another specific case in Section 7.2 because it has characteristics that are important in themselves. Furthermore, our investigation will develop the background we need to solve higher-order differential equations and circuits.

More on Stability, Forced and Natural Responses, and the Dc Steady State

Finally, let's discuss forced and steady-state responses a bit more. We saw that the step response can be written as the sum of forced and natural responses:

$$y(t) = y_f(t) + y_n(t). \tag{7.1-57}$$

If (and only if) the differential equation and the corresponding circuit response are stable, then the natural response $y_n(t)$ decays to zero as $t \to \infty$, thereby leaving only the forced response $y_f(t)$ (which is then approximately equal to the complete response $y(t)$). In this case we call the forced response the dc steady-state response.

Regardless of whether or not the differential equation is stable, we can replace a capacitor by an open circuit and an inductor by a short circuit to compute the forced response because the voltage (or current) forced response is constant. For a circuit with switches, though, we require that the $t \leq 0$ dynamic element circuit response be stable; it is only for such circuits that we can determine the finite initial condition that we need for computing the $t > 0$ response. The $t > 0$ response can then be either stable or unstable and our solution process will succeed.

These ideas, now largely intuitive, will be made mathematically precise in Chapter 10.

Section 7.1 Quiz

Q7.1-1. Find the step and impulse responses for $v(t)$ in the circuit of Figure Q7.1-1. If the current source waveform is $i_s(t) = 8e^{-4t}u(t)$, find the response $v(t)$. Show that $v(t)$ is a stable response. If $i_s(t) = 2u(t)$ A, find the response by sketching the dc steady-state equivalent circuit and using dc analysis techniques.

Figure Q7.1-1

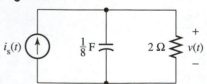

Q7.1-2. Find the voltage $v(t)$ and the current $i(t)$ in the circuit of Figure Q7.1-2.

Figure Q7.1-2

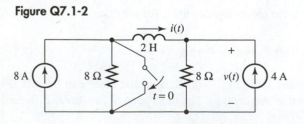

Q7.1-3. If $R_1 = 2\,\Omega$ and $R_2\,\Omega = -4$ in Figure Q7.1-3, find and sketch $v(t)$ for $t > 0$ using operators.[3]

Figure Q7.1-3

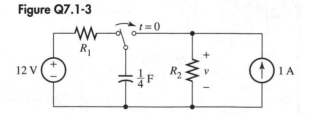

Q7.1-4. Find the forced and natural responses for $v(t)$ in Figure Q7.1-3 for $t > 0$ assuming the same element values as in Question Q7.1-3. Is this response stable? If so, find the dc steady-state value.

Q7.1-5. Let $R_1 = -2\,\Omega$ and $R_2 = 4\,\Omega$ in Figure Q7.1-3. Is the circuit solvable for $v(t)$? Why, or why not?

Q7.1-6. Solve the lowpass differential equation $(dy/dt) + 2y = \cos(\omega t)u(t)$. Identify the forced response $y_f(t)$ and the natural response $y_n(t)$. Sketch $|y_f(t)|_{\max}$ (the maximum value of $y_f(t)$ relative to t) versus ω. Your sketch should show you why we

---

[3] In Chapter 5 we showed that a two-terminal subcircuit containing a dependent source can be made equivalent to a single resistor with a negative value.

have used the term "lowpass" for this differential equation. *Note:* The answer to this question is a bit more involved mathematically than the others. You can probably use the "table form" for integration by parts covered in Appendix C to advantage. You will also need the trigonometric identity $A \cos(x) + B \sin(x) = \sqrt{A^2 + B^2} \cos(\omega t - \tan^{-1}(B/A))$.

7.2 | First-Order Highpass Response

The First-Order Highpass Differential Equation

Figure 7.23 shows a circuit whose voltage response $v(t)$ is not of the lowpass type. We can justify this statement by deriving the differential equation. If we apply one KCL equation at the single essential node, we obtain, in conventional notation for the time derivative,

$$C \frac{d}{dt}[v(t) - v_s(t)] + \frac{v(t)}{R} = 0. \tag{7.2-1}$$

Figure 7.23 A first-order highpass circuit

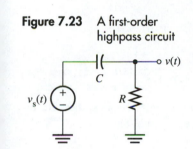

This can be simplified to

$$\frac{dv(t)}{dt} + \frac{v(t)}{RC} = \frac{dv_s(t)}{dt}. \tag{7.2-2}$$

This is not a differential equation of lowpass type because of the derivative term on the right-hand side. We call it a *highpass* differential equation of first order, and the circuit response giving rise to it is called a *highpass* response.[4] Question 7.2-6 in the quiz at the end of this section offers an explanation of the term "highpass."

Solution of the First-Order Highpass Differential Equation

Let's have a look at solving the highpass differential equation. First, though, let's change our notation to more general symbols and use operator notation:

$$py(t) + ay(t) = px(t). \tag{7.2-3}$$

Our first step is to factor the operator terms on the left-hand side to the left and write

$$(p + a)y(t) = px(t). \tag{7.2-4}$$

This has the solution

$$y(t) = \frac{1}{p + a}[px(t)]. \tag{7.2-5}$$

Notice carefully the order of operations in the solution. We must first differentiate $x(t)$— and then apply the first-order solution operator.

We are, however, looking for a general solution process, so let's think about things for a moment. If we write out the preceding solution explicitly, we get

$$y(t) = e^{-at} \int_{-\infty}^{t} e^{a\alpha} x'(\alpha) \, d\alpha, \tag{7.2-6}$$

---

[4] The term highpass *circuit* is also used, but the same circuit can have lowpass responses as well, so this is not good terminology. For instance, the capacitor voltage in Figure 7.23 is a lowpass response.

where we have used the "prime" notation for the derivative of $x(t)$. Now we integrate by parts, letting $u = e^{a\alpha}$ and $dv = x'(\alpha)\,d\alpha$ to get[5]

$$y(t) = e^{-at}\left[e^{at}x(t) - a\int_{-\infty}^{t} e^{a\alpha}x(\alpha)\,d\alpha \right] = x(t) - ae^{-at}\frac{1}{p}[e^{at}x(t)], \quad (7.2\text{-}7)$$

or

$$y(t) = \frac{1}{p+a}px(t) = \left[1 - \frac{a}{p+a} \right]x(t). \quad (7.2\text{-}8)$$

This looks suspiciously like using long division on $p/(p+a)$ without regard for which operator comes first, so let's look at what happens when we do things in the other order. We have

$$p\frac{1}{p+a}x(t) = p\left\{ e^{-at}\frac{1}{p}[e^{at}x(t)] \right\} \quad (7.2\text{-}9)$$

$$= -ae^{-at}\frac{1}{p}[e^{at}x(t)] + e^{-at}p\cdot\frac{1}{p}[e^{at}x(t)]$$

$$= \frac{-a}{p+a}x(t) + x(t) = \left[1 - \frac{a}{p+a} \right]x(t).$$

In the third line, we have merely canceled the p and $1/p$ operators because they are inverses of each other (see footnote 5). Thus, we have the useful result that

$$\frac{p}{p+a} = p\frac{1}{p+a} = \frac{1}{p+a}p = 1 - \frac{a}{p+a}; \quad (7.2\text{-}10)$$

that is, the derivative and first-order solution operators commute with one another. The operator $p/(p+a)$ has an unambiguous definition, and we can manipulate it just like a rational function of a real or complex variable to reduce it to a proper fraction. The constant one in the two-term form in (7.2-10) represents a "direct feed-through" term (it passes the forcing function unhindered), and the second term subtracts a lowpass response—that due to the lowpass operator $a/(p+a)$.

Example 7.7

Find the differential equation for the voltage $v(t)$ in Figure 7.24 and compute the unit impulse and step responses.

Figure 7.24
An example highpass response

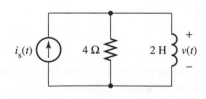

Solution   We first derive the differential equation. We show the circuit prepared for mesh analysis in Figure 7.25, with the inductor represented by its impedance operator. The KVL equation around the essential mesh is

$$2pi(t) + 4[i(t) - i_s(t)] = 0. \quad (7.2\text{-}11)$$

---

[5] We are, as usual, assuming one-sided waveforms, so $x(-\infty) = 0$ and integration by parts works.

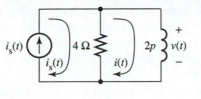

Figure 7.25
The circuit prepared for mesh analysis

Simplifying, we have

$$(p + 2) i(t) = 2i_s(t).$$ (7.2-12)

Thus, the inductor current is a lowpass response, but it is the inductor voltage we are interested in. For this reason, we note that

$$i(t) = \frac{1}{2p} v(t).$$ (7.2-13)

Inserting this in (7.2-12) gives

$$(p + 2) \frac{1}{2p} v(t) = 2i_s(t).$$ (7.2-14)

Multiplying both sides from the left by the operator $1/(p + 2)$, then by $2p$, we get

$$v(t) = \frac{4p}{p + 2} i_s(t).$$ (7.2-15)

We could just as well have noted that

$$(p + 2) \cdot \frac{1}{2p} = p \cdot \frac{1}{2p} + 2 \cdot \frac{1}{2p} = \frac{1}{2p} \cdot (p + 2),$$

so the operator $(p + 2)/2p$ is unambiguously defined. We could have then multiplied both sides by its inverse, $2p/(p + 2)$, to derive (7.2-15).[6] A quicker method, and one we will often use in the future, is to simply use current division on the circuit:

$$i(t) = \frac{4}{2p + 4} i_s(t) = \frac{2}{p + 2} i_s(t).$$ (7.2-16)

The *v-i* relationship of the inductor then gives

$$v(t) = 2pi(t) = \frac{4p}{p + 2} i_s(t) = 4 \left[1 - \frac{2}{p + 2} \right] i_s(t).$$ (7.2-17)

There is an even quicker technique. We just use impedance operators on the two parallel elements:

$$v(t) = \frac{4 \cdot 2p}{2p + 4} i_s(t) = \frac{4p}{p + 2} i_s(t).$$ (7.2-18)

---

[6] Notice that

$$\frac{2p}{p + 2} \cdot \frac{p + 2}{2p} = \frac{1}{p + 2} \cdot 2p \cdot \frac{1}{2p} \cdot (p + 2) = \frac{1}{p + 2} \cdot (p + 2) = 1.$$

We can do this with a clear conscience because we know that first-order differential operators behave just like real or complex constants or variables—and impedances behave like resistances. Are you beginning to see how neatly and efficiently operators solve circuit problems? As long as you keep all operators to the left of the time waveforms, they behave just like resistances (or conductances)!

At any rate, we can now determine the unit impulse response. It is

$$h(t) = \frac{4p}{p+2}\delta(t) = 4\left[1 - \frac{2}{p+2}\right]\delta(t) = 4\delta(t) - 8e^{-2t}u(t). \qquad (7.2\text{-}19)$$

The unit step response, on the other hand, is

$$s(t) = \frac{4p}{p+2}u(t) = 4\left[1 - \frac{2}{p+2}\right]u(t) = 4e^{-2t}u(t). \qquad (7.2\text{-}20)$$

These waveforms are shown in Figure 7.26(a) and (b). Observe that the unit step waveform for our highpass response variable looks like the impulse response for the lowpass response variable[7]—and that the impulse response for our highpass response has an **impulse function** in it.

Figure 7.26
Response waveforms

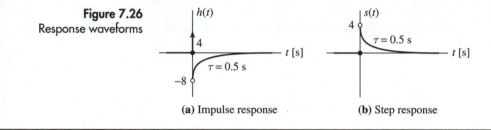

(a) Impulse response   (b) Step response

Example 7.8   Find the voltage $v(t)$ in the switched circuit of Figure 7.27.

Figure 7.27
A switched circuit
example

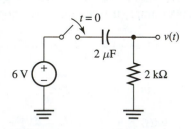

Solution   The circuit for $t \le 0$ is shown in Figure 7.28. The current through the capacitor has been zero since $t = -\infty$, so its voltage must (by causality) be zero. Hence, the circuit is in the dc steady state, as shown, and the voltage $v(t)$ is zero. Now, we imagine the switch to be flipped. Using either initial condition model for the capacitor (the series v-source is equivalent to a short circuit and the parallel i-source to an open circuit), we obtain the $t > 0$ circuit of Figure 7.29. Notice that we have "turned off" the input source for all $t \le 0$ by multiplying it by the unit step function.

---

[7] $s(t) = \dfrac{4p}{p+2}u(t) = \dfrac{4}{p+2}pu(t) = \dfrac{4}{p+2}\delta(t)$, so it is, indeed, the impulse response of a lowpass operator.

Figure 7.28
The $t \leq 0$ circuit

Figure 7.29
The scaled $t > 0$ circuit

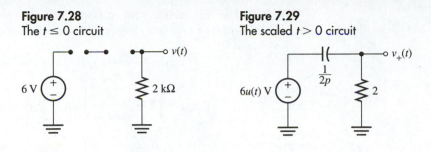

"But, wait a moment!" you might object. "Why is the capacitor operator impedance only $1/2p$ and not $10^6/(2p)$?" The answer is that we have scaled it. Recall that $mA \times k\Omega = V$ results in consistency if we choose the unit of current to be the milliampere and the unit of resistance to be the kilohm. Suppose we do that, and also notice that the time constant of our circuit for $t > 0$ is

$$\tau = RC = 2 \text{ k}\Omega \times 2 \text{ } \mu F = 4 \text{ ms}. \tag{7.2-21}$$

Thus, if we choose the unit of C to be the microfarad (in addition to our choice of $k\Omega$ for R, mA for i, and V for v), we will automatically have agreed to choose the unit of ms for time. Therefore, our circuit in Figure 7.29 has been scaled to mA, $k\Omega$, μF, and ms. If you are still a bit hazy on this, you should reread Section 6.8 in Chapter 6 and review the discussion there on scaling with compatible units.

We can now apply the voltage divider rule to get

$$v_+(t) = \frac{2}{2 + \dfrac{1}{2p}} \times 6u(t) = 6\frac{p}{p + \dfrac{1}{4}}u(t) \tag{7.2-22}$$

$$= 6\frac{1}{p + \dfrac{1}{4}}pu(t) = 6\frac{1}{p + \dfrac{1}{4}}\delta(t) = 6e^{-t/4}u(t).$$

Rather than doing long division on the highpass operator and computing the step response directly, we have been a little more efficient. Because impulse responses are easier to compute than step responses, we have associated the p in the numerator with the $u(t)$ function to produce the delta function. Then we merely computed an impulse response for the lowpass operator that remained after the factoring of the p in the numerator. The waveform is shown in Figure 7.30, where we have used the fact that $v(t) = 0$ for $t \leq 0$. Let's recall that the time constant is the time it takes for an exponential waveform to undergo about 63% of the total change it will experience. Thus, the total change in Figure 7.30 is 6 V, and we therefore see that after 4 ms of time lapse, $v(t)$ will have a value of approximately 2.22 V ($6 - 0.63 \times 6 = 6 - 3.78 = 2.22$).

Figure 7.30
The response waveform

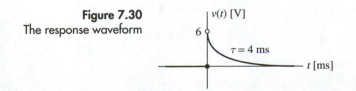

We will close this section with an observation on stability. We have shown that the general solution of the first-order highpass differential equation is

$$y(t) = \frac{p}{p + a}x(t) = \left[1 - \frac{a}{p + a}\right]x(t). \tag{7.2-23}$$

Thus,

$$h(t) = \left[1 - \frac{a}{p + a}\right]\delta(t) = \delta(t) - ae^{-at}u(t). \tag{7.2-24}$$

Because $\delta(t) = 0$ for all $t \neq 0$, we see that the highpass response is stable if $a > 0$, unstable if $a < 0$, and marginally stable if $a = 0$—exactly as is the case for the lowpass response.

Section 7.2 Quiz

Q7.2-1. Show that the two labeled responses of the circuit in Figure Q7.2-1 are stable.

Figure Q7.2-1

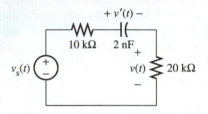

Q7.2-2. Find appropriate units for voltage, current, resistance, and time in order to simplify the numerical computations for the circuit in Figure Q7.2-1.

Q7.2-3. Find the unit step and impulse responses for $v(t)$ and $v'(t)$ in Figure Q7.2-1.

Q7.2-4. Assume that $v_s(t) = 12u(t)$ in Figure Q7.2-1. Find the Thévenin equivalent of the circuit "seen by" the capacitor and use it and the dc steady-state equivalent to compute $v'(t)$ without deriving a differential equation. Then use the substitution theorem (Chapter 3, Section 3.6) and superposition to compute $v(t)$.

Q7.2-5. Based upon your result in Question Q7.2-3, without using the result in Question Q7.2-4, find the response of $v(t)$ to $v_s(t) = 12u(t)$.

Q7.2-6. Solve the highpass differential equation $(dy/dt) + 2y = px(t)$, for $x(t) = \cos(\omega t)u(t)$. Identify $y_f(t)$ and $y_n(t)$. Then plot $|y_f(t)|_{max}$, the maximum value of $|y_f(t)|$ over all time, versus ω. *Note:* This problem is somewhat more involved mathematically than the others. You might find the table form of the integration by parts method of Appendix C helpful. The trigonometric identity

$$A\sin(x) - B\cos(x) = \sqrt{A^2 + B^2}\sin(x - \tan^{-1}(B/A))$$

might also prove to be of use.

7.3 | General First-Order Response

The General First-Order Differential Equation and Its Solution Operator

We are now in a position to analyze any first-order circuit. We will show how by solving the general first-order differential equation[8]

$$\frac{dy(t)}{dt} + ay(t) = b\frac{dx(t)}{dt} + cx(t), \tag{7.3-1}$$

or, in our more convenient operator form,

$$py(t) + ay(t) = bpx(t) + cx(t). \tag{7.3-2}$$

Factoring the operators to the left gives

$$(p + a)y(t) = (bp + c)x(t). \tag{7.3-3}$$

---

[8] The order of any differential equation is the order of the highest-order derivative appearing on either side.

The solution is

$$y(t) = \frac{1}{p + a}(bp + c)x(t) = \frac{bp + c}{p + a}x(t) = H(p)x(t). \qquad (7.3\text{-}4)$$

We observe that our first-order Heaviside, or differential, operator $H(p)$ can be unambiguously written as the rational fraction of p shown in (7.3-4) because, as we showed in the last section, constants and p commute with the first-order lowpass operator $1/(p + a)$. Notice that we can write $H(p)$ as a linear combination of highpass and lowpass operators:

$$H(p) = \frac{bp + c}{p + a} = b\frac{p}{p + a} + c\frac{1}{p + a} = bH_{hp}(p) + cH_{lp}(p). \qquad (7.3\text{-}5)$$

Furthermore, we can use long division to simplify it. We simply write

$$H(p) = \frac{bp + c}{p + a} = b + \frac{c - ab}{p + a}. \qquad (7.3\text{-}6)$$

Impedance and Admittance Operators

Now suppose that our first-order differential equation describes the v-i characteristic of a source-free two-terminal subcircuit such as the one shown in Figure 7.31. If we identify the response $y(t)$ with the voltage $v(t)$ and the forcing function $x(t)$ with the current $i(t)$, we call $H(p)$ the *operator impedance,* or *impedance operator,* of the subcircuit and write the solution of our differential equation in the form

$$v(t) = Z(p)i(t); \qquad (7.3\text{-}7)$$

Figure 7.31 A first-order subcircuit

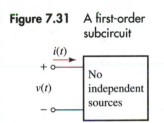

that is, we use the special symbol $Z(p)$ for the Heaviside operator rather than the more general $H(p)$. If, conversely, we associate the response $y(t)$ with the current $i(t)$ and the forcing function $x(t)$ with the voltage $v(t)$, we call $H(p)$ the *operator admittance,* or *admittance operator.* We then use the notation

$$i(t) = Y(p)v(t). \qquad (7.3\text{-}8)$$

The impedance and admittance operators are of course inversely related:

$$Y(p) = Z^{-1}(p) = \frac{1}{Z(p)}. \qquad (7.3\text{-}9)$$

Here is the advantage of our development of impedance and admittance operators. Equations (7.3-7) and (7.3-8) are merely generalized forms of Ohm's law. Thus, impedances and admittances act just like resistances and conductances *as long as we keep them to the left of the time functions upon which they operate.* A careful look at our development of resistive circuits will show that this restriction changes nothing. We developed all of our analysis machinery for resistive circuits based upon three things: Ohm's law, KVL, and KCL. Thus, we now know that all our procedures of resistive circuit analysis continue to hold for circuits with capacitors and inductors if we use operator impedances in the place of resistance.[9] This includes, of course, nodal and mesh analysis, series and parallel equivalent subcircuits, Thévenin and Norton equivalent subcircuits, etc.

A General First-Order Circuit Example

Example 7.9   Find the impulse and step responses for the voltage $v(t)$ shown in the circuit of Figure 7.32.

---

[9] This is true as long as we never—at this point—have to consider operators of higher order than first. (Even if we do, our analysis will still be valid, as we will see in Chapter 9.)

Figure 7.32
An example circuit

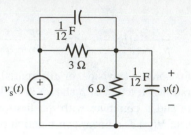

Solution   The circuit is shown with operator impedances in Figure 7.33. Now notice the special form of this circuit, topologically speaking. We have a two-terminal subcircuit consisting of the parallel combination of a 3-Ω resistor and a 1/12-F capacitor at the top and a similar subcircuit consisting of the 6-Ω resistor and its parallel 1/12-F capacitor at the right. We show this in a more abstract form in Figure 7.34. We can compute each of the individual operator impedances by using the parallel equivalent for resistors extended to impedances. Thus,

$$Z_1(p) = \frac{3 \times \dfrac{12}{p}}{3 + \dfrac{12}{p}} = \frac{12}{p + 4} \tag{7.3-10}$$

and

$$Z_2(p) = \frac{6 \times \dfrac{12}{p}}{6 + \dfrac{12}{p}} = \frac{12}{p + 2} \tag{7.3-11}$$

Figure 7.33
Operator impedance form

Figure 7.34
Equivalent with two subcircuits

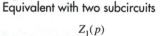
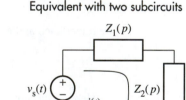
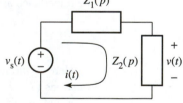

We can now simply use the voltage divider rule:

$$v(t) = \frac{Z_2(p)}{Z_1(p) + Z_2(p)} v_s(t) = \frac{\dfrac{12}{p + 2}}{\dfrac{12}{p + 2} + \dfrac{12}{p + 4}} v_s(t) \tag{7.3-12}$$

$$= \frac{p + 4}{2(p + 3)} v_s(t) = \frac{1}{2}\left[1 + \frac{1}{p + 3}\right] v_s(t).$$

In the last step, we used long division to simplify the operator.
We can now immediately write down the impulse response,

$$h(t) = \frac{1}{2}\left[1 + \frac{1}{p + 3}\right]\delta(t) = \frac{1}{2}\delta(t) + \frac{1}{2}e^{-3t}u(t), \tag{7.3-13}$$

and the step response,

$$s(t) = \frac{1}{2}\left[1 + \frac{1}{p+3}\right]u(t) = \frac{1}{2}u(t) + \frac{1}{6}[1 - e^{-3t}]\,u(t) \qquad (7.3\text{-}14)$$

$$= \left[\frac{2}{3} - \frac{1}{6}e^{-3t}\right]u(t).$$

These waveforms are shown in Figure 7.35(a) and (b).

Figure 7.35
Waveforms for
Example 7.9

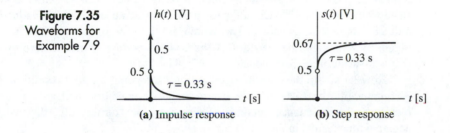

(a) Impulse response   (b) Step response

An Observation on "Fast" and "Slow" Circuit Responses

There is an interesting bit of intuition that we can derive from the last example. If we look closely at the step response, we see that it is a stable one because the time constant is positive and the natural response settles down to zero as t becomes large. More importantly, notice that the dc steady state response is 2/3—the result predicted by replacing the capacitors with their dc steady-state equivalent open circuits and using the voltage divider rule on the resistors alone. Here is something even more interesting: the step response "jumps up" immediately to 1/2, the value predicted by applying the voltage divider rule to only the capacitors with the resistors ignored. The idea is that fast changes are coupled right through the capacitors. The current in a capacitor is proportional to the rate of change of the voltage; therefore, most of the current (in fact, all of it) flows through the capacitors at the "leading edge" of the input step function. This circuit will be analyzed more thoroughly—and the preceding intuitive statements justified analytically—in Section 7.6 (Example 7.23).

A Switched
Circuit Example

Example 7.10   Find the two node voltages $v_1(t)$ and $v_2(t)$ in Figure 7.36.

Figure 7.36
An example circuit

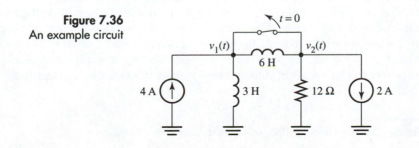

Solution   The $t \leq 0$ circuit is operating in the dc steady state with the equivalent shown in Figure 7.37. Why is it in the dc steady state? The answer also gives our rationale for showing $i_2(t)$ with the value of 0 A. As our circuit is assumed to have been in the "switch closed" situation forever, the voltage across the 6-H inductor has always been zero. The principle of causality we have used so often before (which tells us loosely that nothing can happen without a cause) then tells us that $i_2(t)$ must be zero. Thus, that inductor (at least) must be in the dc steady state and appear as the short circuit we have drawn. If we next "stand on" the other inductor and "look away from it" with the two i-sources deactivated, we will see an equivalent resistance of 12 Ω. Because this is positive, as is the time constant $\tau = L/R = 3/12 = 1/4$ s, we know that the $t < 0$ circuit 3-H inductor current is a stable response and has—at any finite negative value of time—settled down into the dc steady state as shown.[10] KCL at the top node now tells us that the 3-H inductor is carrying a dc current of 2 A, as shown. The switch, by the way, is carrying 2 A from left to right (again by KCL) and the resistor 0 A (because its voltage is held at zero by the short circuit equivalent of the 3-H inductor).

Now let's "flip the switch" open and use the parallel initial condition model for the 3-H inductor (the 6-H inductor has zero initial condition and so can be represented merely by its impedance operator). Doing this and "turning off" the two i-sources for negative time results in the $t > 0$ circuit shown in Figure 7.38.

Figure 7.37
The $t \leq 0$ dc steady-state equivalent

Figure 7.38
The $t > 0$ circuit

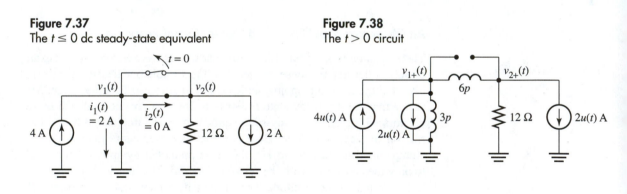

The nodal equations are

$$\frac{v_{1+}(t)}{3p} + \frac{v_{1+}(t) - v_{2+}(t)}{6p} = 4u(t) - 2u(t) \tag{7.3-15}$$

and

$$\frac{v_{2+}(t)}{12} + \frac{v_{2+}(t) - v_{1+}(t)}{6p} = -2u(t). \tag{7.3-16}$$

Before proceeding, a word about notation. Often, as now, when we write nodal equations, it is convenient to place operators involving p directly beneath voltage or current variables. It is to be understood that this is merely compact notation and that such a term really means that we are to apply the reciprocal of that operator to the numerator from the left-hand side. We can simplify our nodal equations by multiplying equation (7.3-15) by $6p$ (from the left, equivalent to differentiating both sides, then multiplying both sides

---

[10] The other inductor is shorted out, so we are dealing with a first-order circuit.

by 6) and equation (7.3-16) by 12p. This gives

$$3v_{1+}(t) - v_{2+}(t) = 12\delta(t) \tag{7.3-17}$$

and

$$-2v_{1+}(t) + (p + 2)v_{2+}(t) = -24\delta(t). \tag{7.3-18}$$

Recall now that we can multiply, divide, add, and subtract the operators p, $1/p$, $(p + a)$, and $1/(p + a)$ as we would real or complex numbers because we have shown that they commute and have inverses just like the latter. We choose here to use a matrix approach, because all the mechanics of matrices (based as they are upon the properties of addition, multiplication, addition, and subtraction of numbers) continue to hold for operators. Thus, we write (7.3-17) and (7.3-18) in the form

$$\begin{bmatrix} 3 & -1 \\ -2 & p + 2 \end{bmatrix} \begin{bmatrix} v_{1+}(t) \\ v_{2+}(t) \end{bmatrix} = \begin{bmatrix} 1 \\ -2 \end{bmatrix} 12\delta(t). \tag{7.3-19}$$

The inverse of a (2×2) matrix is, of course, easy to obtain. We multiply both sides from the left by this inverse to get

$$\begin{bmatrix} v_{1+}(t) \\ v_{2+}(t) \end{bmatrix} = \frac{1}{3p + 4} \begin{bmatrix} p + 2 & 1 \\ 2 & 3 \end{bmatrix} \begin{bmatrix} 1 \\ -2 \end{bmatrix} 12\delta(t) = \begin{bmatrix} \dfrac{4p}{p + \dfrac{4}{3}} \\[3ex] \dfrac{-16}{p + \dfrac{4}{3}} \end{bmatrix} \delta(t) \tag{7.3-20}$$

$$= \begin{bmatrix} \dfrac{4p}{p + \dfrac{4}{3}}\delta(t) \\[3ex] \dfrac{-16}{p + \dfrac{4}{3}}\delta(t) \end{bmatrix} = \begin{bmatrix} \left\{ 4 - \dfrac{\dfrac{16}{3}}{p + \dfrac{4}{3}} \right\}\delta(t) \\[4ex] \dfrac{-16}{p + \dfrac{4}{3}}\delta(t) \end{bmatrix} = \begin{bmatrix} 4\delta(t) - \dfrac{16}{3}e^{-4t/3}u(t) \\[2ex] -16e^{-4t/3}u(t) \end{bmatrix}.$$

These waveforms are only valid for $t > 0$. For this circuit though, we know that both node voltages were really zero for $t \le 0$; thus, (7.3-20) gives the true waveforms for all t.

A Practical Consideration for Switched Inductor Circuits

Let's discuss the waveforms in (7.3-20) at bit. Looking back at Figure 7.38, we see that the positive time circuit has a node that is somewhat special—the only elements connected to it are inductors and current sources. Now KCL demands that the sum of the currents into that node from the sources be equal, at each instant of time, to the current leaving through the inductors. Because these current sources both have step waveforms, the current through the inductors is also a step. But the voltage across an inductor is proportional to the time derivative of its current, in this case resulting in an impulse of voltage. The basic idea is a practical one to remember: attempting to switch or otherwise abruptly change the current in an inductor leads to extremely large voltages.

The Relationship Between Impulse and Step Responses

Now, let's shift gears and look at the general relationship between the impulse response $h(t)$ and the step response $s(t)$ for our general first-order operator. We know that

$$s(t) = H(p)u(t) = H(p)\frac{1}{p}\delta(t) \qquad (7.3\text{-}21)$$

because the integral of the unit impulse function is the unit step function. Writing $H(p)$ explicitly, we have

$$s(t) = \frac{bp + c}{p + a}\frac{1}{p}\delta(t). \qquad (7.3\text{-}22)$$

But we know that the integration operator $1/p$ commutes with the operators $bp + c$ and $1/(p + a)$, so we can bring the $1/p$ operator to the extreme left and write

$$s(t) = \frac{1}{p}\frac{bp + c}{p + a}\delta(t) = \frac{1}{p}h(t). \qquad (7.3\text{-}23)$$

Thus, we have

$$s(t) = \frac{1}{p}h(t), \qquad (7.3\text{-}24)$$

and, by multiplying both sides by p—that is, by differentiating both sides—we have

$$h(t) = ps(t). \qquad (7.3\text{-}25)$$

It is often simpler to compute *one* of these responses (usually the impulse response) and integrate or differentiate to obtain the other. We will illustrate this with an example.

Example 7.11   Find the unit step response for $v(t)$ in Figure 7.39 by first finding the unit impulse response, then integrating. (Scale the elements so that they have nice values for hand calculation.)

Figure 7.39
An example circuit

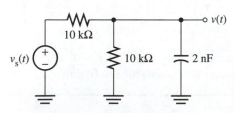

Solution   Before we start, let's think about the scaling issue. If we take the unit of resistance to be the kΩ, the unit of current to be the mA, and the unit of capacitance to be the nF, then the unit of voltage will be the V and the unit of time the μs. (Recall that kΩ $\times$ mA $=$ V and kΩ $\times$ nF $=$ μs.) Furthermore, let's not forget that we can use subcircuit equivalences to simplify a given circuit. Choosing to apply Thévenin's equivalent to the subcircuit to the left of the capacitor, we obtain the equivalent circuit shown in operator form in Figure 7.40. The voltage divider rule now gives

$$v(t) = \frac{\dfrac{1}{2p}}{5 + \dfrac{1}{2p}}0.5v_s(t) = \frac{\dfrac{1}{20}}{p + \dfrac{1}{10}}v_s(t). \qquad (7.3\text{-}26)$$

Figure 7.40
An equivalent circuit

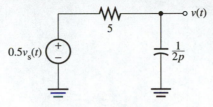

To compute the unit impulse response, we adjust the voltage source to produce the waveform $\delta(t)$, and the voltage $v(t)$ becomes $h(t)$:

$$h(t) = \frac{\dfrac{1}{20}}{p + \dfrac{1}{10}}\delta(t) = \frac{1}{20}e^{-t/10}u(t).$$ (7.3-27)

Integrating to find the unit step response, we have

$$s(t) = \frac{1}{p}h(t) = \frac{1}{20}\left[\int_0^t e^{-\alpha/10}\,d\alpha\right]u(t) = 0.5\,[1 - e^{-t/10}]\,u(t).$$ (7.3-28)

Example of "Charge Trapping" in Switched Capacitor Circuits

Example 7.12 Find the response variables $v(t)$, $i(t)$, $v_x(t)$, and $v_y(t)$ in Figure 7.41.

Figure 7.41
An example circuit

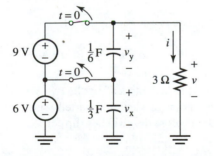

Solution Before $t = 0$, we clearly have $v_x(t) = 6$ V, $v_y(t) = 9$ V, $v(t) = 15$ V, and $i(t) = 5$ A—where all variables are constant (dc). The $t > 0$ circuit is shown in Figure 7.42, where we have used the series initial condition models for the capacitors. Notice that we have decided not to use the usual subscript "+" just to simplify the notation. The variables in Figure 7.42 are only valid, however, for $t > 0$. Thus,

$$i(t) = \frac{15u(t)}{\dfrac{9}{p} + 3} = \frac{5p}{p + 3}u(t) = \frac{5}{p + 3}\delta(t) = 5e^{-3t}u(t)\text{ A}.$$ (7.3-29)

The voltage $v(t)$, therefore, is just

$$v(t) = 3i(t) = 15e^{-3t}u(t)\text{ V}.$$ (7.3-30)

Figure 7.42
The $t > 0$ circuit

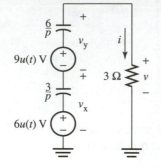

The voltage across the bottom capacitor (remembering the initial condition source!) is

$$v_x(t) = 6u(t) - \frac{3}{p}i(t) = 6u(t) - \frac{15}{p+3}u(t) \qquad (7.3\text{-}31)$$

$$= 6u(t) - 5[1 - e^{-3t}]u(t) = [1 + 5e^{-3t}]u(t).$$

The voltage across the top capacitor is

$$v_y(t) = 9u(t) - \frac{6}{p}i(t) = 9u(t) - \frac{30}{p+3}u(t) \qquad (7.3\text{-}32)$$

$$= 9u(t) - 10[1 - e^{-3t}]u(t) = [-1 + 10e^{-3t}]u(t).$$

Now observe that $v(t)$ and $i(t)$ both approach zero as $t \to \infty$, but $v_x(t)$ and $v_y(t)$ do not! In fact, we have $v_x(t) \to 1$ V and $v_y(t) \to -1$ V as $t \to \infty$. These limiting voltages do, however, add up to zero—as they must to give the proper value for $v(t)$ as $t \to \infty$. We see that there is a net charge stored on the bottom capacitor of

$$q_x = \frac{1}{3} \times (1) = \frac{1}{3} \text{ C} \qquad (7.3\text{-}33)$$

and, on the top one, there is a stored charge of

$$q_y = \frac{1}{6} \times (-1) = -\frac{1}{6} \text{ C}. \qquad (7.3\text{-}34)$$

These charges are said to be "trapped" by the two series capacitors.[11]

From an energy point of view, we started with a total energy stored on the charged capacitors at $t = 0-$ (before the switches flip) of

$$w(0-) = \frac{1}{2} \times \frac{1}{3} \times (6)^2 + \frac{1}{2} \times \frac{1}{6} \times (9)^2 = 12.75 \text{ J}. \qquad (7.3\text{-}35)$$

As $t \to \infty$, though, we have

$$w(\infty) = \frac{1}{2} \times \frac{1}{3} \times (1)^2 + \frac{1}{2} \times \frac{1}{6} \times (-1)^2 = 0.25 \text{ J}. \qquad (7.3\text{-}36)$$

Thus, 12.5 J has been transferred to the resistor and 0.25 J retained by the capacitors. We

---

[11] Remember here that we are assuming only a single type of charge (to free the reader from the necessity of having a background in physics). From the point of view of physics, there is a charge of 1/3 C on the top plate of the bottom capacitor, a charge of $-1/3$ C on the bottom plate of that capacitor, a charge of $-1/6$ C on the top plate of the top capacitor, and a charge of $+1/6$ C on the bottom plate of that capacitor. This means that "charge conservation" (as understood by physicists) is obeyed, though a cursory glance at (7.3-33) and (7.3-34) might make one think otherwise.

will leave it to you to compute the energy delivered to the resistor by integrating the product of $v(t)$ and $i(t)$ from zero to infinity and verify that it is indeed 12.5 J.

Flux Linkage Trapping

There is a phenomenon for switched inductor circuits, called *flux linkage trapping,* that is similar to charge trapping in switched capacitor circuits. We explore this concept in the following example.

Example 7.13   Find the response variables $v(t)$, $i(t)$, $i_x(t)$, and $i_y(t)$ in Figure 7.43.

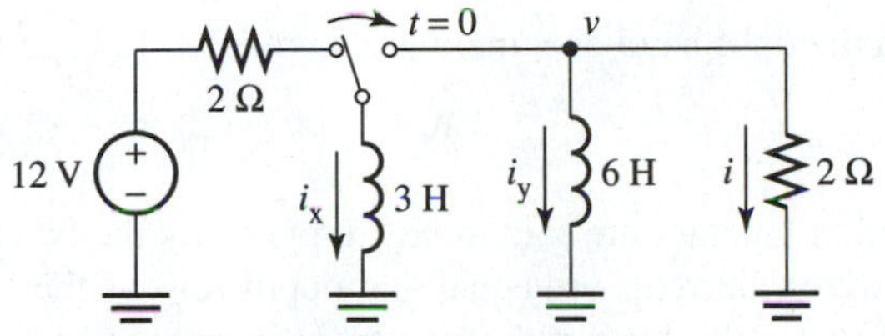

Figure 7.43
An example circuit

Solution   We will leave it to you to show that the two halves of the "$t \leq 0$" circuit are stable and therefore that the responses shown are all dc steady-state variables. The values of these variables are easily computed to be $i_x = 12 \text{ V}/2 \; \Omega = 6 \text{ A}$, $v(t) = 0 \text{ V}$, and $i_y(t) = i(t) = 0 \text{ A}$. The $t > 0$ circuit, therefore, is the one shown in Figure 7.44, where we have used the parallel initial condition models for the inductors. Notice that we have once again decided not to use the usual subscript "+" notation to simplify notation. The variables in Figure 7.44 are only valid, therefore, for positive time. Thus, writing one nodal equation in the unknown variable $v(t)$, we get

$$\frac{v(t)}{3p} + \frac{v(t)}{6p} + \frac{v(t)}{2} = -6u(t). \tag{7.3-37}$$

Solving for the voltage $v(t)$, we get

$$v(t) = \frac{-12p}{p+1}u(t) = \frac{-12}{p+1}\delta(t) = -12e^{-t}u(t) \text{ V}. \tag{7.3-38}$$

The current through the 3-H inductor, therefore, is

$$i_x(t) = \frac{v(t)}{3p} + 6u(t) = \frac{-4}{p+1}u(t) + 6u(t) = [2 + 4e^{-t}]u(t) \text{ A}. \tag{7.3-39}$$

The current through the 6-H inductor is

$$i_y(t) = \frac{v(t)}{6p} = \frac{-2}{p+1}u(t) = -2[1 - e^{-t}]u(t) \text{ A}. \tag{7.3-40}$$

Figure 7.44
The $t > 0$ circuit

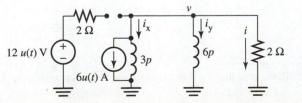

Finally, the current through the 2-Ω resistor is

$$i(t) = \frac{v(t)}{2} = -6e^{-t}u(t) \text{ A}. \tag{7.3-41}$$

(You can check these currents by noting that KCL is satisfied at the top node.)

Now observe that $v(t)$ and $i(t)$ both approach zero as $t \to \infty$, but $i_x(t)$ and $i_y(t)$ do not! In fact, we have $i_x(t) \to 2$ A and $i_y(t) \to -2$ A as $t \to \infty$. These limiting currents do, however, add up to zero—as they must to give the proper value for $i(t)$ as $t \to \infty$. We see that there is a net flux linkage[12] stored in the left-hand inductor of

$$\lambda_x = 3 \times (2) = 6 \text{ Wb} \tag{7.3-42}$$

and, in the right-hand one, there is a stored flux linkage of

$$\lambda_y = 6 \times (-2) = -12 \text{ Wb}. \tag{7.3-43}$$

These flux linkages are said to be "trapped" by the two parallel inductors. Notice that the two inductor currents are equal and opposite—as they must be to make KCL "work" at the top node—so we could consider the current to be "trapped" also.

From an energy point of view, we started with a total energy stored on the inductors at $t = 0-$ (before the switch flips) of

$$w(0-) = \frac{1}{2} \times 3 \times (6)^2 + \frac{1}{2} \times 6 \times (0)^2 = 54 \text{ J}. \tag{7.3-44}$$

As $t \to \infty$, though, we have the following for the energy trapped by the inductors:

$$w(\infty) = \frac{1}{2} \times 3 \times (2)^2 + \frac{1}{2} \times 6 \times (-2)^2 = 18 \text{ J}. \tag{7.3-45}$$

Thus, 36 J has been transferred to the resistor and 18 J retained by the inductors. We will leave it to you to compute the energy delivered to the resistor by integrating the product of $v(t)$ and $i(t)$ from zero to infinity and verify that it is indeed 36 J.

Section 7.3 Quiz

Q7.3-1. Find the unit impulse and step responses for the current $i(t)$ in the circuit in Figure Q7.3-1 by adjusting the current source waveform alternately to $i_s(t) = \delta(t)$ and $i_s(t) = u(t)$.

Q7.3-2. Find the voltage $v(t)$ in the circuit of Figure Q7.3-2.

Figure Q7.3-2

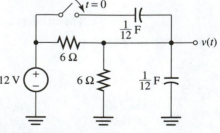

Figure Q7.3-1

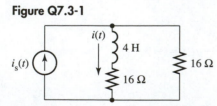

---

[12] Remember that the flux linkage for an inductor is given by $\lambda = Li$.

Q7.3-3. Use the unit impulse response you found in Question Q7.3-1 to find the unit step response by integration.

Q7.3-4. Use the unit step response you found in Question Q7.3-1 to find the unit impulse response by differentiation.

7.4 | Simulation Diagrams

The Basic Concept of a Simulation Diagram

When we analyze a "real" circuit we draw interconnected circuit elements: an inductor symbol stands for the real inductor, the resistor symbol stands for the real resistor, and so on. We can measure the actual voltages across and currents through these elements in the laboratory and compare them with the ones we compute using the mathematics of circuit analysis. Thus, on a "reality scale," the actual circuit components are the most concrete, their symbols on a piece of paper more abstract, and the equations describing them even more abstract. In order to make the mathematical equations more visual, we use the concept of a *block diagram*, which is also called a *system diagram* or a *simulation diagram.* Such a diagram represents the equations, not the physical component directly.

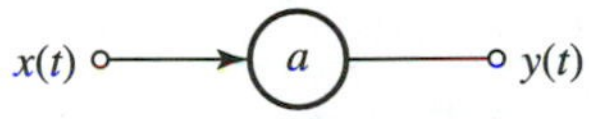

Figure 7.45 shows a *system element* called a scalar multiplier. The defining equation of this element is

Figure 7.45 The scalar multiplier

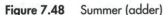
$x(t)$ a $y(t)$

$$y(t) = ax(t). \tag{7.4-1}$$

The parameter a is a constant. It is assumed to be real here, though later on we will allow it to be complex as well. The scalar multiplier merely multiplies the *input* $x(t)$ by the constant a at each instant of time to produce the *output* $y(t)$. It is nondynamic in the sense that, like a resistor, it has no memory. As a matter of fact, if we identify $x(t)$ with the current $i(t)$ in a resistor and $y(t)$ with its voltage, we can *simulate* the resistor as shown in Figure 7.46. Notice that $x(t)$ and $y(t)$ in Figure 7.45 have no dimensions. They are *information signals;* that is, they are abstractions of real signals. In Figure 7.46, however, we have identified them with real signals in a real device. A resistor is *invertible.* This means that we can "turn it around" and solve for the current (as output) in terms of the voltage (as input). In this case, we use the conductance to get the simulation diagram in Figure 7.47.

Figure 7.46
Resistor simulation:
Resistance form

$i(t)$ R $v(t)$

Figure 7.47
Resistor simulation:
Conductance form

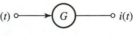
$v(t)$ G $i(t)$

Figure 7.48 Summer (adder)

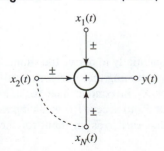

Another nondynamic system element (one that has no memory) is the *summer,* or *adder,* shown in Figure 7.48. It can have as many inputs as desired, and we permit each input to have either a $+$ or $-$ sign affixed. If there is no sign, we assume it to be $+$. The defining characteristic of this element is

$$y(t) = \pm x_1(t) \pm x_2(t) + \cdots \pm x_N(t). \tag{7.4-2}$$

We can use it, among other things, to simulate KVL and KCL. For example, suppose that we have written KVL for a given loop in a circuit as

$$v_1(t) + v_2(t) + v_3(t) = 0. \tag{7.4-3}$$

Figure 7.49   KVL simulation

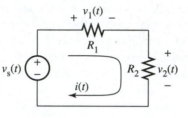

$v_1(t)$

$v_2(t)$

$v_3(t)$

We choose any of the three voltages, say $v_2(t)$, and solve for it:

$$v_2(t) = -v_1(t) - v_3(t).\qquad\qquad(7.4\text{-}4)$$

It then has the simulation diagram shown in Figure 7.49. Notice that it is not unique; we could just as well have solved for either of the other two voltages and treated it as the output, with the other two being the inputs. The resistor simulation was not unique, either, as we saw above: either of the two terminal variables can be selected as the input and the other as the output. The following example demonstrates how one might simulate a circuit.

Example 7.14   Draw a simulation diagram of the simple voltage divider circuit shown in Figure 7.50.

Figure 7.50
An example circuit

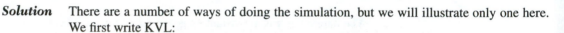

$+\ v_1(t)\ -$

R_1

$v_s(t)$

R_2 $v_2(t)$

$+$

$-$

$i(t)$

Solution   There are a number of ways of doing the simulation, but we will illustrate only one here. We first write KVL:

$$v_s(t) = v_1(t) + v_2(t).\qquad\qquad(7.4\text{-}5)$$

Next, we write the Ohm's law equation for each of the two resistors:

$$v_1(t) = R_1 i(t)\qquad\qquad(7.4\text{-}6)$$

and

$$v_2(t) = R_2 i(t).\qquad\qquad(7.4\text{-}7)$$

Now here is an essential observation: The *source voltage $v_s(t)$ cannot be the output of a system element; it must be an input.* The reason for this is simply that it is an independently specified waveform, and the output of a system element is determined by its inputs. Thus, all independent sources are constrained to be inputs. Therefore, let's solve (7.4-5) by selecting either of the other two voltages as the output. We will arbitrarily pick $v_1(t)$. Then (7.4-5) becomes

$$v_1(t) = v_s(t) - v_2(t).\qquad\qquad(7.4\text{-}8)$$

But now we have two different definitions of $v_1(t)$: it appears as an output in both (7.4-8) and (7.4-6), so let's turn (7.4-6) around and use $v_1(t)$ as an input by putting it in conductance form:

$$i(t) = G_1 v_1(t).\qquad\qquad(7.4\text{-}9)$$

We can now use (7.4-8), then (7.4-9), and finally (7.4-7) sequentially to derive the simulation diagram shown in Figure 7.51. We first drew the adder, assuming that $v_2(t)$ was available from *somewhere,* then drew the scalar multiplier for R_1 in conductance form to generate $i(t)$ as its output, then drew the scalar multiplier for R_2 to accept $i(t)$ as its input. This scalar multiplier then generates $v_2(t)$, which we needed earlier as an input for the

Figure 7.51
Simulation diagram

Figure 7.51
Simulation diagram

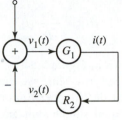

adder. We therefore merely connected the output of the R_2 scalar multiplier to that input and we were finished. Do you see how things work? The source voltage, being an independent variable, is not defined by any element.

As we mentioned at the beginning of our solution of the last example, a simulation diagram is not unique. In our solution, we actually simulated each element, then connected them together. If we are only interested in, say, the single voltage $v_2(t)$, we can apply the voltage divider rule to our circuit in Figure 7.50, to obtain

Figure 7.52 Voltage divider simulation

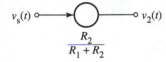

$$v_2(t) = \frac{R_2}{R_1 + R_2} v_s(t). \tag{7.4-10}$$

We can now simulate this equation with the single scalar multiplier shown in Figure 7.52. We have, however, lost the "internal" variables $v_1(t)$ and $i(t)$.

Dynamic System Element: The Differentiator and the Integrator

A resistive circuit can be simulated with only the two elements we have discussed: the scalar multiplier and the summer. Energy storage elements, however, require integrators and/or differentiators. We show the symbol for a differentiator in Figure 7.53(a) and the one for an integrator in Figure 7.53(b). Once again, we remind you that $x(t)$ and $y(t)$ are *information signals*—they have no physical dimensions, unless we wish to make them correspond to physical variables. They merely represent the abstract operations of differentiation and integration. We can think of these signals as arising from a simulation on a digital or analog computer, and the system diagram merely a way of organizing our simulation program.

Figure 7.53
Dynamic system elements

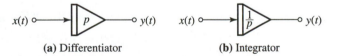

(a) Differentiator   **(b)** Integrator

We can use the differentiator and the integrator to simulate an inductor as follows. The derivative relationship for the inductor is

$$v(t) = Lpi(t). \tag{7.4-11}$$

The corresponding system simulation is shown in Figure 7.54(a). Because the scalar multiplication by L commutes with differentiation, we could just as well have drawn the scalar multiplier at the output of the differentiator rather than the way we have shown it.

Figure 7.54
Inductor simulation

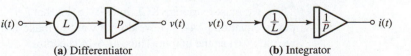

(a) Differentiator   **(b)** Integrator

If we invert (7.4-11), we have

$$i(t) = \frac{1}{Lp}v(t), \tag{7.4-12}$$

with the simulation diagram shown in Figure 7.54(b). Remember that the operator $1/Lp$ means $1/L$ followed by $1/p$, or vice versa. Thus, we can move the $1/L$ scalar multiplier to the output of the integrator because the two operations commute with one another.

The capacitor, of course, has quite similar simulation diagrams. The equation

$$i(t) = Cpv(t) \tag{7.4-13}$$

leads to the simulation diagram in Figure 7.55(a), and its inverse,

$$v(t) = \frac{1}{Cp}i(t), \tag{7.4-14}$$

gives the one in Figure 7.55(b).

Figure 7.55
Capacitor simulation

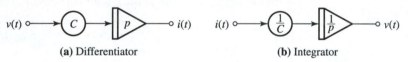

(a) Differentiator   **(b)** Integrator

Example 7.15   Simulate the first-order circuit shown in Figure 7.56.

Figure 7.56
An example first-order
circuit

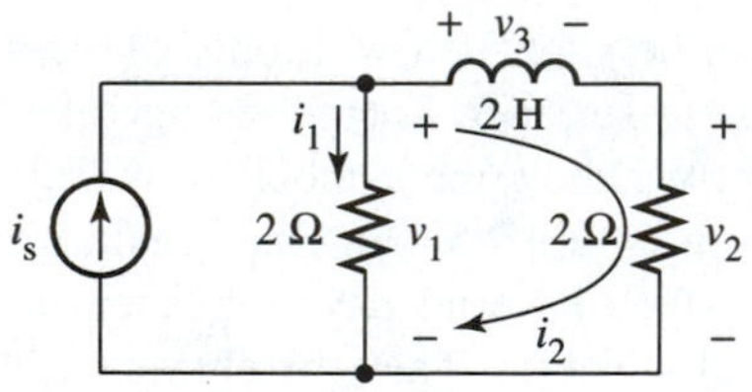

Solution   There are, as we have said, many ways to draw a simulation diagram for a given circuit. One way is as follows. We first write KVL for the essential mesh and one KCL equation for the node at the top left, then write the element equations:

$$v_1(t) = v_2(t) + v_3(t), \tag{7.4-15}$$

$$i_s(t) = i_1(t) + i_2(t), \tag{7.4-16}$$

$$v_1(t) = 2i_1(t), \tag{7.4-17}$$

$$v_2(t) = 2i_2(t), \tag{7.4-18}$$

and $$v_3(t) = 2pi_2(t). \tag{7.4-19}$$

As $v_1(t)$ appears as an output in the first equation, we cannot treat it as the output in any other; thus, we invert equation (7.4-17):

$$i_1(t) = 0.5v_1(t). \tag{7.4-17a}$$

Then, we must rearrange KCL in (7.4-16) so that $i_s(t)$ is not the output. As $i_1(t)$ is the output in equation (7.4-17a), we do not want to make it the output in our KCL equation, so we pick $i_2(t)$ as the output. Thus, (7.4-16) becomes

$$i_2(t) = i_s(t) - i_1(t). \tag{7.4-16a}$$

Figure 7.57
Simulation diagram

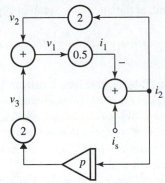

Now we simulate (7.4-15), then (7.4-17a), then (7.4-16a), then (7.4-18), and finally (7.4-19)—in that order. This results in the simulation diagram shown in Figure 7.57. It is not the only one possible, as we observed previously. One could start with any element. The basic idea is to assume that the inputs are available *somewhere,* then work through all the elements. You should always check to make sure that each of the inputs is defined by one *and only one* output, or by an external variable from an independent source.

Differentiators Versus Integrators: A Practical Discussion

There is a feature of the simulation diagram that we derived in the last example that should be discussed. The single dynamic element was a differentiator. One often requires that the simulation diagram not have any differentiators if possible. Here's why. Suppose we apply a signal that has been corrupted with additive noise to a differentiator as in Figure 7.58(a). Here $x(t)$, the signal, represents something of interest to us, whereas the added noise component $n(t)$ is undesired; noise, however, is present in all systems whether we like it or not. For the sake of simplicity, let's assume that $n(t)$ has the form of a sinusoid with amplitude N_o and frequency ω:

$$n(t) = N_o \sin(\omega t)u(t). \tag{7.4-20}$$

The output of the differentiator, then is

$$y(t) = p[x(t) + n(t)] = px(t) + \omega N_o \cos(\omega t)u(t) = y_s(t) + y_n(t), \tag{7.4-21}$$

where

$$y_s(t) = px(t) \tag{7.4-22}$$

is the output due to the *signal,* $x(t)$, and

$$y_n(t) = \omega N_o \cos(\omega t)u(t) \tag{7.4-23}$$

is the output due to the *noise.* Notice that if we allow ω to take on larger and larger values, $y_n(t)$ becomes larger and larger. Thus, we say that *a differentiator enhances high-frequency noise.*

Figure 7.58
Differentiation and integration with noise

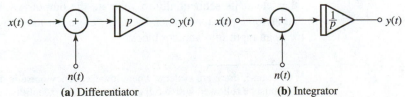

(a) Differentiator   (b) Integrator

Look, on the other hand, at the integrator setup in Figure 7.58(b). We see that

$$y(t) = \frac{1}{p}[x(t) + n(t)] = \frac{1}{p}x(t) - \frac{1}{\omega}N_o \cos(\omega t)u(t) = y_s(t) + y_n(t). \quad (7.4\text{-}24)$$

This looks very much like (7.4-21); there is, however, a very great difference. Now the output due to the noise, $y_n(t)$, is *inversely* proportional to the frequency of the noise. Thus, it decreases as ω increases. *We say that integration reduces high-frequency noise.* It is a fact that many types of systems contain more high-frequency noise than low-frequency noise. *Thus, we generally prefer that any simulation be done with integrators and try our best to eliminate any differentiators.*[13]

If we now glance back at Figure 7.57 in the last example, we see that we have violated our preference there. The question is, can we generate another simulation diagram without a differentiator? We certainly can. What we will do is to invert equation (7.4-19), labeling it as (7.4-19a):

$$i_2(t) = \frac{1}{2p}v_3(t). \quad (7.4\text{-}19a)$$

Because this integration is so important, let's start with it as our most important equation and draw its system simulation, assuming that $v_3(t)$ is available as an input. We then rearrange (7.4-15) to provide $v_3(t)$:

$$v_3(t) = v_1(t) - v_2(t). \quad (7.4\text{-}15a)$$

We see that $v_2(t)$ is derived from $i_2(t)$ by (7.4-18). But what about $v_1(t)$? To compute it, we need $i_1(t)$, which we can get by solving (7.4-16) for $i_1(t)$ in terms of $i_s(t)$ and $i_2(t)$:

$$i_1(t) = i_s(t) - i_2(t). \quad (7.4\text{-}16b)$$

We then use (7.4-17) to obtain $v_1(t)$. The resulting diagram is shown in Figure 7.59.

Figure 7.59
An "integrator-only"
simulation

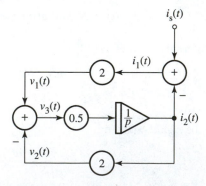

Summary of Simulation Procedures

To summarize, we start with the dynamic element (inductor or capacitor) and simulate it with an integrator. We then manipulate the equations so that the input to that integrator is defined by outputs of other blocks (and/or by the independent source(s)). We then use the other elements sequentially to generate the outputs we need until we are finished. Note that at some point we generally must "close the loop" by using the output of the integrator as an input into another block.

---

[13] Of course, there are systems where low-frequency noise is dominant; thus, this, like any general rule, must be followed with a full awareness of the exceptional cases.

In addition to simulating circuits, we can also use the technique we have developed to simulate differential equations. For instance, suppose we are to simulate the linear, constant-coefficient lowpass differential equation

$$py(t) + ay(t) = x(t) \tag{7.4-25}$$

using an integrator. We first solve for $py(t)$:

$$py(t) = x(t) - ay(t). \tag{7.4-26}$$

Then we multiply both sides by $1/p$ from the left (that is, we integrate both sides). This results in

$$y(t) = \frac{1}{p}[x(t) - ay(t)]. \tag{7.4-27}$$

Thus, we obtain $y(t)$ by integrating the difference between $x(t)$ and a times $y(t)$ itself. We show the result in Figure 7.60(a). Because this will be needed so often, we also show it just as a block by itself in Figure 7.60(b). We will often use the block in order to conserve space on our diagrams.

Figure 7.60
First-order block

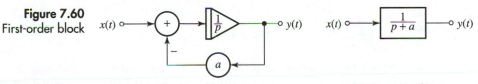

(a) Integrator simulation   (b) First-order block

The general first-order differential equation is given by

$$py(t) + ay(t) = bpx(t) + cx(t). \tag{7.4-28}$$

The solution is

$$y(t) = H(p)x(t), \tag{7.4-29}$$

where

$$H(p) = \frac{bp + c}{p + a}. \tag{7.4-30}$$

By doing long division, we can convert this to the sum of two simpler operators and write

$$y(t) = bx(t) + \frac{c - ab}{p + a}x(t). \tag{7.4-31}$$

We show the simulation diagram in Figure 7.61. The first-order block shown is, of course, only a shorthand symbol for the more complicated block diagram shown in Figure 7.60(a).

Figure 7.61
General first-order
simulation

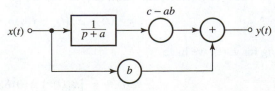

Recursive and Nonrecursive Simulations

The simulation in Figure 7.60(a) and, indeed, those we have developed earlier in Figures 7.51, 7.57, and 7.59 are called *recursive* because a variable is "fed back" to assist in the development of a variable that is ultimately used as an input into itself. But we can generate *nonrecursive* simulations as well: those that are not fed back in this fashion. For instance, let's look more closely at the first-order lowpass operator (the solution operator for the first-order lowpass differential equation):

$$y(t) = \frac{1}{p + a}x(t) = e^{-at}\frac{1}{p}[e^{at}x(t)]. \tag{7.4-32}$$

The last expression means that we multiply the forcing function $x(t)$ by e^{at}, integrate the product, then multiply the result by e^{-at}. We show this in Figure 7.62. Notice that we have multiplied the two exponentials by the unit step function $u(t)$. This is permissible because we are assuming that all waveforms (including the input $x(t)$) are one sided, that is, they are zero for $t \leq 0$. We have also introduced a new system element: the *function multiplier*. This is a block that accepts two time-varying waveforms and multiplies their values together at each point in time to produce its output.

Figure 7.62
Nonrecursive simulation of the first-order solution operator

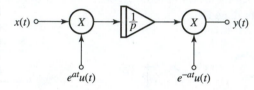

$x(t) \circ\!\!\longrightarrow X \longrightarrow \boxed{\frac{1}{p}} \longrightarrow X \longrightarrow\!\!\circ\; y(t)$

$e^{at}u(t)$ $e^{-at}u(t)$

Simulation of Systems with Initial Conditions

Now let's talk about the simulation of circuits and differential equations on the positive time axis subject to initial conditions. In fact, we will look at initial condition models for the integrator. We know that the input to an integrator is the derivative of the output, so we are to solve the differential equation

$$py(t) = y'(t) = x(t) \tag{7.4-33}$$

on the interval $0 < t < \infty$. We restrict things to the nonnegative time axis by multiplying both sides by the unit step function $u(t)$. Thus,

$$y'(t)u(t) = x(t)u(t). \tag{7.4-34}$$

In the past, we have used the following notation (in general):

$$f_+(t) = f(t)u(t), \tag{7.4-35}$$

so the right-hand side of (7.4-34) is just $x_+(t)$. The left-hand side is a bit trickier. Recall, however, that if we apply the Leibniz rule for differentiating a product, and note that the derivative of the unit step function is simply the unit impulse and that the product of $y(t)$ and the unit impulse can be simplified by the sampling property to $y(0)\delta(t)$, we have

$$py_+(t) = p[y(t)u(t)] = y'(t)u(t) + y(0)\delta(t). \tag{7.4-36}$$

This transforms (7.4-34) into a relationship between one-sided waveforms only: if we solve (7.4-36) for $y'(t)u(t)$ and use it in (7.4-34), we get

$$py_+(t) = x_+(t) + y(0)\delta(t). \tag{7.4-37}$$

Solving for $y_+(t)$, we have

$$y_+(t) = \frac{1}{p}[x_+(t) + y(0)\delta(t)]. \tag{7.4-38}$$

Figure 7.63
Initial condition models
for the integrator

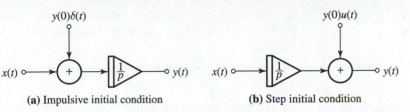

(a) Impulsive initial condition   **(b)** Step initial condition

This gives the initial condition integrator model shown in Figure 7.63(a). If we integrate the delta function term, we will have

$$y_+(t) = \frac{1}{p}x_1(t) + y(0)u(t), \qquad (7.4\text{-}39)$$

whose corresponding system model is shown in Figure 7.63(b). These two simulation diagrams correspond to the two initial condition models for the capacitor and inductor.

Example 7.16 Draw the simulation for the differential equation (holding for $t > 0$)

$$\frac{dy(t)}{dt} + 3y(t) = x(t), \qquad (7.4\text{-}40)$$

subject to the initial condition $y(0) = 2$.

Solution We will outline two approaches to solving this problem. The first is to juggle the differential equation in order to find $y(t)$ as an integral,

$$y(t) = \frac{1}{p}[x(t) - 3y(t)], \qquad (7.4\text{-}41)$$

realize it in the form of a simulation diagram, then merely add the initial condition for $y(t)$ as in Figure 7.63. This gives the configuration (using an impulsive initial condition) of Figure 7.64. We have economized on summers by adding the initial condition in the same adder as the term $x(t) - 3y(t)$. This approach is quite effective as long as the initial condition on the integrator output is specified; in other problems, however, this will not be the case. In that event, the following procedure will work—thus, it is the more general of the two. (It will work for higher-order simulations as well.) We multiply both sides of our differential equation by $u(t)$,

$$y'(t)u(t) + 3y(t)u(t) = x(t)u(t), \qquad (7.4\text{-}42)$$

then use the result in equation (7.4-36) to replace $y'(t)u(t)$ by $py_+(t) - y(0)\delta(t)$. This gives

$$py_+(t) + 3y_+(t) = x_+(t) + 2\delta(t). \qquad (7.4\text{-}43)$$

Figure 7.64
Simulation diagram

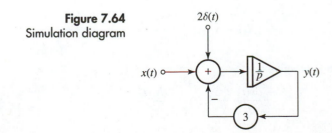

Now we simply treat the initial condition impulse waveform as an added forcing function, solve for $y_+(t)$ in terms of an integral, and draw the resulting diagram. We get Figure 7.64 once more.

Initial Conditions on the Forcing Function

We will now illustrate the method just explained at the end of the preceding example a bit more with the following initial condition problem. Solve

$$\frac{dy(t)}{dt} + ay(t) = \frac{dx(t)}{dt}, \tag{7.4-44}$$

for $t > 0$ subject to the initial condition $y(0) = y_o$. As in the last example (second method), we multiply both sides by $u(t)$, getting

$$y'(t)u(t) + ay(t)u(t) = x'(t)u(t). \tag{7.4-45}$$

Now we use the relationship

$$pf_+(t) = p[f(t)u(t)] = f'(t)u(t) + f(0)\delta(t) \tag{7.4-46}$$

on both sides of (7.4-45). This results in the following one-sided problem:

$$py_+(t) + ay_+(t) = px_+(t) + [y_o - x_o]\,\delta(t), \tag{7.4-47}$$

where x_o is the initial condition for $x(t)$. Solving for $y_+(t)$ as an integral, we obtain

$$y_+(t) = x_+(t) + \frac{1}{p}[-ay_+(t) + [y_o - x_o]\,\delta(t)]. \tag{7.4-48}$$

We interpret this equation in the form of a simulation diagram in Figure 7.65.

Figure 7.65
Simulation diagram

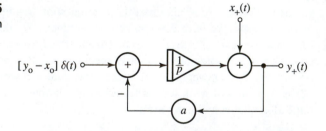

Do you see something curious about this diagram? We need to know not only the initial condition on the output but an initial condition on the input as well. Let's translate this into a more physical context with a circuit problem to show how this might arise. Figure 7.66 shows a relatively simple circuit, which we will analyze to find $v(t)$. We have labeled the single 1/3-farad capacitor with its impedance operator. A straightforward nodal analysis gives

Figure 7.66   An example circuit

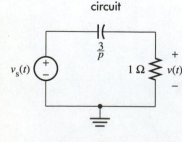

$$\frac{p}{3}[v(t) - v_s(t)] + \frac{v(t)}{1} = 0, \tag{7.4-49}$$

which can be rearranged to give

$$\frac{dv(t)}{dt} + 3v(t) = \frac{dv_s(t)}{dt}. \tag{7.4-50}$$

This is precisely the same differential equation as (7.4-44) with $a = 3$ and with different symbols for the time functions.

But our problem of interpretation above had to do with the initial conditions. If we multiply both sides of (7.4-50) by $u(t)$ and solve as before, we will introduce the necessity for knowing $v_s(0)$—again as before. Rather than doing this, let's inquire how the initial conditions come into existence. We know by now that they often arise because a circuit has switches that are flipped at $t = 0$, thus setting up the $t > 0$ circuit (that is, the one we have just analyzed in Figure 7.66). Thus, let's look at the switched circuit in Figure 7.67. It might not be clear at once that this circuit is closely related to the one in Figure 7.66, but it is, as we will see. Notice that two of the sources are dc.

Figure 7.67
A switched circuit

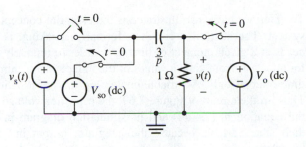

The $t \le 0$ circuit is shown in Figure 7.68. Notice that the voltage across the capacitor is held at $v_c(t) = V_o - V_{so}$ by the two dc v-sources, so the circuit is in the dc steady state as shown. Notice also that the response voltage $v(t)$ is "clamped" at V_o—a specification of the initial condition on $v(t)$, while $v_s(t)$ is replaced (for $t \le 0$) by the constant value V_{so}—its initial condition.

Figure 7.68
The $t \le 0$ circuit

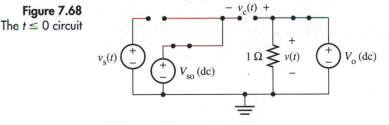

Now let's look at the $t > 0$ circuit shown in Figure 7.69. We have used the parallel initial condition model for the capacitor. The differential equation for the voltage of interest is now

$$\frac{p}{3}[v_+(t) - v_{s+}(t)] + \frac{v_+(t)}{1} = \frac{1}{3}[V_o - V_{so}]\delta(t). \tag{7.4-51}$$

Figure 7.69
The $t > 0$ circuit

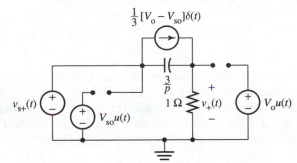

Multiplying by $3/p$ (that is, integrating both sides and then multiplying both sides by 3), we obtain

$$v_+(t) = v_{s+}(t) + \frac{1}{p}[-3v_+(t) + [V_o - V_{so}]\delta(t)]. \tag{7.4-52}$$

Do you see the similarity to equation (7.4-48)? The forcing function has been replaced by $v_{s+}(t)$ and its initial condition by V_{so}; similarly, the response variable $y(t)$ is now $v_+(t)$, and its initial condition is V_o. The whole idea is that the initial voltage condition on the capacitor is the *difference* of two node voltages—one the response variable and the other the forcing function variable.

State and State Variables: A Brief Introduction

The idea we have just illustrated is the essential concept of the *state* of a circuit or other system. The state of a circuit, informally speaking, is the amount of information one needs at a given moment of time t_o in order to uniquely determine any response variable for all $t > t_o$ from known values of the independent sources or forcing functions over that time interval only. A variable containing this information is known as a *state variable*. Thus, in the circuit of Figure 7.69, the capacitor voltage is a state variable. The differential equation in (7.4-44) and its simulation diagram in Figure 7.65 have $y(t) - x(t)$ as their state variable because the integrator output in Figure 7.65 is $y_+(t) - x_+(t)$ and knowledge of $y_0 - x_0$ allows the response $y(t)$ to be determined for all $t > 0$. (If this variable were known at any other time t_o, one could easily show that $y(t)$ could then be determined for all $t > t_o$, knowing $x(t)$ only over the time interval $(t_o, t]$.)

Let's develop just a bit more insight into these new ideas by solving (7.4-44) in a slightly different way. Recalling this differential equation, now numbered (7.4-53),

$$\frac{dy(t)}{dt} + ay(t) = \frac{dx(t)}{dt}, \tag{7.4-53}$$

we make the definition

$$z(t) = y(t) - x(t), \tag{7.4-54}$$

thus transforming it into

$$\frac{dz}{dt} + az = -ax(t). \tag{7.4-55}$$

We can now multiply both sides by the unit step function and let $z_+(t) = z(t)u(t)$ to get

$$pz_+(t) + az_+(t) = -ax_+(t) + z_0\delta(t), \tag{7.4-56}$$

where z_0 is the initial condition on the variable $z(t)$. Solving for $pz_+(t)$ and multiplying both sides by $1/p$ (integrating both sides) we get

$$z_+(t) = \frac{1}{p}[-ax_+(t) + z_0\delta(t) - az_+(t)]. \tag{7.4-57}$$

The corresponding simulation diagram is shown in Figure 7.70. Notice that it is almost the same as the one in Figure 7.65. The only difference is that we have labeled the integrator output with the state variable $z_+(t)$ and "fed it back" to the summer at the integrator input. Thus, we have derived a state variable for the differential equation by a simple linear combination of the input and output variables. Note that it is the initial condition

Figure 7.70
Simulation diagram

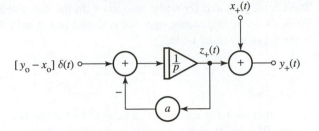

for the integrator by itself, not a simple value of the response variable $y(t)$. This is typical: the integrator outputs are the state variables.

We will not go any further with our discussion of state and state variables. There are many good texts covering this material, and you will undoubtedly encounter it again in more advanced courses, particularly in the study of linear systems.

Section 7.4 Quiz

Q7.4-1. Draw a simulation diagram for the circuit in Figure Q7.4-1.

Figure Q7.4-1

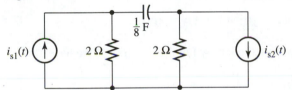

Q7.4-2. Draw a simulation diagram for the initial condition

problem given by the differential equation

$$\frac{dy(t)}{dt} + 5y(t) = 3\frac{dx(t)}{dt} + 2x(t), \qquad t > 0$$

subject to the initial conditions $y(0) = 3$ and $x(0) = 2$.

Q7.4-3. Derive a first-order state variable differential equation for the differential equation in Question Q7.4-2; that is, find a linear combination $z(t)$ of the response variable $y(t)$ and the input variable $x(t)$ such that no derivative of $x(t)$ appears on the right-hand side. What is the initial condition on this state variable corresponding to the initial conditions in Question Q7.4-2?

7.5 | Active First-Order Circuits

Time Domain Analysis of Circuits with Dependent Sources

The analysis of circuits with dependent sources is exactly the same as for those without—providing that the extra steps of taping and untaping are added. Our first example illustrates this. This extended example also shows that, unlike passive circuits, circuits having active elements can be unstable.

Example 7.17

For the circuit shown in Figure 7.71:

a. Find $H(p)$ in the solution $v(t) = H(p)v_s(t)$.

b. For which values of β is this response unstable?

c. Find the step and impulse responses for $\beta = 0, 1, 2$, and 3.

Figure 7.71
An example circuit

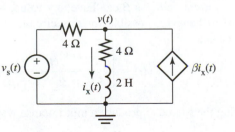

Solution   The circuit prepared for nodal analysis with the dependent source taped and the inductor labeled with its impedance operator is shown in Figure 7.72. We write one nodal equation at the single essential node:

$$\frac{v(t) - v_s(t)}{4} + \frac{v(t)}{2p + 4} = i_c(t). \tag{7.5-1}$$

Notice that we have used a series equivalent for the vertical 4-Ω resistor and the 2-H inductor. We can do this because we now know that its current is equal to $v(t)/(2p + 4)$; that is, the inverse of the operator $2p + 4$ is well defined. Rationalizing by multiplication from the left on both sides by $4(2p + 4) = 8(p + 2)$, then dividing by 2, gives

$$(p + 4)v(t) - (p + 2)v_s(t) = 4(p + 2)i_c(t). \tag{7.5-2}$$

Now we untape the dependent source and express it in terms of the unknown node voltage:

$$i_c(t) = \beta i_x(t) = \beta \frac{v(t)}{2(p + 2)}. \tag{7.5-3}$$

Using this in (7.5-2), we obtain the differential equation

$$[p + 2(2 - \beta)] \, v(t) = (p + 2)v_s(t). \tag{7.5-4}$$

Solving by multiplying both sides from the left by $[p + 2(2 - \beta)]^{-1}$, we have

$$v(t) = \frac{p + 2}{p + 2(2 - \beta)} v_s(t), \tag{7.5-5}$$

so $H(p)$ is the operator

$$H(p) = \frac{p + 2}{p + 2(2 - \beta)}. \tag{7.5-6}$$

Figure 7.72
The circuit prepared for
nodal analysis

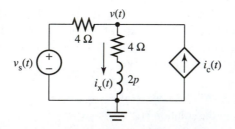

The presence of the dependent source is reflected by the term in the denominator involving β, its current gain. As β can have any value, we will see that it dramatically affects the type of behavior possible for this circuit. Using one step of long division on (7.5-5), we see that

$$v(t) = \left[1 + \frac{2(\beta - 1)}{p + 2(2 - \beta)} \right] v_s(t). \tag{7.5-7}$$

Thus, adjusting the source to produce a unit impulse waveform, we get

$$h(t) = \delta(t) + 2(\beta - 1)e^{-2(2-\beta)t}u(t). \tag{7.5-8}$$

Because $h(t)$ approaches zero as $t \to \infty$ if $\beta < 2$, $v(t)$ is a stable response for this range of values of β; $h(t)$ becomes unbounded if $\beta > 2$ and therefore $v(t)$ is unstable; and $h(t)$ is bounded as $t \to \infty$, but approaches neither zero nor infinity if $\beta = 2$, so $v(t)$ is marginally stable. The step response, assuming that $\beta \neq 2$, is

$$s(t) = u(t) + \frac{\beta - 1}{2 - \beta}[1 - e^{-2(2-\beta)t}]\, u(t) \tag{7.5-9}$$

$$= \frac{1}{2 - \beta}[1 - (\beta - 1)e^{-2(2-\beta)t}]\, u(t).$$

If $\beta = 0$, equation (7.5-8) gives

$$h(t) = \delta(t) - 2e^{-4t}u(t). \tag{7.5-10}$$

The step response, according to equation (7.5-9), is

$$s(t) = \frac{1}{2}[1 + e^{-4t}]\, u(t). \tag{7.5-11}$$

This response is stable because $h(t)$ approaches zero for large values of t. (The occurrence of the impulse at $t = 0$ has no bearing on stability.)

When $\beta = 1$, we have

$$h(t) = \delta(t) \tag{7.5-12}$$

and
$$s(t) = u(t). \tag{7.5-13}$$

This seems to be a special case, and it is. Go back to equation (7.5-5) and investigate it for the condition $\beta = 1$. You will see that the numerator factor of $p + 2$ exactly cancels the denominator operator $p + 2(2 - \beta) = p + 2$. Hence, the output is precisely the same as the input! Again, the response is stable because $h(t)$ becomes (actually is identical to) zero for large values of time.

If $\beta = 2$, the impulse response above in equation (7.5-8) is still valid, and

$$h(t) = \delta(t) + 2u(t). \tag{7.5-14}$$

Now, however, our expression for the step response in (7.5-9) is not valid. Going back to (7.5-7), however, and letting $v_s(t) = u(t)$, we get

$$s(t) = \left[1 + \frac{2}{p}\right]u(t) = (1 + 2t)u(t). \tag{7.5-15}$$

Notice here that in letting $\beta = 2$ we have modified the operator on the bottom in equation (7.5-5) so that it merely represents the integration operator, rather than the first-order lowpass operator. Because $h(t)$ does not go to zero for large values of time, but does not become infinitely large either (that is, it is nonzero but bounded), we see that our response is now marginally stable.

When $\beta = 3$ we have

$$h(t) = \delta(t) + 4e^{2t}u(t) \tag{7.5-16}$$

and
$$s(t) = -[1 - 2e^{2t}]\, u(t). \tag{7.5-17}$$

This represents unstable operation for $h(t)$ becomes unbounded as t gets infinitely large.

In our example here, the current gain β of the CCCS affected only the denominator operator polynomial in $H(p)$; in general, it can affect both numerator and denominator. However, it is only the denominator polynomial operator that affects stability. We will see in succeeding chapters that this is generally true, regardless of the order of the circuit.

Time Domain Analysis of Circuits with Operational Amplifiers

We will now discuss the analysis of first-order circuits containing op amps. Our technique will be based upon the dc methods we established in Chapter 5, Sections 5.4 and 5.6.

Example 7.18(O)

For the op amp circuit in Figure 7.73:

a. Find $H(p)$ in the equation $v_o(t) = H(p)v_s(t)$.

b. Find $h(t)$ and $s(t)$.

c. Discuss the stability of the $v_o(t)$ response as a function of $k = 1 + R_b/R_a$.

d. If $R = 10 \text{ k}\Omega$, $C = 10 \text{ nF}$, and $v_i(t) = 2 \cos(10^4 t)u(t)$ V, find $v_o(t)$ for the two conditions $k = 2$ (say $R_a = R_b = 10 \text{ k}\Omega$) and $k = 3$ (say $R_a = 10 \text{ k}\Omega$ and $R_b = 20 \text{ k}\Omega$).

Figure 7.73
An example circuit

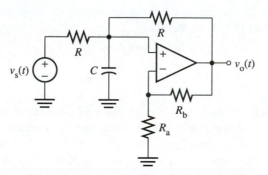

Solution

In Chapter 5 we developed a number of techniques for dealing with op amp circuits. We choose here to recognize the standard op amp subcircuit called a noninverting voltage amplifier. This subcircuit consists of the op amp and the two resistors R_a and R_b. As we showed in Chapter 5, it can be replaced by a VCVS having a voltage gain of $k = 1 + R_b/R_a$ as shown in Figure 7.74. We can analyze this configuration with one nodal equation,

$$\frac{v(t) - v_s(t)}{R} + \frac{v(t) - v_o(t)}{R} + Cpv(t) = 0. \tag{7.5-18}$$

Recall that the output node is a nonessential node and—as we discussed in Chapter 5 does not require an equation. Although we have not explicitly shown this, we are consid-

Figure 7.74
The example circuit prepared for nodal analysis

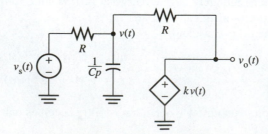

ering the VCVS to have been taped and to be represented by the value $v_o(t)$. We simplify (7.5-18) into the form

$$\left[p + \frac{2}{RC}\right]v(t) - \frac{1}{RC}v_o(t) = \frac{1}{RC}v_s(t). \tag{7.5-19}$$

Now we untape the VCVS and express its value in terms of the unknown node voltage, $v_o(t) = kv(t)$, and insert it in (7.5-19):

$$\left[p + \frac{2 - k}{RC}\right]v(t) = \frac{1}{RC}v_s(t). \tag{7.5-20}$$

But $v(t) = v_o(t)/k$ and it is $v_o(t)$ that we want. Using this in (7.5-20) gives

$$v_o(t) = \frac{\dfrac{k}{RC}}{p + \dfrac{2 - k}{RC}}v_s(t). \tag{7.5-21}$$

Thus, the required operator is

$$H(p) = \frac{\dfrac{k}{RC}}{p + \dfrac{2 - k}{RC}}. \tag{7.5-22}$$

The impulse response, therefore, is

$$h(t) = \frac{\dfrac{k}{RC}}{p + \dfrac{2 - k}{RC}}\delta(t) = \frac{k}{RC}e^{-(2-k)t/RC}u(t), \tag{7.5-23}$$

and the step response is

$$s(t) = \frac{\dfrac{k}{RC}}{p + \dfrac{2 - k}{RC}}u(t) = \frac{k}{2 - k}[1 - e^{-(2-k)t/RC}]u(t). \tag{7.5-24}$$

In computing the step response we have assumed that $k \neq 2$. If $k = 2$, (7.5-24) becomes

$$s(t) = \frac{2}{RCp}u(t) = \frac{2}{RC}tu(t). \tag{7.5-25}$$

As for the stability, we see from (7.5-23) that $h(t)$ approaches zero as t approaches infinity, provided that $k < 2$. It approaches infinity (becomes unbounded) if $k > 2$ and so is unstable. If $k = 2$ exactly, $h(t)$ is a step function of magnitude $2/RC$; hence the response is marginally stable.

Now let's compute the response to the cosine forcing function. When $k = 2$, equation (7.5-21) gives

$$v_o(t) = \frac{2}{RCp}[2\cos(10^4 t)u(t)]. \tag{7.5-26}$$

We know that the use of kΩ for resistance, mA for current, and V for voltage is consistent; furthermore, noting that the product $RC = 10 \text{ k}\Omega \times 10 \text{ nF} = 0.1 \text{ ms}$, we decide to use the millisecond for our unit of time. This transforms (7.5-26) into

$$v_o(t) = \frac{40}{p}[\cos(10t)u(t)] = 4\sin(10t)u(t) \text{ V}. \tag{7.5-27}$$

When $k = 3$, the computation is a bit more difficult. Equation (7.5-21) gives

$$v_o(t) = \frac{30}{p - 10}[2\cos(10t)u(t)] = 60e^{10t}\frac{1}{p}[e^{-10t}\cos(10t)u(t)] \tag{7.5-28}$$

$$= 60e^{10t}\left[\int_0^t e^{-10\alpha}\cos(10\alpha)\,d\alpha\right]u(t).$$

We compute the indefinite integral $F(\alpha)$ by the tabular method (see Appendix C), as in Table 7.1. We then find the definite integral by evaluating $F(t) - F(0)$. This results in the voltage $v_o(t)$ that we are seeking. Thus,

$$F(\alpha) = 0.1e^{-10\alpha}[\sin(10\alpha) - \cos(10\alpha)] - F(\alpha), \tag{7.5-29}$$

and hence

$$F(\alpha) = 0.05e^{-10\alpha}[\sin(10\alpha) - \cos(10\alpha)]. \tag{7.5-30}$$

Therefore,

$$F(t) - F(0) = 0.05e^{-10t}[\sin(10t) - \cos(10t)] + 0.05 \tag{7.5-31}$$

and

$$v_o(t) = 3[\sin(10t) - \cos(10t)]u(t) + 3e^{10t}u(t) \text{ V}. \tag{7.5-32}$$

Table 7.1
Computation of $F(\alpha)$
by tabular method

$e^{-10\alpha}$	$+$	$\cos(10\alpha)$
$-10e^{-10\alpha}$	$-$	$0.1\sin(10\alpha)$
$100e^{-10\alpha}$	$+1/p$	$-0.01\cos(10\alpha)$

Notice an important thing: the natural response $3e^{10t}u(t)$ is increasing in magnitude with time. This means that our response is unstable and there is no steady state. We observe, however, that the *ac forced response* still exists and is given by the sinusoids in (7.5-32), which can be simplified somewhat using the trigonometric identity $A\cos(\theta) - B\sin(\theta) = \sqrt{A^2 + B^2}\cos(\theta + \tan^{-1}(B/A))$. With $A = B = -3$ and $\theta = 10t$, we have

$$v_{of}(t) = 3\sqrt{2}[\cos(10t - 135°)]u(t) \text{ V} \tag{7.5-33}$$

Active Circuit
Design Using
Simulation Diagrams

Our next example presents some ideas about designing first-order active op amp circuits using the simulation diagrams we developed in Section 7.4.

Example 7.19(O)

Suppose a system operator is given by $H(p) = (bp + c)/(p + a)$, with $a = 2 \times 10^4 \text{ s}^{-1}$, $b = 10$, and $c = 10^4 \text{ s}^{-1}$. Design an active circuit to accept an input voltage as the forcing function and produce an output voltage as the response with $H(p)$

as the system operator. Use only resistors, a single capacitor, and two operational amplifiers.

Solution   First, notice that the units we gave for the constants are consistent. Because p is the derivative operator, it has the dimension of inverse time (1/time). Thus, a must have the same dimension because it must add to the differentiation operator. Similarly, because $H(p)$ must multiply (operate on) a voltage to produce a voltage, it must be dimensionless. This implies that b is also dimensionless and c has the same dimension as p and a. We write

$$v_o(t) = H(p)v_i(t) = \frac{bp + c}{p + a}v_i(t) = b\frac{p + c/b}{p + a}v_i(t) = 10\frac{p + 10^3}{p + 2 \times 10^4}v_i(t). \qquad (7.5\text{-}34)$$

Now, have a look at the op amp subcircuit shown in Figure 7.75. The output v_y is related to the input v_x by

$$v_y(t) = \frac{-Z_2(p)}{Z_1(p)}v_x(t) = -\frac{C_1}{C_2}\frac{p + 1/R_1C_1}{p + 1/R_2C_2}v_x(t). \qquad (7.5\text{-}35)$$

This has precisely the same form as equation (7.5-34), except for an overall sign difference. Thus, we can match coefficients, getting

$$C_1/C_2 = 10, \qquad (7.5\text{-}36)$$

$$\frac{1}{R_1C_1} = 10^3, \qquad (7.5\text{-}37)$$

and

$$\frac{1}{R_2C_2} = 2 \times 10^4. \qquad (7.5\text{-}38)$$

Taking the ratio of the last two equations, we get

$$R_1/R_2 = 20. \qquad (7.5\text{-}39)$$

Arbitrarily choosing $R_2 = 10\text{ k}\Omega$ and $C_2 = 1\text{ nF}$, equations (7.5-36) and (7.5-39) tell us that we must choose $R_1 = 200\text{ k}\Omega$ and $C_1 = 10\text{ nF}$.

Figure 7.75
A first-order op amp subcircuit

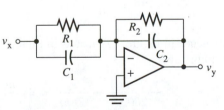

In many applications, such as audio signal processing, an overall sign inversion is of no importance. If this is the case, we are done—merely choosing the appropriate values for the resistors and capacitors and letting $v_x = v_i$ and $v_o = v_y$. Otherwise, we must add

Figure 7.76
Final design

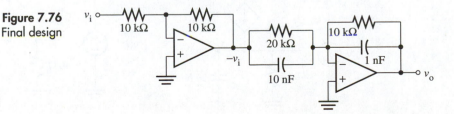

an inverter subcircuit, as shown in Figure 7.76. We have somewhat arbitrarily chosen to make the equal-valued inverter resistor values 10 kΩ each.

Our last topic in this section will be the time domain analysis of first-order circuits with bipolar junction transistors (BJTs).

Example 7.20(T)   Find the unit step response for $v_o(t)$ in the circuit shown in Figure 7.77. Assume that the small signal model for the BJT is valid and note that the circuit shown is already small signal in the sense that any dc biasing sources have been deactivated. Assume that the value of r_π is 2 kΩ and the value of β is 49.

Figure 7.77
A BJT emitter follower
circuit

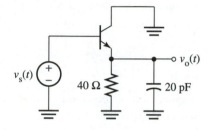

Solution   We merely "unplug" the BJT symbol and "plug in" its hybrid π model. We adjust the input source to be a unit step function in order to compute the step response. The result is shown in Figure 7.78. Before we launch too abruptly into our analysis, let's stop and reflect a bit. The resistance values are widely different, so there seems to be no benefit in changing their units from Ω to something else. Thus, we will keep the units of ohms, amperes, and volts. How about the capacitance? Well, we see that $\Omega \cdot pF = ps$—so let's use the pF as our unit of capacitance and the ps as our unit of time. Thus, we will draw the operator form of our circuit as shown in Figure 7.79. The operator impedance of the parallel equivalent of the 40-Ω resistor and the $1/20p$ capacitive impedance is

$$Z(p) = \frac{40 \times \dfrac{1}{20p}}{40 + \dfrac{1}{20p}} = \frac{40}{800p + 1} = \frac{\dfrac{1}{20}}{p + \dfrac{1}{800}}. \tag{7.5-40}$$

Using this equivalence in Figure 7.79 results in Figure 7.80. Now we can write one nodal equation at the output node:

$$\frac{v_o - u(t)}{2000} + \frac{v_o}{Z(p)} = 49\frac{u(t) - v_o}{2000}. \tag{7.5-41}$$

Figure 7.78
Circuit with the hybrid π equivalent for
the BJT

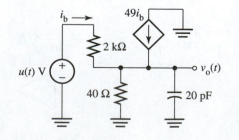

Figure 7.79
The hybrid π equivalent after scaling

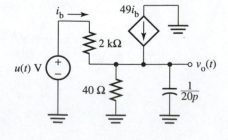

We did not tape the dependent source because of the simplicity of the circuit. Shifting the term on the right side to the left, we get

$$\frac{v_o - u(t)}{40} + \frac{v_o}{Z(p)} = 0. \tag{7.5-42}$$

Multiplying by $40Z(p)$ and rearranging, we obtain

$$[Z(p) + 40]\, v_o(t) = Z(p)u(t). \tag{7.5-43}$$

Solving for $v_o(t)$,

$$v_o(t) = \frac{Z(p)}{Z(p) + 40}u(t) = \frac{\frac{1}{800}}{p + \frac{1}{400}}u(t) = \frac{1}{2}[1 - e^{-t/400}]\,u(t). \tag{7.5-44}$$

Now we recall that our unit of time is the ps, but the time constant in the exponential seems a bit large for comfortable interpretation. Therefore, let's decrease it by changing the unit of time to the nanosecond (1 ns = 1000 ps). This makes the last equation

$$v_o(t) = \frac{\frac{1}{800}}{p + \frac{1}{400}}u(t) = 0.5[1 - e^{-2.5t}]\,u(t), \tag{7.5-45}$$

where t is now in ns. The waveform is sketched in Figure 7.81.

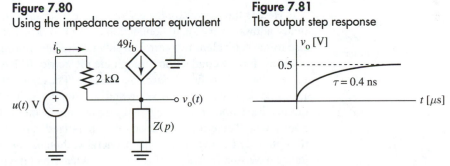

Figure 7.80
Using the impedance operator equivalent

Figure 7.81
The output step response

Section 7.5 Quiz

Q7.5-1.   For the circuit shown in Figure Q7.5-1:

Figure Q7.5-1

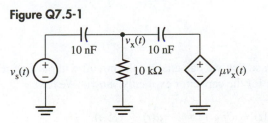

a.   Find the voltage $v_x(t)$ with the voltage gain μ of the VCVS as a parameter.

b.   Find the values of μ that result in stable, unstable, and marginally stable operation.

c.   If $v_s(t) = 10u(t)$ V and $\mu = 0.5$, find $v_x(t)$. (*Hint:* Choose convenient units for R, C, and t.)

7.6 | Practical Circuit Models for Passive Elements

On several occasions we have mentioned that the theory of circuit analysis we are developing pertains to ideal models—and that selecting combinations of these ideal models to represent "real world" devices is in the province of applications courses such as electronic circuits. Here, however, we will depart briefly to consider how this might be done. We devoted an earlier section (Section 5.7 of Chapter 5) to a discussion of diodes and bipolar transistors because approximations must be made to derive even dc steady-state models. We have waited until now to discuss nonideal models for passive elements because it requires that we have some understanding of the time domain (transient) behavior of circuits.

The "Lossy" Capacitor

We will start by considering the capacitor a bit more carefully. In Figure 7.82 we show a somewhat stylized representation of how a capacitor is actually constructed. The metal plates actually make contact with the dielectric, which is another term for an insulator. The metal plates are often of foil and the dielectric is pliable so the capacitor can be folded or rolled into other shapes. On an integrated circuit chip the metal plates are often aluminum or polysilicon and the dielectric is silicon dioxide, a "skin" that is grown on top of the integrated circuit for isolation and protection.

Figure 7.82
Construction of a capacitor

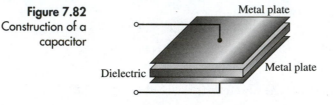

Here is the reason we have shown the construction of the capacitor: the dielectric actually allows a very small conduction current to flow between the two plates, thus giving rise to the equivalent two-terminal subcircuit shown in Figure 7.83. The resistor labeled r accounts for the conduction through the dielectric. It, like any element one does not wish to be present, is called a *parasitic element*. Thus, we refer to it as a parasitic resistance. Its value depends upon the value and way the capacitor is constructed. For a very high-quality capacitor on an integrated circuit chip it might have the value of 10^{15} Ω; for a poor-quality "electrolytic" capacitor (a polarized capacitor used in power supply design) it might only be several kilohms. Generally, though not invariably, larger values of capacitance have smaller values of r. Let's see what effect the parasitic resistance has.

Figure 7.83 A first-order nonideal model for the capacitor

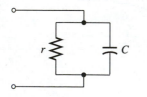

Example 7.21   Find the current $i(t)$ in Figure 7.84.

Figure 7.84
An example circuit

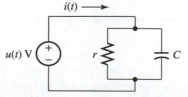

Solution   By now we have practiced enough so that we can analyze the circuit without first redrawing it with the impedance operator for the capacitor explicitly shown. We have

$$i(t) = \frac{1}{r}u(t) + Cpu(t) = \frac{1}{r}u(t) + C\delta(t). \qquad (7.6\text{-}1)$$

Figure 7.85
Current step response
of the nonideal
capacitor

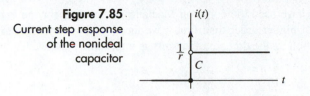

For an ideal capacitor, we would only expect to have the delta function, which delivers a charge of C coulombs to the capacitor instantaneously; the new feature is the (small) step function of current. We show this in Figure 7.85.

Example 7.22   Repeat Example 7.21 for the circuit shown in Figure 7.86; that is, find the step response for the voltage with current source "drive."

Figure 7.86
An example circuit

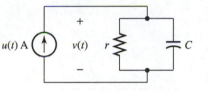

Solution   Here we must work a bit harder:

$$v(t) = \frac{r \times \dfrac{1}{Cp}}{r + \dfrac{1}{Cp}} u(t) = \frac{\dfrac{1}{C}}{p + \dfrac{1}{rC}} u(t) = r[1 - e^{-t/rC}]\, u(t) \text{ V}. \qquad (7.6\text{-}2)$$

This response is plotted in Figure 7.87.

Figure 7.87
Voltage step response
of the nonideal
capacitor

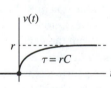

Ideal Approximation
for the Lossy
Capacitor Model

Does this last example disturb you a bit? It should if you have assimilated our earlier discussion of Chapter 6 on the operation of capacitors. "After all," you might say, "I know that the voltage step response of a capacitor to a current source drive is

$$v(t) = \frac{1}{Cp} u(t) = \frac{1}{C} tu(t), \qquad (7.6\text{-}3)$$

Figure 7.88   Voltage step response of the ideal capacitor

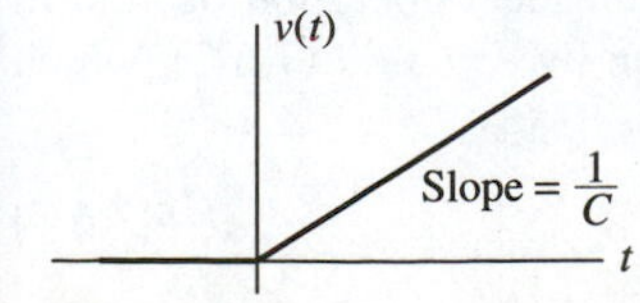

which is the ramp waveform shown in Figure 7.88. This looks nothing like the waveform we have derived in Figure 7.87!"

You are quite correct in this objection. The key to reconciling these facts lies in the idea of "looks like," which depends on the time scale. To further explain this idea, we need the following common approximation of the exponential:

$$e^x = \sum_{n=0}^{\infty} \frac{x^n}{n!} = 1 + x + \frac{x^2}{2} + \frac{x^3}{6} + \cdots \cong 1 + x, \qquad (7.6\text{-}4)$$

provided $|x| \ll 1$. We have used the common Maclaurin expansion for the exponential and dropped all terms of higher order than first—we have, as they say, "linearized" the exponential. The restriction on the domain of validity is important. Now we let

$$x = -t/rC \tag{7.6-5}$$

in equation 7.6-2 for the nonideal step response. This converts it to the form

$$v(t) = r[1 - e^{-t/rC}] \, u(t) \cong r[1 - (1 - t/rC)] \, u(t) = \frac{1}{C} tu(t) \text{ V}, \tag{7.6-6}$$

which is precisely the same as equation (7.6-3) for the ideal capacitor.

What have we learned from this little derivation? Simply that the nonideal capacitor with parasitic resistance behaves exactly like an ideal capacitor *as long as we impose the restriction that*

$$|x| = t/rC \ll 1, \tag{7.6-7}$$

which means that

$$t \ll rC = \tau, \tag{7.6-8}$$

the time constant of the circuit with the parasitic resistance. This means that we must agree not to watch the capacitor for too long and still expect it to behave ideally. Our derivation was only for the voltage step response of the nonideal capacitor with current source drive, but the method works in general. We will leave the verification of this statement as an exercise.

The Nonideal Resistor

Now let's investigate the nonideal behavior of a resistor. Again, the specifics depend upon the particular method of construction, but we will choose here to look at the stylized form shown in Figure 7.89. The block is made of some material such as carbon, that conducts current—but not readily. By adjusting the dimensions, one can fabricate a resistor to have just the "resistance" to current flow that is desired. (Other types of resistors are "wire-wound," "metal film," etc., but the general issues are the same.)

Here is the issue we wish to discuss. Because the material is not an ideal conductor, charge tends to collect on the end surfaces, which thereby act as capacitor plates. Furthermore, there is stray capacitance between the leads. These observations lead to the first-order model for the nonideal resistor shown in Figure 7.90. Does it look familiar? It should, for it is exactly the same as that which we presented for the nonideal capacitor in Figure 7.83! Well, just like it except for one fact: the resistance is now the value desired, and the capacitance is the unwanted parasitic element. Thus, the capacitance is a small value. Notice the notation—in both cases (capacitor model and resistor model) we have used a lowercase letter for the parasitic element value and an uppercase one for the desired element value.

As the same models hold for the two elements, we see that both our preceding examples hold for the resistor as well as for the capacitor. In Example 7.21, the delta function of current would be the parasitic if the model were that of a practical resistor. This means that if we apply a step voltage change to a resistor, we must supply c coulombs of charge immediately to charge the parasitic capacitance—after which ideal operation occurs. In Example 7.22, the voltage response to a current step is given by equation (7.6-2), which applies here with only a slight change in notation:

$$v(t) = R[1 - e^{-t/Rc}] \, u(t) \text{ V}. \tag{7.6-9}$$

Figure 7.89 Physical construction of a resistor

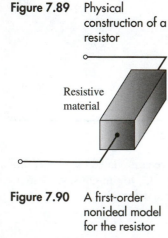

Resistive material

Figure 7.90 A first-order nonideal model for the resistor

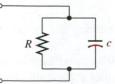

R c

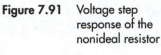

Figure 7.91 Voltage step response of the nonideal resistor

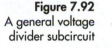

In this case, the time constant will be quite short relative to the time scale of the circuit in which the resistor is used (or the design is quite poor!), and the waveform will appear as shown in Figure 7.91. The "leading edge" of the step is not infinitely steep, but requires a small amount of time to rise to its final (dc steady state) value of R. Do you recall our convention about discontinuous waveforms relative to our open circles and filled dots? This is the reason we have chosen to use these symbols on ideal waveforms: to simply remind ourselves that in the lab we always see waveforms such as the one in Figure 7.91, rather than ideal step functions.

An Example: The Time Response of a "Nonideal" Circuit

The models for the nonideal resistor and the nonideal capacitor are identical except for element values, so we will work only one "more complicated" example that will handle both cases.

Example 7.23

Analyze the nonideal voltage divider shown in Figure 7.92 for the unit step response of $v_o(t)$. Note that if we make $R_1 = R_2 = \infty$ it will become an ideal capacitor divider, whereas if we make $C_1 = C_2 = 0$ it will be an ideal resistor divider.

Figure 7.92
A general voltage divider subcircuit

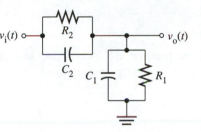

Solution

We first notice that there are two two-terminal subcircuits having identical topology (only the element values are different). Let us first, therefore, note that the impedance of a general parallel subcircuit consisting of a resistor of value R and a capacitor of value C is

$$Z(p) = \frac{R \times \dfrac{1}{Cp}}{R + \dfrac{1}{Cp}} = \frac{R}{RCp + 1}. \qquad (7.6\text{-}10)$$

Using the appropriate subscripts, we see that

$$v_o(t) = \frac{Z_1(p)}{Z_1(p) + Z_2(p)} v_i(t) = \frac{\dfrac{R_1}{R_1 C_1 p + 1}}{\dfrac{R_1}{R_1 C_1 p + 1} + \dfrac{R_2}{R_2 C_2 p + 1}} v_i(t) \qquad (7.6\text{-}11)$$

$$= \frac{R_1(R_2 C_2 p + 1)}{R_1 R_2 [C_1 + C_2] p + R_1 + R_2} v_i(t) = \frac{C_2}{C_1 + C_2} \frac{p + \dfrac{1}{R_2 C_2}}{p + \dfrac{R_1 + R_2}{R_2 R_1(C_1 + C_2)}} v_i(t).$$

We note that the constant in the denominator can be recognized as $1/RC$, where

$$R = \frac{R_1 R_2}{R_1 + R_2} = R_1 \| R_2 \qquad (7.6\text{-}12)$$

and

$$C = C_1 + C_2 = C_1 \| C_2. \qquad (7.6\text{-}13)$$

The notation $\parallel$ means the value of the "parallel combination of"; that is, $R_1 \parallel R_2$ means the value of "R_1 connected in parallel with R_2" and $C_1 \parallel C_2$ means the value of "C_1 connected in parallel with C_2."

We next perform one step of long division on the first-order operator and adjust the input voltage to have a unit step waveform $u(t)$ volts. This results in

$$v_o(t) = \frac{C_2}{C_1 + C_2}\left[1 + \frac{\dfrac{1}{R_2 C_2} - \dfrac{1}{RC}}{p + \dfrac{1}{RC}}\right]u(t) \tag{7.6-14}$$

$$= \frac{C_2}{C_1 + C_2}\left\{1 + \left\{\frac{RC}{R_2 C_2} - 1\right\}[1 - e^{-t/RC}]\right\}u(t).$$

We have used our known step response for the first-order lowpass operator in the last step. Separating the natural and forced responses gives

$$v_o(t) = \frac{R_1}{R_1 + R_2}u(t) - \frac{C_2}{C_1 + C_2} \cdot \frac{R_1 C_1 - R_2 C_2}{[R_1 + R_2]C_2}e^{-t/RC}u(t). \tag{7.6-15}$$

Because R and C are positive, the natural response approaches zero as t becomes infinitely large, leaving the dc steady-state response

$$v_{of}(t) = \frac{R_1}{R_1 + R_2} = F_r. \tag{7.6-16}$$

We have used the symbol F_r for the resistor-only voltage divider ratio. If, on the other hand, we evaluate the response at $t = 0+$ (that is, immediately after the step has occurred), we have

$$v_o(0+) = \frac{C_2}{C_1 + C_2} = F_c. \tag{7.6-17}$$

You can evaluate (7.6-14) at $t = 0$ to verify this. Here we have used the symbol F_c for the capacitor-only voltage divider ratio. Notice that the two divider ratios can be equal, or either can be larger than the other. We show the response for all three cases in Figure 7.93.

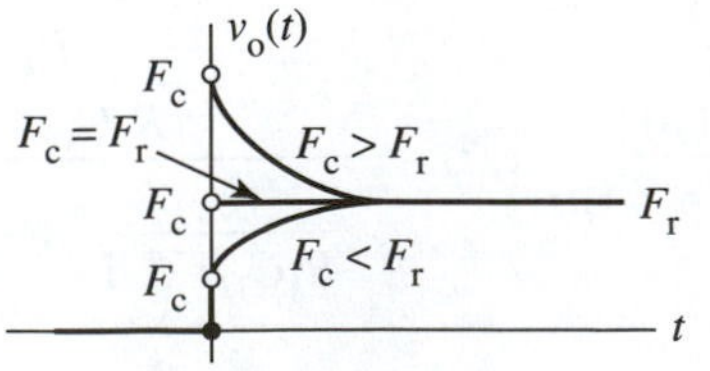

Figure 7.93
Step response of the
general voltage divider

The divider subcircuit in the preceding example is an important one; we should perhaps discuss it a bit more now that we know its step response. First, we note that the leading edge of the step function experiences a voltage division that consists of simply the capacitor divider alone. In the dc steady state, it experiences the resistor divider alone. Each of the RC subcircuits represents either a nonideal resistor or a nonideal capacitor, so we can see the nonideal effects in a voltage divider made from either type of element.

For a capacitive divider, the resistors will have quite large values. As the time constant is equal to the product RC, where R is the resistance of the parallel combination of the two parasitic resistances and C is the parallel capacitance, we see that it will have a relatively large value if both resistances are large. Thus, for intervals of time that are small in comparison with this time constant, the voltage divider effect is given approximately by F_c (using the approximation that $e^x \cong 1$ for small values of x). On the other hand, if the divider is made from physical resistors, the parasitic capacitors will be quite small in value. This means that the time constant will be very small—and the dc steady state occurs very quickly.

What does all this mean? Simply that we must know something about the time scale of the waveforms with which we are concerned in a given circuit. We then make the decision, based on our preceding analysis, as to whether the nonideal effects are important or not.

Finally, we mention one more thing about the preceding example. Often the input of an electronic device can be modeled fairly well as a parallel RC circuit. One example is an oscilloscope. For such devices, one then adds another parallel RC subcircuit in series with the input and adjusts its values so that $F_r = F_c$. Why? Because this means that the step response will be a perfect step (to within the limits of the parallel RC input approximation, anyway).

The "Lossy" Inductor The one passive element we have not discussed is the inductor. The physical construction of this circuit element is sketched in Figure 7.94. As you can see, it has a "core," which is often made of a nonmetallic material. For larger values of inductances, the core material is made of a magnetic material; unfortunately, this results in nonlinear behavior and other undesirable things. This—as well as the fact that inductors are quite bulky—means that they are often considered as undesirable elements, particularly in circuits to be constructed on integrated circuit chips. For this reason, physical inductors are frequently simulated with op amps, resistors, and capacitors. One such simulation was given in Chapter 6, Section 6.7, Example 6.25(O).

Figure 7.94
Physical construction of an inductor

Figure 7.95
First-order model for the nonideal inductor

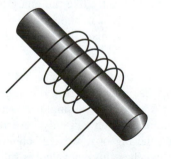

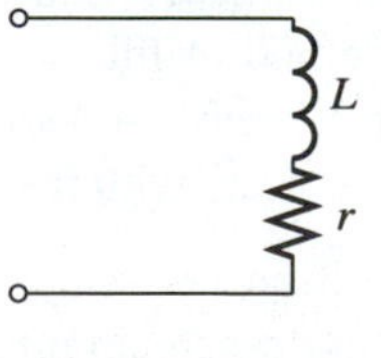

Here we wish merely to investigate first-order nonideal behavior, so we only note that the coil of wire certainly possesses resistance. This observation leads to the equivalent circuit shown in Figure 7.95. We will leave it as an exercise for you to investigate its nonideal behavior as we did earlier for the capacitor. The resistor is the parasitic element, and its value is often quite small—perhaps less than one ohm—but depends upon the value of the inductance and other factors.

Some General Comments on Models Before closing this section, we would like to point out that models for actual physical elements are *never* exact. Thus, one postulates a model made up of one or more ideal circuit elements, checks its behavior against experimental data, and then perhaps goes back and modifies the model and analyzes its behavior again. For instance, let us again consider the

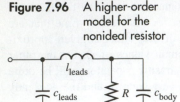

Figure 7.96 A higher-order model for the nonideal resistor

lowly resistor. As a matter of fact, its leads have inductance—which we have until now neglected. To include this effect, we postulate an even more complicated model, such as the one shown in Figure 7.96. The resistor labeled R is the element we desire. The two inductances come from the self-inductance of each lead, the left-hand capacitance comes from the capacitance between them, and the right-hand capacitance is the capacitance between the ends of the resistor bulk material itself.

Are you getting the picture? The resistor in Figure 7.96 has four energy storage elements, and this leads to a v-i characteristic that is a fourth-order differential equation—which we have not yet even discussed. If you have several resistors in a circuit and were to model them like this-----? You would undoubtedly use a computer simulation; all intuition about the effects of the resistance in the circuit would have been lost, or at least strained to the limit. Such models must, however, be dealt with in some fashion at high frequencies and when the current and voltage waveforms vary very rapidly. The choice of which model to use is often more of an intuitive art than a science.

Section 7.6 Quiz

Q7.6-1. Find the voltage step response of the nonideal inductor model given in Figure 7.95.

Q7.6-2. Repeat Question Q7.6-1 for the current step response.

Q7.6-3. Find the approximate time interval over which the step response of the first-order lowpass operator $H(p) = 1/(p + a)$ is well approximated by an ideal integrator $1/p$. (*Hint:* Write the step response explicitly, then use the first-order exponential approximation given by equation (7.6-4).)

7.7 | Equivalent Circuit Analysis for Single Energy Storage Circuits with Dc Sources and Switches

In this section we will develop a method for analyzing first-order circuits having only one energy storage element, dc sources, and switches. Although we consider the procedure we have already developed using operators to be superior because it is a general method, is easy to apply, and proceeds exactly like the analysis of resistive circuits, we present the method of this section because it is outlined (perhaps in less detail than we present here) in many modern texts. Thus, many engineers rely on it and you should be able to communicate with them.

A Quick Stability Test   There is an easy way to check the time constant, and therefore determine the stability, of any first-order circuit that has a single energy storage element. Look, for example, at Figure 7.97. The little box represents a single inductor or a single capacitor and the big box the rest of the circuit, which can only have resistors and/or sources. If we do a Thévenin equivalent on the big box, we will obtain the equivalent circuit shown in Figure 7.98. (If there are switches, this equivalent is valid for a given switch state.)

Figure 7.97
A general single L or C circuit

Figure 7.98
The Thévenin equivalent

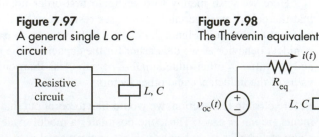

Let us assume first that the storage element is a capacitor. Then one nodal equation, after only a bit of manipulation, gives

$$\left[p + \frac{1}{R_{eq}C} \right] v(t) = \frac{1}{R_{eq}C} v_{oc}(t). \tag{7.7-1}$$

We are assuming that the storage element is an actual capacitor (not an equivalent capacitance and particularly not an equivalent capacitance of an active subcircuit), so C is positive. Hence we only need to check R_{eq}. If its value is positive, then so is the time constant, and the circuit response $v(t)$ will be a stable one. If R_{eq} is negative, then the response $v(t)$ will be unstable. In what follows, we will assume that R_{eq} is a finite, nonzero constant (either positive or negative). Thus, the method we will develop will not work for the exceptional cases $R_{eq} = 0$ (ideal v-source) or $R_{eq} = \infty$ (ideal i-source).

If the storage element is an inductor, we can derive the following differential equation for the current:

$$\left[p + \frac{R_{eq}}{L} \right] i(t) = \frac{1}{L} v_{oc}(t). \tag{7.7-2}$$

If L is positive, we can check on stability, once again, merely by computing the equivalent resistance. If R_{eq} is positive, the response is stable; if it is negative, the response is unstable.

Continuity Conditions

Equations (7.7-1) and (7.7-2) illustrate something quite important. The capacitor voltage—or inductor current—in a circuit having a single energy storage element and in which $R_{eq} \neq 0$ or ∞ is a lowpass response variable. This has an interesting and useful consequence: the capacitor voltage—or inductor current—is a continuous waveform: that is, it has no instantaneous jumps. To see this, we simply rewrite (7.7-1) and (7.7-2) in the generic form

$$\left[p + \frac{1}{\tau} \right] y(t) = x(t), \tag{7.7-3}$$

where $y(t)$ is either the capacitor voltage or the inductor current, as appropriate. We can solve for $py(t)$, which we write in conventional notation as

$$\frac{dy}{dt} = py(t) = x(t) - \frac{1}{\tau} y(t). \tag{7.7-4}$$

We next notice that $y(t)$ must always be finite because the stored energy in the capacitor or inductor is finite, $w_C = (1/2)Cv^2 < \infty$ or $w_L = (1/2)Li^2 < \infty$, respectively, leading to $|v| < \infty$ for the capacitor and $|i| < \infty$ for the inductor. If the right-hand side forcing function $x(t)$ is finite as well, we have (*provided that* $\tau \neq 0$)

$$\left| \frac{dy}{dt} \right| = \left| x(t) - \frac{1}{\tau} y(t) \right| \le |x(t)| + \frac{1}{|\tau|} \cdot |y(t)| < \infty. \tag{7.7-5}$$

Now look at Figure 7.99(a). There we see a waveform that has a discontinuity. What does this imply about its derivative? Because the derivative is the slope of the waveform, which is infinite at $t = t_o$, we see that the derivative, too, is infinite at $t = t_o$. This is shown in Figure 7.99(b). But this is precisely the behavior that is barred by (7.7-5). Thus, we can state the following result: *The capacitor voltage—or inductor current—in a circuit having only a single energy storage element in addition to finite valued independent sources and resistors is always a continuous waveform, provided the time constant τ* ($= R_{eq}C$ *or* L/R_{eq}) *is nonzero*. We will use this result to develop a method of analyzing such circuits using equivalent circuits. We note, however, that many circuits violate this restriction. If the circuit has an active element (dependent source or op amp), the method (with a certain restriction) still works, but the response might be unstable.

Figure 7.99
The derivative of a discontinuous waveform

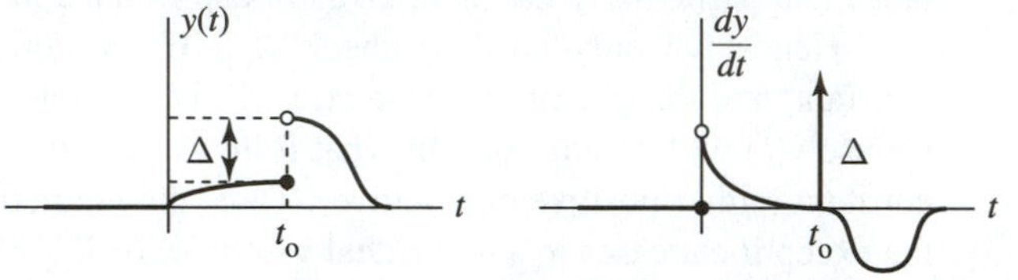

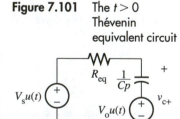

The Equivalent Circuit Method

We will now develop the method. It is limited to single energy storage circuits;[14] more importantly, it is limited to such circuits in which the time constant is nonzero. We show such a circuit in Figure 7.100. We have "extracted" the independent sources and the single energy storage element—which we assume, merely to be specific, is a capacitor. The box remaining is assumed to have only resistors and switches that operate only once at $t = 0$. We assume that the problem is to find either the voltage v between a pair of nodes or the current i in a conductor or element.

Figure 7.100
A dc circuit having only one energy storage element

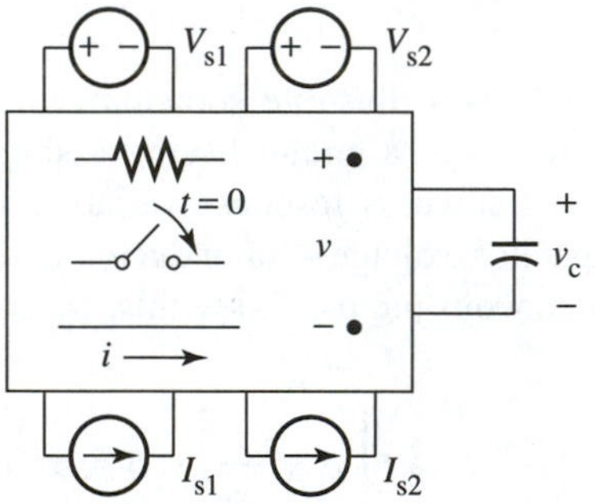

Figure 7.101   The $t > 0$ Thévenin equivalent circuit

Our first step in the development of our method is to replace everything except the capacitor by its $t > 0$ Thévenin equivalent, as shown in Figure 7.101. (Norton would also work.) It is important to recognize that the subcircuit consisting of everything except the capacitor has a purely real Thévenin resistance (there are no other energy storage elements). We note that the switches have all flipped to their alternate position and that the capacitor, in general, has nonzero initial conditions because of circuit action for $t \leq 0$. We assume that the circuit for $t \leq 0$ is stable (this is the restriction referred to above), so this initial condition can be derived as usual from the dc steady-state equivalent (with the capacitor replaced, as usual, by an open circuit). We also assume that $R_{eq} \neq 0$ for $t > 0$. This means that there will be no instantaneous charge transfer.

We can derive the voltage $v_{c+}(t)$ in a number of ways. Choosing voltage division to obtain the voltage across the capacitor element and adding it to the initial condition source voltage, we obtain (after a bit of manipulation that we leave to you)

$$v_{c+}(t) = V_o u(t) + \frac{1/Cp}{R_{eq} + 1/Cp}[V_s u(t) - V_o u(t)] \tag{7.7-6}$$

$$= V_s u(t) + (V_o - V_s)e^{-t/R_{eq}C}u(t).$$

---

[14] Or to circuits containing unswitched series or parallel capacitors or inductors.

Although we have the quantitative response, let's rewrite it for $t > 0$ in the general form

$$v_c(t) = A + Be^{-t/\tau} = y_f(t) + y_n(t), \qquad (7.7\text{-}7)$$

where A and B are constants. We note that the forced response $y_f(t)$ is a constant and the natural response is an exponential with time constant $\tau = R_{eq}C$.

Now let's go back to our original circuit for $t > 0$ shown in Figure 7.100 and note that it is not necessarily the capacitor voltage we want, but any voltage or current in the circuit. We know, however, that all the independent sources are dc. We invoke the substitution theorem in its voltage source form (see Chapter 3, Section 3.6) to replace the capacitor with a v-source having the known value given in equation (7.7-7). Thus, we obtain the equivalent shown in Figure 7.102. If we apply superposition to determine any voltage or current in the circuit—let's call it $y(t)$ to be general—we see that each of the dc sources produces a constant partial response and the capacitor produces a partial response having the form shown in equation (7.7-7). Thus, using A and B again for *general* constant values, we have

$$y(t) = A + Be^{-t/\tau} = y_f(t) + y_n(t). \qquad (7.7\text{-}8)$$

(*A* and *B* have different values in general than they did in (7.7-7)). Again the forced response is a constant and the natural response is an exponential—*with the same time constant as that of the energy storage element response.*

Figure 7.102
Circuit after using the substitution theorem on the capacitor

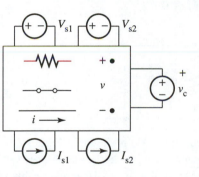

Because the forced response is a constant, we can find $y_f(t) = A$ from the ($t > 0$) dc equivalent circuit. To find B, we note that

$$y(0+) = A + B. \qquad (7.7\text{-}9)$$

Thus, we need only find the value of the desired variable $y(t)$ at the time $t = 0+$ (right after the switch flip). We do this by using the substitution theorem on the capacitor (or inductor), replacing it with a voltage source (or current source) having the value found earlier for $t \le 0$. Knowing A already, we solve (7.7-9) for B. We already know the time constant from $\tau = R_{eq}C$ (or $\tau = L/R_{eq}$) and the form of $y_n(t)$. Here, then, is the method:

The Equivalent Circuit Method

1. Find the dc *steady-state capacitor voltage or inductor current* for $t \le 0$. This, of course, assumes that the $t \le 0$ circuit is stable and therefore that a dc steady state exists.

2. Find the forced response for $y(t)$ in the $t > 0$ circuit by replacing the capacitor by an open circuit or the inductor by a short circuit. This is the constant A in (7.7-8).

3. Knowing that the capacitor voltage or inductor current is continuous (as we showed above), draw the equivalent circuit at $t = 0+$ (just after the switch flip) by using the

substitution theorem to replace the capacitor by a voltage source or the inductor by a current source having a value equal to the initial condition voltage or current you found in step 2. Solve this circuit to obtain the initial condition $y(0+)$ and use it in (7.7-9) to find B.

4. Find R_{eq} for the $t > 0$ circuit by deactivating all the independent sources and analyzing the two-terminal subcircuit formed when the capacitor is removed. Use this value to compute $\tau = R_{eq}C$ or $\tau = L/R_{eq}$, as appropriate.

5. Assemble all the parts of (7.7-8) to determine the response for $t > 0$.

Stability Issues   Let's emphasize one thing: we do not have to assume that the circuit is stable for $t > 0$, only that it is stable for $t \leq 0$. Why? Simply because the forced response part of the capacitor voltage is constant for $t > 0$, and this means that the capacitor is equivalent to an open circuit for that part of the response.[15] (Similar comments hold for inductive circuits.) Most texts do not discuss stability at this point (which you can readily see is a highly important issue!); those that do, insist that the $t > 0$ circuit be stable and that the capacitor voltage settle down into its dc steady-state response. As you can see, however, the forced response is a constant whether or not the circuit is stable. In the case of stability, the dc steady-state and forced responses are one and the same. We call the method we have just developed the *equivalent circuit method*.

Example 7.24   Find $i(t)$ in Figure 7.103 using the equivalent circuit method. Assume dc steady state conditions are valid for $t \leq 0$.

Figure 7.103
An example circuit

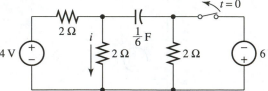

Solution   The equivalent circuit for $t \leq 0$ is shown in Figure 7.104. If we mentally deactivate the v-sources (thus replacing them by short circuits), we can easily compute the equivalent resistance seen by the capacitor. It is $1\,\Omega$. Because this value is positive,[16] the $t \leq 0$ circuit is stable and all variables are in the dc steady state for all finite negative time values. For this reason we have shown the capacitor as an open circuit. It is important to note that we have also defined the capacitor voltage. We could have selected the opposite polarity just as well, but once we have made our choice we must stick to it, for this is the quantity that is "continuous across the switch flip" at $t = 0$. We easily see that $i = 1\,\text{A}$ and $v_c = -8\,\text{V}$.

Figure 7.104
The $t \leq 0$ equivalent circuit

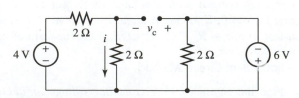

We next "flip the switch," thus removing the right-hand voltage source from the circuit. We draw the resulting circuit in Figure 7.105. We could, of course, solve this circuit

---

[15] Notice that this is a superposition argument.

[16] The right-hand 2-Ω resistor is shorted out by the deactivated 6-V source.

using our operator impedance technique, but instead we intend here to illustrate our equivalent circuit method. We first find the forced response by replacing the capacitor by an open circuit (remember that the forced response is a constant), resulting in the forced response equivalent of Figure 7.106. We quickly analyze this equivalent circuit to find the forced response. It is $i_f(t) = 4 \text{ V}/4 \text{ }\Omega = 1 \text{ A}$. This is only coincidentally (because of the specific circuit topology and the specific circuit response chosen) the same as the dc steady-state value for $t \leq 0$. Our next step is to find the constant B. To do this, we first find $i(0+)$ by replacing the capacitor by a v-source having an initial condition value of -8 V.[17] This gives the $t = 0+$ equivalent circuit of Figure 7.107. We will leave it to you to analyze this simple circuit. The result is $i(0+) = 2 \text{ A}$. We use this in our known response form given by equation (7.7-9), which is equation (7.7-8) evaluated at $t = 0$, to obtain

$$i(0+) = 2 = A + B = 1 + B. \tag{7.7-10}$$

Solving, we get $B = 1$.

Figure 7.105
The $t > 0$ circuit

Figure 7.106
The $t > 0$ forced response equivalent circuit

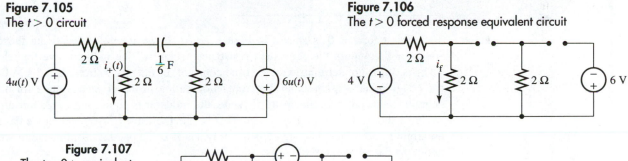

Figure 7.107
The $t = 0+$ equivalent circuit

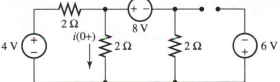

Next, we compute the time constant for the $t > 0$ circuit. Referring to Figure 7.106, we deactivate the two v-sources to find the equivalent resistance "seen" by the capacitor, and find that $R_{eq} = 3 \text{ }\Omega$. Thus, $\tau = R_{eq}C = 1/2 \text{ s}$. Our last step is to merely insert all the now known values in equation (7.7-8):

$$i(t) = 1 + e^{-2t}, \quad t > 0. \tag{7.7-11}$$

For $t \leq 0$, however, we know the value already: it is 1 A. A sketch of $i(t)$ for all values of t is shown in Figure 7.108. Notice that the *current* is discontinuous. It is only the *capacitor voltage* that must be continuous.

Figure 7.108
Response waveform

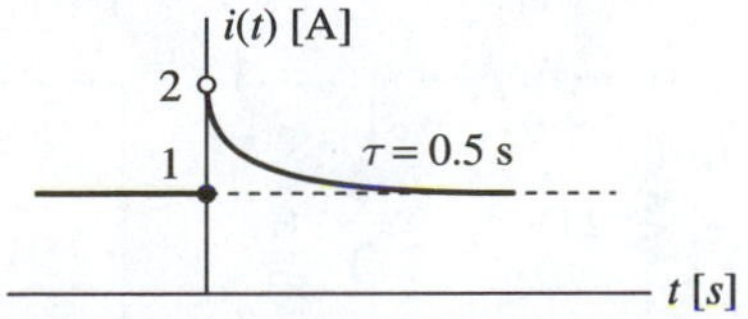

---

[17] Be careful not to confuse this with a dc equivalent circuit. We are using the substitution theorem here.

The main attractive feature of the equivalent circuit method is that one works only with dc resistive circuits—even though the final result is the time domain response. We remind you, though, that it only works for a limited class of circuits, whereas our operator method is not only quite general but results in the need for analyzing fewer circuits—and if you have committed the step response of the first-order operator to memory, it does not require the solution of a differential equation either.

The equivalent circuit method, however, does apply to circuits with a single inductor with all the obvious changes being made: the time constant is $\tau = L/R_{eq}$, and it is the inductor current that is continuous across the switch flip.

Example 7.25   Find the response $i(t)$ in the circuit shown in Figure 7.109.

Figure 7.109
An example circuit

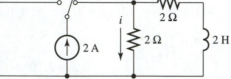

Solution   The circuit for $t \leq 0$ is shown in Figure 7.110. If we mentally deactivate the current source and compute the equivalent resistance "seen by" the inductor, we find that it is 4 Ω. As this is greater than zero, the time constant for the $t \leq 0$ circuit is 0.5 s > 0. Thus, the $t \leq 0$ circuit is stable and therefore is operating in the dc steady state for all finite negative values of t. In the dc steady state, the inductor is equivalent to a short circuit as we have shown; therefore, $i(t) = 1$ A. This is also the value of $i_L(t)$, which we have given an arbitrary reference. We need this quantity, for it is the one bit of information we know about the circuit at $t = 0+$.

Figure 7.110
The $t \leq 0$ equivalent
circuit

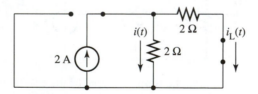

We next "flip the switch," obtaining the $t > 0$ circuit shown in Figure 7.111. We find the forced response by replacing the inductor with a short circuit (after all, the forced response for the inductor current, a lowpass variable, is a constant, so the forced response equivalent of the inductor itself is a short circuit), as shown in Figure 7.112. We quickly conclude that the forced response for our variable of interest is 0 A, so $A = 0$ in the general response equation (7.7-8).

Figure 7.111
The $t > 0$ circuit

Figure 7.112
The $t > 0$ forced response equivalent circuit

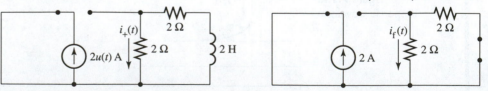

Next, we draw the $t = 0+$ equivalent circuit, using the substitution theorem (in i-source form this time) to replace the inductor by a current source having as its value the inductor current at $t = 0$—which we know, by continuity, to be the same as its dc steady-state value just before the switch flip. This equivalent is shown in Figure 7.113. We see that $i(0+) = -1$ A. In equation (7.7-9) this gives

$$i(0+) = -1 = A + B = 0 + B, \qquad (7.7\text{-}12)$$

for we have already determined that $A = 0$. Thus, $B = -1$.

Figure 7.113
The $t = 0+$ equivalent circuit

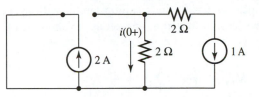

Now, we find the time constant. We easily compute the Thévenin equivalent resistance "seen by the inductor" for $t > 0$ by inspecting, say Figure 7.111. This resistance is $R_{eq} = 4\ \Omega$. This gives $\tau = L/R = 0.5$ s. Putting together the pieces of equation (7.7-8), we have

$$i(t) = A + Be^{-t/\tau} = -e^{-2t}\ \text{A}; \quad t > 0. \qquad (7.7\text{-}13)$$

The forced response is zero and the natural response is also the complete response.

We know, though, that $i(t) = 1$ A for $t \leq 0$. Putting together the $t \leq 0$ and $t > 0$ results, we have the waveform sketched in Figure 7.114.

Figure 7.114
The response waveform

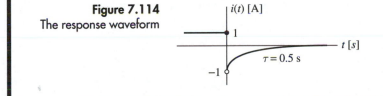

Comparison with the Operator Method

We have presented the equivalent circuit method primarily because you will find it in other texts, ones that do not use operators, and many circuit engineers therefore use this technique. You should be able to communicate with them and understand their solution procedure. Figure 7.115, however, shows a circuit that might *seem* amenable to analysis by the equivalent circuit method. The capacitors are parallel connected for $t > 0$, and we know that two parallel capacitors are equivalent to a single capacitor—or do we? If we use the series v-source initial condition model for each capacitor, we see[18] that it is the *complete models* that are parallel connected, not the capacitors in the model. Of course if we use the parallel i-source model, the capacitor symbols will be parallel connected; however, the current sources will be impulsive—and this violates our constraints that all forcing functions have finite values. The capacitor voltage will change instantaneously.

Figure 7.115
An example circuit that cannot be analyzed by the equivalent circuit method

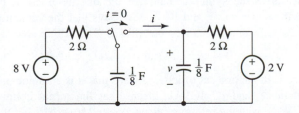

---

[18] In an earlier footnote, we explicitly disallowed series or parallel capacitors that are *switched* together as these are.

We will leave it as an exercise for you to analyze the circuit in Figure 7.115 for both $i(t)$ and $v(t)$ using the operator method. (The voltage is $v(t) = 2$ V for $t \leq 0$ and $2[1 + e^{-2t}]$ for $t > 0$.)

There is another type of circuit that is not analyzable by any technique whatsoever, for it is not well posed. Figure 7.116 shows such a circuit. It is not well posed because it is unstable for $t \leq 0$ and so a dc steady state does not exist. If we reverse the switch so that it opens at $t = 0$, the resulting circuit will be well posed, but unstable for $t > 0$. Again, we leave it to you as an exercise to show that the waveform for $i(t)$ is then given by $i(t) = [1 - e^{t/2}]\, u(t)$ A.

Figure 7.116
An example of an improperly posed circuit

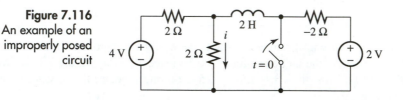

Section 7.7 Quiz

Q7.7-1. Find the node voltage $v(t)$ in Figure Q7.7-1 for all values of t.

Figure Q7.7-1

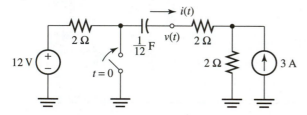

Q7.7-2. Find $i(t)$ and $v(t)$ in the circuit in Figure 7.115 using the operator method.

Q7.7-3. Find $i(t)$ in the circuit in Figure 7.116, with the switch reversed, using the equivalent circuit method.

7.8 | SPICE: Transient Analysis

Capacitor and Inductor Specification in SPICE

The SPICE statement for a capacitor has the form

```
CXXXX---X    NPLUS    NMINUS    CVAL    IC=ICVAL
```

The first letter in the symbolic name means that the element being described is a capacitor. As usual, NPLUS is the symbolic name of the node assumed to be positive, and NMINUS the symbolic name of the one assumed to be negative. CVAL is the value of the capacitance, and IC=ICVAL says that the capacitor has an initial voltage value of ICVAL, with the plus sign on NPLUS. The SPICE statement for an inductor is similar:

```
LXXXX---X    NPLUS    NMINUS    LVAL    IC=ICVAL
```

The L in the first set of letters means that it is an inductance. LVAL is the value of this inductance, and IC=ICVAL means that the initial condition on the inductor current is ICVAL, with the reference arrow pointed from NPLUS through the inductor to NMINUS.

The .TRAN Command

The control statement that tells SPICE to do a time domain, or *transient,* analysis is

```
.TRAN    PINT    FNLTIM    NPINT    STPCEIL    UIC
```

.TRAN tells SPICE to do a transient analysis. PINT means "print interval." If you intend to print out a table of values of voltage or current versus time, this parameter specifies how much time elapses between each printed value. FNLTIM is short for "final time." This is the largest value of time simulated (that is, it is the last time point). NPINT means "no-print interval." In some cases, one is interested in only investigating the steady state, for instance, and does not wish to generate output while the natural response is significant. Thus, although the simulation starts at $t = 0$, no output is generated until $t =$ NPINT. Then output is generated all the way to $t =$ FNLTIM.

The term STPCEIL requires a bit more discussion, and it is quite important. SPICE uses what is called a "variable step size" integration algorithm for solving the circuit differential equation. Such an algorithm automatically selects the next time point on the basis of an estimated accuracy of the result. In some cases, such as when the circuit variables are changing very slowly, this step interval will be so large that the output looks poor. In this case, one can use the parameter STPCEIL to place a maximum limit on the time interval between the computed time points.

The last mnemonic, UIC, means "use initial conditions." If it is not present, PSPICE will ignore all initial conditions, even if they have been specified in the capacitor and/or inductor statements.

The following example will perhaps make things clear.

Example 7.26

Simulate the circuit shown in Figure 7.117 to find the capacitor voltage as a function of time for $t > 0$. Assume that $v(0) = -2$ V.

Figure 7.117
An example circuit

Figure 7.118
The circuit prepared for SPICE

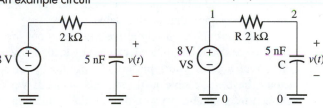

Solution

We label the elements and nodes as shown in Figure 7.118. We translate this circuit into a NETLIST as shown below:

```
****08/23/94 13:30:40***********EVALUATION PSPICE (JAN 1992)*****
 EXAMPLE 1
 ****   CIRCUIT DESCRIPTION
 *****************************************************************
 *******
 ******
 VS   1   0    8
 R    1   2    2K
 C    2   0    5N    IC= -2
 .TRAN  5U    50U   0    1U   UIC
 .PRINT TRAN   V(2,0)
 .END
 ****08/23/94 13:30:40***********EVALUATION PSPICE (JAN 1992)*****
 EXAMPLE 1
 ****   TRANSIENT ANALYSIS          TEMPERATURE= 27.000 DEG C
 *****************************************************************
 ******
 TIME    V(2,0)
 0.000E+00 -2.000E+00
 5.000E-06 1.931E+00
 1.000E-05 4.320E+00
```

```
1.500E-05 5.769E+00
2.000E-05 6.647E+00
2.500E-05 7.180E+00
3.000E-05 7.503E+00
3.500E-05 7.699E+00
4.000E-05 7.817E+00
4.500E-05 7.889E+00
5.000E-05 7.933E+00
   JOB CONCLUDED
   TOTAL JOB TIME      1.35
```

Most of the lines in the input file are self-explanatory, based upon our preceding discussion. The .PRINT statement has an added modifier, TRAN. The reason is quite simple. It is possible to ask SPICE to do a number of different types of simulation, all for one NETLIST. Thus, we must indicate to SPICE which of these simulations is to be printed out. In this case, we have done only one type of simulation, a transient one, and we are asking that it be printed.

Selection of the
Simulation Time

A more complicated question is: "Why have we chosen the values in the .TRAN statement shown in our example above?" Here's the reasoning. Our basic circuit has a time constant of $\tau = RC = 2 \text{ k}\Omega \times 5 \text{ nF} = 10 \ \mu s$. We anticipate that the circuit will have settled down into the dc steady state after $5\tau = 50 \ \mu s$. This is the reason for the final time of 50U. We have somewhat arbitrarily chosen a tenth of the final time as the print interval: 5U. In order to make sure that the results will look acceptable when we plot them, we have chosen to make the step ceiling one-fifth of the print interval. Why? Well, we are "fudging around a bit," and if that value did not work well, we would have performed the simulation again with a smaller value.

Now look more closely at the SPICE output for our simulation in the last example. Do you see that it has, indeed, essentially settled down to its dc value of 8 V by the end of our simulation? If you replace the capacitor by an open circuit (its dc equivalent), you should have no trouble justifying this value. If we felt that our output did not convey enough information, we would rerun the simulation with a smaller print interval to produce more data between the time points.

Our initial computation of the time constant points out a very important issue in time simulation: the time scale. Suppose we had decided to simulate our circuit for only 0.5 microsecond, with an output interval of 0.05 microsecond. Here is the output:

```
****08/23/94 13:49:52**********EVALUATION PSPICE (JAN 1992)*****
  EXAMPLE 1: POOR CHOICE OF SIMULATION TIME
  ****   CIRCUIT DESCRIPTION
********************************************************************
******
VS   1   0    8
R    1   2    2K
C    2   0    5N    IC=-2
.TRAN    0.05U  0.55U   0     0.01U    UIC
.PRINT TRAN    V(2,0)
.END
****08/23/94 13:49:52**********EVALUATION PSPICE (JAN 1992)*****
  EXAMPLE 1
  ****   TRANSIENT ANALYSIS          TEMPERATURE= 27.000 DEG C
********************************************************************
******
  TIME      V(2,0)
 0.000E+00  -2.000E+00
 5.000E-08  -1.950E+00
 1.000E-07  -1.901E+00
```

```
1.500E-07 -1.851E+00
2.000E-07 -1.802E+00
2.500E-07 -1.753E+00
3.000E-07 -1.704E+00
3.500E-07 -1.656E+00
4.000E-07 -1.608E+00
4.500E-07 -1.560E+00
5.000E-07 -1.512E+00
5.500E-07 -1.465E+00
      JOB CONCLUDED
      TOTAL JOB TIME      1.65
```

Do you see the problem? The capacitor voltage has not approached anywhere near the final dc steady-state value of 8 V by the time the simulation ends. We have very grossly underestimated the simulation time.

Suppose we go the other way and make the simulation time 5 seconds. Here is the output:

```
****08/23/94 13:58:56***********EVALUATION PSPICE (JAN 1992)*****
 EXAMPLE 1
 ****  CIRCUIT DESCRIPTION
 *****************************************************************
 ******
 VS    1   0   8
 R    1   2   2K
 C    2   0   5N     IC=-2
 .TRAN    0.5   5    0   0.1   UIC
 .PRINT TRAN    V(2,0)
 .END
****08/23/94 13:58:56***********EVALUATION PSPICE (JAN 1992)*****
 EXAMPLE 1
 ****  TRANSIENT ANALYSIS        TEMPERATURE= 27.000 DEG C
 *****************************************************************
 ******
 TIME      V(2,0)
 0.000E+00  4.887E+00
 5.000E-01  8.000E+00
 1.000E+00  8.000E+00
 1.500E+00  8.000E+00
 2.000E+00  8.000E+00
 2.500E+00  8.000E+00
 3.000E+00  8.000E+00
 3.500E+00  8.000E+00
 4.000E+00  8.000E+00
 4.500E+00  8.000E+00
 5.000E+00  8.000E+00
      JOB CONCLUDED
      TOTAL JOB TIME      1.40
```

The problem here is that essentially all we obtain is the dc steady state. We have lost the transient behavior entirely. Notice, too, that PSPICE has become a bit confused about our choice of time scale—even the initial condition has been "messed up." The moral of the story is simply this: always be sure to use an appropriate time scale.

"But," you might object, "suppose the circuit is so complicated that I do not know the time constant." Actually, it is often quite easy to compute $\tau = R_{th}C$, because one need only compute the Thévenin equivalent resistance. However, your concern is still legitimate because a circuit involving, say, dependent sources or op amps could present a fairly difficult computation for even this item of information. In that case, you make a guess and do the simulation. If you only see the dc steady-state behavior (a constant), you reduce the simulation time and the print interval. If the variables are still changing significantly at the end of the printout, on the other hand, you increase the simulation time.

Finally, a word of warning. If you make the parameter STPCEIL too small, you might have to wait a long time for your simulation to stop. Therefore, you should use this parameter with a bit of caution and forethought.

Plotting   Now let's look at another aspect of the simulation in Example 7.26. We only generated a table of values. Of course, we could graph our waveform from this table, but we suspect that the computer should be able to do this for us—and it can. The answer is the .PLOT statement. It has the form

```
.PLOT  TRAN  V(2,0)
```

Notice that, once more, we must specify that it is a transient plot, as opposed to one of another type. The output of a SPICE run for the last example, with the original values for the simulation time, etc., is

```
****08/23/94 14:11:58***********EVALUATION PSPICE (JAN 1992)*****
 EXAMPLE 1
 ****  CIRCUIT DESCRIPTION
 ****************************************************************
 ******
 VS    1    0    8
 R     1    2    2K
 C     2    0    5N    IC=-2
 .TRAN    5U    50U    0    1U    UIC
 .PLOT    TRAN    V(2,0)
 .END
****08/23/94 14:11:58***********EVALUATION PSPICE (JAN 1992)*****
 EXAMPLE 1
 ****  TRANSIENT ANALYSIS          TEMPERATURE= 27.000 DEG C
 ****************************************************************
 ******
 TIME    V(2,0)
 (*)---------- -5.0000E+00 0.0000E+00 5.0000E+00 1.0000E+01 1.5000E+01
              ---------------------------------------
 0.000E+00 -2.000E+00 .    *  .         .         .         .
 5.000E-06 1.931E+00.       . *         .         .         .
 1.000E-05 4.320E+00.       .       *.  .         .         .
 1.500E-05 5.769E+00.       .         .*         .         .
 2.000E-05 6.647E+00.       .         . *         .         .
 2.500E-05 7.180E+00.       .         .  *         .         .
 3.000E-05 7.503E+00.       .         .   *       .         .
 3.500E-05 7.699E+00.       .         .    *.      .         .
 4.000E-05 7.817E+00.       .         .    *.      .         .
 4.500E-05 7.889E+00.       .         .     *.     .         .
 5.000E-05 7.933E+00.       .         .     *.     .         .
              ---------------------------------------
     JOB CONCLUDED
     TOTAL JOB TIME      1.42
```

You have to read this graph by rotating it through 90 degrees: the time axis is vertical, with time increasing in the downward direction. Each computed point is represented by an asterisk. If you wish, the plot can be of several variables at once; in this event, each is assigned a different symbol (plus sign, etc.). If you leave the .PRINT statement in the SPICE input file, you will generate a table, then the plot (which, as you see, also contains the same information). This is a general principle in SPICE. You can do many things at once—you get everything you ask for.

Does our graph appear to be a bit primitive? It is. As a matter of fact, it is what was once known as a "print-plot." It was a way of generating a plot on a "page printer." A page printer was a printer designed only to print out a basic character set. The spacing across a line was fixed, often to 80 or 130 characters. The spacing between lines was also fixed, and this limited the quality of the graph.

There is a better way. If we include the control statement .PROBE in our input file, the simulation results will be "dumped" into a graphics output file (this will be automatically generated). After the simulation is run—either automatically or upon command, depending upon the way your particular version of SPICE is set up—a graphics post-processor program called PROBE, evoking the image of an oscilloscope probe, is run. This piece of software allows you to make high-quality graphs of waveforms. Here's the way it looks for our last example.

```
****08/23/94 14:23:20***********EVALUATION PSPICE (JAN 1992)*****
  EXAMPLE 1
  ****  CIRCUIT DESCRIPTION
  ****************************************************************
  ******
  VS   1   0   8
  R    1   2   2K
  C    2   0   5N   IC=-2
  .TRAN    5U   50U   0   1U   UIC
  .PROBE
  .END
       JOB CONCLUDED
       TOTAL JOB TIME       1.33
```

It seems that nothing has been produced! When we execute PROBE, however, we obtain the graph shown in Figure 7.119. We have used dashed lines to indicate the value after a single time constant of $\tau = 10$ μs. Noting that the total change in voltage is from -2 V to $+8$ V, a change of $+10$ V, after one time constant the voltage should have changed by about $(1 - e^{-1}) \times 100 = 63\%$ of the total. This is about 6.3 V, and $-2 + 6.3 = 4.3$ V. This checks with the value shown on the graph.

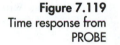

Figure 7.119
Time response from
PROBE

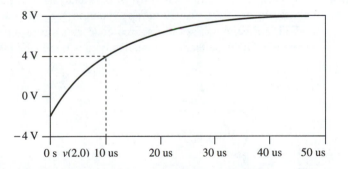

We should mention that quite a number of display options are available within PROBE. We will not discuss them here, however, but will merely note that our command .PROBE saves *all* of the circuit variables for later display. As you can see from the graph, PROBE generates a much higher quality plot than does the outdated print-plot version obtained with .PLOT. From here on, all of our graphs will be presented in the form of .PROBE plots.

Source Waveforms

Now let's discuss the initial condition issue a bit more. The circuit of Example 7.26, Figure 7.117 (the one we have been investigating thus far) represents the "$t > 0$" circuit; that is, it is the circuit after the switch or switches have been flipped into their alternate position. The initial conditions in such a circuit must have been obtained by solving another circuit for $t \leq 0$, this circuit being (for example) a dc steady-state one. It is necessary that the $t \leq 0$ circuit be stable in order for such a steady state to exist.

Suppose, however, that the circuit is not switched. In this case, we are considering a circuit on the entire time axis. Therefore, the various independent sources must have waveform values that are zero for $t \leq 0$. We can specify such waveforms in a number of ways. Here is one:

$$\text{VS} \qquad \text{PWL}(0, 0 \qquad 1\text{E}-9, 1)$$

This specifies that the voltage source named VS (being zero for all $t \leq 0$) starts at $t = 0$ with the value 0 V and increases linearly to 1 V at $t = 1$ ns, thus approximating a unit step function of voltage. In general, the piecewise linear (PWL) source is specified by

$$\text{VS} \qquad \text{PWL (T1, V1} \qquad \text{T2, V2} \qquad \text{---} \qquad \text{TN, VN)}$$

Thus, you only have to specify the "break points," that is, the points where the slope changes. The commas are for readability only; the SPICE compiler simply ignores them, so they are optional.

Special Cases Causing Trouble with SPICE

The next example uses a PWL source in an inductive circuit.

Example 7.27

Simulate the circuit in Figure 7.120 and plot the response voltage across the 200-Ω resistor.

Figure 7.120
An example circuit

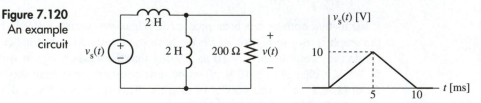

Solution

We show the circuit prepared for SPICE analysis in Figure 7.121. It seems that a natural choice is the PWL waveform specification for $v_s(t)$, so that is what we will use. The NETLIST below shows our source code and the resulting output, obtained in this case from the .OUT file that is automatically generated when the input file is run.

```
****08/23/94 15:22:20***********EVALUATION PSPICE (JAN 1992)*****
  EXAMPLE 2
  ****  CIRCUIT DESCRIPTION
  ***************************************************************
  ******
  VS   1   0   PWL(0,0   5M,10   10M,0   )
  L1   1   2   2
  L2   2   0   2
  R    2   0   200
  .TRAN    0.5M  15M   0   0.1M
  .PROBE
  .END
  ERROR-VOLTAGE SOURCE AND/OR INDUCTOR LOOP INVOLVING VS
  YOU MAY BREAK THE LOOP BY ADDING A SERIES RESISTANCE
```

What is wrong? The output file tells us. SPICE does not allow inductor and/or v-source loops. The reason for this should be apparent if you draw the dc steady-state equivalent circuit. You will then see that the two inductors "short-out" the v-source VS. Thus, if VS were a dc source, the dc steady-state circuit would be inconsistent. As a matter of fact, this state of affairs means that a dc steady state would not exist. SPICE, however, has made a suggestion to us to use a series resistance.

Figure 7.121
The example circuit
prepared for SPICE

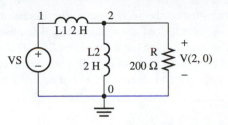

What does SPICE mean by this suggestion? Simply that we add a very small resistance in series with any one of the offending elements that make up the inductor/v-source loop. Choosing to do this with the input v-source, we have the circuit in Figure 7.122 (shown prepared for SPICE analysis). We have added a resistor, called RDUM, with a very small value of 0.01 Ω. (Notice that we have also added a new node, node 4.) How do we know that our value for RDUM is small enough? Although we will not do so here, in practice we would make several runs with successively smaller values of RDUM. When decreasing its value caused no noticeable change in the response, we would stop and settle for the last value. The SPICE NETLIST is shown here.

Figure 7.122
The modified example
circuit prepared for
SPICE

```
****08/23/94 15:48:47***********EVALUATION PSPICE (JAN 1992)*****
  EXAMPLE 2
  ****  CIRCUIT DESCRIPTION
  *****************************************************************
  ******
  VS    4   0   PWL(0,0  5M,10   10M,0  )
  RDUM      4   1    0.01
  L1    1   2   2
  L2    2   0   2
  R     2   0   200
  .TRAN     0.5M     15M     0    0.1M
  .PROBE
  .END

****08/23/94 15:48:47***********EVALUATION PSPICE (JAN 1992)*****
  EXAMPLE 2
  ****  INITIAL TRANSIENT SOLUTION   TEMPERATURE= 27.000 DEG C
  *****************************************************************
  ******
  NODE VOLTAGE  NODE VOLTAGE  NODE VOLTAGE  NODE VOLTAGE
  (  1)  0.000 (  2)  0.0000 (  4)  0.00000
    VOLTAGE SOURCE CURRENTS
    NAME    CURRENT
    VS      0.000E+00
    TOTAL POWER DISSIPATION 0.00E+00 WATTS
      JOB CONCLUDED
      TOTAL JOB TIME     3.68
```

The PROBE output is shown in Figure 7.123. To be consistent with our previously outlined procedure, we actually should have made another run or two and settled on the simulation time that showed the voltage returning essentially to zero by the end of the plot. However, our waveform does show the essential features of the response. We also notice another defect in our plot: we have instructed SPICE to simulate at too few time points to provide a good plot. Notice the straight line segments between points? We should reduce our maximum step size and make another run so that a better plot will be generated.

Figure 7.123
Response of example
circuit

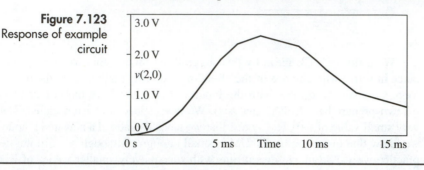

The equivalent situation for a capacitive circuit that causes trouble for SPICE is one for which there is a node to which only capacitors and/or i-sources are connected. The solution in that case is to add a very large resistor in parallel with one of these elements. We will leave this for you to do in quiz question Q7.8-1.

Section 7.8 Quiz

Q7.8-1. Use SPICE to simulate the circuit in Figure Q7.8-1 if $i_s(t)$ is a dc source having a value of 10 mA. Use PROBE to plot the voltage $v(t)$ for $t > 0$. Assume that the initial conditions on the two capacitors are $v_{c1}(0) = 10$ V and $v_{c2}(0) = -2$ V. (*Hint:* Add a 10^9-Ω resistor in parallel with the i-source.)

Figure Q7.8-1

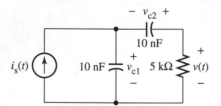

Q7.8-2. If the i-source in the circuit in Figure Q7.8-1 has the waveform shown in Figure Q7.8-2, simulate the circuit and use PROBE to plot the voltage $v(t)$.

Figure Q7.8-2

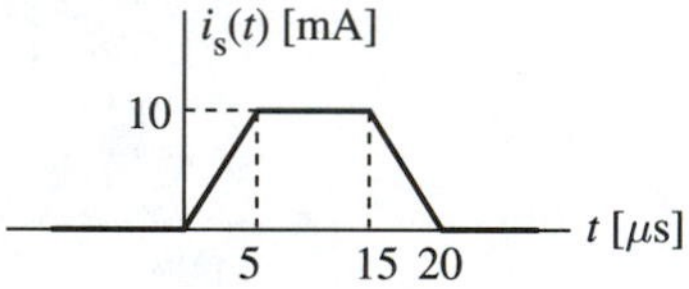

Chapter 7 Summary

This chapter has presented the analysis of first-order circuits: those whose voltage and current waveforms satisfy first-order differential equations. We derived the solution to the first-order *lowpass differential equation* in the form of the first-order *lowpass (or solution) operator.* As we will see in later chapters, this is a very fundamental result upon which the analysis of higher-order circuits (and, hence the solution of higher-order differential equations) is based. We also considered the first-order *highpass differential equation and circuit response.* The *general first-order differential equation (and circuit response)* was solved and the operator in its solution was shown to be a linear combination of the lowpass and highpass operators. We derived our theory for circuits without switches and for which the independent source waveforms were known for all values of time, but which are *one sided;* that is, identically zero for all $t \leq 0$. We then used the initial condition models derived in Chapter 6 and the foregoing theory to analyze first-order *switched circuits.*

We introduced the idea of a *simulation diagram,* the idea being to represent various mathematical operations graphically in order to make the mathematical equations more visual and concrete. Simulation diagrams are a mainstay of system theory and find applications in the design of *control systems* and *active filters,* which was illustrated by a practical design example in the section on time domain response of *active circuits.*

An entire section was devoted to a discussion of *more realistic circuit models* for the basic passive elements: *R, L,* and *C.* The discussion included the issue of *parasitic elements* and showed how actual observed waveforms differ from those obtained using the ideal models for the elements.

The next-to-last section presented a classical method for determining the response of a circuit containing a single energy storage element (or nonswitched series and/or parallel combinations of any number of a single type of such element), dc independent sources, and switches. Though often referred to as the *single time constant* approach, or *the $A + Be^{-t/\tau}$ method,* we call it the *equivalent circuit method* because one solves a problem by analyzing several dc equivalent circuits. It was shown to have limited applicability (relative to the general operator method we previously developed) and to be conceptually more involved.

Transient analysis using SPICE simulation was presented as the last topic. We covered the SPICE statements for specifying capacitors and inductors to the SPICE compiler and also the use of the *.TRAN command.* The issue of *selection of simulation parameters* was discussed, using examples to illustrate the general rules suggested.

Chapter 7 Review Quiz

RQ7.1. Show that the current $i(t)$ in Figure RQ7.1 is a lowpass response and find its value for all t.

Figure RQ7.1

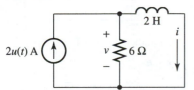

RQ7.2. Repeat Problem RQ7.1 for $v(t)$, showing that it is a highpass response.

RQ7.3. Assume that the switch S is always open in Figure RQ7.2 and find the impulse response for both $v(t)$ and $v_x(t)$. Determine the stability (unstable, stable, marginally stable) for each. Use a convenient set of units.

Figure RQ7.2

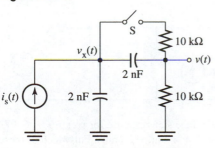

RQ7.4. Assume that $i_s(t) = 4\,\text{mA}$ (dc) in Figure RQ7.2 and that the capacitor response voltages are dc steady-state values for $t \leq 0$. Further, assume that S has been closed since $t = -\infty$, opens at $t = 0$, and remains open forever after. Find $v(t)$ and $v_x(t)$ for all t. (Note that solving for $v_x(t)$ is fairly difficult. Use integration by parts as discussed in Appendix C.)

RQ7.5. Draw a simulation diagram for the circuit in Figure RQ7.1 using only one integrator.

RQ7.6. Assume that the 2-H inductor in Figure RQ7.1 has a series parasitic resistance of $0.1\,\Omega$. Find the response current $i(t)$ and determine how it is different from the response found in Question RQ7.1.

RQ7.7. Find $v(t)$ in Figure RQ7.3 for all t using the equivalent circuit method and a convenient set of units for the element values.

Figure RQ7.3

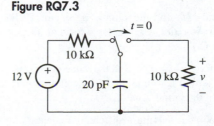

Chapter 7 Problems

Section 7.1 First-Order Lowpass Response

7.1-1. Find and sketch the unit step response $s(t)$ for $i(t)$ in the circuit in Figure P7.1-1. What is the time constant?

Figure P7.1-1

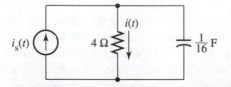

7.1-2. Find and sketch the impulse response $h(t)$ for $i(t)$ in the circuit in Figure P7.1-1.

7.1-3. For the circuit of Figure P7.1-1, find $i(t)$ if $i_s(t) = 4e^{-4t}u(t)$ A.

7.1-4. For the circuit of Figure P7.1-1, find $i(t)$ if $i_s(t) = 4\sqrt{2}\cos(4t)u(t)$ A. (*Hint:* Use the table form of integration by parts in Appendix C.)

7.1-5. Solve the differential equation $(dy/dt) + 2y = t^2u(t)$ using operators. What is the time constant? (*Hint:* Use the table form of integration by parts in Appendix C.)

7.1-6. Solve the differential equation $(dy/dt) + 2y = \cos(2t)u(t)$ using operators. (*Hint:* Use the table form of integration by parts in Appendix C.)

7.1-7. Find the unit step response for $v(t)$ in the circuit in Figure P7.1-2. What is the time constant?

Figure P7.1-2

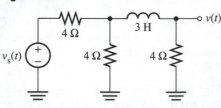

7.1-8. Find the unit impulse response for $v(t)$ in Figure P7.1-2.

7.1-9. Find $v(t)$ for the circuit in Figure P7.1-2 if $v_s(t) = 12\sin(2t)u(t)$ V. (*Hint:* Use the table form of integration by parts in Appendix C.)

7.1-10. If $x(t) = X_o e^{-t/\tau}$, find the value of $x(t)$ at $t = \tau$, $t = 3\tau$, and $t = 5\tau$. At the last time given, how close is $x(t)$ to zero as a percentage of the initial value X_o?

7.1-11. Without solving a differential equation for the circuit response $v(t)$ shown in Figure P7.1-3, determine its "final value." How long does it take to attain 99% of that final value? Answer this question without solving a differential equation.

Figure P7.1-3

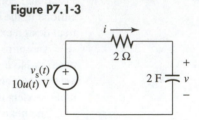

7.1-12. What is the final value of $i(t)$ in the circuit shown in Figure P7.1-3? How long does it take for $i(t)$ to drop to 1% of the initial value? Answer this question without solving a differential equation.

7.1-13. Suppose $v_s(t)$ in Figure P7.1-3 is changed to $v_s(t) = 10[u(t) - u(t -)]$ V, where τ is the circuit time constant. Find and plot the response voltage $v(t)$ and current $i(t)$.

7.1-14. What is the time constant in Figure P7.1-4 if $R = 1\ \Omega$? What is its value if $R = 0.1\ \Omega$? What is its value if $R = 0.01\ \Omega$? For these three values of R, find the total charge q delivered to the capacitor. What is the charge delivered to the capacitor if we allow R to approach zero?

Figure P7.1-4

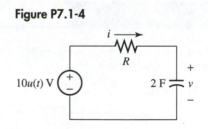

7.1-15. Find $v(t)$ in the circuit of Figure P7.1-5.

Figure P7.1-5

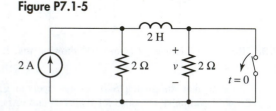

7.1-16. Find $v(t)$ and $i(t)$ in the circuit of Figure P7.1-6.

Figure P7.1-6

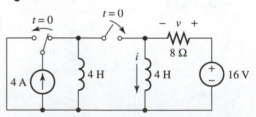

7.1-17. Find the step response for $v(t)$ in Figure P7.1-7 assuming that $R = 2$, -4, and -2 ohms. For which of these resistance values does $v(t)$ achieve a dc steady-state value? Notice that $v(t)$ is not a lowpass response variable, but can be found by the technique discussed in Example 7.6 in Section 7.1.

Figure P7.1-7

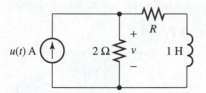

7.1-18. Consider the circuit shown in Figure P7.1-8.
a. Find $i(t)$ for all t if $R = 4\,\Omega$.
b. Find $i(t)$ for all t if $R = -2\,\Omega$.
c. Find $i(t)$ for all t if $R = -4\,\Omega$. (*Hint:* Replace the subcircuit to the right of terminals a and b by its Norton equivalent subcircuit.)

Figure P7.1-8

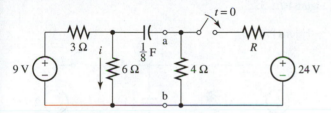

7.1-19. Consider the switched circuit in Figure P7.1-9.
a. Compute the energy stored on both capacitors for $t \le 0$.
b. Compute the current $i(t)$ for all t.
c. Compute the energy absorbed by the resistor for all t. Does your answer depend on the value of R?
d. In the light of your work in parts a, b, and c of this question, answer the question posed in Problem 6.5-7 at the end of Chapter 6.

Figure P7.1-9

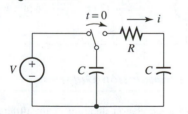

Section 7.2 First-Order Highpass Response

7.2-1. Solve the differential equation $(dy/dt) + 3y = 6(dx/dt)$ for $y(t)$ by solving the equivalent lowpass differential equation $(dy/dt) + 3y = \bar{x}(t)$, where $\bar{x}(t) = 6(dx/dt)$. Do this for two different forcing functions:
a. $x(t) = u(t)$ (that is, find the step response)
b. $x(t) = 2e^{-3t}u(t)$.

7.2-2. Solve Problem 7.2-1 by using long division on the system operator $H(p) = 6p/(p + 3)$.

7.2-3. Find the natural and forced responses for the differential equation in Problem 7.2-1 assuming that $x(t) = 2e^{-t}u(t)$.

7.2-4. Find $v(t)$ for the circuit shown in Figure P7.2-1 assuming that the current source waveform is:
a. $9u(t)$ A
b. $9\delta(t)$ A
c. $90e^{-3t}u(t)$ A
d. $9e^{-2t}u(t)$ A.

Figure P7.2-1

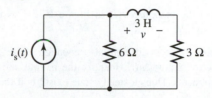

7.2-5. Find $v(t)$ in Figure P7.2-1 if $i_s(t) = I_m \cos(\omega t)u(t)$ A. (*Hint:* Your answer will depend upon the frequency ω as a parameter.)

7.2-6. This problem assumes that you have solved Problem 7.2-5. Using the result from that problem, find the forced and natural responses. What is the resulting forced response if the frequency ω is allowed to approach zero? Infinity? Is the response $v(t)$ highpass, lowpass, or neither?

7.2-7. Find an appropriate system of units for solving the circuit in Figure P7.2-2. Using the scaled circuit, find the response $v(t)$.

Figure P7.2-2

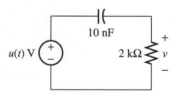

7.2-8. For the circuit shown in Figure P7.2-3, find the range of values of R for which $i(t)$ is a stable response by replacing all the elements to the left of the terminals marked a and b by a Norton equivalent subcircuit. Find the step and impulse responses for $i(t)$ if $R = 4\,\Omega$. Finally, find the step and impulse responses for $i(t)$ if $R = 2\,\Omega$.

Figure P7.2-3

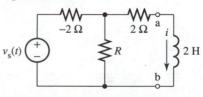

Section 7.3 General First-Order Response

7.3-1. Find the response for $v(t)$ in Figure P7.3-1 (on p. 390) in terms of an operator equation.

Figure P7.3-1

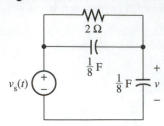

7.3-2. Find the response $v(t)$ in Figure P7.3-1 assuming that $v_s(t) = 4e^{-2t}u(t)$ V.

7.3-3. Solve the differential equation $(dy/dt) + 3y = (dx/dt) + 5x$ for $x(t) = tu(t)$. Identify the natural and forced response components. Does a steady state exist? Is it the dc steady state?

Problems 7.3-4, 7.3-5, and 7.3-6 should be worked as a unit.

7.3-4. Find the impedance operator $Z(p)$ for the two-terminal subcircuit shown in Figure P7.3-2. It is called the "driving point" impedance because (in general) you must "drive it" with a source to determine $Z(p)$. In this case, however, you can simply use series/parallel impedance combinations.

Figure P7.3-2

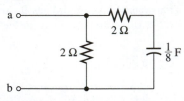

7.3-5. If a current source with a waveform of $i(t) = e^{-4t}u(t)$ A is applied to the subcircuit in Figure P7.3-2 (into terminal a), find the response terminal voltage $v(t)$ assuming the passive sign convention.

7.3-6. Repeat Problem 7.3-5 for $i(t) = e^{-2t}u(t)$ A.

Problems 7.3-7 and 7.3-8 should be worked as a unit.

7.3-7. Find the impulse response $h(t)$ for the response voltage $v(t)$ in Figure P7.3-3.

Figure P7.3-3

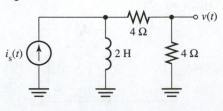

7.3-8.
a. Sketch a graph of $i_s(t) = u(t) - 2u(t - 2) + u(t - 3)$.
b. Use the impulse response you found in Problem 7.3-7 to

find the response $v(t)$ to the current $i_s(t)$ in step a, assuming that the unit of current is the ampere. (*Hint:* Write an analytical expression for $i_s(t)$ as a sum of integrals of impulses.)

7.3-9. Compute the unit step response for $v(t)$ in Figure P7.3-4 and differentiate it to find the unit impulse response.

Figure P7.3-4

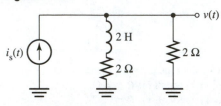

7.3-10. Repeat Problem 7.3-9 in the reverse order: compute the unit impulse response directly and integrate it to find the unit step response.

7.3-11. Find and sketch the current $i(t)$ for the circuit shown in Figure P7.3-5.

Figure P7.3-5

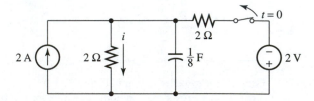

7.3-12. Find and sketch the voltage $v(t)$ for the circuit in Figure P7.3-6.

Figure P7.3-6

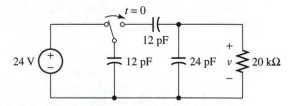

Section 7.4 Simulation Diagrams

7.4-1. Draw a simulation diagram of the differential equation $py + 3y = 3px + 6x$ using summers, scalar multipliers, and a single integrator—but no differentiators. Assume that the forcing function $x(t)$ is defined over the entire time axis and is one sided.

7.4-2. A first-order circuit response has the impulse response $h(t) = 4e^{4t}u(t)$. It is stable, unstable, or marginally stable? Find the differential operator $H(p)$ corresponding to $h(t)$ and draw a simulation diagram using only one integrator.

7.4-3. Repeat Problem 7.4-2 if the step response is $s(t) = 0.5[1 - e^{-4t}]u(t)$.

7.4-4. A *system* is said to be unstable if any one of its response variables is unstable, stable if all of its response variables are stable, and marginally stable if all are stable except for one or more that are marginally stable. Suppose a system has three response variables with corresponding impulse responses $h_1(t) = u(t)$, $h_2(t) = tu(t)$, and $h_3(t) = e^{-2t}u(t)$. Is this system stable, unstable, or marginally stable? Justify your answer. Draw a simulation diagram that represents this system. Use only adders, scalar multipliers, and integrators.

7.4-5. A system response variable has the impulse response $h(t) = t^2e^{-4t}u(t)$. Is it stable, unstable, or marginally stable? Derive the corresponding differential operator $H(p)$ and draw a simulation diagram. (*Hint:* One approach is to differentiate $h(t)$ three times, solve for $h(t)$, then factor. This results in a third-order system, but of a special type: three of first order, one after the other.)

7.4-6. Suppose a system response satisfies the differential equation

$$py(t) + 3y(t) = 3x_1(t) + (4p + 3)x_2(t).$$

Draw a simulation diagram using only a single integrator.

7.4-7. Draw a simulation diagram for the circuit in Figure P7.4-1 with v and i as the output variables. Then assume that $i_s(t) = u(t)$ A and $v_s(t) = 8e^{-2t}u(t)$ V and find $i(t)$ and $v(t)$.

Figure P7.4-1

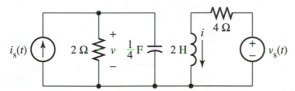

7.4-8. If the initial condition for the circuit shown in Figure P7.4-2 is $i(0) = 2$ A, draw a simulation diagram representing the response $i(t)$ for all $t > 0$.

Figure P7.4-2

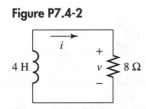

7.4-9. Repeat Problem 7.4-8 under the assumption that $v(0) = 4$ V, rather than $i(0) = 2$ A.

7.4-10. Draw a simulation diagram for the differential equation

$$\frac{dy}{dt} + 2y = 4\frac{dx}{dt} + 8x, \quad t > 0.$$

Assume that $x(t) = 2e^{-2t}$ for $t > 0$ and that $y(0) = 2$. *Warning:* This one is just a bit tricky. Before attempting it, you might wish to reread the material at the end of Section 7.4.

Section 7.5 Active First-Order Circuits

7.5-1. Find the response voltage $v(t)$ in Figure P7.5-1. Use a convenient set of units to simplify your work.

Figure P7.5-1

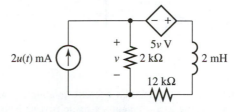

7.5-2. Find the values of g_m in Figure P7.5-2 that result in the response voltage $v(t)$ being stable, unstable, and marginally stable. Assuming that $g_m = 3/8$ S, find the response voltage $v(t)$ assuming that $i_s(t) = 10u(t)$ A.

Figure P7.5-2

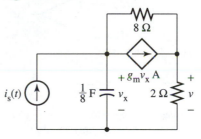

7.5-3(O). Find the differential operator relating the response voltage $v(t)$ to the input voltage $v_s(t)$ in Figure P7.5-3. Use an appropriate set of units. Find the unit step response, then differentiate it to find the unit impulse response. Finally, find $v(t)$ if $v_s(t) = 2e^{-t/80}u(t)$ V, with t in μs.

Figure P7.5-3

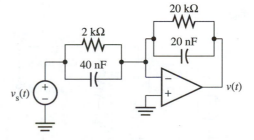

7.5-4(O). Find the differential operator relating $v(t)$ to $v_s(t)$ in Figure P7.5-4, using a convenient set of units. Use it to find the unit impulse and step responses. Then find the response voltage $v(t)$ for $v_s(t) = e^{-t/80}u(t)$ V, with t in microseconds.

Figure P7.5-4

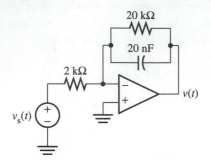

7.5-5(O). Find the differential operator relating $v(t)$ to $v_s(t)$ in Figure P7.5-5, using a convenient set of units. Use it to find the impulse and step responses, then find the $v(t)$ that results if $v_s(t) = e^{-t}u(t)$ V.

Figure P7.5-5

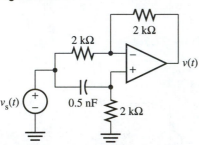

7.5-6(O). Consider the operator $H(p) = 4p/(p + a)$, with $a = 10^3 \text{ s}^{-1}$. Design a circuit with only resistors, capacitors, and op amps to realize this operator.

7.5-7(T). Find the responses $v_x(t)$ and $v_y(t)$ in Figure P7.5-6 for $v_s(t) = 3u(t)$ mV and for the more complex waveform $v_s(t) = 3e^{-t}u(t)$ mV, with t in milliseconds. Use a convenient set of units and solve the problem by replacing the two-terminal circuit to the left of the capacitor (relative to ground) by a Norton equivalent. Note that any dc power supply sources have been deactivated and assume that $r_\pi = 1$ kΩ for the BJT.

Figure P7.5-6

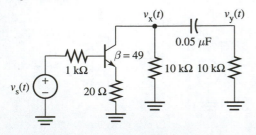

7.5-8(T). Find $v(t)$ for the small signal BJT circuit in Figure P7.5-7 in terms of an operator equation. Assume that $r_\pi = 1$ kΩ. (*Hint:* This is a fairly difficult problem, so be methodical in your solution. First, convert the problem to a convenient set of units. Then use the hybrid π model in the place of the BJT symbol and

apply the current source substitution theorem covered in Section 3.6 of Chapter 3. You can then either write two nodal equations or use the Miller equivalent established in Chapter 6, Section 6.7 (assuming that the voltage v/v_b is very large, with v_b being the voltage from the BJT base contact to ground). What is the $v(t)$ response to a 1-mV step function?

Figure P7.5-7

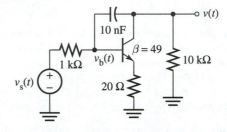

7.5-9(T). The high-frequency hybrid π model for the BJT is shown in Figure P7.5-8. Use it to solve for the impulse response of the short circuit current gain operator $\beta(p)$. This is defined from Figure P7.5-9 in the following manner. Solve for the response current $i_c(t)$, then express $i_c(t)$ in the form $i_c(t) = \beta(p)i_b(t)$.

Figure P7.5-8

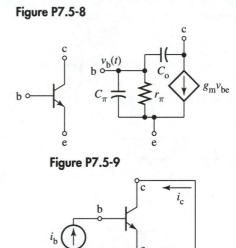

Figure P7.5-9

Section 7.6 Practical Circuit Models for Passive Elements

7.6-1. Find the unit step response for $i(t)$ in Figure P7.6-1 assuming that the inductor is ideal. Then repeat, assuming that it has a series parasitic resistance of 0.01 Ω. Over what time interval is the nonideal result approximately equal to the ideal one?

Figure P7.6-1

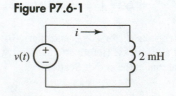

7.6-2(O). Show that the op amp circuit in Figure P7.6-2 is an inverting integrator; that is, the system operator relating $v(t)$ to $v_s(t)$ is

$$H(p) = -\frac{1}{R'Cp},$$

assuming all elements are ideal. Then suppose that the capacitor has a parasitic parallel resistance r_p. Show that the resulting system operator is

$$H(p) = -\frac{r_p/R'}{r_pCp + 1} = \frac{1/R'C}{p + 1/r_pC}.$$

If $r_p = 10\ M\Omega$, $C = 10\ nF$, and $R' = 2\ k\Omega$, find the unit step response. Use a convenient set of units. Find the forced and natural responses.

Figure P7.6-2

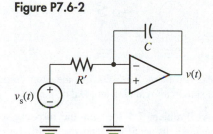

Section 7.7 Equivalent Circuit Analysis for Single Energy Storage Circuits with Dc Sources and Switches

7.7-1. Assuming that the circuit in Figure P7.7-1 is in the dc steady state for all $t < 0$, find the responses $v(t)$ and $i(t)$ for all $t > 0$. Use the equivalent circuit method.

Figure P7.7-1

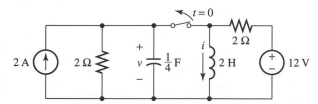

7.7-2. Find $v(t)$ in the circuit shown in Figure P7.7-2 using the equivalent circuit method.

Figure P7.7-2

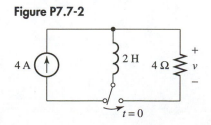

7.7-3. Consider the circuit in Figure P7.7-3.

a. Draw the circuit for $t \le 0$ and determine its stability type.
b. Find the dc steady-state capacitor voltage (if one exists).
c. If you determined that the dc steady-state condition exists in step b, find the value of $i(t)$ for all t. Otherwise, assume that the initial capacitor voltage is zero and find and sketch the waveform $i(t)$ for all $t > 0$. Use the equivalent circuit method.

Figure P7.7-3

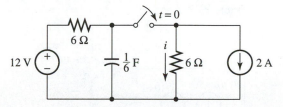

7.7-4. Use the equivalent circuit method to find $v(t)$ and $i(t)$ for all t in the circuit shown in Figure P7.7-4. Use a convenient set of units. Sketch the waveforms.

Figure P7.7-4

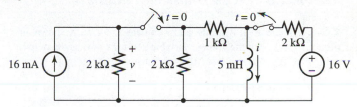

7.7-5. Use the equivalent circuit method to find and sketch $v(t)$ for all t in Figure P7.7-5.

Figure P7.7-5

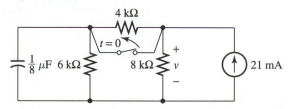

7.7-6(O). Use a convenient set of units and the equivalent circuit method to find the response voltages $v_x(t)$ and $v_y(t)$ for $t > 0$ in the circuit shown in Figure P7.7-6. Assume that the circuit is in

Figure P7.7-6

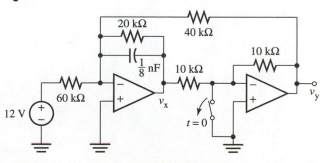

the dc steady state for $t \leq 0$. Sketch the waveforms you have computed.

7.7-7(T). Assuming that dc steady-state conditions hold for the circuit shown in Figure P7.7-7, find and sketch the voltage $v(t)$ for all t using the equivalent circuit technique. (*Hint:* Use the hybrid π small signal model developed in Chapter 5 and note that the circuit shown is already small signal in the sense that any dc sources have been deactivated.)

Figure P7.7-7

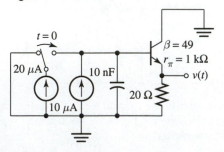

Section 7.8 SPICE: Transient Analysis

7.8-1. Write a SPICE file and simulate the circuit shown in Figure P7.8-1. Plot the voltage variable $v(t)$ for $t > 0$ as the output using PROBE. Assume that $i(0) = 0.5$ mA.

Figure P7.8-1

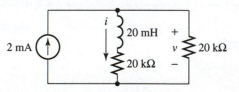

7.8-2. Change the i-source in Problem 7.8-1 to a piecewise linear (PWL) pulse that starts at 0 mA at $t = 0$, rises linearly to 2 mA after 0.5 μs, stays at 2 mA until $t = 1$ μs, then drops linearly to 0 at $t = 1.5$ μs. Repeat the simulation and plot $v(t)$ for $t > 0$ assuming $i(0) = 0.5$ mA.

7.8-3. Simulate the circuit shown in Figure P7.8-2 and use PROBE to plot the response of $v(t)$ for a 1-mA step function input. First write a NETLIST using ideal capacitors. After trying to run SPICE (which will give you an error message), recode and simulate the circuit that results when you use a nonideal model for the capacitors with each having a parallel parasitic resistance of 10 MΩ. See Section 7.6 for details.

Figure P7.8-2

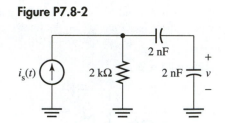

7.8-4. Solve Problem 7.5-5(O) for the step response using SPICE and a subcircuit realizing an ideal op amp.

7.8-5. Solve Problem 7.5-8(T) for the response to a 10-mV step function input using SPICE and the small signal hybrid π model for the BJT.

8 Complex Signals and Systems

Although you have undoubtedly encountered complex numbers before, you were probably introduced to them in a slightly different manner than the one we adopt here: as ordered pairs of real numbers. Using this interpretation, we show that a complex number can be considered as, for example, a pair of conductors carrying two real voltages. In general, we refer to two conductors carrying two real waveforms as a "two-wire signal." We review all of the basic properties of complex numbers that we will need in the next chapter (Chapter 9), where we discuss the time domain solution of higher-order circuits, and a following one (Chapter 11), where we develop special techniques for finding the forced response of circuits whose independent sources are sinusoidal. Section 8.1 defines the set of complex numbers and discusses their elementary properties, emphasizing their interpretation with system diagrams. Section 8.2 then develops the usual rectangular form, $z = x + jy$, and polar form,

$z = (|z|, \theta)$. We show that any complex number z can be written in the Euler form[1] $z = |z|e^{j\theta} = |z|\angle\theta$. In Section 8.4 we show how complex numbers can be used to represent sinusoidal waveforms and give a real interpretation of the complex exponential $z(t) = \bar{Z}e^{j\omega t}$, where $\bar{Z}$ is the phasor associated with the signal. Section 8.5 describes how SPICE can be used to generate and study complex time-varying waveforms. Specifically, it treats the "real world" generation of the complex exponential waveform $e^{-j\omega t}$. Finally, Section 8.6 is a discussion of the relationship between real and complex number representations of physical systems and reality itself.

8.1 | Complex Numbers and Signals

Preliminary Discussion of Complex Numbers

In Part I of the text, we worked with signals (voltage and current source waveforms, for instance) that were real; for the most part, in fact, they were dc signals (constants). Now we will extend our concept of signal to include those that vary with time and have possibly complex values. Recall the way complex numbers were first introduced in your high school algebra text. The basic idea is that, for example, the polynomial equation $x^2 = 1$ has the solutions $x = \pm 1$, whereas the polynomial equation $x^2 = -1$ has none—as long as we expect the answer to be a real number. So what do we do? Why, we *define* the answer for the latter problem to be $\pm\sqrt{-1}$. But merely labeling the answer in this manner is no real help because we must know how to compute it. For example, we can compute $\sqrt{144} = \pm 12$. We have computed the square root in terms of a more fundamental quantity, a simple real number. Now what is $\sqrt{1}$? It is ± 1 of course, and we do not ask for a simpler answer: 1 is a fundamental quantity. In the same manner, we must consider $\sqrt{-1}$ to be a fundamental quantity; but this entails adding such elements to the real number system; that is, expanding our concept of number. We will develop this idea here in a physical context, assuming that we know all the properties of the real number system.

Complex Numbers as "Two-Wire" Signals

We will consider numbers to be signals in a circuit or system with a physical meaning; we assume they are quantities like voltage or current that we can measure. Thus, we will deal with numbers or signals in the form of real number pairs $z = (x, y)$. Such a signal is illustrated in Figure 8.1(a), in which we can think of x and y as a pair of real voltages. For obvious reasons we will refer to such a number pair as a "two-wire" signal. We are, of course, going to call our pair of real numbers a *complex number,* but we will define this idea more formally a bit later. We call x the *real part of z* and y the *imaginary part of z,*

Figure 8.1
The idea of a complex number

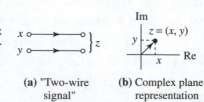

(a) "Two-wire signal"

(b) Complex plane representation

---

[1] The number pair $(|z|, \theta)$ is often defined in math texts to be the polar form. For this reason, we distinguish it from the form $z = |z|e^{j\theta} = |z|\angle\theta$ by referring to the latter as the "Euler form" (as it is so closely related to the classical Euler formula $e^{j\theta} = \cos(\theta) + j\sin(\theta)$). We therefore use the term "polar form" for the former. This is somewhat at variance with other circuits texts.

and write them as

$$x = \text{Re}\{z\} \tag{8.1-1}$$
$$y = \text{Im}\{z\}$$

As we show in Figure 8.1(b), we can plot our number pair in a plane that we will refer to as *the complex plane*. We choose to call the horizontal axis the *real axis* and the vertical one the *imaginary axis*. Both, we hasten to note, are axes of real numbers. We plot $\text{Re}\{z\} = x$ on the horizontal axis and $\text{Im}\{z\} = y$ on the vertical axis. Finally, we observe that we can think of $z = (x, y)$ as either a point in the complex plane or as a vector from the origin to the point (as shown by the arrow in the figure).

The Negative of a Complex Number

In a quite intuitive and straightforward fashion, we define the *negative of z* to be $-z = (-x, -y)$. Notice that the negative sign in front of z does not (as yet, anyway) represent subtraction. The two characters "$-$" and "z" are to be interpreted as a single symbol. The physical interpretation of $-z$ as a two-wire signal and its representation in the complex plane are shown in Figure 8.2(a) and (b). The small circles are merely symbols representing multiplication of the input signal (on the line with the arrowhead) by the real numbers inside. They are, therefore, (real) *multipliers* or *scalar multipliers*.

Figure 8.2
The negative of a complex number

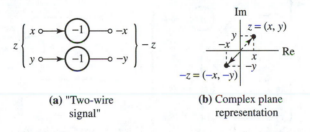

(a) "Two-wire signal"   (b) Complex plane representation

The Conjugate of a Complex Number

The negative of a real number, of course, is a familiar concept. There is an equally simple concept relating to complex numbers, however, which has no counterpart for real numbers. This is the idea of *complex conjugate,* or simply *conjugate*. The conjugate of z is the complex number z^* defined by

$$z^* = (x, -y). \tag{8.1-2}$$

In this case the sign of only the imaginary part of z is changed. The complex conjugate is depicted physically in Figure 8.3.

Figure 8.3
The conjugate of a complex number

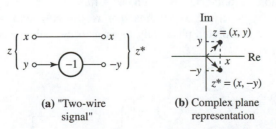

(a) "Two-wire signal"   (b) Complex plane representation

Equality of Two Complex Numbers

We define z_1 and z_2 to be *equal* if their real and imaginary components are separately equal as real numbers:

$$z_1 = z_2 \iff x_1 = x_2 \quad \text{and} \quad y_1 = y_2. \tag{8.1-3}$$

In number pair form, we have $(x_1, y_1) = (x_2, y_2) \iff x_1 = x_2$ and $y_1 = y_2$. (The symbol $\iff$ is to be read "if and only if.")

The Sum and Difference of Two Complex Numbers

Thus far, we have been discussing *unary operations* on complex numbers. These are operations on a single entity. Now we will deal with operations on two at a time: *binary operations*. Thus, let's define $z_1 = (x_1, y_1)$ and $z_2 = (x_2, y_2)$. We will also let $w = (u, v)$ be the result of each of our binary operations. We define the *sum* of z_1 and z_2 to be the complex number that has the sum of the two individual real parts for its real part and the sum of the two individual imaginary parts for its imaginary part:

$$w = z_1 + z_2 = (x_1 + x_2, y_1 + y_2) = (u, v). \tag{8.1-4}$$

In number pair form, $(x_1, y_1) + (x_2, y_2) = (x_1 + x_2, y_1 + y_2) = (u, v)$, or $u = x_1 + x_2$ and $v = y_1 + y_2$. Figure 8.4(a) shows the implementation of a complex summer with real elements (the small circles with plus signs are real summers), and Figure 8.4(b) shows the way we will represent it in complex form. The latter, more compact, notation is the same symbolism as that for a real summer; it should be clear from the context which is meant. The *difference* of z_1 and z_2 is defined to be the sum of z_1 and the negative of z_2:

$$z_1 - z_2 = z_1 + (-z_2) = (x_1 - x_2, y_1 - y_2). \tag{8.1-5}$$

Figure 8.4
The complex summer

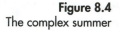
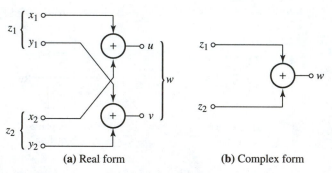

(a) Real form   **(b)** Complex form

Product of Two Complex Numbers

All of our operations on complex numbers, with the sole exception of the conjugate, have so far been straightforward extensions of the corresponding concepts for real numbers. In the case of multiplication, the extension is not obvious; however, it is a necessary one if we wish to extend as many of the properties of real numbers as possible to the complex domain. Mathematically, the *product* of z_1 and z_2 is defined by

$$w = z_1 z_2 = (x_1 x_2 - y_1 y_2, x_1 y_2 + x_2 y_1) = (u, v). \tag{8.1-6}$$

In more explicit terms, with $w = (u, v) = z_1 z_2$, we have $(u, v) = (x_1, y_1)(x_2, y_2) = (x_1 x_2 - y_1 y_2, x_1 y_2 + x_2 y_1)$; thus, $u = x_1 x_2 - y_1 y_2$ and $v = x_1 y_2 + x_2 y_1$. The complex multiplier is interpreted physically in terms of real elements in Figure 8.5(a) and more com-

Figure 8.5
The complex multiplier

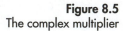
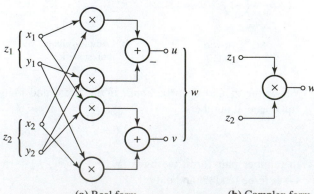

(a) Real form   **(b)** Complex form

pactly in Figure 8.5(b). We will most often use the latter, but when we think of the physical meaning we must always visualize the former.

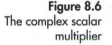

Scalar Multiplication

Each small circle with an "x" inside in Figure 8.5 is a function multiplier; that is, if z_1 and z_2 are functions of time, then the response w is also a function of time—one whose value at any time t is the product of the two values of z_1 and z_2 at the same time. A special (but very important) case is that of the *scalar multiplier,* which we have already mentioned. For the scalar multiplier, one of the two operands is constant, say $a = (\alpha, \beta)$. We have

$$w = az = (\alpha, \beta)(x, y) = (\alpha x - \beta y, \alpha y + \beta x) = (u, v). \qquad (8.1\text{-}7)$$

Equating real and imaginary parts gives the real system diagram of Figure 8.6(a) and the more compact complex form of Figure 8.6(b). The circles labeled α and β in part (a) of the figure are real scalar multipliers, and the one labeled a in part (b) is a complex scalar multiplier.

Figure 8.6
The complex scalar
multiplier

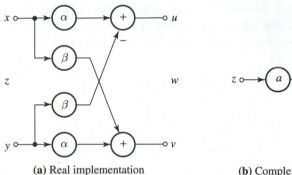

(a) Real implementation (b) Complex form

Now that we have completed our definitions of the basic operations involving number pairs, let's formally define what we mean by a complex number and the set of all complex numbers.

Definition 8.1   The set of all possible real number pairs, $C = \{z = (x, y)\}$, with the preceding definitions of equality, sum, conjugate, and product is called the *set of complex numbers.* A single number pair $z = (x, y)$ in C is referred to as *a complex number.*

Section 8.1 Quiz

Let $z_1 = (1, 1)$, $z_2 = (3, 4)$, $z_3 = (-1, 1)$, and $a = (4, 3)$ for the following questions.

Q8.1-1. Find $z_1 + z_2$ and sketch it in the complex plane.

Q8.1-2. Find $z_1 - z_3$ and sketch it in the complex plane.

Q8.1-3. Find az_2 and sketch it in the complex plane.

Q8.1-4. Find $z_2{}^*$ and sketch it, together with z_2, in the complex plane.

Q8.1-5. Find $z_2 z_2{}^*$ and sketch it in the complex plane.

8.2 | Rectangular and Polar Forms

**Real Numbers as
a Subset of the
Complex Numbers**

We will begin this section by exploring the question of exactly how real signals and real number manipulations relate to those in the complex realm. We will consider all number pairs of the form $(x, 0)$. Suppose $z_1 = (x_1, 0)$ and $z_2 = (x_2, 0)$ in what follows. Then

$$z_1 = z_2 \Longleftrightarrow x_1 = x_2, \qquad (8.2\text{-}1a)$$

$$z_1 + z_2 = (x_1 + x_2, 0), \qquad (8.2\text{-}1b)$$

$$z_1 z_2 = (x_1 x_2, 0). \qquad (8.2\text{-}1c)$$

What does this mean? Simple! If we start with number pairs of the form $(x, 0)$, then none of the usual complex number manipulations will ever produce anything other than another complex number of the same type; furthermore, to obtain this number one need only do real number manipulations on the real components. Thus, the set of all number pairs of the form $(x, 0)$ is identical with the set of all real numbers, except in name, and we will therefore consider these names to be equivalent—that is, we will identify $(x, 0)$ with the real number x:

$$(x, 0) = x. \tag{8.2-2}$$

Figure 8.7 shows the situation for summation. The zero imaginary part is merely "taken along for the ride" when we consider the two numbers to be complex. If we choose to leave it off, that is acceptable, but if we ever have to sum a real number with one that is complex, we must replace the zero imaginary part. By omitting it we save one real summer. The corresponding situation for multiplication is shown in Figure 8.8. Again, one saves elements by merely recognizing that it is unnecessary to multiply numbers by zero and to sum the two resulting zero numbers.

Figure 8.7
A complex summer
with real inputs

Figure 8.8
A complex multiplier
with real inputs

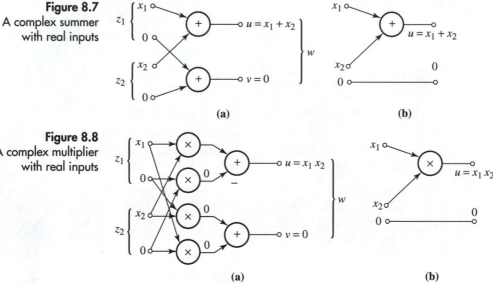

The Imaginary Unit j   There is a special complex number that plays an important role in the representation of such numbers: it is called the *imaginary unit*. We define it by

$$j = (0, 1). \tag{8.2-3}$$

Mathematicians use the symbol i, but we have already used this notation for current. The complex number j has a quite interesting property, namely,

$$j^2 = (0, 1)(0, 1) = (-1, 0) = -1. \tag{8.2-4}$$

This is what we *rigorously mean* by the statement that $j = \sqrt{-1}$: the special number pair $j = (0, 1)$ has the property that its square (using the preceding definition of product) gives a number pair that is equivalent in all respects to the real number -1. We also find that another property is enormously useful. It is spelled out by the following equation:

$$jy = (0, 1)(y, 0) = (0, y). \tag{8.2-5}$$

Thus, multiplication of a real number by j turns it into a *purely imaginary number*. (We will use this term for all complex numbers of the form $(0, y)$.) For this reason, j is often called the *imaginary operator*. It is also referred to as the imaginary unit by analogy with the idea of a unit vector.

We will use this second property in the following way. We write

$$z = (x, y) = (x, 0) + (0, y) = x + jy. \tag{8.2-6}$$

We call this the *rectangular form* of z. In somewhat more general terms,

$$z = \text{Re}\{z\} + j\text{Im}\{z\}. \tag{8.2-7}$$

Thus, the real part appears without a j multiplier, and the imaginary part appears with one. This is a nice, compact representation; by using it, we can avoid the necessity of dealing directly with number pairs. To support this statement, note that

$$z_1 = z_2 \Longleftrightarrow x_1 + jy_1 = x_2 + jy_2 \Longleftrightarrow x_1 = x_2 \quad \text{and} \quad y_1 = y_2. \tag{8.2-8}$$

Thus, we can simply equate the terms without a j and those with a j separately. Furthermore, we have

$$z_1 + z_2 = (x_1 + jy_1) + (x_2 + jy_2) = (x_1 + x_2) + j(y_1 + y_2), \tag{8.2-9}$$

from which we see at once that the real part of the sum is $x_1 + x_2$ and the imaginary part is $y_1 + y_2$. Finally, let us investigate the product. We have

$$z_1 z_2 = (x_1 + jy_1)(x_2 + jy_2) = (x_1 x_2 - y_1 y_2) + j(x_1 y_2 + x_2 y_1). \tag{8.2-10}$$

We have simply applied the familiar algebraic procedure for multiplying two sums, and have recognized that $j^2 = -1$. The result of what we have just shown is that by using the j operator in the rectangular form, we can manipulate complex numbers just like real ones—as long as we recognize that $j^2 = -1$.

A vector interpretation of the rectangular form of z is shown in Figure 8.9. There, we show the real number $1 = 1 + j0$ and the purely imaginary unit $j = 0 + j1$. We see that they act like basis vectors.[2]

Figure 8.9
Representation of z
in terms of basis
vectors 1 and j

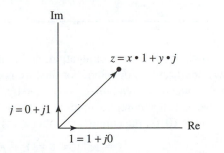

Let's return for a moment to the idea of a scalar multiplier, which we covered in Section 8.1. If a is a real constant, then

$$az = a(x + jy) = ax + jay = (ax, ay). \tag{8.2-11}$$

Thus, we can "multiply in" a real constant.[3] Finally, we note that taking the complex conjugate is very easy: one simply changes the sign of the j term. Thus,

$$z^* = (x + jy)^* = (x, y)^* = (x, -y) = x - jy. \tag{8.2-12}$$

---

[2] That is, like $\hat{i}$ and $\hat{j}$ (or, if you prefer, $\hat{a}_x$ and $\hat{a}_y$).

[3] Thus, we can now interpret $-z$ as $-1 \cdot z$ because $-1(x, y) = (-x, -y) = -z$ by definition.

Polar Coordinates

Figure 8.10   Polar coordinates

As we know from analytic geometry, one can express a number pair in terms of *polar co-ordinates*. We show these two coordinates in Figure 8.10. We designate the *polar angle* by θ and the *magnitude,* or *absolute value,* by $|z|$. Now, suppose we know the complex number z in rectangular form; that is, we know the values of x and y. What are the values of $|z|$ and θ? By Pythagoras's theorem, we know that

$$|z| = \sqrt{x^2 + y^2} = \sqrt{\text{Re}^2\{z\} + \text{Im}^2\{z\}} \tag{8.2-13}$$

and that

$$\theta = \tan^{-1}\left[\frac{y}{x}\right] = \tan^{-1}\left[\frac{\text{Im}\{z\}}{\text{Re}\{z\}}\right]. \tag{8.2-14}$$

Some care must be applied in working with these equations, as we show in the following short (but tricky) example.

Example 8.1   Find the polar coordinates of $z = -1 - j = -1 - j1$.

Solution $|z| = \sqrt{(-1)^2 + (-1)^2} = \sqrt{2}$. No problem. But let's compute the polar angle: $\theta = \tan^{-1}[-1/-1] = \tan^{-1}(1) = \pi/4$ rad. But this is incorrect! Why? Look at Figure 8.11. The actual angle is $5\pi/4$ rad (or $-3\pi/4$ rad of course). The problem is that the inverse tangent has a period of only π rad; thus, there is a possible ambiguity of π radians. Always make a sketch to determine the correct quadrant of the complex plane before deciding upon the angle.

Figure 8.11
Sketch of $z = -1 - j$

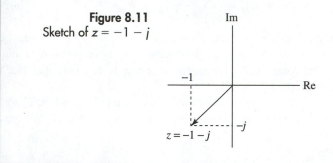

The Inverse of a Complex Number

We have thus far defined negation, conjugation, addition, and multiplication of complex numbers. We have not discussed division. For this, we need the concept of the inverse of a complex number, which we symbolize as z^{-1} or $1/z$. For a given complex number $z = x + jy$, the inverse is that complex number $z^{-1} = u + jv$, which has the property that (assuming $z = (x, y) \neq (0, 0)$, the complex number zero)

$$z^{-1}z = zz^{-1} = 1 = 1 + j0. \tag{8.2-15}$$

In rectangular form, we have

$$(u + jv)(x + jy) = (xu - yv) + j(vx + uy) = 1 + j0. \tag{8.2-16}$$

Equating the real and imaginary parts and solving for u and v results in

$$u = \frac{x}{x^2 + y^2} \tag{8.2-17a}$$

and

$$v = \frac{-y}{x^2 + y^2}. \tag{8.2-17b}$$

Thus,

$$z^{-1} = \frac{1}{z} = \frac{x - jy}{x^2 + y^2} = \frac{z^*}{|z|^2} = \frac{z^*}{zz^*}. \tag{8.2-18}$$

We will take the last expression as our working definition for it is very suggestive notationally: one merely multiplies the top and bottom of the "fraction" $1/z$ by z^* to compute the inverse. Notice that $z = 0$ does not have an inverse because $|0| = 0$ and (8.2-18) is invalid.

We are now in a position to define division. Letting $z_1 = x_1 + jy_1$ and $z_2 = x_2 + jy_2$, we define

$$\frac{z_1}{z_2} = z_2^{-1}z_1 = z_1 z_2^{-1}. \tag{8.2-19}$$

This is the reason for introducing the notation $1/z$ for the inverse: it suggests division of the real number one by z. If we use (8.2-18), we can write

$$\frac{z_1}{z_2} = \frac{z_1 z_2^*}{z_2 z_2^*} = \frac{z_1 z_2^*}{|z_2|^2}, \tag{8.2-20}$$

a process called "rationalizing the denominator."

Example 8.2   Find $(1 + j)/(1 - j)$ in rectangular form.

Solution   We merely note that the conjugate of the denominator is $1 + j$ and multiply it top and bottom:

$$\frac{1 + j}{1 - j} = \frac{(1 + j)(1 + j)}{(1 - j)(1 + j)} = \frac{0 + j2}{2} = j. \tag{8.2-21}$$

The result has zero real part—it is purely imaginary.

The Complex Exponential Function and the Euler Formulas   There is a form into which complex numbers can be placed that is very convenient for multiplication and division; furthermore, it serves as the foundation for the ac steady-state circuit analysis techniques to be covered in later chapters. To introduce this form, we need the concept of a complex function of a complex variable. Just as we write $y = f(x)$ for a real function of the real variable x, we can write

$$w = f(z) \tag{8.2-22}$$

for a complex-valued function of the complex variable z. Letting $z = x + jy$, as usual, we see that this equation actually defines two functions of two real variables. In fact, because $f(z) = f(x + jy)$ is complex, we can write

$$w = u(x, y) + jv(x, y), \tag{8.2-23}$$

where $u(x, y)$ is the real part of w and $v(x, y)$ is its imaginary part. Both are functions of the two real variables x and y. A basic problem in complex variable theory is this: given our stock of standard functions of a real variable, how do we extend[4] them to become complex functions of a complex variable so that they retain all of their ordinary properties

---

[4] The process is called *analytic continuation*. See any standard text on complex analysis for further details.

when $y = \text{Im}\{z\} = 0$; that is, when z is real? We will not be concerned with many such functions, but one is extremely important.

Definition 8.2   If $z = x + jy$, the *complex exponential* function is defined by

$$e^z = e^x \cos(y) + je^x \sin(y). \qquad (8.2\text{-}24)$$

Does this reduce to the usual exponential function when $y = \text{Im}\{z\} = 0$? Yes it does, for $\cos(0) = 1$ and $\sin(0) = 0$, and a more advanced look at the problem shows that (8.2-24) is the only "nice" function that has this property. The special case in which $x = \text{Re}\{z\} = 0$ is known as *Euler's formula:*

$$e^{jy} = \cos(y) + j\sin(y). \qquad (8.2\text{-}25)$$

Noticing that the cosine is an even function and the sine is an odd function, we can readily add and subtract (8.2-25) and the equivalent expression for e^{-jy} to show that

$$\cos(y) = \frac{e^{jy} + e^{-jy}}{2} \qquad (8.2\text{-}26a)$$

and

$$\sin(y) = \frac{e^{jy} - e^{-jy}}{2j}. \qquad (8.2\text{-}26b)$$

This expresses the real cosine and sine functions in terms of the complex exponential. Their value will prove to be very high in our following work. Because they are so useful, we will emphasize the preceding three relations to call your attention to them when you are leafing through the chapter:

Euler's Formulas $e^{jy} = \cos(y) + j\sin(y)$

$$\cos(y) = \frac{e^{jy} + e^{-jy}}{2}$$

$$\sin(y) = \frac{e^{jy} - e^{-jy}}{2j}$$

The Euler Form of a Complex Number

We can immediately put Euler's formula to work to develop an alternate form for complex numbers. Refer to the polar coordinates for the complex number z shown in Figure 8.10 and repeated here for convenience as Figure 8.12. If we apply Euler's formula to z, we obtain

$$z = x + jy = |z|\cos(\theta) + j|z|\sin(\theta) = |z|[\cos(\theta) + j\sin(\theta)] = |z|e^{j\theta}. \qquad (8.2\text{-}27)$$

Notice that we have applied Euler's formula with the real number θ replacing y in our original discussion of the formula. We will refer to the preceding expression for z in terms of its polar coordinates,

$$z = |z|e^{j\theta}, \qquad (8.2\text{-}28)$$

Figure 8.12   Polar coordinates

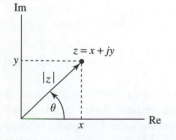

as the *Euler form* for z.[5] The quiz exercises that follow ask you to practice converting back and forth between rectangular and Euler form. For circuit analysis, it is helpful to become very familiar with this process.

---

[5] Many texts use the term *polar form*, but this is ambiguous because ($|z|$, θ)—as a number pair— is often called the polar form or polar representation of z.

The Euler form is quite useful in multiplication and division. For instance, if $z_1 = |z_1|e^{j\theta_1}$ and $z_2 = |z_2|e^{j\theta_2}$ are two complex numbers given in Euler form, then

$$z_1 z_2 = |z_1||z_2|e^{j(\theta_1 + \theta_2)} \qquad (8.2\text{-}29a)$$

and

$$\frac{z_1}{z_2} = \frac{|z_1|\,e^{j\theta_1}}{|z_2|\,e^{j\theta_2}} = \frac{|z_1|}{|z_2|}\,e^{j(\theta_1 - \theta_2)}. \qquad (8.2\text{-}29b)$$

We have assumed here that the usual law of exponents works for the complex exponential, which is true, although we will not prove it.

Section 8.2 Quiz

Q8.2-1. If $z_1 = 1 + j$, $z_2 = -1 + j$, $z_3 = 3 + j4$, $z_4 = 4 + j3$, and $z_5 = (1/2) + j(\sqrt{3}/2)$, compute (using the rectangular form exclusively)

a. $z_1 + z_2$ **b.** $z_1 - z_2$ **c.** $|z_3|$ **d.** $z_3 z_4$
e. $|z_5|$ **f.** z_1/z_2 **g.** $z_1 z_2{}^*$.

Q8.2-2. Convert each of the complex numbers z_1 through z_5 in Problem Q8.2-1 to Euler form.

Q8.2-3. Using the Euler form, find each of the following complex numbers and express your answer in rectangular form.

    a. z_2/z_1 **b.** $z_1 z_2$ **c.** $z_3/z_3{}^*$ **d.** $(1/z_5)$.

8.3 | Complex Signals and Systems

Differentiation and Integration of Complex Signals and Other Real Operators

Let's think a bit about complex signals on the system level. We might ask, for example, what is the meaning of $pz(t)$, where $z(t) = x(t) + jy(t)$ is a complex function of time? It is a matter of *definition*. We define

$$pz(t) = (px(t), py(t)) = px(t) + jpy(t); \qquad (8.3\text{-}1)$$

that is, we define the derivative of a complex waveform to be that complex signal whose real part is the derivative of the real part of the original function and whose imaginary part is the derivative of the original imaginary part. Figure 8.13 illustrates this definition with a system diagram in which $w(t) = u(t) + jv(t) = pz(t)$. Similarly, we define the integral to be

$$\frac{1}{p}z(t) = \left(\frac{1}{p}x(t), \frac{1}{p}y(t)\right) = \frac{1}{p}x(t) + j\,\frac{1}{p}y(t). \qquad (8.3\text{-}2)$$

The system diagram is shown in Figure 8.14, with $w(t) = (u(t), v(t)) = (1/p)\,z(t)$.

Figure 8.13
Differentiation of a complex signal

Figure 8.14
Integration of a complex signal

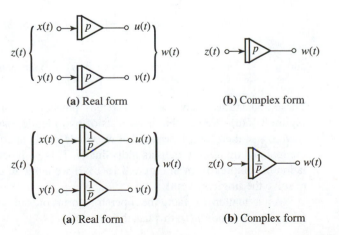

(a) Real form   (b) Complex form

(a) Real form   (b) Complex form

Now let us generalize. If $H(p)$ is a differential operator with real coefficients, one that we will therefore call a *real operator*, we define

$$H(p)z(t) = H(p)(x(t), y(t)) = (H(p)x(t), H(p)y(t)) = H(p)x(t) + jH(p)y(t). \qquad (8.3\text{-}3)$$

This is shown in the system diagram of Figure 8.15, with $w(t) = (u(t), v(t)) = H(p)z(t)$.

Figure 8.15
A general real
differential operation
on a complex signal

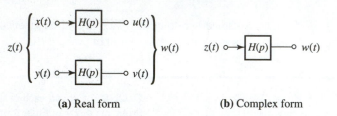

              (a) Real form   **(b)** Complex form

Real Differential Equations with Complex Forcing Functions

We now give an example of a real system with a complex forcing function. (Be careful to note that here we are using $u(t)$ in two senses: as the unit step function and as the real part of w.)

Example 8.3   Solve the differential equation $(dw/dt) + 3w = 6e^{j3t}u(t)$.

Solution   By "real differential equation" we mean one with real coefficients. As the forcing function is complex, we expect this to be the case for the solution also; for this reason, we have used the symbol w. In terms of real functions, $w(t) = u(t) + jv(t)$. In operator notation, using Euler's formula on the complex exponential forcing function, we have $p[u(t) + jv(t)] + 3[u(t) + jv(t)] = 6\cos(3t)u(t) + j6\sin(3t)u(t)$. Applying the definition of equality of two complex numbers, we merely equate the real terms and the "j terms" individually. Thus, we have

$$pu(t) + 3u(t) = 6\cos(3t)u(t) \qquad (8.3\text{-}4a)$$

and $\qquad\qquad\qquad pv(t) + 3v(t) = 6\sin(3t)u(t). \qquad (8.3\text{-}4b)$

The important issue here is that a single first-order differential equation with a complex forcing function is equivalent to two first-order differential equations with real forcing functions—but they are identical except for the forcing functions. If we solve for $pu(t)$ and $pv(t)$, and then multiply both sides of the resulting equations by the integration operator $1/p$, we will have

$$u(t) = \frac{1}{p}[-3u(t) + 6\cos(3t)u(t)] \qquad (8.3\text{-}5a)$$

and $\qquad\qquad\qquad v(t) = \frac{1}{p}[-3v(t) + 6\sin(3t)u(t)]. \qquad (8.3\text{-}5b)$

These equations can be interpreted physically in the simulation diagram of Figure 8.16(a). We see that it consists of two identical systems driven by different input waveforms; therefore, it seems that we would duplicate quite a bit of effort in solving for the real and imaginary outputs individually. This is, indeed, the case. We show the simulation in complex form in Figure 8.16(b). If we agree to use complex math, we need only to solve the single system!

As a matter of fact, the operator form of our original differential equation is $(p + 3)w(t) = 6e^{j3t}u(t)$; this has the solution

Figure 8.16
Simulation diagrams:
Real and complex
implementations

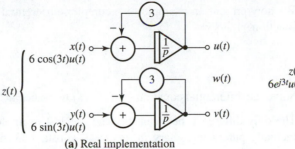

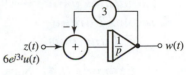

(a) Real implementation

(b) Complex form

$$w(t) = \frac{1}{p+3}[6e^{j3t}u(t)] = e^{-3t}\frac{1}{p}[6e^{3t}e^{j3t}u(t)] \tag{8.3-6}$$

$$= \frac{2}{1+j}e^{j3t}u(t) - \frac{2}{1+j}e^{-3t}u(t)$$

If we want our solution in rectangular form (that is, in the form $w(t) = u(t) + jv(t)$), we can use just a bit of the complex algebra we practiced in Section 8.1 along with Euler's formula to put (8.3-6) in the form

$$w(t) = [\sqrt{2}\cos(3t - \pi/4) - e^{-3t}]\,u(t) + j[\sqrt{2}\sin(3t - \pi/4) + e^{-3t}]u(t). \tag{8.3-7}$$

We will leave the steps in the verification to you; it is good practice for the type of analysis we will be doing in later chapters on ac steady-state methods.

Complex Operators   Example 8.3 was a real system (that is, a real operator—one with real coefficients) with a complex forcing function. It might surprise you to learn that we can also consider physical systems that have complex operators; nevertheless, this is done quite often in circuits and systems work. Such operators express important systems that have physical reality, so they can be considered just as "real" as those represented by real operators. Let's see how we might define a complex operator. We follow the same pattern we used in defining complex multiplication. If $H(p) = (H_r(p), H_i(p))$, where $H_r(p)$ and $H_i(p)$ are real operators, we say that $H(p)$ is a complex operator and define $H_r(p) = \mathrm{Re}\{H(p)\}$ and $H_i(p) = \mathrm{Im}\{H(p)\}$. We then define

$$w(t) = u(t) + jv(t) = H(p)z(t) = (H_r(p), H_i(p))(x(t), y(t)) \tag{8.3-8}$$

$$= [H_r(p) + jH_i(p)][x(t) + jy(t)]$$

$$= [H_r(p)x(t) - H_i(p)y(t)] + j[H_r(p)y(t) + H_i(p)x(t)].$$

This equation leads to the situation shown in Figure 8.17. Notice that it is simply the generalization of a complex scalar multiplier.

Figure 8.17
A complex system

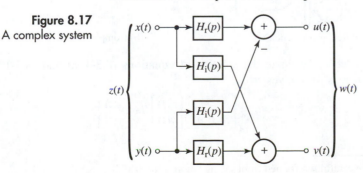

*Complex Differential
Equations with
Complex Forcing
Functions*

We now present an example of a complex differential equation with a complex forcing function.

Example 8.4

Solve the differential equation $(dw/dt) + j3w = 6e^{j3t}u(t)$.

Solution

The only difference between this and our last example is the appearance of a complex (actually purely imaginary in this case) constant $j3$ in the place of the real constant 3. Again letting $w(t) = u(t) + jv(t)$ and $z(t) = x(t) + jy(t) = 6e^{j3t}u(t)$, we have

$$p[u(t) + jv(t)] + j3[u(t) + jv(t)] = x(t) + jy(t), \qquad (8.3\text{-}9)$$

or—equating real and imaginary parts—

$$pu(t) - 3v(t) = x(t) \qquad (8.3\text{-}10a)$$

$$3u(t) + pv(t) = y(t). \qquad (8.3\text{-}10b)$$

Notice that this is quite a bit more complicated[6] than our previous example because the two differential equations are *coupled!* That is, each involves both $u(t)$ and $v(t)$. This means that the two are much more difficult to solve. We will explore this in a bit more detail shortly, but for now we will simulate our differential equation by solving equation (8.3-10a) for $u(t)$ and multiplying the result by $1/p$ from the left—then repeating the process for $v(t)$ in equation (8.3-10b). This results in the equations

$$u(t) = \frac{1}{p}[3v(t) + x(t)] \qquad (8.3\text{-}11a)$$

$$v(t) = \frac{1}{p}[-3u(t) + y(t)]. \qquad (8.3\text{-}11b)$$

Figure 8.18(a) shows a system diagram that realizes these equations and Figure 8.18(b) the corresponding complex form. The key thing to notice is that the imaginary constant $j3$ has *coupled* the real and imaginary "channels." They are no longer independent of one another as in Example 8.3, Figure 8.16.

Figure 8.18
Simulation diagrams:
Real and complex
implementations

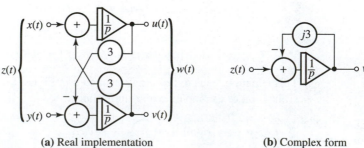

(a) Real implementation   (b) Complex form

Now let's investigate the solution of equations (8.3-10a) and (8.3-10b). Rewriting them in matrix form gives

$$\begin{bmatrix} p & -3 \\ 3 & p \end{bmatrix} \begin{bmatrix} u(t) \\ v(t) \end{bmatrix} = \begin{bmatrix} x(t) \\ y(t) \end{bmatrix}. \qquad (8.3\text{-}12)$$

---

[6] We have carefully refrained from saying "more complex" !

We solve this equation by multiplying both sides (from the left) by the inverse of the 2×2 operator matrix on the left-hand side of the equation, resulting in

$$\begin{bmatrix} u(t) \\ v(t) \end{bmatrix} = \frac{1}{p^2 + 9} \begin{bmatrix} p & +3 \\ -3 & p \end{bmatrix} \begin{bmatrix} x(t) \\ y(t) \end{bmatrix}. \tag{8.3-13}$$

If we solve for $u(t)$ and $v(t)$, we get

$$u(t) = \frac{p}{p^2 + 9} x(t) + \frac{3}{p^2 + 9} y(t) \tag{8.3-14a}$$

and

$$v(t) = \frac{-3}{p^2 + 9} x(t) + \frac{p}{p^2 + 9} y(t). \tag{8.3-14b}$$

We do not as yet know how to interpret the operators on the right-hand side of (8.3-14), though we will develop a complete procedure for their solution in a later chapter. However, we note that if we multiply both sides of each by the operator $p^2 + 9$ and express the result in conventional form, we will have the following two *uncoupled* differential equations:

$$\frac{d^2 u}{dt^2} + 9u = \frac{dx}{dt} + 3y \tag{8.3-15a}$$

and

$$\frac{d^2 v}{dt^2} + 9v = -3x + \frac{dy}{dt}. \tag{8.3-15b}$$

We now observe that even though these two differential equations are uncoupled, they are of *second order*. A first-order differential equation with complex coefficients has resulted in two real ones of second order. Shouldn't it be simpler to solve our original problem in complex form?

The answer is a definite yes! We merely rewrite our differential equation in operator form: $pw + j3w = 6e^{j3t}u(t)$. The solution is

$$w(t) = \frac{1}{p + j3}[6e^{j3t}u(t)] = e^{-j3t}\frac{1}{p}[6e^{j3t}e^{j3t}u(t)] \tag{8.3-16}$$

$$= \frac{1}{j}e^{-j3t}[e^{j6t} - 1]u(t) = \frac{e^{j3t} - e^{-j3t}}{j}u(t) = 2\sin(3t)u(t).$$

This is much simpler. Rather than solving two coupled second-order differential equations, we simply solve one complex differential equation of the first order using our usual methods. What does the solution mean? If you refer back to Figure 8.18(a) and perform the indicated operations, you will see that the imaginary channel output $v(t)$ is identically zero and the real output $u(t)$ is $2\sin(3t)u(t)$.

Stability of Complex Operators

Our final topic in this section will be the important one of stability. In Chapter 7 we defined stability and found conditions under which a given first-order (real) operator is stable. Here, we will extend these ideas to complex systems. Suppose a given system is represented by the general first-order operator shown in Figure 8.19; we will suppose here that any or all of the constant parameters are complex. We inquire as to the conditions for stable, unstable, and marginally stable operation. We first simplify the operator by one step of long division:

Figure 8.19   A complex first-order system

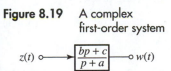

$z(t) \circ\!\!\longrightarrow\!\!\boxed{\dfrac{bp + c}{p + a}}\!\!\longrightarrow\!\circ\, w(t)$

$$H(p) = \frac{bp + c}{p + a} = k_0 + \frac{k_1}{p + a}, \tag{8.3-17}$$

where $k_0 = b$ and $k_1 = c - ab$. The impulse response is

$$h(t) = H(p)\delta(t) = k_0\delta(t) + \frac{k_1}{p + a}\delta(t) = k_0\delta(t) + k_1e^{-at}u(t). \quad (8.3\text{-}18)$$

Stability only concerns the behavior of the system for large values of time; thus, we ignore the $k_0\delta(t)$ part of the response and concentrate on the exponential term. Because a is complex, we write it in the rectangular form

$$a = a_r + ja_i, \quad (8.3\text{-}19)$$

where a_r and a_i are real numbers (the real and imaginary parts, respectively, of a). Thus, we have

$$e^{-at}u(t) = e^{-(a_r+ja_i)t}u(t) = e^{-a_rt}[\cos(a_it) - j\sin(a_it)]u(t). \quad (8.3\text{-}20)$$

We have used Euler's formula to obtain this last result. Now we recall that

$$|\cos(a_it) + j\sin(a_it)| = \sqrt{\cos^2(a_it) + \sin^2(a_it)} = 1. \quad (8.3\text{-}21)$$

Thus, we have

$$|e^{-at}| = |e^{-(a_r+ja_i)t}| = e^{-a_rt}|\cos(a_it) - j\sin(a_it)| = e^{-a_rt}. \quad (8.3\text{-}22)$$

Thus, we finally see that if $a_r > 0$, the impulse response approaches zero for large values of time and the operator is, therefore, stable; if $a_r < 0$, the impulse response becomes unbounded as $t \to \infty$ and the response is unstable; and if $a_r = 0$, the response is bounded for large values of time but does not approach zero, thus implying that the system is marginally stable. Note that for the last case, the response does not approach a constant, as in the case of a real first-order system, but in fact it is *oscillatory*, being composed of sinusoids. The magnitude, however, does approach a constant as time becomes infinitely large. Notice that it is the *real part* of the constant a that determines stability. The imaginary part has no effect upon this aspect of system behavior. It merely determines the frequency of the sinusoidal components of the response.

Section 8.3 Quiz

Q8.3-1. Check the stability of the differential equation $(dw/dt) + 4w = z(t)$.

Q8.3-2. Check the stability of the differential equation $(dw/dt) - jw = z(t)$.

Q8.3-3. Consider the differential equation $j3(dw/dt) + 6w = 3z(t)$. Check its stability, draw a real implementation of its simulation diagram, and find the response for $z(t) = 4\,e^{j2t}u(t)$.

8.4 | Sinusoids, Complex Exponentials, and Phasors

The Fundamental Property

We will be working quite often in the rest of the text with sinusoids. Unfortunately, they are somewhat involved to deal with, mainly because of the following property:

$$p\cos(\omega t) = -\omega\sin(\omega t) \quad (8.4\text{-}1)$$

$$p^2\cos(\omega t) = -\omega^2\cos(\omega t).$$

This means that it takes two derivatives to reproduce a sinusoid of the same type. We note, however, that

$$pe^{j\omega t} = j\omega e^{j\omega t}. \quad (8.4\text{-}2)$$

The exponential function requires only one derivative to reproduce itself. As we will see, this has a remarkable effect upon the amount of work one must do in solving many circuit problems.

The Complex Extension of a Sinusoid

The preceding fact leads to the following speculation. Could we not represent a sinusoid, say a cosine, as an exponential? The answer is, indeed, yes. We use the Euler form with $\theta = \omega t$ and write

$$e^{j\omega t} = \cos(\omega t) + j\sin(\omega t). \tag{8.4-3}$$

We merely add an imaginary component $\sin(\omega t)$, thus rendering the signal complex, and thereby obtain an exponential with a purely imaginary exponent. Going the other way, we can extract the original sinusoid by means of the equation

$$\cos(\omega t) = \text{Re}\{e^{j\omega t}\}. \tag{8.4-4}$$

In physical terms, this just means that we drop the imaginary part. We could, of course, do the same thing with the sine, but it has become conventional in circuit analysis to work with cosine functions, rather than with sines. The physical interpretation of this process of forming the *complex extension* of the cosine waveform in terms of a voltage source is shown in Figure 8.20. This is primarily a conceptual process, and for practical work we would simply redraw the original voltage source and label it with a complex value.

Figure 8.20
The complex extension of a sinusoid

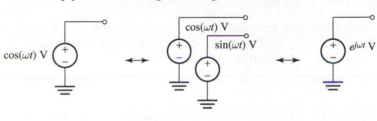

Not all sine waves have unit amplitude and zero phase. In order to incorporate more general sinusoids into our methods, we will use the complex scalar multiplier shown in Figure 8.21. Such complex multipliers are so important that we will give them a formal definition.

Figure 8.21
A phasor

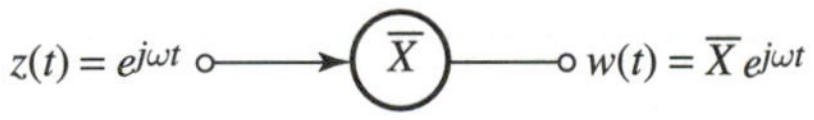

$$z(t) = e^{j\omega t} \circ\!\longrightarrow\!\boxed{\overline{X}}\!\longrightarrow\!\circ\ w(t) = \overline{X}\,e^{j\omega t}$$

Definition 8.3   If $\overline{X}\,e^{j\omega t}$ is any complex exponential with purely imaginary exponent, we will define the complex multiplier $\overline{X}$ to be a *phasor*.

Now suppose we start with a real sinusoidal signal with a general amplitude X_m and a general phase ϕ, that is, $x(t) = X_m \cos(\omega t + \phi)$. We perform the complex extension by adding an imaginary component $y(t) = X_m \sin(\omega t + \phi)$, with the same amplitude and the same phase. This gives the complex exponential

$$w(t) = X_m \cos(\omega t + \phi) + jX_m \sin(\omega t + \phi) = X_m\, e^{j(\omega t + \phi)} = \overline{X}e^{j\omega t} \tag{8.4-5}$$

where

$$\overline{X} = X_m\, e^{j\phi} \tag{8.4-6}$$

is the phasor (expressed in Euler form) *associated with* $x(t) = X_m \cos(\omega t + \phi)$. Expressed in rectangular form, it is

$$\overline{X} = X_m \cos(\phi) + jX_m \sin(\phi). \tag{8.4-7}$$

In deriving this, we have simply applied Euler's formula to $e^{j\phi}$.

The Phasor as a Rotating Vector

A picture of $w(t)$ in the complex plane is shown in Figure 8.22(a) and one of the phasor itself in Figure 8.22(b). Observe that $\overline{X}$ is the value of $w(t)$ at $t = 0$: $w(0)$. We see that the original cosine function is the horizontal projection of the rotating vector $w(t)$. (Note that the angle $\omega t + \phi$ increases linearly with time.) Now let us make an important observation: if the frequency of the sinusoid is known, then *the phasor completely determines the sinusoid, and vice versa*. In fact, the conversion from one to the other is a very simple process, as we will now illustrate with some examples.

Figure 8.22
The complex extension and the associated phasor

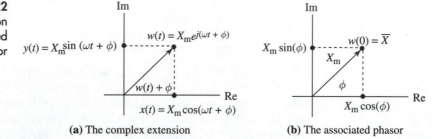

(a) The complex extension   **(b)** The associated phasor

Example 8.5   Let $\overline{V} = 1 + j$ be the phasor associated with a given sinusoidal voltage. Assuming that the frequency of this sinusoid is $\omega = 1000$ rad/sec, find the mathematical representation for the sinusoid as a function of time.

Solution   We first place the phasor in Euler form: $\overline{V} = 1 + j = \sqrt{2}\, e^{j\pi/4}$. (A simple sketch will help in verifying the angle.) We then immediately write down the sinusoid: $v(t) = \sqrt{2}\, \cos(1000t + \pi/4)$ V.[7]

Example 8.6   If $i(t) = 10\cos(50t + 30°)$ is a sinusoidal current waveform, find its associated phasor in rectangular form.

Solution   We merely pick off the magnitude and the angle of the phasor from the given sinusoid; then the phasor in Euler form is $\overline{I} = 10e^{j30°}$.[8] In rectangular form, using Euler's formula, we have $\overline{I} = 10\cos(30°) + j10\sin(30°) = 5\sqrt{3} + j5$.

Example 8.7   If $\overline{X} = 4 - j3$ is the phasor associated with a sinusoid with a radian frequency of $\omega = 10^6$ rad/sec, find $x(t)$.

Solution   We must first put the given phasor into Euler form. This results in $\overline{X} = 5e^{-j36.9°}$. Thus, $x(t) = 5\cos(10^6 t - 36.9°)$.

Example 8.8   If $v(t) = 20\sin(1000t)$, find its phasor in rectangular form.

Solution   There is a bit of a tricky situation here! We have expressed $v(t)$ as a *sine*, rather than as a *cosine*. Therefore, we must first convert it to the latter form: $v(t) = 20\cos(1000t - 90°)$. Then we have $\overline{V} = 20e^{-j90°} = -j20$. To verify this last equality, just use Euler's formula and note that $\cos(-90°) = 0$ and that $\sin(-90°) = -1$.

---

[7] *Conceptually,* we have $v(t) = \mathrm{Re}\{\overline{V}e^{j\omega t}\} = \mathrm{Re}\{V_m e^{j\phi}\, e^{j\omega t}\} = \mathrm{Re}\{V_m e^{j(\omega t + \phi)}\} = V_m \cos(\omega t + \phi)$.

[8] We will express the polar angle either in radians or in degrees. If we were very careful, we would note that the entire argument of the cosine function is $\omega t + \phi$—so ϕ should, strictly speaking, be in radians.

Section 8.4 Quiz

Q8.4-1.   Sketch the following phasors in the complex plane and convert them to Euler form: $1 + j$, $1 - j$, $-1 + j$, and $-1 - j$.

Q8.4-2.   Repeat Question Q8.4-1 for: $3 + j4$ and $4 + j3$.

Q8.4-3.   Repeat Question Q8.4-1 for: $0.5 + j0.5\sqrt{3}$ and $0.5\sqrt{3} + j0.5$.

Q8.4-4.   If $v(t) = 3\cos(100t) - 4\sin(100t)$, find the phasor of each component sinusoid and add to find the phasor for $v(t)$ in polar form. Using this result, write an expression for $v(t)$ as a single sinusoid.

8.5 | SPICE: Complex Signals and Circuits—An Op Amp Example

We will use SPICE in this section to underscore the point that complex signals and circuits can certainly be regarded as real physical systems. We will specifically investigate the circuit shown in Figure 8.23. This circuit has real components, and we could analyze it (using techniques that will be developed in Chapter 9 for higher-order circuits) as a second-order system. Here, however, we wish to show that it can be considered as a complex first-order active circuit. Thus, we use the notation $x(t)$ for the real part of a complex waveform $z(t)$ and $y(t)$ for its imaginary part:

$$z(t) = x(t) + jy(t) = (x(t), y(t)). \tag{8.5-1}$$

What we will do is compute the response $z(t)$ for all $t > 0$ under the assumption that $x(0) = 1$ V and $y(0) = 0$ V. This is the same as requiring that $z(0) = 1 + j0 = (1, 0)$ V.

Figure 8.23
An example complex
circuit

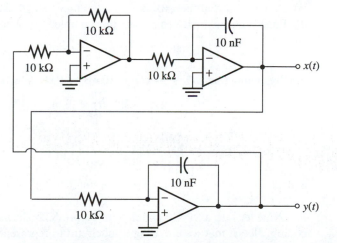

If we consider the op amp inverter-integrator at the top to represent the "real channel" of our circuit and the op amp integrator at the bottom to be the "imaginary channel" and notice that all resistors have a common value $R = 10\ k\Omega$ and the capacitors a common value $C = 10$ nF, we can readily derive the relationships

$$RCpx(t) = y(t) \tag{8.5-2}$$

and

$$RCpy(t) = -x(t). \tag{8.5-3}$$

The initial condition, of course, is $x(0) = 1$ V and $y(0) = 0$ V. We can, therefore, draw the simulation diagram of Figure 8.24 (where we have used the step function form for establishment of the initial condition on the "real" integrator, so that $u(t)$ refers to the unit step function).

Figure 8.24
Simulation diagram

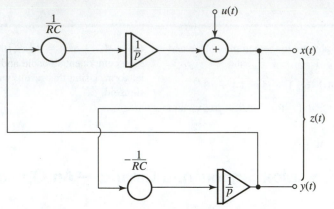

By multiplying equation (8.5-3) by j and adding it to (8.5-2), we obtain

$$RCp[x(t) + jy(t)] = -j[x(t) + jy(t)], \qquad (8.5\text{-}4)$$

or

$$pz(t) + \frac{z(t)}{-jRC} = 0. \qquad (8.5\text{-}5)$$

Equation (8.5-5) is nothing more than a first-order differential equation in the complex variable $z(t)$ with *an imaginary time constant,* $\tau = -jRC$ and initial condition $z(0) = 1 + j0$. Therefore, it has the solution

$$z(t) = z(0)e^{-t/\tau} = e^{-jt/RC} \text{ V}; \quad t > 0. \qquad (8.5\text{-}6)$$

We have recognized that $z(0)$ is the real number 1 and that $1/\tau = 1/(-jRC) = j1/RC$. Using Euler's formula, we can rewrite this complex exponential in the form

$$z(t) = \cos(t/RC) - j\sin(t/RC) \text{ V}; \quad t > 0. \qquad (8.5\text{-}7)$$

Finally, noting that $RC = 10 \text{ k}\Omega \times 10 \text{ nF} = 100 \text{ }\mu\text{s} = 0.1 \text{ ms}$, we can write it as

$$z(t) = \cos(10^4 t) - j\sin(10^4 t) \text{ V}; \quad t > 0. \qquad (8.5\text{-}8)$$

The real and imaginary parts are

$$x(t) = \cos(10^4 t) \text{ V}; \quad t > 0 \qquad (8.5\text{-}9)$$

and

$$y(t) = -\sin(10^4 t) \text{ V}; \quad t > 0. \qquad (8.5\text{-}10)$$

Now let's do a SPICE simulation. Figure 8.25 shows our circuit prepared for SPICE coding. Notice that we are using a subcircuit to represent the op amps, which we assume are ideal. The SPICE file itself is as follows:

```
SECTION 8.5 COMPLEX SPICE EXAMPLE 1
R1    1      2     10K
R2    2      3     10K
R3    3      4     10K
R4    5      6     10K

C1    5      4     10N    IC=1V    ;IC IS STEP INITIAL CONDITION
C2    6      1     10N             ;N IS NANOFARADS
X1    0      2     3      IOA      ;IDEAL OP AMP SUBCIRCUIT
X2    0      4     5      IOA      ;IDEAL OP AMP SUBCIRCUIT
X3    0      6     1      IOA      ;IDEAL OP AMP SUBCIRCUIT
```

```
.SUBCKT    IOA      PLUS      MINUS     OUT   ;DEFINES SUBCIRCUIT
RHI  PLUS        MINUS     1E9
EOUT OUT         0        PLUS      MINUS     1E9
.ENDS

.TRAN        0.05E-3    1.25E-3   0        0.05E-3   UIC   ;APP. 2 PDS
.PROBE
.END
```

The results are plotted using PROBE in Figure 8.26, verifying our equations (8.5-9) and (8.5-10).

Figure 8.25
The example circuit prepared for SPICE coding

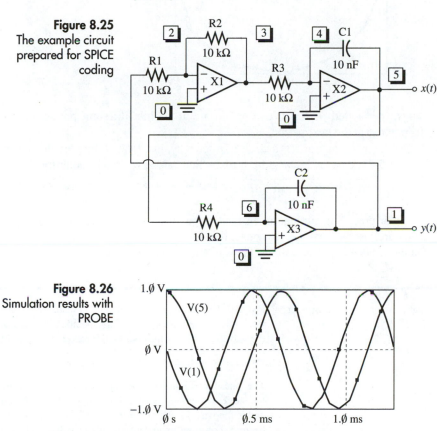

Figure 8.26
Simulation results with PROBE

Although we have not—and will not—discuss PROBE in a lot of detail, we would like to point out that it has provisions for displaying data in a number of formats. This allows us to plot a voltage or current variable along the horizontal axis instead of time. If we elect to plot V(5) (that is, $x(t)$) horizontally and V(1) (that is, $y(t)$) vertically, we obtain a plot of the complex waveform $z(t) = x(t) + jy(t)$ with t as a parameter. That is, for each time point PROBE plots one point in the z plane. This is repeated for each time point in the simulation. The result is shown in Figure 8.27. It is a polygonal approximation to a circle, rather than a perfect circle because PROBE draws a straight line approximation to the curve between two simulation time points. (Refer back to Figure 8.26.) Using more points results in a better circle. Furthermore, your curve might look more like an ellipse (on some computers) unless you make certain that the vertical and horizontal scales are exactly the same. We ran our SPICE simulation and PROBE plot on a Macintosh IIci, so we had only to resize the graph window using a mouse to make the circle symmetric. Notice that several periods are plotted and the traces do not exactly overlap. The reason for this is that our simulation interval did not divide the period an integral number of times.

Figure 8.27
A plot of $z(t)$ in the
complex z plane

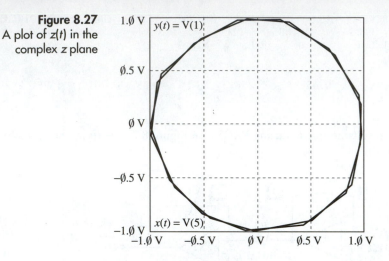

Thus, the plotted points in one period are at slightly different points on the waveform from those in the next period.

Why is our plot of $z(t)$ a circle? To see why, just look at the polar form for $z(t)$. Equation (8.5-7) gives the rectangular form, which can be written in polar form as

$$z(t) = \sqrt{\cos^2(t/RC) + \sin^2(t/RC)}e^{-j\tan^{-1}[\sin(t/RC)/\cos(t/RC)]} = e^{-jt/RC} \text{ V}; \quad t > 0. \quad (8.5\text{-}11)$$

Our preceding simulation will, we hope, convince you that complex signals and circuits are just as "real" as real ones; that is, there is a physical interpretation of complex quantities that is just as valid as the way we interpret real ones.

Section 8.5 Quiz

Q8.5-1. Derive the differential equations for $x(t)$ and $y(t)$ in Figure Q8.5-1.

Figure Q8.5-1

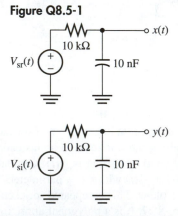

Q8.5-2. Draw a simulation diagram for $z(t) = x(t) + jy(t)$ using the results of Question Q8.5-1.

Q8.5-3. Change the differential equations you found in Question Q8.5-1 into a single one with a complex dependent variable $z(t) = x(t) + jy(t)$ and solve it. Find the real and imaginary part waveforms $x(t)$ and $y(t)$. Assume that $v_{sr}(t) = 4\cos(10^4 t)u(t)$ V and that $v_{si}(t) = -4\sin(10^4 t)u(t)$ V.

Q8.5-4. Write a SPICE file and simulate the circuit in Figure Q8.5-1. Plot your results for $z(t)$ in the complex z plane using

PROBE. Use the general "damped" sinusoid waveform specification shown in Figure Q8.5-2. The specification of this function as the waveform of a voltage source, for example, would be

```
VS   N+   N−   SIN(VO, VA, F, TD, A, PHASE).
```

The phase in Figure Q8.5-2 is zero merely for drawing convenience. For a pure sinusoid, the damping factor A would be zero, and if the waveform starts at $t = 0$, the delay time TD would also be zero. For a pure sinusoid, the offset voltage VO would be zero. Be careful to notice that this is a sine function—not a cosine; thus, a cosine with zero phase would have PHASE = 90 (the PHASE parameter is in degrees). The parameter F is the (Hertz) frequency, that is, $F = 1/T = \omega/2\pi$. TF is the "final time" of the simulation.

Figure Q8.5-2

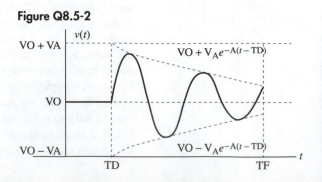

8.6 | Real and Complex Numbers and Reality

This section has no equations, examples, exercises, or problems; yet, in the author's opinion it covers ideas that—when made a part of an engineer's outlook on his or her discipline—will prove to be quite valuable.

In the early days of science fiction, back when the genre was written in the "pulps" (magazines with rough paper), the typical story had a bug-eyed monster, a space ship or two, a ray gun, and a strong-jawed male hero who arrived just in time to save a sexy maiden from the clutches of the bug-eyed monster. And often there was an additional character: a mathematical whiz kid who sat down with pencil and paper and played around with some equations until he found a new physical phenomenon, a way of penetrating the advanced electronic defense systems of the bug-eyed monster's space ship.

Science fiction has come a long way since those days, and is now considered to be a worthwhile genre for books and films that are considered to be more literary. Yet there is a fundamental idea—a misconception—in the above scenario that is still widespread, not only in science fiction, but even in technical books on such topics as circuit theory. It is the confusion of mathematical manipulations with reality. One cannot control reality by the manipulation of equations, nor do the equations tell us any new truth about reality. Mathematics starts with a set of fundamental axioms, assumptions about the objects or entities it deals with, and derives results from these assumptions based upon the formal rules of logic. If we assume that axioms A and B are valid, then proposition C about the fundamental objects follows inescapably. *But mathematics never asks questions about the basic truth of axioms A and B!* Their truth or falsity can only be checked by running laboratory experiments and checking the results against what we have predicted by assuming their truth. If our experimental results check, fine; otherwise, we must modify our assumptions and do our analysis all over from scratch.

In fact, this is proper engineering practice. One uses very simple models to analyze a proposed circuit to predict its behavior using simple element models, then checks on their accuracy with an experiment.[9] If the results do not agree closely enough, one brings out the pencil and paper anew and analyzes the circuit once again with more complicated models for the elements and then performs the experiment yet again. This process continues until the results agree to the accuracy desired. Thus, the mathematics of circuit analysis only permits one to manipulate the basic assumptions or axioms (in the present case, the axioms are the element models and the axioms of circuit theory) and infer what the results should be—*provided the assumptions are correct.*

How does all of this fit in with our chapter on complex numbers and signals? The reason we have included it can be found in many introductory texts on circuit analysis. Many insert a disclaimer, something to the effect that "Complex signals are never found in real physical circuits or systems; thus, we are merely using complex mathematics to simplify the analysis of such systems." This statement is categorically untrue! In fact, the entire question of whether complex signals exist in physical circuits and systems is not properly posed. It is the circuit or the signal that has physical reality—not the mathematics.[10]

Let's look at an example. When was the last time you analyzed a physical system and used the actual value of the real number $\sqrt{2}$? Never! The value of $\sqrt{2}$ as a decimal number

---

[9] Computer simulation is rapidly taking the place of physical laboratory experiments because the mathematical models used by the computer often are a more accurate reflection of the elements on a microscopic integrated circuit chip than are the discrete components used to construct larger-scale laboratory models.

[10] The author is grateful to Professor Gary Ford at the University of California at Davis for bringing this important fact to his attention.

can only be expressed with an infinite number of digits, as is true of all irrational numbers. Thus, we use the symbol for $\sqrt{2}$ but we never compute with it as an irrational number; when we must actually use a value, we truncate it to a finite number of digits. This, however, converts it to an approximate rational number. So, we ask: Do irrational numbers have physical reality? Who cares? If they simplify the mathematics (and they do), we will simply use them as a tool.

It is the same with complex numbers. If it suits our needs to conceive a signal on two wires as a complex time function and to agree to certain ways of adding it with another, multiplying it times another, and so on, we will do so. As it happens, there are many physical systems that contain such signals and for which it is more convenient to use complex, rather than real, mathematics. The field of digital signal processing, that is, the use of computer hardware to manipulate signals, has recognized the value of complex hardware for quite some time. Strangely, the field of analog signal processing (of which circuit analysis is a fundamental topic, along with analog electronics) has only recently begun to incorporate such an idea—at least on a hardware basis—in the design of electronic filters.

In summary, one should always remain aware that mathematics is a tool; as such, it possesses no physical reality in and of itself. To think of complex signals and systems as having actual physical reality makes as much sense as insisting that all signals and systems be real. Although real numbers have become rather ingrained (because they have been in use in commerce since Phoenician cargo ships plied the waters of the Mediterranean), it is now quite convenient to assume the physical reality of complex entities— and we will do so with impunity! That said, however, we should also point out that the *main thrust of this text is the analysis of real circuits; thus, we will primarily use the idea of complex signals as a tool for simplifying the analysis of real circuits.*

Chapter 8 Summary

This chapter has presented the basic concepts of *complex numbers,* their meaning, and their manipulation. Though this material permeates almost every analytically oriented subject an electrical engineering student takes, it is often presented somewhat in the form of a *fait accompli.* The idea of the *j operator* is merely presented as a property: $j^2 = -1$. (Mathematics texts use *i* rather than *j*.) We, however, have presented the idea of a complex number as a *two-wire* (or *two-channel*) signal—one that can be constructed and measured in the laboratory. This provides a physical feel for the subject that is lacking in the more conventional approach.

That said, we stress that our rationale has been to provide the mathematical apparatus for analyzing *real circuits and systems.* We have shown in this chapter that *complex circuits and systems* do exist (at least, in the sense that they have as much physical reality as "real" circuits and systems), but it is to the problem of analyzing circuits of the real variety that the mathematics of complex numbers will be applied.

After defining the basic operations on complex numbers as number pairs, we defined the unit *j* as a number pair, $j = (0, 1)$, and showed that all complex numbers of the form $(x, 0)$ (called *purely real* numbers) have the property that adding two of them or multiplying two of them results in another purely real number. This means that they have all the same properties as real numbers of the more ordinary variety. For this reason, we agreed to merely drop the zero imaginary part and write $(x, 0)$ as x. Furthermore, we showed that multiplication by *j* turns a purely real number into one that is purely imaginary: $jy = (0, 1)(y, 0) = (0, y)$. This allowed us to write any complex number in *rectangular form: $z = (x, y) = x + jy$.*

We next talked about the general problem of extending all the usual panoply of real functions of a real variable into the complex domain, and we mentioned that the *complex exponential* is defined as $e^z = e^x \cos(y) + je^x \sin(y)$ for a number of valid reasons, one of which is that it reduces to e^x when $y = \text{Im}\{z\} = 0$. When one evaluates it for purely imaginary values ($z = jy$), however, one obtains $e^{jy} = \cos(y) + j\sin(y)$. This formula, as well as those closely related to it, are quite fundamental for the analysis of circuits and systems for their sinusoidal forced response. We used it to define the *Euler form* of a complex number by $z = |z|e^{j\theta}$. We agreed to the conventional shorthand notation $z = |z|\angle\theta$, where $\angle\theta$ means $e^{j\theta}$. (This is often referred to in other texts as the *polar form,* but this term is also used on occasion to mean the ordered pair $(|z|, \theta)$.)

We next showed that one can consider the cosine function to be the real part of the complex exponential $e^{j\omega t}$, with the physical interpretation being a two-wire signal, one channel of which has the original waveform $\cos(\omega t)$ and the other of which has the waveform $\sin(\omega t)$. We then considered signals of the form $V_m \cos(\omega t + \theta)$ and showed that if one "adds another wire"—an "imaginary channel"—carrying $V_m \sin(\omega t + \theta)$, the complex equivalent carries the waveform $\overline{V}e^{j\omega t}$, where $\overline{V} = V_m e^{j\theta}$. Notice that this is a complex constant in Euler form, which can be converted to rectangular form by the equation $\overline{V} = V_m \cos(\theta) + jV_m \sin(\theta)$. This complex constant is called a phasor. Its physical meaning is merely that of a complex multiplier that multiplies $e^{j\omega t}$ by the complex constant $\overline{V}$ to change it to the more general form just cited.

We discussed *real and complex operators* (including the real operations of differentiation and integration), and showed how they can be simulated by means of system diagrams. One of the most fundamental results, one that you will use in many courses and in many places later in this text, is that of stability. The cornerstone of stability is composed of the properties of the single complex system operator $H(p) = 1/(p + a)$. We thoroughly investigated the stability of this operator for all complex values of the parameter a.

Finally, we gave a brief example in SPICE to show how the idea of a complex circuit can be made physically meaningful.

Chapter 8 Review Quiz

RQ8.1. If $z_1 = (2, 3)$ and $z_2 = (4, 5)$, compute $z_1 + z_2, z_1 z_2$, and $z_1 z_2{}^*$ using number pairs.

RQ8.2. Make a sketch of each of the complex numbers (z_1, z_2 and the results of the indicated operations) in Question RQ8.1.

RQ8.3. Write each of the complex numbers in Question RQ8.1 in rectangular form $z = x + jy$.

RQ8.4. Find the polar coordinates $(|z|, \theta)$ for the following complex numbers: $z_1 = -3 - j4, z_2 = -4 + j3$, and $z_3 = 1 - j$.

RQ8.5. Write each of the complex numbers in Question RQ8.4 in Euler form, $z = |z|e^{j\theta} = |z|\angle\theta$.

RQ8.6. Solve the equation $z^2 = -1 - j$ for z in rectangular and Euler forms.

RQ8.7. Solve the differential equation $pz + jz = x(t)$ for z assuming that $x(t) = u(t)$; that is, find the unit step response of the differential equation. Express your answer in both rectangular and Euler form.

RQ8.8. Draw a system simulation diagram for the differential equation in Question RQ8.7.

RQ8.9. Determine the stability type for the differential equation in Question RQ8.7.

Chapter 8 Problems

Section 8.1 Complex Numbers and Signals

8.1-1. Solve the equation $z^2 = -1$ using $z = (x, y)$ and the basic definitions.

8.1-2. Repeat Problem 8.1-1 for the equation $z^2 = -j$.

8.1-3. Repeat Problem 8.1-1 for the equation $z^2 + z + 1 = 0$. (*Hint:* Note that $y \neq 0$ because the equation $x^2 + x + 1 \neq 0$ for any real number x.)

8.1-4. Let $z_1 = (-1, -1)$, $z_2 = (1, 1)$, $z_3 = (3, -4)$, and $z_4 = (-\sqrt{3}, 1)$. Perform the indicated operations and sketch the answer in the complex plane. Use *number pairs*.

   a. $z_1 + z_2$
   b. $z_1 z_2$
   c. $z_3 + z_3{}^*$
   d. $z_1 [z_2 + z_3] z_4{}^*$.

Section 8.2 Rectangular and Polar Forms

8.2-1. Express each of the complex numbers in Problem 8.1-4 in the form $z = x + jy$ and repeat Problem 8.1-4.

8.2-2. Find the Euler form for the complex numbers z_1, z_2, z_3, and z_4 in Problem 8.1-4.

8.2-3. Using the Euler form for each of the complex numbers z_1, z_2, z_3, and z_4 in Problem 8.1-4, find the following quantities and express your answer in rectangular form: $z = x + jy$.

   a. $z_1 z_3$
   b. z_1/z_2
   c. $z_3 z_3{}^*$
   d. z_3/z_4
   e. $z_3 z_4/z_1 z_2{}^*$.

8.2-4. Using the Euler form, find the solution to $z^3 = -1$. Note that the polar angles of two equal complex numbers can differ by an integral multiple of 2π radians.

8.2-5. Repeat Problem 8.2-4 for $z^5 = -32$.

Section 8.3 Complex Signals and Systems

8.3-1. Solve the differential equation $(dw/dt) + 4w = z(t)$ for $z(t) = 4e^{(-4+j4)t}u(t)$, expressing your answer in rectangular form: $w(t) = u(t) + jv(t)$. Draw the simulation diagram in real form.

8.3-2. Investigate the stability of the differential equation $(dw/dt) + (a + j3)w = (dz/dt)$ for various values of the real constant a; that is, find the range of values of a such that the differential equation is stable, unstable, and marginally stable. Draw a simulation diagram using no differentiators.

Section 8.4 Sinusoids, Complex Exponentials, and Phasors

8.4-1. Assume that $\overline{V} = 4 - j3$ is the phasor associated with a real sinusoid of frequency 10 rad/s. What is the waveform $v(t)$?

8.4-2. Assume that $\overline{V} = 3 + j4$ is the phasor associated with a real sinusoid of frequency 10 rad/s. What is the waveform $v(t)$?

8.4-3. Let $i(t) = 20 \cos(4t - 225°)$ A. What is the (rectangular form) phasor associated with this waveform?

8.4-4. Let $i(t) = 20 \cos(4t + 135°)$ A. What is the (rectangular form) phasor associated with this waveform? How is it related to the phasor in Problem 8.4-3?

8.4-5. Figure P8.4-1(a) shows a closed surface S penetrated by four conductors carrying the sinusoidal waveforms shown. Figure P8.4-1(b) shows the same surface with the currents represented by

Figure P8.4-1

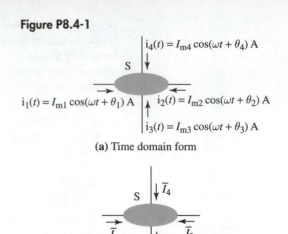

(a) Time domain form

(b) Phasor domain form

their phasors. Show that KCL, $i_1(t) + i_2(t) + i_3(t) + i_4(t) = 0$, in the time domain implies that KCL is also valid in phasor form: $\overline{I}_1 + \overline{I}_2 + \overline{I}_3 + \overline{I}_4 = \overline{0}$, where $\overline{I}_1 = I_{m1}e^{j\theta_1}$, $\overline{I}_2 = I_{m2}e^{j\theta_2}$, $\overline{I}_3 = I_{m3}e^{j\theta_3}$, $\overline{I}_4 = I_{m4}e^{j\theta_4}$, and $\overline{0} = 0 + j0$ is the complex number (phasor) zero. (*Hint:* Note first that the sum of the cosine waveforms is zero for all values of t. This means that it is also zero if we shift time to $t - \pi/(2\omega)$, which changes each of the cosines to a sine. Then add the sum of the cosine waveforms to j times the sum of the sine waveforms and express each term as a complex number in Euler form—then cancel the common $e^{j\omega t}$ factor. This gives the desired phasor equation.)

8.4-6. Figure P8.4-2(a) shows two paths between the same pair of nodes with elements carrying the sinusoidal waveforms shown. Figure P8.4-2(b) shows the same pair of paths and the same elements with the element voltages represented by their phasors. Show that KVL, $v_1(t) + v_2(t) = v_3(t) + v_4(t) = 0$, in the time domain implies that KVL is also valid in phasor form: $\overline{V}_1 + \overline{V}_2 = \overline{V}_3 + \overline{V}_4$, if $\overline{V}_1 = V_{m1}e^{j\theta_1}$, $\overline{V}_2 = V_{m2}e^{j\theta_2}$, $\overline{V}_3 = V_{m3}e^{j\theta_3}$, $\overline{V}_4 = V_{m4}e^{j\theta_4}$. (*Hint:* Use the same procedure outlined in the hint for Problem 8.4-5.)

Figure P8.4-2a

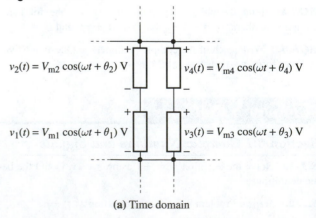

(a) Time domain

Figure P8.4-2b

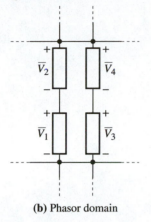

(b) Phasor domain

Section 8.5 SPICE: Complex Signals and Circuits

8.5-1. Figure P8.5-1(a) shows two real circuits coupled by two CCVSs. By deriving two differential equations, one for $i_r(t)$ and one for $i_i(t)$, show that these two coupled circuits are equivalent to the single complex circuit shown in Figure P8.5-1(b); thus, show that the two dependent sources are equivalent to an imaginary resistor with a value of $j4\Omega$.

8.5-2. Find the impulse response of the complex circuit in Figure P8.5-1(b); that is, assume that $v_r(t) = \delta(t)$ and $v_i(t) = 0$ (equivalent to making $\tilde{v}_s(t) = \delta(t) + j0 = \delta(t)$), and find $i(t)$. Use your solution to find $i_r(t)$ and $i_i(t)$.

8.5-3. Assume that $v_{sr}(t) = 15\cos(4t)u(t)$ V and $v_{si}(t) = -15\sin(4t)u(t)$ V; that is, $v_s(t) = 15e^{-j4t}u(t)$ V in Figure P8.5-1, and find $\tilde{i}(t)$. Use your result to find $i_r(t)$ and $i_i(t)$.

8.5-4. Write a SPICE file and simulate the circuit in Figure P8.5-1 assuming that $v_{sr}(t)$ and $v_{si}(t)$ have the waveforms given in Question 8.5-3. Use the SIN function specification given in the review quiz at the end of Section 8.5 to generate the voltage source waveforms. Use PROBE to plot $v_r(t)$ and $v_s(t)$ versus time in your circuit simulation, then use it to produce a complex plane plot of $\tilde{i}(t)$ with $i_r(t)$ plotted on the horizontal axis and $i_i(t)$ on the vertical axis and t as a parameter.

Figure P8.5-1

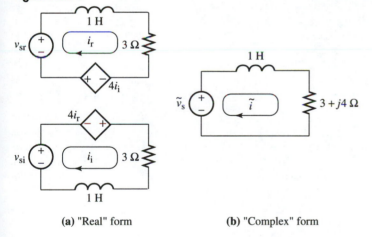

(a) "Real" form   **(b) "Complex" form**

9

Time Response of Higher-Order Circuits

This chapter develops analysis procedures for higher-order circuits and systems. It starts by developing the solution of higher-order differential equations based upon the first-order lowpass operator. It then considers the special properties of second-order circuits and systems. In particular, the concept of damping is developed and illustrated by analogy with a mechanical system: the automobile shock absorber system. Next, a very powerful tool—the partial fraction expansion (PFE)—is developed. It is explained in several steps: first for operators whose characteristic polynomial has simple roots, then for those having multiple roots. The special case of complex conjugate roots is treated in detail. Finally, a method is developed for analyzing essentially any circuit using strictly algebraic methods based upon the partial fraction expansion. For readers who have previously come into contact with the Laplace transform, we point out that the methods developed in this chapter are (mechanically) identical to those commonly associated with that technique—although a transform is never taken. The final section of the chapter deals with the application of SPICE to finding the response of higher-order circuits and systems by using system simulation diagrams.

The prerequisites for this chapter are Chapter 7 on first-order circuit response and Chapter 8 on complex variables and waveforms.

9.1 | Solution of Higher-Order Circuits and Differential Equations

Circuits Described by Second-Order Differential Equations

The first-order solution operator, which we have thoroughly covered in Chapter 7, is fundamental; as we will now see, the solution of higher-order circuits and differential equations is based upon it. To see how such higher-order differential equations arise, look at the circuit shown in Figure 9.1. If we apply current division, we obtain the following equation for the indicated current response:

$$i(t) = \frac{2p + \dfrac{8}{p}}{6 + 2p + \dfrac{8}{p}} i_s(t) = \frac{p^2 + 4}{p^2 + 3p + 4} i_s(t). \tag{9.1-1}$$

Cross-multiplying by the denominator, we obtain $[p^2 + 3p + 4]i(t) = [p^2 + 4]i_s(t)$, or in more conventional notation,

$$\frac{d^2i}{dt^2} + 3\frac{di}{dt} + 4i = \frac{d^2i_s}{dt^2} + 4i_s. \tag{9.1-2}$$

As an exercise, you might check this result with a single mesh equation.

Figure 9.1
An example circuit

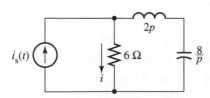

Second-Order Operator Polynomials: Factorization

Now that we have seen how such differential equations arise in a circuits context, let's change our notation to a more general one. We will investigate the solution of

$$\frac{d^2y}{dt^2} + A\frac{dy}{dt} + By = x(t), \tag{9.1-3}$$

or, in operator form,

$$[p^2 + Ap + B]\, y(t) = x(t). \tag{9.1-4}$$

We will investigate the effects of derivatives in the forcing function term later. We define the *characteristic polynomial* by

$$D(z) = z^2 + Az + B; \tag{9.1-5}$$

that is, we form a polynomial in the real or complex variable z with each power of z being the corresponding order of the derivative on the left-hand side of our differential equation. One of the most important results of algebra, you will recall, is that such a polynomial can be written as a product of linear factors, with the number of such factors being the same as the degree of the polynomial. In our case, we can write

$$D(z) = (z + a)(z + b) = (z + b)(z + a). \tag{9.1-6}$$

The order of the factors is unimportant, that is, they *commute*. The *roots* of the characteristic polynomial are the values of z for which $D(z) = 0$, namely $-a$ and $-b$. Furthermore, $a + b = A$ and $ab = B$.

As it happens, we can easily see that the differential operator on the left-hand side of (9.1-4) also satisfies (9.1-6). In fact, using simple properties of the differentiation operator, we have, for any differentiable time function $y(t)$,

$$(p + a)(p + b)y = (p + a)(py + by) = p^2y + bpy + apy + aby \tag{9.1-7}$$
$$= [p^2 + (a + b)p + ab]y = D(p)y,$$

and the same result holds for $(p + b)(p + a)y$. Thus,

$$D(p) = (p + a)(p + b) = (p + b)(p + a); \tag{9.1-8}$$

that is, the operator $D(p)$ can be factored exactly like the characteristic polynomial $D(z)$ into the product of two linear operators that commute.

Second-Order Solution Operators

Let's apply this result to our differential equation (9.1-4) by writing

$$(p + b)(p + a)y = x(t). \tag{9.1-9}$$

If we multiply both sides from the left by the inverse of the operator $(p + b)$, we have

$$(p + a)y = \frac{1}{p + b}x(t). \tag{9.1-10}$$

If we then multiply both sides by the inverse of $(p + a)$, we obtain the solution:

$$y(t) = \frac{1}{p + a} \cdot \frac{1}{p + b}x(t). \tag{9.1-11}$$

Thus, we only need to apply two first-order lowpass operators in succession. We can, of course, write our differential equation in the form

$$(p + a)(p + b)y = x(t) \tag{9.1-12}$$

and apply the two first-order operators in the reverse order to that we have just shown to obtain the solution in the form

$$y(t) = \frac{1}{p + b} \cdot \frac{1}{p + a}x(t). \tag{9.1-13}$$

Thus, not only have we solved our differential equation, but we also have the important fact that any two first-order lowpass operators commute with one another:

$$\frac{1}{D(p)} = \frac{1}{p^2 + Ap + B} = \frac{1}{p + b} \cdot \frac{1}{p + a} = \frac{1}{p + a} \cdot \frac{1}{p + b}. \tag{9.1-14}$$

We can represent these results in the form of system diagrams as shown in Figure 9.2. Although the overall responses are the same, we note that $w(t) \neq v(t)$ in general.

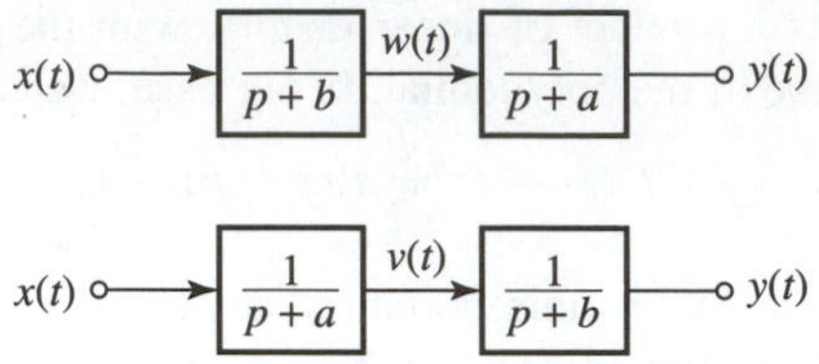

Figure 9.2
Simulation diagrams
for the second-order
solution operator

We now present a second-order example.

Example 9.1   Solve the differential equation $(d^2y/dt^2) + 3(dy/dt) + 2y = x(t)$, with the forcing function $x(t) = e^{-t}u(t)$.

Solution   The characteristic polynomial is $D(z) = z^2 + 3z + 2 = (z + 1)(z + 2)$, so our differential equation can be written $(p + 1)(p + 2)y = e^{-t}u(t)$. The solution is, therefore,

$$y(t) = \frac{1}{p + 2} \frac{1}{p + 1} [e^{-t}u(t)] = \frac{1}{p + 2}\left[e^{-t}\frac{1}{p} [e^t\, e^{-t}u(t)] \right]$$

$$= \frac{1}{p + 2} [te^{-t}u(t)] = e^{-2t}\frac{1}{p} [e^{2t}\, te^{-t}u(t)] = e^{-2t}\frac{1}{p} [te^t u(t)] \qquad (9.1\text{-}15)$$

$$= e^{-2t}\left[\int_0^t \alpha e^\alpha \, d\alpha \right] u(t) = e^{-2t}[(t - 1)\, e^t + 1]\, u(t) = [(t - 1)\, e^{-t} + e^{-2t}]\, u(t).$$

We recognized the fact that $(1/p)\, u(t) = tu(t)$ in going from the first line to the second and used the table method for integration by parts (see Appendix C) in going from the second line to the third. If you have trouble seeing where the 1 comes from, just recall that we are evaluating the definite integral from 0 to t (when $t > 0$), so we must subtract the indefinite integral evaluated at the zero lower limit.

The General Case:
Heaviside Operator
Solutions

Now let's extend what we have learned to more general differential equations. The most general linear, constant-coefficient differential equation can be written as

$$a_n\frac{d^n y}{dt^n} + a_{n-1}\frac{d^{n-1}y}{dt^{n-1}} + \cdots + a_0 y = b_m\frac{d^m x}{dt^m} + b_{m-1}\frac{d^{m-1}x}{dt^{m-1}} + \cdots + b_0 x. \qquad (9.1\text{-}16)$$

If we convert it to operator form, we will have

$$[a_n p^n + a_{n-1}p^{n-1} + \cdots + a_0]y(t) = [b_m p^m + b_{m-1}p^{m-1} + \cdots + b_0]x(t), \qquad (9.1\text{-}17)$$

or in factored form

$$a_n(p + r_1)(p + r_2) \cdots (p + r_n)\, y(t) = b_m(p + s_1)(p + s_2) \cdots (p + s_m)\, x(t). \qquad (9.1\text{-}18)$$

We now multiply by $1/a_n$ and the inverses of the various $(p + r_k)$ factors, one at a time, to obtain the solution:

$$y(t) = \frac{b_m(p + s_1)(p + s_2) \cdots (p + s_m)}{a_n(p + r_1)(p + r_2) \cdots (p + r_n)}\, x(t) = H(p)x(t), \qquad (9.1\text{-}19)$$

where
$$H(p) = \frac{b_m(p + s_1)(p + s_2) \cdots (p + s_m)}{a_n(p + r_1)(p + r_2) \cdots (p + r_n)} \qquad (9.1\text{-}20)$$

is called the *system operator* or the *Heaviside operator* for the differential equation or a *(higher-order) differential operator*. We note that we can apply the numerator operators and the reciprocals of the denominator operators in any order, so $H(p)$ has an unambiguous meaning. This permits us to write the system operator directly from the original differential equation (9.1-16) or (9.1-17) as

$$H(p) = \frac{b_m p^m + b_{m-1} p^{m-1} + \cdots + b_0}{a_n p^n + a_{n-1} p^{n-1} + \cdots + a_0}. \qquad (9.1\text{-}21)$$

We say that $H(p)$ is *proper* if $m \leq n - 1$; otherwise we call it *improper*. If $m \geq n + 1$, we say that it is *singular*. We also define its *order* to be the larger of m and n:

$$\text{order}[H(p)] = \max(m, n). \qquad (9.1\text{-}22)$$

Cascade Simulation of the General Solution Operator

We can draw a simulation diagram for the differential equation or, equivalently, for the solution in equation (9.1-19) in the form shown in Figure 9.3. It is called a *cascade representation* because of its form: one first-order system followed by another like a cascade of waterfalls in a river. Now, in fact, this simulation diagram can be simplified. As we have pointed out, the various first-order systems commute with one another; hence, let us—as far as possible—combine each $p + s_k$ factor with one of the form $1/(p + r_k)$, obtaining

$$H_k(p) = \frac{p + s_k}{p + r_k} = 1 + \frac{s_k - r_k}{p + r_k}. \qquad (9.1\text{-}23)$$

We have performed one long division step, which is of course perfectly permissible for we have shown now that operators behave just like real or complex numbers. The simulation diagram for $H_k(p)$ is shown in Figure 9.4. We have simplified our simulation because $p + s_k$, by itself, requires a differentiator, a summer, and a scalar multiplier. We have therefore eliminated the differentiator, a move that we have already decided is desirable from a noise standpoint as well as in terms of a simple savings in components. If the degree of the numerator of $H(p)$ is larger than its denominator, $m > n$ (that is, the operator is singular), we will not be able to combine all of the $p + s_k$ factors, and therefore one or more differentiators *will* be required.

Figure 9.3
Cascade simulation of $H(p)$

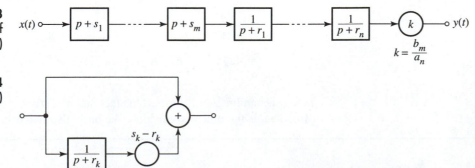

Figure 9.4
Simulation of $H_k(p)$

Operator Impedance and Admittance: The General Case

If the forcing function $x(t)$ is the current into one terminal of a two-terminal subcircuit or element, such as the one shown in Figure 9.5, and $y(t)$ is the voltage across its terminals (assuming the passive sign convention, of course), we call the system operator an *impedance* operator. If $x(t)$ is the voltage and $y(t)$ the current, we say that it is an *admittance* operator. These definitions are the same as the ones we presented earlier in the text for first-order operators, only now we see that they hold for differential operators of all orders.

Figure 9.5
Impedance and
admittance operators

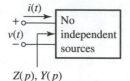

$Z(p), Y(p)$

We now present a circuit example using impedance operators.

Example 9.2   Find the system function for the response $v(t)$ in the second-order circuit shown in Figure 9.6 and, assuming that $R = 4/3 \ \Omega$, find its unit step response.

Figure 9.6
An example circuit

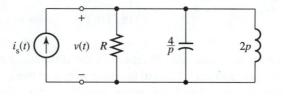

Solution   The system operator we are looking for is an impedance operator—the impedance operator of the RLC subcircuit. Thus,

$$Z(p) = \frac{1}{Y(p)} = \frac{1}{\dfrac{1}{R} + \dfrac{p}{4} + \dfrac{1}{2p}} = \frac{4Rp}{Rp^2 + 4p + 2R}. \tag{9.1-24}$$

The desired system operator is, therefore,

$$Z(p) = \frac{4p}{p^2 + \dfrac{4}{R}p + 2}. \tag{9.1-25}$$

Letting $R = 4/3 \ \Omega$ and $i_s(t) = u(t)$ (to compute the unit step response), we obtain

$$s(t) = Z(p)u(t) = \frac{4p}{p^2 + 3p + 2}u(t) = \frac{4}{(p + 1)(p + 2)}\delta(t) \tag{9.1-26}$$

$$= \frac{4}{p + 1}\left[e^{-2t}\frac{1}{p}[e^{2t}\delta(t)]\right] = \frac{4}{p + 1}[e^{-2t}u(t)]$$

$$= 4e^{-t}\frac{1}{p}[e^t e^{-2t}u(t)] = 4e^{-t}[1 - e^{-t}]u(t) = 4[e^{-t} - e^{-2t}]u(t).$$

In the second line, we have used the fact that $e^{2t}\delta(t) = e^0\delta(t) = \delta(t)$ and that its integral, of course, is the unit step function. We have also noted in the first line that $pu(t) = \delta(t)$.

The Relationship Between Unit Step Response and Unit Impulse Response

We would now like to generalize an important relationship between the impulse and step responses that was given in Chapter 7 for first-order operators. The unit impulse response is defined by

$$h(t) = H(p)\delta(t). \tag{9.1-27}$$

Let's look at what we obtain by integrating both sides:

$$\frac{1}{p}h(t) = \frac{1}{p}H(p)\delta(t) = H(p)\frac{1}{p}\delta(t) = H(p)u(t). \qquad (9.1\text{-}28)$$

We have used the fact that all differential operators commute with each other, as we have shown. We have also noted that the integral of the unit impulse is the unit step. Thus, because the last term in (9.1-28) is the step response, we have the important relationship

$$s(t) = \frac{1}{p}h(t); \qquad (9.1\text{-}29)$$

that is, the integral of the impulse response is the step response. Equation (9.1-29) also implies

$$h(t) = ps(t). \qquad (9.1\text{-}30)$$

As an example, we note that the unit impulse response of the first-order lowpass operator is $e^{-at}u(t)$. Its running integral is $1/p[e^{-at}u(t)] = 1/a[1 - e^{-at}]\,u(t)$, which we know is the unit step response of this operator.

Extension of Dc Analysis Methods to the Time Domain Using Operators

This is perhaps a good point to make a fundamental observation: all of our passive two-terminal circuit elements (R, L, and C), as well as two-terminal subcircuits containing these elements and having no independent sources, obey a generalized form of Ohm's law:

$$v(t) = Z(p)i(t). \qquad (9.1\text{-}31)$$

The only way in which this equation differs from the resistive form is the additional constraint that $Z(p)$ must be kept on the left of $i(t)$. Now if we reflect back upon the development of our resistive circuit analysis techniques, we see that they follow logically from only three basic principles: Ohm's law, KVL, and KCL. Furthermore, the added restriction of always writing $Z(p)$ on the left of $i(t)$—or $Y(p)$ on the left of $v(t)$—*does not change our results*. Thus, we can legitimately use all of these techniques for *RLC* circuits or subcircuits: linearity, superposition, mesh analysis, nodal analysis, Thévenin and Norton equivalent subcircuits, and the substitution theorem—to name just a few. We give two examples in the following sections.

A Superposition Example

Example 9.3   Find the differential equation for $v(t)$ in Figure 9.7 using superposition.

Figure 9.7
An example circuit

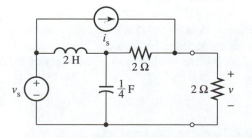

Solution   We first deactivate the current source and obtain the operator form shown in Figure 9.8. A small amount of algebra, using voltage and current division, gives

$$v_1(t) = \frac{1}{p^2 + p + 2}v_s(t). \qquad (9.1\text{-}32)$$

Next, we deactivate the v-source and find the partial response due to the i-source acting alone. The associated partial circuit is shown in Figure 9.9. Again, application of current and voltage division and some care in recognizing series and parallel elements results in

$$v_2(t) = \frac{p^2 + 2p + 2}{p^2 + p + 2} i_s(t). \tag{9.1-33}$$

Summing the two partial responses, we have

$$v(t) = \frac{1}{p^2 + p + 2} v_s(t) + \frac{p^2 + 2p + 2}{p^2 + p + 2} i_s(t). \tag{9.1-34}$$

Notice that this expression has the form

$$v(t) = H_1(p)v_s(t) + H_2(p)i_s(t), \tag{9.1-35}$$

with obvious choices for the two operators. Thus, we see that in a circuit with multiple sources there is one system operator for each independent source—for each response variable chosen. Furthermore, from the form of the expression in (9.1-35) one would think perhaps that all of the system operators have the same denominator polynomial, that is, the same characteristic polynomial. This is actually true, except for possible cancellation of factors between the numerator of a given operator and the denominator.

Figure 9.8
First partial circuit

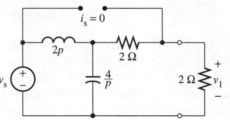

Figure 9.9
Second partial circuit

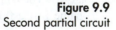
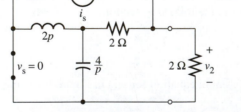

The expression in (9.1-35) would suffice to obtain a solution if we were given the waveforms for the independent sources; however, our task was that of deriving the differential equation. We can do this by cross-multiplying by the denominator operator polynomial, giving

$$[p^2 + p + 2]v(t) = v_s(t) + [p^2 + 2p + 2]i_s(t). \tag{9.1-36}$$

In conventional notation, this becomes

$$\frac{d^2v}{dt^2} + \frac{dv}{dt} + 2v = \frac{d^2i_s}{dt^2} + 2\frac{di_s}{dt} + 2i_s + v_s. \tag{9.1-37}$$

A Time Domain Thévenin Equivalent Example

Example 9.4   Derive the differential equation for $v(t)$ in the circuit of Example 9.3 using a Thévenin equivalent for all elements except the 2-Ω resistor across which $v(t)$ is defined.

Solution   Figure 9.10 shows the subcircuit obtained by removing the resistor in question. We have attached a test i-source for the purpose of determining the general v-i characteristic and have expressed the inductor and capacitor in terms of their impedance operators. We have also anticipated performing nodal analysis by selecting a ground reference and labeling the nonreference node voltages. At this point we notice a somewhat unusual feature of the circuit that permits us to simplify things: by applying KCL at the top terminal of the subcircuit, we see that the current through the resistor is a known quantity. Therefore, we can apply the substitution theorem to replace it with a current source having the known value $i + i_s$. Then, writing one nodal equation at the node marked with the unknown node voltage v_x, we have

$$\frac{p}{4}v_x + \frac{v_x - v_s}{2p} = i_s + i. \tag{9.1-38}$$

Solving, we get

$$v_x = \frac{2}{p^2 + 2}v_s + \frac{4p}{p^2 + 2}i_s + \frac{4p}{p^2 + 2}i. \tag{9.1-39}$$

Now we see that we must add the voltage across the horizontal 2-Ω resistor to get v:

$$v = v_x + 2(i + i_s) = \frac{2}{p^2 + 2}v_s + \frac{2p^2 + 4p + 4}{p^2 + 2}i_s + \frac{2p^2 + 4p + 4}{p^2 + 2}i. \tag{9.1-40}$$

The operator multiplying i is the Thévenin impedance, and the remaining terms form the open circuit voltage. We show the Thévenin equivalent in Figure 9.11 (on p. 432) with the resistor replaced. We can now apply the voltage divider rule to obtain v:

$$v(t) = \frac{2}{2 + \dfrac{2p^2 + 4p + 4}{p^2 + 2}} \times \frac{2v_s(t) + (2p^2 + 4p + 4)i_s(t)}{p^2 + 2} \tag{9.1-41}$$

$$= \frac{v_s(t) + (p^2 + 2p + 2)i_s(t)}{p^2 + p + 2}.$$

Cross-multiplying and reverting to standard notation now results in equation (9.1-37) at the end of Example 9.3.

Figure 9.10
Testing the subcircuit with an i-source

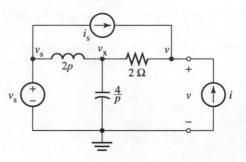

Figure 9.11
The Thévenin
equivalent

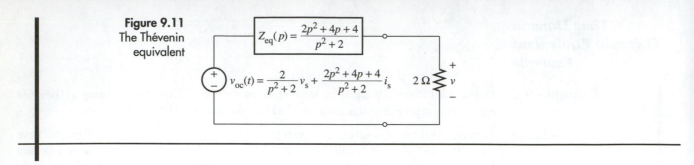

$$Z_{eq}(p) = \frac{2p^2 + 4p + 4}{p^2 + 2}$$

$$v_{oc}(t) = \frac{2}{p^2 + 2}v_s + \frac{2p^2 + 4p + 4}{p^2 + 2}i_s \qquad 2\,\Omega \quad v$$

Section 9.1 Quiz

Q9.1-1. Draw a simulation diagram for the differential equation

$$\frac{d^3y}{dt^3} + 4\frac{d^2y}{dt^2} + 5\frac{dy}{dt} + 2y = \frac{dx}{dt} + ax.$$

To what value(s) should a be adjusted to simplify the simulation? (*Hint:* One root of the characteristic polynomial is -1.)

Q9.1-2. Solve the differential equation

$$\frac{d^2y}{dt^2} + 5\frac{dy}{dt} + 6y = e^{-2t}u(t).$$

Q9.1-3. Find the unit step response for the indicated variable of the circuit in Figure Q9.1-1 using operator impedances and voltage division. Then differentiate it to find the impulse response.

Figure Q9.1-1

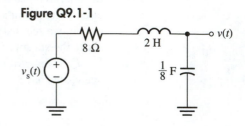

Q9.1-4. For the circuit in Figure Q9.1-1, sketch the Norton equivalent of the subcircuit with the capacitor removed and determine its parameters. Use this equivalent to find the differential equation for $v(t)$.

9.2 | Properties of Second-Order Circuits and Systems: Damping

Second-Order Circuits and Systems: Response Modes

We will now focus our attention on the specific properties of second-order circuits and systems; more specifically, we will investigate the unit impulse response of differential equations whose characteristic polynomial is of second order. The following differential equation is sufficiently general for our purposes:

$$\frac{d^2y(t)}{dt^2} + A\frac{dy(t)}{dt} + By(t) = x(t), \qquad (9.2\text{-}1)$$

where A and B are real parameters. We will write it in factored operator form as

$$(p + a)(p + b)y(t) = x(t). \qquad (9.2\text{-}2)$$

As we will see in later sections of this chapter, the right-hand side will only affect amplitudes of certain terms in our solution and dictate whether any impulses are present in the response; thus, we will compute the unit impulse response as a function of the parameters a and b. We set $x(t) = \delta(t)$ and compute the general form of the unit impulse response:

$$h(t) = \frac{1}{p + b}\frac{1}{p + a}\delta(t) = e^{-bt}\frac{1}{p}\left[e^{bt}e^{-at}\frac{1}{p}\left[e^{at}\delta(t)\right]\right] \qquad (9.2\text{-}3)$$

$$= e^{-bt}\frac{1}{p}\left[e^{(b-a)t}u(t)\right].$$

Here, we have used the facts that $e^{at}\delta(t) = e^0\delta(t) = \delta(t)$ and that $1/p\ \delta(t) = u(t)$. Before going any further, we see that we must make an assumption on the parameters a and b in order to compute the second integral. This leads to three modes of response.

Mode 1 (a, b, real and not equal)

In this case, we have

$$h(t) = \frac{1}{b-a}e^{-bt}[e^{(b-a)t} - 1]\,u(t) = \frac{1}{b-a}[e^{-at} - e^{-bt}]\,u(t). \qquad (9.2\text{-}4)$$

If we assume that a and b are both positive real numbers, we obtain the results shown in Figure 9.12. If either a or b is negative, the plot starts at zero and increases without limit as $t \to \infty$, as shown in Figure 9.13. If both parameters a and b are positive, the response is stable (approaches zero for large t) and if either (or both) is negative, the response is unstable. We can say that the response in (9.2-4) "has two time constants," $\tau_1 = 1/a$ and $\tau_2 = 1/b$.

Figure 9.12   Impulse response (real distinct positive values of a and b)

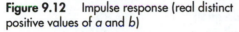
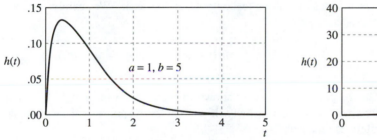

Figure 9.13   Impulse response (one parameter negative)

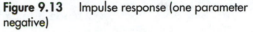
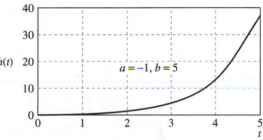

Mode 2: (a and b complex conjugates)

There is another possible situation that can occur when our two parameters are unequal: they can be complex. In this case, they must be *complex conjugates* for they are roots of the characteristic polynomial

$$D(z) = z^2 + Az + B = (z + a)(z + b), \qquad (9.2\text{-}5)$$

which has real coefficients. In this case, we can write

$$a = \sigma_0 - j\omega_0 \qquad (9.2\text{-}6a)$$

and

$$b = \sigma_0 + j\omega_0, \qquad (9.2\text{-}6b)$$

with $\omega_0 \neq 0$. The impulse response in (9.2-4) can then be written in the form

$$h(t) = \frac{1}{j2\omega_0}[e^{-(\sigma_0 - j\omega_0)t} - e^{-(\sigma_0 + j\omega_0)t}]u(t) = \frac{1}{\omega_0}e^{-\sigma_0 t}\frac{e^{j\omega_0 t} - e^{-j\omega_0 t}}{j2}u(t) \qquad (9.2\text{-}7)$$

$$= \frac{1}{\omega_0}e^{-\sigma_0 t}\sin(\omega_0 t)u(t).$$

This waveform is plotted in Figure 9.14 (for $\sigma_0 > 0$). Notice that it is quite unlike the real parameter case, so we cannot say that it "has two time constants" or that it is a "two time constant response"; rather, we say that it is "oscillatory" or that it "rings." The latter term arises because the same response waveform describes the motion of the surface of a bell when struck with a hammer. In Figure 9.14, we have assumed that the value of ω_0 is fairly large relative to σ_0. If, on the other hand, it is not so much larger, we will obtain a somewhat different plot, as shown in Figure 9.15. The parameter ω_0 is a measure of how fast the response oscillates and σ_0 is a measure of how fast it "decays," or goes to zero. Because σ_0 is larger in Figure 9.15, the response "damps out" faster than in Figure 9.14.

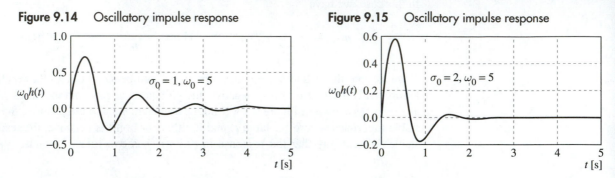

Figure 9.14   Oscillatory impulse response

$\sigma_0 = 1, \omega_0 = 5$

$\omega_0 h(t)$

Figure 9.15   Oscillatory impulse response

$\sigma_0 = 2, \omega_0 = 5$

$\omega_0 h(t)$

Mode 3 (a and b real and equal)

The response is easier to compute in this case. We merely use the right-hand side of (9.2-3) to write

$$h(t) = e^{-at} \frac{1}{p} [u(t)] = te^{-at}u(t). \tag{9.2-8}$$

This response waveform is shown in Figure 9.16 for $a = 5$. If a is smaller, the decay to zero will occur more slowly. If $a < 0$, the response will be unstable, increasing indefinitely as time gets large.

Figure 9.16
Impulse response with real, equal parameters

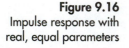

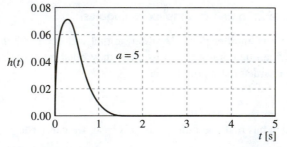

$a = 5$

$h(t)$

Physical Interpretation of Second-Order Response Modes: Damping

The foregoing discussion is a mathematical one: we have merely been analyzing the impulse response of a differential equation. Now, let's look at the origin of the term "damping" a little more—in a physical sense. In fact, we will first discuss a mechanical system with which you are most certainly familiar: the suspension system of an automobile. We have shown (in a quite stylized way) a car traveling along a smooth-surfaced street in Figure 9.17. At some point, however, it runs over a very small, but sharp, rock in the road. The rock acts on a wheel like an impulse of upward force.

Figure 9.17
Impulse response of an automobile suspension system

Each wheel is connected to the frame of the car through a spring and, in parallel with that, a shock absorber or dashpot (which is also referred to as a "damper"). We show the situation in Figure 9.18. On the left, we see a schematic diagram of the interconnections: the coil represents a spring, the circle a wheel, and the symbol to the right of the spring a damper. The spring has a "spring constant" K and the wheel a mass M, as shown. The operation of the damper is as follows. There is a fluid inside a cylinder containing a piston. The piston is connected to a rod that is, in turn, connected to the frame of the car. The cylinder is connected rigidly to the wheel axle. When a bump occurs, the wheel presses upward on the spring and dashpot. The spring pushes back against the force applied by the wheel with a strength of Ky, where y is the displacement of the wheel axle in an upward direction. At the same time, the cylinder moves upward relative to the piston in the damper. The fluid in the cylinder must move past the piston—between it and the cylinder wall. This, too, results in a force downward against the wheel; now, however, the force is proportional to the velocity of displacement of the piston (and, therefore, the wheel) due to the viscosity of the fluid. This component of force, therefore, is given by $B \, dy(t)/dt$, where B is the "damping coefficient."

Figure 9.18
An automobile
suspension system

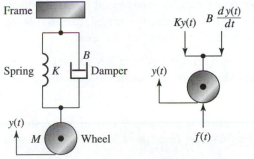

The net result is illustrated by the "free body diagram" shown to the right in the figure. The force upward is given the symbol $f(t)$, and we know from Newton's laws that the resultant force is equal to the mass of the wheel times the acceleration:

$$f(t) - Ky(t) - B\frac{dy(t)}{dt} = M\frac{d^2y(t)}{dt^2}. \tag{9.2-9}$$

In operator form, we have

$$\left(p^2 + \frac{B}{M}p + \frac{K}{M}\right)y(t) = f(t)/M = x(t). \tag{9.2-10}$$

This, of course, is the second-order differential equation we have just been studying. Thus, if we consider the rock in the road as applying an impulsive force, we need only look at the characteristic polynomial

$$D(z) = z^2 + \frac{B}{M}z + \frac{K}{M} = (z + a)(z + b). \tag{9.2-11}$$

The parameters a and b will be real and unequal, real and equal, or complex conjugates accordingly as the discriminant

$$\Delta = \left[\frac{B}{M}\right]^2 - 4\frac{K}{M} \begin{cases} > 0 \\ = 0, \\ < 0 \end{cases} \tag{9.2-12}$$

which leads to

$$B \begin{cases} > 2\sqrt{KM} \\ = 2\sqrt{KM}. \\ < 2\sqrt{KM} \end{cases} \qquad (9.2\text{-}13)$$

Thus, the motion of the wheel in response to the rock (the impulse response) will be a two time constant response (which we will call *overdamped*) if the "greater than" relation holds; it will be of the form Ate^{-at} (which we will call *critically damped*) if the equality holds; and it will be oscillatory (which we will call *underdamped*) if the "less than" relation holds.

If you have ever ridden in a large truck you have experienced the overdamped phenomenon. In this case, it is not the passengers' comfort that is of importance—merely that the wheels not move up and down, for this leads to a loss of traction and steering problems. In a large, "softly sprung" car, the suspension is typically underdamped. The wheels move up and down a lot and the ride is very soft, but the steering is poor because the tires lose contact with the surface of the road. If you have ever ridden in an expensive sports car, you will have experienced a compromise between steering and comfort. The suspension system is critically damped; thus, the passenger has as comfortable a ride as is consistent with the performance of the car.

A Second-Order Circuit Example: Damping Modes

Damping is a term applicable to any second-order differential equation, so we see that circuits can also be under, over, or critically damped.

Example 9.5   Find the values of R in Figure 9.19 that will result in under, over, and critical damping. Find the unit impulse responses for $R = 1$, 5, and 6 ohms.

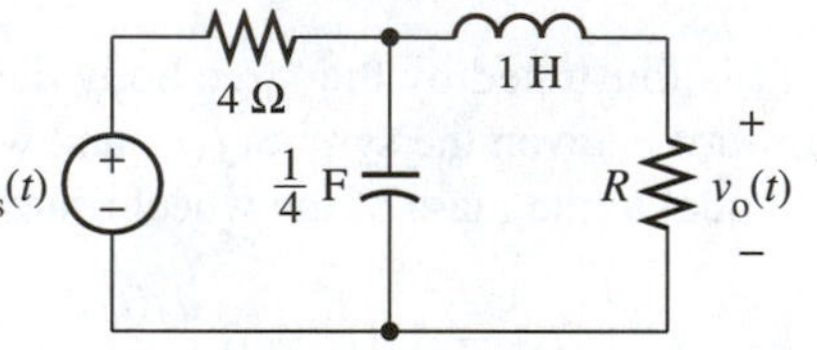

Figure 9.19
An example circuit

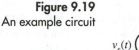

Solution   We show the circuit prepared for analysis in Figure 9.20. We have used impedance operators for the inductor and the capacitor. We have also chosen a reference node and will perform nodal analysis (a somewhat arbitrary choice of method). The single nodal equation required is

$$\frac{v(t) - v_s(t)}{4} + \frac{p}{4}\, v(t) + \frac{v(t)}{p + R} = 0. \qquad (9.2\text{-}14)$$

We remind you that although we have written $p + R$ directly beneath $v(t)$, it actually means that the operator $1/(p + R)$ is to be applied to $v(t)$ from the left. Now we multiply both sides of (9.2-14) from the left by $4(p + R)$; that is, we multiply by R, differentiate, add, and then multiply by 4. This results in

$$[p^2 + (R + 1)p + (R + 4)]\, v(t) = (p + R)\, v_s(t). \qquad (9.2\text{-}15)$$

The solution for $v(t)$ is

$$v(t) = \frac{p + R}{p^2 + (R + 1)p + (R + 4)}\, v_s(t). \qquad (9.2\text{-}16)$$

However, what we want is $v_0(t)$. This quantity can be determined from $v(t)$ by the voltage divider rule:

$$v_o(t) = \frac{R}{p + R}\, v(t). \qquad (9.2\text{-}17)$$

Thus,

$$v_o(t) = \frac{R}{p^2 + (R + 1)p + (R + 4)}\, v_s(t). \qquad (9.2\text{-}18)$$

Figure 9.20
The example circuit prepared for analysis

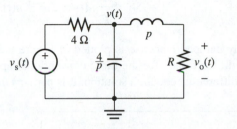

By simply replacing the denominator polynomial operator by the same polynomial in the real or complex variable z, we obtain the characteristic polynomial:

$$D(z) = z^2 + (R + 1)z + (R + 4). \qquad (9.2\text{-}19)$$

The discriminant of this polynomial is

$$\Delta = (R + 1)^2 - 4(R + 4). \qquad (9.2\text{-}20)$$

If $R = 5$, then $\Delta = 0$ and we have critical damping (real, equal roots). If $R < 5$, then $\Delta < 0$ and we have an underdamped circuit (complex conjugate roots), and if $R > 5$, then $\Delta > 0$ and the circuit is overdamped (real, unequal roots).

Now let's assume that $R = 1\ \Omega$. The impulse response is

$$h(t) = \frac{1}{p^2 + 2p + 5}\, \delta(t) = \frac{1}{(p + 1 + j2)(p + 1 - j2)}\, \delta(t) \qquad (9.2\text{-}21)$$

$$= e^{-(1+j2)t}\, \frac{1}{p}\left[e^{j4t}\, \frac{1}{p}\, [e^{(1-j2)t}\delta(t)] \right]$$

$$= e^{-(1+j2)t}\, \frac{1}{p}\, [e^{j4t}\, u(t)] = e^{-(1+j2)t}\, \frac{1}{j4}\, [e^{j4t} - 1]\, u(t)$$

$$= \frac{1}{j4}\, e^{-t}\, [e^{j2t} - e^{-j2t}]\, u(t) = 0.5e^{-t}\sin(2t)u(t).$$

The response looks qualitatively something like that in Figure 9.14 or 9.15; that is, it is oscillatory. However, we show a plot (for $t > 0$) in Figure 9.21.

Figure 9.21
Underdamped impulse response

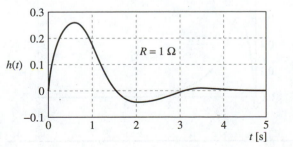

When $R = 5\,\Omega$ the circuit is critically damped. The impulse response is

$$h(t) = \frac{5}{p^2 + 6p + 9}\,\delta(t) = \frac{5}{(p+3)^2}\,\delta(t) \tag{9.2-22}$$

$$= 5e^{-3t}\frac{1}{p}\left[e^{3t}e^{-3t}\frac{1}{p}\left[e^{3t}\delta(t)\right]\right]$$

$$= 5e^{-3t}\frac{1}{p}\,u(t) = 5te^{-3t}\,u(t).$$

This critically damped response is plotted in Figure 9.22. We should point out that this case is "the difficult one" for the more conventional approach (that is, without operator calculus) to differential equation solution. It is perhaps the easiest for us.

Figure 9.22
Critically damped
impulse response

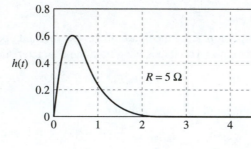

The last case to be considered is overdamped. If $R = 6\,\Omega$, the impulse response is

$$h(t) = \frac{6}{p^2 + 7p + 10}\,\delta(t) = \frac{6}{(p+2)(p+5)}\,\delta(t) \tag{9.2-23}$$

$$= 6e^{-2t}\frac{1}{p}\left[e^{2t}e^{-5t}\frac{1}{p}\left[e^{5t}\delta(t)\right]\right]$$

$$= 6e^{-2t}\frac{1}{p}\left[e^{-3t}\,u(t)\right] = -2e^{-2t}\left[e^{-3t} - 1\right]u(t)$$

$$= 2\left[e^{-2t} - e^{-5t}\right]u(t).$$

This response, shown in Figure 9.23, does not look appreciably different from that of Figure 9.22; in fact, it is not, for the corresponding values of R are fairly close together percentagewise. They are similar only qualitatively because the overdamped response has two real, but *unequal,* parameters. A more careful study would show that $h(t)$ reaches its peak value fastest, but without overshoot (another term for ringing), for critical damping.

Figure 9.23
Overdamped impulse
response

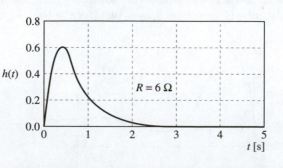

Section 9.2 Quiz

Q9.2-1. For the circuit in Figure Q9.2-1, find the nonnegative value of R for critical damping.

Q9.2-2. Find the unit step response for $v(t)$ in the circuit in Figure Q9.2-1 for $R = 1\ \text{k}\Omega$ and $R = 4\ \text{k}\Omega$. (*Hint:* Use units of mA, kΩ, V, nF, and mH—this implies that the unit of time is the μs. RC and L/R both have units of μs in that case.)

Figure Q9.2-1

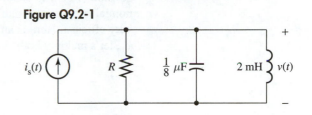

9.3 | Partial Fraction Expansion: Simple Roots

Partial Fraction Expansion: Simple Factors

The method of circuit analysis that we outlined in Section 9.1 is general: it will work with any forcing function $x(t)$. The sequential integrations involved, however, can become a bit complicated, particularly for higher-order systems. We will, therefore, now develop a procedure that is quite a bit simpler. In this section we will present the simplest case; in later sections we will elaborate upon it for more complicated ones.

Let's approach our subject by means of a simple real rational function. We will let $Q(z) = (cz + d)/((z + a)(z + b))$, where z is a real or complex variable. We claim that

$$Q(z) = \frac{cz + d}{(z + a)(z + b)} = \frac{k_1}{z + a} + \frac{k_2}{z + b}, \tag{9.3-1}$$

where k_1 and k_2 are constants to be determined. To investigate the truth of this claim, let's multiply both sides by the denominator of $Q(z)$. This gives

$$(z + a)(z + b)Q(z) = cz + d = k_1(z + b) + k_2(z + a) \tag{9.3-2}$$
$$= (k_1 + k_2)z + (k_1 b + k_2 a).$$

If we can show that this equation is valid, then our assertion will be true. As the powers of z are linearly independent, we can equate their coefficients on both sides,[1] giving the matrix equation

$$\begin{bmatrix} 1 & 1 \\ b & a \end{bmatrix} \begin{bmatrix} k_1 \\ k_2 \end{bmatrix} = \begin{bmatrix} c \\ d \end{bmatrix}. \tag{9.3-3}$$

We solve by multiplying both sides by the inverse of the (2 × 2) matrix on the left-hand side of this equation, obtaining

$$\begin{bmatrix} k_1 \\ k_2 \end{bmatrix} = \frac{1}{a - b} \begin{bmatrix} a & -1 \\ -b & 1 \end{bmatrix} \begin{bmatrix} c \\ d \end{bmatrix} = \begin{bmatrix} \dfrac{ac - d}{a - b} \\ \dfrac{d - bc}{a - b} \end{bmatrix}. \tag{9.3-4}$$

This gives the solution—*but only if $a \neq b$!*

---

[1] Alternatively, let z become very large. Then the linear terms on both sides of (9.3-2) are much greater than the constant terms, and we can equate the coefficients of the linear terms. Then, because these two linear terms are equal, we can cancel them from both sides and set the constant terms equal to each other.

A Simpler Method   At this point we have shown that the preceding expansion, called the *partial fraction expansion (PFE)*, is valid, provided that the denominator factors of the denominator polynomial are distinct. Our procedure of solving simultaneous equations, however, is not very efficient. Here is an equivalent procedure that is much easier and faster. To compute k_1, let's multiply both sides of (9.3-1) by $(z + a)$:

$$(z + a)Q(z) = \frac{cz + d}{z + b} = k_1 + k_2 \frac{z + a}{z + b}. \tag{9.3-5}$$

Now let z have the particular value $-a$. Assuming that $b \neq a$, the second term on the right vanishes and we are left with k_1 all by itself:

$$k_1 = [(z + a)Q(z)]_{z=-a} = \left[\frac{cz + d}{z + b} \right]_{z=-a} = \frac{d - ca}{b - a}. \tag{9.3-6}$$

We can, of course, compute k_2 in exactly the same fashion by exchanging a and b in the last equation. By the way, note that $(z + a)Q(z)$ is not zero when $z = -a$ because we cancel the $(z + a)$ term with the same factor in the denominator of $Q(z)$ before letting $z = -a$.

Example 9.6   Find the partial fraction expansion of

$$Q(z) = \frac{3z + 4}{z^2 + 3z + 2}.$$

Solution   First, we write the PFE with the denominator in factored form:

$$\frac{3z + 4}{(z + 1)(z + 2)} = \frac{k_1}{z + 1} + \frac{k_2}{z + 2}. \tag{9.3-7}$$

Then we compute k_1 and k_2, yielding

$$k_1 = [(z + 1)Q(z)]_{z=-1} = \left[\frac{3z + 4}{z + 2} \right]_{z=-1} = 1 \tag{9.3-8a}$$

and

$$k_2 = [(z + 2)Q(z)]_{z=-2} = \left[\frac{3z + 4}{z + 1} \right]_{z=-2} = 2. \tag{9.3-8b}$$

Therefore,

$$\frac{3z + 4}{(z + 1)(z + 2)} = \frac{1}{z + 1} + \frac{2}{z + 2}. \tag{9.3-9}$$

The Degree Condition   The partial fraction expansion clearly simplifies a rational function, but there are a couple of restrictions. We have already seen that we must require that the denominator factors be different. To get an idea for the second restriction, let's look at $Q(z) = (cz^2 + dz + e)/((z + a)(z + b))$. Assuming that the same PFE works, we write

$$Q(z) = \frac{cz^2 + dz + e}{(z + a)(z + b)} = \frac{k_1}{z + a} + \frac{k_2}{z + b}. \tag{9.3-10}$$

Now let's try to pursue the same procedure that we used before to demonstrate the validity of this equation. We multiply both sides by the denominator, getting

$$(z + a)(z + b)Q(z) = cz^2 + dz + e = k_1(z + b) = k_2(z + a) \tag{9.3-11}$$
$$= (k_1 + k_2)z + (k_1 b + k_2 a).$$

When we try to match coefficients now, we find that there are not enough undetermined constants on the right-hand side! Thus, our procedure fails unless $c = 0$. The numerator of $Q(z)$ must be at least one degree lower than the denominator for the process to work.

Application to Circuit Analysis

How does all of this apply to our circuit analysis problem? It works like this. We have already seen that rational differential operators have all the same properties as rational functions of a real or complex variable. Thus, our PFE will work just as well for a differential operator. Exactly the same type of reasoning we have just gone through will work, in general, to establish the validity of the following theorem.

Theorem 9.1 Let

$$H(p) = \frac{N(p)}{D(p)} = \frac{N(p)}{(p + a_1)(p + a_2) \cdots (p + a_n)}$$

be any rational system operator with distinct values for the a_k's (that is, the roots of the characteristic polynomial $D(z)$ are distinct) and with the degree of $N(p)$ being strictly less than the degree of $D(p)$,[2] that is, $\deg\{N(p)\} \leq n - 1$. Then

$$H(p) = \sum_{i=1}^{n} \frac{k_i}{p + a_i} = \frac{k_1}{p + a_1} + \frac{k_2}{p + a_2} + \cdots + \frac{k_n}{p + a_n}, \qquad (9.3\text{-}12)$$

where
$$k_i = [(p + a_i)H(p)]_{p \leftarrow -a_i}. \qquad (9.3\text{-}13)$$

(Note that $p \leftarrow -a_i$ means to replace p everywhere it occurs by $-a_i$.)

Before applying the PFE to an operator example, we remind you that for the first-order lowpass operator $H(p) = k/(p + a)$, the unit impulse response is $h(t) = H(p)\,\delta(t) = ke^{-at}u(t)$, and the unit step response is $s(t) = (1/p)\,h(t) = (k/a)\,[1 - e^{-at}]u(t)$. As we have succeeded in representing a more complex, higher-order operator as the sum of first-order operators, we will find these results quite useful. We will find impulse and step responses often in our examples because they are simple to work with as well as being useful results themselves. As we will later show, most practical circuit responses can be computed as the unit impulse response of *some* operator.

Example 9.7 Use the PFE to find the impulse response of the differential equation

$$\frac{d^2y}{dt^2} + 5\frac{dy}{dt} + 6y = x(t). \qquad (9.3\text{-}14)$$

Solution The system operator is

$$H(p) = \frac{1}{p^2 + 5p + 6} = \frac{1}{(p + 2)(p + 3)}.$$

If we apply the PFE,[3] we obtain $H(p) = k_1/(p + 2) + k_2/(p + 3)$. But note that $k_1 = [(p + 2)H(p)]_{p \leftarrow -2} = 1$ and, similarly, $k_2 = [(p + 3)H(p)]_{p \leftarrow -3} = -1$. Thus,

---

[2] In other words, $H(p)$ is a *proper* operator (see Section 9.1).

[3] After first verifying, of course, that the operator is proper and that the roots of the characteristic polynomial are distinct.

$H(p) = 1/(p + 2) - 1/(p + 3)$. Hence, the unit impulse response is

$$h(t) = H(p)\delta(t) = \frac{1}{p + 2}\,\delta(t) - \frac{1}{p + 3}\,\delta(t) = e^{-2t}\,u(t) - e^{-3t}\,u(t)$$

$$= [e^{-2t} - e^{-3t}]u(t).$$

If $H(p)$ has simple denominator factors, that is, nonrepeated roots for its characteristic polynomial, then we know how to proceed; however, let us suppose that the degree of the numerator does not meet the condition of our PFE theorem (Theorem 9.1) in the sense that $H(p)$ is not proper. As it happens, there is a simple solution to this problem. We will show how it works in the next example.

Example 9.8   Find the unit impulse response of the following differential equation:

$$\frac{d^2y}{dt^2} + 5\frac{dy}{dt} + 6y = \frac{d^3x}{dt^3} + 7\frac{d^2x}{dt^2} + 19\frac{dx}{dt} + 19x. \qquad (9.3\text{-}15)$$

Solution   The system operator is

$$H(p) = \frac{p^3 + 7p^2 + 19p + 19}{p^2 + 5p + 6},$$

by inspection. We note that it is not proper because the degree of the numerator is larger than $2 - 1 = 1$. How can we solve this problem? We can solve it simply by using long division to reduce the degree of the numerator until the remainder is a proper operator. Thus, after two steps, we have

$$H(p) = p + 2 + \frac{3p + 7}{(p + 2)(p + 3)} \qquad (9.3\text{-}16)$$

We now apply the PFE to the proper remainder operator to give

$$H(p) = p + 2 + \frac{1}{p + 2} + \frac{2}{p + 3}. \qquad (9.3\text{-}17)$$

The impulse response is, then $h(t) = H(p)\delta(t) = \delta'(t) + 2\delta(t) + [e^{-2t} + 2e^{-3t}]\,u(t)$.

Example 9.9   Find the unit step response of the circuit shown in Figure 9.24.

Figure 9.24
An example circuit

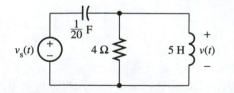

Solution   We require a voltage, and nodal analysis will require $N - 1 - N_V = 3 - 1 - 1 = 1$ equation, whereas mesh analysis would require $B - N + 1 - N_I = 4 - 3 + 1 - 0 = 2$ equations—so let's do nodal analysis. The circuit prepared for nodal analysis is shown in Figure 9.25. We have chosen the bottom node as the reference and have used operator impedances for the capacitor and the inductor.

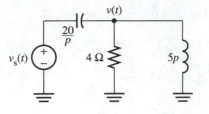

Figure 9.25
The example circuit
prepared for nodal
analysis

The nodal equation at the essential node is

$$\left[\frac{p}{20} + \frac{1}{4} + \frac{1}{5p}\right] v(t) = \frac{p}{20} v_s(t). \tag{9.3-18}$$

Rationalizing and solving for $v(t)$, we get

$$v(t) = \frac{p^2}{p^2 + 5p + 4} v_s(t). \tag{9.3-19}$$

Thus, $H(p) = p^2/(p^2 + 5p + 4)$, which is improper. We do one step of long division, then perform a PFE to get

$$H(p) = 1 + \frac{\frac{1}{3}}{p + 1} - \frac{\frac{16}{3}}{p + 4}. \tag{9.3-20}$$

The unit step response is $s(t) = H(p)u(t)$, or

$$s(t) = \left[1 + \frac{1}{3}[1 - e^{-t}] - \frac{4}{3}[1 - e^{-4t}]\right] u(t). \tag{9.3-21}$$

This can be simplified to $s(t) = (1/3)\,[4e^{-4t} - e^{-t}]\,u(t)$. This is a fairly interesting response waveform, so we have plotted it in Figure 9.26. The sign reversal is caused by the fact that the positive exponential term has a much shorter time constant than does the negative one. Thus, when t is very small, the positive exponential term—being larger—"wins," whereas the opposite is true for larger values of t. Finally, however, both approach zero.

Figure 9.26
Step response

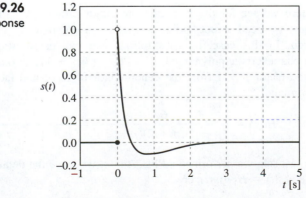

Example 9.10   Find the unit step response for $v(t)$ in Figure 9.27.

Figure 9.27
An example circuit

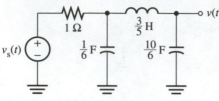

Solution   We will use impedance operators and nodal analysis. The circuit prepared for nodal analysis is shown in Figure 9.28.

Figure 9.28
The example circuit prepared for analysis

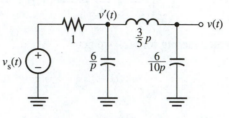

The nodal equations in matrix form, which we will leave to you to verify, are

$$
\begin{bmatrix}
1 + \dfrac{p}{6} + \dfrac{5}{3p} & -\dfrac{5}{3p} \\[3mm]
-\dfrac{5}{3p} & \dfrac{10p}{6} + \dfrac{5}{3p}
\end{bmatrix}
\begin{bmatrix} v' \\ v \end{bmatrix} =
\begin{bmatrix} v_s \\ 0 \end{bmatrix}.
\tag{9.3-22}
$$

If we multiply both rows by $6p$ and invert the resulting (2×2) coefficient matrix, we will get

$$
\begin{bmatrix} v' \\ v \end{bmatrix} = \frac{1}{10p\,[p^3 + 6p^2 + 11p + 6]}
\begin{bmatrix} 10p^2 + 10 & 10 \\ 10 & p^2 + 6p + 10 \end{bmatrix}
\begin{bmatrix} 6pv_s \\ 0 \end{bmatrix}.
\tag{9.3-23}
$$

Solving for $v(t)$ gives

$$
v(t) = \frac{6}{p^3 + 6p^2 + 11p + 6}\, v_s(t),
\tag{9.3-24}
$$

so

$$
H(p) = \frac{6}{p^3 + 6p^2 + 11p + 6}.
\tag{9.3-25}
$$

Here we see for the first time the major difficulty in the analysis of higher-order systems: we need a root-finding algorithm. A computer or calculator will generally be required. In this case, by substitution, you can easily show that there is one root of the characteristic polynomial $D(z) = z^3 + 6z^2 + 11z + 6$ at $z = -1$. We can thus divide the factor $p + 1$ from the denominator of $H(p)$ and factor the resulting quadratic polynomial:

$$
H(p) = \frac{6}{(p + 1)(p + 2)(p + 3)} = \frac{3}{p + 1} - \frac{6}{p + 2} + \frac{3}{p + 3}.
\tag{9.3-26}
$$

In the last step we performed a PFE. Again, we leave the details to you. The unit step response is $s(t) = H(p)u(t)$, or

$$
s(t) = [1 - 3e^{-t} + 3e^{-2t} - e^{-3t}]u(t).
\tag{9.3-27}
$$

Example 9.11   Find the unit impulse response for the current $i(t)$ in Figure 9.29.

Figure 9.29
An example circuit

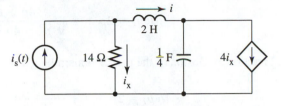

Solution   Using our usual procedure of determining the complexity of analysis, we find that two nodal equations are required, whereas only one mesh equation is necessary; thus, we will do mesh analysis. If we test the circuit by temporarily deactivating all sources, we find that the two end meshes are nonessential. Thus, we will write KVL for the inner mesh. We have prepared the circuit for mesh analysis in Figure 9.30. We have taped the CCCS and have labeled the inductor and the capacitor with their operator impedances. We have chosen the mesh currents in the two end meshes to coincide with the current sources defining them. The KVL equation around the center mesh is

$$2pi + \frac{4}{p}(i - i_c) + 14(i - i_s) = 0. \tag{9.3-28}$$

If we now untape the dependent source and express its value in terms of the known and unknown mesh currents, using $i_c = 4i_x = 4(i_s - i)$, we will obtain

$$\left[2p + \frac{4}{p} + 14\right]i - \frac{4}{p}(4)[i_s - i] = 14i_s. \tag{9.3-29}$$

Solving, we obtain

$$i(t) = \frac{7p + 8}{p^2 + 7p + 10} i_s(t). \tag{9.3-30}$$

Thus, the system operator is

$$H(p) = \frac{7p + 8}{p^2 + 7p + 10} = \frac{-2}{p + 2} + \frac{9}{p + 5} \tag{9.3-31}$$

by PFE. The unit impulse response is, therefore,

$$h(t) = H(p)\,\delta(t) = [-2e^{-2t} + 9e^{-5t}]\,u(t). \tag{9.3-32}$$

Figure 9.30
The example circuit
prepared for mesh
analysis

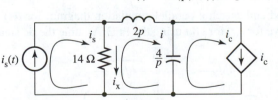

Simulation of Higher-Order Differential Equations and Operators

Before we leave the subject of partial fraction expansion with distinct roots of the characteristic polynomial, let's discuss simulation for just a moment. To make our discussion concrete, let's assume that

$$H(p) = \frac{2p^3 + 12p^2 + 26p + 18}{p^2 + 4p + 3}, \tag{9.3-33}$$

a singular operator. Because $H(p)$ is improper, we divide twice to make *the remainder proper*:

$$H(p) = 2p + 4 + \frac{4p + 6}{p^2 + 4p + 3}. \qquad (9.3\text{-}34)$$

Partial fraction expansion of the proper remainder gives

$$H(p) = 2p + 4 + \frac{1}{p + 1} + \frac{3}{p + 3}. \qquad (9.3\text{-}35)$$

For any input $x(t)$, we have $y(t) = H(p)x(t)$, or

$$y(t) = H(p)x(t) = 2px(t) + 4x(t) + \frac{1}{p + 1}x(t) + \frac{3}{p + 3}x(t). \qquad (9.3\text{-}36)$$

Thus, the overall response is the sum of four constituent signals. This leads to the simulation diagram shown in Figure 9.31. We refer to it as a *parallel realization* for an obvious reason: the signal paths are parallel to each other.

Figure 9.31
Parallel simulation
diagram for $H(p)$

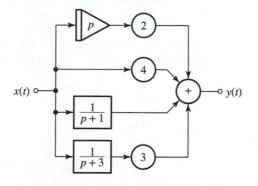

We can draw several conclusions by observing the form of our parallel realization. If $H(p)$ is proper (that is, the degree of its numerator is strictly less than that of its denominator), then all signals will pass through first-order blocks, which do not require differentiators for realization. If $H(p)$ is improper (the degree of its numerator is at least equal to that of its denominator), there will be a direct feedforward path through a scalar multiplier (which can be of unity value). Finally, if $H(p)$ is singular (the degree of the numerator is strictly greater than that of the denominator), there will be one or more differentiators in feedforward paths. We will now provide a circuit example for parallel realization.

Example 9.12   Find and sketch a parallel simulation diagram for $i(t)$ for all $t > 0$ in Figure 9.32. Then solve for $i(t)$ for all t using the PFE. Assume the dc steady state for $t \le 0$.

Figure 9.32
An example circuit

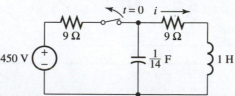

Solution   Though there are many ways of simulating a circuit (including drawing a block for each element and then using KVL and KCL to connect them), we will analyze the $t > 0$ circuit by first deriving a differential equation for i. We find the initial conditions from the dc steady-state circuit for $t \le 0$ shown in Figure 9.33. Then we draw the equivalent circuit that holds for $t > 0$ using the initial condition models for the energy storage elements as in Figure 9.34. Then we derive the differential equation by writing one mesh equation. This equation takes the form

$$\left[\frac{14}{p} + 9 + p\right] i(t) = 225u(t) + 25\delta(t) = 25(p + 9)u(t). \qquad (9.3\text{-}37)$$

Multiplying both sides by p and solving, we get

$$i(t) = \frac{25(p + 9)}{p^2 + 9p + 14}\delta(t) = H(p)x(t), \qquad (9.3\text{-}38)$$

where $x(t) = \delta(t)$ (note that $pu(t) = \delta(t)$) and $H(p) = (25(p + 9))/(p^2 + 9p + 14)$ is the system operator. Performing the PFE, we obtain

$$H(p) = \frac{25(p + 9)}{(p + 2)(p + 7)} = \frac{35}{p + 2} - \frac{10}{p + 7}. \qquad (9.3\text{-}39)$$

Multiplying $H(p)$ times $x(t)$ gives the simulation diagram shown in Figure 9.35. We can solve to obtain

$$i(t) = H(p)x(t) = \left[\frac{35}{p + 2} - \frac{10}{p + 7}\right]\delta(t) = 5[7e^{-2t} - 2e^{-7t}]\,u(t). \qquad (9.3\text{-}40)$$

As we have said, this is only one of many ways to do the simulation. For instance, we could have treated each of the initial condition sources as a separate input variable.

Figure 9.33
The $t \le 0$ steady-state equivalent circuit

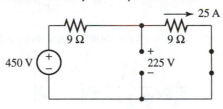

Figure 9.34
The $t > 0$ circuit

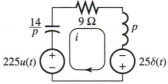

Figure 9.35
Simulation diagram

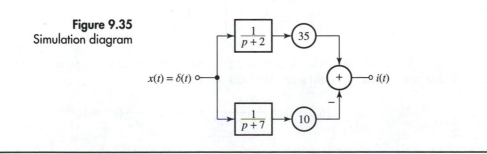

Section 9.3 Quiz

Q9.3-1. Perform a partial fraction expansion on

$$H(p) = \frac{p^2 + 4p}{p^2 + 5p + 6}$$

and find its unit step response.

Q9.3-2. Find $H(p)$, then perform a PFE to find the response to $x(t) = e^{-t}u(t)$ for the differential equation

$$\frac{d^2y}{dt^2} + 8\frac{dy}{dt} + 7y = 36\frac{d^2x}{dt^2}.$$

Q9.3-3. Find the unit step response for the voltage $v(t)$ indicated in Figure Q9.3-1 using partial fraction expansion techniques and draw the simulation diagram.

Figure Q9.3-1

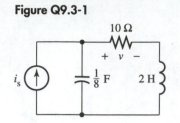

9.4 | Partial Fraction Expansion: Multiple Roots

As we mentioned in the last section, our method of partial fraction expansion will not work unless the roots of the characteristic polynomial are distinct (that is, all roots and factors are simple). We will now modify the PFE process to accommodate multiple roots. The procedure does become more complicated, but more complicated problems require more complicated solutions. The PFE that we derive will still be much more powerful than the cascade, or sequential integration, procedure of Section 9.1.

An Explicit Form for the Operator $H(p) = 1/(p + a)^m$

Let's start by looking at the multiple root[4] case more carefully. Suppose our operator is $H(p) = 1/(p + a)^m$. If we recognize that this operator is the sequential application of the first-order operator $1/(p + a)$ m times, we can write

$$H(p) = \frac{1}{(p + a)^m}\, x(t) = \frac{1}{(p + a)^{m-1}}\left[\frac{1}{p + a}\, x(t)\right]$$

$$= \frac{1}{(p + a)^{m-1}}\left[e^{-at}\frac{1}{p}\left[e^{at}x(t)\right]\right]. \tag{9.4-1}$$

But if $m \geq 2$ (which must be the case if there are multiple roots), we can do this same thing again, getting

$$H(p) = \frac{1}{(p + a)^{m-2}}\left[e^{-at}\frac{1}{p}\left[e^{at}e^{-at}\frac{1}{p}\left[e^{at}x(t)\right]\right]\right] \tag{9.4-2}$$

$$= \frac{1}{(p + a)^{m-2}}\left[e^{-at}\frac{1}{p^2}\left[e^{at}x(t)\right]\right].$$

We can repeat this same process again and again, reducing the power of the operator outside the brackets until it reaches zero. We then have

$$H(p)x(t) = \frac{1}{(p + a)^m}\, x(t) = e^{-at}\frac{1}{p^m}\left[e^{at}x(t)\right]. \tag{9.4-3}$$

Therefore, we see that the mth-order operator with equal characteristic polynomial roots is exactly the same in form as the first-order operator—except that there is an m-fold integration, rather than only one.

---

[4] We remind you that the characteristic polynomial is the denominator of $H(p)$ with p replaced by the real or complex variable z. Thus, $D(z) = (z + a)^m$. The "roots" we are referring to are the roots of $D(z)$.

Example 9.13   Find the unit impulse response of $H(p) = 1/(p + 2)^2$.

Solution   We simply write

$$h(t) = H(p)\,\delta(t) = \frac{1}{(p + 2)^2}\,\delta(t) = e^{-2t}\frac{1}{p^2}[e^{2t}\,\delta(t)] = te^{-2t}\,u(t). \qquad (9.4\text{-}4)$$

In the last step we used the fact that $e^{2t}\delta(t) = e^0\delta(t) = \delta(t)$ and that $(1/p^2)\,\delta(t) = (1/p)\,u(t) = tu(t)$.

Because many of our examples will treat impulse responses[5] we will now present the general form for the unit impulse response of $H(p) = 1/(p + a)^m$. It is

$$\frac{1}{(p + a)^m}\,\delta(t) = e^{-at}\frac{1}{p^m}[e^{at}\,\delta(t)] = e^{-at}\frac{1}{p^m}\,\delta(t) = \frac{t^{m-1}}{(m - 1)!}e^{-at}\,u(t). \qquad (9.4\text{-}5)$$

We have here used the very important result that

$$\frac{1}{p^m}\,\delta(t) = \frac{1}{p^{m-1}}\frac{1}{p}\,\delta(t) = \frac{1}{p^{m-1}}\,u(t) = \frac{t^{m-1}}{(m - 1)!}\,u(t), \qquad (9.4\text{-}6)$$

which you should verify by performing several integrations and generalizing. Using this result, let's get an idea as to how we can develop our PFE for multiple roots with an example.

Example 9.14   Find the impulse response of $H(p) = (p + 1)/(p + 2)^2$.

Solution   The denominator has two factors: $(p + 2)$ and $(p + 2)$. They are, however, not distinct—as we require for the PFE. How do we correct the situation? Let's factor out one and perform the PFE on the operator that remains:

$$H(p) = \frac{1}{p + 2}\left[\frac{p + 1}{p + 2}\right] = \frac{1}{p + 2}\left[1 - \frac{1}{p + 2}\right] = \frac{1}{p + 2} - \frac{1}{(p + 2)^2}. \qquad (9.4\text{-}7)$$

We noted that the operator inside the first set of brackets is not proper and did one long division step, then brought the outside operator back inside the brackets by multiplication. But, you might ask, have we really accomplished anything? After all, we still have a higher-order operator to deal with! The answer, of course, is yes. We did not know how to compute the response of an operator with a general numerator; we do, however, know how to compute the response of any operator of the form $k/(p + a)^m$ by our foregoing result. Thus,

$$h(t) = H(p)\delta(t) = \left[\frac{1}{p + 2} - \frac{1}{(p + 2)^2}\right]\delta(t) \qquad (9.4\text{-}8)$$

$$= e^{-2t}\,u(t) - te^{-2t}\,u(t) = (1 - t)\,e^{-2t}\,u(t).$$

This example is the prototype of the more general development to follow.

---

[5] We will show later in this chapter that most circuit analysis problems can be reduced to the computation of an impulse response.

The PFE:
Multiple Roots

Now suppose that $H(p)$ is a general system operator that we assume to be proper—for otherwise we simply do long division steps first to make it proper—and whose characteristic polynomial $D(z)$ has, among others, a multiple root at $-a$. We can then write

$$H(p) = \frac{F(p)}{(p + a)^m}, \qquad (9.4\text{-}9)$$

where $F(p)$ is a rational fraction having as its numerator the numerator of $H(p)$ and as its denominator all of the factors of $H(p)$ corresponding to the other roots of the characteristic polynomial. Let's now write

$$H(p) = \frac{1}{(p + a)^{m-1}}\left[\frac{F(p)}{p + a}\right] = \frac{1}{(p + a)^{m-1}}\left[\frac{k_m}{p + a} + F_1(p)\right], \qquad (9.4\text{-}10)$$

where k_m is a constant and $F_1(p)$ is the operator consisting of all terms in the PFE of $F(p)/(p + a)$ other than $k_m/(p + a)$. We now bring the factor outside the brackets back in to obtain

$$H(p) = \frac{k_m}{(p + a)^m} + \frac{F_1(p)}{(p + a)^{m-1}}. \qquad (9.4\text{-}11)$$

The second operator is one that has all the same denominator factors as $H(p)$, but the one corresponding to the root at $-a$ has a multiplicity *one less*. This drop in multiplicity is important because it means that we can repeat the procedure until the multiplicity drops to zero. This results in

$$H(p) = \frac{k_m}{(p + a)^m} + \frac{k_{m-1}}{(p + a)^{m-1}} + \cdots + \frac{k_{m-i}}{(p + a)^{m-i}} \qquad (9.4\text{-}12)$$

$$+ \cdots + \frac{k_1}{p + a} + F_m(p).$$

The operator $F_m(p)$ contains all the terms in the PFE other than those explicitly shown.

Example 9.15   Perform a PFE on

$$H(p) = \frac{5p^3 + 31p^2 + 57p + 27}{(p + 2)^3(p + 1)}.$$

Solution   We specified the denominator of $H(p)$ in factored form to avoid numerical root-finding complications; in practical applications, this step would be necessary. We first factor out two of the multiple factors and obtain, after multiplying the remaining factors and doing two long division steps,

$$H(p) = \frac{1}{(p + 2)^2}\left[\frac{5p^3 + 31p^2 + 57p + 27}{p^2 + 3p + 2}\right] \qquad (9.4\text{-}13)$$

$$= \frac{1}{(p + 2)^2}\left[5p + 16 - \frac{p + 5}{(p + 2)(p + 1)}\right].$$

We can now do a conventional PFE on the last term, for it only involves simple denominator factors. Thus,

$$H(p) = \frac{1}{(p + 2)^2}\left[5p + 16 + \frac{3}{p + 2} - \frac{4}{p + 1}\right]. \qquad (9.4\text{-}14)$$

We now bring *one* of the multiple factors back inside the brackets:

$$H(p) = \frac{1}{p+2} \left[\frac{5p}{p+2} + \frac{16}{p+2} + \frac{3}{(p+2)^2} - \frac{4}{(p+2)(p+1)} \right]. \quad (9.4\text{-}15)$$

Again, we do a PFE on the last term and long division on the first. This gives

$$H(p) = \frac{1}{p+2} \left[5 - \frac{10}{p+2} + \frac{16}{p+2} + \frac{3}{(p+2)^2} + \frac{4}{p+2} - \frac{4}{p+1} \right]. \quad (9.4\text{-}16)$$

Now let's gather like terms and bring the last of the multiple factors back inside, yielding

$$H(p) = \frac{5}{p+2} + \frac{10}{(p+2)^2} + \frac{3}{(p+2)^3} - \frac{4}{(p+2)(p+1)}. \quad (9.4\text{-}17)$$

We are nearly finished! One last PFE (on the last term) does the trick! Performing this step and combining the last $p + 2$ term with the first one gives

$$H(p) = \frac{9}{p+2} + \frac{10}{(p+2)^2} + \frac{3}{(p+2)^3} - \frac{4}{p+1}. \quad (9.4\text{-}18)$$

The Derivative Formula for the PFE Coefficients

The procedure we have just outlined is simple, though perhaps somewhat lengthy. There is another way of performing the PFE for multiple roots that is more compact. To develop this procedure, let's start with the general form (9.4-12), repeated here for convenience:

$$H(p) = \frac{k_m}{(p+a)^m} + \frac{k_{m-1}}{(p+a)^{m-1}} + \cdots + \frac{k_{m-i}}{(p+a)^{m-i}} + \cdots \quad (9.4\text{-}19)$$

$$+ \frac{k_1}{p+a} + F_m(p).$$

The problem is only to evaluate the constants. To do this, let's first multiply both sides by the factor $(p + a)^m$. Thus, we have

$$(p+a)^m H(p) = k_m + k_{m-1}(p+a) + \cdots + k_{m-i}(p+a)^i \quad (9.4\text{-}20)$$

$$+ \cdots + k_1(p+a)^{m-1} + F_m(p)(p+a)^m.$$

If we replace p by $-a$ everywhere in this equation, we see that all terms on the right-hand side vanish except for the first. This gives k_m:

$$k_m = [(p+a)^m H(p)]_{p \leftarrow -a}. \quad (9.4\text{-}21)$$

The symbol $p \leftarrow -a$ means "replace p at each occurrence by $-a$." Now suppose we differentiate[6] both sides of (9.4-20) with respect to p. The derivative of a constant is zero, so the first term on the right-hand side vanishes. All of the terms other than the one involving k_{m-1} originally had powers of $p + a$ greater than one; thus, they will still have multiplicative powers of this factor after differentiation. If we substitute $-a$ for p, all of these

---

[6] This is a purely formal process. If it disturbs you, just look at $H(z)$—the system operator with p replaced everywhere by the real or complex variable z. The general form of expansion is still valid, and one can now differentiate with respect to z. This is perfectly rigorous and it results in exactly the same formulas as those above. For this reason we have chosen to merely symbolically differentiate with respect to p so that we do not waste time in converting everything to a function of z.

terms will therefore vanish. This gives

$$k_{m-1} = \frac{d}{dp}[(p + a)^m H(p)]_{p \leftarrow -a}.$$ (9.4-22)

Note carefully that the replacement of p by $-a$ occurs *after* the differentiation.

We can repeat this process with higher orders of differentiation to compute the remaining coefficients. The general term on the right-hand side, after differentiation i times (and substituting $-a$ for p) is $k_{m-i} \cdot i!$. All of the lower-order terms vanish upon differentiation, and the higher-order ones do so also after substituting $-a$ for p. Thus,

$$k_{m-i} = \frac{1}{i!}\frac{d^i}{dp^i}[(p + a)^m H(p)]_{p \leftarrow -a}; \quad i = 0, 1, \ldots, m - 1.$$ (9.4-23)

The form of this result might appear somewhat involved, but it is quite effective—as we will see in the next example. In practice, usually only a few differentiations are required.

Example 9.16   Find the impulse response of the circuit in Figure 9.36 relative to the voltage variable indicated.

Figure 9.36
An example circuit

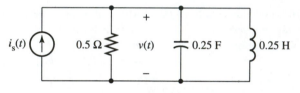

Solution   Rather than performing nodal analysis, let's use our knowledge of impedances. We first note that the system function is an operator impedance $Z(p)$. We show this conceptually by the box in Figure 9.37(a), that is, as a two-terminal subcircuit driven by the independent current source. In Figure 9.37(b) we indicate explicitly that the resistor, the inductor, and the capacitor make up the subcircuit. Noting that these elements are connected in parallel and that admittances add for parallel elements, we can write

$$Z(p) = \frac{1}{Y(p)} = \frac{1}{2 + \dfrac{p}{4} + \dfrac{4}{p}} = \frac{4p}{p^2 + 8p + 16} = \frac{4p}{(p + 4)^2}.$$ (9.4-24)

The PFE takes the form

$$\frac{4p}{(p + 4)^2} = \frac{k_2}{(p + 4)^2} + \frac{k_1}{p + 4}.$$ (9.4-25)

The coefficient k_2 can be computed as

$$k_2 = \left[(p + 4)^2 \frac{4p}{(p + 4)^2}\right]_{p \leftarrow -4} = 4(-4) = -16.$$ (9.4-26)

The remaining constant can be computed with a single derivative:

$$k_1 = \frac{d}{dp}\left[(p + 4)^2 \frac{4p}{(p + 4)^2}\right]_{p \leftarrow -4} = \frac{d}{dp}[4p]_{p \leftarrow -4} = 4.$$ (9.4-27)

Thus, we have

$$h(t) = H(p)\,\delta(t) = \left[\frac{-16}{(p+4)^2} + \frac{4}{p+4} \right] \delta(t) \tag{9.4-28}$$

$$= -16te^{-4t}\,u(t) + 4e^{-4t}\,u(t) = 4(1 - 4t)\,e^{-4t}\,u(t)\ \text{V}.$$

Figure 9.37
The system operator as
an impedance

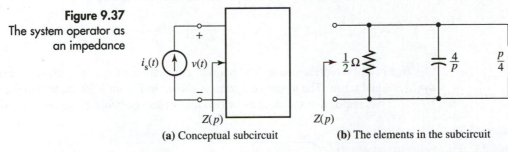

$Z(p)$ $Z(p)$

(a) Conceptual subcircuit   **(b)** The elements in the subcircuit

We call to your attention that the second-order response in the last example is critically damped. The next one illustrates this type of response a bit more.

Example 9.17   Find the value of R in Figure 9.38 for critical damping.

Figure 9.38
An example circuit

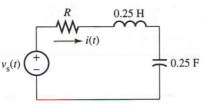

Solution   Here we desire the operator admittance of the series RLC subcircuit because $i(t) = Y(p)\,v_s(t)$; however, because the impedances of series elements sum, we have

$$Y(p) = \frac{1}{Z(p)} = \frac{1}{R + \dfrac{4}{p} + \dfrac{p}{4}} = \frac{4p}{p^2 + 4Rp + 16}. \tag{9.4-29}$$

Writing $Y(p) = N(p)/D(p)$, we have $N(p) = 4p$ and $D(p) = p^2 + 4Rp + 16$; hence, the characteristic polynomial for $Y(p)$ is $D(z) = z^2 + 4Rz + 16$. It has roots at

$$z_{1,2} = \frac{-4R \pm \sqrt{16R^2 - 64}}{2} = -2R \pm 2\sqrt{R^2 - 4}, \tag{9.4-30}$$

so the roots are real and equal when $R^2 = 4$, or $R = 2\ \Omega$. We exclude the case of negative values because we are assuming that the resistor is a physical resistor, not the equivalent resistance of a two-terminal subcircuit (which *could* exhibit negative resistance).

Simulation of
Operators: The
Multiple Root Case

We discussed the parallel realization of an operator in the last section—under the assumption, of course, that the roots of the characteristic polynomial were distinct. Now we will look at the modifications that occur when there are multiple roots. In order to keep our discussion manageable, we will work with a specific example. This example, however, will illustrate all of the essential concepts.

Suppose that we are to simulate the operator

$$H(p) = \frac{p^4 + 8p^3 + 28p^2 + 41p + 23}{(p + 1)^2(p + 2)}. \tag{9.4-31}$$

We will leave it to you to verify that the partial fraction expansion is

$$H(p) = p + 4 + \frac{3}{(p + 1)^2} + \frac{2}{p + 1} + \frac{5}{p + 2}. \tag{9.4-32}$$

$H(p)$ can be represented by the parallel connection of systems whose operators are the individual terms. The system diagram is shown in Figure 9.39. Note that the operator $1/(p + 1)^2$ is realized as the cascade of two first-order operators of the form $1/(p + 1)$.

Figure 9.39
Simulation of an operator with multiple characteristic polynomial roots

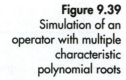

Section 9.4 Quiz

Q9.4-1. Do a PFE on

$$H(p) = \frac{p^2 + 4p + 6}{(p + 1)^3}$$

using both methods studied in this section.

Q9.4-2. Do a PFE on

$$H(p) = \frac{2p^2 + 6p + 6}{(p + 1)^2(p + 2)}$$

using both methods studied in this section.

Q9.4-3. Find R in Figure Q9.4-1 for critical damping and compute the unit impulse response. Then sketch the simulation diagram.

Figure Q9.4-1

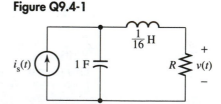

9.5 | Complex Conjugate Roots

The basic ideas of partial fraction expansion have all now been covered: the characteristic polynomial has roots that are either simple or multiple, so the preceding two sections have been exhaustive.[7] We can perform the PFE on any system function we encounter. Some elaboration is helpful, though, if any of the roots are complex. We illustrate this issue with the following example.

---

[7] But, we hope, not exhausting!

Example 9.18   Find the unit impulse response of the circuit shown in Figure 9.40 relative to the indicated response variable $v(t)$.

Figure 9.40
An example circuit

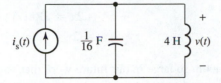

Solution   The response is the voltage across the same pair of terminals as those to which the current source is attached, so the system function is the operator impedance $Z(p)$ of the subcircuit shown in Figure 9.41. The inductor and capacitor are connected in parallel, so we can use the parallel impedance formula:

$$Z(p) = \frac{\dfrac{16}{p} \times 4p}{\dfrac{16}{p} + 4p} = \frac{16p}{p^2 + 4}. \qquad (9.5\text{-}1)$$

If we perform a partial fraction expansion, we will have

$$Z(p) = \frac{16p}{p^2 + 4} = \frac{k_1}{p + j2} + \frac{k_2}{p - j2}. \qquad (9.5\text{-}2)$$

We compute the constants in the PFE according to

$$k_1 = [(p + j2)Z(p)]_{p \leftarrow -j2} = \left[(p + j2) \frac{16p}{(p + j2)(p - j2)} \right]_{p \leftarrow -j2} \qquad (9.5\text{-}3)$$

$$= \frac{16(-j2)}{-j4} = 8$$

and $$k_2 = [(p - j2)Z(p)]_{p \leftarrow j2} = \left[(p - j2) \frac{16p}{(p + j2)(p - j2)} \right]_{p \leftarrow j2} \qquad (9.5\text{-}4)$$

$$= \frac{16(j2)}{j4} = 8.$$

Thus, the impulse response is

$$h(t) = Z(p)\,\delta(t) = \left[\frac{8}{p + j2}\delta(t) + \frac{8}{p - j2}\delta(t) \right] \qquad (9.5\text{-}5)$$

$$= 8\,e^{-j2t}\,u(t) + 8\,e^{j2t}\,u(t).$$

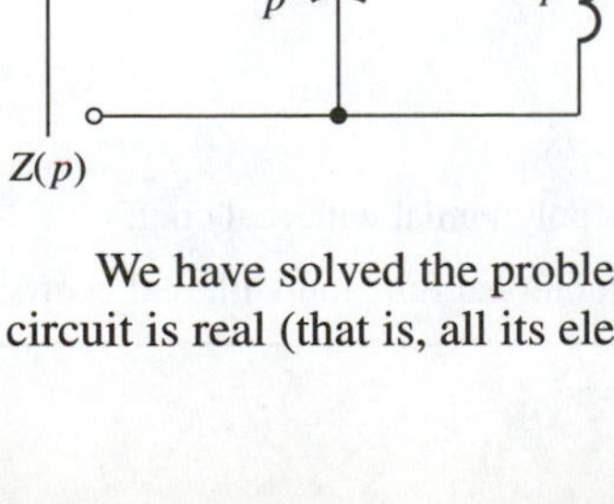

Figure 9.41
LC subcircuit

$Z(p)$

We have solved the problem, but it really isn't a very satisfactory solution. Why? Our circuit is real (that is, all its element values are real) and so is the input—so why does the response appear to be complex? In fact, it is not. The two terms in (9.5-5) have a very particular form. If we recall that the complex conjugate of any complex number $z = x + jy$ is defined by $z* = x - jy$, we immediately have the following two relations:

$$z + z* = 2x = 2\text{Re}\{z\} \qquad\qquad (9.5\text{-}6)$$
$$z - z* = j2y = j2\text{Im}\{z\}.$$

These equations loom so large in our future work that you should file them in your memory now. Applying them to (9.5-5) with $z = e^{j2t}$, we obtain the perfectly real result

$$h(t) = 16 \cos(2t)u(t). \qquad\qquad (9.5\text{-}7)$$

Circuit Response: Complex Root Case

We will now generalize upon the last example, showing that circuits—and therefore their system operators—having real parameters and real forcing functions have real responses, even if the roots of the characteristic polynomial are complex. Toward this end, we present a few properties of complex conjugates in Table 9.1. We leave the verification of properties 1, 2, and 3 to you. You need only compute both sides of each identity (letting $z = x + jy$, $z_1 = x_1 + jy_1$, and $z_2 = x_2 + jy_2$) and place them in, say, rectangular form to confirm their validity.

Property 4 follows immediately from the linearity properties of the differentiation operator: $p^m(z) = p^m(x + jy) = p^m x + jp^m y$. Thus, $[p^m z]* = p^m x - jp^m y = p^m(x - jy) = p^m(z*)$. (See Chapter 8 if you need a bit of a review on complex numbers.) Let's illustrate property 5 with a concrete example that generalizes immediately to prove the result. Let $Q(z) = 2z^2 + 3z + 2$. Then $Q*(z) = [2z^2 + 3z + 2]*$. We can use property 1 to write this as $Q*(z) = [2z^2]* + [3z]* + [2]*$; then, using property 2 and the fact that the conjugate of a real number is the same real number, we can express this in the form $Q*(z) = 2[z*]^2 + 3[z*] + 2 = Q(z*)$. You see how it goes. This thinking works for the general case. Coupling property 5 with property 3, we now see that property 6 is true.

Let's put these properties to use. Suppose that $H(p)$ is a real system operator of the form

$$H(p) = \frac{N(p)}{D(p)}, \qquad\qquad (9.5\text{-}8)$$

where $N(p)$ and $D(p)$ are polynomials in p with real coefficients. The characteristic polynomial is, then $D(z)$. If $D(z)$ has a simple complex root at $-a$, we will have

$$D(-a) = 0. \qquad\qquad (9.5\text{-}9)$$

But this implies that

$$D(-a*) = D*(-a) = 0* = 0 \qquad\qquad (9.5\text{-}10)$$

Table 9.1
Some properties of
complex conjugates

1. $(z_1 + z_2)* = z_1^* + z_2^*$

2. $(z_1 z_2)* = z_1^* z_2^*$

3. $\left[\dfrac{z_1}{z_2}\right]^* = \dfrac{z_1^*}{z_2^*}$

4. $(p^m z)^* = p^m(z^*)$

5. $Q^*(z) = Q(z^*)$; $Q(z)$ a polynomial with real coeffs.

6. $H^*(z) = H(z^*)$; $H(z)$ a rational function with real coeffs.

also; thus, if $-a$ is a root, so is its conjugate $-a^*$. Therefore, we can partially factor $D(z)$:

$$D(z) = (z + a)(z + a^*)D_0(z), \tag{9.5-11}$$

where $D_0(z)$ has no roots at either $-a$ or $-a^*$. Our system operator can then be written as

$$H(p) = \frac{N(p)}{(p + a)(p + a^*)D_0(p)}, \tag{9.5-12}$$

where $N(p)$ and $D_0(p)$ are polynomial operators with real coefficients.

If we now do a PFE, we can write $H(p)$ in the form

$$H(p) = \frac{N(p)}{(p + a)(p + a^*)D_0(p)} = \frac{k_1}{p + a} + \frac{k_2}{p + a^*} + F(p), \tag{9.5-13}$$

where $F(p)$ does not involve the complex conjugate roots at $-a$ and $-a^*$. We will evaluate the constants k_1 and k_2 and show that they are related in a special way—the recognition of which halves our computational load. First, k_1:

$$k_1 = [(p + a)H(p)]_{p \leftarrow -a} = \left[\frac{N(p)}{(p + a^*)D_0(p)} \right]_{p \leftarrow -a} = \frac{N(-a)}{(-a + a^*)D_0(-a)}. \tag{9.5-14}$$

Now let's look at k_2:

$$k_2 = [(p + a^*)H(p)]_{p \leftarrow -a^*} \left[\frac{N(p)}{(p + a)D_0(p)} \right]_{p \leftarrow -a^*} = \frac{N(-a^*)}{(-a^* + a)D_0(-a^*)}. \tag{9.5-15}$$

If we use our table of properties of complex conjugates, we will see at once that

$$k_2 = k_1^*. \tag{9.5-16}$$

The importance of this property cannot be stressed enough. It saves half the work in doing the PFE. As we will now see, it also saves half the labor in computing the time response.

The response of a system having the system operator $H(p)$ of equation (9.5-13) will be

$$y(t) = H(p)x(t) = \frac{k_1}{p + a}x(t) + \frac{k_1^*}{p + a^*}x(t) + F(p)x(t). \tag{9.5-17}$$

We notice, however, that the second term is the complex conjugate of the first—*provided that $x(t)$ is a real time function*. In this case, we can use the relation $z + z^* = 2\text{Re}\{z\}$ with $z = (k_1/(p + a))x(t)$ to write the response as

$$y(t) = 2\text{Re}\left[\frac{k_1}{p + a}x(t) \right] + F(p)x(t). \tag{9.5-18}$$

We have again saved half the labor in computing that part of the response due to the complex conjugate pair of roots. Let's put our theory to use, now, with an example.

Example 9.19 Find the unit impulse response of the differential equation

$$\frac{d^2y}{dt^2} + 4\frac{dy}{dt} + 13y = 2\frac{d^2x}{dt^2} + 8\frac{dx}{dt} + 44x. \tag{9.5-19}$$

Solution   The characteristic polynomial is $D(z) = z^2 + 4z + 13 = (z + 2)^2 + 3^2$, which has roots at $z_{1,2} = -2 \pm j3$. Thus, we write the system operator by inspection:

$$H(p) = \frac{2p^2 + 8p + 44}{p^2 + 4p + 13} = \frac{2p^2 + 8p + 44}{(p + 2 + j3)(p + 2 - j3)}. \qquad (9.5\text{-}20)$$

We must be careful here because the numerator has the same degree as the denominator. Thus, one step of long division is required. This gives

$$H(p) = \frac{2p^2 + 8p + 44}{p^2 + 4p + 13} = 2 + \frac{18}{(p + 2 + j3)(p + 2 - j3)} \qquad (9.5\text{-}21)$$

$$= 2 + \frac{k}{p + 2 + j3} + \frac{k^*}{p + 2 - j3},$$

where we have made use of our knowledge of the conjugate relationship of the constants in the PFE. We need only to compute k:

$$k = \left[(p + 2 + j3) \frac{18}{(p + 2 + j3)(p + 2 - j3)} \right]_{p \leftarrow -2 - j3} = \frac{18}{-j6} = j3. \qquad (9.5\text{-}22)$$

As $k^* = -j3$, we have (with $x(t) = \delta(t)$)

$$h(t) = 2\delta(t) + 2\mathrm{Re}\left\{ \frac{j3}{p + 2 + j3} \delta(t) \right\} = 2\delta(t) + 2\mathrm{Re}\{j3\, e^{-(2+j3)t}u(t)\} \qquad (9.5\text{-}23)$$

$$= 2\delta(t) + 2e^{-2t}\mathrm{Re}\{j3[\cos(3t) - j\sin(3t)]\}\, u(t) = 2\delta(t) + 6e^{-2t}\sin(3t)u(t).$$

The PFE: General Case — Complex, Multiple Roots

Let's go on now to the general case — that for which the characteristic polynomial has multiple complex conjugate roots in its characteristic polynomial. In this case, the form of the system operator is

$$H(p) = \frac{N(p)}{(p + a)^m(p + a^*)^m D_0(p)}, \qquad (9.5\text{-}24)$$

where $D_0(p)$ has no factors involving a or a^*. Now, we know from our results in Section 9.4 that the PFE has the form

$$H(p) = \frac{k_m}{(p + a)^m} + \frac{k_{m-1}}{(p + a)^{m-1}} + \cdots + \frac{k_{m-i}}{(p + a)^{m-i}} + \cdots + \frac{k_1}{p + a} \qquad (9.5\text{-}25)$$

$$+ \frac{k'_m}{(p + a^*)^m} + \frac{k'_{m-1}}{(p + a^*)^{m-1}} + \cdots + \frac{k'_{m-i}}{(p + a^*)^{m-i}} + \cdots + \frac{k'_1}{p + a^*} + F(p),$$

where the operator $F(p)$ consists of all terms in the PFE other than those involving either $p + a$ or $p + a^*$. It is a fact that should be intuitively plausible, but which we will defer to the end of chapter problems for proof, that the primed coefficients are the complex conjugates of the unprimed ones:

$$k'_{m-i} = k^*_{m-i}, \quad i = 0, 1, \ldots, m - 1. \qquad (9.5\text{-}26)$$

The response to any (real) input, then, is

$$y(t) = H(p)x(t) = 2\mathrm{Re}\left\{ \left[\frac{k_m}{(p + a)^m} + \cdots + \frac{k_1}{p + a} \right] x(t) \right\} + F(p)x(t). \qquad (9.5\text{-}27)$$

The following example should convince you that such a situation is not just a mathematical curiosity, but can occur in a physical circuit.

Example 9.20   Find the unit impulse response for $v(t)$ in Figure 9.42.

Figure 9.42
An example circuit

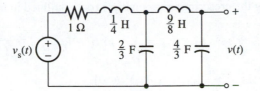

Solution   Let's use impedances and admittances to solve this problem. The circuit is shown in Figure 9.43 with impedance operators for the elements and an auxiliary impedance defined, along with a ground reference and an auxiliary node voltage v'. We will first compute the indicated impedance $Z(p)$ and then use it in the voltage divider rule to find $v'(t)$. It is

$$Z(p) = \frac{\dfrac{3}{2p} \times \left[\dfrac{9}{8}p + \dfrac{3}{4p}\right]}{\dfrac{3}{2p} + \dfrac{9}{8}p + \dfrac{3}{4p}} = \frac{3p^2 + 2}{2p^3 + 4p}. \tag{9.5-28}$$

Now we compute the auxiliary node voltage $v'(t)$ using voltage division:

$$v'(t) = \frac{Z(p)}{1 + \dfrac{p}{4} + Z(p)} v_s(t) = \frac{3p^2 + 2}{\dfrac{1}{2}p^4 + 2p^3 + 4p^2 + 4p + 2} v_s(t). \tag{9.5-29}$$

Then we apply voltage division once more, this time to the rightmost inductor and capacitor:

$$v(t) = \frac{\dfrac{3}{4p}}{\dfrac{9}{8}p + \dfrac{3}{4p}} v'(t) = \frac{2}{3p^2 + 2} \times \frac{3p^2 + 2}{\dfrac{1}{2}p^4 + 2p^3 + 4p^2 + 4p + 2} v_s(t) \tag{9.5-30}$$

$$= \frac{4}{p^4 + 4p^3 + 8p^2 + 8p + 4} v_s(t).$$

Thus,

$$H(p) = \frac{4}{p^4 + 4p^3 + 8p^2 + 8p + 4} = \frac{4}{[(p + 1)^2 + 1]^2} \tag{9.5-31}$$

$$= \frac{4}{(p + 1 + j)^2 (p + 1 - j)^2}.$$

Figure 9.43
The example circuit
prepared for analysis

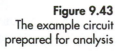
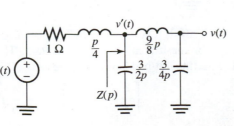

The PFE has the form

$$H(p) = \frac{k_2}{(p + 1 + j)^2} + \frac{k_1}{p + 1 + j} + \frac{k_2^*}{(p + 1 - j)^2} + \frac{k_1^*}{p + 1 - j}. \qquad (9.5\text{-}32)$$

Following the prescription for multiple roots outlined in the last section, we find that

$$k_2 = [(p + 1 + j)^2 H(p)]_{p \leftarrow -1-j} = \frac{4}{(-j2)^2} = -1 \qquad (9.5\text{-}33)$$

and

$$k_1 = \frac{d}{dp}[(p + 1 + j)^2 H(p)]_{p \leftarrow -1-j} \qquad (9.5\text{-}34)$$

$$= \frac{d}{dp}\left[\frac{4}{(p + 1 - j)^2}\right]_{p \leftarrow -1-j} = \left[\frac{-8}{(p + 1 - j)^3}\right]_{p \leftarrow -1-j} = \frac{-8}{(-j2)^3} = j.$$

Thus,

$$h(t) = 2\mathrm{Re}\left\{\left[\frac{-1}{(p + 1 + j)^2} + \frac{j}{p + 1 + j}\right]\delta(t)\right\} \qquad (9.5\text{-}35)$$

$$= 2\mathrm{Re}\,\{-te^{-(1+j)t}u(t) + je^{-(1+j)t}u(t)\}.$$

Here, we used the formula we derived at the beginning of this section for the impulse response of a system operator of the form $1/(p + a)^m$ with $a = 1 + j$ and $m = 2$. Finally,

$$h(t) = 2e^{-t}\mathrm{Re}\{-t\cos(t) + jt\sin(t) + j\cos(t) + \sin(t)\}\,u(t) \qquad (9.5\text{-}36)$$

$$= 2e^{-t}[-t\cos(t) + \sin(t)]\,u(t).$$

A Rationale for Using the Impulse Response

We have thus far concentrated on the impulse response of various circuits for a couple of reasons. As we pointed out earlier, this is perhaps the simplest possible input; thus, it allows us to concentrate upon our techniques of circuit analysis without becoming embroiled in more complicated integrations. The second reason will be developed in Section 9.6: virtually any waveform of interest in the analysis of circuits and systems can itself be represented as the impulse response of a differential operator. In the example that follows, however, we illustrate the computation of the response to a more general forcing function.

Example 9.21   Find the response of the circuit in Figure 9.44 if $i_s(t) = 3e^{-t}u(t)$.

Figure 9.44
An example circuit

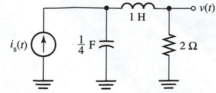

Solution   Because mesh analysis is so straightforward, let's use it on this example. We show the circuit prepared for mesh analysis in Figure 9.45. The single KVL equation around the right-hand mesh (the left-hand mesh is nonessential) is

$$\left[\frac{4}{p} + p + 2\right]i(t) - \frac{4}{p}i_s(t) = 0, \qquad (9.5\text{-}37)$$

or

$$v(t) = 2i(t) = \frac{8}{p^2 + 2p + 4} \, i_s(t), \tag{9.5-38}$$

so

$$H(p) = \frac{8}{p^2 + 2p + 4} = \frac{8}{(p + 1 + j\sqrt{3})(p + 1 - j\sqrt{3})} \tag{9.5-39}$$

$$= \frac{k}{p + 1 + j\sqrt{3}} + \frac{k^*}{p + 1 - j\sqrt{3}},$$

where

$$k = [(p + 1 + j\sqrt{3}) \, H(p)]_{p \leftarrow -1 - j\sqrt{3}} = \frac{8}{-j2\sqrt{3}} = j \frac{4}{\sqrt{3}}. \tag{9.5-40}$$

The desired response, then, is

$$v(t) = 2\text{Re}\left\{ \frac{j\frac{4}{\sqrt{3}}}{p + 1 + j\sqrt{3}} 3e^{-t} u(t) \right\} = 8\sqrt{3}\,\text{Re}\left\{ je^{-(1+j\sqrt{3})t} \frac{1}{p} [e^{(1+j\sqrt{3})t}e^{-t} u(t)] \right\} \tag{9.5-41}$$

$$= 8\sqrt{3}\,\text{Re}\left\{ je^{-(1+j\sqrt{3})t} \frac{1}{j\sqrt{3}} [e^{j\sqrt{3}t} - 1] \right\} u(t) = 8e^{-t}[1 - \cos(\sqrt{3}t)] \, u(t).$$

This waveform is sketched in Figure 9.46.

Figure 9.45
The example circuit prepared for mesh analysis

Figure 9.46
Response waveform

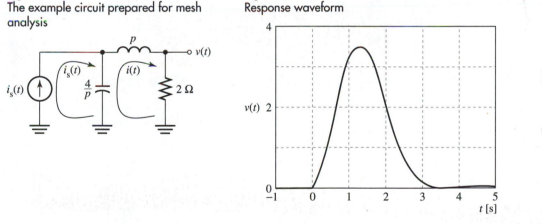

We now provide a final example.

Example 9.22 Find and sketch the unit step response of the second-order circuit drawn in Figure 9.47 (using three ground symbols for the bottom node in that figure) for $R = 6\,\Omega$, $5\,\Omega$, and $3\,\Omega$.

Figure 9.47
An example circuit

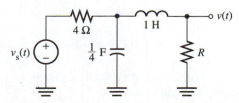

Solution   This is the same circuit that we used to discuss the damping properties of second-order systems in Section 9.2. We computed the unit impulse response there, and you might wish to compare it with the results we obtain here for the step response. We will leave it to you to show (using nodal, mesh, or series-parallel impedance/admittance analysis), or by simply referring back to Section 9.2, that

$$H(p) = \frac{R}{p^2 + (R + 1)p + (R + 4)} . \tag{9.5-42}$$

Case a: For $R = 6\ \Omega$, we have (leaving the details of the PFE up to you)

$$H(p) = \frac{6}{p^2 + 7p + 10} = \frac{6}{(p + 2)(p + 5)} = \frac{2}{p + 2} - \frac{2}{p + 5} . \tag{9.5-43}$$

The step response, therefore, is

$$s(t) = H(p)u(t) = \frac{1}{5}[3 - 5e^{-2t} + 2e^{-5t}]\, u(t). \tag{9.5-44}$$

Case b: For $R = 5\ \Omega$,

$$H(p) = \frac{5}{p^2 + 6p + 9} = \frac{5}{(p + 3)^2} ,$$

so $$s(t) = 5e^{-3t} \frac{1}{p^2}\, [e^{3t}u(t)] = \frac{5}{9}\, [1 - (3t + 1)e^{-3t}]\, u(t). \tag{9.5-45}$$

Again, we have left the details of integration up to you.

Case c: For $R = 1\ \Omega$,

$$H(p) = \frac{1}{p^2 + 2p + 5} = \frac{1}{(p + 1 + j2)(p + 1 - j2)} \tag{9.5-46}$$

$$= \frac{j\dfrac{1}{4}}{p + 1 + j2} - \frac{j\dfrac{1}{4}}{p + 1 - j2} .$$

The step response is

$$s(t) = 2\mathrm{Re}\left\{ \frac{j\dfrac{1}{4}}{p + 1 + j2} u(t) \right\} = \frac{1}{2}\, \mathrm{Re}\left\{ \frac{j}{1 + j2}[1 - e^{-(1+j2)t}]\, u(t) \right\}, \tag{9.5-47}$$

which, after just a bit of complex algebra, simplifies to

$$s(t) = \frac{1}{10}[2 - e^{-t}[2\cos(2t) + \sin(2t)]]\, u(t). \tag{9.5-48}$$

Case a is overdamped, Case b is critically damped, and Case c is underdamped. Plots of all three are shown in Figure 9.48.

Figure 9.48
Step response
waveforms

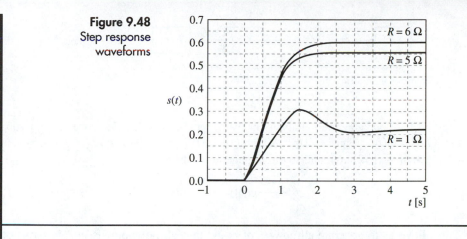

If you have mastered the complex number manipulations presented in this section, you will be well prepared to solve general time response problems as well as ac forced response problems, to which we will devote quite a bit of attention in succeeding chapters.

Section 9.5 Quiz

Q9.5-1.   Perform a partial fraction expansion on

$$H(p) = \frac{12}{p^2 + 4p + 13}$$

and find the unit impulse and step responses.

Q9.5-2.   Find $H(p)$ for the circuit in Figure Q9.5-1. Then find the unit impulse response $h(t)$ and the unit step response $s(t)$.

Figure Q9.5-1

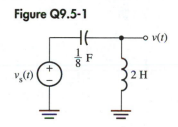

9.6 | An Algebraic Circuit Analysis Method

Time Domain Analysis
Without Integration:
Elementary Waveforms

Our technique for circuit analysis thus far requires integration in the final step—except for the unit impulse and step responses, which we agreed to commit to memory. As it happens, though, all of the forcing functions normally encountered in circuits are elementary: linear combinations of exponentials, sinusoids, and products of these waveforms. For such forcing functions (that is, waveforms of the independent sources) it is possible to reduce our procedure to one that is strictly algebraic. *No integrations will be required!*

We will define an *elementary waveform* to be one that can be written as the impulse response of a differential system, or operator, as shown in Figure 9.49(a). If we then apply such a signal to a circuit or system, we will have the cascade representation of Figure 9.49(b). It is then only necessary to find the impulse response of the entire equivalent system. In equation form, this means that if

$$x(t) = X(p)\delta(t), \tag{9.6-1}$$

then

$$y(t) = H(p)x(t) = H(p)X(p)\delta(t). \tag{9.6-2}$$

Thus, if we know the impulse responses of just a very few elementary types of operator we can perform a PFE and find the response $y(t)$ *entirely with algebraic operations.*

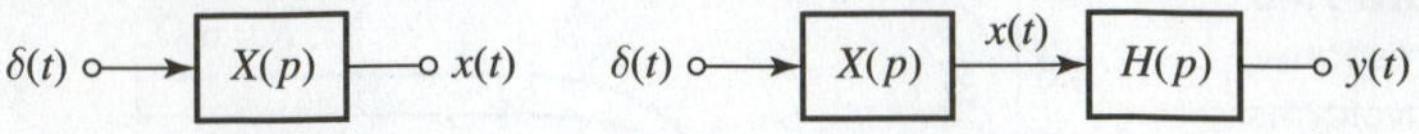

Figure 9.49
Response of a
differential system to an
elementary forcing
function

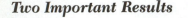

(a) Generation of an
elementary waveform

(b) Cascade representation

Two Important Results   We will base our algebraic analysis on two important results. First, we remind you that $(1/p)\,\delta(t) = u(t)$. Thus, $(1/p^2)\,\delta(t) = (1/p)\,u(t) = tu(t)$; $(1/p^3)\,\delta(t) = (1/p^2)\,u(t) = (1/p)\,[tu(t)] = (t^2/2)\,u(t)$; and so on. In general, we have

$$\frac{t^m}{m!}u(t) = \frac{1}{p^{m+1}}\delta(t). \tag{9.6-3}$$

This allows us to generate all powers of t, and thus all polynomials in t, in terms of the impulse response of simple differential systems. Our second important result above is the following:

Theorem 9.2 (The   If $H(p)$ is a rational differential operator (a Heaviside, or system, operator), then for
Shift Theorem)   any real or complex number r

$$e^{-rt}\,H(p)\,[e^{rt}\,x(t)] = H(p + r)\,x(t). \tag{9.6-4}$$

Proof   The PFE for $H(p)$ shows that it can be written as the linear combination of a constant, powers of p, that is p^m, and powers[8] of the first-order solution operator, $1/(p + a)^m$.

For the constant, the proof is quite clear because the exponentials cancel each other and $H(p + r) = H(p) = $ constant. Thus, suppose that $H(p) = p^m$. Then

$$e^{-rt}\,p^m\,[e^{rt}\,x(t)] = e^{-rt}\,p^{m-1}\,p\,[e^{rt}\,x(t)] \tag{9.6-5}$$

$$e^{-rt}\,p^{m-1}\,[e^{rt}\,px(t) + re^{rt}\,x(t)] = e^{-rt}\,p^{m-1}\,[e^{rt}\,(p + r)\,x(t)].$$

We have reduced the power of the differentiation operator by one and have replaced $x(t)$ with $(p + r)x(t)$. If we repeat the same argument $m - 1$ more times, the exponentials will cancel in the last step and we will have

$$e^{-rt}\,p^m\,[e^{rt}\,x(t)] = (p + r)^m\,x(t) \tag{9.6-6}$$

Finally, let $H(p) = 1/(p + a)^m$. Then

$$e^{-rt}\,\frac{1}{(p + a)^m}[e^{rt}\,x(t)] = e^{-rt}\,e^{-at}\,\frac{1}{p^m}[e^{at}\,e^{rt}\,x(t)], \tag{9.6-7}$$

by our earlier result for this operator. Combining the exponentials now gives the result.

We summarize two foundational results in Table 9.2. Our entire method of analysis depends upon these two results and the partial fraction expansion, so in the event of a memory failure for any result that we now derive, you can return to these two facts and

---

[8] If $H(p)$ is a proper operator, then the constant and powers of p do not appear; if $H(p)$ is improper but nonsingular, the constant appears, but the powers of p do not.

Table 9.2
Two important
operator formulas

1. $\dfrac{t^m}{m!} u(t) = \dfrac{1}{p^{m+1}} \delta(t)$

2. $e^{-rt} H(p)[e^{rt} x(t)] = H(p + r)x(t)$

rederive it. Before proceeding, however, let's note the particular form taken on by item 2 in Table 9.2 when $x(t) = \delta(t)$:

$$e^{-rt} H(p)[e^{rt} \delta(t)] = e^{-rt} H(p)[\delta(t)] = e^{-rt} h(t), \qquad (9.6\text{-}8)$$

where $h(t) = H(p)\delta(t)$ is the impulse response of $H(p)$. Thus, we have

$$H(p + r)\delta(t) = e^{-rt} h(t). \qquad (9.6\text{-}9)$$

A Table of Impulse
Response
Representations

In each of the following examples we will derive the impulse response representation of an elementary waveform. They are all essential to the algebraic analysis to be developed, so we will summarize them in a table for your convenience in using them later.

Example 9.23   If $x(t) = e^{-at} u(t)$, write $x(t)$ in the operator form $x(t) = X(p)\delta(t)$.

Solution   We simply write

$$e^{-at} u(t) = e^{-at} \frac{1}{p} \delta(t) = e^{-at} \frac{1}{p}[e^{at} \delta(t)] = \frac{1}{p + a} \delta(t), \qquad (9.6\text{-}10)$$

so

$$e^{-at} u(t) = \frac{1}{p + a} \delta(t). \qquad (9.6\text{-}11)$$

Example 9.24   Write $x(t) = \cos(\omega_0 t) u(t)$, in the operator form $x(t) = X(p)\delta(t)$.

Solution   We first use one form of Euler's formula: $\cos(\omega_0 t) = (e^{j\omega_0 t} + e^{-j\omega_0 t})/2$, then apply the result derived in Example 9.23, that is, equation (9.6-11):

$$x(t) = \frac{e^{j\omega_0 t} + e^{-j\omega_0 t}}{2} u(t) = \frac{1}{2}[e^{j\omega_0 t} u(t) + e^{-j\omega_0 t} u(t)]$$

$$= \frac{1}{2}\left[\frac{1}{p - j\omega_0} + \frac{1}{p + j\omega_0}\right]\delta(t) = \frac{p}{p^2 + \omega_0^2} \delta(t).$$

Thus,
$$\cos(\omega_0 t) u(t) = \frac{p}{p^2 + \omega_0^2} \delta(t). \qquad (9.6\text{-}12)$$

Example 9.25   Write $x(t) = \sin(\omega_0 t) u(t)$ in the operator form $x(t) = X(p)\delta(t)$.

Solution   An alternate form of Euler's formula is $\sin(\omega_0 t) = (e^{j\omega_0 t} - e^{-j\omega_0 t})/j2$. This gives, when we apply the result derived in Example 9.23,

$$x(t) = \frac{e^{j\omega_0 t} - e^{-j\omega_0 t}}{j2} u(t) = \frac{1}{j2} [e^{j\omega_0 t} u(t) - e^{-j\omega_0 t} u(t)]$$

$$= \frac{1}{j2} \left[\frac{1}{p - j\omega_0} - \frac{1}{p + j\omega_0} \right] \delta(t) = \frac{\omega_0}{p^2 + \omega_0^2} \delta(t),$$

or
$$\sin(\omega_0 t) u(t) = \frac{\omega_0}{p^2 + \omega_0^2} \delta(t). \qquad (9.6\text{-}13)$$

Example 9.26　Find the operator forms for the waveforms $x(t) = e^{-at} \cos(\omega_0 t) u(t)$ and $x(t) = e^{-at} \sin(\omega_0 t) u(t)$.

Solution　After doing Examples 9.24 and 9.25 this one is easy. For the cosine, we write

$$x(t) = e^{-at} \frac{p}{p^2 + \omega_0^2} \delta(t) = e^{-at} \frac{p}{p^2 + \omega_0^2} [e^{at} \delta(t)] \qquad (9.6\text{-}14)$$

and apply the shift theorem, to obtain

$$e^{-at} \cos(\omega_0 t) u(t) = \frac{(p + a)}{(p + a)^2 + \omega_0^2} \delta(t). \qquad (9.6\text{-}15)$$

Similarly, for the sine function we get

$$e^{-at} \sin(\omega_0 t) u(t) = \frac{\omega_0}{(p + a)^2 + \omega_0^2} \delta(t). \qquad (9.6\text{-}16)$$

Example 9.27　Write $x(t) = (t^m/m!) \, e^{-at} u(t)$ in the operator form $x(t) = X(p) \, \delta(t)$.

Solution　We already know that $(t^m/m!) \, u(t) = 1/(p^{m+1}) \, \delta(t)$, so we simply use the shift theorem to get

$$\frac{t^m}{m!} e^{-at} u(t) = \frac{1}{(p + a)^{m+1}} \delta(t). \qquad (9.6\text{-}17)$$

This, of course, is of great interest for this $x(t)$ is the impulse response of the basic operator in the PFE.

We summarize the results of Examples 9.23 through 9.27 in Table 9.3.

With Table 9.3 and the partial fraction expansion, we can solve most typical problems in circuit analysis algebraically. Thus, the importance of the algebraic method should not be underestimated. It will save an enormous amount of work in the solution of circuits problems. Examples 9.28 through 9.30 illustrate this.

Table 9.3
Impulse response representations

$x(t)$	$X(p)$
1. $e^{-at}u(t)$	$\dfrac{1}{p+a}$
2. $\dfrac{t^m}{m!}e^{-at}u(t)$	$\dfrac{1}{(p+a)^{m+1}}$
3. $\cos(\omega_0 t)u(t)$	$\dfrac{p}{p^2+\omega_0^2}$
4. $\sin(\omega_0 t)u(t)$	$\dfrac{\omega_0}{p^2+\omega_0^2}$
5. $e^{-at}\cos(\omega_0 t)u(t)$	$\dfrac{p+a}{(p+a)^2+\omega_0^2}$
6. $e^{-at}\sin(\omega_0 t)u(t)$	$\dfrac{\omega_0}{(p+a)^2+\omega_0^2}$

Example 9.28   Solve the differential equation

$$\frac{d^2y}{dt^2} + 2\frac{dy}{dt} + y = x(t) \tag{9.6-18}$$

with the forcing function $x(t) = 4\cos(t)u(t)$.

Solution   Expressing the differential equation and the forcing function in operator form, we have

$$(p+1)^2 y(t) = \frac{4p}{p^2+1}\delta(t). \tag{9.6-19}$$

Therefore, the solution is

$$y(t) = \frac{4p}{(p+1)^2(p^2+1)}\delta(t). \tag{9.6-20}$$

Thus,

$$Y(p) = \frac{4p}{(p+1)^2(p^2+1)} = \frac{1}{p+1}\left[\frac{4p}{(p+1)(p^2+1)}\right] \tag{9.6-21}$$

$$= \frac{1}{p+1}\left[\frac{-2}{p+1} + \frac{1+j}{p+j} + \frac{1-j}{p-j}\right]$$

$$= \frac{-2}{(p+1)^2} + \frac{2(p+1)}{(p+1)(p^2+1)} = \frac{-2}{(p+1)^2} + \frac{2}{p^2+1}.$$

Notice that the numerator and denominator factors $(p+1)$ cancel in the last line. Finally, we have

$$y(t) = Y(p)\,\delta(t) = -2te^{-t}u(t) + 2\sin(t)u(t). \tag{9.6-22}$$

A Special PFE Technique: Complex Roots with Real Algebra

Before closing this highly important section we will present one more procedure that reduces the computational load. In the case of *simple* complex conjugate factors in the characteristic polynomial, it is possible to work entirely with real algebra. The key idea is that the PFE terms for complex conjugate roots can be combined into a single one with a second-order denominator and a first-order numerator—both with real coefficients. To be specific, the simple conjugate roots in the characteristic polynomial give rise to the factors $k/(p + a) + k^*/(p + a^*)$ in the PFE. But we can combine them to give

$$\frac{(k + k^*)p + (ka^* + k^*a)}{p^2 + (a + a^*)p + aa^*} = \frac{Ap + B}{p^2 + rp + q},$$ (9.6-23)

where all the coefficients are real. The only problem now is to solve for A and B. We will illustrate the procedure with two examples. The first example demonstrates the algebraic method for solving operator equations.

Example 9.29   If $H(p) = 4/(p + 2)$ and $x(t) = 5e^{-t}\cos(2t)u(t)$ for a given system, find the system output $y(t) = H(p)x(t)$.

Solution   First, we use Table 9.3, item 5, to write

$$x(t) = \frac{5(p + 1)}{(p + 1)^2 + 4}\delta(t),$$ (9.6-24)

so the response is

$$y(t) = H(p)\,x(t) = \frac{20(p + 1)}{(p + 2)[(p + 1)^2 + 4]}\delta(t) = Y(p)\delta(t).$$ (9.6-25)

Thus, the characteristic polynomial $D(z) = (z + 2)[(z + 1)^2 + 4]$ has a real root at $z = -2$ and complex conjugate roots at $z = -1 \pm j2$. We will, however, use the real form that we have just derived for the complex conjugate factors. Thus, the PFE has the form

$$Y(p) = \frac{k}{p + 2} + \frac{Ap + B}{p^2 + 2p + 5}.$$ (9.6-26)

We evaluate k in the usual manner:

$$k = [(p + 2)Y(p)]_{p \leftarrow -2} = -4.$$ (9.6-27)

We evaluate A and B in a slightly different manner. We multiply both sides of equation (9.6-26) by the entire denominator of $Y(p)$, giving

$$20(p + 1) = -4(p^2 + 2p + 5) + (Ap + B)(p + 2).$$ (9.6-28)

Then we equate coefficients of each power of p. Thus,

$$p^2: 0 = -4 + A$$ (9.6-29a)

$$p^1: 20 = -8 + (2A + B)$$ (9.6-29b)

$$p^0: 20 = -20 + 2B.$$ (9.6-29c)

The first of these equations gives $A = 4$ and the last gives $B = 20$. We can use the second equation to check our results. With these values of A and B, it does reduce to an identity.

Thus, the response of our system is

$$y(t) = \frac{-4}{p+2}\delta(t) + \frac{4p+20}{(p+1)^2+2^2}\delta(t). \qquad (9.6\text{-}30)$$

But now what? Well, we see that the second operator looks almost like the sum of a sine and a cosine multiplied by an exponential (see Table 9.3, items 5 and 6). We can express it in this form by writing

$$y(t) = \frac{-4}{p+2}\delta(t) + 4\frac{(p+1)}{(p+1)^2+2^2}\delta(t) + 8\frac{2}{(p+1)^2+2^2}\delta(t) \qquad (9.6\text{-}31)$$

$$= -4e^{-2t}u(t) + 4e^{-t}\cos(2t)u(t) + 8e^{-t}\sin(2t)u(t).$$

The next example illustrates the algebraic method for finding circuit responses.

Example 9.30   Find the system operator for the indicated response of the circuit shown in Figure 9.50 in terms of R. Then, letting $R = 12\ \Omega$, find the response to the independent voltage source waveform $v_s(t) = 8e^{-t}u(t)$.

Figure 9.50
An example circuit

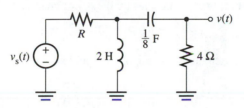

Solution   We will use impedances and voltage division in our solution. The circuit is shown in Figure 9.51 with impedance operators for the dynamic elements and an auxiliary impedance and an auxiliary node voltage defined. Now

$$Z(p) = \frac{2p\left(\dfrac{8}{p}+4\right)}{2p+\dfrac{8}{p}+4} = \frac{4p(p+2)}{p^2+2p+4}. \qquad (9.6\text{-}32)$$

Using voltage division to determine $v'(t)$, we have

$$v'(t) = \frac{Z(p)}{R+Z(p)}v_s(t) = \frac{4p(p+2)}{(R+4)p^2+(2R+8)p+4R}v_s(t), \qquad (9.6\text{-}33)$$

and, using voltage division once more,

$$v(t) = \frac{4}{4+\dfrac{8}{p}}v'(t) = \frac{4p^2}{(R+4)p^2+(2R+8)p+4R}v_s(t). \qquad (9.6\text{-}34)$$

Thus,

$$H(p) = \frac{4p^2}{(R+4)p^2+(2R+8)p+4R}. \qquad (9.6\text{-}35)$$

We now let $R = 12\ \Omega$ and $v_s(t) = 8e^{-t}u(t)$. Then

$$v(t) = V(p)\,\delta(t) = H(p)\,V_s(p)\delta(t), \qquad (9.6\text{-}36)$$

Figure 9.51
The example circuit
prepared for analysis

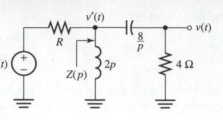

where
$$v_s(t) = 8e^{-t}u(t) = V_s(p)\delta(t) = \frac{8}{p+1}\delta(t) \tag{9.6-37}$$

and
$$H(p) = \frac{4p^2}{16p^2 + 32p + 48} = \frac{\frac{1}{4}p^2}{p^2 + 2p + 3}. \tag{9.6-38}$$

Thus,
$$V(p) = \frac{2p^2}{(p+1)[(p+1)^2+2]} = \frac{k_1}{(p+1)} + \frac{k_2 p + k_3}{(p+1)^2 + 2}. \tag{9.6-39}$$

Now
$$k_1 = [(p+1)V(p)]_{p \leftarrow -1} = 1, \tag{9.6-40}$$

and, after multiplying both sides by the denominator of $V(p)$, we have

$$2p^2 = 1\,[(p+1)^2 + 2] + (k_2 p + k_3)(p+1). \tag{9.6-41}$$

Matching coefficients, we obtain $k_2 = 1$ and $k_3 = -3$. Thus,

$$v(t) = V(p)\delta(t) = \left[\frac{1}{p+1} + \frac{p+1}{(p+1)^2 + 2} - \frac{3}{(p+1)^2 + 2}\right]\delta(t) \tag{9.6-42}$$

$$= e^{-t}u(t) + e^{-t}\left[\cos(\sqrt{2}t) + \frac{3}{\sqrt{2}}\sin(\sqrt{2}t)\right]u(t).$$

***The Algebraic
Method: Summary***

To summarize our last two examples, we evaluate all coefficients associated with the real denominator factors by our previously developed methods; for the simple complex conjugate factors, we write them together as a fraction with a quadratic denominator and a linear numerator factor, all of whose coefficients are real. We then multiply both sides of our PFE by the entire denominator and equate the coefficients of various powers of p to solve for the unknown coefficients. We then "massage" the fraction corresponding to the complex factors into the form of a sine and a cosine in operator form. Thus, we never have to perform complex arithmetic! This method, coupled with Table 9.3, results in a very efficient method for working circuits problems.

Section 9.6 Quiz

Q9.6-1. If $x(t) = 6t^2 \cos(4t)u(t) + e^{-t}\sin(6t)u(t)$, write $x(t)$ in the form $x(t) = X(p)\delta(t)$, where $X(p)$ is a rational differential operator.

Q9.6-2. Solve the differential equation $\dfrac{dy}{dt} + 2y = (t^2 + t + 1)e^{-2t}u(t)$.

Q9.6-3. Find the response voltage $v(t)$ in the circuit shown in Figure Q9.6-1 with $v_s(t) = 12\cos(2t)u(t)$ V.

Figure Q9.6-1

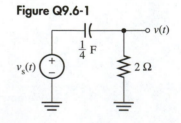

9.7 | Multiple Sources and Initial Conditions

If a circuit has more than one independent source, the analysis proceeds quite similarly to that for circuits we have been considering, which, for the most part, had a single independent source. We must now deal, however, with more than one system operator.

Solution of Circuits with Multiple Sources

Example 9.31   Find the current $i(t)$ in Figure 9.52 as a linear combination of the two independent source waveforms.

Figure 9.52
An example circuit

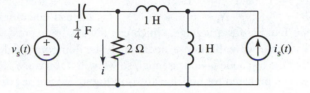

Solution   This circuit might, at first glance, appear to be of third order; actually it is only of second order because of the particular arrangement of the two inductors and the current source. Let's recall the Norton equivalent circuit and derive it for the subcircuit consisting of the two inductors and the current source. The result is shown in Figure 9.53. For this equivalent circuit one nodal equation suffices:

$$\frac{p}{4}(v - v_s) + \frac{v}{2} + \frac{v}{2p} = \frac{1}{2}i_s, \tag{9.7-1}$$

or

$$(p^2 + 2p + 2)v(t) = p^2 v_s(t) + 2pi_s(t). \tag{9.7-2}$$

Solving explicitly for the resistor current $i(t) = v(t)/2$, we get

$$i(t) = H_1(p) v_s(t) + H_2(p) i_s(t), \tag{9.7-3}$$

where

$$H_1(p) = \frac{p^2}{2(p^2 + 2p + 2)} \tag{9.7-4}$$

and

$$H_2(p) = \frac{p}{p^2 + 2p + 2}. \tag{9.7-5}$$

The conclusion of this example is that one has a separate system operator that connects each response variable (i here) with each independent source variable (v_s and i_s).

Figure 9.53
After "Nortonization"

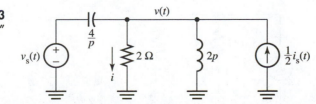

In the preceding example, the denominators of the two system operators were identical. This is often true, though not necessarily so, as we illustrate in the next example.

Example 9.32   Find the response $v(t)$ of the circuit shown in Figure 9.54.

Figure 9.54
An example circuit

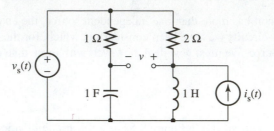

Solution   If we count branches, nodes, v-sources, and i-sources, we will find that $B = 6, N = 4,$ $N_V = 1,$ and $N_I = 1.$ Thus, $N - 1 - N_V = 4 - 1 - 1 = 2$ nodal equations and $B - N + 1 - N_I = 6 - 4 + 1 - 1 = 2$ mesh equations will be required. Because we are looking for a voltage, which can always be expressed as the difference of two node voltages, we will choose nodal analysis. Temporarily deactivating the circuit shows us that the top node is nonessential, so we will not write a KCL equation for it. The circuit is shown prepared for nodal analysis in Figure 9.55. The two nodal equations are

$$pv_1 + \frac{v_1 - v_s}{1} = 0 \tag{9.7-6}$$

and

$$\frac{v_2}{p} + \frac{v_2 - v_s}{2} = i_s. \tag{9.7-7}$$

Solving simultaneously, we get

$$v_1 = \frac{1}{p + 1} v_s \tag{9.7-8}$$

and

$$v_2 = \frac{pv_s + 2pi_s}{p + 2}. \tag{9.7-9}$$

The voltage we are looking for is $v = v_2 - v_1$, or

$$v(t) = \frac{p^2 - 2}{(p + 1)(p + 2)} v_s(t) + \frac{2p}{p + 2} i_s(t). \tag{9.7-10}$$

The two system operators have different denominators. Why? If you sketch the deactivated circuit, you will see that the top node, the nonessential one, disappears into the ground reference. This splits the circuits, effectively, into two separate parts. The right-hand node voltage is affected by both sources, but the left-hand one is only affected by the voltage source.

Figure 9.55
The example circuit prepared for nodal analysis

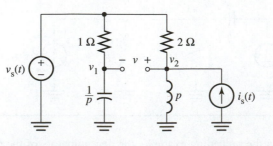

Circuits with Initial Conditions

We have analyzed a couple of circuits with initial conditions in this chapter already, but let's review a bit. As we showed in Chapter 6, Section 6.5, we can analyze such circuits by forcing all waveforms to be zero for $t \leq 0$ and using the initial condition models for the inductor and the capacitor shown in Figure 9.56. In this figure, the plus subscripts refer to the one-sided versions of the corresponding waveforms; in general, if $x(t)$ is a waveform unknown for $t \leq 0$, we write

$$x_+(t) = x(t)u(t). \qquad (9.7-11)$$

Let us agree, however, to drop the subscript when it is clear that we are working with one-sided signals. This will simplify our notation.

Figure 9.56
Series and parallel
initial condition models
for the inductor and the
capacitor

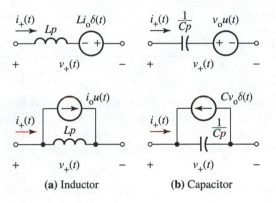

(a) Inductor   **(b)** Capacitor

Example 9.33

Find the response voltage $v(t)$ for $t > 0$ for the circuit shown in Figure 9.57, assuming that $i_L(0) = 2$ A and $v_C(0) = 12$ V.

Figure 9.57
An example circuit

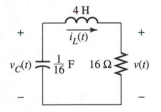

Solution

Solving this circuit is a simple matter of "plugging in" the equivalent condition models for the capacitor and the inductor. The result is shown in Figure 9.58. We quickly compute the voltage $v(t)$ by using the voltage divider rule and noting that $\delta(t) = pu(t)$:

$$v(t) = \frac{16}{16 + 4p + \dfrac{16}{p}}[12u(t) + 8\delta(t)] = \frac{16(2p + 3)}{(p + 2)^2}\delta(t) \qquad (9.7\text{-}12)$$

$$= \left[\frac{k_2}{(p + 2)^2} + \frac{k_1}{p + 2}\right]\delta(t).$$

The constants are

$$k_2 = [16(2p + 3)]_{p \leftarrow -2} = -16 \qquad (9.7\text{-}13)$$

and

$$k_1 = \frac{d}{dp}[16(2p + 3)]_{p \leftarrow -2} = 32. \qquad (9.7\text{-}14)$$

Thus, $v(t) = -16te^{-2t}u(t) + 32e^{-2t}u(t) = 16(2 - t)e^{-2t}u(t).$ (9.7-15)

We note, however, that our solution is not really valid for $t < 0$ because we know nothing about that time interval; thus, we simply write

$$v(t) = 16(2 - t)e^{-2t}, \quad t > 0.$$ (9.7-16)

Figure 9.58
Initial condition model

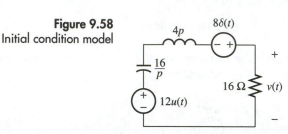

Switched Circuits   If our circuit to be analyzed has one or more switches, the analysis proceeds in exactly the same manner—*provided that the capacitor voltages and inductor currents are known quantities prior to the switching action.* A typical situation is when the circuit is in the dc steady state for $t \leq 0$. We illustrate with an example.

Example 9.34   Find the voltage $v(t)$ in Figure 9.59 for all t.

Figure 9.59
An example circuit

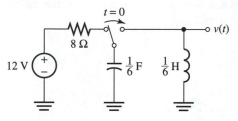

Solution   Here it makes sense to ask for the solution for all time because we *do* know something about the circuit for $t \leq 0$. As a matter of fact, we know that the capacitor voltage is $v_c(t) = 12$ V (positive at the top). To see this, we note that there is a first-order RC circuit on the left whose time constant is $\tau = RC = 4/3$ s > 0. Thus, we know that it is stable and consequently is in the dc steady state for $t \leq 0$. The equivalent circuit holding for all $t \leq 0$ is shown in Figure 9.60. By looking at this figure, we know that the inductor current is zero for negative time because nothing is connected to it. Hence, applying causality, its voltage is also zero. Using these values in the initial condition models for the capacitor and the inductor results in the equivalent circuit shown in Figure 9.61 for $t > 0$. We simply apply the voltage divider rule, obtaining

$$v(t) = \frac{\dfrac{p}{6}}{\dfrac{p}{6} + \dfrac{6}{p}}12u(t) = \frac{12p}{p^2 + 36}\delta(t) = 12\cos(6t)u(t) \text{ V}.$$ (9.7-17)

Here we have used the fact that $u(t) = 1/p\ \delta(t)$. As we know independently from the circuit for $t \leq 0$ that $v(t) = 0$ for that time interval, the unit step function multiplier in the present situation is actually correct for all time.

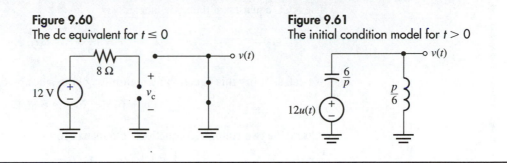

Figure 9.60
The dc equivalent for $t \leq 0$

Figure 9.61
The initial condition model for $t > 0$

Now let's work a much more complicated example. The more dynamic elements a circuit has, the more initial conditions it has, and the more complicated the analysis will be.

Example 9.35   Solve the circuit in Figure 9.62 for the voltage $v(t)$ for $t > 0$ assuming that $v_s(t) = 6e^{-2t}$ for $t > 0$, $v_C(0) = 2$ V, and $i_L(0) = 1$ A.

Figure 9.62
An example circuit

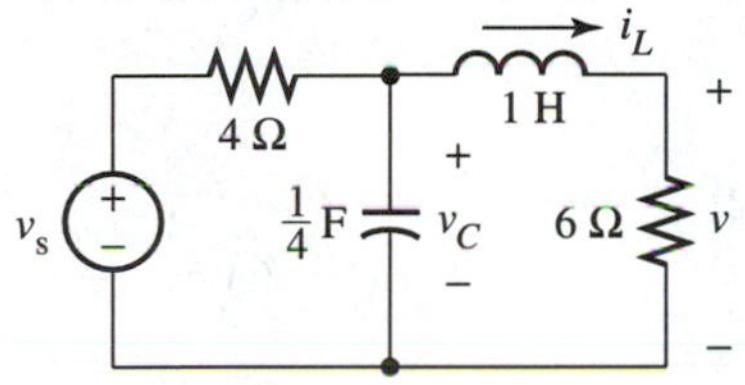

Solution   Here, we are given the initial conditions without having to solve a circuit prior to $t = 0$. For $t > 0$, we will perform nodal analysis and so will use the parallel initial condition models for both capacitor and inductor.[9] The $t > 0$ circuit is shown in Figure 9.63. Note that this equivalent circuit is only valid for $t > 0$ for we do not know anything at all about it for $t < 0$—except for the initial conditions. The KCL equation at the node labeled v_C is

$$\frac{v_C - v_s}{4} + \frac{p}{4}v_C + \frac{v_C - v}{p} = -u(t) + \frac{1}{2}\delta(t). \qquad (9.7\text{-}18)$$

We can clear fractions by multiplying both sides by $4p$:

$$(p^2 + p + 4)v_C - 4v = 2p\,\delta(t) - 4\delta(t) + p\,v_s. \qquad (9.7\text{-}19)$$

Figure 9.63
The equivalent model with initial conditions

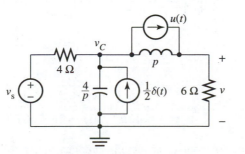

---

[9] We could perhaps make a case just as easily for using series models because they introduce only nonessential nodes, which do not require KCL equations.

The KCL equation at the node labeled v is

$$\frac{v}{6} + \frac{v - v_C}{p} = u(t). \tag{9.7-20}$$

We can simplify this equation by multiplying both sides by $6p$, giving

$$(p + 6) v - 6v_C = 6 \, \delta(t). \tag{9.7-21}$$

Putting the two simplified equations into matrix form, we have

$$\begin{bmatrix} p^2 + p + 4 & -4 \\ -6 & p + 6 \end{bmatrix} \begin{bmatrix} v_C \\ v \end{bmatrix} = \begin{bmatrix} 2p\delta(t) - 4\delta(t) + pv_s(t) \\ 6\delta(t) \end{bmatrix}. \tag{9.7-22}$$

Multiplying by the inverse of the (2×2) matrix on the left, we obtain

$$\begin{bmatrix} v_C(t) \\ v(t) \end{bmatrix} = \frac{1}{p(p^2 + 7p + 10)} \begin{bmatrix} p + 6 & 4 \\ 6 & p^2 + p + 4 \end{bmatrix} \begin{bmatrix} 2p\delta(t) - 4\delta(t) + pv_s(t) \\ 6\delta(t) \end{bmatrix}. \tag{9.7-23}$$

Thus,

$$v(t) = \frac{6(p + 3)}{p^2 + 7p + 10} \delta(t) + \frac{6}{p^2 + 7p + 10} v_s(t). \tag{9.7-24}$$

We know also that

$$v_s(t) = 6e^{-2t}u(t) = \frac{6}{p + 2}\delta(t). \tag{9.7-25}$$

Remember that we are multiplying all time functions by $u(t)$. Thus

$$v(t) = \frac{6(p + 3)}{p^2 + 7p + 10} \delta(t) + \frac{36}{(p + 2)^2(p + 5)} \delta(t) = \frac{6(p^2 + 5p + 12)}{(p + 2)^2(p + 5)} \delta(t) \tag{9.7-26}$$

$$= \left[\frac{k_2}{(p + 2)^2} + \frac{k_1}{p + 2} + \frac{k_3}{p + 5} \right]\delta(t).$$

We find that

$$k_2 = \left[\frac{6(p^2 + 5p + 12)}{p + 5} \right]_{p \leftarrow -2} = 12, \tag{9.7-27}$$

$$k_1 = \frac{d}{dp} \left[\frac{6(p^2 + 5p + 12)}{p + 5} \right]_{p \leftarrow -2} = 6 \frac{d}{dp} \left[p + \frac{12}{p + 5} \right]_{p \leftarrow -2} \tag{9.7-28}$$

$$= 6 \left[1 - \frac{12}{(p + 5)^2} \right]_{p \leftarrow -2} = -2,$$

and

$$k_3 = 6 \left[\frac{p^2 + 5p + 12}{(p + 2)^2} \right]_{p \leftarrow -5} = 8. \tag{9.7-29}$$

Thus, $v(t) = 12te^{-2t}u(t) - 2e^{-2t}u(t) + 8e^{-5t}u(t). \tag{9.7-30}$

This solution, of course, artificially predicts that $v(t) = 0$ for $t \leq 0$. We simply throw away the $u(t)$ multiplier and write the final answer as

$$v(t) = (12t - 2)e^{-2t} + 8e^{-5t}; \quad t > 0. \tag{9.7-31}$$

**Solution of
Differential Equations
with Prescribed
Initial Conditions**

We have now covered our topic of multiple sources and initial conditions rather thoroughly through the examples we have worked; however, there is one more topic that we should perhaps mention. Sometimes a differential equation is to be solved subject to known conditions on the response and one or more of its derivatives at $t = 0$. In this case, we simply note that from $y_+(t) = y(t)u(t)$ we derive the following equations:

$$py_+(t) = \frac{d}{dt}[y(t)u(t)] = \frac{dy}{dt}u(t) + y(0)\delta(t) \qquad (9.7\text{-}32)$$

and

$$p^2 y_+(t) = \frac{d}{dt}\left[\frac{dy}{dt}u(t) + y(0)\delta(t)\right] = \frac{d^2y}{dt^2}u(t) + y'(0)\delta(t) + y(0)\,p\,\delta(t). \qquad (9.7\text{-}33)$$

Similar equations hold for higher-order derivatives. We have used the p notation for one-sided functions and derivatives of impulses, conventional notation for the derivatives of the two-sided time function $y(t)$, and primes for derivatives evaluated at $t = 0$. We will illustrate how to use these equations in the next example.

Example 9.36   Solve the differential equation

$$\frac{d^2y}{dt^2} + 2\frac{dy}{dt} + y = 1; \quad t > 0 \qquad (9.7\text{-}34)$$

for $y(t)$ subject to the initial conditions $y_0 = y(0) = 2$ and $y_0' = y'(0) = 1$.

Solution   We will first multiply both sides of our differential equation by $u(t)$, thus obtaining a relationship that holds for all time:

$$\frac{d^2y}{dt^2}u(t) + 2\frac{dy}{dt}u(t) + yu(t) = u(t). \qquad (9.7\text{-}35)$$

Now we solve equations (9.7-32) and (9.7-33) for the various derivatives multiplied by the unit step function and use them in equation 9.7-35, giving

$$[p^2 y_+(t) - y_0'\delta(t) - y_0 p\,\delta(t)] + 2[py_+(t) - y_0\delta(t)] + y_+(t) = u(t). \qquad (9.7\text{-}36)$$

Rearranging, and noting that $u(t) = (1/p)\,\delta(t)$, we have

$$y_+(t) = \frac{y_0 p^2 + (2y_0 + y_0')p + 1}{p(p+1)^2}\delta(t) = \frac{2p^2 + 5p + 1}{p(p+1)^2}\delta(t). \qquad (9.7\text{-}37)$$

We now do a PFE in the usual manner, to obtain

$$y_+(t) = \left[\frac{1}{p} + \frac{2}{(p+1)^2} + \frac{1}{p+1}\right]\delta(t) = [1 + (2t+1)e^{-t}]\,u(t). \qquad (9.7\text{-}38)$$

Finally, noting that our solution is not necessarily valid for $t \leq 0$, we throw away the step function multiplier and write

$$y(t) = 1 + (2t+1)e^{-t}; \quad t > 0. \qquad (9.7\text{-}39)$$

In the preceding example we note that one can check the solution by substituting $t = 0$, then taking the derivative and evaluating it at $t = 0$.

Section 9.7 Quiz

Q9.7-1. Solve for the current $i(t)$, $t > 0$, in Figure Q9.7-1 for $v_C(0) = 8$ V. Assume that $i(t)$ has been zero for all $t < 0$. (There is a circuit, not shown, that has established the initial conditions.)

Figure Q9.7-1

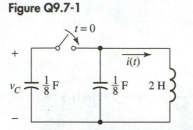

Q9.7-2. Solve the differential equation

$$\frac{d^2y}{dt^2} + 2\frac{dy}{dt} + 2y = e^{-t}; \quad t > 0$$

for $t > 0$ subject to the initial conditions $y_0 = 2$ and $y_0' = 1$.

9.8 | SPICE: System Simulation (Behavioral Modeling)

We have already covered the simulation of time-domain circuits using SPICE. The simulation of higher-order circuits is no different than for first-order ones—although it becomes harder to predict the simulation parameters because higher-order circuits are not necessarily describable in terms of time constants. Therefore, we will explore a new aspect of SPICE in this section: the simulation of circuits and differential equations using the system diagram. This is referred to in a SPICE context as *behavioral modeling*. Not all SPICE versions have this capability, so we will concentrate exclusively on PSPICE—which does.

Circuit Simulation:
The Scalar Multiplier

Let's begin with what is perhaps the most basic system element: the scalar multiplier. We remind you of its symbol in Figure 9.64(a). In SPICE, we will think of it as a *subcircuit* having input and output variables that are both voltages; that is, $x(t)$ is a voltage signal and so is $y(t)$. The constant a is, therefore, a unitless multiplier. We will simulate the scalar multiplier with the PSPICE subcircuit shown in Figure 9.64(b). The resistor RIN is included to satisfy the SPICE constraint that there be no floating nodes. The gain of EOUT is to be equal to a. Here is the simulation code.

```
.SUBCKT   SCALAR  IN   OUT   PARAMS:A=1
RIN  IN   0  1
EOUT   OUT  0  VALUE={A*V(IN)}
.ENDS
```

We discussed the PARAMS option in Section 4.7 of Chapter 4. The parameter a is set to unity as a default value: if we do not "pass" a parameter, this statement causes it to be unity. The inclusion of the phrase VALUE = {a*V(IN)} is one we have not discussed yet. It is quite straightforward. It merely allows us to use the parameter a as the value of voltage gain. The noun VALUE is a keyword in SPICE, not one we have dreamed up, and { } denotes that the symbols inside the braces form an expression that SPICE must compute.

Figure 9.64
The scalar multiplier

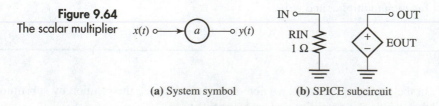

(a) System symbol   **(b)** SPICE subcircuit

Now let's think of a circuit using the subcircuit SCALAR. One possibility is shown in Figure 9.65, with the SPICE code shown below.

```
SCALAR MULTIPLIER TEST FILE
VS  10   0  1
XMULT1  10  34   SCALAR   PARAMS:A=4
RLOAD  34  0  1
.LIB  "DAVIS.LIB"
.DC  VS  -1  1  0.1
.PROBE
.END
```

In keeping with SPICE requirements, we have given the subcircuit a name starting with the letter X. In this case, we have called it XMULT1 to denote the fact that it is a multiplier (we think that perhaps XSCALAR1 might cause some confusion with the subcircuit itself). The verb .LIB "DAVIS.LIB" tells SPICE to look in the currrent directory (for DOS computers), or the current folder (for Macintoshes) and open the file DAVIS.LIB. We have coded and stored our preceding subcircuit code in that file. We have decided to do a dc sweep of the source voltage VS from -1 volt to $+1$ volt so that we can check the slope of the resulting graph of output versus input, which should be the parameter a (in this case 4). Remember that although the .DC statement causes the source VS to be "swept" we still must give it a nominal value in the NETLIST.

Figure 9.65
Test circuit for SCALAR

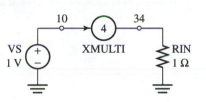

The SPICE simulation results are shown as "Probe output" in Figure 9.66 (notice that the slope of the output (v(34)) is 4, as we expected).

Figure 9.66
Probe output for dc
sweep

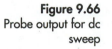

Circuit Simulation:
The Summer

Next, let's look at how we might simulate a summer. We show the system symbol in Figure 9.67(a) and the SPICE subcircuit in Figure 9.67(b). We have decided to allow each input to have a weight, thus combining the functions of scalar multiplier and summer into one unit. The added code in doing so is very slight and it adds quite a bit of flexibility to our simulation capability. We have chosen to use three inputs, but more or fewer could be used. If we do not require the third input, we merely force it to zero. If more inputs are needed, we must change our subcircuit SPICE code, which is as follows:

```
.SUBCKT  SUMMER  IN1  IN2  IN3  OUT  PARAMS:A1=1,A2=1,A3=1
RIN1  IN1  0  1
RIN2  IN2  0  1
RIN3  IN3  0  1
EOUT  OUT  0  VALUE={A1*V(IN1)+A2*V(IN2)+A3*V(IN3)}
.ENDS
```

Figure 9.67
SPICE subcircuit for the
weighted summer

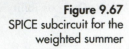
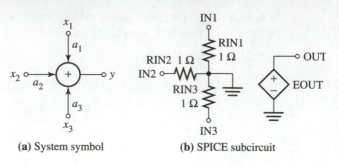

(a) System symbol   **(b)** SPICE subcircuit

A simple circuit for evaluating SUMMER is shown in Figure 9.68. Now is a good time to note that our simulations are a bit different from the simulation diagrams we have drawn earlier in the text. We must apply voltage sources for our input signals (rather than strictly mathematical ones, such as x's and y's) and must include a load resistor to eliminate "floating nodes." Note that the boxed numbers in Figure 9.68 refer to nodes and the "nonboxed" numbers to weights. The SPICE output file is as follows:

```
***03/21/95 20:47:34***********EVALUATION PSPICE (JULY 1993)****
*********SUMMER TEST FILE
****  CIRCUIT DESCRIPTION
****************************************************************
******
VS1   1   0   1
VS2   2   0   2
VS3   3   0   3
XSUM1  1   2   3   4  SUMMER  PARAMS:A1=2,A2=3,A3=4
RLOAD   4   0   1
.LIB "DAVIS.LIB"
.DC  VS1  1   1   1
.PRINTDC   V(4)
.END
****03/21/95 20:47:34***********EVALUATION PSPICE (JULY 1993)****
SUMMER TEST FILE
**** DC TRANSFER CURVES       TEMPERATURE=27.000 DEG C
****************************************************************
******
 VS1          V(4)
 1.000E+00   2.000E+01
     JOB CONCLUDED
     TOTAL JOB TIME      1.20
```

Note that we expect $v(4) = 2 \times 1 + 3 \times 2 + 4 \times 3 = 20$ V, which is verified by our output.

Figure 9.68
Test circuit for SUMMER

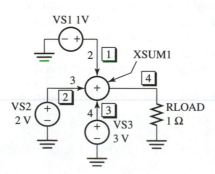

Circuit Simulation:
The Integrator

The next item we will simulate is the integrator, which we will allow to have a constant gain term a, as shown in Figure 9.69(a). The SPICE subcircuit is shown in Figure 9.69(b). Our simulation procedure will hinge upon a new SPICE statement that allows us to specify the dependent source with a voltage gain that is an operator. It is

```
EOUT  NPLUS  NMINUS  LAPLACE {INPUTVARIABLE}={EXPRESSION}
```

NPLUS and NMINUS are the nodes between which the dependent source is connected, as usual. EXPRESSION defines the operator that will operate upon the input variable INPUTVARIABLE (a voltage) to produce the value of the dependent source. EXPRESSION is written as usual as a rational differential operator, but there is a slight difference in notation: we must use the letter s instead of the letter p. The reason for this is that SPICE uses the concept of "Laplace transform," rather than that of "differential operator." As we will show in Chapter 13, however, the mathematical machinery of the Laplace transform is exactly the same as that of differential operators; therefore, we will agree to use the p notation for our operators, but to convert all the p's to s's when we code our operators in SPICE. The SPICE NETLIST for the integrator, then, is as follows:

```
.SUBCKT INT   IN   OUT   PARAMS:A=1
RIN  IN  0  1
EOUT  OUT  0  LAPLACE  {V(IN)}={A/S}
.ENDS
```

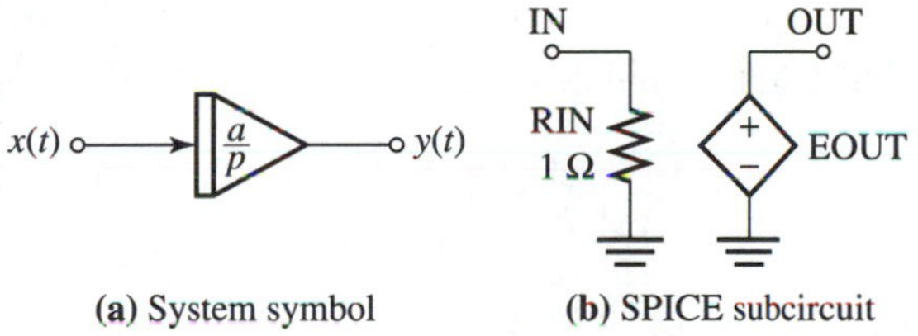

Figure 9.69
The integrator (with gain)

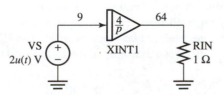

(a) System symbol   (b) SPICE subcircuit

In order to test our subcircuit, we will use the simple test circuit shown in Figure 9.70. We apply a step function having a value of 2 to an integrator having a gain of 4. The corresponding SPICE NETLIST is

```
INTEGRATOR TEST FILE
VS  9  0  PWL(0,0  1E-8,2)
XINT1  9  64  INT  PARAMS:A=4
RLOAD  64  0  1
.LIB  "DAVIS.LIB"
.TRAN  0.1  5  0 0.1
.PROBE
.END
```

Figure 9.70
A test circuit for INT

The PROBE output is shown in Figure 9.71. We can check it by merely performing the required integration:

$$y(t) = \frac{4}{p}[2u(t)] = \frac{8}{p}u(t) = 8tu(t).$$

The slope should be 8 (which it is), and at $t = 5$ s the equation should have the value $8 \times 5 = 40$ (which it does.)

Figure 9.71
PROBE output

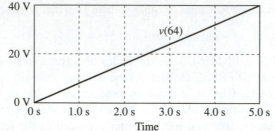

Circuit Simulation:
The Differentiator and
Differential Equations

We have previously discussed the undesirability of differentiators if they can possibly be avoided, and PSPICE is no exception to this rule. It simply does not allow differentiators to be modeled adequately. You are welcome to try—simply use the integrator subcircuit we have just presented, replacing the expression a/s with a*s in the EOUT source and changing labels where appropriate, and try to simulate such inputs as a ramp or an exponential. You will find that the results are not at all good. For this reason, we omit the differentiator.

Instead, let's show that our other system blocks suffice to simulate a differential equation.

Example 9.37   Use PSPICE to simulate the differential equation

$$\frac{d^2y}{dt^2} + 4\frac{dy}{dt} + 13y = x(t), \tag{9.8-1}$$

with $x(t) = 26u(t)$.

Solution   First, we express the equation in operator form:

$$(p^2 + 4p + 13)y(t) = x(t). \tag{9.8-2}$$

Next, we solve for $p^2y(t)$ and then multiply both sides by $1/p^2$, to obtain

$$y(t) = \frac{1}{p}\left[-4y(t) + \frac{1}{p}[x(t) - 13y(t)]\right]. \tag{9.8-3}$$

This is represented by the simulation diagram shown in Figure 9.72. (Note that this is only one of many techniques we could use to obtain the simulation.) We have chosen to use unity gain integrators (so will use the default values for the SPICE subcircuit INT). Notice that we did not have to add an extra load resistor to any of our subcircuits because each output is connected to at least one input of another subcircuit—which contains an internal resistor connected to that input terminal. The NETLIST for doing the simulation using the subcircuits we have developed (and stored in DAVIS.LIB) is

```
SECOND ORDER  DIFFERENTIAL EQUATION EXAMPLE
VS      2  0     PWL(0,0  1E-8,26)
XSUM1   1  2     0 3    SUMMER      PARAMS:A1=-13
XINT1   3  4     INT
XSUM2   1  4     0 5    SUMMER      PARAMS:A1=-4
XINT2   5  1     INT
.LIB     "DAVIS.LIB"
```

```
.TRAN    5   5
.PROBE
.END
```

We show the Probe output in Figure 9.73. As we can quickly verify by solving the differential equation by hand, the response is underdamped (with complex conjugate roots).

Figure 9.72
The simulation diagram

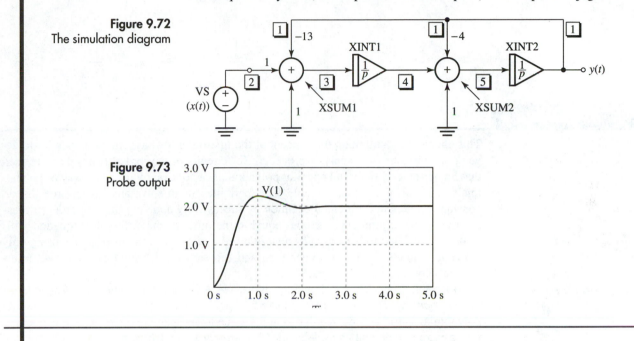

Figure 9.73
Probe output

Library of Subcircuits for Behavioral Simulation

Finally, for your reference, we list the content of DAVIS.LIB, the library version of the subcircuits we have developed in this section. Note that you must either place this library in the directory (folder for Macintosh computers) that contains the PSPICE executable file or provide a complete path describing its location.

```
*****************************DAVIS.LIB***************************
*******THIS IS A LIBRARY OF SUBCIRCUITS FOR SIMULATING CIRCUITS,
*******SYSTEMS, AND DIFFERENTIAL EQUATIONS VIA SIMULATION
*******DIAGRAMS
****************************************************************
*******SCALAR MULTIPLIER*****
.SUBCKT      SCALAR   IN    OUT    PARAMS:A=1
RIN          IN    0    1
EOUT    OUT   0    VALUE={A*V(IN)}
.ENDS
*******SUMMER*****************
.SUBCKT SUMMER    IN1   IN2    IN3    OUT   PARAMS:A1=1,A2=1,A3=1
R1   IN1  0    1
R2   IN2  0    1
R3   IN3  0    1
EOUT  OUT   0   VALUE={A1*V(IN1)+A2*V(IN2)+A3*V(IN3)]
.ENDS
******INTEGRATOR**************
.SUBCKT   INT   IN    OUT  PARAMS:A=1
RIN          IN    0    1
EOUT       OUT   0    LAPLACE{V(IN)]={A/S}
.ENDS
```

Section 9.8 Quiz

Q9.8-1. Write a differential equation for the response $v(t)$ of the circuit in Figure Q9.8-1, draw a simulation diagram for it, then run SPICE to determine the response to the input $v_S(t) = 8e^{-t}u(t)$ V and $R = 4 \; \Omega$.

Figure Q9.8-1

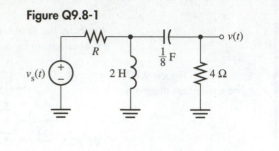

Chapter 9 Summary

This chapter has built upon the concept of the first-order lowpass solution operator to develop a comprehensive analysis procedure for circuits of any order. It has demonstrated that higher-order differential operator polynomials factor exactly the same way as polynomials in a real or complex variable. This permitted us to solve any higher-order linear, constant-coefficient differential equation by successively applying the first-order lowpass solution operator a number of times equal to the order of the differential equation. This procedure also led to the *cascade simulation* of a circuit or system response using only scalar multipliers, summers, integrators, and (possibly) differentiators—the last occurring only for *singular operators*.

In connection with the simulation issue, we defined a *differential (or Heaviside, or system) operator* $H(p) = N(p)/D(p)$ to be *proper* if the degree of the numerator operator polynomial $N(p)$ has a degree that is at least one lower than the degree of the denominator operator polynomial $D(p)$. We called it *improper* if the degree of $N(p)$ is at least equal to that of $D(p)$, and *singular* if the degree of the former is actually *greater* than that of the latter.

We next showed that the *partial fraction expansion (PFE)* is valid for such differential operators and discussed procedures for finding this expansion. In particular, we developed procedures for the case of real and unequal roots of the *characteristic polynomial $D(z)$ (which is $D(p)$ with p replaced by the real or complex variable z).* We discussed the case of both proper and improper operators and showed that if $H(p)$ is improper, one must first go through a *preamble* consisting of a finite number of steps of long division. We developed two different methods for dealing with the case in which the characteristic polynomial has *multiple roots,* and derived an explicit formula for the coefficients in the PFE expansion that holds for all cases. We also developed explicit forms and methods of dealing with the *complex conjugate root case* using strictly real algebra.

As a culmination of the partial fraction expansion discussion, we developed a technique of solving linear differential equations with constant coefficients and the circuit responses corresponding to such differential equations in a *purely algebraic manner.* This procedure is based upon the representation of waveforms as impulse responses of simple differential operators. We labeled such time functions *elementary waveforms.* For those having a prior exposure to the Laplace transform, we noted that the mathematical machinery that we have developed is identical to that associated with the transform—but without the necessity of computing the improper integral by which the Laplace transform is defined.

We discussed the issue of circuits with multiple inputs and for which a number of circuit variables are desired as responses. This represented a brief introduction to the theory of *multiple-input multiple-output (MIMO) systems.* In this case, the differential operator becomes a matrix of scalar operators.

We discussed the issue of solving differential equations and circuits with *prescribed initial conditions*. We used the technique of *truncation*, that is forcing all waveforms to be identically zero for all $t \le 0$, representing the effects of nonzero waveforms for $t \le 0$ with corresponding initial conditions. We noted that this leads in a natural way to the twin concepts of *state* and *state variables*.

Finally, we discussed the use of the *behavioral modeling* capability of SPICE to generate system simulation diagram blocks.

Chapter 9 Review Quiz

RQ9.1. Find the differential equation in operator form for $v(t)$ in Figure RQ9.1. (Do not solve.) (*Hint:* Determine whether mesh or nodal analysis is simpler, then use one of these methods.)

Figure RQ9.1

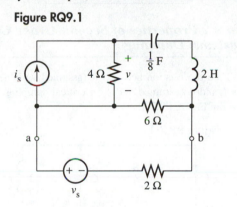

RQ9.2. Solve the differential equation

$$\frac{d^2y}{dt^2} - 3\frac{dy}{dt} - 4y = x(t)$$

if $x(t) = 2u(t)$.

RQ9.3. Draw a cascade simulation diagram for $y(t)$ in Question RQ9.2 using first-order lowpass operator blocks (as well as scalar multipliers and summers).

RQ9.4. Find and sketch the Thévenin equivalent of the subcircuit above terminals a and b in the circuit shown in Figure RQ9.1.

RQ9.5. Find the nonnegative value of R in Figure RQ9.2 required for critical damping.

Figure RQ9.2

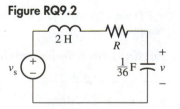

RQ9.6. Solve for $v(t)$ in Figure RQ9.2 if $R = 20 \, \Omega$ and $v_S(t) = 16u(t)$ V.

RQ9.7. Answer Question RQ9.6 for $R = 12 \, \Omega$ and $v_S(t) = u(t)$V.

RQ9.8. Draw a parallel simulation diagram for the differential equation in Question RQ9.2.

RQ9.9. Derive the differential equation for $v(t)$ in the circuit shown in Figure RQ9.3. Then solve for $v(t)$ assuming that $i_S(t) = u(t)$ A and $R = 1 \, \Omega$.

Figure RQ9.3

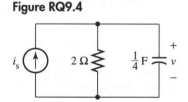

RQ9.10. Repeat Question RQ9.9 for $R = 6 \, \Omega$ and $i_S(t) = 5u(t)$ A.

RQ9.11. Answer Question RQ9.9 for $R = 5 \, \Omega$ and $i_S(t) = 5u(t)$ A.

RQ9.12. Find the response voltage $v(t)$ in Figure RQ9.4 if $i_S(t) = \cos(4t)u(t)$ A.

Figure RQ9.4

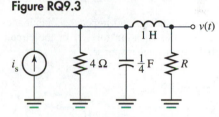

RQ9.13. Find the voltage $v(t)$ in Figure RQ9.5 for all values of t. (*Hint:* Notice that this circuit actually consists of two circuits prior to $t = 0$, each of which is stable and therefore operating in the dc steady state.)

Figure RQ9.5

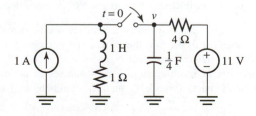

Chapter 9 Problems

Section 9.1 Solution of Higher-Order Circuits and Differential Equations

9.1-1. Solve the differential equation

$$\frac{d^2y}{dt^2} + a\frac{dy}{dt} + 4y = x(t),$$

if $x(t) = \delta(t)$ for the following values of the parameter a:
a. $a = 5$ **b.** $a = 4$ **c.** $a = 2$

9.1-2. Solve the differential equation

$$\frac{d^2y}{dt^2} + a\frac{dy}{dt} + 9y = x(t),$$

if $x(t) = \delta(t)$, for the following values of the parameter a:
a. $a = 10$ **b.** $a = 6$ **c.** $a = 4$

9.1-3. Use mesh analysis and impedance/admittance operators to find the differential equation for $v(t)$ in the circuit of Figure P9.1-1 and find the unit step response.

Figure P9.1-1

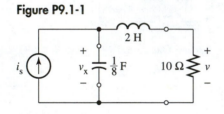

9.1-4. Use nodal analysis and impedance/admittance operators to find the differential equation for $v_X(t)$ in the circuit of Figure P9.1-1 and find the unit impulse response.

9.1-5. Remove the 10-Ω resistor in the circuit for Figure P9.1-1 and, using impedance and/or admittance operators, find the parameters $Z_{eq}(p)$ and $v_{oc}(t)$ for the Thévenin equivalent of the resulting two-terminal subcircuit. Then replace the 10-Ω resistor and use the Thévenin equivalent to find the differential equation for $v(t)$.

9.1-6. Repeat Problem 9.1-5, except remove the capacitor, find the Thévenin equivalent parameters, then replace it and find the differential equation for $v_X(t)$.

9.1-7. Repeat Problem 9.1-5 using the Norton equivalent.

9.1-8. Repeat Problem 9.1-6 using the Norton equivalent.

9.1-9. Using impedance and/or admittance operators and either mesh or nodal analysis, find the differential equation for $v(t)$ in Figure P9.1-2. (*Hint:* Recall that $N_{me} = B - N + 1 - N_I$, where N_{me} = number of mesh equations, N = number of nodes, B = number of elements (branches), and N_I = number of i-sources.

Recall also that $N_{ne} = N - 1 - N_v$, where N_{ne} = number of nodal equations and N_v = number of v-sources.)

Figure P9.1-2

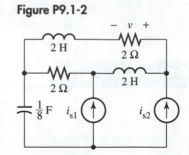

Section 9.2 Properties of Second-Order Circuits and Systems: Damping

9.2.1. Find the relation between R (assumed variable) and the variables L and C (assumed fixed) for critical damping in the circuit shown in Figure P9.2-1.

Figure P9.2-1

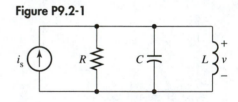

9.2-2. For the circuit in Figure P9.2-1, assume that $L = 2$ mH, and $C = 0.5$ nF. Find the impulse response for the variable $v(t)$ if $R = 1$ kΩ, 800 Ω, and 2 kΩ. Use a consistent set of units.

9.2-3. Find the nonnegative value of R for critical damping of the response $v(t)$ in the circuit shown in Figure P9.2-2.

Figure P9.2-2

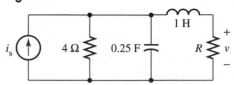

9.2-4. Let $R = 5\ \Omega$ and $i_S(t) = e^{-3t}u(t)$ A in the circuit of Figure P9.2-2 and compute $v(t)$.

Section 9.3 Partial Fraction Expansion: Simple Roots

9.3-1. Find the solution to

$$\frac{d^2y}{dt^2} + 5\frac{dy}{dt} + 6y = 12\ u(t)$$

using the PFE.

9.3-2. Find the solution to

$$\frac{d^2y}{dt^2} + 7\frac{dy}{dt} + 10y = 8e^{-t}u(t)$$

using the PFE.

9.3-3. Find the unit impulse response of

$$\frac{d^2y}{dt^2} + 6\frac{dy}{dt} + 8y = 2\frac{dx}{dt} + 6x$$

using the PFE.

9.3-4. Find the unit step response of

$$\frac{d^3y}{dt^3} + 7\frac{d^2y}{dt^2} + 14\frac{dy}{dt} + 8y = 6\frac{dx}{dt} + 18x$$

using the PFE. (*Hint:* One factor of the characteristic polynomial $D(z)$ is $(z + 1)$.)

9.3-5. Find the solution to

$$\frac{d^2y}{dt^2} + 3\frac{dy}{dt} + 2y = P_1(t)$$

using the PFE, where $P_T(t)$ is the unit pulse function of duration T (with $T = 1$ s) sketched in Figure P9.3-1.

Figure P9.3-1

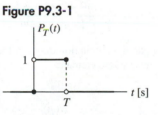

9.3-6. Find the value of $v(t)$ in Figure P9.3-2 for all values of t.

Figure P9.3-2

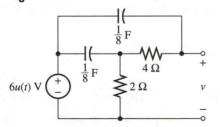

9.3-7. Find the value of $v(t)$ in Figure P9.3-3 for all values of t.

Figure P9.3-3

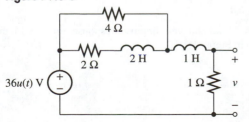

9.3-8. Find the value of $i(t)$ in Figure P9.3-4 for $t > 0$ if $i(0) = 4$ A and $v(0) = 8$ V.

Figure P9.3-4

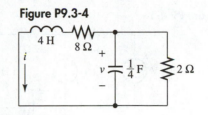

9.3-9. Find the current $i(t)$ in Figure P9.3-5 for all values of t.

Figure P9.3-5

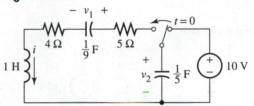

Section 9.4 Partial Fraction Expansion: Multiple Roots

9.4-1. Solve the differential equation

$$\frac{d^2y}{dt^2} + 2\frac{dy}{dt} + y = x(t)$$

assuming that $x(t) = u(t)$.

9.4-2. Solve the differential equation

$$\frac{d^2y}{dt^2} + 2\frac{dy}{dt} + y = x(t)$$

assuming that $x(t) = te^{-t}u(t)$.

9.4-3. Solve the differential equation

$$\frac{d^2y}{dt^2} + 6\frac{dy}{dt} + 9y = x(t)$$

assuming that $x(t) = \frac{1}{2}t^2e^{-3t}u(t)$.

9.4-4. Solve the differential equation

$$\frac{d^2y}{dt^2} + 8\frac{dy}{dt} + 16y = \frac{dx}{dt} + 4x$$

if $x(t) = te^{-4t}u(t)$.

9.4-5. Draw a system simulation diagram for $y(t)$ in Problem 9.4-1.

9.4-6. Draw a system simulation diagram for $y(t)$ in Problem 9.4-3.

9.4-7. Draw a system simulation diagram for $y(t)$ in Problem 9.4-4.

9.4-8. Find $v(t)$ for all values of t for the circuit shown in Figure P9.4-1.

Figure P9.4-1

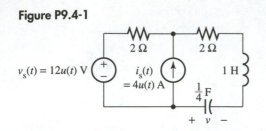

9.4-9. Find $v(t)$ for all values of t for the circuit shown in Figure P9.4-2.

Figure P9.4-2

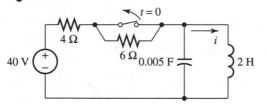

9.4-10. Find $i(t)$ for all values of t for the circuit shown in Figure P9.4-3. Assume dc steady-state conditions for all $t \le 0$.

Figure P9.4-3

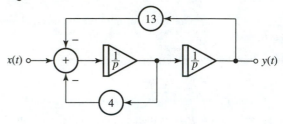

9.4-11. Draw a simulation diagram for $i(t)$, $t > 0$, in the circuit shown in Figure P9.4-3.

Section 9.5 Complex Conjugate Roots

9.5-1. Find the unit impulse response of the differential equation

$$\frac{d^2y}{dt^2} + \omega_0^2 y = \omega_0 x(t).$$

9.5-2. Find the unit impulse response of the differential equation

$$\frac{d^2y}{dt^2} + \omega_0^2 y = \omega_0 \frac{dx}{dt}.$$

9.5-3. Solve the differential equation

$$\frac{d^2y}{dt^2} + 8\frac{dy}{dt} + 20y = 80\frac{dx}{dt} + 160x$$

if $x(t) = u(t)$.

9.5-4. If $v(0) = 6$ V and $i(0) = 1$ A in Figure P9.5-1, find $i(t)$ for all $t > 0$.

Figure P9.5-1

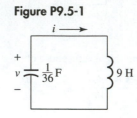

9.5-5. Find the unit step response for the output $y(t)$ of the system whose simulation diagram appears in Figure P9.5-2.

Figure P9.5-2

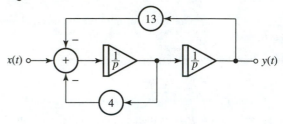

9.5-6. Find the unit impulse response of

$$H(p) = \frac{6p^2 + 38p + 52}{(p + 1)(p^2 + 4p + 13)}.$$

9.5-7. Draw a parallel simulation diagram for $H(p)$ in Problem 9.5-6 using only real system elements.

9.5-8. Let

$$H(p) = \frac{N(p)}{(p + a)^m(p + a^*)^m D_0(p)},$$

where the coefficients of $N(p)$ and $D_0(p)$ are real and the latter has no denominator factors of the form $(p + a)$ or $(p + a^*)$. In the PFE

$$H(p) = \frac{k_m}{(p + a)^m} + \cdots + \frac{k_{m-i}}{(p + a)^{m-i}} + \cdots + \frac{k_1}{p + a}$$
$$+ \frac{k'_m}{(p + a^*)^m} + \cdots + \frac{k'_{m-i}}{(p + a^*)^{m-i}} + \cdots + \frac{k'_1}{p + a^*} + F(p),$$

where $F(p)$ consists of the terms in the PFE due to $D_0(p)$, show that the primed constants are the complex conjugates of the unprimed ones. (*Hint:* Write the explicit expression for k_{m-i}, conjugate it to obtain k^*_{m-i}, and show that the result is the same as that for k'_{m-i}.)

9.5-9. Find the unit impulse response for $v(t)$ in the circuit shown in Figure P9.5-3.

Figure P9.5-3

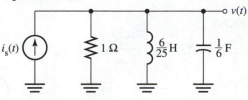

9.5-10. Find the unit impulse response for $v(t)$ in Figure P9.5-4.

Figure P9.5-4

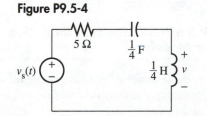

Section 9.6 An Algebraic Circuit Analysis Method

9.6-1. Draw a simulation diagram and find the response for the differential equation

$$\frac{d^2y}{dt^2} + 6\frac{dy}{dt} + 8y = x(t)$$

assuming that $x(t) = 4e^{-2t}u(t)$.

9.6-2. Draw a simulation diagram and find the response for the differential equation

$$\frac{d^2y}{dt^2} + 3\frac{dy}{dt} + 2y = \frac{dx}{dt}$$

assuming that $x(t) = e^{-t}u(t)$.

9.6-3. Find the step response for the circuit variable $i(t)$ shown in Figure P9.6-1 if $i_s(t) = 9e^{-t}u(t)$.

Figure P9.6-1

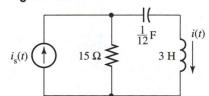

9.6-4. Find $i(t)$ in Figure P9.6-2 if $v_s(t) = 4e^{-t}u(t)$ V and $R = 4\ \Omega$.

Figure P9.6-2

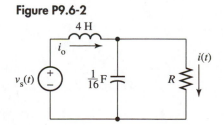

9.6-5. Find the response $i_0(t)$ for the circuit in Figure P9.6-2 with $R = 16/5\ \Omega$ and $v_s(t) = 8u(t)$ V.

9.6-6. Find the response $v(t)$ for the circuit shown in Figure P9.6-3 if $i_s(t) = (25/8)\cos(4t)u(t)$ A.

Figure P9.6-3

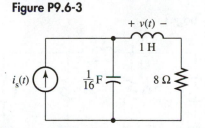

9.6-7. Find the response $v(t)$ in Figure P9.6-4 assuming that $i_s(t) = \cos(5t)u(t)$ V.

Figure P9.6-4

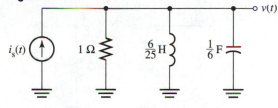

9.6-8. If $i_s(t) = 4\cos(2t)u(t)$ A in Figure P9.6-5, find $i(t)$.

Figure P9.6-5

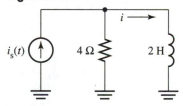

9.6-9. Find $y(t)$ for the differential equation

$$\frac{d^2y}{dt^2} + 4\frac{dy}{dt} + 4y = e^{-2t}u(t).$$

9.6-10. If $v_s(t) = 5e^{-3t/2}u(t)$ V in the circuit shown in Figure P9.6-6, find $v(t)$.

Figure P9.6-6

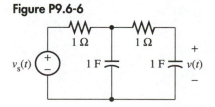

Section 9.7 Multiple Sources and Initial Conditions

9.7-1. Find $i(t)$ in terms of $i_{s1}(t)$ and $i_{s2}(t)$ for the circuit shown in Figure P9.7-1.

Figure P9.7-1

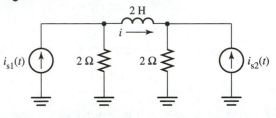

9.7-2. Find $v(t)$ in Figure P9.7-2 for all values of t.

Figure P9.7-2

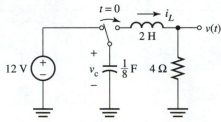

9.7-3. Find $i(t)$ in Figure P9.7-3 for all values of t.

Figure P9.7-3

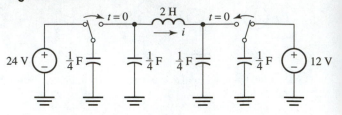

Section 9.8 SPICE: System Simulation (Behavior Modeling)

9.8-1. Find the response of the differential equation in Problem 9.3-1 using simulation diagrams and PSPICE.

9.8-2. Find the response of the differential equation in Problem 9.4-2 using simulation diagrams and PSPICE. (*Hint:* Use the PWL waveform specification.)

9.8-3. Find the response of the differential equation in Problem 9.5-3 using simulation diagrams and PSPICE.

10 Stability and Forced Response

In this chapter we focus on the ideas of stability and response decomposition. We present the idea of the pole-zero diagram, a graphical technique for determining the stability properties of a given circuit response. We give the definitions of stable, unstable, and marginally stable responses and present an algorithm for determining the type of stability without factoring a polynomial: the Routh algorithm. We show that any response can be written as the sum of two component responses, the forced and natural responses. We define the steady-state response (for exponential forcing functions and those that can be expressed in some manner as an exponential—for instance, dc and sinusoidal waveforms) to be the forced response *if the response is a stable one*. It is only for stable circuit responses that the term steady-state response makes sense. We also briefly treat the idea of a switched circuit being well posed; we define a circuit to be well posed if it is stable for $t \leq 0$ so that a steady state exists. Finally, we present several case studies of the stability and transient analysis of active circuits of a fairly complex nature. (Examples flagged with an (O) pertain to operational amplifiers.)

The prerequisite for this chapter is the material in Chapter 9.

10.1 | System Functions and Pole-Zero Diagrams

At this point we have the tools to find the time response of any circuit we encounter, with any independent source waveform. Sometimes, though, the solution can be quite complicated. If we are interested in only one or several features or aspects of the circuit response, rather than in the complete response, we might choose to employ methods that give us these aspects, rather than going through a complete solution process. We will develop an important method here for doing precisely this.

Response Decomposition and the System Function H(s) Suppose, as we show in Figure 10.1(a), that we are interested in how a circuit or system behaves in response to an exponential input:

$$x(t) = e^{st}u(t), \tag{10.1-1}$$

where

$$s = \sigma + j\omega \tag{10.1-2}$$

is any real or complex number (real if $\omega = 0$, purely imaginary if $\sigma = 0$) and is called the *complex frequency* of the exponential. From our work in the last chapter, we know that we can represent the exponential waveform as the impulse response of a differential system. We show this in Figure 10.1(b). Assuming that s is such that $p - s$ does not duplicate any of the factors in $H(p)$,[1] we can write the response $y(t)$ in terms of a PFE as

$$y(t) = \frac{H(p)}{p - s}\delta(t) = \frac{k}{p - s}\delta(t) + Y_n(p)\delta(t). \tag{10.1-3}$$

The operator $Y_n(p)$ is that part of the PFE resulting from $H(p)$, itself. We can evaluate k as usual:

$$k = \left[(p - s)\frac{H(p)}{p - s}\right]_{p \leftarrow s} = H(s). \tag{10.1-4}$$

This is a remarkable result! It says that the response to a one-sided exponential waveform is the sum of two terms,

$$y(t) = H(s)e^{st}u(t) + y_n(t). \tag{10.1-5}$$

Because the first term has the same form as the input (remember, $H(s)$ is a constant with respect to time), we call it the *exponential forced response:*

$$y_f(t) = H(s)e^{st}u(t). \tag{10.1-6}$$

The second term, $y_n(t)$, is called the *natural response* because it comes from the system operator, itself[2]—it represents the manner in which the system "wants" to respond "natu-

Figure 10.1
Exponential response of a differential system

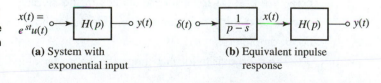

(a) System with exponential input

(b) Equivalent inpulse response

---

[1] That is, s does not coincide with any of the roots of the characteristic polynomial of $H(p)$.

[2] Notice that the natural response has the same form as the impulse response because it contains precisely the same terms (because of the denominator factors in $H(p)$).

rally." Thus, we can write the *complete response* as the sum of the forced and natural responses,

$$y(t) = y_f(t) + y_n(t). \tag{10.1-7}$$

The quantity $H(s)$ is the system operator with each symbol p replaced by the real or complex variable s:

$$H(s) = [H(p)]_{p \leftarrow s}. \tag{10.1-8}$$

It is called the *system function*. Notice that it is *not* an operator, but a rational function of the real or complex variable s. Thus, there is a one-to-one correspondence between the set of all differential operators (or system operators) and the set of all system functions. Anything we can do with one, we can do with the other—as far as exponential forced response is concerned.

Examples: Exponential Forced Response

Example 10.1 For the differential equation

$$\frac{d^2y}{dt^2} + 2\frac{dy}{dt} + 2y = x(t), \tag{10.1-9}$$

find the forced response to the forcing function $x(t) = 4e^{-t}u(t)$.

Solution The system function can be read directly from the differential equation:

$$H(s) = \frac{1}{s^2 + 2s + 2} = \frac{1}{(s + 1)^2 + 1}; \tag{10.1-10}$$

furthermore, $x(t)$ is a constant times e^{st}, with $s = -1$. Thus, the forced response is

$$y_f(t) = [4H(s)e^{st}u(t)]_{s=-1} = 4H(-1)e^{-t}u(t) = 4e^{-t}u(t). \tag{10.1-11}$$

The preceding example was an easy "plug-in." Because we did not ask for the complete response, and because the forcing function was an exponential, our labor was quite reduced.

Example 10.2 Find the forced response of the operator

$$H(p) = \frac{p + 1}{p + 2} \tag{10.1-12}$$

to the waveform $x(t) = 2e^{-t}u(t)$ and to $x(t) = 2e^{-2t}u(t)$.

Solution By inspection, we write the system function

$$H(s) = \frac{s + 1}{s + 2}. \tag{10.1-13}$$

Thus, for the first input waveform, $s = -1$ and $H(-1) = 0$; hence, the forced response is identically zero. For the second input waveform, $s = -2$ and $H(-2) = \infty$! Thus, the forced response seems to be infinite. You will recall that in deriving the exponential forced response, however, we assumed that the denominator factor introduced by the exponential was not the same as one already in $H(p)$—but this means that the exponent should not be -2 in our current example. If we convert $H(s)$ back to its operator form and solve for y in the usual way, we find that the complete response is actually $2(1 - t)e^{-2t}u(t)$. Thus, we should interpret $H(-2) = \infty$ as indicating a special case in the response computation.

Poles and Zeros

Based on the last example, we introduce the following important definition and illustrate it with an example.

Definition 10.1

The *zeros* of $H(s)$ are those values of s for which $H(s) = 0$; the *poles* of $H(s)$ are those values for which $H(s) = \infty$.

Example 10.3

Find the poles and zeros of

$$H(s) = 10\frac{(s + 1)^2(s + 2)(s + 4)^3}{s(s + 3)^2(s + 5)}. \tag{10.1-14}$$

(We have given $H(s)$ in factored form for convenience.)

Solution   The *finite* values of s that cause $H(s)$ to have the value zero are clearly at $s = -1$, -2, and -4. We say that the zeros at -1 and -4 are *multiple*, or that there are *two zeros at* $s = -1$ and *three zeros at* $s = -4$. Alternately, we say that the zero at $s = -1$ is of *second order*, and the one at $s = -4$ is of *third order*. We say that the zero at $s = -2$ is *simple*, or that it is of *first order*. Similar comments hold for the *finite* poles, of which there are two at $s = -3$, one at $s = -5$, and one at $s = 0$.

The only remaining question concerns the behavior of $H(s)$ as s becomes very large. We see that $H(s) \sim 10s^6/s^4 = 10s^2$ as $s \to \infty$; thus, we say that $H(s)$ has a second-order pole at $s = \infty$.[3] If the degree of the denominator were greater than that of the numerator, $H(s)$ would have one or more zeros at $s = \infty$.

The Pole-Zero Diagram (PZD)

It is easy to see that the finite zeros and poles completely determine $H(s)$, except for a scale factor. Therefore a graph showing the finite poles and zeros would be useful. Such a picture is provided by the *pole-zero diagram*, or *PZD*, which is simply a plot of all the finite poles and zeros in the complex s plane using the symbols x for a pole and o for a zero. A PZD for the system function in the last example is shown in Figure 10.2. It is con-

Figure 10.2

The pole-zero diagram for

$$H(s) = 10\frac{(s + 1)^2(s + 2)(s + 4)^3}{s(s + 3)^2(s + 5)}$$

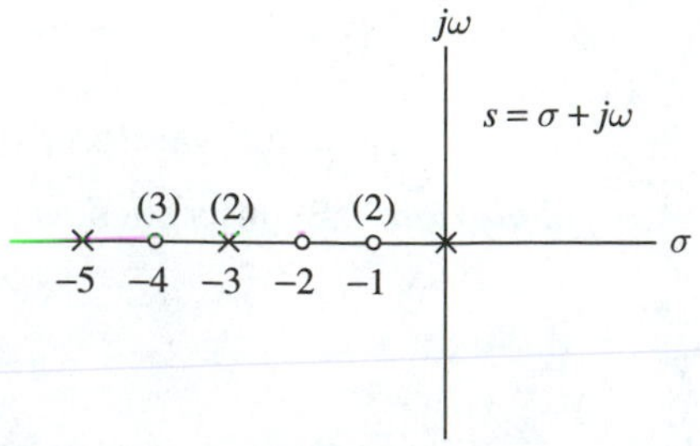

---

[3] We write $F(s) \sim G(s)$ if $\lim [F(s)/G(s)] = 1$ as $s \to \infty$, and say that $F(s)$ is "asymptotically equal to" $G(s)$.

ventional to indicate the multiplicity in parentheses for higher-order poles and zeros. In the example shown all of the poles and zeros are real; this, however, does not always have to be the case, as the next example shows.

Example 10.4   Draw the PZD for

$$H(s) = 10\frac{s(s-2)}{s^2 + 2s + 2}. \qquad (10.1\text{-}15)$$

Does it have a pole or a zero at $s = \infty$? If so, what is its order?

Solution   The finite zeros of $H(s)$ are at 0 and $+2$ on the real axis. The poles are given by the zeros of the *characteristic polynomial* (the roots of the *characteristic equation*), which we now write in terms of the complex variable s (rather than z) as

$$D(s) = s^2 + 2s + 2. \qquad (10.1\text{-}16)$$

Thus, there are simple poles at $s = -1 + j$ and $s = -1 - j$. The PZD is sketched in Figure 10.3. In order to answer the question regarding the presence or absence of poles and/or zeros at infinity, we must make the usual asymptotic approximation. Letting s become very large, we ignore all powers of s but the highest in both numerator and denominator, obtaining

$$H(s) \sim 10\frac{s^2}{s^2} \longrightarrow 10. \qquad (10.1\text{-}17)$$

Thus, there are no poles or zeros at $s = \infty$. Notice carefully that this information is already contained within the finite poles and zeros for we do not need the scale factor to investigate the poles or zeros at $s = \infty$.

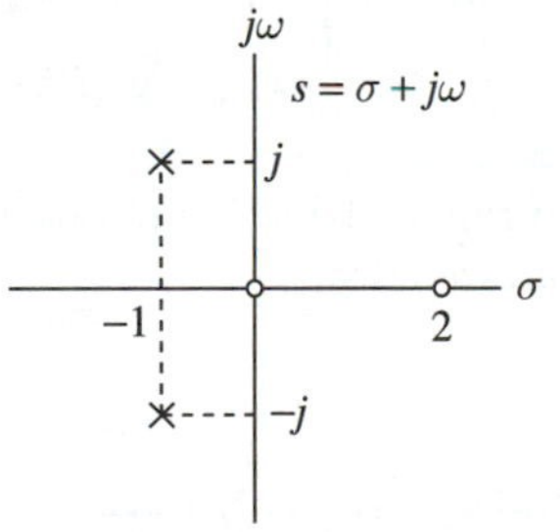

Figure 10.3

The pole-zero diagram for
$$H(s) = 10\frac{s(s-2)}{s^2 + 2s + 2}$$

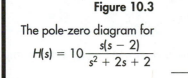

Complex Impedance and Admittance

As we show schematically in Figure 10.4, if the stimulus for a subcircuit is considered to be the current or voltage at a given pair of terminals, and the response the other variable at that same terminal pair, the system operator becomes either an impedance operator or an admittance operator, respectively. The same identification holds for the exponential forced response of these variables. If the current has the form $i(t) = e^{st}u(t)$, then the *complex impedance* is defined by

$$v_f(t) = Z(s)e^{st}u(t). \qquad (10.1\text{-}18)$$

Observe carefully that it is only the forced response for the voltage that appears in this equation; if one attempts to use the complete response, the result will be erroneous. An

Figure 10.4 Complex impedance and admittance

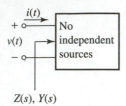

$Z(s)$, $Y(s)$

alternate way of writing (10.1-18) is

$$Z(s) = \left[\frac{v_f(t)}{i(t)} \right]_{i(t)=e^{st}u(t)} \tag{10.1-19}$$

Either of the two preceding equations can be used as the definition for the complex impedance, which is often shortened to simply "impedance."

If the voltage is known to have the form $v(t) = e^{st}u(t)$, then the *complex admittance* is defined by the equation

$$i_f(t) = Y(s)e^{st}u(t), \tag{10.1-20}$$

or by

$$Y(s) = \left[\frac{i_f(t)}{v(t)} \right]_{v(t)=e^{st}u(t)}. \tag{10.1-21}$$

It should be clear that if either the impedance or the admittance is not identically zero, then one is the inverse of the other—and knowledge that one terminal variable is exponential leads to the assurance that the other is exponential also, at least as far as the forced response component is concerned, because the current and voltage are related by a linear, constant-coefficient differential equation.

Exponential Forced Response Models for R, L, and C

The simplest examples of impedances and admittances are for resistors, inductors, and capacitors. The impedance of a resistor is clearly its resistance (as one can see immediately upon application of equation (10.1-19)). We know that the operator impedance of an inductor is $Z(p) = Lp$, so its complex impedance is $Z(s) = Ls$. Similarly, the impedance of a capacitor follows from the form of its operator: $Z(p) = 1/Cp$; thus, for a capacitor, $Z(s) = 1/Cs$. We exhibit all three of these results as the following set of equations:

$$Z(s) = \begin{cases} R; & \text{resistor} \\ Ls; & \text{inductor} \\ \dfrac{1}{Cs}; & \text{capacitor} \end{cases} \tag{10.1-22}$$

and show them graphically in Figure 10.5. Let's look now at some more general impedances by means of a couple of examples.

Figure 10.5
Exponential forced response models for
R, L, and C

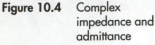

Example 10.5

Find the complex impedance of the series LC subcircuit shown in Figure 10.6 and sketch its PZD.

Figure 10.6
A series LC circuit

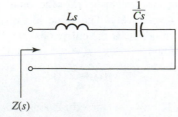

$Z(s)$

Solution   We already know that operator impedances in series add; hence, so do complex impedances. Thus,

$$Z(s) = Ls + \frac{1}{Cs} = L\frac{s^2 + \dfrac{1}{LC}}{s} = L\frac{s^2 + \omega_0^2}{s} \qquad (10.1\text{-}23)$$

where

$$\omega_0 = \frac{1}{\sqrt{LC}} \qquad (10.1\text{-}24)$$

is called the *series resonant frequency*. (We will discuss this concept at greater length in
Chapter 12.) There is one finite pole at $s = 0$ and two complex conjugate finite zeros at
the purely imaginary values $s = \pm j\omega_0$. The PZD is shown in Figure 10.7. We point out
an unusual feature of $Z(s)$, one that is true for any impedance that has finite zeros:
$Z(\pm j\omega_0) = 0$. For complex exponential excitation with an exponent equal to either of
these two purely imaginary values, the series LC subcircuit is equivalent to a short circuit!
This is not a complete equivalence, however, because a short circuit requires the voltage
to be zero voltage for *any* excitation.

Figure 10.7

The PZD for
$Z(s) = L\dfrac{s^2 + \omega_0^2}{s}$

Example 10.6   Find the complex impedance for the parallel LC subcircuit shown in Figure 10.8 and
sketch its PZD.

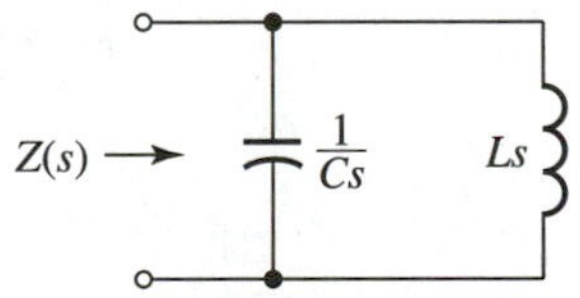

Figure 10.8
The parallel LC
subcircuit

Solution   Again, we know how to combine impedance operators that are connected in parallel: we
add the admittances, then invert—or multiply the impedances and divide by their sum.
These manipulations hold also for complex impedances; hence,

$$Z(s) = \frac{\dfrac{1}{Cs} \times Ls}{\dfrac{1}{Cs} + Ls} = \frac{1}{C}\frac{s}{s^2 + \dfrac{1}{LC}} = \frac{1}{C}\frac{s}{s^2 + \omega_0^2}, \qquad (10.1\text{-}25)$$

where

$$\omega_0 = \frac{1}{\sqrt{LC}} \qquad (10.1\text{-}26)$$

is called the *parallel resonant frequency*. The impedance is simply the reciprocal of that for the series LC circuit, providing we let L be replaced by the reciprocal of C and vice versa. There is a single finite zero at $s = 0$ and two purely imaginary complex conjugate poles at $s = \pm j\omega_0$. The PZD is shown in Figure 10.9. Because there are two poles at $\pm j\omega_0$, the impedance is infinite there; thus, for exponential excitation with either of these two exponents, the circuit behaves like an open circuit. We hasten to point out, though, that this is not an exact equivalence because an open circuit constrains the terminal current to be zero for all excitation waveforms.

Figure 10.9
The PZD for
$$Z(s) = \frac{1}{C} \frac{s}{s^2 + \omega_0^2}$$

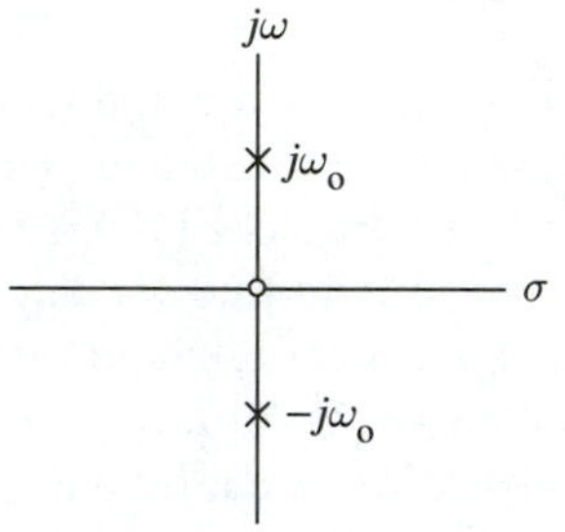

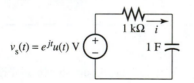

Section 10.1 Quiz

Q10.1-1. Find the forced response for $i(t)$ in Figure Q10.1-1.

Figure Q10.1-1

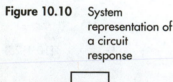

$v_s(t) = e^{jt}u(t)$ V

Q10.1-2. Find the forced response of the differential equation
$$\frac{d^2y}{dt^2} + 5\frac{dy}{dt} + 6y = 2x(t)$$
to the forcing functions $x(t) = u(t)$ and $x(t) = e^{-t}u(t)$.

Q10.1-3. Plot the PZD for the system operator
$$H(p) = \frac{5(p - 2)}{p(p^2 + 2p + 2)}.$$

10.2 | Stability

The Definition of Stability

Now let's discuss stability in terms of the system function $H(s)$. We will focus our attention on a circuit having a single independent source and upon a particular response voltage or current. We represent the situation as shown in Figure 10.10. The operator $H(p)$ connects the input $x(t)$ (the independent source waveform) with the desired response variable $y(t)$ by means of the equation $y(t) = H(p)x(t)$. If we adjust the source waveform to be the unit impulse, the response variable $y(t)$ will be the unit impulse response $h(t)$:

$$h(t) = H(p)\delta(t). \tag{10.2-1}$$

Figure 10.10   System representation of a circuit response

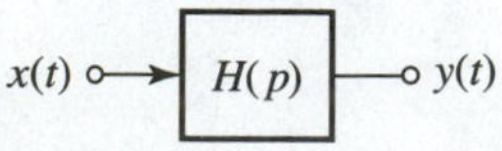

$x(t) \longrightarrow \boxed{H(p)} \longrightarrow y(t)$

We then have the following definition, which we presented earlier in an informal manner, but which we now give here more formally.

Definition 10.2   A system operator is *stable* if its unit impulse response[4] approaches zero for large values of t. It is called *unstable* if the unit impulse response is unbounded as t becomes infinitely large. It is called *marginally stable* if it is neither stable nor unstable; that is, if $h(t)$ does not approach zero as t becomes infinitely large, but does remain bounded.

More generally, one can have several (or many) voltage and current responses in a circuit; likewise, a circuit can have more than one independent source. Thus, we say that a *circuit* is stable if every possible operator connecting a source with a response variable is stable. If at least one is unstable, we call the circuit unstable. Finally, if at least one operator is marginally stable—but none is unstable—we say the circuit is marginally stable.

In the last section we developed the idea of a system function and its PZD. As it turns out, these tools give us a very powerful way to check the stability of an operator. For instance, suppose

$$H(p) = \frac{p^3 + 6p^2 + 15p + 16}{p^2 + 4p^2 + 4} = p + 2 + \frac{3}{p + 2} + \frac{2}{(p + 2)^2}. \qquad (10.2\text{-}2)$$

Then the unit impulse response is $h(t) = p\delta(t) + 2\delta(t) + 3e^{-2t}u(t) + 2te^{-2t}u(t)$. We see that the "improper part," that is, the part that arises from the long division steps (or, equivalently, from the "excess degree" of the numerator) consists in general of an impulse function and perhaps one or more of its derivatives. *Such waveforms are concentrated around $t = 0$, whereas stability is a property of the behavior of the impulse response for large values of time.* Thus, we need only investigate the proper part: that part of the PFE that arises from the factors of the characteristic polynomial. In fact, from (10.2-2), we see that it is the denominator factors of $H(p)$ that determine stability.

Stability Versus Pole Location   Here is the idea that makes the system function and the PZD so important for stability studies. The system function $H(s)$ is identical in form to the system operator $H(p)$; hence, its poles completely determine stability—the denominator factors of $H(p)$ are identical in form to the denominator factors of $H(s)$.

Example 10.7   Investigate the stability of the complex first-order system operator

$$H(p) = \frac{1}{p - s_0}, \qquad (10.2\text{-}3)$$

where

$$s_0 = \sigma_0 + j\omega_0. \qquad (10.2\text{-}4)$$

Solution   The system function is

$$H(s) = \frac{1}{s - s_0}, \qquad (10.2\text{-}5)$$

which has a simple (or first-order) pole at $s = s_0$. The PZD is shown in Figure 10.11. We have assumed in our drawing that σ_0 is negative and ω_0 is positive, but we will allow s_0 to conceptually move about to any point in the s plane and investigate the stability as a func-

---

[4] Taking into account footnote 2 (page 492) the phrase "natural response" can be substituted everywhere in this definition for the phrase "unit impulse response."

tion of the pole position. Stability depends on the size of the unit impulse response, so we look at its magnitude for various values of s_0. We have

$$\left| h(t) \right| = \left| e^{s_0 t} u(t) \right| = \left| e^{\sigma_0 t} e^{j\omega_0 t} \right| u(t) = e^{\sigma_0 t} \left| e^{j\omega_0 t} \right| u(t). \tag{10.2-6}$$

Here we have used the fact that the real exponential $e^{\sigma_0 t}$ is nonnegative, as is $u(t)$. What about the exponential with the complex exponent? Using Euler's formula, we find that its magnitude is

$$\left| e^{j\omega_0 t} \right| = \sqrt{\cos^2(\omega_0 t) + \sin^2(\omega_0 t)} = 1. \tag{10.2-7}$$

Thus, (10.2-6) becomes

$$\left| h(t) \right| = e^{\sigma_0 t} u(t). \tag{10.2-8}$$

Figure 10.12 shows a time plot of $\left| h(t) \right|$ for three different values of σ_0: positive, negative, and zero. *If you glance at the PZD in Figure 10.11, you will see that the operator is stable if the pole s_0 is in the open left half-plane (as shown in that figure), unstable if it is in the open right half-plane, and marginally stable if it is on the $j\omega$ axis.*

Figure 10.11
The PZD

Figure 10.12
Impulse response magnitude

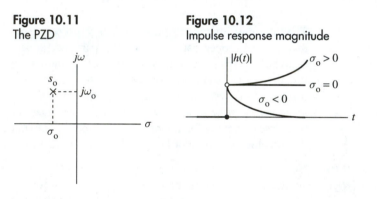

Determination of Stability Type from the PZD

Now let's investigate the stability of a general rational system operator. We will suppose that $H(s)$ has a pole of order m at $s_0 = \sigma_0 + j\omega_0$; thus, it can be written in the form

$$H(s) = \frac{N(s)}{(s - s_0)^m D_0(s)}. \tag{10.2-9}$$

The corresponding system operator is

$$H(p) = \frac{N(p)}{(p - s_0)^m D_0(p)}. \tag{10.2-10}$$

The PFE of $H(p)$ is

$$H(p) = \frac{k_1}{p - s_0} + \frac{k_2}{(p - s_0)^2} + \cdots + \frac{k_i}{(p - s_0)^i} + \cdots + \frac{k_m}{(p - s_0)^m} + H_0(p), \tag{10.2-11}$$

where $H_0(p)$ consists of the terms due to the factors of $D_0(p)$, that is, of all the poles of $H(s)$ other than the one at s_0. We will look at the time function corresponding to the ith term in the impulse response $h(t) = H(p)\delta(t)$, which is

$$h_i(t) = \frac{k_i}{(p - s_0)^i} \delta(t) = k_i \frac{t^{i-1}}{(i-1)!} e^{s_0 t} u(t) = k_i \frac{t^{i-1}}{(i-1)!} e^{\sigma_0 t} e^{j\omega_0 t} u(t). \quad (10.2\text{-}12)$$

Taking the magnitude, we have

$$|h_i(t)| = |k_i| \frac{t^{i-1}}{(i-1)!} e^{\sigma_0 t} u(t). \quad (10.2\text{-}13)$$

Our stability check comes from equation (10.2-13). There are three separate cases:
Case 1 ($\sigma_0 > 0$): In this case, it is clear that $h_i(t)$ is unbounded as $t \to \infty$ regardless of the value of i. As the entire impulse response is made up of the sum of such terms, the system operator is unstable.[5] We note that $\sigma_0 > 0$ represents a pole in the right half-plane (RHP). Thus, *if H(s) has at least one finite pole in the RHP, H(p) is unstable.*
Case 2 ($\sigma_0 < 0$): Here, even though t^{i-1} in (10.2-13) tends to infinity by itself (for $i \geq 2$), the exponential term goes to zero faster—as is easily shown by an application of L'Hôpital's rule. *If H(s) has all its finite poles in the left half-plane (LHP), H(p) is stable.*
Case 3 ($\sigma_0 = 0$): In this case, we can write

$$|h_i(t)| = |k_i| \frac{t^{i-1}}{(i-1)!} u(t). \quad (10.2\text{-}14)$$

Clearly, if $i \geq 2$, the waveform will become infinitely large, implying instability. If, however, the pole at s_0 is simple, then $i = m = 1$, and it remains bounded but nonzero. Thus, *if there are one or more finite poles on the imaginary axis of the s plane that are simple, and any others are in the LHP, the response will be marginally stable. If any of these $j\omega$ axis poles have multiplicity greater than one, the response will be unstable.*

   We can summarize our discussion with a drawing of the complex plane showing the RHP, LHP, and imaginary (or $j\omega$) axis as in Figure 10.13. The regions of stability, marginal stability, and instability are shown shaded. Note that it is only the location of the finite poles (and their multiplicity, if they are on the $j\omega$ axis) that is important. *The zeros have no effect on stability—nor do poles at $s = \infty$.* Because there is so much similarity between finite poles and zeros on the PZD, there is a tendency to forget the fact that only the finite poles influence stability.

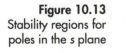

Figure 10.13
Stability regions for poles in the s plane

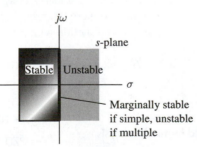

Example 10.8   Investigate the stability of the voltage response $v(t)$ for the circuit drawn in Figure 10.14 using the system function and PZD technique just developed. Then use $H(p)$ to compute the unit impulse response and verify the result you have obtained.

---

[5] Two different such terms cannot cancel each other because of a sign difference if the corresponding poles are different or the corresponding orders are different. We will not pause to prove this statement, but it should be intuitively acceptable.

Figure 10.14
An example circuit

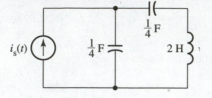

Solution We show the circuit prepared for mesh analysis in Figure 10.15. The source has been adjusted so that it produces a complex exponential waveform $e^{st}u(t)$. As each circuit variable is of the form $y(t) = H(p)i_s(t) = H(p)e^{st}u(t)$, for some system operator $H(p)$, the forced response for that variable is $y_f(t) = H(s)e^{st}u(t)$; hence, we can use complex impedances for the elements in computing the forced response. The single mesh equation required is

$$\frac{4}{s}[i_f(t) - i_s(t)] + \frac{4}{s}i_f(t) + 2si_f(t) = 0. \tag{10.2-15}$$

Simplifying, we have

$$(s^2 + 4)i_f(t) = 2i_s(t), \tag{10.2-16}$$

or

$$i_f(t) = \frac{2}{s^2 + 4}i_s(t). \tag{10.2-17}$$

Finally, using the inductor relationship for the voltage, we get

$$v_f(t) = 2si_f(t) = \frac{4s}{s^2 + 4}i_s(t). \tag{10.2-18}$$

The system function is

$$H(s) = \frac{4s}{s^2 + 4}. \tag{10.2-19}$$

The single finite zero is at $s = 0$, and the two finite poles are at $s = \pm j2$. The PZD is shown in Figure 10.16. The operator is clearly marginally stable because the two finite poles on the $j\omega$ axis are simple and there are none in the RHP. Because the operator is itself exactly the same as the system function with p replacing s everywhere, we have

$$h(t) = H(p)\delta(t) = \frac{4p}{p^2 + 4}\delta(t) = 4\cos(2t)u(t). \tag{10.2-20}$$

Figure 10.15
The circuit prepared for mesh analysis

Figure 10.16
The PZD for $H(s) = \dfrac{4s}{s^2 + 4}$

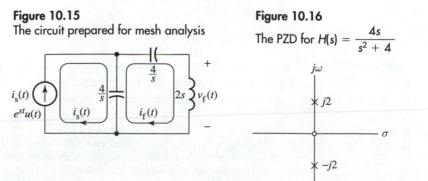

Clearly, this impulse response does not approach zero or become unbounded for large t. Thus, it is marginally stable, as predicted.

Summary:
Determination
of Stability

Let's summarize our technique for determining the stability of a given response variable. We derive the exponential forced response for that variable—in the process determining the appropriate $H(s)$. Because $H(s)$ is a rational function of the complex variable s, its finite poles are the roots of its denominator polynomial. We must, therefore, factor this polynomial to determine its roots. This can be quite a numerical chore, particularly for higher-order polynomials. Furthermore, we do not actually need such precise information. All we need to know is whether or not there are roots in the right half-plane. If there are, we know the circuit is unstable. If there are none, we need to know whether or not there are $j\omega$ axis roots. If there are none, we know the circuit is stable. If there are $j\omega$ axis roots, we must check the multiplicity of those roots. If they are all simple, the circuit is marginally stable. If there is at least one of higher order than one, the system is unstable.

The Routh
Algorithm: Stability
Determination
Without Factoring

It is fortunate that there is an algorithm, simple to apply, that determines just this information without the need for factorization. It is called the Routh algorithm or the Routh criterion (for stability). The proof is beyond the level of this text, so we will merely show how to apply it.

To understand the Routh algorithm we need the ideas of *even* and *odd* polynomials. $E(s) = s^4 + 2s^2 + 5$ is an even polynomial because

$$E(-s) = E(s). \tag{10.2-21}$$

This equation implies that only even powers are present in the polynomial, and vice versa.[6] On the other hand, $O(s) = 3s^5 + 2s^3 + s$ is an odd polynomial because

$$O(-s) = -O(s). \tag{10.2-22}$$

An odd polynomial has only odd powers of s.

The Routh criterion relies upon the following obvious fact: each polynomial can be written ("decomposed" into) the sum of an even polynomial and an odd one. If the original polynomial $P(s)$ is of odd degree $2n + 1$, we have

$$P(s) = O_{2n+1}(s) + E_{2n}(s), \tag{10.2-23}$$

where $O_{2n+1}(s)$ is an odd polynomial of degree $2n + 1$ and $E_{2n}(s)$ is an even polynomial of degree at most $2n$ (if its leading coefficient is zero, it will be of lower order). Here is an example:

$$P(s) = s^5 + 2s^4 + s^3 + 2s^2 + s + 2 = (s^5 + s^3 + s) + (2s^4 + 2s^2 + 2). \tag{10.2-24}$$

If $P(s)$ is of even degree $2n$, we can write

$$P(s) = E_{2n}(s) + O_{2n-1}(s), \tag{10.2-25}$$

where $E_{2n}(s)$ is an even polynomial of degree $2n$, and $O_{2n-1}(s)$ is an odd one of degree at most $2n - 1$. For example, we can write

$$P(s) = 4s^4 + 2s^2 + s + 4 = (4s^4 + 2s^2 + 4) + (0s^3 + s). \tag{10.2-26}$$

---

[6] This is not "immediately obvious," but should be intuitively easy to accept.

Now we can demonstrate the Routh algorithm. We will work with a specific example—one for which we know the root distribution, so that we can check our results. We will choose the polynomial

$$P(s) = (s + 1)(s - 2)(s + 3)(s - 4) = s^4 - 2s^3 - 13s^2 + 14s + 24 \quad (10.2\text{-}27)$$
$$= (s^4 - 13s^2 + 24) + (-2s^3 + 14s) = E_4(s) + O_3(s).$$

There are clearly two roots in the RHP and two in the LHP, with none on the $j\omega$ axis. We form the Routh array shown in Table 10.1 by inserting the coefficients of $E_4(s)$ in descending order from left to right in the first row and those of $O_3(s)$ in the second row. We already know that the coefficients in the first row form an even polynomial of order 4 and those in the second row form an odd polynomial of order 3. That is the significance of the powers of s to the left of each row. The Routh algorithm successively generates an even polynomial of order 2, an odd one of order 1, and finally an even one of order 0 (a constant). Here's how the algorithm works. Assuming that a_{ij} is the entry in row i and column j of the array, we generate the elements in row $i + 1$ by this formula:

$$a_{i+1,j} = \frac{a_{i,1}a_{i-1,j+1} - a_{i-1,1}a_{i,j+1}}{a_{i,1}}, \quad j = 1, 2, \text{etc.} \quad (10.2\text{-}28)$$

Here are how the computations go for the s^2 row:

$$a_{3,1} = \frac{a_{2,1}a_{1,2} - a_{1,1}a_{2,2}}{a_{2,1}} = \frac{(-2) \times (-13) - (1) \times (14)}{-2} = -6, \quad (10.2\text{-}29)$$

$$a_{3,2} = \frac{a_{2,1}a_{1,3} - a_{1,1}a_{2,3}}{a_{2,1}} = \frac{(-2) \times (24) - (1) \times (0)}{-2} = 24, \quad (10.2\text{-}30)$$

$$a_{3,3} = \frac{a_{2,1}a_{1,4} - a_{1,1}a_{2,4}}{a_{2,1}} = \frac{(-2) \times (0) - (1) \times (0)}{-2} = 0. \quad (10.2\text{-}31)$$

Note: In applying this algorithm, we assume that all array entries "to the right" of the actual entries are zero. We will leave it to you to compute the other entries and verify that they are the ones given in Table 10.1.

Table 10.1
Routh array for
equation (10.2-27)

s^4	1	-13	24
s^3	-2	14	0
s^2	-6	24	0
s^1	6	0	0
s^0	24	0	0

Now here is the payoff: we look at the first column. *The number of sign transitions (+ to − or − to +) is equal to the number of roots in the RHP.* In our case, there are two. There is a transition from + to − from the first row to the second and one from − to + between the third and the fourth. Thus, there are two roots in the RHP.

There are a couple of ways the algorithm can "go wrong," but both are correctable. For instance, look at the polynomial

$$P(s) = 2s^4 + 2s^3 + 3s^2 + 3s + 2 = (2s^4 + 3s^2 + 2) + (2s^3 + 3s) \quad (10.2\text{-}32)$$
$$= E^4(s) + O_3(s).$$

The array is shown in Table 10.2. As a zero appears in the (3, 1) position, we cannot per-form any more computations because our algorithm says we must divide by it. Now a the-oretical investigation shows that this (a zero in the first column, but the row is not identically zero) occurs whenever there are an even number of roots symmetrically lo-cated relative to the origin. Here's how we solve the problem: we go back to the original polynomial and add another root by multiplying by a factor $s + a$, where a is an arbitrary *positive number*. This added root is in the LHP and therefore does not affect the number of roots in the RHP or on the $j\omega$ axis—and the Routh algorithm is only concerned with these. Taking the factor to be $s + 1$ (a root at $s = -1$), we have the modified polynomial

$$P(s) = 2s^5 + 4s^4 + 5s^3 + 6s^2 + 5s + 2 \tag{10.2-33}$$
$$= (2s^5 + 5s^3 + 5s) + (4s^4 + 6s^2 + 2) = O_5(s) + E_4(s).$$

The Routh array is shown in Table 10.3. There are clearly two roots in the RHP in this modified polynomial and thus two in the original one as well.

Table 10.2
Routh array for equation (10.2-32)

s^4	2	3	2
s^3	2	3	0
s^2	0	2	0
s^1	?	?	?
s^0	?	?	?

Table 10.3
Routh array for equation (10.2-33)

s^5	2	5	5
s^4	4	6	2
s^3	2	4	0
s^2	-2	2	0
s^1	6	0	0
s^0	2	0	0

When an entire row consists of zero entries the situation is a bit different. The last nonzero row, in this case, corresponds to an even or odd polynomial that is a common factor of the even and odd polynomials in the original decomposition. The roots of this polynomial are then roots of the original one. Here's an example:

$$P(s) = s^4 - 3s^3 + 6s^2 - 12s + 8 \tag{10.2-34}$$
$$= (s^4 + 6s^2 + 8) + (-3s^3 - 12s) = E_4(s) + O_3(s).$$

The Routh array is shown in Table 10.4. We are once more unable to continue. The theory behind the Routh algorithm tells us that the even polynomial of degree 2 in the row im-mediately above the zero row is a common factor of the original even and odd polynomi-als; hence, its roots are also roots of the original polynomial. Because the degree is only 2, we can actually find these roots. The polynomial itself is $2s^2 + 8$, and its roots are at $\pm j2$. There are, therefore, two simple roots on the $j\omega$ axis. Thus, if there had been no sign variations in the first column prior to the zero row, there would have been no roots in the RHP and only two simple ones on the $j\omega$ axis; hence, the system from which the polyno-

Table 10.4
Routh array for
equation (10.2-34)

s^4	1	6	8
s^3	-3	-12	0
s^2	2	8	0
s^1	0	0	0
s^0	?	?	?

mial arose would have been marginally stable. In our present example, however, we already have two sign variations—one from the first to second rows and one from the second to the third. Thus, we know that we also have two roots in the RHP. In fact, in applying the Routh algorithm we can actually stop whenever we detect a sign variation because that single transition tells us that the response is unstable.

If there had been no sign variations above our all-zero row, however, we would have been forced to continue, for there could be one or more sign transitions below that row. *Here's what we do to "patch up" the algorithm: we compute the derivative of the polynomial in the last nonzero row (the derivative of an even polynomial is odd, whereas the derivative of an odd polynomial is even). In this case, the derivative of $2s^2 + 8$ is $4s$. We replace the zeros in the all-zero row with the coefficients of this polynomial and complete the algorithm.* The modified Routh array for our example is shown in Table 10.5. We see that there are two RHP roots. In general, this process always gives us the number of RHP roots. If the last polynomial before the all-zero row were of high order, we would have to test it to determine its root locations. This will not often occur for circuits that are sufficiently simple for hand analysis, so we will defer this aspect of the Routh algorithm to the problems at the end of this chapter.

Table 10.5
Modified Routh array

s^4	1	6	8
s^3	-3	-12	0
s^2	2	8	0
s^1	4	0	0
s^0	8	0	0

A Circuit Example: Stability

Now let's tackle a circuit, the one shown in Figure 10.17. Suppose we vary μ through all nonnegative values. Which values result in stable operation, which in marginally stable operation, and which in instability? We will take the indicated current $i(t)$ as the response and, of course, $i_s(t)$ as the input.

Figure 10.17
An example circuit

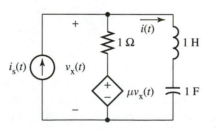

We show the circuit in Figure 10.18 prepared for mesh analysis with the inductor and the capacitor represented by their complex impedances and the current source adjusted to put out a complex exponential. We have also taped the v-source and defined the mesh currents in that figure. We can write one mesh equation to find the desired forced response current $i_f(t)$:

$$(s + 1/s + 1)i_f(t) - 1i_s(t) = v_c(t). \tag{10.2-35}$$

Untaping, we have (using KVL on the inductor and capacitor)

$$v_c(t) = \mu v_x(t) = \mu(s + 1/s)i_f(t). \tag{10.2-36}$$

Figure 10.18
Forced response equivalent circuit in complex impedance form

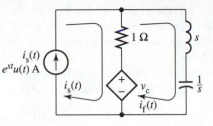

Solving these two equations simultaneously, we obtain

$$(1 - \mu)s^2 + s + 1 - \mu]i_f(t) = si_s(t). \tag{10.2-37}$$

Thus, the system function for the response variable $i(t)$ is

$$H(s) = \frac{s}{(1 - \mu)s^2 + s + 1 - \mu}. \tag{10.2-38}$$

Mechanically, the computation of the system function $H(s)$ is precisely the same as computation of the system operator $H(p)$.

At this point we can begin our stability study. The Routh array for the denominator polynomial is shown in Table 10.6. It is quite simple, as you can see, and tells us that there are two roots in the RHP if $\mu > 1$ (so the response is unstable) and none if $\mu < 1$ (it is stable). In the exceptional case in which $\mu = 1$, we merely note that (10.2-38) becomes $H(s) = 1$. There are no poles and no zeros, and the response current is identical to the source current. Hence the response is stable (the unit impulse response is a unit impulse, which is zero for $t > 0$). It is not marginally stable for any finite value of μ.

Table 10.6
Routh Array

s^2	$1-\mu$	$1-\mu$	0
s^1	1	0	0
s^0	$1-\mu$	0	0

Section 10.2 Quiz

Q10.2-1. Use the Routh algorithm to determine the stability type for the system whose characteristic polynomial is $P(s) = s^4 - 2s^3 - 7s^2 + 8s + 12$.

Q10.2-2. Find the values of $\beta \geq 0$ for which the response indicated in Figure Q10.2-1 is stable, marginally stable, and unstable.

Q10.2-3. Draw the PZD for the response $i(t)$ in Figure Q10.2-1. Assume that $\beta = 2$.

Figure Q10.2-1

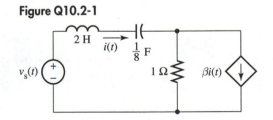

10.3 | Forced and Steady-State Response

The Concept of Steady-State Behavior

Now that we have developed the idea of stability thoroughly, we are in a position to discuss another important concept: *steady state*. The phrase *"steady state"* is intended to suggest that the behavior is unchanging: something is going on that does not change radically with time. If a response is constant, dc, then this certainly fits the description. However, a sinusoidal response does, too—even though it changes over one period, it repeats this same behavior over and over with time. In fact, any periodic response can be considered to be a steady-state behavior.

Let's look at Figure 10.19 for review. It is Figure 10.1 of Section 10.1, but is repeated here for ease of reference. We know from the analysis in Section 10.1 that

$$y(t) = H(s)e^{st}u(t) + y_n(t) = y_f(t) + y_n(t); \qquad (10.3\text{-}1)$$

that is, the total response is the sum of the forced response $y_f(t)$ and the natural response $y_n(t)$. The system function $H(s)$ is obtained from the operator $H(p)$ by replacing the symbol p by s everywhere it occurs. Thus, if $H(p)$ is known, the forced response is easy to find. The natural response, however, must be computed by doing a complete partial fraction expansion on $H(p)/(p - s)$. The response due to the denominator factor $p - s$ gives the forced response component in (10.3-1), and the remaining denominator factors (those of $H(p)$ itself) give the natural response component.

Figure 10.19
Exponential response
of a differential system

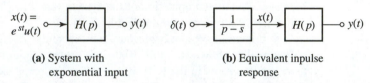

(a) System with   (b) Equivalent impulse
exponential input   response

The Common Form for the Natural and Impulse Responses

As it turns out, the natural response is related in a "natural" way to the impulse response. In fact, if you look at the defining equation for the impulse response,

$$h(t) = H(p)\delta(t), \qquad (10.3\text{-}2)$$

you will see that the *form* of $h(t)$ is the same as that of $y_n(t)$. The reason is simple: both waveforms can be obtained by doing a partial fraction expansion and both arise from the denominator factors of $H(p)$ only. Thus, all of the exponential terms in the PFE for $h(t)$ will be the same as those in the PFE for $y_n(t)$. The constants will differ in general, but the exponents will be the same—and it is these exponents that determine stability. Thus, if the response $y(t)$ (or, what is the same thing, the operator $H(p)$) is stable, then both $h(t)$ and $y_n(t)$ will approach zero for large values of time. This leaves only the forced response $y_f(t)$. In this case, we say that the circuit response is *steady state*.

Definition 10.3   If $y(t) = H(p)x(t)$ is a stable response variable, then we define the *steady-state* response to be the forced response $y_f(t)$.

For large values of time, therefore, the complete response is approximately the same as the forced response, and we say that steady state has occurred. If this is true for *all* the voltage and current variables in a circuit, we say that it is "operating in the steady state." We now present a steady-state response example.

Example 10.9   Suppose that a circuit variable $y(t)$ obeys the operator equation

$$y(t) = H(p)x(t) = \frac{24(p - 3)}{(p + 1)(p + 2)}. \qquad (10.3\text{-}3)$$

a.   Determine the stability of $y(t)$ from $H(s)$.

b.   Find the forced response to $x(t) = 2e^{2t}u(t)$.

c. Find the natural response.

d. Find the impulse response.

Solution   We immediately see that the system function is

$$H(s) = \frac{24(s - 3)}{(s + 1)(s + 2)}. \tag{10.3-4}$$

The PZD is shown in Figure 10.20. $H(s)$ has one right half-plane zero and two left half-plane poles. As all the poles are in the left half-plane, the response $y(t)$ is stable. The forced response to $x(t) = 2e^{2t}u(t)$ can be found at once by evaluating $H(s)$ at $s = 2$:

$$y_f(t) = H(2)2e^{2t}u(t) = 2 \times \frac{24(2 - 3)}{(2 + 1)(2 + 2)}e^{2t}u(t) = -4e^{2t}u(t). \tag{10.3-5}$$

Because the response is a stable one, we know that (10.3-5) also gives us the exponential steady-state behavior as t becomes large. We will, however, find the natural and impulse responses, as required in steps c and d to verify this.

Figure 10.20
The PZD

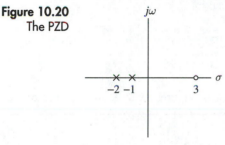

To compute the natural response, we must do a PFE on $H(p)x(t)$:

$$y(t) = H(p)x(t) = \frac{24(p - 3)}{(p + 1)(p + 2)} \times \frac{2}{p - 2}\delta(t) \tag{10.3-6}$$

$$= \left[\frac{-4}{p - 2} + \frac{64}{p + 1} - \frac{60}{p + 2}\right]\delta(t)$$

$$= -4e^{2t}u(t) + 64e^{-t}u(t) - 60e^{-2t}u(t).$$

Here, we have expressed the exponential input in the form

$$x(t) = \frac{2}{p - 2}\delta(t). \tag{10.3-7}$$

The very first term in (10.3-6) is the forced response, as we have already computed. The last two terms form the natural response, which clearly vanishes as $t \rightarrow \infty$.

The unit impulse response is

$$h(t) = H(p)\delta(t) = \frac{24(p - 3)}{(p + 1)(p + 2)}\delta(t) = \left[\frac{-96}{p + 1} + \frac{120}{p + 2}\right]\delta(t) \tag{10.3-8}$$

$$= -96e^{-t}u(t) + 120e^{-2t}u(t).$$

As you can see, the *form* of $h(t)$ is the same as the natural response component in (10.3-6).

Steady-State Response: The System Function Method

The foregoing discussion and example lead to a simple method for finding the exponential steady-state response—provided the associated operator is stable. We merely check the poles of $H(s)$. If they are all in the left half-plane, we know our response is a stable one and that the forced response component dominates for large values of time. Then we merely use $H(s)$ to compute the forced response.

Complex Impedance and Admittance

We remind you that if $x(t)$ is a current and $y(t)$ is a voltage relative to the pair of terminals in a two-terminal network, then we call $H(s)$ the (complex) impedance $Z(s)$, and if $x(t)$ is a voltage and $y(t)$ a current, we call $H(s)$ the (complex) admittance $Y(s)$. If the independent source(s) are exponentials of the form $e^{st}u(t)$ (same s for each such source), we can use the element impedances (derived in Section 10.1 and shown in Figure 10.21 for reference) to compute the desired forced response. We remind you, however, that *these models are only valid for determining the forced component of a response variable.*

Figure 10.21
Exponential forced response models for R, L, and C

Example 10.10

Determine whether the circuit in Figure 10.22 is stable, unstable, or marginally stable. Find the forced response for $v(t)$ if $i_s(t) = 4e^t u(t)$ A.

Figure 10.22
An example circuit

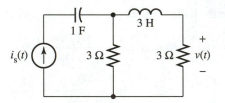

Solution

We carefully note that we are not required to find the complete response. Thus, we will simply work with system functions. We show the circuit with the current source adjusted to supply $e^{st}u(t)$ A and the energy storage elements labeled with their complex impedances in Figure 10.23. We have added the subscript f to $v(t)$ to emphasize the fact that our circuit is only valid for forced responses. Furthermore, we have labeled the other element voltages as well, for we must check them *all* in order to determine the stability properties of the *circuit*. We remind you (as per the comment after Definition 10.2 in Section 10.2) that a *circuit* is stable only if *all* circuit variables are stable, unstable if any one is

Figure 10.23
The forced response equivalent

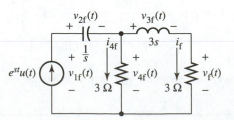

unstable, and marginally stable if one or more are marginally stable and all others are stable.

We can use the current divider rule, together with the impedances shown, to determine the following values:

$$v_{1f}(t) = v_{2f}(t) + v_{4f}(t) = \frac{1}{s}e^{st}u(t) + \frac{3(s+1)}{s+2}e^{st}u(t) \qquad (10.3\text{-}9\text{a})$$

$$= \frac{3s^2 + 4s + 2}{s(s+2)}e^{st}u(t) = H_1(s)e^{st}u(t),$$

$$v_{2f}(t) = \frac{1}{s}e^{st}u(t) = H_2(s)e^{st}u(t), \qquad (10.3\text{-}9\text{b})$$

$$v_{3f}(t) = 3s \times \frac{3}{3s+6}e^{st}u(t) = \frac{3s}{s+2}e^{st}u(t) = H_3(s)e^{st}u(t), \qquad (10.3\text{-}9\text{c})$$

$$v_{4f}(t) = 3 \times \frac{3s+3}{3s+6}e^{st}u(t) = \frac{3(s+1)}{s+2}e^{st}u(t) = H_4(s)e^{st}u(t), \qquad (10.3\text{-}9\text{d})$$

$$i_{4f}(t) = \frac{v_{4f}(t)}{3} = \frac{s+1}{s+2}e^{st}u(t) = H'_4(s)e^{st}u(t), \qquad (10.3\text{-}9\text{e})$$

$$v_f(t) = 3 \times \frac{3}{3s+6}e^{st}u(t) = \frac{3}{s+2}e^{st}u(t) = H(s)e^{st}u(t), \qquad (10.3\text{-}9\text{f})$$

and

$$i_f(t) = \frac{v_f(t)}{3} = \frac{1}{s+2}e^{st}u(t) = H'(s)e^{st}u(t). \qquad (10.3\text{-}9\text{g})$$

We see that all system functions are stable (finite poles in the left half-plane only), except for $H_1(s)$ and $H_2(s)$, both of which have a simple pole at $s = 0$. Thus, these last two are only marginally stable, and hence our circuit is marginally stable.

The forced response $v_f(t)$ when $i_s(t) = 4e^t u(t)$ A can be found by using $s = 1$ in (10.3-9f) and scaling it up by 4. This gives

$$v_f(t) = 4e^t u(t). \qquad (10.3\text{-}10)$$

Because the circuit responses for $v_1(t)$ and $v_2(t)$ are only marginally stable, the exponential steady state does not exist for these variables; however, it does exist for the others.

Dc Steady-State Response

Exponential steady state response is not so useful in itself; its usefulness is related to the fact that we can specialize it to include dc and ac steady-state behavior—which are both *extremely* useful. To obtain a step forcing function, we merely specialize the exponential waveform by making $s = 0$. Thus, we have

$$x(t) = Ae^{0t}u(t) = Au(t). \qquad (10.3\text{-}11)$$

The forced response to this waveform, therefore, is

$$y_f(t) = H(0)\,x(t) = H(0)Ae^{0t}u(t) = H(0)Au(t). \qquad (10.3\text{-}12)$$

We have already observed that steady state is a concept applying only for large values of time; therefore, in the steady-state case, the $u(t)$ multiplier is irrelevant, and we see that the forced response is actually a constant (dc). We note, by the way, that the forced response always exists. The steady state only occurs, however, if the associated circuit variable is stable. If we make $s = 0$ in our exponential forced response models in Figure 10.21, we obtain the dc steady-state models in Figure 10.24. For the inductor, we have $Ls = L \cdot 0 = 0$ (a short circuit), and for the capacitor, we have $1/Cs = 1/(C \cdot 0) = \infty$ (an open circuit). We have already used these models earlier in the text, but they were derived in an ad hoc fashion under the assumption that all currents and voltages were constant. Now we know precisely the condition that must hold for these models to be valid: the circuit must be stable (at least the inductor currents and capacitor voltages must be stable responses) and have only dc sources.

Figure 10.24
Dc steady-state models
for R, L, and C

Ac Steady State:
Gain and Phase of
a Circuit or System

Now let's suppose that the forcing function $x(t)$ is an ac waveform and write it as

$$x(t) = X_m \cos(\omega t + \theta)u(t) = \frac{X_m}{2}e^{j(\omega t + \theta)}u(t) + \frac{X_m}{2}e^{-j(\omega t + \theta)}u(t) \qquad (10.3\text{-}13)$$

$$= \frac{X_m e^{j\theta}}{2}e^{j\omega t}u(t) + \frac{X_m e^{-j\theta}}{2}e^{-j\omega t}u(t).$$

We have used one form of Euler's formula that was derived in Chapter 8. This means that our cosine waveform is the sum of two exponentials. One has $s = j\omega$ and the other has $s = -j\omega$. We know how to find the forced response to each exponential—we merely evaluate the system function $H(s)$ at the appropriate value and multiply it times the input exponential. We then use superposition and add them to give the total forced response:

$$y_f(t) = \frac{X_m e^{j\theta}}{2}H(j\omega)e^{j\omega t}u(t) + \frac{X_m e^{-j\theta}}{2}H(-j\omega)e^{-j\omega t}u(t). \qquad (10.3\text{-}14)$$

Now let's observe that, for each value of ω, the quantity $H(j\omega)$ is a complex number; therefore, we can put it into Euler form:

$$H(j\omega) = A(\omega)e^{j\phi(\omega)}. \qquad (10.3\text{-}15)$$

$A(\omega)$ is called the *gain* and $\phi(\omega)$ is called the *phase*, or the *phase shift*. We recall now the properties of real functions of complex numbers (those with real coefficients) given in Table 9.1 in Section 9.5. Specifically, we note that if $H(s)$ has real coefficients, then

$$H(-j\omega) = H^*(j\omega), \qquad (10.3\text{-}16)$$

where the asterisk denotes the complex conjugate. Using this in (10.3-15), we see that

$$A(-\omega)e^{j\phi(-\omega)} = A(\omega)e^{-j\phi(\omega)}. \qquad (10.3\text{-}17)$$

We can equate the magnitudes and angles of the complex numbers on each side to get

$$A(-\omega) = A(\omega) \qquad (10.3\text{-}18a)$$

and

$$\phi(-\omega) = -\phi(\omega). \qquad (10.3\text{-}18b)$$

The gain is even and the phase is odd as a function of ω. Using this in (10.3-14), we have

$$y_f(t) = A(\omega)X_m \frac{e^{j(\omega t+\theta+\phi(\omega))} + e^{-j(\omega t+\theta+\phi(\omega))}}{2} u(t) \qquad (10.3\text{-}19)$$

$$= A(\omega)X_m \cos(\omega t + \theta + \phi(\omega))u(t).$$

The main point is simply this: the ac forced response — and hence the ac steady-state response if the response is stable — is a sinusoid with the same frequency as the input, but with a different amplitude and phase, in general. The result in (10.3-19) is highly important for the analysis of ac steady-state circuits. We now present an ac steady-state example.

Example 10.11   Determine if the circuit response $v(t)$ in Figure 10.25 is stable. If it is, compute the ac steady-state response.

Figure 10.25
An example circuit

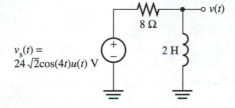

$v_s(t) =$
$24\sqrt{2}\cos(4t)u(t)$ V

Solution   The circuit is shown in complex impedance form in Figure 10.26. We have altered the source to have the form $e^{st}u(t)$ to determine $H(s)$. A quick application of voltage division gives

$$v_f(t) = \frac{2s}{2s+8} e^{st}u(t) = \frac{s}{s+4} e^{st}u(t). \qquad (10.3\text{-}20)$$

The system function associated with our desired response voltage, therefore, is

$$H(s) = \frac{s}{s+4}. \qquad (10.3\text{-}21)$$

It has a single zero at the origin and a single pole at $s = -4$; the PZD is shown in Figure 10.27. Our desired voltage is, therefore, a stable response variable (notice that the *zero* at $s = 0$ does not imply marginal stability!). The input frequency is $\omega = 4$ rad/s, so we evaluate $H(s)$ at $s = j4$. Thus,

$$H(j4) = \frac{j4}{j4+4} = \frac{j}{1+j} = \frac{1\angle 90°}{\sqrt{2}\angle 45°} = \frac{1}{\sqrt{2}}\angle 45°. \qquad (10.3\text{-}22)$$

Figure 10.26
The exponential forced response equivalent

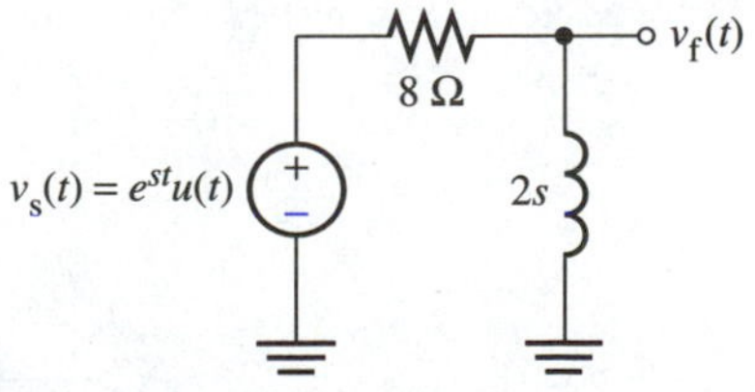

$v_s(t) = e^{st}u(t)$

Figure 10.27
The PZD

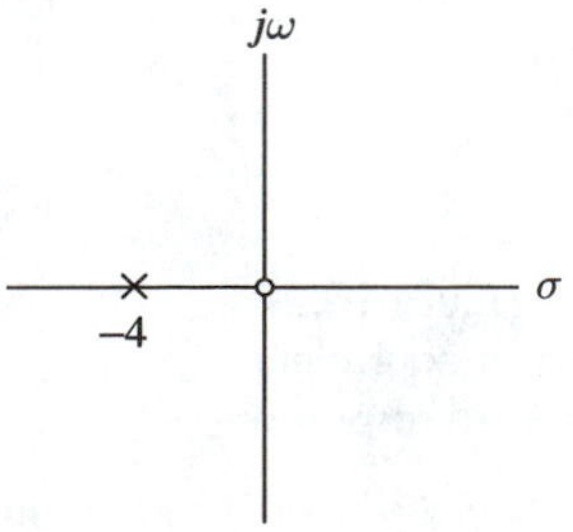

Therefore, $A(4) = 1/\sqrt{2}$ and $\phi(4) = 45°$. Using (10.3-19), we merely write down the sinusoidal (ac) forced response:

$$v_f(t) = 24 \cos(4t + 45°) \text{ V}.\qquad(10.3\text{-}23)$$

Ac Forced Response:
The Phasor Equivalent

Now let us derive models for our dynamic elements for the ac forced response (and, of course, the ac steady state for stable circuits). We rewrite equation (10.3-14) as

$$y_f(t) = \frac{X_m e^{j\theta}}{2} H(j\omega) e^{j\omega t} u(t) + \frac{X_m e^{-j\theta}}{2} H(-j\omega) e^{-j\omega t} u(t)\qquad(10.3\text{-}24)$$

$$= \text{Re}\{H(j\omega) X_m e^{j(\omega t + \theta)} u(t)\} = \text{Re}\{H(j\omega)\tilde{x}(t)\},$$

where

$$\tilde{x}(t) = X_m e^{j(\omega t + \theta)} u(t) = \overline{X} e^{j\omega t} u(t).\qquad(10.3\text{-}25)$$

The complex constant $\overline{X} = X_m e^{j\theta}$ is called the *phasor* associated with $x(t) = \cos(\omega t + \theta)u(t)$. We now define the complex forced response by

$$\tilde{y}_f(t) = H(j\omega)\,\tilde{x}(t)\qquad(10.3\text{-}26)$$

Figure 10.28 Complex
response system
diagram

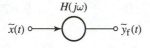

$$H(j\omega)$$

$$\tilde{x}(t) \circ\!\!\longrightarrow\!\!\bigcirc\!\!\longrightarrow\!\!\circ \tilde{y}_f(t)$$

and observe that we need only take the real part of this waveform to find the response we are actually seeking, as we see from (10.3-24). We can visualize the process in terms of a system diagram as in Figure 10.28. We merely evaluate the system function $H(s)$ at the purely imaginary value $s = j\omega$ (where ω is the radian frequency of the input sinusoid) and multiply it by the complex input waveform to obtain the complex form of the forced response.

As a matter of fact, we can simplify things even further. We note that $\tilde{x}(t) = \overline{X}e^{j\omega t}u(t)$ has the exponential factor $e^{j\omega t}u(t)$ — and so does the complex forced response:

$$\tilde{y}_f(t) = H(j\omega)\tilde{x}(t) = H(j\omega)\overline{X}e^{j\omega t}u(t) = \overline{Y}e^{j\omega t}u(t).\qquad(10.3\text{-}27)$$

We can factor out common terms from this equation to give us an equation involving only the phasors:

$$\overline{Y} = H(j\omega)\overline{X}.\qquad(10.3\text{-}28)$$

This is a very simple end result. We merely multiply the input phasor by the system function evaluated at $s = j\omega$ to obtain the output phasor. We conceptually multiply this result by $e^{j\omega t}u(t)$ and take the real part to obtain the (real) ac forced response we are seeking. In practice, one merely writes down the input phasor, solves for the output phasor (converting it to polar form if necessary), then writes down the ac forced response. This is shown in Figure 10.29.

Figure 10.29
The phasor equivalent
for ac forced response

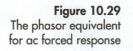

$$x(t)$$
$$= X_m \cos(\omega t + \theta)$$

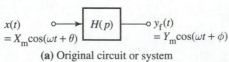

(a) Original circuit or system

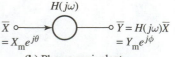

$$H(j\omega)$$

(b) Phasor equivalent

For our three basic circuit elements R, L, and C, the appropriate system operator is the impedance (or admittance). In terms of impedance operators, these are

$$Z(p) = R, \tag{10.3-29}$$

$$Z(p) = Lp, \tag{10.3-30}$$

and $$Z(p) = 1/Cp. \tag{10.3-31}$$

We need only to replace p by $j\omega$ to obtain the *ac forced response impedance:*

$$Z(j\omega) = R, \tag{10.3-32}$$

$$Z(j\omega) = j\omega L, \tag{10.3-33}$$

and $$Z(j\omega) = 1/j\omega C. \tag{10.3-34}$$

The ac forced response impedances are shown in Figure 10.30. We will now use these equivalents to rework Example 10.11 in a much simpler manner.

Figure 10.30
Ac forced response models for R, L, and C

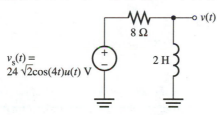

Example 10.12 Find the ac forced response of the circuit of Example 10.11 using the ac forced response models we have just derived. (The circuit is repeated for convenience of reference here as Figure 10.31.)

Figure 10.31
An example circuit

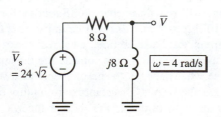

Solution The first step in our solution is to convert the entire circuit to phasor form, replacing the voltage source with one having the complex constant (phasor) $\overline{V}_s = 24\sqrt{2}$ as its value and replacing the resistor and the inductor by their ac forced response equivalents. The resulting *phasor equivalent circuit* is shown in Figure 10.32. Notice that we have placed the radian frequency $\omega = 4$ rad/s in a box for future reference. The phasor equivalent is only valid at that frequency, and when we converted to phasor form we lost all reference to the input frequency. For this reason, we suggest noting it as shown. We can now derive the response phasor by simple voltage division:

$$\overline{V} = \frac{j8}{8 + j8}\overline{V}_s = \frac{j24\sqrt{2}}{1 + j} = 24e^{j45°}. \tag{10.3-35}$$

Figure 10.32
The phasor equivalent

We can now at once write down the ac forced response, using the frequency we have noted on the phasor equivalent circuit. It is

$$v_f(t) = 24 \cos(4t + 45°) \text{ V}. \tag{10.3-36}$$

If you compare the result of this last example with that derived in Example 10.11 you will see that they are the same. For more involved ac forced response problems, as you will see in Part III, the phasor equivalent circuit offers additional conceptual insight into the behavior of circuits, as well as being computationally effective. *We note once more in closing this section, that the ac forced response always exists for any well-posed circuit; the ac steady state, on the other hand, is only meaningful for stable circuits.*

Section 10.3 Quiz

Q10.3-1. Determine the type of stability of the response $i(t)$ in Figure Q10.3-1.

Figure Q10.3-1

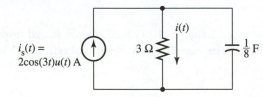

$i_s(t) = 2\cos(3t)u(t) \text{ A}$

Q10.3-2. Compute the ac forced response for $i(t)$ in Figure Q10.3-1 using the ratio $H(s) = \left[\dfrac{i_f(t)}{i_s(t)} \right]_{i_s(t) = e^{st}}$.

Q10.3-3. Compute the ac forced response for $i(t)$ in Figure Q10.3-1 using the phasor equivalent.

10.4 | Scaling

Thus far, we have primarily been working with "unrealistic" numbers; that is, resistor values have been on the order of one ohm, inductors one henry, and capacitors one farad (or a simple fraction thereof). In fact, particularly in electronics work, one often encounters resistances in the several kilohm range, inductors of microhenry or millihenry value, and capacitors with values of perhaps microfarads or nanofarads. Here we will show that these can be converted to the former "unrealistic" values by *scaling,* thus making them easier to calculate.

We will look first at impedance scaling. The basic equation is

$$v(t) = Z(p)i(t) = aZ(p) \times \frac{i(t)}{a} = \frac{Z(p)}{1/a} \times \frac{i(t)}{a}. \tag{10.4-1}$$

There are two ways of looking at this equation. The first says that if we multiply all impedances by a and divide all currents by a, then the voltages will all remain unchanged. The second says that if we let the unit for current be a and the unit for impedance be $1/a$, the unit of voltage will remain unchanged. Suppose, for instance, that we let $a = 10^{-3}$. Then when $i = 10^{-3} \text{ A} = 1 \text{ mA}$, i/a will be unity. This means that we are using the mA as the unit of current. At the same time, each resistor of 1-kilohm resistance will have a value of unity. This means that we are considering the unit of resistance to be the kilohm.

Now let's see what happens to the elements themselves under this transformation. Figure 10.33 shows the results. Resistances are multiplied by a, thus changing their unit to $1/a$. Inductors are transformed in the same manner. Capacitor impedance, however, is proportional to the inverse of the capacitance, thus capacitance is divided by a. The transformed capacitor, therefore, has the new unit a.

Figure 10.33
Element transformation
under impedance
scaling

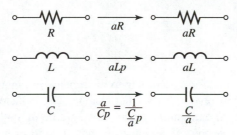

Example 10.13   Impedance scale the circuit in Figure 10.34 so that the resistor has a value of 4 ohms.

Figure 10.34
An example circuit

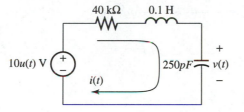

Solution   We wish to multiply the resistor by 10^{-4}, to make it have a value of 4 Ω. This means that our scale factor $a = 10^{-4}$. The new inductor will have a value of $aL = 10^{-5}$ and the capacitor, $c/A = 25 \times 10^{-7}$, as shown in Figure 10.35. Note that the response voltage $v(t)$ will be unchanged, but the current will be scaled to a unit of $a = 0.1$ mA. It is more intuitive to use the first interpretation in (10.4-1) for the elements and say that the new resistance is 4 Ω, the new inductance is 10 μH, and the new capacitance is 2.5 μF, as we have shown in Figure 10.35.

Figure 10.35
The impedance scaled
circuit

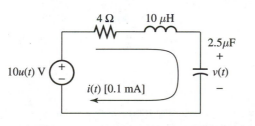

Even after impedance scaling, the circuit in the last example hardly has "nice" values. To achieve them we will develop another scaling procedure, one involving time. We agree to make the new unit for time, let us say, b. Then we have

$$t_n = t/b. \tag{10.4-2}$$

(Thus, when $t = b$ seconds, $t_n = 1$.) We can, of course, invert this to write

$$t = bt_n. \tag{10.4-3}$$

This allows us to go in the reverse direction, that is, to unscale the circuit and its response. The effect of a time scaling on the response of a circuit can be determined by its effect on the differentiation operator:

$$p = \frac{d}{dt} = \frac{1}{b}\frac{d}{d(t/b)} = \frac{1}{b}\frac{d}{dt_n} = \frac{1}{b}p_n.$$

Thus,

$$p = \frac{1}{b}p_n \qquad (10.4\text{-}4)$$

and

$$p_n = bp. \qquad (10.4\text{-}5)$$

The elements transform according to the scheme in Figure 10.36. As the resistor relationship does not involve p, it does not change. The inductor impedance, however, is Lp. If we divide L by b and multiply p by b, we are doing nothing to the impedance itself, but are agreeing to express L in units of b and t in units of b seconds as well. The impedance of the capacitor is $1/Cp$, so C only appears in the product Cp; thus, capacitance transforms exactly like inductance.

Figure 10.36
Element transformation
under time scaling:
$t = bt_n$

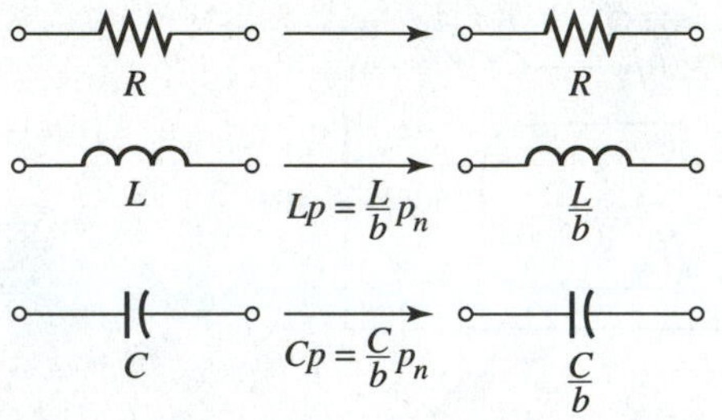

R		R
L	$Lp = \frac{L}{b}p_n$	$\frac{L}{b}$
C	$Cp = \frac{C}{b}p_n$	$\frac{C}{b}$

Example 10.14   Time scale the impedance-scaled circuit that was the *result* in Example 10.13 to a time unit of 0.1 μs.

Solution   Here, we let $b = 10^{-5}$. This gives the circuit shown in Figure 10.37. The 10-μH inductor is transformed to 1 H and the 2.5-μF capacitor to 0.25 F, while the resistor has remained invariant. Now, noting that the scaled inductor has an impedance of p_n, and the scaled capacitor an impedance of $4/p_n$, we obtain the response voltage

$$v_n(t_n) = \frac{4/p_n}{4 + p_n + 4/p_n}u(t_n) = \frac{4}{p_n^2 + 4p_n + 4}\delta(t_n) \qquad (10.4\text{-}6)$$

$$= \frac{4}{(p_n + 2)^2}\delta(t_n) = 4t_n e^{-2t_n}u(t_n).$$

It is crucial here to note that the response voltage is actually a different time function than the one we are seeking. Thus, we must use $t_n = t/b$ to compute $v(t)$:

$$v(t) = v_n(t/b) = v_n(10^5 t) = 4 \times 10^5 t e^{-2\times10^5 t}u(t). \qquad (10.4\text{-}7)$$

Notice that $u(10^5 t) = u(t)$ because the unit step is unity when the *argument* is positive, and multiplication by a positive constant has no effect on the positivity.

Figure 10.37
The time-scaled circuit

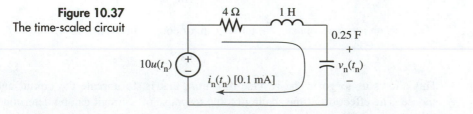

Section 10.4 Quiz

Q10.4-1. Impedance and time scale the circuit in Figure Q10.4-1 so that R_1 has a value of 4 Ω and so that both the inductor and the capacitor have "nice" values. Then, using this scaled circuit, find $v(t)$.

Figure Q10.4-1

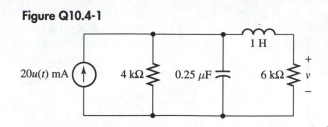

10.5 | Stability and Transient Analysis of Active Circuits

We will now present a series of examples of analysis of circuits with active devices. Each is actually a "case study," for each solution is fairly involved. It is safe to say that these examples represent perhaps the pinnacle of circuit analysis considered in this text. If you read and understand each of them, you should have no trouble performing a transient or stability analysis of any circuit you encounter either in later courses or in engineering practice.

Example: Time Domain Circuit Analysis

Example 10.15   For the circuit in Figure 10.38, find $i(t)$ for $C = 1/12$ F, $C = 1/16$ F, and $C = 1/32$ F.

Figure 10.38
An example circuit

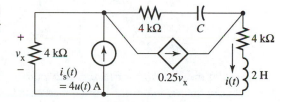

Solution   Let's recall our results from Part I on complexity. If we replace the capacitor and its series resistor and the inductor and its series resistor with equivalent impedances, we see that there will be $N = 3$ nodes, $B = 5$ branches (two of them being equivalent impedance elements), $N_V = 0$ v-sources, and $N_I = 2$ i-sources. We will be required to write $N_{ne} = N - 1 - N_V = 2$ nodal equations or $N_{me} = B - N + 1 - N_I = 1$ mesh equation. Therefore, we will opt for mesh analysis. We show the circuit prepared for mesh analysis in Figure 10.39. We have used operator impedances for the inductor and the capacitor. Furthermore, we have decided to use the consistent units of mA, kΩ, V, µF, and H. (Notice that $RC = $ kΩ $\times$ µF $=$ ms and that $L/R = $ H/kΩ $=$ ms.) Thus, we have impedance and time scaled by $a = b = 10^{-3}$. By merely changing units we have accomplished this in a bit more informal a manner.

Figure 10.39
The example circuit prepared for mesh analysis

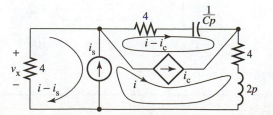

Notice that only one supermesh equation is required: testing the circuit by deactivating both the current sources shows that there is one supermesh (the outside loop) and no essential meshes. We have defined the independent mesh current so that it is the same as the variable we are seeking and have expressed the other two mesh currents in terms of it using the current source constraints. It is also important to observe that we have taped the dependent current source. The single supermesh equation is

$$4(i - i_s) + (4 + 1/Cp)(i - i_c) + (4 + 2p)i = 0, \tag{10.5-1}$$

or

$$(12 + 2p + 1/Cp)i - (4 + 1/Cp)i_c = 4i_s. \tag{10.5-2}$$

Untaping, we have the dependent source constraint

$$i_c = 0.25v_x = 0.25 \times 4(i_s\text{-}i) = i_s\text{-}i. \tag{10.5-3}$$

(Be careful of the sign here!) Using this in (10.5-2) gives

$$\left[16 + \frac{2}{Cp} + 2p\right]i = (8 + \frac{1}{Cp})i_s. \tag{10.5-4}$$

Thus, we have

$$H(p) = 4\frac{p + \dfrac{1}{8C}}{p^2 + 8p + \dfrac{1}{C}}. \tag{10.5-5}$$

Assuming now that $C = 1/12\ \mu F$ and noting that $4u(t) = (4/p)\delta(t)$, we have

$$i(t) = H(p) \times \frac{4}{p}\delta(t) = \frac{16p + 24}{p(p^2 + 8p + 12)}\delta(t) \tag{10.5-6}$$

$$= \left[\frac{2}{p} + \frac{1}{p + 2} + \frac{-3}{p + 6}\right]\delta(t) = [2 + e^{-2t} - 3e^{-6t}]u(t)\ \text{mA}.$$

We have noted that the denominator factors into $p(p + 2)(p + 6)$ and have performed a PFE. This is clearly the second-order overdamped case because the roots of the characteristic polynomial (the denominator of H(s)) are real and unequal.

If we choose $C = 1/16\ \mu F$, we see that

$$i(t) = H(p) \times \frac{4}{p}\delta(t) = \frac{4(4p + 8)}{p(p + 4)^2}\delta(t) \tag{10.5-7}$$

$$= \left[\frac{2}{p} + \frac{-2}{p + 4} + \frac{8}{(p + 4)^2}\right]\delta(t) = 2[1 + (4t - 1)e^{-4t}]u(t)\ \text{mA}.$$

The response component due to the repeated second order factor is clearly critically damped.

Finally, we pick $C = 1/32\ \mu F$. For this value, we have

$$i(t) = H(p) \times \frac{4}{p}\delta(t) = \frac{16(p + 4)}{p[(p + 4)^2 + 4^2]} \tag{10.5-8}$$

$$= \left[\frac{2}{p} - 2\frac{p + 4}{(p + 4)^2 + 4^2} + 2\frac{4}{(p + 4)^2 + 4^2}\right]\delta(t)$$

$$= 2[1 - e^{-4t}[\cos(4t) + \sin(4(t)]]u(t),$$

an underdamped response.

A Thévenin-Norton Example: Operator Form

The next example illustrates the fact that Thévenin/Norton equivalents are sound and useful techniques to apply in the operator world.

Example 10.16

Replace the subcircuit to the left of terminals a and b in Figure 10.40 by its Norton equivalent subcircuit, then find $v(t)$ for $L = 1/4$ H, 2/9 H, and 1/9 H.

Figure 10.40
An example circuit

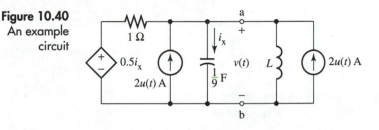

Solution

We show the two-terminal subcircuit with the terminals a and b in Figure 10.41. We have taped the dependent source and applied a test voltage source. We look for a relationship in the form $i(t) = Y(p)v(t) - i_{sc}(t)$. A KCL equation at the top node gives:

$$i(t) = \frac{p}{9}v(t) + \frac{v(t) - v_c(t)}{1} - i_s(t) = \left[\frac{p}{9} + 1\right]v(t) - v_c(t) - i_s(t). \qquad (10.5\text{-}9)$$

Notice that we have labeled the independent current source $i_s(t)$ merely for compactness of notation. Untaping the dependent source, we find that

$$v_c(t) = 0.5i_x(t) = 0.5 \times \frac{p}{9}v(t) = \frac{p}{18}v(t). \qquad (10.5\text{-}10)$$

Using this in equation (10.5-9) and replacing $i_s(t)$ by its known value, we get

$$i(t) = \frac{p + 18}{18}v(t) - 2u(t). \qquad (10.5\text{-}11)$$

This describes the Norton equivalent subcircuit drawn in Figure 10.42, where we have replaced the external components across terminals a and b. The two-terminal subcircuit shown as the box represents the Norton admittance operator $Y(p)$. We can now write one nodal equation to obtain $v(t)$:

$$\left[\frac{p + 18}{18} + \frac{1}{Lp}\right]v(t) = 4u(t), \qquad (10.5\text{-}12)$$

Figure 10.41
Testing the subcircuit

Figure 10.42
The Norton equivalent

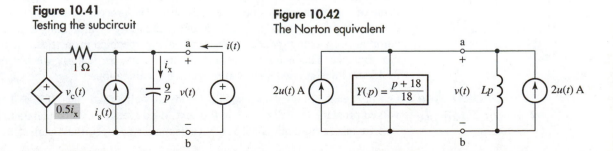

or
$$v(t) = \frac{72Lp}{Lp^2 + 18Lp + 18}u(t) = \frac{72}{p^2 + 18p + \dfrac{18}{L}}\delta(t). \qquad (10.5\text{-}13)$$

(We have used the fact here that $pu(t) = \delta(t)$.)

Letting $L = 1/4$ H, we get

$$v(t) = \frac{72}{p^2 + 18p + 72}\delta(t) = \frac{72}{(p + 6)(p + 12)}\delta(t) \qquad (10.5\text{-}14)$$

$$= \left[\frac{12}{p + 6} - \frac{12}{p + 12}\right]\delta(t) = 12[e^{-6t} - e^{-12t}]u(t) \text{ V.}$$

This is clearly an overdamped response.

When $L = 2/9$ H, we have

$$v(t) = \frac{72}{p^2 + 18p + 81}\delta(t) = \frac{72}{(p + 9)^2}\delta(t) = 72te^{-9t}u(t) \text{ V,} \qquad (10.5\text{-}15)$$

a critically damped situation.

Finally, letting $L = 1/9$ H, we obtain

$$v(t) = \frac{72}{p^2 + 18p + 162}\delta(t) = 8\frac{9}{(p + 9)^2 + 9^2}\delta(t) = 8e^{-9t}\sin(9t)u(t) \text{ V.} \qquad (10.5\text{-}16)$$

This is an underdamped type of response.

We have raised an interesting issue in the last example. When we computed the Norton admittance $Y(p)$ we showed it merely as a box with $Y(p)$ labeled inside. In this case, it turns out that we can do much better. If we write

$$Y(p) = \frac{p + 18}{18} = \frac{1}{18}p + 1 = C_{eq}p + G_{eq}, \qquad (10.5\text{-}17)$$

we can make the identifications

$$C_{eq} = \frac{1}{18}\text{F,} \qquad (10.5\text{-}18a)$$

$$G_{eq} = \frac{1}{R_{eq}} = 1 \text{ S.} \qquad (10.5\text{-}18b)$$

Figure 10.43 Synthesized subcircuit

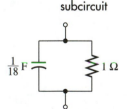

Because the admittances of parallel elements add we obtain the two-terminal subcircuit shown in Figure 10.43. In this case, it is a simple two-element subcircuit with a positive value of capacitance and a positive resistance. The process of making this identification is called *network synthesis* and consists of procedurally manipulating a given impedance or admittance so as to produce a subcircuit having specified properties, such as positivity of elements, etc. This is a major area of circuit theory, particularly when it is applied to the problem of designing a filter to meet certain specifications.

A Switched Circuit Example

Our next example involves a stability study as well as switches. It is perhaps the most involved problem we will pursue in this text. If you feel that you understand each and every step we take, congratulations! You have become an adept circuit analyst!

Example 10.17   For the circuit in Figure 10.44,

    a.   Find the nonnegative values of μ such that the response $v(t)$ is stable for $t \leq 0$.

    b.   Find $v(t)$ for all time under the assumption that $\mu = 0.5$.

Figure 10.44
An example circuit

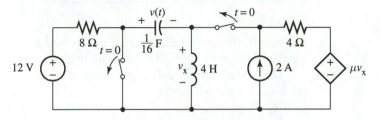

Solution   The circuit, with general source values, for $t \leq 0$ is shown in Figure 10.45. In the past, we have usually made the assumption that the circuit is operating in the dc steady state; here, however, we are going to investigate the very existence of such a condition. Furthermore, since we will ultimately compute a complete response, we will do most of our analysis using differential operators. Notice, also, that we have taped the dependent source. We can write a single nodal equation at the essential node, whose voltage is $v_x(t)$:

$$\frac{p}{16}v_x(t) + \frac{v_x(t)}{4p} + \frac{v_x(t) - v_c(t)}{4} = i_s(t). \tag{10.5-19}$$

If we multiply both sides (from the left!) by $16p$, we obtain

$$(p^2 + 4p + 4)v_x(t) - 4pv_c(t) = 16pi_s(t). \tag{10.5-20}$$

Untaping the dependent source gives

$$v_c(t) = \mu v_x(t). \tag{10.5-21}$$

Thus, (10.5-20) becomes

$$v_x(t) = \frac{16p}{p^2 + 4(1 - \mu)p + 4}i_s(t). \tag{10.5-22}$$

We now observe from Figure 10.45 that $v(t)$ is the negative of $v_x(t)$; hence

$$v(t) = \frac{-16p}{p^2 + 4(1 - \mu)p + 4}i_s(t). \tag{10.5-23}$$

In order to investigate stability, we change $H(p)$ (the coefficient of $i_s(t)$ in (10.5-23)) to $H(s)$, the system function:

$$H(s) = \frac{-16s}{s^2 + 4(1 - \mu)s + 4}. \tag{10.5-24}$$

Figure 10.45
The $t \leq 0$ circuit

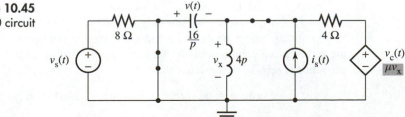

Table 10.7
Routh array for
equation (10.5-24)

s^2	1	4	0
s^1	$4(1-\mu)$	0	0
s^0	4	0	0

The Routh array for the denominator polynomial is shown in Table 10.7. We see that our response is unstable if $\mu > 1$ and stable if $\mu < 1$. If $\mu = 1$, the second row has only zeros. This means that the roots of the polynomial in the first row are also roots of the original denominator polynomial (in this case, all the roots). Thus, we solve for the roots of $s^2 + 4$. These are $s = \pm j2$. They are simple poles on the imaginary axis, therefore, our response is marginally stable if $\mu = 1$.

Now that we have completed step a, we proceed to step b. Here, $\mu = 0.5$, which is in the stable range (for the $t \leq 0$ circuit) of values for μ. Thus, for $\mu = 0.5$ a dc steady-state condition does exist for $v(t)$ (in the $t \leq 0$ circuit). This means that the capacitor is equivalent to an open circuit for finite values of $t \leq 0$. A glance at Figure 10.45 and an application of Ohm's law for the inductor gives the inductor current (downward) $i(t)$:

$$i(t) = \frac{v_x(t)}{4p}.$$ (10.5-25)

Using this result in (10.5-22) we get

$$i(t) = \frac{4}{p^2 + 4(1 - \mu)p + 4}i_s(t).$$ (10.5-26)

Because the denominator is precisely the same as that for $v(t)$, we see that the $i(t)$ response has the same stability properties as does $v(t)$. Therefore, we can assume that the dc steady state exists for both the capacitor and inductor for any finite negative value of t if $\mu = 0.5$. In this case, the $t \leq 0$ circuit can be modeled as in Figure 10.46. The analysis is easy—much more so than our stability analysis. We see at once that $v(t) = 0$ V and $i(t) = 2$ A for $t \leq 0$.

Figure 10.46
The $t \leq 0$ dc steady-
state equivalent

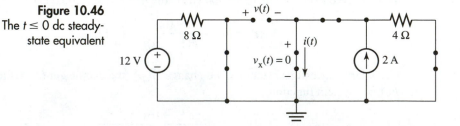

Next, we "flip the switches," producing the circuit shown in Figure 10.47. It is split into two halves; however, we are only interested in the left-hand part. We write a single mesh equation using the mesh current $i(t)$ shown in the figure. (Notice that we have used the series initial condition model for the inductor and noted that the initial voltage on the capacitor is zero.) The mesh equation is

$$(4p + 8 + 16/p)i(t) = 12u(t) + 8\delta(t) = 4(3/p + 2)\delta(t).$$ (10.5-27)

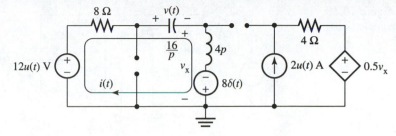

Figure 10.47
The $t > 0$ circuit

We note, however, that our variable of interest is $v(t)$. It is related to the mesh current by

$$v(t) = \frac{16}{p}i(t) = \frac{16}{p}\frac{4(2p+3)}{4p(p+2+4/p)}\delta(t). \qquad (10.5\text{-}28)$$

Together, the two preceding equations give

$$v(t) = \frac{16(2p+3)}{p(p^2+2p+4)}\delta(t) \qquad (10.5\text{-}29)$$

To get this form, we divided both sides of (10.5-27) by 4, multiplied from the left by p, (recognizing that $u(t) = 1/p\delta(t)$), and multiplied the result from the left on both sides by $16/p$. We see that the corresponding $H(s)$ has complex conjugate poles at $-1 \pm j\sqrt{3}$. Because both are in the open left half-plane, $v(t)$ is a stable response. Doing a PFE, whose details we leave to you, we get

$$v(t) = 4\left[\frac{3}{p} + \frac{-3(p+1) + \dfrac{5}{\sqrt{3}}\sqrt{3}}{(p+1)^2 + 3}\right]\delta(t) \qquad (10.5\text{-}30)$$

$$= 4\left[3 + e^{-t}\left[-3\cos(\sqrt{3}t) + \frac{5}{\sqrt{3}}\sin(\sqrt{3}t)\right]\right]u(t).$$

The preceding example illustrates several things. First, not all circuits with dc independent sources have a steady-state behavior (the $t \le 0$ circuit was unstable for $\mu > 1$). It is only the stable ones that do; furthermore, stability must be checked for each and every energy storage element in order to determine the existence (or lack thereof) of such a steady state. A great deal of study has been directed toward making such a complete stability study more efficient. The resulting theory is called *state variable analysis*. We will not discuss this body of knowledge here, however. A circuit with switches that flip once actually consists of two circuits, and each can have any of the three types of stability. For the $t \le 0$ circuit, a problem or circuit is "well posed" if and only if it is stable. The $t > 0$ circuit can then possess any stability property and the problem does not fail to be well posed.

A Circuit Example
with Scaling

Example 10.18(O) For which value(s) of $k = (1 + R_b/R_a)$ is the output voltage $v_o(t)$ in Figure 10.48 stable, unstable, and marginally stable? Find the response $v_o(t)$ for $k = 3$.

Figure 10.48
An example circuit

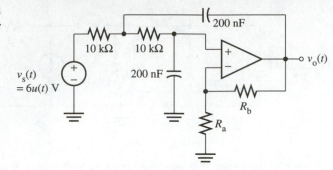

Solution   Let's first perform impedance scaling. Because we are interested in a voltage (v_o), this does not affect our result at all. It would appear that $a = 10^{-4}$ would be a convenient number because this makes all resistor values 1 Ω. The capacitors are divided by 10^{-4}, meaning that they have the common value 2 mF. Now suppose that we choose the unit of time to be the millisecond. This corresponds to choosing a normalized time:

$$t_n = t/b, \tag{10.5-31}$$

with $b = 1$ ms $= 10^{-3}$ s. Therefore, we must divide both capacitors by $b = 10^{-3}$, that is, we must multiply them by 1000. This makes both capacitor values 2 F. The resulting circuit is shown in Figure 10.49. Notice that the response is in terms of normalized time, that is, $t_n = 1$ s corresponds to a real time of 1 ms. Also, we point out that $u(t_n) = u(t/b) = u(t)$ because t and t_n are positive, zero, and negative simultaneously.

Figure 10.49
An example circuit

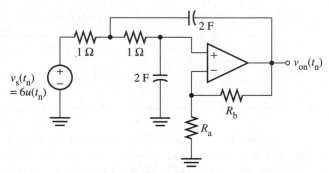

With these scaling changes, we can now do our analysis. We first note that the op amp, in conjunction with R_a and R_b, forms a noninverting voltage amplifier subcircuit with a voltage gain of $k = (1 + R_b/R_a)$. Using the VCVS equivalent for this subcircuit gives the scaled operator equivalent shown in Figure 10.50. Considering the VCVS to be

Figure 10.50
Preparation for
analysis

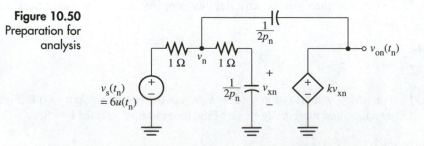

taped with the value v_o and treating the series 2-F capacitor and 1-Ω resistor as a single entity, we obtain

$$\frac{v_n(t_n) - v_s(t_n)}{1} + \frac{v_n(t_n)}{1 + 1/2p_n} + \frac{v_n(t_n) - v_{on}(t_n)}{1/2p_n} = 0. \tag{10.5-32}$$

This simplifies after both sides are multiplied by $2p_n + 1$ from the left to:

$$(4p_n^2 + 6p_n + 1)v_n(t_n) - 2p_n(2p_n + 1)v_{on}(t_n) = (2p_n + 1)v_s(t_n). \tag{10.5-33}$$

Untaping the VCVS and expressing its value in terms of the unknown node voltage v_n, we have

$$v_{on}(t_n) = kv_{xn}(t_n) = k \times \frac{1/2p_n}{1 + 1/2p_n} v_n(t_n) = \frac{k}{2p_n + 1} v_n(t_n). \tag{10.5-34}$$

Using this in (10.5-33) gives

$$v_n(t_n) = \frac{2p_n + 1}{4p_n^2 + 2(3 - k)p_n + 1} v_s(t_n). \tag{10.5-35}$$

Finally, we solve for the output voltage of the op amp:

$$v_{on}(t_n) = \frac{k}{2p_n + 1} v_n(t_n) = \frac{k}{4p_n^2 + 2(3 - k)p_n + 1} v_s(t_n). \tag{10.5-36}$$

Now it should be clear that time scaling (multiplication of t by a positive real constant) has no effect on stability. Thus, we will check stability by forming the equivalent system function by replacing p_n by s in (10.5-36). This gives

$$H(s) = \frac{k}{4s^2 + 2(3 - k)s + 1}. \tag{10.5-37}$$

The Routh array is shown in Table 10.8. Our response is stable if $k < 3$ and unstable if $k > 3$. If $k = 3$, the second row has all zeros and we compute the poles from the polynomial in the first row: $4s^2 + 1$. The roots are at $s = \pm j0.5$. Because these are simple and purely imaginary, our response is marginally stable for $k = 3$.

Table 10.8
Routh array for
equation (10.5-37)

s^2	4	1	0
s^1	2(3−k)	0	0
s^0	1	0	0

We will now compute the voltage response $v_{on}(t_n)$ for $k = 3$, which we now know to be the marginally stable situation. We have

$$v_{on}(t_n) = \frac{3}{4p_n^2 + 1} v_s(t_n) = \frac{3}{4p_n^2 + 1} \times 6u(t_n) = \frac{9/2}{p_n^2 + \frac{1}{4}} \times \frac{1}{p_n} \delta(t_n) \tag{10.5-38}$$

$$= \left(\frac{18}{p_n} - \frac{18}{p_n^2 + \frac{1}{4}} \right) \delta(t_n) = 18[1 - 2\sin(t_n/2)]u(t_n) \text{ V}.$$

We now recognize that $t_n = t/b = 1000t$. This transforms (10.5-38) into

$$v_o(t) = v_{on}(1000t) = 18[1 - 2\sin(500t)]u(t) \text{ V}. \tag{10.5-39}$$

Here, we have used the fact that

$$u(1000t) = u(t). \tag{10.5-40}$$

The value of the step function depends only on the *sign* of its argument, and the sign of $1000t$ is the same as the sign of t. The practical situation is this: we have suddenly applied a dc voltage at $t = 0$, and the response is a never-ending sinusoid. In this case we call the circuit a sinusoidal "oscillator"—one having a frequency of 500 rad/s.

A Design Example (with Operational Amplifiers)

Our next example treats a design problem using simulation diagrams and op amps.

Example 10.19(O)

Design a circuit using op amps, resistors, and capacitors that will exhibit the system operator

$$H(p) = \frac{16p}{p^2 + 8p + 16}. \tag{10.5-41}$$

Solution

Let us choose the symbols $x(t)$ as the input to our system and $y(t)$ as the response. We can then write

$$y(t) = H(p)x(t) = \frac{16p}{p^2 + 8p + 16}x(t). \tag{10.5-42}$$

If we solve for $p^2 y(t)$, we will have

$$p^2 y(t) = 16px(t) - 8py(t) - 16y(t). \tag{10.5-43}$$

Multiplying both sides from the left by $1/p^2$, we get

$$y(t) = \frac{1}{p}\left[16x(t) - 8y(t) - \frac{16}{p}y(t)\right]. \tag{10.5-44}$$

This expresses $y(t)$ in terms of integrations, scalar multiplications, and summations. (If one solves for the highest derivative of $y(t)$, $p^m y(t)$, then multiplies by $1/p^m$, one will always obtain $y(t)$ in terms of integrations. Thus, we are exhibiting a general method.) We want to implement this equation with op amp integrators, which we know are always of the sign inverting type. Thus, let's change the signs to obtain inverting integrations and make them have "gains" that are different:

$$y(t) = \frac{-8}{p}\left[y(t) - 2x(t) - \left[\frac{-2}{p}[y(t)]\right]\right]. \tag{10.5-45}$$

If you follow this equation from left to right, you will be able to see that Figure 10.51 (from right to left) is a valid simulation diagram. Now let's decide what we need to implement this diagram: two inverting scalar multipliers (the signs are negative), two inverting integrators, and one three-input summer. The inverting scalar multiplier can be

Figure 10.51
System simulation

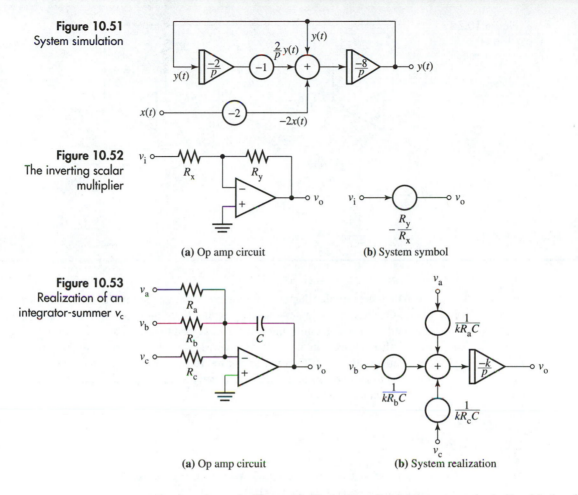

Figure 10.52
The inverting scalar
multiplier

(a) Op amp circuit   **(b)** System symbol

Figure 10.53
Realization of an
integrator-summer v_c

(a) Op amp circuit   **(b)** System realization

realized as shown in Figure 10.52. An inverting integrator can be built with the circuit elements shown in Figure 10.53(a). In actuality, this is a *weighted* inverting integrator because we can assign a different weight to each input variable. Thus, it actually implements an inverting integrator and three scalar multipliers as shown in Figure 10.53(b). We can assign any gain constant k to the integrator and "take it out" in the scalar multipliers, as shown in part (b) of the figure. This often allows us to make one or more of the scalar multipliers have unit value, thus eliminating a block.

Using these building blocks, we derive the active op amp circuit shown in Figure 10.54. The variables $x(t)$ and $y(t)$ are to be interpreted as voltages. We have labeled each of the nodes except for the virtual grounds at the op amp inputs. We see that we must make the resistance ratio R_2/R_1 equal to 2 and the ratio R_5/R_4 equal to 1. We can therefore, quite arbitrarily, choose $R_1 = R_4 = R_5 = 10$ kΩ and $R_2 = 20$ kΩ. As we have mentioned previously, op amps can generally supply currents only in the milliampere range, and this sets a lower limit on the resistor values. On the other hand, nonideal op amp characteristics set an upper limit of perhaps several megohms. (We will leave a discussion of this to a later course in electronics.) We require the op amp inverting integrator on the left to have a system operator of $-2/p$, so we make $1/R_3C_1 = 2$. If we (again quite arbitrarily) choose $C_1 = 1$ μF, then we must set $R_3 = 500$ kΩ. Finally, we consider the output integrator-summer on the right. The system functions relating the node labeled $-2x(t)$, the node labeled $y(t)$, and the node labeled $(2/p)y(t)$ to the op amp output $y(t)$ must be set to $-8/p$ (refer to Figure 10.51), so we make $R_6 = R_7 = R_8 = R$. If we assume that $C_2 = 1$ μF, then $R = 125$ kΩ.

Figure 10.54
The active op amp
realization

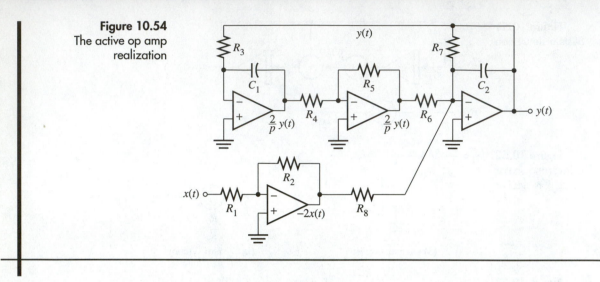

The last example illustrated a very common method in the design of electronic active filters. One first draws a simulation diagram, then implements it with op amp integrators, summers, and scalar multipliers. We should point out that such a design is not unique—many others are possible. In fact, we will leave it to you to show that an op amp can be eliminated in the circuit shown in Figure 10.54. (*Hint:* Consider moving the input variable $x(t)$ to another op amp input.) The next example shows how to analyze a circuit of order higher than second.

A Higher-Order Circuit Example

Example 10.20   Do a mesh analysis on the circuit shown in Figure 10.55. Then find the mesh currents when $v_{s1}(t) = u(t)$ V, while $v_{s2}(t) = v_{s3}(t) = 0$ V.

Figure 10.55
A higher-order circuit

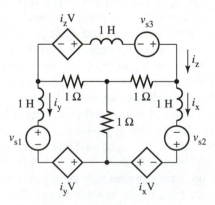

Solution   The circuit is shown prepared for mesh analysis in Figure 10.56. We have taped the dependent sources and used operator notation. We can now write the matrix mesh equations by inspection:

$$\begin{bmatrix} p+2 & -1 & -1 \\ -1 & p+2 & -1 \\ -1 & -1 & p+2 \end{bmatrix} \begin{bmatrix} i_1 \\ i_2 \\ i_3 \end{bmatrix} = \begin{bmatrix} v_{s1} - v_{c1} \\ v_{s2} + v_{c2} \\ v_{s3} + v_{c3} \end{bmatrix}. \qquad (10.5\text{-}46)$$

Figure 10.56
The circuit prepared for mesh analysis

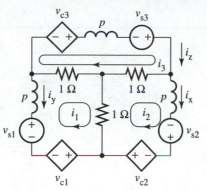

Now we untape the dependent sources, express them in terms of the mesh currents, and transpose them to the other side. This results in

$$
\begin{bmatrix} p+1 & -1 & -1 \\ -1 & p+1 & -1 \\ -1 & -1 & p+1 \end{bmatrix} \begin{bmatrix} i_1 \\ i_2 \\ i_3 \end{bmatrix} = \begin{bmatrix} v_{s1} \\ v_{s2} \\ v_{s3} \end{bmatrix}. \tag{10.5-47}
$$

If we multiply both sides by the inverse[7] of the p matrix on the left (the mesh impedance matrix), we will obtain

$$
\begin{bmatrix} i_1 \\ i_2 \\ i_3 \end{bmatrix} = \begin{bmatrix} \dfrac{p}{(p-1)(p+2)} & \dfrac{1}{(p-1)(p+2)} & \dfrac{1}{(p-1)(p+2)} \\ \dfrac{1}{(p-1)(p+2)} & \dfrac{p}{(p-1)(p+2)} & \dfrac{1}{(p-1)(p+2)} \\ \dfrac{1}{(p-1)(p+2)} & \dfrac{1}{(p-1)(p+2)} & \dfrac{p}{(p-1)(p+2)} \end{bmatrix} \begin{bmatrix} v_{s1} \\ v_{s2} \\ v_{s3} \end{bmatrix}. \tag{10.5-48}
$$

With the waveforms specified, we have

$$
\begin{bmatrix} i_1 \\ i_2 \\ i_3 \end{bmatrix} = \begin{bmatrix} \dfrac{p}{(p-1)(p+2)} & \dfrac{1}{(p-1)(p+2)} & \dfrac{1}{(p-1)(p+2)} \\ \dfrac{1}{(p-1)(p+2)} & \dfrac{p}{(p-1)(p+2)} & \dfrac{1}{(p-1)(p+2)} \\ \dfrac{1}{(p-1)(p+2)} & \dfrac{1}{(p-1)(p+2)} & \dfrac{p}{(p-1)(p+2)} \end{bmatrix} \begin{bmatrix} u(t) \\ 0 \\ 0 \end{bmatrix} \tag{10.5-49}
$$

$$
= \begin{bmatrix} \dfrac{1}{(p-1)(p+2)} \\ \dfrac{1}{p(p-1)(p+2)} \\ \dfrac{1}{(p-1)(p+2)} \end{bmatrix} \delta(t) = \begin{bmatrix} \dfrac{e^t - e^{-2t}}{3} \\ \dfrac{-3 + 2e^t + e^{-2t}}{6} \\ \dfrac{-3 + 2e^t + e^{-2t}}{6} \end{bmatrix} u(t).
$$

We have used the fact that $u(t) = 1/p\,\delta(t)$.

---

[7] In the explicit form given in Appendix A, $Z^{-1}(p) = \mathrm{adj}(Z(p))/\det(Z(p))$.

Section 10.5 Quiz

Q10.5-1.   Find the value(s) of g_m in the active circuit of Figure Q10.5-1 such that $v_x(t)$ is a stable response.

Figure Q10.5-2

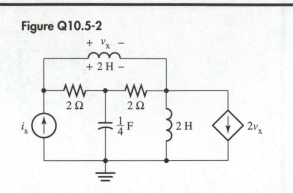

Figure Q10.5-1

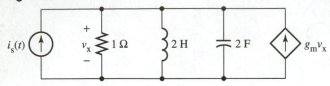

Q10.5-2.   Write the p operator matrix solution for the node voltages in Figure Q10.5-2.

Chapter 10 Summary

We have discussed a number of important ideas in this chapter—not only for circuit analysis, but for courses to follow. One of the most important of these is the idea of *forced and natural response* components of the total response of a differential equation or a circuit. The forced response is that part of the total response that "looks like" the forcing function. For instance, if the forcing function is dc (constant), then the *dc forced response* is the constant part of the response; if the forcing function is a sinusoid, then the "pure sinusoid" component of the response is the *ac forced response*. A generalization of both is the *exponential forced response*. This is the result of adjusting the forcing function to be an exponential of the form $e^{st}u(t)$, where $s = \sigma + j\omega$ is called the *complex frequency*.

We showed that the exponential forced response of a differential equation (or a given circuit response) satisfying the operator equation $y(t) = H(p)x(t)$ is $H(s)e^{st}u(t)$; that is, it is a constant (relative to t) times the forcing function. This constant $H(s)$ is called the *system function*. Note that it depends on the complex frequency parameter s.

We showed that the system function carries a lot of information. We used it to define the concept of *poles and zeros*—values of s that cause $H(s)$ to be either infinite or zero, respectively. We showed how to plot them in the complex s plane, calling the resulting plot the *pole-zero diagram (or PZD)*.

We pointed out that the poles of $H(s)$ determine the system *stability* and defined three modes of stability: *stable, unstable,* and *marginally stable*. This information is carried graphically and obviously in the PZD; however, constructing the PZD requires factoring a polynomial—which is difficult if its degree is large. Therefore, we presented a table method—called the *Routh algorithm*—that does not require factorization.

We defined the *steady-state response* to be the same as the forced response for stable system responses. If the response is stable, then the complete response is approximately the same as the forced response for large values of time because the *natural response* component approaches zero with increasing values of time for stable responses.

We discussed *magnitude (or impedance) scaling* and *frequency scaling* and how these operations can be used to simplify the numbers appearing in a hand analysis.

We ended the chapter with a number of fairly comprehensive case studies of various types of active circuits, examining their responses from a stability point of view. In doing so, we covered both circuits without switches and those that are switched.

Chapter 10 Review Quiz

RQ10.1.   Consider the differential equation

$$\frac{d^3y}{dt^3} + 5\frac{d^2y}{dt^2} + 6\frac{dy}{dt} = 20\frac{dx}{dt} + 10x.$$

Find the system function $H(s)$ and plot the pole-zero diagram (PZD). What type of stability is exhibited by $y(t)$?

RQ10.2.   Answer Question RQ10.1 for the differential equation

$$\frac{d^2y}{dt^2} + 4\frac{dy}{dt} + 4y = \frac{d^2x}{dt^2} + \frac{dx}{dt}.$$

RQ10.3.   Answer Question RQ10.1 for the differential equation

$$\frac{dy}{dt} + 3y = \frac{d^2x}{dt^2} - 2\frac{dx}{dt}.$$

RQ10.4.   Answer Question RQ10.1 for the differential equation

$$\frac{dy}{dt} + 3y = \frac{d^2x}{dt^2} - 2\frac{dx}{dt}.$$

RQ10.5.   If $H(s) = [4s(s + 2)]/(s^3 + 2s^2 - 5s - 6)$, plot the PZD and determine the type of stability.

RQ10.6.   If $P(s) = s^4 + s^3 - 7s^2 - s + 6$, find the number of roots in the right half-plane.

RQ10.7.   Draw the exponential forced response equivalent for the circuit in Figure RQ10.1 (that is, assume that $i_g(t) = I_g e^{st}u(t)$ A and $v_g(t) = V_g e^{st}u(t)$ V, where I_g and V_g are constants, and use the forced response equivalents for the passive elements).

Figure RQ10.1

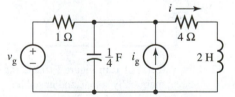

RQ10.8.   Draw the dc forced response equivalent for the circuit in Figure RQ10.1.

RQ10.9.   Find the system function

$$H_1(s) = \frac{i_f(t)}{i_g(t)}\bigg]_{i_g(t)=e^{st},\,v_g(t)=0}$$

and draw the PZD for the circuit in Figure RQ10.1.

RQ10.10.   Find the system function

$$H_2(s) = \frac{i_f(t)}{v_g(t)}\bigg]_{v_g(t)=e^{st},\,i_g(t)=0}$$

and draw the PZD for the circuit in Figure RQ10.1.

RQ10.11.   Find the forced response $i_f(t)$ in Figure RQ10.1 due to $i_g(t) = 2e^{-5t}u(t)$ A with $v_g(t) = 0$.

RQ10.12.   Find and sketch the dc forced response equivalent for the circuit in Figure RQ10.1 and solve for the dc forced response $i(t)$ with $v_g = 10$ V and $i_g = 5$ A.

RQ10.13.   Draw the phasor equivalent circuit for the network in Figure RQ10.2 assuming that $v_g(t) = 24\sqrt{2}\cos(3t)u(t)$ V and solve for the ac forced response $i_f(t)$.

Figure RQ10.2

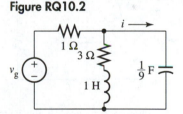

RQ10.14.   Impedance and time scale the circuit in Figure RQ10.1 so that the resistors have values of 10 kΩ and 40 kΩ and the capacitor has a value of 0.25 μF.

RQ10.15.   Impedance and time scale the circuit in Figure RQ10.2 so that the resistors have values of 10 kΩ and 30 kΩ and the capacitor has a value of 1/9 μF.

Chapter 10 Problems

Section 10.1 System Functions and Pole-Zero Diagrams

10.1-1.   If $(d^2y/dt^2) + (3dy/dt) + 2y = x(t)$, find the system function $H(s)$. If $x(t) = 3e^{-t/2}u(t)$, find the forced response $y_f(t)$.

10.1-2.   If $(d^2y/dt^2) + (4dy/dt) + 13y = (dx/dt) + x$, find the system function $H(s)$. If $x(t) = 18e^{-2t}u(t)$, find the forced response $y_f(t)$.

10.1-3.   If $(d^2y/dt^2) + (4dy/dt) + 4y = (dx/dt) + x$, find the system function $H(s)$. If $x(t) = 4e^{-3t}u(t)$, find the forced response $y_f(t)$.

10.1-4.   Let $H(s) = 10\,(s^2 + 4s + 9)/[(s + 2)^2(s^2 + 5s + 6)]$. Find the poles and zeros and plot the PZD.

10.1-5.   Let $H(s) = 20\,[s^2(s + 3)]/(s^2 + 6s + 25)$. Find poles and zeros and plot the PZD.

10.1-6. For the circuit shown in Figure P10.1-1:
   a. Find $Z(s)$.
   b. Use $Z(s)$ to find a differential equation for $v(t)$ in terms of $i(t)$.
   c. Find the poles and zeros of $Z(s)$ and plot the PZD.
   d. Assume that $i(t) = 4e^{-t}u(t)$ A and find the forced response for $v(t)$.

Figure P10.1-1

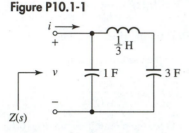

10.1-7. For the circuit shown in Figure P10.1-2:
   a. Find $Y(s)$.
   b. Use $Y(s)$ to find a differential equation for $i(t)$ in terms of $v(t)$.
   c. Find the poles and zeros of $Y(s)$ and plot the PZD.
   d. Assume that $v(t) = 4e^{-t}\cos(\sqrt{2}\,t)u(t)$ V and find the forced response for $i(t)$.

Figure P10.1-2

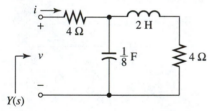

10.1-8. Find the impulse response for the voltage $v(t)$ in the circuit of Example 10.5.

10.1-9. Find the impulse response for the current $i(t)$ in the circuit of Example 10.6.

Section 10.2 Stability

10.2-1. Consider the differential equation

$$\frac{d^2y}{dt^2} + \mu\frac{dy}{dt} + 4y = \frac{dx}{dt} + x.$$

For which values of μ is the response $y(t)$ stable, unstable, and marginally stable?

10.2-2. Consider the differential equation

$$\frac{d^2y}{dt^2} + 4\frac{dy}{dt} + \mu y = \frac{dx}{dt} + x.$$

For which values of μ is the response $y(t)$ stable, unstable, and marginally stable?

10.2-3. Consider the system function

$$H(s) = 10\frac{s-1}{s^2+4s+4}.$$

Plot the PZD. Is the corresponding forced response stable, marginally stable, or unstable?

10.2-4. Consider the system function

$$H(s) = -10\frac{s-1}{s^2+s-2}.$$

Plot the PZD. Is the corresponding forced response stable, marginally stable, or unstable?

10.2-5. Consider the system function

$$H(s) = 10\frac{s^2+s-2}{s(s^2+5s+6)}.$$

Plot the PZD. Is the corresponding forced response stable, marginally stable, or unstable?

10.2-6. Consider the system function

$$H(s) = 10\frac{s^2+s-2}{s^2(s^2+5s+6)}.$$

Plot the PZD. Is the corresponding forced response stable, marginally stable, or unstable?

10.2-7. Find the value(s) of μ in Figure P10.2-1 for which the response $v(t)$ is stable, marginally stable, and unstable.

Figure P10.2-1

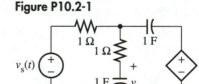

10.2-8. Find the value(s) of μ in Figure P10.2-2 for which the response $v(t)$ is stable, marginally stable, and unstable.

Figure P10.2-2

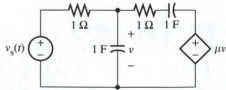

10.2-9. Find the number of roots of $P(s) = s^4 + s^3 + s^2 + 2s + 3$ in the RHP, LHP, and on the imaginary axis.

10.2-10. Find the number of roots of $P(s) = s^3 + 6s^2 + 11s + 6$ in the RHP, LHP, and on the imaginary axis.

10.2-11. Find the number of roots of $P(s) = s^4 + 3s^3 + 6s^2 + 12s + 8$ in the RHP, LHP, and on the imaginary axis.

10.2-12. Find the number of roots of $P(s) = s^6 + s^5 + 2s^4 + 2s^3 + 3s^2 + s + 1$ in the RHP, LHP, and on the imaginary axis.

10.2-13. Consider the system simulation diagram in Figure P10.2-3. Find the range of values of μ such that the response $y(t)$ is stable, marginally stable, and unstable.

Figure P10.2-3

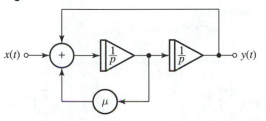

Section 10.3 Forced and Steady-State Response

10.3-1. Given the differential equation

$$\frac{d^2y}{dt^2} - 2\frac{dy}{dt} - 3y = \frac{dx}{dt} - x$$

with forcing function $x(t) = 6e^{2t}u(t)$, find the forced response $y_f(t)$. Is $y(t)$ a stable response? If so, determine the steady-state response.

10.3-2. Given the differential equation

$$\frac{d^2y}{dt^2} + 3\frac{dy}{dt} + 2y = x(t)$$

with forcing function $x(t) = 2e^{-3t}u(t)$, find the forced response $y_f(t)$. Is $y(t)$ a stable response? If so, determine the steady-state response.

10.3-3. Given the differential equation $(dy/dt) + 2y = x(t)$ with forcing function $x(t) = 4\cos(2t)u(t)$, find the forced response $y_f(t)$. Is $y(t)$ a stable response? If so, determine the steady-state response.

10.3-4. Given the differential equation

$$\frac{d^2y}{dt^2} + 4\frac{dy}{dt} + 4y = x(t)$$

with forcing function $x(t) = 4\cos(2t)u(t)$, find the forced response $y_f(t)$. Is $y(t)$ a stable response? If so, determine the steady-state response.

10.3-5. If $g_m = 1$ S, $R = 2$ Ω, and $v_g(t) = 4\cos(2t)u(t)$ in Figure P10.3-1, find the forced response for $v(t)$ using phasor methods.

Figure P10.3-1

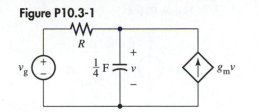

10.3-6. Repeat Problem 10.3-5 for $g_m = 0.5$ S and $R = 2$ Ω.

10.3-7. Repeat Problem 10.3-5 for $g_m = 0.25$ S and $R = 2$ Ω.

10.3-8. If $R = 8$ Ω and $i_g(t) = \cos(2t)u(t)$ A, draw the phasor equivalent for the circuit in Figure P10.3-2 and find the forced response for $v(t)$. Is $v(t)$ a stable response? If so, find the steady-state response if $i_g(t) = 12e^{2t}u(t)$ A.

Figure P10.3-2

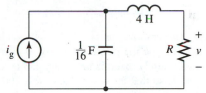

10.3-9. Repeat the first half (ac forced response) of Problem 10.3-8 for $R = 16$ Ω and $i_g(t) = \cos(2t)u(t)$ A.

10.3-10. Repeat the first half (ac forced response) of Problem 10.3-8 for $R = 20$ Ω and $i_g(t) = 10\cos(2t)u(t)$ A.

10.3-11. If $R = 8$ Ω in Figure P10.3-2 and $i_g(t) = 2u(t)$ A, determine whether or not a dc steady-state response exists for $v(t)$ and draw the dc forced response equivalent circuit.

Section 10.4 Scaling

10.4-1. Impedance scale the circuit in Figure P10.4-1 so that $R = 6$ Ω. Find the *scaled* values of V_s, I, and V.

Figure P10.4-1

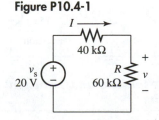

10.4-2. Time scale the differential equation $(d^2y/dt^2) + 10^4y = u(t)$ so that the unit of time is 0.1 millisecond and write down the resulting differential equation.

10.4-3. Impedance and time scale the circuit in Figure P10.4-2 so that $R = 4$ Ω and $C = 1/8$ F. Use the resulting scaled circuit to find $v(t)$.

Figure P10.4-2

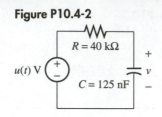

10.4-4. Impedance and time scale the circuit in Figure P10.4-3 so that $R = 1\ \Omega$ and $C = 0.25$ F. Then use the resulting scaled circuit to find $v(t)$.

Figure P10.4-3

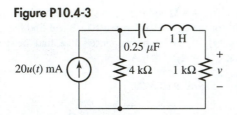

10.4-5. The resonant frequency of the subcircuit in Figure P10.4-4 is defined to be the positive value of ω (denoted by ω_0) for which $Z(j\omega)$ is purely real (the imaginary part is zero). Impedance and time scale so that the new value of R is 100 kΩ and the new value of the resonant frequency, ω_0', is 10^6 rad/s.

Figure P10.4-4

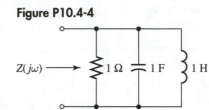

10.4-6. Impedance and time scale the circuit in Figure P10.4-5 so that $R = 40$ kΩ and $C = 1/4$ nF.

Figure P10.4-5

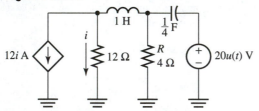

Section 10.5 Stability and Transient Analysis of Active Circuits

10.5-1. Find the voltage $v(t)$ in Figure P10.5-1 for
- **a.** $i_s(t) = 6u(t)$ mA and
- **b.** $i_s(t) = 6\delta(t)$ mA.

(Though this is not an active circuit, it is used as a building block in such circuits.)

Figure P10.5-1

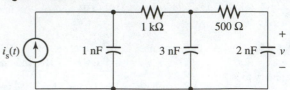

10.5-2. Find the voltage $v(t)$ in Figure P10.5-2 assuming that $v_{g1}(t) = v_{g2}(t) = 3u(t)$ V and that the value of $i_s(t)$ is the unit step function $u(t)$ A. (*Hint:* Decide first which is simpler, mesh or nodal analysis.)

Figure P10.5-2

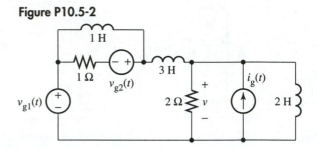

10.5-3. Find the voltage $v(t)$ in Figure P10.5-3 assuming that $i_g(t) = 4u(t)$ A and $v_g(t) = 2u(t)$ V.

Figure P10.5-3

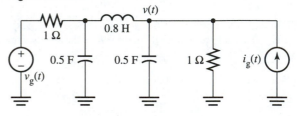

10.5-4. Find the unit impulse response for the voltage $v(t)$ in Figure P10.5-4.

Figure P10.5-4

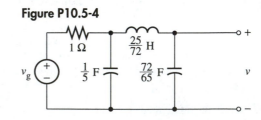

Part III

FREQUENCY DOMAIN ANALYSIS

11

Phasor Analysis of Ac Circuits

This chapter presents methods of analysis for finding the forced response of circuits containing sinusoidal independent sources. Here we assume that all circuits are "stable," that is, that any "turn-on" transients die out with time. In this case, the circuit is said to be operating in the "ac steady state," with the forced response being the dominant effect. This means that all voltages and currents are sinusoidal, with differing amplitudes and phases. Thus, the ac steady-state analysis problem consists in finding all of these amplitudes and phases. Section 11.1 presents the idea of a phasor in such a fashion that one can read the material in this chapter without having covered Part II on time domain analysis. If this material *has* been covered, then Section 11.1 should be considered as a brief refresher on phasors—and perhaps the different approach will enhance your understanding of this topic. Section 11.2 then presents examples of nodal and mesh analysis, equivalent circuits, and so on. It illustrates the point we make in Section 11.1 that all elements obey Ohm's law in phasor form and that KVL and KCL continue to hold for phasors, so all of the analysis techniques previously presented in the text continue to hold. Section 11.3 discusses the idea of average (as opposed to instantaneous) power, power factor, and real, reactive, and apparent powers. The idea of complex power is presented in Section 11.4, and Tellegen's theorem[1] is used to discuss conservation of complex power. A number of examples are worked in detail to provide a solid understanding of this important topic. Finally, SPICE analysis of circuits in the ac steady state is covered in Section 11.5. A side issue that will prove useful in Chapter 12 on frequency response is the exploration of the power and energy relations for parallel and series tuned circuits at resonance.

11.1 | The Phasor Equivalent Circuit

We will now develop an analysis method that makes determination of the forced response of circuits straightforward and algorithmic. Although we have developed this idea in some theoretical detail in Section 10.3 of Chapter 10, we will repeat it here—in a different manner. If you have already read the material just mentioned, simply treat the next few pages as a refresher to provide a different insight into ac forced response. If you have not, this section will develop the methodology you need without relying on that material.

Ac Forced Response   First, precisely what do we mean by "ac forced response?" Simply this. Suppose that we have constructed a circuit and "turned it on," that is, we have made one or more independent sources in the circuit nonzero. We will further assume that each and every one of these sources has a waveform that is sinusoidal at a single common frequency. Each resulting circuit response voltage and current has two parts: one is a "transient" resulting from the initial charging of the dynamic elements (L or C) in the circuit, and the other is a sinusoid. The latter part is called the ac forced response. If the circuit is stable—that is, if the transient part approaches zero as time increases—the "complete response" (the sum of the transient and forced responses) is approximately the same as the forced component for large values of time. Figure 11.1 illustrates the basic idea. After a turn-on transient of about 0.3 or 0.4 second, our waveform $x(t)$ has "settled down" into the ac steady state consisting of a sinusoid of about four units in amplitude and a period of about 0.2 seconds, corresponding to a radian frequency of about $2\pi/0.2$ s $= 10\pi$ radians/second.

Figure 11.1
Sinusoidal complete response

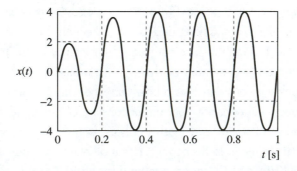

We will ignore the turn-on transient and concentrate on the ac forced response component; thus, we will simply assume that all waveforms are sinusoidal and extend them backwards to $-\infty$, resulting in the graph of Figure 11.2. (We have shown only the seg-

---

[1] Stated and proved in Appendix B.

ment from $t = -1$ to $t = +1$ in the drawing.) Thus, we will be dealing with two-sided periodic waveforms in the form of sinusoids. Because the turn-on transient is no longer important, we will not distinguish between the forced and ac steady-state responses—though we stress the fact that the forced response exists independently of the stability of the circuit, whereas the ac steady-state response only exists for stable ones.

Figure 11.2
The ac forced response

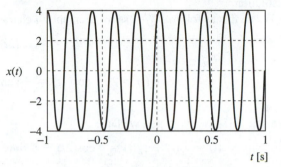

The General Sinusoid   Because we will be dealing so much with sinusoids, let's investigate them a bit more. Figure 11.3(a) shows the graph of two periods of $x(t) = 3 \cos(4\pi t)$, and Figure 11.3(b) shows two periods of $y(t) = 3 \sin(4\pi t)$. As it happens, we can describe either one—and an entire host of other sinusoids as well—by using the *general sinusoid*

$$f(t) = A \cos(\omega t + \theta). \tag{11.1-1}$$

If we use the trigonometric identity

$$\cos(x + y) = \cos(x) \cos(y) - \sin(x) \sin(y), \tag{11.1-2}$$

with $x = \omega t$ and $\theta = y = -90°$, equation (11.1-1) becomes

$$f(t) = A \sin(\omega t). \tag{11.1-3}$$

(Notice that choosing $\theta = y = +90°$ results in $f(t) = -A \sin(\omega t)$.) The trigonometric identity in (11.1-2) is one with which you should become very familiar for we will be using it on a number of occasions in this chapter.

Figure 11.3
Cosine and sine
waveforms

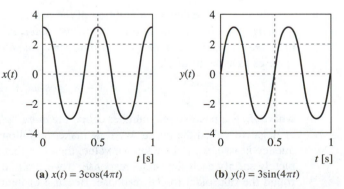

(a) $x(t) = 3\cos(4\pi t)$ (b) $y(t) = 3\sin(4\pi t)$

Ac Forced Response of   Now let's look a bit at the forced response of a circuit. A general method of analysis
an Example Circuit   would be to derive a differential equation for a given response variable and then use conventional differential equation solution techniques to determine the forced response (it is called the "particular integral" in many math texts). We will, however, adopt a different approach—one that provides more physical insight into circuit behavior. To help moti-

vate our discussion, we will investigate the simple circuit shown in Figure 11.4. Although we are developing a method of analyzing such a circuit for its ac forced response, we will for now "beg the question" and simply show that

$$i(t) = \cos(2t) \text{ A} \tag{11.1-4}$$

is a solution.

We do this as follows. We use the element v-i characteristics to find their voltages:

$$v_R(t) = 4i(t) = 4 \cos(2t) \text{ V} \tag{11.1-5}$$

and

$$v_L(t) = 2\frac{d}{dt} \cos(2t) = -4 \sin(2t) \text{ V}. \tag{11.1-6}$$

If we sum these two voltages, we have

$$v_s(t) = v_L(t) + v_R(t) = 4 \cos(2t) - 4 \sin(2t) \tag{11.1-7}$$

$$= \sqrt{4^2 + 4^2} \left[\frac{4}{\sqrt{4^2 + 4^2}} \cos(2t) - \frac{4}{\sqrt{4^2 + 4^2}} \sin(2t) \right]$$

$$= 4\sqrt{2} \left[\frac{1}{\sqrt{2}} \cos(2t) - \frac{1}{\sqrt{2}} \sin(2t) \right] = 4\sqrt{2} \cos(2t + 45°).$$

In the last line, we used the trigonometric identity in (11.1-2) with $x = 2t$ and $y = 45°$. Our computed voltage agrees with the one given in Figure 11.4; that is, our assumed mesh current satisfies the resistor and inductor element relationships with KVL as well. Thus, it is a solution for our circuit.

Figure 11.4
An example circuit

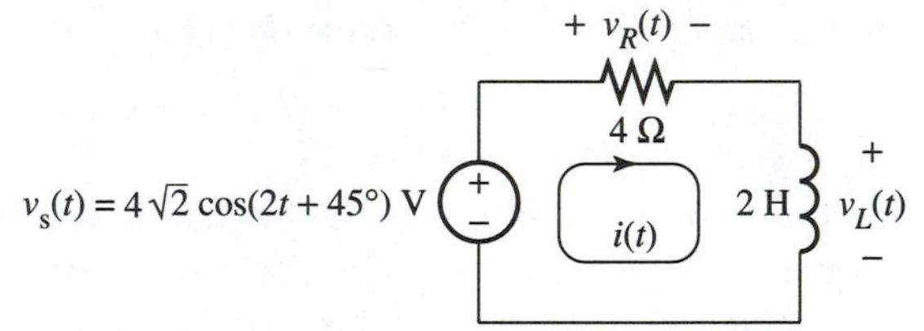

$$v_s(t) = 4\sqrt{2} \cos(2t + 45°) \text{ V}$$

$$+ v_R(t) -$$

$$4 \, \Omega$$

$$i(t)$$

$$2 \text{ H} \quad v_L(t)$$

Phasors   The thing to notice about our solution of the preceding example circuit is this: *the forced response for any variable in a circuit whose independent sources are all sinusoidal with the same frequency is a sinusoid at that same frequency*. This means that all circuit responses have the form

$$x(t) = X_m \cos(\omega t + \theta), \tag{11.1-8}$$

where X_m is a positive real number called the *amplitude* and θ is an angle, either positive or negative, called the *phase*. We now make the following crucial observation: if the frequency is assumed to be a known value, then two real numbers X_m and θ suffice to completely specify each sinusoidal response. If we can find these two constants, then we have found the sinusoidal forced response. Because of their importance, we give this number pair a name.

Definition 11.1   If $x(t) = X_m \cos(\omega t + \theta)$, then the pair of real numbers (X_m, θ) is called the *(polar form of the) phasor associated with $x(t)$.*

At this point, let's recall that a real number pair can be considered to be a single complex number provided that we agree to use the definitions for addition, multiplication, and complex conjugation given in Chapter 8.[2] Let's do this and plot our phasor in the complex plane as in Figure 11.5. We use a bar above a letter to denote the fact that it is a phasor quantity. Recalling the Euler form for complex numbers developed in Chapter 8, we can write our phasor as

Figure 11.5 The phasor in the complex plane

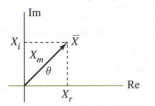

$$\overline{X} = X_m e^{j\theta} = X_m \angle \theta. \tag{11.1-9}$$

We remind you that $\angle \theta$ is shorthand notation for $e^{j\theta}$. Alternatively, in rectangular form, we have

$$\overline{X} = X_r + jX_i = X_m \cos(\theta) + jX_m \sin(\theta). \tag{11.1-10}$$

$X_r = X_m \cos(\theta)$ is called the *real part* of $\overline{X}$ and $X_i = X_m \sin(\theta)$ is called the *imaginary part* of $\overline{X}$. The latter notation is unfortunate because X_i, like X_r, is a real number.

Basic Phasor Examples

Although we have agreed to impose the structure of complex arithmetic on the quantities we call phasors, we have yet to show that this step leads to a worthwhile analysis procedure. We will do this shortly, but let's pause to gain a bit of practice in manipulating phasors before doing so.

Example 11.1   If a voltage in a circuit has the form $v(t) = 10 \cos(2t - 45°)$ V, find the corresponding phasor in rectangular form.

Solution   We can immediately write the Euler form by inspection:

$$\overline{V} = 10\angle{-45°} \text{ V.} \tag{11.1-11}$$

Notice that we have used the same symbol as that for the time-varying voltage, only in uppercase and with an overbar. We will use this notation consistently in what follows. We next simply use Euler's formula to write

$$\overline{V} = 10 \cos(45°) - j10 \sin(45°) = \frac{10}{\sqrt{2}} - j\frac{10}{\sqrt{2}} = 5\sqrt{2} - j5\sqrt{2} \text{ V.} \tag{11.1-12}$$

The unit for a phasor is the same as the unit for the time quantity it represents.

Example 11.2   If $\overline{I} = 6 + j8$ A and $\omega = 5$ rad/s, find $i(t)$.

Solution   Our phasor $\overline{I}$ is given in rectangular form, so we must convert it to Euler form. We do so in the following way:

$$\overline{I} = 6 + j8 = 2(3 + j4) = 2(5\angle53.1°) = 10\angle53.1° \text{ A.} \tag{11.1-13}$$

We can now simply write the time-varying form for $i(t)$:

$$i(t) = 10 \cos(5t + 53.1°) \text{ A.} \tag{11.1-14}$$

We just identify the magnitude of the complex phasor with the amplitude of the sinusoid, and its angle with the phase of the latter.

---

[2] It might be a good idea here to go back to Chapter 8 and review the basics of complex numbers.

Cosine Phasors   Let's discuss the preceding examples a bit more to establish a convention. We note that we have used cosine functions in each case rather than sine functions. This is a convention, but one we will stick to: *each and every phasor represents a cosine function of time (not a sine function).* We now provide an example to illustrate this.

Example 11.3   Let $x(t) = 4\sqrt{2} \sin(3t + 45°)$. Find the phasor $\overline{X}$ in rectangular form.

Solution   Here we must perform the preliminary step of expressing $x(t)$ as a cosine function. Using the trigonometric identity $\cos(x + y) = \cos(x)\cos(y) - \sin(x)\sin(y)$ with $x = 3t$ and $y = -90°$, we have

$$x(t) = 4\sqrt{2} \cos(3t - 45°). \tag{11.1-15}$$

(Look back at equation 11.1-2 and the discussion corresponding to it.) The Euler form for our phasor now is

$$\overline{X} = 4\sqrt{2} \angle -45°. \tag{11.1-16}$$

We next convert to rectangular form:

$$\overline{X} = 4\sqrt{2} \cos(-45°) + j4\sqrt{2} \sin(-45°) = 4 - j4. \tag{11.1-17}$$

Observe here that the cosine is an even function and the sine is an odd one.

A Possible Pitfall When Working with Phasors   One problem that arises in working with phasors is that angles can be expressed in two ways (as either positive or negative) and are always ambiguous to within multiples of 360°. Another problem can occur when converting from rectangular to Euler form. If care is not taken, a false answer is possible. Consider the problem of finding the Euler form of the phasor $\overline{X} = -4 - j4$. We compute the magnitude as usual to be $X_m = \sqrt{(-4)^2 + (-4)^2} = 4\sqrt{2}$. No problem. But let's compute the polar angle. It is given by $\theta = \tan^{-1}[-4/-4] = \tan^{-1}(1) = 45°$. But this is wrong! The problem is that the tangent has period 180° (π radians), unlike the sine and cosine whose periods are both 360° (2π radians). The best way to guard against such difficulties is to make a sketch, as we have done in Figure 11.6. In this case, the angle is 225° (or $-135°$).

Figure 11.6
Finding the correct quadrant

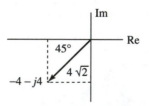

A Rationale for Phasors: The System Function   Now let's support our decision to use complex numbers to express sinusoids. We will abstract the problem as shown in Figure 11.7. In Figure 11.7(a) we show a circuit, represented by a rectangle, having only one independent source whose waveform $x(t)$ is sinusoidal as shown with amplitude X_m and phase θ. (The extension to circuits having more than one independent source is straightforward, but it makes the discussion simpler to assume only one.) We select any voltage or current variable in the circuit and call it $y(t)$. From our earlier discussion we know that this response will be sinusoidal, having some amplitude Y_m and phase β as shown in the figure. If we can find the amplitude and phase of $y(t)$ we will have succeeded in our analysis. The circuit with its time-varying sources is referred to as the "time domain form."

To solve for our circuit response, we represent the input and response waveforms by their phasors as shown in Figure 11.7(b). Now consider this: the ratio

$$\frac{\overline{Y}}{\overline{X}} = \frac{Y_m \angle \beta}{X_m \angle \theta} = \frac{Y_m}{X_m} \angle \phi, \qquad (11.1\text{-}18)$$

where

$$\phi = \beta - \theta \qquad (11.1\text{-}19)$$

is simply a complex constant. We refer to it as a *system function,* and write it (in polar form) as

$$H(\omega) = |H(\omega)| \angle \phi(\omega). \qquad (11.1\text{-}20)$$

By expressing it as a function of ω we are allowing for the fact that the system function might (and it generally will) change if we alter the frequency. Once we have found $H(\omega)$, we *use it* in our analysis procedure as follows:

$$\overline{Y} = H(\omega)\overline{X}. \qquad (11.1\text{-}21)$$

We merely multiply the magnitudes of $H(\omega)$ and $\overline{X}$ and add their angles to find the polar form of $\overline{Y}$.

Figure 11.7
The idea of a phasor
domain representation

$x(t) = X_m \cos(\omega t + \theta)$ | Circuit | $y(t) = Y_m \cos(\omega t + \beta)$

(a) Time domain

$\overline{X} = X_m \angle \theta$ — $H(\omega)$ — $\overline{Y} = Y_m \angle \beta$

(b) Phasor domain

Complex Impedance and Admittance

Now let's have a look at the element or, more generally, the two-terminal subcircuit shown in Figure 11.8(a). We are assuming that the two-terminal subcircuit has no independent sources.[3] The phasor representation is shown in Figure 11.8(b). We define the *impedance* of the subcircuit to be the system function relative to $i(t)$ as the input and $v(t)$ as the output (relative to Figure 11.7):

$$Z(j\omega) = \frac{\overline{V}}{\overline{I}} = |Z(j\omega)| \angle \phi. \qquad (11.1\text{-}22)$$

Here we see that $|Z(j\omega)| = V_m/I_m$ and $\phi = \beta - \theta$.[4] If we reverse the situation and consider $v(t)$ to be the input and $i(t)$ as the output (relative to Figure 11.7), we call the system function the *admittance* of the subcircuit and write

$$Y(j\omega) = \frac{\overline{I}}{\overline{V}} = \frac{1}{Z(j\omega)} = \frac{I_m}{V_m} \angle -\phi. \qquad (11.1\text{-}23)$$

---

[3] Strictly speaking, this is not a necessary assumption, provided that all the internal independent sources are sinusoidal at the same frequency ω. The concept of impedance, however, would be somewhat artificial.

[4] We are including j as a multiplier for ω in the impedance and admittance expressions by convention; however, for simplicity, we are suppressing the argument entirely from the phase angle θ.

Figure 11.8
Impedance and
admittance

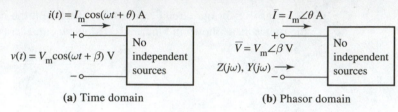

(a) Time domain   (b) Phasor domain

The unit of impedance is the ohm and that of admittance the siemens. Thus, impedance is the generalization of resistance, and admittance that of conductance.

Complex Impedance
Models for R, L, and C

To make our definitions of impedance and admittance pay off, we will now see what forms they have for resistors, inductors, and capacitors. We will start with the resistor, shown in Figure 11.9(a). We write the time domain equation in the form

$$v(t) = V_m \cos(\omega t + \beta) = Ri(t) = RI_m \cos(\omega t + \theta). \tag{11.1-24}$$

Next we simply equate magnitudes and phases:

$$V_m = RI_m \tag{11.1-25}$$

and

$$\beta = \theta. \tag{11.1-26}$$

The impedance is

$$Z(j\omega) = \frac{\overline{V}}{\overline{I}} = \frac{RI_m \angle \theta}{I_m \angle \theta} = R. \tag{11.1-27}$$

Thus, Ohm's law continues to hold for phasors:

$$\overline{V} = R\overline{I}. \tag{11.1-28}$$

The phasor representation is shown in Figure 11.9(b).

Figure 11.9
Impedance of resistor

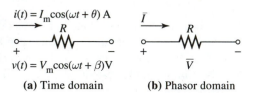

(a) Time domain   (b) Phasor domain

The time domain form for the inductor is shown in Figure 11.10(a). The time domain relationship is

$$v(t) = V_m \cos(\omega t + \beta) = L\frac{di}{dt} = -\omega L I_m \sin(\omega t + \theta) \tag{11.1-29}$$

$$= \omega L I_m \cos(\omega t + \theta + 90°).$$

Figure 11.10
Impedance of inductor

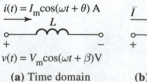

(a) Time domain   (b) Phasor domain

(Look back at equation (11.1-2) if necessary to justify the last step.) Equating magnitudes and angles, we obtain

$$V_m = \omega L I_m \tag{11.1-30}$$

and

$$\beta = \theta + 90°. \tag{11.1-31}$$

The impedance of the inductor, therefore, is

$$Z(j\omega) = \frac{\overline{V}}{\overline{I}} = \frac{\omega L I_m \angle(\theta + 90°)}{I_m \angle \theta} = \omega L \angle 90° = j\omega L. \tag{11.1-32}$$

In the last step, we have simply converted from Euler to rectangular form. We now see that the inductor, too, obeys Ohm's law:

$$\overline{V} = j\omega L \overline{I}. \tag{11.1-33}$$

The only difference is that the "resistance"—the impedance—is purely imaginary. The resulting phasor representation for the inductor is shown in Figure 11.10(b).

The time domain form for the capacitor is shown in Figure 11.11(a). The time domain relationship is

$$i(t) = I_m \cos(\omega t + \theta) = C\frac{dv}{dt} = -\omega C V_m \sin(\omega t + \beta) \tag{11.1-34}$$

$$= \omega C V_m \cos(\omega t + \beta + 90°).$$

Equating magnitudes and angles, we obtain

$$I_m = \omega C V_m \tag{11.1-35}$$

and

$$\beta = \theta - 90°. \tag{11.1-36}$$

The impedance of the capacitor therefore, is

$$Z(j\omega) = \frac{\overline{V}}{\overline{I}} = \frac{V_m \angle(\theta - 90°)}{\omega C V_m \angle \theta} = \frac{1}{\omega C} \angle -90° = -j\frac{1}{\omega C} = \frac{1}{j\omega C}. \tag{11.1-37}$$

In the last step, we converted from polar to rectangular form and noted that $1/j = -j$ (you should verify this). We now see that the capacitor, like the inductor and the resistor, obeys Ohm's law:

$$\overline{V} = \frac{1}{j\omega C}\overline{I}. \tag{11.1-38}$$

Its phasor equivalent model is shown in Figure 11.11(b).

Figure 11.11
Impedance of capacitor

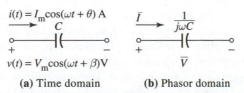

(a) Time domain   (b) Phasor domain

KVL and KCL in Phasor Form   There is one last property of phasors that we must develop before we proceed to the analysis of circuits using phasors. Consider the problem of summing two sinusoids with different amplitudes and phases:

$$x(t) = x_1(t) + x_2(t) = A\cos(\omega t + \theta) + B\cos(\omega t + \phi) \tag{11.1-39}$$

$$= A\cos(\omega t)\cos(\theta) - A\sin(\omega t)\sin(\theta) + B\cos(\omega t)\cos(\phi) - B\sin(\omega t)\sin(\phi)$$

$$= [A\cos(\theta) + B\cos(\phi)]\cos(\omega t) - [A\sin(\theta) + B\sin(\phi)]\sin(\omega t)$$

$$= A_0\cos(\omega t) - B_0\sin(\omega t) = \sqrt{A_0^2 + B_0^2}\left[\frac{A_0}{\sqrt{A_0^2 + B_0^2}}\cos(\omega t) - \frac{B_0}{\sqrt{A_0^2 + B_0^2}}\sin(\omega t)\right]$$

$$= \sqrt{A_0^2 + B_0^2}\cos(\omega t + \beta).$$

Here, $A_0 = A\cos(\theta) + B\cos(\phi)$, $B_0 = A\sin(\theta) + B\sin(\phi)$, and $\beta = \tan^{-1}[B_0/A_0]$. Notice that we have used the trigonometric identity in equation (11.1-2) once again, grouped the cosine and sine terms, and then multiplied and divided by the radical shown. This converts the coefficient of $\cos(\omega t)$ into $\cos(\beta)$ and the coefficient of $\sin(\omega t)$ into $\sin(\beta)$, where β is as previously defined. We then read the trigonometric identity (11.1-2) backwards to convert the result into a cosine waveform.

Now let's see what happens if we represent each sinusoid by its phasor and add the result. We have

$$\overline{X} = \overline{X}_1 + \overline{X}_2 = A\angle\theta + B\angle\phi \tag{11.1-40}$$

$$= A\cos(\theta) + jA\sin(\theta) + B\cos(\phi) + jB\sin(\phi)$$

$$= [A\cos(\theta) + B\cos(\phi)] + j[A\sin(\theta) + B\sin(\phi)]$$

$$= A_0 + jB_0 = \sqrt{A_0^2 + B_0^2}\angle\tan^{-1}[B_0/A_0].$$

But this is simply the phasor associated with the result in (11.1-39). Thus, we see that we can find the phasor for a sum of sinusoids by summing the individual phasors using complex arithmetic. In fact, a simple induction argument shows that this works for the sum of any number of sinusoids.

Now suppose the $x(t)$'s are voltages and we are applying KVL. Then the first equality in equation (11.1-39) becomes

$$v(t) = v_1(t) + v_2(t), \tag{11.1-41}$$

with the corresponding phasor equation being

$$\overline{V} = \overline{V}_1 + \overline{V}_2. \tag{11.1-42}$$

This clearly generalizes to any number of voltages—and to currents in any number, as well. Thus, we can make the following statement: *if KVL is satisfied by any number of sinusoidal voltages (or KCL by any number of sinusoidal currents) having a common frequency, then the corresponding KVL (or KCL) equation is valid for the phasor form of this equation.*

Here is the net result of what we have accomplished. We now know that each of the elements R, L, and C obeys Ohm's law, provided we use the impedance of the appropriate element in the place of resistance. Furthermore, we now know that both KVL and KCL hold for phasor voltages and currents. A bit of reflection will reveal that all of our dc analysis techniques—superposition, Thévenin and Norton equivalents, nodal and mesh analysis, etc.—are all based upon only these facts and linearity. *Thus, all of our dc analysis techniques continue to hold for ac forced response with impedances replacing resistances and phasors replacing time-varying voltages and currents.* We illustrate this with the next example.

Example 11.4   Solve for the forced response of the currents and voltages for the circuit shown in Figure 11.12 using phasor techniques.

Figure 11.12
An example circuit

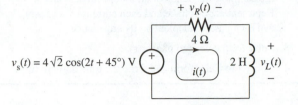

$$v_s(t) = 4\sqrt{2}\cos(2t + 45°) \text{ V}$$

Solution   We convert all of the voltages and currents to phasors and represent the inductor and the resistor by their impedances as shown in Figure 11.13. The resistor and the inductor are connected in series, so we do as we would with resistors: we simply add their impedances, then use Ohm's law to find the phasor current:

$$\bar{I} = \frac{\bar{V}}{Z} = \frac{4\sqrt{2}\angle 45°}{4 + j4} = \frac{4\sqrt{2}\angle 45°}{4\sqrt{2}\angle 45°} = 1\angle 0° \text{ A}. \tag{11.1-43}$$

Thus, $i(t) = \cos(2t)$ A. We easily find the inductor and resistor voltages using the voltage divider rule:

$$\bar{V}_L = \frac{j4}{4 + j4}\bar{V}_s = \frac{j}{1 + j}4\sqrt{2}\angle 45° = \frac{1\angle 90° \times 4\sqrt{2}\angle 45°}{\sqrt{2}\angle 45°} = 4\angle 90° \text{ V} \tag{11.1-44}$$

and

$$\bar{V}_R = \frac{4}{4 + j4}\bar{V}_s = \frac{1}{1 + j}4\sqrt{2}\angle 45° = \frac{4\sqrt{2}\angle 45°}{\sqrt{2}\angle 45°} = 4\angle 0° \text{ V}. \tag{11.1-45}$$

Therefore, we see that $v_L(t) = 4\cos(2t + 90°)$ V and $v_R(t) = 4\cos(2t)$ V.

Figure 11.13
The phasor equivalent circuit

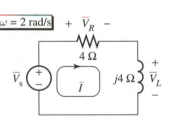

The Phasor Method: Summary   Notice that when we converted to the phasor equivalent circuit shown in Figure 11.13 we made a note of the frequency in a small box. This is the reason: at a given frequency, the impedances, voltages, and currents are all complex constants. Therefore, the value of ω does not appear anywhere on this diagram—or in our analysis procedure. The last step, however, consists of writing down the time domain form of the corresponding voltage or current. For this we need the frequency, and having it clearly visible on the diagram makes it instantly accessible. There is one facet of phasor analysis that creates some problems for beginners: confusing impedances with phasors. Both are complex numbers, but only phasors are representative of sinusoidal time-varying waveforms. The impedance (or admittance) is simply a complex constant that takes the place of resistance in dc circuit analysis. To help you to learn the various steps in phasor analysis, we list them in Table 11.1. To illustrate the importance of the first step, we will work one more simple example.

Table 11.1
The phasor method

1. Convert all sine functions to cosines (if necessary).

2. Draw the phasor equivalent circuit, making a note of the common frequency of all independent sources. Represent each voltage and each current by a phasor and each passive element by its impedance.

3. Solve for the desired phasor(s).

4. Convert to Euler form and write the time domain form.

Example 11.5   Solve for the forced response of the voltage $v(t)$ in the circuit shown in Figure 11.14 using phasor techniques. Assume that $i_s(t) = 10 \sin(3t)$ A.

Figure 11.14
An example circuit

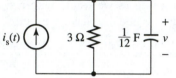

Solution   We first convert the current source to cosine form, changing it to

$$i_s(t) = 10 \cos(3t - 90°) \text{ A.} \tag{11.1-46}$$

We then convert voltages and currents to phasors and the passive elements to impedances, resulting in the phasor equivalent shown in Figure 11.15 with

$$\bar{I}_s = 10\angle{-90°} = -j10 \text{ A.} \tag{11.1-47}$$

The two passive elements are connected in parallel, with an equivalent impedance of

$$Z = \frac{3 \times (-j4)}{3 - j4} = \frac{12\angle{-90°}}{5\angle{-53.1°}} = 2.4\angle{-36.9°} \text{ Ω.} \tag{11.1-48}$$

Thus, the phasor voltage is

$$\bar{V} = Z\bar{I} = 2.4\angle{-36.9°} \times 10\angle{-90°} = 24\angle{-126.9°} \text{ V.} \tag{11.1-49}$$

The corresponding time-varying sinusoidal waveform is

$$v(t) = 24 \cos(3t - 126.9°) \text{ V.} \tag{11.1-50}$$

Figure 11.15
The phasor equivalent circuit

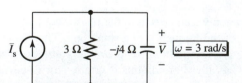

Section 11.1 Quiz

Q11.1-1.   If $i(t) = 8 \sin(100t)$ A, find the associated phasor $\bar{I}$ in both Euler and rectangular forms.

Q11.1-2.   If $\bar{V} = -30 - j40$ V is the phasor associated with a sinusoidal voltage $v(t)$ and the frequency ω is known to be 377 rad/s, find $v(t)$.

Q11.1-3.   Draw the phasor equivalent for the circuit shown in Figure Q11.1-1.

Figure Q11.1-1

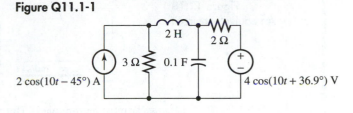

Q11.1-4.   If $x(t) = 5 \cos(5t - 36.9°)$ and $y(t) = 10 \sin(5t)$, compute:

    a. $x(t) + y(t)$ using trigonometric identities
    b. $x(t) + y(t)$ using phasors
    c. the rectangular form of the phasor representing dx/dt.
Express each of your answers in cosine form.

11.2 | Methods of Ac Circuit Analysis

Ac Circuit
Analysis Examples

This section will consist mostly of examples because all of our circuit analysis methods—nodal, mesh, Thévenin, Norton, etc.—continue to apply for complex impedances and phasors. We will, however, present a couple of new ideas between the examples.

Example 11.6   Find the forced response for the circuit shown in Figure 11.16.

Figure 11.16
An example circuit

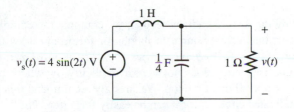

Solution   We show the circuit in phasor form in Figure 11.17. Notice that our first step was to convert the sine function to a cosine; only then did we convert the voltage source to phasor form. There are many ways to solve this circuit, but—remembering that *all of our dc analysis techniques continue to hold for the phasor equivalent*—we choose to find the equivalent impedance of the capacitor and its parallel resistor:

$$Z = \frac{1 \times (-j2)}{1 + (-j2)} = \frac{-j2}{1 - j2} \times \frac{1 + j2}{1 + j2} = \frac{4 - j2}{5} = 0.8 - j0.4 \ \Omega. \quad (11.2\text{-}1)$$

Using this equivalent impedance, we can redraw the circuit in the equivalent form shown in Figure 11.18. Next, we simply use the voltage divider rule (in phasor form) to compute the phasor associated with the voltage we are looking for:

$$\overline{V} = \frac{Z}{Z + j2} \overline{V}_s = \frac{0.8 - j0.4}{0.8 - j0.4 + j2} \times (-j4) = \frac{-1.6 - j3.2}{0.8 + j1.6} = -2 + j0 \ \text{V}. \quad (11.2\text{-}2)$$

Finally, we must convert to Euler form (noticing that the negative sign in front of our an-

Figure 11.17
The phasor equivalent

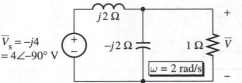

Figure 11.18
An equivalent circuit

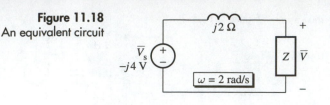

swer must be removed). But we know that $-2 = 2\angle\pm180°$. Arbitrarily choosing the negative sign, we write

$$v(t) = 2\cos(2t - 180°) \text{ V}. \tag{11.2-3}$$

Example 11.7   Find the forced response for $i(t)$ in the circuit shown in Figure 11.19, working with element impedances and/or admittances.

Figure 11.19
An example circuit

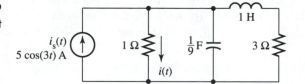

Solution   Noting that $\omega = 3$ radians/second, we compute the complex impedance of each element, convert the current source to its phasor form, and draw the *phasor equivalent circuit* in Figure 11.20. *This equivalent, we stress, is only valid at the single frequency of 3 rad/s.* As this frequency does not appear explicitly anywhere on the circuit, we have made a note of it in a box at the top. We strongly recommend this practice, particularly with more complicated circuits, for it can be easily forgotten. The 1-Ω resistor is connected in parallel with the capacitor and the series combination of the inductor and the 3-Ω resistor, so we compute the impedance of the subcircuit made up of all the passive elements to be

$$Z(j3) = \frac{1}{1 + \dfrac{1}{-j3} + \dfrac{1}{3 + j3}} = \frac{3(1-j)}{4 - j3} \ \Omega. \tag{11.2-4}$$

The current through the 1-Ω resistor is the same, by Ohm's law, as the voltage across it. Thus,

$$\bar{I} = \frac{3(1-j)}{4 - j3} \times 5\angle0° = \frac{3\sqrt{2}\angle-45°}{5\angle-36.9°} \times 5\angle0° = 3\sqrt{2}\angle-8.1° \text{ A}. \tag{11.2-5}$$

Thus, the time waveform for the forced response of the current $i(t)$ is

$$i(t) = 3\sqrt{2}\cos(3t - 8.1°) \text{ A}. \tag{11.2-6}$$

Figure 11.20
The phasor equivalent

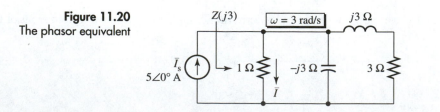

Example 11.8   Use nodal analysis to determine the forced response for $v(t)$ in Figure 11.21.

Figure 11.21
An example circuit

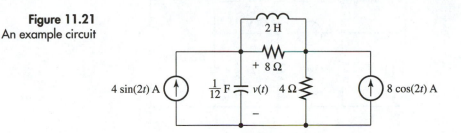

Solution   By now, the process should be familiar: we represent the sinusoidal current sources with their phasors and convert all passive elements to impedances. This results in the circuit shown in Figure 11.22. We have chosen the bottom node as the reference and have labeled the two nonreference nodes with symbols for unknown phasor voltages. The equation at the node labeled $\overline{V}$ is

$$\frac{\overline{V} - \overline{V}_a}{j4} + \frac{\overline{V} - \overline{V}_a}{8} + \frac{\overline{V}}{-j6} = 4\angle -90° = -j4, \qquad (11.2\text{-}7)$$

and that at the node labeled $\overline{V}_a$ is

$$\frac{\overline{V}_a - \overline{V}}{j4} + \frac{\overline{V}_a - \overline{V}}{8} + \frac{\overline{V}_a}{4} = 8\angle 0° = 8. \qquad (11.2\text{-}8)$$

If we rationalize by multiplying the first equation by $j24$ and the second by $j8$, then assemble the two equations into matrix form, we have

$$\begin{bmatrix} (2 + j3) & -(6 + j3) \\ -(2 + j) & (2 + j3) \end{bmatrix} \begin{bmatrix} \overline{V} \\ \overline{V}_a \end{bmatrix} = \begin{bmatrix} 96 \\ j64 \end{bmatrix}. \qquad (11.2\text{-}9)$$

We solve by inverting the (2×2) coefficient matrix to get $\overline{V} = 48\angle -90°$ and $\overline{V}_a = 16\angle -90°$. Thus, the time response we are seeking is $v(t) = 48 \cos(2t - 90°)$ V $= -48 \sin(2t)$ V.

Figure 11.22
The phasor equivalent circuit

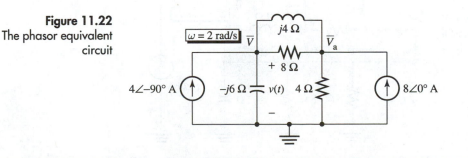

Example 11.9   Find the forced response for the current $i(t)$ in Figure 11.23, using either mesh or nodal analysis—whichever is simpler.

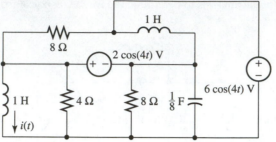
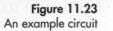

Solution If we count the nodes, the branches, the number of voltage sources, and the number of current sources and compute the required number of mesh equations, we will find that the number of mesh equations $= N_{\text{me}} = B - N + 1 - N_I = 8 - 4 + 1 - 0 = 5!$ We can reduce this number to 3 by noticing that the 4-Ω resistor and 1-H inductor are in parallel, as are the 8-Ω resistor and 1/8-F capacitor. Using equivalent parallel impedances for these two subcircuits would reduce the number of "elements" by 2, hence the number of required mesh equations would be reduced to 3. The number of nodal equations required, though, is $N_{\text{ne}} = N - 1 - N_V = 4 - 1 - 2 = 1!$ Clearly, we wish to perform nodal analysis.

Choosing the ground reference at the bottom (notice that this grounds one of the voltage sources, though one cannot ground both—which would, of course, be desirable) results in the phasor equivalent circuit shown in Figure 11.24. As we have indicated by shading, there is one supernode. The only other nonreference node is nonessential. The single equation at the supernode is, therefore,

$$\frac{\overline{V}}{j4} + \frac{\overline{V}}{4} + \frac{\overline{V} - 2}{8} + \frac{\overline{V} - 2}{-j2} + \frac{\overline{V} - 6}{8} + \frac{\overline{V} - 2 - 6}{j4} = 0. \qquad (11.2\text{-}10a)$$

The solution is $\overline{V} = 2 - j2 = 2\sqrt{2}\angle{-45°}$, so

$$\overline{I} = \frac{\overline{V}}{j4} = \frac{2\sqrt{2}\angle{-45°}}{4\angle 90°} = \frac{1}{\sqrt{2}}\angle{-135°}. \qquad (11.2\text{-}10b)$$

Hence, the time variation is $i(t) = 1/\sqrt{2}\,\cos(4t - 135°)$ A.

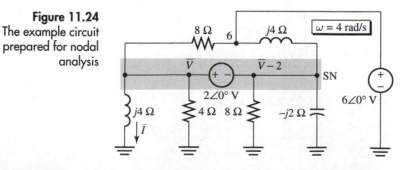

Example 11.10 Find the forced response for $i_x(t)$ in Figure 11.25 using either nodal or mesh analysis— whichever is simpler.

Figure 11.25
An example circuit

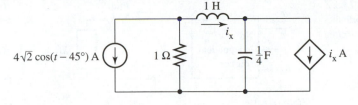

Solution Counting branches, nodes, and sources, we find that $B = 5$, $N = 3$, $N_v = 0$, and $N_i = 2$. Therefore, we would have to write $N - 1 - N_v = 2$ nodal equations and $B - N + 1 - N = 1$ mesh equation; therefore, mesh analysis is simpler. The phasor equivalent circuit with the mesh currents defined is shown in Figure 11.26. Note that we have taped the CCCS and have used the "natural" directions for the two outside mesh currents—those that coincide with the current source definitions.

There is only one mesh equation—that around the central mesh. The two end meshes are nonessential, for they disappear if we test the circuit by deactivating the sources. Thus, we have

$$1(\bar{I} + 4 - j4) + j\bar{I} + (-j4)(\bar{I} - \bar{I}_c) = 0. \tag{11.2-11}$$

We now untape the CCCS and note that $\bar{I}_c = \bar{I}$. The last term vanishes, so

$$\bar{I} = \frac{-4(1 - j)}{1 + j} = -4\angle -90° = j4. \tag{11.2-12}$$

Thus, $i_x(t) = i(t) = 4\cos(t + 90°)\text{ A} = -4\sin(t)\text{ A}$.

Figure 11.26
The phasor equivalent circuit

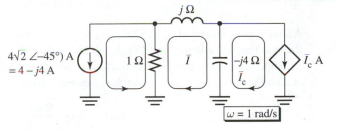

Series and Parallel Equivalent Subcircuits

The preceding examples form the prototype for all sinusoidal forced response circuit calculations. Notice that the radian frequency only entered in computing the element impedances, then again in the final step of converting back to the real time domain. It was for this reason that we labeled the phasor equivalent circuit with the frequency in a box at the top. The phasor equivalent by itself does not contain any frequency information.

Let's now investigate the idea of an equivalent subcircuit a bit more. Figure 11.27(a) shows a two-terminal passive subcircuit in phasor form. Let us suppose that we have computed or measured the impedance to be $Z(j\omega) = R(\omega) + jX(\omega)$; that is,

$$R(\omega) = \text{Re}\{Z(j\omega)\} \tag{11.2-13}$$

and

$$X(\omega) = \text{Im}\{Z(j\omega)\}. \tag{11.2-14}$$

We call $R(\omega)$ the *resistance* and $R(\omega)$ the *reactance* of the subcircuit. Note that, unlike the resistance of a resistive subcircuit, per se, the resistance of a circuit with inductors and/or capacitors can be a function of frequency, as can the reactance. The reactance is a mea-

Figure 11.27
Series and parallel
equivalent subcircuits

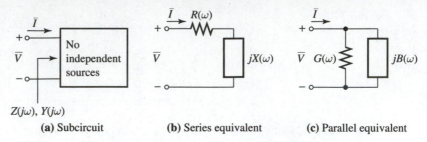

(a) Subcircuit   (b) Series equivalent   (c) Parallel equivalent

sure of the effect of these inductances and capacitances. Because $Z(j\omega) = R(\omega) + jX(\omega)$, and because we know that the impedances of elements connected in series add, we have the *series equivalent* subcircuit shown in Figure 11.27(b).

Now let's compute the admittance:

$$Y(j\omega) = \frac{1}{Z(j\omega)} = \frac{1}{R(\omega) + jX(\omega)} = \frac{1}{R(\omega) + jX(\omega)} \times \frac{R(\omega) - jX(\omega)}{R(\omega) - jX(\omega)} \quad (11.2\text{-}15)$$

$$= \frac{R(\omega)}{R^2(\omega) + X^2(\omega)} + j\frac{-X(\omega)}{R^2(\omega) + X^2(\omega)} = G(\omega) + jB(\omega).$$

We call $G(\omega)$ the *conductance* and $B(\omega)$ the *susceptance* of the subcircuit. Thus, we define

$$G(\omega) = \text{Re}\{Y(j\omega)\} \quad (11.2\text{-}16)$$

and

$$B(\omega) = \text{Im}\{Y(j\omega)\}. \quad (11.2\text{-}17)$$

Admittances connected in parallel sum to produce the equivalent admittance, thus we see that the parallel subcircuit of Figure 11.27(c) also is a valid equivalent for our two-terminal subcircuit. Note that $G(\omega)$ and $B(\omega)$ in that diagram have the unit S (siemens). We can equate the terms in (11.2-15) and thus express the conversion formulas as follows:

$$G(\omega) = \frac{R(\omega)}{R^2(\omega) + X^2(\omega)} \quad (11.2\text{-}18)$$

and

$$B(\omega) = \frac{-X(\omega)}{R^2(\omega) + X^2(\omega)}. \quad (11.2\text{-}19)$$

Example 11.11   Find the equivalent parallel subcircuit for the two-element subcircuit shown in Figure 11.28(a) that will be valid at $\omega = 2$ rad/s.

Figure 11.28
An example subcircuit

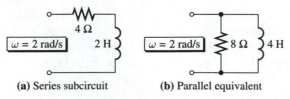

(a) Series subcircuit   (b) Parallel equivalent

Solution   The impedance of this subcircuit is $Z(j2) = 4 + j4\ \Omega$. If we compute the admittance, we will get $Y(j2) = 1/(4 + j4) = (4 - j4)/(4^2 + 4^2) = 1/8 - j\,1/8$ S. But we note that this represents a resistor having a conductance of 1/8 S (a resistance of 8 Ω) connected in parallel with an inductance having an admittance of $1/j\omega L = -j\,1/8$ S (which, with

$\omega = 2$ rad/s, can be solved to give $L = 4$ H). This equivalent subcircuit is shown in Figure 11.28(b).

The last example was an important one, but we must add a warning. *The equivalence we have shown between elements (as opposed to simply drawing boxes labeled with $R(\omega)$ and $X(\omega)$) is only valid for sinusoidal excitation at the frequency of $\omega = 2$ rad/s.* Let us illustrate this idea by changing the frequency to $\omega = 4$ rad/s. In this case, the impedance is $Z = 4 + j8$ Ω, $Y = 1/Z = 0.05 - j0.1$ S. But the latter is the admittance of the parallel connection of a 20-Ω resistor and a 2.5-H inductor.

Inductive and Capacitive Subcircuits   The three passive elements themselves form two-terminal subcircuits and, as such, have complex impedances. For the resistor, $Z(j\omega) = R + j0$, so its resistance is R and its reactance is zero as we would expect. The inductor has $Z(j\omega) = 0 + j\omega L$—its resistance is zero and its reactance is ωL (note that the reactance does not include the j!). Finally, the capacitor has $Z(j\omega) = 1/j\omega C = 0 + j(-1/\omega C)$, so its resistance is zero and its reactance is $-1/\omega C$. Thus, the resistor is purely *resistive,* and both the inductor and the capacitor are purely *reactive.* Now notice carefully that the inductor has a positive reactance and the capacitor a negative reactance; thus, for *any* passive two-terminal network with impedance $Z(j\omega) = R + jX(\omega)$, we say that it is *inductive* if $X(\omega) > 0$ and *capacitive* if $X(\omega) < 0$. Because $X(\omega)$ is a function of ω, a general two-terminal subcircuit can be inductive at one frequency and capacitive at another.

Series and Parallel Resonance   As an example, consider the subcircuit in Figure 11.29(a). Its complex impedance is $Z(j\omega) = (1/j\omega C) + j\omega L = j[\omega L - (1/\omega C)]$, which is zero when $\omega = \omega_0 = 1/\sqrt{LC}$. This phenomenon is called *series resonance* and ω_0 is called the *series resonant frequency.* For frequencies larger than this value, the reactance is positive and the subcircuit is inductive; for lower values of ω, the reactance is negative and so the subcircuit is capacitive. The parallel LC subcircuit in Figure 11.29(b) has an impedance of

$$Z(j\omega) = \frac{\dfrac{1}{j\omega C} \times j\omega L}{\dfrac{1}{j\omega C} + j\omega L} = j\frac{\omega L}{1 - \omega^2 LC}.$$

The impedance is infinite at $\omega_0 = 1/\sqrt{LC}$. Here ω_0 is called the *parallel resonant frequency,* and the phenomenon itself is called *parallel resonance.* Notice that the parallel subcircuit is inductive for $\omega < \omega_0$ and capacitive for larger values of ω. Notice also that the series subcircuit has zero impedance at $\omega_0 = 1/\sqrt{LC}$ and therefore is equivalent to a short circuit at that frequency; the parallel subcircuit, however, has infinite impedance at that frequency and so is equivalent to an open circuit there.

Figure 11.29
Two important LC
subcircuits

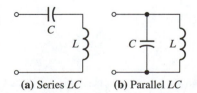

(a) Series LC (b) Parallel LC

Superposition for Ac Forced Response   When a circuit has two or more independent sources, some care must be taken because *a phasor represents a sinusoidal voltage or current at one single frequency. Thus, it is only if all the independent sources in a circuit are sinusoidal and have the same frequency that* *we can draw a single phasor equivalent.* If the circuit has independent sources with sinusoidal waveforms at different frequencies, we must use superposition—deactivating all sources but one at a time—and draw a phasor equivalent for each of these single source partial circuits; then, we convert each solution phasor back to the time domain. When we have found the *time domain* waveform for each partial response, we add them using superposition.

Example 11.12   Suppose the voltage sources in Figure 11.30 are adjusted to $v_{s1}(t) = 10\cos(3t)$ V and $v_{s2}(t) = 10\cos(4t)$ V. Find the resulting forced response for $v(t)$. Repeat the solution if both are adjusted to be the same: $v_{s1}(t) = v_{s2}(t) = 10\cos(3t)$ V.

Figure 11.30
An example circuit

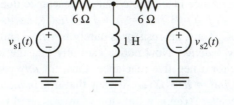

Solution   In the first situation the sources are at different frequencies, so we are forced to use superposition. To see why on an intuitive basis, consider this: when the two sources have different frequencies, which frequency do we use in computing the inductor's impedance? There must be a separate circuit and a separate set of impedances for each frequency. Thus, deactivating the two sources one at a time produces the two partial response equivalent circuits shown in Figure 11.31. We will leave it to you as an exercise to verify that $\overline{V}_1 = (5/\sqrt{2})\angle 45°$, or in the time domain, $v_1(t) = (5/\sqrt{2})\cos(3t + 45°)$ V and that $\overline{V}_2 = 4\angle 36.9°$, or $v_2(t) = 4\cos(4t + 36.9°)$ V. Thus, the actual voltage is the sum: $v(t) = (5/\sqrt{2})\cos(3t + 45°) + 4\cos(4t + 36.9°)$ V. When both sources are adjusted to the identical values of $10\cos(3t)$ V, however, we can draw a single phasor equivalent incorporating both sources, as shown in Figure 11.32. Again, one can use many techniques, but we will use nodal analysis. The single nodal equation is

$$\frac{\overline{V} - 10}{6} + \frac{\overline{V} - 10}{6} + \frac{\overline{V}}{j3} = 0. \tag{11.2-20a}$$

Solving, we get

$$\overline{V} = \frac{10j}{1 + j} = \frac{10\angle 90°}{\sqrt{2}\angle 45°} = 5\sqrt{2}\angle 45°, \tag{11.2-20b}$$

so that $v(t) = 5\sqrt{2}\cos(3t + 45°)$ V.

Figure 11.31
The two partial circuits
for superposition

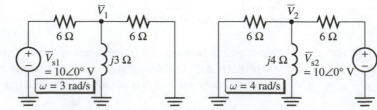

Figure 11.32
The single phasor
equivalent when both
sources have the same
frequency

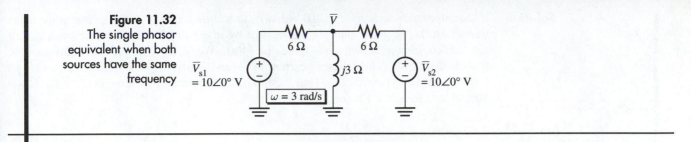

Impedance Scaling

Thus far, we have been working with "unrealistic" values: 1 F, 1 H, and 1 Ω, for instance. These might be somewhat reasonable in the generation, transmission, and distribution of electrical power. However, many applications of sinusoidal forced response are in the area of electronics and control, where much smaller values of capacitance and inductance and larger values of resistance are appropriate. Thus, we will consider how to deal efficiently with such realistic values.[5]

Ohm's law for impedance and phasors is

$$\overline{V} = Z(j\omega)\,\overline{I}. \tag{11.2-21}$$

We can *impedance scale* by the real constant a by noting that

$$\overline{V} = [aZ(j\omega)]\left[\frac{\overline{I}}{a}\right] = Z'(j\omega)\,\overline{I}'. \tag{11.2-22}$$

This is equivalent to a change of units: the unit of current is now a amperes and the unit of impedance is $1/a$ ohms. The unit of voltage V remains invariant. An important special case occurs when a is a power of 10. For instance, if $a = 10^{-3}$, then current is measured in mA and impedance in kΩ.

How does this impedance scaling affect the three passive circuit elements? Well, because a resistor has $Z(j\omega) = R$, we have $Z'(j\omega) = aR$; because an inductor has $Z(j\omega) = j\omega L$, we have $Z'(j\omega) = j\omega(aL)$. Thus all resistances and inductances are multiplied by a. The capacitor, on the other hand, has $Z(j\omega) = 1/j\omega C$, so $Z'(j\omega) = 1/j\omega(C/a)$; hence, all capacitances are divided by a. We show all of these results in Table 11.2.

Table 11.2
Impedance scaling
of *RLC*

Element	Scaled Value
Resistor	$R' = aR$
Inductor	$L' = aL$
Capacitor	$C' = C/a$

Example 11.13

Impedance scale the circuit in Figure 11.33 and find the actual value of the forced response for the current $i(t)$.

Figure 11.33
An example circuit

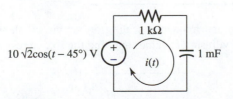

---

[5] Much of this material on scaling repeats material found elsewhere in the text in a different context.

Solution   If we impedance scale by $a = 10^{-3}$ (that is, to a current unit of 1 mA and an impedance unit of 1 kΩ), we must multiply the resistor by a and divide the capacitor by a, giving the scaled circuit shown in Figure 11.34(a). We draw the phasor equivalent in Figure 11.34(b). The phasor current $\bar{I}'$ can easily be computed to be $\bar{I}' = (10 - j10)/(1 - j) = 10\angle 0°$. Thus, the nonscaled current phasor is $\bar{I} = 10^{-3}\bar{I}' = 10\angle 0°$ mA, so $i(t) = 10 \cos(t)$ mA.

Figure 11.34
Impedance scaling

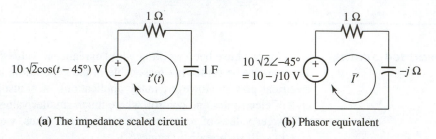

(a) The impedance scaled circuit   (b) Phasor equivalent

Frequency Scaling   In addition to impedance scaling, we can perform another type of scaling to simplify element values: frequency scaling. We agree to change the unit of ω by writing

$$\omega' = b\omega. \tag{11.2-23}$$

Now the phasor voltage and phasor current in

$$\bar{V} = Z(j\omega)\bar{I} \tag{11.2-24}$$

are constants; thus, Ohm's law changes with frequency scaling as follows:

$$\bar{V} = Z(j\omega)\bar{I} = Z\left(j\frac{\omega'}{b}\right)\bar{I}. \tag{11.2-25}$$

When $\omega' = b$ rad/s, the scaled impedance has the same value as the unscaled version at $\omega = 1$ rad/s. A resistor, whose impedance is independent of frequency, obviously does not change in value with frequency scaling; the inductor and the capacitor, however, undergo alterations. For the inductor, $Z(j\omega) = j\omega L = j\omega'(L/b)$, so we must divide the inductance by b; for the capacitor, $Z(j\omega) = 1/j\omega C = 1/(j\omega'(C/b))$, so we must also divide the capacitance by b. These results are shown in Table 11.3.

Table 11.3
Frequency scaling of
R, L, and C

Element	Scaled Value
Resistor	$R' = R$
Inductor	$L' = L/b$
Capacitor	$C' = C/b$

Example 11.14   Do impedance and frequency scaling on the circuit shown in Figure 11.35 to produce a scaled circuit with simple values and find the forced response for $i(t)$.

Figure 11.35
An example circuit

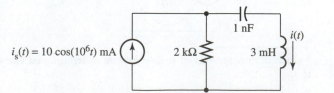

Solution   First, let's frequency scale. The argument of the current source sinusoid is 10^6 rad/s, so we will try $b = 10^{-6}$. This will scale the capacitance to $C' = 1$ nF/$10^{-6} = 1$ mF and the inductance to $L' = 3$ mH/$10^{-6} = 3000$ H. This inductance value does not look so good, but we must still impedance scale. Our frequency-scaled circuit is shown in Figure 11.36. Notice that the current source has a scaled frequency of 1 rad/s (which is equivalent to changing the unit of frequency to 10^6 rad/s), but its unit of magnitude is still the mA. We have retained the same notation for our desired current $i(t)$, though we know that its behavior has changed (in a known way). Now let's impedance scale. The resistance value is in kΩ, so we will try $a = 10^{-3}$. This gives a scaled resistance of 2 Ω, a scaled capacitance value of 1 mF/$10^{-3} = 1$ F, and a scaled inductance value of $10^{-3} \times 3000$ H $= 3$ H. The resulting circuit after both impedance and frequency scaling is shown in Figure 11.37. Notice that the unit of current is now the mA—an A in our scaled circuit. We can draw the phasor equivalent as in Figure 11.38 (the frequency now is 1 rad/s, remember). A quick application of current division gives

$$\bar{I} = \frac{2}{2 - j + j3} \times 10\angle 0° = \frac{10\angle 0°}{1 + j} = \frac{10\angle 0°}{\sqrt{2}\angle 45°} = 5\sqrt{2}\angle -45°. \qquad (11.2\text{-}26)$$

Thus, for our *scaled* circuit, $i(t) = 5\sqrt{2}\cos(t - 45°)$ A; however, our scaling has changed the unit of frequency to 10^6 rad/s and the unit of current to the mA, so the response for the *unscaled* circuit is $i(t) = 5\sqrt{2}\cos(10^6 t - 45°t)$ mA.

Figure 11.36
Circuit after frequency scaling

Figure 11.37
Circuit after frequency and impedance scaling

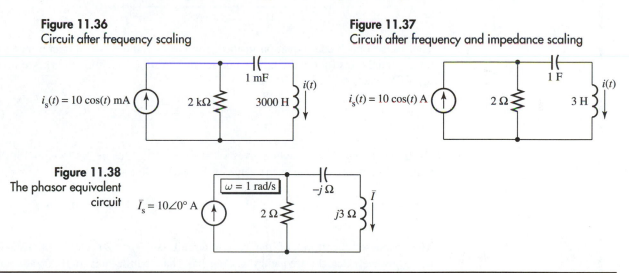

Figure 11.38
The phasor equivalent circuit

Example 11.15   Find the forced response for the voltage $v(t)$ in Figure 11.39.

Figure 11.39
An example circuit

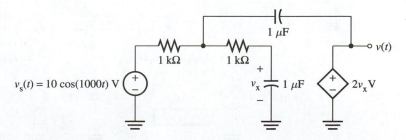

Solution   If we impedance scale by $a = 10^{-3}$, the resistors will both be reduced to 1 Ω; if we frequency scale the resulting circuit by $b = 10^{-3}$ also, we will have the equivalent circuit shown in Figure 11.40. (Note that the capacitor is divided twice by 10^{-3}, the resistor only gets multiplied once.) The input voltage source is now normalized in frequency to 1 rad/s, but recall that neither impedance nor frequency scaling alters the unit of voltage, the volt. Another issue we should address is the VCVS. Because the voltage gain is unitless, it is not modified in either scaling step; neither would a CCCS. A VCCS or CCVS would, however, become altered in the impedance scaling step because the gain units are those of conductance and resistance, respectively.

Figure 11.40
The impedance and frequency scaled equivalent

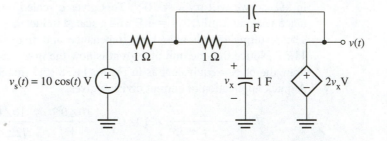

The phasor equivalent of our normalized circuit is shown in Figure 11.41. Performing the usual steps of nodal analysis (taping the VCVS, etc.) gives

$$\frac{\overline{V}' - \overline{V}_s}{1} + \frac{\overline{V}' - \overline{V}_c}{-j} + \frac{\overline{V}'}{1 - j} = 0. \qquad (11.2\text{-}27)$$

This is the only nodal equation needed, for the remaining two nonreference nodes are nonessential ones (not counting the one inside the series RC subcircuit). We now untape the VCVS and see that

$$\overline{V}_c = 2\overline{V}_x = 2\,\frac{-j}{1 - j}\,\overline{V}'. \qquad (11.2\text{-}28)$$

Using this in our nodal equation gives $\overline{V}' = (1 - j)\,\overline{V}_s$. But we are looking for the output voltage, which is the same as the value of the VCVS; thus,

$$\overline{V} = \overline{V}_c = 2\overline{V}_x = \frac{-j2}{1 - j} \times (1 - j)\,\overline{V}_s = -j2 \times 10\angle 0° = 20\angle -90°. \qquad (11.2\text{-}29)$$

The response, therefore, of the scaled circuit is $v(t) = 20\cos(t - 90°)$ V. We must note, however, that the frequency scaling has changed the unit to 1000 rad/s. Thus, the response of the original circuit is $v(t) = 20\cos(1000t - 90°)$ V. Notice that the voltages in the scaled and unscaled circuits are identical, unlike the impedances and currents.

Figure 11.41
The phasor equivalent

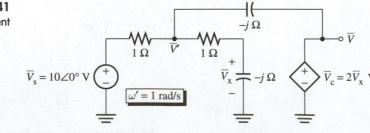

Thévenin and Norton Equivalents: Phasor Form

We have, at this point, covered a number of examples using nodal and mesh analysis. We note, however, that all of our other techniques of circuit analysis continue to hold. Specifically, the Thévenin and Norton equivalent subcircuits continue to be valid. The next example explores this aspect of ac steady-state analysis.

Example 11.16

Find the Thévenin and Norton phasor equivalents for the subcircuit shown in Figure 11.42.

Figure 11.42
An example subcircuit

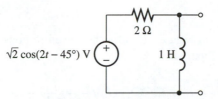

Solution

We have drawn the phasor equivalent circuit in Figure 11.43. Remember from our previous work that one option is to find two of three things: the open circuit voltage, the short circuit current, and the equivalent impedance of the deactivated subcircuit. Choosing the first and the last of these, we use the voltage divider rule to find the open circuit voltage,

$$\overline{V}_{oc} = \frac{j2}{2 + j2} \times \sqrt{2}\angle -45° = 1\angle 0° \text{ V}, \tag{11.2-30}$$

and the equivalent impedance,

$$Z = \frac{2 \times j2}{2 + j2} = \frac{2\angle 90°}{\sqrt{2}\angle 45°} = \sqrt{2}\angle 45° = 1 + j. \tag{11.2-31}$$

We can draw the Thévenin equivalent as in Figure 11.44(a), or—recognizing that *at the given frequency the j-Ω impedance is equivalent to a 0.5-H inductor*—as in Figure 11.44(b). To get the Norton equivalent, we need only find the short circuit current, which we know is the ratio of open circuit voltage to equivalent impedance:

$$\overline{I}_{sc} = \frac{\overline{V}_{oc}}{Z(j2)} = \frac{1\angle 0°}{\sqrt{2}\angle 45°} = \frac{1}{\sqrt{2}}\angle -45° \text{ A}. \tag{11.2-32}$$

The Norton equivalent is the parallel connection of a current source having this value and the equivalent impedance, as we show in Figure 11.45(a) on p. 564; in Figure 11.45(b) we show the interpretation of the equivalent impedance as a series *RL* circuit, as in the Thévenin equivalent. We remind you, however, that we can also use the equivalent parallel model for the equivalent impedance, as we show in Figure 11.45(c). This follows from $Y = 1/Z = (1/2) - j(1/2)$ S, which represents the parallel connection of a 2-Ω resistor and a *j*2-Ω inductor.

Figure 11.43
The phasor equivalent circuit

Figure 11.44
The Thévenin equivalent circuit

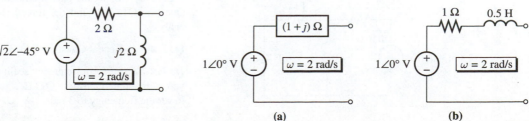

(a)   (b)

Figure 11.45
The Norton equivalent circuit

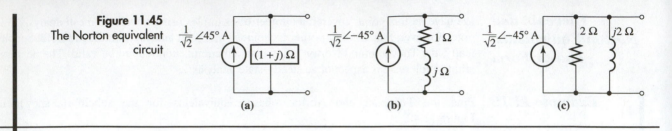

(a)   (b)   (c)

Example 11-17   Solve the circuit in Figure 11.46 for the forced response $i(t)$ using Thévenin and/or Norton equivalent subcircuits.

Figure 11.46
An example circuit

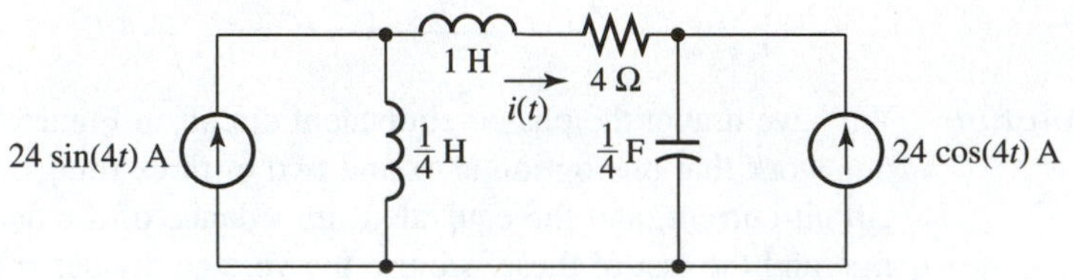

Solution   The phasor equivalent circuit is shown in Figure 11.47. We see that the two current sources have elements connected in parallel with them; hence, we can combine these and derive the Thévenin equivalent subcircuit for each. This results in the equivalent circuit shown in Figure 11.48, where we have combined the two series Thévenin impedances ($+j\,\Omega$ and $-j\,\Omega$) with the $j4$-Ω inductive impedance and the 4-Ω resistive impedance. This gives $\bar{I} = (24(1 - j))/(4(1 + j)) = 6\angle-90°$ A. Thus, $i(t) = 6\cos(4t - 90°)$ A.

Figure 11.47
The phasor equivalent

Figure 11.48
After Thévenin equivalents

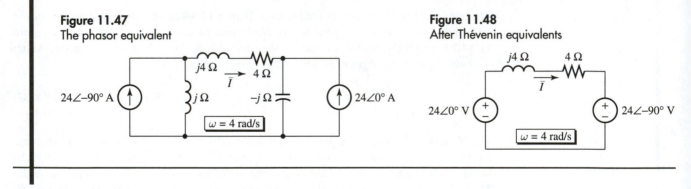

Section 11.2 Quiz

Q11.2-1. Draw the phasor equivalent circuit for the example shown in Figure Q11.2-1.

Figure Q11.2-1

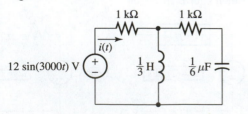

Q11.2-2. Draw the time domain equivalent circuit for the example in Figure Q11.2-1 after it has been impedance scaled by $a = 10^{-3}$ and frequency scaled by $b = 10^{-3}$.

Q11.2-3. Using the phasor domain impedance scaled circuit of Question Q11.2-2, find the forced response for $i(t)$ using nodal analysis.

Q11.2-4. Repeat Question Q11.2-3 using mesh analysis.

Q11.2-5. Perform frequency scaling on the time domain circuit you derived in Question Q11.2-2 (after impedance scaling) to lower the frequency to 3 rad/s.

11.3 | Average Power in the Sinusoidal Steady State

Averages Electrical energy is widely distributed in ac form; therefore, it is essential that we investigate the nature of ac power more deeply. We will start by recalling the notion of an average. Suppose we are interested in a waveform such as the one sketched in Figure 11.49(a) that has finite duration. We ask how it behaves "on the average." The mathematically precise definition of this average (or mean) behavior is as follows: the average of $x_T(t)$ *over the interval [0, T]* is that constant $\langle x_T(t) \rangle$ that encloses exactly the same area as $x_T(t)$. We have used the subscript T to emphasize the fact that we are restricting ourselves to a finite interval of duration T. Mathematically, area is simply the definite integral, so we can equate areas:

$$A = \int_0^T x(t)\, dt = T \langle x_T(t) \rangle. \tag{11.3-1}$$

Because $\langle x_T(t) \rangle$ is a constant, its enclosed area is that of a rectangle with height $\langle x_T(t) \rangle$ and length T. Thus, we can solve for the value of this constant:

$$\langle x_T(t) \rangle = \frac{1}{T} \int_0^T x(t)\, dt. \tag{11.3-2}$$

Figure 11.49
The average over an interval

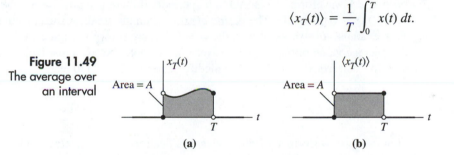

Unfortunately, things are not always this simple. Often we are confronted with a signal—often one-sided in physical applications—that extends to $t = \infty$; that is, it continues for an indefinite time. We show such a signal in Figure 11.50(a). In this case, we merely compute the average over the finite interval [0, T] and ask for its limiting behavior as the interval length becomes infinitely large. We show this in Figure 11.50(b). We truncate $x(t)$ so that it is zero outside the finite interval [0, T] and compute the average of the resulting waveform $x_T(t)$ over the finite interval [0, T], then take the limit as $T \to \infty$. Thus, the average of $x(t)$ is defined to be

$$\langle x(t) \rangle = \lim_{T \to \infty} \langle x_T(t) \rangle = \lim_{T \to \infty} \frac{1}{T} \int_0^T x(t)\, dt. \tag{11.3-3}$$

Figure 11.50
The average over an infinite interval

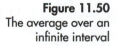
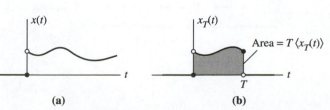

Notice carefully the distinction between "average over an interval" and "average." The latter is for an infinite time interval, even if the waveform is only nonzero for a finite amount of time. Thus, the averages of the waveforms in Figure 11.51(a) and (b) are zero because their areas are finite (for the exponential, too, because it approaches zero so rapidly as $t \to \infty$), and the division by T forces them to zero in the limit. More mathematically, we can assume that $T > 1$ for the pulse waveform in Figure 11.51(a) for we are going to let it become infinitely large anyway. We can then note that the area enclosed by $x(t)$ is unity, so $\langle x(t) \rangle = \lim_{T \to \infty} 1/T = 0$. For the exponential, we note that $\int_0^T e^{-at} dt = (1/a)[1 - e^{-aT}] \to (1/a)$ as $T \to \infty$. Therefore, after dividing by T and letting it approach infinity, we obtain an average of zero. The average of the unit step function in Figure 11.51(c) is easy, for its integral from 0 to T is simply T; dividing by T gives unity for any positive value of T. This remains true in the limit, so the average value is unity.

Figure 11.51
Three examples of averages

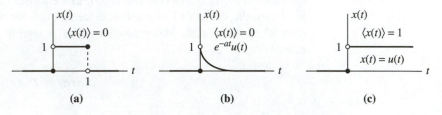

(a)   (b)   (c)

From the foregoing discussion, we see that a waveform must possess infinite area in order to have a nonzero average value. A periodic waveform is one that can have this property. A waveform $x(t)$ is said to be periodic if there is a positive number T_0 such that $x(t + T_0) = x(t)$ for all t. The *period of $x(t)$* is then the smallest such positive value of T_0. Four examples of two-sided periodic waveforms are shown in Figure 11.52. In computing the average value of a two-sided waveform, we note that we must include areas on both sides of the vertical axis. Thus, we modify equation (11.3-3) to read

$$\langle x(t) \rangle = \lim_{T \to \infty} \frac{1}{2T} \int_{-T}^{T} x(t) \, dt. \tag{11.3-4}$$

Now we can perform an important simplification for the average value of a periodic waveform. We simply let our time interval $2T$ be an integral number of periods on both sides of the vertical axis; that is, $T = 2nT_0$. Thus $\int_{-nT_0}^{nT_0} x(t) \, dt = 2n \int_0^{T_0} x(t) \, dt$ (just glance at the waveforms in Figure 11.52 and sum the areas for $2n$ periods symmetric with respect to the vertical axis) and we can write $(1/2T) \int_{-T}^{T} x(t) \, dt = (2n/2nT_0) \int_0^{T_0} x(t) \, dt = (1/T_0) \int_0^{T_0} x(t) \, dt$. Thus, the limit is

$$\langle x(t) \rangle = \frac{1}{T_0} \int_0^{T_0} x(t) \, dt. \tag{11.3-5}$$

Figure 11.52
Examples of averages of periodic waveforms

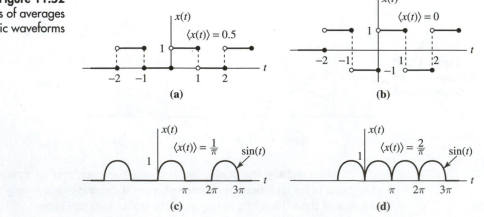

(a)   (b)

(c)   (d)

This means that we can find the area enclosed by one period of the waveform and divide by the period to find the average value. Thus, in Figure 11.52(a), the period is 2 and the area enclosed is 1; hence, the average is 0.5. For the waveform in part (b) of the figure, the period is still 2, but now the area enclosed is zero (note that the area below the horizontal axis is considered to be negative); hence, the average value is 0. We will leave the integration for the remaining two up to you, simply noting that the period of the waveform in part (c) of the figure is 2π, whereas that of the one in part (d) is π. Thus, the area enclosed during one period is the same for each, but the period of the latter is half that of the former; thus, the average value of the second is twice that of the former. The basic integration one must perform is of the sine from 0 to π in order to determine the area enclosed. The waveform in part (c) of the figure, incidentally, is called a "half-wave rectified sinusoid," and the one in part (d) a "full-wave rectified sinusoid." They are both waveforms one can observe in dc power supplies for electronic apparatus: electronic circuitry that accepts the alternating voltage from the power mains and produces constant (or dc) voltages for operating solid-state circuits.

Averages of Sinusoids and Related Waveforms

We have covered the basic ideas of averages because one must, on occasion, deal with such general waveforms; *for sinusoidal waveforms,*[6] *however, the situation becomes much simpler.* We first note that the process of "taking the average" is an operator. It accepts a waveform and produces a constant: the average value. In fact, this operator is linear, for it is easy to show using the basic properties of the integral that

$$\langle ax(t) + by(t) \rangle = a\langle x(t) \rangle + b\langle y(t) \rangle \qquad (11.3\text{-}6)$$

for any two waveforms $x(t)$ and $y(t)$ and any pair of numbers a and b. We also note that the average of a constant is the constant itself, that is,

$$\langle k \rangle = k. \qquad (11.3\text{-}7)$$

Figure 11.53   Graph of a portion of a general sinusoid

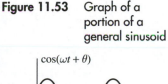

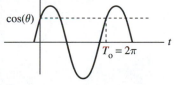

A glance at the graph of a portion of the waveform $\cos(\omega t + \theta)$ shown in Figure 11.53 convinces us that its average is zero. The area above the time axis is positive and that below, negative. Thus, the areas above and below the horizontal axis having equal magnitude, the net area over one full period is zero. Because $\sin(\omega t + \theta) = \cos(\omega t + \theta - 90°)$, we see that its average is also zero; in fact, the average of any sinusoid—with any frequency and with any phase—is zero.

Another waveform whose average we will encounter many times in this chapter is the square of the general sinusoid, one period of which is shown in Figure 11.54. We have drawn it with a phase angle of zero for convenience of representation. On a strictly intuitive basis, we would expect the average value to be 0.5 because the waveform is symmetric with respect to a horizontal line at that value. Clearly the waveform is nonnegative because it is a squared function of time.

Figure 11.54   Graph of one period of the square of a general sinusoid

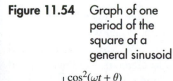

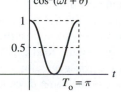

More mathematically, we will use the following trigonometric identity:

$$\cos^2(x) = \frac{1 + \cos(2x)}{2}. \qquad (11.3\text{-}8)$$

We will, in the future, also occasionally need the similar identity for the sine function,

$$\sin^2(x) = \frac{1 - \cos(2x)}{2}. \qquad (11.3\text{-}9)$$

---

[6] Note that the waveforms in Figure 11.52(c) and (d) are *not* sinusoids, but *parts* of sinusoids extended periodically.

Thus, for the general time-varying sinusoid, we have

$$\cos^2(\omega t + \theta) = \frac{1 + \cos[2(\omega t + \theta)]}{2}. \tag{11.3-10}$$

We can now apply the property of linearity, expressed above as (11.3-6) and our knowledge of the average of a constant and of a general sinusoid as follows:

$$\langle\cos^2(\omega t + \theta)\rangle = \left\langle\frac{1 + \cos[2(\omega t + \theta)]}{2}\right\rangle \tag{11.3-11}$$

$$= \left\langle\frac{1}{2}\right\rangle + \frac{1}{2}\langle\cos[2(\omega t + \theta)]\rangle = \frac{1}{2}.$$

We remind you here that $\cos[2(\omega t + \theta)] = \cos(2\omega + 2\theta)$, which is a general sinusoid with a radian frequency of 2ω and a phase of 2θ. As the identity for the square of the sine waveform is the same as that for the cosine except for the sign (see equations (11.3-8) and (11.3-9)), we see that $\langle\sin^2(\omega t + \theta)\rangle = 1/2$ also. The five results we have just derived are very fundamental; in fact, they will — in themselves — be all the mathematical average results we will require for our entire study of sinusoidal steady-state power! For this reason, we summarize them in Table 11.4.

Table 11.4
Averages of sinusoids and related waveforms

$x(t)$	$<x(t)>$
1. k	k
2. $\cos(\omega t + \theta)$	0
3. $\sin(\omega t + \theta)$	0
4. $\cos^2(\omega t + \theta)$	1/2
5. $\sin^2(\omega t + \theta)$	1/2

The Sinusoidal Average Power Absorbed by a Load

Now let's apply our ideas about averages to determine the average sinusoidal power absorbed by a general two-terminal load, sketched in Figure 11.55. When we are discussing power, we use the term "load" synonymously with "subcircuit." We assume that the load we are considering consists of any combination of the circuit elements we have studied thus far: resistors, inductors, capacitors, dependent sources, and/or independent sources. We assume, however, that each of the independent sources (if there are any) is sinusoidal at the frequency ω. The question we pose is this: What is the *average* power absorbed by the load? This quantity can be either positive or negative; if negative, the load is actually delivering power to the external world. As an example, an ac electric generator will deliver electrical power if mechanical torque is applied to its shaft in the same direction as the shaft rotation and will absorb power if the torque is in the opposite direction (it is then running as a motor).

Figure 11.55   A general two-terminal load

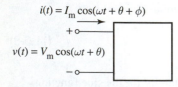

$i(t) = I_m \cos(\omega t + \theta + \phi)$

$v(t) = V_m \cos(\omega t + \theta)$

In our analysis, we will allow the voltage to have the general phase angle θ. If the entire circuit into which the load is connected, including the load itself, is stable, then after some time has passed the current will be a sinusoidal waveform — the circuit will be operating in the ac steady state. In this case the current will have some phase angle, which we have shown as $\theta + \phi$; thus, ϕ is zero when the current phase is exactly the same as that of the voltage. If $\phi > 0$, we say the current *leads* the voltage; if $\phi < 0$, we say the current *lags* the voltage. The reason for this can be seen in Figure 11.56, which shows one

Figure 11.56
Illustration of a
lagging current

$v(t) = V_m\cos(\omega t)$ $i(t) = I_m \cos(\omega t - 90°)$

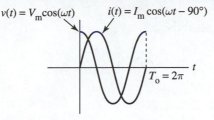

$T_0 = 2\pi$

complete period of both $v(t)$ and $i(t)$ for the case in which the current lags[7] the voltage
with a phase angle of $\phi = -90°$. (It is assumed there, for simplicity, that $\theta = 0°$.) The
crest, or peak, of the current comes after—or lags—the crest of the voltage waveform.

   We can use any analysis technique we like, the most efficient being phasor analysis,
to find the current $i(t)$ and the voltage $v(t)$ for any given set of sinusoidal waveforms of
the independent sources in the circuit. We will assume here that this analysis step has
been done.

   The *instantaneous* power absorbed by the load is

$$P(t) = v(t)i(t) = V_mI_m \cos(\omega t + \theta) \cos(\omega t + \theta + \phi). \tag{11.3-12}$$

The *average* power is the average value of the instantaneous power, or

$$P = \langle P(t) \rangle = \langle V_mI_m \cos(\omega t + \theta) \cos(\omega t + \theta + \phi) \rangle. \tag{11.3-13}$$

Letting $x = \omega t + \theta + \phi$ and $y = \omega t + \theta$, we apply the trigonometric identity

$$\cos(x) \cos(y) = \frac{1}{2} [\cos(x - y) + \cos(x + y)] \tag{11.3-14}$$

to get

$$P = \frac{1}{2}V_mI_m \langle \cos(\phi) + \cos[2(\omega t + \theta) + \phi] \rangle, \tag{11.3-15}$$

or—recognizing once again that the averaging operator is linear, that the average of the
constant $\cos(\phi)$ is simply that constant, and that the average of the time-varying cosine
term is zero—we obtain what is perhaps the most important result in the theory of ac
power:

$$P = \frac{1}{2}V_mI_m \cos(\phi). \tag{11.3-16}$$

The *constant* $\cos(\phi)$ is called the *power factor*, and is usually written *PF*; the adjective
leading or *lagging* is added to indicate the sign of ϕ, which does not affect the value
of *PF*:

$$PF = \cos(\phi), \text{ leading or lagging}. \tag{11.3-17}$$

Example 11.18   Find the power absorbed by the load in Figure 11.57 for values of the voltage source
phase angle $\alpha = 0°, 90°,$ and $-90°$.

---

[7] Note that there is the always-annoying and ever-present ambiguity of angle here. A current *lead-ing* the voltage by ϕ could also be interpreted as *lagging* the voltage by $360° - \phi$ and vice versa. It
is common, however, to always assume that ϕ is less than 180° in magnitude. Thus, in the case
shown in Figure 11.56, ϕ is interpreted to have the value $-90°$, rather than $+270°$, and *arbitrarily*
for $\phi = \pm180°$.

Figure 11.57
An example load

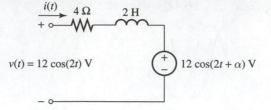

Solution   The basic problem is to find the current $i(t)$, which, to be consistent with our preceding results, must be defined consistently with the passive sign convention as shown in the figure. More specifically, we must only find I_m and ϕ, the angle by which the current leads the voltage. To do this, we convert to phasor form, as shown in Figure 11.58. Then a simple application of KVL and Ohm's law gives

$$\bar{I} = \frac{12\angle 0° - 12\angle \alpha°}{4 + j4} = \frac{12\angle 0° - 12\angle \alpha°}{4\sqrt{2}\angle 45°}. \tag{11.3-18a}$$

For $\alpha = 0°, \bar{I} = 0$; for $\alpha = 90°$,

$$\bar{I} = \frac{12\angle 0° - 12\angle 90°}{4\sqrt{2}\angle 45°} \frac{12 - j12}{4\sqrt{2}\angle 45°} = \frac{12\sqrt{2}\angle -45°}{4\sqrt{2}\angle 45°} = 3\angle -90°; \tag{11.3-18b}$$

and for $\alpha = -90°$,

$$\bar{I} = \frac{12\angle 0° - 12\angle -90°}{4\sqrt{2}\angle 45°} \frac{12 + j12}{4\sqrt{2}\angle 45°} = \frac{12\sqrt{2}\angle + 45°}{4\sqrt{2}\angle 45°} = 3\angle 0°. \tag{11.3-18c}$$

Thus, in the first case $I_m = 0$, so $P = 0$. In the second case, $I_m = 3$ A and $\phi = -90°$; hence, $P = (1/2)\, V_m I_m \cos(\phi) = (1/2) \times 12$ V $\times 3$ A $\cos(-90°) = 0$ W. Notice that the average power is zero even though neither the current nor the voltage is zero. In the last case, $I_m = 3$ A also, and $\phi = 0°$; thus, $P = (1/2) \times 12$ V $\times 3$ A $\cos(0°) = 18$ W.

Figure 11.58
Phasor form

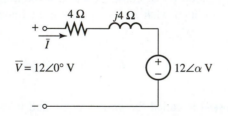

The Sinusoidal   Now let's look at the power consumed by our three basic two-terminal elements: the re-
Average Power   sistor, the capacitor, and the inductor—shown in phasor form in Figure 11.59. We
Absorbed by the   assume that $\bar{V} = V_m\angle\theta$ and that $\bar{I} = I_m\angle(\theta + \phi)$. For the resistor, $Z(j\omega) = R$, so $\phi = 0$
Passive Elements and $I_m = V_m/R$; hence, $P = 1/2\, V_m I_m \cos(0°) = V_m^2/(2R)$. Note that $\cos(0°) = 1$, so the
                      power factor is unity. For the inductor, $Z(j\omega) = j\omega L = \omega L\angle 90°$; hence, $\bar{I} = \bar{V}/Z =$

Figure 11.59
Power relations for the
basic elements

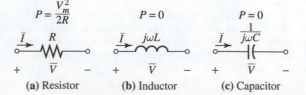

(a) Resistor   (b) Inductor   (c) Capacitor

$(V_m \angle \theta)/(\omega L \angle 90°) = (V_m/\omega L)\angle(\theta - 90°)$. Thus, $\phi = -90°$ and $PF = \cos(-90°) = 0$ and so the average power is zero. The same is true for the capacitor, except that $\phi = +90°$. We will leave this for you to show.

Parallel Resonance: Power Relations

Let's look at this a bit more closely in the time domain. Figure 11.60 shows a circuit that is simple to solve, but that will serve to illustrate the points we wish to make. We will assume that $v(t) = V_m \cos(\omega t)u(t)$ V and look at all the ac steady-state currents indicated in the circuit diagram and the power absorbed by each of the three passive elements. We will use phasor techniques to solve for the forced response. The phasor resistor current is $\bar{I}_R = (V_m \angle 0°)/R = V_m/R$, so the ac steady-state response is

$$i_R(t) = \frac{v(t)}{R} = \frac{V_m}{R} \cos(\omega t) \text{ A}. \tag{11.3-19}$$

The phasor inductor current is $\bar{I}_L = V_m/j\omega L = (V_m/\omega L)\angle -90°$, so the steady-state response is

$$i_L(t) = \frac{V_m}{\omega L} \sin(\omega t) \text{ A}. \tag{11.3-20}$$

The phasor capacitor current is $\bar{I}_c = j\omega C V_m = \omega C V_m \angle +90°$, so

$$i_C(t) = -\omega C V_m \sin(\omega t) \text{ A}. \tag{11.3-21}$$

These three currents are plotted in Figure 11.61 for one period of positive time. Though their amplitudes are different in general, there is a remarkable relationship among their phases. The resistor current is exactly in step ($\phi = 0°$) with the applied voltage, whereas the inductor and capacitor currents are 90° out of step with the voltage; furthermore, the signs of the inductor and capacitor currents are opposite of one another. This implies that the capacitor supplies some of the current for the inductor during one half-period, and that the opposite is true during the next half-period.

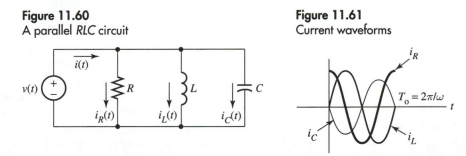

Figure 11.60
A parallel *RLC* circuit

Figure 11.61
Current waveforms

Let's now look at the *instantaneous* steady-state power absorbed by each element. For the resistor, we have

$$P_R(t) = v(t)i_R(t) = \frac{V_m^2}{R} \cos^2(\omega t)u(t) = \frac{V_m^2}{2R}[1 + \cos(2\omega t)]u(t) \text{ W}. \tag{11.3-22}$$

There is a time-varying part and a constant part. When we compute the average, we find that the average of the time-varying part is zero; thus,

$$P_R = \langle P_R(t) \rangle = \frac{V_m^2}{2R} \text{ W}. \tag{11.3-23}$$

A graph of $P_R(t)$ is shown in Figure 11.62 for one period of positive time. Notice that it is

Figure 11.62
Resistor instantaneous
power

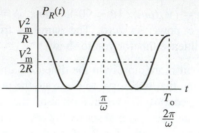

always positive—the resistor only absorbs power. Also, notice that the period of the power waveform is one-half the period of the voltage (and current).

Because the inductor and capacitor currents are so similar, let's look at the associated powers absorbed together, on the same plot. The ac steady-state inductor power is

$$P_L(t) = v(t)i_L(t) = \frac{V_m^2}{\omega L} \cos(\omega t) \sin(\omega t) = \frac{V_m^2}{2\omega L} \sin(2\omega t) \text{ W.} \qquad (11.3\text{-}24)$$

The instantaneous steady-state power absorbed by the capacitor is

$$P_C(t) = v(t)i_C(t) = -\omega C V_m^2 \cos(\omega t) \sin(\omega t) = -\frac{1}{2}\omega C V_m^2 \sin(2\omega t) \text{ W.} \qquad (11.3\text{-}25)$$

Figure 11.63   Instantaneous
power
waveforms

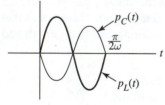

Plots of these two interesting results are shown in Figure 11.63. Power flows from the capacitor to the inductor during one half-period and conversely during the next. The two maximum values are different, in general, with the difference being supplied by the voltage source. But let's look at the special case in which the two maxima are equal; in this case, we set $(1/2)\omega C V_m^2 = V_m^2/2\omega L$. If we think of keeping the inductance and the capacitance values fixed and varying the frequency, this will occur when

$$\omega = \frac{1}{\sqrt{LC}}, \qquad (11.3\text{-}26)$$

a condition called *parallel resonance*. The adjective "parallel" is used because the circuit is a parallel one, and "resonance" refers to a similar phenomenon that occurs in the theory of acoustics—when an opera star hits a high note, for instance, and shatters a wine glass. The key thing is this: at parallel resonance, the peak energy stored by the capacitor is equal and opposite to that stored by the inductor. The inductor and capacitor swap this energy back and forth, and the voltage source must supply only the energy absorbed by the resistor. The average value of energy absorbed by each of these elements (L and C) is zero. If you use condition (11.3-26) in equations (11.3-20) and (11.3-21) you can easily show that at resonance the inductor and capacitor currents are equal and opposite and, therefore, that the current $i(t)$ from the voltage source only supplies current to the resistor.

Rms Values   Now that we have discussed the time relationships for resistor, inductor, and capacitor currents and instantaneous powers, let's consider how effective ac voltages and currents are in delivering average power to appliances like electric stoves, heaters, and refrigerators. In Figure 11.64 we show a conceptual experiment. In part (a) of the figure, we show

Figure 11.64
The concept of rms
voltage and current

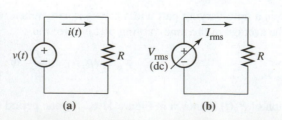

a general time-varying voltage source connected to a resistive load. In part (b) we show a dc supply connected to the same load. The arrow indicates that it is adjustable, and we will adjust it so that for some value, which we will call V_{rms}, the average power absorbed by the resistor is exactly the same as that absorbed in part(a) of the figure. Let's find the value of V_{rms}. Equating the two average powers, we get

$$P = \langle v(t)i(t) \rangle = \left\langle v(t)\frac{v(t)}{R} \right\rangle = \frac{1}{R}\langle v(t)^2 \rangle = \langle V_{rms}I_{rms} \rangle = \frac{V_{rms}^2}{R}, \qquad (11.3\text{-}27)$$

where we have used the fact that the average of the constant $V_{rms}I_{rms}$ is simply that constant itself, as well as the fact that $I_{rms} = V_{rms}/R$. Thus, solving for V_{rms}, we have

$$V_{rms} = \sqrt{\langle v(t)^2 \rangle}. \qquad (11.3\text{-}28)$$

We call this quantity the *root-mean-square* value of $v(t)$ because one must square it, then compute the mean (the average) and then extract the (square) root. We usually shorten this term to "rms." This is the reason for the subscript rms on the dc voltage; note carefully that the rms value itself is a constant, whereas $v(t)$ is a time-varying waveform. Thus, the rms value is a scalar measure of how effective the waveform is in delivering average power to a load. For this reason, rms values are often referred to as *effective* values. We will leave it to you to show that the same relationship holds for the currents $i(t)$ and I_{rms}:

$$I_{rms} = \sqrt{\langle i(t)^2 \rangle}. \qquad (11.3\text{-}29)$$

In the special case in which the voltage $v(t)$ is a sinusoid, we can write $v(t) = V_m \cos(\omega t + \phi)$. We can quickly compute $\langle v^2(t) \rangle = V_m^2/2$, so

$$V_{rms} = \frac{V_m}{\sqrt{2}}. \qquad (11.3\text{-}30)$$

Similarly,

$$I_{rms} = \frac{I_m}{\sqrt{2}}. \qquad (11.3\text{-}31)$$

We note that *these results are only for sinusoids*. A novice circuit analyst often tries to apply this result to nonsinusoidal waveforms such as those in Figure 11.65. For a general waveform, however, one must square the function, average it, and then extract the square root of the result. Squaring the waveform in Figure 11.65(a), for instance, gives exactly the same waveform because $1^2 = 1$ and $0^2 = 0$ (it only takes on these two values). Thus,

Figure 11.65
Examples of rms values of one-sided periodic waveforms

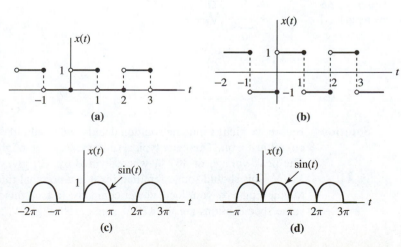

(a)

(b)

(c)

(d)

the average of the squared waveform is 1/2. Taking the square root, we obtain $X_{rms} = 1/\sqrt{2}$. Because $(-1)^2 = 1$, the square of the waveform in Figure 11.65(b) becomes the constant function of unit value, which has the average value of 1. Thus, extracting the square root gives $X_{rms} = 1$. The nonzero portion of the waveform in Figure 11.65(c) is the sine function, so we compute

$$\langle x^2(t) \rangle = \frac{1}{2\pi} \int_0^\pi \sin^2(t)\, dt = \frac{1}{2\pi} \int_0^\pi \frac{1 - \cos(2t)}{2}\, dt = \frac{1}{4}.$$

Hence, $X_{rms} = 1/2$. The period of the last waveform, in Figure 11.65(d), is one-half of the previous one; other than that, the computation is the same. Thus $X_{rms} = 1/\sqrt{2}$. We note that these waveforms are the same as those for which we earlier computed the average values.

We warn you about one point: in the expression $x(t) = X_m \cos(\omega t + \theta)$, the value X_m is *not* an rms value—it is simply a value that happens to be the *peak* (or largest) value of the sinusoid. It is *related* to the rms value by means of the equation $V_{rms} = V_m/\sqrt{2}$. The two quantities have entirely different physical meanings.

Power in Rms Terms   The rms value of a sinusoid finds wide application in power computations for a seemingly trivial reason. If we reexamine equation (11.3-16) in terms of rms values, we see that we can write

$$P = \frac{1}{2} V_m I_m \cos(\phi) = \frac{V_m}{\sqrt{2}} \frac{I_m}{\sqrt{2}} \cos(\phi) = V_{rms} I_{rms} \cos(\phi) \qquad (11.3\text{-}32)$$

$$= V_{rms} I_{rms} \text{PF}.$$

Thus, we can drop the factor of 1/2. When one makes many such computations, this can be a decided benefit. Additionally, it makes working with phasors easier. Recall that the sinusoidal voltage $v(t) = V_m \cos(\omega t + \theta)$, for example, has the phasor $\overline{V} = V_m \angle \theta°$. *We will now agree to scale all phasors by the factor of $1/\sqrt{2}$ and thus write $\overline{V} = V_{rms} \angle \theta°$.* We still have the same form of Ohm's law, $\overline{V} = Z(j\omega)\,\overline{I}$ ($\overline{V}$ and $\overline{I}$ each are divided by $\sqrt{2}$), and we can compute the power using equation (11.3-32), where $|\overline{V}| = V_{rms}$.

Example 11.19   For the two-terminal load shown in Figure 11.66, find the rms values of the terminal voltage, the resistor voltage, the capacitor voltage, and the terminal current. Also compute the power factor and the average power absorbed by the load.

Figure 11.66
An example load

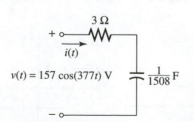

$v(t) = 157 \cos(377t)$ V

Solution   Before we plunge into the solution details, we would like to discuss the values in this subcircuit just a bit. These are typical of the values one might expect in realistic applications. The peak voltage of 157 V, when divided by $\sqrt{2}$, gives an rms value of approximately 110 V. This should sound familiar, for it is a nominal rating for the rms voltage delivered to your home by your local power company. In fact, this value is not very well controlled; some specifications are for 120 V, some are for 117 V, and still others more realistically specify a range: 110–120 Vrms. The radian frequency of 337 rad/s corresponds to $377/2\pi = 60$ Hz. This frequency is maintained very accurately by the power company, so accurately, in fact, that many electric clocks use the power line frequency as a time reference.

Now let's get to the calculations. The phasor equivalent (using rms phasors for the voltages and currents, as we have agreed to do) is shown in Figure 11.67. The rms value of the current is $\bar{I} = (110\angle 0°)/(3 - j4) = (110\angle 0°)/(5\angle -53.1°) = 22\angle 53.1°$ A (rms). The resistor voltage is, therefore, $\bar{V}_R = 66\angle 53.1°$ V (rms), and the capacitor voltage is obtained as the product: $\bar{V}_C = (-j4) \times 22\angle 53.1° = 4\angle -90° \times 22\angle 53.1° = 88\angle -36.9°$ V(rms).

Let us examine what we have here before we proceed. Why do the capacitor voltage and the resistor voltage not sum up to give the terminal voltage? After all, $66 + 88 = 154 \neq 110$! Well, we are perhaps being a bit facile, for one cannot simply add the magnitudes. We remind you that we must add the two voltages as a *phasor sum:* $66\angle 53.1° + 88\angle -36.9° = 13.2\,[3 + j4] + 17.6\,[4 - j3] = 110 + j0$. The angle of the current is 53.1°, positive, so PF = cos(53.1°) = 0.4 leading. The average power absorbed by the load is $P = V_{rms}I_{rms} \times$ PF, or $P = 110 \times 22 \times 0.4 = 968$ W.

Figure 11.67
Phasor equivalent

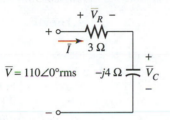

Apparent Power

Often the power absorbed by a load is more than 1000 W, in which case a more convenient unit is the *kilowatt,* kW. We will use this unit on occasion. Additionally, we define the *apparent power* to be the product $V_{rms}I_{rms}$, without the power factor multiplier. It has the same unit as that of power, but we will distinguish it from the true power by calling it the *voltampere* (VA) or *kilovoltampere* (kVA).

Passive Loads: Series and Parallel

You might have noticed that our last example considered a fairly specialized circuit, but one that is highly important. It contained no sources. Thus, it is representative of any situation in which a subcircuit consumes power, but is not capable of delivering any net power back to the source driving it. This is the case with all home appliances, for instance. Let's investigate this situation more closely and use some of the concepts we developed in the last section.

Figure 11.68(a) shows such a passive two-terminal load. As there are no independent sources inside, it can be completely described by means of its terminal—or *driving point*—impedance and/or admittance. Letting $Z = R + jX$, we obtain the series equiva-

Figure 11.68
A passive load and two equivalent subcircuits

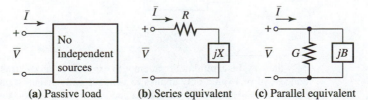

(a) Passive load   **(b)** Series equivalent   **(c)** Parallel equivalent

lent subcircuit shown in part (b) of the figure. Inverting it, we obtain the admittance $Y = G + jB$ and the parallel equivalent subcircuit shown in part (c) of the figure. We remind you that R (the resistance), X (the reactance), G (the conductance), and B (the susceptance) are all functions of ω; furthermore, it is not true that $G = 1/R$ in general. In fact, $G = R/(R^2 + X^2)$ and $B = -X/(R^2 + X^2)$.

As it happens, admittance and the parallel model are somewhat more convenient to use than impedance and the series equivalent. To take parallelism first, we point out that the power company delivers electrical power to your house via its wall receptacles, which are wired in parallel. Furthermore, to a good first approximation, the Thévenin equivalent circuit looking into any receptacle is a good voltage source (very low impedance). Thus, this voltage source and all the appliances in your house can be considered to be a parallel circuit.

Now for the admittance issue. Because the voltage at each of the receptacles in your house is connected to a voltage source, their terminal voltages are independent of whatever is "plugged in." If we plug in a heavy load, that is, one that absorbs lots of power, a large current will be "drawn" from the receptacle. If the load is light, a small current will be drawn. The point is, the voltage is fixed and the current is determined by the characteristics of the load. Thus, it is more natural to use Ohm's law in admittance form:

$$\bar{I} = Y(j\omega)\bar{V}. \tag{11.3-33}$$

There is an even more important reason for using this form. Do you recall the way we determine whether the current leads or lags the voltage? It is by the sign of the phase of the current. Now suppose that $\bar{V} = V_{rms}\angle\theta$ and $\bar{Y} = |Y(j\omega)|\angle\phi$.[8] Then we can compute the current as $\bar{I} = \bar{Y}\ \bar{V} = [|Y(j\omega)|\angle\phi][V_{rms}\angle\theta] = |Y(j\omega)|V_{rms}\angle(\theta + \phi)$. Thus, $I_{rms} = |Y(j\omega)|V_{rms}$ and *the angle by which the current leads the voltage—that is, the power factor angle—is ϕ, the angle of the admittance $Y(j\omega)$.* Let's apply this idea with an example.

Example 11-20   A passive electrical load connected to a 100-V rms ac power line draws an apparent power of 100 kVA at a power factor of 0.8 lagging. Find the rms load current and draw the parallel and the series equivalent subcircuits.

Solution   We can conceptualize the load by the box shown in Figure 11.69, where we have assumed that the angle of the voltage is zero. The apparent power is specified to be 100 kVA, therefore, the *magnitude* of the current is $I_{rms} = 100\,\text{kVA}/100\,\text{V} = 1000\,\text{A}$. The basic question is, what is the angle of Y? We have this information in the power factor: PF $= \cos(\phi) = 0.8$ lagging. Because the PF is lagging, the angle must be negative. Thus, $\phi = \cos^{-1}(0.8) = -36.9°$. Furthermore, as we have just seen, the power factor angle is the same as the angle of the admittance. Thus, $Y = \bar{I}/\bar{V} = (1000\angle\theta)/(100\angle 0°) = 10\angle -36.9°$ S. We now know the angle of the current: it is the same as the angle of the admittance, because the angle of the voltage is zero. If we express the admittance in rec-

Figure 11.69
An example load

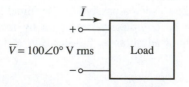

$\bar{V} = 100\angle 0°$ V rms   Load

---

[8] By the way, be careful not to attempt to express $|Y|$ as an "rms value." This is meaningless, for the concept of rms value only applies to voltages and currents—much the same as the idea of phasor only applies to sinusoidal waveforms and not to complex constants such as Z and Y.

tangular form, we have $Y = 10 \cos(-36.9°) + j10 \sin(-36.9°) = 8 - j6$ S. This corresponds to a parallel resistance of $0.125\ \Omega$ and a parallel reactance of $-0.166\ \Omega$, as shown in Figure 11.70. Inverting the Euler form of admittance gives $Z = 1/Y = 1/(10\angle -36.9°) = 0.1\angle 36.9°\ \Omega = 0.08 + j0.06\ \Omega$. The resulting series equivalent is drawn in Figure 11.71.

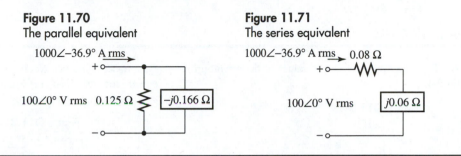

Figure 11.70
The parallel equivalent

Figure 11.71
The series equivalent

In the last example, we could have represented the reactive element in each equivalent circuit by an inductor or a capacitor of known value if we had known the frequency.

Superposition for Power   We have used superposition on quite a few occasions, both as a computational tool and to derive other theoretical results. We note here, however, that *superposition does not hold, in general, for power*. To see this, just recall that the average power dissipated in a resistor is given by

$$P = \langle v(t)i(t)\rangle = R\langle i^2(t)\rangle = G\langle v^2(t)\rangle, \qquad (11.3\text{-}34)$$

so the power is the average of the *square* of the voltage or the current, which is not a linear function of the independent source values.

Recalling superposition, let's assume that $x_1(t)$ is the partial response of a resistor variable (either voltage or current) to a given independent source alone and that $x_2(t)$ is the partial response due to another. Then the superposition principle tells us that the total response (voltage or current) is given by $x(t) = x_1(t) + x_2(t)$. This means that the average power absorbed by the given resistor is proportional to the square of x ($x = v$ or i):

$$\begin{aligned} P &= k\langle x^2(t)\rangle = k\langle[x_1(t) + x_2(t)]^2\rangle = k\langle x_1^2(t)\rangle + k\langle x_2^2(t)\rangle + 2k\langle x_1(t)x_2(t)\rangle \qquad (11.3\text{-}35) \\ &= P_1 + P_2 + 2k\langle x_1(t)x_2(t)\rangle. \end{aligned}$$

Here, P_1 is the average power due to the first source acting alone, P_2 is the average power due to the second acting alone, and k is a constant. Thus, $P \neq P_1 + P_2$ in general. If, however, $\langle x_1(t)x_2(t)\rangle = 0$, then superposition *does* hold for power—and there are a couple of important cases in which this is true. Suppose, for example, that the two sources are sinusoidal at *different frequencies*. Then we have

$$\langle x_1(t)x_2(t)\rangle = \langle X_{m1}\cos(\omega_1 t)X_{m2}\cos(\omega_2 t)\rangle \qquad (11.3\text{-}36)$$

$$= \frac{1}{2}X_{m1}X_{m2}\langle\cos[(\omega_1 - \omega_2)t] + \cos[(\omega_1 + \omega_2)t]\rangle$$

$$= \frac{1}{2}X_{m1}X_{m2}\langle\cos[(\omega_1 - \omega_2)t]\rangle + \frac{1}{2}X_{m1}X_{m2}\langle\cos[(\omega_1 + \omega_2)t]\rangle = 0,$$

because the average of each sinusoidal term is zero individually. If, however, $\omega_1 = \omega_2$, the difference term is merely $\cos(0) = 1$—whose average is unity—and the term in question is not zero. Thus, ac average powers add for sinusoidal waveforms at different frequencies.

Another special case occurs when one signal is dc, say $x_1(t) = k$ (a constant), whereas the other is ac, say $x_2(t) = X_m \cos(\omega t)$. Then we have

$$\langle x_1(t)x_2(t)\rangle = \langle k \cos(\omega t)\rangle = k\langle\cos(\omega t)\rangle = 0. \qquad (11.3\text{-}37)$$

Thus, *mixed dc and ac powers obey the superposition principle.*

Section 11.3 Quiz

Q11.3-1.   Find the average and rms values of the waveform in Figure Q11.3-1.

Figure Q11.3-1

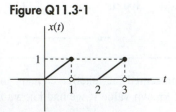

Q11.3-2.   Find the average power and the PF of the load in Figure Q11.3-2 for $\phi = -60°$.

Figure Q11.3-2

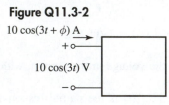

Q11.3-3.   Find the average power, power factor, and apparent power for the load in Figure Q11.3-3 at a frequency of $\omega = 2\text{rad/s}$. Also find and draw the series equivalent.

Figure Q11.3-3

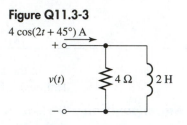

11.4 | Complex Power

Motivation for the Definition of Complex Power

We will now investigate more fully the relationship between apparent and average power. To do this, we remind you that we are dealing with a subcircuit such as the one shown in Figure 11.72, which can possibly have independent sources. We assume the current to be leading the voltage by a phase angle ϕ. The instantaneous power absorbed by the load is

$$p(t) = v(t)i(t) = V_m I_m \cos(\omega t + \theta) \cos(\omega t + \theta + \phi) \qquad (11.4\text{-}1)$$

$$= V_m I_m \frac{\cos(\phi) + \cos[2(\omega t + \theta) + \phi]}{2}$$

$$= V_{rms} I_{rms} \cos(\phi) + V_{rms} I_{rms} \cos[2(\omega t + \theta) + \phi].$$

We have used the trigonometric identity $\cos(x) \cos(y) = (\cos(x - y) + \cos(x + y))/2$. Equation (11.4-1) provides a lot of information. Let's see what kind and how much.

Figure 11.72
A general two-terminal load

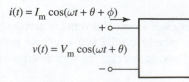

$i(t) = I_m \cos(\omega t + \theta + \phi)$

$v(t) = V_m \cos(\omega t + \theta)$

Recall that we defined the *apparent power* to be the product of (rms) voltage and current:

$$A = V_{rms}I_{rms}. \tag{11.4-2}$$

For convenience, we have given the apparent power the symbol A. We call its unit the *voltampere* (VA) to distinguish it from that of the *average power P*,

$$P = A \cos(\phi) = V_{rms}I_{rms} \cos(\phi), \tag{11.4-3}$$

which has the unit *watt* (W). We often use the more convenient units of kVA and kW for these two quantities. The *power factor* is the value

$$PF = \cos(\phi), \tag{11.4-4}$$

to which we add the adjective *leading* if $\phi > 0$ and *lagging* if $\phi < 0$. Thus, we can write (11.4-1) more compactly as

$$p(t) = P + A \cos[2(\omega t + \theta) + \phi]. \tag{11.4-5}$$

Figure 11.73 The instantaneous absorbed power

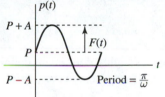

Thus, P represents the constant part of the instantaneous power, and the apparent power is the amplitude of the fluctuation, or time-varying part. Noting that $P \leq A$ (because A is multiplied by $PF = \cos(\phi)$ to get P), we see that the peak amplitude of the fluctuating part is no less than the constant part. We show the general situation in Figure 11.73. Note that the period is one-half the period of either the voltage or the current. Also notice the geometric interpretation of the time-varying component, which we have called $F(t)$.

Let's look a bit more closely at this term. If it becomes more negative than P, the subcircuit is giving some instantaneous power back to the rest of the circuit in which it is imbedded. We can simplify $F(t)$ by using the trigonometric identity $\cos(x + y) = \cos(x) \cos(y) - \sin(x) \sin(y)$:

$$F(t) = A \cos[2(\omega t + \theta) + \phi] \tag{11.4-6}$$
$$= \underbrace{A \cos(\phi)}_{P} \cos[2(\omega t + \theta)] - \underbrace{A \sin(\phi)}_{Q} \sin[2(\omega t + \theta)].$$

We have noted in this equation the term corresponding to the average power P, and we have defined a new quantity, the *reactive power Q*:

$$Q = A \sin(\phi) = V_{rms}I_{rms} \sin(\phi). \tag{11.4-7}$$

If ϕ is in the first or fourth quadrants, that is, $-\pi/2 \leq \phi \leq \pi/2$, the average power is nonnegative; however, the reactive power is negative if $\phi < 0$ and positive when $\phi > 0$. We give the reactive power the unit of *voltampere-reactive,* or VAR (or kVAR), to distinguish it from the unit of average power and the unit of apparent power. We summarize the three quantities in Table 11.5.

Table 11.5
Types of power

Parameter	Definition	Unit
1. Average power	$P = V_{rms}I_{rms} \cos(\phi)$	W, kW
2. Reactive power	$Q = V_{rms}I_{rms} \sin(\phi)$	VAR, kVAR
2. Apparent power	$A = V_{rms}I_{rms}$	VA, kVA
4. Power factor	$PF = \cos(\phi)$	——

We can write (11.4-6) for the time-varying component of absorbed power more compactly as

$$F(t) = P\cos[2(\omega t + \theta)] - Q\sin[2(\omega t + \theta)]. \qquad (11.4\text{-}8)$$

Does this suggest an even more compact way of writing $F(t)$? It will if you remember that the real part of a product of two complex numbers is the product of the two real parts *minus* the product of the two imaginary parts. Thus, we write

$$F(t) = \text{Re}\{(P + jQ)e^{j2(\omega t + \theta)}\}. \qquad (11.4\text{-}9)$$

This small equation opens a realm of new possibilities for us—it allows us to do all our power computations strictly in the complex phasor domain! Because the term $P + jQ$ appears so prominently, we will define it formally in the following section.

The Definition of Complex Power

Definition 11.2   The *complex power* is defined by

$$S = P + jQ = V_{\text{rms}}I_{\text{rms}}\cos(\phi) + jV_{\text{rms}}I_{\text{rms}}\sin(\phi). \qquad (11.4\text{-}10)$$

We will now spend some little time interpreting and using this highly convenient definition. First, we notice that (11.4-10) is the rectangular form of S; we can, however, write it in Euler form as

Figure 11.74   The complex power S

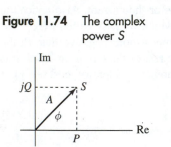

$$S = Ae^{j\phi} = V_{\text{rms}}I_{\text{rms}}e^{j\phi}. \qquad (11.4\text{-}11)$$

Figure 11.74 shows a sketch of this complex constant in the complex plane. S can be thought of as a vector whose magnitude is the apparent power A, whose polar angle is the phase angle ϕ by which the current leads the voltage, whose real part is the average power, and whose imaginary part is the reactive power. That's a lot of information in one symbol!

We can obtain a convenient expression for the complex power in terms of the terminal voltage and current phasors of the load by writing[9]

Table 11.6
Complex power relations

$$S = V_{\text{rms}}I_{\text{rms}}e^{j\phi} = V_{\text{rms}}e^{-j\theta}I_{\text{rms}}e^{j(\theta + \phi)} = \overline{V}{}^*\overline{I}. \qquad (11.4\text{-}12)$$

Thus, we can write

$$P = \text{Re}\{S\} = \text{Re}\{\overline{V}{}^*\overline{I}\} \qquad (11.4\text{-}13)$$

and

$$Q = \text{Im}\{S\} = \text{Im}\{\overline{V}{}^*\overline{I}\}. \qquad (11.4\text{-}14)$$

We summarize these important relations in Table 11.6, along with the facts that

$$A = |S| \qquad (11.4\text{-}15)$$

Quantity
1. $S = \overline{V}{}^*\,\overline{I}$
2. $P = \text{Re}\{S\}$
3. $Q = \text{Im}\{S\}$
4. $A = |S|$
5. $p(t) = P + \text{Re}\{Se^{j2(\omega t + \theta)}\}$

---

[9] Most texts use the equation $S = \overline{V}\overline{I}{}^*$ as a definition; however, this leads to a somewhat awkward situation in which S appears in the fourth quadrant (at an angle of $-\phi$) for a *leading* power factor. Thus, we have chosen the definition that leads to equation (11.4-12). There should be no confusion because the average power is the same regardless of which variable is conjugated, though the sign of Q is different. You should be aware of this difference, however, when reading other circuits texts.

and that the following compact expression (see (11.4-9)) for the instantaneous power holds:

$$p(t) = P + \text{Re}\,\{Se^{j2(\omega t + \theta)}\}. \qquad (11.4\text{-}16)$$

A Comprehensive Example

Example 11.21   For the load in Figure 11.75, find A, P, Q, S, and PF; then sketch S and $p(t)$ for the following values of ϕ: $-90°$, $-45°$, $0°$, $+45°$, and $+90°$.

Figure 11.75
An example two-terminal load

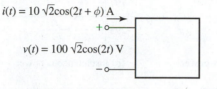

$$i(t) = 10\sqrt{2}\cos(2t + \phi)\ \text{A}$$
$$v(t) = 100\sqrt{2}\cos(2t)\ \text{V}$$

Solution   This example is something of an exercise; it will be easy and straightforward, but the results will illustrate some important points. Thus, we will work through it carefully and in detail. We show the load in Figure 11.76 with its terminal variables in phasor form. Notice that we are using rms phasors and that $\theta = 0°$.

Figure 11.76
Phasor form

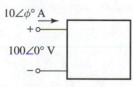

$$10\angle\phi°\ \text{A}$$
$$100\angle 0°\ \text{V}$$

Case a ($\phi = -90°$): We have $S = \overline{V}{}^*\,\overline{I} = 1000\angle{-90°} = -j1000$ VA, so $A = 1000$ VA $= 1$ kVA. Additionally, we have $PF = \cos(-90°) = 0$ (lagging), $P = A \times PF = 0$, and $Q = A\sin(\phi) = -1000$ VAR, or -1 kVAR. A sketch of S in the complex plane is shown in Figure 11.77(a).

Figure 11.77
Complex and instantaneous powers for $\phi = -90°$

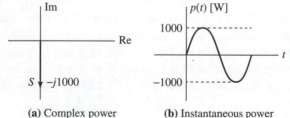

(a) Complex power   **(b)** Instantaneous power

Because $P(t) = P + \text{Re}\{Se^{j2(\omega t + \theta)}\} = 0 + \text{Re}\{-j1000e^{j4t}\} = \text{Re}\{1000e^{j(4t-90°)}\} = 1000\cos(4t - 90°) = 1000\sin(4t)$ VA, we have the sketch shown in Figure 11.77(b). Note that $P(t)$ has the unit W (watts), just as does the average power.

Case b ($\phi = -45°$): Here we have $S = (100\angle 0°)^*(10\angle{-45°}) = 1000\angle{-45°} = 500\sqrt{2} - j500\sqrt{2}$ VA. The power factor is $PF = \cos(-45°) = 0.707$ (lagging), the apparent power is given by $A = |S| = 1000$ VA (as in Case a), and the reactive power is $Q = -500\sqrt{2}$ VAR. The instantaneous power is $P(t) = 500\sqrt{2} + \text{Re}\{1000e^{-j45°}e^{j4t}\} = 500\sqrt{2} + 1000\cos(4t - 45°)$ W. Sketches of S and $P(t)$ are shown in Figure 11.78. Notice that the apparent power is the same as for Case a, but, by swinging S over to a smaller angle than 90° in absolute value, we have increased its horizontal component from 0 to $500\sqrt{2} = 707$ W. Notice, though, that the instantaneous power plot tells us that we are still giving back some energy to the driving source (or external circuit, in general)—the maximum rate of flow of this returned energy is the peak negative power of -293 W. Notice, too, that the phase angle of the instantaneous power has changed, though this is not usually important in applications.

Figure 11.78
Complex and
instantaneous powers
for $\phi = -45°$

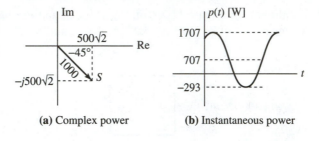

(a) Complex power   (b) Instantaneous power

Case c ($\phi = 0°$):   Here $S = (100\angle 0°)^*(10\angle 0°) = 1000\angle 0° = 1000$ VA. The power factor is $PF = \cos(0°) = 1.0$ (neither leading nor lagging), the apparent power is $|S| = 1000$ VA (as in Cases a and b), and the reactive power is $Q = 0$ VAR. The instantaneous power is $P(t) = 1000 + \text{Re}\{1000e^{-j0°}e^{j4t}\} = 1000 + 1000\cos(4t)$ W. The sketches of S and $P(t)$ are shown in Figure 11.79. Notice that no energy is delivered back to the driving source (or exterior circuit) in this case; however, the instantaneous power still fluctuates symmetrically about the average value.

Figure 11.79
Complex and
instantaneous powers
for $\phi = 0°$

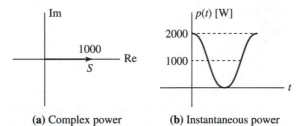

(a) Complex power   (b) Instantaneous power

Case d ($\phi = +45°$):   Here we have $S = (100\angle 0°)^*(10\angle 45°) = 1000\angle 45° = 500\sqrt{2} + j500\sqrt{2}$ VA. The power factor is $PF = \cos(45°) = 0.707$ (leading), the apparent power is $|S| = 1000$ VA (as in each other case), and the reactive power is $Q = 500\sqrt{2}$ VAR. The instantaneous power is $P(t) = 500\sqrt{2} + \text{Re}\{1000e^{j45°}\,e^{j4t}\} = 500\sqrt{2} + 1000\cos(4t + 45°)$ W. The sketches of S and $p(t)$ are shown in Figure 11.80. They are exactly the same as Case b, in which $\phi = -45°$, except for the sign of the angle of S, the sign of the reactive power Q, and the fact that the instantaneous power is shifted in phase.

Figure 11.80
Complex and
instantaneous powers
for $\phi = 45°$

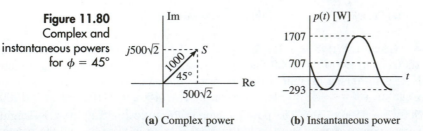

(a) Complex power   (b) Instantaneous power

Case e ($\phi = +90°$):   We have $S = \overline{V}^*\,\overline{I} = 1000\angle 90° = j1000$ VA, so $A = 1000$ VA $=$ 1 kVA. Additionally, we have $PF = \cos(90°) = 0$ leading, $P = A \times PF = 0$, and $Q = A\sin(\phi) = 1000$ VAR, or 1 kVAR. The sketch of S in the complex plane is shown in Figure 11.81(a). Because $P(t) = P + \text{Re}\{Se^{j2(\omega t+\theta)}\} = 0 + \text{Re}\{j1000e^{j4t}\} = \text{Re}\{1000e^{j(4t+90°)}\} = 1000\cos(4t + 90°) = -1000\sin(4t)$ W, we have the sketch shown in Figure 11.81(b). Notice that the only difference relative to Case a ($\phi = -90°$) is the sign of the reactive power and the signs of the various angles—including the phase shift on the instantaneous power.

Figure 11.81
Complex and
instantaneous powers
for $\phi = 90°$

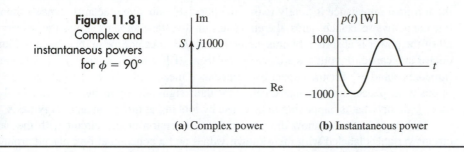

(a) Complex power   (b) Instantaneous power

**An Example:
Series Resonance**

The next example will illustrate an actual circuit, as opposed to a general load with the voltage and current specified.

Example 11.22

Find P, Q, S, and PF for the two-terminal load illustrated in Figure 11.82 for three cases:
(a) $R = 10 \; \Omega$, $C = 1/24$ F; (b) $R = 10 \; \Omega$, $C = 1/4$ F; (c) $R = 0 \; \Omega$, $C = 1/2$ F.

Figure 11.82
An example load

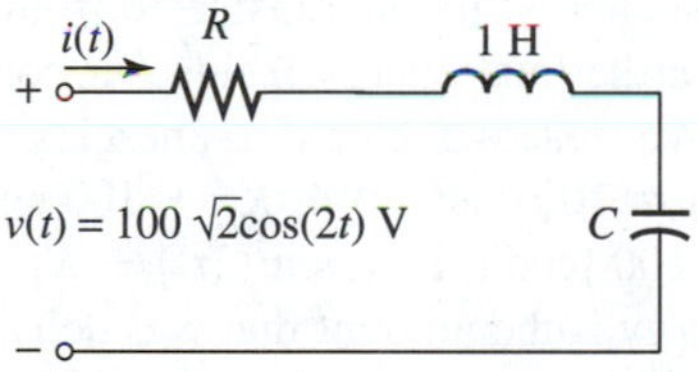

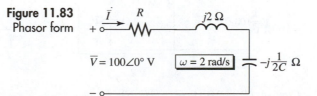

$v(t) = 100\sqrt{2}\cos(2t)$ V

Solution

The phasor form of the subcircuit is given in Figure 11.83. Because it is a series subcircuit, it is easy to first find the impedance.

$$Z = R + j\left[2 - \frac{1}{2C}\right].$$

We now proceed on a case-by-case basis.

Figure 11.83
Phasor form

$\bar{I}$ R $j2 \; \Omega$

$\bar{V} = 100\angle0°$ V $\boxed{\omega = 2 \text{ rad/s}}$ $-j\frac{1}{2C} \; \Omega$

Figure 11.84
The complex power for
$C = \dfrac{1}{24}$ F

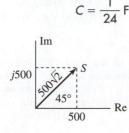

Case a ($R = 10 \; \Omega$, $C = 1/24$ F): In this case, $Z = 10 - j10 = 10\sqrt{2}\angle-45°$, so the admittance is $Y = (1/(10\sqrt{2}))\angle45°$. The current, then, is $\bar{I} = Y\bar{V} = 5\sqrt{2}\angle45°$ A. (Remember that by our earlier convention we converted the peak voltage to rms when we drew the phasor equivalent circuit in Figure 11.83!) Thus, $S = \bar{V}*\bar{I} = 500\sqrt{2}\angle45°$ VA. In rectangular form this becomes $S = 500 + j500$ VA. We show the complex plane sketch in Figure 11.84. The apparent power is $A = |S| = 500\sqrt{2}$ VA and the power factor is $PF = \cos(45°) = 0.707$ (leading). We also have $P = 500$ W and $Q = 500$ VAR.

Figure 11.85
The complex power

for $C = \dfrac{1}{4}$ F

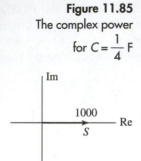

Case b ($R = 10\ \Omega$, $C = 1/4$ F): In this case, $Z = 10 + j0 = 10\angle 0°$, so the admittance is $Y = 0.1\angle 0°$. The current, then, is $\bar{I} = Y\bar{V} = 10\angle 0°$ A. Thus, $S = 1000\angle 0°$ VA. In rectangular form, this becomes $S = 1000 + j0$ VA. We show the complex plane sketch in Figure 11.85. The apparent power is $A = |S| = 1000$ VA and the power factor is $PF = \cos(0°) = 1.0$. We also have $P = 1$ kW and $Q = 0$ VAR.

Let's look a bit more at the power and energy relationships for this case. Notice that the terminal impedance is purely real—the positive inductive reactance cancels the negative capacitive reactance. Thus, the entire terminal voltage appears across the resistor, and all of the power is absorbed by this element. But this does not mean that the individual inductor and capacitor voltages are zero—far from it! The presence of $+j$ for the inductive impedance and $-j$ for the capacitive impedance means that their two voltages are 180° opposite in phase, but this just means that their signs are opposite. We will shed a bit more light on what is happening in this case by looking at things on an energy basis.

In Figure 11.86 we show the phasor domain version of our circuit with the various circuit variables labeled with their known values and a conceptual box placed around the subcircuit consisting of the inductor and the capacitor. We see that $\bar{V}_L = 20\angle 90°$ and $\bar{V}_C = 20\angle -90°$ (rms, remember!). Thus, $v_L(t) = 20\sqrt{2}\cos(2t + 90°) = -20\sqrt{2}\sin(2t)$ V and $v_C(t) = +20\sqrt{2}\sin(2t)$ V. The inductor current is $i(t) = 10\sqrt{2}\cos(2t)$ A. Recalling that the power absorbed is $P(t) = dw/dt = d/dt\,[v(t)i(t)]$, we integrate[10] to find the energy stored on the inductor: $w_L(t) = (1/2)\,Li^2(t) + K_L = 100\cos^2(2t) + K_L$ J. Similarly, $w_C(t) = (1/2)\,C\,v_C^2(t) + K_C = 100\sin^2(2t) + K_C$ J. These two waveforms are plotted in Figure 11.87 (for $K_L = K_C = 0$). Note that they are equal in magnitude, but different in phase by 180°; thus, the energy first is transferred from the inductor to the capacitor, then vice versa. We note that the total energy inside our conceptual box in Figure 11.86 must be the sum of the two reactive element energies. Hence, we have the following: $w_{\text{tot}}(t) = w_L(t) + w_C(t) = 100\cos^2(2t) + K_L + 100\sin^2(2t) + K_C$. Combining the sinusoidal terms, $w_{\text{tot}}(t) = 100\,[\cos^2(2t) + \sin^2(2t)] + K_L + K_C = 100J + K_L + K_C$, a constant. This value of energy is the amount that was delivered to the box before steady-state conditions were reached. This case illustrates the idea of *series resonance,* and the mathematics is quite similar to that for *parallel resonance* studied in Section 11.3.

Figure 11.86

Phasor form $C = \dfrac{1}{4}$ F

Figure 11.87

Energy relations for the reactive elements

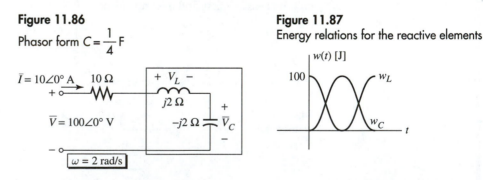

Case c ($R = 10\ \Omega$, $C = 1/2$ F): In this case, $Z = j = 1\angle 90°$, so the admittance is $Y = 1/j = 1\angle -90°$. The current, then, is $\bar{I} = Y\bar{V} = 100\angle -90°$ A. Thus, $S = \bar{V}*\bar{I} = 10{,}000\angle -90°$ VA. In rectangular form, this becomes $S = -j10{,}000$ VA. We show the

---

[10] We are dealing with two-sided waveforms in this chapter, so we must include an arbitrary constant of integration. For a more thorough investigation that would include the value of this constant, we would assume a sinusoidal input that was "turned on" at $t = 0$, compute the energy as a function of time, then let time become large. This would permit determination of the constant of integration.

Figure 11.88
The complex power for
$R = 0\,\Omega$ and $C = \dfrac{1}{2}$ F

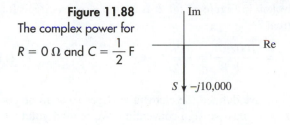

complex plane sketch in Figure 11.88. The apparent power is $A = |S| = 10$ kVA, and the power factor is PF $= \cos(-90°) = 0$ (lagging). We also clearly have $P = 0$ W and $Q = -10$ kVAR.

Conservation of Complex Power

Now let us use Tellegen's theorem, which is proved in Appendix B. It states that if $v_1, v_2, \ldots v_N$ are branch currents for a given circuit with N elements and if $i_1, i_2, \ldots i_N$ are branch (element) voltages of the same circuit, or of any other circuit with the same topology (interconnections), then

$$\sum_{k=1}^{N} v_k(t) i_k(t) = 0. \tag{11.4-17}$$

The only assumptions upon which this result is based are Kirchhoff's laws; we only assume in the derivation that the voltages obey KVL and the currents obey KCL. We know that KVL and KCL continue to hold for the phasor equivalent; furthermore, KVL and KCL are invariant under conjugation (e.g., if the sum of phasor voltages is zero, then so is the sum of their conjugates). This implies that

$$\sum_{k=0}^{N} S_k = \sum_{k=0}^{N} \overline{V}_k^* \overline{I}_k = 0. \tag{11.4-18}$$

This extremely important result is known as *conservation of complex power*. It says that the sum of the complex powers absorbed by all of the elements in a circuit is zero. Taking the real part, we see that it also implies

$$\sum_{k=0}^{N} P_k = 0; \tag{11.4-19}$$

that is, the average power, too, is conserved[11]—a result that, by averaging, also follows from the time domain version of Tellegen's theorem.

Example 11.23

Compute the complex power absorbed by each element of the circuit in Figure 11.89 and demonstrate directly that the total is zero.

Figure 11.89
An example circuit

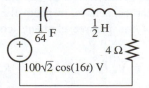

$\frac{1}{64}$ F $\frac{1}{2}$ H

$4\,\Omega$

$100\sqrt{2} \cos(16t)$ V

---

[11] If we take the imaginary part of (11.4-18), we see that the reactive power is conserved as well.

Solution   We first draw the phasor equivalent in Figure 11.90. It is a simple series circuit for which we readily find the phasor current $\bar{I}$:

$$\bar{I} = \frac{100\angle 0°}{4 + j4} = \frac{25}{\sqrt{2}}\angle -45°. \tag{11.4-20}$$

As there is only one current, we have defined the element voltages so as to be consistent with the direction of $\bar{I}$ relative to the passive sign convention. We remind you that we are using rms phasors, so we have multiplied the peak value of the voltage source sinusoid by $1/\sqrt{2}$. The element voltages are $\bar{V}_C = 50\sqrt{2}\angle -135°$ V, $\bar{V}_L = 100\sqrt{2}\angle 45°$ V, $\bar{V}_R = 50\sqrt{2}\angle -45°$ V, and $\bar{V}_S = 100\angle 0°$ V. We now compute the complex power absorbed by each of the elements, $S_n = \bar{V}_n^* \bar{I}_n$ (here we will index the elements by C, L, R, and s for "source"). The complex power *absorbed* by the source is

$$S_s = 100\angle 0° \times \left[-\frac{25}{\sqrt{2}}\angle -45° \right] = 100\angle 0° \times \left[\frac{25}{\sqrt{2}}\angle 135° \right] \tag{11.4-21}$$

$$= 1250\sqrt{2}\angle 135° = -1250 + j1250 \text{ VA}.$$

To find the complex power absorbed by each of the other elements, we first conjugate the voltage and then multiply it by the current. Thus, $S_C = \bar{V}_C^* \bar{I} = 50\sqrt{2}\angle 135° \times (25/(\sqrt{2}\angle -45°)) = 1250\angle 90° = j1250$ VA. Similarly, we have $S_L = 100\sqrt{2}\angle -45° \times (25/\sqrt{2})\angle -45° = 2500\angle -90° = -j2500$ VA. The complex power absorbed by the resistor is $S_R = 50\sqrt{2}\angle 45° \times 25/\sqrt{2}\angle -45° = 1250\angle 0° = 1250$ VA. Summing all of the complex powers gives the complex number $0 + j0 = 0$.

Figure 11.90
The phasor equivalent

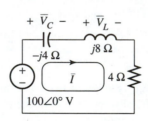

Where the Real and Reactive Powers Go in RLC Circuits   Figure 11.91 shows an *RLC* circuit in which we have conceptually "pulled out" all of the elements from a box representing their interconnections. We assume for simplicity that there is a single sinusoidal source. Then conservation of complex power says that[12]

$$S_{\text{source}} = \sum_{n=1}^{N} S_{R_n} + \sum_{n=1}^{M} S_{L_n} + \sum_{n=1}^{K} S_{C_n}, \tag{11.4-22}$$

where S_{source} is the complex power *delivered* by the source and N, M, and K are the numbers of resistors, inductors, and capacitors, respectively. Letting phasor voltages and currents be defined for each of the passive elements consistently with the passive sign convention, we find that:

$$S_{R_n} = \bar{V}_n^* \bar{I}_n = [R_n \bar{I}_n^*] \bar{I}_n = R_n |\bar{I}_n|^2, \tag{11.4-23}$$

$$S_{L_n} = \bar{V}_n^* \bar{I}_n = [-j\omega L_n \bar{I}_n^*] \bar{I}_n = -j\omega L_n |\bar{I}_n|^2, \tag{11.4-24}$$

---

[12] Notice that the complex power *absorbed* by the source is $\bar{V}^*[-\bar{I}] = -S_{\text{source}}$. Moving it to the left side results in equation (11.4-22). The generalization of our analysis to include more than one independent source is straightforward.

Figure 11.91
An *RLC* circuit in the ac
steady state

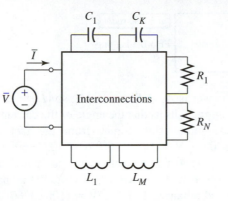

and
$$S_{C_n} = \overline{V}_n^* \overline{I}_n = \overline{V}_n^* [j\omega C_n \overline{V}_n] = j\omega C_n |\overline{V}_n|^2. \qquad (11.4\text{-}25)$$

Thus, summing over all the resistors, inductors, and capacitors individually, we find that

$$S_{\text{source}} = P + jQ, \qquad (11.4\text{-}26)$$

where
$$P = \sum_{n=1}^{N} S_{R_n} = \sum_{n=1}^{N} P_{R_n} = \sum_{n=1}^{N} R_n |\overline{I}_n|^2 \qquad (11.4\text{-}27)$$

and
$$Q = j\omega \left[\sum_{n=1}^{N} C_n |\overline{V}_n|^2 - \sum_{n=1}^{M} L_n |\overline{I}_n|^2 \right]. \qquad (11.4\text{-}28)$$

Because *S*, in addition to being the complex power generated by the source, is the total complex power absorbed by the rest of the circuit, we see that *the average power P (the real part of the total absorbed complex power) is the average power absorbed by all the resistors; similarly, Q (the imaginary part of the total absorbed complex power) is the total reactive power absorbed by the dynamic elements—the capacitors and the resistors. If there are no inductors or capacitors, the reactive power absorbed is zero; if there are no resistors, there is no real average power absorbed. Thus, we now know where the real and imaginary components of complex power "go." The real power goes to the resistors, and the reactive power goes to the reactive elements. If there are only resistors, we call the circuit "resistive"; if there are none (that is, there are only inductors and/or capacitors), we call it "reactive" or "lossless."*

A Passivity Condition on Impedance and Admittance

Our preceding observations place a restriction on the driving point admittance "seen" by the independent source. If all the resistors are positive, then Re{S_{source}} is positive also. If we plot S_{source} in the complex plane, as in Figure 11.92, we see that its polar angle ϕ must lie between $-\pi/2$ and $+\pi/2$ radians—or between $-90°$ and $+90°$. But recall that the complex power is given by

$$S_{\text{source}} = \overline{V}^* \overline{I} = Y(j\omega) |\overline{V}|^2. \qquad (11.4\text{-}29)$$

Figure 11.92 The complex power delivered by the source

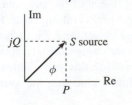

Thus, the angle of the admittance is the same as the angle of the complex power; hence, the angle of the admittance is constrained to lie between $\pm 90°$ also. Finally, the angle of the impedance $Z(j\omega)$ is the negative of the angle of the admittance because the impedance is the reciprocal of the admittance, so the angle of the impedance is also constrained to lie between $\pm 90°$. This means that Re{Y} and Re{Z} are both nonnegative. In this case, because power cannot be supplied by the subcircuit, we say that it is "passive."

Example 11.24

Figure 11.93 shows a two-terminal passive load that draws an apparent power of 10 kVA at a power factor of 0.5 lagging from a 100-Vrms line whose frequency is 100 rad/s. Find the average power absorbed and draw a parallel equivalent circuit for the load.

Figure 11.93
An example load

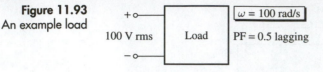

Solution   The apparent power is $A = |S| = |\overline{V}||\overline{I}|$, so $|\overline{I}| = 10$ kVA/100 V $= 100$ A. The power factor of 0.5 lagging tells us that the angle of the current relative to the voltage is given by $\phi = -60°$ (the minus sign comes from the fact that the PF is lagging). Thus, $S = \overline{V}^*\overline{I} = 100\angle0°$ V $\times 100\angle-60°$ A $= 10\angle-60°$ kVA $= 5-j5\sqrt{3}$ kVA. We show a sketch in Figure 11.94. We see that the average power absorbed by the load is 5 kW, and the reactive power is $-5\sqrt{3}$ kVAR. To draw the parallel equivalent subcircuit, we compute the admittance Y: $Y = \overline{I}/\overline{V} = (100\angle-60°)/(100\angle0°) = 1\angle-60° = (1/2) - j(\sqrt{3}/2)$ S (note that the S here means siemens, the unit of admittance, the reciprocal of ohms). Because admittances add for parallel circuits, we have the parallel equivalent circuit shown in Figure 11.95.

Figure 11.94
The complex power

Figure 11.95
The parallel equivalent circuit

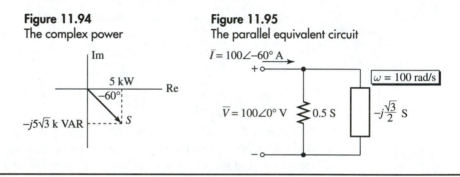

Power Factor Correction   The angle of the load in the last example is rather large. This implies that there is a sizable reactive component of power. What does the presence of this reactive component mean? Simply this. The power company installs a meter at each drop[13] that measures the energy used by that customer during a given time period. Now energy is proportional to the average power P—the reactive power Q does not contribute anything to energy usage, but it must be supplied and then received by the power company. This means that the apparent power can be much larger than the real average power, and—for a given voltage—larger apparent power means larger current. Thus, the power company must install large conductors, transformers, and other equipment to handle this large current, all without being rewarded by the customer. Therefore, it penalizes customers whose power factors are low by an increased rate structure. Thus, it is to the user's benefit to maintain a power factor close to unity.

Let's think about how we can increase the power factor in our last example. By looking at the parallel equivalent in Figure 11.95, we see that the 0.5-S resistor (a 2-Ω resistance) is directly across the line. Therefore, its power is independent of the susceptance. We have already seen this in our discussion of complex power conservation. In an *RLC* circuit, the resistors alone are responsible for the real average power, and the inductors and capacitors for the reactive component. Hmmm! Suppose we change the susceptance without modifying the resistance in our parallel circuit. We could in this way change the angle ϕ of the admittance—and thus the power factor—without changing the average power.

---

[13] The electrical lines going into your house from the "mains" (the transmission cables) are called a "drop" because they allow power to be "dropped" from the mains.

How can we do this? Well, we see that the susceptance of our load is negative; thus, the reactance is positive and so the load is "inductive." Does this suggest anything? A capacitance has positive susceptance (negative reactance); that is, its susceptance is opposite in sign to that of the load! Furthermore, the complex power absorbed by a capacitor is purely imaginary, so an added capacitor would consume no real power. How do we connect it? The only place we can logically place it is across the terminals of the load. (Remember that our parallel subcircuit in Figure 11.95 is only an *equivalent* circuit, so we will assume that only the terminals are accessible.) This leads to the *power factor correction* scheme shown in Figure 11.96.

Figure 11.96
Power factor correction

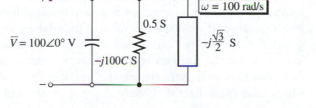

Recalling that the admittance of the capacitor is $j\omega C = j100C$ S, we can compute the modified load admittance: $Y' = j100C + 0.5 - j\sqrt{3}/2 = 0.5 + j[100C - (\sqrt{3}/2)]$ S. Now what value of capacitance is required to adjust *PF* to unity? Clearly, it is the value that makes the imaginary part of the admittance equal to zero. Thus, $C = \sqrt{3}/200 = 8.66$ mF. We will leave it to you to show that the overall load current is now $50\angle 0°$ A, half of the original.

As it happens, larger-valued capacitors are more expensive than their smaller-valued counterparts, so one does not always correct the power factor to unity; however, one can specify the power factor, compute the required angle, and from there compute the required value of capacitance to achieve that specified value of power factor.

Maximum Power Transfer: Conjugate Matching

Our final topic in this section is maximum power transfer. Recall that we addressed this question earlier in our discussion of Thévenin and Norton equivalent circuits for the case in which the given subcircuit was purely resistive. We now generalize.

Thus, consider the circuit shown in Figure 11.97(a). We suppose that there is a load impedance, specified by Z, to which we are to deliver power. The box represents the remainder of the circuit, which we assume to be operating in the ac steady state. In Figure 11.97(b) we show our circuit with the box replaced by its Thévenin equivalent. The load voltage is $\overline{V} = Z/(Z + Z_{eq}) \overline{V}_{oc}$, and the current is $\overline{I} = \overline{V}_{oc}/(Z + Z_{eq})$. Thus, the complex power absorbed by the load is $S = \overline{V}^* \overline{I} = (Z^*/|Z + Z_{eq}|^2)|\overline{V}_{oc}|^2$. If we take the real part, we get the average power P:

$$P = \text{Re}\{S\} = \frac{R}{|Z + Z_{eq}|^2}|\overline{V}_{oc}|^2 = \frac{R}{(R + R_{eq})^2 + (X + X_{eq})^2}|\overline{V}_{oc}|^2, \qquad (11.4\text{-}30)$$

where we have written $Z = R + jX$ and $Z_{eq} = R_{eq} + jX_{eq}$.

Figure 11.97
Maximum power transfer

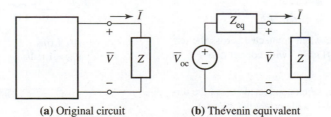

(a) Original circuit   (b) Thévenin equivalent

We can now formulate two different practical problems. One is to vary the load impedance while keeping the Thévenin equivalent fixed and determine the value of Z that results in maximum absorbed power in the load; the other is to keep the load fixed and vary the Thévenin equivalent impedance (while maintaining constant Thévenin equivalent voltage) and ask the question: What value of Z_{eq} results in maximum dissipated power in the load? In either case, we see upon examining equation (11.4-30) that we should make $X = -X_{eq}$. In this case, we have

$$P = \frac{R}{(R + R_{eq})^2}\, |\overline{V}_{oc}|^2. \tag{11.4-31}$$

From this point on, our solution depends upon which question we ask. Taking the second question first, suppose we first keep the load fixed and vary the Thévenin equivalent impedance. If we allow *any* value of impedance, we should clearly make $R = -R_{eq}$, giving $P = \infty$! This is clearly somewhat absurd (at least for passive circuits) because we have seen that the real part of the resistance is never negative for *RLC* circuits (that is, those that have only these elements). Thus, if we vary R_{eq} over only nonnegative values, which value maximizes P? Clearly, $R_{eq} = 0$. Hence, our answer to the second question is

$$Z_{eq} = -jX; \tag{11.4-32}$$

that is, we "tune out" the load reactance with a reactance of the opposite sign.

To answer our first question, we must work a bit harder. If we vary R through all non-negative values, what value maximizes the expression in (11.4-31)? To find this value, let's simplify the equation by dividing both top and bottom by R. This gives

$$P = \frac{|\overline{V}_{oc}|^2}{\left(\sqrt{R} + \dfrac{R_{eq}}{\sqrt{R}}\right)^2}. \tag{11.4-33}$$

Here is our approach. We simply note that *minimizing* $\sqrt{R} + R_{eq}/\sqrt{R}$ maximizes P. Let's define $x = \sqrt{R}$ and minimize $f(x) = x + R_{eq}/x$. This function has a minimum at $x = \sqrt{R_{eq}}$, which you can verify by setting the first derivative to zero. Computing the second derivative and evaluating it at $x = \sqrt{R_{eq}}$ gives a positive result, thus showing that our value is, indeed, a relative minimum. Thus, we must set $\sqrt{R} = \sqrt{R_{eq}}$, which implies that we must make $R = R_{eq}$. Therefore, we must make

$$Z_{eq} = R - jX = Z^*. \tag{11.4-34}$$

This is referred to as the *conjugate matched,* or simply *matched,* condition.

Section 11.4 Quiz

Q11.4-1. For the subcircuit shown in Figure Q11.4-1, find the following:

    a. Y *b.* I_{rms} *c.* S *d.* PF.

Q11.4-2. Find an element that, when connected across the terminals of the load in Figure Q11.4-1, will correct the *PF* to $0.866 = \sqrt{3}/2$ lagging.

Q11.4-3. Find an element that, when connected across the terminals of the load in Figure Q11.4-1, will correct the *PF* to 0.

Figure Q11.4-1

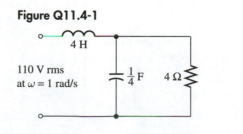

110 V rms
at $\omega = 1$ rad/s

4 H $\frac{1}{4}$ F   4 Ω

Q11.4-4. Find the power absorbed by the resistor in Figure Q11.4-1 and the peak energies stored in the inductor and the capacitor.

Q11.4-5. If the subcircuit in Figure Q11.4-1 is a load connected to the source shown in Figure Q11.4-2, find the value of X_a for maximum power delivered to the load. What is that maximum power?

Figure Q11.4-2

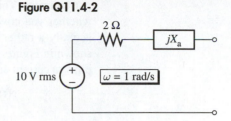

11.5 | SPICE Analysis of Circuits in the Ac Steady State

SPICE is quite convenient to use in analyzing circuits in the ac steady state by simulation. We will discuss here how to set up a circuit file for doing such an analysis. First, however, let's use the capabilities of SPICE that we have already discussed to illustrate the approach of a circuit to ac steady-state conditions. Figure 11.98 offers a nice, simple example for this. More specifically, we will compute the response $v(t)$ to an input $i_s(t)$ that consists of a sinusoid at a frequency of 2 rad/s which is "turned on" at $t = 0$—and we will do this for $k = -1$.

Figure 11.98
An example circuit

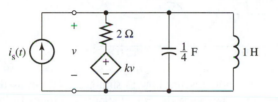

The circuit is shown prepared for SPICE simulation in Figure 11.99. The SPICE file itself is as follows (for $k = -1$):

```
FIRST EXAMPLE
IS  0  1  SIN(0, 1, {2/6.2832}, 0, 0)  ; HERTZ FREQUENCY!
ES  2  0  1  0    -1
R   1  2  2
C   1  0  0.25
L   1  0  1
.PROBE
.TRAN  0.01  10  0  0.01
.END
```

The time response obtained from SPICE (using Probe) is shown in Figure 11.100. Notice that the ac steady-state condition has been reached by the end of the first period of the in-

Figure 11.99
The example circuit prepared for SPICE analysis

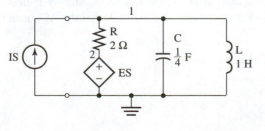

Figure 11.100
Response for $k = -1$

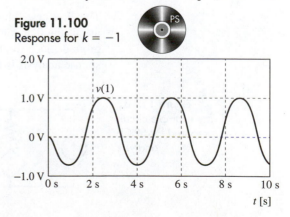

put sinusoid. The only possibly new aspect of our SPICE NETLIST (depending upon whether you covered the material on time domain analysis) is the SIN function. This is actually a rather general waveform available in SPICE that simulates the time function shown in Figure 11.101. (We have drawn the figure for a voltage, but everything there applies to current as well.) It simulates the waveform

$$v(t) = \text{VO} + \text{VA}e^{-\alpha(t-\text{TD})}\sin[2\pi\text{FREQ}(t - \text{TD}) + \text{PHASE}]; \qquad (11.5\text{-}1)$$

for TD $\leq t \leq$ TSTOP.

The SPICE syntax (for a voltage source named VG) is

$$\text{VG} \quad \text{N+} \quad \text{N-} \quad \text{SIN(VO, VA, FREQ, TD, ALPHA, PHASE)} \qquad (11.5\text{-}2)$$

where N+ and N− are, as usual, the positive and negative nodes (respectively) of the voltage source, SIN specifies this particular waveform, FREQ is the sinusoidal frequency in hertz, and the other parameters are shown in Figure 11.101 and/or equation (11.5-1).

Figure 11.101
General damped
sinusoidal waveform

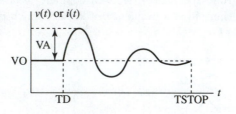

We will now discuss how to simulate a circuit assumed to be in the ac steady state. We will once again use the circuit shown in Figure 11.98 and the SPICE prepared circuit in Figure 11.99. We show the latter again here for ease of reference as Figure 11.102. This time, we would like to find all of the circuit voltages and currents for the ac forced response. Because the topology is the same as before, so is the NETLIST. This time, however, we change our control statements slightly:

```
SECOND EXAMPLE
IS 0  1   AC 1  0   ; NOTE THAT PHASE HAS UNITS OF DEGREES
ES 2  0   1 0 -1
R  1  2   2
C  1  0   0.25
L  1  0   1
.AC LIN  1   0.3183   0.3183   ;FREQUENCY IN HERTZ!
.PRINT  AC  VM(1)    VP(1)  VM(2)   VP(2)   IM(R)   IP(R)
+IM(C) IP(C) IM(L)   IP(L)
.END
```

The first new thing in this listing is the AC specification for the current source IS; as usual, the 0 and 1 right after IS specify that the current source is drawing current from ground (node 0) and pushing it into node 1. The AC simply means that an ac forced response analysis is to be done. Finally, the 1 and 0 after the AC mean that the amplitude of

Figure 11.102
The example circuit
prepared for SPICE
analysis

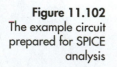
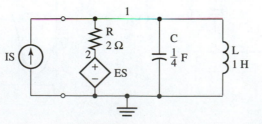

the current source is 1 ampere and that the phase is zero degrees—in that order. There is no need to specify the frequency, for this is done by the .AC control statement. This has the general form

```
.AC ATYPE NP FSTART FSTOP
```

The parameter ATYPE specifies the type of "frequency sweep": LIN, OCT, or DEC. LIN means that the number of frequency points will be equally spaced, OCT means that NP simulations will be done over each frequency range of 2:1 between FSTART and FSTOP, and DEC means that each such frequency range will have a 10:1 spread. This distinction is not important for single frequency analysis, but will be used later in the next chapter on frequency response. In our example, we specified NP = 1 for a single frequency analysis. FSTART is the first frequency at which analysis will occur—SPICE does an ac analysis at a frequency FSTART, then increments the frequency to another value depending upon ATYPE. If the new frequency is greater than FSTOP, no more analyses are performed. If it is less than FSTOP, another analysis is done. It is important to note that FSTART and FSTOP must be in hertz frequency, that is in radian frequency divided by 2π. Thus, our preceding analysis specified a frequency of 0.3183, which is the frequency in hertz corresponding to a radian frequency of 2 rad/s. We gave both FSTART and FSTOP this value, which resulted in analysis being performed at this one frequency. The .AC control statement sets each of the independent sinusoidal sources in the circuit to the proper frequency, but uses the amplitude and phase specified for each in the NETLIST.

We used the .PRINT statement to output our circuit variables. This has been discussed before in connection with the .DC statement. However, the way we specify ac variables is new. The symbol VM means "voltage magnitude," and VP means "voltage phase," and similarly for currents. Otherwise, the output variables are specified as always. If you want the phasor representing any voltage or current in rectangular form, you should use VR for the real part of the voltage V and VI for its imaginary part. As a matter of fact, if you simply specify (for example) V(1), SPICE will automatically output V(1) in rectangular form; that is, it will give you VR(1) and VI(1).

Finally, we would like to mention the small + sign in the statement just before the .END. It is a "continuation" statement. If a given statement is too long to fit on one line, you merely start the next line with a + in the first column, then continue with the rest of the statement.

The output from a SPICE run for the circuit we have been discussing (slightly cleaned up to save space) is:

```
****07/03/95 09:27:21***********EVALUATION PSPICE (JULY 1993)****
SECOND EXAMPLE
**** CIRCUIT DESCRIPTION
*****************************************************************
******
IS 0 1  AC  1  0
ES 2 0  1  0  -1
R  1 2  2
C  1 0  0.25
L  1 0  1
.AC LIN  1  0.3183 0.3183
.PRINT AC  VM(1) VP(1) VM(2) VP(2) IM(R) IP(R)
+ IM(C) IP(C) IM(L) IP(L)
.END
****07/03/95 09:27:21***********EVALUATION PSPICE (JULY 1993)****
SECOND EXAMPLE
**** SMALL SIGNAL BIAS SOLUTION      TEMPERATURE= 27.000 DEG C
```

```
****************************************************************
******
NODE VOLTAGE   NODE    VOLTAGE   NODE    VOLTAGE   NODE
VOLTAGE
( 1) 0.0000 ( 2) 0.0000
 VOLTAGE SOURCE CURRENTS
 NAME    CURRENT
 TOTAL POWER DISSIPATION  0.00E+00  WATTS
****07/03/95 09:27:21***********EVALUATION PSPICE (JULY 1993)****
SECOND EXAMPLE
**** AC ANALYSIS                TEMPERATURE= 27.000 DEG C
****************************************************************
******
 FREQ     VM(1)     VP(1)     VM(2)    VP(2)     IM(R)
 3.183E-01  1.000E+00  1.780E-03 1.000E+00  -1.800E+02  1.000E+00
****07/03/95  09:27:21***********EVALUATION PSPICE (JULY 1993)****
SECOND EXAMPLE
**** AC ANALYSIS                 TEMPERATURE= 27.000 DEG C
****************************************************************
******
 FREQ  IP(R)   IM(C)    IP(C)     IM(L)     IP(L)
3.183E-01  1.780E-03 5.000E-01    9.000E+01    5.000E-01 -9.000E+01
     JOB CONCLUDED
     TOTAL JOB TIME        .48
```

We would like to point out a couple of things about this SPICE output. If you have a look at the resistor and its series VCVS in Figure 11.102, you will see that it forms a sub-circuit that is equivalent to a 1-Ω resistor. To see why, look at Figure 11.103. We "bent" the subcircuit around a bit, but you should have no trouble seeing that the top terminal is that attached to node 1 in the circuit of Figure 11.102 and that the bottom is the one connected to the ground node, node 0. We will leave it to you to show that this subcircuit is equivalent to a resistance of value

$$R_{\text{eq}} = \frac{R}{1 - \mu}. \tag{11.5-3}$$

In our circuit, $R = 2\ \Omega$ and $\mu = -1$, so $R_{\text{eq}} = 1\ \Omega$. This means that we can draw the equivalent circuit shown in Figure 11.104 to represent the behavior of all elements except those in the subcircuit. We will leave it to you to show that $\overline{V} = 1\angle 0°\ \text{V}$, $\overline{I}_R = 1\angle 0°\ \text{A}$, $\overline{I}_C = 0.5\angle 90°\ \text{A}$, and $\overline{I}_L = 0.5\angle -90°\ \text{A}$—thereby verifying our SPICE simulation. Notice that our circuit is *parallel resonant*, with the capacitive and inductive impedances canceling each other out so that the parallel *LC* subcircuit appears as an open circuit to the rest of the elements. Note, however, that their currents are not zero; they are, rather, the same in magnitude and 180° different in phase. This verifies our earlier statement that the two elements provided exactly the amount of current required by the other at resonance.

Figure 11.103
Resistive subcircuit

Figure 11.104
The phasor form of the equivalent circuit

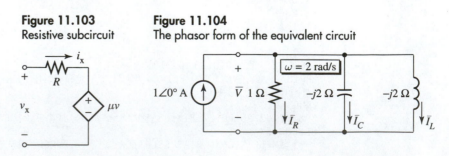

Section 11.5 Quiz

Q11.5-1.   Simulate the circuit in Question Q11.1-3 of the Quiz
for Section 11.1.

Chapter 11 Summary

This chapter has developed efficient methods of analyzing ac circuits: those that operate
in the *ac steady state*. A circuit is said to be operating in the ac steady state when all voltages and currents are sinusoidal. The chapter opens with a discussion of how a circuit approaches ac steady-state behavior. Under such conditions, we do not have to concern
ourselves with the fact that, in practice, all waveforms are one sided (that is, they are
turned on at some point in time); rather, we can extend our sinusoidal waveforms back
to $t = -\infty$ in a periodic manner and simply deal with waveforms of the form
$x(t) = X_m \cos(\omega t + \theta)$. X_m is called the *amplitude* of the sinusoid, and θ is called its
phase.

We observe that determination of the amplitude and phase (assuming that the frequency is known) is sufficient to completely determine the sinusoid. Therefore, the ac
steady-state analysis problem boils down to determining these two quantities.

Each sinusoid is described by two real quantities, so we grouped them together into a
single complex number of the form $\overline{X} = X_m \angle \theta = X_m e^{j\theta}$, which is the *Euler form* of the
complex number (as defined in Chapter 8). We defined this complex number to be the
phasor associated with the sinusoid.

Next, we showed that if all the independent sources in a circuit are sinusoidal
at the same frequency ω, then each phasor response voltage and current response variable
is related to each of the independent sources (circuit inputs) by a constant (complex)
multiplier called the *system function* $H(j\omega)$ (which depends upon the frequency
in general). From this general concept, we defined the *complex impedance* $Z(j\omega)$
to be the system function for a two-terminal element or subcircuit with the current treated
as the input and the voltage as the response. Conversely, if the voltage is the input
and the current the response, we called our system function the *complex admittance,*
$Y(j\omega)$. We noted that the complex impedance of the resistor is R, of the inductor $j\omega L$, and
of the capacitor $1/j\omega C = -j1/\omega C$. Thus, for each of these three elements, one
can write the voltage-current phasor relationship as $\overline{V} = Z(j\omega)\overline{I}$. This is simply
Ohm's law for phasors. We next showed that *KVL and KCL continue to hold for the phasors associated with sinusoidal waveforms obeying these laws*. Then we argued that
all our dc analysis techniques were constructed on the sole basis of these
three things: KVL, KCL, and Ohm's law. This, we observed, means that all the
circuit analysis techniques we previously developed for dc circuits continue to hold for
phasors.

Following this reasoning, we developed the *phasor equivalent* for a circuit operating
in the sinusoidal steady state: one in which the passive elements are represented by their
complex impedances and the independent source waveforms by their equivalent phasors.

Following this general theoretical development, we devoted an entire section to
working examples. Our computational techniques were simply those of dc circuits; the
only added labor was caused by the fact that we had to manipulate complex numbers
rather than real ones.

We next discussed the idea of *average* and presented a specialized and efficient
bracket notation for the process of computing the average of a waveform. We showed that
the average of a constant is a constant, that the average of a sinusoid is zero, that the aver-

age of a squared sinusoid[14] is 1/2, and that the averaging operator is linear. These facts suffice for all applications of phasor analysis to the problem of computing ac steady-state power.

We computed the average power for a general load (one that can contain independent sources as well as passive elements, provided that all such sources operate at the same frequency as the one driving the load) and showed that it is

$$P = \frac{1}{2}V_m I_m \cos(\phi),$$

where V_m and I_m are the maximum voltage and current values, respectively. We have already noted the origin of the "extra factor" of 1/2. The number $\cos(\phi)$ is called the *power factor (PF)*. The angle ϕ is called the *power factor angle,* and it is the angle by which the current "leads" the voltage. We say that the load has a "leading" power factor if $\phi > 0$ and has a "lagging" power factor if $\phi < 0$. If $\phi = 0$, then $PF = 1$ and the load is said to have a "unity" power factor.

We derived *series* and *parallel equivalents* for the *passive load:* one that does not contain any independent sources. We then defined *equivalent resistance* and *reactance* as the real and imaginary parts, respectively, of the load impedance. In like manner, we defined *conductance* and *susceptance* as the real and imaginary parts, respectively, of the load admittance. We noted that all four parameters can vary with frequency. Thus, the resistance and conductance are not exactly like the similar concepts for dc circuits; furthermore, in general, the conductance is not the reciprocal of the resistance. Using the preceding concepts, we defined a load to be *inductive* if its equivalent reactance is positive (and its susceptance is negative) and to be capacitive if its equivalent reactance is negative (and its susceptance is positive). We noted that it is called *resistive* if the reactance (and susceptance) is zero.

We next defined the *root-mean-square, or rms,* equivalent of voltage and current waveforms. The rms value of a voltage or current waveform is merely the dc voltage or current that delivers the same power to a resistive load as the given waveform. We showed that there is a "square root of 2" relation between maximum and rms values if the waveform is sinusoidal. We also noted that, in terms of rms voltage and current, the power is given by

$$P = V_{rms} I_{rms} PF;$$

that is, our extra factor of 1/2 disappears. This motivated us to agree to consider all phasors in power-related problems to be *rms phasors*. This means that an extra square root of 2 must be supplied whenever one converts a phasor to sinusoidal time domain form.

We spent some time discussing power flow to and from a load driven by a sinusoidal source. In doing so, we introduced three quantities: *apparent power* (in voltamperes, or VA, or kVA), *real power* (a new name for our old average power in watts, or W, or kW), and *reactive power* (in voltamperes reactive, or VAR, or kVAR). We defined the *complex power* to be the complex number whose real part is the real power and whose imaginary part is the reactive power. We also noted that the magnitude of the complex power is the apparent power and its angle is the power factor angle. We used these concepts to discuss the process of *power factor compensation.*

Our last section in this chapter discussed using SPICE to simulate circuits operating in the ac steady state.

---

[14] This result is quite important in that it accounts for an "extra factor" of 0.5 (relative to dc values) in computations of many ac quantities.

Chapter 11 Review Quiz

RQ11.1. If $v(t) = 6 \cos(10t) - 8 \sin(10t)$, what is the associated phasor $\overline{V}$ in Euler form? In rectangular form?

RQ11.2. A current $i(t)$ is represented by the phasor $\overline{I} = -1/2 - j\sqrt{3}/A$. Assuming that its frequency is 4 rad/s, what is $i(t)$?

RQ11.3. Two circuit variables $x(t)$ and $y(t)$ (each either a voltage or a current) are related as shown in Figure RQ11.1. What is the value of ω and what is the corresponding transfer function $H(j\omega)$ in Euler form? What is $H(j\omega)$ in rectangular form?

Figure RQ11.1

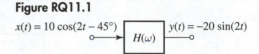

$x(t) = 10 \cos(2t - 45°)$ → $H(\omega)$ → $y(t) = -20 \sin(2t)$

RQ11.4. Consider the two-terminal subcircuit shown in Figure RQ11.2. Suppose that $v(t) = 12 \cos(\omega t)$ and that when $\omega = 3$ rad/s, $i(t) = 3\sqrt{2} \cos(3t - 45°)$ A. What are the values of R and L?

Figure RQ11.2

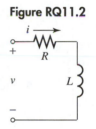

RQ11.5. If $i(t) = 2 \cos(2t)$ A in the circuit shown in Figure RQ11.3, use voltage and current division and/or series and parallel equivalents and/or Norton/Thévenin equivalents to find the ac forced response for $v(t)$.

Figure RQ11.3

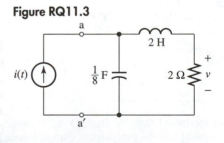

RQ11.6. Find the answer to Question RQ11.5 using nodal analysis.

RQ11.7. Repeat Question RQ11.6 using mesh analysis.

RQ11.8. Find the series equivalent subcircuit for the elements to the right of terminals a and a' in Figure RQ11.3 at a frequency of 1 rad/s. (Give specific element values.)

RQ11.9. Find the parallel equivalent subcircuit for the elements to the right of terminals a and a' in Figure RQ11.3 at a frequency of 1 rad/s. (Give specific element values.)

RQ11.10. Impedance and frequency scale the circuit of Figure RQ11.3, with the current source waveform given in Question RQ11.5 so that $R = 2$ kΩ and the impedance seen looking to the right into terminals a and a' is purely real at 2 krad/s.

RQ11.11. Consider the subcircuit of Figure RQ11.2 with $R = 3\ \Omega$ and $L = 1$ H. Find the complex power S delivered by a voltage source with the waveform $v(t) = 10 \cos(4t)$ V placed across its terminals. Sketch S as a vector in the complex plane.

RQ11.12. Consider the subcircuit of Figure RQ11.2 with $R = 3\ \Omega$ and $L = 1$ H. Find the average power P delivered by a voltage source with the waveform $v(t) = 10 \cos(4t)$ V placed across its terminals.

RQ11.13. Consider the subcircuit of Figure RQ11.2 with $R = 3\ \Omega$ and $L = 1$ H. Find the reactive power Q delivered by a voltage source with the waveform $v(t) = 10 \cos(4t)$ V placed across its terminals.

RQ11.14. Consider the subcircuit of Figure RQ11.2 with $R = 3\ \Omega$ and $L = 1$ H. Find the apparent power $A = |S|$ delivered by a voltage source with the waveform $v(t) = 10 \cos(4t)$ V placed across its terminals.

RQ11.15. What reactive element must be added across the terminals of the subcircuit in Figure RQ11.2, assuming that a voltage source with the waveform $v(t) = 10 \cos(4t)$ V is placed across these terminals, to correct the power factor of the "load" to unity? To 0.8 lagging? To 0.6 leading?

Chapter 11 Problems

Review Drill on Complex Numbers and Phasors

11.0-1. The complex numbers in parts (a) through (h) represent current or voltage phasors for a circuit. Convert those in Euler form to rectangular form and those in rectangular form to Euler form. In parts (i) through (m) carry out the indicated operations and express the result in Euler form. (The mathematical manipulations in this problem are typical of those needed to solve subsequent problems.)

a. $10\angle 30°$	**b.** $6\angle 150°$
c. $-8\angle 60°$	**d.** $20\angle -225°$
e. $-5 + j12$	**f.** $16 - j12$

g. $-1 - j15$ **h.** $-10 + j0.5$

i. $10\angle 135° - 12\angle -60° + 6\angle 120°$

j. $\dfrac{(1 + j)^2}{1 - j}$ **k.** $\dfrac{2\angle 30°}{j10} + \dfrac{1}{-8 + j6}$

l. $\dfrac{(3 - j8)^2(-1 + j2)}{j(-1 - j)^2}$

m. $\left[\dfrac{3 - j}{1 + j} - \dfrac{j2}{2 + j3}\right]\dfrac{j2}{9 - j12}$

11.0-2. Find the following quantities. (This is a more difficult problem than 11.0-1.)

a. $j^{1/3}$ **b.** j^9

c. e^j **d.** j^j

e. $|(3 - j4)(10\angle 20°)e^{j\pi/8}|$

f. $\mathrm{Re}\{w\}$ and $\mathrm{Im}\{w\}$ if $w = \dfrac{z^2}{|z|^2}$

11.0-3. Show the regions in the complex plane $(z = x + jy)$ described by the following inequalities:

a. $|z - 3| < 2$ **b.** $\mathrm{Im}\{z\} < 4$ **c.** $\mathrm{Re}\left\{\dfrac{1}{z}\right\} > \dfrac{1}{4}$

11.0-4. Express each of the following functions as a single cosine with a phase angle:

a. $-5\cos(3t) + 3\sin(3t)$
b. $2\cos(2t + 45°) - 2\sin(2t)$
c. $3\sin(2t - 120°) - 4\cos(2t)$

Section 11.1 The Phasor Equivalent Circuit

11.1-1. For the subcircuit shown in Figure P11.1-1, $i_1(t) = 2\cos(100t)$ A, $i_2(t) = 2\sqrt{2}\cos(100t - 45°)$ A, and $i_3(t) = 5\cos(100t + 90°)$ A. First, use appropriate trigonometric identities to find an expression for $i(t)$ that has the form $i(t) = A\cos(100t + \theta°)$. Second, convert each cosine to a phasor rectangular form, add, convert to Euler form, and then express as a cosine function of time. Do the two procedures give the same result?

Figure P11.1-1

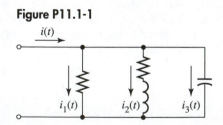

11.1-2. Figure P11.1-2 shows a simple four-terminal subcircuit. Assuming that $i_1(t) = 4\cos(3t)$ A, $i_2(t) = 6\cos(3t - 45°)$ A, $i_3(t) = 4\cos(3t - 180°)$ A, and $i_4(t) = 6\sin(3t - 135°)$ A,

a. Show that $i_1(t) + i_2(t) + i_3(t) + i_4(t) = 0$ for all t.

b. Draw the phasor equivalent and show that $\bar{I}_1 + \bar{I}_2 + \bar{I}_3 + \bar{I}_4 = \bar{0} = 0$. (Note that $\bar{0}$ is the complex constant $(0, 0) = 0 + j0$.)

Figure P11.1-2

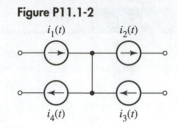

11.1-3. Use the phasor method to find the ac forced response indicated for each simple circuit in Figure P11.1-3. (Note that the $u(t)$ is actually irrelevant for forced response; furthermore, none of these circuits is absolutely stable, so the ac steady state does not exist. The last remark is for those having covered Part II).

Figure P11.1-3

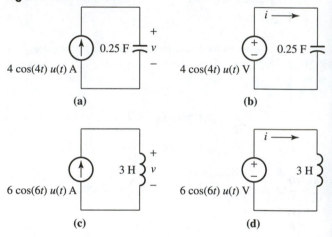

(a)   **(b)**

(c)   **(d)**

11.1-4. Find the ac forced response for the current $i(t)$ in the circuit shown in Figure P11.1-4 assuming $v(t) = 10\sqrt{2}\cos(2t)u(t)$ V. What is the system function $H(j\omega) = \bar{I}/\bar{V}_s$?

Figure P11.1-4

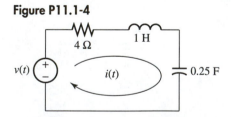

11.1-5. Find the forced response for $i_x(t)$ in the circuit of Figure P11.1-5 assuming that the independent current source has the waveform $i_s(t) = 2\cos(8t)$ A. What is the system function $H(j\omega) = \bar{I}_x/\bar{I}_s$?

Figure P11.1-5

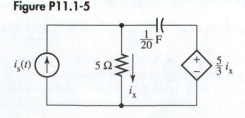

11.1-6. Find the ac forced response for the current $i(t)$ and the voltage $v(t)$ in Figure P11.1-6. Assume that $i_s(t) = 5\cos(2\sqrt{3}t)$ V. What is the system function $H(j\omega) = \bar{I}/\bar{I}_s$?

Figure P11.1-6

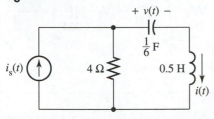

11.1-7. Find the forced response for the current $i(t)$ in Figure P11.1-7.

Figure P11.1-7

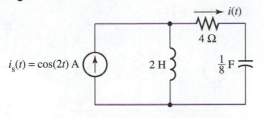

Section 11.2 Methods of Ac Circuit Analysis

11.2-1. Use nodal analysis to find the ac forced response for $v(t)$ in Figure P11.2-1. Assume that $i_s(t) = 4\sqrt{2}\cos(2t)$ A.

Figure P11.2-1

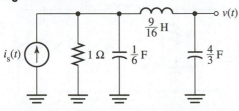

11.2-2. Solve Problem 11.2-1 using mesh analysis. Do not simplify by using series/parallel equivalents.

11.2-3. Use Thévenin/Norton, series-parallel equivalents, voltage division, and/or current division to solve Problem 11.2-1.

11.2-4. Find the forced response for $v(t)$ in Figure P11.2-2. Assume that $i_{s1}(t) = \sin(2t)$ A and that $i_{s2}(t) = \cos(2t)$ A. What are the system functions $H_1(j\omega) = [\bar{V}/\bar{I}_{s1}]_{\bar{I}_{s2}=0}$ and $H_2(j\omega) = [\bar{V}/\bar{I}_{s2}]_{\bar{I}_{s1}=0}$? (By superposition, $\bar{V} = H_1(j\omega)\bar{I}_{s1} + H_2(j\omega)\bar{I}_{s2}$.)

Figure P11.2-2

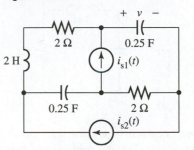

11.2-5. Find the forced response for $v(t)$ in Figure P11.2-3, assuming that $v_s(t) = 10\sqrt{2}\cos(2t - 45°)$ V.

Figure P11.2-3

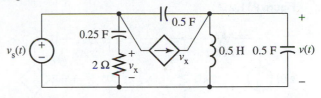

11.2-6. Find the forced response for $i(t)$ in Figure P11.2-4 using mesh analysis and assuming $i_s(t) = 6\cos(2t)$ A.

Figure P11.2-4

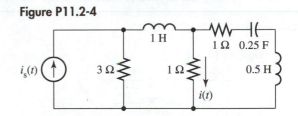

11.2-7. Find the forced response for $v(t)$ in Figure P11.2-5 using nodal analysis. Assume that $v_{s1}(t) = v_{s2}(t) = 12\cos(4t)$ V and $i_s(t) = 2\sin(4t)$ A.

Figure P11.2-5

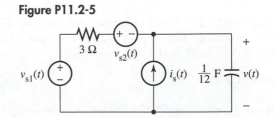

11.2-8. Draw the series and parallel equivalent circuits, using individual elements, of the subcircuit shown in Figure P11.2-6 at a frequency of $\omega = 2$ rad/s.

Figure P11.2-6

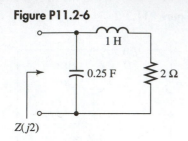

$Z(j2)$

11.2-9. Frequency and impedance scale the circuit in Figure P11.2-7 to obtain "nice" values, then use any method you like to find the forced response of $i(t)$. Be careful to note that the CCVS changes with impedance scaling because its gain constant is trans-resistance. Assume that $v_s(t) = 4\cos(5000t)$ V.

Figure P11.2-7

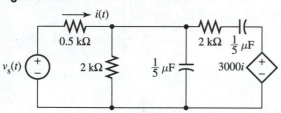

11.2-10. Frequency and impedance scale the circuit in Figure P11.2-8 for "nice" values, then find the forced response for $v(t)$. Assume that $v_s(t) = 2\cos(2 \times 10^4 t)$ V.

Figure P11.2-8

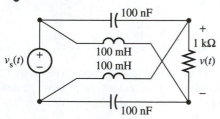

11.2-11. Find the forced response for $v(t)$ in Figure P11.2-9, assuming that $i_{s1}(t) = 3\cos(2t)$ A and $i_{s2}(t) = 10\cos(4t + 30°)$ A. (*Hint:* Use superposition.)

Figure P11.2-9

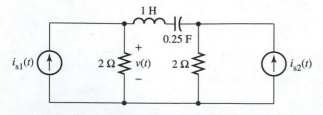

11.2-12. Assuming that $i_{s1}(t) = 4\cos(2t)$ A, $i_{s2}(t) = 6\cos(2t)$ A, and $i_{s3}(t) = 6\cos(2t)$ A, find the forced response for $v(t)$ in Figure P11.2-10.

Figure P11.2-10

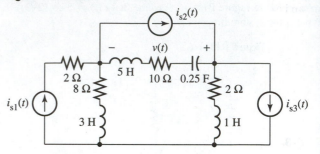

11.2-13. Solve the circuit in Figure P11.2-11 for the forced response of $v(t)$ by replacing the subcircuit to the left of terminals aa′ by its Thévenin equivalent. The independent current source waveform is $i_s(t) = 3\sqrt{2}\cos(2t + 45°)$ A.

Figure P11.2-11

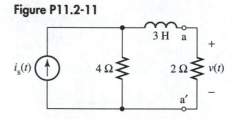

11.2-14. Find the forced response for the voltage $v(t)$ in Figure P11.2-12 by replacing the subcircuit to the left of the terminals aa′ by its Norton equivalent. Assume that $i_s(t) = 3\sin(4t)$ A and $\beta = 1$. *Warning:* Be very careful here because there is a possibly unexpected result!

Figure P11.2-12

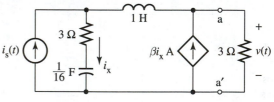

Section 11.3 Average Power in the Sinusoidal Steady State

11.3-1. Find the average value of the periodic waveform in Figure P11.3-1.

Figure P11.3-1

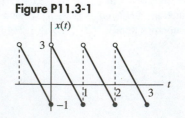

11.3-2. Repeat Problem 11.3-1 for the periodic waveform shown in Figure P11.3-2.

Figure P11.3-2

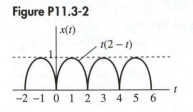

11.3-3. Find the power factor and the average power absorbed by the load shown in Figure P11.3-3. Assume that $i_s(t) = \sin(4t)$ A and $v(t) = 12 \sin(4t)$ V.

Figure P11.3-3

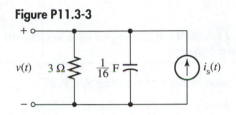

11.3-4. Find the rms value of $x(t)$ in Figure P11.3-1.

11.3-5. Find the rms value of $x(t)$ in Figure P11.3-2.

11.3-6. Find the parallel and series equivalents for the subcircuit in Figure P11.3-4 at $\omega = 4$ rad/s.

Figure P11.3-4

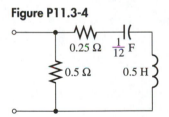

11.3-7. Find the average power absorbed by each element in Figure P11.3-5 and compute their sum. Assume that $v(t) = 2\sqrt{2} \cos(2t)$ V. Find the two-element series equivalent subcircuit for the subcircuit consisting of all the elements except the voltage source. Find the power absorbed by each of the two equivalent elements.

Figure P11.3-5

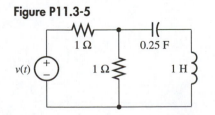

11.3-8. Find the rms value of $i(t)$ in Figure P11.3-6. Assume that $v(t) = 16\sqrt{2} \cos(2t)$ V. Find the two-element parallel equivalent subcircuit for the subcircuit consisting of all the elements except the voltage source. Find the power absorbed by each of the two equivalent elements.

Figure P11.3-6

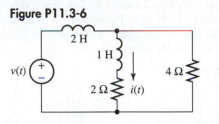

Section 11.4 Complex Power

11.4-1. For the circuit shown in Figure P11.4-1, $i(t) = 10\sqrt{2} \cos(\omega t)$ mA.

    a. Find the value of ω for which the driving point admittance $Y(j\omega)$ has an angle of 53.1°.

    b. For $\omega = 1000$ rad/s, find the complex power absorbed by each element, the peak energies stored by the capacitor and the inductor, and the average power dissipated in the resistor.

Figure P11.4-1

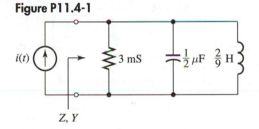

Z, Y

11.4-2. Verify the conservation of complex power for the circuit that is shown in Figure P11.4-2. Assume that $v_s(t) = 10\sqrt{2} \cos(4t + 53.1°)$ V. (Lengthy, but not hard!)

Figure P11.4-2

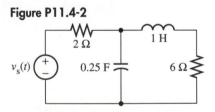

11.4-3. Assume that $\omega = 1000$ rad/s in the circuit of Problem 11.4-1. What is the power factor of the load? What element, when placed across the terminals, results in a unity power factor?

11.4-4. Replace the current source in Problem 11.4-1 with the subcircuit shown in Figure P11.4-3. Find R (≥ 0) and X for maximum power dissipated in the 3-mS resistor ($\omega = 1000$ rad/s).

Figure P11.4-3

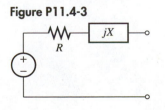

11.4-5. If $v_s(t) = 8\sqrt{2}\cos(2t)$ V, find the average power absorbed by the 2-Ω resistor in Figure P11.4-4. Note that the current source is dc. Assume the circuit is stable—and therefore both ac and dc steady-state conditions exist.

Figure P11.4-4

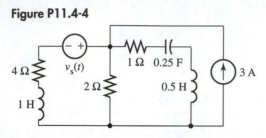

11.4-6. Find the reactive element to be placed in parallel with the independent voltage source in Figure P11.4-5 to correct the power factor of the remainder of the circuit to 0.8 lagging. $v(t)$ is sinusoidal with a frequency of 10 rad/s.

Figure P11.4-5

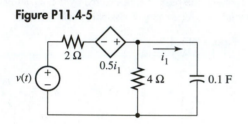

11.4-7. Find the real, reactive, apparent, and complex powers delivered by the source in Figure P11.4-6. Assume that $v(t) = 10\cos(2t)$ V.

Figure P11.4-6

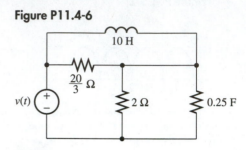

Section 11.5 SPICE Analysis of Circuits in the Ac Steady State

11.5-1. Solve Problem 11.1-5 using SPICE.

11.5-2. Solve Problem 11.2-4 using SPICE.

12

Frequency Response and Filtering

This chapter starts with an introduction to a new way of looking at the signals present in circuits: the frequency domain viewpoint. The concept of representing a signal as the weighted sum of sinusoids or complex exponentials is presented in a qualitative manner—deferring the rigorous mathematics until a later chapter. Several examples are presented showing how knowledge of the ac forced response of a circuit at any arbitrary frequency can be used to find the response to each of the individual sinusoids making up the given signal and how superposition can then be used to find the response to the waveform itself. Several practical examples are given, one of which consists of the removal of a sinusoidal interfering noise signal from a desired sinusoidal signal. A standard classification of filter types based upon gain approximation is introduced, and the effects of phase distortion are discussed. The problem of tuning a radio or television receiver is treated and related to the ideal parallel tuned circuit.

The quality factor (Q) is then defined for any two-terminal element or subcircuit and used to derive the series-parallel transformation for a lossy inductor. Then the non-ideal parallel tuned circuit is carefully analyzed to determine the frequency response, the peaking frequency, and the bandwidth. A short section on finding the time response of the parallel tuned circuit to a gated sinusoid at the tuning frequency and the relationship between this time response and the pole-zero diagram is presented for those who have previously covered the material in Part II on time domain analysis techniques. Following this, the parallel-series conversion of a lossy capacitor is derived, and the lossy series tuned circuit is analyzed. The next section contains a discussion of Bode plots; in particular, a fast "back-of-the-envelope" technique for rapidly sketching the linearized gain and phase plots is presented that works for any order, as long as all pole and zero factors are real. The chapter then concludes with sections on the frequency response of active circuits and the determination of the frequency response of circuits using SPICE. (Examples flagged with an (O) or (T) pertain to operational amplifiers and transistors, respectively.)

The prerequisite for this chapter (with the exception of Section 12.4, which requires a knowledge of time domain analysis) is a knowledge of the ac forced response analysis presented in Chapter 11.

12.1 | The Frequency Domain Point of View

How to Read This Section

This section will develop a new way of looking at circuits and other linear systems. We call it the *frequency domain.* Just as there is an instrument called the oscilloscope that one can use to obtain a picture of the time behavior point of a given waveform, there is another—the *spectrum analyzer*—that provides another type of picture: the behavior of the waveform in terms of its *frequency content.* Before we can make this term precise, we must consider the basic definition of frequency a bit more closely. But first, we will make a suggestion about how to read this section. In the background, we will be calling upon some rather involved mathematics (that is, the author did when writing this section); however, at the moment you should treat this as irrelevant and only concentrate on the "special effects." That is, you should focus on the pictures we plot and the basic ideas behind the equations we write. You should not, at this point, be concerned about the theoretical justification for either. The next few chapters will develop all of the ideas we will present in more mathematical detail, so you should concentrate here upon seeing the "big picture."

On the Definition of Frequency

Let's have a look at Figure 12.1. There we show a simple *RC* circuit driven by an independent voltage source. We will suppose that the independent source, whose waveform is labeled $v_i(t)$ (the i for "input") has the waveform shown in Figure 12.2. It is a triangular waveform having a peak amplitude of 1 V and a "period" of 2 seconds. It is one sided, that is, it is identically zero for $t \leq 0$ and "turns on" at $t = 0$. Thus, we see that the waveform is not *truly* periodic because that would require that it be nonzero and repetitive for $t \leq 0$ as well as for $t > 0$.

Figure 12.1
An example circuit

Figure 12.2
The "input" waveform

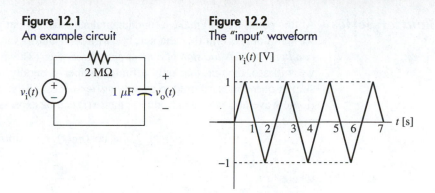

First, we ask, "What is the frequency of $v_i(t)$?" Heretofore, we have defined the frequency *only for a sinusoid* so the question merits study. Thus, if

$$v_i(t) = V_{im} \cos(\omega t) \text{ V},\tag{12.1-1}$$

a truly periodic, two-sided sine wave, we would say that its radian frequency is ω rad/s and that its frequency in hertz is $f = \omega/2\pi$. But the waveform in Figure 12.2 is not a sinusoid, yet it *seems* to be somewhat periodic and we somehow feel that its "frequency" should be $(1/2)s^{-1}$, or 0.5 Hz. How do we reconcile our intuition with the fact that it is not a sinusoid?

Before we discuss our answer to this question, let's investigate the response of our circuit in Figure 12.1, $v_o(t)$, to the waveform in Figure 12.2. This is shown in Figure 12.3. (The author used PSPICE to determine this waveform, then cleaned it up a bit using a graphics program.)

Figure 12.3
The response of the example circuit

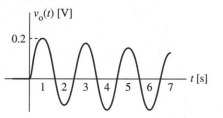

There are several things we can observe about this waveform. First, we see that it, like the input, is zero for $t \leq 0$. This is the consequence of a fundamental property of all physical systems: they are *causal,* that is, they do not react before they are acted upon. The second thing we see is that there is a "transient" part, exactly as there was for the sinusoidal response that we investigated in Chapter 11. Also, just as was the case for the sinusoidal input, this response "settles down" into steady-state behavior after perhaps two periods or so of the input waveform. After that, the output waveform is periodic, with the same period as that of the input, namely 2 seconds. We see, however, that the response looks quite different from the input. In fact, it looks almost sinusoidal (it isn't, not quite!), whereas the input is a triangular waveform. If we are only interested in the steady-state response, we can simply ignore the transient part and extend $v_o(t)$ (and $v_i(t)$ too, of course) backward periodically in time and simply imagine that all waveforms of interest are truly periodic, with the same period. We will do that here and we will see that this assumption leads to a quite powerful way of looking at waveforms and the effects they experience upon passing through various types of circuits.

Fourier Series   At this point, we will make a fundamental statement about periodic waveforms—one that we will not prove in this chapter, but will defer until a later one: *any periodic waveform can be written as the sum of a number of sinusoids.* (The number, theoretically, is infinite; we will see, though, that only a finite number generally must be used in practice.) The sum in question is referred to as the *Fourier series* for the waveform. Thus, if $x(t)$ is a periodic waveform whose period is T, then $x(t)$ can be expressed as

$$x(t) = \sum_{n=0}^{\infty} [a_n \cos(n\omega_0 t) + b_n \sin(n\omega_0 t)], \qquad (12.1\text{-}2)$$

where

$$\omega_0 = = \frac{2\pi}{T} \qquad (12.1\text{-}3)$$

and the a_n's and b_n's are specific constants whose values depend upon $x(t)$. The constant ω_0 is called the *fundamental frequency,* and $n\omega_0$ is called, for each $n \geq 1$, the *nth harmonic.*

Here, now, is the answer to the question we posed earlier about frequency: a periodic waveform has an infinite number of frequencies—all harmonically related to the fundamental, which is what we agreed earlier was the intuitive definition of frequency.

To be specific, let's look at the *periodic extension* of $v_i(t)$ in our previous example (that is, we simply repeat periods of $v_i(t)$ "backwards" in time so that it is nonzero there and looks exactly like its waveform for positive t values).[1] The result is shown in Figure 12.4 for a few periods. Though it is in actuality a different waveform, we will call it $v_i(t)$ simply to keep the notation from becoming too complicated, with the understanding that our analysis will only be valid for the steady-state behavior of the actual waveforms involved.

Figure 12.4
Periodic extension of
the input waveform

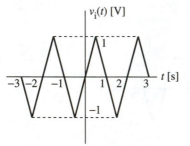

A more involved mathematical analysis (to be covered in a later chapter) shows that the Fourier series of $v_i(t)$ is

$$x(t) = \sum_{k=0}^{\infty} \frac{8}{(2k + 1)^2 \pi^2} \sin[(2k + 1)\pi t]. \qquad (12.1\text{-}4)$$

We have reindexed and let $n = 2k + 1$ for convenience. From equations (12.1-2) and (12.1-3), we see that $a_n = 0$ for all n, $b_n = 0$ for even n, $b_n = 8/n^2\pi^2$ for odd n, and $\omega_0 = \pi$ rad/s. In more detailed form, we have

$$v_i(t) = 0.811 \sin(\pi t) + 0.090 \sin(3\pi t) + 0.032 \sin(5\pi t) + \cdots. \qquad (12.1\text{-}5)$$

---

[1] Technically, this is called the *odd extension* because it is an odd function of t. One can also form an *even extension in a similar manner.*

The three dots at the end imply that there are more terms, and there always will be if we specifically only write out a finite number of them. Thus, equation (12.1-4) carries more information than does equation (12.1-5). Notice, however, that the amplitudes of the sinusoidal terms are decreasing rapidly with the number n; that is, with the number of harmonics included. Also note that only sine terms appear, and only the odd harmonics at that. There are no cosines (this is typical of a waveform with "odd" symmetry).

Now let's look at the result of truncating the sum in (12.1-4) after a given number of terms, that is, of including only a finite but given number of harmonics in addition to the fundamental. For reference purposes, we show one period of the input waveform in Figure 12.5. Figure 12.6 shows the fundamental component by itself ($n = 1$). Notice that the maximum value is about equal to our computed value in equation (12.1-4), at least to the accuracy of our graph. (We did our Fourier computations and plots using MATLAB.)

Figure 12.5
A triangular waveform

Figure 12.6
The fundamental component

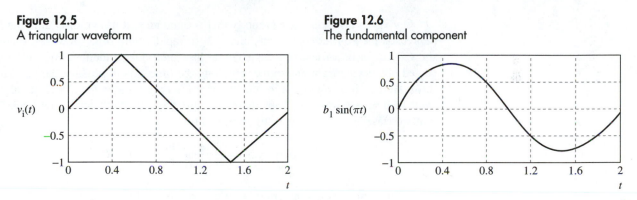

Figure 12.7 shows what happens when we include the fundamental and third harmonic terms in our Fourier series (that is, the first two nonzero terms in (12.1-4)). As you can see, this is beginning to look quite a bit like the original waveform. If we include three nonzero terms (that is, the fundamental, third, and fifth harmonics), we obtain the waveform shown in Figure 12.8. It is an even better approximation to the original. We show the sum of the first four nonzero harmonics (first, third, fifth, and seventh) in Figure 12.9. Finally, we show the sum of the first five nonzero harmonics (first, third, fifth, seventh, and ninth) in Figure 12.10. In our figures we have used a synonym for the fundamental component: the *first harmonic*. We could, of course, keep going and obtain better and better approximations until we were satisfied with the accuracy or until our patience wore out. Fortunately, the patience of a computer does not become frazzled; thus, we could simply program it to add up a given number of terms and plot the result without investigating each sum.

Figure 12.7
The sum of the first and third harmonics

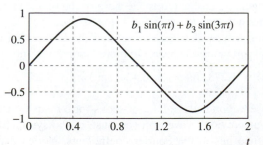

$$b_1 \sin(\pi t) + b_3 \sin(3\pi t)$$

Figure 12.8
The sum of the first, third, and fifth harmonics

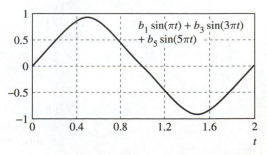

$$b_1 \sin(\pi t) + b_3 \sin(3\pi t) + b_5 \sin(5\pi t)$$

Figure 12.9
The sum of the first, third, fifth, and seventh harmonics

Figure 12.10
The sum of the first, third, fifth, seventh, and ninth harmonics

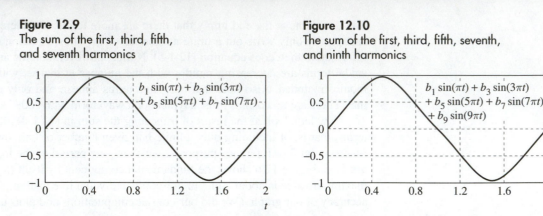

Our $v_i(t)$ waveform has a property that is quite nice: it has no discontinuities. This is the reason the Fourier coefficients diminish so rapidly with time. However, Figure 12.11 shows one period of a waveform $x(t)$ that has jump discontinuities. It is, of course, repeated both to the right and to the left along the time axis to produce the periodic waveform. We call it a *square wave,* and say that the *duty cycle* is 50%. In general, the duty cycle of a square wave (perhaps more appropriately called a "rectangular wave," though "square wave" is most often used) is defined by

$$\text{duty cycle} = \frac{\text{"high time"}}{\text{period}} \times 100\%. \qquad (12.1\text{-}6)$$

If the "high time" is half a period, as is the case here, the duty cycle is 50%, and we say that the square wave is symmetrical.

Figure 12.11
One period of a 50% duty cycle square wave

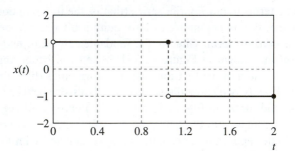
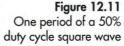

The Fourier series of the waveform in Figure 12.11 is

$$x(t) = \sum_{k=1}^{\infty} \frac{4}{(2k-1)\pi} \sin(k\pi t). \qquad (12.1\text{-}7)$$

Thus, there are again no cosine terms (because of the odd symmetry of $x(t)$) and only the odd harmonics are nonzero. (This is due to the 50% duty cycle. Note that this is a special way of expressing a type of symmetry around the half-period time point that both this waveform and our previous one have.) In more "long-hand" notation, (12.1-7) reads

$$x(t) = 1.273 \sin(\pi t) + 0.424 \sin(3\pi t) + 0.255 \sin(5\pi t) + \cdots. \qquad (12.1\text{-}8)$$

We make the observation that the Fourier series for our square wave is precisely the same *in form* as for our previous waveform (because of the symmetries we have mentioned). The only difference lies in the Fourier coefficients. Notice that the fundamental has a larger peak value than does $x(t)$ itself, and the coefficients diminish less rapidly than do those associated with our earlier waveform.

Figure 12.12 shows the fundamental component, the first harmonic. Figure 12.13 shows what happens when we evaluate (12.1-7) using only the first two nonzero terms. Notice that the negative peak of the third harmonic, the sine wave at the frequency 3π rad/s, pulls down the peak value of the fundamental and brings it closer to a peak value of unity, the peak value of our original waveform. Adding one more nonzero harmonic gives the result shown in Figure 12.14. Notice that we are beginning to better approximate our square wave, but the approximation is still not terribly good. In fact, adding one additional term only results in the approximation shown in Figure 12.15. This is graphical evidence of our previous comment: the discontinuity causes the Fourier series to converge more slowly. We now jump to the case in which we include the fundamental and the next 10 nonzero harmonics. We show the result in Figure 12.16. Notice that although the approximation is quite good where the waveform is continuous, it is not very good right at the discontinuity. In fact, a deeper investigation than we will make in this text shows

Figure 12.12
The fundamental component of $x(t)$

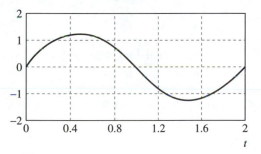

Figure 12.13
Fundamental and third harmonic

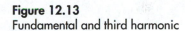
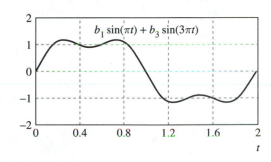

Figure 12.14
The sum of the first, third, and fifth harmonics

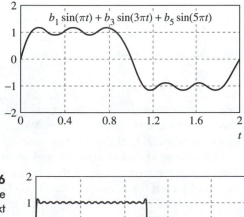

Figure 12.15
The sum of the first, third, fifth, and seventh harmonics

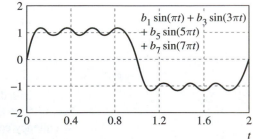

Figure 12.16
The sum of the fundamental and next ten nonzero harmonics

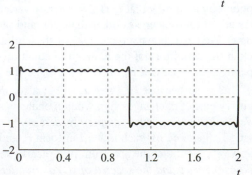

that the Fourier series converges to the average values of the left-hand limit and the right-hand limit at each point. Thus, at the discontinuity points, the series converges to zero. What's more, the finite approximations undershoot and overshoot the constant values by about 10% regardless of how many terms we include. This is referred to as the *Gibbs phenomenon.*

The Complex Form of the Fourier Series: Amplitude and Phase Spectra

We can place the Fourier series in equation (12.1-2) in a more convenient form for the following discussion by splitting the sum into two parts, one involving only the cosine terms and the other involving only the sine terms, splitting off the $n = 0$ term, and using Euler's formula on the sine and cosine functions. This results in

$$x(t) = a_0 + \sum_{n=0}^{\infty} a_n \cos(n\omega_0 t) + \sum_{n=0}^{\infty} b_n \sin(n\omega_0 t) \tag{12.1-9}$$

$$= a_0 + \sum_{n=0}^{\infty} a_n \frac{e^{jn\omega_0 t} + e^{-jn\omega_0 t}}{2} + \sum_{n=0}^{\infty} b_n \frac{e^{jn\omega_0 t} - e^{-jn\omega_0 t}}{2j}$$

$$= a_0 + \sum_{n=0}^{\infty} \frac{1}{2}(a_n - jb_n)e^{jn\omega_0 t} + \sum_{n=0}^{\infty} \frac{1}{2}(a_n + jb_n)e^{-jn\omega_0 t}$$

$$= \sum_{n=-\infty}^{\infty} c_n e^{jn\omega_0 t}.$$

In going from the second line to the third, we simply regrouped terms and brought the j in the second summation "upstairs." In going from the third line to the last, we changed dummy summation variables from n to $-n$ in the sum over negative values, then incorporated the constant a_0 term. In this last transformation, we have defined

$$c_n = \begin{cases} \dfrac{a_n - jb_n}{2}; & n > 0 \\ a_0; & n = 0. \\ \dfrac{a_n + jb_n}{2}; & n < 0 \end{cases} \tag{12.1-10}$$

The last expression in (12.1-9) is called the *complex form* of the Fourier series. Notice that the Fourier coefficients in (12.1-10) have the property that $c_{-n} = c_n^*$.

Because each constant c_n is complex, we can write it as

$$c_n = |c_n| e^{j\phi_n}. \tag{12.1-11}$$

$|c_n|$ is called the *amplitude spectrum* of the waveform $x(t)$ (of which the c_n are the Fourier coefficients), and the ϕ_n are called its *phase spectrum.* We can get a very good idea of the frequency content by assuming that ω is a frequency variable and that the frequency content of $x(t)$ is zero at all frequencies other than $n\omega_0$ and plotting the amplitude and phase spectra versus ω. We have shown a typical set of these plots in Figure 12.17(a) and (b). If you look back at the definitions of the c_n given in equation (12.1-10), you will see that the amplitude spectrum is even, that is, $|c_{-n}| = |c_n|$, and the phase spectrum is odd, that is, $\phi_{-n} = -\phi_n$. This is so because $|c_n| = \sqrt{a_n^2 + b_n^2}$; also, $\phi_n = \tan^{-1}[(\text{Im}\{c_n\})/(\text{Re}\{c_n\})]$, and the imaginary part of c_n changes sign when n changes sign. It is important to note that *the amplitude and phase spectra completely determine x(t), and vice versa.* In other words, knowing $|c_n|$ and ϕ_n for each n, we can form c_n and then sum the Fourier series in (12.1-9) to find $x(t)$. Note that the spectrum plot clearly shows the value of ω_0. Thus, *we refer to the Fourier series and the spectra of x(t) as frequency domain descriptions of x(t).*

Figure 12.17
The amplitude and
phase spectra

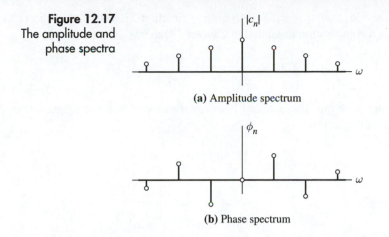

(a) Amplitude spectrum

(b) Phase spectrum

They are just as valid, and sometimes more informative, than a plot or an equation for $x(t)$ itself.

The Spectrum of a Nonperiodic Waveform: The Fourier Transform

Not all waveforms, of course, are periodic. When $x(t)$ is not periodic, it turns out that we can still obtain a frequency domain description. In fact, the more involved theory to be presented later in this text (in Chapter 14) will demonstrate that virtually any nonperiodic waveform encountered in circuits and systems analysis work can be represented by its *Fourier integral,* or inverse *Fourier transform:*

$$x(t) = \frac{1}{2\pi} \int_{-\infty}^{\infty} X(\omega)e^{j\omega t}\, d\omega. \tag{12.1-12}$$

If you remember that an integral is merely the limit of a sum, you should be able to see the analogy between (12.1-12) and (12.1-2)—the latter being the definition of the Fourier series—at once. However, it might be helpful to show the analogy just a bit more explicitly.[2] Thus, suppose we (intuitively) approximate the integral in (12.1-12) by a Riemann sum, taking the increments of frequency to be the uniform value $\Delta\omega = \omega_0$. Then we will have

$$x(t) = \frac{1}{2\pi} \int_{-\infty}^{\infty} X(\omega)e^{j\omega t}\, d\omega \cong \sum_{n=-\infty}^{\infty} \frac{X(n\omega_0)}{2\pi} e^{jn\omega_0 t}\, \omega_0 \tag{12.1-13}$$

$$= \sum_{n=-\infty}^{\infty} \frac{X(n\omega_0)\omega_0}{2\pi} e^{jn\omega_0 t}.$$

If you compare this equation with the last expression in (12.1-9), you will see that this is nothing more than the complex Fourier series with

$$c_n = \frac{X(n\omega_0)\omega_0}{2\pi}. \tag{12.1-14}$$

Let's look at an example. Figure 12.18 shows a sketch of the waveform

$$x(t) = e^{-at}u(t), \tag{12.1-15}$$

where a is a positive real parameter. The graph is shown for three different values of a:

---

[2] The integral in (12.1-12) is actually an improper one, so it is actually the limit of integrals over finite time intervals.

$a = a_1$, $a = a_2$, and $a = a_3$. The frequency function $X(\omega)$ used above in (12.1-12) turns out (after a computation discussed in Chapter 14) to be

$$X(\omega) = \frac{1}{a + j\omega} = \frac{1}{\sqrt{a^2 + \omega^2}} e^{-j\tan^{-1}(\omega/a)}. \tag{12.1-16}$$

By analogy with our discussion of Fourier series, we define the magnitude of $X(\omega)$ to be the amplitude spectrum,

$$|X(\omega)| = \frac{1}{\sqrt{a^2 + \omega^2}}, \tag{12.1-17}$$

and its angle,

$$\phi(\omega) = -\tan^{-1}\left[\frac{\omega}{a}\right], \tag{12.1-18}$$

to be the phase spectrum of $x(t)$. The amplitude spectrum for our example is plotted in Figure 12.19(a) and its phase spectrum in Figure 12.19(b). Notice that there is an inverse relationship between a signal and its amplitude spectrum. In our example, for instance, $x(t)$ changes the most rapidly with time for the largest value of a, that is, for $a = a_3$; the amplitude spectrum is "spread out" the most for this case; thus, *a large amplitude spectrum at large values of frequencies means that the time waveform changes rapidly.* Conversely, if the $x(t)$ waveform changes slowly, that is, $a = a_1$ (for instance), the amplitude spectrum is small for large values of frequency—it is concentrated around the vertical axis.

Figure 12.18
The time waveform

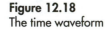

Figure 12.19
The spectra of the exponential waveform

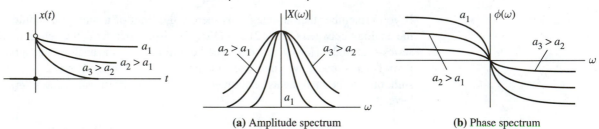

(a) Amplitude spectrum   (b) Phase spectrum

Figure 12.20   An example circuit

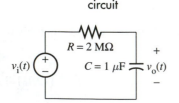

The discussion we have just given of the Fourier integral and its relation to the Fourier series is, of necessity, quite brief and it glosses over a number of important theoretical issues. The main thing to glean from it and our presentation of the Fourier series is that essentially any time-varying waveform can be represented as a weighted sum of sinusoids (or, equivalently, of complex exponentials).

Now let's return to the circuit with which we opened this section, repeated here for ease of reference as Figure 12.20. We will show analytically that the characteristics of the circuit, specifically its ac steady-state transfer function, tell us how the response waveform "looks" in the frequency domain.

Thus, suppose that the independent voltage source has the waveform shown in Figure 12.2, repeated here as Figure 12.21, and that we are to find the *steady-state* response for the circuit variable $v_o(t)$. We first form the periodic extension of the waveform as shown in Figure 12.22 (we will continue to call it $v_i(t)$ for convenience). Now, let's approximate the Fourier series of this waveform as we did in equation (12.1-5), repeated here for ease of reference as equation (12.1-19):

$$v_i(t) = 0.811 \sin(\pi t) + 0.090 \sin(3\pi t) + 0.032 \sin(5\pi t) + \cdots. \tag{12.1-19}$$

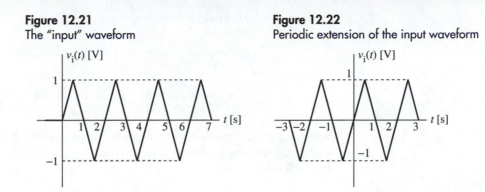

Figure 12.21
The "input" waveform

Figure 12.22
Periodic extension of the input waveform

Let's find the forced response of our circuit that relates $v_o(t)$ to the independent source voltage $v_i(t)$, which we assume to be adjusted to produce a sinusoid at frequency ω, that is,

$$v_i(t) = V_{im}\cos(\omega t + \theta).\qquad(12.1\text{-}20)$$

Next, we represent it by its phasor, as shown in the phasor equivalent circuit in Figure 12.23. Notice that we have expressed the capacitor impedance in terms of the general frequency variable ω, rather than evaluating it at a fixed frequency as we did in our single frequency analysis methods developed in Chapter 11. We can now simply apply the voltage divider rule using impedances to get

$$\overline{V}_o = \frac{1/j\omega C}{R + 1/j\omega C}\overline{V}_i = \frac{1}{1 + j\omega RC}V_{im}\angle\theta°\qquad(12.1\text{-}21)$$

$$= \frac{V_{im}\angle\theta°}{\sqrt{1 + (\omega RC)^2}\,\angle\tan^{-1}(\omega RC)},$$

from which we get

$$\overline{V}_o = \frac{V_{im}}{\sqrt{1 + (\omega RC)^2}}\angle(\theta - \tan^{-1}(\omega RC)).\qquad(12.1\text{-}22)$$

Figure 12.23
The phasor equivalent

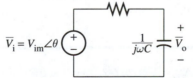

$\overline{V}_i = V_{im}\angle\theta$

$\dfrac{1}{j\omega C}$

$\overset{+}{\underset{-}{\overline{V}_o}}$

Now that we have found the phasor for the forced response, let's go back to equation (12.1-2) and express it in terms of cosines. (Remember, a phasor always represents a cosine, never a sine.) This gives

$$v_i(t) = 0.811\cos(\pi t - 90°) + 0.090\cos(3\pi t - 90°)\qquad(12.1\text{-}23)$$
$$+\ 0.032\cos(5\pi t - 90°) + \cdots.$$

We note that this is the sum of cosine functions at the odd harmonic frequencies $\omega_0 = \pi, 3\omega_0 = 3\pi$, etc. *We can therefore find the forced response due to each, then sum these forced responses using superposition to get the steady-state response for $v_o(t)$.* The computation for the fundamental frequency goes like this:

$$\overline{V}_{o1} = \frac{V_{1m}}{\sqrt{1 + (\omega_0 RC)^2}} \angle(-90° - \tan^{-1}(\omega_0 RC)) \qquad (12.1\text{-}24)$$

$$= \frac{0.811}{\sqrt{1 + (\pi \times 2)^2}} \angle(-90° - \tan^{-1}(2\pi)) = 0.128\angle -90°.$$

Here, V_{1m} is the amplitude of the fundamental term in (12.1-23). This gives the phasor $\overline{V}_{o1}$ associated with the fundamental sinusoidal component of the response. Going back to the time domain, we have

$$v_{o1}(t) = 0.1275 \sin(\pi t). \qquad (12.1\text{-}25)$$

Doing the same thing for each of the component waveforms, we come up with

$$v_o(t) = 0.1275 \sin(\pi t) + 0.0048 \sin(\pi t) + 0.0010 \sin(\pi t) + \cdots. \qquad (12.1\text{-}26)$$

Notice that the fundamental is much larger than any of the higher-order harmonics (by a factor of perhaps 20 or more). This means that the steady-state waveform is approximately a sinusoid at the fundamental frequency—a fact that we noted at the beginning of this section. If you glance back at Figure 12.3, which consists of the computed complete response, you will see that the *steady-state* peak amplitude of the waveform plotted there is therefore correct to within the accuracy of our plot.

We will close this section by noting that we *could* have computed the spectrum of the original one-sided waveform in Figure 12.2 and used it to compute the complete response of the output voltage. However, the problem is complicated by the fact that we would have been forced to multiply our Fourier series, the general form given in (12.1-2) and the specific form for our example waveform in (12.1-4), by a unit step function. Then we would have been forced to compute the complete response for each "turned-on" sinusoidal waveform before using superposition to assemble them to find the final answer. Thus, *the concepts of the Fourier series, Fourier transform, and frequency spectrum are all most aptly suited for the discussion of concepts relating to the steady state only.*

Summary   The main point of this section has been to demonstrate in a largely intuitive manner that there is a frequency domain point of view that, in many situations, is more convenient to use than a time domain equation or plot. We will use this idea often and intensively in the remaining sections of this chapter.

Section 12.1 Quiz

Q12.1-1.   Using a computer software package (say MATLAB) or a graphing calculator, find and plot the time waveform corresponding to the frequency spectra given by $a_0 = 0.5$, $a_n = 0$ for $n \geq 1$, and $b_n = 1/n$ for all n. Assume that the fundamental frequency is $\omega_0 = 2\pi$ rad/s.

Q12.1-2.   For the frequency spectra in Question Q12.1-1, find the corresponding complex Fourier coefficients c_n and plot the amplitude and phase spectra using graph paper, a computer plotting package, or a graphing calculator.

12.2 | The Definition and Use of Frequency Response

In the last section we developed the idea of the frequency domain point of view; in this section, we will illustrate how this idea is used. Generally, we will explore the idea of *frequency response*. Specifically, we will consider one important application: the *filtering* of signals to remove noise or interference.

Ac Forced Response: A Review

Let's begin by recalling how one computes the forced response (steady-state response for stable circuits—and we will assume that all of our circuits are stable). To fix ideas, look at Figure 12.24, which consists of a general circuit or linear system having no internal independent sources, with a sinusoidal input (or excitation) waveform $x(t)$ and an output (or response) waveform $y(t)$. We apply the ac forced response methods we developed in Chapter 11 by looking at the phasor equivalent shown in Figure 12.25. The input signal is represented by its phasor, having amplitude (or magnitude) X_m and angle (or phase) θ. As we know, the forced response phasor will have amplitude Y_m and angle $\theta + \phi$, both of which are to be determined by our analysis. Note that we have chosen the angle of the input as the reference angle and have expressed the angle of the response relative to it by using the additive term $\theta + \phi$. Following the methods developed in Chapter 11, we define the system function in the box in Figure 12.25 as

$$H(\omega) = \frac{\overline{Y}}{\overline{X}} = \frac{Y_m\angle(\theta + \phi)}{X_m\angle\theta} = \frac{Y_m}{X_m}\angle\phi = |H(\omega)|\angle\phi(\omega). \qquad (12.2\text{-}1)$$

Figure 12.24
The ac steady-state response of a general circuit or other linear system

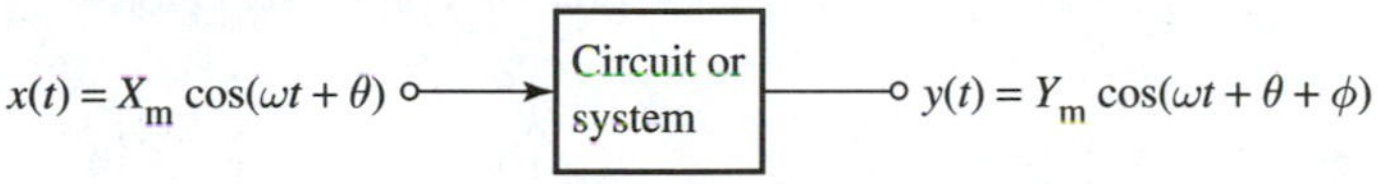

$$x(t) = X_m \cos(\omega t + \theta) \circ\longrightarrow \boxed{\begin{array}{c}\text{Circuit or}\\\text{system}\end{array}} \longrightarrow\circ\; y(t) = Y_m \cos(\omega t + \theta + \phi)$$

Figure 12.25
The phasor equivalent

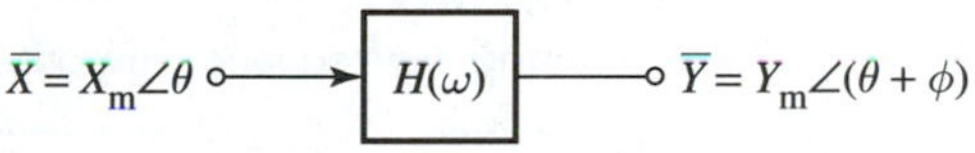

$$\overline{X} = X_m\angle\theta \;\circ\longrightarrow\; \boxed{H(\omega)} \;\longrightarrow\circ\; \overline{Y} = Y_m\angle(\theta + \phi)$$

We have included the frequency ω as an argument for the system function because, in general, the ratio of the output phasor to the input phasor is frequency dependent. If we know $H(\omega)$ for each value of ω, we can determine the ac forced response $y(t)$ for any input sinusoid $x(t)$. Therefore, $H(\omega)$ contains all the information we need to know about the circuit—provided we know its value for each value of ω—and we note that, for each ω, $H(\omega)$ is a *complex* number. Thus, we simply write down the phasor for the input, then multiply it by $H(\omega)$ to determine the response phasor:

$$\overline{Y} = H(\omega)\overline{X}. \qquad (12.2\text{-}2)$$

We recall at this point that the spectrum of any waveform consists of plots of the amplitude and phase as a function of ω. We adapt this idea for the system function $H(\omega)$. We define

$$|H(\omega)| = A(\omega) \qquad (12.2\text{-}3)$$

to be the circuit or system *gain,* and

$$\angle H(\omega) = \phi(\omega) \qquad (12.2\text{-}4)$$

to be the *phase shift.* Notice that both A and ϕ are functions of ω because this is true of the system function $H(\omega)$. Thus, we see that the amplitude of the output is given by

$$|Y(\omega)| = |H(\omega)\overline{X}| = A(\omega)|\overline{X}| = A(\omega)X_m| \qquad (12.2\text{-}5)$$

and its phase by

$$\angle Y(\omega) = \phi(\omega) + \theta. \qquad (12.2\text{-}6)$$

Thus, we multiply the input amplitude by the system gain and add its phase angle to the system phase shift to obtain the output amplitude and phase angle, respectively.

Frequency Response: Gain and Phase

Now let's recall what we did in Section 12.1. We developed the idea that any waveform can be expressed as a weighted sum of pure sinusoids (or complex exponentials); furthermore, the response of a linear circuit or system to that waveform can be determined by finding the response for each component (fundamental and harmonics) and then adding them up using superposition.

This is a highly significant result! To see just what it implies, look at Figure 12.26. There we have shown a test source (a laboratory signal generator) applying a cosine waveform with unit amplitude and general frequency ω to our system under test. We imagine the ac steady-state response to be measured at a given frequency ω, then the input signal to be changed to a new frequency $\omega + \Delta\omega$ and the response measured once again. By observing the steady-state output signal, we can determine the value of the gain $A(\omega)$ and the phase shift $\phi(\omega)$ at each frequency. We can then plot the *frequency response,* that is, the gain and the phase versus frequency, as shown in Figure 12.27.[3] There are, in fact, laboratory instruments that automatically vary the frequency, determine the gain and phase shift at each frequency, and plot the results. They are called *gain and phase sets.*

Figure 12.26
Measuring the frequency response

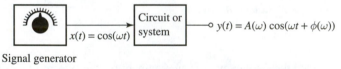

Signal generator

Figure 12.27
The gain and phase shift plots

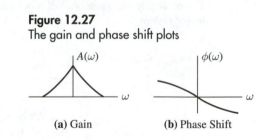

(a) Gain   **(b)** Phase Shift

Here is the significance of what we have just accomplished: *our gain and phase determination allows us to compute the spectrum of the output of our system for any input waveform.* We know that the overall response of our circuit or system is the superposition of its response to individual sinusoids, so we simply multiply our gain function by the amplitude of the sinusoidal component of the input at each frequency and add our phase shift function at each frequency with the phase angle of the input sinusoidal component at that frequency. Thus, the gain and phase shift tell us all we wish to know about our system.

A Practical Example: Removal of "Noise" by Filtering

Let's look at a practical example. Suppose a sinusoidal signal is transmitted over a noisy channel that adds an interfering signal; specifically, assume the "noisy" signal is

$$x(t) = \cos(2\pi t) + 0.5 \cos(200\pi t), \tag{12.2-7}$$

where the $\cos(2\pi t)$ term is the desired signal and the $0.5 \cos(200\pi t)$ term is the additive noise. A plot of one period of the noisy signal is shown in Figure 12.28. We notice that the noise has a frequency that is 100 times that of our desired signal. Does this suggest a possible way of removing the noise? If we could come up with a "filter" that would pass the sinusoid at the frequency of 2π radians per second and block the one at 200π radians per second, we would have achieved our objective.

Let's look at the simple *RC* circuit shown in Figure 12.29. We will assume that the noisy signal is available as a voltage source having the prescribed $x(t)$ as its waveform

Figure 12.28 A noisy waveform

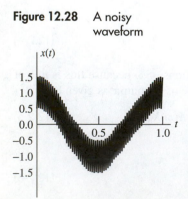

_____

[3] In our more theoretical development of the Fourier series and the Fourier transform in Chapter 14, we will show that the gain is an even function of ω and the phase an odd function, as illustrated in Figure 12.27.

and we will consider the output voltage to be the response of our system. Suppose that we run a test by applying a sinusoidal source at the input and adjusting it so that its amplitude is one volt and its frequency is variable. We can use sinusoidal steady-state techniques to compute the response. For this reason, we show the ac steady-state equivalent circuit in Figure 12.30. We assume that the input signal has zero phase, compute the phasor response, then convert to the time domain. The phasor response is

$$\bar{Y} = \frac{\frac{1}{j\omega C}}{R + \frac{1}{j\omega C}} 1\angle 0° = \frac{1}{1 + j\omega RC} = \frac{1}{\sqrt{1 + (\omega RC)^2}} \angle -\tan^{-1}[\omega RC]. \qquad (12.2\text{-}8)$$

Thus, we see that

$$y(t) = \frac{1}{\sqrt{1 + [\omega RC]^2}} \cos[\omega t - \tan^{-1}[\omega RC]] \qquad (12.2\text{-}9)$$

$$= A(\omega) \cos(\omega t + \phi(\omega)).$$

Thus,

$$A(\omega) = \frac{1}{\sqrt{1 + (\omega RC)^2}} \qquad (12.2\text{-}10)$$

and

$$\phi(\omega) = -\tan^{-1}(\omega RC). \qquad (12.2\text{-}11)$$

We see, therefore, that the phasor response to a unit amplitude, zero phase cosine is, in general, the system function $H(\omega)$.[4] The magnitude of the response phasor is the gain, and the angle is the phase shift. The gain of our particular circuit is plotted in Figure 12.31. From this graph, you can see why we refer to it as a lowpass filter. (Recall that we used this term in Chapter 7 for the fundamental operator upon which all of our time domain

Figure 12.29
A first-order lowpass filter

Figure 12.30
Ac steady-state response

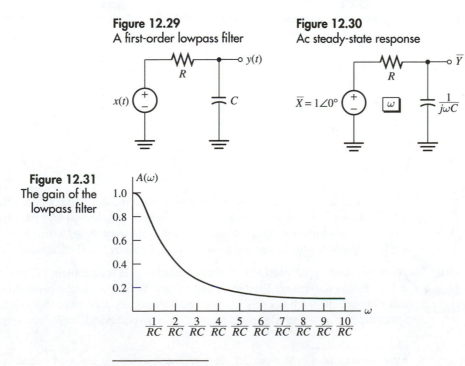

Figure 12.31
The gain of the lowpass filter

---

[4] This is merely a restatement of equation (12.2-2).

analysis was based.) Higher-frequency components in any input signal will be attenuated much more than those at low frequencies. Furthermore, if we make the time constant RC larger, we reduce the high-frequency gain, and if we make the time constant smaller, we increase the high-frequency gain.

Noticing again that the noise in our signal is a sinusoid with a frequency 100 times that of the signal, we consider making the RC time constant such that 2π rad/s $\ll 1/RC \ll 200\pi$ radians per second. Thus, let's see what happens if we adjust it so that

$$\frac{1}{RC} = 20\pi \, \text{rad/s}. \tag{12.2-12}$$

This is a factor of 10 higher than our signal and a similar factor of 10 lower than the noise.

Knowing the response to each sinusoid, we can use superposition to compute the output signal resulting from our original noisy signal of equation (12.2-7) as the input waveform. It is

$$y(t) = A(2\pi) \cos(2\pi t + \phi(2\pi)) + A(200\pi) \cos(200\pi t + \phi(200\pi)) \tag{12.2-13}$$
$$= 0.995 \cos(2\pi t - 5.7°) + 0.05 \cos(200\pi t - 84.3°).$$

The signal component has been almost unchanged (amplitude reduced by about one-half of one percent and the phase shifted by only $-5.7°$), but the noise waveform has been attenuated by a factor of 100 (the phase shift is irrelevant because the signal is noise anyway)! A plot of one period of the response is shown in Figure 12.32. Notice how remarkably clean this signal is! There is essentially no trace of the contaminating noise. Also, carefully reconsider how we have used our frequency domain point of view to come up with the circuit that did the job we wanted it to do.

Figure 12.32
Response of the lowpass filter with $1/RC = 20\pi$

Figure 12.33
A first-order highpass filter

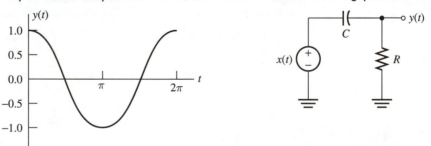

We will leave it as an exercise for you to design the *highpass* filter shown in Figure 12.33 (the same circuit as the one we have worked with, except that the capacitor and resistor have been interchanged) so that the high-frequency signal, considered now to be the signal, is passed, and the low-frequency signal, considered to be the noise, is blocked.

The "Tuning" Problem: Bandpass Filters

Another useful application of frequency response is to the problem of "tuning" a radio receiver. Radio stations are limited by federal law to have a certain bandwidth; that is, they can only occupy so much space in the radio frequency spectrum. Figure 12.34 shows the spectrum[5] of a waveform consisting of two hypothetical radio stations, KOKA and

---

[5] Per our discussion in Section 12.1, the amplitude spectrum is an even function of frequency, and its phase is odd; therefore, we have plotted only the positive frequency half.

Figure 12.34
Spectrum of two
radio stations

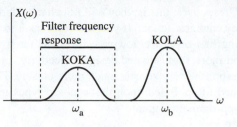

KOLA. This plot represents the amplitude spectrum of the very weak signal that is picked up by the antenna of our radio receiver. Notice that the spectra of the two stations do not overlap; furthermore, the signal strength of the one we desire (say KOKA) is weaker than the other (KOLA). The problem is this: if we do not select one station and reject the other, we will hear a babble from both stations in our speaker. Our solution is to pass the composite signal through a *bandpass* filter,[6] ideally having the rectangular shape shown in the figure. Thus we can pass the station having the frequency components we desire and reject those we do not want. This process is known as *tuning*.

Band Reject (Notch) Filters

Figure 12.35 shows another typical situation to which we can apply the ideas of frequency response and filtering. It is the spectrum of, let us say, the audio signal in a power amplifier in an audio sound system. It is unfortunately true that an amplifier with a very high gain also picks up unwanted disturbances, and one of the most common interfering signals comes from the ac power cables behind the walls in the room in which the amplifier is located. This interference is in the form of a sinusoidal signal whose frequency is 60 Hz, or 120π radians per second, which causes a hum in the speaker. Now, assuming that the audio signal components in a frequency range close to the interfering frequency of 120π rad/s are not very important to the intelligibility of the waveform, we can pass the composite waveform through a *band reject,* or *notch,* filter having the frequency response shown. This eliminates only the interfering waveform and passes our desired audio signal relatively unaffected, providing the "notch" is very narrow.

Figure 12.35
Spectrum of an audio
signal with 60-Hz
interference

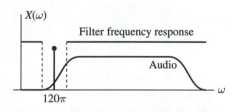

Summary: Filter Types

Let's summarize our discussion. There are a number of standard filter frequency response types that can be applied in a host of practical situations. They are called *lowpass, highpass, bandpass,* and *band reject (or notch)* filters. They have, respectively, the gain versus frequency responses and the acronyms shown in Figures 12.36(a) through (d), respectively. We hasten to note that these are *ideal* filter types. Actual filters will not have these

Figure 12.36
Four standard
filter types

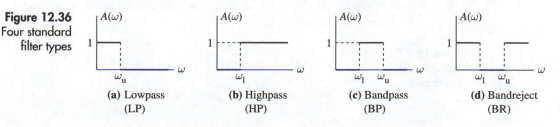

---

[6] There are usually other stations below KOKA in frequency, so we could not use a lowpass filter.

"brick wall" responses; that is, they will not change abruptly from one value to another as the frequency changes. In fact, if you will refer back to Figure 12.31, you will see that the corresponding first-order *RC* lowpass filter was far from being ideal. With more circuit elements and more sophisticated design procedures, one can approximate the ideal filter frequency response characteristic much more closely. There are catalogs containing tables of predesigned filters based upon the standard types in Figure 12.36 and some standard method of approximating them.

The Effects of Phase Shift

Looking at the Figure 13.36 once again, we note that we have defined our ideal responses in terms of the gain only. We have not mentioned phase. Why? The answer lies deeply imbedded in the history of electrical engineering. The telephone system was the first major application of filtering techniques in electrical engineering, and the truth is that the human ear is not sensitive to phase distortion. That is, any two audio signals having the same amplitude spectrum sound very similar, regardless of how their phase spectra might differ. Add this information to the fact that our standard filter types were developed for the telephone system, and you can readily see why we have not mentioned the phase shift characteristics. They were simply not important when the standard filter types were developed.

Phase characteristics *are* quite important in other applications, such as television. What one does for such applications is compare two designs with similar gain characteristics in terms of the resulting phase characteristics. But this brings up the question, "What is a good phase characteristic?" We will discuss this issue in the remainder of this section.

Let's first of all consider a *band-limited* signal[7] such as the one shown in Figure 12.37 being applied to a circuit or other linear time-invariant system with frequency response function $H(\omega)$. Now let's assume that over the frequency range for which the input signal spectrum is nonzero, the *gain is constant* with value A_o and the *phase is linear* with the form $\phi(\omega) = -\omega\tau$ (τ a constant). If the input is a cosine waveform,

$$x(t) = X_m \cos(\omega t + \theta),\qquad (12.2\text{-}14)$$

we see that the response will be

$$y(t) = A_o X_m \cos(\omega t + \theta - \omega\tau) = A_o X_m \cos(\omega[t - \tau] + \theta) = A_o x(t - \tau).\quad (12.2\text{-}15)$$

As we have discussed in Section 12.1, a more general signal $x(t)$ can be considered to be made up of the sum of a number of such cosine waveforms as in (12.2-14). By superposition, the spectrum of the response waveform $y(t)$ is the sum of the individual responses as given by (12.2-15); thus, $y(t)$ is merely the input waveform $x(t)$ multiplied by the real constant factor A_o and delayed by τ seconds:

$$y(t) = A_o x(t - \tau).\qquad (12.2\text{-}16)$$

Our conclusion is simply this: for distortionless filtering of a band-limited waveform, it is desirable that the filter gain be constant and its phase to be linear over the frequency range of that waveform.

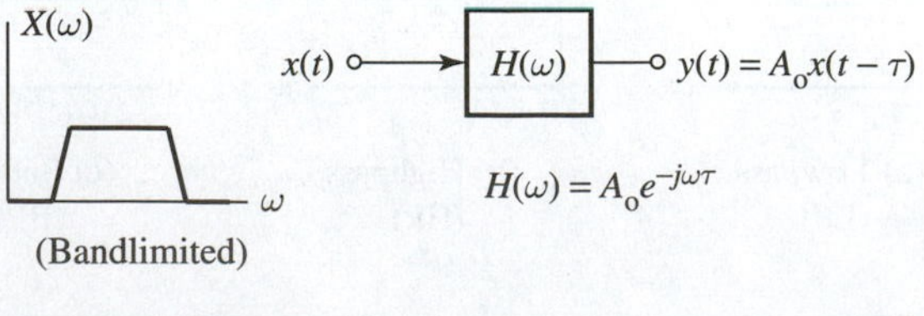

Figure 12.37
The effects of constant gain and linear phase

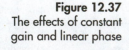

$X(\omega)$

(Bandlimited)

$x(t) \longrightarrow \boxed{H(\omega)} \longrightarrow y(t) = A_o x(t - \tau)$

$H(\omega) = A_o e^{-j\omega\tau}$

---

[7] That is, a signal that is zero except over a finite range of frequencies.

The Allpass Filter

There is one more standard filter type whose gain is a bit surprising. It is shown (for positive frequencies) in Figure 12.38(a). Of what value is such a constant gain circuit? Simply this. It has a phase characteristic, such as the one shown in Figure 12.38(b), that can be adjusted to compensate for nonlinear phase distortion in another filter without affecting the gain. Such a filter is termed an *allpass* filter.

Figure 12.38
The allpass filter

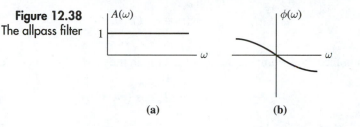

(a)   (b)

In conclusion, we remind you that our time domain analysis in Part II was based upon the time response of a first-order lowpass filter to a general time-varying input signal; thus, you can see that it is a very fundamental configuration indeed.

Section 12.2 Quiz

Q12.2-1. Find the response of the first-order highpass filter in Figure 12.33 to the signal in equation (12.2-7) and find an *RC* product that will effectively eliminate the low-frequency sinusoid and pass the higher-frequency one relatively unaffected. (The practical interpretation here is that the higher-frequency sinusoid is the signal and the lower-frequency sinusoid the interference.)

12.3 | The Parallel Tuned Circuit

In the last section we discussed the tuning problem. This involved the passing of a specific range of frequencies and the rejection of all others; that is, the application of a bandpass filter. The word "tuning" is a musical term referring to the adjustment of an instrument for the proper frequency of operation, and the adjustment of a radio receiver is similar. The simplest and most classical example of a bandpass filter is the *LC tuned circuit,* which we will discuss in this section. At the outset, we observe that we are primarily concerned with the frequency response, that is, the ac forced response with the frequency as a parameter to be varied. We will address the time response of the parallel tuned circuit in Section 12.4.

The Quality Factor Q

Let us start by defining a quantity that measures the quality of an energy storage element. Such an element *should* only store energy and not dissipate any, as is the case for ideal inductors and capacitors. Considering the ac forced response of the two-terminal element (or subcircuit) drawn in Figure 12.39, we define the *quality factor,* or *Q,* by the equation

Figure 12.39 The ac forced response of a circuit element

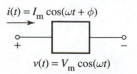

$i(t) = I_m \cos(\omega t + \phi)$

$v(t) = V_m \cos(\omega t)$

$$Q = 2\pi \frac{\text{peak energy stored}}{\text{energy dissipated in one period}} = 2\pi \frac{w_P}{w_D}. \qquad (12.3\text{-}1)$$

In general, we will pick either the voltage or the current to have zero phase angle as a reference and ϕ to be the angle of the other variable, though this is not essential. However, we must remember that our definition of Q is in terms of ac forced response quantities. It can be *used* in other ways, but it is *defined* in terms of the ac forced response.

Q of a Lossy Inductor

Now, let's apply our definition to a practical case: the "lossy" inductor. As we have discussed earlier, inductors are constructed in the form of a coil of wire having finite resistance. Thus, a fairly practical model of a realistic inductor consists of an ideal inductor in

Figure 12.40   A lossy inductor

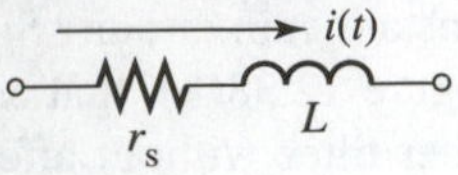

series with a small resistance r_s, as we show in Figure 12.40. We will assume that the current $i(t)$ is sinusoidal, that is, of the form

$$i(t) = I_m \cos(\omega t + \phi). \tag{12.3-2}$$

Recalling that the energy stored in an inductor is given by

$$w(t) = \frac{1}{2}Li^2(t), \tag{12.3-3}$$

we can immediately compute the stored energy in the ac steady state to be

$$w(t) = \frac{1}{2}LI_m^2 \cos^2(\omega t + \phi). \tag{12.3-4}$$

This means that the *peak* value of stored energy—the value we need for equation (12.3-1)—is

$$w_P = \frac{1}{2}LI_m^2. \tag{12.3-5}$$

The ideal inductor element is responsible for all the energy stored in our simple circuit. On the other hand, the only element that dissipates (or absorbs) energy is the resistor. Recall that energy is the integral of power, so the energy dissipated over one full period is

$$W_D = \int_0^T P(t)\, dt = T \cdot \frac{1}{T}\int_0^T P(t)\, dt = T\langle P(t)\rangle \tag{12.3-6}$$

$$= T\langle r_s I_m^2 \cos^2(\omega t + \phi)\rangle = \frac{1}{2}I_m^2 r_s T.$$

Therefore, the Q of our lossy inductor is

$$Q_L = 2\pi\frac{w_P}{w_D} = 2\pi\frac{\frac{1}{2}LI_m^2}{\frac{1}{2}I_m^2 r_s T} = \frac{\omega L}{r_s}. \tag{12.3-7}$$

We have used the fact here that $\omega = 2\pi f = 2\pi/T$, and have placed the subscript L on Q to denote the fact that it is a quantity associated with an inductor.

Example 12.1   A lossy inductor has a series winding resistance of 10 Ω and a nominal value of 10 mH. Find the quality factor at a frequency of 100 krad/s.

Solution   We simply apply (12.3-7):

$$Q_L = \frac{\omega L}{r_s} = \frac{10^5 \times 10^{-2}}{10} = 100. \tag{12.3-8}$$

This, by the way, is a reasonably good inductor Q.

Q of a Lossy Capacitor

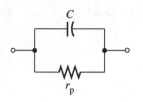

Figure 12.41 A lossy capacitor

Let's now investigate the Q for a lossy capacitor. A capacitor is constructed in the form of parallel metal plates (perhaps rolled up or folded in the final construction phase) separated by some sort of dielectric. Thus, we see that if the dielectric has finite resistance (and it always will!) the lossy capacitor can be quite well approximated by the equivalent circuit shown in Figure 12.41. Assuming that the terminal voltage is sinusoidal, that is, of the form

$$v(t) = V_m \cos(\omega t + \phi), \tag{12.3-9}$$

we see that the energy stored on the ideal capacitor element as a function of time is

$$w(t) = \frac{1}{2} Cv^2(t) = \frac{1}{2} CV_m^2 \cos^2(\omega t + \phi). \tag{12.3-10}$$

The *peak* energy stored is, therefore,

$$w_P = \frac{1}{2} CV_m^2, \tag{12.3-11}$$

and the energy dissipated in one period—that is, in the resistor—is, as for the inductor,

$$w_D = T\langle P(t) \rangle = T\langle \frac{1}{r_p} V_m^2 \cos^2(\omega t + \phi) \rangle = \frac{TV_m^2}{2r_p}. \tag{12.3-12}$$

Taking the ratio of these two quantities and multiplying by 2π results in

$$Q_C = \omega r_p C. \tag{12.3-13}$$

We have used the subscript C on Q to denote the fact that it is the Q associated with a capacitor. We have observed that the ideal capacitor element is responsible for the energy storage, whereas the ideal resistor element is solely responsible for the energy dissipation.

Example 12.2 A lossy capacitor has a parallel dielectric resistance of 10 MΩ and a nominal value of 10 nF. Find the quality factor at a frequency of 100 krad/s.

Solution We simply apply (12.3-13)

$$Q_C = \omega r_p C = 10^5 \times 10^7 \times 10^{-8} = 10,000! \tag{12.3-14}$$

Capacitors typically have much better Q's than do inductors.

Series-to-Parallel Transformation for a Lossy Inductor

We can use the idea of Q to streamline an idea we developed in Chapter 11: series-parallel equivalence. To be specific, let's look once more at the lossy inductor of Figure 12.40, repeated here for ease of reference as Figure 12.42(a). We will show that, over a restricted range of frequencies at least, it is equivalent to the subcircuit in Figure 12.42(b). To do

Figure 12.42 Series-parallel equivalence

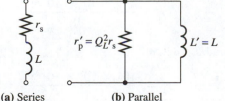

$$r_p' = Q_L^2 r_s \qquad L' = L$$

(a) Series subcircuit   **(b) Parallel** subcircuit

this, let's compute the admittance of the series subcircuit:

$$Y(j\omega) = \frac{1}{r_s + j\omega L} = \frac{r_s - j\omega L}{r_s^2 + (\omega L)^2} = \frac{\dfrac{1}{r_s} - j\dfrac{Q_L}{r_s}}{1 + Q_L^2}. \tag{12.3-15}$$

Now, let us assume that the inductor Q is large—much larger than unity, in fact. Then (because $Q_L^2 \gg 1$) we can write

$$Y(j\omega) \cong \frac{1}{Q_L^2 r_s} - j\frac{1}{Q_L r_s} \tag{12.3-16}$$

$$= \frac{1}{Q_L^2 r_s} + \frac{1}{j\dfrac{\omega L}{r_s} r_s} = \frac{1}{Q_L^2 r_s} + \frac{1}{j\omega L}.$$

If we let

$$r_p' = Q_L^2 r_s \tag{12.3-17}$$

and

$$L' = L, \tag{12.3-18}$$

we will have the equivalent subcircuit of Figure 12.42(b).

The Narrow Band Approximation

We must now address one issue that has perhaps been covered up by our symbols, but it will be quite important in our development of the tuned circuit. The equivalence that we have just derived is only valid as long as the frequency variation is small. The reason for this is simply that $Q_L = \omega L/r_s$ is a function of frequency; hence, our r_p' cannot be a simple resistance because it varies with frequency. If we concentrate on a specific frequency, however, and are only interested in small percentage changes relative to the *center frequency*, we can assume that Q_L is a constant whose value is that assumed at the center frequency. This is called the *narrow band approximation*. The next example explores this idea a bit more concretely.

Example 12.3

Find the parallel equivalent circuit for the lossy inductor of Example 12.1 and find the percent by which r_p' varies over a frequency range of 99 krad/s to 101 krad/s.

Solution

We have already computed the inductor Q to be 100 at 100 krad/s. Assuming that Q_L is much greater than one, we therefore have

$$r_p'(\omega) = Q_L^2 r_s = \left[\frac{\omega L}{r_s}\right]^2 r_s = \frac{(\omega L)^2}{r_s}. \tag{12.3-19}$$

Thus, at $\omega = 100$ krad/s, we have

$$r_p' = \frac{(10^5 \times 10^{-2})^2}{10} = 100 \text{ k}\Omega. \tag{12.3-20}$$

At $\omega = 99$ krad/s, we have

$$r_p' = \frac{(99 \times 10^3 \times 10^{-2})^2}{10} = 98.01 \text{ k}\Omega. \tag{12.3-21}$$

At $\omega = 101$ krad/s, we have

$$r_p' = \frac{(101 \times 10^3 \times 10^{-2})^2}{10} = 102.01 \text{ k}\Omega. \tag{12.3-22}$$

Thus, we see that a variation of $\pm 1\%$ in the frequency results in only a $\pm 2\%$ variation in the resulting parallel resistance. Hence, it is safe to assume (at least over that frequency range) that it is constant at 100 kΩ.

The Lossless Tuned Circuit: Resonance

There is a similar parallel-to-series transformation for the capacitor, but we will leave this for a later section. Instead, we will move on to a study of tuned circuits. We begin by looking at the ideal lossless *LC* parallel circuit in Figure 12.43. We assume that the driving source is a sinusoidal current source and that we are to compute the ac forced response for the voltage as a function of the frequency parameter ω—that is, we are to find the frequency response for the voltage $v(t)$. To do this, we redraw the circuit in the phasor form shown in Figure 12.44. Notice that the frequency ω appears in the element impedances, so we do not have to make a special note of its value on the phasor circuit diagram—as we did in Chapter 11 when considering single frequency analysis with a numerical value for ω. The impedance function for the two-terminal subcircuit is the system function of interest. It is

$$Z(j\omega) = \frac{j\omega L \times \dfrac{1}{j\omega C}}{j\omega L + \dfrac{1}{j\omega C}} = \frac{j\omega L}{1 - \omega^2 LC} = jX(\omega). \tag{12.3-23}$$

This result means that the impedance is always purely imaginary. $X(\omega)$ is the reactance function. This reactance (that is, the impedance without the j multiplier) is plotted versus ω in Figure 12.45. Notice that the response goes to infinity at $\omega = \omega_0$, where

$$\omega_0 = \frac{1}{\sqrt{LC}}. \tag{12.3-24}$$

This phenomenon is called *resonance* and the frequency ω_0 is called the *resonant frequency*.

Figure 12.43
A lossless *LC* parallel tuned circuit

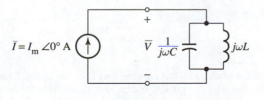

Figure 12.44
The phasor equivalent

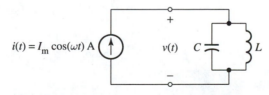

Figure 12.45
The reactance function

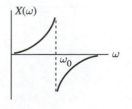

The Lossy Tuned Circuit

Now let's move along[8] to a more practical circuit, but one constructed on the same basic principles. We simply suppose that the inductor and the capacitor both have finite Q—that both are lossy. This gives the equivalent circuit shown in Figure 12.46. If we perform our series-to-parallel transformation on the lossy inductor, we will obtain the equivalent subcircuit shown in Figure 12.47. We note that the parallel resistor is a composite of the loss resistance r_p of the capacitor, the (transformed) parallel equivalent resistance r_p' of the inductor, and any source resistance that might be present (think of the Norton equivalent for the driving source). We will make the narrow band assumption: that this resistance is a constant over the frequency range of interest. The analysis procedure that we are now going to develop will work for any circuit of the form shown in the figure, but the most practical origin of such a subcircuit is from the one shown in Figure 12.46.

Figure 12.46
An *LC* tuned circuit with lossy elements

Figure 12.47
The narrow band equivalent tuned circuit

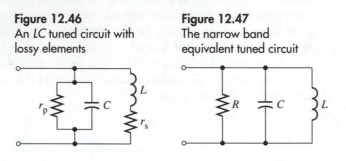

Our <u>first</u> objective will be to find the Q of our subcircuit at the resonant frequency $\omega_0 = 1/\sqrt{LC}$. Suppose we therefore hook up a test current source to our subcircuit and adjust it to be a sinusoid at this special frequency. We have done this in Figure 12.48. It is more convenient to work with phasors, so let's convert it to phasor form as in Figure 12.49. At the resonant frequency $\omega_0 = 1/\sqrt{LC}$, we know that the part of the subcircuit consisting of the capacitor and the inductor presents an infinite impedance, that is $Z'(j\omega_0) = \infty$, to the rest of the circuit because this set of elements is exactly that which we have analyzed earlier. Thus, we see that the impedance at the two terminals of the subcircuit is

$$Z(j\omega_0) = R. \tag{12.3-25}$$

Thus, we can easily determine the phasor terminal voltage:

$$\overline{V} = R\overline{I} = RI_m\angle 0^\circ. \tag{12.3-26}$$

Figure 12.48
Testing the Q of the tuned circuit

Figure 12.49
Phasor form

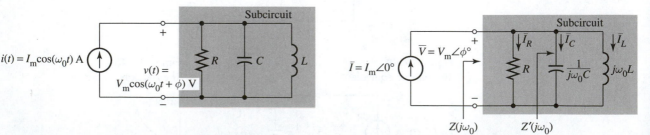

---

[8] A note to the reader who has covered Part II of the text: the preceding subcircuit, the ideal parallel tuned circuit, is only marginally stable. This means that the ac steady-state response does not exist. The ac forced response, however, does exist regardless of stability. The subcircuit about to be considered—the lossy tuned circuit—*does* have an ac steady-state response because it is stable.

In other words,

$$V_m = RI_m. \tag{12.3-27}$$

Let's compute the peak energy stored by our subcircuit. We know that the energy stored on the capacitor as a function of time is

$$w_C(t) = \frac{1}{2}CV_m^2 \cos^2(\omega_0 t). \tag{12.3-28}$$

To find the energy stored by the inductor, we must first determine the inductor current. The phasor form of this current is

$$\bar{I}_L = \frac{\bar{V}}{j\omega_0 L} = \frac{V_m}{j\omega_0 L} = \frac{V_m}{\omega_0 L}\angle -90°. \tag{12.3-29}$$

Therefore, in the time domain, we have

$$i_L(t) = \frac{V_m}{\omega_0 L}\cos(\omega_0 t - 90°) = \frac{V_m}{\omega_0 L}\sin(\omega_0 t). \tag{12.3-30}$$

The energy stored in the inductor is, thus,

$$w_L(t) = \frac{1}{2}L\left[\frac{V_m}{\omega_0 L}\sin(\omega_0 t)\right]^2 = \frac{1}{2}CV_m^2 \sin^2(\omega_0 t). \tag{12.3-31}$$

(To obtain the last expression, we substituted $1/\sqrt{LC}$ for ω_0 and canceiled some terms.) We now see that the peak energy stored in the inductor is the same as that stored in the capacitor—and these are the only two elements capable of storing energy. The only difference between the two is this: the waveform for the inductor is a sine-squared function, rather than a cosine-squared one. This means that when the capacitor is storing its maximum energy, the inductor is storing no energy, and vice versa. The energy is, therefore, being swapped back and forth between the capacitor and the inductor, and none is coming from the source. To illustrate this point, we show graphs of the two stored energies in Figure 12.50. Notice that both axes have been normalized (the time axis is the variable $\omega_0 t$ and the vertical axis is $w(t)/0.5CV_m^2$) and that the two waveforms are 180° out of phase with each other. The key point we need to make, however, is simply that the peak energy stored inside our subcircuit is

$$w_P = \frac{1}{2}CV_m^2. \tag{12.3-32}$$

Figure 12.50
Stored energies in the dynamic elements

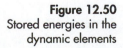

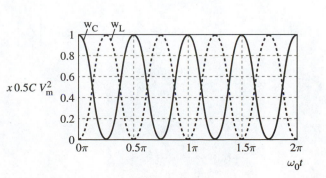

Because the impedance $Z'(j\omega_0) = \infty$ the current into the parallel LC combination is zero; hence, the terminal current of our subcircuit only feeds the resistor. Therefore, the energy absorbed by the subcircuit in one period is the same as the energy absorbed by the resistor in that period. Thus,

$$w_D = \frac{1}{2}\frac{V_m^2}{R}T_0. \tag{12.3-33}$$

The subscript on the period indicates that it is the period at the special frequency ω_0. Thus, the Q of the subcircuit at $\omega = \omega_0$, which we will call Q_0, is

$$Q_0 = 2\pi\frac{w_P}{w_D} = 2\pi\frac{\frac{1}{2}CV_m^2}{\frac{1}{2}\frac{V_m^2}{R}T_0} = \omega_0 RC. \tag{12.3-34}$$

Frequency Response: Fractional Frequency Deviation

Now that we have computed this important quantity, let's return to our parallel tuned circuit and compute the impedance at its terminals as a function of the general frequency ω. To do so, we redraw the circuit in the phasor domain in Figure 12.51. For this parallel circuit we have

$$Z(j\omega) = \cfrac{1}{\cfrac{1}{R} + j\omega C + \cfrac{1}{j\omega L}} = \cfrac{1}{\cfrac{1}{R} + C\left[j\omega + \cfrac{\omega_0^2}{j\omega}\right]} \tag{12.3-35}$$

$$= \cfrac{1}{\cfrac{1}{R} + j\omega_0 C\left[\cfrac{\omega}{\omega_0} - \cfrac{\omega_0}{\omega}\right]} = \cfrac{R}{1 + j\omega_0 RC\left[\cfrac{\omega}{\omega_0} - \cfrac{\omega_0}{\omega}\right]}$$

$$= \cfrac{R}{1 + jQ_0\left[\cfrac{\omega}{\omega_0} - \cfrac{\omega_0}{\omega}\right]}.$$

This last expression is a standard form for the tuned circuit impedance, and the expression in the denominator brackets is called the *fractional frequency deviation*. Thus, considering the terminal current as the input to our system and the voltage as the response, the gain is the magnitude of $Z(j\omega)$ and the phase shift is its angle:

$$A(\omega) = |Z(j\omega)| = \cfrac{R}{\sqrt{1 + Q_0^2\left[\cfrac{\omega}{\omega_0} - \cfrac{\omega_0}{\omega}\right]^2}} \tag{12.3-36}$$

and

$$\phi(\omega) = \angle Z(j\omega) = -\tan^{-1}\left[Q_0\left[\cfrac{\omega}{\omega_0} - \cfrac{\omega_0}{\omega}\right]\right]. \tag{12.3-37}$$

Figure 12.51
Finding the impedance
of the tuned circuit

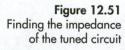
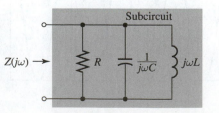

The gain function normalized to R is plotted in Figure 12.52. As you can see, the peak value occurs at $\omega = \omega_0$, the resonant frequency. We have plotted our response, called the *resonance curve,* for two different values of Q_0. Note that the larger the value of Q_0, the sharper is the curve, or, as we say, the more *selective* is the tuned circuit. Thus, if we were tuning in a radio station, we would need a large Q_0 if there were another station very close in frequency and only a small one if the interfering station were very far away in frequency. Notice that the curve is that of a bandpass filter, but it is certainly not ideal because the gain is not flat in the *passband,* that is, over the band of frequencies close to ω_0. In fact, we observe that there is a trade-off: as we make the tuned circuit more selective, we also attenuate our desired signal more at frequencies different from ω_0. It is for this reason that a considerable amount of effort has been expended in developing the branch of applied circuit theory called filter design—simply to design a bandpass filter that is flat in the passband, yet whose response drops rapidly for frequencies beyond those we define as our passband. The simple tuned circuit is, however, significantly important, primarily because of its simplicity and the ease with which it can be tuned. One needs only to vary one parameter, either the capacitance or the inductance, to vary the center frequency ω_0.

Figure 12.52
Resonance curve

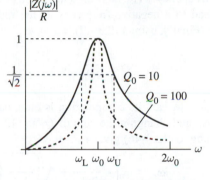

Bandwidth   Let us define the term *passband* more carefully. We will say that it is the range of frequencies for which the normalized gain is greater than $1/\sqrt{2}$. We see from Figure 12.52 that there are two frequencies, one above ω_0 and the other below it, at which the response drops to this value. We call these the *upper cutoff frequency* and the *lower cutoff frequency,* respectively. (These are shown for $Q_0 = 10$ in Figure 12.52.) We define the *bandwidth* of the tuned circuit by

$$B = \omega_U - \omega_L. \tag{12.3-38}$$

Let's compute this bandwidth. Checking equation (12.3-36) we see that at the upper and lower cutoff frequencies, we must have

$$Q_0^2 \left[\frac{\omega}{\omega_0} - \frac{\omega_0}{\omega} \right]^2 = 1. \tag{12.3-39}$$

Because ω_U is greater than ω_0 and ω_L is less than ω_0, we see that

$$Q_0 \left[\frac{\omega_U}{\omega_0} - \frac{\omega_0}{\omega_U} \right] = +1 \tag{12.3-40}$$

and

$$Q_0 \left[\frac{\omega_L}{\omega_0} - \frac{\omega_0}{\omega_L} \right] = -1. \tag{12.3-41}$$

We will solve first for the upper cutoff frequency by letting

$$\frac{\omega_U}{\omega_0} = x. \tag{12.3-42}$$

Thus, (12.3-40) becomes

$$x - \frac{1}{x} = \frac{1}{Q_0}, \tag{12.3-43}$$

or

$$x^2 - \frac{1}{Q_0}x - 1 = 0. \tag{12.3-44}$$

The solution is

$$x = \frac{1}{2Q_0} \pm \sqrt{\left[\frac{1}{2Q_0}\right]^2 + 1}. \tag{12.3-45}$$

Which sign do we accept and which do we reject? (After all, there can be only one positive upper cutoff frequency!) The answer is the positive sign. Why? Because choosing the negative would lead to a negative frequency as the radical term is larger in magnitude than the first term. Finally, using (12.3-42), we have

$$\omega_U = \omega_0\left[\frac{1}{2Q_0} + \sqrt{\left[\frac{1}{2Q_0}\right]^2 + 1}\right]. \tag{12.3-46}$$

What about ω_L? In fact, its computation is quite easy now. We simply note that exchanging ω_L for ω_U and changing the sign of Q_0 in (12.3-40) results in (12.3-41); hence, doing the same thing to (12.3-46) gives ω_L:

$$\omega_L = \omega_0\left[-\frac{1}{2Q_0} + \sqrt{\left[\frac{1}{2Q_0}\right]^2 + 1}\right]. \tag{12.3-47}$$

Notice that these two values are not arithmetically symmetric relative to ω_0; in fact, they are geometrically symmetric. We leave it to you to verify this by showing that the square root of their product is ω_0:

$$\sqrt{\omega_U\omega_L} = \omega_0. \tag{12.3-48}$$

The bandwidth, however, is given by the simple expression

$$B = \omega_U - \omega_L = \frac{\omega_0}{Q_0}. \tag{12.3-49}$$

Does this give you a hint as to how one might measure ω_0 and Q_0 in the laboratory? Looking back at our resonance curve in Figure 12.52, we see that we merely search for the peak frequency, which is ω_0. Next, we find the upper cutoff frequency by noting the frequency at which the response has dropped by about 30% (that is to $1/\sqrt{2} = 0.707$) and identifying it as ω_U. Finally, we find ω_L in a similar manner and subtract them to find the bandwidth. Then we compute Q_0 from (12.3-49).

Phase Shift   Although the magnitude or gain characteristic is enough for many applications, we have noted that the phase shift is important for others. Thus, it is of interest to see how the phase shift varies with frequency. We have plotted $\phi(\omega)$ as given by equation (12.3-37) in

Figure 12.53
The phase shift of the
tuned circuit

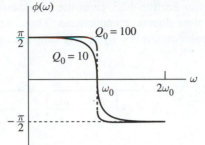

Figure 12.53 for two values of Q_0. Observe that the phase shift is zero at resonance and approaches $\pm\pi/2$ rad for frequencies well below and well above the resonant frequency. Before leaving the steady state frequency response characteristics of the parallel tuned circuit, we should perhaps define resonance more carefully: *the resonant frequency of a two-terminal subcircuit is the frequency at which its impedance is purely resistive.* A look at equation (12.3-35) shows immediately that this frequency is ω_0, which is consistent with the term we have been using for this frequency.

Section 12.3 Quiz

Q12.3-1. If the parallel resistance of the lossy capacitor in Figure Q12.3-1 is 10 MΩ, the series resistance of the lossy inductor is 100 Ω, $C = 10$ nF, and $L = 10$ mH, use the narrow band approximation to:

  a. Find the values of ω_0 and Q_0,
  b. Find the values of ω_U, ω_L, and B,
  c. Find the frequencies at which $\phi = \pm45°$.

Then find the percentage error in making the narrow band as-

sumption by computing the effective parallel resistance for the entire tuned circuit at ω_0, ω_U, and ω_L.

Figure Q12.3-1

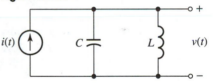

12.4 | Time Response of the Parallel Tuned Circuit

The Time Response to a "Gated" Sinusoid

We have not completely finished our investigation of the parallel tuned circuit because the steady-state frequency response is not the only item of concern. We are also interested in how quickly the circuit responds in time. For ease of reference, we repeat the circuit diagram[9] for the current source-driven parallel tuned circuit in Figure 12.54. Because a tuned circuit only passes signals in the vicinity of $\omega = \omega_0$, let's apply a sinusoid at that frequency and "turn it on" at $t = 0$:

$$i(t) = \cos(\omega_0 t)u(t). \tag{12.4-1}$$

We can represent it as an impulse response of the form

$$i(t) = \frac{p}{p^2 + \omega_0^2}\delta(t). \tag{12.4-2}$$

---

[9] We use operator impedances for the inductor and capacitor. We will be analyzing this circuit for its time response, so you must be familiar with the contents of Part II in order to follow our development. Alternatively, if you have covered Chapter 13, simply replace p by s, replace $\delta(t)$ by 1, and use Laplace transforms.

Equation (12.3-35) in Section 12.3 gives the impedance of our tuned circuit, but in an awkward form for time domain computations. Thus, we first put it into another standard form as a ratio of polynomials in p by simply substituting p for $j\omega$ and rearranging:

$$Z(p) = \frac{R}{1 + Q_0\left[\dfrac{p}{\omega_0} + \dfrac{\omega_0}{p}\right]} = \frac{R\dfrac{\omega_0}{Q_0}p}{p^2 + \dfrac{\omega_0}{Q_0}p + \omega_0^2}. \tag{12.4-3}$$

Thus,

$$v(t) = Z(p)i(t) = \frac{R\dfrac{\omega_0}{Q_0}p^2}{[p^2 + \omega_0^2]\left[p^2 + \dfrac{\omega_0}{Q_0}p + \omega_0^2\right]}\delta(t). \tag{12.4-4}$$

Expanding in partial fractions (using real arithmetic), we have

$$v(t) = \frac{R\dfrac{\omega_0}{Q_0}p^2}{[p^2 + \omega_0^2]\left[p^2 + \dfrac{\omega_0}{Q_0}p + \omega_0^2\right]}\delta(t) \tag{12.4-5}$$

$$= \left[\frac{Ap + B}{p^2 + \omega_0^2} + \frac{Cp + D}{p^2 + \dfrac{\omega_0}{Q_0}p + \omega_0^2}\right]\delta(t).$$

Cross-multiplying and equating coefficients, we obtain (after a bit of algebra left to you) $B = D = 0, A = R$, and $C = -R$. Thus, we have

$$v(t) = \left\{\frac{Rp}{p^2 + \omega_0^2} - \frac{Rp}{p^2 + \dfrac{\omega_0}{Q_0}p + \omega_0^2}\right\}\delta(t). \tag{12.4-6}$$

To make the algebra simpler, we recall that ω_0/Q_0 was identified to be the bandwidth B, so we will use this symbol. We will also introduce the *damped natural frequency*

$$\omega_d = \omega_0\sqrt{1 - \left[\frac{1}{2Q_0}\right]^2}. \tag{12.4-7}$$

Using these symbols and completing the square on the denominator of the second term in equation (12.4-6), we get

$$v(t) = \left\{\frac{Rp}{p^2 + \omega_0^2} - \frac{R\left[p + \dfrac{B}{2}\right] - R\dfrac{B}{2}}{\left[p + \dfrac{B}{2}\right]^2 + \omega_d^2}\right\}\delta(t) \tag{12.4-8}$$

$$= \left\{\frac{Rp}{p^2 + \omega_0^2} - \frac{R\left[p + \dfrac{B}{2}\right]}{\left[p + \dfrac{B}{2}\right]^2 + \omega_d^2} + \frac{R}{\sqrt{(2Q_0)^2 - 1}}\frac{\omega_d}{\left[p + \dfrac{\omega_0}{2Q_0}\right]^2 + \omega_d^2}\right\}\delta(t).$$

Thus, assuming that $Q_0 > 0.5$ (so that the roots of the denominator in the second two

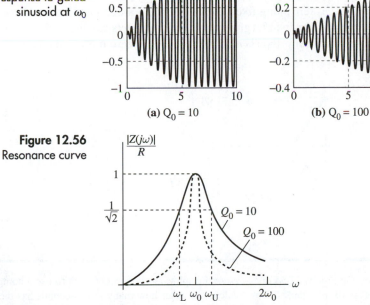

Figure 12.54 The parallel tuned circuit

i(t)= I_m cos(ω_0 t)u(t) A
+
v(t)
–
R
1/Cp
Lp

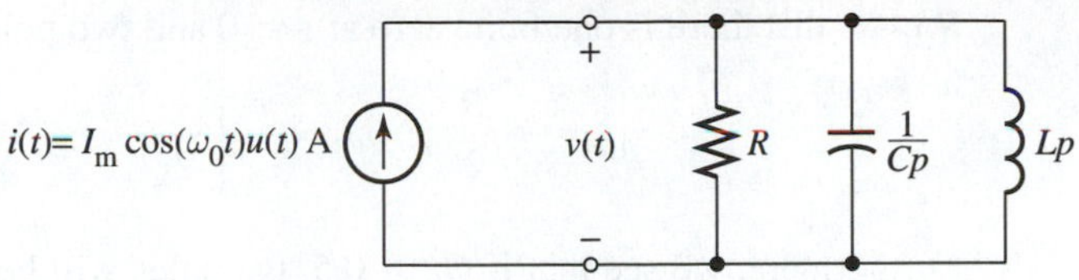

terms of (12.4-8) are complex), we get

$$v(t) = R\cos(v_0 t)u(t) - Re^{-Bt/2}\left\{\cos[\omega_d t] - \frac{1}{\sqrt{(2Q_0)^2 - 1}}\sin[\omega_d t]\right\}u(t). \quad (12.4\text{-}9)$$

This response is graphed in Figure 12.55 for the two values of Q_0 for which the frequency response was plotted in Figure 12.52 of Section 12.3, repeated here for ease of reference as Figure 12.56. The center frequency for the plots in Figure 12.55 was assumed to be $\omega_0 = 10$ rad/s and R was assumed to be 1 Ω (both, merely for numerical convenience). Notice that the response for $Q_0 = 10$ has essentially reached its steady state amplitude of 1 V after 10 seconds have elapsed, but the higher Q_0 response has attained only about 40% of that value at the end of the same time interval. The conclusion is that the higher the value of Q_0, that is, the higher the selectivity, the longer it will take the circuit to settle down into the steady state—the response time is slower for higher values of Q_0.

Figure 12.55 Time response to gated sinusoid at ω_0

1
0.5
0
–0.5
–1
0
5
10
(a) Q_0 = 10
0.4
0.2
0
–0.2
–0.4
0
5
10
(b) Q_0 = 100

Figure 12.56 Resonance curve

|Z(jω)|/R
1
1/√2
Q_0 = 10
Q_0 = 100
ω
ω_L ω_0 ω_U
2ω_0

The Pole-Zero Diagram for the Parallel Tuned Circuit

There is a strong correspondence between the time and frequency responses of the tuned circuit and the location of the poles of $Z(s)$ in the complex plane. If we look back at equation (12.4-3) and simply replace p with s, we obtain

$$Z(s) = \frac{R}{1 + Q_0\left[\dfrac{s}{\omega_0} + \dfrac{\omega_0}{s}\right]} = \frac{R\dfrac{\omega_0}{Q_0}s}{s^2 + \dfrac{\omega_0}{Q_0}s + \omega_0^2} \quad (12.4\text{-}10)$$

$$= \frac{RBs}{s^2 + Bs + \omega_0^2} = \frac{RBs}{(s + B/2)^2 + \omega_d^2}.$$

We see that there is one finite zero at $s = 0$ and two poles located at

$$s_{1,2} = -\frac{B}{2} \pm j\omega_\mathrm{d} = \omega_0\left[\frac{-1}{2Q_0} \pm j\sqrt{1 - \left[\frac{1}{2Q_0}\right]^2}\right]. \qquad (12.4\text{-}11)$$

Furthermore, we see that if $Q_0 > 0.5$, the poles will be complex. This case of high Q_0 is the one of interest. The PZD (pole-zero diagram) is plotted in Figure 12.57. We observe that the poles form a complex conjugate pair and that their magnitude is

$$|s_{1,2}| = \omega_0 \qquad (12.4\text{-}12)$$

independent of the value of Q_0. The resonant frequency ω_0 is also known as the *natural frequency* and $\omega_\mathrm{d} = \omega_0\sqrt{1 - 1/(2Q_0)^2}$ as the *damped natural frequency*. Notice that as Q_0 is allowed to increase, the poles swing around toward the imaginary axis, and the damped natural frequency approaches the natural frequency more and more closely. The angle β is a measure of the degree by which the poles deviate from the $j\omega$ axis; in fact, we have

$$\tan(\beta) = \frac{\omega_0\sqrt{1 - \left[\frac{1}{(2Q_0)^2}\right]}}{\dfrac{\omega_0}{2Q_0}} = \sqrt{(2Q_0)^2 - 1}. \qquad (12.4\text{-}13)$$

When $Q_0 = 0.5$, we see that $\beta = 0$; as $Q_0 \to \infty$, $\beta \to \tan^{-1}(\infty) = \pi/2$. As Q_0 becomes larger, with ω_0 remaining fixed, the poles swing around and approach the $j\omega$ axis. At the same time, the bandwidth (given by $B = \omega_0/Q_0$) gets smaller and smaller, and the natural response (whose envelope is $e^{-[\omega_0/2Q_0]t}$) decays to zero more and more slowly.

Figure 12.57
The PZD of $Z(s)$

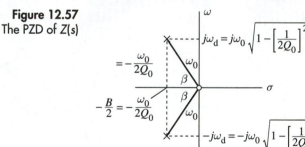

Section 12.4 Quiz

Q12.4-1. Consider the parallel tuned circuit in Figure 12.54 of this section. Assume that $R = 100\ \mathrm{k\Omega}$, $C = 10$ nF, and $L = 10\mathrm{mH}$.

  a. Draw the PZD, giving all numerical values.
  b. Find and plot the response waveform for $v(t)$ if $i(t) = 10 \sin(10^5 t)u(t)\ \mu A$.

Note: For plotting, you might wish to use a computer or graphing calculator. After how many microseconds has the response settled down into the ac steady state?

12.5 | The Series Tuned Circuit and Second-Order Passive Filters

The Lossless Series Tuned Circuit

Figure 12.58 A series tuned circuit

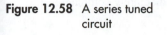

Figure 12.58 shows an ideal *series tuned circuit*. Its behavior is very much like that of the parallel tuned circuit that we investigated in the last section, except that all the properties of the latter which we considered on an impedance basis hold for this circuit on an admittance basis. We will analyze this circuit and its more realistic lossy counterpart in this section, then we will show how knowledge of the series and parallel tuned circuits allows us to work with several practical second-order filter circuits.

The admittance of the series tuned circuit in Figure 12.58 is

$$Y(j\omega) = \frac{1}{Z(j\omega)} = \frac{1}{j\omega L + \dfrac{1}{j\omega C}} = \frac{1}{L}\frac{j\omega}{(j\omega)^2 + \omega_0^2} = j\frac{\omega/L}{\omega_0^2 - \omega^2} = jB(\omega) \quad \text{(12.5-1)}$$

Figure 12.59 The susceptance as a function of frequency

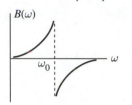

where

$$\omega_0 = \frac{1}{\sqrt{LC}}, \quad \text{(12.5-2)}$$

and $B(\omega)$ is the susceptance. A sketch of $B(\omega)$ versus ω is shown in Figure 12.59. Notice that the susceptance approaches infinity at $\omega = \omega_0$; this means that it is equivalent to a short circuit at ω_0. Thus, we see that the admittance of our series tuned circuit behaves precisely like the impedance of the parallel tuned circuit in the last section.

The Parallel-to-Series Transformation for the Lossy Capacitor

Before we consider the lossy version of the series tuned circuit, look at Figure 12.60(a), which shows a lossy capacitor. The loss can be modeled very well by the parallel resistance shown, which is the resistance of the dielectric of the capacitor. The admittance is

$$Y(j\omega) = j\omega C + \frac{1}{r_p}. \quad \text{(12.5-3)}$$

Figure 12.60 Series equivalent of a lossy capacitor

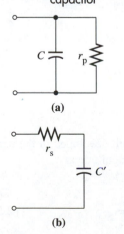

Inverting to obtain the impedance, we get

$$Z(j\omega) = \frac{r_p}{1 + j\omega r_p C} = \frac{r_p}{1 + jQ_C} = \frac{r_p - jr_p Q_C}{1 + Q_C^2}. \quad \text{(12.5-4)}$$

If we assume that $Q_C \gg 1$ (virtually always a good assumption for a capacitor!), then

$$Z(j\omega) = \frac{r_p - jr_p Q_C}{1 + Q_C^2} = \frac{r_p}{1 + Q_C^2} - j\frac{r_p Q_C}{1 + Q_C^2} \quad \text{(12.5-5)}$$

$$\cong \frac{r_p}{Q_C^2} + \frac{r_p}{j\omega r_p C} = \frac{r_p}{Q_C^2} + \frac{1}{j\omega C}.$$

This gives the series equivalent shown in Figure 12.60(b) with

$$r_s = \frac{r_p}{Q_C^2} \quad \text{and} \quad C' = C. \quad \text{(12.5-6)}$$

The Q of the Lossy Series Tuned Circuit at Resonance

Look, once again, at the series tuned circuit in Figure 12.58; this time, however, assume that both the inductor and the capacitor are lossy. This results in the equivalent shown in Figure 12.61. Any driving source resistance (considered as a Thévenin equivalent) can be

Figure 12.61 A lossy series tuned circuit

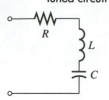

included in the series resistance R along with the series inductor loss resistance and the transformed capacitor loss resistance. Let's apply a sinusoidal test voltage source to our subcircuit, as shown in Figure 12.62, and test the Q_0, that is, the Q of the entire subcircuit at the frequency ω_0. We see right away that as $j\omega_0 L + (1/j\omega_0 C) = 0$, we have $Z'(j\omega_0) = 1/Y'(j\omega_0) = 0$, so $Z(j\omega_0) = R$. Thus, the current is in phase with the source voltage; furthermore, the voltage drop across the inductor/capacitor series combination is zero because its series impedance is zero. Hence, we can immediately compute

$$Q_0 = 2\pi \times \frac{w_P}{w_D} = 2\pi \times \frac{\frac{1}{2}L\frac{V_m^2}{R^2}}{\frac{1}{2}\frac{V_m^2}{R}T_0} = \frac{\omega_0 L}{R}; \qquad (12.5\text{-}7)$$

where T_0 is the period of a sinusoid at the resonant frequency ω_0. Although we will not go into the details, it is easy to show (just as we did in the last section) that the peak energies stored by the inductor and the capacitor are equal to one another—and therefore to the peak energy stored by the subcircuit itself.

Figure 12.62
Testing the Q at the resonant frequency

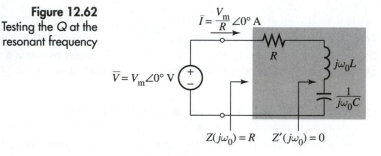

Frequency Response   Now look back at Figure 12.61 and compute the impedance. It is

$$Z(j\omega) = R + j\omega L + \frac{1}{j\omega C} = R\left[1 + \frac{\omega_0 L}{R}\left[\frac{j\omega}{\omega_0} + \frac{\omega_0}{j\omega}\right]\right], \qquad (12.5\text{-}8)$$

so the admittance is

$$Y(j\omega) = \frac{\frac{1}{R}}{1 + jQ_0\left[\frac{\omega}{\omega_0} - \frac{\omega_0}{\omega}\right]}. \qquad (12.5\text{-}9)$$

If you compare this with the $Z(j\omega)$ for the parallel tuned circuit developed in the preceding section, you will see that they are identical in form. The magnitude is

$$|Y(j\omega)| = \frac{\frac{1}{R}}{\sqrt{1 + Q_0^2\left[\frac{\omega}{\omega_0} - \frac{\omega_0}{\omega}\right]^2}}, \qquad (12.5\text{-}10)$$

and the phase shift is

$$\angle Y(j\omega) = -\tan^{-1}\left[Q_0\left[\frac{\omega}{\omega_0} - \frac{\omega_0}{\omega}\right]\right]. \qquad (12.5\text{-}11)$$

This behavior is sketched for reference in Figure 12.63 (Y normalized to $1/R$).

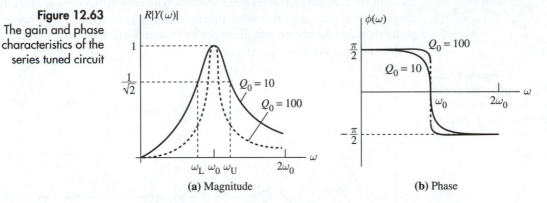

Figure 12.63
The gain and phase characteristics of the series tuned circuit

(a) Magnitude (b) Phase

Filtering Examples   The following example investigates a second-order bandpass filter.

Example 12.4   Find the voltage transfer function of the circuit shown in Figure 12.64. Plot its gain and phase characteristics.

Figure 12.64
A second-order bandpass filter

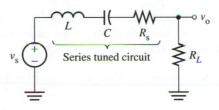

Series tuned circuit

Solution   Notice that there is a series tuned circuit between the voltage source (the filter input) and the output terminal. Calling its impedance $Z(j\omega)$, we have[10]

$$H(\omega) = \frac{\overline{V}_o(\omega)}{\overline{V}_s(\omega)} = \frac{R_L}{Z(j\omega) + R_L} = \frac{R_L Y(j\omega)}{1 + R_L Y(j\omega)}. \qquad (12.5\text{-}12)$$

Now we already know the functional form of $Y(j\omega)$ for it is given by (12.5-9). Using this in equation (12.5-12), we obtain

$$H(\omega) = \frac{R_L/R_s}{1 + R_L/R_s + Q_0\left[\dfrac{j\omega}{\omega_0} + \dfrac{\omega_0}{j\omega}\right]} = \frac{H_0}{1 + jQ_0'\left[\dfrac{\omega}{\omega_0} - \dfrac{\omega_0}{\omega}\right]}, \qquad (12.5\text{-}13)$$

where

$$H_0 = \frac{R_L}{R_s + R_L} \qquad (12.5\text{-}14)$$

and

$$Q_0' = \frac{R_s}{R_s + R_L} Q_0. \qquad (12.5\text{-}15)$$

Thus, our voltage transfer function is exactly the same in form as the admittance of the series tuned circuit or the impedance of the parallel tuned circuit. There are only two differences: Q_0 has been decreased by the resistance ratio given in (12.5-15), and the gain constant H_0 is a voltage gain, not an admittance or impedance. The gain and phase varia-

---

[10] Recall that we have agreed to suppress the j in system functions such as $H(\omega)$, but to retain it for impedance and admittance functions—simply because this is conventional in the literature.

tions with frequency are shown in Figure 12.65. Note that they are identical in form with those of the parallel and series tuned circuits. The only differences are that the gain function is now a voltage gain, a more practical quantity for applications, rather than an impedance or an admittance, and the upper and lower 3-dB frequencies are changed because of the modified value of Q_0.

Figure 12.65
The gain and phase characteristics of the second-order bandpass filter

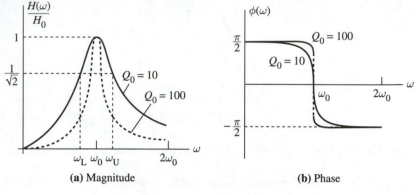

(a) Magnitude

(b) Phase

As you can see by the previous example, we are able to solve quite complicated circuits by recognizing the basic elements in the parallel and/or series tuned circuits. Our next example discusses a second-order notch filter.

Example 12.5

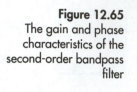

Find the voltage gain transfer function of the circuit in Figure 12.66 and plot its gain and phase versus ω.

Figure 12.66
A second-order notch filter

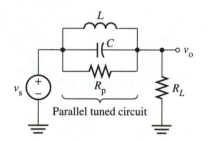

Solution   We start by recognizing that a major component of this circuit is a parallel tuned circuit, whose impedance we know. Thus, we can write the voltage gain transfer function as

$$H(\omega) = \frac{\overline{V}_o}{\overline{V}_s} = \frac{R_L}{Z(j\omega) + R_L} = \frac{R_L}{\dfrac{R_p}{1 + jQ_o\left[\dfrac{\omega}{\omega_o} - \dfrac{\omega_o}{\omega}\right]} + R_L} \quad (12.5\text{-}16)$$

$$= H_o \frac{1 + jQ_o\left[\dfrac{\omega}{\omega_o} - \dfrac{\omega_o}{\omega}\right]}{1 + jQ_o'\left[\dfrac{\omega}{\omega_o} - \dfrac{\omega_o}{\omega}\right]},$$

where

$$H_o = \frac{R_L}{R_L + R_p} \quad (12.5\text{-}17)$$

and

$$Q_o' = \frac{R_L}{R_p + R_L} Q_o. \quad (12.5\text{-}18)$$

This is somewhat different from the transfer function of a tuned circuit itself, so let's investigate it. Converting to Euler form by taking the magnitude and angle, we have the following expressions:

$$A(\omega) = |H(\omega)| = H_\text{o}\sqrt{\frac{1 + Q_\text{o}^2\left[\dfrac{\omega}{\omega_\text{o}} - \dfrac{\omega_\text{o}}{\omega}\right]^2}{1 + Q_\text{o}'^2\left[\dfrac{\omega}{\omega_\text{o}} - \dfrac{\omega_\text{o}}{\omega}\right]^2}}$$ (12.5-19)

and

$$\phi(\omega) = \angle H(\omega) = \tan^{-1}\left[Q_\text{o}\left[\frac{\omega}{\omega_\text{o}} - \frac{\omega_\text{o}}{\omega}\right]\right] - \tan^{-1}\left[Q_\text{o}'\left[\frac{\omega}{\omega_\text{o}} - \frac{\omega_\text{o}}{\omega}\right]\right].$$ (12.5-20)

Figure 12.67 shows a plot of these frequency functions. We have used 0.1 for the H_o value and 20 for Q_o, thus making $Q_\text{o}' = 2$. We have assumed that $\omega_\text{o} = 1$. Although we used a computer to plot Figure 12.67, we can make a few observations about the shape very easily by looking at the circuit. The impedance of the inductor is $j\omega L$ and that of the capacitor is $1/j\omega C$. Thus, at dc ($\omega = 0$) the inductor is a short circuit—this effectively connects the input directly to the output. At infinite frequency ($\omega = \infty$), the capacitor is similarly a short circuit. Thus, we know that the gain is unity and the phase shift zero at these two frequencies. Finally, we see that the inductor and the capacitor are parallel connected; this subcircuit has an effective impedance of infinity at ω_o. Thus, it is equivalent to an open circuit at that frequency. The gain, therefore, is $H_\text{o} = 0.1$ and the phase once more is zero. These points correspond to our plots. Notice that a good notch filter (this one is not terribly good!) would have a large value of R_p relative to the value of R_L and a high Q_o. Thus, the notch would be very narrow and very deep (approximately zero at ω_o).

Figure 12.67
Gain and phase of the example notch filter

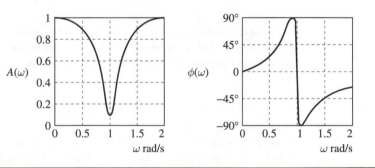

Canonical Forms for Second-Order Filters

Before we end this section, we would like to present a set of standard forms—a synonym is *canonical forms*—for the three most basic standard second-order filter types: lowpass, bandpass, and highpass. They are

$$H_\text{LP}(\omega) = \frac{\omega_\text{o}^2}{(j\omega)^2 + \dfrac{\omega_\text{o}}{Q_\text{o}}(j\omega) + \omega_\text{o}^2},$$ (12.5-21)

$$H_\text{BP}(\omega) = \frac{\dfrac{\omega_\text{o}}{Q_\text{o}}(j\omega)}{(j\omega)^2 + \dfrac{\omega_\text{o}}{Q_\text{o}}(j\omega) + \omega_\text{o}^2},$$ (12.5-22)

and

$$H_{HP}(\omega) = \frac{(j\omega)^2}{(j\omega)^2 + \dfrac{\omega_0}{Q_0}(j\omega) + \omega_0^2}. \tag{12.5-23}$$

There are referred to as standard forms because they all have the same denominator; furthermore, the gains are unity at dc, ω_0, and infinite frequency, respectively. We can relate these expressions to those for tuned circuit impedances and/or admittances by restructuring the denominator polynomial:

$$D(j\omega) = (j\omega)^2 + \frac{\omega_0}{Q_0}(j\omega) + \omega_0^2 = \frac{\omega_0}{Q_0}(j\omega)\left[1 + jQ_0\left[\frac{\omega}{\omega_0} - \frac{\omega_0}{\omega}\right]\right]. \tag{12.5-24}$$

The second factor is one that we easily recognize to be the denominator of our tuned circuit admittance or impedance. Using this form for $D(s)$ we can rewrite equations (12.5-21) through (12.5-23) as

$$H_{LP}(\omega) = \frac{Q_0\omega_0}{j\omega\left[1 + jQ_0\left[\dfrac{\omega}{\omega_0} - \dfrac{\omega_0}{\omega}\right]\right]}, \tag{12.5-25}$$

$$H_{BP}(\omega) = \frac{1}{1 + jQ_0\left[\dfrac{\omega}{\omega_0} - \dfrac{\omega_0}{\omega}\right]}, \tag{12.5-26}$$

and

$$H_{HP}(\omega) = \frac{\dfrac{Q_0}{\omega_0}(j\omega)}{1 + jQ_0\left[\dfrac{\omega}{\omega_0} - \dfrac{\omega_0}{\omega}\right]}. \tag{12.5-27}$$

These forms clearly show their relationship to parallel and series tuned circuits. For example, we see that as $\omega \to 0$, $H_{LP}(\omega) \to 1$; furthermore, as $\omega \to \infty$, $H_{LP}(\omega) \to 0$. Thus, this transfer function is indeed that of a lowpass filter. Similar considerations hold for the other two.

Filter Design Using the Canonical Forms

Let's see how these forms can be useful. Let's look at the series ac steady-state equivalent circuit shown in Figure 12.68. If there are exactly one resistor, one capacitor, and one inductor in this circuit, we see that if we look into the input terminal with nothing connected at the output, then it is a series tuned circuit and its admittance (the reciprocal of the sum of the three impedances) is given by (12.5-9). Where should we place each element? The answer depends upon the filter type desired. Using (12.5-9) and writing

$$H(\omega) = \frac{Z_3(j\omega)}{Z_1(j\omega) + Z_2(j\omega) + Z_3(j\omega)} = \frac{Z_3(j\omega)/R}{1 + jQ_0\left[\dfrac{\omega}{\omega_0} - \dfrac{\omega_0}{\omega}\right]}, \tag{12.5-28}$$

we see that we can get any of the three basic types. For a lowpass configuration, we merely choose the capacitor for Z_3. Thus, we have

$$H(\omega) = \frac{\dfrac{1}{RC(j\omega)}}{1 + jQ_0\left[\dfrac{\omega}{\omega_0} - \dfrac{\omega_0}{\omega}\right]}, \tag{12.5-29}$$

so, if we compare this result with equation (12.5-25), we must have

$$Q_o \omega_o = \frac{\omega_o L}{R} \qquad \omega_o = \frac{1}{RC}.$$ (12.5-30)

Thus, we see that we have the correct numerator for realizing the standard lowpass filter regardless of the value of the capacitor. The circuit is shown in Figure 12.69 in time domain form. We will leave it to you to explore the other configurations.

Figure 12.68
A general three-element series-type circuit

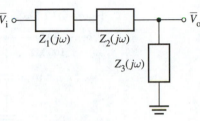

Figure 12.69
A second-order lowpass filter based on the series tuned circuit

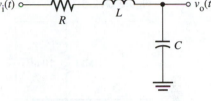

A Resonance Example

Before leaving the topic of resonance, we remind you of the general definition. A two-terminal circuit is said to be *resonant* at any nonzero, finite frequency[11] for which the ac steady-state impedance is purely real. This does not always occur where an inductor and a capacitor produce equal and opposite reactances, as shown in the next example.

Example 12.6 Find the resonant frequency(ies) of the circuit in Figure 12.70.

Figure 12.70
An example circuit

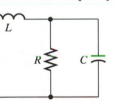

Solution The ac steady-state impedance is

$$Z(j\omega) = j\omega L + \frac{R \times \dfrac{1}{j\omega C}}{R + \dfrac{1}{j\omega C}} = j\omega L + \frac{R}{1 + j\omega RC}$$ (12.5-31)

$$= j\omega L + \frac{R(1 - j\omega RC)}{1 + (\omega RC)^2} = \frac{R}{1 + (\omega RC)^2} + j\omega \left[L - \frac{R^2 C}{1 + (\omega RC)^2} \right].$$

Setting the imaginary part to zero gives $\omega = 0$ and

$$\omega_o = \frac{1}{\sqrt{LC}} \sqrt{1 - \frac{L}{R^2 C}}.$$ (12.5-32)

The second radical represents a correction factor. If $R \rightarrow \infty$, we see that the resonant frequency approaches that of an ideal series tuned circuit. Note that we must discard

---

[11] All *RLC* circuits exhibit purely real impedances at zero and infinite frequency because each capacitor and each inductor is either an open circuit or a short circuit at these frequencies.

$\omega = 0$ as a candidate because all *RLC* circuits have real impedance at zero and infinite frequency, as we have already observed. Notice also that if $Y(j\omega)$ is purely real at any frequency, then the same statement is true of the $Z(j\omega)$, and vice versa.

Section 12.5 Quiz

Q12.5-1. Find the series equivalent circuit of the lossy capacitor in Figure Q12.5-1 at $\omega = 100$ krad/s. Repeat this computation for $\omega = 10$ krad/s.

Q12.5-3. For the circuit in Figure Q12.5-3,
 a. Classify it (LP, BP, HP, or notch).
 b. Find the peak value of voltage gain and the bandwidth.

Figure Q12.5-1

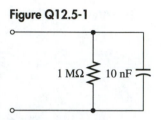

Figure Q12.5-3

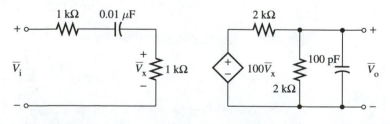

Q12.5-2. Find the resonant frequency of the circuit in Figure Q12.5-2.

Figure Q12.5-2

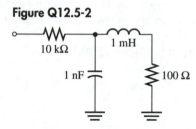

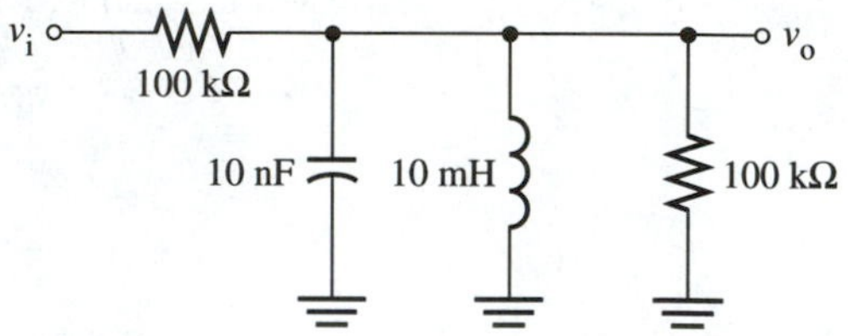

12.6 | Bode Plots

We will now discuss some of the practical issues surrounding the idea of frequency response. What we have in mind is the development of skill in rapidly sketching the frequency response (gain and phase) in a "back-of-the-envelope" manner.[12] Let's start with a practical example to fix ideas.

Example 12.7   Find the system function $H(\omega) = \overline{V}_o/\overline{V}_i$ for the circuit in Figure 12.71.

Figure 12.71
An example circuit

[Circuit diagram: Figure 12.71 with input $\overline{V}_i$, 1 kΩ resistor, 0.01 μF capacitor, $\overline{V}_x$ across 1 kΩ, dependent source $100\overline{V}_x$, 2 kΩ, 2 kΩ, 100 pF capacitor, and output $\overline{V}_o$]

---
[12] This term is intended to conjure up a picture of two engineers discussing a design over coffee. One searches for writing material to illustrate a point and pulls out an old letter, then jots down equations and frequency response sketches on its reverse side. The phrase "pushing the envelope," which evokes an image of the math becoming too complicated to fit on the back of the envelope, has come to mean the creation of a more complex and higher performance circuit or system.

Solution

This circuit is a fairly respectable model for a voltage amplifier, one such as the "preamp" in a stereo system. As we are interested in the frequency response, we will use the phasor equivalent circuit shown in Figure 12.72. The unit of impedance is the ohm, so if we scale the unit of impedance to be 1 kΩ and the unit of current to be 1 mA, the unit of voltage will remain 1 V; furthermore, the system function $H(\omega) = \overline{V}_o/\overline{V}_i$ will remain unchanged as well. Two applications of the voltage divider rule yield, after a bit of algebraic simplification and "unscaling" which we will leave to you,

$$H(\omega) = \frac{\overline{V}_o}{\overline{V}_i} = \frac{25 \times 10^7 (j\omega)}{(j\omega + 5 \times 10^4)(j\omega + 10^7)}. \tag{12.6-1}$$

We will be using this result in the discussion to follow, so we will write it in symbols for greater compactness of notation:

$$H(\omega) = \frac{K(j\omega)}{(j\omega + p_1)(j\omega + p_2)}. \tag{12.6-2}$$

Figure 12.72
The phasor equivalent circuit

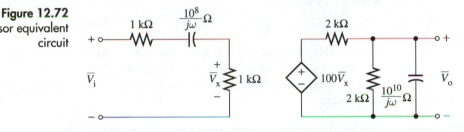

Ac Forced Response
Form of the
System Function

By generalizing upon the preceding example, we see that the system function of any response variable of a circuit constructed from our standard supply of elements, namely R, L, C, and dependent sources, has the form

$$H(\omega) = K \frac{(j\omega + z_1)\,(j\omega + z_2)\,\cdots\,(j\omega + z_m)}{(j\omega + p_1)\,(j\omega + p_2)\,\cdots\,(j\omega + p_n)}. \tag{12.6-3}$$

We call the factors $(j\omega + z_1)$, $(j\omega + z_2)$, . . ., $(j\omega + z_m)$ the *zero factors* of $H(\omega)$ and the factors $(j\omega + p_1)$, $(j\omega + p_2)$, . . ., $(j\omega + p_n)$ the *pole factors* of $H(\omega)$.[13] We call K the *scale factor*. Finally, we note that any finite zero or pole factor can be *repeated*; that is, we might have $z_k = z_{k+1} = \cdots = z_{k+q}$, in which case we say that the zero factor $(j\omega + z_k)^q$ *has order q*, and similarly for a pole factor. In Example 12.7-1 there is one zero factor $j\omega$ (that is, the corresponding z_k is 0). There are two finite pole factors: $(j\omega + 5 \times 10^4)$ and $(j\omega + 10^7)$. The scale factor is $K = 25 \times 10^7$. These zero and pole factors are all *simple,* which is a synonym for the expression "of order one." If we plot the gain frequency response of a circuit such as the one in our last example, we will obtain a graph with the general shape shown in Figure 12.73. It is zero at dc ($\omega = 0$) and also at infinite frequency ($\omega = \infty$). The response is constant over a very wide frequency range. The corresponding value of gain over this frequency range is called the "midband gain," A_{mb}.

---

[13] If you have covered Part II on time domain analysis, you will have learned that the poles and zeros (that is, the constants $-z_1$, . . . , $-z_m$ and $-p_1$, . . . , $-p_n$) are useful in determining the stability of the circuit response. If you have not, just accept the terms "zero factors" and "pole factors" here as definitions.

Figure 12.73
A typical gain
frequency response

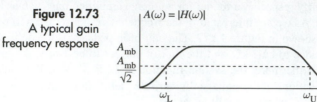

Decibels   Now here is a problem that often occurs. We investigate an experimental plot of $A(\omega)$ taken in the lab and discover that $A_{mb} = 100$. We notice something strange about the plot in the low frequency range where $A(\omega)$ has a value of approximately one. As we wish to investigate this effect more closely, we adjust the scale of our plot so that a gain of one corresponds to a height of, say, 1 cm. If we wish, however, to closely investigate the shape of the plot in the midband range as well—where $A(\omega)$ is 100—we find that our plot must be at least 100 cm tall (that is, about 254 inches or 21 feet)! This would be a very unwieldy plot to handle. For this reason, we perform a nonlinear transformation on our scale by setting

$$A_{dB}(\omega) = 20 \log A(\omega). \qquad (12.6\text{-}4)$$

The logarithm is to the base 10, not to the base e. (We use "ln" for the latter.)

There is a historical reason for the factor of 20. The *bel* (named for Alexander Graham Bell, inventor of the telephone) was first defined as the unit of the logarithm of a *power* ratio:

$$\text{power ratio in bels} = \log \left[\frac{P_2}{P_1} \right]. \qquad (12.6\text{-}5)$$

It was then discovered that for most applications this unit was too large; thus, the *decibel* became the standard. This was defined by the equation

$$\text{power ratio in decibels} = 10 \log \left[\frac{P_2}{P_1} \right]. \qquad (12.6\text{-}6)$$

Table 12.1
Common dB values

A_{dB}	B
0.001	−60
0.01	−40
0.1	−20
1/2	−6
$1/\sqrt{2}$	−3
1	0
$\sqrt{2}$	+3
2	+6
10	+20
100	+40
1000	+60

The prefix "deci" means that this unit is one-tenth of a bel (when the log changes by unity, the entire expression changes by 10 units). This is related to our definition in (12.6-4) as follows. Assuming that our powers are carried in sinusoidal voltage waveforms measured across resistors, we note that $P = (1/2) V_{rms}^2/R$. Thus, *if both powers are measured relative to the same resistance value,* then

$$\text{power ratio in decibels} = 20 \log \left[\frac{V_2}{V_1} \right], \qquad (12.6\text{-}7)$$

rms values typically being understood.[14] This definition in terms of voltage has today come to mean equation (12.6-7), regardless of resistance levels. Thus, we have equation (12.6-4).

There are a number of common values that you should memorize. They will occur again and again, particularly if you decide to specialize in analog electronics. We present them to you in Table 12.1. Notice that a gain of 0 dB does not mean that the response is zero—it means that it has exactly the same magnitude as the input. Observe also the special values $\sqrt{2}$, $1/\sqrt{2}$, 2, and 1/2: the former pair have dB values of ± 3 and the latter, ± 6. By inspecting the table carefully, you will see that a gain of $1/x$ has the same magnitude

---

[14] Because $V_{max} = \sqrt{2}\, V_{rms}$, though, it is immaterial whether rms or peak values are used.

in dB as that of x. The only difference is their signs, which are reversed relative to one another. Finally, note that doubling or halving the gain corresponds to adding or subtracting 6 dB, whereas multiplying or dividing it by a factor of 10 corresponds to adding or subtracting 20 dB.

The next example aids in developing facility in the manipulation of gains in decibels.

Example 12.8   Find the values of gain corresponding to 26 dB, -100 dB, and 34 dB.

Solution   We recall that $\log(xy) = \log(x) + \log(y)$; thus, adding dB values corresponds to multiplying the actual quantities. Hence, we can write 26 dB = 20 dB + 6 dB. We now note that 20 dB corresponds to a gain of 10 and 6 dB to a gain of 2. Thus, 26 dB corresponds to a gain of 20. We can write -100 dB = $20 \log(x)$ and solve for x: $x = 10^{-100/20} = 10^{-5}$. Finally, we can write 34 dB = 40 dB $-$ 6 dB; thus, the corresponding gain is $100 \times 0.5 = 50$.

In order to gain facility with dB manipulation you should learn the logarithms of a few basic positive integers: $\log(2) = 0.301$, $\log(3) = 0.477$, $\log(5) = 0.699$, and $\log(7) = 0.845$. With these, you can learn to calculate simple dB values in your head. All other values follow easily. For instance, $4 = 2^2$, so $\log(4) = 2 \log(2) = 0.602$. In the same manner, $\log(9) = \log(3^2) = 2 \log(3) = 0.954$. A typical dB computation, then, might run as follows. What is the dB value of a gain of 30? It is the dB value of 3 plus the dB value of 10. But the former is $20 \log(3) = 20 \times 0.477 = 9.54$ and the latter is 20. Thus, our required value is 29.54. Isn't it simple?

The Logarithmic Frequency Scale

Now look back at Figure 12.73 and look at the frequencies ω_U and ω_L. Suppose we find in our lab experiment that $\omega_U = 10^5$ rad/s and $\omega_L = 10$ rad/s. Suppose we want to investigate both regions in the frequency plot carefully, so we make 10 rad/s correspond to 1 cm of horizontal length on our plot. We find that in order to include $\omega_U = 10^5$ rad/s we must make the plot $10^4 = 10,000$ cm long (that is, 100 m or about 300 feet). This is indeed a long chart! For this reason, we perform a logarithmic transformation on the frequency axis:

$$\omega' = \log(\omega). \tag{12.6-8}$$

Notice that we do not have the factor of 20 present in this case as we do for the gain.

We show the situation in Figure 12.74. We note that multiplying any given frequency value by 10 results in adding one unit of log frequency; we call such a frequency change a *decade*. Dividing a given frequency by a factor of 10 results in *subtracting* one unit of log frequency. Similarly, multiplying or dividing by a factor of 2 results in adding or subtracting 0.3 unit of log frequency. We call such a frequency change an *octave*. This name would seem to indicate a frequency change of *eight* times, rather than two, but this is not the case. The name comes from music, wherein there are eight notes in any frequency increment of a factor of 2.

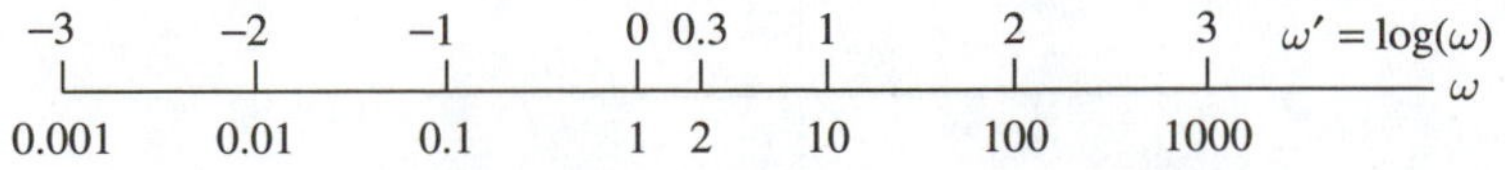

Figure 12.74
The logarithmic frequency axis

Bode Plots   We are now in a position to develop the idea of a *Bode plot*,[15] which is a plot of the gain $A(\omega)$ versus $\log(\omega)$ and/or of the phase $\phi(\omega)$ versus $\log(\omega)$. We start by recalling that the

---

[15] Pronounced "Bode-uh," after its inventor Hendrik Bode. You will, however, often hear it pronounced (incorrectly) "Bodee."

general expression for the system function of a circuit with R, L, C, and possibly dependent sources (including op amps) is

$$H(\omega) = K\frac{(j\omega + z_1)(j\omega + z_2)\cdots(j\omega + z_m)}{(j\omega + p_1)(j\omega + p_2)\cdots(j\omega + p_n)}. \tag{12.6-9}$$

We then take the absolute value and note that the absolute value of a product is the product of the absolute values and that the absolute value of a ratio is the ratio of the absolute values:

$$A(\omega) = |H(\omega)| = |K|\frac{|j\omega + z_1||j\omega + z_2|\cdots|j\omega + z_m|}{|j\omega + p_1||j\omega + p_2|\cdots|j\omega + p_n|} \tag{12.6-10}$$

Finally, we take the base 10 logarithm and multiply by 20 to get the dB value:

$$\begin{aligned} A_{dB}(\omega) = {} &20\log|K| + 20\log|j\omega + z_1| + \cdots + 20\log|j\omega + z_m| \\ &-20\log|j\omega + p_1| - 20\log|j\omega + p_2| - \cdots - 20\log|j\omega + p_n|. \end{aligned} \tag{12.6-11}$$

For the phase, we merely start with (12.6-9) and compute angles, noting that the polar angle of a product is the sum of the polar angles of the elements in the product, and the polar angle of a ratio is the difference between the numerator and denominator angles:

$$\begin{aligned} \phi(\omega) = {} &\angle K + \tan^{-1}\left[\frac{\omega}{z_1}\right] + \tan^{-1}\left[\frac{\omega}{z_2}\right] + \cdots + \tan^{-1}\left[\frac{\omega}{z_m}\right] \\ &-\tan^{-1}\left[\frac{\omega}{p_1}\right] - \tan^{-1}\left[\frac{\omega}{p_2}\right] - \cdots - \tan^{-1}\left[\frac{\omega}{p_n}\right]. \end{aligned} \tag{12.6-12}$$

Note that the angle of k must be included for it is $\pm 180°$ if k is negative.

Now we can formulate our strategy. We will compute and sketch each factor, then we will simply add them up point by point. If we desire a highly accurate plot, we merely use a programmable calculator or computer to do the plotting for us. Our interest here, however, is in developing quick approximate methods for sketching the Bode plot. Thus, let's investigate the factors in (12.6-11) one at a time. We will develop the log magnitude plot first, and consider phase plots later. We will assume that all the zero and pole constants (z_k and p_k, respectively) are real.

The Scale Factor

We look first at the scale factor, $20\log|K|$. We have sketched this factor in Figure 12.75. The solid line represents a scale factor with magnitude greater than unity (positive dB value) and the dotted line one with a magnitude less than unity. We remind you that the horizontal axis is *labeled* with values of ω, but distances are *plotted* in terms of $\log(\omega)$. Thus, the origin (zero distance) is at $\omega = 1$ rad/s. Most practical electronic circuits do not have interesting behavior in the frequency range below about 10 rad/s; other circuits and

Figure 12.75
The scale factor

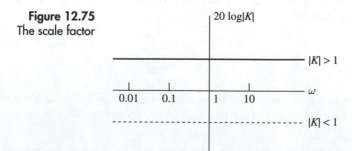

systems can. For instance, the feedback control system that positions the thruster engines in a giant booster rocket might have signals varying at this rate. For this reason, we must be quite general in allowing all frequencies to be represented.

The General Factor   Now suppose that we have $z_1 = z_2 = \cdots = z_q = a$ or $p_1 = p_2 = \cdots = p_q = a$ in our general system function of equation (12.6-9); that is q finite zeros or q finite pole constants are equal. In this case, there is a zero or a pole factor of order q. Thus, we will consider factors of the form

$$F(\omega) = (j\omega + a)^q. \qquad (12.6\text{-}13)$$

If the order of the zero or pole factor is unity, then $q = 1$; otherwise, it will be larger than 1. If the $F(\omega)$ term is a qth-order zero factor, it will appear in the numerator. If it is a pole factor, it will appear in the denominator.

Before we consider a factor such as (12.6-13) with a nonzero value of a, we will look at the case for which $a = 0$. In this event, we will say that there is a zero or pole factor of order q at $\omega = 0$. This is a special case, because dc does not appear anywhere on our plot. We remind you that $\log(0) = -\infty$, so dc (or zero frequency) is at an infinite distance to the left of the origin of our Bode plot. Now, let's consider

$$F(\omega) = (j\omega)^q. \qquad (12.6\text{-}14)$$

We start by taking the absolute value, to obtain

$$|F(\omega)| = |(j\omega)^q| = |j\omega|^q = \omega^q. \qquad (12.6\text{-}15)$$

We have used the fact that the absolute value of a product is the product of the absolute values in the second equality and the fact that $|j| = 1$ in the third. Finally, we take the logarithm and multiply by 20:

$$|F(\omega)|_{\text{dB}} = 20\log|F(\omega)| = 20q\log(\omega). \qquad (12.6\text{-}16)$$

We have plotted this factor in Figure 12.76(a), which represents a qth-order zero factor at $\omega = 0$. As one must subtract the factor if it represents a pole, we have merely changed the sign in order to obtain the plot in Figure 12.76(b). A 10-fold increase or decrease in frequency results in a $20q$-dB increase or decrease in the factor; this is a slope of $\pm 20q$ dB/decade. We will leave it to you to convince yourself that it is also $\pm 6q$ dB/octave.

Figure 12.76
A multiple zero or pole factor at dc ($\omega = 0$)

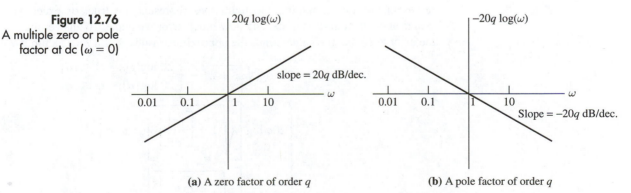

(a) A zero factor of order q (b) A pole factor of order q

Suppose, now, that $a \neq 0$; that is, our zero or pole factor is of order q, and the associated zero or pole constant is nonzero. Let's follow the same set of steps. First, we replace s by $j\omega$, obtaining

$$F(\omega) = (j\omega + a)^q. \qquad (12.6\text{-}17)$$

Next, we take the absolute value:

$$|F(\omega)| = |(j\omega + a)^q| = |j\omega + a|^q. \qquad (12.6\text{-}18)$$

Finally, taking the logarithm, we have

$$|F(\omega)|_{dB} = 20\log|(j\omega + a)^q| = 20q\log|j\omega + a| = 20q\log\sqrt{\omega^2 + a^2} \qquad (12.6\text{-}19)$$
$$= 10q\log[\omega^2 + a^2].$$

At this point, we recognize that plotting the log term is not so easy! For this reason, we resort to an approximation. We note that

$$|F(\omega)|_{dB} \cong \begin{cases} 20q\log(\omega); & \omega \gg |a| \\ 20q\log|a|; & \omega \ll |a| \end{cases}. \qquad (12.6\text{-}20)$$

The Straight-Line Approximation

Now here is the fundamental approximation involved in constructing a *linearized Bode gain plot:* we assume that the approximation holds everywhere; that is, the factor is equal to $20q\log(\omega)$ for $\omega \geq |a|$ and to $20q\log|a|$ for $\omega \leq |a|$. Note that we have used magnitude signs around the parameter a for it could possibly be negative. The resulting plot is shown for a zero factor in Figure 12.77(a) and for a pole factor in Figure 12.77(b).

Figure 12.77
A multiple zero or pole factor $(j\omega + a)^q$

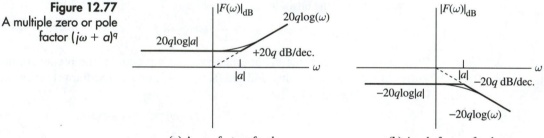

(a) A zero factor of order q (b) A pole factor of order q

The Approximation Error

Let's discuss this plot a bit. First, we observe that if the line with constant slope of $\pm 20q$ dB/decade is extended back past the "break frequency" $|a|$, it crosses through the origin because $20q\log(1) = 0$ dB. The value of $|a|$ can be either greater or less than unity, so the constant segment can be either above or below the horizontal axis. The heavy lines represent the approximation; the light lines represent the true, or exact, values. This brings about the question of exactly how much error we make. To answer this, let's define the error to be the true value minus the approximate value. For a zero factor, it will be

$$e(\omega) = 20\log(\sqrt{\omega^2 + a^2} - \begin{cases} 20\log(\omega); & \omega \geq |a| \\ 20\log|a|; & \omega \leq |a| \end{cases}$$

$$= \begin{cases} 20q\log\left[\sqrt{1 + \left[\dfrac{|a|}{\omega}\right]^2}\right]; & \omega \geq |a| \\[4ex] 20q\log\left[\sqrt{1 + \left[\dfrac{\omega}{|a|}\right]^2}\right]; & \omega \leq |a| \end{cases} \qquad (12.6\text{-}21)$$

If we evaluate this expression at the *break frequency* $|a|$, we obtain

$$e(|a|) = 3q \text{ dB}. \qquad (12.6\text{-}22)$$

The plus sign will hold for a zero factor and a negative sign will take its place for a pole factor. The true plot is above the approximate one by $3q$ dB at a zero break frequency and is below it by $3q$ dB at a pole break frequency. An octave above and below the break frequency, we have

$$e(2|a|) = e(0.5|a|) = 0.97q \text{ dB} \cong q \text{ dB}, \tag{12.6-23}$$

whereas at a decade above and below we have

$$e(10|a|) = 0.1|a| = 0.04q \text{ dB} \cong 0 \text{ dB}, \tag{12.6-24}$$

for at least moderately small orders of q.

We now give an example showing how to sketch a Bode gain plot.

Example 12.9 Suppose that a circuit or system has the system function

$$H(s) = \frac{10^5 s}{(s + 10)(s + 1000)^2}. \tag{12.6-25}$$

Sketch the linearized Bode gain plot.

Solution The scale factor is $K = 10^5$, or 100 dB—a constant. It is plotted in Figure 12.78(a). There is a simple zero factor at $\omega = 0$. The corresponding factor is plotted in Figure 12.78(b). There is a simple pole factor $(\omega + 10)$, with the corresponding Bode factor plotted in Figure 12.78(c); as you can see, the break frequency is 10 rad/s. Finally, in Figure 12.78(d), we have plotted the Bode factor corresponding to the pole factor $(\omega + 1000)^2$ of order 2. In this last plot, we have not preserved the same scale as for the others because this would lead to mechanical problems in presenting our plot. Thus, we have not shown the dotted line extending back through the origin, and the -120-dB constant gain value has been "squeezed" so that the plot is not too low. However, we see that now we must simply add all four plots at each frequency point in order to obtain the overall linearized Bode gain plot, which we show in Figure 12.78(e). Notice that we have

Figure 12.78
The linearized Bode gain plot

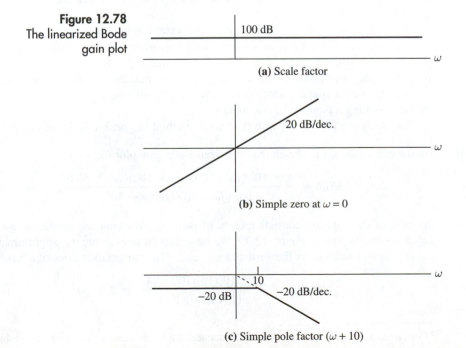

(a) Scale factor

(b) Simple zero at $\omega = 0$

(c) Simple pole factor $(\omega + 10)$

Figure 12.78 (Cont)
The linearized Bode
gain plot

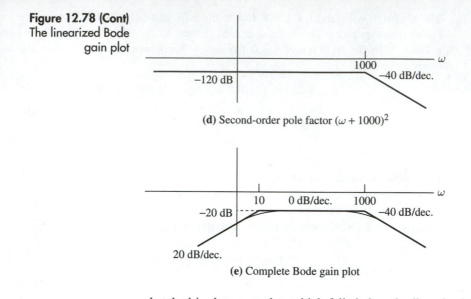

(d) Second-order pole factor $(\omega + 1000)^2$

(e) Complete Bode gain plot

sketched in the exact plot, which falls below the linearized approximation, because both finite nonzero break frequencies were due to poles. What is the error at these break frequencies? *Because these break frequencies are widely separated,* the error at each is almost that computed for each in isolation, i.e., -3 dB at 10 rad/s and -6 dB at 1000 rad/s.

The Quick Method

Now that we have worked through a Bode gain plot in somewhat painstaking detail, we can reveal some good news: there is a much faster way of doing this type of plot! Let's recap our procedure. If you think about how you would have drawn Figure 12.78(e) had you actually been wielding the pencil, you would probably agree that you would have started at very low values of frequency with a sketch upward having the slope dictated by the order of the zero factor at $\omega = 0$. Then you would make the slope break upward or downward by an amount $\pm 20q$ dB/dec at each zero or pole break frequency[16] due to a zero or pole of order q.

Here is the intuitive way to think of this process. Imagine that there is a small demon inside our system (with a frequency meter) who monitors the input frequency and compares it with all of the break frequencies. When the input frequency is below a given break frequency, our pint-sized demon flips a switch making that factor equal to its constant value; when it is above that break frequency, the demon flips the switch in the other direction, making it equal to $\pm 20q \log(\omega)$.

We codify our discussion with the "quick method" algorithm in Table 12.2.

Example 12.10

Use the quick method to sketch the linearized Bode gain plot for

$$H(\omega) = \frac{4 \times 10^5 \, (j\omega + 10)(j\omega + 100)(j\omega + 500)}{(j\omega)^2(j\omega + 1000)(j\omega + 10^4)}. \tag{12.6-26}$$

Solution

The bulk of the solution consists merely of steps in sketching the graph, so we merely sketch the final shape in Figure 12.79. We have chosen to evaluate the approximate magnitude at $\omega = 1$ rad/s to fix the vertical axis scale. The computation goes like this:

$$|H(1)|_{app} = \frac{4 \times 10^5(10)(100)(500)}{(1)^2(1000)(10^4)} = 2 \times 10^4, \tag{12.6-27}$$

---

[16] For this reason, we use the term "break frequency" for the constant in a zero or pole factor.

Table 12.2	1. Draw a frequency axis.
The "quick method" for Bode gain plots	2. Evaluate the approximate magnitude at one single value of frequency to fix the vertical scale (typically at a low value of frequency, perhaps $\omega = 1$ rad/s) and draw the vertical axis. Note that "approximate" means to use the approximation for each of the factors.
	3. Label all the finite break frequencies with an o for a zero and an x for a pole. If the associated zero or pole is of order higher than one, label its order in parentheses.
	4. Start the Bode plot at low values of frequency with a slope of $\pm 20q$ dB/dec., where q is the order of the zero or pole at $s = 0$, if any. If there are none, $q = 0$ and the initial slope is zero.
	5. Imagine the frequency to increase slowly. Continue drawing the plot, causing the slope to break upward by $+20q$ dB/dec. at each break frequency associated with a zero of order q and downward by $-20q$ dB/dec. at each break frequency associated with a pole of order q. Continue this until the last break frequency has been included.

which we easily compute—using our quick techniques for dB computations—to be 86 dB. We can similarly compute the dB value at any frequency we desire; thus, we are not terribly concerned whether or not our overall plot is to scale. We note further that the exact plot would be 3 dB above our linearized one at the corner frequency at 10 rad/s and 3 dB below at 10^4 rad/s. The errors at the other break frequencies cannot be determined so easily because they are not widely separated from their nearest neighbors.

Figure 12.79
The linearized Bode gain plot—quick method

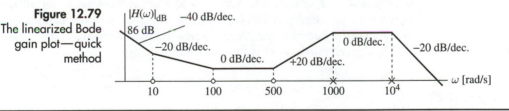

The Bode Phase Plot   Now let's see what we can do with the Bode phase plot. The scale factor can be either positive or negative; thus its phase angle is either 0 or $\pm 180°$. (Remember that there is an ambiguity of 360°, or 2π radians, in the phase.) We have plotted this factor in Figure 12.80, choosing $-180°$ for the case in which $K < 0$.

Figure 12.80
Phase angle of the scale factor

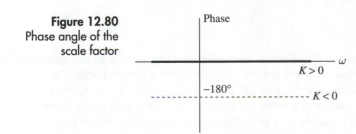

For the qth-order zero or pole at $s = 0$, we note that

$$s^q \longrightarrow (j\omega)^q \longrightarrow [e^{jq90°}\omega^q] \longrightarrow \angle = 90°q. \tag{12.6-28}$$

The arrows indicate steps in the process: first, we replace s by $j\omega$, then we compute the angle. We show the result in Figure 12.81. The solid line represents the phase due to a zero of order q and the dotted one that of a pole of order q.

Figure 12.81
Phase of zero or pole
at $s = 0$ of order q

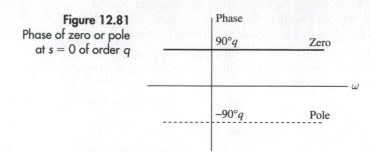

The Straight-Line Phase Approximation

The preceding two factors are quite easy to deal with. Somewhat more complicated is the qth-order zero or pole factor $(j\omega + a)^q$. The computation in this case proceeds as follows:

$$(j\omega + a)^q \longrightarrow \left[\sqrt{\omega^2 + a^2}\, e^{j\tan^{-1}(\omega/a)} \right]^q \longrightarrow \angle = q\tan^{-1}(\omega/a). \qquad (12.6\text{-}29)$$

For a zero factor, this is the phase; for a pole factor, of course, one takes the negative. This gives the plot shown in Figure 12.82 (for a zero factor in which a is positive). The smooth curve is the exact plot, and the broken straight line is the approximation we will make. We approximate the phase by zero for all frequencies a decade or more below the break frequency, by $90q$ degrees for all frequencies a decade or more above the break frequency, and by a straight line having slope $45q$ degrees/decade for all frequencies between these two limits. A careful analysis, which we will not perform, shows that the maximum error occurs at the "decade below and decade above" points and has the value $\pm 5.7q$ degrees and the error is zero at $\omega = a$.

Figure 12.82
Phase angle of a factor
due to a zero factor
$(j\omega + a)^q$

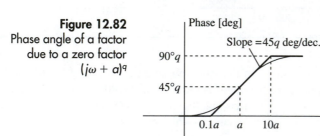

A Caveat About Signs in the Phase Plot

Before we continue, let's discuss something that might seem to be a fine point but, in fact, is not. Notice that we have not put absolute value signs around a. The reason for this is that its sign affects the sign of the phase because $\tan^{-1}(-x) = -\tan^{-1}(x)$. Rather than drawing three additional graphs, we summarize the situation in the following manner: if a is positive and it occurs in a zero (numerator) factor, the phase is positive as shown in Figure 12.82; if a is negative and occurs in a zero factor, the phase is the negative of that shown; if a is positive and occurs in a pole (denominator) factor, the phase is the negative of that shown; and if a is negative and occurs in a pole factor, the phase is positive as shown in the figure. We note that the last case is seldom of interest, for it corresponds to a pole in the right half of the s plane. For this reason, such a circuit or system is unstable.

The other three cases can all occur because stable circuits and systems can have *zeros* in the right half of the *s* plane.[17]

The Phase Plot as an Equivalent Gain Plot

Now refer once more to the straight-line approximation shown in Figure 12.82. *We see that the shape is exactly the same as that for the linearized Bode gain plot of a system function having a zero factor* $(\omega + 0.1a)$ *and a pole factor* $(\omega + 10a)$. *The only difference is in the slope: it is 45q degrees/decade, rather than 20q decibels/decade.* We show this in Figure 12.83(a). In any of the cases in which the parameter *a* is negative, the approximation is similar to that of the approximate Bode gain plot having the preceding zero and pole factors reversed; that is, with the pole factor $(\omega + 0.1a)$ and the zero factor $(\omega + 10a)$. Now we can state the quick method for sketching the approximate phase plot. We simply replace each finite zero and each finite pole by the appropriate pole/zero pair a decade below and above the original, then we draw a gain plot for the new transfer function. Further, *a zero or pole factor at the origin gives constant phase versus frequency.* The only other difference in our plotting procedure is that we evaluate phases, rather than gains, and compute slopes in the appropriate increments of ±45 degrees per decade, rather than in increments of ±20 dB/decade.

Figure 12.83
Equivalent gain plots
for a zero factor
$(j\omega + a)^q$

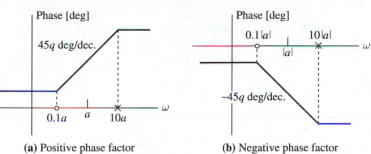

(a) Positive phase factor   **(b)** Negative phase factor

Example 12.11

Sketch the linearized Bode phase plot for

$$H(\omega) = \frac{10^5(j\omega)}{(j\omega + 10)(j\omega + 1000)^2}. \qquad (12.6\text{-}30)$$

Solution

This is the same system function for which we sketched the linearized gain plot in Example 12.9. We note that both the finite pole constants are positive, so we replace each by a pole a decade in frequency below and a zero a decade above the original. This gives

$$H(\omega) = \frac{10^5(j\omega)(j\omega + 100)(j\omega + 10^4)^2}{(j\omega + 1)(j\omega + 100)^2} = \frac{10^5(j\omega)(j\omega + 10^4)^2}{(j\omega + 1)(j\omega + 100)}. \qquad (12.6\text{-}31)$$

Notice the pole-zero factor cancellation at $\omega = 100$ rad/s. Of course the size of the scale factor is immaterial, though its sign is important. At very low values of frequency, the $(j\omega)$ factor dominates; hence, the low frequency phase is $+90°$. This permits us to fix the scale on the vertical axis. We then sketch a gain plot for (12.6-31), with all slopes in increments of ±45 degrees/decade. The result is given in Figure 12.84. We emphasize the fact that the single factor $(j\omega)$, representing a zero at $\omega = 0$, acts exactly like the scale factor as far as phase is concerned: the scale factor itself contributes either $0°$ or $\pm 180°$,

---

[17] If you have not read the material in Part II on time domain analysis, this three-sentence explanation will perhaps not make sense to you. Just interpret it as meaning that such circuits or systems are usually undesirable and do not often occur.

depending on its sign, and a single zero factor at $\omega = 0$ contributes $+90°$. The finite nonzero poles and zeros act like *pairs* of poles and zeros.

Figure 12.84
The quick phase
plot technique

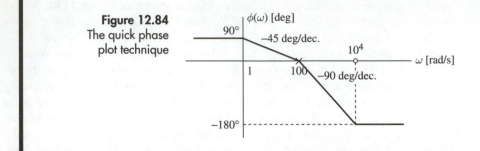

Example 12.12   Find the transfer function and sketch the linearized Bode gain and phase plots for the circuit in Figure 12.85.

Figure 12.85
An example circuit

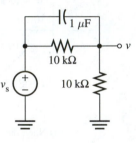

Solution   We let the common resistor value be R and the capacitor value be C. Working with the phasor domain equivalent shown in Figure 12.86, we use the voltage divider rule, then insert the values of R and C, to easily compute

$$H(\omega) = \frac{j\omega + 100}{j\omega + 200}. \tag{12.6-32}$$

The gain at low frequencies is 1/2, or -6 dB. There is one finite zero factor ($j\omega + 100$) and one finite pole factor ($j\omega + 200$). This gives the linear gain plot shown in Figure 12.87. Notice that a slope of 20 dB/decade is the same as a slope of 6 dB/octave, so the fi-

Figure 12.86
The phasor equivalent

Figure 12.87
The linearized Bode gain plot

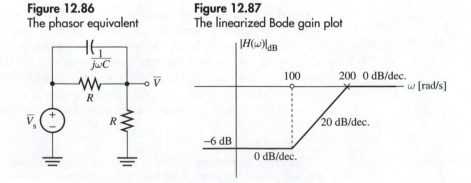

nal, high-frequency value of gain is zero dB. This checks with the circuit, because the capacitor is a short circuit for high frequencies and an open circuit for low frequencies.

For the linearized phase plot, we replace the zero factor at 100 with a zero factor at 10 and a pole factor at 1000, and the pole factor at 200 with a pole factor at 20 and a zero factor at 2000. This gives the phase plot shown in Figure 12.88. Notice that the scale factor is positive, and there is neither a pole factor nor a zero factor at $\omega = 0$; hence, the phase plot starts with zero degree at low frequencies. Now recalling that a frequency increase of an octave adds 0.3 unit to $\log(\omega)$, we see that our linearized plot predicts that between 20 and 1000 radians per second the phase will be approximately $0.3 \times 45 = 13.5°$. The error in *any* of our linearized approximations is, however, very poor when the break frequencies are not widely separated. On the other hand, it is not really that bad in this example. The actual phase can be easily computed at, say, 500 rad/s to be 12.5°—only a degree off. The actual value is 5.6° at 20 rad/s and 18.4° at 1000 rad/s.

Figure 12.88
The linearized Bode
phase plot

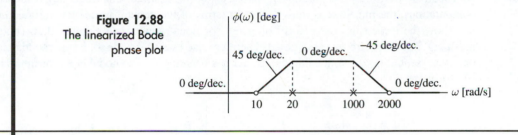

Summary: Bode Plots for Complex Poles and Zeros

We stress that we have not described a quick sketching procedure for complex poles and zeros. The reason for this is simply that complex poles and zeros make the procedure quite complicated—there is not really a simple way of sketching them. It is our opinion that it is better in such cases to use a calculator or a computer to construct the Bode plot.

Section 12.6 Quiz

Q12.6-1.   Sketch the linearized Bode gain and phase plots for the system function

$$H(\omega) = 8 \times 10^8 \frac{(j\omega + 2)(j\omega + 100)}{j\omega(j\omega + 10)(j\omega + 2000)(j\omega + 10^4)}.$$

Q12.6-2.   Sketch the linearized Bode gain and phase plots for the circuit shown in Figure Q12.6-1.

Figure Q12.6-1

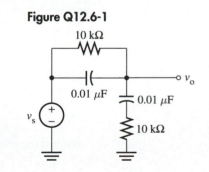

12.7 | Frequency Response of Active Circuits

We will now investigate the frequency response of circuits having active elements: dependent sources, operational amplifiers, and transistors. The analysis procedure is quite analogous to the dc analysis of such circuits with capacitors (and inductors, if any) described by their impedances.

An Active Circuit Example

Example 12.13   Determine the ac forced response transfer function for the circuit response v shown in Figure 12.89 and sketch the linearized gain and phase Bode plots.

Figure 12.89
An example active circuit

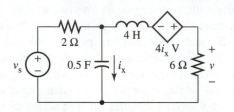

Solution   As usual, we first draw the phasor equivalent using impedances. For frequency response computations, the impedances must be left in terms of the general frequency variable ω. We show the circuit in phasor form prepared for nodal analysis in Figure 12.90 (both mesh and nodal require two equations). There are one essential node, one supernode, and one nonessential node. Notice that the CCVS has been taped. The nodal equations in matrix form are

$$\begin{bmatrix} \dfrac{1}{6} + \dfrac{1}{j4\omega} & -\dfrac{1}{j4\omega} \\[3mm] -\dfrac{1}{j4\omega} & \dfrac{1}{2} + \dfrac{j\omega}{2} + \dfrac{1}{j4\omega} \end{bmatrix} \begin{bmatrix} \overline{V} \\[2mm] \overline{V}_x \end{bmatrix} = \begin{bmatrix} \dfrac{\overline{V}_c}{j4\omega} \\[3mm] \dfrac{\overline{V}_s}{2} - \dfrac{\overline{V}_c}{j4\omega} \end{bmatrix}. \qquad (12.7\text{-}1)$$

We now untape the CCVS to obtain

$$\overline{V}_c = 4\overline{I}_x = 4\left[\frac{j\omega}{2}\overline{V}_x\right] = j2\omega\overline{V}_x. \qquad (12.7\text{-}2)$$

Using this in (12.7-1) and transposing the unknown terms to the left-hand side gives

$$\begin{bmatrix} \dfrac{1}{6} + \dfrac{1}{j4\omega} & -\left[\dfrac{1}{2} + \dfrac{1}{j4\omega}\right] \\[3mm] -\dfrac{1}{j4\omega} & 1 + \dfrac{j\omega}{2} + \dfrac{1}{j4\omega} \end{bmatrix} \begin{bmatrix} \overline{V} \\[2mm] \overline{V}_x \end{bmatrix} = \begin{bmatrix} 0 \\[2mm] \dfrac{\overline{V}_s}{2} \end{bmatrix}. \qquad (12.7\text{-}3)$$

We will leave it to you to solve this matrix equation for $\overline{V}$ (perhaps using the explicit formula for the inverse of a (2×2) matrix or Cramer's rule). The result is

$$\overline{V} = \frac{3(1 + j2\omega)}{(4 - 2\omega^2) + j7\omega}\overline{V}_s. \qquad (12.7\text{-}4)$$

Figure 12.90
The example circuit in phasor form

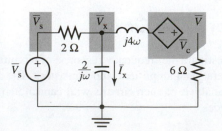

Thus, the system function we are seeking is

$$H(\omega) = \overline{V}/\overline{V}_s = \frac{3(1 + j2\omega)}{(4 - 2\omega^2) + j7\omega}. \tag{12.7-5}$$

In order to sketch the Bode plots, we must have the system function $H(\omega)$ as an explicit function of the variable $j\omega$ in factored form. We thus manipulate (12.7-5) into the form

$$H(\omega) = \frac{3 + 6(j\omega)}{4 + 7(j\omega) + 2(j\omega)^2} = \frac{3(j\omega + 0.5)}{(j\omega + 0.72)(j\omega + 2.78)}. \tag{12.7-6}$$

We first divided the top and bottom by 2 to make the leading coefficient in the denominator unity, then factored the denominator polynomial. We see that there is one zero factor, two pole factors, and a scale factor of 3. We evaluate the approximate gain at $\omega = 0.1$ rad/s (arbitrarily chosen to be less than the lowest break frequency) as follows:

$$|H(0.1)|_{app} = \left|\frac{3(j0.1 + 0.5)}{(j0.1 + 0.72)(j0.1 + 2.78)}\right|_{app} = \frac{3 \times 0.5}{0.72 \times 2.78} = 0.75. \tag{12.7-7}$$

We first inserted the value of ω, then made the approximation on each factor. Now we convert the approximate gain at $\omega = 0.1$ rad/s to dB:

$$|H(0.1)|_{dB} = 20 \log(0.75) = -2.5 \text{ dB}. \tag{12.7-8}$$

The linearized Bode gain plot is shown in Figure 12.91. We will leave it to you to show that the "midband gain" is 0.67 dB as noted on the graph.

Figure 12.91
The linearized Bode gain plot

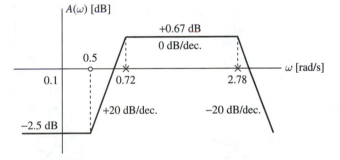

Now let's compute the linearized Bode phase plot. To do so, we first note that the scale factor is positive and that there are no poles or zeros at dc ($\omega = 0$ rad/s). This means that the phase plot starts out at low frequencies with zero phase angle. Next, we replace the zero factor ($j\omega + 0.5$) with a zero factor ($j\omega + 0.05$) and a pole factor ($j\omega + 5$), the pole factor ($j\omega + 0.72$) with a pole factor ($j\omega + 0.72$) and a zero factor ($j\omega + 7.2$), and the pole factor ($j\omega + 2.78$) with a pole factor ($j\omega + 0.278$) and a zero factor ($j\omega + 27.8$). This gives us the formal transfer function

$$H'(\omega) = \frac{3(j\omega + 0.05)(j\omega + 0.72)(j\omega + 27.8)}{(j\omega + 5)(j\omega + 0.072)(j\omega + 0.278)}. \tag{12.7-9}$$

We show the linearized phase plot in Figure 12.92. Remember that we start the graph as just mentioned (with zero slope), and all other slopes are in increments of $\pm 45°$ per decade.

Figure 12.92
The linearized Bode
phase plot

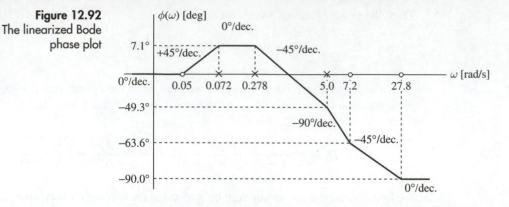

For the sake of comparison, we offer the "exact" (nonlinearized) Bode gain plot in Figure 12.93. "Exact" means that a computer was used to plot equation (12.7-6). The "exact" phase plot is shown in Figure 12.94. Notice that these exact plots do not convey as much information that one can ascertain in a quick glance as do the linearized plots because they do not show the sharp corners as do the latter. This might be a somewhat surprising comment, but if you think just a moment about the design process, you will see that the linearized plots show the break frequencies at a glance—and these are directly related to our circuit parameters. (Neither one of the linearized plots is drawn to scale.)

Figure 12.93
"Exact" Bode gain plot

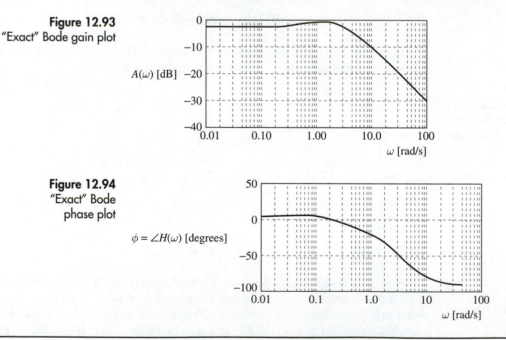

Figure 12.94
"Exact" Bode
phase plot

**A Comment
on Notation**

We have analyzed our circuits for their ac forced response—both in this chapter and in the preceding one—in terms of the variable $j\omega$. Each inductor obeys Ohm's law with the impedance $j\omega L$ replacing the resistance of the resistor; similarly, each capacitor also obeys Ohm's law with the impedance $1/j\omega C$ replacing the resistance of the resistor. Ohm's law, coupled with KCL and KVL, suffices for our derivations of circuit response. Can you see how these analysis procedures relate to those for dc circuits? We are performing exactly the same algebraic operations. The only difference is that in ac forced response analysis we must use complex algebra, rather than real algebra as we did for dc analysis. Finally, a comment to the reader who has covered Part II: if you analyze a circuit using the operator p instead of $j\omega$, you will derive an operator solution that gives the ac forced response transfer function *if you simply replace p by $j\omega$.*

Operational Amplifier Examples

Example 12.14(O)

Determine the voltage gain transfer function and sketch its linearized Bode plots for the simple op amp integrator subcircuit shown in Figure 12.95.

Figure 12.95
An op amp integrator

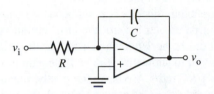

Solution

We add a sinusoidal test source at the general frequency ω and use the phasor equivalent for the resulting circuit as in Figure 12.96. Now just how should we analyze this circuit? There are a number of options. We could replace the op amp by its nullor equivalent (as explained in Chapter 5) and do nodal or mesh analysis, or we could simply use the general equivalent shown in Figure 12.97. We have already analyzed this subcircuit for its equivalent, with resistors in the place of the two impedances shown here, in Example 5.11 of Chapter 5. The equivalent subcircuit for the resistive case was shown in Figure 5.54. It is shown here with impedances replacing the resistances in Figure 12.98. We stress that the output voltage is connected to the negative terminal of the VCVS; we could just as well have connected it to the positive terminal and affixed a negative sign to the voltage gain.

Figure 12.96
The phasor equivalent

Figure 12.97
The general inverting configuration

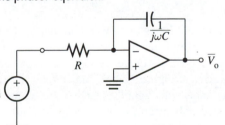

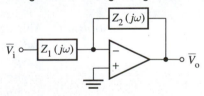

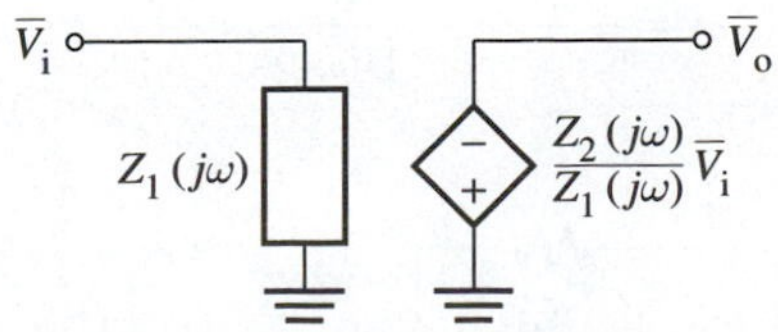

Figure 12.98
The ac forced response equivalent

If we use the equivalent in Figure 12.98 with $Z_1(j\omega) = R$ and $Z_2(j\omega) = 1/j\omega C$, we can simply write the voltage gain by inspection:

$$H(\omega) = \frac{\overline{V}_o}{\overline{V}_i} = -\frac{1/j\omega C}{R} = -\frac{1}{j\omega RC} = j\frac{1}{\omega RC} = \frac{1}{\omega RC}\angle 90°. \qquad (12.7\text{-}10)$$

Figure 12.99
Bode gain plot

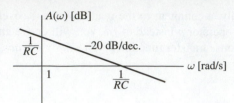

We note that the phase is constant with respect to frequency, so we will not bother to plot it. The gain, however, is a straight line with a slope of -20 dB/decade, as shown in Figure 12.99. We note that this plot is exact because no approximation is involved for factors whose constant parameter is zero. We also remind you that dc ($\omega = 0$) is an infinite distance to the left on the horizontal axis and that the origin of the plot is at $\omega = 1$ rad/s. We will leave it to you to show that at $\omega = 1$ the gain is $1/RC$ and the frequency at which the gain is 0 dB is $1/RC$ rad/s. We have, incidently, pursued a common course that might be confusing if we did not point it out. Although the vertical scale is in dB, we have labeled the gain at $\omega = 1$ rad/s in "non-dB" terms. This is a common practice for values that have obvious physical and intuitive meaning. In our example, RC is the "integration time constant."

Example 12.15(O)

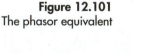

Find the voltage gain transfer function for the simple op amp differentiator circuit shown in Figure 12.100 and sketch the linearized Bode plot.

Figure 12.100
An op amp
differentiator

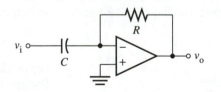

Solution   Although this circuit looks remarkably similar to the one in the last example, it differs in that the capacitor and the resistor have been exchanged. We add a sinusoidal voltage source having a general frequency ω to the input and draw the resulting ac forced response equivalent in Figure 12.101. This circuit has the topology of the general inverting configuration shown in Figure 12.97, so the equivalent shown in Figure 12.98 is also valid—we need only swap the impedances $Z_1(j\omega)$ and $Z_2(j\omega)$ relative to the last example. This gives the voltage gain transfer function

$$H(\omega) = \frac{\overline{V}_o}{\overline{V}_i} = -\frac{R}{1/j\omega C} = -j\omega RC = \omega RC \angle -90°. \qquad (12.7\text{-}11)$$

Figure 12.101
The phasor equivalent

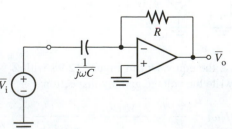

Figure 12.102
Bode gain plot

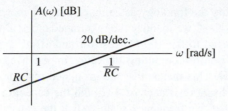

The phase is again constant (this time at $-90°$), so we will not plot it. The gain function is linearly rising with ω, as shown in Figure 12.102. Remember that the vertical axis is in dB, so the vertical intercept is actually at $20 \log(RC)$. In most situations RC is less than unity, so we have plotted it as a negative value on the vertical axis. We will leave it to you to show that this quantity is correct and to derive the value of the "crossover frequency," that is, the value of ω for which the gain curve crosses the 0-dB axis (the horizontal axis), and show that it is, indeed, $1/RC$.

A Brief Review of the Noninverting Operational Amplifier Topology

Before we tackle another example, let's pause for a moment and consider the general ac forced response of the subcircuit shown in phasor form in Figure 12.103. This is the topology that was explored in Chapter 5, Section 5.4 in Example 5.14 (resistive case). There we showed that the input current into the positive input terminal of the op amp was zero—as it is here—and that the equivalent "looking" between this terminal and ground was an open circuit. Looking back into the output terminal, between that terminal and ground, we see (in electrical terms) a VCVS whose voltage gain is one plus the ratio of the "feedback" resistance value to the value of the grounded resistance. Translating these results into impedance/phasor terminology gives the equivalent circuit shown in Figure 12.104.

Figure 12.103
The noninverting topology

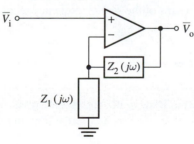

Figure 12.104
VCVS equivalent

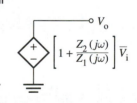

Example 12.16(O) Find the voltage gain transfer function for the subcircuit shown in Figure 12.105 and sketch the linearized Bode gain and phase plots.

Figure 12.105
An example circuit

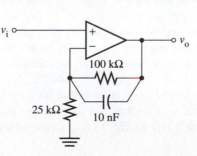

Solution   We first identify the topology as being of the sort shown in Figure 12.103: the noninverting topology. Next, we identify the two impedances: $Z_1(j\omega)$ corresponds to the 25-kΩ resistor and $Z_2(j\omega)$ to the 100-kΩ resistor and 10-nF capacitor connected in parallel with one another. Before computing the latter impedance, let's agree to impedance scale to units of 25 kΩ. This means that the grounded resistor has a resistance of 1 Ω, the feedback resistor has a resistance of 4 Ω, and the capacitor has a value of $25 \times 10^3 \times 10^{-8}$, or 250 μF. Next, we frequency scale so that the capacitor has a scaled value of 1 F. This requires that we change our frequency variable to $\omega' = (250 \times 10^{-6})\omega = \omega/4000$. The impedance and frequency scaled (time domain) subcircuit is shown in Figure 12.106. Because this subcircuit is so simple we will not explicitly draw the phasor equivalent, though you should do so if you have trouble following our derivation.

Figure 12.106
The impedance and
frequency scaled
subcircuit

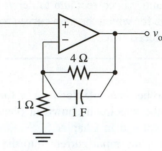

Now let's compute the ac forced response impedance $Z_2(j\omega')$:

$$Z_2(j\omega') = \frac{4 \times \dfrac{1}{j\omega'}}{4 + \dfrac{1}{j\omega'}} = \frac{1}{j\omega' + 0.25}. \tag{12.7-12}$$

Next, we use the equivalent circuit in Figure 12.104 to write the ac forced response voltage gain function in terms of the scaled frequency ω' as

$$H(\omega') = \frac{\overline{V}_o}{\overline{V}_i} = 1 + \frac{\dfrac{1}{j\omega' + 0.25}}{1} = \frac{j\omega' + 1.25}{j\omega' + 0.25}. \tag{12.7-13}$$

Before we sketch the Bode gain and phase plots, let's unscale the frequency with $\omega' = \omega/4000$ (noting that $H(\omega')$ is a different function from $H(\omega)$):

$$H(\omega) = H(\omega/4000) = \frac{j\omega + 5000}{j\omega + 1000}. \tag{12.7-14}$$

In case scaling is still a bit awkward for you, let's discuss it a little more. When we impedance or frequency scale, all functions of frequency are changed to new ones. If we were to use different notation for these functions after each scaling operation the notation would become cumbersome. So we simply write, for instance, $H(\omega')$ rather than, say, $H'(\omega')$. If we were to have done so, then unscaling would result in the equation of transformation $H(\omega) = H'(\omega')|_{\omega' = \omega/4000} = H'(\omega/4000)$. We hope these comments have helped and not confused you. The main point is that we do not simply substitute blindly, but recognize that the actual transfer function results when the scaled one has its frequency variable replaced by one that is a factor of four thousand times larger.

The Bode gain plot is shown in Figure 12.107. Note here that the 14 for the gain at frequencies below 250 rad/s is in dB, unlike our previous two examples. It is computed as follows. We select an arbitrary frequency below 1000 rad/s and compute the value of the linearized approximation of gain at that frequency. At $\omega = 1$ rad/s, we have

$$|H(j1)|_{app} = \left[\frac{j1 + 5000}{j1 + 1000}\right]_{app} = \frac{5000}{1000} = 5. \tag{12.7-15}$$

Next, we convert this value to dB by taking the base 10 log, then multiplying it by 20:

$$A_{dB}(1) = 20 \log(5) = 14 \text{ dB}. \tag{12.7-16}$$

We will leave it to you to verify that the approximate gain at frequencies above 5000 rad/s is 0 dB, as shown in the figure.

Figure 12.107
The linearized Bode gain plot

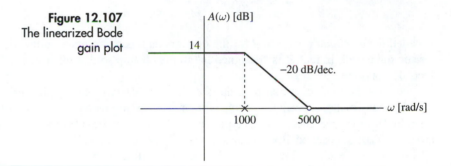

To compute the phase plot, we replace the pole at 1000 rad/s with a pole at 100 rad/s and a zero at 10,000 rad/s. Likewise, we replace the zero at 5000 rad/s with a zero at 500 rad/s and a pole at 50 krad/s. As there is no pole or zero factor at $\omega = 0$ and the scale factor is positive, the phase plot starts at 0°. The result is shown in Figure 12.108. We leave it to you once again to show that the phase at high frequencies is 0 degree, as we have shown in our plot.

Figure 12.108
The linearized Bode phase plot

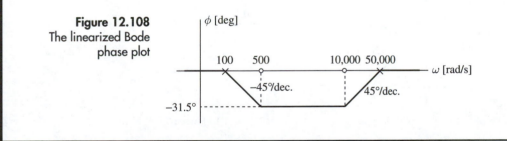

Frequency of the Operational Amplifier Itself

In our work thus far in this section with operational amplifiers we have assumed that the op amp itself was ideal. We will now explore one type of nonideality: the case where the op amp has a frequency response that is a function of frequency. Though there are other nonidealities in a practical op amp, most of them are better explored in a course in electronics; the frequency response of the nonideal op amp is something we are equipped to tackle right now.

Figure 12.109(a) shows the various op amp voltages for ease of reference. We note that there is always a ground predefined in a circuit involving an op amp and that all voltages are relative to that ground reference. The current into each of the input terminals is

Figure 12.109
Frequency response
model of the
nonideal op amp

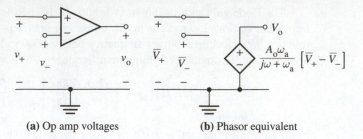

(a) Op amp voltages   (b) Phasor equivalent

always zero; therefore, it is equivalent to the open circuit shown in Figure 12.109(b), which itself is a phasor equivalent circuit for the op amp that takes into consideration the finite frequency response of the device.[18] The gain function

$$H(\omega) = \frac{A_o \omega_a}{j\omega + \omega_a} \qquad (12.7\text{-}17)$$

is called the *dominant pole model* for the op amp, and many op amps follow this characteristic quite well. In fact, it is "designed in" to the op amp so that other nondesirable effects do not occur.

Perhaps the most common op amp is called the 741 type; that is the serial number used by many manufacturers.[19] For this type of op amp, to give you a general idea of the magnitudes of the parameters, A_o is typically on the order of 100,000 and $f_a (= \omega_a/2\pi)$ is around 10 Hz. Linearized Bode gain and phase plots of (12.7-17) are shown in Figure 12.110. At frequencies below ω_a, the ideal approximation seems to be quite good, for it depends upon the gain being infinite, and, at least for "reasonable" values for the external components in a circuit using the op amp, 100,000 seems close enough. However, because ω_a is so small, we see that if our signals are not dc, we must use the dominant pole approximation for even small values of ω. We will investigate the effects of this frequency response characteristic on several simple circuits in the following development.

Figure 12.110
The linearized plot *s*
for the voltage gain
of an op amp

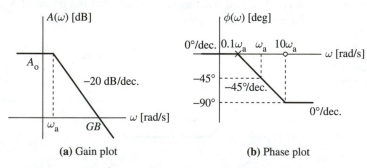

(a) Gain plot   (b) Phase plot

Before continuing, we note that the *unity gain crossover frequency* is denoted by the symbol *GB* in Figure 12.110. This is the frequency where the basic op amp voltage gain is zero dB, or unity. We will pause for a moment to see why it has been given that symbol. To do so, let's rewrite the voltage gain transfer function $H(\omega)$ in Euler form:

$$H(\omega) = \frac{A_o \omega_a}{j\omega + \omega_a} = \frac{A_o}{1 + j\omega/\omega_a} = \frac{A_o}{\sqrt{1 + (\omega/\omega_a)^2}} \angle -\tan^{-1}(\omega/\omega_a). \quad (12.7\text{-}18)$$

---

[18] Refer back to Section 5.5 of Chapter 5 and look at Figure 5.73. The "dynamic" op amp model in that figure corresponds to the frequency domain model in Figure 12.109(b).

[19] A prefix μA in front of the 741 indicates a device manufactured by Fairchild Corporation, a prefix LM denotes that it was fabricated by National Semiconductor, and so on.

Therefore, the Bode gain function is

$$A(\omega) = \frac{A_o}{\sqrt{1 + (\omega/\omega_a)^2}}.$$ (12.7-19)

The basic Bode linearization approximation is simply that at frequencies less than ω_a we ignore the term ω/ω_a in the denominator, and for frequencies above ω_a we ignore the 1. Thus, the -20 dB per decade "rolloff" (that is, the straight line with negative slope) has the "non-dB" equation

$$A(\omega) = \frac{A_o}{\omega/\omega_a} = \frac{A_o \omega_a}{\omega}.$$ (12.7-20)

Now the definition of GB is that frequency at which $A(\omega) = 1$ (0 dB), and we see at once from (12.7-20) that this frequency is given by $A_o \omega_a$. Thus,

$$GB = A_o \omega_a.$$ (12.7-21)

Because A_o is the low-frequency (or dc) voltage gain and ω_a is the 3-dB bandwidth,[20] GB is the *gain bandwidth product,* that is, the product of the dc gain and the 3-dB bandwidth. Because A_o is unitless, however, GB has the unit of rad/s. Sometimes one uses the hertz form

$$GB_{Hz} = A_o f_a = A_o \omega_a / 2\pi.$$ (12.7-22)

If you look back at the typical values we gave a bit earlier for the 741 type of op amp, you will see that its GB_{Hz} is about 1 MHz.

Example 12.17(O)   Find the voltage gain transfer function for the noninverting op amp subcircuit shown in Figure 12.111 and sketch the linearized Bode gain and phase plots. Assume that the op amp is well represented by its dominant pole model.

Figure 12.111
The noninverting
topology

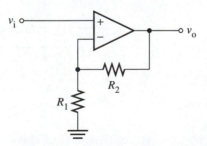

Solution   First, let's make a simple but important point: this subcircuit *is not described by its ideal voltage gain of* $1 + R_2/R_1$. This is a mistake often made by a neophyte circuit analyst because he or she has been working with ideal models for so long. This is a new type of problem—a whole new "ball game." Though the current into the input terminals is zero, the voltage between them is not zero any longer. Thus, we will "unplug" the op amp symbol in Figure 12.111 and substitute the model given in Figure 12.109. This results in the (phasor domain) configuration shown in Figure 12.112. Note that we have added a sinu-

---

[20] To see this, just recall that the actual gain is 3 dB below the linearized approximation at $\omega = \omega_a$, or just insert ω_a into (12.7-19) and then convert to dB.

Figure 12.112
The example subcircuit
being tested with a
sinusoidal voltage
source with the
dominant pole model
for the op amp

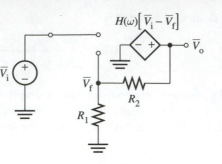

soidal voltage source at the input in order to determine the voltage gain. The analysis can be performed by simple KVL, KCL, and voltage division applications. Thus, we have

$$\overline{V}_o = H(\omega)[\overline{V}_i - \overline{V}_f] = H(\omega)[\overline{V}_i - F\overline{V}_o], \tag{12.7-23}$$

where

$$F = \frac{R_2}{R_1 + R_2} \tag{12.7-24}$$

is the voltage divider ratio of the two resistors in the "feedback" path. Solving (12.7-23) for $\overline{V}_o$, we get

$$\overline{V}_o = \frac{H(\omega)}{1 + FH(\omega)}\overline{V}_i. \tag{12.7-25}$$

We next insert the explicit equation for $H(\omega)$ given by equation (12.7-17). This results in

$$\overline{V}_o = \frac{A_o\omega_a}{j\omega + \omega_a + FA_o\omega_a}\overline{V}_i = \frac{A_o\omega_a}{j\omega + (1 + FA_o)\omega_a}\overline{V}_i. \tag{12.7-26}$$

We write this in terms of the "closed loop" voltage gain transfer function $G(\omega)$, as follows:

$$G(\omega) = \frac{\overline{V}_o}{\overline{V}_i} = \frac{\dfrac{A_o}{1 + FA_o}(1 + FA_o)\omega_a}{j\omega + (1 + FA_o)\omega_a} = \frac{G_o\omega'_a}{j\omega + \omega'_a}, \tag{12.7-27}$$

where

$$G_o = \frac{A_o}{1 + FA_o} \tag{12.7-28}$$

is the dc "closed loop gain," that is, $G(\omega)$ evaluated at $\omega = 0$, and

$$\omega'_a = (1 + FA_o)\omega_a. \tag{12.7-29}$$

Notice that (12.7-27)—the voltage gain for the complete circuit—is exactly the same in form as that for the op amp itself; however, the various parameters are modified.

Figure 12.113
Linearized gain and
phase plots

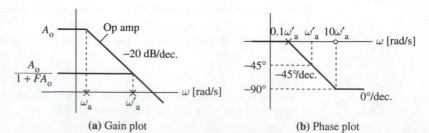

(a) Gain plot   (b) Phase plot

We show the linearized Bode gain and phase plots in Figure 12.113(a) and (b). We have superimposed a plot of the "open loop" gain (the op amp gain) over the gain plot of the complete circuit for the sake of comparison. The question is: Is this plot really correct?

We will investigate by asking for the frequency at which the op amp gain linearized approximation has dropped to the dc gain of the entire circuit. Thus, we must solve the equation

$$\frac{A_o \omega_a}{\omega} = \frac{A_o}{1 + FA_o}. \tag{12.7-30}$$

But the solution of this equation is

$$\omega = (1 + FA_o)\omega_a = \omega'_a. \tag{12.7-31}$$

Therefore, our plot is correct. Observe that the 3-dB bandwidth of the complete circuit has been expanded by the factor $1 + FA_o$, which is approximately FA_o unless F is tremendously small (because A_o is something on the order of 10^5)! Using this approximation in equation (12.7-31) we get

$$\omega'_a \cong FA_0\omega_a = FGB; \tag{12.7-32}$$

that is, the "closed loop" 3-dB bandwidth is about equal to the feedback ratio F multiplied by the gain bandwidth product of the op amp.

Now we ask: What is the gain bandwidth product of the subcircuit in Figure 12.111? To find the answer, we simply compute it:

$$GB' = G_o\omega'_a = \frac{A_o}{1 + FA_o}(1 + FA_o)\omega_a = A_o\omega_a = GB. \tag{12.7-33}$$

In words, we say that the "gain bandwidth product is conserved." That is, the gain bandwidth of the subcircuit in Figure 12.111 is the same as that of the op amp itself. This means that if we design the subcircuit for a dc gain that is 0.01 times the dc gain of the op amp, the bandwidth of the subcircuit will be 100 times as large as that of the op amp itself. It is for this reason that op amp circuits can possess a reasonable bandwidth even though that of the op amp itself is so small.

We will do the same analysis now for the inverting circuit, and we will find that the gain bandwidth is not conserved.

Example 12.18(O)   Find the voltage gain transfer function for the inverting op amp subcircuit shown in Figure 12.114 and sketch the linearized Bode gain and phase plots. Assume that the op amp is well represented by its dominant pole model.

Figure 12.114
The inverting topology

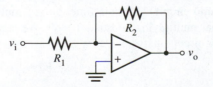

Solution   As in Example 12.17, we insert the dominant pole model for the op amp, attach a sinusoidal voltage source at the input, and use phasors. This gives us the equivalent circuit shown in Figure 12.115. Notice that the negative terminal of the VCVS is connected to the output terminal because the positive input terminal is the one that is grounded. Just

Figure 12.115
The phasor equivalent

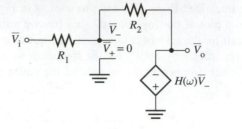

for the sake of variety (though we could simply use, say, superposition), let's do a nodal analysis. The single nodal equation required is at the inverting input terminal of the op amp:

$$\frac{\overline{V}_- - \overline{V}_i}{R_1} + \frac{\overline{V}_- + H(\omega)\overline{V}_-}{R_2} = 0.$$ (12.7-34)

Solving for $\overline{V}_-$, we obtain

$$\overline{V}_- = \frac{\dfrac{R_2}{R_1 + R_2}}{1 + \dfrac{R_1}{R_1 + R_2}H(\omega)}\overline{V}_i.$$ (12.7-35)

However, we want the overall output phasor $\overline{V}_o$. To obtain this quantity, we just notice that it is the negative of $H(\omega)$ times the voltage $\overline{V}_-$. To simplify things, we define the "feedback factor" F by

$$F = \frac{R_1}{R_1 + R_2},$$ (12.7-36)

just as for the noninverting circuit of the last example. We note that

$$1 - F = \frac{R_2}{R_1 + R_2}.$$ (12.7-37)

This gives $$\overline{V}_o = -\frac{(1 - F)H(\omega)}{1 + FH(\omega)}\overline{V}_i.$$ (12.7-38)

Thus, the voltage gain function we are looking for is the factor multiplying the input phasor in (12.7-38). To put it into explicit form we use the dominant pole equation for $H(\omega)$:

$$G(\omega) = -\frac{(1 - F)A_o\,\omega_a}{j\omega + (1 + FA_o)\omega_a}.$$ (12.7-39)

Now let's compare this result with the one we obtained in the last example for the noninverting op amp circuit. This gain function is simply a scaled version of the former because of the real multiplying constant $1 - F$. We see that the 3-dB bandwidth is the same as the earlier circuit, that is,

$$\omega_a' = (1 + F)\,\omega_a.$$ (12.7-40)

If we compute the dc gain by evaluating (12.7-39) at $\omega = 0$, we obtain

$$G_o = -\frac{(1 - F)A_o}{1 + FA_o}.$$ (12.7-41)

There are two differences between this gain and the one we found earlier for the noninverting circuit: there is an extra negative sign and an extra constant multiplier of $1 - F$ (which you can see from (12.7-37) is less than one). This means that the gain bandwidth of the overall inverting circuit is given by

$$GB' = |G_o|\omega_a' = (1 - F)\frac{A_o}{1 + FA_o}(1 + FA_o)\omega_a = (1 - F)A_o\omega_a \quad (12.7\text{-}42)$$

$$= (1 - F)GB.$$

Notice that we are using the *absolute value* of the gain here. The negative sign will affect the phase response, but we will always use "gain bandwidth product" in a positive sense. The net result is this: for the inverting op amp circuit topology, the gain bandwidth of the circuit is not the same as that of the op amp. It has been shrunk by a factor of $1 - F$. Thus, the *gain bandwidth is not conserved* for the inverting topology!

   We show the linearized Bode gain and phase plots in Figure 12.116. Note that although the 3-dB bandwidths of the noninverting and inverting circuits are identical, the -20 dB/decade rolloff does not match that of the op amp in this case because of the lower dc gain. The gain bandwidth product is clearly less than that of the op amp itself. One note about the phase plot: the negative sign can be considered to add either $+180°$ or $-180°$ to the phase—which, otherwise, is identical to that of the noninverting circuit. We have chosen to add $+180°$ because that choice yields a better (more compact) plot. Notice, once more, that our plots are not to scale. This is typical of our "back-of-the-envelope" approach.

Figure 12.116
Linearized gain and phase plots

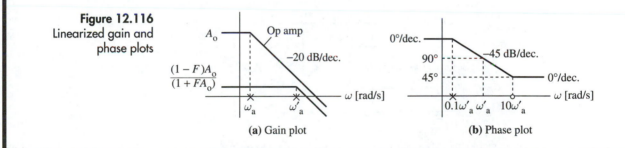

(a) Gain plot   **(b)** Phase plot

BJT Frequency Response

Now let's consider the frequency response of the bipolar junction transistor. A quite common model for the BJT that includes frequency effects is actually a refinement of the *hybrid π model* that we introduced in Chapter 5. It is obtained by adding two capacitances, as shown in Figure 12.117. As we pointed out in Chapter 5, the base resistance r_π and the transconductance g_m are both functions of the dc bias current of the transistor. The capacitor c_π that we have just added is a function of this current as well, and the capacitor c_o is a function of the dc bias voltage between the collector and the base. We will simply provide

Figure 12.117
The small signal high-frequency hybrid π model for the BJT

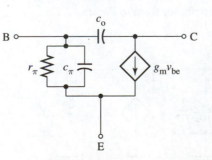

values for these elements when necessary and leave a more thorough exploration to a course in physical electronics. We now present an example for the bipolar junction transistor.

Example 12.19(T)   Find the voltage gain transfer function for the small signal BJT circuit shown in Figure 12.118 and sketch the linearized Bode gain and phase plots. Assume that the BJT is well represented by its high frequency hybrid π model with $r_\pi = 1\ k\Omega$, $c_\pi = 10\ pF$, $c_0 = 0.1\ pF$, and $g_m = 20\ mS$.

Figure 12.118
An example BJT circuit

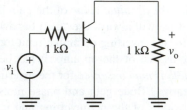

Solution   We simply "unplug" the BJT symbol and "plug in" the high-frequency small signal hybrid π model, resulting in the equivalent circuit shown in Figure 12.119. We are remaining for the moment in the time domain for we want to discuss the component values a bit before going to the phasor form. We call your attention, however, to the fact that we have gone to a single subscript notation for v_{be}. The emitter terminal is grounded, so v_{be} is the same as v_b.

Figure 12.119
The hybrid π
equivalent

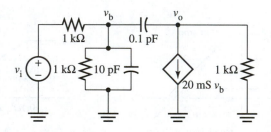

First, let's make the resistances have "nice" values by impedance scaling down by 1000. This makes each of the resistors have a value of $1\ \Omega$, and the transconductance, which was 20 mS, now becomes 20 S. We must increase the size of each of the capacitances by a factor of 1000, recalling that increasing the capacitance lowers the impedance. Thus, we change c_π from 10 pF to 10 nF and c_0 from 0.1 pF to 0.1 nF. Next, let's frequency scale downward by a factor of 10^9 (equivalent to making the unit of frequency the gigahertz). This again increases the capacitance values, this time by a multiplicative factor of 10^9 (note that increasing the capacitance is equivalent to lowering the frequency). Thus, the frequency scaled capacitor values are 10 F and 0.1 F for c_π and c_0, respectively. The resistors are not affected. The resulting impedance and frequency scaled circuit is shown in phasor form in Figure 12.120.

Figure 12.120
The phasor form
after scaling

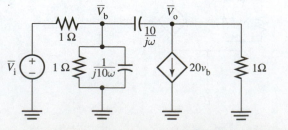

Now let's do a nodal analysis. The equations are

$$\frac{\overline{V}_b - \overline{V}_i}{1} + \frac{\overline{V}_b}{1} + 10(j\omega)\,\overline{V}_b + \frac{j\omega}{10}(\overline{V}_b - \overline{V}_o) = 0 \tag{12.7-43}$$

and

$$\frac{\overline{V}_o}{1} + \frac{j\omega}{10}(\overline{V}_o - \overline{V}_b) = -20\overline{V}_b. \tag{12.7-44}$$

We will leave it to you to solve these simultaneous equations for $\overline{V}_o$ in terms of $\overline{V}_i$ and then compute the ratio to verify that the voltage gain transfer function is

$$H(\omega) = \frac{\overline{V}_o}{\overline{V}_i} = \frac{0.1\,(j\omega - 200)}{(j\omega)^2 + 12.3\,(j\omega) + 2}. \tag{12.7-45}$$

We must be careful to remember that the unit of frequency is the gigaradian/second, but that impedance scaling has not affected the voltage transfer function.

In order to determine the linearized Bode plots we must factor the denominator polynomial. This, in general, requires a calculator or a computer. For the second-order one in this case, however, we can use the quadratic formula or simply complete the squares to obtain the roots, and hence the denominator factors. The resulting voltage gain transfer function is

$$H(\omega) = \frac{0.1(j\omega - 200)}{(j\omega + 0.17)(j\omega + 12.13)}. \tag{12.7-46}$$

The negative zero factor parameter does not affect the Bode gain plot, as we mentioned in the last section. We use our "quick plot" method to sketch the graph as in Figure 12.121.

We must be a bit careful in dealing with the phase plot because of the negative parameter (-200) in the numerator zero factor. We first note that for positive ω, the complex number representing this factor lies in the second quadrant, and as ω increases, the angle decreases from $+180°$ down to $+90°$ smoothly. Thus, we start the phase plot with an an-

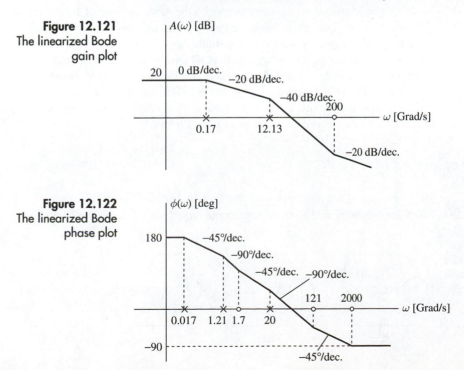

Figure 12.121
The linearized Bode gain plot

Figure 12.122
The linearized Bode phase plot

gle of $+180°$ and then simply treat the zero factor exactly like a pole factor (changing the sign of the parameter from -200 to $+200$). Recall that we represent each pole factor by a pole a decade below the original and a zero a decade above, then do a "gain plot" with slopes at multiples of $45°$/decade. The plot is shown in Figure 12.122.

Section 12.7 Quiz

Q12.7.1. Find the voltage gain transfer function for the circuit shown in Figure Q12.7-1 and sketch the linearized Bode gain and phase plots.

Figure Q12.7-1

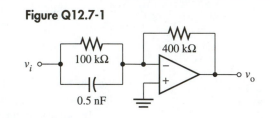

12.8 | SPICE: Frequency Response

The .AC Statement: A Review

Although we covered the .AC control statement in the last chapter, let's review it and extend it a bit. The general format is

```
.AC    ATYPE    NPTS    FSTART    FSTOP
```

The verb .AC tells SPICE that we would like to do an ac steady-state analysis (the circuit must be stable so that this is the same as the ac forced response). The adjective ATYPE should be one of the three key words: LIN, DEC, or OCT. LIN means that the simulation will be done over a "linear" frequency interval; that is, the frequencies at which the simulation is performed will be equally spaced. If one specifies DEC, the analysis frequencies will be evenly spaced in log frequency with the same number of frequency points in each decade. The OCT adjective is similar, except that the analysis frequencies will be evenly spaced in log frequency with the same number of frequency points in each octave. NPTS is the total number of frequency points at which the analysis will be done if LIN is specified and the number per decade or octave if DEC or OCT, respectively, is specified. The analysis starts with the frequency set to FSTART. An analysis is performed there and the frequency is incremented and compared with FSTOP. If the frequency is greater than FSTOP, no analysis is performed and execution stops.

Frequency Response of R, L, and C

Example 12.20   Perform a SPICE analysis on the example circuit shown in Figure 12.123.

Figure 12.123
An example circuit

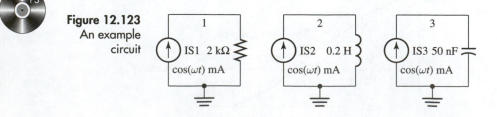

Solution   Note that, in reality, our example actually consists of three independent circuits having a common ground reference. This is a common "dodge" for obtaining results from several closely related circuits simultaneously. The sources are sinusoidal at a general frequency of ω rad/s. The component values have been selected so that the magnitude of the impedance of each is 2 kΩ at $\omega = 10^4$ rad/s. Thus, we will simulate our circuit at several frequency points between, say, 8 and 12 krad/s. The SPICE code is:

```
EXAMPLE 1: RLC IMPEDANCES
IS1 0    1      AC   1E-3
IS2 0    2      AC   1E-3
IS3 0    3      AC   1E-3
R   1    0      2E3
L   2    0      0.2
C   3    0      50E-9
RHI 3    0      1E9
.AC LIN  200    1273 1910   ;NOTE: HERTZ FREQ
.PROBE
.END
```

Notice that we have not specified the frequency of the three ac sources, for this is computed automatically by SPICE when it encountereds the .AC statement. We have specified 200 frequency points at which analysis is to occur between 1273 Hz (about 8 krad/s) and 1910 Hz (about 12 krad/s) and have decided to use the high-quality graphics output by specifying PROBE.

The PSPICE output file is as follows:

```
****07/24/95 09:29:51***********EVALUATION PSPICE (JULY 1993)
*************
EXAMPLE 1: RLC IMPEDANCES
****   CIRCUIT DESCRIPTION
******************************************************************
IS1  0   1  AC   1E-3
IS2  0   2  AC   1E-3
IS3  0   3  AC   1E-3
R    1   0  2E3
L    2   0  0.2
C    3   0  50E-9
RHI  3   0   1E9
.AC  LIN 200  1273  1910   ;NOTE: HERTZ FREQ
.PROBE
.END
****07/24/95 09:29:51***********EVALUATION PSPICE (JULY 1993)
EXAMPLE 1: RLC IMPEDANCES
****  SMALL SIGNAL BIAS SOLUTION            TEMPERATURE= 27.000
DEG C
******************************************************************
NODE   VOLTAGE  NODE   VOLTAGE     NODE    VOLTAGE
( 1)0.0000 ( 2) 0.0000 ( 3) 1.101E-27
VOLTAGE SOURCE CURRENTS
NAME    CURRENT
TOTAL POWER DISSIPATION  0.00E+00 WATTS
  JOB CONCLUDED
   TOTAL JOB TIME     2.58
```

A few notes are in order. First, why have we specified RHI (1 gigaohm) in our NETLIST? We have done this because SPICE does not allow what is referred to as a "floating node," and node 3 is one such because it has attached to it only a capacitor and a current source. In SPICE, a floating node is one that is not connected to ground at dc. SPICE always performs a dc analysis (that is, with $\omega = 0$) before doing the frequency response analysis. At dc, the 50-nF capacitor is an open circuit, and this effectively means

Figure 12.124
Probe output

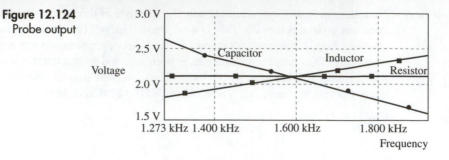

that the current source IS3 is attempting to force 1 mA of current into an open circuit. This is a violation of KCL.

The Probe graphics output is shown in Figure 12.124. The trace labeled "resistor" is the voltage of node 1, the one labeled "inductor" that of node 2, and the one labeled "capacitor" that of node 3. Observe that all have the same value of 2 V at $f = 1.59$ kHz, which corresponds to 10 krad/s. We should also mention that we have plotted the *magnitude* of these node voltages (each is a complex quantity).

SPICE Analysis of a Tuned Circuit

Example 12.21   Perform a SPICE analysis on the parallel tuned circuit shown in Figure 12.125 and plot the magnitude and phase of $v(t)$ for the following values of R: 100 kΩ, and 1 MΩ. Assume that $C = 5$ nF and $L = 2$ H.

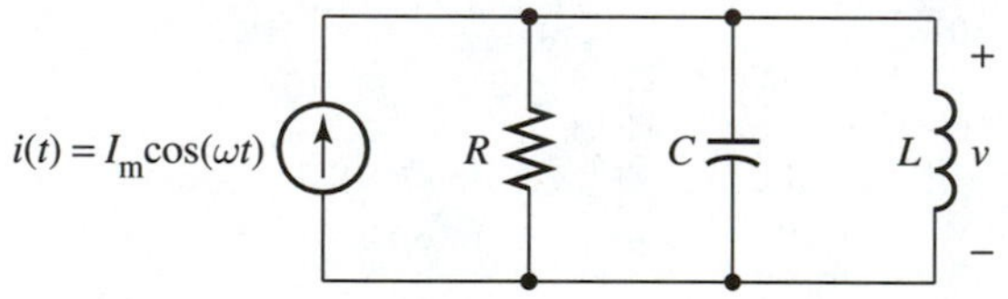

Figure 12.125
An example circuit

$$i(t) = I_m \cos(\omega t) \qquad R \qquad C \qquad L \quad v$$

Solution   The parallel resonant frequency is given by

$$\omega_o = \frac{1}{\sqrt{LC}} = \frac{1}{\sqrt{2 \times 5 \times 10^{-9}}} = 10 \text{ krad/s.} \tag{12.8-1}$$

The quality factor at resonance (which determines the bandwidth) is

$$Q_o = \omega_o RC = 10^4 \times R \times 5 \times 10^{-9} = 5 \times 10^{-5} R. \tag{12.8-2}$$

Therefore, when $R = 100$ kΩ, $Q_o = 5$, and when $R = 1$ MΩ, $Q_o = 50$. The bandwidth is

$$B = \frac{\omega_o}{Q_o} = \frac{10^4}{Q_o}. \tag{12.8-3}$$

Thus, when $R = 100$ kΩ, $B = 2$ krad/s, and when $R = 1$ MΩ, $B = 200$ rad/s. These preliminary computations will give us some idea of how to set the simulation parameters for SPICE. The circuit is shown prepared for SPICE analysis in Figure 12.126. Because the resistance, along with the current source value, sets the peak voltage, we will choose the current source to have a value of 10 μA. This will set the peak voltage to be 1 V when $R = 100$ kΩ and 10 V when $R = 1$ MΩ. We will use a couple of new statements that we will discuss after we present the resulting output. The SPICE file is as follows:

```
EXAMPLE 2A: PARALLEL TUNED CIRCUIT
IS 0   1     AC         10E-6
R  1   0     RMOD       1
L  1   0     2
C  1   0     5E-9
.MODEL RMOD          RES(R=1E5)
.STEP      RES     RMOD(R)   LIST 1E5, 1E6
.AC    DEC 400     0.159E3   15.9E3              ;HERTZ FREQ!
.PROBE
.END
```

Let's first run SPICE and finish the example, then we will discuss the new statement .MODEL appearing in the SPICE file. The PROBE graphics output is shown in Figure 12.127. As we would expect, the tuned circuit response is much sharper for $R = 1\ \text{M}\Omega$ (with $Q_o = 50$) than for $R = 100\ \text{k}\Omega$ (with $Q_o = 5$).

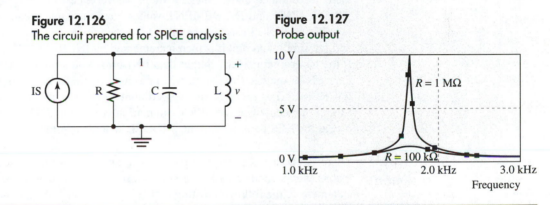

Figure 12.126
The circuit prepared for SPICE analysis

Figure 12.127
Probe output

SPICE Models Now let's discuss the new statements. The term RMOD in the statement specifying the resistor is new. Each element in SPICE has a model associated with it that contains all of the parameters defining that element. In SPICE, a resistor has certain nonideal facets of behavior, such as the coefficient of variation of resistance with temperature. It also has a resistance *multiplier* as one of its parameters. When one specifies Rxxx, a name beginning with R and having any number of additional characters to the right, one is telling SPICE to look for a .MODEL statement somewhere in the file that specifies certain parameters in the internal model for that resistor. (This can be done with the other elements, too, such as Lxxx for an inductor.) We have named our model here RMOD, short for "resistor model." In this model statement, which occurs in our file right after the line that specifies the capacitor, we start with the key verb .MODEL. This tells the SPICE compiler that this line specifies a model. The name of the model is then listed as RMOD. This is followed by the parameter we wish to specify. In this case, it is RES for the resistance parameter. Although the following parentheses are not necessary, we have inserted them for readability. The SPICE compiler simply ignores them, but inside them we have specified $R = 1E5$. The letter R corresponds to the name we have given the single resistor in the circuit. When the SPICE file is run, it takes the value we specified in the earlier line defining R and *multiplies* it by the R value found in the .MODEL statement. Thus, the RES parameter is a multiplier.

The .STEP Statement If we had not included the next statement, the .STEP command, SPICE would have assigned a resistance value of 1*1E5 to R for the simulation run. The .STEP command, however, is one that allows us to step parameters through a set of values. In our file, it has the form

```
.STEP   STEPVAR   MODELNAME   LIST   VAL1, VAL2, . . . , VALN
```

STEPVAR is the name of the variable to be incremented. If it is the name of a parameter in a model, we must give the model name after the variable name as shown. The key word LIST means that a list of values is to follow. In the preceding example, STEPVAR is RES, the resistance parameter in our resistor model; MODELNAME is RMOD, the name we have given our particular resistor model; VAL1 is 1E5 (100 kΩ); and VAL2 is 1E6 (1 megohm).

Another form for the .STEP command is as follows:

```
.STEP   STEPTYPE   STEPVAR   START   END   INCREMENT
```

STEPTYPE can be LIN, DEC, or OCT—with the same meanings as for the .AC statement we outlined above. STEPVAR is the variable to be stepped, or incremented. The START, END, and INCREMENT values have the same meanings as for the .AC command: if STEPTYPE is chosen to be LIN, then the named parameter is first set to START and the simulation run; it is then incremented by INCREMENT and compared with END. If the named parameter is larger than END, no simulation is performed. If it is no greater than END, another simulation will be run. If STEPTYPE is chosen to be DEC or OCT, INCREMENT is set to be the desired number of simulation points per decade or octave, as appropriate. When the SPICE simulation is run, the .STEP value that is computed *replaces* the RES parameter in the .MODEL statement—the latter is simply ignored.

The .PARAM Statement

There is another way to vary parameters, one we have discussed earlier in the text, but that is worth another look. We simply use the concept of a "parameter." The .PARAM statement in the following listing defines a parameter RVAL, and the following statement defining the resistor R uses it. Note that the braces are necessary, and that one could use a more complicated mathematical expression inside them if so desired. The .STEP statement then takes a slightly different form. In this case, the key word .PARAM takes the place of a model name and indicates that it is a user-defined parameter that is being varied. Here is the listing:

```
EXAMPLE 2B: PARALLEL TUNED CIRCUIT
IS    0    1    AC         10E-6
.PARAM    RVAL = 100E3
R     1    0    {RVAL}
L     1    0    2
C     1    0    5E-9
.STEP    PARAM    RVAL    LIST   1E5, 1E6
.AC DEC   400   0.159E3   15.9E3            ;HERTZ FREQ!
.PROBE
.END
```

The Probe output is identical to that shown in Figure 12.127 and so it is not shown.

A Strategic Approach to Determining the Frequency Sweep Parameters

We should perhaps caution you about one thing. The choice of values for the starting frequency, the stopping frequency, and the increment in a frequency response simulation is often an intuitive thing. In practice, one simulates a circuit because it is quite complicated and therefore not a lot of previous knowledge is available about that response. Thus, one usually makes a run over a broad range of frequencies with a course frequency increment and gets a general "feel" for the actual range over which interesting behavior occurs—

then refines the frequency sweep parameters. This is often done several times before a good choice is made.

In the case of a high-Q tuned circuit, it is quite possible to miss the peak value by taking the frequency increment too large, so one should be able to predict, to at least a crude approximation, where this peak will occur and call to mind the basic behavior of this type of phenomenon when doing the iterations.

Section 12.8 Quiz

Q12.8-1. Run a frequency response (gain and phase) for the circuit in Example 12.12 in Section 12.6 and use Probe to produce a graphics plot. Note that the syntax used in Probe is the same as that which we have covered in Section 11.5 of Chapter 11. Here, you should use VP for "voltage phase," etc.

Chapter 12 Summary

This chapter started with an overview in Section 12.1 (a presentation of the results without the mathematically rigorous development that will follow in Chapter 14) of the topic of *Fourier analysis*. In essence, this mathematical discipline demonstrates that a waveform has an alternate description: the *frequency domain* description. The basic idea is that a waveform can be written as a linear combination of sinusoids at different frequencies. This permits us to use the techniques we developed in Chapter 11, phasor analysis, to determine the response of a given circuit to each of the sinusoidal components and then use superposition to add up these responses to give us the complete response of the circuit. Rather than being a computational tool, this process allows us to visualize the operation of a circuit in a way that is distinctively different from the way we have previously envisioned the operation of a circuit in time. We simply consider the way the circuit modifies each frequency component of the input waveform. In Section 12.2, we used this newly introduced viewpoint to define the basic filter types: *lowpass, highpass, bandpass, band reject* (or *notch*), and *allpass*.

In Section 12.3, we developed the idea of the *quality factor,* or Q, of any two-terminal element or subcircuit. This is a measure of how well the element or subcircuit stores energy as opposed to dissipating it. We developed expressions for the Q of a lossy inductor and a lossy capacitor. We introduced the idea of series-to-parallel transformation of a lossy inductor into an equivalent parallel model. We also discussed the idea of *narrow band approximation,* which means that the Q (which varies with frequency) is assumed to be constant over the frequency range of interest.

We next applied our new viewpoint and definitions in Section 12.3 with a discussion and analysis of the *parallel tuned circuit,* which is a basic second-order bandpass filter. This subcircuit is the basic mechanism for tuning a radio receiver to a desired station so that the station is passed, relatively unchanged, and those at other frequencies are rejected.

We followed this discussion with an examination in Section 12.4 of how the *time domain response* of a parallel tuned circuit is influenced by its Q. We showed via the pole-zero diagram that the time response was inversely related to the bandwidth and showed how the poles of the tuned circuit vary as a function of the Q.

In Section 12.5, we considered the *series tuned circuit* and showed that the mathematics of its analysis was precisely the same as that which we developed earlier in Section 12.3 for the parallel tuned circuit—with impedance being replaced by admittance and current by voltage (and vice versa). We then presented *canonical* (or standard) forms for the ac forced response system function of lowpass, bandpass, and highpass second-order filters. We used them to design several types of second-order passive filters.

In Section 12.6 we discussed the twin ideas of Bode gain and phase plots, paying particular attention to the straight-line approximations of these plots for both gain and phase. We presented a quick, "back-of-the-envelope" method for generating these important design aids.

We next discussed the frequency response of active circuits in Section 12.7. This was primarily a set of case studies of various types of dependent source, op amp, and BJT circuits in the frequency domain.

Finally, we discussed SPICE simulation of circuits for their ac forced response, including the ideas of SPICE models, parameters, and step statements.

Chapter 12 Review Quiz

RQ12.1. If a voltage waveform $v(t)$ is known to have a period of 2 μs, what are the frequencies present in its (real, sine and cosine) Fourier series?

RQ12.2. The current waveform $i(t)$ in the circuit shown in Figure RQ12.1 is described by the equation $i(t) = 2 + \cos(4t) + \cos(8t)$ V. What is the voltage waveform $v(t)$?

Figure RQ12.1

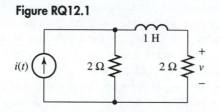

RQ12.3. Find the parallel equivalent of the "lossy" inductor in Figure RQ12.2 at $\omega = 100$ krad/s.

Figure RQ12.2

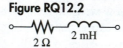

RQ12.4. The Q of the inductor in Figure RQ12.3 is 200, and the other elements are ideal. Find the resonant frequency ω_o, Q of the tuned circuit at resonance (Q_0), the upper and lower cutoff frequencies ω_U and ω_L, and the bandwidth B.

Figure RQ12.3

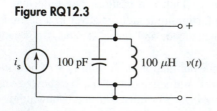

RQ12.5. If $i(t) = 2\cos(\omega t)$ A in the circuit shown in Figure RQ12.4, find the frequency at which the voltage between terminals a and a' has a phase angle of 0°.

Figure RQ12.4

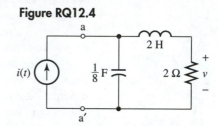

RQ12.6. Find the series equivalent of the "lossy" capacitor in Figure RQ12.5 at $\omega = 100$ krad/s.

Figure RQ12.5

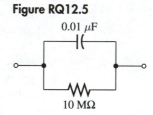

RQ12.7. For the circuit in Figure RQ12.6, $v(t) = 4\cos(10^5 t)$ V. Find the frequency at which $i(t)$ is a maximum. Then find the frequencies where the magnitude of $i(t)$ has dropped to a factor of $1/\sqrt{2}$ times this maximum value. What is the bandwidth B?

Figure RQ12.6

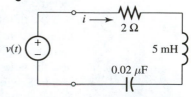

RQ12.8. Sketch a linearized Bode gain plot for the transfer function

$$H(\omega) = 200 \frac{(j\omega)^3(j\omega + 100)}{(j\omega + 5)(j\omega + 10)^2(j\omega + 15)(j\omega + 1000)}.$$

RQ12.9. Sketch a linearized Bode phase plot for the transfer function in Question RQ12.8.

RQ12.10. Find the voltage gain transfer function and sketch linearized Bode gain and phase plots for the active circuit shown in Figure RQ12.7.

Figure RQ12.7

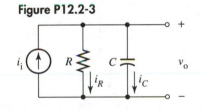

Chapter 12 Problems

Section 12.1 The Frequency Domain Point of View

12.1-1. Assume that the Fourier series of a given periodic waveform has $a_0 = 0$, $b_n = 2/(n\pi)$ for $n = $ odd (that is, $n = 1, 3, 5,...$), and $b_n = -4/(n\pi)$ for n even and nonzero (that is, $n = 2, 4, 6,...$). Sum the Fourier series for the first five nonzero terms and plot the result assuming that the period is 1 s. (Use a graphing calculator or a computer.)

12.1-2. Assume that the waveform of the Problem 12.1-1 is available as the waveform of the voltage source in the circuit of Figure 12.20 in the text. Find the response voltage $v_o(t)$ to the first five nonzero terms of the Fourier series.

Section 12.2 The Definition and Use of Frequency Response

12.2-1. Find and sketch the gain and phase response for the circuit in Figure P12.2-1.

Figure P12.2-1

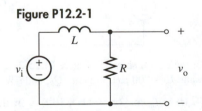

12.2-2. Repeat Problem 12.2-1 for the circuit in Figure P12.2-2.

Figure P12.2-2

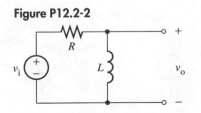

12.2-3. Repeat Problem 12.2-1 for the circuit in Figure P12.2-3 with i_R as the output.

Figure P12.2-3

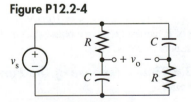

12.2-4. Repeat Problem 12.2-1 for the circuit in Figure P12.2-3 with i_C as the output.

12.2-5. Repeat Problem 12.2-1 for the circuit in Figure P12.2-3 with v_o as the output.

12.2-6. Find and sketch the gain and phase response for the circuit variable v_o in Figure P12.2-4. What type of filter is this?

Figure P12.2-4

12.2-7. A waveform is given by $x(t) = s(t) + n(t)$, where $s(t)$ is the signal and $n(t)$ is added noise. Suppose $s(t) = 10 \cos(2\pi \times 10^4 t)$ and $n(t) = 5 \sin(2\pi \times 10t)$.
  a. Plot $x(t)$.
  b. Assuming $x(t)$ is available as a current source waveform, design a first-order RC filter to remove the noise.
  c. Find and plot the filter output. *Note:* A computer is useful for plotting.

Section 12.3 The Parallel Tuned Circuit

12.3-1. Find and sketch the reactance function $X(\omega)$ for the ideal tuned circuit in Figure P12.3-1.

Figure P12.3-1

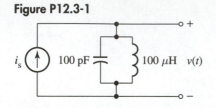

12.3-2. Assuming an ideal inductor and an ideal capacitor (their resistances have been included in the value of R) in Figure P12.3-2,

   a. Find the resonant frequency and the quality factor of the three element subcircuit at the resonant frequency.

   b. Find the 3-dB bandwidth.

Figure P12.3-2

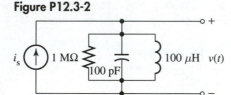

12.3-3. The inductor and the capacitor in Figure P12.3-3 are ideal.

   a. Find the value of R to give a resonant frequency quality factor of 200.

   b. Find the resulting bandwidth.

Figure P12.3-3

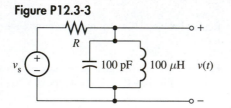

12.3-4. If $v_s(t) = 10 \cos(2\pi \times 4 \times 10^4 t)$ V in Figure P12.3-3, find the value of R required to give a phase shift of $+45°$ in the ac steady-state response voltage $v(t)$.

Section 12.4 Time Response of the Parallel Tuned Circuit

12.4-1. Plot the pole-zero diagram (PZD) for $H(s) = V_o(s)/V_i(s)$ for the circuit in Figure P12.3-3 using the value of R that gives $Q_0 = 100$.

12.4-2. Find and plot the response voltage $v(t)$ as a function of time for the tuned circuit in Figure P12.3-2 for $R = 100$ kΩ if $i_s(t) = \cos(10^7 t)u(t)$ μA.

12.4-3. Repeat Problem 12.4-2 for the circuit in Figure P12.3-3 assuming that $v_s(t) = \cos(10^7 t)u(t)$ V.

12.4-4. Repeat Problem 12.4-2 for the circuit in Figure P12.3-3 assuming $v_s(t) = u(t)$ V.

Section 12.5 The Series Tuned Circuit and Second-Order Passive Filters

12.5-1. Find the series equivalent for the lossy capacitor shown in Figure P12.5-1 at a frequency of 100 kHz. Find the percentage error at 99 and 110 kHz if the series resistance is approximated by a constant whose value is that found at 100 kHz.

Figure P12.5-1

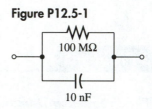

12.5-2. Find and plot the ac steady-state energy stored in the capacitor, $w_C(t)$, and in the inductor, $w_L(t)$, at the resonant frequency in Figure P12.5-2. Assume that $v_s(t) = V_m \cos(\omega t)$ V.

Figure P12.5-2

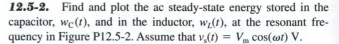

12.5-3. Find and plot the variation of susceptance $B(\omega)$ of the circuit in Figure P12.5-3 versus ω.

Figure P12.5-3

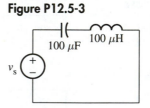

12.5-4. For the circuit shown in Figure P12.5-2, let $R = 0.1$ Ω, $C = 100$ μF, and $L = 100$ μH. Find ω_0, Q_0, and BW.

12.5-5. If $i_s(t) = I_m \cos(2\pi \times 11 \times 10^4 t)$ A in the circuit of Figure P12.5-4, find the value of R required to give a phase shift of $-45°$ in $i(t)$.

Figure P12.5-4

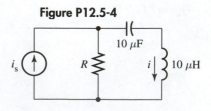

12.5-6. Consider the standard *LP* filter function:

$$H(\omega) = \frac{\omega_0^2}{(j\omega)^2 + \dfrac{\omega_0}{Q_0} j\omega + \omega_0^2}.$$

Show that $A(\omega)$ has a relative maximum (peak) if $Q_0 > 1/\sqrt{2}$ and find that relative maximum and the frequency where it occurs. (*Hint:* Note that maximizing $A^2(\omega)$ is the same as maximizing $A(\omega)$. Write $A^2(\omega) = 1/f(\omega)$ and *minimize* $f(\omega)$.) Assuming that peaking occurs, plot $A(\omega)$ for $Q_0 = 10$.

12.5-7. Using the method developed in the text, sketch the circuit diagram of a highpass filter based on the series tuned circuit.

12.5-8. Repeat Problem 12.5-6 for the highpass filter transfer function. (*Hint:* Let $\omega' = 1/\omega$ first.)

12.5-9. Find $H(\omega) = \dfrac{\overline{V}_o}{\overline{V}_i}$ for the circuit in Figure P12.5-5. What type of filter is it?

Figure P12.5-5

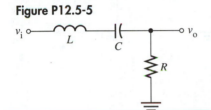

12.5-10. Repeat Problem 12.5-9 for the circuit in Figure P12.5-6.

Figure P12.5-6

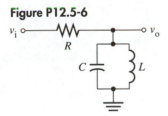

12.5-11. Show that $H_{LP}(\omega) + H_{BP}(\omega) + H_{HP}(\omega) = 1$ for all values of ω, providing each of these transfer functions is in standard form (as defined in the text, Section 12.5).

Section 12.6 Bode Plots

12.6-1. Convert the following gains to dB values: 36, 155, 98, 10^4.

12.6-2. Convert the following dB values to plain gains: 800, -40, 36, -80.

12.6-3. If $\omega_1 = 156$ rad/s and $\omega_2 = 15{,}600$ rad/s, how many decades apart are the two frequencies? If $\omega_1 = 1.56$ rad/s and $\omega_2 = 156$ rad/s, how many decades apart are they?

12.6-4. If $\omega_1 = 84$ rad/s and $\omega_2 = 21$ rad/s, how many octaves are there between ω_1 and ω_2? If $\omega_1 = 84$ krad/s and $\omega_2 = 21000$ rad/s?

12.6-5. Let $H(\omega) = \dfrac{K}{(j\omega + a)(j\omega + b)}$, with a and b positive real numbers. Sketch the linearized Bode gain and phase plots and find the *exact* upper 3-dB frequency. Show that $\omega_{3\,\text{dB}} \cong a$ if $b \gg a$.

12.6-6. Repeat Problem 12.6-5 for $H(\omega) = K\dfrac{j\omega + b}{j\omega + a}$, with $b > a$.

12.6-7. Sketch the linearized gain and phase Bode plots for the transfer function

$$H(\omega) = 10^6 \frac{(j\omega + 10)(j\omega + 80)}{(j\omega)^2(j\omega + 500)(j\omega + 10^4)}.$$

12.6-8. Repeat Problem 12.6-7 for

$$H(\omega) = -10^6 \frac{(j\omega)^2(j\omega + 800)}{(j\omega + 10)(j\omega + 100)(j\omega + 10^4)^2}.$$

Be careful with the negative sign.

Section 12.7 Frequency Response of Active Circuits

12.7-1(O). Find the voltage transfer function and use it to sketch the linearized Bode gain and phase plots for the circuit in Figure P12.7-1.

Figure P12.7-1

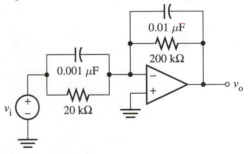

12.7-2(O). Find the voltage transfer function and use it to sketch the linearized Bode gain and phase plots for the circuit in Figure P12.7-2.

Figure P12.7-2

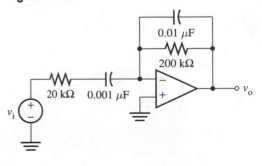

12.7-3. Find the voltage transfer function and use it to sketch the linearized Bode gain and phase plots for the circuit in Figure P12.7-3.

Figure P12.7-3

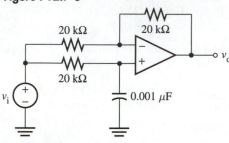

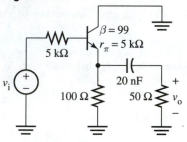

12.7-4(T). Find the voltage transfer function and use it to sketch the linearized Bode gain and phase plots for the circuit in Figure P12.7-4. Use the small signal model for the BJT.

Figure P12.7-4

Section 12.8 SPICE: Frequency Response

12.8-1. Write a SPICE program to solve Problem 12.3-2 using Probe graphics.

12.8-2. Repeat Problem 12.8-1 for Problem 12.3-3.

12.8-3. Repeat Problem 12.8-1 for Problem 12.5-5.

12.8-4(O). Write a SPICE file and simulate the circuit shown in Figure P12.8-1 for $R = 1.5$ kΩ, 1.9 kΩ, and 1.99 kΩ. Plot the frequency response, both gain and phase, using Probe graphics.

Figure P12.8-1

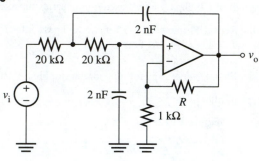

13 Laplace Transform Analysis of Circuits

This chapter opens with a brief historical sketch of Oliver Heaviside, the Victorian electrical scientist and engineer who introduced the idea of circuit analysis using differential operators. It points out that it is only a historical accident, one due to the personality of Heaviside, that—along with the work of Bromwich and others—diverted the engineering profession into the use of the Laplace transform. The literature of the last half-century has been written with Laplace transforms, so the reader must absorb its terminology, although the analytical machinery is precisely the same as that of the Heaviside operators covered in Part II of this text. Section 13.0 offers a very concise and condensed summary of the Laplace transform method (and its relationship to the operator method of Part II) *for those who have already covered operators in depth.* The rest of the chapter is a more leisurely development for those who have not. It, therefore, repeats much of the material of Part II, but in a Laplace transform context.

Section 13.1 gives the definition of the Laplace transform in a physically motivating manner as the steady-state response—if one exists—of a simple first-order system. It briefly discusses convergence and derives the transform of the unit step function (along with its region of convergence, or ROC), develops the very important linearity property, and presents the idea of a time-limited function. It is shown that time-limited waveforms have the entire s plane as their region of convergence. Section 13.1 also derives most of the basic transforms that will be used later in the chapter, as well as most of the basic properties of the transform (time shift, frequency shift, scaling, and linearity); these two sets of facts will later be used to do most Laplace analysis problems without computation of the defining integral; it is the *operational* properties that are of importance in circuit applications. Discussion here is restricted to "ordinary" waveforms—those that are piecewise continuous and whose derivative of any order has the same property.

Section 13.2 offers a fairly rigorous, yet elementary, presentation of the theory of generalized functions, one that is more thorough than that presented in earlier chapters and more thorough than that of most existing texts. It is shown that the inclusion of impulse functions and their derivatives makes the Fundamental Theorem of Calculus "work," in the sense that the integral of the derivative is the original function. This has important implications in the development of the all-important time differentiation property for Laplace transforms, which is the basis for all of its operational aspects.

Section 13.3 develops the time differentiation property in some detail and presents lots of examples of its use to find Laplace transforms and, conversely, to find generalized derivatives by inversion of the Laplace transform. This section also develops the frequency differentiation property.

Section 13.4 derives the *partial fraction expansion,* which is more properly referred to as the *Heaviside expansion theorem* because Heaviside was the one who first derived this result.

Section 13.5 considers the solution of differential equations using the Laplace transform. It also discusses the important concepts of *stability, response decomposition* (into *forced* and *natural* components and into *zero state* and *zero input* parts), and differential equations with prescribed *initial conditions*. Additionally, the idea of a *system function* is presented, along with the associated idea of a *pole-zero diagram*. It is shown that a steady-state response (dc or ac) only exists for stable responses; the forced response always exists, whether or not the response is stable, as long as the forcing function has no poles that are shared with the system function.

After defining *complex impedance* and *admittance,* Section 13.6 applies the preceding material to the analysis of linear time-invariant circuits: those without switches. The *damping properties of second-order systems* are discussed, the *Thévenin/Norton equivalents in the s domain* are illustrated, and the procedures of *impedance and frequency scaling* are covered.

Section 13.7 discusses the analysis of circuits containing switches. A physical discussion is presented to justify the inclusion of any impulses associated with the switching process in the $t > 0$ circuit. The $t \leq 0$ circuit—the circuit holding before the switches flip at $t = 0$—is discussed rather thoroughly. It is demonstrated that only those $t \leq 0$ circuits that have stable responses for the capacitor voltages and inductor currents are *well posed*, though a circuit may be unstable for $t > 0$ and yet be well posed.

Section 13.8 serves two purposes: it demonstrates that the Laplace transform approach presented in this chapter and the Heaviside operator approach developed in Part II are *completely equivalent*, and it develops the idea of time domain convolution as the inverse transform of the product of two *s*-domain functions.

The material in this chapter serves as a general method of circuit analysis, as is the operator method of Part II, and the student is given the choice between the two. In the remainder of the text, the Laplace notation (*s*, rather than *p*) is used for the most part in deference to the current literature. When we wish to emphasize the time domain aspects of a problem, however, we will still use *p* in preference to d/dt and $1/p$ rather than $\int_{-\infty}^{t} (\ ) \, d\alpha$. If the reader wishes, he or she may (in future chapters) simply treat *s* wherever it occurs as a differential operator if that treatment is preferred.

In order to read this chapter, one needs the material in Part I on dc circuit analysis techniques plus the more basic material in Chapter 6 concerning the operation of capacitors and inductors. For those who have covered Part II on operator techniques, much of this chapter will be repetitive; however, they will find here a comparison of the two methods and will acquire the Laplace transform language commonly used in most of the engineering and scientific literature.

13.0 | The Laplace Transform: Historical Perspective

A Bit of History: Oliver Heaviside

It is time for a bit of history in the development of circuit analysis techniques. The time domain analysis approach we presented in Part II was initially developed by Oliver Heaviside, a British self-educated pioneer in electrical science. He originated the idea of applying operators to the solution of circuit analysis problems as a way of speeding up the process and making it more efficient. Until perhaps the 1920s this was the approach used in most texts on circuit analysis.

Unfortunately, Heaviside was an eccentric individual. He was unschooled in rigorous mathematics, did not prove his assertions, and in fact insulted in print those who did not accept his methods. This led to a heated debate between Heaviside and mathematicians at Cambridge University.[1] (Perhaps it would be accurate to say that the debate was actually between Oliver and the *rest* of the scientific and engineering world!)

---

[1] See the delightfully written book by P. J. Nahin, *Oliver Heaviside: Sage in Solitude,* IEEE Press, 1987, for a very readable account of this fascinating Victorian engineer and electrical scientist.

As a result of this debate, a British mathematician named Bromwich developed a rigorous basis for Heaviside's operational calculus based upon the contour integration of complex functions. Instead of using differential operators, which are intuitive to apply directly in the time domain, he transformed the problem into one involving the analytical manipulation of complex functions. The mathematical quantity upon which his method in essence relied is called the Laplace transform. Because the Laplace transform depends upon the convergence of an improper integral, it is less general than the Heaviside method; furthermore, as we just mentioned, it converts the actual problem into another with an entirely different set of symbols and terms.

Why, then, are we devoting an entire chapter to the subject of the Laplace transform? We do so because immediately after Bromwich's work, the engineering community took the Laplace transform to its bosom. Because of the history of the Heaviside method and engineers' increasing desire for rigor, essentially all of the engineering texts on circuit analysis quickly adopted the language of the Laplace transform. To be fair, we must also point out that the earlier texts on Heaviside operator techniques were, for the most part, uniformly poor in their presentation because their authors did not really understand the theoretical underpinnings of the subject.

Thus, for the past half-century, essentially all of the engineering literature has been written in Laplace transform language, with all its defects and cumbersome theoretical justification, even though during this period the Heaviside approach has been placed on a solid theoretical foundation. For this reason, the reader must understand the Laplace transform in order to read existing literature and to communicate with his or her fellow engineers. We will, however, show in a later section of this chapter that all of the Laplace transform machinery we develop is exactly the same in form (with only a change of symbolism from s to p) to that of the Heaviside operator approach, which we have already extensively developed in Part II. Thereafter, in the remainder of the text, if you have covered Part II and prefer the operator approach (as the author hopes you will), you need only substitute p for s in all the formulas we derive and consider all functions of s to be operators in the variable p.

The Purpose of This Section

The following pages of this section are devoted to a "quick fix" on Laplace methods *for the reader who has covered the operator techniques in Part II thoroughly* (though they would clearly benefit more by reading the entire chapter). *Others should skip this material.* The development here is primarily oriented toward developing the basic facts of the Laplace transform and how the Laplace technique relates to the operator method. Therefore, it will be somewhat "telegraphic." We will assume that all waveforms are one sided; that is, they are identically zero for all $t \le 0$.

Definition of the Laplace Transform

Definition 13.1   The Laplace transform of a time waveform $x(t)$ is defined by

$$\pounds\{x(t)\} = X(s) = \int_{-\infty}^{\infty} x(t)e^{-st}\, dt. \qquad (13.0\text{-}1)$$

The parameter $s = \sigma + j\omega$ is called the *complex frequency*. Note that our definition introduces some notation for the transform of $x(t)$. Note also that the Laplace transform is an improper integral (with attendant convergence problems).

Three Fundamental
Properties
Table 13.1 shows the three most fundamental properties of the Laplace transform.[2] We will now develop them sequentially. For the time differentiation property, we simply write down the definition of the Laplace transform of the derivative of $x(t)$, which we will assume has the Laplace transform $X(s)$ (as per Definition 13.1). Thus,

$$\pounds\{px(t)\} = \int_{-\infty}^{\infty} x'(t)e^{-st}\,dt = [x(t)e^{-st}]_{-\infty}^{\infty} + s\int_{-\infty}^{\infty} x(t)e^{-st}\,dt. \qquad (13.0\text{-}2)$$

We have used integration by parts, letting $u = e^{-st}$ and $dv = x'(t)\,dt$. Now, we notice that the lower limit on the integrated term is zero because $x(t)$ is zero for all $t \leq 0$. Furthermore, we will assume that $x(t)e^{-st} \to 0$ as $t \to \infty$—at least for some value of s. Finally, we recognize that the integral in the last term of equation (13.0-2) is just the Laplace transform of $x(t)$. This gives the result shown as Property 1 in Table 13.1.

Table 13.1
Basic properties

Property 1: Time differentiation	$\pounds\{px(t)\} = sX(s)$
Property 2: Time integration	$\pounds\{\frac{1}{p}x(t)\} = \frac{1}{s}X(s)$
Property 3: Frequency shift	$\pounds\{e^{-at}x(t)\} = X(s + a)$

We now offer a brief critique of our derivation. It assumes that integration by parts works; that is, the conditions necessary for the application of this technique are met by our waveform $x(t)$. In fact, a great many waveforms of interest in the circuit applications of the Laplace transform do not—yet most circuits texts assume that they do. As you will see, Property 1 is the foundation upon which the whole structure of the Laplace transform method is erected—and it is somewhat shaky. The resolution of this problem leads one in a natural way to the use of operators, but that is not our aim in this brief section, so we will simply accept the property as being proved. Notice, however, that our requirement that $x(t)e^{-st}$ approach zero for large values of t is actually met by most engineering waveforms. For your future reference, we mention in passing that such waveforms are said to be "of exponential order." The fact that the preceding limit exists for only some values of s, in general, leads to the idea of a region of convergence (ROC) of the Laplace transform. You can easily find these ideas discussed in more depth in many texts and in the remaining sections of this chapter.

To prove Property 2, the time integration property, we just apply the time differentiation property to the function

$$y(t) = \frac{1}{p}x(t). \qquad (13.0\text{-}3)$$

Thus, we have $py(t) = x(t)$, so Property 1 gives

$$sY(s) = X(s). \qquad (13.0\text{-}4)$$

Solving for $Y(s) = \pounds\left\{\frac{1}{p}x(t)\right\} = X(s)/s$ gives Property 2 in Table 13.1.

---

[2] In addition, of course, to *linearity,* which is an immediate consequence of the definition.

The proof of Property 3 is a straightforward computation:

$$\pounds\{e^{-at}x(t)\} = \int_{-\infty}^{\infty} e^{-at}x(t)e^{-st}\,dt = \int_{-\infty}^{\infty} x(t)e^{-(s+a)t}\,dt = X(s + a). \qquad (13.0\text{-}5)$$

In the last step, we have just recognized that the integral is the Laplace transform of $x(t)$ with s replaced by $s + a$. Because the complex frequency s is "shifted" by a, we call it the frequency shift property.

Five Fundamental Laplace Transforms

The transforms of only a few basic waveforms, those shown in Table 13.2, suffice for most applications of the Laplace transform to the analysis of circuits. To derive them, we start with the transform of the unit impulse function:

$$\pounds\{\delta(t)\} = \int_{-\infty}^{\infty} \delta(t)e^{-st}\,dt = \int_{-\infty}^{\infty} \delta(t)\,dt = 1. \qquad (13.0\text{-}6)$$

Next, we use the time integration property to see that

$$\pounds\{u(t)\} = \pounds\left\{\frac{1}{p}\delta(t)\right\} = \frac{1}{s} \times 1 = \frac{1}{s}. \qquad (13.0\text{-}7)$$

Then, we apply this same property n times to the result in equation (13.0-7), to obtain

$$\pounds\left\{\frac{t^n}{n!}u(t)\right\} = \pounds\left\{\frac{1}{p^n}u(t)\right\} = \frac{1}{s^n} \times \frac{1}{s} = \frac{1}{s^{n+1}}. \qquad (13.0\text{-}8)$$

This gives us transform 4 in our table. Finally, we can obtain transforms 3 and 5 in our table by applying the frequency shift property to transforms 2 and 4.

Table 13.2
Basic transforms

$x(t)$	$X(s)$
1. $\delta(t)$	1
2. $u(t)$	$1/s$
3. $e^{-at}u(t)$	$1/(s + a)$
4. $(t^n/n!)u(t)$	$1/s^{n+1}$
5. $(t^n/n!)e^{-at}u(t)$	$1/(s+a)^{n+1}$

The Relationship Between Laplace Transforms and Differential Operators

Now recall our definition of an *elementary function:* one that can be expressed as the impulse response of a differential operator, that is, in the form

$$x(t) = X(p)\delta(t), \qquad (13.0\text{-}9)$$

where $X(p)$ is a rational fraction in the operator p. Furthermore, $X(p)$ can be expanded using a partial fraction expansion. This expresses $x(t)$ in the form of a sum, the individual terms being of the form $kp^n\delta(t)$, $k\delta(t)$, and $k/(p + a)^n\,\delta(t)$. Thus, we can compute the Laplace transform of $x(t)$ by computing a number of terms, each of which has one of the preceding three forms. Therefore, let's look at the Laplace transforms of these three basic waveforms.

First, we see by n applications of the time differentiation property that

$$\pounds\{p^n\delta(t)\} = s^n \times 1 = s^n. \qquad (13.0\text{-}10)$$

Transform 1 in Table 13.2 shows us that

$$\pounds\{k\delta(t)\} = k \times 1 = k. \tag{13.0-11}$$

Finally, using transform 4 in Table 13.2 (with n replaced by $n - 1$) followed by an application of the frequency shift property of Table 13.1, we get

$$\pounds\left\{\frac{k}{(p+a)^n}\,\delta(t)\right\} = \pounds\left\{k\frac{t^{n-1}}{(n-1)!}e^{-at}\delta(t)\right\} = \frac{k}{(s+a)^n}. \tag{13.0-12}$$

Here is the bottom line: if we expand $x(t)$ in a partial fraction expansion and then compute the Laplace transform, we will generate exactly the same terms, except that p will be replaced at each occurrence with s and the $\delta(t)$ term will be dropped. If we "read the partial fraction expansion in the reverse order," we will see that the Laplace transform of $x(t)$ is $X(p)$ with p replaced at each occurrence by s; that is,

$$\pounds\{x(t)\} = \pounds\{X(p)\delta(t)\} = X(s). \tag{13.0-13}$$

Our development has actually shown us quite a bit more than the result in (13.0-13). It has shown us that solution of a problem using the Laplace transform is identical in mechanics with solving the same problem using differential operators. As the following example will demonstrate, it is the time differentiation property of the Laplace transform that is of the most fundamental importance for applications to circuit analysis.

Example 13.1   Using Laplace transform techniques, solve the differential equation

$$\frac{d^2y}{dt^2} + 3\frac{dy}{dt} + 2y = x(t), \tag{13.0-14}$$

assuming that $x(t) = e^{-t}u(t)$.

Solution   We take the Laplace transform of both sides of (13.0-14), using the property of linearity and the time differentiation property to obtain

$$s^2Y(s) + 3sY(s) + 2Y(s) = X(s) = \frac{1}{s+1}. \tag{13.0-15}$$

We have also used the known transform of $x(t) = e^{-t}u(t)$, which is entry 3 of Table 13.2 with $a = 1$. Solving for $Y(s)$, we get

$$Y(s) = \frac{1}{(s+1)^2(s+2)}. \tag{13.0-16}$$

Now, we know that the partial fraction expansion is valid for Laplace transforms, so we can expand $Y(s)$ (leaving the details to you) as

$$Y(s) = \frac{-1}{s+1} + \frac{1}{(s+1)^2} + \frac{1}{s+2}. \tag{13.0-17}$$

Our next step is to take the inverse transform of each term (that is, find the time waveform having that s function as its transform). We use Tables 13.1 and 13.2:

$$y(t) = \pounds^{-1}\{Y(s)\} = \pounds^{-1}\left\{\frac{-1}{s+1}\right\} + \pounds^{-1}\left\{\frac{1}{(s+1)^2}\right\} + \pounds^{-1}\left\{\frac{1}{s+2}\right\} \tag{13.0-18}$$

$$= -e^{-t}u(t) + te^{-t}u(t) + e^{-2t}u(t) = [(t-1)e^{-t} + e^{-2t}]u(t).$$

We should comment on several issues related to the preceding example. We are assuming that the inverse $\pounds^{-1}\{X(s)\}$ of the Laplace transform $\pounds\{x(t)\}$ of any waveform $x(t)$ exists, is unique, and is a linear operator. These, however, are theoretical concerns that can be justified with more mathematical effort (and more caveats and discussions of limitations on the waveforms to be transformed). However, the procedure seems to be intuitive, and it should be easy enough for you to accept these statements at face value—after having covered the operator material in Part II. The next example discusses application of the Laplace transform to solve initial condition problems.

Using the Laplace Transform to Solve Initial Condition Problems

Example 13.2　Using Laplace transform techniques, solve the differential equation

$$\frac{d^2y}{dt^2} + 3\frac{dy}{dt} + 2y = 12, \quad t > 0, \tag{13.0-19}$$

subject to the initial conditions $y(0) = 3$ and $y'(0) = 2$.

Solution　We proceed exactly as with the operator approach, first multiplying both sides by $u(t)$, to obtain

$$y''(t)u(t) + 3y'(t)u(t) + 2y(t)u(t) = 12u(t). \tag{13.0-20}$$

Next, we recall the following identities (obtained by application of the Leibniz rule and the sampling property of the unit impulse function):

$$p\{y_+(t)\} = p\{y(t)u(t)\} = y'(t)u(t) + y(0)\delta(t), \tag{13.0-21}$$

or

$$y'(t)u(t) = py_+(t) - y(0)\delta(t), \tag{13.0-22}$$

and

$$p^2\{y_+(t)\} = p\{py_+(t)\} = p\{y'(t)u(t) + y(0)\delta(t)\} \tag{13.0-23}$$
$$= y''(t)u(t) + y'(0)\delta(t) + y(0)p\delta(t),$$

or

$$y''(t)u(t) = p^2y_+(t) - y'(0)\delta(t) - y(0)p\delta(t). \tag{13.0-24}$$

We use (13.0-22) and (13.0-24), substituting into our original differential equation in (13.0-19), to obtain

$$[p^2y_+ - 2\delta(t) - 3p\delta(t)] + 3[py_+ - 3\delta(t)] + 2y_+ = 12u(t). \tag{13.0-25}$$

Moving the singularity functions to the right-hand side, we see that they appear as additional forcing functions:

$$p^2y_+ + 3py_+ + 2y_+ = 12u(t) + 11\delta(t) + 3p\delta(t). \tag{13.0-26}$$

Taking the Laplace transform of both sides and solving for $Y_+(s) = \pounds\{y_+(t)\}$, we have

$$Y_+(s) = \frac{3s^2 + 11s + 12}{s(s + 1)(s + 2)} = \frac{6}{s} - \frac{4}{s + 1} + \frac{1}{s + 2}. \tag{13.0-27}$$

Here, we have used the known transforms (see Table 13.2) $\pounds\{u(t)\} = 1/s$ and $\pounds\{\delta(t)\} = 1$. Furthermore, we have once again left to you the details of the partial fraction expansion. Recognizing the three terms as the transforms of corresponding time waveforms, we have

$$y_+(t) = [6 - 4e^{-t} + e^{-2t}]u(t). \tag{13.0-28}$$

Recall, however, that our solution is only valid for $t > 0$—*because we do not even know that the differential equation is valid for negative values of t.* Hence, we write our solution in the form

$$y(t) = 6 - 4e^{-t} + e^{-2t}, \quad t > 0. \tag{13.0-29}$$

You can check for yourself that $y(0) = 3$ and $y'(0) = 2$, as required.

Comments on Solution of Initial Condition Problems with the Laplace Transform

If you have taken a course in differential equation theory from a nonoperator point of view, you might feel that our solution of the preceding example is more involved than the "conventional" technique. *In fact, the "conventional" technique is built upon the known solution of particular forms of differential equations (that is, one must recall these forms by rote); furthermore, the evaluation of the unknown constants that occur in the "conventional" solution from the given initial conditions is much more difficult. In the operator approach, the initial conditions are incorporated right at the outset and arbitrary constants do not appear.* Moreover, if the formulas for the one-sided derivatives in equations (13.0-22) and (13.0-24) are committed to memory, the procedure becomes even more straightforward than the conventional approach.

Our final topic in this section will be the initial condition models for the dynamic elements—inductors and capacitors—in their Laplace transform forms. Thus, recall the two time-domain initial condition models shown in Figure 13.1(a) and (b). For the capacitor, we can write KVL in the form

$$v_+(t) = \frac{1}{Cp}i_+(t) + v(0)u(t). \tag{13.0-30}$$

Taking the Laplace transform of both sides, and applying linearity and the time integration property, we get

$$V_+(s) = \frac{1}{Cs}I_+(s) + \frac{v(0)}{s}, \tag{13.0-31}$$

which is just an s domain KVL equation representing the equivalent shown in Figure 13.1(c). Take note of the following fact: the voltage $v(0)$ in the term $v(0)/s$ is the *time domain* initial condition, even though equation (13.0-31) is an s domain entity. We leave the verification of the equivalent for the inductor shown in Figure 13.1(d) to you.

Figure 13.1
Initial condition model for the inductor in the time and complex frequency domains

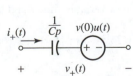

(a) Capacitor: time domain

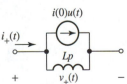

(b) Inductor: time domain

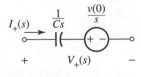

(c) Capacitor: s domain

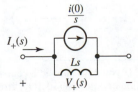

(d) Inductor: s domain

Section 13.0 Quiz

Q13.0-1. Solve the differential equation

$$\frac{d^2y}{dt^2} + 4\frac{dy}{dt} + 13y = e^{-2t}u(t).$$

Q13.0-2. Solve the differential equation

$$\frac{d^2y}{dt^2} + 6\frac{dy}{dt} + 25y = e^{-3t}, \quad \text{for } t > 0,$$

subject to the initial conditions $y(0) = 2$ and $y'(0) = -2$.

Q13.0-3. For the circuit shown in Figure Q13.0-1, find the voltage $v(t)$ for all $t > 0$. Assume that the circuit is operating in the dc steady state for all $t \leq 0$.

Figure Q13.0-1

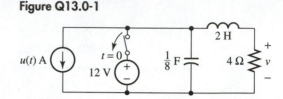

13.1 | Definition and Basic Properties

Physical Motivation for the Laplace Transform

Let's start by investigating the problem of finding the response $y(t)$ of the first-order linear system in Figure 13.2 for any input waveform $x(t)$. We have labeled the integrator with the symbol $1/p$ to denote integration and the input as the derivative of the output waveform using the "prime" notation for differentiation. The summer relationship is

$$y'(t) = x(t) + sy(t). \tag{13.1-1}$$

Moving the term involving $y(t)$ on the right side to the left, we obtain

$$y'(t) - sy(t) = x(t), \tag{13.1-2}$$

which is a differential equation for $y(t)$ with the input $x(t)$ as the forcing function.[3] Now let's solve it by noting that

$$\frac{d}{dt}[e^{-st}y(t)] = e^{-st}y'(t) - se^{-st}y(t) = e^{-st}[y'(t) - sy(t)]; \tag{13.1-3}$$

but if $y(t)$ is to be the solution to (13.1-2), the bracketed term in (13.1-3) must be equal to $x(t)$ (see equation (13.1-2)). Thus, we can use this fact in (13.1-3) to write

$$\frac{d}{dt}[e^{-st}y(t)] = e^{-st}x(t). \tag{13.1-4}$$

Integrating both sides from T to a variable time t and taking the limit as $T \rightarrow -\infty$, we get

$$e^{-st}y(t) - \lim_{T\to-\infty} e^{-sT}y(T) = \lim_{T\to-\infty}\int_T^t x(\alpha)e^{-s\alpha}\, d\alpha = \int_{-\infty}^t x(\alpha)e^{-s\alpha}\, d\alpha. \tag{13.1-5}$$

Figure 13.2
A first-order system

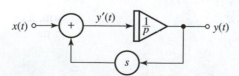

---

[3] If you have not had a course in differential equations, don't be alarmed. We will develop the solution of equation (13.1-1) "from scratch."

Now if the limit on the leftmost side of this expression approaches zero as T approaches $-\infty$ and if the integral on the rightmost side converges at the lower limit (remember that we are dealing with an improper integral here), then we have

$$e^{-st}y(t) = \int_{-\infty}^{t} x(\alpha)e^{-s\alpha}\,d\alpha. \tag{13.1-6}$$

Next we multiply both sides by e^{st} to derive the response waveform $y(t)$:

$$y(t) = \left[\int_{-\infty}^{t} x(\alpha)e^{-s\alpha}\,d\alpha\right]e^{st} = X(s, t)e^{st}. \tag{13.1-7}$$

If the bracketed integral converges at $t = +\infty$ (that is, $X(s, t)$ approaches a finite limit as $t \to +\infty$), then we define it to be the Laplace transform of the input waveform $x(t)$.

Before continuing, we point out that we have used a number of ifs, ands, and thens; this underscores the restrictions we must place on our waveforms. We will discuss these restrictions in the next section and, in the one following that, we will extend the definition of Laplace transform so that most of these restrictions are eliminated; it should be clear, however, that the Laplace transform approach is less general than the operator method outlined in Part II. We will now provide a formal definition of the Laplace transform.

Definition 13.2   The Laplace transform of the waveform $x(t)$ is defined to be

$$X(s) = \lim_{t \to \infty} \int_{-\infty}^{t} x(\alpha)e^{-s\alpha}\,d\alpha = \int_{-\infty}^{\infty} x(t)e^{-st}\,dt, \tag{13.1-8}$$

provided that the integral in question converges, where

$$s = \sigma + j\omega \tag{13.1-9}$$

is called the *complex frequency*.

Notice that we have shifted to an integration variable of t in the improper integral because the upper and lower limits are both constant, rather than being functions of t.

The "One-Sidedness" Assumption

It should be clear from the definition that the Laplace transform is, in theory at least, a quite complicated construct. We will, however, immediately make an assumption that will make our mathematical task a lot simpler. *We will, unless otherwise stated, assume that all forcing functions (system inputs, independent source waveforms in our circuits, etc.) are one sided; that is, they are all identically zero for $t \leq 0$.* This means that all responses are one sided also because a system cannot respond before a stimulus occurs (physical systems are causal). In this case, the only condition that must be met is that the improper integral must be convergent at its upper limit. We will explore this a bit more in a forthcoming example.

The Laplace Transform as the Response of a Spectrum Analyzer

Now that we have defined the Laplace transform, let's go back to equation (13.1-7) for a moment and look only at the bracketed term: the time-varying function whose limit, if one exists, is the Laplace transform of $x(t)$. We show in Figure 13.3 how an instrument can be constructed to find the value of this Laplace transform for each value of s. We have multiplied the exponential function by the unit step to make it zero for $t \leq 0$. This is valid, that is, $x(t)e^{-st} = x(t)e^{-st}u(t)$, because $x(t)$ is assumed to be one sided. After multiplying $x(t)$ by the exponential having the complex exponent $-st = -(\sigma + j\omega)t$, we pass the resulting waveform through an integrator. The output of this integrator will clearly be

Figure 13.3
The spectrum analyzer

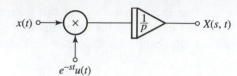

a function of both t and s, $X(s, t)$.[4] If $X(s, t)$ approaches a finite (complex) number as $t \to \infty$, then that value will be the Laplace transform of $x(t)$. We call this system a spectrum analyzer for reasons that we will explore in Chapter 14.

Example 13.3   Find the response of the spectrum analyzer in Figure 13.3 to the input waveform $x(t) = u(t)$ and use it to find the Laplace transform of $u(t)$.

Solution

$$X(s, t) = \int_{-\infty}^{t} u(\alpha)e^{-s\alpha}\, d\alpha = \left[\int_{0}^{t} e^{-s\alpha}\, d\alpha\right]u(t) = \frac{1 - e^{-st}}{s}u(t). \qquad (13.1\text{-}10)$$

Letting $t \to \infty$ and assuming that $\sigma = \mathrm{Re}\{s\} > 0$, one gets $X(s, t) \to X(s) = 1/s$.

Notation   There are a number of notational schemes present in the circuits and systems literature concerning Laplace transforms. We will use each of them at some point in our future work, so we will present them here as a collection for future reference:

$$X(s) = \int_{-\infty}^{\infty} x(t)e^{-st}\, dt, \qquad (13.1\text{-}11)$$

$$X(s) = £\{x(t)\}, \qquad (13.1\text{-}12)$$

$$x(t) \longleftrightarrow X(s), \qquad (13.1\text{-}13)$$

and $$x(t) = £^{-1}\{X(s)\}. \qquad (13.1\text{-}14)$$

Notice that when we are discussing a time function $x(t)$ symbolically, we refer to its Laplace transform with the uppercase form of the same letter. The symbol £ means "Laplace transform of," so (13.1-12) should be read as "$X(s)$ is the Laplace transform of $x(t)$." Equation (13.1-13) should be read "$x(t)$ and $X(s)$ are a Laplace transform pair," and equation (13.1-14) should be read "$x(t)$ is the inverse Laplace transform of $X(s)$." The issue of the existence of an inverse transform lies at the heart of our development, so we will be discussing this at much greater length in a later section. Equations (13.1-12) and (13.1-13) are both synonyms for equation (13.1-11): they mean the same thing. If $x(t)$ happens to be a particular function such as $\cos(t)u(t)$, the uppercase-lowercase distinction is irrelevant—one does not write $\mathrm{COS}(s)$ as the Laplace transform of $\cos(t)$. However, we *will* write such expressions as

$$£\{\cos(2t)u(t)\} = \frac{s}{s^2 + 4}, \qquad (13.1\text{-}15)$$

which (as you will soon see) is a valid equation connecting this particular time waveform with its transform.

---

[4] In the literature, $X(s, t)$ is referred to as the "running" or the "short-time" Laplace transform.

Linearity

The Laplace transform is a linear operation. That is, if $x_1(t)$ and $x_2(t)$ are two wave-forms having Laplace transforms, then

$$\mathcal{L}\{ax_1(t) + bx_2(t)\} = a\mathcal{L}\{x_1(t)\} + b\mathcal{L}\{x_2(t)\}, \qquad (13.1\text{-}16)$$

where a and b are arbitrary constants.

We will not bother with a proof of this property, for it follows directly from the fact, covered in all calculus courses, that an integral is a linear operator acting on its integrand. Now, let's work an example. It should be considered as highly important, however, and not merely as a specific example of limited application.

An Example: The Region of Convergence

Example 13.4   Find the Laplace transform of the waveform

$$x(t) = e^{-at}u(t), \qquad (13.1\text{-}17)$$

where

$$a = a_R + ja_I \qquad (13.1\text{-}18)$$

is an arbitrary complex constant.

Solution   We simply do what the Laplace transform requires: compute the defining integral over a finite time interval, then let the interval length approach infinity:

$$X(s) = \mathcal{L}\{e^{-at}u(t)\} = \lim_{T\to\infty}\int_0^T e^{-at}e^{-st}\, dt = \lim_{T\to\infty}\frac{1 - e^{-(s+a)T}}{s + a}. \qquad (13.1\text{-}19)$$

Notice that we have used a lower limit of zero because the function to be integrated is zero for nonpositive values of t. We must find the limit in question. The s in the denominator and the 1 in the numerator are constants as far as the limit is concerned, so we must only find the limit of the exponential. We first observe that

$$e^{-(s+a)t} = e^{-(\sigma + a_R)t}\, e^{-j(\omega + a_I)t}; \qquad (13.1\text{-}20)$$

furthermore, we note that

$$\left|e^{-j(\omega + a_I)t}\right| = \sqrt{\cos^2[(\omega + a_I)t] + \sin^2[(\omega + a_I)t]} = 1. \qquad (13.1\text{-}21)$$

Using the fact that the absolute value of a product of two complex numbers is the product of the absolute values of those numbers, we have

$$\left|e^{-(s+a)t}\right| = \left|e^{-(\sigma + a_R)t}\, e^{-j(\omega + a_I)t}\right| = \left|e^{-(\sigma + a_R)t}\right|\left|e^{-j(\omega + a_I)t}\right| = e^{-(\sigma + a_R)t}. \qquad (13.1\text{-}22)$$

In the last equality, we have used the fact that e raised to a real power is always positive.

Now we can state our result based on the last term in equation (13.1-22). If $\sigma + a_R > 0$, then $e^{-(s+a)t}$ will approach zero as $t \to \infty$; if $\sigma + a_R = 0$, that function will oscillate (you should use the Euler form to convince yourself that this is true—it is the complex sum of a cosine and a sine); and if $\sigma + a_R < 0$, that function will grow bigger and bigger as $t \to \infty$. Thus, it is only if $\sigma > -a_R$, or in terms of s and a, if $\text{Re}\{s\} > -\text{Re}\{a\}$, that the Laplace transform of the exponential waveform in (13.1-17) exists. We call this set of s values, thought of as a region in the (complex frequency) s plane, the *region of convergence* of the Laplace transform and abbreviate it as ROC.

Returning to equation (13.1-19), we see that

$$X(s) = \pounds\{e^{-at}u(t)\} = \frac{1}{s + a} \tag{13.1-23}$$

with an ROC of $\text{Re}\{s\} > -\text{Re}\{a\}$. The ROC is shown in Figure 13.4 for $a_R = \text{Re}\{a\} > 0$.

Figure 13.4
Region of convergence

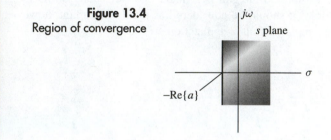

The Laplace Transform of Time-Limited Waveforms

Example 13.5   Find the Laplace transform of the *unit pulse function of duration T* shown in Figure 13.5.

Figure 13.5
The unit pulse function
of duration T

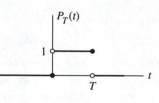

Solution   Analytically, we can write this waveform in the form

$$P_T(t) = \begin{cases} 0; & t \le 0 \\ 1; & 0 < t \le T. \\ 0; & t > T \end{cases} \tag{13.1-24}$$

We will, however, not try to write an analytic expression and "turn the crank," as is so often the course neophyte circuit analysts attempt to follow. We will, rather, notice that the Laplace transform is a definite integral whose interval of integration is the entire real line and just restrict the interval of integration to those values for which the integrand is not zero; that is, from 0 to T. Thus, we have

$$\pounds\{P_T(t)\} = \int_0^T e^{-st}\,dt = \frac{1 - e^{-sT}}{s}. \tag{13.1-25}$$

The last example was an easy computation, wasn't it? This simplicity is reflected in the following general statement: *if a waveform is time limited, then the ROC of its Laplace transform is the entire s plane, that is, the Laplace integral converges for any finite values of s.* By "time limited," we mean that the waveform is nonzero over only a time interval of finite length. In general, if this interval were, say, $T_1 \leq t \leq T_2$, then we would replace the lower limit in (13.1-24) by T_1 and the upper limit by T_2. As both are constants, however, there is no convergence to consider and therefore the ROC is the entire s plane.[5] (In general, time-limited functions are not necessarily one sided; we will, however, assume they *are* unless otherwise stated.)

Next, we will derive some of the most basic transforms and properties of the Laplace transform. These properties and transforms do not require many assumptions at all on the waveforms to be transformed, but in the next section we will discuss others that do—and we will develop the theory of generalized functions to help us in this task. It is surprising that such a small number of transforms and properties are necessary in order to solve any circuit we will encounter *without directly evaluating the Laplace integral itself.*

Laplace Transform of Some Elementary Waveforms

Example 13.6 Find $\mathcal{L}\{u(t)\}$.

Solution We observe that the waveform in question is the same as the one whose transform is given in equation (13.1-23) with $a = 0$. Thus, we can immediately write

$$\mathcal{L}\{u(t)\} = \frac{1}{s}; \quad \text{ROC}: \text{Re}\{s\} > 0. \tag{13.1-26}$$

Note that $\text{Re}\{0\} = 0$.

Example 13.7 Find $\mathcal{L}\{e^{j\omega_0 t}u(t)\}$.

Solution Again, the waveform in question is the same as the one whose transform is given in equation (13.1-23) with $a = -j\omega_0$ this time. Thus, we can again write

$$\mathcal{L}\{e^{j\omega_0 t}u(t)\} = \frac{1}{s - j\omega_0}; \quad \text{ROC}: \text{Re}\{s\} > 0. \tag{13.1-27}$$

Note that $\text{Re}\{j\omega_0\} = 0$.

The waveform in the last example was complex and, for the most part, we will be concerned only with real time functions; however, the next two examples will show the value of that transform in determining the transforms of two very important real ones.

Example 13.8 Find $\mathcal{L}\{\cos(\omega_0 t)u(t)\}$.

Solution We first recall Euler's formula. It has several versions, but the one in which we will be interested is

$$\cos(\omega_0 t) = \frac{e^{j\omega_0 t} + e^{-j\omega_0 t}}{2}. \tag{13.1-28}$$

---

[5] You can use L'Hôpital's rule, or just evaluate the integral in (13.1-25) with $s = 0$, to show that $\mathcal{L}\{P_T(t)\} = T$ at $s = 0$.

Next we use the linearity of the Laplace transform along with the result of Example 13.7, then the same result with ω_0 replaced by $-\omega_0$ to obtain

$$\mathcal{L}\{\cos(\omega_0 t)u(t)\} = \frac{\mathcal{L}\{e^{j\omega_0 t}u(t)\} + \mathcal{L}\{e^{-j\omega_0 t}u(t)\}}{2} = \frac{\dfrac{1}{s - j\omega_0} + \dfrac{1}{s + j\omega_0}}{2} \qquad (13.1\text{-}29)$$

$$= \frac{s}{s^2 + \omega_0^2}.$$

The ROC is $\text{Re}\{s\} > 0$ because $a = -j\omega_0$ in the first exponential waveform and $a = j\omega_0$ in the second.

Example 13.9

Find $\mathcal{L}\{\sin(\omega_0 t)u(t)\}$.

Solution   We again use Euler's formula, this time for the sine function:

$$\sin(\omega_0 t) = \frac{e^{j\omega_0 t} - e^{-j\omega_0 t}}{2j}. \qquad (13.1\text{-}30)$$

We then follow the same derivation as that of Example 13.8, except for a negative sign in the numerator and a j operator in the denominator. The result is

$$\mathcal{L}\{\sin(\omega_0 t)u(t)\} = \frac{\mathcal{L}\{e^{j\omega_0 t}u(t)\} - \mathcal{L}\{e^{-j\omega_0 t}u(t)\}}{2j} = \frac{\dfrac{1}{s - j\omega_0} - \dfrac{1}{s + j\omega_0}}{2j} \qquad (13.1\text{-}31)$$

$$= \frac{\omega_0}{s^2 + \omega_0^2}.$$

The ROC is once again $\text{Re}\{s\} > 0$ because $a = -j\omega_0$ in the first exponential waveform and $a = j\omega_0$ in the second.

A Small Memory Aid

One of the problems encountered by many beginners in Laplace transform theory applications is a difficulty of remembering the distinction between the sine and cosine transforms. They are the same except for the numerator: the cosine transform has an s, whereas the sine transform has a constant equal to the square root of the constant denominator term. To distinguish them, just remember that "co" means "goes with," and co-sine "goes with" the s! Notice, by the way, that we have derived four Laplace transform pairs without directly evaluating the integral.

Time Shift and Frequency Shift Properties

Theorem 13.2 (Time Shift Property)

Let $x(t) \leftrightarrow X(s)$, that is, let $x(t)$ and $X(s)$ be a Laplace transform pair. Then

$$\mathcal{L}\{x(t - a)\} = e^{-as}X(s) \qquad (13.1\text{-}32)$$

and the region of convergence of $x(t - a)$ is the same as that of $x(t)$.

Proof   The proof goes quite easily by simply computing the Laplace transform integral and changing the variable of integration:

$$\mathcal{L}\{x(t - a)\} = \int_{-\infty}^{\infty} x(t - a)e^{-st}\, dt \tag{13.1-33}$$

$$= \int_{a}^{\infty} x(\beta)e^{-s(\beta + a)}\, d\beta = e^{-as} \int_{a}^{\infty} x(\beta)e^{-s\beta}\, d\beta.$$

We used the fact that $x(t - a)$ is zero for $t \le a$ (because $x(t)$ is zero for $t \le 0$) and changed variables to $\beta = t - a$.[6] But note that the last integral is simply the Laplace transform of $x(t)$ with the dummy variable of integration being β rather than t. The region of convergence of $\mathcal{L}\{x(t - a)\}$ is the same as that of $\mathcal{L}\{x(t)\}$. This proves Theorem 13.2. ∎

Example 13.10   Find $\mathcal{L}\{e^{-3t}u(t - 2)\}$.

Solution   We let $x(t) = e^{-3t}u(t - 2) = e^{-6}e^{-3(t-2)}u(t - 2) = e^{-6}y(t - 2)$, where we have defined $y(t) = e^{-3t}u(t)$. (Note that $y(t)$ is zero for $t \le 0$.) But we already know $Y(s)$ from our basic example in the last section: $Y(s) = 1/(s + 3)$, with an ROC of $\text{Re}\{s\} > -3$. Next, we use linearity to factor out the e^{-6} from the Laplace transform of $x(t)$ and use Theorem 13.2 (time shift) to get $X(s) = e^{-6}e^{-2s}/(s + 3)$, with an ROC of $\text{Re}\{s\} > -3$.

Theorem 13.3 (Frequency Shift Property)   Let $x(t) \leftrightarrow X(s)$. Then

$$\mathcal{L}\{e^{-at}x(t)\} = X(s + a). \tag{13.1-34}$$

Furthermore, if the ROC of $\mathcal{L}\{x(t)\}$ is $\text{Re}\{s\} > \alpha$, then the ROC of $\mathcal{L}\{e^{-at}x(t)\}$ is $\text{Re}\{s\} > \alpha - \text{Re}\{a\}$.

Proof   We simply compute the required transform:

$$\mathcal{L}\{e^{-at}x(t)\} = \int_{-\infty}^{\infty} e^{-at}x(t)e^{-st}\, dt = \int_{-\infty}^{\infty} x(t)e^{-(s+a)t}\, dt = X(s + a). \tag{13.1-35}$$

The ROC part is easy; just note that it is $\text{Re}\{s + a\} > \alpha$, so $\text{Re}\{s\} > \alpha - \text{Re}\{a\}$. ∎

Example 13.11   Find $x(t) = \mathcal{L}^{-1}\left\{ \dfrac{4}{s^2 + 4s + 8} \right\}$.

Solution   We will do the inverse transform in steps. First, we complete the squares on the denominator polynomial, giving

$$X(s) = \frac{4}{s^2 + 4s + 8} = \frac{4}{(s + 2)^2 + 4}. \tag{13.1-36}$$

Now we let $Y(s) = \dfrac{4}{s^2 + 4}$, which has the inverse transform $y(t) = 2\sin(2t)u(t)$ (see

---

[6] We remind you once again that all our waveforms are assumed to be one sided unless specifically stated otherwise. One-sidedness is highly important in this property—without it, the property is not true.

Example 13.9). But the transform we are working with is $X(s) = Y(s + 2)$, so we use the frequency shift property:

$$x(t) = \pounds^{-1}\{Y(s + 2)\} = 2e^{-2t}\sin(2t)u(t). \qquad (13.1\text{-}37)$$

Using Transform Properties to Compute Laplace Transforms

We would like to call your attention to our procedure for using the Laplace transform properties to solve problems. We do not simply start with a given function and "crank out" a chain of equalities. What we are doing is expressing a given, more complicated, function in terms of simpler, less complicated ones—then applying the various properties to them. So always ask yourself the question, "Is there a related, but simpler, waveform whose transform or inverse transform I already know?" Then convert the original one into the simpler one (or vice versa) using the various operations occurring in the properties, such as time shift, multiplication by a scalar or exponential, etc.

Scaling Property

**Theorem 13.4
(Scaling Property)**   Let $x(t) \leftrightarrow X(s)$ and let a be a positive real number. Then

$$\pounds\{x(at)\} = \frac{1}{a}X(s/a). \qquad (13.1\text{-}38)$$

Proof   We just write the definition of the Laplace transform of $x(at)$ and change variables to $\beta = at$:

$$\pounds\{x(at)\} = \int_{-\infty}^{\infty} x(at)e^{-st}\,dt = \frac{1}{a}\int_{-\infty}^{\infty} x(\beta)e^{\frac{s}{a}\beta}\,d\beta = \frac{1}{a}X(s/a). \qquad (13.1\text{-}39) \quad \blacksquare$$

Example 13.12   If $x(t) = 4e^{-1000t}\cos(2000t)u(t)$, find $X(s)$.

Solution   Suppose we let $a = 10^{-3}$ and let

$$y(t) = x(at) = x(10^{-3}t) = 4e^{-t}\cos(2t)u(t). \qquad (13.1\text{-}40)$$

Notice here that $u(at) = u(t)$ if $a > 0$. Now the Laplace transform of $y(t)$ is

$$Y(s) = \frac{4(s + 1)}{(s + 1)^2 + 4} = \frac{4s + 4}{s^2 + 2s + 5}. \qquad (13.1\text{-}41)$$

But

$$Y(s) = \pounds\{x(at)\} = \frac{1}{a}X(s/a), \qquad (13.1\text{-}42)$$

so

$$X(s) = aY(as) = 10^{-3}\frac{4(10^{-3}s) + 4}{(10^{-3}s)^2 + 2(10^{-3}s) + 5} \qquad (13.1\text{-}43)$$

$$= \frac{4(s + 1000)}{s^2 + 2000s + 5 \times 10^6}.$$

The scaling property can clearly be used to avoid working with "nasty" values until the final step.

We end this section with two tables. Table 13.3 is a listing of the basic one-sided transforms we have developed thus far. Table 13.4 is a listing of the basic properties we have proved for one-sided waveforms. We call these transforms and properties basic because, after we have added just two more transforms and two more properties, we will be able to use them to solve virtually any problem we encounter in circuit analysis without directly evaluating the defining integral. Though we did not *need* to make this assumption in deriving all properties in the table, you should assume that all are for *one-sided wave-forms*—and you should commit them to memory. In order to conserve space we have not listed the regions of convergence. As a matter of fact, it is the "operational properties" in which we are most interested because these are the ones we will most often use to solve circuits problems.

Table 13.3
Basic transforms

$x(t)$	$X(s)$
1. $u(t)$	$1/s$
2. $e^{-at}u(t)$	$\dfrac{1}{s+a}$
3. $\cos(\omega_0 t)u(t)$	$\dfrac{s}{s^2 + \omega_0^2}$
4. $\sin(\omega_0 t)u(t)$	$\dfrac{\omega_0}{s^2 + \omega_0^2}$

Table 13.4
Basic properties

$x(t)$	$X(s)$
1. $ax(t) + by(t)$	$aX(s) + bY(s)$
2. $x(t - a)$	$e^{-as}X(s)$
3. $e^{-at}x(t)$	$X(s + a)$
4. $x(at); a > 0$	$\dfrac{1}{a}X(s/a)$

Section 13.1 Quiz

Q13.1-1. Draw the "real form" of the Laplace analyzer shown in Figure 13.3. You might wish to return to Chapter 8 and review complex operations in order to do this problem. Let $x(t)$ be a general complex waveform with real part $x_R(t)$ and imaginary part $x_I(t)$; similarly, let $y(t) = y_R(t) + jy_I(t)$, and note that $s = \sigma + j\omega$.

Q13.1-2. By explicitly evaluating the defining integral in Definition 13.2, equation (13.1-8), find the Laplace transforms of the following waveforms and sketch their ROCs in the complex s plane:
 a. $x(t) = e^{-2t}u(t)$
 b. $x(t) = e^{2t}u(t)$
 c. $x(t) = e^{j3t}u(t)$
 d. $x(t) = e^{(2+j3)t}u(t)$.

Q13.1-3. Answer Question Q13.1-1 assuming that $x(t) = 20e^{-t}u(t)$. Assume that the complex frequency in this case is $s = -4 + j4$ and compute the constant value to which the output $X(s, t)$ converges. Express your answer in both Euler form and rectangular form.

Q13.1-4. If $x(t) = e^{-3t}u(t - 1)$, find $X(s)$ without evaluating the defining integral.

Q13.1-5. If $x(t) = e^{-2t}\cos(3t)u(t)$, find $X(s)$ without evaluating the defining integral.

Q13.1-6. If $x(t) = e^{-2000t}u(t)$, find $X(s)$ using the scaling property.

13.2 | Generalized Functions and Their Transforms

Time Differentiation: The Need for Generalized Functions

In the last section we developed several basic properties of the Laplace transform of ordinary functions. In order to make further progress, we must develop one—the time differentiation property—that is not basic at all; in fact, it is somewhat involved to state and prove. We must do so, however, because it is the basis for all applications of the Laplace transform. We will postpone a discussion of this property until the next section, for we must first develop the idea of generalized functions and singularity functions. What is essential for the time differentiation property to work is actually quite obvious: the derivative of each time function must exist and be Laplace transformable. Unfortunately, this is not true of many of the ordinary functions, and we must therefore generalize our idea of function so that all have derivatives. We will start out with an example to illustrate our comments.

Example 13.13   Find and sketch the derivative of $x(t) = (2t + 3)u(t)$, then compute the running integral of this derivative, $\int_{-\infty}^{t} x'(\alpha)\, d\alpha$. (We remind you here that α, though representing time, must be assigned a different symbol than the upper limit t).

Solution   We first sketch the function $x(t)$ itself and show it as Figure 13.6(a). It has a discontinuity at $t = 0$, but otherwise is continuous everywhere. In part (b) of the figure, we show a practical laboratory realization of this signal. Notice that it is continuous *everywhere* because of the finite response-time limitations of our instruments. Thus, we will always assume that our idealized time waveforms are "left continuous," taking a finite time to make the transition from one value to another. For this reason, we use a small dark dot to symbolize the fact that the associated value is actually achieved and the small open circle to denote the fact that a finite (though often very small) time is required to make the transition to the associated value.

Figure 13.6
An example waveform

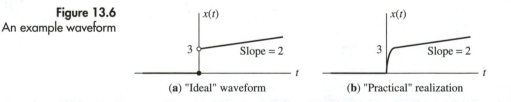

(a) "Ideal" waveform   (b) "Practical" realization

We note that our waveform has another very important property: it is identically zero for $t \leq 0$. As mentioned earlier, we call functions having this property *one sided*. Because the derivative is just the slope of the waveform, we see that it is zero for all $t \leq 0$ and 2 for all $t > 0$. In analytical terms,

$$x'(t) = 2u(t), \qquad (13.2\text{-}1)$$

where $u(t)$ is the unit step function (zero for all $t \leq 0$ and 1 for all $t > 0$). This function is graphed in Figure 13.7(a). We have changed the independent variable to α for we intend to integrate this function to the upper limit t, as shown in the figure. The integral is additive and its value from $-\infty$ to zero is zero, so the integral from $-\infty$ to any value of $t > 0$ is the shaded rectangular area shown. Therefore, we see that the desired integral is zero for all nonpositive values of t and $2t$ for $t > 0$. Analytically, we can write

$$\int_{-\infty}^{t} x'(\alpha)\, d\alpha = 2tu(t). \tag{13.2-2}$$

This function is sketched in Figure 13.7(b).

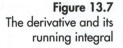

Figure 13.7
The derivative and its
running integral

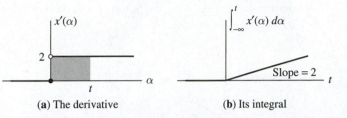

(**a**) The derivative   (**b**) Its integral

Now that we have performed the operations we were asked to do, let's discuss their significance. The bottom line is this: we have lost the discontinuity at $t = 0$! The original waveform is reproduced except for this feature. In terms of mathematical formulas, we see that we can write the original function in the form

$$x(t) = (2t + 3)u(t) = 2tu(t) + 3u(t) = x_c(t) + x_j(t), \tag{13.2-3}$$

where $x_c(t) = 2tu(t)$ is the "continuous part" of $x(t)$ and $x_j(t)$ its "jump function" part.

These two parts are sketched in Figure 13.8. There are two important things to notice about these waveforms. First, the derivative of $x_j(t)$ is zero everywhere except at $t = 0$, at which point it is undefined. Thus, when we differentiate, we lose the jump function part, as we have observed (the step function gives the jump discontinuity at $t = 0$). Furthermore, if we compare $x_c(t)$ with our result in Figure 13.8(b), we see that when we differentiate and then integrate, we recover the continuous part waveform $x_c(t)$ only—the jump function part having been lost in the differentiation process.

Figure 13.8
The continuous
function-jump function
decomposition

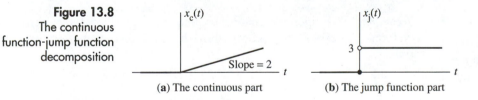

(**a**) The continuous part   (**b**) The jump function part

*More on Jump
Functions*

The preceding example is an important one upon which we will generalize. First, let's discuss the idea of a jump discontinuity. Figure 13.9 shows a waveform that is continuous everywhere except at $t = t_0$. At this value of t, the right-hand limit $x(t_0+)$ and the left-hand limit $x(t_0-)$ both exist and are finite.[7] We define the jump of $x(t)$ at t_0 to be

$$\Delta = x(t_0+) - x(t_0-). \tag{13.2-4}$$

We define the "jump function" component of $x(t)$ by

$$x_j(t) = \Delta u(t - t_0). \tag{13.2-5}$$

---

[7] We remind you that we are using solid and open circles to remind us that laboratory waveforms have finite transition times; thus, even the idealized ones are considered to be left-continuous.

If we subtract this waveform from $x(t)$, we will have the "continuous function" part:

$$x_c(t) = x(t) - x_j(t). \tag{13.2-6}$$

These component waveforms are shown in Figure 13.10. As you can see, these waveforms can, in practice, be drawn immediately from the original waveform. One just thinks of tracing the original waveform, but refusing to lift the pencil at a jump, thereby tracing out the continuous part. The jump part is derived by merely sketching a step function at the proper point in time.

Figure 13.9
A waveform with a
jump discontinuity

Figure 13.10
The continuous and
jump components

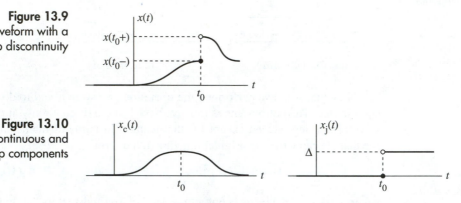

Piecewise Continuous Waveforms

We will insist that all of our waveforms have only jump discontinuities; further, we will demand that they not occur too close together. More precisely, we require that they be *piecewise continuous*. A piecewise continuous waveform is one that is continuous everywhere except at a discrete set of points at which it has only jump discontinuities — and there can be only a finite number of such jumps in each time interval of finite length. (We will continue to require that all waveforms be one sided, as well.)

Decomposition of
a Waveform into
Continuous and
Jump Functions

We can generalize upon the continuous-jump decomposition for piecewise continuous waveforms. We explore this process through an example, then write down the general formulas.

Example 13.14   Let $x(t) = \cos(t)[u(t) - u(t - 2\pi)]$. Find $x_c(t)$ and $x_j(t)$.

Solution   The waveform is sketched in Figure 13.11. It has a positive jump of unit amplitude at $t = 0$ and a negative one of equal magnitude at $t = 2\pi$. The jump function is

$$x_j(t) = u(t) - u(t - 2\pi), \tag{13.2-7}$$

because this is just the function whose subtraction from $x(t)$ is required to "lower" it so that it starts and finishes on the t axis, thus making the remainder continuous. This waveform is shown in Figure 13.12(a). The continuous part is

$$x_c(t) = [\cos(t) - 1][u(t) - u(t) - 2\pi)], \tag{13.2-8}$$

which is the original waveform lowered by one over the range $0 < t \le 2\pi$. It is shown in Figure 13.12(b). Clearly $x_c(t)$ is continuous, for we have removed both the jump discontinuities; furthermore, as you can verify yourself, the sum of $x_c(t)$ and $x_j(t)$ is $x(t)$.

Figure 13.11
An example waveform

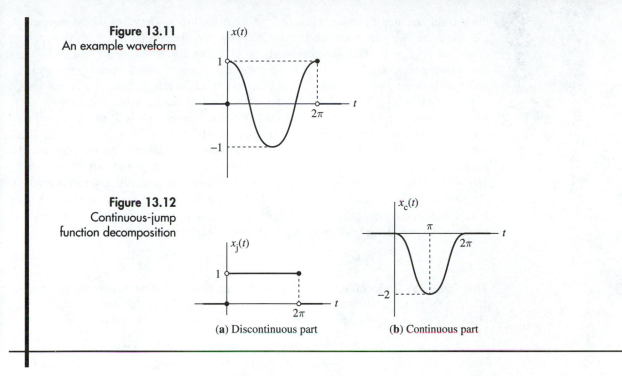

Figure 13.12
Continuous-jump
function decomposition

(a) Discontinuous part   **(b)** Continuous part

Based upon our experience with the waveform in the preceding example, we see that we can write any piecewise continuous waveform as

$$x(t) = x_c(t) + x_j(t), \tag{13.2-9}$$

where the jump function part is given by

$$x_j(t) = \sum_{i=0}^{\infty} \Delta_i u(t - t_i). \tag{13.2-10}$$

Here, Δ_i is the jump in $x(t)$ at $t = t_i$. We are using the fact that $x(t)$ is one sided, so that there are no jumps for negative values of t. Furthermore, we are ordering the time points at which jump discontinuities occur so that $t_0 < t_1 < t_2 < \ldots$. Often we will have the first jump at the origin: $t_0 = 0$. We are allowing for the possibility of an infinite number of jumps in $x(t)$ by using an upper limit of infinity in our sum; however, this limit would be replaced by a finite number if there were only a finite number of jumps. In either case, only the jumps to the left of t affect the value of $x_j(t)$ at any time point t, for those to the right give zero value for the associated step functions. Now that we have an explicit expression for the jump function, we can simply subtract it from the original function to obtain the continuous part:

$$x_c(t) = x(t) - x_j(t). \tag{13.2-11}$$

Example 13.15   Let $x(t) = 4tu(t) - 2tu(t - 1) - 2(t - 1)u(t - 2) - 2u(t - 4)$. Find and sketch the continuous and the jump function parts of this piecewise continuous waveform.

Solution   The function itself is sketched in Figure 13.13. Here's how we did it. We noticed that all of the unit step functions are zero for $t \leq 0$. Between 0 and 1, only the first one is nonzero—and over that interval the function is $x(t) = 4t$, which we have sketched in the graph as a straight line going upward with a slope of $+4$. What happens at $t = 1$? Why, the second unit step function "turns on," going from 0 to 1. Therefore, between 1 and 2, the function we are sketching is given by the sum of the first two terms: $x(t) = 4t - 2t = 2t$. This function starts at $x(1+) = 2 \times 1 = 2$. Note that $x(1+)$ has the intuitive meaning "at $t = 1$ plus a very small amount," which means evaluation of $x(t)$ with both the first two terms "turned on" at $t = 1$. Therefore, there is a discontinuity, with the function jumping downward by 2 units, as we have shown. The function $2t$ is valid between 1 and 2, and has an upward slope of 2 units. At $t = 2$, the third unit step function "turns on," and results in a function of $x(t) = 4t - 2t - 2(t - 1) = 2$ (constant) between 2 and 4. At $t = 4$, the fourth unit step function "turns on," and the mathematical expression for our waveform is the sum of all four terms: $x(t) = 4t - 2t - 2(t - 1) - 2 = 0$. That is, the waveform is identically zero for $t > 4$.

What are the jumps? This is clear from inspection of the graph of $x(t)$ in Figure 13.13: there is a downward jump of 2 units at $t = 1$, another downward jump of 2 units at $t = 2$, and a final downward jump of 2 units at $t = 4$. Thus, the jump function is just

$$x_j(t) = -2u(t - 1) - 2u(t - 2) - 2u(t - 4). \tag{13.2-12}$$

This waveform is plotted in Figure 13.14. We can now determine the continuous part of $x(t)$ merely by subtracting $x_j(t)$ from $x(t)$. We can, of course, do this analytically by using (13.2-12) and the expression for $x(t)$ in the problem statement. This gives

$$x_c(t) = x(t) - x_j(t) = \{4tu(t) - 2tu(t - 1) - 2(t - 1)u(t - 2) \tag{13.2-13}$$
$$- 2u(t - 4)\} - \{-2u(t - 1) - 2u(t - 2) - 2u(t - 4)\}$$
$$= 4tu(t) - 2(t - 1)u(t - 1) - 2(t - 2)u(t - 2).$$

Notice that the last jump is a "pure jump" in $x(t)$; therefore, it does not appear in the continuous part, which is sketched in Figure 13.15. The sketch itself can be done much more quickly than referring to (13.2-13) by simply looking at Figure 13.13 and imagining yourself to be tracing the curve. Instead of lifting your pencil and moving it down abruptly at each jump, you merely keep it going, except at the new slope.

Figure 13.13
An example waveform

Figure 13.14
The jump function

Figure 13.15
The continuous part

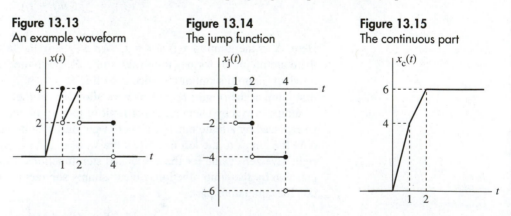

We would like to point out one thing: neither the continuous part of $x(t)$ nor its jump function is zero for $t > 4$, although $x(t)$ itself has this property.

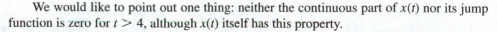

If we were to differentiate the $x(t)$ waveform in Figure 13.13, we would lose the jump function part in Figure 13.14 and would retain only the derivative of the continuous part in Figure 13.15 (in this case, the derivative of $x_c(t)$ would be a pure jump function).

A Fundamental Signal Set for Circuit and System Analysis: K_∞ Waveforms

Our circuit analysis procedures result in derivative relationships. If you have covered Part II on time domain analysis techniques, you will be aware that, in general, one must solve nth-order differential equations. This means that we would like to be able to take an arbitrary number of derivatives of all waveforms we encounter. We will assure ourselves of this by assuming that all of the ordinary waveforms we encounter fall into the following category.

Definition 13.3

Suppose $x(t)$ is a piecewise continuous function; furthermore, assume that for each n its nth derivative $x^{(n)}(t)$ exists and is piecewise continuous. Then we will say that $x(t)$ is a K_∞ waveform and will denote the set of all such functions by the symbol K_∞ itself.

The reason for the subscript ∞ is simply to remind us that an infinite number of derivatives exist. Because C_∞ is used widely in mathematics literature to denote the set of all waveforms that have an infinite number of derivatives that are continuous everywhere, we have selected K (it has the same sound as the c in "continuous") to remind us that our waveforms are continuous "almost everywhere." We stress once again that we are also assuming that all of our waveforms are one sided, that is, they are identically zero for all nonpositive values of time. This is an important property, and one that will make our Laplace transform methods work as we wish them to.

The Fundamental Theorem of Calculus

The time has come now for some more discussion of basic calculus ideas. In your calculus course, you encountered the *Fundamental Theorem of Calculus,* which we will shorten to FTC. This result comes in many forms; which one you have met depends upon the author of your particular calculus text. What it says, in effect, is that the integral of the derivative of a waveform is the same as the waveform itself. But we have just seen that this is not true for functions that have discontinuities. In fact, for K_∞ waveforms, we have just shown by example that one loses the jump function part when one differentiates. Thus, for our waveforms (K_∞), we have the following form of the FTC:

$$\int_{-\infty}^{t} x'(\alpha)\, d\alpha + x_j(t) = x(t); \qquad (13.2\text{-}14)$$

that is, we have to add the jump function back in "by hand." We must remember all the jump amplitudes and occurrence times as we do our operations. This can become quite tedious and cumbersome, so what we will do is this: we will generalize our idea of derivative so that $x_j(t)$ has a derivative and so that the integral of this derivative is the original function itself. Such generalized derivatives will automatically carry the jump information symbolically. This will allow us to use the original form of the FTC in its more compact form

$$\int_{-\infty}^{t} px(\alpha)\, d\alpha = x(t), \qquad (13.2\text{-}15)$$

where we have used the symbol p for our generalized derivative.

Singularities and Generalized Functions

Our generalized differentiation process will center upon the concept of an nth-order *singularity,* which we will symbolize as $A\delta^{(n)}(t - t_o)$ and sketch as in Figure 13.16. The constant A will be called the *strength* (or *amplitude*) of the singularity and the time t_o will be referred to as its *occurrence time*. Often we will simply refer to, say, a "third-order singularity at t_o," with the "at t_o" meaning "with an occurrence time t_o."

Figure 13.16   Symbol for an
                    *n*th-order
                    singularity

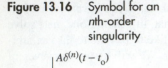

Here's the significance of our singularity. Consider it to be a "sticky pad" note attached to the time axis of an ordinary waveform to remind us, in effect, that after we integrate the given waveform $n + 1$ times,[8] we "add in" a jump discontinuity of height A at the time t_0. We will call our original function (before integration) with the "sticky pad" note attached a *generalized function,* and write it as

$$x(t) = x_0(t) + A\delta^{(n)}(t - t_0). \qquad (13.2\text{-}16)$$

The subscript o refers to the fact that $x_0(t)$ is an "ordinary" function; that is, it is a K_∞ waveform. The plus sign between $x_0(t)$ and the singularity must be interpreted correctly: it means exactly the same as the sticky pad note—it does not mean to add anything to the ordinary function numerically. To keep things clear, we will therefore define the singularity to have the *value* zero for each and every point in time a, writing

$$A\delta^{(n)}(a - t_0) = 0. \qquad (13.2\text{-}17)$$

This means that if we evaluate our generalized function in (13.2-16) at any arbitrary time point $t = a$ (even if $a = t_0$!), we get

$$x(a) = x_0(a) + A\delta^{(n)}(a - t_0) = x_0(a). \qquad (13.2\text{-}18)$$

(Notice the heavy base line we have drawn in Figure 13.16 to emphasize the zero value of our singularity.) Thus, "adding" our sticky pad note has not modified the value of the ordinary function; it has only reminded us to add in the proper jump after we integrate $n + 1$ times.

A New Notation for
Generalized
Differentiation
and Integration

Now let's discuss notation. As we have said, we will use the symbol p to denote generalized differentiation, which we will now proceed to define. We will do this in several steps. First, we define

$$p\{A\delta^{(n)}(t - t_0)\} = A\delta^{(n+1)}(t - t_0) \qquad (13.2\text{-}19)$$

and

$$\frac{1}{p}\{A\delta^{(n)}(t - t_0)\} = A\delta^{(n-1)}(t - t_0). \qquad (13.2\text{-}20)$$

Thus, differentiation (multiplication from the left by p) increases the order of the singularity by one, and integration (multiplication from the left by $1/p$) decreases it by one. The connection between singularities and ordinary functions occurs when $n = 0$; in this case, we write $A\delta^{(0)}(t - t_0)$ simply as $A\delta(t - t_0)$ and express its significance as a "producer of jump discontinuities" with the equations

$$p\{Au(t - t_0)\} = A\delta(t - t_0) \qquad (13.2\text{-}21)$$

and

$$\frac{1}{p}\{A\delta(t - t_0)\} = Au(t - t_0). \qquad (13.2\text{-}22)$$

We now define the derivative and integral of a generalized function by means of the equations

$$px(t) = px_0(t) + p\{A\delta^{(n)}(t - t_0)\} = \frac{dx_0(t)}{dt} + A\delta^{(n+1)}(t - t_0) \qquad (13.2\text{-}23)$$

---

[8] $n + 1$ is used, rather than n, because $n = 0$ will refer to the zeroth-order singularity $A\delta(t - t_0)$— which must be integrated once to produce a discontinuity.

and

$$\frac{1}{p}x(t) = \frac{1}{p}x_o(t) + \frac{1}{p}\{A\delta^{(n)}(t - t_o)\} = \int_{-\infty}^{t} x_o(\alpha)\, d\alpha + A\delta^{(n-1)}(t - t_o). \quad (13.2\text{-}24)$$

We extend this notation in general to any sum of ordinary functions and singularities. That is, the generalized differentiation and integration operators are additive with respect to the sum of ordinary functions and singularities; furthermore, they become the usual operations when applied to ordinary functions.

Example 13.16 Using singularities, find and sketch the first and second generalized derivatives of the waveform $x(t) = (2t + 3)u(t)$ discussed in Example 13.13. Then find the second generalized integral of $p^2 x(t)$ and show that it is the same as the original function.

Solution Recall that the continuous-jump function decomposition of $x(t)$ is

$$x(t) = x_c(t) + x_j(t) = 2tu(t) + 3u(t). \quad (13.2\text{-}25)$$

Thus, we have

$$px(t) = p\{2tu(t)\} + p\{3u(t)\} = 2u(t) + 3\delta(t). \quad (13.2\text{-}26)$$

Applying our generalized differentiation operator once more, we have

$$p^2x(t) = p\{2u(t)\} + p\{3\delta(t)\} = 2\delta(t) + 3\delta'(t), \quad (13.2\text{-}27)$$

where we have written $\delta'(t)$ instead of $\delta^{(1)}(t)$. We will often use the "prime" notation for the first few generalized derivatives of $u(t)$. The two generalized derivatives of $x(t)$ are shown in Figure 13.17. We have only shown one singularity symbol for $p^2 x(t)$ because the two singularities occur at the same time. Now we apply one generalized integration operation to $p^2 x(t)$, obtaining

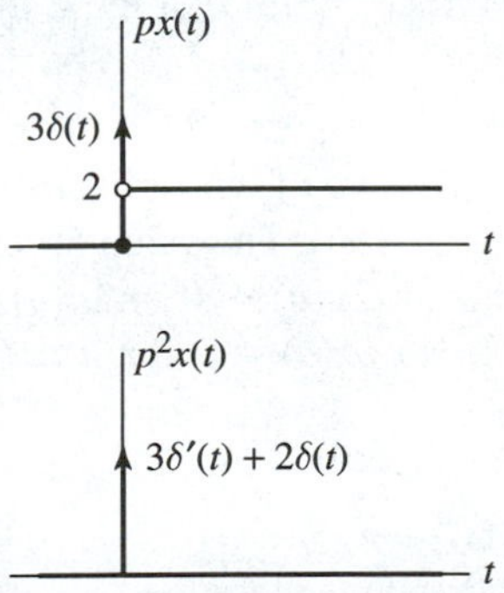

$$\frac{1}{p}\{p^2x(t)\} = 2\frac{1}{p}\{\delta(t)\} + 3\frac{1}{p}\{\delta'(t)\} = 2u(t) + 3\delta(t) = px(t). \quad (13.2\text{-}28)$$

A second application gives

$$\frac{1}{p^2}\{p^2x(t)\} = \frac{1}{p}\{px(t)\} = 2\frac{1}{p}u(t) + 3\frac{1}{p}\delta(t) = 2tu(t) + 3u(t) = x(t). \quad (13.2\text{-}29)$$

Thus, we see that generalized integration reverses the effect of generalized differentiation: the operators p and $1/p$ are inverses of one another. Therefore, we have

$$p\frac{1}{p} = \frac{1}{p}p = 1. \quad (13.2\text{-}30)$$

Figure 13.17 The first two generalized derivatives

Generalized Functions: Formal Definition

Do you see how easy our introduction of singularities as sticky pad notes and the resulting idea of generalized functions have made the process of integration and differentiation? Thus far, our definition of generalized function has been on a highly intuitive level, but we will now make it more mathematically precise. First, a preliminary idea.

Definition 13.4   A *singularity function* is a linear combination of singularities, all of whose orders are less than some fixed finite least integer N called the *order* of the singularity function. In other words, a singularity function can be written in the form

$$x_s(t) = \sum_{i=0}^{\infty} k_i \delta^{(n_i)}(t - t_i), \qquad (13.2\text{-}31)$$

where each n_i satisfies the inequality $n_i \leq N$.

The sum in (13.2-31) might or might not have an infinite number of terms; we stress, however, that it is *not* a sum of real numbers. The *value* of each singularity is zero at each instant of time and *we therefore define the value of $x_s(t)$ to also be zero at each instant of time*. Thus, the sum in question is more like a set union. Again, simply think of it as a sequence of sticky pad note tags along the time axis telling us to add an appropriate step function (jump discontinuity) of height k_i at each occurrence time t_i after $n_i + 1$ integrations.

Definition 13.5   A *generalized function* $x(t)$ is the sum of an ordinary (K_∞) waveform $x_o(t)$ and a singularity function $x_s(t)$,

$$x(t) = x_o(t) + x_s(t) = x_o(t) + \sum_{i=0}^{\infty} k_i \delta^{(n_i)}(t - t_i), \qquad (13.2\text{-}32)$$

where the "addition" of $x_s(t)$ to $x_o(t)$ has the same meaning as just discussed relative to the singularity function definition. Furthermore, the *value* of $x(t)$ is defined by

$$x(a) = x_o(a) + x_s(a) = x_o(a) \qquad (13.2\text{-}33)$$

for each and every value $t = a$—including any of the occurrence times t_i of the singularities.

In Example 13.16, for instance, the first generalized derivative of our ordinary waveform $x(t) = (2t + 3)u(t)$ was computed to be $px(t) = 2u(t) + 3\delta(t)$. The value of $px(t)$ at any arbitrary time instant $t = a$ is $px(a) = 2u(a) + 3\delta(a) = 2u(a) + 0 = 2u(a)$. The step function is, of course, zero if $a \leq 0$ and 1 if $a > 0$, so $px(a) = 0$ for $a \leq 0$ and $px(a) = 2$ in the event that $a > 0$.

Definition 13.6   The derivative of a generalized function $x(t) = x_o(t) + x_s(t)$ is defined to be

$$px(t) = px_o(t) + px_s(t) = x_o'(t) + \sum_{i=0}^{\infty} k_i \delta^{(n_i+1)}(t - t_i). \qquad (13.2\text{-}34)$$

The (running, or indefinite) integral of $x(t)$ is defined to be

$$\frac{1}{p}x(t) = \frac{1}{p}x_o(t) + \frac{1}{p}x_s(t) = \int_{-\infty}^{t} x_o(\alpha)\, d\alpha + \sum_{i=0}^{\infty} k_i \delta^{(n_i-1)}(t - t_i). \qquad (13.2\text{-}35)$$

A couple of comments about Definition 13.6 are perhaps in order. First, we point out that we are using the notation

$$\delta^{(-1)}(t - t_i) = \frac{1}{p}\delta(t - t_i) = u(t - t_i). \qquad (13.2\text{-}36)$$

Thus, in (13.2-34), the $\delta^{(0)}(t - t_i)$ terms come from the step functions in $x_o(t)$ (that is, from the jump function component). Furthermore, in (13.2-35), when we integrate a $\delta(t - t_i)$ term, we turn it into a step function, which gets transferred to the ordinary component as part of its jump function. These two operations link ordinary functions to singularity functions. For this reason, they form the heart of our procedure in that they make computation of the derivative of a discontinuous function possible.

The Definite
Integral of a
Generalized Function

We will need the idea of a definite integral of a generalized function as well as that of the indefinite, or running, integral we have just discussed. Here is the definition.

Definition 13.7   Let $x(t)$ be a generalized function. Then, assuming that $b > a$, we define the definite integral of $x(t)$ between the limits a and b by

$$\int_a^b x(t)\,dt = \left[\frac{1}{p}x(t)\right]_a^b = \frac{1}{p}x(b) - \frac{1}{p}x(a). \qquad (13.2\text{-}37)$$

The symbols on the right mean the values of the generalized running integral at the upper and lower limits, respectively, not integration of the constants $x(b)$ and $x(a)$.

For example, suppose $x(t) = (2t + 3)u(t)$ as in our earlier examples. Then, noting that the generalized derivative is $px(t) = 2u(t) + 3\delta(t)$, we can write, for example,

$$\int_{-2}^3 [2u(t) + 3\delta(t)]\,dt = [(2t + 3)u(t)]_{-2}^3 = 9u(3) - (-u(-2)) = 9. \qquad (13.2\text{-}38)$$

This should come as something of a relief, for it means that *we can manipulate generalized functions using the FTC with impunity.*

Generalized Functions
as the mth Derivative
of a K_∞ Waveform

One interesting aspect of our definition of generalized functions is this: each generalized function $x(t)$ is the mth-order derivative of an ordinary (K_∞) waveform. That is, there is an ordinary function $y(t)$ and an integer m such that

$$x(t) = p^m y(t). \qquad (13.2\text{-}39)$$

This is quite easy to see if you just look at Definition 13.6 and use the fact that the order of each and every singularity is bounded by the same integer N. Thus, integration N times results in an ordinary function $y(t)$; then differentiating the same number of times gives $x(t)$ once more. Here we are using the obvious definitions of higher-order differentiation and integration given inductively by

$$p^m x(t) = p\{p^{m-1}x(t)\} \qquad (13.2\text{-}40)$$

and

$$\frac{1}{p^m}x(t) = \frac{1}{p}\left\{\frac{1}{p^{m-1}}x(t)\right\}. \qquad (13.2\text{-}41)$$

In the example we have already used several times, the generalized function $2\delta(t) + 3\delta'(t)$ is the second generalized derivative of the ordinary function $(2t + 3)u(t)$.

This result will be useful in the next section when we prove some of the more involved properties of Laplace transforms.

The Sampling Property of Impulse Functions

We are almost ready to define the Laplace transform of a generalized function; however, we recall that this involves computing the product of $x(t)$, the waveform to be transformed, and the exponential function e^{-st}. Therefore, we must first define such a product. To do so, let's first look at an ordinary function $\phi(t)$ that is continuous at $t = t_o$ and its product with the unit step function at $t = t_o$—shown in Figure 13.18. If we compute the ordinary derivative, we will obtain the identically zero function for $t < t_o$ and $\phi'(t)$ for $t > t_o$—that is, $\phi'(t)u(t - t_o)$. The derivative will not exist at $t = t_o$, but this is unimportant (we "fill in" its value there using left continuity). More important is the information about the time point t_o that is provided by the generalized derivative sketched at the bottom of Figure 13.18. Analytically, the jump at $t = t_o$ "adds" a first-order singularity:

Figure 13.18 The derivative of the product of an ordinary function and a step function

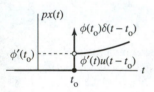

$$p\{\phi(t)u(t - t_o)\} = \phi'(t)u(t - t_o) + \phi(t_o)\delta(t - t_o). \tag{13.2-42}$$

Now, let us *assume* that the Leibniz rule (about the derivative of a product of two functions) is valid for generalized derivatives. This gives

$$p\{\phi(t)u(t - t_o)\} = \phi'(t)u(t - t_o) + \phi(t)\delta(t - t_o) \tag{13.2-43}$$

because the derivative of the unit step function is the first-order singularity $\delta(t - t_o)$.[9] Setting (13.2-42) and (13.2-43) equal gives

$$\phi(t)\delta(t - t_o) = \phi(t_o)\delta(t - t_o). \tag{13.2-44}$$

We observe that this is a *definition,* though one that follows logically from the assumption that the product in question obeys the Leibniz rule. Equation (13.2-44) is called the *sampling property* of the unit impulse function, though this definition is often given to a simple consequence: if we integrate both sides of (13.2-44) from $-\infty$ to $+\infty$ using our definition of definite integral given in Definition 13.7, we will get

$$\int_{-\infty}^{\infty} \phi(t)\delta(t - t_o)\,dt = \int_{-\infty}^{\infty} \phi(t_o)\delta(t - t_o)\,dt = \phi(t_o)[u(t - t_o)]_{-\infty}^{\infty} = \phi(t_o). \tag{13.2-45}$$

The Product of an Ordinary Function and a Singularity

If you differentiate both sides of (13.2-44) and assume the Leibniz rule to be satisfied, you can obtain the required definition for $\phi(t)\delta'(t - t_o)$. Continuing this same process for higher-order derivatives of the unit impulse results in the following definition. Though somewhat complicated, it is highly important. The inductive proof that it follows logically from (13.2-44) above and the Leibniz rule is given in Appendix D.

Definition 13.8 If $\phi(t)$ is a time function whose nth derivative $\phi^{(n)}(t)$ is continuous at $t = t_o$, then the product of $\phi(t)$ and the nth-order singularity $\delta^{(n)}(t - t_o)$ is defined by

$$\phi(t)\delta^{(n)}(t - t_o) = \sum_{k=0}^{n} \binom{n}{k}(-1)^k \phi^{(k)}(t_o)\delta^{(n-k)}(t - t_o). \tag{13.2-46}$$

---

[9] We will refer to $\delta(t - t_o)$ as a unit impulse at $t = t_o$ in keeping with tradition. Then we will use the notation $\delta^{(n)}(t - t_o) = p^n\delta(t - t_o)$.

We can consider this to be a sampling property of the nth-order singularity function. It is often expressed in terms of a definite integral, as follows:

$$\int_{-\infty}^{\infty} \phi(t)\delta^{(n)}(t - t_o)\,dt = \sum_{k=0}^{n}\binom{n}{k}(-1)^k\phi^{(k)}(t_o)\int_{-\infty}^{\infty}\delta^{(n-k)}(t - t_o)\,dt \qquad (13.2\text{-}47)$$

$$= \sum_{k=0}^{n}\binom{n}{k}(-1)^k\phi^{(k)}(t_o)[\delta^{(n-k-1)}(t - t_o)]_{-\infty}^{\infty}$$

$$= (-1)^n\phi^{(n)}(t_o).$$

We have used here the fact that the value of the integrated singularity $\delta^{(n-k-1)}(t - t_o)$ is zero except for $k = n$—in which case $\delta^{(n-k-1)}(t - t_o) = u(t - t_o)$. Finally, $u(\infty) = 1$, and $u(-\infty) = 0$. We will now use this result in computing the Laplace transform of a generalized function.

The Laplace Transform of a Generalized Function

First, we note that the Laplace transform of a singularity of order n is

$$\int_{-\infty}^{\infty}\delta^{(n)}(t - t_o)e^{-st}\,dt = (-1)^n\frac{d^n}{dt^n}[e^{-st}]_{t=t_o} = s^n e^{-st_o}. \qquad (13.2\text{-}48)$$

We use this, along with the assumption of linearity, to define the Laplace transform of an arbitrary generalized function.

Definition 13.9   The Laplace transform of a generalized function $x(t)$, written in the form given by Definition 13.5, is defined by

$$\pounds\{x(t)\} = \pounds\{x_o(t)\} + \pounds\{x_s(t)\} = X_o(s) + \sum_{i=0}^{\infty}k_i s^{n_i}e^{-s(t-t_i)}, \qquad (13.2\text{-}49)$$

providing that the Laplace transform $X_o(s)$ exists and the series in question converges.

We note here that, unlike the case for generalized functions of time, the sum in question is an ordinary one: for each value of s it is a sum of numbers. As such, it can be an infinite series if an infinite number of terms are present.

Example 13.17   Find the Laplace transform of the generalized function

$$x(t) = e^{-at}u(t) + \sum_{i=0}^{\infty}\delta(t - iT). \qquad (13.2\text{-}50)$$

Solution   We merely apply the definition:

$$X(s) = \frac{1}{s + a} + \sum_{i=0}^{\infty}e^{-iTs} = \frac{1}{s + a} + \frac{1}{1 - e^{-sT}}. \qquad (13.2\text{-}51)$$

The first term is the Laplace transform of the ordinary function $x_o(t) = e^{-at}u(t)$, which converges if $\sigma = \text{Re}\{s\} > 0$. The second is the sum of the geometric series

$$\sum_{i=0}^{\infty}a^i = \frac{1}{1 - a}, \qquad (13.2\text{-}52)$$

with $a = e^{-sT}$, which converges to the expression on the right if $|a| = e^{-\sigma T} < 1$—or $\sigma > 0$. Thus, if $\sigma = \text{Re}\{s\} > 0$, then the Laplace transform of $x(t)$ converges and has the value on the right-hand side of (13.2-51).

Properties of Laplace Transforms of Generalized Functions

We must now address one final topic before we close this section: the properties we derived in Section 13.1 for ordinary functions. We will format them as theorems. Linearity was at the very heart of our definitions in this section, so it should be clear that the linearity property continues to hold for generalized functions. We will, therefore, not go through a proof. Instead, we will accept it as being quite clear.

Theorem 13.5 (Linearity) For any two generalized functions $x(t)$ and $y(t)$ and any two scalars a and b, we have

$$\pounds\{ax(t) + by(t)\} = a\pounds\{x(t)\} + b\pounds\{y(t)\}. \tag{13.2-53}$$

We can easily see that the time shift property holds for singularities because this is the content of equation (13.2-48).

Theorem 13.6 (Time Shift) Let $x(t)$ be an arbitrary generalized function with transform $X(s)$. Then

$$\pounds\{x(t - a)\} = e^{-as}X(s). \tag{13.2-54}$$

Proof We note that replacing t by $t - a$ is the same as replacing t_o by $t_o + a$ in equation (13.2-48), to obtain

$$\pounds\{\delta^{(n)}(t - a - t_o)\} = \int_{-\infty}^{\infty} \delta^{(n)}(t - a - t_o)e^{-st}\,dt \tag{13.2-55}$$

$$= (-1)^n\frac{d^n}{dt^n}[e^{-st}]_{t=t_o+a} = e^{-as}s^n e^{-st_o}$$

$$= e^{-as}\pounds\{\delta^{(n)}(t - t_o)\}.$$

Thus, the time shift property holds for singularities. We can factor out the common exponential term e^{-as} from the singularity sum in equation (13.2-49) of Definition 13.9. Then, as we already know that our property is valid for the ordinary waveform $x_o(t)$, we see that it is valid for the generalized function $x(t)$ as well. ∎

Theorem 13.7 (Frequency Shift) Let $x(t)$ be an arbitrary generalized function with Laplace transform $X(s)$ and let a be an arbitrary real number. Then

$$\pounds\{e^{-at}x(t)\} = X(s + a). \tag{13.2-56}$$

Proof We just use equation (13.2-47) with $\phi(t) = e^{-(s+a)t}$. This gives

$$\pounds\{e^{-at}\delta^{(n)}(t - t_o)\} = \int_{-\infty}^{\infty} \delta^{(n)}(t - t_o)e^{-(s+a)t}\,dt \tag{13.2-57}$$

$$= (-1)^n\frac{d^n}{dt^n}[e^{-(s+a)t}]_{t=t_o} = (s + a)^n e^{-(s+a)t_o}$$

$$= \pounds\{\delta^{(n)}(t - t_o)\}_{s \leftarrow s+a}.$$

Therefore, our property has been proved for singularities—and by extension—for the singularity function component $x_s(t)$ (for we can factor the e^{-as} term from the sum defining $X_s(s)$). We already know that the property holds for ordinary functions $x_o(t)$, so we have proved our property for generalized functions. ■

The only remaining property for us to verify is the scaling property. To do so, we must consider the idea a bit more carefully than we did in the last section. In the scaling process, we agree to transform our time variable from t to t' by letting $t' = at$ (with $a > 0$). This means that this substitution must be done *at each and every occurrence* of t in any formula. Thus, we would write

$$x_a(t') = x(at) \tag{13.2-58}$$

for each occurrence of $x(t)$ and

$$p_a^n = \frac{d^n}{d(at)^n} = a^{-n}\frac{d^n}{dt^n} = a^{-n}p^n \tag{13.2-59}$$

for each occurrence of the nth-order time differentiation operator. We have written this operator explicitly for the ordinary derivative to explain our motivation; *we will, however, merely take this as part of our definition of the generalized operator p^n—an added property to be assumed as part of its definition so as to make it consistent with the ordinary derivative.* Notice carefully that p_a^n is the nth derivative with respect to t'.

Theorem 13.8 (Scaling) Let $x(t)$ be an arbitrary generalized function with Laplace transform $X(s)$ and let $t' = at$, where a is any positive real number. Then

$$\pounds\{x_a(t')\} = \frac{1}{a}X(s/a). \tag{13.2-60}$$

Proof We already know that the property is valid for $x_o(t)$, the ordinary function part of $x(t)$. Thus, if we let $x(t)$ be an nth-order singularity at $t = t_0$, we have

$$\pounds\{\delta^{(n)}(t' - t_i)\} = \pounds\{p_a^n\delta(t' - t_i)\} = a^{-n}\pounds\{p^n\delta(at - t_i). \tag{13.2-61}$$

To compute the Laplace transform of $\delta(t' - t_i)$, we note that

$$u(t' - t_i) = u(at - t_i) = u(a[t - t_i/a]) = u(t - t_i/a). \tag{13.2-62}$$

The last step comes from the fact that the sign of the argument is unaffected. Next, we recall that

$$\delta(t' - t_i) = p_a u(t' - t_i) = a^{-1}pu(at - t_i) = a^{-1}pu(t - t_i/a) \tag{13.2-63}$$
$$= a^{-1}\delta(t - t_i/a).$$

Using this result in (13.2-61) gives

$$\pounds\{\delta^{(n)}(t' - t_i)\} = a^{-(n+1)}\pounds\{\delta(t - t_i/a) = \frac{1}{a}(s/a)^n e^{-(s/a)t_i}. \tag{13.2-64}$$

This proves our property for singularities and, by extension, to $x(t)$ itself. ■

Section 13.2 Quiz

Q13.2-1. Sketch the first and second *ordinary* derivatives of the waveform shown in Figure Q13.2-1, then integrate the $x'(t)$ waveform and sketch the result. What would be the result of integrating $x''(t)$ twice?

Figure Q13.2-1

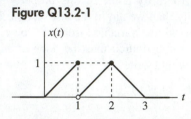

Q13.2-2. Answer Question Q13.2-1 using generalized functions; that is, sketch the first and second generalized derivatives, then integrate $p^2 x(t)$ twice to recover $x(t)$. Do the latter part analytically.

Q13.2-3. Find the *generalized* derivative of the waveform shown in Figure Q13.2-2.

Figure Q13.2-2

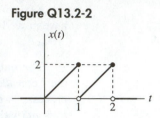

Q13.2-4. Consider the ordinary waveform $x(t)$ in Figure Q13.2-1. Is the generalized function $x(t)\delta(t - 2)$ well defined? Is $x(t)\delta'(t - 2)$ well defined? Is $x(t)\delta(t - 1)$ well defined?

Q13.2-5. Find the definite integral of the generalized function $\cos(2t)\delta'(t)$ over the time interval $[-1, 1]$. Find the integral over the interval $(-\infty, \infty)$ using equation (13.2-47).

Q13.2-6. Find the Laplace transform of $x(t)$ in Figure Q13.2-2.

Q13.2-7. Find the Laplace transform of the generalized function $x(t) = 3u(t - 1) + \delta''(t - 3)$.

13.3 | The Time Differentiation Property

We will now derive the important time differentiation property of the Laplace transform, then show how to apply it in computing transforms. First, though, we need a bit more theory.

Integration by Parts for Generalized Functions

Many of the operations one has to perform in circuits and systems work rely upon the *integration by parts* formula of integral calculus. We will show here that it works for generalized functions. We consider an arbitrary generalized function, which as we know can be written

$$x(t) = x_0(t) + x_s(t), \tag{13.3-1}$$

where $x_0(t)$ is an ordinary (K_∞) waveform and $x_s(t)$ is a singularity function, which we recall is a linear combination of singularities, the order of each being no greater than some common integer N. We will let $\phi(t)$ be an arbitrary ordinary function whose $(N + 1)$-th derivative exists and is continuous at the occurrence time of each singularity in $x_s(t)$.[10] Because differentiation raises each singularity order by one, we see that the two products $\{px(t)\}\phi(t)$ and $x(t)p\phi(t)$ are well defined. Then, as the Leibniz rule is applicable to the product of $x(t)$ and $\phi(t)$, we see that

$$p\{x(t)\phi(t)\} = \{px(t)\}\phi(t) + x(t)p\phi(t). \tag{13.3-2}$$

---

[10] If $x(t)$ is an ordinary waveform, so that the singularity function is not present, we only require that $\phi(t)$ be continuous at each jump in $x(t)$. This is consistent, though, if we interpret a step function as being a singularity of order -1: $u(t - t_i) = \delta^{(-1)}(t - t_i)$.

(Notice that $p\phi(t) = \phi'(t)$, the ordinary derivative.) Because our restrictions on $\phi(t)$ imply that the product $x(t)\phi(t)$ is a well defined generalized function for which the

FTC works, that is $\dfrac{1}{p}[p\{x(t)\phi(t)\}] = x(t)\phi(t)$, we see that

$$x(t)\phi(t) = \frac{1}{p}[\{px(t)\}\phi(t)] + \frac{1}{p}[x(t)p\phi(t)] \qquad (13.3\text{-}3)$$

or

$$\frac{1}{p}[\{px(t)\}\phi(t)] = x(t)\phi(t) - \frac{1}{p}[x(t)p\phi(t)]. \qquad (13.3\text{-}4)$$

Equation (13.3-4) is our indefinite integral form of the integration by parts formula; however, if we evaluate it between fixed limits, we will have the definite integral form:

$$\int_a^b [px(t)]\phi(t)\,dt = x(t)\phi(t)]_a^b - \int_a^b x(t)p\phi(t)\,dt. \qquad (13.3\text{-}5)$$

We need one more concept before we can tackle the time differentiation property, one which assures us that a given ordinary function has a Laplace transform.

Definition 13.10   Let $x_o(t)$ be an ordinary waveform and assume that there exist two real numbers M and σ_o such that

$$|x_o(t)| \le Me^{\sigma_o \tau} \qquad (13.3\text{-}6)$$

for all values of t. Then we say that $x_o(t)$ is *of exponential order* and write "$x_o(t)$ is EO(M, σ_o)."

If $x_o(t)$ is EO(M, σ_o), then its Laplace transform exists and has a region of convergence (ROC) given by $\sigma = \mathrm{Re}\{s\} > \sigma_o$. To see this, look at

$$\left| \int_0^T x_o(t)e^{-st}\,dt \right| \le \int_0^T |x_o(t)|\,e^{-\sigma t}\,dt \le \int_0^T Me^{\sigma_o t}e^{-\sigma t}\,dt = M\,\frac{1 - e^{-(\sigma - \sigma_o)T}}{\sigma - \sigma_o}. \qquad (13.3\text{-}7)$$

Letting $T \to \infty$, we see that

$$\left| \int_0^\infty x_o(t)e^{-st}\,dt \right| \le \frac{M}{\sigma - \sigma_o}, \qquad (13.3\text{-}8)$$

provided of course that $\sigma > \sigma_o$. This says that the improper Laplace integral converges absolutely and hence it converges.[11] Thus, the Laplace transform of a function of exponential order always exists.

---

[11] It might be helpful to thumb back through your calculus text material on improper integrals and review this result.

The Time
Differentiation
Property

Theorem 13.9   Assume that $x(t)$ is a generalized function whose ordinary function $x_o(t)$ is of expo-
(Time Differentiation   nential order and for which the transform of its singularity function $x_s(t)$ exists (the as-
Property)   sociated sum converges). Then the Laplace transform of the generalized derivative of
$x(t)$ exists and is given by

$$\pounds\{px(t)\} = sX(s), \qquad (13.3\text{-}9)$$

where $X(s) = \pounds\{x(t)\}$, as usual.

Proof   We let $x(t)$ be replaced by $x_o(t)$ and $\phi(t) = e^{-st}$ in our integration by parts formula, equation (13.3-5), with the upper and lower limits given by $a = -\infty$ and $b = \infty$, respectively.
This gives

$$\int_{-\infty}^{\infty} px_o(t)e^{-st}\,dt = x_o(t)e^{-st}]_{-\infty}^{\infty} + s\int_{-\infty}^{\infty} x_o(t)e^{-st}\,dt. \qquad (13.3\text{-}10)$$

Now if $\mathrm{Re}\{s\} = \sigma$ is large enough, the integrated product on the right-hand side approaches zero at the upper limit. Furthermore, as we are assuming that all of our signal
waveforms are one sided, $x(t) = 0$ for all $t \le 0$, so this term is zero at the lower limit
also. Finally, we notice that the integral on the left-hand side is the Laplace transform of
the generalized derivative $px(t)$ and the one on the right-hand side is the Laplace transform of $x(t)$ itself. This proves equation (13.3-9) for ordinary functions.

For singularities, the proof is immediate, because of the form of the transform. For a
singularity of order n, we know that $\pounds\{\delta^{(n)}(t - t_i)\} = s^n e^{-st_i}$. Thus, because the derivative
of the nth-order singularity is a singularity of order $n + 1$, we see that

$$\pounds\{p\delta^{(n)}(t - t_i)\} = \pounds\{\delta^{(n+1)}(t - t_i)\} = s^{n+1}e^{-st_i} = s\pounds\{\delta^{(n)}(t - t_i)\}. \qquad (13.3\text{-}11)$$

We can then just factor the common s term out of each factor in $x_s(t)$ to show that the result holds for $x_s(t)$ as well. Finally, because $x(t)$ is the sum of $x_o(t)$ and $x_s(t)$, the property
holds for $x(t)$ also.   ∎

Notice that the time differentiation property generalizes immediately to

$$\pounds\{p^n x(t)\} = s^n X(s). \qquad (13.3\text{-}12)$$

We now provide some examples using the time differentiation property.

Example 13.18   Find the Laplace transform of the waveform shown in Figure 13.19.

Figure 13.19
An example waveform

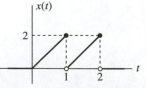

Solution   Using graphical differentiation, we determine the first derivative, which is sketched in
Figure 13.20. The open dot on the time axis at $t = 2$ and the closed one just above it

merely indicate that, in practice, a finite amount of time is required to make the transition from one to the other—and the same applies for the two at $t = 0$. We then compute the second derivative as shown in Figure 13.21. Notice that there are actually two singularity functions at $t = 2$: one is the derivative of the one shown in Figure 13.20, and the other appears as a result of the downward jump in $px(t)$. The solid base line serves to remind us that the *value* of the second derivative is zero at each and every time point. We can write the second derivative analytically in the form

$$p^2x(t) = 2\delta(t) - 2\delta'(t - 1) - 2\delta(t - 2) - 2\delta'(t - 2). \qquad (13.3\text{-}13)$$

We now compute the Laplace transform of both sides, using the time differentiation property on $p^2x(t)$:

$$s^2X(s) = 2 - 2se^{-s} - 2e^{-2s} - 2se^{-2s}. \qquad (13.3\text{-}14)$$

Solving for $X(s)$, our desired Laplace transform, we get

$$X(s) = \frac{2(1 - se^{-s} - (s + 1)e^{-2s})}{s^2}. \qquad (13.3\text{-}15)$$

Figure 13.20
First derivative waveform

Figure 13.21
Second derivative waveform

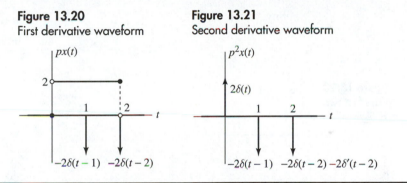

Can you see how easy the concept of generalized functions, coupled with the time differentiation property, makes the computation of Laplace transforms? Let's work another example that shows how to use these ideas to compute generalized derivatives with the Laplace transform. We will work "backwards" relative to the last example.

Example 13.19   If $x(t) = 8\sin(2t)u(t)$, find $p^3x(t)$ using the Laplace transform.

Solution   We apply the time differentiation property three times, along with the known transform of the sinusoid. This results in

$$\pounds\{p^3x(t)\} = s^3X(s) = 8s^3\frac{2}{s^2 + 4} = \frac{16s^3}{s^2 + 4}. \qquad (13.3\text{-}16)$$

Noticing that this function of s is improper—the degree of the numerator polynomial is higher than that of the denominator—we perform one step of long division to get

$$\pounds\{p^3x(t)\} = \frac{16s^3}{s^2 + 4} = 16s - \frac{64s}{s^2 + 4} = \pounds\{16\delta'(t) - 64\cos(2t)u(t)\}. \qquad (13.3\text{-}17)$$

Taking the inverse transform of both sides, we have

$$p^3x(t) = 16\delta'(t) - 64\cos(2t)u(t). \qquad (13.3\text{-}18)$$

Example 13.20   Compute the Laplace transform of the waveform in Figure 13.22.

Figure 13.22
An example waveform

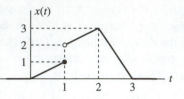

Solution   We have sketched the first and second derivatives in Figure 13.23 using the graphical method of slopes and the properties of singularity functions. Analytically, we can write

$$p^2 x(t) = \delta(t) + \delta'(t-1) - 4\delta(t-2) + 3\delta(t-3). \qquad (13.3\text{-}19)$$

Taking the Laplace transform of both sides of (13.3-19) using the time differentiation property now gives

$$s^2 X(s) = 1 + se^{-s} - 4e^{-2s} + 3e^{-3s}. \qquad (13.3\text{-}20)$$

Solving, we have

$$X(s) = \frac{1 + se^{-s} - 4e^{-2s} + 3e^{-3s}}{s^2}. \qquad (13.3\text{-}21)$$

Figure 13.23
The first two
generalized derivatives

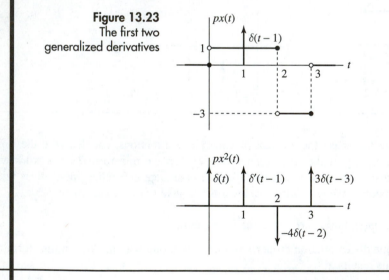

As we have shown, generalized integration exactly "undoes" generalized differentiation, so we see at once that

$$\frac{1}{p}\{px(t)\} = x(t). \qquad (13.3\text{-}22)$$

If we let $y(t) = \dfrac{1}{p}x(t)$, we see that $x(t) = py(t)$ and thus $X(s) = sY(s)$; therefore

$$\pounds\left\{\frac{1}{p}x(t)\right\} = Y(s) = \frac{1}{s}X(s). \qquad (13.3\text{-}23)$$

This is often referred to as the "time integration property." We will not do this, however, because it is only the time differentiation property thinly disguised. Note that

$$\pounds\left\{\frac{1}{p^n}x(t)\right\} = \frac{1}{s^n}X(s). \tag{13.3-24}$$

Example 13.21   Extend the time differentiation property to find $\pounds\{(t^n/n!)u(t)\}$.

Solution   We just write

$$\frac{t^n}{n!}u(t) = \frac{1}{p}\left\{\frac{t^{n-1}}{(n-1)!}u(t)\right\} = \frac{1}{p^2}\left\{\frac{t^{n-2}}{(n-2)!}u(t)\right\} = \cdots = \frac{1}{p^n}u(t). \tag{13.3-25}$$

Then we just apply the generalized form of the time integration property in equation (13.3-23) n times successively to get

$$\pounds\left\{\frac{t^n}{n!}u(t)\right\} = \pounds\left\{\frac{1}{p^n}u(t)\right\} = \frac{1}{s^n}\pounds\{u(t)\} = \frac{1}{s^{n+1}}. \tag{13.3-26}$$

The preceding result is a quite useful one in solving circuits problems with the Laplace transform. The next one is also frequently of use, though we will not prove it rigorously. It involves the concept of generalized limit, which we will not develop here for reasons of space, but we will state it and (we hope) make it plausible.

Theorem 13.10 (Frequency Differentiation Property)   Let $x(t)$ be a generalized function with transform $X(s)$ and assume that the derivative with respect to s, namely $X'(s)$, exists. Then we have

$$\pounds\{-tx(t)\} = X(s). \tag{13.3-27}$$

Proof   (Plausibility Discussion Only) Notice that the derivative of $X(s)$ can be written as a limit:

$$X'(s) = \lim_{\epsilon \to 0}\frac{X(s) - X(s - \epsilon)}{\epsilon} = \lim_{\epsilon \to 0}\frac{1 - e^{-\epsilon s}}{\epsilon}X(s) \tag{13.3-28}$$

$$= \lim_{\epsilon \to 0}\pounds\left\{\frac{1 - e^{-\epsilon t}}{\epsilon}x(t)\right\}.$$

We now use L'Hôpital's rule to see that

$$\lim_{\epsilon \to 0}\left\{\frac{1 - e^{-\epsilon t}}{\epsilon}\right\} = -t. \tag{13.3-29}$$

Thus, assuming that we can interchange the limit and the Laplace transform operation (and this is where the weakness of our "proof" lies), we see that equation (13.3-28) becomes equation (13.3-27), which is what we were trying to prove. ∎

Example 13.22   Find the Laplace transform of $te^{-t}u(t)$ using the frequency differentiation property.

Solution   A sketch of $x(t) = te^{-t}u(t)$ is shown in Figure 13.24. This waveform is rather tedious to plot by hand, so we used MATLAB to plot it for us. Notice that the graph is only for $t > 0$, though it is understood that it is identically zero for negative values of time. We know the transform of $e^{-t}u(t)$; it is

$$\pounds\{e^{-t}u(t)\} = \frac{1}{s+1}. \tag{13.3-30}$$

We can now use the frequency differentiation property to write

$$\pounds\{te^{-t}u(t)\} = -\frac{d}{ds}\left\{\frac{1}{s+1}\right\} = \frac{1}{(s+1)^2}. \tag{13.3-31}$$

This result can also be derived using other properties we developed in Section 13.1 of this chapter. You should repeat this example using some of them for practice.

Figure 13.24
Plot of the example
waveform

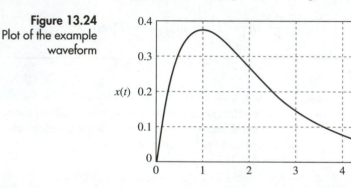

Tables of Basic
Transforms and
Properties

At this point we have developed all of the theory of the Laplace transform we will need to analyze circuits. The following sections of this chapter will develop the Laplace transform method of circuit analysis, and you should be very comfortable finding Laplace transforms and manipulating them before you continue. For your ease of reference, we will now present our final two tables: one of basic transforms (Table 13.5) and the other of basic properties of the Laplace transform (Table 13.6). *Both now hold for all generalized functions.* They should be committed to memory for future application.

Table 13.5
Six basic transforms

$x(t)$	$X(s)$
1. $\delta(t)$	1
2. $u(t)$	$1/s$
3. $\dfrac{t^n}{n!}u(t)$	$1/s^{n+1}$
4. $e^{-at}u(t)$	$1/(s+a)$
5. $\sin(\omega_0 t)u(t)$	$\omega_0/(s^2+\omega_0^2)$
6. $\cos(\omega_0 t)u(t)$	$s/(s^2+\omega_0^2)$

Table 13.6
Six basic properties

$x(t)$	$X(s)$
1. $ax(t) + by(t)$	$aX(s) + bY(s)$
2. $x(t-a)$	$e^{-as}X(s)$
3. $e^{-at}x(t)$	$X(s+a)$
4. $x(at); a>0$	$\dfrac{1}{a}X(s/a)$
5. $px(t)$	$sX(s)$
6. $-tx(t)$	$dX(s)/ds$

Section 13.3 Quiz

Q13.3-1. If $x(t) = t^2 e^{-3t} u(t-1)$, find $X(s)$ without evaluating the defining integral.

Q13.3-2. If $x(t) = e^{-3t} \cos(4t) u(t)$, find $X(s)$ without evaluating the defining integral.

Q13.3-3. If $X(s) = \dfrac{1}{s^2(s+2)}$, find $x(t)$ using transform properties.

Q13.3-4. Find the Laplace transform of the time function $x(t) = 4e^{-2t}u(t) + 3e^{4t}u(-t)$.

Q13.3-5. Use the time differentiation property (backward) to compute $\mathcal{L}^{-1}\{4/s^2\}$.

Q13.3-6. Use the time differentiation property to compute $p^3 x(t)$ if $x(t) = e^{-2t}\cos(3t)u(t)$.

Q13.3-7. If $x(t)$ is the waveform shown in Figure Q13.3-1, find $X(s)$ using the time differentiation property.

Figure Q13.3-1

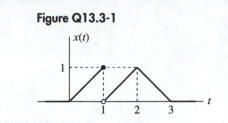

Q13.3-8. Check your answer to Question Q13.3-2 by computing the defining integral for the Laplace transform.

Q13.3-9. If $X(s) = \dfrac{s^3}{(s+1)(s+2)}$, find $x(t)$ using Laplace transform properties.

Q13.3-10. If $x(t) = e^{-2t}u(t) - 2\cos(2\pi t)u(t-1)$, use the time differentiation property to compute the first and second generalized derivatives $px(t)$ and $p^2 x(t)$.

Q13.3-11. Find the Laplace transform of $x(t)$ in Question Q13.3-10.

13.4 | The Partial Fraction Expansion

Until now we have not developed a general technique for determining the time function $x(t)$ from the Laplace transform $X(s)$; our procedure has been something of a hit-and-miss application of the basic transform properties. We will now remedy this lack by developing a methodical procedure for inverting the Laplace transform. To be sure, it works only for functions $X(s)$ that are rational, that is, which are ratios of polynomials in s. This is not a serious shortcoming, however, because most of the circuit applications of the Laplace transform concern such functions.

An Example: The Partial Fraction Expansion for Simple Factors

Example 13.23 Let $X(s) = \dfrac{s}{(s+1)(s+2)}$. Find $x(t) = \mathcal{L}^{-1}\{X(s)\}$.

Solution Let's assume that $X(s)$ can be written in terms of simpler functions of s as follows:

$$X(s) = \frac{s}{(s+1)(s+2)} = \frac{k_1}{s+1} + \frac{k_2}{s+2}, \qquad (13.4\text{-}1)$$

where k_1 and k_2 are constants to be determined. We can certainly do this if we can find values of the constants for which (13.4-1) is an identity. We will check this by multiplying both sides by the denominator of $X(s)$, resulting in the equation

$$s = k_1(s+2) + k_2(s+1) = (k_1 + k_2)s + (2k_1 + k_2). \qquad (13.4\text{-}2)$$

If this equation is satisfied for all s by some set of values for k_1 and k_2, then (13.4-1) will be satisfied also. But if we are to satisfy (13.4-2) for all values of s, then we can, say, let $s = 0$. This gives $2k_1 + k_2 = 0$. Then if we use this and let $s = 1$, for instance, we get $3k_1 + 2k_2 = 1$. Solving these two equations simultaneously gives $k_1 = -1$ and $k_2 = 2$. These values then satisfy both (13.4-2) and (13.4-1), and we can write

$$X(s) = \frac{-1}{s + 1} + \frac{2}{s + 2}. \tag{13.4-3}$$

We know the time functions having these two terms as their transforms and, using the linearity of the inverse transform, we simply write down the solution:

$$x(t) = \pounds^{-1}\{X(s)\} = \pounds^{-1}\left\{\frac{-1}{s + 1}\right\} + \pounds^{-1}\left\{\frac{2}{s + 2}\right\} = [2e^{-2t} - e^{-t}]u(t). \tag{13.4-4}$$

An Easier Method   The procedure in the preceding example is not as efficient as we might like. After all, if there are two factors in the denominator of $X(s)$, there will be two equations to solve in two unknown coefficients. If there are three factors, three equations will result, and so on. Solving large systems of equations such as this can be cumbersome and time consuming, so let's look for another method of evaluating the constants.

Let's have a look once again at equation (13.4-1) and this time multiply both sides by the first factor, thus canceling it in two places—on the left-hand side of the equality and on the right-hand side in the term involving k_1:

$$(s + 1)X(s) = \frac{s}{s + 2} = k_1 + \frac{k_2(s + 1)}{s + 2}. \tag{13.4-5}$$

Now, let's let s take on the value -1. This results in the equation

$$[(s + 1)X(s)]_{s=-1} = \left[\frac{s}{s + 2}\right]_{s=-1} = k_1 + \left[\frac{k_2(s + 1)}{s + 2}\right]_{s=-1}. \tag{13.4-6}$$

It is important to notice that we canceled the $s + 1$ term in the expression $(s + 1)X(s)$ before we let $s = -1$; otherwise, the left-hand side seems to result in zero, which is not correct. Evaluating both sides results in $k_1 = -1$ because the second term vanishes due to the factor $(s + 1)$ in its numerator. We will leave it to you to use this same method to determine the value of k_2. With a small amount of practice, you will be able to do this procedure mentally and thus save time and paper.

Things That Can Go Wrong

There are a couple of things that can go wrong with this process. Here's one. Suppose that we are to determine the inverse Laplace transform of

$$X(s) = \frac{s}{(s + 1)^2}. \tag{13.4-7}$$

If we try the same expansion as before, we get

$$X(s) = \frac{s}{(s + 1)^2} = \frac{k_1}{s + 1} + \frac{k_2}{s + 1}. \tag{13.4-8}$$

This clearly will not work because the right-hand side is just $\dfrac{k_1 + k_2}{s + 1}$; therefore, we can never find k_1 and k_2 to make equation (13.4-8) valid. Thus, we must demand that each of the denominator factors be different from each of the others. In this case, we call them *simple denominator factors* or *distinct denominator factors* or *nonrepeated denominator factors*.

Example 13.24 Let $X(s) = (6s^2 + 23s + 21)/((s + 1)(s + 2)(s + 3))$. Find $x(t) = \mathcal{L}^{-1}\{X(s)\}$.

Solution We first check whether or not the denominator factors are distinct (different from each other). They clearly are, and thus we can write

$$X(s) = \frac{6s^2 + 23s + 21}{(s + 1)(s + 2)(s + 3)} = \frac{k_1}{s + 1} + \frac{k_2}{s + 2} + \frac{k_3}{s + 3}. \qquad (13.4\text{-}9)$$

Using our quick method, we successively compute:

$$k_1 = \left[(s + 1)\frac{6s^2 + 23s + 21}{(s + 1)(s + 2)(s + 3)} \right]_{s = -1} = \frac{6(-1)^2 + 23(-1) + 21}{(-1 + 2)(-1 + 3)} = 2, \qquad (13.4\text{-}10)$$

$$k_2 = \left[(s + 2)\frac{6s^2 + 23s + 21}{(s + 1)(s + 2)(s + 3)} \right]_{s = -2} = \frac{6(-2)^2 + 23(-2) + 21}{(-2 + 1)(-2 + 3)} = 1, \qquad (13.4\text{-}11)$$

and

$$k_3 = \left[(s + 3)\frac{6s^2 + 23s + 21}{(s + 1)(s + 2)(s + 3)} \right]_{s = -3} = \frac{6(-3)^2 + 23(-3) + 21}{(-3 + 1)(-3 + 2)} = 3. \qquad (13.4\text{-}12)$$

This gives

$$X(s) = \frac{6s^2 + 23s + 21}{(s + 1)(s + 2)(s + 3)} = \frac{2}{s + 1} + \frac{1}{s + 2} + \frac{3}{s + 3} \qquad (13.4\text{-}13)$$

and

$$x(t) = [2e^{-t} + e^{-2t} + 3e^{-3t}]u(t). \qquad (13.4\text{-}14)$$

There is a second problem that can arise in this procedure. Suppose we modify the numerator of $X(s)$ in Example 13.23, equation (13.4-1), so that the new problem is this: find k_1 and k_2 such that

$$X(s) = \frac{s^2}{(s + 1)(s + 2)} = \frac{k_1}{s + 1} + \frac{k_2}{s + 2}. \qquad (13.4\text{-}15)$$

Multiplying both sides by the denominator now yields

$$s^2 = k_1(s + 2) + k_2(s + 1) = (k_1 + k_2)s + (2k_1 + k_2). \qquad (13.4\text{-}16)$$

Do you see the problem here? There is no way we can select k_1 and k_2 to satisfy this equation because the polynomial on the right-hand side is only of degree 1, whereas the one on the left-hand side is of degree 2. In fact, this is a general result: *the degree of the numerator polynomial of X(s) must be strictly less than the degree of the denominator polynomial.* To see this, just notice that when we multiply both sides of our expansion by the denominator, exactly one factor is canceled in each right-hand side term. This lowers the degree of the denominator polynomial by one. *The right-hand side will always be of degree one lower than the denominator—and that degree must be matched by the numerator polynomial of X(s),* which forms the left-hand side of the resulting expression. Here's an example of what can be done to remedy this defect.

Example 13.25   Let $X(s) = (2s^4 + 15s^3 + 46s^2 + 68s + 39)/((s + 1)(s + 2)(s + 3))$.   Find $x(t) = £^{-1}\{X(s)\}$.

Solution   We first check the degrees of the numerator and denominator. The numerator is of degree 4, and the denominator degree is 3. This means that our expansion process will not work as is. So what do we do? The question we must ask now is this: Is there a way of expressing $X(s)$ in some fashion as a sum of functions of s whose transform we know— or know how to compute? Well, we know that we can reduce the degree of the numerator by long division, so let's do this until its degree is less than the degree of the denominator. This requires two long division steps, after which we can write

$$X(s) = \frac{2s^4 + 15s^3 + 46s^2 + 68s + 39}{(s + 1)(s + 2)(s + 3)} \qquad (13.4\text{-}17)$$

$$= 2s + 3 + \frac{6s^2 + 23s + 21}{(s + 1)(s + 2)(s + 3)}.$$

We next do our expansion on the last term because it satisfies the degree condition. In this example, we see that this last term is the $X(s)$ of Example 13.24, so we will simply write down this expansion (which we have already determined). This gives

$$X(s) = \frac{2s^4 + 15s^3 + 46s^2 + 68s + 39}{(s + 1)(s + 2)(s + 3)} \qquad (13.4\text{-}18)$$

$$= 2s + 3 + \frac{2}{s + 1} + \frac{1}{s + 2} + \frac{3}{s + 3}.$$

Therefore, we get (using our table of basic transforms and properties)

$$x(t) = 2\delta'(t) + 3\delta(t) + [2e^{-t} + e^{-2t} + 3e^{-3t}]u(t). \qquad (13.4\text{-}19)$$

The Partial Fraction Expansion

We have, at this point, proved the following general theorem covering the above procedure, which we call the *partial fraction expansion* (PFE).

Theorem 13.11 (Partial Fraction Expansion)   Let $X(s)$ be a function of s that can be written in the form

$$X(s) = \frac{N(s)}{(s + a_1)(s + a_2) \cdots (s + a_n)}, \qquad (13.4\text{-}20)$$

where all the a_i's have different values and the degree of $N(s)$ is no more than $n - 1$. Then $X(s)$ can be written in the form

$$X(s) = \frac{k_1}{s + a_1} + \frac{k_2}{s + a_2} + \cdots + \frac{k_i}{s + a_i} + \cdots + \frac{k_n}{s + a_n} \qquad (13.4\text{-}21)$$

and the constants k_i are given by

$$k_i = [(s + a_i)X(s)]_{s=-a_i}.$$ (13.4-22)

We again issue the warning that the denominator factor $(s + a_i)$ must be canceled in (13.4-22) before $-a_i$ is substituted for s. Finally, we note that if the degree condition is not met, one can perform long division steps until the remainder does satisfy that condition.

A Common Mistake and an Easy Correction

We will now cover a point that is not strictly essential from a logical point of view. It is, rather, motivated by the author's experience over a number of years grading students' papers. One of the most common mistakes is the failure to check the degree condition. Here's the situation. After working through a long process of deriving the various coefficients, the student suddenly recognizes his or her failure to check the degree condition—then starts all over from scratch.

Fortunately, one can save work. Here's how. Let's choose the $X(s)$ in Example 13.25, so that we can work with concrete numbers and recall the result of performing two long division steps to reduce the numerator degree. With general coefficients for the remainder PFE expansion, we have

$$X(s) = \frac{2s^4 + 15s^3 + 46s^2 + 68s + 39}{(s + 1)(s + 2)(s + 3)}$$ (13.4-23)

$$= 2s + 3 + \frac{k_1}{s + 1} + \frac{k_2}{s + 2} + \frac{k_3}{s + 3}.$$

Now suppose that we blindly do what seems to come naturally, that is multiply both sides by, say, $s + 1$ and let $s = -1$. This gives

$$k_1 \overset{?}{=} \left[(s + 1) \frac{2s^4 + 15s^3 + 46s^2 + 68s + 39}{(s + 1)(s + 2)(s + 3)} \right]_{-1}$$ (13.4-24)

$$= \frac{2(-1)^4 + 15(-1)^3 + 46(-1)^2 + 68(-1) + 39}{(-1 + 2)(-1 + 3)} = 2.$$

We have placed a question mark above the first equality sign because we have not justified the procedure (unless the degree reduction is first done by long division). We notice, however, that our result is correct! This will always be the case. Here's why. Just look at what happens to the "extra degree terms" $2s$ and 3 in (13.4-23) when we multiply both sides by $(s + 1)$ and let $s = -1$. They become

$$[2s(s + 1)]_{s=-1} + [3(s + 1)]_{s=-1} = 0 + 0 = 0!$$ (13.4-25)

Thus, if we have forgotten about the degree condition, we can still use the coefficients we have already computed! We need only go back to the original rational function and do the long division steps to find the extra terms.[12]

---

[12] Come now, tell the truth. You *did* think that professors (and textbook authors) are never concerned about the plight of their students, didn't you?

An Example: Repeated Factors

Example 13.26   Let $X(s) = (s^2 + 6s + 11)/(s + 2)^3$. Find $x(t) = \mathcal{L}^{-1}\{X(s)\}$.

Solution   We cannot use the PFE here because the three factors of $(s + 2)$ are repeated; that is, they are not distinct. What do we do? We could use various properties, such as frequency shifting, differentiation, etc., to solve our problem, but what we are doing here is looking for a general method. So let's ask ourselves, "What is the problem?" The problem, of course, is the duplicated denominator factors. But suppose we get to the heart of the difficulty by factoring out the *excess* factors—in this case $(s + 2)^2$. We get

$$X(s) = \frac{1}{(s + 2)^2}\left[\frac{s^2 + 6s + 11}{s + 2}\right]. \tag{13.4-26}$$

There is only one denominator factor in the term inside the brackets, but the degree condition is not met by this rational fraction. Therefore, we do what we normally do—we perform long division until the remainder does meet that condition. This results in

$$X(s) = \frac{1}{(s + 2)^2}\left[s + \frac{4s + 11}{s + 2}\right] = \frac{1}{(s + 2)^2}\left[s + 4 + \frac{3}{s + 2}\right]. \tag{13.4-27}$$

We next multiply our "outside factor" back in, to obtain

$$X(s) = \frac{s}{(s + 2)^2} + \frac{4}{(s + 2)^2} + \frac{3}{(s + 2)^3}. \tag{13.4-28}$$

We see that we have simple constants divided by powers of $(s + 2)$ except for the first term. But we can do the same thing with it, namely, factor out one of the denominator factors, obtaining

$$\frac{s}{(s + 2)^2} = \frac{1}{(s + 2)}\left[\frac{s}{s + 2}\right] = \frac{1}{(s + 2)}\left[1 - \frac{2}{s + 2}\right] \tag{13.4-29}$$

$$= \frac{1}{(s + 2)} - \frac{2}{(s + 2)^2}.$$

Using this in (13.4-28) results in

$$X(s) = \frac{1}{s + 2} + \frac{2}{(s + 2)^2} + \frac{3}{(s + 2)^3}. \tag{13.4-30}$$

We have succeeded in expressing $X(s)$ as the sum of powers of $1/(s + 2)$ from unity through the highest power present in the original form of $X(s)$. We need only determine the inverse transform of each term. We know the inverse transform of $1/(s + 2)$ for it is entry 4 in Table 13.5 of Section 13.3: our table of basic transforms.

We can derive the second inverse transform in the following way. We just let

$$Y(s) = \frac{2}{(s + 2)^2} \tag{13.4-31}$$

and let $$Z(s) = Y(s - 2) = \frac{2}{s^2}. \tag{13.4-32}$$

This means that (see entry 3 in Table 13.5 of Section 13.3)

$$z(t) = 2tu(t). \tag{13.4-33}$$

But if we solve (13.4-32) in reverse for $Y(s) = Z(s + 2)$, we know by the frequency shift property (entry 3 in Table 13.6 of Section 13.3) that

$$y(t) = 2te^{-2t}u(t). \tag{13.4-34}$$

Performing a similar procedure for the last term in equation (13.4-30), which we leave as an exercise for you to do, we find that

$$x(t) = (1 + 2t + 1.5t^2)e^{-2t}u(t). \tag{13.4-35}$$

Repeated Factors: A General Formula for the PFE Coefficients

Let's see if we can go from the specific example we have just worked to a general result about the PFE of $X(s)$ functions with repeated denominator factors. Suppose that

$$X(s) = \frac{N(s)}{(s + a)^m D_0(s)}, \tag{13.4-36}$$

where the degree of $N(s)$ is strictly less than the degree of the entire denominator and $D_0(s)$ is a polynomial containing all the denominator factors different from $(s + a)$. We can factor out $m - 1$ of these factors and do a PFE on what's left, obtaining

$$X(s) = \frac{N(s)}{(s + a)^m D_0(s)} = \frac{1}{(s + a)^{m-1}} \left[\frac{N(s)}{(s + a)D_0(s)} \right] \tag{13.4-37}$$

$$= \frac{1}{(s + a)^{m-1}} \left[\frac{k_m}{(s + a)} + R_1(s) \right].$$

The coefficient k_m is obtained in the usual way, and $R_1(s)$ is the rest of the PFE—that is, the sum of all the terms due to factors in $D_0(s)$. Next, we multiply the "excess factors" back in, obtaining

$$X(s) = \frac{k_m}{(s + a)^m} + \frac{1}{(s + a)^{m-1}} R_1(s). \tag{13.4-38}$$

What have we accomplished? Plenty, though it might not be immediately evident. We have gained one term because of the repeated factors in our PFE and only need to do the PFE on the second term, in which this factor has one fewer repetitions than in the original. In fact, we can continue exactly this same process until we have

$$X(s) = \frac{k_m}{(s + a)^m} + \frac{k_{m-1}}{(s + a)^{m-1}} + \cdots + \frac{k_{m-i}}{(s + a)^{m-i}} \tag{13.4-39}$$

$$+ \cdots + \frac{k_1}{(s + a)} + F(s).$$

The function $F(s)$ is that part of the PFE due to all of the factors in $D_0(s)$, that is, all of the denominator factors in $X(s)$ different from $(s + a)^m$. Here's how we compute the coefficients. We multiply both sides of (13.4-39) by $(s + a)^m$, to obtain

$$(s + a)^m X(s) = k_m + (s + a)k_{m-1} + \cdots + (s + a)^i k_{m-i} \tag{13.4-40a}$$

$$+ \cdots + k_1(s + a)^{m-1} + F(s)(s + a)^m.$$

If we just evaluate both sides at $s = -a$, we have

$$K_m = [(s + a)^m X(s)]_{s=-a}. \tag{13.4-40b}$$

If we differentiate with respect to s one time, we see that the k_m constant term on the right vanishes. Then, when we evaluate at $s = -a$, all of the other terms vanish because of the presence of some positive power of the factor $(s + a)$—except the one involving k_{m-1}. We continue in this way, noting that differentiation "pulls down" the integers in the exponents, obtaining the general term:

$$K_{m-i} = \frac{1}{i!} \frac{d^i}{ds^i} [(s + a)^m X(s)]_{s=-a}. \tag{13.4-41}$$

Example 13.27   Find $\pounds^{-1}\{(s + 2)/(s(s + 1)^2(s + 3))\}$.

Solution   We let $X(s)$ be the given s domain function. We notice that the degree condition is met, but there is a double factor of the form $(s + 1)$. We will solve this problem by using equation (13.4-41). Thus, we write the PFE for $X(s)$ in the form

$$X(s) = \frac{s + 2}{s(s + 1)^2(s + 3)} = \frac{k_2}{(s + 1)^2} + \frac{k_1}{s + 1} + \frac{k_3}{s} + \frac{k_4}{s + 3}. \tag{13.4-42}$$

It might seem that we have named our coefficients rather oddly, but we have labeled them so that the repeated coefficient factors come first and in an order that is consistent with (13.4-39) and (13.4-41). This is not essential, it just aids in keeping things straight. Now if you look at equation (13.4-40), you will see that the highest-order coefficient of the repeated factor is determined exactly like those for nonrepeated factors—the only difference is that we must multiply by $(s + a)^m$, the "complete" factor in the denominator. Therefore, let's compute it and the simple factor coefficients. (The author has found it to be effective to postpone the more difficult work until the end.) Thus, we have

$$k_2 = \left[(s + 1)^2 \frac{s + 2}{s(s + 1)^2(s + 3)} \right]_{s=-1} = \frac{(-1 + 2)}{(-1)(-1 + 3)} = -\frac{1}{2}, \tag{13.4-43}$$

$$k_3 = \left[s \frac{s + 2}{s(s + 1)^2(s + 3)} \right]_{s=0} = \frac{(0 + 2)}{(0 + 1)^2(0 + 3)} = \frac{2}{3}, \tag{13.4-44}$$

and $$k_4 = \left[(s + 3) \frac{s + 2}{s(s + 1)^2(s + 3)} \right]_{s=-3} = \frac{(-3 + 2)}{(-3)(-3 + 1)^2} = \frac{1}{12}. \tag{13.4-45}$$

Now, let's apply (13.4-41) to obtain k_1. Here, $m = 2$ and $i = 1$. This gives the formula

$$k_1 = \frac{1}{1!} \frac{d}{ds} \left[(s + 1)^2 \frac{s + 2}{s(s + 1)^2(s + 3)} \right]_{s=-1} = \frac{d}{ds} \left[\frac{s + 2}{s^2 + 3s} \right]_{s=-1} \tag{13.4-46}$$

$$= \left[\frac{(s^2 + 3s) \times 1 - (s + 2)(2s + 3)}{(s^2 + 3s)^2} \right]_{s=-1} = \left[\frac{-s^2 - 4s - 6}{s^2(s + 3)^2} \right]_{s=-1}$$

$$= \frac{-[(-1)^2 + 4(-1) + 6]}{(-1)^2(-1 + 3)^2} = -\frac{3}{4}.$$

This results in

$$X(s) = \frac{-\dfrac{1}{2}}{(s + 1)^2} + \frac{-\dfrac{3}{4}}{s + 1} + \frac{\dfrac{2}{3}}{s} + \frac{\dfrac{1}{12}}{s + 3}. \tag{13.4-47}$$

We next take the inverse transform term by term using our tables of basic transforms and properties (Tables 13.5 and 13.6):

$$x(t) = \left[\frac{2}{3} + \frac{1}{12}e^{-3t} - \frac{1}{4}(2t + 3)e^{-t} \right]u(t). \qquad (13.4\text{-}48)$$

An Important Transform

Let's pause for a moment to derive an important transform that does not appear in our table of *basic* transforms, yet does often find application in the partial fraction expansion. We ask for the inverse Laplace transform of

$$X(s) = \frac{1}{(s + a)^m}. \qquad (13.4\text{-}49)$$

To compute it, we first let

$$Y(s) = X(s - a) = \frac{1}{s^m}. \qquad (13.4\text{-}50)$$

This implies that

$$y(t) = \mathcal{L}^{-1}\{1/s^m\} = \frac{t^{m-1}}{(m - 1)!}u(t). \qquad (13.4\text{-}51)$$

Solving (13.4-50) for $X(s)$, we have

$$X(s) = Y(s + a). \qquad (13.4\text{-}52)$$

Thus, by the frequency shift property, we get

$$\boxed{x(t) = \mathcal{L}^{-1}\left\{ \frac{1}{(s + a)^m} \right\} = e^{-at}y(t) = \frac{t^{m-1}}{(m - 1)!}e^{-at}u(t).} \qquad (13.4\text{-}53)$$

Because this transform is of so much use, we have put a box around it for ease of reference.

Our discussion of the partial fraction expansion is now complete logically because the only types of denominator factors one can have for rational $X(s)$ are either simple or repeated, and if the degree condition is not met, one only encounters additional terms consisting of either a constant or powers of s—and these are transforms of impulses or derivatives of impulses. If factors are in complex conjugate pairs, however, one is forced to perform complex arithmetic, as the next example shows.

Complex Factors

Example 13.28   Find $\mathcal{L}^{-1}[s/(s^2 + 2s + 5)]$.

Solution   We let $X(s)$ be the given s domain function. We see that the degree condition is satisfied. The next step is to factor the denominator. The factored form for $X(s)$, along with the PFE,[13] is

$$X(s) = \frac{s}{(s + 1 + j2)(s + 1 - j2)} = \frac{k_1}{s + 1 + j2} + \frac{k_2}{s + 1 - j2}. \qquad (13.4\text{-}54)$$

---

[13] In practical situations, one must have provision for factoring polynomials numerically—particularly when the polynomials are of high order. In this text, however, such factoring will not be necessary unless we wish to stress this particular issue.

We find the coefficients in the usual way, albeit with complex algebra required:

$$k_1 = \left[\frac{(s + 1 + j2)s}{(s + 1 + j2)(s + 1 - j2)} \right]_{s = -1 - j2} = \left[\frac{s}{s + 1 - j2} \right]_{s = -1 - j2} \qquad (13.4\text{-}55)$$

$$= \frac{-1 - j2}{-1 - j2 + 1 - j2} = \frac{2 - j1}{4}$$

and

$$k_2 = \left[\frac{(s + 1 - j2)s}{(s + 1 + j2)(s + 1 - j2)} \right]_{s = -1 + j2} = \left[\frac{s}{s + 1 + j2} \right]_{s = -1 + j2} \qquad (13.4\text{-}56)$$

$$= \frac{-1 + j2}{-1 + j2 + 1 + j2} = \frac{2 + j1}{4}.$$

Thus, we find

$$X(s) = \frac{\dfrac{2 - j1}{4}}{s + 1 + j2} + \frac{\dfrac{2 + j1}{4}}{s + 1 - j2}. \qquad (13.4\text{-}57)$$

The inverse transform is

$$x(t) = \left[\frac{2 - j1}{4} e^{(-1-j2)t} + \frac{2 + j1}{4} e^{(-1+j2)t} \right] u(t). \qquad (13.4\text{-}58)$$

Now this form does not make us very happy for several reasons. It is complicated, and that's one reason—but there's another reason. All of the coefficients in $X(s)$ are real, so we would somehow expect that $x(t)$ would be real also. As it turns out, this expectation is indeed true. To see that it is, factor out the common factor e^{-t}, then use Euler's formula to put (13.4-58) in the real form

$$x(t) = \frac{1}{2}[2 \cos(2t) - \sin(2t)]e^{-t}u(t). \qquad (13.4\text{-}59)$$

(We have left a bit of complex algebra to you, so expect to do a couple of lines of work in going from (13.4-58) to (13.4-59).)

Complex Factors with Real Algebra

There is a way of avoiding the complex algebra involved in our computations with complex factors that we will now develop. It is the *method* here that is of importance, so do not try to memorize the *formula* that we will develop. Simply go through the following line of reasoning each time you encounter complex factors.

We will start with an s-domain function of the form

$$X(s) = \frac{N(s)}{(s + s_0)(s + s_0^*)D_0(s)}, \qquad (13.4\text{-}60)$$

where

$$s_0 = \sigma_0 + j\omega_0 \qquad (13.4\text{-}61)$$

and where $D_0(s)$ is a polynomial containing all of the denominator factors of $X(s)$ other than the two involving s_0. We write the PFE, as follows:

$$X(s) = \frac{N(s)}{(s + s_0)(s + s_0^*)D_0(s)} = \frac{k_1}{s + s_0} + \frac{k_2}{s + s_0^*} + R(s). \qquad (13.4\text{-}62)$$

$R(s)$ is the rest of the terms in the PFE; that is, those due to factors other than those involving s_0. We next compute k_1. It is

$$k_1 = \left[(s + s_0) \frac{N(s)}{(s + s_0)(s + s_0^*)D_0(s)} \right]_{s=-s_0} = \frac{N(-s_0)}{(-s_0 + s_0^*)D_0(-s_0)}. \quad (13.4\text{-}63)$$

The parameter k_2 takes the form

$$k_2 = \left[(s + s_0^*) \frac{N(s)}{(s + s_0)(s + s_0^*)D_0(s)} \right]_{s=-s_0^*} = \frac{N(-s_0^*)}{(-s_0^* + s_0)D_0(-s_0^*)}. \quad (13.4\text{-}64)$$

Next, we will need to make use of some basic properties of complex numbers. These were discussed in Chapter 9, Section 9.5, Table 9.1. We show this table below as Table 13.7 for ease of reference with s as the complex variable argument of the various expressions and functions shown there. We will use entries 5 and 6 on the last expression in equation (13.4-64). This results in

$$k_2 = \frac{N^*(-s_0)}{(-s_0 + s_0^*)^* D_0^*(-s_0)} = \left[\frac{N(-s_0)}{(-s_0 + s_0^*)D_0(-s_0)} \right]^* = k_1^*. \quad (13.4\text{-}65)$$

If we were to stop at this point, we would be "ahead of the game," because we would have saved half the work. We need only compute k_1, then take its complex conjugate to find k_2. Look back at Example 13.28, equation (13.4-55) and (13.4-56) to see that this result is verified in practice. But we can do better. Here's how.

We rewrite equation (13.4-62), the PFE, with the preceding result incorporated:

$$X(s) = \frac{N(s)}{(s + s_0)(s + s_0^*)D_0(s)} = \frac{k}{s + s_0} + \frac{k^*}{s + s_0^*} + R(s). \quad (13.4\text{-}66)$$

We have changed notation from k_1 to simply k to simplify our work. We combine the two terms involving s_0 in (13.4-66) to obtain

$$X(s) = \frac{(k + k^*)s + (ks_0^* + k^*s_0)}{(s + s_0)(s + s_0^*)} + R(s) = \frac{As + B}{s^2 + as + b} + R(s). \quad (13.4\text{-}67)$$

The coefficients A and B are real because each is the sum of a complex number and its complex conjugate. Why? If you let $z = x + jy$, then $z^* = x - jy$, and $z + z^* = 2x = 2\text{Re}\{z\}$. This suffices for the constant A if we let $z = k$. Letting $z = ks_0^*$ in the second numerator expression, we see that $B = ks_0^* + k^*s_0 = ks_0^* + (ks_0^*)^*$ (see entry 2 in Table 13.7), so B is real too. The same holds for a and b in the denominator because $a = s_0 + s_0^* = 2\sigma_0$ and the parameter b is given by $s_0 s_0^* = |s_0|^2$.

Table 13.7
Properties of
complex numbers

1. $(s_1 + s_2)^* = s_1^* + s_2^*$

2. $(s_1 s_2)^* = s_1^* s_2^*$

3. $\left[\dfrac{s_1}{s_2} \right]^* = \dfrac{s_1^*}{s_2^*}$

4. $(p^m s)^* = p^m(s^*)$

5. $Q^*(s) = Q(s^*) \leftrightarrow$ polynomial, real coeffs.

6. $H^*(s) = H(s^*) \leftrightarrow$ rational runction, real coeffs.

Now remember that we have cautioned you not to memorize these formulas! Here's why. The important thing is the resulting expression in (13.4-67). In solving a problem, we merely multiply both sides of (13.4-67) by the complete denominator and match coefficients to determine the A and B coefficients. Rather than writing out a long expression, we will simply work an example. The ideas will be clearer that way.

Example 13.29   Find $\mathcal{L}^{-1}[(20(s^2 + 6))/(s(s^2 + 2s + 5))]$.

Solution   Letting $X(s)$ be the given s-domain function, we write the form of the PFE for a denominator consisting of one factor of s and another that is of second order with real coefficients:

$$X(s) = \frac{20(s^2 + 6)}{s(s^2 + 2s + 5)} = \frac{k_0}{s} + \frac{As + B}{s^2 + 2s + 5}. \qquad (13.4\text{-}68)$$

Here is a practical tip: if we simply multiply both sides by the complete denominator and match coefficients, we will have to solve three equations in three unknowns to determine the values of k_0, A, and B. It is much easier, however, to evaluate k_0 (or, in general, any coefficients due to other simple factors) by the usual method—then to match coefficients for A and B. Thus, for the simple factor s, we have

$$k_0 = \left[s \frac{20(s^2 + 6)}{s(s^2 + 2s + 5)} \right]_{s=0} = 24. \qquad (13.4\text{-}69)$$

Now we multiply both sides of (13.4-68) by the complete denominator using the given value of k_0. This results in

$$20s^2 + 120 = 24(s^2 + 2s + 5) + s(As + B). \qquad (13.4\text{-}70)$$

We next match coefficients of the various powers of s. We show this in Table 13.8. We have already computed the value of k_0, so the last row is redundant; however, it provides a check on our work. If the two entries are not equal, we have made an earlier mistake somewhere. We see immediately that $A = -4$ and $B = -48$. Using these values in equation (13.4-68) gives

$$X(s) = \frac{24}{s} - 4\frac{s + 12}{s^2 + 2s + 5}. \qquad (13.4\text{-}71)$$

Now what? Our tables of basic properties and transforms come to the rescue, but only if we do just a bit of "massaging" on the quadratic term. We rearrange it in the following way, called "completing the square" in high school algebra:

$$X(s) = \frac{24}{s} - 4\frac{(s + 1) + 11}{(s + 1)^2 + 2^2} \qquad (13.4\text{-}72)$$

$$= \frac{24}{s} - 4\frac{s + 1}{(s + 1)^2 + s^2} - 22\frac{2}{(s + 1)^2 + 2^2}.$$

In addition to completing the square, we have also arranged things a bit in the numerator so that the two last terms are recognizable from our tables. What are they? Well, they are "almost" the transforms of the cosine and sine, entries 5 and 6 in Table 13.5 (the table of basic transforms) in Section 13.3, but there is one slight difference: instead of s, in each occurrence of s we have the expression $s + 1$. Now do you remember the frequency shift property? It's entry number 3 in Table 13.6 (basic transform properties) in Section 13.3.

This means that we multiply the sine and cosine functions by e^{-t}. After we recognize these facts, we can write the time function we are looking for:

$$x(t) = 24u(t) - 2e^{-t}[2 \cos(2t) + 11 \sin(2t)]u(t). \qquad (13.4\text{-}73)$$

Table 13.8
Coefficient matching

s^n	LHS	RHS
s^2	20	$24 + A$
s^1	0	$48 + B$
s^0	120	$120 + 0$

Summary   The main thing to remember about inversion of Laplace transforms having complex conjugate coefficients is the logical procedure. We first wrote the PFE with the term corresponding to the pair of complex conjugate factors in terms of a quadratic denominator (the two complex factors being multiplied together) and a first-order numerator of first degree with real coefficients. We then evaluated all of the coefficients corresponding to the other terms in order to reduce the work in later coefficient matching. We next multiplied both sides of our PFE by the complete denominator of $X(s)$ and matched coefficients. If there is only one quadratic expression, there will be only two equations in two unknowns; if there are two quadratics, there will be four equations in four unknowns, and so on. Notice that we have not covered multiple complex conjugate factors, for they seldom occur. If and when they do, it is probably just as easy to use our general results for PFEs and do the complex algebra. We should point out that the procedure we have just developed for complex factors also works for *any* pair of factors—they do not have to be complex. Thus, if you prefer to do coefficient matching, you can. It just happens that the quick method we have already outlined is probably more efficient. The best policy is probably to use a combination of approaches that you can alter at will to tackle any specific type of problem.

Section 13.4 Quiz

Q13.4-1.   Find $x(t) = \pounds^{-1}\left\{\dfrac{7s^2 + 33s + 20}{s(s^2 + 7s + 10)}\right\}$.

Q13.4-2.   Find $x(t) = \pounds^{-1}\left\{\dfrac{2s^3 + 4s^2 + 12s + 8}{s(s + 2)}\right\}$.

Q13.4-3.   Find $x(t) = \pounds^{-1}\left\{\dfrac{s^4 + 4s^3 + 6s^2 + 8s + 4}{s(s + 2)^2}\right\}$.

Q13.4-4.   Find $x(t) = \pounds^{-1}\left\{\dfrac{5s^2 + 14s + 26}{(s + 2)(s^2 + 4s + 13)}\right\}$.

13.5 | Laplace Transform Solution of Differential Equations

We will now apply what we have learned about Laplace transforms to the solution of differential equations and to several ideas resulting from the nature of the solution, such as stability, forced and natural responses, and the idea of steady state. We will start with an example. We will use the differential operator notation discussed in Section 13.2; that is, we will use p^n for d^n/dt^n and $1/p^n$ for the n-fold integration with limits $-\infty$ and t. We will at first assume that all waveforms are one sided (that is, identically zero for $t \le 0$); later on in this section, however, we will discuss how to handle waveforms that do not satisfy this condition.

We will consider two basic types of differential equation problem. In the first, the differential equation is assumed to hold for all t, $-\infty < t < \infty$, but the forcing function is assumed to be one sided, that is, identically zero for all $t \leq 0$. As for the second, the differential equation will be assumed to hold only for $t > 0$, but initial conditions will be given at $t = 0$.

Differential Equations on $(-\infty, \infty)$ with One-Sided Forcing Functions

We will deal at length with differential equations defined over the entire time line, $-\infty < x < \infty$, with one-sided forcing function waveforms. We will start with an example.

Example 13.30

Solve the differential equation

$$p^2 y(t) + 3py(t) + 2y(t) = px(t) + 3x(t), \tag{13.5-1}$$

for **a.** $x(t) = \delta(t)$, **b.** $x(t) = u(t)$, **c.** $x(t) = e^{-3t}u(t)$, and **d.** $x(t) = e^{-t}u(t)$.

Solution

We start by taking the Laplace transform of both sides of (13.5-1), using the linearity property:

$$\pounds\{p^2 y(t)\} + 3\pounds\{py(t)\} + 2\pounds\{y(t)\} = \pounds\{px(t)\} + 3\pounds\{x(t)\}. \tag{13.5-2}$$

Next, we use the time differentiation property to obtain

$$(s^2 + 3s + 2)Y(s) = (s + 3)X(s). \tag{13.5-3}$$

Solving for $Y(s)$ gives

$$Y(s) = \frac{s + 3}{s^2 + 3s + 2}X(s). \tag{13.5-4}$$

We will pause here to briefly note that this is a *general solution* procedure for our differential equation because it is valid for any forcing function $x(t)$. We need only transform $x(t)$, multiply by the s-domain fraction shown in (13.5-4) to find $Y(s)$, then use the PFE on $Y(s)$ and compute the inverse transform to find $y(t)$.

a. $x(t) = \delta(t)$: Here we have $X(s) = \pounds\{\delta(t)\} = 1$, so

$$Y(s) = \frac{s + 3}{s^2 + 3s + 2} = \frac{s + 3}{(s + 1)(s + 2)} = \frac{2}{s + 1} - \frac{1}{s + 2}. \tag{13.5-5}$$

(We have left the details of computing the coefficients in the PFE to you.) This gives

$$y(t) = h(t) = [2e^{-t} - e^{-2t}]u(t). \tag{13.5-6}$$

Notice that the presence of $u(t)$ is necessary; we are assuming that all of our operations involved in solving for $y(t)$ are *causal;* that is, the resulting waveform is one sided if the forcing function is. As our $x(t)$ is one sided, $y(t)$ must be one sided also. Note also that we have defined our response as being $h(t)$. We will always use this symbol for the *unit impulse response* (often shortened simply to "impulse response," with "unit" being understood), which is the response of a differential equation when the forcing function is the unit step.

b. $x(t) = u(t)$: Because $X(s) = \pounds\{u(t)\} = 1/s$, we have

$$Y(s) = \frac{s + 3}{s^2 + 3s + 2} \times \frac{1}{s} = \frac{s + 3}{s(s + 1)(s + 2)} = \frac{\frac{3}{2}}{s} - \frac{2}{s + 1} + \frac{\frac{1}{2}}{s + 2}. \tag{13.5-7}$$

The associated time response is

$$y(t) = s(t) = \frac{1}{2}[3 - 4e^{-t} + e^{-2t}]u(t). \tag{13.5-8}$$

Again, be aware of the $u(t)$ multiplier. We have defined the response $y(t)$, when the forcing function $x(t)$ is the unit step function, with the symbol $s(t)$—which we will always use to mean the *unit step response*. Like the impulse response, it is often abbreviated simply to "step response," with "unit" being understood.

c. $x(t) = e^{-3t}u(t)$: Here $X(s) = \pounds\{e^{-3t}u(t)\} = 1/(s + 3)$, so

$$Y(s) = \frac{\cancel{s + 3}}{s^2 + 3s + 2} \times \frac{1}{\cancel{s + 3}} = \frac{1}{(s + 1)(s + 2)} = \frac{1}{(s + 1)} - \frac{1}{(s + 2)} \tag{13.5-9}$$

and

$$y(t) = [e^{-t} - e^{-2t}]u(t). \tag{13.5-10}$$

Notice the cancellation of the $s + 3$ factor between the numerator of the multiplying rational fraction and the transform of $x(t)$. This is called "pole-zero cancellation" for reasons we will explain later on in this section.

d. $x(t) = e^{-t}u(t)$: The Laplace transform of the forcing function now is $X(s) = 1/(s + 1)$. This results in

$$Y(s) = \frac{s + 3}{s^2 + 3s + 2} \times \frac{1}{s + 1} = \frac{s + 3}{(s + 1)^2(s + 2)} \tag{13.5-11}$$

$$= \frac{-1}{s + 1} + \frac{2}{(s + 1)^2} + \frac{1}{s + 2}$$

and

$$y(t) = [(2t - 1)e^{-t} + e^{-2t}]u(t). \tag{13.5-12}$$

Again, we have left the PFE details to you. We suggest that you evaluate the coefficient of $1/(s + 2)$ and of $1/(s + 1)^2$ in the usual fashion—that is, by multiplying by $(s + 2)$ and $(s + 1)^2$, then letting $s = -2$ and $s = -1$, respectively—then letting the coefficient of $1/(s + 1)$ be unknown, multiplying both sides by the entire denominator of the left-hand side and matching coefficients. The special case here is this: the denominator factor $(s + 1)$ appears in both $X(s)$ and in the rational fraction multiplier; thus, it is of multiplicity two in $Y(s)$.

The System Function for a Differential Equation

The rational fraction $(s + 3)/(s^2 + 3s + 2)$ in the last example was quite useful, wasn't it? It did not change when we went from one $X(s)$ to another. This seems to indicate that we should always compute it for a given differential equation with the understanding that it will give us the solution for any forcing function. All we must do is compute the Laplace transform of the forcing function, then multiply it by the rational fraction to find the Laplace transform of the desired response, $Y(s)$. This is a general solution method, so let's consider the most general linear, constant-coefficient differential equation:

$$p^n y(t) + a_{n-1}p^{n-1}y(t) + \cdots + a_1 py(t) + a_0 y(t) \tag{13.5-13}$$
$$= b_m p^m x(t) + b_{m-1}p^{m-1}x(t) + \cdots + b_1 px(t) + b_0 x(t).$$

We take the Laplace transform of both sides and solve for $Y(s)$, as in Example 13.30:

$$Y(s) = \frac{b_m s^m + b_{m-1}s^{m-1} + \cdots + b_1 s + b_0}{s^n + a_{n-1}s^{n-1} + \cdots + a_1 s + a_0}X(s). \tag{13.5-14}$$

We call the rational fraction of s that multiplies $X(s)$ the *system function* of the given differential equation. Thus, we have the general result that

$$Y(s) = H(s)X(s), \tag{13.5-15}$$

where $$H(s) = \frac{Y(s)}{X(s)} = \frac{b_m s^m + b_{m-1} s^{m-1} + \cdots + b_1 s + b_0}{s^n + a_{n-1} s^{n-1} + \cdots + a_1 s + a_0}. \tag{13.5-16}$$

We advise you to study this carefully because many, at first exposure, think that the fraction is "upside down" because the a coefficients are on the left side of the differential equation and the b coefficients on the right. If you follow the reasoning we used in getting to (13.5-14), then to (13.5-15), and hence to (13.5-16), however, you should not have this problem.

If we set the forcing function $x(t)$ to be the unit impulse function, then $X(s) = 1$

and $$h(t) = \mathcal{L}^{-1}\{H(s)\} = \mathcal{L}^{-1}\left\{\frac{b_m s^m + b_{m-1} s^{m-1} + \cdots + b_1 s + b_0}{s^n + a_{n-1} s^{n-1} + \cdots + a_1 s + a_0}\right\}. \tag{13.5-17}$$

This means that our notation for the system function has been selected in a way that is consistent with our Laplace transform naming convention: uppercase letters for the transform, lowercase of the same letter for the inverse transform.

What about the step response? Well, $X(s) = \mathcal{L}\{u(t)\} = 1/s$ for the step function, so we see that

$$s(t) = \mathcal{L}^{-1}\{H(s) \times 1/s\} = \frac{1}{p}\{h(t)\} \tag{13.5-18}$$

because multiplication of the transform of $h(t)$, that is $H(s)$, by $1/s$ is equivalent to integration in the time domain. Thus, the step response is the integral of the impulse response, and, conversely, the impulse response is the derivative of the step response.

Summary of the System Function Method

We would like to make an important observation about what we have accomplished. We have exhibited a solution that is valid for any linear, constant-coefficient differential equation and for any forcing function for which differentiation and integration are defined any number of times; in short, we can let the forcing function be any generalized function. Here is an example dealing with singularity forcing functions.

Example 13.31

Find the response $y(t)$ for the differential equation

$$p^2 y(t) + 4y(t) = px(t) \tag{13.5-19}$$

when the forcing function is given by $x(t) = 4\delta(t) - 4\delta'(t-2)$.

Solution

We first find the system function $H(s)$. If you followed our development above carefully, you will see that we can simply write it by inspection, converting all "powers of p" to powers of s and using the right-hand side coefficients in the numerator of $H(s)$ and the left-hand side coefficients in the denominator. Thus, we have

$$H(s) = \frac{Y(s)}{X(s)} = \frac{s}{s^2 + 4}. \tag{13.5-20}$$

Next, we find the Laplace transform of $x(t)$:

$$X(s) = 4 - 4se^{-2s}. \tag{13.5-21}$$

We now write the Laplace transform relation for $Y(s)$:

$$Y(s) = H(s)X(s) = 4\frac{s}{s^2 + 4} - 4e^{-2s}\frac{s^2}{s^2 + 4}. \tag{13.5-22}$$

We follow this by recognizing the first rational fraction as a standard transform in our table of basic transforms—it is a cosine—and expanding the second rational function in a PFE, which is actually only one step of long division, giving

$$\frac{s^2}{s^2 + 4} = 1 - \frac{4}{s^2 + 4}. \tag{13.5-23}$$

Rewriting (13.5-22) now, using (13.5-23), we obtain

$$Y(s) = 4\frac{s}{s^2 + 4} - 4e^{-2s} + 8e^{-2s}\frac{2}{s^2 + 4}. \tag{13.5-24}$$

(We have strategically left a factor of 2 "upstairs" in the last fraction so that we can recognize it from our table of basic transforms.) Thus, we have

$$Y(s) = 4\frac{s}{s^2 + 4} - 4e^{-2s} + 8e^{-2s}\frac{2}{s^2 + 4}. \tag{13.5-25}$$

Taking the inverse transform of each term using linearity and the frequency shift property, we obtain the time function we are looking for, namely

$$y(t) = 4\cos(2t)u(t) - 4\delta(t - 2) + 8\sin(2[t - 2])u(t - 2). \tag{13.5-26}$$

The preceding example has, we hope, convinced you that the system function is an extremely powerful tool in solving differential equations.

Poles and Zeros: The Pole-Zero Diagram (PZD)

Let's have a look at another important aspect of the system function by recalling the Fundamental Theorem of Algebra, which states that each polynomial of order n has n factors, either real or complex, some of which might be repeated (or multiple). Thus, if we factor out the leading numerator coefficient in equation (13.5-16), which defines the system function $H(s)$, we can factor both the numerator and the denominator and write $H(s)$ in the form

$$H(s) = \frac{Y(s)}{X(s)} = \frac{b_m s^m + b_{m-1}s^{m-1} + \cdots + b_1 s + b_0}{s^n + a_{n-1}s^{n-1} + \cdots + a_1 s + a_0} \tag{13.5-27}$$

$$= b_m \frac{(s + z_1)(s + z_2)\cdots(s + z_m)}{(s + p_1)(s + p_2)\cdots(s + p_n)}.$$

The negatives of the numerator coefficient $-z_i$ ($i = 1, 2, \ldots, m$) and the denominator coefficients $-p_i$ ($i = 1, 2, \ldots, n$) are called the *finite zeros* and *finite poles*, respectively, of $H(s)$. (Note that p_i is *not* a differential operator, but a constant whose value we find by factoring the denominator polynomial.) We call $-z_i$ a zero because $H(-z_i) = 0$ and, similarly, we call $-p_i$ a pole because $H(-p_i) = \infty$. The latter term is used because if you think of $|H(s)|$ as the height of a circus tent in the process of being erected, you will see that the tent drags on the ground at some places (the height is zero), and at others its height becomes very large because of the tent poles.

Figure 13.25 A pole-zero diagram (PZD)

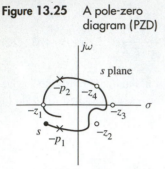

Except for the constant b_m, which we call the *scale factor*, the finite poles and zeros completely define $H(s)$. For this reason, we will plot them in the complex s plane, as shown in Figure 13.25. We call this plot a *pole-zero diagram, (PZD)*. We use open circles to represent zeros and crosses (x's) to represent poles. We have also shown a small dark dot in the drawing to represent a general point s that we are imagining to move along the path shown by the solid line to various parts of the s plane. When the test point s moves through a zero (as it has in our drawing at $s = -z_1$ and at $s = -z_4$), the system function goes to zero. When it moves through a pole (as it has at $s = -p_1$ and at $s = -p_2$), the system function becomes infinitely large. We stress that the small black dot and its dark line trajectory are *not* parts of the PZD, but only items we have added for explanatory purposes. If a pole or a zero has a multiplicity m greater than one, we place the expression (m) near the cross or circle to denote that multiplicity.

Example 13.32 Plot the PZD for the system function

$$H(s) = 10\frac{s + 3}{(s + 1)(s^2 + 2s + 5)}. \tag{13.5-28}$$

Solution The denominator is not completely factored, so we must finish the procedure, obtaining

$$H(s) = 10\frac{s + 3}{(s + 1)(s + 1 + j2)(s + 1 - j2)}. \tag{13.5-29}$$

Thus, we have a zero at $s = -3$ (because of the numerator factor) and poles at $s = -1$ and at $s = -1 - j2$ and $s = -1 + j2$ (because of the denominator factors). We show these poles and zeros in Figure 13.26.

Figure 13.26
The pole-zero diagram

(PZD) for $H(s) =$

$$10\frac{s + 3}{(s + 1)(s^2 + 2s + 5)}$$

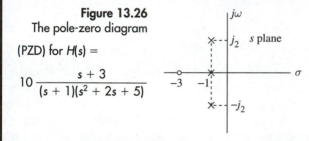

Forced and Natural Responses

We will now demonstrate that the system function $H(s)$ and its PZD are useful tools in answering certain qualitative questions about the solution $y(t)$ of a given differential equation without actually computing that solution. The questions we are referring to are ones regarding *stability*. Before discussing this important topic, though, we will talk about another facet of a differential equation, namely, its *forced response* and its *natural response*.

We will start with the system function solution to a given differential equation and perform a PFE in a special way; specifically, we first gather together all terms that appear in the denominator of $X(s)$, then those that appear in $H(s)$:

$$Y(s) = H(s)X(s) = Y_f(s) + Y_n(s). \tag{13.5-30}$$

Thus, $Y_f(s)$ is the sum of all the PFE terms coming from the denominator of $X(s)$, and $Y_n(s)$ contains all those coming from the denominator of $H(s)$. Now, *assuming that $H(s)$ and $X(s)$ have no common finite poles*, we see that the poles of $X(s)$ determine the form of

$Y_f(s)$, and the poles of $H(s)$ determine the form of $Y_n(s)$. We then write

$$y(t) = y_f(t) + y_n(t). \tag{13.5-31}$$

Why have we said that the poles of $H(s)$ determine the *form* of $y_n(t)$, and the poles of $X(s)$ determine the *form* of $y_f(t)$? Because the zeros influence the values of the *constants* in the PFE, but the poles determine the *form* of the response (such a response will contain a term of the form $e^{-at}u(t)$ if $s = a$ is a pole of order 1 or $\dfrac{t^{m-1}}{(m-1)!}e^{-at}u(t)$ if $s = a$ is a pole of order m).

Example 13.33   Find the forced and natural responses for the differential equation

$$\frac{d^2y}{dt^2} + 3\frac{dy}{dt} + 2y = 4\frac{dx}{dt} + 12x, \tag{13.5-32}$$

assuming that $x(t) = 4u(t)$.

Solution   We take the Laplace transform of both sides and solve, obtaining

$$Y(s) = \frac{4s + 12}{s^2 + 3s + 2}X(s) = \frac{16(s + 3)}{s(s + 1)(s + 2)} \tag{13.5-33}$$

$$= \frac{24}{s} - \frac{32}{s + 1} + \frac{8}{s + 2}.$$

Thus, the forced response transform is $Y_f(s) = 24/s$, so the forced response itself is

$$y_f(t) = 24u(t). \tag{13.5-34}$$

The natural response transform is $Y_n(s) = -8[4/(s + 1) - 1/(s + 2)]$, so the natural response is

$$y_n(t) = -8[4e^{-t} - e^{-2t}]u(t). \tag{13.5-35}$$

Stability of a Differential Equation

Now suppose that $x(t) = \delta(t)$. In this case, $X(s) = 1$ *has no poles*. Therefore, all the poles in $Y(s)$ come from the poles of $H(s)$, and *the response $y(t)$ only consists of a natural response. We can therefore identify $y_n(t)$ with the impulse response $h(t)$.*

$$y_n(t) = h(t), \quad provided \ x(t) = \delta(t). \tag{13.5-36}$$

We say that a differential equation is *stable* if its natural response approaches zero as time becomes infinitely large. In mathematical symbols,

$$y_n(t) \longrightarrow 0 \quad \text{as } t \longrightarrow \infty. \tag{13.5-37}$$

From (13.5-36) we see that we only have to find the impulse response $h(t)$ and check to see whether or not

$$h(t) \longrightarrow 0 \quad \text{as } t \longrightarrow \infty \tag{13.5-38}$$

to determine stability. If, on the other hand $y_n(t)$ becomes unbounded as $t \to \infty$ (or, equivalently, the impulse response $h(t)$ becomes unbounded as $t \to \infty$), we say that the differential equation is *unstable*. Finally, if the differential equation is neither unstable nor stable, we say that it is *marginally stable*. Equivalently, we see that a marginally stable differential equation is one for which the natural response (and the impulse response) does not approach zero or become unbounded as $t \to \infty$. Thus, it must either approach a nonzero constant or oscillate up and down indefinitely as time gets larger and larger. We now provide a stability example.

Example 13.34   Check the stability properties of a system function having an mth-order pole at $s = s_0$ and no zeros, where

$$s_0 = \sigma_0 + j\omega_0. \tag{13.5-39}$$

Solution   Such a system function has the form

$$H(s) = \frac{k}{(s - s_0)^m}, \tag{13.5-40}$$

where k is a constant. The impulse response is

$$h(t) = k\frac{t^{m-1}}{(m - 1)!}e^{s_0 t}u(t). \tag{13.5-41}$$

We can check the stability by looking at the absolute value of $h(t)$:

$$|h(t)| = |k|\frac{t^{m-1}}{(m - 1)!}e^{\sigma_0 t}u(t). \tag{13.5-42}$$

(Remember that

$$|e^{j\omega_0 t}| = \sqrt{\cos^2(\omega_0 t) + \sin^2(\omega_0 t)} = 1.) \tag{13.5-43}$$

We now see that if $\sigma_0 > 0$, the exponential term in (13.5-42), for any value of m, will approach infinity as t becomes large; it will approach zero as t becomes large if $\sigma_0 < 0$ (for any value of m). Finally, we see that if $\sigma_0 = 0$, there will be two cases. If $m = 1$, the magnitude of $h(t)$ will be constant for all positive time, whereas $h(t)$ itself is given by

$$h(t) = ke^{j\omega_0 t}u(t) = k[\cos(\omega_0 t) + j\sin(\omega_0 t)]u(t). \tag{13.5-44}$$

If $m > 1$, the power of t in (13.5-42) will make $h(t)$ again become unbounded as $t \to \infty$.

In summary, if $\text{Re}\{s_0\} = \sigma_0 > 0$, $H(s)$ will be unstable; if $\text{Re}\{s_0\} = \sigma_0 < 0$, $H(s)$ will be stable; and if $\text{Re}\{s_0\} = \sigma_0 = 0$, $H(s)$ will be unstable if $m > 1$ and marginally stable if $m = 1$. We show this in the form of a PZD in Figure 13.27. If the pole is in the LHP (left-half plane) the differential equation is stable and if it is in the RHP (right-half plane) the differential equation is unstable. If it is on the imaginary axis it is on the boundary of stability: if the multiplicity of the pole is one, the differential equation is stable; if the multiplicity of the pole is more than one, the differential equation is unstable.

Figure 13.27
Stability regions for a single pole

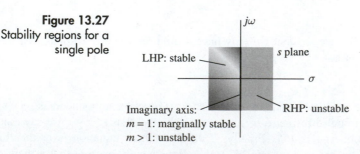

Determination of Stability from the PZD

Now consider an arbitrary differential equation that has a rational system function $H(s)$. If the degree, m, of the numerator polynomial is greater than that of the denominator, n, one first does long division, getting the following form:

$$H(s) = \sum_{i=0}^{m-n} k_i s^i + H_o(s). \tag{13.5-45}$$

The corresponding time domain impulse response is

$$h(t) = \sum_{i=0}^{m-n} k_i \delta^i(t) + h_o(t), \tag{13.5-46}$$

where $h_o(t)$ is an ordinary waveform whose form is determined only by the poles of $H(s)$. Because stability only concerns the behavior of $h(t)$ for large values of t, we can disregard the impulses and derivatives of impulses at $t = 0$. $H_o(s)$ in (13.5-45) can be expanded using the PFE, and each term will be of the form $h_i(t) = \dfrac{k_i}{(s - s_i)^{m_i}}$. For a pole at $s = 0$, $s_i = 0$. If a pole at s_i is simple, then the order $m_i = 1$.

Here is the result of our development. *If any one pole of $H(s)$ is in the RHP, the entire differential equation will be unstable* because the term $h_i(t)$ due to that pole will force the entire impulse response $h(t)$ to become infinite as t becomes large. On the other hand, *if all of the poles of $H(s)$ are in the LHP,* all of the $h_i(t)$ terms will approach zero as time becomes large, and *the differential equation will be stable. If all of the poles are in the LHP with the exception of one or more on the imaginary axis, we must check the multiplicities of the latter. If all are simple, the differential equation is marginally stable; if one or more have a higher multiplicity, then the differential equation is unstable.* (If you are a bit concerned as to whether or not two or more terms could approach infinity, yet have opposing signs and thus cancel each other, rest assured that this cannot happen because the individual terms are linearly independent. We will leave this for you to show.)

Example 13.35   Check the stability properties of the differential equation

$$p^3 y(t) + 4p^2 y(t) + 13 p y(t) = 2px(t) + 4x(t) \tag{13.5-47}$$

without solving it.

Solution   We write the system function $H(s)$ by inspection of the given differential equation. It is

$$H(s) = \frac{2(s + 2)}{s(s^2 + 4s + 13)}. \tag{13.5-48}$$

(Notice that the denominator is merely the left-hand side of the original differential equation with the p's changed to s's and the $y(t)$ term removed, whereas the numerator is the right-hand side with the $x(t)$ removed and the p's changed to s's.) If we factor the quadratic in the denominator, we have

$$H(s) = \frac{2(s + 2)}{s((s + 2)^2 + 3^2)} = \frac{2(s + 2)}{s(s + 2 - j3)(s + 2 + j3)}. \tag{13.5-49}$$

The zeros are those values of s making the numerator zero; the poles are the values of s making the denominator zero. The PZD is shown in Figure 13.28. There are three poles: one at $s = -2 + j3$, one at $s = -2 - j3$ (the conjugate of the first one), and one at $s = 0$. The two complex poles are "stable poles" because they are in the LHP. The one at

$s = 0$, however, is on the imaginary axis. Because it is simple (has multiplicity one), it is a "marginally stable pole." Therefore, the differential equation itself is marginally stable. Notice carefully that the zero (at $s = -2$, in this case) does not enter into the determination of stability. For example, it could be at $s = +2$ and the differential equation would still be marginally stable. *It is only the locations of the poles that determine stability.*

Figure 13.28
The pole-zero diagram
(PZD) for $H(s) =$

$$\frac{2(s + 2)}{s(s + 2 - j3)(s + 2 + j3)}$$

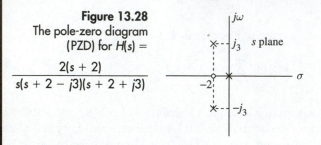

Example 13.36   Find the impulse response of the differential equation

$$p^3 y(t) + 4p^2 y(t) + 13py(t) = 2px(t) + 4x(t), \tag{13.5-50}$$

and thus show directly that it satisfies the definition of marginal stability.

Solution   We let the forcing function be a unit impulse, $x(t) = \delta(t)$. Because this waveform has a Laplace transform of one, the PFE of $Y(s) = H(s)X(s) = H(s)$ is

$$H(s) = \frac{2(s + 2)}{s(s^2 + 4s + 13)} = \frac{4/13}{s} + \frac{As + B}{s^2 + 4s + 13}. \tag{13.5-51}$$

Notice that we have already determined the coefficient of $1/s$ in the PFE because it is so simple to find. Multiplying both sides by the denominator of $H(s)$, we have

$$2s + 4 = \frac{4}{13}(s^2 + 4s + 13) + s(As + B). \tag{13.5-52}$$

Solving by matching coefficients of powers of s, we find that $A = -4/13$ and $B = 10/13$. In (13.5-51), these results give

$$H(s) = \frac{\frac{4}{13}}{s} - \frac{2}{13}\frac{2s - 5}{s^2 + 4s + 13} = \frac{\frac{4}{13}}{s} - \frac{2}{13}\frac{2(s + 2) - 9}{(s + 2)^2 + 9} \tag{13.5-53}$$

$$= \frac{\frac{4}{13}}{s} - \frac{4}{13}\frac{s + 2}{(s + 2)^2 + 9} + \frac{6}{13}\frac{3}{(s + 2)^2 + 9}.$$

Using the frequency shift property and known transforms, we obtain

$$h(t) = \frac{1}{13}[4 - e^{-2t}(4\cos(3t) - 6\sin(3t))]u(t). \tag{13.5-54}$$

As $t \to \infty$, we see that $h(t) \to 4/13$, a constant. This is consistent with our earlier determination (on the basis of the PZD) that the differential equation was marginally stable.

The Concept of **The forced and natural response decomposition of the solution of a differential equation
Steady State always works—provided that the poles of $X(s)$ do not replicate any in $H(s)$. The idea of a
steady state, however, is only valid for stable differential equations. Let's look at the most
important types of steady-state behavior.**

Dc Steady State

First, suppose that the forcing function of a given differential equation is of the form
$x(t) = ku(t)$, where k is an arbitrary constant. Then we can write

$$Y(s) = H(s)X(s) = H(s) \times \frac{k}{s} = \frac{H(0)k}{s} + Y_n(s). \qquad (13.5\text{-}55)$$

We have noted that the Laplace transform of $ku(t)$ is k/s and have expanded $Y(s)$ in a PFE.
The term due to the input pole at $s = 0$ is $H(0)k/s$ and is, therefore, the Laplace transform
of the forced response. Assuming that $H(s)$ has no poles at dc, we can then identify
$Y_n(s)$—the rest of the PFE—as the terms due to the poles of $H(s)$. Inverting the Laplace
transform, then, we have

$$y(t) = H(0)ku(t) + y_n(t). \qquad (13.5\text{-}56)$$

Now, if the differential equation is stable (no poles of $H(s)$ in the RHP or on the $j\omega$ axis),
then we know that $y_n(t)$, the natural response, decays to zero as time increases. Thus, for
large values of t, we have

$$y(t) = H(0)k = y_{dc} = y_{ss}; \qquad (13.5\text{-}57)$$

that is, the forced response—which we now refer to as the "dc forced response"—is the
complete (or total) response for large values of t. In this case, we call it the *dc steady
state* response. As long as $H(s)$ has no poles at $s = 0$, the forced response always exists.
The dc steady state, however, only exists for stable responses. We have suppressed the
$u(t)$ multiplier because we are typically only concerned about these responses for large
positive values of t; furthermore, as they are constant for positive t, we have suppressed
the "(t)" argument symbols.

Example 13.37 Show that the differential equation

$$p^2 y(t) + 5py(t) + 6y(t) = 2px(t) - x(t) \qquad (13.5\text{-}58)$$

is stable, and find the dc forced (and steady-state) response to $x(t) = 12u(t)$.

Solution The system function is

$$H(s) = \frac{2s - 1}{s^2 + 5s + 6} = \frac{2(s - 0.5)}{(s + 2)(s + 3)}. \qquad (13.5\text{-}59)$$

The PZD is shown in Figure 13.29 on p. 746. Both of the finite poles are in the LHP;
therefore, the differential equation is stable. (The zero at $s = +0.5$ has no bearing on stability.) Because $H(0) = -1/6$, the dc forced response is

$$y_{dc} = H(0) \times 12 = -2. \qquad (13.5\text{-}60)$$

Because the differential equation is stable, this is also the value of the dc steady-state
response.

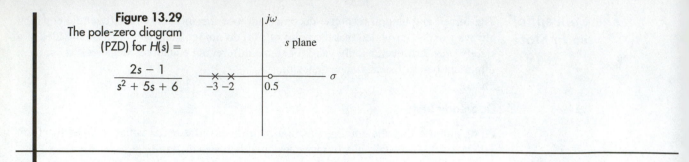

Figure 13.29
The pole-zero diagram
(PZD) for $H(s) =$

$$\frac{2s - 1}{s^2 + 5s + 6}$$

Ac Steady State

Another useful type of forced response occurs when the forcing function is a pure sinusoid, that is

$$x(t) = X_m \cos(\omega_0 t)\, u(t),\tag{13.5-61}$$

whose Laplace transform is

$$X(s) = \frac{sX_m}{s^2 + \omega_0^2}.\tag{13.5-62}$$

The s-domain response of the differential equation under study is, therefore,

$$Y(s) = H(s)X(s) = \frac{sH(s)X_m}{s^2 + \omega_0^2} = \frac{sH(s)X_m}{(s + j\omega_0)(s - j\omega_0)}.\tag{13.5-63}$$

Notice here that $H(s)$ is not a polynomial—it, too, has poles and zeros. The poles of $X(s)$ are simple and on the $j\omega$ axis, and these poles are lumped together with those of $H(s)$ in determining the poles of $Y(s)$. We now assume that the system function for the $H(s)$ of our differential equation has no poles at $s = \pm j\omega_0$. The PFE of $Y(s)$ then has the form

$$Y(s) = \frac{k}{s - j\omega_0} + \frac{k^*}{s + j\omega_0} + Y_n(s).\tag{13.5-64}$$

(Recall that the PFE coefficients for complex conjugate poles are, themselves, complex conjugates.) The value of the constant k is

$$k = [(s - j\omega_0)Y(s)]_{s = j\omega_0} = \left[(s - j\omega_0)\frac{sH(s)X_m}{(s - j\omega_0)(s + j\omega_0)}\right]_{s = j\omega_0} = \frac{1}{2}H(j\omega_0)X_m.\tag{13.5-65}$$

The forced response is the inverse Laplace transform of the first two terms in (13.5-64), that is,

$$y_f(t) = \frac{1}{2}H(j\omega_0)X_m e^{j\omega_0 t}u(t) + \frac{1}{2}H^*(j\omega_0)X_m e^{-j\omega_0 t}u(t)\tag{13.5-66}$$

$$= \text{Re}\{H(j\omega_0)e^{j\omega_0 t}\}X_m u(t).$$

Here we have used the following properties of complex numbers: if z_1 and z_2 are any two complex numbers, then $(z_1 z_2)^* = z_1^* z_2^*$, $(az)^* = az^*$ (if a is real), and $z + z^* = 2\text{Re}\{z\}$. Furthermore, we observe that $[e^{-j\omega_0 t}]^* = e^{j\omega_0 t}$.

Next, let's note that $H(j\omega_0)$ is a complex number, so it can be written in Euler form as

$$H(j\omega_0) = A(\omega_0)e^{j\phi(\omega_0)},\tag{13.5-67}$$

where
$$A(\omega_0) = |H(j\omega_0)| \tag{13.5-68}$$

is called the *gain* and

$$\phi(\omega_0) = \angle H(j\omega_0) \tag{13.5-69}$$

is called the *phase*. Notice that these are properties of the differential equation only, and do not depend upon the forcing function. If, however, we use (13.5-67) in (13.5-66), we can write

$$y_f(t) = \text{Re}\{A(\omega_0)e^{j[\omega_0 t + \phi(\omega_0)]}\}X_m u(t) = A(\omega_0)X_m \cos[\omega_0 t + \phi(\omega_0)]u(t). \tag{13.5-70}$$

This is called the *ac forced response*. Notice that in doing the PFE (13.5-64) we did not assume that the differential equation was stable—only that $H(s)$ had no poles at $s = \pm j\omega_0$. If, however, the differential equation is stable, the natural response (that is, the inverse transform of the s-domain function $Y_n(s)$ in (13.5-64), $y_n(t) = \pounds^{-1}\{Y_n(s)\}$) approaches zero for large values of time. In this case, we also call $y_f(t)$ in (13.5-70) the *ac steady-state response*. The forced response always exists if $H(s)$ does not contain the poles of the forcing function, but the steady state only exists for stable differential equations.

Example 13.38   Show that the differential equation

$$py(t) + 2y(t) = x(t) \tag{13.5-71}$$

is stable, and find the ac forced (and steady-state) response to $x(t) = 12 \cos(2t)u(t)$.

Solution   The system function is

$$H(s) = \frac{1}{s + 2}. \tag{13.5-72}$$

Because this $H(s)$ is so simple we will not show the PZD; we simply note that the one finite pole is at $s = -2$, hence the differential equation is stable. Next, we compute

$$H(j2) = \frac{1}{j2 + 2} = \frac{1}{2\sqrt{2}\angle 45°} = \frac{1}{2\sqrt{2}}\angle -45°. \tag{13.5-73}$$

Thus,
$$A(2) = \frac{1}{2\sqrt{2}} \tag{13.5-74}$$

and
$$\phi(2) = -45°. \tag{13.5-75}$$

Therefore,
$$y_f(t) = A(2)12 \cos(2t + \phi(2)) = 3\sqrt{2} \cos(2t - 45°). \tag{13.5-76}$$

We will often drop the u(t) multiplier (as we did in (13.5-76)) when working with the ac steady-state response, for we are assuming that a long enough time has passed for the natural response to have decayed to zero; hence, the u(t) is a bit redundant. When we discuss forced response, however, we will at times retain the u(t).

Initial Condition Problems   The differential equations we have been investigating up until now have all been defined on the entire time axis $-\infty < t < +\infty$, and all the waveforms have been assumed to be one sided, that is, they have been assumed to be identically zero for all $t \leq 0$. We will now discuss another type of problem: the *initial condition problem.* Given a differential equation and forcing function, both of which are only known to hold for $t > 0$, find the response for all $t > 0$. The difference is that we no longer know the forcing function for $t \leq 0$. In fact, we do not even know that the differential equation is valid for $t \leq 0$! We will have to be given additional information, though, as we will soon see.

Our analysis will rely on waveforms that are truncated for $t \leq 0$. One such is shown in Figure 13.30. The light line is the graph of a waveform $f(t)$, which is possibly nonzero for $t \leq 0$. We will only watch it, however, for $t > 0$. Because our Laplace transform theory is built up for waveforms that are zero for $t \leq 0$, we will simply "turn it off" for $t \leq 0$. This gives us the waveform shown with the heavy line. It is the same as $f(t)$ for $t > 0$, but is identically zero for $t \leq 0$. We use the notation $f_+(t)$ for this waveform. Mathematically, the relation between $f(t)$ and $f_+(t)$ can be written

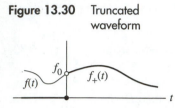

Figure 13.30   Truncated waveform

$$f_+(t) = f(t)u(t), \tag{13.5-77}$$

the $u(t)$ multiplier acting as a time-sensitive switch.

Next, suppose we take the time derivative (using generalized differentiation, as discussed in Section 13.2). We see that

$$pf_+(t) = \{pf(t)\}u(t) + f_0\delta(t). \tag{13.5-78}$$

We are adding an assumption: the original $f(t)$ is continuous at $t = 0$ so the derivative of $f_+(t)$ is well defined.[14] This is certainly allowable, for we only know $f(t)$ for $t > 0$; hence, we can make any assumptions we like about it for $t \leq 0$. Notice that this adds an artificially produced delta function at $t = 0$. If we take the second derivative of $f_+(t)$ using (13.5-78) (assuming that the ordinary derivative $f'(t)$ is well defined at $t = 0$), we get

$$p^2f_+(t) = \{p^2f(t)\}u(t) + f_0'\delta(t) + f_0\delta'(t). \tag{13.5-79}$$

We have now added an extraneous derivative of an impulse as well as an extraneous impulse. You see how this goes; we could continue for any order derivative we like, resulting in

$$p^nf_+(t) = \{p^nf(t)\}u(t) + f_0^{(n-1)}\delta(t) + f_0^{(n-2)}\delta'(t) + \cdots + f_0\delta^{(n-1)}(t). \tag{13.5-80}$$

As we do not know $f(t)$ or any of its derivatives for $t \leq 0$, we do not know their values at $t = 0$; hence, these must be supplied in the problem statement or from continuity considerations. Let's see how we incorporate these *initial conditions* with an example.

Example 13.39   Solve the differential equation

$$py(t) + y(t) = px(t) + 2x(t) \tag{13.5-81}$$

for all $t > 0$ if $x(t) = 2$ for $t > 0$, assuming that the initial condition is $y_0 = 2$.

Solution   The right-hand side of our differential equation is known for all positive values of t, so we can rewrite (13.5-81) in the form

$$py(t) + y(t) = f(t), \tag{13.5-82}$$

where $$f(t) = px(t) + 2x(t) = p\{2\} + 2\{2\} = 4 \tag{13.5-83}$$

---

[14] If there are any singularities or discontinuities in our original $f(t)$ at $t = 0$ we do not know about them; furthermore, all of our waveforms must be continuous for a small enough neighborhood of $t = 0$. This is one of the basic assumptions we made in our discussion of generalized functions in Section 13.2.

for all $t > 0$. Now let's multiply both sides of our differential equation by $u(t)$, to obtain

$$[py(t)]u(t) + y(t)u(t) = f(t)u(t) = 4u(t). \tag{13.5-84}$$

We note that $f(t)u(t) = f_+(t) = 4u(t)$ and $y(t)u(t) = y_+(t)$; then, using (13.5-78), we have

$$py_+(t) - y_0\delta(t) + y_+(t) = f_+(t) = 4u(t), \tag{13.5-85}$$

which becomes (after rearranging)

$$py_+(t) + y_+(t) = 4u(t) + y_0\delta(t). \tag{13.5-86}$$

The initial condition has simply become an extra impulsive forcing function. We are given the value of y_0, so using it and taking the Laplace transform, (13.5-86) becomes

$$Y_+(s) = \frac{2s + 4}{s(s + 1)} = \frac{4}{s} - \frac{2}{s + 1}. \tag{13.5-87}$$

Inverting $Y_+(s)$, we obtain

$$y_+(t) = [4 - 2e^{-t}]u(t). \tag{13.5-88}$$

This solution is not valid for $t \leq 0$, however, so we remove the $u(t)$ multiplier and simply write

$$y(t) = 4 - 2e^{-t}; \quad t > 0. \tag{13.5-89}$$

Notice that $y(0+) = y(0) = 2$, as required.

The General Solution for the Initial Condition Problem

Here is the general method we have established in the preceding example. If the differential equation to be solved on $t > 0$ is

$$p^n y(t) + a_{n-1}p^{n-1}y(t) + \cdots + a_1 py(t) + a_0 y(t) \tag{13.5-90}$$
$$= b_m p^m x(t) + b_{m-1}p^{m-1}x(t) + \cdots + b_1 px(t) + b_0 x(t),$$

we define

$$f(t) = b_m p^m x(t) + b_{m-1}p^{m-1}x(t) + \cdots + b_1 px(t) + b_0 x(t). \tag{13.5-91}$$

We explicitly compute the derivatives and form $f(t)$, a function that is known for all $t > 0$. Next, we use $f(t)$ in (13.5-90) and multiply both sides by the unit step function $u(t)$, obtaining

$$[p^n y(t)]u(t) + a_{n-1}[p^{n-1}y(t)]u(t) + \cdots + a_1[py(t)]u(t) + a_0 y(t)u(t) = f_+(t). \tag{13.5-92}$$

We solve for each of the terms on the left-hand side using (13.5-80) to express $p^n y(t)$ in terms of $p^n y_+(t)$. For each $i = 1, 2, \ldots, n$, we have (using $y(t)$ rather than $f(t)$)

$$\{p^i y(t)\}u(t) = p^i y_+(t) - y_0^{(i-1)}\delta(t) - y_0^{(i-2)}\delta'(t) - \cdots - y_0\delta^{(i-1)}(t). \tag{13.5-93}$$

We will not present the general formula, which is quite involved notationally; it is the procedure, however, that is essential. In this way, because $i = n$ is the largest value of the index in (13.5-92), we see that we must independently be given n initial conditions: $y_0, y_0', \ldots, y_0^{(n-1)}$. Once we insert these into our differential equation (13.5-90), we take the Laplace transform, do the PFE, and invert the result to find the solution.

Example 13.40   Solve the differential equation

$$p^2 y(t) + 4py(t) + 4y(t) = px(t) + 3x(t) \qquad (13.5\text{-}94)$$

for all $t > 0$ if $x(t) = e^{-t}$ for $t > 0$, assuming that general values of the initial conditions y_0 and y_0' are given.

Solution   The right-hand side of our differential equation is known for all positive values of t, so we can compute it directly:

$$f(t) = pe^{-t} + 3e^{-t} = 2e^{-t}. \qquad (13.5\text{-}95)$$

Using this in (13.5-94) and multiplying both sides by $u(t)$ gives

$$[p^2 y(t)]u(t) + 4[py(t)]u(t) + 4y(t)u(t) = 2e^{-t}u(t). \qquad (13.5\text{-}96)$$

Inserting the expressions for the various derivatives in terms of one-sided waveforms and the initial conditions results in

$$[p^2 y_+(t) - y_0'\delta(t) - y_0\delta'(t)] + 4[py_+(t) - y_0\delta(t)] + 4y_+(t) = 2e^{-t}u(t). \qquad (13.5\text{-}97)$$

Taking the Laplace transform of both sides and rearranging, we have

$$(s^2 + 4s + 4)Y_+(s) = \frac{2}{s+1} + y_0' + 4y_0 + y_0 s. \qquad (13.5\text{-}98)$$

Solving for $Y_+(s)$, we obtain

$$Y_+(s) = \frac{2}{(s+1)(s^2 + 4s + 4)} + \frac{y_0 s + y_0' + 4y_0}{s^2 + 4s + 4}. \qquad (13.5\text{-}99)$$

Now we observe that there is a pole of multiplicity two in each of the two right-hand side Laplace transforms. Expanding in a PFE (the details of which we leave to you), we get

$$Y_+(s) = \frac{2}{s+1} + \frac{y_0 - 2}{s+2} + \frac{y_0' + 2y_0 - 2}{(s+2)^2}. \qquad (13.5\text{-}100)$$

(We suggest that you evaluate the $s + 1$ and $(s + 2)^2$ coefficients in the usual way—by multiplying both sides by the appropriate denominator and evaluating at the associated value of s—then multiply both sides by the complete denominator of $Y_+(s)$ and match coefficients to obtain the constant in the $s + 2$ fraction.) Inverting, we get

$$y(t) = 2[e^{-t} - (t + 1)e^{-2t}] + [y_0 e^{-2t} + (y_0' + 2y_0)te^{-2t}]. \qquad (13.5\text{-}101)$$

We have perhaps been just a bit terse, for what we did was simply write the expression for $t > 0$ without the $u(t)$ multiplier. We are actually solving for $y_+(t)$—with the $u(t)$ multiplier—then dropping it, recognizing that the solution is not valid for $t \le 0$.

Now let's explore some more how the initial conditions appear in the solution. Glance back at equation (13.5-99). There you will see that the transform of the one-sided response function $y_+(t)$ was expressed as the sum of two components. This is a general result, as you can easily see. We can place the Laplace transform of any differential equation in the form

$$D(s)Y_+(s) = F_+(s) + Y_0(s), \qquad (13.5\text{-}102)$$

where $D(s)$ is merely the left-hand side of the differential equation with the powers of p replaced by powers of s and factored, $F_+(s)$ is the (known) transform of the (one-sided) right-hand side, and $Y_0(s)$ is the transform of all the initial condition impulse functions and derivatives of impulse functions transferred to the right-hand side of the equation. Solving, we have

$$Y_+(s) = \frac{F_+(s)}{D(s)} + \frac{Y_0(s)}{D(s)} = Y_{zs}(s) + Y_{zi}(s). \tag{13.5-103}$$

Equation (13.5-99) of the preceding example is in this form. In the time domain, we have

$$y(t) = y_{zs}(t) + y_{zi}(t). \tag{13.5-104}$$

The function $y_{zs}(t)$ is called the *zero state response* and the function $y_{zi}(t)$ the *zero input response*. The initial conditions are called the *initial state* of the differential equation. If the initial state is zero (e.g., $y_0 = y_0' = 0$) then the zero state response is the actual response; conversely, if we make $x(t) = 0$ (for all $t > 0$), the zero input response is the actual response. It depends upon the initial conditions. Notice that the poles of $Y_{zs}(s)$ are the poles of $F_+(s)$ and the zeros of $D(s)$, whereas the poles of $Y_{zi}(s)$ are only the zeros of $D(s)$. Though similar to the forced/natural response decomposition already discussed, the zero state/zero input response is not the same thing. Typically, we do not include the $u(t)$ function in these waveforms because they are valid only for $t > 0$.

Example 13.41 Find the zero state and zero input responses for

$$p^2 y(t) + 5py(t) + 6y(t) = 6px(t) + 24x(t). \tag{13.5-105}$$

Assume that $x(t) = e^{-2t}$ for $t > 0$.

Solution The right-hand side is given by

$$f(t) = 6pe^{-2t} + 24e^{-2t} = 12e^{-2t}. \tag{13.5-106}$$

Using this in (13.5-105) and multiplying both sides by $u(t)$ gives

$$[p^2 y(t)]u(t) + 5[py(t)]u(t) + 6y(t)u(t) = 12e^{-2t}u(t). \tag{13.5-107}$$

Inserting the expressions for the various derivatives in terms of one-sided waveforms and the initial conditions results in

$$[p^2 y_+(t) - y_0'\delta(t) - y_0\delta'(t)] + 5[py_+(t) - y_0\delta(t)] + 6y_+(t) = 12e^{-2t}u(t). \tag{13.5-108}$$

Taking the Laplace transform of both sides and rearranging, we have

$$(s^2 + 5s + 6)Y_+(s) = \frac{12}{s + 2} + y_0 s + y_0' + 5y_0. \tag{13.5-109}$$

Solving for $Y_+(s)$,

$$Y_+(s) = \frac{12}{(s + 2)(s^2 + 5s + 6)} + \frac{y_0 s + y_0' + 5y_0}{s^2 + 5s + 6}. \tag{13.5-110}$$

Rather than solving for the complete response this time, we will simply identify the zero state and zero input terms and invert them individually. Thus, we have

$$Y_{zs}(s) = \frac{12}{(s+2)(s^2+5s+6)} = \frac{12}{(s+2)^2(s+3)} \qquad (13.5\text{-}111)$$

$$= \frac{-12}{s+2} + \frac{12}{(s+2)^2} + \frac{12}{s+3}$$

because it is that part of $Y_+(s)$ that does not depend upon the initial conditions (the initial state). Again, we have left the details of the PFE up to you. Inverting, we get

$$y_{zs}(t) = 12[(t-1)e^{-2t} + e^{-3t}]; \quad t > 0. \qquad (13.5\text{-}112)$$

The term in $Y_+(s)$ corresponding to the zero input response is

$$Y_{zi}(s) = \frac{y_0s + y_0' + 5y_0}{s^2+5s+6} = \frac{y_0s + y_0' + 5y_0}{(s+2)(s+3)} = \frac{y_0' + 3y_0}{s+2} - \frac{y_0' + 2y_0}{s+3}. \qquad (13.5\text{-}113)$$

The time domain waveform is given by

$$y_{zi}(t) = (y_0' + 3y_0)e^{-2t} - (y_0' + 2y_0)e^{-3t}; \quad t > 0. \qquad (13.5\text{-}114)$$

Section 13.5 Quiz

All of these questions pertain to the following differential equation:

$$p^3y(t) + 6p^2y(t) + 11py(t) + 6y(t) = 4p^2x(t) + 4px(t) + 2x(t).$$

Q13.5-1. Find the system function $H(s) = Y(s)/X(s)$ and then draw the PZD. (*Hint:* One root of the polynomial $s^3 + 6s^2 + 11s + 6$ is $s = -1$; that is, one of its factors is $(s+1)$). Determine the stability characteristic of the differential equation.

Q13.5-2. Find the unit step and unit impulse responses.

Q13.5-3. Find the responses to:
  a. $x(t) = e^{-0.5t}u(t)$
  b. $x(t) = e^{-2t}u(t)$
  c. $x(t) = 12u(t)$
  d. $x(t) = 12\cos(2t)u(t)$.

Note that these responses might not be "nice"; that is, the numbers will not all be integers or even simple fractions.

Q13.5-4. Assume that the differential equation above only holds for $t > 0$ and let the forcing function $x(t)$ be given by $x(t) = tu(t)$. Find the zero input and zero state responses.

13.6 | Laplace Transform Analysis of Linear Time-Invariant Circuits

We will now begin to apply the Laplace transform theory we have thus far developed in this chapter. Specifically, we will discuss the analysis of linear time-invariant circuits, which, for all practical purposes here, means that our circuits will have no switches. Switched circuits will be discussed in Section 13.7.

Impedance and Admittance for a Two-Terminal Subcircuit

We start by considering a general two-terminal element or subcircuit such as the one shown in Figure 13.31. If it is an element, we assume that it is a passive one; that is, we assume that it is a resistor, inductor, or capacitor. If it is a subcircuit, we assume that it has no independent sources, though we will allow it to have dependent sources. We now use the following fact:[15] the terminal voltage $v(t)$ and terminal current $i(t)$ are related by a

---

[15] This follows from the operator theory developed in Part II; however, it will also follow from the general analysis techniques to be presented here. (Our reasoning will not be circular, because the general fact will follow from the fact that the basic R, L, and C elements all have this property.)

Figure 13.31 A two-terminal element or subcircuit

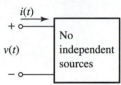

linear, constant-coefficient differential equation:

$$p^n v(t) + a_{n-1} p^{n-1} v(t) + \cdots + a_1 pv(t) + a_0 v(t) \tag{13.6-1}$$
$$= b_m p^m i(t) + b_{m-1} p^{m-1} i(t) + \cdots + b_1 pi(t) + b_0 i(t).$$

We can treat (13.6-1) as being a differential equation giving the response $v(t)$ to a known waveform for $i(t)$—or we can read it backwards and assume that $v(t)$ is known, with $i(t)$ being the response function to be found.

Selecting the voltage as the response function, we take the transform of (13.6-1) and solve for $V(s)$, obtaining

$$V(s) = H(s)I(s), \tag{13.6-2}$$

where

$$H(s) = \frac{b_m s^m + b_{m-1} s^{m-1} + \cdots + b_1 s + b_0}{s^n + a_{n-1} s^{n-1} + \cdots + a_1 s + a_0} \tag{13.6-3}$$

is the system function. In our special case, that is with $v(t)$ being the response and $i(t)$ being the forcing function, we refer to $H(s)$ as the *complex impedance* (often shortened to simply *impedance*) and change our notation from $H(s)$ to $Z(s)$. Thus, we write

$$Z(s) = H(s) = \frac{V(s)}{I(s)}. \tag{13.6-4}$$

In the future, we will use $H(s)$ to represent the ratio of any two transforms, one of which is the response and the other the forcing function, and the notation $Z(s)$ will apply only to the special case we are considering.

If we choose to read (13.6-1) backwards, that is, we select $i(t)$ as the response and $v(t)$ as the forcing function, we have

$$\frac{I(s)}{V(s)} = \frac{s^n + a_{n-1} s^{n-1} + \cdots + a_1 s + a_0}{b_m s^m + b_{m-1} s^{m-1} + \cdots + b_1 s + b_0} = \frac{1}{Z(s)} = Y(s). \tag{13.6-5}$$

We call $Y(s) = 1/Z(s)$ the *complex admittance* (often shortened to simply *admittance*). Sometimes the phrase "driving point" is applied to $Z(s)$ and $Y(s)$ because other system functions can be the ratio of voltage to current or current to voltage, with the variables defined in different parts of the circuit or subcircuit under consideration. In that situation, the system function is often referred to as a *transfer impedance* or a *transfer admittance*. Thus, the adjective "driving point" will serve to stress the fact that the voltage and current are both being considered at a given pair of terminals. We will often not use this expression, though, if the interpretation is obvious.

For the important special case in which the two-terminal subcircuit is a resistor, inductor, or capacitor, the impedance and admittance assume simple forms. For a resistor, for instance, we have

$$v(t) = Ri(t). \tag{13.6-6}$$

This is a degenerate case of our differential equation (13.6-1) for which $n = m = 0$, $a_0 = 1$, and $b_0 = R$. Thus, we have

$$V(s) = RI(s). \tag{13.6-7}$$

The impedance is

$$Z(s) = \frac{V(s)}{I(s)} = R \tag{13.6-8}$$

and the admittance is

$$Y(s) = \frac{I(s)}{V(s)} = 1/R = G. \tag{13.6-9}$$

The inductor obeys the relationship[16]

$$v(t) = Lpi(t). \tag{13.6-10}$$

This is a special case of our differential equation (13.6-1) for which $n = 0$ and $m = 1$, with $a_0 = 1$, $b_1 = L$, and $b_0 = 0$. Taking the Laplace transform of both sides gives

$$V(s) = LsI(s). \tag{13.6-11}$$

Thus, the impedance of the inductor is

$$Z(s) = \frac{V(s)}{I(s)} = Ls \tag{13.6-12}$$

and its admittance is

$$Y(s) = \frac{I(s)}{V(s)} = \frac{1}{Z(s)} = \frac{1}{Ls}. \tag{13.6-13}$$

Last (though certainly not the least in importance) comes the capacitor. Its defining time-domain relationship is

$$i(t) = Cpv(t). \tag{13.6-14}$$

Again, this is a special case of our differential equation (13.6-1) "read backwards" with $n = 1$, $m = 0$, $a_1 = C$, $a_0 = 0$, and $b_0 = 1$. In the s domain, this becomes

$$I(s) = CsV(s), \tag{13.6-15}$$

with the impedance being

$$Z(s) = \frac{V(s)}{I(s)} = \frac{1}{Cs} \tag{13.6-16}$$

and the admittance being

$$Y(s) = \frac{I(s)}{V(s)} = \frac{1}{Z(s)} = Cs. \tag{13.6-17}$$

Notice that all three of our basic elements satisfy the s-domain relationship

$$V(s) = Z(s)I(s), \tag{13.6-18}$$

which is Ohm's law with the complex impedance $Z(s)$ replacing the real constant R. We will therefore draw the s-domain equivalents of our elements as in Figure 13.32 with the impedance given next to the element.

---

[16] We are extending the relationship to include generalized differentiation. This is an added assumption about the inductor v-i characteristic.

Figure 13.32
s-Domain impedance
models for *R*, *L*, and *C*

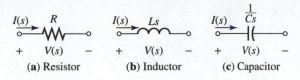

(a) Resistor (b) Inductor (c) Capacitor

***Kirchhoff's Laws
in the s Domain***

Here is another important fact: Kirchhoff's laws continue to hold when voltages and currents are interpreted as Laplace transforms. To see this, just recall that KVL and KCL have the form

$$x_1(t) + x_2(t) + \cdots + x_N(t) = 0, \qquad (13.6\text{-}19)$$

where the $x(t)$'s are voltages for KVL and currents for KCL. Taking the Laplace transform and using the linearity property, we have

$$X_1(s) + X_2(s) + \cdots + X_N(s) = 0. \qquad (13.6\text{-}20)$$

But this is just KVL or KCL, as appropriate, phrased in terms of transforms of voltages or currents, as appropriate.

***The s-Domain
Equivalent Circuit***

Why are these facts so important? Simply because all of our analysis machinery developed in Part I on dc circuits was built upon Ohm's law, KVL, and KCL. Therefore, all of these analysis techniques: nodal and mesh analysis, Thévenin and Norton equivalent subcircuits, superposition, etc., continue to hold for the *s*-domain equivalent circuit (the circuit with all currents and voltages represented as functions of *s* and with the resistors, inductors, and capacitors represented by their complex impedances).

Example 13.42

Find the impulse and step responses of the circuit in Figure 13.33.

Figure 13.33
An example circuit

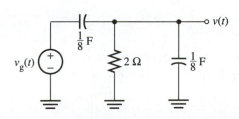

Solution

Our first step is to draw the *s*-domain equivalent circuit. This is shown in Figure 13.34. Notice that the capacitors are represented by their impedances, both being $8/s$, and the voltages in terms of their transforms. We can choose any of our standard techniques to analyze this circuit, but a quite simple approach is to recognize that the 2-Ω resistor and the right-hand capacitor are connected in parallel and that the resulting two-terminal equivalent subcircuit is connected in series with the remaining capacitor. In connection with this, note that we are (by convention) assuming that nothing else is connected to the terminal whose voltage is $V(s)$ relative to ground. The impedance of the parallel combination is

$$Z(s) = \frac{2 \times 8/s}{2 + 8/s} = \frac{8}{s + 4}. \qquad (13.6\text{-}21)$$

We show the resulting equivalent circuit in Figure 13.35. We can use the voltage divider rule to obtain the voltage $V(s)$:

$$V(s) = \frac{Z(s)}{Z(s) + 8/s}V_g(s) = \frac{\dfrac{8}{s+4}}{\dfrac{8}{s+4} + 8/s}V_g(s) = \frac{s}{2(s+2)}V_g(s). \qquad (13.6\text{-}22)$$

To find the unit impulse response, we simply adjust $v_g(t)$ so that its waveform is the unit impulse function; hence, we have $V_g(s) = 1$ and

$$h(t) = v(t) = \pounds^{-1}\left\{\frac{s}{2(s+2)}\right\} = \frac{1}{2}\pounds^{-1}\left\{1 - \frac{2}{s+2}\right\} \qquad (13.6\text{-}23)$$

$$= \frac{1}{2}\delta(t) - e^{-2t}u(t).$$

Figure 13.34
The s-domain equivalent circuit

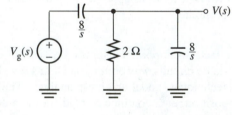

Figure 13.35
Using the parallel equivalent subcircuit

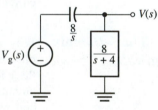

The unit step response can be found by integrating $h(t)$; however, it is probably more straightforward to simply let $v_g(t) = u(t)$, with the transform $V_g(s) = 1/s$, and do the computation as above. Thus, we now have

$$V(s) = \frac{s}{2(s+2)} \times \frac{1}{s} = \frac{1}{2(s+2)}. \qquad (13.6\text{-}24)$$

Notice the pole-zero cancellation; that is, the cancellation of a zero of $H(s) = s/[2(s+2)]$ with a pole of the input $V_g(s)$. Now we simply invert (13.6-24) to obtain

$$s(t) = v(t) = \pounds^{-1}\left\{\frac{1}{2(s+2)}\right\} = \frac{1}{2}e^{-2t}u(t). \qquad (13.6\text{-}25)$$

We now present an example of nodal analysis in the s-domain.

Example 13.43   Find the s-domain responses $V(s)$ and $I(s)$ corresponding to the voltage $v(t)$ and the current $i(t)$ shown in Figure 13.36 in terms of the transforms of the independent source waveforms $V_{g1}(s)$ and $V_{g2}(s)$. Use nodal analysis.

Figure 13.36
An example circuit

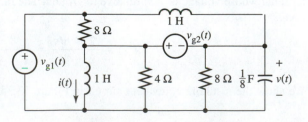

Solution   The s-domain equivalent circuit is shown in Figure 13.37. Not only have we expressed each of the dynamic elements in terms of its s-domain impedance, we have chosen a ground reference and labeled the nodes with their voltages as usual for nodal analysis. Notice that there is one supernode and one nonessential node. The single nodal equation required is

$$\frac{V(s)}{8/s} + \frac{V(s)}{8} + \frac{V(s) + V_{g2}(s)}{4} + \frac{V(s) + V_{g2}(s)}{s} \tag{13.6-26}$$

$$+ \frac{V(s) + V_{g2}(s) - V_{g1}(s)}{8} + \frac{V(s) - V_{g1}(s)}{s} = 0.$$

After a bit of algebra, we obtain

$$V(s) = \frac{s + 8}{s^2 + 4s + 16}V_{g1}(s) - \frac{3s + 8}{s^2 + 4s + 16}V_{g2}(s). \tag{13.6-27}$$

We can now solve for $I(s)$ in terms of the node voltages. This yields

$$I(s) = \frac{V(s) + V_{g2}(s)}{s} = \frac{s + 8}{s(s^2 + 4s + 16)}V_{g1}(s) + \frac{s^2 + s + 8}{s(s^2 + 4s + 16)}V_{g2}(s). \tag{13.6-28}$$

If we write our s-domain solutions (13.6-27) and (13.6-28) in matrix form, we obtain the more compact expression

$$\begin{bmatrix} V(s) \\ I(s) \end{bmatrix} = \begin{bmatrix} \dfrac{s + 8}{s^2 + 4s + 16} & -\dfrac{3s + 8}{s^2 + 4s + 16} \\ \dfrac{s + 8}{s(s^2 + 4s + 16)} & \dfrac{s^2 + s + 8}{s(s^2 + 4s + 16)} \end{bmatrix} \begin{bmatrix} V_{g1}(s) \\ V_{g2}(s) \end{bmatrix}. \tag{13.6-29}$$

Figure 13.37
The s-domain circuit prepared for nodal analysis

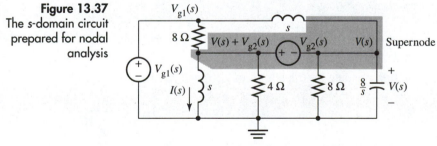

The System Function: Matrix Form   By placing our solution to the last example in matrix form we have illustrated a very important fact: there are quite a number of different system functions associated with a given circuit—one for each independent source/response variable pair. As we had two independent sources in the last circuit and were interested in two response variables, we had four system functions. We can therefore write our solution in (13.6-29) in the form

$$\begin{bmatrix} Y_1(s) \\ Y_2(s) \end{bmatrix} = \begin{bmatrix} H_{11}(s) & H_{12}(s) \\ H_{21}(s) & H_{22}(s) \end{bmatrix} \begin{bmatrix} X_1(s) \\ X_2(s) \end{bmatrix}, \tag{13.6-30}$$

where $X_1(s)$ and $X_2(s)$ are the two source transforms and $Y_1(s)$ and $Y_2(s)$ are our two desired responses. $H_{11}(s)$ is then the system function connecting the input variable $X_1(s)$ to the response variable $Y_1(s)$, $H_{12}(s)$ is the system function connecting the input variable $X_2(s)$ to the response variable $Y_1(s)$, and so on. In general, we can write

$$\overline{Y}(s) = H(s)\overline{X}(s), \tag{13.6-31}$$

where $\overline{Y}(s)$ is an ($m \times 1$) column matrix of response transforms, $\overline{X}(s)$ is an ($n \times 1$) column matrix of source transforms, and $H(s)$ is an ($m \times n$) rectangular matrix of system functions. Note that it does not have to be square.

We make the observation, by the way, that if there is only one source and only one response variable, nodal analysis shows that the response variable $Y(s)$ has the form

$$Y(s) = H(s)X(s) = \frac{b_m s^m + b_{m-1}s^{m-1} + \cdots + b_1 s + b_0}{s^n + a_{n-1}s^{n-1} + \cdots + a_1 s + a_0}X(s). \tag{13.6-32}$$

Cross-multiplying by the denominator, we get

$$(s^n + a_{n-1}s^{n-1} + \cdots + a_1 s + a_0)Y(s) \tag{13.6-33}$$
$$= (b_m s^m + b_{m-1}s^{m-1} + \cdots + b_1 s + b_0)X(s).$$

If we multiply out the terms and invert the Laplace transform, we get

$$p^n y(t) + a_{n-1}p^{n-1}y(t) + \cdots + a_1 py(t) + a_0 y(t) \tag{13.6-34}$$
$$= b_m p^m x(t) + b_{m-1}p^{m-1}x(t) + \cdots + b_1 px(t) + b_0 x(t).$$

This proves our assertion at the beginning of this section that a response variable was linked to a given forcing function by a linear, constant coefficient differential equation because in deriving (13.6-34) we have only used nodal analysis (based on Ohm's law, KVL, and KCL only) and the s-domain characteristics of the resistor, the inductor, and the capacitor—which we independently showed obeyed Ohm's law in terms of s-domain impedances.

Example 13.44   Assume that $v_{g1}(t) = 6\cos(4t)u(t)$ V and $v_{g2}(t) = 2\cos(4t)u(t)$ V in the circuit of Example 13.43 (Figure 13.36). Find $i(t)$.

Solution   We first compute the transforms of $v_{g1}(t)$ and $v_{g2}(t)$:

$$V_{g1}(s) = \pounds\{6\cos(4t)u(t)\} = \frac{6s}{s^2 + 16} \tag{13.6-35}$$

and

$$V_{g2}(s) = \pounds\{2\cos(4t)u(t)\} = \frac{2s}{s^2 + 16}. \tag{13.6-36}$$

Using these transforms in equation (13.6-28), which we derived in Example 13.43 for $I(s)$, we get

$$I(s) = \frac{s + 8}{s(s^2 + 4s + 16)} \times \frac{6s}{s^2 + 16} + \frac{s^2 + s + 8}{s(s^2 + 4s + 16)} \times \frac{2s}{s^2 + 16} \tag{13.6-37}$$
$$= \frac{2(s^2 + 4s + 32)}{(s^2 + 16)(s^2 + 4s + 16)}.$$

Notice that there is a pole-zero cancellation that removes the s factor in the numerator and the denominator. We see that there are four simple poles of $I(s)$ at $s = \pm j4$ and

$s = -2 \pm j2\sqrt{3}$. Those due to $H(s)$ are both in the left half-plane, therefore our response is stable.[17] Choosing to do the PFE with real algebra, we write

$$I(s) = \frac{2(s^2 + 4s + 32)}{(s^2 + 16)(s^2 + 4s + 16)} = \frac{As + B}{s^2 + 16} + \frac{Cs + D}{s^2 + 4s + 16}. \qquad (13.6\text{-}38)$$

We then multiply both sides by the denominator of $I(s)$ and match coefficients. This gives

$$I(s) = -\frac{1}{2}\frac{s - 4}{s^2 + 16} + \frac{1}{2}\frac{s + 4}{s^2 + 4s + 16} \qquad (13.6\text{-}39)$$

$$= -\frac{1}{2}\frac{s}{s^2 + 16} + \frac{1}{2}\frac{4}{s^2 + 16} + \frac{1}{2}\frac{s + 2}{(s + 2)^2 + 12} + \frac{1}{2\sqrt{3}}\frac{2\sqrt{3}}{(s + 2)^2 + 12}.$$

We have "massaged" the result just a bit so that we could recognize standard transforms. Doing so gives

$$i(t) = -\frac{1}{2}[\cos(4t) - \sin(4t)]u(t) + \frac{1}{2}e^{-2t}[\cos(2\sqrt{3}t) + \frac{1}{\sqrt{3}}\sin(2\sqrt{3}t)]u(t). \qquad (13.6\text{-}40)$$

Notice that the first term is the ac forced response and the second term is the natural response. Because $i(t)$ is a stable response, the natural response decays to zero as time increases, leaving only the forced response. We then say that the circuit response is in the ac steady state.

Mesh Analysis in the s Domain

Our next example illustrates the application of mesh analysis in the s domain; in addition, we use it to illustrate the properties of the time response of second-order networks. The example is lengthy, but detailed.

Example 13.45

Using mesh analysis, find the transfer function $H(s)$ relating the response $v(t)$ to $i_g(t)$ for the second-order circuit in Figure 13.38 with R as a parameter. Then find the unit impulse and step responses for **a.** $R = 10\ \Omega$, **b.** $R = 8\ \Omega$, and **c.** $R = 4\ \Omega$.

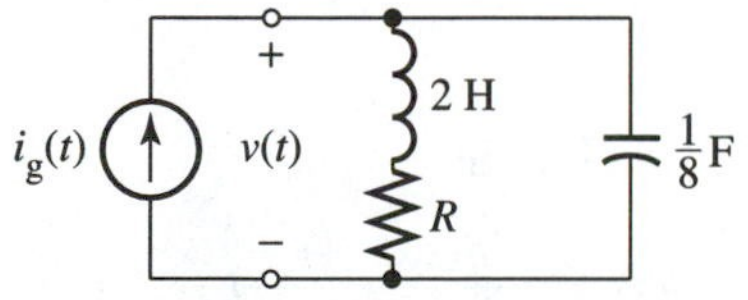

Figure 13.38
An example circuit

Solution

We show the transformed equivalent circuit in Figure 13.39 prepared for mesh analysis. The left-hand is nonessential, so the single mesh equation required is the one for the essential mesh on the right. It is

$$(2s + R)[I(s) - I_g(s)] + \frac{8}{s}I(s) = 0. \qquad (13.6\text{-}41)$$

Solving for $I(s)$, we have

$$I(s) = \frac{s(2s + R)}{2s^2 + Rs + 8}I_g(s) = \frac{s(s + R/2)}{s^2 + \dfrac{R}{2}s + 4}I_g(s). \qquad (13.6\text{-}42)$$

---

[17] Our discussion in connection with equations (13.6-30) and (13.6-31) implies that we have to consider stability for each and every input-response variable pair.

Figure 13.39
The s-domain equivalent prepared for mesh analysis

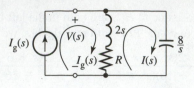

Next, we note that $V(s) = \dfrac{8}{s} I(s)$, so

$$V(s) = \frac{8}{s} \frac{s(s + R/2)}{s^2 + \dfrac{R}{2}s + 4} I_g(s) = \frac{8(s + R/2)}{s^2 + \dfrac{R}{2}s + 4} I_g(s). \tag{13.6-43}$$

a. $R = 10\ \Omega$: If $R = 10\ \Omega$ we have

$$V(s) = \frac{8(s + 5)}{s^2 + 5s + 4} I_g(s) = \frac{8(s + 5)}{(s + 1)(s + 4)} I_g(s). \tag{13.6-44}$$

For the impulse response, $I_g(s) = 1$ and (leaving the PFE for you to verify)

$$H(s) = V(s) = \frac{8(s + 5)}{(s + 1)(s + 4)} = \frac{\dfrac{32}{3}}{s + 1} - \frac{\dfrac{8}{3}}{s + 4}; \tag{13.6-45}$$

thus, the impulse response is

$$h(t) = \frac{8}{3}[4e^{-t} - e^{-4t}]u(t). \tag{13.6-46}$$

To compute the step response we let $V_g(s) = 1/s$, the transform of the unit step function:

$$S(s) = V(s) = H(s) \times \frac{1}{s} = \frac{8(s + 5)}{s(s + 1)(s + 4)} = \frac{10}{s} - \frac{\dfrac{32}{3}}{s + 1} + \frac{\dfrac{2}{3}}{s + 4}, \tag{13.6-47}$$

so

$$s(t) = \frac{2}{3}[15 - 16e^{-t} + e^{-4t}]u(t). \tag{13.6-48}$$

b. $R = 8\ \Omega$: If $R = 8\ \Omega$ we have

$$V(s) = \frac{8(s + 4)}{s^2 + 4s + 4} I_g(s) = \frac{8(s + 4)}{(s + 2)^2} I_g(s). \tag{13.6-49}$$

For the impulse response, $I_g(s) = 1$ and (leaving the PFE for you to verify)

$$H(s) = V(s) = \frac{8(s + 4)}{(s + 2)^2} = \frac{8}{s + 2} + \frac{16}{(s + 2)^2}; \tag{13.6-50}$$

thus, the impulse response is

$$h(t) = 8(2t + 1)e^{-2t}u(t). \tag{13.6-51}$$

To compute the step response we let $V_g(s) = 1/s$ and compute

$$S(s) = V(s) = H(s) \times \frac{1}{s} = \frac{8(s + 4)}{s(s + 2)^2} = \frac{8}{s} - \frac{8}{s + 2} - \frac{8}{(s + 2)^2}, \tag{13.6-52}$$

so

$$s(t) = 8[1 - (t + 1)e^{-2t}]u(t). \tag{13.6-53}$$

c. $R = 4\ \Omega$: If $R = 4\ \Omega$ we have

$$V(s) = \frac{8(s + 2)}{s^2 + 2s + 4}I_g(s) = \frac{8(s + 2)}{(s + 1)^2 + 3}I_g(s). \qquad (13.6\text{-}54)$$

For the impulse response, $I_g(s) = 1$ and (again leaving the algebra for you to verify)

$$H(s) = V(s) = \frac{8(s + 2)}{(s + 1)^2 + 3} = \frac{8(s + 1)}{(s + 1)^2 + 3} + \frac{8}{\sqrt{3}}\frac{\sqrt{3}}{(s + 1)^2 + 3}; \qquad (13.6\text{-}55)$$

thus, the impulse response is

$$h(t) = 8e^{-t}[\cos(\sqrt{3}t) + \frac{1}{\sqrt{3}}\sin(\sqrt{3}t)]u(t). \qquad (13.6\text{-}56)$$

To compute the step response we let $V_g(s) = 1/s$ and compute

$$S(s) = H(s) \times \frac{1}{s} = \frac{8(s + 2)}{s(s^2 + 2s + 4)} = \frac{4}{s} - \frac{4s}{s^2 + 2s + 4} \qquad (13.6\text{-}57)$$

$$= \frac{4}{s} - 4\left[\frac{s + 1}{(s + 1)^2 + 3} - \frac{1}{\sqrt{3}}\frac{\sqrt{3}}{(s + 1)^2 + 3}\right],$$

so

$$s(t) = 4\left[1 - e^{-t}\left\{\cos(\sqrt{3}t) - \frac{1}{\sqrt{3}}\sin(\sqrt{3}t)\right\}\right]u(t). \qquad (13.6\text{-}58)$$

We show a sketch of the impulse responses for the three different resistance values in Figure 13.40. For $R = 10\ \Omega$, the response is sluggish and decays to zero quite slowly; for $R = 3\ \Omega$, it decays faster. For neither of these two cases, though, is there "undershoot," that is, the response never goes below its final value of zero. For $R = 4\ \Omega$, however, undershoot does occur: the impulse response oscillates positively and negatively while the peak amplitude decays toward zero. By analogy with a mechanical system, such as an automobile suspension system, the three cases are called overdamped ($R = 10\ \Omega$, real and unequal poles of $H(s)$), critically damped ($R = 8\ \Omega$, real and equal poles of $H(s)$), and underdamped ($R = 4\ \Omega$, complex conjugate poles of $H(s)$). The idea is that resistance in an electrical circuit acts much like a "damper" (shock absorber) in a mechanical system. This is discussed more thoroughly in Chapter 9, Section 9.2, if you want to explore the mechanical ideas a bit more or refresh your memory about them.

Figure 13.40
Impulse response—
under, over, and
critically damped

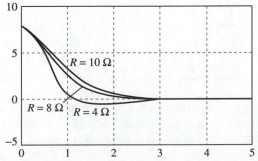
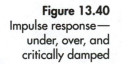

In order to help you in coordinating the time responses in Figure 13.40 with the pole locations, we offer the PZD plots of them all in Figure 13.41. We remind you that the small parentheses by the pole symbol in Figure 13.41(b) refer to the order of that pole.

Figure 13.41
Pole-zero diagrams for
the three cases of
damping

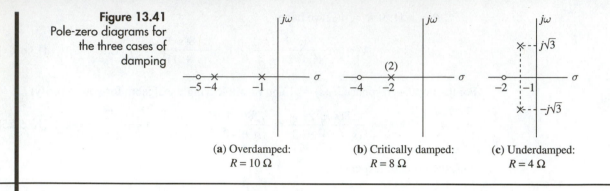

(a) Overdamped:
$R = 10\ \Omega$

(b) Critically damped:
$R = 8\ \Omega$

(c) Underdamped:
$R = 4\ \Omega$

***Thévenin/Norton
Equivalents in
the s Domain***

We have already used series and parallel equivalent circuits, substituting complex impedances in our current work for pure resistances in our earlier dc analysis. Now we will
have a look at the Thévenin/Norton equivalents in the *s* domain.

Example 13.46

Replace all of the elements to the left of terminals a and b in Figure 13.42 by a Thévenin
equivalent and use it to find the system function $H(s) = V(s)/V_g(s)$.

Figure 13.42
An example circuit

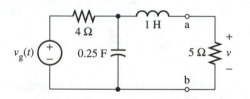

Solution

We remove the 5-Ω resistor and draw the *s*-domain equivalent for the subcircuit remaining. This results in Figure 13.43. The terminals a and b are open circuited, so the current
in the inductor is zero; hence, there is no voltage drop across it and the remaining elements are series connected. Thus, because $V(s)$ in this case is the open circuit voltage, we
have

$$V_{oc}(s) = \frac{\dfrac{4}{s}}{4 + \dfrac{4}{s}} V_g(s) = \frac{1}{s+1} V_g(s). \tag{13.6-59}$$

Figure 13.43
The s-domain subcircuit

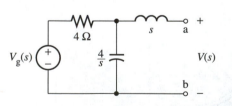

To find the Thévenin equivalent impedance, we deactivate the independent voltage
source, thus getting the deactivated subcircuit shown in Figure 13.44. Now we need only
use parallel-series impedance equivalents to get

$$Z_{eq}(s) = s + \frac{4 \times \dfrac{4}{s}}{4 + \dfrac{4}{s}} = s + \frac{4}{s+1} = \frac{s^2 + s + 4}{s+1}. \tag{13.6-60}$$

Figure 13.44
The deactivated
subcircuit

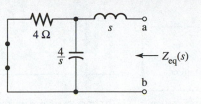

The resulting Thévenin equivalent for the subcircuit is shown in Figure 13.45 with the 5-Ω resistor replaced. It is now a simple matter to compute the desired system function using voltage division. We first compute $V(s)$,

$$V(s) = \frac{5}{5 + Z_{eq}(s)} V_{oc}(s) = \frac{5}{5 + \dfrac{s^2 + s + 4}{s + 1}} \times \frac{V_g(s)}{s + 1} \qquad (13.6\text{-}61)$$

$$= \frac{5}{s^2 + 6s + 9} V_g(s),$$

and then recognize that the s-domain function multiplying $V_g(s)$ is the desired system function. (Caution: Notice that it is not the factor multiplying $V_{oc}(s)$!) Thus,

$$H(s) = \frac{V(s)}{V_g(s)} = \frac{5}{s^2 + 6s + 9}. \qquad (13.6\text{-}62)$$

Figure 13.45
Thévenin equivalent

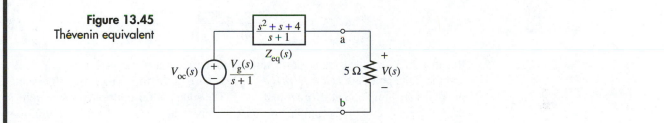

Poles and Zeros of Z(s) and Y(s): Physical Meaning

We will now discuss the physical meaning of poles and zeros of impedance and admittance functions. With this in mind, look at Figure 13.46. There we have drawn a box representing a two-terminal subcircuit having no independent sources and to which we have applied an independent test source. We do not commit ourselves at the outset as to its type, i-source or v-source. We will allow the subcircuit to possibly have dependent sources.

Figure 13.46
An *RLC* subcircuit

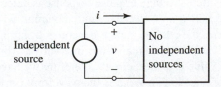

Now suppose we elect to apply a current source. Then we can draw the s-domain equivalent as in Figure 13.47. There we have also drawn an approximation to the delta function that we will use for the following explanation. The current source, in practical terms, delivers a very strong pulse of current to the subcircuit. It lasts for a very brief period of time, then returns to zero. A zero-valued current source is equivalent to an open circuit, therefore we see that the positive time equivalent of Figure 13.48 is valid. Our question is this: What is the resulting terminal voltage for $t > 0$?

Figure 13.47
Current source drive

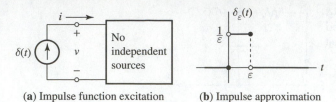

(a) Impulse function excitation   (b) Impulse approximation

Figure 13.48   Positive time equivalent

$i(t) = 0$

```
+
v(t)        No
            independent
−           sources
```

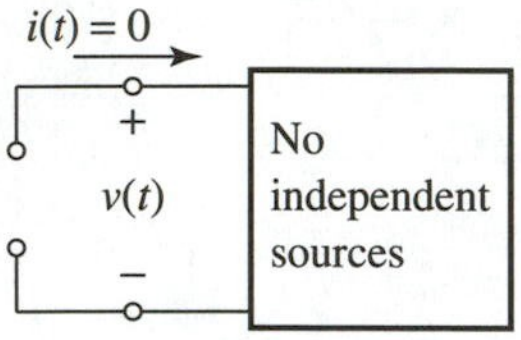

To answer this, let's turn to the s domain. We know that the network can be described by its driving point impedance.[18] Thus, we can write

$$V(s) = Z(s)\pounds\{\delta(t)\} = Z(s). \qquad (13.6\text{-}63)$$

In the time domain, we have

$$v(t) = \pounds^{-1}\{Z(s)\} = z(t) = h(t). \qquad (13.6\text{-}64)$$

We have used a couple of different symbols here. The small $h(t)$ means that we are looking at the impulse response, whereas the small $z(t)$ is the lowercase equivalent of its transform $Z(s)$, with the uppercase of the same letter. We will use $z(t)$ here for we are considering only the impulse response of a driving point impedance.

And what is $z(t)$? We have seen that, in general, $Z(s)$ can be written in terms of its finite zeros and finite poles as

$$Z(s) = K\frac{(s + z_1)(s + z_2)\cdots(s + z_m)}{(s + p_1)(s + p_2)\cdots(s + p_n)}. \qquad (13.6\text{-}65)$$

Recall that K is called the scale factor, the finite zeros are at $s = -z_i$ ($i = 1, 2, \ldots m$), and the finite poles are at $s = -p_i$ ($i = 1, 2, \ldots, n$). In the following, we will assume that the poles are simple; that is, $p_i \neq p_j$ if $i \neq j$. The situation for multiple poles is quite similar, but we will choose the simple pole case merely to be specific. We use the PFE to write

$$Z(s) = \frac{k_1}{s + p_1} + \frac{k_2}{s + p_2} + \cdots + \frac{k_n}{s + p_n}; \qquad (13.6\text{-}66)$$

therefore, the open circuit voltage function will have the form

$$z(t) = [k_1 e^{-p_1 t} + k_2 e^{-p_2 t} + \cdots + k_n e^{-p_n t}]u(t). \qquad (13.6\text{-}67)$$

The constants in the exponents of the exponentials are the poles of $Z(s)$. What influence do the zeros have? They merely affect the values of the constants k_i in the PFE. Thus, the form of the open circuit voltage impulse response is determined by the poles of $Z(s)$.

Let's change things now and drive our subcircuit with a voltage source, as shown in Figure 13.49. Here, the *voltage* source generates a sharp, high-amplitude pulse of a very short duration. Thus, after a small time interval the voltage returns to zero. A voltage source with zero value is a short circuit, therefore we have the positive time equivalent of Figure 13.50. Our question is: What is the waveform for the short circuit current for positive values of t? (Notice that the short circuit current is *into* the top terminal because we chose it that way in defining $Z(s)$ originally.)

We are now considering the terminal voltage to be the forcing function and the current the response. As we discussed at the beginning of this section, we will "read" the differential equation connecting the voltage to the current backward. This is equivalent to

---

[18] The reason for the term "driving point" is now clear: we are "driving" the subcircuit with an externally applied independent source.

Figure 13.49
Voltage source drive

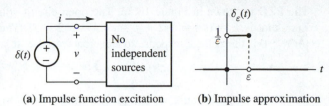

(a) Impulse function excitation   (b) Impulse approximation

Figure 13.50   Positive time equivalent

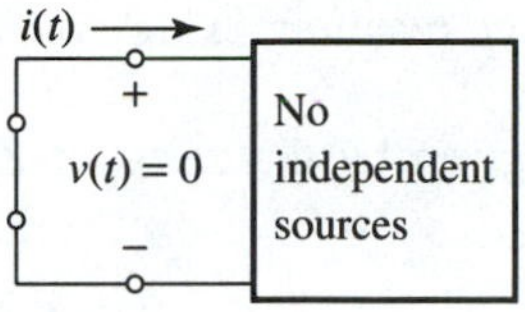

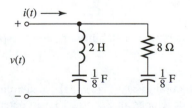

using the admittance, rather than the impedance. Thus, in the s domain we can write

$$I(s) = Y(s)V(s) = Y(s) = \frac{1}{Z(s)} = \frac{1}{K}\frac{(s + p_1)(s + p_2)\cdots(s + p_n)}{(s + z_1)(s + z_2)\cdots(s + z_m)}. \quad (13.6\text{-}68)$$

Follow through this equation carefully to avoid confusion. We have written the admittance equation for current in terms of voltage. Next, we have used $V(s) = 1$ because $v(t)$ has been adjusted to be the unit impulse function, whose transform is unity. Finally, we have used equation (13.6-65) to express the admittance in terms of *the zeros and poles of $Z(s)$*. The italicized phrase is very important. It is tempting to confuse the zeros and poles of $Z(s)$ with those of $Y(s)$. *We have decided to use only the zeros and poles of $Z(s)$, notationally. We must remember that the zeros of $Y(s)$ are the poles of $Z(s)$ and the poles of $Y(s)$ are the zeros of $Z(s)$.*

Next we do the PFE (assuming once again for convenience that all of the zeros of $Z(s)$ are simple). This gives

$$I(s) = Y(s) = \frac{q_1}{s + z_1} + \frac{q_2}{s + z_2} + \cdots + \frac{q_m}{s + z_m}. \quad (13.6\text{-}69)$$

The positive time short circuit current now is $i(t) = \mathcal{L}^{-1}\{Y(s)\} = y(t)$:

$$y(t) = [q_1e^{-z_1t} + q_2e^{-z_2t} + \cdots + q_me^{-z_mt}]u(t). \quad (13.6\text{-}70)$$

The positive time short circuit current due to an impulsive voltage source drive is, therefore, determined in form by the zeros of $Z(s)$. The poles merely affect the constants q_i. The following example should help to put these observations into perspective.

Example 13.47   Find the form of the open circuit voltage and short circuit current (for positive time) for the subcircuit shown in Figure 13.51.

Figure 13.51
An example subcircuit

$i(t) \longrightarrow$

$+\ \circ$

$v(t)$

2 H   8 Ω

$\frac{1}{8}$ F $\frac{1}{8}$ F

$-\ \circ$

Solution   The s-domain equivalent for our subcircuit is shown in Figure 13.52. The driving point impedance is

$$Z(s) = \frac{\left(2s + \dfrac{8}{s}\right)\left(8 + \dfrac{8}{s}\right)}{\left(2s + \dfrac{8}{s}\right) + \left(8 + \dfrac{8}{s}\right)} = \frac{8(s^3 + s^2 + 4s + 4)}{s(s^2 + 4s + 8)} \quad (13.6\text{-}71)$$

$$= \frac{8(s + 1)(s^2 + 4)}{s(s + 2 + j2)(s + 2 - j2)}.$$

The PZD is shown in Figure 13.53. The open circuit voltage is specified by the poles of $Z(s)$, that is, $s = 0$, $s = -2 + j2$, and $s = -2 - j2$. Therefore, the open circuit voltage for $t > 0$ is the voltage response to a current impulse. It has the form

$$z(t) = k_1 + k_2 e^{(-2+j2)t} + k_3 e^{(-2-j2)t}. \tag{13.6-72}$$

By using Euler's formula, we can place it in the real form

$$z(t) = k_1 + k_2 e^{-2t} \cos(2t) + k_3 e^{-2t} \sin(2t). \tag{13.6-73}$$

The constants in (13.6-73) are, in general, different from those in (13.6-72). To determine them, we would have to perform a PFE on $Z(s)$, then invert the resulting expression one term at a time.

The positive time form for the short circuit current is given by the zeros of $Z(s)$. Thus, we have

$$y(t) = k_1 e^{-t} + k_2 e^{j2t} + k_3 e^{-j2t} \tag{13.6-74a}$$

or, in real form,

$$y(t) = k_1 e^{-t} + k_2 \cos(2t) + k_3 \sin(2t). \tag{13.6-74b}$$

Again we point out that our constants are "general"; that is, those in (13.6-72), (13.6-73), (13.6-74a), and (13.6-74b) are all nonspecific, and if we were to perform the PFE on either $Z(s)$ or $Y(s)$ we could find their values. The point of this example, though, is that the *form* of $z(t)$ and $y(t)$ can be determined *without* doing the PFE.

Figure 13.52
The s-domain equivalent

Figure 13.53
The PZD

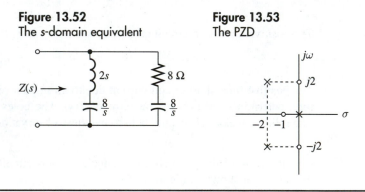

Example: Circuit Stability

We now turn to a consideration of stability of circuit response. The next example is such a stability study.

Example 13.48

Derive the system function $H(s) = V(s)/I_g(s)$ for the circuit shown in Figure 13.54 and use the stability tests developed in Section 13.5 to determine the stability of the voltage response $v(t)$ for $g_m = 0.5$ and $g_m = 1.25$.

Figure 13.54
An example circuit

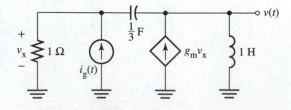

Solution   The s-domain equivalent for our subcircuit is shown in Figure 13.55. We have noted that whereas nodal analysis would require two KCL equations, mesh analysis—because of the presence of the two current sources—requires only one KVL equation. We have therefore decided to do mesh analysis; this entails defining mesh currents as we have done in the figure. Notice that we have also taped the dependent source, thus temporarily turning it into an independent one. Furthermore, we have assigned the single unknown mesh current to the rightmost mesh because it is the most directly related to the response voltage $V(s)$ that we want. The single supermesh equation is

$$1[I(s) - I_c(s) - I_g(s)] + \frac{3}{s}[I(s) - I_c(s)] + sI(s) = 0, \tag{13.6-75}$$

which simplifies to

$$(s^2 + s + 3)I(s) - (s + 3)I_c(s) = sI_g(s). \tag{13.6-76}$$

Now we *untape* the dependent source and express its value in terms of the unknown, $I(s)$. This gives the implicit equation

$$I_c(s) = g_m V_x(s) = g_m(-1)[I(s) - I_c(s) - I_g(s)]. \tag{13.6-77}$$

The solution of (13.6-77) for $I_c(s)$—assuming that $g_m \neq 1$ (certainly true for our given values of $g_m = 0.5$ and 1.25—is

$$I_c(s) = \frac{-g_m}{1 - g_m}I(s) + \frac{g_m}{1 - g_m}I_g(s). \tag{13.6-78}$$

Using this in (13.6-76) and simplifying gives

$$I(s) = \frac{s + 3g_m}{(1 - g_m)s^2 + s + 3}I_g(s). \tag{13.6-79}$$

Our desired response $V(s)$ is the voltage across the inductor, so we have

$$V(s) = sI(s) = \frac{s(s + 3g_m)}{(1 - g_m)s^2 + s + 3}I_g(s). \tag{13.6-80}$$

The system function relating $V(s)$ to the independent source transform $I_g(s)$, then, is

$$H(s) = \frac{s(s + 3g_m)}{(1 - g_m)s^2 + s + 3}. \tag{13.6-81}$$

Figure 13.55
The s-domain equivalent (prepared for mesh analysis)

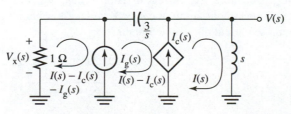

So much for the circuit analysis phase of our work. Now that we have $H(s)$, we need only check its poles to determine stability—or lack thereof. Using the quadratic formula, we see that these poles are at

$$p_{1,2} = \frac{-1 \pm \sqrt{1 - 12(1 - g_m)}}{2(1 - g_m)} = \frac{1}{2(1 - g_m)}[-1 \pm \sqrt{12g_m - 11}]. \tag{13.6-82}$$

In the first case we are considering, $g_m = 0.5$. This gives poles at

$$p_{1,2} = -1 \pm j\sqrt{5}. \tag{13.6-83}$$

Because the real part of each of the poles is negative, both are in the left half-plane; hence, our circuit response is stable. In the second case, $g_m = 1.25$ and the poles are at

$$p_{1,2} = -2, +6. \tag{13.6-84}$$

Both poles are real, but one has a positive real part and hence is in the right half-plane. This makes our circuit response unstable. Notice, by the way, that the impulse response has the form

$$h(t) = k_1 e^{-t} \cos(\sqrt{5}t) + k_2 e^{-t} \sin(\sqrt{5}t) \tag{13.6-85}$$

when $g_m = 0.5$ and

$$h(t) = k_1 e^{6t} + k_2 e^{-2t} \tag{13.6-86}$$

when $g_m = 1.25$. We have used the real form for $h(t)$ in (13.6-85). In the case we are considering, the response is a voltage and the input a current, but the $H(s)$ we derived is not a "driving point" impedance because the voltage is not that across the two terminals of the subcircuit into which the independent current source is "looking."

Example: Ac Forced Response

In the next example we show how the Laplace transform is related to the ac forced response.

Example 13.49

Derive the system function $H(s) = I(s)/V_g(s)$ for the circuit shown in Figure 13.56. Then find and sketch the magnitude and phase of the sinusoidal forced response component of the current $i(t)$ assuming that $v_g(t) = V_m \cos(\omega t) u(t)$.[19]

Figure 13.56
An example circuit

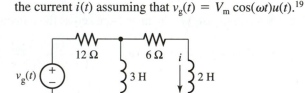

Solution

The s-domain equivalent for our subcircuit is shown in Figure 13.57. We choose here to use impedances and current division. Thus, we compute the total impedance "seen" by the source, divide it into the source voltage to obtain the current in the 12-Ω resistor, and then use current division to obtain $I(s)$:

$$I(s) = \frac{3s}{6 + 5s} \times \frac{V_g(s)}{12 + \dfrac{3s(6 + 2s)}{3s + 6 + 2s}} = \frac{s}{2(s + 1)(s + 12)} V_g(s). \tag{13.6-87}$$

We see that the system function relating $I(s)$ to $V_g(s)$ is

$$H(s) = Y(s) = \frac{s}{2(s + 1)(s + 12)}. \tag{13.6-88}$$

---

[19] Equivalent, if $V_m = 1$, to the gain and phase Bode plots (see Chapters 11 and 12).

Both its poles are in the left half-plane (at $s = -1$ and $s = -12$), so the response $i(t)$ is stable; hence, the ac forced response becomes the ac steady-state response for large values of time.

Figure 13.57
The s-domain equivalent

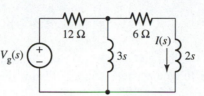

According to the theory we developed in Section 13.5, we need to evaluate $H(s)$ at $s = j\omega$ and take its magnitude and angle. Thus, we have

$$H(j\omega) = \frac{j\omega}{2(j\omega + 1)(j\omega + 12)} \tag{13.6-89}$$

$$= \frac{\omega \angle 90°}{2\sqrt{(\omega^2 + 1)(\omega^2 + 144)}\angle(\tan^{-1}(\omega) + \tan^{-1}(\omega/12))}.$$

We can now simply pick off the magnitude,

$$A(\omega) = |H(j\omega)| = \frac{\omega}{2\sqrt{(\omega^2 + 1)(\omega^2 + 144)}}, \tag{13.6-90}$$

and phase,

$$\phi(\omega) = 90° - \tan^{-1}(\omega) - \tan^{-1}(\omega/12). \tag{13.6-91}$$

The ac forced response theory we developed in the last section applies, and we can write

$$i_f(t) = A(\omega)V_m \cos(\omega t + \phi(\omega)) \tag{13.6-92}$$

$$= \frac{\omega}{2\sqrt{(\omega^2 + 1)(\omega^2 + 144)}}V_m \cos[\omega t + 90° - \tan^{-1}(\omega) - \tan^{-1}(\omega/12)].$$

The amplitude and phase are plotted versus ω in Figures 13.58 and 13.59.[20]

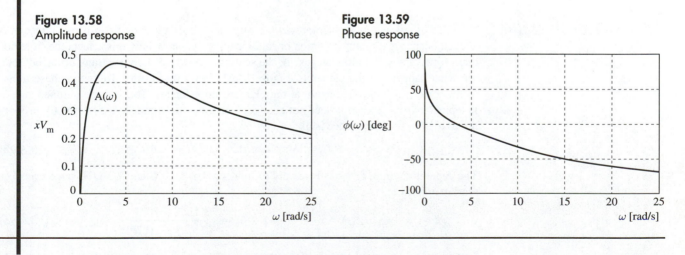

Figure 13.58
Amplitude response

Figure 13.59
Phase response

---

[20] The (linearized) Bode plot, alluded to in a previous footnote, is a *rapid* way of obtaining these plots.

Impedance Scaling   Our last topic in this section is that of scaling. We have already discussed scaling within the strict confines of Laplace transform theory, but now we will look at how it is applied in circuit analysis. First, let's look at *impedance scaling*. Ohm's law in the *s* domain takes the form

$$V(s) = Z(s)I(s) = [aZ(s)]\left[\frac{I(s)}{a}\right]. \tag{13.6-93}$$

Here, we have multiplied and divided the right-hand side of this equation by the scalar a, which we assume to be a positive number. What is the physical meaning? Just this. Suppose we make $a = 0.001 = 10^{-3}$, for example. Then if $Z(s)$ happens to be 1 kΩ at a particular value of s, $aZ(s)$ will be 1 Ω—a nicer value. But, at the same time, when $I(s) = 1$ mA, $I(s)/a = 1$ A—again a nicer value. Furthermore, the voltage $V(s)$ is unaffected. Thus, (13.6-93) has the following meaning: if we agree to take $1/a$ Ω as our unit of impedance and a A as our unit of current, the unit of voltage remains as 1 V. Of course, we can select any other value of a we choose to simplify our work. Figure 13.60 shows the effect on the passive element values. To see how this works, just note that $aZ(s) = aR$ for the resistor, $aZ(s) = (aL)s$ for the inductor, and $aZ(s) = \dfrac{1}{(C/a)s}$ for the capacitor. To remember this, just think of a as a number bigger than one. Multiplying impedances by a causes resistors and inductors to increase in size, whereas it causes capacitors to decrease. The important issue in this: resistors and inductors have impedances that are directly proportional to the element values, whereas the impedance of a capacitor is inversely proportional to its element value.

Figure 13.60
Impedance scaling

Figure 13.61
Frequency scaling

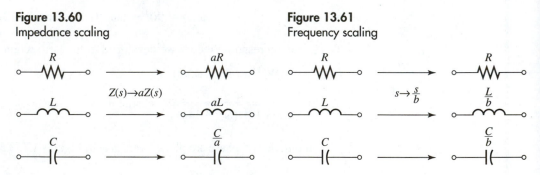

Frequency Scaling   Now, let's think about *frequency scaling*. Suppose we divide s by the positive constant b at each occurrence in any equation or relationship. We know how this affects the time domain form of any function because of the scaling property of Laplace transforms, but let's see what happens to the elements. Figure 13.61 shows the results. To justify these transformations, just look once again at the element impedances. The resistor's impedance is independent of s, and so it does not change. The inductor's impedance is directly proportional to s, so the inductance is divided by b:

$$Z(s) = Ls \rightarrow Z(s/b) = L(s/b) = (L/b)s. \tag{13.6-94}$$

The capacitor's impedance is inversely proportional to both C and s, so the same transformation applies to it:

$$Z(s) = \frac{1}{Cs} \longrightarrow Z(s/b) = \frac{1}{C(s/b)} = \frac{1}{(C/b)s}. \tag{13.6-95}$$

The idea is this: apply impedance and/or frequency scaling to make the element values "nice," then do the circuit analysis for the desired transformed variable. Finally, apply the frequency scaling property to find the time domain equivalent. We illustrate this process in our final example.

An Example: Scaling

Example 13.50 Use impedance and frequency scaling on the circuit shown in Figure 13.62 to make the elements have "nice" values. Then find $i(t)$ using the frequency scaling property of the Laplace transform.

Figure 13.62
An example circuit

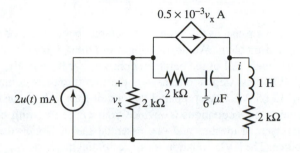

Solution Let us say right off that the process of finding proper scaling factors is something of an art, rather than a science. Therefore, our solution will not be unique, and sometimes one simply cannot come up with scaling parameters that will convert all element values to "nice" ones. In this event, though, it usually means that the design has a large "element spread," which is actually a poorly designed circuit. But that is a topic for a more advanced course.

Our example has element values that are quite amenable to being scaled. Let's first choose $a = 10^{-3}$ and do impedance scaling. The $2u(t)$-mA current source value changes to $2u(t)$ A. (Note that this is equivalent to choosing the mA as the unit, but that is a different point of view. We are simply saying that the new value is $2u(t)$ A.) All of the resistors now have values of $2\ \Omega$, but the voltages are unaffected. Finally, we must consider the dependent source. It is of the VCCS variety, so its gain constant is a conductance—with units of siemens = 1/ohms. Thus, we must *divide* its value by our impedance scaling factor of $a = 10^{-3}$. We also divide the capacitance value by a. The resulting impedance scaled circuit is shown in Figure 13.63.

Figure 13.63
After impedance scaling

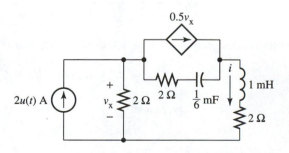

Next, let's recall that if we divide s by the constant b, the resistors are unaffected and the inductor and capacitor values are also divided by b. What is a good value for b? It should be 10^{-3}! (Because the capacitor and inductor both have "milli values.") This transforms the capacitance to 1/6 F and the inductor to 1 H (its original value, by the way). We could have chosen a value of b to make the capacitance value an integer, but then the inductor would have had a fractional value. The circuit after both impedance and frequency scaling is shown in Figure 13.64. The dependent source parameter remains unchanged because it is a reciprocal resistance.

Figure 13.64
After impedance and
frequency scaling

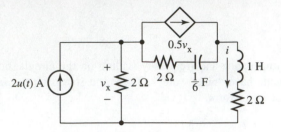

Figure 13.64
After impedance and
frequency scaling

Let's pause at this point for a simple, but crucial, observation. We are *not* solving the original problem. We are solving a different, but *related* problem. After we solve it, we must "undo" all of the parameter changes we have made.

Our next step is to transform the circuit to the s domain, with the result shown in Figure 13.65. We have prepared the circuit for mesh analysis because nodal analysis would require more equations (convince yourself of the truth of this statement). There are two nonessential meshes and one essential mesh. Notice that we have taped the dependent source. The KVL equation for the essential mesh is

$$2\left[I(s) - \frac{2}{s}\right] + \left(2 + \frac{6}{s}\right)[I(s) - I_c(s)] + (s + 2)I(s) = 0. \qquad (13.6\text{-}96)$$

Simplifying, we have

$$\left(6 + \frac{6}{s} + s\right)I(s) - \left(2 + \frac{6}{s}\right)I_c(s) = \frac{4}{s}. \qquad (13.6\text{-}97)$$

Next, we untape the dependent source and express its value in terms of the unknown mesh current $I(s)$:

$$I_c(s) = 0.5 \times (2)\left[\frac{2}{s} - I(s)\right] = \frac{2}{s} - I(s). \qquad (13.6\text{-}98)$$

Inserting this in (13.6-97) results in

$$I(s) = \frac{8s + 12}{s(s + 2)(s + 6)} = \frac{1}{s} + \frac{\frac{1}{2}}{s + 2} - \frac{\frac{3}{2}}{s + 6}. \qquad (13.6\text{-}99)$$

(As usual in this section, we have left the PFE details up to you.) The solution is

$$i(t) = \frac{1}{2}[2 + e^{-2t} - 3e^{-6t}]u(t) \text{ A}. \qquad (13.6\text{-}100)$$

Figure 13.65
The s-domain
equivalent prepared for
mesh analysis

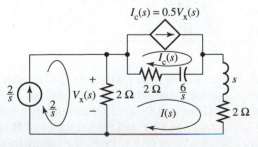

Remember our earlier warning that we were not solving the original problem, but a related problem? Here's where the comment becomes important. Equation (13.6-99) is not actually $I(s)$; rather, it is $(1/a)I(s/b) = 1000I(1000s)$ because of impedance and frequency scaling. We see that we must multiply our $i(t)$ in (13.6-100) by $a = 10^{-3}$ to obtain the actual current level, thereby reversing our impedance scaling. However, this only means that we must change the A on the right-hand side to mA. Easy enough, but what about the frequency scaling? Here's what we do. The frequency scaling property of the Laplace transform is given as entry 4 of Table 13.6 of Section 13.3. Using b, rather than a as in that table, we have

$$i(bt) \longleftrightarrow \frac{1}{b}I(s/b). \qquad (13.6\text{-}101)$$

But our b value was 10^{-3}, so our Laplace transform was actually the right-hand side of (13.6-101), and we see that equation (13.6-100) is actually $i(bt) = i(10^{-3}t)$. Hence, we must substitute $1000t$ for each occurrence of t in (13.6-100). Thus, our final answer is

$$i(t) = \frac{1}{2}[2 + e^{-2000t} - 3e^{-6000t}]u(t) \text{ mA.} \qquad (13.6\text{-}102)$$

If you have trouble with this more-or-less intuitive approach, let's dot all the mathematical i's and cross all the mathematical t's. In doing the impedance and frequency scaling, what we have arrived at in equation (13.6-99) is $(1/ab)I(s/b)$, which corresponds to the time function $(1/a)i(bt) = i_s(t)$, where the subscript means "scaled." Solving this equation, we get $i(t) = ai_s(t/b) = 10^{-3}i_s(1000t)$ A.

Section 13.6 Quiz

Questions Q13.6-1 through Q13.6-7 refer to the circuit shown in Figure Q13.6-1.

Figure Q13.6-1

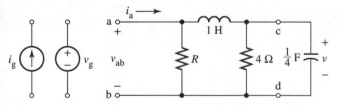

Q13.6-1. Assuming that the current source is connected to terminals a and b, use mesh analysis to find the value of $R \geq 0$ required for
   a. overdamped,
   b. underdamped, and
   c. critically damped responses for $v_{ab}(t)$.

Q13.6-2. Assuming that $R = 1\ \Omega$, find $Z_{eq}(s)$ "looking into" terminals a and b using series and parallel equivalent impedances.

Q13.6-3. What is the form of the open circuit voltage response $v_{ab}(t)$ for $t > 0$ and $R = 1\ \Omega$?

Q13.6-4. What is the form of the short circuit current response $i_b(t)$ for $t > 0$ and $R = 1\ \Omega$?

Q13.6-5. Assuming that the current source is connected to ter-

minals a and b, find the step and impulse responses for $v(t)$ using nodal analysis for
   a. $R = 4\ \Omega$,
   b. $R = 3\ \Omega$, and
   c. $R = 1\ \Omega$.

Q13.6-6. Determine the stability characteristics of the response current $i(t)$ with the voltage source connected to terminals a and b. Draw the PZD of the system function associated with the response $v(t)$ and forcing function $v_g(t)$.

Q13.6-7. Find the Norton equivalent circuit "looking into" terminals c and d (away from the capacitor) with the current source connected to terminals a and b and use it to deduce the system function $H(s) = V(s)/I_g(s)$. Assume that $R = 3\ \Omega$.

Q13.6-8. Perform impedance and frequency scaling on the circuit in Figure Q13.6-2 to transform the element values to "nice" ones. Then use your scaled circuit and the frequency scaling property of the Laplace transform to find the response voltage $v(t)$ if $v_g(t) = 6e^{-40,000t}\cos(80,000t)u(t)$ V.

Figure Q13.6-2

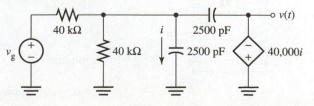

13.7 | Switched Circuits with Dc Sources

Time-Varying
Differential Equations

We will now investigate a special, but highly important, type of circuit—one that requires a bit more machinery than we developed in the last chapter. Mathematically, we will treat circuits whose variables $y(t)$ satisfy what is called a linear time-varying differential equation. Another term is "linear variable-coefficient differential equation." Here is a linear variable-coefficient differential equation of first order:

$$\frac{dy(t)}{dt} + a(t)y(t) = x(t). \tag{13.7-1}$$

Such equations are, in general, quite difficult to solve. The Laplace transform does not work as usual because

$$£\{a(t)y(t)\} \neq a(t)£\{y(t)\}; \tag{13.7-2}$$

for this reason, we cannot simply apply the Laplace transform to both sides and use linearity.

Switched Circuits

What we *will* do is restrict our analysis to circuits in which the coefficient variation is of a special type: it will always have one value for all $t \leq 0$ and another for all $t > 0$. But even this restriction does not quite suffice to allow us to solve (13.7-1) in its most general form. We will, however, take another approach. We will recognize that circuits must obey certain additional physical constraints and *we will assume that all circuits with which we deal have only dc values for their independent sources and switches that operate only once at $t = 0$.* With these assumptions, we can develop a procedure that always works. We will investigate the example circuit in Figure 13.66.

Figure 13.66
An example circuit

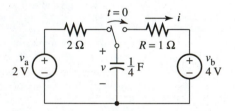

The switch is assumed to have been in its leftmost position since $t = -\infty$, flips in the direction of the arrow (instantaneously) at $t = 0$, and stays in the resulting position until $t = +\infty$. Thus, we must in actuality solve two circuits: one that we will refer to as the $t \leq 0$ circuit and another that we will call the $t > 0$ circuit. Our problem will be to find the voltage $v(t)$ and the current $i(t)$ for all time.

Let's have a look at the $t \leq 0$ circuit first. Because the switch is merely an open circuit with respect to the right half of the circuit, we see that $i(t)$ is identically zero for $t \leq 0$ and we will simply not draw that part of the circuit. The left half is shown in Figure 13.67. We can write a differential equation for $v(t)$ using the upper right-hand node with a reference at the bottom:

Figure 13.67   The $t \leq 0$ circuit

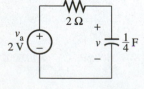

$$\frac{1}{4}pv(t) + \frac{v(t) - v_a(t)}{2} = 0, \tag{13.7-3}$$

which simplifies to

$$pv(t) + 2v(t) = 2v_a(t). \tag{13.7-4}$$

Our standard Laplace solution procedure would be to take the Laplace transform of both sides and solve for $V(s)$. Let's see what happens if we try to do this. We get

$$sV(s) + 2V(s) = 2\pounds\{v_a(t)\}. \tag{13.7-5}$$

This is fine so far, but now let's see about computing $V_a(s)$. We write

$$V_a(s) = \pounds\{v_a(t)\} = \int_{-\infty}^{\infty} 2e^{-st}\, dt = -\frac{2}{s}\, e^{-st}]_{-\infty}^{\infty} = -\frac{2}{s}\, e^{-\sigma t}e^{-j\omega t}\,]_{-\infty}^{\infty}. \tag{13.7-6}$$

The problem is the computation of the upper and lower limits. If we assume $\sigma > 0$, we see that $e^{-\sigma t}$, becomes infinite as $t \to -\infty$, and if we assume $\sigma < 0$, $e^{-\sigma t} \to \infty$ as $t \to +\infty$. If we consider the case in which $\sigma = 0$, we see that $e^{-\sigma t} = 1$ and the remaining term $e^{-j\omega t}$, being (by Euler's formula) the sum of a sine and a cosine, merely oscillates periodically as t becomes infinite with either sign. The Laplace transform of a constant does not exist! And this defeats our method.

But we do not give up, for we suspect that the circuit does have a solution and it is merely our mathematics that needs a bit of modification. Here's how we will solve the problem. We reflect that the assumption of an eternally constant signal is a bit at variance with the physical facts, for we must turn the voltage source on at some time.[21] Let's go along with this observation and assume that our voltage source was turned on at $t = -T$ seconds; that is, let's assume that $v_a(t) = 2u(t + T)$, which we have plotted in Figure 13.68. To verify that this is, indeed, a picture of our $v_a(t)$, just remember that the unit step function is unity when its *argument* (in this case $t + T$) is greater than zero and zero when its argument is less than or equal to zero. Its Laplace transform is

$$V_a(s) = \pounds\{v_a(t)\} = \int_{-T}^{\infty} 2e^{-st}\, dt = -\frac{2}{s}\, e^{-st}]_{-T}^{\infty} = \frac{2}{s}\, e^{sT}. \tag{13.7-7}$$

In computing the last term, we have assumed that $\sigma > 0$, and this implies that the exponential is zero at the upper limit. The lower limit is now finite, and no trouble is encountered in computing it.

Now that we have computed $V_a(s)$, let's use it in (13.7-5). This gives

$$sV(s) + 2V(s) = \frac{4}{s}e^{sT}. \tag{13.7-8}$$

We can now solve for $V(s)$:

$$V(s) = \frac{4}{s(s+2)}e^{sT}. \tag{13.7-9}$$

Doing a PFE on the rational fraction on the right-hand side of this equation, we get

$$V(s) = 2\left[\frac{1}{s} - \frac{1}{s+2}\right]e^{sT}. \tag{13.7-10}$$

The inverse transform of the s-domain function inside the brackets is $[1 - e^{-2t}]u(t)$, and after we apply the time shift property of the Laplace transform,[22] we get

$$v(t) = 2[1 - e^{-2(t+T)}]u(t + T). \tag{13.7-11}$$

Figure 13.68   A "turned-on" source waveform

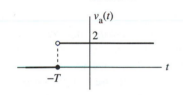

---

[21] This is the reason for our earlier insistence that all time waveforms be one sided.

[22] Although our previous theory is only phrased in terms of one-sided signals, the time shift property holds for two-sided transforms as well. To see this, go back and look over the derivation in Section 13.1.

Figure 13.69 The turned-on response waveform

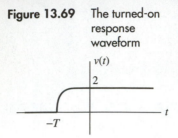

A sketch of this waveform is given in Figure 13.69. Suppose that we now let $T \to \infty$, that is, we turn on the waveform earlier and earlier and thus let the "turn-on time" approach $-\infty$. We see from Figure 13.69 and equation (13.7-11) that $v(t)$ approaches its dc steady-state value of 2 V as $T \to \infty$—*for any finite value of t.*

Although we have been working with a specific example, we easily see that this argument can be used to prove the following general statement: *if all of the* $t \leq 0$ *circuit response variables (voltages and currents) are stable and if all of its independent sources are dc (constant), then the circuit is operating in the dc steady state for any finite value of negative time,* $t \leq 0$. This is an important result, for it permits us to solve the $t \leq 0$ circuit by simply drawing its dc steady-state equivalent. At this point, we recall that a capacitor has the defining v-i relationship $i(t) = Cpv(t)$. In the dc steady state, all variables are constant, so $pv(t) = 0$. Because $i(t)$ is then identically zero for $t \leq 0$, we obtain the dc steady-state equivalent circuit for the capacitor shown in Figure 13.70. If we apply this equivalent to our specific example circuit, we have the $t \leq 0$ equivalent shown in Figure 13.71. We emphasize that we could not use this equivalent circuit if the $t \leq 0$ circuit were not stable. The problem would simply not be well posed. If the $t \leq 0$ circuit were unstable, $v(t)$ would be infinite for any specific finite negative value of t we chose. If the $t \leq 0$ circuit were only marginally stable, the value would depend upon the "turn-on" time, a result that would not be physically meaningful. Because our specific circuit is stable, though, we see that $v(t) = 2$ V for any $t \leq 0$.

Figure 13.70
Dc equivalent circuit model for a capacitor

Let's now return to our original circuit in Figure 13.66 and imagine that we use our fingers and flip the switch in the direction of the arrow. Let's stop the motion of the switch in the middle of its travel—in "open air" as it were. Thus, we are considering the circuit of Figure 13.72 in which we have "frozen" the switch in the straight-up position. Now just before we moved the switch, the current in the 2-Ω resistor was zero (the voltage across it was $2\,V - 2\,V = 0$) and an amount of charge $q = Cv = 0.25\,F \times 2\,V = 0.5\,C$ (the C on the right standing for "coulombs") was stored on the capacitor. When we move the switch away from its position of contact with the left-hand half of the circuit, this charge is trapped and certainly does not "spray off into open air. Thus, we see that we have also trapped a voltage of 2 V across the capacitor. That settled, let's continue our flipping of the switch so that it moves in the direction of the arrow, makes contact with the right-hand half of the circuit, and remains in contact with it until $t = \infty$.

Figure 13.71 The $t \leq 0$ dc equivalent

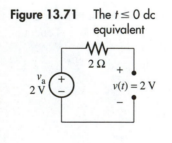

Figure 13.72
The "frozen" circuit

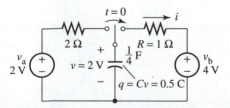

Before we draw the resulting circuit, let's take some time out to discuss the v-i characteristic of the capacitor for positive time. We show a capacitor symbol again at the left-hand side of Figure 13.73, along with passive sign convention definitions of current and voltage. We recall that the defining characteristic for the capacitor is

$$q(t) = Cv(t), \tag{13.7-12}$$

where $q(t)$ is the total charge that has flowed into the capacitor in the direction of the current arrow between $-\infty$ and t, and is therefore the total charge accumulated on the capacitor at time t. Let's rearrange the equation, writing

$$v(t) = \frac{q(t)}{C} = \frac{1}{C}\int_{-\infty}^{t} i(\alpha)\, d\alpha = \frac{1}{C}\int_{0}^{t} i(\alpha)\, d\alpha + \frac{q(0)}{C} \qquad (13.7\text{-}13)$$

$$= \frac{1}{C}\int_{0}^{t} i(\alpha)\, d\alpha + v(0).$$

We have expressed the charge that has flowed into the capacitor since $-\infty$ as the integral of the current (going into the capacitor—not $i(t)$ in our example circuit), then split the integral into two terms: one is over the range $(-\infty, 0]$ and the other over the interval $(0, t]$. The former is just the charge $q(0)$, and $q(0)/C$ is simply the voltage across the capacitor at $t = 0$. Next, let's multiply both sides of (13.7-13) by the unit step function $u(t)$, to obtain

$$v(t)u(t) = \left[\frac{1}{C}\int_{0}^{t} i(\alpha)\, d\alpha\right]u(t) + v(0)u(t). \qquad (13.7\text{-}14)$$

Figure 13.73
An initial condition model for a capacitor

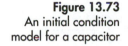

Let's pause for a moment to consider what we have done. In Figure 13.74 we show a general waveform $f(t)$ being multiplied by the unit step function, thus "turning it off" for negative values of time. For such a waveform, you can readily check that the following relationship holds:

Figure 13.74   A truncated waveform

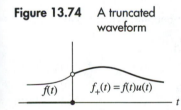
$$\left[\int_{0}^{t} f(\alpha)\, d\alpha\right]u(t) = \int_{-\infty}^{t} f_{+}(\alpha)\, d\alpha = \frac{1}{p}f_{+}(t). \qquad (13.7\text{-}15)$$

Because $f_{+}(t)$ is identically zero for $t \leq 0$, the integral from $-\infty$ to 0 contributes nothing. We will use this on equation (13.7-14), thus obtaining

$$v_{+}(t) = \left[\frac{1}{C}\int_{-\infty}^{t} i_{+}(\alpha)\, d\alpha\right] + v(0)u(t) = \frac{1}{Cp}i_{+}(t) + v(0)u(t). \qquad (13.7\text{-}16)$$

As the one-sided voltage $v_{+}(t)$ is the sum of two voltages, we see that each represents a series connected element. The term $(1/Cp)i_{+}(t)$ can be thought of as the voltage across a capacitor with zero initial charge, and the second term as an "initial condition" voltage source representing the effect of the stored charge at $t = 0$. The resulting equivalent model is shown in Figure 13.73 on the right-hand side. The past history of the capacitor is entirely accounted for by the series voltage source.

Figure 13.75   The $t > 0$ circuit

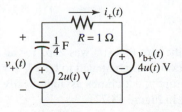

Let's return to our "frozen" circuit in Figure 13.72 now and continue with our switch flip. We move the switch all the way to the right and leave it there forever after. Using our newly derived initial condition model for the capacitor, we obtain the $t > 0$ circuit of Figure 13.75. We are not interested in the variables associated with the elements in the left-hand part any longer, so we have not drawn them in. We have done several things here. First, of course, we have truncated the capacitor variables, making them identically zero for $t \leq 0$. We have also done the same thing for the remaining circuit variables. Because the rest of the circuit consists of only a resistor and an independent source, it has no memory. The operation of this part of the circuit is not influenced at all by what happened for $t \leq 0$. The only memory in the circuit is that of the capacitor. This means that we can truncate all the independent sources in such a circuit without affecting the values of any of its variables on the interval $(0, \infty)$. We point out, by the way, that $i_+(t)$ here is the truncated version of the originally defined current in our example circuit—the negative of the capacitor current for $t > 0$. Finally—and this is a very important observation—the capacitor voltage $v_+(t)$ is the voltage across the *entire equivalent model for the capacitor;* as such, it includes the initial condition source as well as the symbol for the initially uncharged capacitor.

At this point, we have a circuit of the type that we have been analyzing with the Laplace transform. We simply draw the s-domain equivalent for the $t \geq 0$ circuit, as shown in Figure 13.76, and analyze it as usual. The current transform is

Figure 13.76 The s-domain equivalent

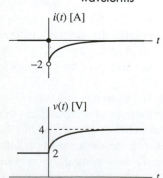

$$I_+(s) = \frac{\dfrac{2}{s} - \dfrac{4}{s}}{\dfrac{4}{s} + R} = -\frac{\dfrac{2}{R}}{s + \dfrac{4}{R}} \tag{13.7-17}$$

and the transform of the capacitor voltage is

$$V_+(s) = \frac{4}{s} + RI_+(s) = \frac{4}{s} - \frac{2R}{Rs + 4} = \frac{4}{s} - \frac{2}{s + \dfrac{4}{R}}. \tag{13.7-18}$$

We have expressed both in terms of R for reasons to soon become evident. We compute the inverse transforms easily and find that

$$i_+(t) = -\frac{2}{R}e^{-4t/R}u(t) \tag{13.7-19}$$

and

$$v_+(t) = 2[2 - e^{-4t/R}]u(t). \tag{13.7-20}$$

Inserting our specified value of $R = 1 \, \Omega$, we get

$$i_+(t) = -2e^{-4t}u(t) \tag{13.7-21}$$

and

$$v_+(t) = 2[2 - e^{-4t}]u(t). \tag{13.7-22}$$

Figure 13.77 The response waveforms

At this stage, we must stop and think back over our solution procedure. We have used truncated waveforms to find the desired time responses for $t > 0$, but *our solution is not valid for $t \leq 0$.* We must go back to the $t \leq 0$ circuit and determine our values for that time interval from it. Doing so, we get

$$i(t) = -2e^{-4t}u(t) \, \text{A} \tag{13.7-23}$$

and

$$v(t) = \begin{cases} 2; & t \leq 0 \\ 2[2 - e^{-4t}]; & t > 0 \end{cases}. \tag{13.7-24}$$

The $i(t)$ waveform is the same as that for $i_+(t)$ due to the fact that $i(t)$ was zero for all $t \leq 0$ because of the open switch. The voltage $v(t)$, however, was computed to be 2 V for $t \leq 0$. We see from (13.7-24) that it does not change abruptly, but climbs asymptotically to 4 V as t becomes large. The two waveforms are shown in Figure 13.77.

Analysis Procedure for Switched Circuits

Our procedure for solving circuits with switches, dc sources, and capacitors should now be clear: we first check the $t \leq 0$ circuit for stability. If it is stable, then we merely use the dc steady-state equivalent (with all capacitors open circuited) to find the capacitor voltage and any desired response variables. (We need the capacitor voltage for our initial condition model for the capacitor.) All of these variables will have constant values. We next "flip" the switch (or switches if there are more than one) and replace each capacitor symbol by its initial condition equivalent—carefully noting that the physical capacitor voltage is the voltage across the new capacitor symbol and the initial condition v-source. We solve the $t > 0$ circuit using Laplace transform techniques. Finally, we "piece together" the two different parts of the solution to obtain the actual waveform(s) on the entire time line $(-\infty, \infty)$.

The Limiting Case: Impulses in Switched Circuits

Let's now consider the solutions for our example circuit in a bit more detail. We purposely derived these responses in terms of a general resistance value R for the analysis we are about to perform. In Figure 13.78, we have plotted these solutions for $R = 1\ \Omega$ and $R = 0.1\ \Omega$. (Note that they are not to scale.) As R becomes smaller and smaller, we see that the magnitude of the current waveform becomes larger and larger, but decays faster and faster to zero. The voltage waveform, on the other hand, has a rise time that gets shorter and shorter. We suspect that for small, but finite values of R, the current waveform approximates an impulse function—and the voltage waveform certainly approximates a step function.

Figure 13.78
Response waveforms with R as a parameter

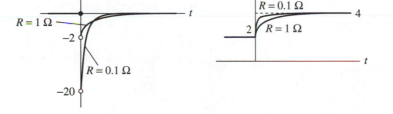

Let's check our guess. We go back to the Laplace transform of $i(t)$ (which, you recall, is the same as that of $i_+(t)$) and take the limit as $R \to 0$. This limit is

$$\lim_{R \to 0} I_+(s) = -\lim_{R \to 0} \frac{2}{Rs + 4} = -\frac{1}{2}. \qquad (13.7\text{-}25)$$

Recall, however, that this says that the limit of $i_+(t)$ is an impulse at $t = 0$ with a strength of -0.5:

$$\lim_{R \to 0} i_+(t) = -0.5\delta(t). \qquad (13.7\text{-}26)$$

We will leave the computation of the corresponding generalized limit of the voltage waveform $v_+(t)$ to you. It is clear, however, that it approaches a step function.

Here's the point of our argument. Many (indeed most) circuit analysis texts insist that you learn and apply the following rule: "a capacitor voltage is always a continuous function of time." They then apply this to solving a circuit differential equation for $t > 0$. We have just seen that this old adage is simply not always true; as a matter of fact, the basic operating principle of modern switched capacitor circuits violates it. This is shown in the next example. What we have just seen, however, is that we can use the $t \leq 0$ value of capacitor voltage as the value of an initial condition source. Then, *any impulses that occur in the circuit do so immediately after the switch flip*. This is a form of the principle of causality, which we have used formally several times before and informally on many occasions.

Example: A Switched
Capacitor Circuit

Example 13.51   Find the node voltages $v_1(t)$ and $v_2(t)$ for all time in the circuit shown in Figure 13.79.

Figure 13.79
A switched capacitor
example circuit

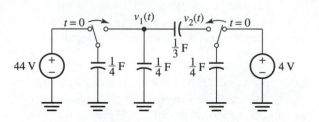

Solution   For our example circuit, we see that the capacitor voltages are known for all $t \leq 0$ because they are connected in parallel with voltage sources or are uncharged. Furthermore, because there are no independent sources in the remainder of the circuit, we can use the causality principle to see that $v_1(t)$ and $v_2(t)$ are zero for all $t \leq 0$. Thus, we need only draw the circuit for $t > 0$ using the initial condition models for the two capacitors that were connected across v-sources for $t \leq 0$. This results in Figure 13.80. (Note that we have not used subscript plus signs because the variables are identically zero already.) Our next step is to draw the transformed circuit, which is shown in Figure 13.81. We can analyze this simple circuit with one nodal equation:

$$\frac{V_1(s) - \dfrac{44}{s}}{\dfrac{4}{s}} + \frac{V_1(s)}{\dfrac{4}{s}} + \frac{V_1(s) - \dfrac{4}{s}}{\dfrac{3}{s} + \dfrac{4}{s}} = 0. \tag{13.7-27}$$

The solution is

$$V_1(s) = \frac{18}{s}. \tag{13.7-28}$$

Using voltage division, we find that

$$V_2(s) = \frac{\dfrac{4}{s}}{\dfrac{3}{s} + \dfrac{4}{s}} \left[\frac{18}{s} - \frac{4}{s} \right] + \frac{4}{s} = \frac{12}{s}. \tag{13.7-29}$$

Thus, we have $v_1(t) = 18u(t)$ V and $v_2(t) = 12u(t)$ V.

Figure 13.80
The $t > 0$ circuit

Figure 13.81
The s-domain equivalent

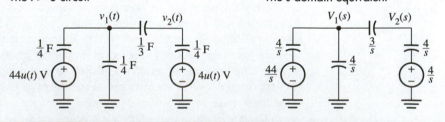

Switched Inductor
Circuits

Practical circuits can, of course, have inductors as well as capacitors. We have developed the procedure for the capacitor thoroughly, so we will not go into a lot of detail with the inductor. We merely show its symbol and terminal variables at the left of Figure 13.82. We recall that the defining characteristic for the inductor is one involving flux linkage and current. Recall that flux linkage is the time integral of the voltage,

$$\lambda(t) = \frac{1}{p}v(t) = \int_{-\infty}^{t} v(\alpha)\, d\alpha. \tag{13.7-30}$$

In terms of flux linkage, then, the inductor's defining relationship is

$$\lambda(t) = Li(t). \tag{13.7-31}$$

If we exchange charge for flux linkage, voltage for current, and capacitance for inductance, we see that the defining capacitor relationship is exactly the same in form as that for the inductor.

Figure 13.82
An initial condition
model for an inductor

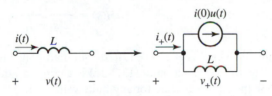

We rewrite (13.7-31) by using (13.7-30) and splitting the integral into two parts:

$$i(t) = \frac{\lambda(t)}{L} = \frac{1}{L}\int_{-\infty}^{t} v(\alpha)\, d\alpha = \frac{1}{L}\int_{0}^{t} v(\alpha)\, d\alpha + \frac{1}{L}\int_{-\infty}^{0} v(\alpha)\, d\alpha \tag{13.7-32}$$

$$= \frac{1}{L}\int_{0}^{t} v(\alpha)\, d\alpha + \frac{\lambda(0)}{L} = \frac{1}{L}\int_{0}^{t} v(\alpha)\, d\alpha + i(0).$$

We now multiply both sides by $u(t)$, to obtain

$$i_{+}(t) = \left[\frac{1}{L}\int_{0}^{t} v(\alpha)\, d\alpha\right]u(t) + i(0)u(t) = \frac{1}{L}\int_{-\infty}^{t} v_{+}(\alpha)\, d\alpha + i(0)u(t). \tag{13.7-33}$$

$$i_{+}(t) = \frac{1}{Lp}i_{+}(t) + i(0)u(t). \tag{13.7-34}$$

This is the one-sided relationship we will use. Because it expresses the inductor current in as the sum of two currents, we see that it is a KCL equation for two elements connected in parallel. The resulting equivalent subcircuit that gives the same KCL equation is shown to the right in Figure 13.82.

Now let's discuss switching in a circuit containing an inductor—because an inductor behaves quite differently from a capacitor. Figure 13.83 shows a simple circuit that contains all the elements necessary for such a discussion. We will find the values of $i(t)$ and $v(t)$ for all time.

Figure 13.83
A switched inductor
circuit

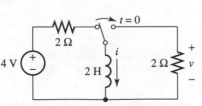

At the outset, we can see from Figure 13.83 that the current through the right-hand resistor is zero for all $t \leq 0$ because it is in series with an open switch. Therefore, by Ohm's law, the voltage $v(t)$ is identically zero for all $t \leq 0$. The part of the $t \leq 0$ circuit that is of more interest (in that it contains the inductor) is shown in Figure 13.84. We can analyze it immediately using mesh analysis to obtain a differential equation for $i(t)$:

Figure 13.84   The $t \leq 0$ circuit

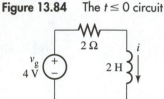

$$2pi(t) + 2i(t) = v_g(t), \tag{13.7-35}$$

where we have generalized the voltage source value to $v_g(t)$, as we did in our initial discussion of capacitor circuits with switches. After taking the Laplace transform, we find the response current $I(s)$ to be

$$I(s) = \frac{1}{2(s+1)} V_g(s). \tag{13.7-36}$$

The system function, therefore, is

$$H(s) = \frac{1}{2(s+1)}. \tag{13.7-37}$$

Its single finite pole is at $s = -1$, in the left half-plane; hence, the response current $i(t)$ is stable. The forcing function in (13.7-35) is a constant; therefore, the response current $i(t)$ will be in the dc steady state for any finite value of $t \leq 0$. Thus, we can replace the inductor symbol in Figure 13.84 with its dc equivalent—a short circuit[23]—as shown in Figure 13.85. Clearly, $i(t) = 2$ A for $t \leq 0$.

Figure 13.85   The dc equivalent for $t \leq 0$

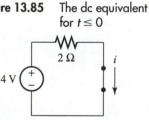

Now, let's flip the switch and draw the right half of the $t > 0$ circuit using the initial condition model for the inductor. This equivalent circuit is shown in Figure 13.86. We have stretched the circuit in the vertical direction a bit merely so that our reference arrow for $i_+(t)$ would fit. Notice, however, that $i_+(t)$ is the *total* current into the *initial condition equivalent model* for the inductor—not the current into the inductor *symbol*. The s-domain equivalent for this circuit is shown in Figure 13.87. We have added a ground reference symbol at the bottom for the purpose of nodal analysis. (You should verify our analysis by using some other method, such as mesh analysis.) We have also labeled the single resistor with the letter R because we will do a symbolic analysis with the idea of altering its value to see its effect on the voltage and current waveforms.

Figure 13.86
The $t > 0$ circuit

Figure 13.87
The s-domain equivalent ($t > 0$)

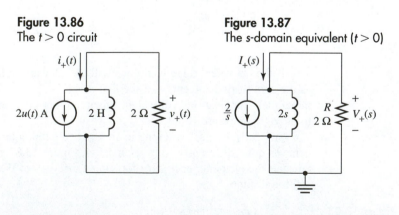

---

[23] We remind you that all current and voltage variables are constant in a circuit operating in the dc steady state. The inductor relationship is $v(t) = Lpi(t) = 0$, so the inductor is equivalent to a short circuit.

The single nodal equation required is

$$\frac{V_+(s)}{R} + \frac{V_+(s)}{2s} = -\frac{2}{s}. \tag{13.7-38}$$

The resulting s-domain voltage is

$$V_+(s) = -\frac{2R}{s + \dfrac{R}{2}}. \tag{13.7-39}$$

The current is

$$I_+(s) = -\frac{V_+(s)}{R} = \frac{2}{s + \dfrac{R}{2}}. \tag{13.7-40}$$

The corresponding time domain waveforms are

$$v_+(t) = -2\mathrm{Re}^{-Rt/2}u(t) \text{ V} \tag{13.7-41}$$

and

$$i_+(t) = 2e^{-Rt/2}u(t) \text{ A}. \tag{13.7-42}$$

In particular, when $R = 2\ \Omega$, we have

$$v_+(t) = -4e^{-t}u(t) \text{ V} \tag{13.7-43}$$

and

$$i_+(t) = 2e^{-t}u(t) \text{ A}. \tag{13.7-44}$$

Our problem has been solved (if we recall the known values for $t \leq 0$), but our objective goes a bit further. We want to get some idea of what happens when the resistor value R goes to infinity. Looking at (13.7-41) and (13.7-42), we see that $i_+(t)$ approaches a constant value of zero ($e^{-\infty t} = 0$) and we suspect that $v_+(t)$ approaches an impulse. To verify this, we simply compute the limit of $V_+(s)$ as $R \to \infty$. From (13.7-39), we see that

$$\lim_{R\to\infty} V_+(s) = -\lim_{R\to\infty}\frac{2R}{s + \dfrac{R}{2}} = -\lim_{R\to\infty}\frac{4}{\dfrac{2s}{R} + 1} = -4. \tag{13.7-45}$$

Thus, we have

$$\lim_{R\to\infty} v_+(t) = -4\delta(t). \tag{13.7-46}$$

A Practical Caveat   Now don't let this result go past without considering it for a moment! An extremely high voltage, as we all know, causes the air (normally considered to be an insulator) to break down, and arcing will occur. Actually, this puts a rather negative light on the situation. In a more positive vein, it is this same principle that causes the spark plugs to fire in your car: the current through an inductor is interrupted suddenly, and the resulting extremely high voltage causes arcing in the air gap of the spark plug. For this practical reason, older circuits texts often drew the switched inductor circuit (Figure 13.83) as in Figure 13.88.

Figure 13.88
An older switch
convention

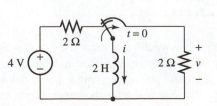

Figure 13.89 The s-domain equivalent with $R = \infty$

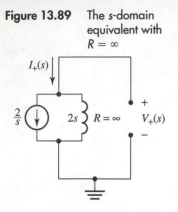

The switch is called a "make-before-break" type; that is, the switch moves in the direction of the arrow and *makes* contact with the right-hand switch conductor *before* it *breaks* contact with the one on the left-hand side (the switch conductors being shown by the dark arcs). In this manner, the inductor never sees an open circuit(as long as the resistor on the right is finite). This convention is no longer used because it is hard to draw and it clutters a circuit diagram too much. For this reason, modern circuits literature uses the symbols we have previously used, along with the understanding that the switching is "instantaneous"; that is, no time is required for the switching action. Then, any voltage impulses are assumed to occur after the switching operation has been completed.

To illustrate our last point, let's look at our circuit above for $t > 0$ (Figure 13.87) with a resistor value of infinity. The transformed version of this circuit is shown in Figure 13.89. KCL shows that $I_+(s) = 0$ and so $V_+(s) = -2/s \times 2s = -4$; hence, $v_+(t) = -4\delta(t)$ V. Our method will handle impulsive voltages and currents in a circuit— those impulses understood to occur immediately after any switching action.

Another Switched Circuit Example

Example 13.52 Find $v_1(t)$ and $v_2(t)$ in the circuit shown in Figure 13.90 for all time if the circuit is known to be operating in the dc steady state for all negative values of t.

Figure 13.90 An example circuit

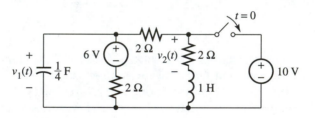

Solution The information that the circuit is operating in the dc steady state for negative values of time saves us a lot of work. If we did not know this, we would have to solve for the capacitor voltage and the inductor current with general source waveforms for the two voltage sources and check to see if both were stable responses. Then and only then could we assume the dc steady state in order to compute the values of these variables for $t \leq 0$. As things stand, though, we simply draw the dc steady-state equivalent for $t \leq 0$, as we have done in Figure 13.91. Notice that we have defined the current i through the inductor because we will need it in our initial condition model that holds for $t > 0$, and similarly for the capacitor. Our circuit is now easy to analyze. Using basic circuit laws we find that $v_1(t) = 4$ V and $i(t) = 1$ A. We also see that $v_2(t) = 2$ V.

Figure 13.91 The dc steady-state equivalent for $t \leq 0$

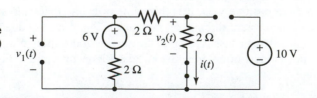

Next, we "flip the switch" and draw the $t > 0$ s-domain equivalent circuit using the initial condition models for the capacitor and the inductor. This results in Figure 13.92. We have added a ground reference symbol for nodal analysis. Only one equation is required for $V_{1+}(s)$, that one being for the supernode at the upper left corner of the circuit.

Thus, we have

$$\frac{v_{1+}(s) - \dfrac{4}{s}}{\dfrac{4}{s}} + \frac{v_{1+}(s) - \dfrac{6}{s}}{2} + \frac{V_{1+}(s) - \dfrac{10}{s}}{2} = 0. \tag{13.7-47}$$

Solving, we get
$$V_{1+}(s) = \frac{4(s + 8)}{s(s + 4)} = \frac{8}{s} - \frac{4}{s + 4}. \tag{13.7-48}$$

(As usual in this section, we have left the PFE for you to verify.) The corresponding time domain voltage, then, is

$$v_{1+}(t) = 4[2 - e^{-4t}]u(t) \text{ V}. \tag{13.7-49}$$

Figure 13.92
The s-domain
equivalent for $t \geq 0$

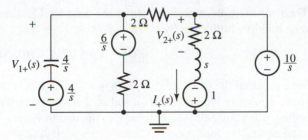

We must be careful as usual not to interpret our formula in (13.7-49) as being valid for $t \leq 0$. In fact, it is not. We have already seen that $v_1(t) = 4$ V for $t \leq 0$. Thus, we can write

$$v_1(t) = \begin{cases} 4; & t \leq 0 \\ 8 - 4e^{-4t}; & t > 0 \end{cases} \text{V}. \tag{13.7-50}$$

We can now easily apply KVL and Ohm's law to derive the current $I_+(s)$:

$$I_+(s) = \frac{\dfrac{10}{s} + 1}{s + 2} = \frac{s + 10}{s(s + 2)} = \frac{5}{s} - \frac{4}{s + 2}. \tag{13.7-51}$$

Thus, we have
$$i_+(t) = [5 - 4e^{-2t}]u(t). \tag{13.7-52}$$

Hence, our other voltage to be found is easily computed from Ohm's law:

$$v_{2+}(t) = 2[5 - 4e^{-2t}]u(t). \tag{13.7-53}$$

Putting this result together with our known value for $t \leq 0$, we have

$$v_2(t) = \begin{cases} 2; & t \leq 0 \\ 10 - 8e^{-2t}; & t > 0 \end{cases} \text{V}. \tag{13.7-54}$$

Circuit Degeneracy   We note in passing that our solution in the preceding example showed that our desired response was only of first order—even though the circuit had two energy elements. Why? Just go back to Figure 13.92 and notice that if we deactivate the rightmost source (the 10-V one), the top node to which is connected the series combination of the right-hand inductor (and its initial condition voltage source, of course) and the rightmost 2-Ω resistor is shorted to ground. This effectively "decouples" this simple circuit from the rest of the network. This is something that often happens in applications, and thus you should be aware of it.

Impulsive Initial Condition Models for L and C

As a final topic in this section, we will derive two additional initial condition equivalent models: one will be for the capacitor and one for the inductor. Taking the capacitor first, we recall its initial condition model in Figure 13.93(a). Because the two elements are series connected, we will henceforth refer to it as the "series initial condition model for the capacitor." Here's how we derive the second model—the one in Figure 13.93(b)—from the first. We simply write the KVL equation for the series model,

$$v_+(t) = \frac{1}{Cp}i_+(t) + v(0)u(t). \tag{13.7-55}$$

Then we multiply both sides (from the left) by Cp; that is, we differentiate both sides, then multiply both sides by C. This gives

$$i_+(t) = Cpv_+(t) - Cv(0)\delta(t). \tag{13.7-56}$$

We have used the fact here that $pu(t) = \delta(t)$. But this is the KCL equation for the two elements shown in Figure 13.93(b), and so that configuration is equivalent, as far as external voltages and currents are concerned, to the one in Figure 13.93(a). Therefore, it is a legitimate equivalent for the original capacitor with its initial condition.

Figure 13.93
The two initial condition models for a capacitor

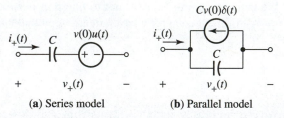

(a) Series model   (b) Parallel model

We can perform a similar manipulation on the (parallel) initial condition model for the inductor shown in Figure 13.94(a). We merely write the KCL equation for this model,

$$i_+(t) = \frac{1}{Lp}v_+(t) + i(0)u(t), \tag{13.7-57}$$

and solve for $v_+(t)$, to obtain

$$v_+(t) = Lpi_+(t) - Li(0)\delta(t). \tag{13.7-58}$$

But this is merely the KVL equation for the series circuit shown in Figure 13.94(b). Be sure to note the direction of the voltage source.

Figure 13.94
The two initial condition models for an inductor

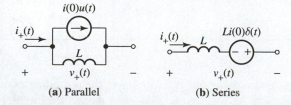

(a) Parallel   (b) Series

One way to remember the references in the (b) parts of the two figures is this: Just imagine the capacitor to be open circuited and the inductor to be short circuited. The capacitor current source would then be in a direction (and with just the right waveform) to establish the known initial voltage. The same would be true for the inductor, with the inductor voltage source being in the right direction (and with just the right waveform) to establish the known initial current. (Remember that the integral of an impulse function is a step function.)

A Final Switched Circuit Example

Example 13.53   Find $v_x(t)$ in the circuit of Figure 13.95 for all t if $g_m = 1$ S (siemens). Comment upon the solution to be expected if the switch operation is reversed (that is, if the switch is closed, rather than opened, at $t = 0$).

Figure 13.95
An example circuit

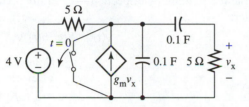

Solution   We will be a bit brief with our $t \leq 0$ circuit analysis, for the dependent current source is shorted out and this "uncouples" the voltage source and its series 5-Ω resistor from the rest of the circuit. Thus, as there is (effectively) no independent source in the circuit to the right of the switch, we know from causality that both capacitor voltages are zero, as is the voltage $v_x(t)$. This means that our initial condition models for the two capacitors will be very simple. We can either consider them to be the series models with zero-valued initial condition voltage source or the parallel models with an initial condition current source value of zero. In either case, the initial condition models reduce to the two-sided linear, time-invariant ones we discussed in Section 13.6. We will not use plus subscripts on our variables in this example because their values are zero for $t \leq 0$.

The $t > 0$ circuit is shown in Figure 13.96 (with the switch open and with $g_m = 1s$). The capacitors, as we have mentioned, have zero initial conditions and so have no initial condition sources. Figure 13.97 shows the s-domain equivalent circuit. We have added a ground reference in anticipation of nodal analysis, which—if we combine the two series elements on the right-hand side—will only require one equation. Notice that we have taped the dependent source, calling its transformed value $I_c(s)$. The nodal equation, then, is

$$\frac{V_a(s) - \dfrac{4}{s}}{5} + \frac{V_a(s)}{\dfrac{10}{s}} + \frac{V_a(s)}{5 + \dfrac{10}{s}} = I_c(s). \tag{13.7-59}$$

Simplifying, we get

$$(s^2 + 6s + 4)V_a(s) - 8 - \frac{16}{s} = 10(s + 2)I_c(s). \tag{13.7-60}$$

Untaping, we have

$$I_c(s) = V(s) = \frac{5}{5 + \dfrac{10}{s}} V_a(s). \tag{13.7-61}$$

Using this in (13.7-60) results in

$$V_a(s) = \frac{8(s+2)}{s(s^2 - 4s + 4)}. \tag{13.7-62}$$

$V_a(s)$ is not, of course, the variable of interest; we can, however, immediately compute the one that is by means of (13.7-61) or—better—by simple inspection of the circuit using voltage division. We obtain

$$V(s) = \frac{5}{5 + \dfrac{10}{s}} V_a(s) = \frac{s}{s+2} V_a(s) = \frac{8}{s^2 - 4s + 4} = \frac{8}{(s-2)^2}. \tag{13.7-63}$$

We can write the solution by inspection—carefully taking note of the negative sign in the denominator:

$$v(t) = 8te^{2t}u(t) \text{ V}. \tag{13.7-64}$$

Figure 13.96
The $t > 0$ circuit

Figure 13.97
s-Domain equivalent of the $t > 0$ circuit

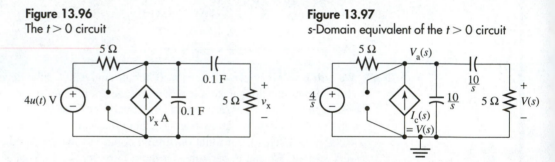

Because $v(t)$ is continuously increasing, we see that our $t > 0$ circuit is unstable; hence, a dc steady state does not exist. This has a very important bearing on our answer to the second part of the question posed in this example, asking us to discuss the solution if the switch action is reversed. This means that the $t > 0$ circuit we just solved now becomes the $t \leq 0$ circuit. It is unstable, so we know that a dc steady-state condition does not exist. In fact, we see that the problem will simply not be well posed if the switching action is reversed because the capacitor voltages will be infinite for any finite $t \leq 0$. This is a situation that is physically not possible because (typically) the dependent source would no longer behave as a nice, linear voltage-controlled current source. Instead, it would "saturate" at some finite constant value, which would force us to reconsider its model and perhaps solve a new circuit with an entirely different structure (and much more involved and difficult methods than we cover in this text).

Section 13.7 Quiz

Q13.7-1. Find the voltage $v(t)$ for all values of t in the circuit shown in Figure Q13.7-1.

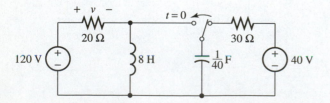

Q13.7-2. For the circuit shown in Figure Q13.7-2:

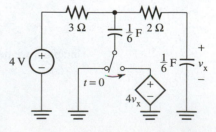

a. Discuss the stability characteristics of the circuit for $t \leq 0$. If that circuit is stable, then find $v_x(t)$ for all t.

b. Reverse the switching action—that is, let the switch flip from right to left at $t = 0$ and stay there forever after. Repeat step a of this question.

13.8 | Operators, Laplace Transforms, and Convolution

The Laplace Transform Related to the Operator Method

In Part II of this text we developed the theory of circuit analysis using operators; in this chapter we have done the same thing using the Laplace transform. A natural question is: How are these two approaches related? As a matter of fact, we claimed in the prefatory material to this chapter that the operator approach contained all of the same mathematical machinery as that of the Laplace transform—except that the operator approach is more general, in that it applies to a wider class of time functions. We will now justify this statement.

The basic problem in circuit analysis—indeed, in the analysis of many types of continuous-time linear time-invariant systems—is the solution of differential equations of the form

$$p^n y(t) + a_{n-1}p^{n-1}y(t) + \cdots + a_1 py(t) + a_0 y(t) \qquad (13.8\text{-}1)$$
$$= b_m p^m x(t) + b_{m-1}p^{m-1}x(t) + \cdots + b_1 px(t) + b_0 x(t).$$

Let's look at both the operator solution and the Laplace transform solution of (13.8-1).

Operator Solution

In Part II we showed that the solution is given by the operator equation

$$y(t) = H(p)x(t), \qquad (13.8\text{-}2)$$

where $H(p)$ is the differential (or Heaviside) operator

$$H(p) = \frac{b_m p^m + b_{m-1}p^{m-1} + \cdots + b_1 p + b_0}{p^n + a_{n-1}p^{n-1} + \cdots + a_1 p + a_0}. \qquad (13.8\text{-}3)$$

It was Heaviside who showed that the partial fraction expansion (PFE) holds for $H(p)$:

$$H(p) = \sum_{i=0}^{m-n} q_i p^i + \sum_{i=1}^{N}\sum_{j=1}^{m_i} \frac{q_{ij}}{(p + a_i)^j}. \qquad (13.8\text{-}4)$$

This is just a condensed, symbolic form of the PFE we developed in Part II in detail. If the degree of the numerator operator in $H(p)$, namely m, is strictly less than that of the denominator operator, n, we agree that the first sum in (13.8-4) is simply not present. We refer to that sum as the *improper part* of the operator $H(p)$ and to the second (double) sum as the *proper part* of $H(p)$. N is the number of distinct (different) denominator factors of the form $(p + a_i)$ and m_i the multiplicity of that factor; that is, there is a factor of the form $(p + a_i)^{m_i}$ for each $i = 1, 2, \ldots, N$. The constants q_i in the improper part are determined by long division and the constants q_{ij} in the proper part by the formula

$$q_{ij} = \frac{1}{(m_i - j)!} \frac{d^{m_i-j}}{dz^{m_i-j}}[(z + a_i)^{m_i}H(z)]_{z=-a_i}. \qquad (13.8\text{-}5)$$

(In Part II, we simply used a formal differentiation with respect to p; however, it is perhaps a bit more rigorous and comfortable to change the p to an arbitrary complex variable z and do the PFE on the resulting $H(z)$. This PFE is then also valid for the Heaviside operator $H(p)$.) Knowing each of the constants, we can now write the general solution to (13.8-1) as

$$y(t) = \sum_{i=0}^{m-n} q_i p^i x(t) + \sum_{i=1}^{N}\sum_{j=1}^{m_i} \frac{q_{ij}}{(p + a_i)^j}\, x(t). \qquad (13.8\text{-}6)$$

In performing the operations indicated in this equation, we use the relation

$$\frac{1}{(p + a)^m}\, x(t) = e^{-at}\, \frac{1}{p^m}\, [e^{at}x(t)], \qquad (13.8\text{-}7)$$

which we derived in Chapter 9, Section 9.4, for any positive integer m. (In (13.8-6), the parameter a would be replaced with a_i and m with j.)

Now let's turn to the Laplace transform solution of (13.8-1).

Laplace Transform Solution

We compute the Laplace transform of both sides of (13.8-1), and this leads to the solution

$$Y(s) = H(s)X(s), \qquad (13.8\text{-}8)$$

where
$$H(s) = \frac{b_m s^m + b_{m-1}s^{m-1} + \cdots + b_1 s + b_0}{s^n + a_{n-1}s^{n-1} + \cdots + a_1 s + a_0}. \qquad (13.8\text{-}9)$$

Furthermore, we showed that the same PFE as that for the Heaviside operator holds:

$$H(s) = \sum_{i=0}^{m-n} q_i s^i + \sum_{i=1}^{N}\sum_{j=1}^{m_i} \frac{q_{ij}}{(s + a_i)^j}, \qquad (13.8\text{-}10)$$

where m, n, N, and m_i are exactly the same parameters as for the Heaviside expansion. The constants, too, are as before. The q_i in the first sum are given by long division of the numerator of $H(s)$ by its denominator and the q_{ij} in the second sum by

$$q_{ij} = \frac{1}{(m_i - j)!}\, \frac{d^{m_i-j}}{ds^{m_i-j}}[(s + a_i)^{m_i}H(s)]_{s=-a_i}. \qquad (13.8\text{-}11)$$

These constants are clearly the same as those in the operator approach; the only difference is that we are using the complex variable s in (13.8-11) in the place of z in (13.8-5). On the basis of (13.8-8) and (13.8-10), we see that the transform $Y(s)$ of the solution to (13.8-1) is

$$Y(s) = \sum_{i=0}^{m-n} q_i s^i X(s) + \sum_{i=1}^{N}\sum_{j=1}^{m_i} \frac{q_{ij}}{(s + a_i)^j}\, X(s). \qquad (13.8\text{-}12)$$

This equation is precisely the same in form as (13.8-6), but (13.8-12) is a Laplace transform relationship, whereas (13.8-6) is a time domain equation. To compare the two time domain solutions, we must compute the inverse transform of (13.8-12).

We first recall that

$$p^i x(t) \longleftrightarrow s^i X(s), \qquad (13.8\text{-}13)$$

that is, multiplication by s in the s domain is equivalent to differentiation in the time domain. This means that the improper part sums of (13.8-6) and the inverse transform of (13.8-12) are identical.

Computing the proper part sums in our answer is only slightly more complicated. Let's define

$$Q(s) = \frac{1}{(s + a)^m} X(s). \tag{13.8-14}$$

Then, applying the frequency shift property of the Laplace transform, we have

$$Q(s - a) = \frac{1}{s^m} X(s - a) = R(s), \tag{13.8-15}$$

where we have defined a new s-domain function $R(s)$ to mean the middle term in (13.8-15). We can now apply the inverse of the time differentiation property (time integration property) and the frequency shift property of Laplace transforms to compute $r(t)$:

$$r(t) = \frac{1}{p^m} [e^{at} x(t)]. \tag{13.8-16}$$

We have recognized that $X(s - a)$ is the transform of the term in the brackets in (13.8-16) and then applied the inverse of the time differentiation property.

Now let's go back to (13.8-15) and solve for $Q(s)$ in terms of $R(s)$:

$$Q(s) = R(s + a). \tag{13.8-17}$$

In the time domain, this relation becomes

$$q(t) = e^{-at} r(t). \tag{13.8-18}$$

Combining (13.8-18) with (13.8-16), we obtain

$$q(t) = e^{-at} \frac{1}{p^m} [e^{at} x(t)] = \frac{1}{(p + a)^m} x(t). \tag{13.8-19}$$

We have recognized here that the expression immediately after the first equality is the same as the one after the last one—according to equation (13.8-7).

If we now look at the proper part sum in (13.8-12) term by term, letting $a = a_i$ and $m = j$, we see that the inverse transform of (13.8-12) is exactly the same as our operator solution in (13.8-6). This means that the two approaches give exactly the same result with the same "operational machinery." The only difference is that in the Laplace approach one puts in the extra step of computing the Laplace transformation of each time function—after which one works with functions of a complex variable rather than with operators.

Because the two approaches are exactly the same "mechanically," we will use the Laplace *notation* for much of our work in the rest of the text, with the understanding that if you wish you may simply substitute s at each occurrence by p and interpret all of our equations as operators in the time domain. In some cases, it is handy to be able to go in the other direction also: if we wish to discuss a particular manipulation as an operation on complex numbers, we have perfect freedom to merely replace p at each occurrence by s, perform the operation desired, then convert back to the time domain.

Convolution[24]   Now that we have settled the question of operators versus Laplace transforms definitively, we will discuss another concept that proves to be useful in a number of situations in circuit analysis and in the analysis of more general systems as well.

 We start our discussion by observing that the inverse transform of the product of two functions of s is not the product of the two time functions that are the inverse transforms of the individual functions of s. Mathematically, we write this observation as

$$\pounds^{-1}\{X(s)Y(s)\} \neq \pounds^{-1}\{X(s)\} \cdot \pounds^{-1}\{Y(s)\}. \qquad (13.8\text{-}20)$$

You can see this quite easily by looking at an example. Suppose that $X(s) = Y(s) = 1/s$. Then $x(t) = y(t) = u(t)$. But $u(t)$ is identically zero for all $t \leq 0$ and identically one for all $t > 0$. Because $0 \times 0 = 0$ and $1 \times 1 = 1$, we see that $x(t)y(t) = u(t)u(t) = u(t)$. Thus, we see that, in the symbolism of (13.8-20), $\pounds^{-1}\{X(s)\} \times \pounds^{-1}\{Y(s)\} = x(t)y(t) = u(t)u(t) = u(t)$; however, we also have $\pounds^{-}\{X(s)Y(s)\} = \pounds^{-1}\{1/s^2\} = tu(t) \neq x(t)y(t)$.

 The question: Just what sort of time function is $\pounds^{-1}\{X(s)Y(s)\}$? We know that it represents a time function (perhaps a generalized function) of some form, which we will name in the following definition.

Definition 13.11   The *convolution* of two time functions $x(t)$ and $y(t)$, having Laplace transforms $X(s)$ and $Y(s)$, respectively, is defined to be the inverse transform of the product of $X(s)$ and $Y(s)$, written

$$x(t) \otimes (t) = \pounds^{-1}\{X(s)Y(s)\}. \qquad (13.8\text{-}21)$$

Note that we are defining the symbol $\otimes$ by equation (13.8-21). There is an integral form for the convolution of two time functions that is derived in the following theorem.

Theorem 13.12   Let $x(t)$ and $y(t)$ be arbitrary waveforms. Then

$$x(t) \otimes y(t) = \int_{-\infty}^{\infty} x(\alpha)y(t - \alpha) \, d\alpha. \qquad (13.8\text{-}22)$$

Proof   We just write the equation

$$X(s)Y(s) = \left[\int_{-\infty}^{\infty} x(\alpha)e^{-s\alpha} \, d\alpha\right]Y(s) = \int_{-\infty}^{\infty} x(\alpha)[Y(s)e^{-s\alpha}] \, d\alpha. \qquad (13.8\text{-}23)$$

Here, we have used α rather than t as the time variable in the integral expression for the Laplace transform of $X(s)$ and have brought the function $Y(s)$ inside the integral sign. Now we recognize that $Y(s)e^{-s\alpha} = \pounds\{y(t - \alpha)\}$ and write

$$X(s)Y(s) = \int_{-\infty}^{\infty} x(\alpha)\pounds\{y(t - \alpha)\} \, d\alpha = \int_{-\infty}^{\infty} x(\alpha)\left[\int_{-\infty}^{\infty} y(t - \alpha)e^{-st} \, dt\right] d\alpha. \qquad (13.8\text{-}24)$$

We have simply written out the definition of the Laplace transform of $y(t - \alpha)$ in terms of an improper integral. We now propose to exchange the order of the two integrations, the one with respect to α and the one with respect to t. We will not derive the conditions

---

[24] Not to be confused with the similar, but much less technical word, *convulsion* (as many students do)!

under which this is permissible mathematically, for they are quite difficult. We simply ask you to accept the fact that it is permissible. Performing the exchange, we write

$$X(s)Y(s) = \int_{-\infty}^{\infty} \left[\int_{-\infty}^{\infty} x(\alpha)y(t - \alpha) \, d\alpha \right] e^{-st} \, dt. \tag{13.8-25}$$

But if we consider the bracket integral simply as a time function, we see that the outside integral is the Laplace transform of that function. Hence, we can write[25]

$$x(t) \otimes y(t) = \pounds^{-1}\{X(s)Y(s)\} = \int_{-\infty}^{\infty} x(\alpha)y(t - \alpha) \, d\alpha, \tag{13.8-26}$$

which is what we were to prove.　■

Example 13.54　Find the convolution of $x(t) = u(t)$ and $y(t) = u(t)$ by direct computation of the integral expression of Theorem 13.12.

Solution　If we look at (13.8-22) carefully, we see that we are supposed to change the variable of $x(t)$ to α, thus forming $x(\alpha)$, and of $y(t)$ to $t - \alpha$, thus forming $y(t - \alpha)$. We have selected very simple functions for our first example to illustrate this process. We show $x(\alpha)$ at the top of Figure 13.98. The waveform $y(t - \alpha)$ is shown immediately below $x(\alpha)$. We stress that α is the "running variable" of integration and t can be considered as a constant relative to the integration process. We pick a value of t, then form $x(\alpha)$ and $y(t - \alpha)$ and multiply these two functions of α together point by point to form the product shown at the bottom of Figure 13.98. Finally, we integrate this product waveform. The integral is the area shown shaded in the figure.

Figure 13.98
Graphical illustration of convolution

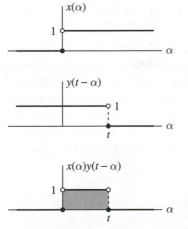

Before proceeding, let's discuss the formation of $y(t - \alpha)$ a bit more. If α is very large and negative, then $t - \alpha$ is very large and positive. We have "turned time around" in the sense that increasing α decreases the argument of $y(t - \alpha)$. This argument is zero when $t - \alpha = 0$, or when $\alpha = t$. For larger values of α, the function argument is negative. This is the type of reasoning you should go through as you compute and sketch this waveform.

---

[25] We are relying here on the fact that if two transforms are equal, then so are their inverse transforms—which requires that certain mathematical conditions hold. Again, we simply assume that they do.

Next, let's note that even though t is a constant in the integration process, we are free to choose it to be any specific value—and we must know the result of the integration process for *all* values of t. Looking at our specific example, we see that if $t \leq 0$, the function $y(t - \alpha)$ is zero for all $\alpha > 0$—but this is precisely where $x(\alpha)$ is nonzero. Hence, the product will be identically zero for all values of α and its integral will certainly be zero. It might help to draw a sketch to see this.

If $t > 0$, on the other hand, the two waveforms in the product overlap and the situation is as shown in Figure 13.98. The area is that of the shaded region shown in that figure, and this area has the value of t. Thus, we see that the convolution process gives

$$x(t) \otimes y(t) = u(t) \otimes u(t) = tu(t). \tag{13.8-27}$$

The graph of this waveform is shown in Figure 13.99.

Figure 13.99
The result of the convolution

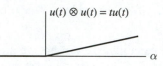

We can check our computation by applying the basic definition. We know that the Laplace transform of $u(t)$ is $1/s$, so $\mathcal{L}\{u(t) \otimes u(t)\} = 1/s \times 1/s = 1/s^2$. But the inverse transform of $1/s^2$ is $tu(t)$—the same waveform that we just computed from the integral representation.

Example 13.55   Find the convolution of $x(t) = e^{-t}u(t)$ and $y(t) = 2e^{-2t}u(t)$ by direct computation of the integral expression of Theorem 13.12.

Solution   We show the waveforms $x(\alpha)$ and $y(t - \alpha)$ in Figure 13.100. Once more, you should go through these sketches carefully to be sure you know how they were formed. Again, we see that if $t \leq 0$, the two waveforms at the top do not overlap and so their product waveform in α will be zero. If $t > 0$, there is overlap as shown at the bottom of the figure. Unlike the case with the unit step functions we worked with in Example 13.54, we cannot simply compute the area using areas of known geometric figures. Now we must actually compute an integral, but our sketches have shown what the limits of integration are. We

Figure 13.100
Another example of convolution

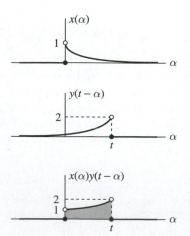

have

$$x(t) \otimes y(t) = \int_0^t e^{-\alpha} 2e^{-2(t-\alpha)} \, d\alpha = 2e^{-2t} \int_0^t e^{\alpha} \, d\alpha = 2e^{-2t}[e^t - 1] \tag{13.8-28}$$

$$= 2[e^{-t} - e^{-2t}].$$

"Turning it off" for $t \le 0$ by multiplication by the unit step function gives

$$e^{-t}(t)u(t) \otimes 2e^{-2t}u(t) = 2[e^{-t} - e^{-2t}]u(t). \tag{13.8-29}$$

A graph of this waveform is shown (for positive t only) in Figure 13.101.

Figure 13.101
The result of the convolution

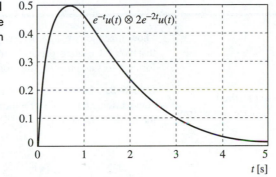

We can once again check our computation by applying the basic definition. We know that the Laplace transform of $e^{-t}u(t)$ is $1/(s+1)$ and that of $2e^{-2t}u(t)$ is $2/(s+2)$, so

$$\pounds\{e^{-t}u(t) \otimes 2e^{-2t}u(t)\} = \frac{1}{(s+1)} \times \frac{2}{(s+2)} \tag{13.8-30}$$

$$= \frac{2}{(s+1)(s+2)} = \frac{2}{s+1} - \frac{2}{s+2}.$$

Here we have used the PFE. If we take the inverse transform, we get (13.8-29), which is a positive check on our direct time-domain computation.

Before we close this brief section, important as it is, we would like to develop just a few more properties of the convolution operation. First, we can change variables by letting $\beta = t - \alpha$ in the integral representation,

$$x(t) \otimes y(t) = \int_{-\infty}^{\infty} x(\alpha)y(t-\alpha) \, d\alpha = \int_{-\infty}^{\infty} y(\beta)x(t-\beta) \, d\beta, \tag{13.8-31}$$

showing in this manner that convolution is a commutative operation:

$$x(t) \otimes y(t) = y(t) \otimes x(t). \tag{13.8-32}$$

We also observe that

$$p^k\{x(t) \otimes y(t)\} = p^j x(t) \otimes p^{k-j} y(t) \tag{13.8-33}$$

for any $j = 0, 1, \ldots, k$ because

$$p^k\{x(t) \otimes y(t)\} \longleftrightarrow s^k X(s) Y(s) = s^j X(s) s^{k-j} Y(s) \longleftrightarrow p^j x(t) \otimes p^{k-j} y(t). \qquad (13.8\text{-}34)$$

This means that if $x(t)$ is a generalized function of order n, that is, $x(t) = p^n x_0(t)$, and $y(t)$ one of order m, that is, $y(t) = p^m y_0(t)$, where both $x_0(t)$ and $y_0(t)$ are K_∞ waveforms, then

$$x(t) \otimes y(t) = p^{n+m} x_0(t) \otimes y_0(t) = p^{n+m} \int_{-\infty}^{\infty} x_0(\alpha) y_0(t - \alpha) \, d\alpha. \qquad (13.8\text{-}35)$$

We will not, however, go into the topic of convolution at any greater depth, deferring it to a more advanced follow-on linear systems course.

Section 13.8 Quiz

Q13.8-1.   Consider the differential equation

$$\frac{d^2 y}{dt^2} + 4 \frac{dy}{dt} + 13y = \frac{dx}{dt} + x.$$

a.   Find the step response using the operator method.
b.   Find the step response using the Laplace transform method.

Show that the two approaches give the same result and observe that you have done precisely the same operations (except transforming to s and inverse transforming to t in step b).

Q13.8-2.   Let $x(t) = u(t) - u(t - 2)$ and $y(t) = 2u(t)$. Find the convolution $x(t) \otimes y(t)$ graphically, then check your result using the Laplace transform.

Chapter 13 Summary

The initial section of this chapter was numbered Section 13.0 because it offered a "quick fix" on the Laplace transform for those who have covered Part II on operator methods. It merely demonstrated quickly that the machinery of the Laplace transform is identical to that of the operator method—except that an improper integral must be computed with the former. The remainder of the chapter was written for those who have elected not to cover the operator method (or to cover both techniques). It is a much more leisurely development of the ideas contained in Section 13.0. However, even if you have not studied the operator methods of Part II, it would be worth your time to read Section 13.0 quickly—it presents a bit of history of both the operator methods and the Laplace transform; furthermore, a quick reading would give you a good overview of this chapter.

Section 13.1 provided motivation for studying the Laplace transform in terms of a *spectrum analyzer* and formally defined the Laplace transform itself as an *improper integral.* As you are aware from your calculus courses, an improper integral carries the idea of *convergence,* and this in turn leads to the idea of a *region of convergence (ROC)* for the transform. The "free parameter" in the Laplace transform (and the one to which the term region of convergence refers) is called the *complex frequency* $s = \sigma + j\omega$. Section 13.1 derived the *transforms of several basic waveforms* and developed several *fundamental properties* of the transform itself.

Section 13.2 developed the idea of *singularity* and *generalized function.* The reason for this is twofold. First, the operations used in the analysis of circuits and other systems require that the *Fundamental Theorem of Calculus (FTC)* work; unfortunately, not all waveforms commonly encountered satisfy the conditions of this theorem. By allowing our waveforms to possess singularities (impulses and their derivatives) this problem was solved. All waveforms then possess the property that the running integral of their derivatives (as well as the derivatives of their running integrals) are the same as the waveforms themselves. Thus, for this class of waveforms, integration and differentiation are inverse operations. Second, we required such generalized functions to make the *time differentiation property* valid.

This important property was covered in detail in Section 13.3. There it was shown that one can employ it very effectively to compute Laplace transforms without computing the defining integral. A converse property, the *frequency differentiation property,* was also covered in this section. Here, too, we presented two fundamental tables. One contains six basic transforms, and the other six basic transform properties. These two tables, along with the *partial fraction expansion (PFE)* developed in Section 13.4, sufficed to permit the computation of any waveform encountered in the circuit analysis we developed later in this chapter.

Section 13.4 derived the partial fraction expansion, which expresses the Laplace transform of a given waveform as a sum of much simpler transforms. This, together with the two tables previously mentioned, offers a complete technique for solving circuits for their time responses.

Section 13.5 discussed the solution of two types of differential equation problem. The first consisted of a differential equation that is known to be valid over the entire time line $-\infty < t < +\infty$, but that has *one-sided* forcing functions; those that are identically zero over the time interval $-\infty < t \leq 0$. The second type of problem consisted of a differential equation that is known to hold only for positive time, $t > 0$, but for which *initial conditions* are given at $t = 0$. We referred to this second type as an *initial condition problem*. It was cast into the same framework as the first by *truncating* the forcing functions so that they were one sided; this process introduced initial conditions in the form of impulse functions and their derivatives. In this section we presented the idea of a *system function,* of *poles* and *zeros,* and of the *pole-zero diagram (PZD)*. We also discussed the decomposition of a solution of a differential equation (and, consequently, of the response of a circuit) into *forced* and *natural* responses, as well as *zero state* and *zero input* responses. Here, too, we developed the idea of *stability* of a differential equation (or circuit response), defining *stable, unstable,* and *marginally stable* operation. We also introduced the idea of *steady state,* which only occurs for stable responses. We then discussed two special, but important, types of steady-state behavior: *dc and ac.*

Section 13.6 then defined *complex impedance* and *admittance* as special types of system function associated with the differential equation expressing the relationship between terminal current and voltage for a two-terminal element or subcircuit. It then developed the properties of second-order circuits in some detail and presented the concept of *damping: underdamping, overdamping,* and *critical* damping. It was argued in this section that because Ohm's law and Kirchhoff's laws hold for impedances and transformed variables (voltages and currents), respectively, all procedures developed in Part I for dc circuits continued to hold: nodal and mesh analysis, superposition, etc. In particular, the *Thévenin and Norton equivalents in the s domain* were presented and used to solve particular example circuits. The techniques of *impedance scaling* and *frequency scaling* were presented in a Laplace transform (or *s*-domain) context.

Section 13.7 discussed the analysis of *circuits containing switches* using *initial condition models* for the dynamic elements (L and C). A critical discussion was given of conditions on the $t \leq 0$ circuit for which the circuit problem is *well posed;* that is, the problem for which steady-state conditions exist prior to $t = 0$. It was argued that only those circuits for which the $t \leq 0$ equivalent is stable are well posed; the circuit holding for $t > 0$ may, however, be stable, unstable, or marginally stable (and still have a well-defined solution).

Section 13.8 served two primary purposes: it demonstrated that the Laplace transform approach is equivalent to the operator method developed in Part II (for those who cover both Part II and Part III), and it introduced the idea of *convolution*. There was a (more leisurely) repetition of material on the Laplace/operator comparison relative to the "quick fix" given in Section 13.0. Convolution was defined to be the inverse transform of two s-domain functions, and several examples were worked out in detail.

Chapter 13 Review Quiz

RQ13.1. Find the Laplace transform of $x(t) = t^2 e^{-3t} u(t - 2)$.

RQ13.2. Find the inverse Laplace transform of

$$X(s) = \frac{10(s + 1)e^{-2s}}{s(s^2 + 5s + 6)}.$$

RQ13.3. Find and sketch the running integral of $x(t) = u(t) + 2u(t - 1) - 3u(t - 2)$.

RQ13.4. Find and sketch the first and second generalized time derivatives of the waveform in Question RQ13.3.

RQ13.5. Find the Laplace transform of the waveform in Question RQ13.3.

RQ13.6. Find and sketch the continuous and jump components of the waveform $x(t) = 2 \cos(2t)u(t) - 4u(t - \pi)$.

RQ13.7. Find the Laplace transform of the waveform in Question RQ13.6.

RQ13.8. Find the unit step and unit impulse responses of the differential equation

$$p^2 y(t) + 4py(t) + 13y(t) = 26x(t)$$

RQ13.9. Find the response of the differential equation

$$\frac{d^2 y}{dt^2} - 8\frac{dy}{dt} + 16y = t$$

for $t > 0$ subject to the initial conditions $y(0) = 2$ and $y'(0) = 4$.

RQ13.10. Find the system function $H(s)$, plot the PZD, and determine the stability type of the response $y(t)$ for the following differential equation for $\mu = 0, -1, 4,$ and 8.

$$p^2 y(t) + (4 - \mu)py(t) + 4y(t) = px(t) - x(t).$$

RQ13.11. Find the unit step and unit impulse responses of $v(t)$ for the circuit shown in Figure RQ13.1.

Figure RQ13.1

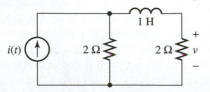

RQ13.12. The current waveform $i(t)$ in the circuit shown in Figure RQ13.1 is described by the equation

$i(t) = [2 + \cos(4t) + \cos(8t)]u(t)$ V. What is the voltage waveform $v(t)$?

RQ13.13. Find the voltage between terminals a and b in Figure RQ13.2 if $i(t) = \delta(t)$.

Figure RQ13.2

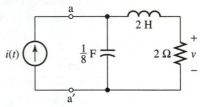

RQ13.14. Replace the i-source in Figure RQ13.2 with a voltage source (+ on terminal a) and find the current $i(t)$ into terminal a if the voltage waveform is $\delta(t)$ V.

RQ13.15. Using scaling techniques, find the current $i(t)$ in Figure RQ13.3 for all values of time if:
    a. $R = 5 \text{ k}\Omega, L = 4$ H
    b. $R = 1 \text{ k}\Omega, L = 4$ H
    c. $R = 1 \text{ k}\Omega, L = 0.5$ H

Figure RQ13.3

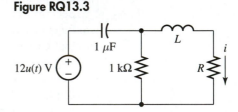

RQ13.16. Find the voltage $v(t)$ in the circuit of Figure RQ13.4 for all values of t. *Note:* You must check the stability of the inductor current and the capacitor voltage for $t \le 0$.

Figure RQ13.4

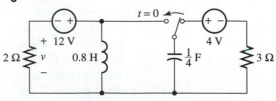

Chapter 13 Problems

Section 13.1 Definition and Basic Properties of the Laplace Transform

13.1-1. Consider the first-order system shown in Figure P13.1-1. Let $x(t) = u(t)$ and $s = 3 + j4$. Find the response $y(t)$. Does the Laplace transform $\mathcal{L}\{x(t)\}$ exist for these parameter values? If so, what is it?

Figure P13.1-1

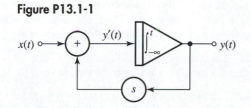

13.1-2. Repeat Problem 13.1-1 assuming that $x(t) = e^{-2t}u(t)$ and that $s = -2$.

13.1-3. Consider the spectrum analyzer shown in Figure P13.1-2. Let $x(t) = e^{-2t}u(t)$ and $s = j\omega$. Does the Laplace transform $\mathcal{L}\{x(t)\}$ exist? If so, what is it? (The Laplace transform evaluated at $s = j\omega$ is called the *Fourier transform* of $x(t)$. We will study it in Chapter 14.)

Figure P13.1-2

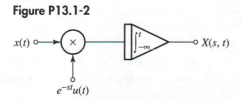

13.1-4. Using the defining improper integral, find the Laplace transform of the waveform shown in Figure P13.1-3. What is the ROC (region of convergence)?

Figure P13.1-3

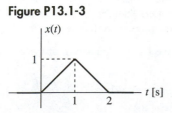

13.1-5. Using the defining improper integral, find the Laplace transform of the waveform shown in Figure P13.1-4. What is the ROC (region of convergence)?

Figure P13.1-4

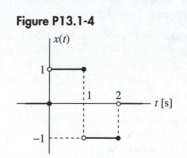

13.1-6. Using the defining improper integral, find the Laplace transform of the waveform $x(t) = e^{2t}u(t)$. What is the ROC (region of convergence)?

13.1-7. Using the defining improper integral, find the Laplace transform of the waveform $x(t) = te^{-2t}u(t)$. What is the ROC (region of convergence)?

13.1-8. If $x(t) = t^3 \cos(2t)u(t)$, find $X(s)$. (*Hint:* Use Euler's formula on $\cos(2t)$ and the frequency shift property.)

13.1-9. If $x(t) = e^{-2t}\sin(2t)u(t)$, find $X(s)$. (*Hint:* Use the frequency shift property.)

13.1-10. If $x(t) = e^{-2(t-3)}u(t-3)$, find $X(s)$. (*Hint:* Use the time shift property.)

13.1-11. If $x(t) = t^4 e^{2t}u(t)$, find $X(s)$. (*Hint:* Use the frequency shift property.)

13.1-12. If $x(t)$ has the waveform shown in Figure P13.1-5, find $X(s)$ using the defining integral.

Figure P13.1-5

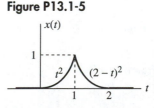

13.1-13. Find $x(t)$ if

$$X(s) = \frac{s + 2}{s^2 + 4s + 13}.$$

(*Hint:* Complete the square for the denominator polynomial.)

Section 13.2 Generalized Functions and Their Transforms

13.2-1. Compute the running integral of the time function $x(t) = 4e^{-2t}u(t)$ and sketch the resulting waveform. Then derive and sketch the second running integral of the same waveform.

13.2-2. Compute and sketch—using graphical methods—the (ordinary) time derivative of the waveform in Problem 13.1-4, then compute and sketch the running integral of the derivative. Repeat using the generalized derivative.

13.2-3. Compute and sketch—using graphical methods—the (ordinary) time derivative of the waveform in Problem 13.1-5, then compute and sketch the running integral of this derivative. How does it compare with the original $x(t)$? Repeat using the generalized derivative.

13.2-4. Compute and sketch the generalized derivative of the waveform in Problem 13.1-6, then compute and sketch the running integral of this derivative.

13.2-5. Compute and sketch the generalized derivative of the waveform in Problem 13.1-7, then compute and sketch the running integral of this derivative.

13.2-6. Compute and sketch the first two generalized derivatives of the waveform in Problem 13.1-12 and compute and sketch the running integrals of the generalized second derivative and the generalized first derivative.

13.2-7. Find the continuous component and the jump function for the waveform $x(t)$ shown in Figure P13.2-1. Then sketch the generalized derivative of $x(t)$.

Figure P13.2-1

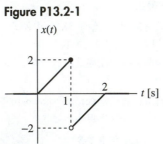

13.2-8. Repeat Problem 13.2-7 for the waveform shown in Figure P13.2-2.

Figure P13.2-2

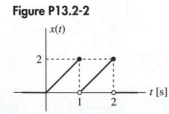

13.2-9. Repeat Problem 13.2-8 assuming that the waveform repeats forever to the right of $t = 2$ s, rather than dropping to zero and staying there as shown in Figure P13.2-2.

13.2-10. Repeat Problem 13.2-7 for the waveform shown in Figure P13.2-3. Note that it repeats forever to the right.

Figure P13.2-3

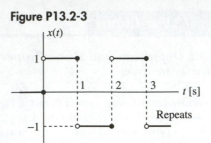

13.2-11. Sketch the waveform of $x(t) = 4\delta(t) - 2u(t - 1) + (2(t - 2)u(t - 2))$. Then compute and sketch the first generalized derivative and the running integral of $x(t)$.

13.2-12. Let $x(t) = u(t) - 3\delta'(t - 2) + e^{2t}\delta(t - 1)$. Compute the definite integral of $x(t)$ over the time interval $[0, 3]$.

13.2-13. Let $x(t) = 4t^3 e^{-3t} u(t)$ and $y(t) = \delta''(t - 1)$. Find the definite integral over $(-\infty, \infty)$ of the product $x(t)y(t)$.

Section 13.3 The Time Differentiation Property

13.3-1. Use the time differentiation property developed in this section (for generalized functions) to find the Laplace transform of the waveform shown in Figure P13.2-1. (*Hint:* Take two generalized derivatives, then find the Laplace transform of $p^2 x(t)$.)

13.3-2. Find the Laplace transform of the waveform in Figure P13.2-2 using the time differentiation property developed in this section for generalized functions. (*Hint:* Differentiate twice, then find the Laplace transform of $p^2 x(t)$.)

13.3-3. Find $x(t)$ if $X(s) = 4/(s^2(s + 2))$. (*Hint:* Use a basic transform and the time differentiation property twice (backward).)

13.3-4. Find $x(t)$ if $X(s) = 4 + s + 3/(s + 2)^2$. (*Hint:* Use transform properties.)

13.3-5. If $x(t) = 4\delta(t) - 3\delta(t - 2)$, find $X(s)$.

13.3-6. Find the Laplace transform of the waveform shown in Figure P13.2-3. (*Hint:* Take the generalized derivative and use the time differentiation property of this section.[26]

13.3-7. Find the Laplace transform of the waveform in Figure P13.3-1.

Figure P13.3-1

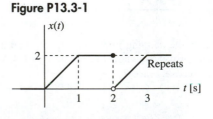

---

[26] For such periodic functions you need to know the sum of the *geometric series:*

$$\sum_{N=0}^{\infty} a^n = \frac{1}{1 - a}; \text{ if } |a| < 1.$$

13.3-8. Find the Laplace transform of the waveform shown in Figure P13.3-2. (*Hint:* Take the generalized derivative and use the time differentiation property.)

Figure P13.3-2

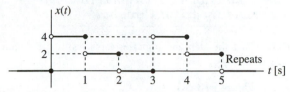

13.3-9. Solve the following differential equation by taking the Laplace transform of both sides, solving for $Y(s)$, and then finding the inverse transform using Laplace transform properties:

$$\frac{d^2y}{dt^2} + 4\frac{dy}{dt} + 4y = e^{-2t}u(t).$$

13.3-10. Find $x(t)$ if $X(s) = (1 - e^{-2s})/s^2$. (*Hint:* Use a basic transform and the frequency shift property.)

13.3-11. Find $x(t)$ if $X(s) = e^{-2s}/(s + 2)^2$. (*Hint:* Use a basic transform and the frequency shift property.)

13.3-12. Find the inverse Laplace transform of the following functions of s using properties:

a. $X(s) = \dfrac{10(s + 1)}{s^2 + 2s + 1}$ **b.** $\dfrac{e^{-s}}{s + 1}$

c. $\dfrac{1 - e^{-2s}}{s}$ **d.** $\dfrac{1 - e^{-s}}{s + 2}$

e. $X(s) = \dfrac{(s + 1)e^{-s}}{s(s + 2)}$

13.3-13. Using the Laplace transform properties presented in this section, find the transforms of the following waveforms:

a. $x(t) = [t - 1 + e^{-(t-1)}]u(t - 1)$
b. $x(t) = [t - 2 + e^{-(t-2)}]u(t - 1)$
c. $x(t) = 2te^{-3t}u(t - 1)$
d. $x(t) = 2te^{-3t}\cos(4t)u(t - 2)$
e. $x(t) = 2\delta''(t) + 4\delta'(t - 1) - 3\delta(t - 2) + 4u(t - 3)$

13.3-14. Using Laplace transform properties, find the inverse Laplace transform of the following functions of s:

a. $X(s) = 4s^2 + 3s + 2 + \dfrac{1}{s + 2}$

b. $X(s) = \dfrac{s^3 + 2s^2 + 3s + 4}{s^2}$

c. $X(s) = \dfrac{s^2e^{-2s} + 2s + 4e^{-3s}}{s}$

d. $X(s) = \dfrac{s^2}{s + 2}e^{-2s}$

e. $X(s) = \dfrac{24s}{s^2 + 4s + 13}$

13.3-15. Let $x(t)$ be the staircase waveform defined on each time interval $(nT, (n + 1)T]$,[27] by $x(t) = n + 1$ for each $n = 0, 1, 2, \ldots$ and zero for all $t \le 0$. Find $X(s)$.

Section 13.4 The Partial Fraction Expansion

13.4-1. Find the time function $x(t)$ if

$$X(s) = \frac{8(s + 5)(s + 12)}{s(s + 8)(s + 6)}.$$

13.4-2. Find the time function $x(t)$ if

$$X(s) = \frac{12(s + 3)}{s(s^2 + 5s + 4)}.$$

13.4-3. Find the time function $x(t)$ if

$$X(s) = \frac{12(s + 3)}{s^2(s + 2)}.$$

13.4-4. Find the time function $x(t)$ if

$$X(s) = \frac{12(s + 1)}{s^2(s + 2)}.$$

13.4-5. Find the time function $x(t)$ if

$$X(s) = \frac{18(s + 3)}{(s + 1)^2(s + 4)}.$$

13.4-6. Find the time function $x(t)$ if

$$X(s) = \frac{25(s + 35)}{(s + 6)(s^2 + 6s + 25)}.$$

13.4-7. Find the time function $x(t)$ if

$$X(s) = \frac{24(s^2 + 1)}{s(s + 2)(s^2 + 2s + 2)}.$$

Section 13.5 Laplace Transform Solution of Differential Equations

13.5-1. Find the unit impulse and step responses of the following differential equations:

a. $p^2y(t) + 2py(t) + 2y(t) = x(t)$
b. $2p^2y(t) + 5py(t) = x(t)$
c. $p^2y(t) - 16y(t) = x(t)$
d. $p^2y(t) - 8y(t) = x(t)$
e. $p^2y(t) + 9y(t) = x(t)$
f. $4p^2y(t) + y(t) = x(t)$
g. $p^2y(t) - 3py(t) + 2y(t) = x(t)$
h. $p^2y(t) - py(t) - 6y(t) = x(t)$
i. $\dfrac{d^2y}{dt^2} + 8\dfrac{dy}{dt} + 16y = x(t)$
j. $\dfrac{d^2y}{dt^2} - 10\dfrac{dy}{dt} + 25y = x(t)$

---

[27] Note that $(a, b]$ means $a < t \le b$.

k. $p^2y(t) + 3py(t) - 5y(t) = x(t)$
l. $8p^2y(t) + 2py(t) - y(t) = x(t)$
m. $p^3y(t) - 4p^2y(t) - 5py(t) = x(t)$
n. $p^3y(t) - y(t) = x(t)$
o. $p^3y(t) - 5p^2y(t) + 3py(t) + 9y(t) = x(t)$
p. $p^3y(t) - p^2y(t) - 4y(t) = x(t)$

13.5-2. Find the response $y(t)$ of the differential equation in Problem 13.5-1, step c, if the forcing function is $x(t) = u(t) + \delta(t) + \delta'(t-1)$.

13.5-3. Solve each of the following differential equations:
a. $p^2y(t) + 4py(t) + 4y(t) = (2t + 6)u(t)$
b. $p^3y(t) + p^2y(t) = 8t^2u(t)$
c. $p^2y(t) - py(t) - 12y(t) = e^{4t}u(t)$
d. $p^2y(t) - 2py(t) - 3y(t) = (4e^t - 9)u(t)$
e. $p^2y(t) + 6py(t) + 8y(t) = (3e^{-2t} + 2t)u(t)$
f. $p^2y(t) + 25y(t) = 6\sin(t)u(t)$
g. $p^2y(t) + 4y(t) = [4\cos(t) + 3\sin(t)]u(t)$
h. $p^2y(t) - 2py(t) + 5y(t) = e^t\sin(t)u(t)$
i. $p^2y(t) + py(t) + y(t) = t\sin(t)u(t)$
j. $p^3y(t) + 8p^2y(t) = [-6t^2 + 9t + 2]u(t)$

13.5-4. Find the system function $H(s) = Y(s)/X(s)$ for each of the following differential equations and plot the PZD. Also, determine the stability characteristic of each.
a. $p^2y(t) + py(t) = 10px(t) + 20x(t)$
b. $p^2y(t) + 3py(t) + 2y(t) = 4p^2x(t) + 12px(t)$
c. $p^2y(t) + 2py(t) + 5y(t) = 8px(t) - 32x(t)$
d. $p^2y(t) - 2py(t) - 3y(t) = (4e^t - 9)u(t)$
e. $p^2y(t) - 4y(t) = 4px(t) + 4x(t)$
f. $p^2y(t) + 4py(t) + 3y(t) = p^3x(t) + p^2x(t) - 2x(t)$
g. $p^3y(t) + p^2y(t) + 4py(t) + 4y(t) = 2p^2x(t) - 2x(t)$
(*Note:* One pole at $s = -1$)
h. $p^4y(t) + 4p^3y(t) + 6p^2y(t) + 4py(t) + y(t)$
$= p^3x(t) - 3p^2x(t) + 3px(t) - x(t)$
(multiple poles at $s = -1$, zeros at $s = 1$)

13.5-5. Solve each of the following differential equations for $t > 0$ subject to the specified initial conditions:
a. $p^2y(t) + 16y(t) = 0; y(0) = 2, y'(0) = -2$
b. $p^2y(t) - 2py(t) + y(t) = 0; y(0) = 5, y'(0) = 10$
c. $p^2y(t) + 6py(t) + 5y(t) = 0; y(0) = 0, y'(0) = 3$
d. $p^2y(t) - 64y(t) = 16; y(0) = 1, y'(0) = 0$
e. $p^2y(t) - 5py(t) = t - 2; y(0) = 0, y'(0) = 2$
f. $p^2y(t) + 5py(t) - 6y(t) = 10e^{2t}; y(0) = 1, y'(0) = 1$
g. $p^2y(t) + py(t) = 8\cos(2t) - 4\sin(t);$
$y(0) = -1, y'(0) = 0$

13.5-6. Find the natural and forced responses for the following differential equations and comment on the existence of a steady state (if one exists and, if so, what type).
a. $py(t) + 2y(t) = 4x(t); x(t) = 2u(t)$
b. $p^2y(t) + 3py(t) + 2y(t) = px(t) + 3x(t); x(t) = 4u(t)$
c. $py(t) + 2y(t) = 4x(t); x(t) = 6\cos(2t)u(t)$
d. $p^2y(t) + py(t) - 6y(t) = 12x(t); x(t) = 2u(t)$
e. $p^2y(t) - 4py(t) + 4y(t) = x(t); x(t) = \cos(2t)u(t)$

13.5-7. Find the zero input and zero state responses for the dif-

ferential equations in Problem 13.5-6, assuming that the differential equation only holds for $t > 0$—and removing the $u(t)$ functions on the right-hand side. For instance, replace $2u(t)$ in part a with the constant 2.

Section 13.6 Laplace Transform Analysis of Linear Time-Invariant Circuits

13.6-1. Find the step and impulse responses of the circuit in Figure P13.6-1.

Figure P13.6-1

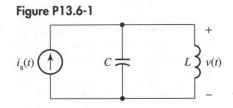

13.6-2. Find the step and impulse responses of the circuit in Figure P13.6-2.

Figure P13.6-2

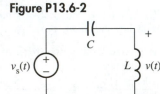

13.6-3. Consider the circuit in Figure P13.6-3.
a. For what value of R is the response $v(t)$ critically damped? Underdamped? Overdamped? Find the appropriate system function $H(s)$ and draw a PZD for each of the damping cases, locating the poles appropriately.
b. For the value of R that gives critical damping, find the step and impulse responses for $v(t)$.

Figure P13.6-3

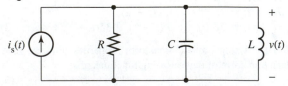

13.6-4. Assuming that $L = 1/13$ H, $C = 1$ F, and $R = 1/4$ Ω in the circuit shown in Figure P13.6-3, find the step and impulse responses.

13.6-5. Repeat Problem 13.6-3 for the circuit in Figure P13.6-4.

Figure P13.6-4

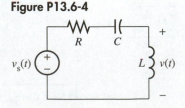

13.6-6. For $R = 4\,\Omega$, $C = 1/13$ F, and $L = 1$ H, find the step and impulse responses for the circuit in Figure P13.6-4.

13.6-7. Find the voltage $v(t)$ in Figure P13.6-5.

Figure P13.6-5

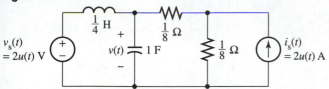

13.6-8. Find the s-domain Thévenin equivalent for the two-terminal subcircuit "seen by the current source" in Figure P13.6-5. Then use it to find the i-source voltage (+ on top).

13.6-9. Find the s-domain Norton equivalent for the two-terminal subcircuit "seen by the voltage source" in Figure P13.6-5. Then use it to find the v-source current (out of + terminal).

13.6-10. Find the voltage $v(t)$ in Figure P13.6-5 if $v_s(t) = 2\cos(2t)u(t)$ V and $i_s(t) = 2\cos(2t)u(t)$ A.

13.6-11. Consider the two-terminal subcircuit shown in Figure P13.6-6. Assume that a current source is attached to terminals a and b to supply $i(t)$. Find the open circuit voltage $v_{oc}(t)$ for $t > 0$.

Figure P13.6-6

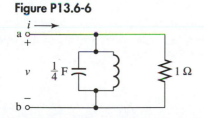

13.6-12. Change the resistor value in Figure P13.6-6 from 1 Ω to 2 Ω and repeat Problem 13.6-11.

13.6-13. Change the resistor value in Figure P13.6-6 from 1 Ω to 1.2 Ω and repeat Problem 13.6-11.

13.6-14. Repeat Problem 13.6-11 for the circuit shown in Figure P13.6-7 with $R = 10\,\Omega$, $8\,\Omega$, and $4\,\Omega$. This time, however, assume that a voltage source is attached to terminals a and b, and find the short circuit current i_{sc} for all $t > 0$.

Figure P13.6-7

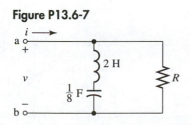

13.6-15. Consider the circuit shown in Figure P13.6-8.
   a. Impedance scale the circuit so that the resistors have values of 1 Ω.
   b. Frequency scale the result of step a so that the capacitors have values of 1 F.

c. Derive the system function $H(s) = V(s)/I_g(s)$ for the scaled circuit.
d. Discuss the stability of the response $v(t)$ for $\mu = 5, 3, 2,$ and 1 using the system function you found in step c.
e. Find the (unscaled) step response for $v(t)$ under the condition $\mu = 1$. Find the dc forced response.
f. Find the (unscaled) ac forced response if $\mu = 2$ and $i_g(t) = \cos(1000t)u(t)$ mA.

Figure P13.6-8

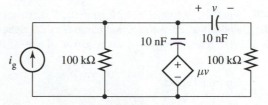

13.6-16. Determine the range of values of μ (both + and −) for which $v_x(t)$ is a stable response in the circuit shown in Figure P13.6-9.

Figure P13.6-9

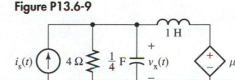
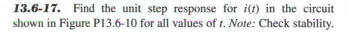

13.6-17. Find the unit step response for $i(t)$ in the circuit shown in Figure P13.6-10 for all values of t. *Note:* Check stability.

Figure P13.6-10

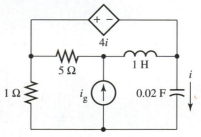

Section 13.7 Switched Circuits with Dc Sources

13.7-1. Find $v(t)$ in Figure P13.7-1. *Note:* Causality implies that $v(t) = 0$ for all $t \leq 0$.

Figure P13.7-1

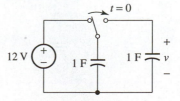

13.7-2. Find the current $i(t)$ in Figure P13.7-2.

Figure P13.7-2

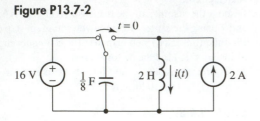

13.7-3. Find the current $i(t)$ in Figure P13.7-3. (Notice that causality implies that all voltages and currents to the right of the switch are zero for all $t \leq 0$—so we do not need any dc steady-state assumption.)

Figure P13.7-3

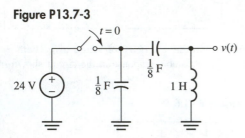

13.7-4. Find $v(t)$ in Figure P13.7-4. Note: You must check the $t \leq 0$ circuit to make sure that the capacitor voltage and inductor current are stable responses in order to be assured that these elements are in the dc steady state for $t \leq 0$. If one or both responses are, in fact, unstable, make the assumption that dc steady state exists anyway in order to solve for $v(t)$.

Figure P13.7-4

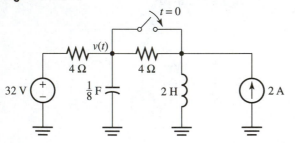

13.7-5. Find $i(t)$ in Figure P13.7-5. Assume dc steady state prior to $t = 0$.

Figure P13.7-5

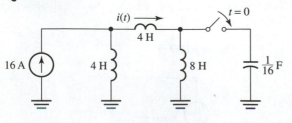

13.7-6. Find $v(t)$ in Figure P13.7-6. The switch is open for $t < 0$. It closes at $t = 0$, opens 1/4 s later, closes 1/4 s later, and so on, repetitively. Find $v(t)$ for all t. *Note:* This is a rather difficult problem!

Figure P13.7-6

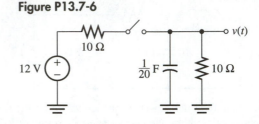

13.7-7. Find the response voltage $v(t)$ for the circuit in Figure P13.7-7. *Note:* You must check stability for $t \leq 0$. Remember that any impulses at $t = 0$ are considered (in accordance with the principle of causality) to occur right after the switch flips.

Figure P13.7-7

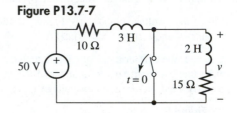

13.7-8. Find the response current $i(t)$ in the circuit of Figure P13.7-8. *Note:* You must check stability of the capacitor voltage and inductor current for $t \leq 0$.

Figure P13.7-8

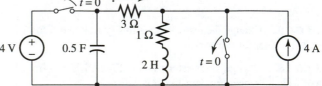

13.7-9. Determine the stability of the capacitor voltages in the $t \leq 0$ circuit in Figure P13.7-9 for $\mu = 1, 2$, and 3. Then for each value of μ for which the $t \leq 0$ circuit is stable, determine the voltage $v(t)$ for all values of t.

Figure P13.7-9

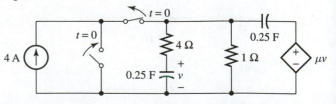

13.7-10. Apply impedance and frequency scaling to the circuit in Figure P13.7-10 to give "nice" values for the elements. Then find $v(t)$ for all values of time. *Note:* You must check the $t \leq 0$ circuit for stability.

Figure P13.7-10

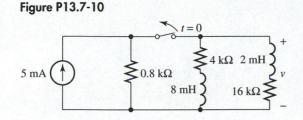

13.7-11. Find the response voltage $v(t)$ for the circuit in Figure P13.7-11 for all t and sketch the response waveform. *Note:* You must check stability.

Figure P13.7-11

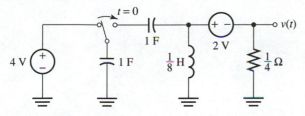

13.7-12. Find the voltage $v(t)$ for all t in the circuit of Figure P13.7-12. *Note:* You must check stability.

Figure P13.7-12

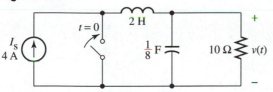

Section 13.8 Operators, Laplace Transforms, and Convolution

13.8-1. Compute and sketch the time convolution of the waveforms shown in Figure P13.8-1 using the integral expression for convolution.

Figure P13.8-1

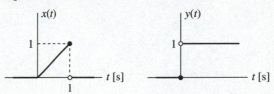

13.8-2. Using the running integral expression for convolution, compute and sketch the convolution of the waveforms shown in Figure P13.8-2.

Figure P13.8-2

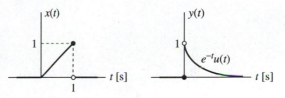

13.8-3. Repeat Problem 13.8-1 using the Laplace transform.

13.8-4. Repeat Problem 13.8-2 using the Laplace transform.

13.8-5. Use the convolution integral to find $e^{-t}u(t) \otimes e^{-t}u(t)$.

13.8-6. Use the convolution integral to find $e^{-t}u(t) \otimes \{u(t) - u(t - 1)\}$.

13.8-7. Repeat Problem 13.8-5 using the Laplace transform.

13.8-8. Repeat Problem 13.8-6 using the Laplace transform.

14 The Fourier Transform

The Fourier transform is defined in Section 14.1 as the Laplace transform evaluated for $s = j\omega$; that is, "on the $j\omega$ axis." This definition is then extended to one-sided functions for which the Laplace integral does not converge for $\sigma = 0$ by introducing the "one-sided convergence factor" e^{-at}. This forces the Laplace transform to converge for $s = j\omega$, and the Fourier transform is then defined as the limit of this value as a approaches zero through positive values. The point is then made that—if one is dealing with stable circuits or systems and is only interested in steady-state behavior—one can extend waveforms backward to $t = -\infty$ and consider two-sided waveforms. A "two-sided convergence factor" $e^{-a|t|}$ is then used to force the Laplace transform to converge on the $j\omega$ axis, and the Fourier transform is defined as the limit of this value as a approaches zero through positive values. This last definition then subsumes the earlier two as special cases.

In Section 14.2 the question of "inverting" the Fourier transform is taken up. The "inversion integral," or "inverse Fourier transform," is defined and applied. The question of convergence for improper integrals is discussed and several definitions of such integrals are explored with respect to convergence. The one chosen is normally associated with Poisson, who first initiated the use of two-sided convergence factors for the study of the Fourier transform. This definition is easier to work with than the one used by Dirichlet, the one usually used in circuits and systems texts.

Section 14.3 discusses the frequency spectrum. It introduces a system of notation that allows the treatment of discrete and continuous spectra on a unified basis and defines the "instantaneous power spectrum" and the "energy density spectrum."

Section 14.4 begins by reconsidering the six properties of the Laplace transform that were presented in Chapter 13 as fundamental, developing their forms when $s = j\omega$. It then continues by developing properties of the Fourier transform that are peculiar to it alone (not shared with the Laplace transform) because of the fact that the frequency variable is purely imaginary.

Section 14.5 develops the Fourier series from the Fourier transform. It begins by showing that the Fourier series for a periodic waveform is a "row of impulses," and inverts the transform to obtain the usual Fourier series. The coefficients are shown to be the Fourier transform of one period of the periodic function sampled at integral multiples of $\omega_0 = 2\pi/T$, where T is the period. This has the advantage of directly demonstrating that the properties of the Fourier coefficients follow directly from those of the Fourier transform.

Finally, Section 14.6 discusses an issue that plagues many students: the question of "negative frequency." The concepts of analytical signal and Hilbert transform are discussed, and the reader is presented with a choice: accept either complex time signals or negative frequencies in the Fourier transform. Although complex signals can be generated in the laboratory (as was shown in Chapter 8), it is argued that one must realize such signals on "two wires," rather than on one—so it seems reasonable to accept the concept of negative frequency and keep the signals real. Finally, with the analytic signal and Hilbert transform already developed, the ideas of AM and SSB radio are discussed. These concepts will prove useful to the reader in follow-on courses in communications and electronics.

The background required for this chapter is the more basic material in Chapter 13 (Sections 13.1–13.3) because we base our definition of the Fourier transform on that of the Laplace transform.

14.1 | The Fourier Transform

The Fourier Transform: Definition

Figure 14.1 shows a simple system called a *spectrum analyzer,* which we discussed in Chapter 13. There the exponential had the more general exponent $s = \sigma + j\omega$. Here, we have let $\sigma = 0$; therefore the exponent is $j\omega$. In Chapter 13 we defined the limiting value of the time response, that is,

$$\lim_{t \to \infty} X(s, t) = X(s) = \mathcal{L}\{x(t)\}, \tag{14.1-1}$$

of this system to be the Laplace transform of $x(t)$—when the limit in question exists. Here, we define the same limit—when $\sigma = 0$—to be the Fourier transform of $x(t)$.

Figure 14.1
The spectrum analyzer

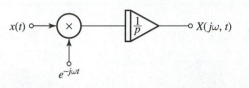

$$x(t) \circlearrowright \times \longrightarrow \boxed{\frac{1}{p}} \longrightarrow \circ\, X(j\omega, t)$$

$$e^{-j\omega t}$$

Definition 14.1   The *Fourier transform* of a waveform $x(t)$ is the Laplace transform of $x(t)$ evaluated on the imaginary axis,

$$X(\omega) = [X(s)]_{s=j\omega}, \tag{14.1-2}$$

provided that this quantity is well defined.

Perhaps a word about notation is in order. It might seem more logical to use the j explicitly in the argument, that is, $X(j\omega)$, because this emphasizes that the Fourier transform is merely the Laplace transform evaluated at purely imaginary values of s. As a matter of fact, some texts and other technical works do use this notation. It has the disadvantage, however, of being somewhat more unwieldy to write; therefore, we will drop the j when writing the Fourier transform of a time function. If we write the definition out explicitly, we will have

$$X(\omega) = \int_{-\infty}^{\infty} x(t)e^{-j\omega t}\, dt, \tag{14.1-3}$$

providing that this improper integral converges.[1] Alternate notational schemes that we will occasionally use for the Fourier transform are:

$$\mathcal{F}\{x(t)\} = X(\omega), \tag{14.1-4}$$

$$x(t) = \mathcal{F}^{-1}\{X(\omega)\},$$

$$x(t) \longleftrightarrow X(\omega).$$

We read the first as "the Fourier transform of $x(t)$ is $X(\omega)$," the second as "$x(t)$ is the inverse Fourier transform of $X(\omega)$," and the third as "$x(t)$ and $X(\omega)$ are a Fourier transform pair."

Examples: Fourier Transforms of Time-Limited Waveforms

Example 14.1   Find $\mathcal{F}\{\delta(t)\}$.

---

[1] We will deal with nonconvergence later in this section. This turns out to be a fairly severe restriction.

Solution   This one is quite easy and straightforward. We simply write

$$\mathscr{F}\{\delta(t)\} = \int_{-\infty}^{\infty} \delta(t)e^{-j\omega t}\, dt = 1. \tag{14.1-5}$$

We have used the sampling property of the delta function and the fact that $e^{j0} = 1$.

Example 14.2   Find $\mathscr{F}\{\delta(t-a)\}$.

Solution   Again, the sampling property suffices:

$$\mathscr{F}\{\delta(t-a)\} = \int_{-\infty}^{\infty} \delta(t-a)e^{-j\omega t}\, dt = e^{-ja\omega}. \tag{14.1-6}$$

Example 14.3   Find $\mathscr{F}\{P_T(t)\}$.

Solution   The unit pulse function of duration T, $P_T(t)$, is drawn in Figure 14.2(a). In Figure 14.2(b) we have sketched its derivative, for this is perhaps the easiest route to computing the Fourier transform. At this point we make the observation that as the Fourier transform is merely the Laplace transform evaluated for $s = j\omega$, all the properties of the latter hold for the Fourier transform. Therefore, we can invoke the time differentiation property of the Laplace transform, obtaining

$$s\pounds\{P_T(t)\} = 1 - e^{-sT}. \tag{14.1-7}$$

Thus,[2]

$$\pounds\{P_T(t)\} = \frac{1 - e^{-sT}}{s}. \tag{14.1-8}$$

Next, we evaluate our result for $s = j\omega$ to obtain the desired Fourier transform:

$$\mathscr{F}\{P_T(t)\} = \frac{1 - e^{-j\omega T}}{j\omega} = e^{-j\omega T/2}\frac{e^{+j\omega T/2} - e^{-j\omega T/2}}{j\omega} \tag{14.1-9}$$

$$= 2Te^{-j\omega T/2}\frac{e^{+j\omega T/2} - e^{-j\omega T/2}}{j2\omega T} = Te^{-j\omega T/2}\frac{\sin\left(\dfrac{\omega T}{2}\right)}{\dfrac{\omega T}{2}}.$$

We factored $e^{-j\omega T/2}$ from the numerator, multiplied and divided by $2T$, and then applied Euler's formula for the sine function.

Figure 14.2
The unit pulse
waveform of duration
T and its derivative

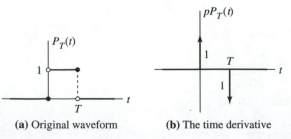

(a) Original waveform   **(b)** The time derivative

---

[2] Notice that the transform has the finite value T as $s = 0$ (apply L'Hôpital's rule).

Example 14.4 Find $X(\omega)$ for the waveform $x(t)$ shown in Figure 14.3.

Figure 14.3
A triangular pulse
waveform

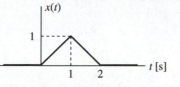

Solution The first two derivatives of $x(t)$ are shown in Figures 14.4(a) and (b), respectively. The computation of the Laplace transform of the second derivative is quite straightforward. We remind you of our notation for singularity functions: a number Δ at the side of an arrow means that the associated impulse function has area Δ. That is, the arrow symbol at t_0 with the number Δ at its side means that there is an impulse at $t = t_0$ with an area of Δ. We also remind you that if we point the arrow downward, this means that the area is negative. Thus, we first compute the Laplace transform of $p^2x(t)$ to be

$$s^2X(s) = \mathcal{L}\{p^2x(t)\} = 1 - 2e^{-s} + e^{-2s}. \qquad (14.1\text{-}10)$$

Then we have

$$X(\omega) = [X(s)]_{s=j\omega} = \frac{1 - 2e^{-j\omega} + e^{-j2\omega}}{(j\omega)^2} = \left(\frac{1 - e^{-j\omega}}{j\omega}\right)^2. \qquad (14.1\text{-}11)$$

What could be simpler?

Figure 14.4
First two derivatives of
the example waveform

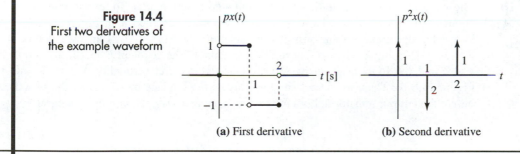

(a) First derivative (b) Second derivative

Example: Fourier Transform of a Non-Time-Limited Waveform

All of the time waveforms until now in this section have been *time limited,* that is, they were zero for sufficiently large values of t. Let's look now at one that is *not.*

Example 14.5 If $x(t) = e^{-at}u(t)$, where a is a positive real number, find $X(\omega)$.

Solution Just to make the situation clear, we have plotted $x(t)$ in Figure 14.5. It is one sided and decays to zero as $t \to \infty$, but it only *approaches* zero—it is not identically zero for large values of t. However, we know that the Laplace transform of this waveform is[3]

$$X(s) = \mathcal{L}\{e^{-at}u(t)\} = \frac{1}{s + a}. \qquad (14.1\text{-}12)$$

---

[3] Compute the integral or simply thumb back through Chapter 13 if you are not sure about this result.

The region of convergence (ROC), we remind you, is $\sigma = \text{Re}\{s\} > -a$. Thus, we can evaluate $X(s)$ on the imaginary axis to obtain

$$X(\omega) = \frac{1}{j\omega + a}.$$
(14.1-13)

Figure 14.5
The one-sided
exponential waveform

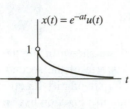
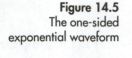

Let's recapitulate a bit. We have worked several examples of the Fourier transforms of time-limited functions. We have just analyzed one that was not time limited, but that did approach zero sufficiently fast as $t \to \infty$ for the integral to converge when $s = j\omega$ (another way of saying that the ROC of the Laplace transform includes the imaginary axis in the s plane). We now ask what happens if this is not true. Suppose the waveform does not approach zero sufficiently fast—or perhaps does not approach zero at all as $t \to \infty$. What then? We will answer this question by considering a particular example, though it is the prototype for all waveforms having this behavior. In the process, we will extend our definition of the Fourier transform to include such functions.

Convergence of the
Fourier Integral: An
Important Example

The waveform we will investigate is the unit step function, $u(t)$. To refresh your memory about the behavior of this basic waveform, we have sketched it in Figure 14.6(a). We know that the Laplace transform of this waveform is $X(s) = 1/s$. Can we evaluate this for $s = j\omega$? It might seem so, but in fact we cannot. The problem is that it has a pole at $s = 0$ and the ROC is given by $\sigma = \text{Re}\{s\} > 0$, a strict inequality. We show this in Figure 14.6(b). Thus, we cannot simply evaluate the Laplace transform on the $j\omega$ axis to obtain the Fourier transform. Looking at things another way, the improper integral

$$\int_{-\infty}^{\infty} u(t)e^{-j\omega t}\, dt = \left.\frac{e^{-j\omega t}}{-j\omega}\right]_{-\infty}^{\infty}$$
(14.1-14)

does not exist as an ordinary limit. We will leave its computation as a *generalized limit* for the end of chapter problems.

Figure 14.6
The unit step function
and the PZD of its
Laplace transform

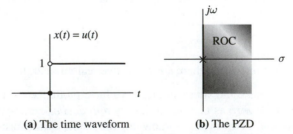

(a) The time waveform   **(b)** The PZD

Here is what we will do.[4] We observe that the decaying exponential waveform that we worked with in Example 14.5 approaches the unit step function if we allow the parameter a to approach zero. We show this as the waveform $x_a(t)$ in Figure 14.7(a). If we restrict a to have only positive real values, the ROC is the region shown in Figure 14.7(b).

---

[4] This example was suggested to the author by a colleague, Michael O'Flynn.

We will refer to e^{-at} as a *convergence factor* because it is merely introduced into the problem to ensure convergence. The Laplace transform of the unit step function with our "convergence factor" thrown in is equation (14.1-12), repeated here for ease of reference with $x(t)$ replaced by $x_a(t)$:

$$X_a(s) = \pounds\{e^{-at}u(t)\} = \frac{1}{s+a}, \qquad (14.1\text{-}15)$$

which has a pole at $s = -a$. This, as usual, is shown by an x in our s-plane plot in Figure 14.7(b). For each positive value of a, the region of convergence includes the $j\omega$ axis, as shown in the figure. Thus, we can evaluate the Laplace transform on the $j\omega$ axis for each positive, nonzero real value of a to get

$$X_a(\omega) = [X_a(s)]_{s=j\omega} = \pounds\{x_a(t)\}_{s=j\omega} = \frac{1}{j\omega + a}. \qquad (14.1\text{-}16)$$

Figure 14.7
The behavior of the convergence factor and its ROC

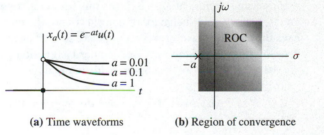

(a) Time waveforms   (b) Region of convergence

If we write out the process just explained as an integral, we see that

$$X_a(\omega) = \int_{-\infty}^{\infty} e^{-at}u(t)e^{-j\omega t}\,dt = \frac{1}{j\omega + a}. \qquad (14.1\text{-}17)$$

Our problem now is to resolve the question: What is the limiting behavior of $X_a(\omega)$ as a approaches zero through positive values? To do this, let's rationalize the complex fraction in (14.1-17), putting it in rectangular form:

$$X_a(\omega) = \frac{1}{j\omega + a} = \frac{a - j\omega}{a^2 + \omega^2} = F_a(\omega) + jG_a(\omega). \qquad (14.1\text{-}18)$$

We will investigate the behavior of the real part function $F_a(\omega)$ and the imaginary part function $G_a(\omega)$ individually as $a \to 0+$.

Taking the imaginary part first, we can write

$$G_a(\omega) = \frac{-\omega}{a^2 + \omega^2} \longrightarrow \begin{cases} -\dfrac{1}{\omega}; & \omega \neq 0 \\ 0; & \omega = 0 \end{cases}. \qquad (14.1\text{-}19)$$

This function is sketched in Figure 14.8. It is a rather well-behaved waveform, having the value 0 at $\omega = 0$ and being a hyperbola elsewhere. We now note an important fact: finite values of a function at isolated points[5] do not affect the value of the integral of that func-

---

[5] A single point, a finite number of points, or an infinite number of points of which only a finite number lie in any interval of finite length are typical examples of what we mean by "isolated points."

Figure 14.8 The imaginary part

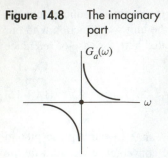

tion—and, as we will see in the next section, the time function can be expressed in terms of its transform by *means* of an integral. Thus, we will simply treat all waveforms that differ by finite values on a set of isolated points as equivalent. Hence, we can write

$$\lim_{a \to 0+} G_a(\omega) = -\frac{1}{\omega}. \tag{14.1-20}$$

Let's look at the real part function. Here we have

$$F_a(\omega) = \frac{a}{a^2 + \omega^2} \longrightarrow \begin{cases} \infty; & \omega = 0 \\ 0; & \omega \neq 0 \end{cases}. \tag{14.1-21}$$

Figure 14.9 The real part

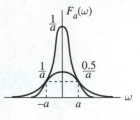

We show a sketch of $F_a(\omega)$ for two different positive values of a in Figure 14.9. The peak occurs at $\omega = 0$, where the value is $1/a$. At $\omega = a$, the value is $0.5/a$. Thus, the smaller value of a corresponds to the "taller" and "skinnier" sketch. Now, even though $F_a(\omega)$ approaches zero everywhere except on an isolated set (the single point $\omega = 0$ in this case), it approaches *infinity* at $\omega = 0$. Does this behavior suggest something? It should—this is precisely the type of behavior of our old friend, the impulse function. We suspect that perhaps the limiting waveform *is* an impulse, so let's check our suspicions. Recalling that the defining property of the impulse singularity function is that the limit of the running integral be a step function, we compute this integral:

$$\int_{-\infty}^{\omega} F_a(\alpha)\, d\alpha = \int_{-\infty}^{\omega} \frac{a}{a^2 + \alpha^2}\, d\alpha = \frac{\pi}{2} + \tan^{-1}\left[\frac{\omega}{a}\right] \longrightarrow \pi u(\omega). \tag{14.1-22}$$

Figure 14.10 The running integral of the real part

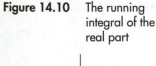
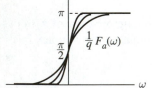

In order to verify this integral, either look it up in a table of integrals or, preferably, do the integration using the trigonometric substitution $\alpha = a \tan(\phi)$. In Figure 14.10 we show this integral for several successively smaller values of a (the ones with smaller values of a vary the most rapidly). In Figure 14.10, we have used the notation

$$\frac{1}{q} F_a(\omega) = \int_{-\infty}^{\omega} F_a(\alpha)\, d\alpha \tag{14.1-23}$$

to mean integration with respect to ω. The key result we have just shown is that

$$\lim_{a \to 0+} \frac{1}{q} F_a(\omega) = \pi u(\omega). \tag{14.1-24}$$

We note that we are no longer demanding that $u(\omega)$ be continuous from the left, but are considering any waveform that is zero for $\omega < 0$ and 1 for $\omega > 0$ to be the unit step function. As we mentioned previously, it is only the integral properties of a waveform that will be of interest; hence, we will consider any set of waveforms that differ only at discontinuity points (by finite values) to be equivalent. We have shown, therefore, that

$$\lim_{a \to 0+} F_a(\omega) = \pi \delta(\omega). \tag{14.1-25}$$

Putting our two results together, we see that

$$\lim_{a \to 0+} \pounds\{u(t)\}_{s=j\omega} = \pi \delta(\omega) + \frac{1}{j\omega}. \tag{14.1-26}$$

We will take this limit to be the Fourier transform, whose definition we will now generalize to include such waveforms that do not decay to zero as time becomes large.

**Definition of the
Fourier Transform:
One-Sided Waveforms**

Definition 14.2   Let $x(t)$ be a one-sided waveform for which the improper integral defining the Laplace transform does not necessarily converge for $s = j\omega$. We define its Fourier transform to be

$$X(\omega) = \lim_{a \to 0+} [X(s + a)]_{s=j\omega} = \lim_{a \to 0+} \pounds\{e^{-at}x(t)\}_{s=j\omega} \qquad (14.1\text{-}27)$$

$$= \lim_{a \to 0+} \mathscr{F}\{e^{-at}x(t)\}_{s=j\omega} = \lim_{a \to 0+} \int_{-\infty}^{\infty} e^{-at}x(t)e^{-j\omega t}\, dt.$$

Each of these forms is equivalent to all the others. We have just shown therefore that

$$\mathscr{F}\{u(t)\} = \pi\delta(\omega) + \frac{1}{j\omega}. \qquad (14.1\text{-}28)$$

The first equivalence in our definition is the frequency shift property of Laplace transforms (entry 3, Table 13.6 in Section 13.3 of Chapter 13). This property will provide us with an effective working tool for computing Fourier transforms.

Example 14.6   Find the Fourier transform of $x(t) = \cos(\omega_0 t)u(t)$.

Solution   Let's use the properties of the Laplace transform and our previous result on the Fourier transform of the unit step function. We use Euler's formula to express the cosine function in terms of exponentials:

$$x(t) = \frac{e^{j\omega_0 t} + e^{-j\omega_0 t}}{2}u(t). \qquad (14.1\text{-}29)$$

Thus,

$$X(s) = \frac{1}{2}\left[\frac{1}{s - j\omega_0} + \frac{1}{s + j\omega_0}\right]; \qquad (14.1\text{-}30)$$

hence, we have

$$[X(s + a)]_{s=j\omega} = \frac{1}{2}\left[\frac{1}{j\omega + a - j\omega_0} + \frac{1}{j\omega + a + j\omega_0}\right] \qquad (14.1\text{-}31)$$

$$= \frac{1}{2}\left[\frac{1}{a + j(\omega - \omega_0)} + \frac{1}{a + j(\omega + \omega_0)}\right].$$

We have, however, already shown above that

$$\lim_{a \to 0+} \frac{1}{j\omega + a} = \pi\delta(\omega) + \frac{1}{j\omega}. \qquad (14.1\text{-}32)$$

Each of the terms in equation (14.1-31) is merely a frequency shifted version of (14.1-32); hence,

$$\mathscr{F}\{\cos(\omega_0 t)u(t)\} = \frac{\pi}{2}\delta(\omega - \omega_0) + \frac{\pi}{2}\delta(\omega + \omega_0) + j\frac{\omega}{\omega_0^2 - \omega^2}. \qquad (14.1\text{-}33)$$

Example 14.7   Find the Fourier transform of $x(t) = \sin(\omega_0 t)u(t)$.

Solution   Because this example is so similar to the last one, we can do it quite quickly. We just note that Euler's formula for the sine function is the same as that for the cosine, except for the fact that the two exponentials are subtracted, and the result divided by $2j$ rather than by only 2. Thus, we have

$$[X(s + a)]_{s=j\omega} = \frac{1}{2j}\left[\frac{1}{j\omega + a - j\omega_0} - \frac{1}{j\omega + a + j\omega_0}\right] \tag{14.1-34}$$

$$= \frac{1}{2j}\left[\frac{1}{a + j(\omega - \omega_0)} - \frac{1}{a + j(\omega + \omega_0)}\right].$$

Hence,

$$F\{\sin(\omega_0 t)u(t)\} = -j\frac{\pi}{2}\delta(\omega - \omega_0) + j\frac{\pi}{2}\delta(\omega + \omega_0) + \frac{\omega_0}{\omega_0^2 - \omega^2}. \tag{14.1-35}$$

To recapitulate, if a waveform is time limited or, in general, has a convergent Fourier transform integral, we merely compute the Laplace transform and evaluate it on the $j\omega$ axis to determine the Fourier transform. If the time waveform is not time limited (but is one sided as per our usual assumption) and the improper integral does not converge, we multiply the waveform by the convergence factor e^{-at}, where a is an arbitrary positive real number, take the Laplace transform, evaluate it on the $j\omega$ axis, and then compute the limit as $a \to 0+$ to determine the Fourier transform.

Two-Sided Waveforms   We now investigate the Fourier transform of two-sided waveforms, those that are not necessarily zero for $t \leq 0$. Let's start our development by looking at an example. Figure 14.11 shows a simple circuit with a sinusoidal forcing function. We remind you that the terminal labeled with $v_i(t)$ is an *implied source;* that is, it represents a voltage source applied from that terminal to the ground reference. We will leave it to you to analyze the circuit, using either Heaviside operator or Laplace transform techniques (they are really the same, as we noted in Chapter 13), and thereby show that the response is

$$v_o(t) = [\cos(2t) + \sin(2t)]u(t) - e^{-2t}u(t) \text{ V.} \tag{14.1-36}$$

The natural response is

$$v_{on}(t) = -e^{-2t}u(t), \tag{14.1-37}$$

and the forced response is

$$v_{of}(t) = [\cos(2t) + \sin(2t)]u(t) = \sqrt{2}\cos(2t - 45°). \tag{14.1-38}$$

We have used the trigonometric identity $\cos(x)\cos(y) + \sin(x)\sin(y) = \cos(x - y)$ in (14.1-38) with $x = 2t$ and $y = 45°$ (whose cosine and sine are both $1/\sqrt{2}$). Our circuit is

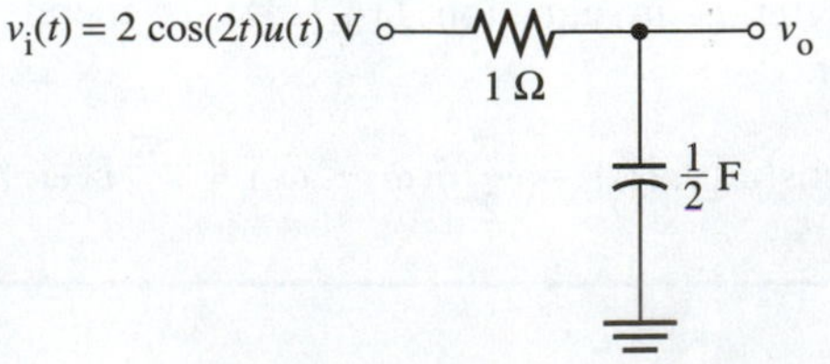

Figure 14.11
An example circuit

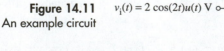

$v_i(t) = 2\cos(2t)u(t)$ V $1\,\Omega$ $\circ\, v_o$ $\frac{1}{2}$ F

clearly stable because the natural response decays to zero with time. Thus, the forced response is also the steady-state response; that is, it is approximately equal to the complete response for large t.

Let's look a bit more closely at this response. In Figure 14.12 we have plotted the complete response for several periods of the sinusoidal component (we used MATLAB to plot the waveform). Now here is the key thing to notice: after a certain time, the natural response has decayed to insignificance and the entire waveform consists of only the forced component. If we are only interested in this component, *we can ignore the initial part of the waveform and imagine that the forced response is extended backward, not only to $t = 0$, but to all negative values of time as well. Thus, for stable systems operating in the steady state (and only for them) we will consider all waveforms to be two sided.*

Figure 14.12
Time response

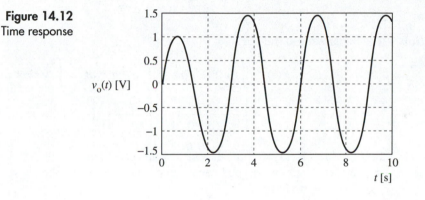

$v_o(t)$ [V]

Example: Fourier Transform of a Two-Sided (but Time-Limited) Waveform

Example 14.8   Compute the Fourier transform of the waveform $x(t) = P_{2T}(t + T)$.

Solution   We have sketched this two-sided, but time-limited, waveform in Figure 14.13. We have shown it as continuous from the left at each point of discontinuity, but our result for the Fourier transform will be valid regardless of how we define the waveform at these points. We compute the Laplace transform and evaluate it at $s = j\omega$:

$$X(s) = \mathcal{L}\{P_{2T}(t + T)\} = \int_{-T}^{T} 1e^{-st}\, dt \tag{14.1-39}$$

$$= \frac{e^{sT} - e^{-sT}}{s} = \frac{e^{j\omega T} - e^{-j\omega T}}{j\omega} = 2T\frac{\sin(\omega T)}{\omega T}.$$

Figure 14.13
A shifted pulse function

$x(t) = P_{2T}(t + T)$

Clearly, no modifications in our procedure are necessary for time-limited waveforms—even if they are not zero for $t \leq 0$; however, not all waveforms are so nice to work with. Some waveforms of interest, particularly in the solution of circuits in the ac steady state, are considered to be two sided and are not time limited. One such waveform is considered in the next example.

Example: Fourier Transform of a Non-Time-Limited Two-Sided Waveform

Example 14.9   Find the Fourier transform of

$$x(t) = e^{-a|t|}. \tag{14.1-40}$$

Solution   The Laplace transform is

$$X(s) = \int_{-\infty}^{\infty} e^{-a|t|} e^{-st} \, dt = \int_{0}^{\infty} e^{-at} e^{-st} \, dt + \int_{-\infty}^{0} e^{at} e^{-st} \, dt \tag{14.1-41}$$

$$= \left[\frac{e^{-(s+a)t}}{-(s+a)} \right]_{0}^{\infty} + \left[\frac{e^{(a-s)t}}{(a-s)} \right]_{-\infty}^{0}.$$

Now if $\sigma = \mathrm{Re}\{s\} > -a$, the first term approaches zero at the upper limit; if $\sigma = \mathrm{Re}\{s\} < a$, the second one approaches zero at the lower limit. Thus, if $-a < \sigma < a$, we have

$$X(s) = \frac{1}{a + s} + \frac{1}{a - s} = \frac{2a}{a^2 - s^2}. \tag{14.1-42}$$

Evaluating it at $s = j\omega$ gives the Fourier transform:

$$X(\omega) = \left[\frac{2a}{a^2 - s^2} \right]_{s=j\omega} = \frac{2a}{a^2 + \omega^2}. \tag{14.1-43}$$

The time waveform, Fourier transform, and ROC of $X(s)$ are plotted in Figure 14.14. We were able to plot the Fourier transform without difficulty because it is purely real. We will be using this particular waveform and its transform often in the following pages.

Figure 14.14
The example waveform and its Fourier transform

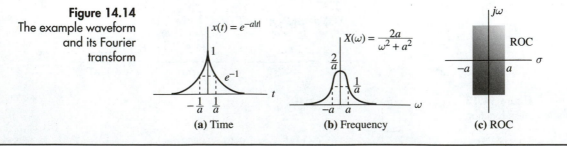

(a) Time   (b) Frequency   (c) ROC

The waveform in the preceding example is one of only very few truly two-sided waveforms that have Fourier transforms in the conventional sense (that is, without impulses in the transform). As we will see, it can be used to advantage in computing other transforms.

**The Fourier Transform
of a Constant**

Let's consider the computation of what is perhaps the simplest two-sided function of all: the dc waveform having the value unity for all time. Let's start by attempting to compute the Laplace transform directly. Thus,

$$\pounds\{1\} = \int_{-\infty}^{\infty} 1 \cdot e^{-st}\, dt = \lim_{T \to \infty} \frac{e^{sT} - e^{-sT}}{s}. \tag{14.1-44}$$

Unfortunately, however, this limit does not exist in the conventional sense. If $\sigma = \text{Re}\{s\}$ is greater than zero, the first term in the numerator increases without bound. If the opposite is true, that is, σ is less than zero, the second term has the same behavior. If $\sigma = 0$, we have

$$\pounds\{1\} = \frac{e^{j\omega T} - e^{-j\omega T}}{j\omega} = 2\frac{\sin(\omega T)}{\omega}. \tag{14.1-45}$$

The evaluation of the limit of this expression as $T \to \infty$ is somewhat difficult; in fact, we cannot do it justice within the limits of this text.[6] For finite T, it is called the *sinc function*.

**A Two-Sided
Convergence Factor**

As it happens, there is another way to approach the computation of this transform that is a simple extension of the way we computed the transform of the unit step function. Recall that we introduced the convergence factor e^{-at}, computed the resulting Laplace transform, and then took the limit as a approached zero through positive values. Let's do the same thing, but extend it to cover two-sided waveforms. A glance at Figure 14.15 should convince you that $e^{-a|t|}$ approaches our dc waveform of unit value as the parameter a approaches zero. We will refer to this waveform as a *two-sided convergence factor*. Now recall our result in Example 14.9, equation (14.1-3). If we define $x_a(t) = e^{-a|t|}$, we will have

$$X_a(\omega) = \frac{2a}{\omega^2 + a^2}. \tag{14.1-46}$$

But (see equations (14.1-24) and (14.1-25)), we know that

$$\lim_{a \to 0+} X_a(\omega) = 2\pi\delta(\omega). \tag{14.1-47}$$

Therefore, we define the Fourier transform of the two-sided waveform $x(t) = 1$ to have the Fourier transform $2\pi\delta(\omega)$, or $\mathcal{F}\{1\} = \pounds\{1\}_{s=j\omega} = 2\pi\delta(\omega)$, or

$$1 \longleftrightarrow 2\pi\delta(\omega). \tag{14.1-48}$$

Let's generalize on this result with the following definition.

Figure 14.15
The two-sided
convergence factor

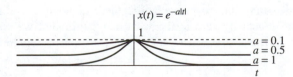

---

[6] It can be shown (with a fair amount of work) that it does exist as a generalized limit and has the value $2\pi\delta(\omega)$.

The Fourier Transform: General Definition

Definition 14.3   If $x(t)$ is a two-sided waveform that is not time limited, we define its Fourier transform to be

$$F\{x(t)\} = \lim_{a \to 0+} \pounds\{e^{-a|t|}x(t)\}_{s=j\omega} = \lim_{a \to 0+} \int_{-\infty}^{\infty} e^{-a|t|}x(t)e^{-j\omega t}\, dt. \qquad (14.1\text{-}49)$$

Looking back at Definition 14.1, we see that this is a generalization of it to cover two-sided waveforms that are not time limited, just as Definition 14.2 is a generalization for one-sided waveforms that are not time limited. In fact, we can take (14.1-49) to be the basic definition now and specialize it to the previous two cases. If $x(t)$ is one sided, equation (14.1-49) becomes (14.1-27); if $x(t)$ is time limited or approaches zero fast enough (as, for instance, is the case with $e^{-at}u(t)$ and $e^{-a|t|}$), it becomes (14.1-2). We will take the pragmatic attitude, therefore, that (14.1-49) defines the Fourier transform and we will then include or not include the appropriate convergence factor as needed.

Example 14.10   Find the Fourier transform of the *signum function*,

$$x(t) = \text{sgn}(t) = \begin{cases} 1; & t > 0 \\ 0; & t = 0. \\ -1; & t < 0 \end{cases} \qquad (14.1\text{-}50)$$

Solution   A sketch of the signum function is shown in Figure 14.16. Note that the definition of $x(t)$ at $t = 0$ has no effect on the transform; however, we will specify it to be zero because that is perhaps the most common definition. Now we use (14.1-49) to compute

$$X_a(\omega) = \lim_{a \to 0+} \int_{-\infty}^{\infty} e^{-a|t|}\,\text{sgn}(t)e^{j\omega t}\, dt \qquad (14.1\text{-}51)$$

$$= \lim_{a \to 0+} \left[\int_0^{\infty} e^{-at}\cdot 1\cdot e^{-j\omega t}\, dt + \int_{-\infty}^0 e^{at}(-1)e^{-j\omega t}\, dt \right]$$

$$= \lim_{a \to 0+} \left[\frac{1}{a + j\omega} - \frac{1}{a - j\omega} \right]$$

$$= \left[\pi\delta(\omega) + \frac{1}{j\omega} \right] - \left[\pi\delta(\omega) - \frac{1}{j\omega} \right] = \frac{2}{j\omega}.$$

Notice that, due to cancellation, there is no impulse at $\omega = 0$. Notice also that the limit is actually zero at the value $\omega = 0$, though we will not stress this and will simply write $2/j\omega$. An alternative to the foregoing method is to evaluate

Figure 14.16
The signum function

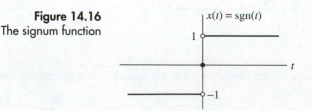

$x(t) = \text{sgn}(t)$

$1/(a + j\omega) - 1/(a - j\omega) = -j2\omega/(a^2 + \omega^2)$ and observe that its limit is 0 for $\omega = 0$ and $2/j\omega$ for $\omega \neq 0$.

Example 14.11   Find the Fourier transform of $x(t) = \cos(\omega_0 t)$.

Solution   Observe that this waveform is the two-sided version of a waveform we have already transformed, $\cos(\omega_0 t)u(t)$. We recall here that

$$\pounds\{e^{-a|t|}\cos(\omega_0 t)\} = \pounds\left\{ \frac{e^{j\omega_0 t} + e^{-j\omega_0 t}}{2} e^{-a|t|} \right\} \tag{14.1-52}$$

$$= \frac{a}{(\omega - \omega_0)^2 + a^2} + \frac{a}{(\omega + \omega_0)^2 + a^2}.$$

These two expressions are merely frequency-translated versions of a waveform whose limit we have already discussed relative to equations (14.1-46) and (14.1-47). Thus, letting $a \rightarrow 0+$, we get

$$\pounds\{e^{-a|t|}\cos(\omega_0 t)\} = \pi\delta(\omega - \omega_0) + \pi\delta(\omega + \omega_0). \tag{14.1-53}$$

Example 14.12   Find the Fourier transform of $x(\ ) = \sin(\omega_0 t)$.

Solution   We follow the same pattern, using Euler's formula for the sine function. Now, however, we divide by $2j$ (rather than 2) and subtract the two terms. The result is

$$\pounds\{e^{-a|t|}\sin(\omega_0 t)\} = -j\pi\delta(\omega - \omega_0) + j\pi\delta(\omega + \omega_0). \tag{14.1-54}$$

We conclude this section with Table 14.1 to summarize our results. Notice that the first five transforms in Table 14.1 are one sided, and the last three are two sided. We should probably also alert you to the fact that our definition of the Fourier transform of a waveform differs from the one most often seen in introductory circuits texts. It is based upon a particular notion of convergence of an improper integral, which we will discuss in the next section.

Table 14.1
Summary of
Fourier transforms

$x(t)$	$X(\omega)$
1. $\delta(t - a)$	$e^{-j\omega a}$
2. $e^{-at}u(t)$	$1/(j\omega + a)$
3. $u(t)$	$\pi\delta(\omega) + 1/j\omega$
4. $\cos(\omega_0 t)u(t)$	$\dfrac{\pi}{2}\delta(\omega - \omega_0) + \dfrac{\pi}{2}\delta(\omega + \omega_0) + \dfrac{j\omega}{\omega_0^2 - \omega^2}$
5. $\sin(\omega_0 t)u(t)$	$-j\dfrac{\pi}{2}\delta(\omega - \omega_0) + j\dfrac{\pi}{2}\delta(\omega + \omega_0) + \dfrac{\omega_0}{\omega_0^2 - \omega^2}$
6. 1	$2\pi\delta(\omega)$
7. $\cos(\omega_0 t)$	$\pi\delta(\omega - \omega_0) + \pi\delta(\omega + \omega_0)$
8. $\sin(\omega_0 t)$	$-j\pi\delta(\omega - \omega_0) + j\pi\delta(\omega + \omega_0)$

Section 14.1 Quiz

Q14.1-1. Find the Fourier transform of the waveform shown in Figure Q14.1-1. (*Hint:* Take two generalized derivatives and use the time differentiation property of the Laplace transform.)

Figure Q14.1-1

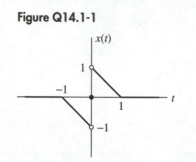

Q14.1-2. Find the Fourier transform of the waveform $x(t) = e^{-2t}\cos(3t)u(t)$. (*Hint:* Express $\cos(3t)$ in terms of exponentials using Euler's formula, then apply the frequency shift property of the Laplace transform.

Q14.1-3. Find the Fourier transform of the waveform $x(t) = 2 + \cos(2t) + \sin(3t)$. (*Hint:* Just use the linearity property of Laplace transforms and the known transforms in Table 14.1.

14.2 | The Inverse Fourier Transform

Improper Integrals

In the last section we defined and worked with the Fourier transform. It is called a "transform" because, like the Laplace transform, it transforms one function into another. In fact, it operates on a time waveform to produce a function of frequency—a different variable. We observed in the course of our development that the Fourier transform

$$\mathscr{F}\{x(t)\} = \int_{-\infty}^{\infty} x(t)e^{-j\omega t}\, dt \tag{14.2-1}$$

is an improper integral. As such, it generally involves taking a limit. If $x(t)$ is time limited, none is in fact required. If not, the limit must actually be computed.

Let's review the idea behind improper integrals. The (two-sided) improper integral of an arbitrary time function $f(t)$ is defined to be the limit

$$I = \lim_{A,B\to\infty} \int_{-A}^{B} f(t)\, dt \tag{14.2-2}$$

whenever this limits exists. Note that A and B are permitted to approach infinity independently. Many time functions of interest in circuits and systems analysis fail, unfortunately, to have convergent improper integrals.[7] The condition that A and B be allowed to approach infinity independently is too strong a requirement.

Another definition of the improper integral is called the *Cauchy principal value*:

$$I = \lim_{A\to\infty} \int_{-A}^{A} f(t)\, dt. \tag{14.2-3}$$

In this case, both limits approach infinity at the same rate. This permits more waveforms to have improper integrals. To gain an idea of what is happening, let's consider the computation of the integral of the signum function, whose Fourier transform was computed in

---

[7] Note, for example, the waveform $x(t) = 1$ for all t—the constant 1 (as discussed in Section 14.1)!

Example 14.10 in the last section.

$$\text{sgn}(t) = \begin{cases} 1; & t > 0 \\ 0; & t = 0. \\ -1; & t < 0 \end{cases} \tag{14.2-4}$$

A sketch of this waveform is shown in Figure 14.17. The value at $t = 0$ is, of course, not important because it does not affect the value of the integral; we have, however, defined it to be zero to be consistent with common practice.

Figure 14.17
The signum function

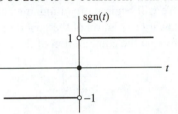

Now let's look at how to compute the improper integral.[8] We show the two interpretations graphically in Figure 14.18. Figure 14.18(a) shows, via shading, the computation of the (more stringent) improper integral, and Figure 14.18(b) shows the computation of the (less stringent) Cauchy principal value. If we keep A fixed and allow B to become infinitely large (the more stringent case), the integral approaches infinity. If we then allow A to approach infinity, the integral remains constant at $+\infty$. If we reverse the procedure, however, allowing A to approach infinity first, the integral value is $-\infty$. Allowing B to approach infinity does not alter this value. Thus, we have computed two different values, depending on the way we allow A and B to approach infinity. Clearly, no unique limit exists.

On the other hand, if we consider the Cauchy principal value illustrated in Figure 14.18(b), we see that allowing both points to recede to infinity at the same rate causes the integral to exist and have the value zero.

Figure 14.18
Two interpretations of the improper integral

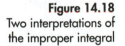

(a)   (b)

The Poisson Integral   The result of our argument is simply this: an improper integral involves a limiting operation, and the existence of the integral depends in general on the way we compute the limit. For the computation of Fourier transforms, we have seen that a third way of interpreting the integral has value. We define the (improper) *Poisson integral*[9] of any waveform $f(t)$ to be[10]

$$\int_{-\infty}^{\infty} f(t) \, dt = \lim_{a \to 0+} \int_{-\infty}^{\infty} e^{-a|t|} f(t) \, dt. \tag{14.2-5}$$

---

[8] Remember that we are *not* computing a transform, only an integral.

[9] Pronounced "Pwa-sanh", with the n not articulated (a nasal sound).

[10] Good references for this definition are Seeley, R. T., *An Introduction to Fourier Series and Integrals,* W. A. Benjamin, New York, 1966; and Rees, Shah, and Stanojevic, *Theory and Application of Fourier Analysis,* Marcel Dekker, New York, 1981. The former small text is worth reading in its entirety. The latter calls our type of convergence *Abel convergence*, rather than Poisson convergence.

Thus, we compute the integral as the limit of a family of improper integrals (depending on the parameter a), each of which exists in the strict sense.[11] The study of convergence of Fourier transforms when integrals are interpreted in the Cauchy principal value sense is associated with the work of the mathematician Dirichlet. It was the French mathematician Poisson, however, who studied the convergence of Fourier transforms when integrals are interpreted in the sense of (14.2-5). This is the approach we will use.

A Fundamental "Transform Pair"   We will now investigate how to "undo" the Fourier transform. That is, once we have a frequency function $X(\omega)$, we would like to go the other way and derive the corresponding time waveform $x(t)$. We will begin with a fundamental transform pair, and our analysis of it will point the way toward a solution of the general problem.

Let's look, therefore, at the computation of the integral

$$\frac{1}{2\pi}\int_{-\infty}^{\infty} e^{j\omega t}\, d\omega. \tag{14.2-6}$$

Because t is a parameter, the result will be a time function. Let's look at each of the three different ways of interpreting this integral. Using the strict definition, we have

$$\frac{1}{2\pi}\int_{-\infty}^{\infty} e^{j\omega t}\, d\omega = \lim_{A,B\to\infty}\frac{1}{2\pi}\int_{-A}^{B} e^{j\omega t}\, d\omega = \lim_{A,B\to\infty}\frac{e^{jBt}-e^{-jAt}}{j2\pi t}, \tag{14.2-7}$$

which clearly does not exist. The Cauchy principal value, on the other hand, is

$$\frac{1}{2\pi}\int_{-\infty}^{\infty} e^{j\omega t}\, d\omega = \lim_{A\to\infty}\frac{1}{2\pi}\int_{-A}^{A} e^{j\omega t}\, d\omega = \lim_{A\to\infty}\frac{e^{jAt}-e^{-jAt}}{j2\pi t} = \lim_{A\to\infty}\frac{\sin(At)}{\pi t}. \tag{14.2-8}$$

A plot of this waveform is shown in Figure 14.19. We see that it is zero at all multiples of π/A, has relative maxima between these values that get smaller with increasing t, and approaches A/π as $t\to 0$ (apply L'Hôpital's rule to verify this). This waveform is a slight modification of what is called the *sinc* function:

$$\mathrm{sinc}(x) = \frac{\sin(x)}{x} \tag{14.2-9}$$

for any variable x. The sinc(x) function appears in many places in transform theory.

Figure 14.19
The sinc function

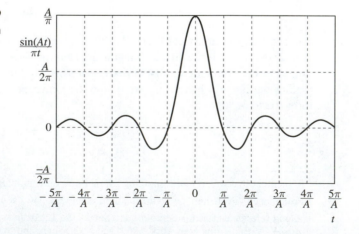

We might suspect that as $A \to \infty$ this waveform approaches an impulse function. For instance, we see that the height at $t = 0$ approaches infinity. Furthermore, if we "rough out" the area of the "principal lobe"—the part between $-\pi/A$ and π/A—by approximating it with a rectangle, we find that it is $A/\pi \times 2\pi/A = 2$, a constant. As a matter of fact, our conjecture is true, but the mathematical derivation is quite difficult. For this reason, we will use our third interpretation of the improper integral: the Poisson integral. We are now in a position to compute a formula that you should *memorize*:

$$\frac{1}{2\pi} \int_{-\infty}^{\infty} e^{j\omega t}\, d\omega = \lim_{a\to 0+} \frac{1}{2\pi} \int_{-\infty}^{\infty} e^{-a|\omega|} e^{j\omega t}\, d\omega = \lim_{a\to 0+} \frac{a}{\pi(a^2 + t^2)} = \delta(t). \qquad (14.2\text{-}10)$$

In the last step, we used our known result derived in the last section (equation (14.1-25)) with t replacing ω in that result. This is clearly a lot easier, and it is for this reason that we adopt the Poisson definition. To summarize, we have shown that with a suitable definition of the improper integral—and the Poisson version is the only one of the three we have discussed that works well—we have

$$\frac{1}{2\pi} \int_{-\infty}^{\infty} e^{j\omega t}\, d\omega = \delta(t). \qquad (14.2\text{-}11)$$

We will now use this result to derive a far-reaching result in Fourier transform theory, known either as the *inverse Fourier transform* or the *inversion integral*.

Theorem 14.1   If

$$X(\omega) = \mathcal{F}\{x(t)\} = \int_{-\infty}^{\infty} x(t) e^{-j\omega t}\, dt \qquad (14.2\text{-}12)$$

is the Fourier transform of $x(t)$, then

$$x(t) = \mathcal{F}^{-1}\{X(\omega)\} = \frac{1}{2\pi} \int_{-\infty}^{\infty} X(\omega) e^{j\omega t}\, d\omega. \qquad (14.2\text{-}13)$$

Proof   With our immediately preceding result, the proof is quite easy. We write

$$\frac{1}{2\pi} \int_{-\infty}^{\infty} X(\omega) e^{j\omega t}\, d\omega = \frac{1}{2\pi} \int_{-\infty}^{\infty} \left[\int_{-\infty}^{\infty} x(\alpha) e^{-j\omega\alpha}\, d\alpha \right] e^{j\omega t}\, d\omega \qquad (14.2\text{-}14)$$

$$= \int_{-\infty}^{\infty} x(\alpha) \left[\int_{-\infty}^{\infty} e^{j\omega(t-\alpha)}\, d\omega \right] d\alpha \longrightarrow \int_{-\infty}^{\infty} x(\alpha)\delta(t - \alpha)\, d\alpha = x(t)$$

at all points of continuity. Notice that we have used the variable α in the Fourier integral representation of $X(\omega)$ because t has already been used in the original exponential. In the second step we interchanged the order of integration, then noted that the inner integral defines the unit impulse function. Finally, we observe that we are considering a waveform to be uniquely defined if we know its value at all points of continuity.   ∎

We should perhaps further discuss the exchange of integrals in equation (14.2-14) of the foregoing proof. Each of the improper integral signs means "if the improper integral does not exist, then multiply the integrand by a two-sided convergence factor, integrate, and take the limit." We have chosen not to belabor this issue here for it is notationally a bit complex, but will instead present several examples.

Example 14.13 Find the inverse Fourier transform of $X(\omega) = P_{2\omega_0}(\omega + \omega_0)$.

Solution We remind you that the unit pulse waveform of duration $2\omega_0$ (translated to the left by ω_0, to be sure) has the appearance sketched in Figure 14.20. The computation is quite simple. We merely write

$$\frac{1}{2\pi}\int_{-\infty}^{\infty} P_{2\omega_0}(\omega - \omega_0)e^{j\omega t}\,d\omega = \frac{1}{2\pi}\int_{-\omega_0}^{\omega_0} e^{j\omega t}\,d\omega \qquad (14.2\text{-}15)$$

$$= \frac{e^{j\omega_0 t} - e^{-j\omega_0 t}}{j2\pi t} = \frac{\sin(\omega_0 t)}{\pi t}.$$

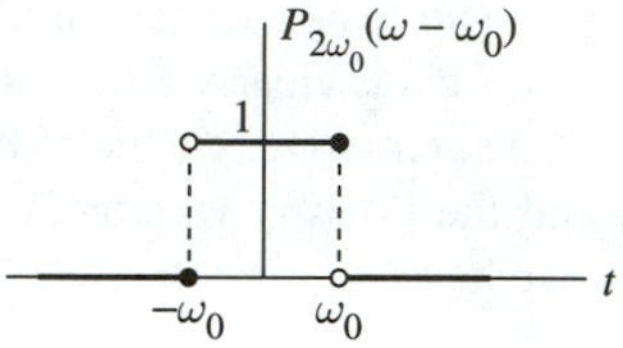

Figure 14.20
A pulse function

Example 14.14 Find the inverse Fourier transform of $X(\omega) = 1$.

Solution We have already done the computation, for equation (14.2-11) shows that

$$x(t) = \mathscr{F}^{-1}\{1\} = \frac{1}{2\pi}\int_{-\infty}^{\infty} e^{j\omega t}\,d\omega = \delta(t). \qquad (14.2\text{-}16)$$

Example 14.15 Find the inverse Fourier transform of $X(\omega) = u(\omega)$.

Solution We must use a convergence factor for this integral; because $X(\omega)$ is zero for negative values of ω, however, this integral is one sided:

$$x(t) = \mathscr{F}^{-1}\{u(\omega)\} = \lim_{a\to 0+} \frac{1}{2\pi}\int_{-\infty}^{\infty} e^{-a|\omega|}u(\omega)e^{j\omega t}\,d\omega \qquad (14.2\text{-}17)$$

$$= \lim_{a\to 0+} \frac{1}{2\pi}\int_{0}^{\infty} e^{-a\omega}e^{j\omega t}\,d\omega = \lim_{a\to 0+} \frac{1}{2\pi(a - jt)}$$

$$= \lim_{a\to 0+} \frac{a + jt}{2\pi(a^2 + t^2)} = \frac{1}{2}\delta(t) + j\frac{1}{2\pi t}.$$

Example 14.16 Find the inverse Fourier transform of $X(\omega) = \text{sgn}(\omega)$.

Solution Again, we use a convergence factor, this time a two-sided one:

$$x(t) = \mathscr{F}^{-1}\{\text{sgn}(\omega)\} = \lim_{a\to 0+} \frac{1}{2\pi}\int_{-\infty}^{\infty} e^{-a|\omega|}\text{sgn}(\omega)e^{j\omega t}\,d\omega \qquad (14.2\text{-}18)$$

$$= \lim_{a\to 0+} \frac{1}{2\pi}\left[\int_{0}^{\infty} e^{-a\omega}e^{j\omega t}\,d\omega - \int_{-\infty}^{0} e^{a\omega}e^{j\omega t}\,d\omega\right]$$

$$= \lim_{a\to 0+}\left[\frac{1}{2\pi(a - jt)} - \frac{1}{2\pi(a + jt)}\right] = \lim_{a\to 0+}\frac{jt}{\pi(a^2 + t^2)} = j\frac{1}{\pi t}.$$

Section 14.2 Quiz

Q14.2-1. Find the inverse Fourier transform of the frequency function shown in Figure Q14.2-1 by direct integration of the inverse transform integral.

Figure Q14.2-1

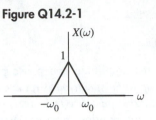

Q14.2-2. If $X(\omega) = e^{-k\omega}u(\omega)$, with k a positive real number, find $x(t)$.

Q14.2-3. If $X(\omega) = e^{-k|\omega|}$, with k a positive real number, find $x(t)$.

14.3 | Spectral Analysis

The Frequency Content of a Waveform

We have pointed out on a number of occasions that the frequency domain is a different way of looking at a signal and at the manner in which a circuit or system affects that signal. We are now in a position to be more precise. If we recall that an integral is merely the limiting case of a sum, we see that the inversion integral developed in Section 14.2 can be approximated by

$$x(t) \cong \sum_{k=-N}^{N} X(\omega_k)e^{j\omega_k t}\frac{\Delta\omega_k}{2\pi}, \tag{14.3-1}$$

where $\Delta\omega_k$ is an incremental frequency interval length, taken as small as one likes to obtain a required precision of approximation, and N is an integer taken large enough so that all frequency components $X(\omega_k)$ of appreciable size are included. Thus, we see that an arbitrary time function can be expressed as a linear combination of complex exponential waveforms. By using Euler's formula, we see that it can also be represented as a linear combination of sine and cosine waveforms. The point is that any arbitrary waveform can be considered to be constructed from a (perhaps large) number of sinusoids. In this section, we will explore this notion a bit further by developing procedures for plotting this "frequency content," that is, of graphing the strength of the sinusoid (or complex exponential) at a given frequency.

The Spectrum of a Waveform: Magnitude and Phase Spectra

Recall that for each value of ω the Fourier transform $X(\omega)$ of a time waveform $x(t)$ is a complex number. It can therefore be placed in Euler form. Thus, we have

$$X(\omega) = A(\omega)e^{j\theta(\omega)}, \tag{14.3-2}$$

where

$$A(\omega) = |X(\omega)| \tag{14.3-3}$$

is the *magnitude,* or *amplitude,* and

$$\theta(\omega) = \angle X(\omega) \tag{14.3-4}$$

is the *angle,* or *phase.* We define a plot of $A(\omega)$ versus ω to be the *amplitude spectrum* and a plot of $\theta(\omega)$ versus ω to be the *phase spectrum.* Together, they are referred to simply as *the spectrum of x(t).* Typical amplitude and phase spectra (spectra is the plural of spectrum) are shown in Figures 14.21(a) and (b), respectively.

Figure 14.21
Typical amplitude and
phase spectra

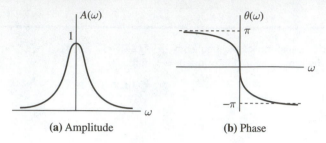

(a) Amplitude   (b) Phase

Do you notice anything special about the amplitude and phase plots? They have properties that *are* somewhat special. *The amplitude spectrum is an even function of frequency, and the phase spectrum is an odd function* (if $x(t)$ is real)—which we will shortly demonstrate. Because we are treating the Fourier transform as a special case of the Laplace transform, *all of the properties of the Laplace transform continue to hold for the Fourier transform; therefore, we will retain the property numbering system of Table 13.6 in Chapter 13.* There are a few properties that hold for the Fourier transform and not for the Laplace transform because $s = j\omega$ is purely imaginary in the Fourier transform. There were six properties of the Laplace transform in Table 13.6 of Chapter 13, so we will refer to those numbers for those properties that the Fourier transform shares with the Laplace transform. We will indicate the special properties of the Fourier transform by prefixing them with an $\mathcal{F}$.

Conjugate Symmetry of the Fourier Transform

Theorem 14.2 ($\mathcal{F}$-Conjugate Symmetry Property)

If $x(t)$ is real, then

$$X(-\omega) = X^*(\omega). \qquad (14.3\text{-}5)$$

Therefore, the amplitude of the Fourier transform of a real waveform is even and the phase is odd. That is,

$$A(-\omega) = A(\omega) \qquad (14.3\text{-}6)$$

and

$$\theta(-\omega) = -\theta(\omega). \qquad (14.3\text{-}7)$$

Proof We will prove our statement in the following way. Writing the Fourier transform definition, we have

$$X(-\omega) = \int_{-\infty}^{\infty} x(t)e^{-j(-\omega)t}\, dt = \int_{-\infty}^{\infty} [x(t)e^{-j\omega t}]^*\, dt \qquad (14.3\text{-}8)$$

$$= \left[\int_{-\infty}^{\infty} x(t)e^{-j\omega t}\, dt\right]^* = X^*(\omega);$$

however, if we equate the two extreme terms and use the polar form $X(\omega) = A(\omega)e^{j\theta(\omega)}$, we will get

$$X^*(\omega) = A(\omega)e^{-j\theta(\omega)} = X(-\omega) = A(-\omega)e^{-j\theta(-\omega)}, \qquad (14.3\text{-}9)$$

so we have equations (14.3-5, 6, and 7). (We have just noted that conjugating a complex number leaves its magnitude unchanged and alters the sign of its angle. Equating magnitudes and angles gives equations (14.3-6) and (14.3-7).) ∎

Example 14.17 Find and plot the spectrum of

$$x(t) = e^{-at}u(t), \tag{14.3-10}$$

where a is a positive real number.

Solution The Fourier transform, which we have already computed, is

$$X(\omega) = \frac{1}{a + j\omega}. \tag{14.3-11}$$

The magnitude of this complex function of ω is

$$|X(\omega)| = \frac{1}{\sqrt{a^2 + \omega^2}} \tag{14.3-12}$$

and the angle is

$$\theta(\omega) = \angle X(\omega) = -\tan^{-1}\left[\frac{\omega}{a}\right]. \tag{14.3-13}$$

The amplitude and phase spectra are shown in Figure 14.22.

Figure 14.22
Spectrum of example

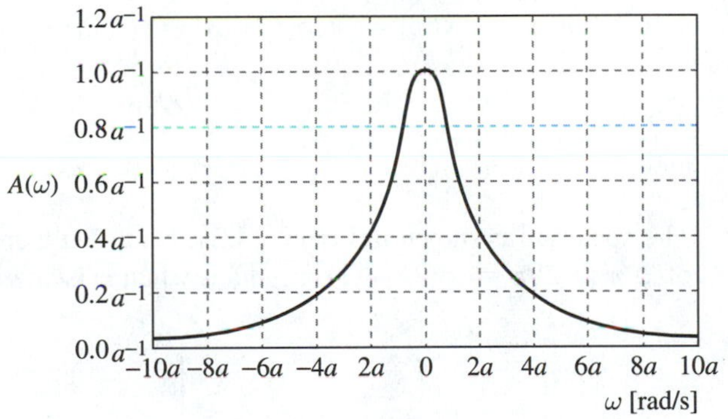

(a) Amplitude spectrum

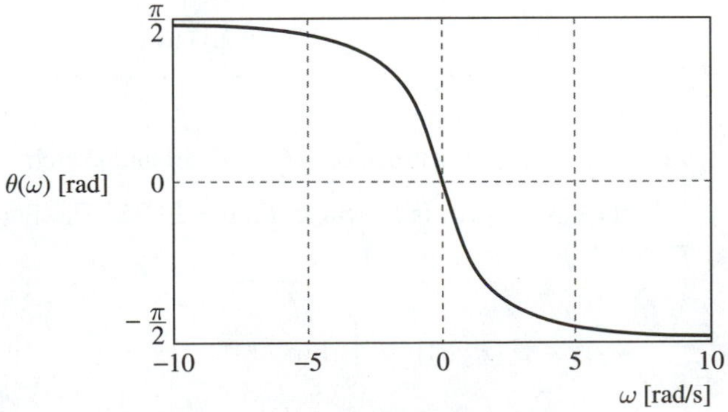

(b) Phase spectrum

Ambiguity of Angle Although this was not apparent in the preceding example, there is an ambiguity in the phase spectrum because an angle is only determined to within 2π radians (or 360°). To help in seeing this, we show the value $X(\omega)$ of the Fourier transform of a time function

$x(t)$ for a given value of ω in Figure 14.23. It is a complex number, and we have identified the amplitude $A(\omega)$ and the phase (angle) $\theta(\omega)$. Suppose we think of increasing the angle while holding $A(\omega)$ constant. It should be clear that once we have increased the angle by 2π radians, we are back where we started—that is, the angle is only uniquely defined to within 2π radians (or 360°). Furthermore, a given angle θ is equally well specified as the angle $\theta - 360°$. These points are shown in the next two examples.

Figure 14.23
Ambiguity of phase

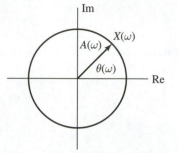

Example 14.18   Find the Fourier transform of $x(t) = \delta(t - a)$ and sketch its spectrum.

Solution   The Fourier transform is

$$X(\omega) = \mathscr{F}\{\delta(t - a)\} = \int_{-\infty}^{\infty} \delta(t - a)e^{-j\omega t}\, dt = e^{-j\omega a}. \qquad (14.3\text{-}14)$$

Thus, $$A(\omega) = 1 \qquad (14.3\text{-}15)$$

and $$\theta(\omega) = -\omega a. \qquad (14.3\text{-}16)$$

The spectrum is shown in Figure 14.24. Note that there is actually an ambiguity in phase of $\pm\pi \pm 2k\pi = \pm(2k + 1)\pi$ radians when ω is any integer multiple of π/a rad/s.

Figure 14.24
The spectrum of
$x(t) = \delta(t - a)$

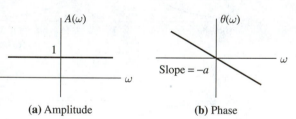

(a) Amplitude   **(b)** Phase

Example 14.19   Find the Fourier transform of $x(t) = P_T(t)$ and sketch its spectrum.

Solution   The time waveform is shown in Figure 14.25. The Fourier transform is given by the expression

$$X(\omega) = \mathscr{F}\{P_T(t)\} = \int_0^T 1\cdot e^{-j\omega t}\, dt = \frac{1 - e^{-j\omega T}}{j\omega} = Te^{-j\omega T/2}\frac{\sin\left(\dfrac{\omega T}{2}\right)}{\left(\dfrac{\omega T}{2}\right)}. \qquad (14.3\text{-}17)$$

(We have factored $e^{-j\omega T/2}$ and used Euler's formula.) The amplitude is

$$A(\omega) = T\left|\frac{\sin\left(\dfrac{\omega T}{2}\right)}{\left(\dfrac{\omega T}{2}\right)}\right| = T\frac{\left|\sin\left(\dfrac{\omega T}{2}\right)\right|}{\left(\dfrac{\omega T}{2}\right)} \qquad (14.3\text{-}18)$$

Figure 14.25 An example waveform

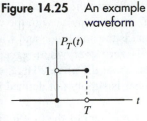

and the phase is

$$\theta(\omega) = \begin{cases} \dfrac{-\omega T}{2}; & \sin\left(\dfrac{\omega T}{2}\right) > 0 \\ \dfrac{-\omega T}{2} \pm \pi; & \sin\left(\dfrac{\omega T}{2}\right) < 0 \end{cases} \quad (14.3\text{-}19)$$

The amplitude plot is shown in Figure 14.26.[12] The phase plot shown in Figure 14.27 is more involved conceptually, so let's discuss it a bit. The $\sin(\omega T/2)/(\omega T/2)$ factor is either positive or negative.[13] If positive, it does not affect the phase; if negative, it adds either $+\pi$ radians or $-\pi$ radians. If you think this over carefully, you will see that this is essentially the same as the ambiguity we referred to earlier—going π radians in one direction back in Figure 14.23 is equivalent to going π radians in the other. Thus, each time the $\sin(\omega T/2)/(\omega T/2)$ factor changes sign, we will have an additive ambiguity of $\pm \pi$ radians in the phase angle. We have decided in Figure 14.27 to add π radians in each case. This is customary because it results in a space savings for the graph.

Figure 14.26
The amplitude spectrum

Figure 14.27
The phase spectrum

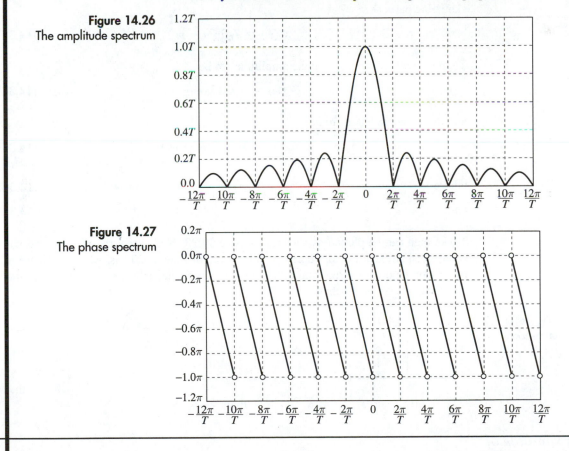

Spectra Containing Impulses

Our preceding program for drawing the spectrum of a waveform is fine as long as the Fourier transform has no impulses. But if it does have impulses, there is a problem. Suppose, for instance, that our transform turns out to be

$$X(\omega) = X_0\delta(\omega - \omega_0), \quad (14.3\text{-}20)$$

---

[12] Use L'Hôpital's rule to show that it has the value T at $\omega = 0$.

[13] Or zero—in which case, we remind you, the phase is undefined.

where X_0 is a complex number. How do we plot the amplitude and the phase? Well, we could perhaps plot the transform itself and mark the "area" of the impulse with the complex number X_0, but this is really not very satisfactory. What is the amplitude and what is the phase? We could very well say that the amplitude is zero everywhere except at $\omega = \omega_0$, at which frequency it has the value $|X_0|$; but what about the phase? Here, there is a genuine problem because the angle of a complex number is simply not defined if its amplitude is zero.

Here's what we will do. We will define a finite-height equivalent of the unit impulse function called the *unit indicator function*:

$$\delta_o(\omega) = \begin{cases} 1; & \omega = 0 \\ 0; & \omega \neq 0 \end{cases}. \tag{14.3-21}$$

Because the notation is so similar to that of the unit impulse, you must be careful to always notice the subscript "o," meaning that it is of finite height. We will now simply replace our actual Fourier transform $X_0\delta(\omega - \omega_0)$ with one of finite height at ω_0 for sketching purposes:

$$X_d(\omega) = X_0\delta_o(\omega - \omega_0). \tag{14.3-22}$$

Then we will define the "discrete amplitude" to be

$$A_d(\omega) = A(\omega_0)\delta_o(\omega - \omega_0) \tag{14.3-23}$$

and the "discrete phase" to be

$$\theta_d(\omega) = \theta(\omega_0)\delta_o(\omega - \omega_0), \tag{14.3-24}$$

where

$$A(\omega_0) = |X_0| \tag{14.3-25}$$

and

$$\theta(\omega_0) = \angle X_0. \tag{14.3-26}$$

These then can be plotted as shown in Figure 14.28. Notice that the vertical line with a small open circle at the top represents the indicator function. Thus, when and if we must plot a continuous spectrum and a discrete one simultaneously, we will be able to distinguish the two types of behavior.

Figure 14.28
A discrete spectrum plot

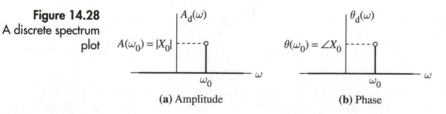

(a) Amplitude　　　　　　　　　　　(b) Phase

Examples of Waveforms with Discrete Spectra

Example 14.20　Plot the spectrum of $x(t) = \cos(\omega_0 t)$.

Solution　We have already computed the Fourier transform. It is

$$X(\omega) = \pi\delta(\omega - \omega_0) + \pi\delta(\omega + \omega_0). \tag{14.3-27}$$

The coefficients of the delta functions are positive real numbers, so we simply replace the impulse functions with unit indicator functions to derive the discrete amplitude spectrum:

$$A_d(\omega) = \pi\delta_o(\omega - \omega_0) + \pi\delta_o(\omega + \omega_0). \tag{14.3-28}$$

As $X(\omega)$ is purely real, the phase is identically zero, so there is nothing interesting to plot. The discrete spectrum is shown in Figure 14.29.

Figure 14.29
The discrete amplitude
spectrum

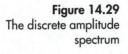

Example 14.21   Plot the spectrum of $x(t) = \sin(\omega_0 t)$.

Solution   The transform of the sine function is

$$X(\omega) = -j\pi\delta(\omega - \omega_0) + j\pi\delta(\omega + \omega_0). \tag{14.3-29}$$

The discrete amplitude spectrum is obtained by taking the magnitude of the coefficients of the impulses:

$$A_d(\omega) = \pi\delta_o(\omega - \omega_0) + \pi\delta_o(\omega + \omega_0). \tag{14.3-30}$$

This is the same as the amplitude spectrum for the cosine function. The phase, however, is now

$$\theta_d(\omega) = -\frac{\pi}{2}\delta_o(\omega - \omega_0) + \frac{\pi}{2}\delta_o(\omega + \omega_0). \tag{14.3-31}$$

The amplitude and phase spectra are plotted in Figure 14.30.

Figure 14.30
The discrete spectrum

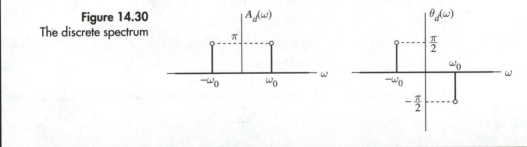

Example 14.22   Plot the time variation and then find the spectrum of the waveform

$$x(t) = 1 + 0.3\cos(\pi t) + 0.4\sin(\pi t) \tag{14.3-32}$$

$$+ \frac{1}{3\sqrt{2}}\cos(3\pi t) + \frac{1}{3\sqrt{2}}\sin(3\pi t).$$

Solution   This is a somewhat practical waveform—one with which we might have to deal in practice. Its graph (for $t > 0$) is plotted in Figure 14.31 (we plotted it using MATLAB). Now this is a quite complicated time waveform as you can see. Let's compute the Fourier transform and look at the spectrum. The Fourier transform is

$$X(\omega) = \mathscr{F}\{1\} + 0.3\mathscr{F}\{\cos(\pi t)\} + 0.4\mathscr{F}\{\sin(\pi t)\} \qquad (14.3\text{-}33)$$

$$+ \frac{1}{3\sqrt{2}}\mathscr{F}\{\cos(3\pi t)\} + \frac{1}{3\sqrt{2}}\mathscr{F}\{\sin(3\pi t)\}$$

$$= 2\pi\delta(\omega) + 0.3[\pi\delta(\omega - \pi) + \pi\delta(\omega + \pi)]$$

$$+ 0.4[-j\pi\delta(\omega - \pi) + j\pi\delta(\omega + \pi)]$$

$$+ \frac{1}{3\sqrt{2}}[\pi\delta(\omega - 3\pi) + \pi\delta(\omega + 3\pi)]$$

$$+ \frac{1}{3\sqrt{2}}[-j\pi\delta(\omega - 3\pi) + j\pi\delta(\omega + 3)].$$

Collecting terms at the same frequency, we have

$$X(\omega) = 2\pi\delta(\omega) + (0.3 - j0.4)\pi\delta(\omega - \pi) + (0.3 + j0.4)\pi\delta(\omega + \pi) \qquad (14.3\text{-}34)$$

$$+ \frac{1}{3\sqrt{2}}(1 - j)\pi\delta(\omega - 3\pi) + \frac{1}{3\sqrt{2}}(1 + j)\pi\delta(\omega + 3\pi).$$

Computing the magnitude of each of the complex numbers and replacing the impulse functions with indicator functions at the corresponding frequencies, we obtain the discrete amplitude spectrum function

$$A_d(\omega) = 2\pi\delta_o(\omega) + \frac{\pi}{2}\delta_o(\omega - \pi) + \frac{\pi}{2}\delta_o(\omega + \pi) + \frac{\pi}{3}\delta_o(\omega - 3\pi) \qquad (14.3\text{-}35)$$

$$+ \frac{\pi}{3}\delta_o(\omega + 3\pi),$$

which we have plotted in Figure 14.32(a). We observe that it is, indeed, an even function of ω, which we would expect, as $x(t)$ is real. If we compute the polar angle of each of the components, we obtain the phase spectrum

$$\theta_d(\omega) = -53.1°\delta_o(\omega - \pi) + 53.1°\delta_o(\omega + \pi) - 45°\delta_o(\omega - 3\pi) \qquad (14.3\text{-}36)$$

$$+ 45°\delta_o(\omega + 3\pi).$$

This spectrum is shown in Figure 14.32(b); as we would expect, it is an odd function.

Figure 14.31
An example time
waveform

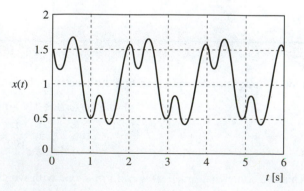

Figure 14.32
The spectrum of the
example waveform

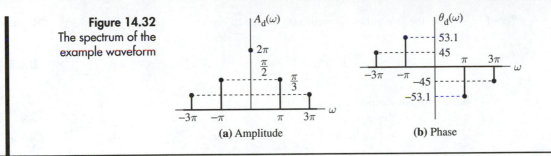

(a) Amplitude   (b) Phase

The conclusion we reach from the last example, which we feel is rather obvious at this point, is the following: *for many waveforms, a frequency domain description offers more information than does a time domain representation.* Looking at it another way, pictorially, which graphs give more information—the one in Figure 14.31 or those in Figure 14.32? The answer is, we feel, evident. At least for waveforms with "discrete spectra," that is, whose Fourier transforms have impulses, the frequency domain offers more information.

Frequency Domain Convolution: The Fourier Transform of a Product

At this stage, we understand that the Fourier transform tells us "how much sinusoid" there is in the waveform at a given frequency ω. We might, however, also be interested in how much *power* there is in a given waveform at a given frequency ω. For an answer, we must develop a bit more theory.

Theorem 14.3   The Fourier transform of the product of two waveforms $x(t)$ and $y(t)$ is equal to the convolution of their Fourier transforms divided by 2π:

$$\mathcal{F}\{x(t)y(t)\} = \frac{1}{2\pi}X(\omega) \otimes Y(\omega).\qquad(14.3\text{-}37)$$

Proof   If you are a bit shaky about convolution, now is the time to review this topic in Chapter 13. We simply compute the transform of the product, then use the inverse transform representation for one of these waveforms. This gives us

$$\mathcal{F}\{x(t)y(t)\} = \int_{-\infty}^{\infty} x(t)y(t)e^{-j\omega t}\,dt = \int_{-\infty}^{\infty}\left[\frac{1}{2\pi}\int_{-\infty}^{\infty}X(\alpha)e^{j\alpha t}\,d\alpha\right]y(t)e^{-j\omega t}\,dt\qquad(14.3\text{-}38)$$

$$= \frac{1}{2\pi}\int_{-\infty}^{\infty}X(\alpha)\left[\int_{-\infty}^{\infty}y(t)e^{-j(\omega-\alpha)t}\,dt\right]d\alpha = \frac{1}{2\pi}\int_{-\infty}^{\infty}X(\alpha)Y(\omega-\alpha)\,d\alpha$$

$$= \frac{1}{2\pi}X(\omega) \otimes Y(\omega).$$

We interchanged the order of integration in going from the first line of the proof to the second and then regrouped terms.   ∎

Parseval's Theorem

Theorem 14.4 (Parseval's Theorem)   If $x(t)$ is a real waveform, then

$$\int_{-\infty}^{\infty} x^2(t)\,dt = \frac{1}{2\pi}\int_{-\infty}^{\infty}|X(\omega)|^2\,d\omega.\qquad(14.3\text{-}39)$$

Proof   We use the last property with $y(t) = x(t)$ to compute the Fourier transform of $x^2(t)$. We obtain

$$\int_{-\infty}^{\infty} x^2(t)e^{-j\omega t}\, dt = \frac{1}{2\pi}\int_{-\infty}^{\infty} X(\alpha)X(\omega - \alpha)\, d\alpha. \tag{14.3-40}$$

If we evaluate both sides at $\omega = 0$ we obtain

$$\int_{-\infty}^{\infty} x^2(t)\, dt = \frac{1}{2\pi}\int_{-\infty}^{\infty} X(\alpha)X(-\alpha)\, d\alpha \tag{14.3-41}$$

$$= \frac{1}{2\pi}\int_{-\infty}^{\infty} X(\alpha)X^*(\alpha)\, d\alpha = \frac{1}{2\pi}\int_{-\infty}^{\infty} |X(\alpha)|^2\, d\alpha.$$

But this is merely (14.3-39) with α as the variable of integration on the right rather than ω. (Recall that, if $x(t)$ is real, $X(-\omega) = X^*(\omega)$ and $X(\omega)X^*(\omega) = |X(\omega)|^2$.)   ∎

The Power Spectrum

Figure 14.33   The power in a waveform

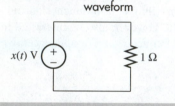

We will use the preceding two properties to develop a workable definition for the *power spectrum* of an arbitrary waveform $x(t)$. Look at Figure 14.33. There we have a voltage source whose waveform is adjusted to be $x(t)$ connected in parallel with a 1-Ω resistor. The instantaneous power, as we well know, is given by

$$P(t) = x^2(t). \tag{14.3-42}$$

Its Fourier transform is given by (14.3-40), and we will define this quantity to be the *instantaneous power spectrum*[14] of $x(t)$.

Definition 14.4   The *instantaneous power spectrum* of an arbitrary waveform is defined[15] to be the Fourier transform of the instantaneous power that it would dissipate in a 1-Ω resistor if it were a voltage or current waveform:

$$S(\omega) = \mathcal{F}\{P(t)\} = \int_{-\infty}^{\infty} x^2(t)e^{-j\omega t}\, dt = \frac{1}{2\pi}X(\omega) \otimes X(\omega). \tag{14.3-43}$$

Notice that the physical dimension of $S(\omega)$ is power $\times$ time = energy. Furthermore, one can recover the instantaneous power from $S(\omega)$ by inverse transformation:

$$P(t) = \mathcal{F}^{-1}\{S(\omega)\} = \frac{1}{2\pi}\int_{-\infty}^{\infty} S(\omega)e^{j\omega t}\, d\omega. \tag{14.3-44}$$

The Energy Density Spectrum

From the definition of $S(\omega)$ with Parseval's theorem, equation (14.3-39), we have

$$S(0) = \int_{-\infty}^{\infty} x^2(t)\, dt = \frac{1}{2\pi}\int_{-\infty}^{\infty} |X(\omega)|^2\, d\omega. \tag{14.3-45}$$

We see that $S(0)$ is the total energy in $x(t)$. This motivates the following definition.

Definition 14.5   The *energy density spectrum* of an arbitrary waveform $x(t)$ is the quantity

$$E(\omega) = \frac{|X(\omega)|^2}{2\pi}. \tag{14.3-46}$$

---

[14] Other texts present a slightly different definition of "power spectrum"; we will therefore use the adjective "instantaneous" for ours.

[15] The idea of a *plot* of magnitude and phase is inherent in our definition.

Observe that the dimension of $E(\omega)$ is, indeed, energy per rad/s if $x^2(t)$ has the dimension of power.

Before we apply this definition, let's use the following example to develop a useful property of the convolution of two impulse functions.

Example 14.23   Show that the convolution of two impulses is given by the formula

$$\delta(\omega - a) \otimes \delta(\omega - b) = \delta(\omega - [a + b]). \tag{14.3-47}$$

Solution   The easiest and shortest proof perhaps is to recall that the Laplace transform of the convolution of two time functions is the product of the two transforms. We then prove (14.3-47) with ω replaced by t. We recall that $\mathfrak{L}\{\delta(t - a)\} = e^{-as}$ and $\mathfrak{L}\{\delta(t - b)\} = e^{-bs}$; thus,

$$\mathfrak{L}\{\delta(t - a) \otimes \delta(t - b)\} = e^{-as}e^{-bs} = e^{-(a+b)s} = \mathfrak{L}\{\delta(t - [a + b])\}. \tag{14.3-48}$$

Taking the inverse Laplace transform of the terms at the extreme ends of this chain of equalities and replacing t with ω results in (14.3-47).

Power Spectrum Examples

Example 14.24   Find the instantaneous power spectrum of the waveform $x(t) = 1$.

Solution   The Fourier transform of $x(t)$ is

$$X(\omega) = 2\pi\delta(\omega), \tag{14.3-49}$$

as we have already shown. Thus,

$$S(\omega) = \frac{1}{2\pi}X(\omega) \otimes X(\omega) = \frac{1}{2\pi}[2\pi\delta(\omega) \otimes 2\pi\delta(\omega)] = 2\pi\delta(\omega). \tag{14.3-50}$$

Figure 14.34   Discrete power spectrum of example waveform

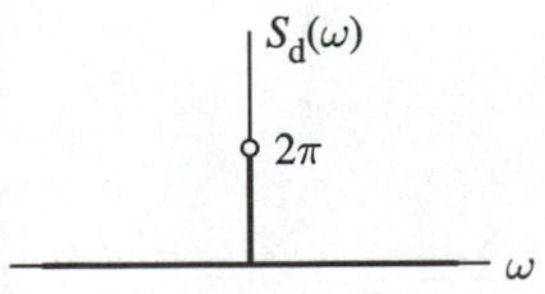

Because this is a "line" or "discrete" spectrum, we will use the indicator function to write

$$S_d(\omega) = 2\pi\delta_o(\omega). \tag{14.3-51}$$

A plot of $S_d(\omega)$ (the subscript standing for the first letter of the word discrete) is shown in Figure 14.34. Notice, by the way, that the energy density spectrum does not exist. $|X(\omega)|^2$ is the square of an impulse function—which does not make sense mathematically. Why? Because the total energy in the constant dc waveform $x(t) = 1$ is infinite; thus, the concept of energy density is meaningless. It is only for finite energy waveforms that this concept is well defined.

Example 14.25   Find the instantaneous power spectrum of the waveform $x(t) = \cos(\omega_0 t)$.

Solution   The Fourier transform is

$$X(\omega) = \mathfrak{F}\{\cos(\omega_0 t)\} = \pi\delta(\omega - \omega_0) + \pi\delta(\omega + \omega_0). \tag{14.3-52}$$

Thus, the instantaneous power spectrum is

$$S(\omega) = \frac{1}{2\pi}X(\omega) \otimes X(\omega) \tag{14.3-53}$$

$$= \frac{1}{2\pi}[\pi\delta(\omega - \omega_0) + \pi\delta(\omega + \omega_0)] \otimes [\pi\delta(\omega - \omega_0) + \pi\delta(\omega + \omega_0)]$$

$$= \frac{\pi}{2}\delta(\omega - \omega_0) \otimes \delta(\omega - \omega_0) + \frac{\pi}{2}\delta(\omega - \omega_0) \otimes \delta(\omega + \omega_0)$$

$$+ \frac{\pi}{2}\delta(\omega + \omega_0) \otimes \delta(\omega - \omega_0) + \frac{\pi}{2}\delta(\omega + \omega_0) \otimes \delta(\omega + \omega_0)$$

$$= \frac{\pi}{2}\delta(\omega - 2\omega_0) + \pi\delta(\omega) + \frac{\pi}{2}\delta(\omega + 2\omega_0).$$

As we will show in the next section, any even waveform has a Fourier transform that is real; thus, as $x^2(t)$ is even here, $S(\omega)$ is real. Hence, we can ignore the phase spectrum and concentrate on the amplitude spectrum, which is the discrete frequency function

$$S_d(\omega) = \frac{\pi}{2}\delta_o(\omega - 2\omega_0) + \pi\delta_0(\omega) + \frac{\pi}{2}\delta_0(\omega + 2\omega_0). \tag{14.3-54}$$

Figure 14.35 Discrete power spectrum of example waveform

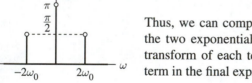

This frequency function is plotted in Figure 14.35. Again, we observe that the energy density spectrum does not exist because the total energy in our waveform is infinite. However, we can write the following expression for the instantaneous power:

$$P(t) = x^2(t) = \cos^2(t) = \frac{1 + \cos(2t)}{2} = \frac{1}{2} + \frac{1}{4}e^{j2t} + \frac{1}{4}e^{-j2t}. \tag{14.3-55}$$

Thus, we can compute the average power to be the constant term, 1/2, because each of the two exponential waveforms has an average value of zero. Furthermore, the Fourier transform of each term in (14.3-55) corresponds to one term in (14.3-54)—and to one term in the final expression in (14.3-53).

Example 14.26   Find the instantaneous power spectrum of $x(t) = e^{-2|t|}$.

Solution   In this case, a direct approach is perhaps appropriate. We simply compute the power waveform to be

$$P(t) = x^2(t) = e^{-4|t|}, \tag{14.3-56}$$

and its transform is

$$S(\omega) = \mathcal{F}\{P(t)\} = \mathcal{F}\{e^{-4|t|}\} = \frac{8}{\omega^2 + 16}. \tag{14.3-57}$$

The total energy in the waveform is the finite value $S(0) = 1/2$. Thus, the energy density spectrum does exist in this case. In fact, the Fourier transform of $x(t)$ is

$$X(\omega) = \mathcal{F}\{e^{-2|t|}\} = \frac{4}{\omega^2 + 4}. \tag{14.3-58}$$

Thus, the energy density spectrum is

$$E(\omega) = \frac{|X(\omega)|^2}{2\pi} = \frac{8}{\pi[\omega^2 + 4]^2}. \tag{14.3-59}$$

A Concluding Observation

Before concluding this section, let us note an important basic fact. We can write the instantaneous power of an arbitrary waveform in the form

$$P(t) = P_{av} + \Delta P(t), \tag{14.3-60}$$

where P_{av} is the average power in $x(t)$ (a constant) and the average value of $\Delta P(t)$ is zero. Taking the Fourier transform, we get

$$S(\omega) = 2\pi P_{av}\delta(\omega) + S_\Delta(\omega). \tag{14.3-61}$$

The frequency function $S_\Delta(\omega)$ has no impulses at $\omega = 0$; hence, the average power is simply the strength of the impulse (if any) in $S(\omega)$ at $\omega = 0$ divided by 2π. For instance, in Example 14.25, the strength of the impulse at dc ($\omega = 0$) was π, and this implies that the average power in $x(t)$ is $\pi/2\pi = 0.5$. We can check this by direct computation: the instantaneous power is $x^2(t) = \cos^2(\omega_0 t) = \frac{1}{2} + \frac{1}{2}\cos(2\omega_0 t)$, so the average power is 0.5.

Section 14.3 Quiz

Q14.3-1. If $x(t)$ has the waveform shown in Figure Q14.3-1, find and sketch its spectrum.

Figure Q14.3-1

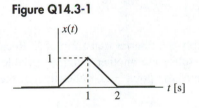

Q14.3-2. If $x(t) = 10\cos(10t)$, find the spectrum and the instantaneous power spectrum. Find the average power by identifying a term in the instantaneous power spectrum.

Q14.3-3. If $x(t) = e^{-2|t|}$, find the power and energy density spectra.

14.4 Properties of the Fourier Transform

The direct computation of the Fourier integral can sometimes be difficult in that the evaluation of a limit is required. In this section we will develop a number of properties of the Fourier transform that will permit us (at least very often) to compute the transform without actually calculating the integral; rather, we will use the properties and other, already known, transforms. Furthermore, the properties provide insight into the nature of the frequency domain itself. In what follows, we will always assume that

$$x(t) \longleftrightarrow X(\omega), \tag{14.4-1}$$

that is, as usual, that $x(t)$ and $X(\omega)$ form a Fourier transform pair. We remind you that we will continue to designate properties 1 through 6 of the Laplace transform (see Chapter 13, Table 13.6) as properties 1 through 6 of the Fourier transform, taking note of the fact that the latter transform is simply the former transform evaluated in some manner on the

$j\omega$ axis. Those properties peculiar to the Fourier transform arise from the fact that the s variable becomes $j\omega$ in the Fourier transform; that is, the real part of the s variable is zero.

The Symmetry Property of the Fourier Transform

We have already derived one property that is peculiar to the Fourier transform: $\mathcal{F}$-conjugate symmetry. Here's another important one.

Theorem 14.5 ($\mathcal{F}$-Time-Frequency Symmetry Property)

$$X(t) \longleftrightarrow 2\pi x(-\omega). \qquad (14.4\text{-}2)$$

Proof The proof is actually simple. The only difficult part is keeping symbols straight. We will go carefully through each step. First, we write the equation for the inverse Fourier transform:

$$x(t) = \frac{1}{2\pi} \int_{-\infty}^{\infty} X(\omega) e^{j\omega t} \, d\omega. \qquad (14.4\text{-}3)$$

Next, we multiply both sides by 2π and rename both independent variables:

$$2\pi x(\alpha) = \int_{-\infty}^{\infty} X(\beta) e^{j\beta\alpha} \, d\beta. \qquad (14.4\text{-}4)$$

The equation remains true in these new, but temporary, variables. Now, let's evaluate this equation at $-\alpha$, rather than at α:

$$2\pi x(-\alpha) = \int_{-\infty}^{\infty} X(\beta) e^{-j\beta\alpha} \, d\beta. \qquad (14.4\text{-}5)$$

We now observe that the right side is simply the Fourier transform of X. Recalling that β is a dummy variable of integration, we substitute another dummy variable—call it t. Then, we decide to evaluate the equation at the point $-\omega$, rather than at the point $-\alpha$. This results in

$$2\pi x(-\omega) = \int_{-\infty}^{\infty} X(t) e^{-j\omega t} \, dt. \qquad (14.4\text{-}6)$$

We see that $2\pi x(-\omega)$ is the Fourier transform of $X(t)$. ∎

Frequency Domain Impulse Functions

The preceding result is a powerful one. We will see just how powerful it is by working several examples using it; first, however, let's have a look at the unit impulse function in the frequency domain a bit more closely. We will show that it is an even function; that is, that $\delta(-\omega) = \delta(\omega)$. This is easy to see by applying the definition that $\delta(\omega)$ is any sequence of waveforms whose integral tends in the limit to the unit step function. Thus, we write the following (and change variables, letting $\beta = -\alpha$)

$$\int_{-\infty}^{\omega} \delta(-\alpha) \, d\alpha = \int_{-\omega}^{\infty} \delta(\beta) \, d\beta = \int_{-\infty}^{\infty} \delta(\beta) \, d\beta - \int_{-\infty}^{-\omega} \delta(\beta) \, d\beta = 1 - u(-\omega) = u(\omega). \qquad (14.4\text{-}7)$$

Here we are not concerned with the fact that $1 - u(-\omega)$ gives the "wrong value" at $\omega = 0$, because in Fourier transform theory we treat two waveforms that differ by a finite value at only one point as being equivalent. A sketch of $1 - u(-\omega)$ is shown in Figure 14.36. Thus, we have shown that

$$\delta(-\omega) = \delta(\omega), \qquad (14.4\text{-}8)$$

that is, that the **unit impulse is an even function.**

Figure 14.36
A sketch of two step functions

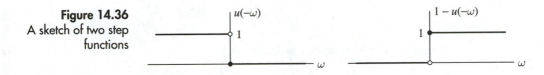

Example 14.27   Use the $\mathcal{F}$-symmetry property to show that $1 \leftrightarrow 2\pi\delta(\omega)$.

Solution   This one is very easy. We simply note that $\delta(t) \leftrightarrow 1$ and apply the symmetry property and use the "evenness" of $\delta(\omega)$. Thus,

$$1 \longleftrightarrow 2\pi\delta(-\omega) = 2\pi\delta(\omega). \qquad (14.4\text{-}9)$$

Some discussion of this last example might be appropriate. We expended a great amount of effort earlier in this chapter to derive this transform; now, we see that it is almost a triviality. Why didn't we simply use this method in the first place? Here is the reason. The symmetry property is based on the inversion integral, which, in turn, relies upon the fact that the unit impulse in ω has an inverse transform of the constant time function having a value of one. Therefore, in a sense, this example is specious for it is based upon circular logic. However, it does illustrate the effectiveness of the symmetry property.

Example 14.28   Show that

$$\frac{1}{\pi} \frac{a}{a^2 + t^2} \longleftrightarrow e^{-a|\omega|}, \qquad (14.4\text{-}10)$$

where a is a positive real number. The time function in this Fourier transform pair is called a *Lorentzian pulse*. (It is a useful approximation to the read back pulse coming from a computer magnetic data storage disk.)

Solution   We use the following transform pair, which we have already derived:

$$e^{-a|t|} \longleftrightarrow \frac{2a}{s^2 + \omega^2}. \qquad (14.4\text{-}11)$$

The symmetry property implies that

$$\frac{2a}{a^2 + t^2} \longleftrightarrow 2\pi e^{-a|\omega|}. \qquad (14.4\text{-}12)$$

(Notice here that $|-\omega| = |\omega|$.) Dividing both sides by 2π (permissible because the Fourier transform is a linear operator) results in our conclusion.

Example 14.29   Show that

$$\delta(t) + \frac{1}{j\pi t} \longleftrightarrow 2u(-\omega). \tag{14.4-13}$$

Solution   We first recall the transform pair (see Section 14.1, equation (14.1-28) or entry 3 in Table 14.1):

$$u(t) \longleftrightarrow \pi\delta(\omega) + \frac{1}{j\omega}. \tag{14.4-14}$$

Applying the $\mathcal{F}$-symmetry property gives

$$\pi\delta(t) + \frac{1}{jt} \longleftrightarrow 2\pi u(-\omega). \tag{14.4-15}$$

Dividing both sides by π gives the desired result.

Fourier Transform Properties   We will now review the properties of the Laplace transform, which we already know hold for the Fourier transform, just to refresh our memory and to express them in the forms containing $j\omega$, rather than s. As for the Laplace transform, we will not investigate the linearity property because it follows at once from the definition of an integral.

Time Shift Property (Table 14.4, Entry 2)

$$x(t - a) \longleftrightarrow e^{-j\omega a}X(\omega). \tag{14.4-16}$$

Proof   We simply write the definition of the Fourier transform to obtain

$$\mathcal{F}\{x(t - a)\} = \int_{-\infty}^{\infty} x(t - a)e^{-j\omega t}\,dt = e^{-j\omega a}\int_{-\infty}^{\infty} x(\beta)e^{-j\omega\beta}\,d\beta. \tag{14.4-17}$$

We have just changed variables from t to $\beta = t - a$. ∎

Example 14.30   Show that $\delta(t - a) \longleftrightarrow e^{-j\omega a}$.

Solution   This one is really obvious, but to be methodical we note that

$$\delta(t) \longleftrightarrow 1, \tag{14.4-18}$$

so the time shift property gives

$$\delta(t - a) \longleftrightarrow e^{-j\omega a}. \tag{14.4-19}$$

Frequency Shift Property (Table 14.4, Entry 3)   This is often called the *modulation property* for Fourier transforms. It states that

$$e^{-j\omega_0 t}x(t) \longleftrightarrow X(\omega + \omega_0). \tag{14.4-20}$$

Proof   Again, our procedure is merely to compute the transform:

$$\mathcal{F}\{e^{-j\omega_0 t}x(t)\} = \int_{-\infty}^{\infty} e^{-j\omega_0 t}x(t)e^{-j\omega t}\,dt = \int_{-\infty}^{\infty} x(t)e^{-j(\omega+\omega_0)t}\,dt = X(\omega + \omega_0). \tag{14.4-21}$$ ∎

We will term the next two properties "subproperties" because they follow so directly from the frequency shift (modulation) property.

Subproperty (a)

$$\mathcal{F}\{\cos(\omega_0 t)x(t)\} = \frac{X(\omega - \omega_0) + X(\omega + \omega_0)}{2}.$$ (14.4-22)

Proof We use Euler's relation for the cosine and the frequency shift property to get

$$\mathcal{F}\{\cos(\omega_0 t)x(t)\} = \mathcal{F}\left\{\frac{e^{j\omega_0 t} + e^{-j\omega_0 t}}{2}x(t)\right\}$$ (14.4-23)

$$= \frac{\mathcal{F}\{e^{j\omega_0 t}x(t)\} + \mathcal{F}\{e^{-j\omega_0 t}x(t)\}}{2} = \frac{X(\omega - \omega_0) + X(\omega + \omega_0)}{2}. \quad\blacksquare$$

Subproperty (b)

$$\mathcal{F}\{\sin(\omega_0 t)x(t)\} = \frac{X(\omega - \omega_0) - X(\omega + \omega_0)}{j2}.$$ (14.4-24)

Proof We use Euler's relation for the sine and the frequency shift property to get

$$\mathcal{F}\{\sin(\omega_0 t)x(t)\} = \mathcal{F}\left\{\frac{e^{j\omega_0 t} - e^{-j\omega_0 t}}{j2}x(t)\right\}$$ (14.4-25)

$$= \frac{\mathcal{F}\{e^{j\omega_0 t}x(t)\} - \mathcal{F}\{e^{-j\omega_0 t}x(t)\}}{j2} = \frac{X(\omega - \omega_0) - X(\omega + \omega_0)}{j2}. \quad\blacksquare$$

Example 14.31 Find $\mathcal{F}\{\cos(\omega_0 t)\cos(\omega_m t)\}$ and sketch the spectrum, assuming that $\omega_m \ll \omega_0$.

Solution We first recall that

$$\cos(\omega_m t) \longleftrightarrow \pi\delta(\omega - \omega_m) + \pi\delta(\omega + \omega_m).$$ (14.4-26)

Using this for $X(\omega)$ in subproperty (a) of the frequency shift property, we get

$$\cos(\omega_0 t)\cos(\omega_m t) \longleftrightarrow \frac{\pi}{2}\delta(\omega - \omega_0 - \omega_m) + \frac{\pi}{2}\delta(\omega - \omega_0 + \omega_m) \quad (14.4\text{-}27)$$

$$+ \frac{\pi}{2}\delta(\omega + \omega_0 - \omega_m) + \frac{\pi}{2}\delta(\omega + \omega_0 + \omega_m).$$

The result is real, so we can plot the spectrum directly, rather than going to the indicator function. Thus, we have the transform shown in Figure 14.37. The term "modulation" arose because that term is used for the process of multiplying two waveforms together in communication networks.

Figure 14.37
Fourier transform of example waveform

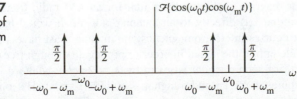

Example 14.32 (AM Radio)

A time waveform $x(t)$ has the frequency spectrum shown in Figure 14.38(a). It is applied to the "amplitude modulator" in Figure 14.38(b). Find and sketch the spectrum of the output signal $y(t)$ assuming that $\omega_m \ll \omega_0$.

Figure 14.38
An amplitude modulator

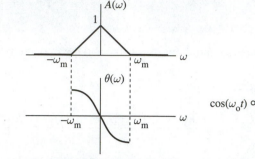

(a) Spectrum of modulating waveform   (b) AM modulator

Solution   We note that

$$y(t) = x(t)\cos(\omega_0 t) \qquad (14.4\text{-}28)$$

and apply subproperty (a) of the frequency shift (modulation) property. This gives

$$Y(\omega) = \frac{X(\omega - \omega_0) + X(\omega + \omega_0)}{2}. \qquad (14.4\text{-}29)$$

Now, if $\omega_m \ll \omega_0$, one has the spectrum in Figure 14.39 (otherwise, the right-hand side might have negative frequency components that could even overlap the left-hand side). Notice that both $x(t)$ and $y(t)$, being real signals, have amplitude spectra that are even and phase spectra that are odd; furthermore, notice that we have not defined the phase in the regions where the amplitude is zero. As we have noted before, the angle of a complex number is not defined if its magnitude is zero.

Figure 14.39
The spectrum of the AM signal

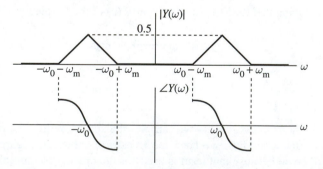

The reason we have labeled this example "AM radio" is simply because this system block diagram forms the basic unit for an AM radio transmitter. If one were to connect the signal $x(t)$ directly to an antenna, not much would happen. As you will discover in your courses in electromagnetics, an antenna must have a length commensurate with the wavelength of the signal in order to propagate through free space to a receiver. The wavelength of such a signal is given by $\lambda = 2\pi c/\omega$, where c is the velocity of light. If one computes λ for an audio signal, say with $f = 10$ kHz, one obtains $\lambda = 2\pi \times 3 \times 10^8$ [m/s]/$(2\pi \times 10^4$ [rad/s]) $= 3 \times 10^4$ [m]. This is an extremely long antenna! By using a modulator, however, one can shift the frequency spectrum up into a range for which the antenna is short. For standard AM broadcast, the frequency is typically 1 MHz. Thus, one has $\lambda = 2\pi \times 3 \times 10^8 \text{ [m/s]}/(2\pi \times 10^6 \text{ [rad/s]}) = 300 \text{ [m]}$, a much more practical value.[16]

Scaling Property (Table 14.4, Entry 4)

If a is any positive real constant, then

$$\mathcal{F}\{x(at)\} = \frac{1}{a}X(\omega/a). \tag{14.4-30}$$

Proof   Once more, our procedure is merely to compute the transform and make the change of variables $\alpha = at$. This gives

$$\mathcal{F}\{x(at)\} = \int_{-\infty}^{\infty} x(at)e^{-j\omega t}\, dt = \frac{1}{a}\int_{-\infty}^{\infty} x(\alpha)e^{-j\omega\alpha/a}\, d\alpha = \frac{1}{a}X(\omega/a). \tag{14.4-31}$$

∎

Of course we cannot forget the property that was all-important for the Laplace transform—and will also be seen to be useful for the Fourier transform—time differentiation.

Time Differentiation Property (Table 14.4, Entry 5)

$$px(t) \longleftrightarrow j\omega X(\omega). \tag{14.4-32}$$

Proof   We write the time function $x(t)$ in terms of $X(\omega)$ using the inversion integral:

$$x(t) = \frac{1}{2\pi}\int_{-\infty}^{\infty} X(\omega)e^{j\omega t}\, d\omega. \tag{14.4-33}$$

Now notice that t is merely a parameter in the integral, which is computed relative to the variable ω. Conceptually one fixes t, then computes the integral to find $x(t)$. One can then change t to a new value and recompute. If we take the time derivative of $x(t)$, we get

$$px(t) = \frac{1}{2\pi}p\int_{-\infty}^{\infty} X(\omega)e^{j\omega t}\, d\omega. \tag{14.4-34}$$

At this point, we will make an assumption in order to complete the proof: the order of time differentiation and frequency integration can be interchanged. Such an exchange is dubious within the context of ordinary functions, but is perfectly legitimate within the theory of generalized functions. Unfortunately, the proof of this is too involved for presentation in this text—so it is one of the things that you will have to "take on faith." Doing the interchange, then, we have

$$px(t) = \frac{1}{2\pi}\int_{-\infty}^{\infty} \frac{\partial}{\partial\tau}[X(\omega)e^{j\omega t}]\, d\omega = \frac{1}{2\pi}\int_{-\infty}^{\infty} j\omega X(\omega)e^{j\omega t}\, d\omega. \tag{14.4-35}$$

---

[16] In fact, broadcast band receivers use only fractional wavelength antennas and are quite inefficient. Shortwave transmitters and receivers use more optimum antenna arrangements.

But this implies that the Fourier transform of $px(t)$ is $j\omega X(\omega)$, which is exactly what we were trying to prove.   ∎

Example 14.33   Show that the time differentiation property works for $1 \leftrightarrow 2\pi\delta(\omega)$.

Solution   Taking the derivative of $x(t) = 1$ gives $px(t) = 0$; we also see that in the frequency domain we have

$$j\omega \cdot 2\pi\delta(\omega) = j0 \cdot \delta(\omega) = 0. \qquad (14.4\text{-}36)$$

We have used the sampling property for impulses to evaluate ω at $\omega = 0$.

In the last chapter we pointed out that integration was the inverse of differentiation—with no additive constant so long as all waveforms were considered to be one-sided, that is, identically zero for $t \leq 0$. In this case, there was no need for a separate time integration property. Now, however, we are dealing with transforms in the variable ω, and we are not making any assumption about one-sidedness. Therefore, we will consider integration in terms of *indefinite integration;* in other words, we will permit an arbitrary additive constant of integration. This leads to a new property for Fourier transforms.

Theorem 14.6
($\mathcal{F}$-Indefinite Time Integration Property)   If $py(t) = x(t)$, then

$$y(t) \longleftrightarrow \frac{X(\omega)}{j\omega} + k\delta(\omega), \qquad (14.4\text{-}37)$$

where k is an arbitrary constant.

Proof   We first observe that if k_0 is any arbitrary constant, then

$$p[y(t) + k_0] = x(t). \qquad (14.4\text{-}38)$$

Using the time differentiation property (equation (14.4-32)), we have

$$j\omega[Y(\omega) + k_0 2\pi\delta(\omega)] = X(\omega) \qquad (14.4\text{-}39)$$

because the transform of unity is $2\pi\delta(\omega)$. Thus, dividing by $j\omega$, we get

$$Y(\omega) = \frac{X(\omega)}{j\omega} - k_0 2\pi\delta(\omega). \qquad (14.4\text{-}40)$$

Defining $-2\pi k_0 = k$ results in equation (14.4-37).   ∎

The preceding property assumes that we know nothing more about $y(t)$ than the fact that its derivative is $x(t)$. If we know something more, we can find the constant k. Often, we know or require that the solution of $py(t) = x(t)$ approach zero as $t \to \pm\infty$. In this case we have

$$Y(\omega) = \int_{-\infty}^{\infty} y(t)e^{-j\omega t}\, dt = \left[y(t)\frac{e^{-j\omega t}}{-j\omega}\right]_{-\infty}^{\infty} + \frac{1}{j\omega}\int_{-\infty}^{\infty} x(t)e^{-j\omega t}\, dt, \qquad (14.4\text{-}41)$$

where we have used integration by parts. But the first term on the extreme right-hand side of this equation is zero by our assumption. Thus, we have

$$Y(\omega) = \frac{X(\omega)}{j\omega},$$

(14.4-42)

and our arbitrary constant has been forced to zero.

Now let's work several examples to get a feel for the practical side of this result. It is often used in calculations, sometimes erroneously (forgetting the constant).

Example 14.34 Let $y(t) = P_{2T}(t + T)$. Find $Y(\omega)$.

Solution We have sketched the time waveform for $y(t)$ in Figure 14.40 to refresh your memory about the time variation of this waveform. We remind you that we have agreed not to worry about the values at points of discontinuity, for these do not affect the transform. Therefore, we will for the most part omit the dots denoting left continuity from here on. The derivative of this function, which we will call $x(t)$, is

$$py(t) = \delta(t + T) - \delta(t - T) = x(t).$$

(14.4-43)

To verify this, just remember that the area of the impulse function in each case is the height of the jump in $y(t)$ at that point. The derivative is shown in Figure 14.41. Its Fourier transform is

$$X(\omega) = \mathcal{F}\{\delta(t + T) - \delta(t - T) = e^{j\omega T} - e^{-j\omega T}.$$

(14.4-44)

Noting that our original function $y(t)$ has the property that $y(\pm\infty) = 0$, we have

$$y(\omega) = \frac{e^{j\omega T} - e^{-j\omega T}}{j\omega} = 2\frac{\sin(\omega T)}{\omega} = 2T\,\text{sinc}(\omega T).$$

(14.4-45)

Figure 14.40
An example waveform

Figure 14.41
The derivative

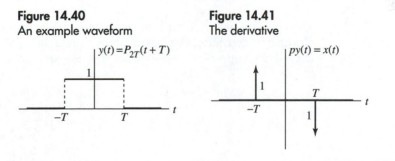

Example 14.35 Find the Fourier transform of $y(t)$ shown in Figure 14.42.

Figure 14.42
An example waveform

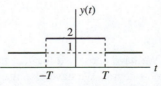

Solution It is easy to see how this waveform is related to the one in the last example. In fact, it is exactly the same except that a constant of unity is added at each point. Thus, the time derivative is precisely the same (shown in Figure 14.41). How do we find the additive constant k? We simply write

$$y(t) = 1 + P_{2T}(t + T).\qquad(14.4\text{-}46)$$

Hence, as we already know the transform of $P_{2T}(t + T)$, we have

$$Y(\omega) = 2\pi\delta(\omega) + 2T\operatorname{sinc}(\omega T).\qquad(14.4\text{-}47)$$

The point of the last two examples is this: the arbitrary additive constant present in an antiderivative is equivalent to an arbitrary added multiple of $\delta(\omega)$ in the frequency domain. This multiple must be found by independent means—just like an arbitrary constant of integration. We will now rederive an important transform pair using the integration property for which it is a bit more difficult to determine the value of the constant (or multiple). (We have already derived this transform pair in Example 14.16.

Example 14.36   Find $\mathscr{F}\{\operatorname{sgn}(t)\}$.

Solution   We show the signum function again for convenience of reference in Figure 14.43(a). We show its derivative in Figure 14.43(b). Remembering that the strength of the impulse is equal to the jump at the discontinuity gives

$$py(t) = 2\delta(t) = x(t).\qquad(14.4\text{-}48)$$

By our integration property, we know that

$$Y(\omega) = \frac{F\{2\delta(t)\}}{j\omega} + k\delta(\omega) = \frac{2}{j\omega} + k\delta(\omega).\qquad(14.4\text{-}49)$$

The only question is: What is the value of k? We must use some other information about $y(t)$ than its derivative. One possibility is to note that $\operatorname{sgn}(t)$ is an odd function; hence, for any value t, $\operatorname{sgn}(-t) = -\operatorname{sgn}(t)$. Using the inversion integral, we find that

$$0 = \operatorname{sgn}(-t) + \operatorname{sgn}(t) = \frac{1}{2\pi}\int_{-\infty}^{\infty}\left[\frac{2}{j\omega} + k\delta(\omega)\right]e^{-j\omega t}\,d\omega\qquad(14.4\text{-}50)$$

$$+ \frac{1}{2\pi}\int_{-\infty}^{\infty}\left[\frac{2}{j\omega} + k\delta(\omega)\right]e^{j\omega t}\,d\omega$$

$$= \frac{1}{\pi}\int_{-\infty}^{\infty}\left[\frac{2}{j\omega} + k\delta(\omega)\right]\cos(\omega t)\,d\omega$$

$$= \frac{2}{\pi}\int_{-\infty}^{\infty}\frac{\cos(\omega t)}{j\omega}\,d\omega + \frac{k}{\pi}\int_{-\infty}^{\infty}\cos(\omega t)\delta(\omega)\,d\omega$$

$$= \frac{k}{\pi}.$$

This implies that $k = 0$. We have here used the fact that $\cos(\omega t)/\omega$ is an odd function of ω, hence has a zero-valued integral. We have also used the sampling property of the impulse function, noting that $\cos(\omega t)\delta(\omega) = \cos(0)\delta(\omega) = \delta(\omega)$. Furthermore, the integral of the delta function is unity. Therefore, we have shown that

$$\mathscr{F}\{\operatorname{sgn}(t)\} = \frac{2}{j\omega}.\qquad(14.4\text{-}51)$$

Figure 14.43
The signum function
and its derivative

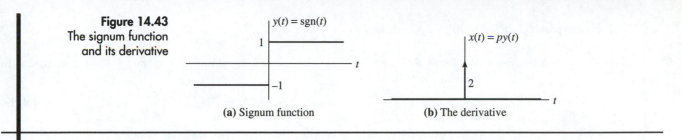

(a) Signum function   (b) The derivative

Example 14.37   Find $\mathcal{F}\{u(t)\}$ using the result expressed by equation (14.4-51).

Solution   Based upon the last example, this one is easy. We merely note that if we shift the signum function upward by one unit we will get twice the unit step function. Thus

$$u(t) = \frac{1}{2}[1 + \text{sgn}(t)] = \frac{1}{2} + \frac{1}{2}\,\text{sgn}(t). \tag{14.4-52}$$

Hence,
$$\mathcal{F}\{u(t)\} = \mathcal{F}\left[\frac{1}{2} + \frac{1}{2}\,\text{sgn}(t)\right] = \pi\delta(\omega) + \frac{1}{j\omega}. \tag{14.4-53}$$

The results of the last two examples are very useful transforms to know. We warn you, however, about a confusing item of notation. Following our convention, we will sometimes write

$$U(\omega) = \pi\delta(\omega) + \frac{1}{j\omega}. \tag{14.4-54}$$

The uppercase letter U, however, does not signify the unit step function—it is the *transform* of the unit step function:

$$U(\omega) = \mathcal{F}\{u(t)\}. \tag{14.4-55}$$

When we want to use the unit step function in the frequency domain, we will consistently use the symbol $u(\omega)$ (lowercase). In pencil and paper writing, though, it is difficult to distinguish one from the other. We recommend that you use some notation of your own choosing to distinguish them.

The next property is, again, a Laplace result that we rederive for the sake of completeness and because the imaginary operator j appears in this form and a sign is different.

Frequency Differentiation Property (Table 14.4, Entry 6)

$$tx(t) \longleftrightarrow jqX(\omega), \tag{14.4-56}$$

where

$$q = \frac{d}{d\omega}. \tag{14.4-57}$$

Proof   We write
$$\mathcal{F}\{tx(t)\} = \int_{-\infty}^{\infty} tx(t)e^{-j\omega t}\,dt = \int_{-\infty}^{\infty} jx(t)\frac{\partial e^{-j\omega t}}{\partial\omega}\,dt \tag{14.4-58}$$

$$= jq\int_{-\infty}^{\infty} x(t)e^{-j\omega t}\,dt = jqX(\omega).$$

We have once more assumed that the order of differentiation with respect to ω can be interchanged with integration with respect to t. ∎

Example 14.38   If $x(t) = \sum\limits_{n=0}^{N} a_n t^n$, find $X(\omega)$.

Solution   We know that $1 \leftrightarrow 2\pi\delta(\omega)$. Thus, $t \leftrightarrow 2\pi(jq)\delta(\omega)$, $t^2 \leftrightarrow 2\pi(jq)^2\delta(\omega)$, and so on. In general, we have

$$t^n \longleftrightarrow 2\pi(jq)^n \delta(\omega). \tag{14.4-59}$$

Thus, using the linearity property of the transform, we have

$$\sum_{n=0}^{N} a_n t^n \longleftrightarrow 2\pi \sum_{n=0}^{N} j^n a_n \delta^{(n)}(\omega), \tag{14.4-60}$$

using the notation $\delta^{(n)}(\omega) = q^n \delta(\omega)$.

Example 14.39   Find the Fourier transform of the waveform in Figure 14.44.

Figure 14.44
An example waveform

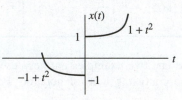

Solution   Admittedly, one does not encounter signals of this form very often, but it does illustrate the use of the result in the last example (based on the frequency differentiation property). We merely note that we can write

$$x(t) = \text{sgn}(t) + t^2. \tag{14.4-61}$$

Thus

$$X(\omega) = \mathcal{F}\{\text{sgn}(t) + t^2\} = \frac{2}{j\omega} + 2\pi j^2 \delta^{(2)}(\omega) = \frac{2}{j\omega} - 2\pi\delta^{(2)}(\omega). \tag{14.4-62}$$

All practical waveforms encountered in circuits and systems work are either bounded as $t \to \infty$ or approach infinity as a power of t. The only notable exception to this statement is the natural response of an unstable circuit or system, which behaves exponentially (with a positive exponent) for large values of time. We are usually not interested, however, in computing the Fourier transform (or spectrum) of signals in such systems.

The next property is one that holds for the Laplace transform, but that we did not derive in Chapter 13. It is much more useful in Fourier analysis.

Theorem 14.7 (Time Reversal Property)

$$x(-t) \longleftrightarrow X(-\omega). \tag{14.4-63}$$

Proof   You should be careful, for it is easy to confuse this property with the conjugate symmetry property of real signals. The proof is easy. Because

$$x(t) = \frac{1}{2\pi} \int_{-\infty}^{\infty} X(\omega) e^{j\omega t} \, dt, \tag{14.4-64}$$

we have

$$x(-t) = \frac{1}{2\pi} \int_{-\infty}^{\infty} X(\omega)e^{-j\omega t}\, dt = \frac{1}{2\pi} \int_{-\infty}^{\infty} X(-\omega)e^{j\omega t}\, dt. \qquad (14.4\text{-}65)$$

But this is merely the same equation, in a lengthier form, as the one we are proving:

$$x(-t) = \mathcal{F}^{-1}\{X(-\omega)\}, \qquad (14.4\text{-}66)$$

or

$$\mathcal{F}\{x(-t)\} = X(-\omega). \qquad (14.4\text{-}67)$$

∎

Examples: Fourier Transform Computation Using Properties

We now have derived most of the manipulative properties for the Fourier transform. For practice in using our just-developed properties, we will rederive several transform pairs that we have already seen.

Example 14.40   Show that

$$e^{j\omega_0 t}u(t) \longleftrightarrow \pi\delta(\omega - \omega_0) + \frac{1}{j(\omega - \omega_0)}. \qquad (14.4\text{-}68)$$

Solution   We use the modulation property and the known transform of the unit step function. Thus, we have

$$u(t) \longleftrightarrow \pi\delta(\omega) + \frac{1}{j\omega}. \qquad (14.4\text{-}69)$$

The modulation property states that

$$e^{j\omega_0 t}x(t) \longleftrightarrow X(\omega - \omega_0). \qquad (14.4\text{-}70)$$

Letting $x(t) = u(t)$, we have the result immediately.

Example 14.41   Show that

$$\cos(\omega_0 t)u(t) \longleftrightarrow \frac{\pi}{2}\delta(\omega - \omega_0) + \frac{\pi}{2}\delta(\omega + \omega_0) + j\frac{\omega}{(\omega_0^2 - \omega^2)}. \qquad (14.4\text{-}71)$$

Solution   We merely note that

$$\cos(\omega_0 t) = \frac{e^{j\omega_0 t} + e^{-j\omega_0 t}}{2} \qquad (14.4\text{-}72)$$

and apply the result derived in the preceding example.

Example 14.42   Show that

$$\sin(\omega_0 t)u(t) \longleftrightarrow -j\frac{\pi}{2}\delta(\omega - \omega_0) + j\frac{\pi}{2}\delta(\omega + \omega_0) + \frac{\omega_0}{(\omega_0^2 - \omega^2)}. \qquad (14.4\text{-}73)$$

Solution   As in the last example, we merely note that

$$\sin(\omega_0 t) = \frac{e^{j\omega_0 t} - e^{-j\omega_0 t}}{2j}, \qquad (14.4\text{-}74)$$

and apply the result derived in the preceding example.

Table 14.2
One-sided transforms

$x(t)$	$X(\omega)$
1. $\delta(t)$	1
2. $u(t)$	$1/j\omega + \pi\delta(\omega)$
3. $\cos(\omega_0 t)u(t)$	$\dfrac{\pi}{2}\delta(\omega - \omega_0) + \dfrac{\pi}{2}\delta(\omega + \omega_0) + j\dfrac{\omega}{(\omega_0^2 - \omega^2)}$
4. $\sin(\omega_0 t)u(t)$	$j\dfrac{\pi}{2}\delta(\omega - \omega_0) + j\dfrac{\pi}{2}\delta(\omega + \omega_0) + \dfrac{\omega_0}{\omega_0^2 - \omega^2}$
5. $e^{-at}u(t)$	$1/(a + j\omega)$

Table 14.3
Two-sided transforms

$x(t)$	$X(\omega)$
1. $\operatorname{sgn}(t)$	$2/j\omega$
2. 1	$2\pi\delta(\omega)$
3. $\cos(\omega_0 t)$	$\pi\delta(\omega - \omega_0) + \pi\delta(\omega + \omega_0)$
4. $\sin(\omega_0 t)$	$-j\pi\delta(\omega - \omega_0) + j\pi\delta(\omega + \omega_0)$
5. $e^{-a|t|}$	$\dfrac{2a}{\omega^2 + a^2}$

Table 14.4
Fourier transform properties

$x(t)$	$X(\omega)$	Name
1. $ax_1(t) + bx_2(t)$	$aX_1(\omega) + bX_2(\omega)$	(Linearity)
2. $x(t - a)$	$e^{-j\omega a}X(\omega)$	(Time Shift)
3. $e^{-j\omega_0 t}x(t)$	$X(\omega + \omega_0)$	(Frequency Shift or Modulation)
(a) $\cos(\omega_0 t)x(t)$	$\dfrac{X(\omega - \omega_0) + X(\omega + \omega_0)}{2}$	
(b) $\sin(\omega_0 t)x(t)$	$\dfrac{X(\omega - \omega_0) - X(\omega + \omega_0)}{j2}$	
4. $x(at)$	$\dfrac{1}{a}X(\omega/a)$	(Scaling)
5. $px(t)$	$j\omega X(\omega)$	(Time Differentiation)
6. $tx(t)$	$jqX(\omega) = jX'(\omega)$	(Freq. Differentiation)
7. $x(t)$ real	$X(-\omega) = X^*(\omega)$	(Conjugate Symmetry)
(a) Magnitude	$|X(-\omega)| = |X(\omega)|$	(Even Symmetry)
(b) Angle	$\angle X(-\omega) = -\angle X(\omega)$	(Odd Symmetry)
8. $X(t)$	$2\pi x(-\omega)$	(Time-Frequency Symmetry)
9. $\dfrac{1}{p}x(t)$	$\dfrac{X(\omega)}{j\omega} + k\delta(\omega)$	(Indef. Time Integration)
10. $x(-t)$	$X(-\omega)$	(Time Reversal)

Summary: Tables of Basic Transforms and Properties

Table 14.2 is a table of one-sided waveforms, Table 14.3 of two-sided functions, and Table 14.4 of basic properties. The first six properties in Table 14.4 are those of the Laplace transform specialized to Fourier transforms (with $s = j\omega$).

Section 14.4 Quiz

Q14.4-1. Find $\mathcal{F}\{10P_2(t - 2)\}$.

Q14.4-2. Find $\mathcal{F}\{e^{-2\pi t}[1 + \delta(t - 4)]\}$.

Q14.4-3. Find $\mathcal{F}\{u(t - 2) - 2u(t - 4) + u(t - 6)\}$ by using the differentiation and integration properties.

Q14.4-4. Find $\mathcal{F}\{tu(t)\}$ using Fourier transform properties.

Q14.4-5. Find the Fourier transform of the waveform shown in Figure Q14.4-1. Sketch its amplitude and phase spectra.

Figure Q14.4-1

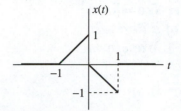

Q14.4-6. Let $x(t) = 4 \sin(2\pi t)[u(t) - u(t - 2)]$. Find $X(\omega) = \mathcal{F}\{x(t)\}$. Sketch its amplitude and phase spectra.

14.5 | Fourier Series

We will now develop the basic ideas behind a very remarkable property of periodic functions: they can be expanded into a series of complex exponentials (or of sinusoids). This idea motivated many of the developments of modern mathematics and sees applications, not only in circuit analysis, but in many other engineering disciplines as well. We start by developing the sampling property of impulse functions a bit more.

A Generalized Sampling Property for Singularity Functions

We will now discuss a relationship that we derived at the end of Chapter 12 for functions of t, but will repeat here in a slightly different way for functions of ω to refresh your memory. It goes like this:

$$\frac{1}{q}[X(\omega)\delta(\omega - a)] = X(a)u(\omega - a), \tag{14.5-1}$$

provided $X(\omega)$ is continuous at $\omega = a$. Notice that we have used the symbol $1/q$ to represent the running integral relative to ω, whereas we will continue to use $1/p$ to represent the time integral. If we evaluate this equation at $\omega = \infty$, noting that $u(\infty) = 1$, we will have

$$\int_{-\infty}^{\infty} X(\omega)\delta(\omega - a)\, d\omega = X(a), \tag{14.5-2}$$

providing, of course, that $X(\omega)$ is continuous at a. If we integrate by parts, an operation that we showed in Chapter 13 to be valid for generalized functions, we will have

$$\int_{-\infty}^{\infty} X(\omega)\delta'(\omega - a)\, d\omega = [X(\omega)\delta(\omega - a)]_{-\infty}^{\infty} - \int_{-\infty}^{\infty} X'(\omega)\delta(\omega - a)\, d\omega = -X'(a). \tag{14.5-3}$$

We have simply noted that the impulse function is zero at both limits. This assumes, of course, that $X'(\omega)$ is continuous at a. Integrating by parts once more, we find that

$$\int_{-\infty}^{\infty} X(\omega)\delta''(\omega - a)\, d\omega = [X(\omega)\delta'(\omega - a)]_{-\infty}^{\infty} - \int_{-\infty}^{\infty} X'(\omega)\delta'(\omega - a)\, d\omega = (-1)^2 X''(a), \tag{14.5-4}$$

assuming that $X''(\omega)$ is continuous at a. The truth of the following equation should now be apparent by induction:

$$\int_{-\infty}^{\infty} X(\omega)\delta^{(n)}(\omega - a)\, d\omega = (-1)^n X^{(n)}(a), \tag{14.5-5}$$

assuming that $X^{(n)}(\omega)$ exists and is continuous at $\omega = a$. We leave the simple inductive proof to you.

Now let's apply this property to the study of periodic functions.

The Fourier Transform of a Periodic Waveform

Definition 14.6   A time function $x(t)$ is periodic with period T if

$$x(t + T) = x(t) \tag{14.5-6}$$

for all values of t, and T is the smallest real number for which this equation is true.

We show the sketch of a stylized periodic time waveform in Figure 14.45. For the sake of brevity, we will use the expression "$x(t)$ is periodic (T)" for the longer phrase "$x(t)$ is a periodic time waveform having period T."

Figure 14.45
A periodic time waveform

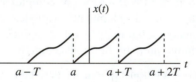

Taking the Fourier transform of equation (14.5-6) and using the time shift property results in

$$e^{j\omega T}X(\omega) = X(\omega). \tag{14.5-7}$$

Let's examine this equation carefully. It is equivalent to

$$[e^{j\omega T} - 1]X(\omega) = 0. \tag{14.5-8}$$

We ask ourselves, "What kind of frequency functions $X(\omega)$ satisfy this equation?" To answer this question, let's observe that $e^{j\omega T} - 1 = 0$ only for

$$\omega = n\omega_0, \tag{14.5-9}$$

where n is any integer $(n = 0, \pm 1, \pm 2, \ldots)$ and

$$\omega_0 = \frac{2\pi}{T}. \tag{14.5-10}$$

Thus, we see that $X(\omega)$ can satisfy our constraining equation only if it is zero at all frequencies not equal to $n\omega_0$. Does this imply that $X(\omega)$ is identically zero?

Not quite, for it can be nonzero at $\omega = n\omega_0$. First, we could have $X(n\omega_0) = k_n$, a finite constant for each value of n. That is, we could have

$$X(\omega) = \sum_{n=-\infty}^{\infty} k_n \delta_0(\omega - n\omega_0), \tag{14.5-11}$$

where $\delta_o(\omega)$ is the indicator function, which has the value one at $\omega = 0$ and is zero everywhere else. This will satisfy equation (14.5-8); however, if we apply the inverse Fourier transform to that equation we will merely get the identity $0 = 0$, rather than (14.5-6) as we desire, because the integral of a function that is nonzero (but of finite value) only at one frequency (or at any sequence of isolated points) is zero.

How about impulse functions at our discrete set of frequencies? Well, for each n, we have

$$[e^{j\omega T} - 1]\delta(\omega - n\omega_0) = [e^{jn\omega_0 T} - 1]\delta(\omega - n\omega_0) = 0 \cdot \delta(\omega - n\omega_0) = 0. \quad (14.5\text{-}12)$$

It seems that impulses work. How about derivatives of impulses? To find out, we will assume that k is a positive integer and use the property we just derived in equation (14.5-5) (letting $X(\omega)$ in the formula be replaced by $[e^{j\omega T} - 1)]$). This results in

$$\int_{-\infty}^{\infty} [e^{j\omega T} - 1]\delta^{(k)}(\omega - n\omega_0)\, d\omega = (-jT)^{(k)} \neq 0. \quad (14.5\text{-}13)$$

But this means that the frequency function inside the integral is nonzero also. Therefore, derivatives of impulses will not work. Thus, we have exhausted all the possible frequency functions that are zero everywhere except at a discrete set of frequencies. In this manner, we see that the only possible function of ω satisfying (14.5-7), and hence (14.5-6), is

$$X(\omega) = \sum_{n=-\infty}^{\infty} 2\pi c_n \delta(\omega - n\omega_0). \quad (14.5\text{-}14)$$

The constants c_n are, at this point, arbitrary. We have multiplied them by 2π in order to simplify our final result. The constants c_n are called the *Fourier coefficients*.

The question now is, what are these constants c_n? To determine them, let's take the inverse Fourier transform of (14.5-14). The result is[17]

$$x(t) = \mathscr{F}^{-1}\{X(\omega)\} = \mathscr{F}^{-1}\left\{ \sum_{n=-\infty}^{\infty} c_n\, 2\pi\delta(\omega - n\omega_0) \right\} \quad (14.5\text{-}15)$$

$$= \sum_{n=-\infty}^{\infty} c_n\, 2\pi \mathscr{F}^{-1}[\delta(\omega - n\omega_0)] = \sum_{n=-\infty}^{\infty} c_n\, e^{jn\omega_0 t}.$$

This is a remarkable result! A periodic function can be expressed as an infinite series of exponential waveforms, whose frequencies are integral multiples of the *fundamental period* $\omega_0 = 2\pi/T$. The frequency $n\omega_0$ is referred to as the *nth harmonic* and the frequency ω_0 itself as either the *first harmonic* or the *fundamental*. The sum in (14.5-15) itself is called the *Fourier series* (representation) of $x(t)$.

But what about the constants? Notice that, for any value of a,

$$\int_a^{a+T} e^{jn\omega_0 t}\, dt = T\delta_0(n); \quad (14.5\text{-}16)$$

that is, the integral is zero except for $n = 0$ and it is equal to T if $n = 0$. Furthermore, the constant a in (14.5-16) is quite arbitrary. To see this, just observe first that if $n = 0$, the equation becomes

$$\int_a^{a+T} 1\, dt = T; \quad (14.5\text{-}17)$$

---

[17] Notice that we are using the linearity of the inverse transform extended to infinite sums, a procedure that is valid for generalized functions.

furthermore, if $n \neq 0$, one has

$$\int_a^{a+T} e^{jn\omega_0 t} \, dt = \left[\frac{e^{jn\omega_0 t}}{jn\omega_0} \right]_a^{a+T} = 0. \tag{14.5-18}$$

Now let's go back to (14.5-15), multiply both sides by $e^{-jm\omega_0 t}$ (m an arbitrary integer) and then integrate both sides from some arbitrary time point a to $a + T$. This results in

$$\int_a^{a+T} x(t)e^{-jm\omega_0 t} \, dt = \int_a^{a+T} \left[\sum_{n=-\infty}^{\infty} c_n e^{j(n-m)\omega_0 t} \right] dt \tag{14.5-19}$$

$$= \sum_{n=-\infty}^{\infty} c_n \int_a^{a+T} e^{j(n-m)\omega_0 t} \, dt = \sum_{n=-\infty}^{\infty} c_n T \delta_0(n - m) = Tc_m.$$

The result of this equation is that we have found the constant c_m. To be compatible notationally with the original series, we simply note that m is any arbitrary integer, so we can write its value for general n:

$$c_n = \frac{1}{T} \int_a^{a+T} x(t)e^{-jn\omega_0 t} \, dt. \tag{14.5-20}$$

Note that c_n will in general be complex—even though $x(t)$ is a real time waveform.

To summarize our development, we see that if $x(t)$ is periodic (T), then it can be written as

$$x(t) = \sum_{n=-\infty}^{\infty} c_n e^{jn\omega_0 t}, \tag{14.5-21}$$

where

$$c_n = \frac{1}{T} \int_a^{a+T} x(t)e^{-jn\omega_0 t} \, dt. \tag{14.5-22}$$

Furthermore, its Fourier transform is

$$X(\omega) = \sum_{n=-\infty}^{\infty} 2\pi c_n \delta(\omega - n\omega_0). \tag{14.5-23}$$

The Singularity Spectrum of a Periodic Waveform

A sketch of $X(\omega)$ is shown in Figure 14.46. It consists of impulses equally spaced ω_0 rad/s apart. The area under each impulse is a complex number. This is perhaps not an intuitive kind of thing—interpreting area as being complex. However, we remind you that the spectrum of an impulsive frequency function is defined in terms of the indicator function. We simply assume for plotting purposes that each of these constants represents the sampled value of a continuous complex frequency function having magnitude $A(\omega)$ and phase $\theta(\omega)$. Thus, the amplitude spectrum is

$$A(\omega) = \sum_{n=-\infty}^{\infty} 2\pi |c_n| \delta_0(\omega - n\omega_0), \tag{14.5-24}$$

and the phase spectrum is

$$\theta(\omega) = \sum_{n=-\infty}^{\infty} 2\pi \angle c_n \, \delta_0(\omega - n\omega_0). \tag{14.5-25}$$

Typical plots are shown in Figure 14.47. Notice that $\theta(0)$ is either 0 (shown) or $\pm 180°$.

Figure 14.46
The Fourier transform
of a periodic time
waveform

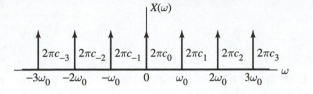

Figure 14.47
The spectrum of a
periodic time waveform

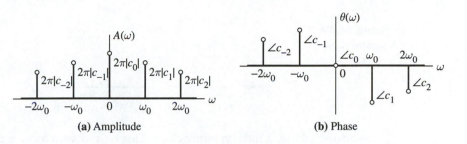

(a) Amplitude   **(b)** Phase

The Fourier coefficients have a simple relationship to the Fourier transform of a simple function related to $x(t)$: the function that is equal to $x(t)$ over a single period and zero elsewhere. We show such a function in Figure 14.48. Its Fourier transform is given by

$$X_0(\omega) = \int_a^{a+T} x_0(t)e^{-j\omega t}\, dt = \int_a^{a+T} x(t)e^{-j\omega t}\, dt \qquad (14.5\text{-}26)$$

because $x_0(t)$ is identical to $x(t)$ over the time interval $[a, a + T]$. If we evaluate this equation at $\omega = n\omega_0$, we obtain

$$c_n = \frac{X_0(n\omega_0)}{T}. \qquad (14.5\text{-}27)$$

This is a very handy result because it says that we must only compute the Fourier transform of a time-limited waveform (which always exists because the Laplace integral converges for any value of s including $s = j\omega$), then evaluate it at the various harmonic frequencies and divide it by the period to obtain the Fourier coefficients. Furthermore, *we now know that all of the Fourier transform properties hold for the Fourier coefficients as well.* For instance, we now see that

$$c_{-n} = c_n^*, \qquad (14.5\text{-}28)$$

which implies that

$$|c_{-n}| = |c_n| \qquad (14.5\text{-}29)$$

and

$$\angle c_{-n} = -\angle c_n. \qquad (14.5\text{-}30)$$

This is illustrated graphically in our typical spectrum sketches in Figure 14.47 because $A(\omega)$ is even and $\theta(\omega)$ is odd.

Figure 14.48
A related, time-limited
waveform

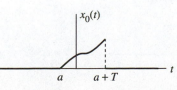

Examples: Fourier Series

Example 14.43   Find the Fourier series representation for the periodic waveform shown in Figure 14.49 and sketch its spectrum.

Figure 14.49
A square wave

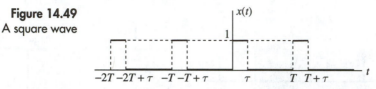

Solution   First, a little on terminology. This type of waveform is called a *square wave,* though the more general term *rectangular wave* might be more appropriate. The *duty cycle* of such a square wave is defined by

$$\text{duty cycle} = \frac{\tau}{T} \times 100\% = \frac{\text{nonzero time}}{\text{period}} \times 100\%. \tag{14.5-31}$$

To compute the Fourier series, we first find the Fourier transform of the one-period equivalent waveform $x_0(t)$ shown in Figure 14.50. This Fourier transform is quite easy to derive. It is, taking $a = 0$ in our general formula for the nth Fourier coefficient,

$$X_0(\omega) = \int_0^\tau 1 e^{-j\omega t}\, dt = \frac{1 - e^{-j\omega\tau}}{j\omega}. \tag{14.5-32}$$

Noting that $\omega_0 T = 2\pi$ (a relationship that we will use again and again), we have

$$c_n = \frac{X_0(n\omega_0)}{T} = \frac{1 - e^{-jn\omega_0\tau}}{jn\omega_0 T} = \frac{[1 - \cos(n\omega_0\tau)] + j\sin(n\omega_0\tau)}{jn2\pi}, \tag{14.5-33}$$

assuming that $n \neq 0$. If $n = 0$ (that is, $\omega = n\omega_0 = 0$), the integration formula in (14.5-32) does not hold. We must integrate afresh:

$$c_0 = \frac{X_0(0)}{T} = \frac{1}{T}\int_a^{a+T} x_0(t)\, dt = \frac{1}{T}\int_0^\tau 1\, dt = \frac{\tau}{T}, \tag{14.5-34}$$

the fractional duty cycle (before multiplying by 100%). The first integral in this equation is a quite important one because one often has to do a special integration for the value $n = 0$ (the *dc value*). Note that it is simply the area under the $x_0(t)$ graph divided by the period and—as such—is often easy to evaluate by inspection, as it is in this case.

Figure 14.50
A single period equivalent

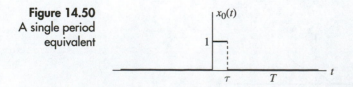

In order to sketch the spectrum, we must compute $|c_n|$ and $\angle c_n$. For $n = 0$, these values are τ/T and 0, respectively. For $n \neq 0$, we have[18]

$$|c_n| = \frac{1}{n2\pi}\sqrt{[1 - \cos(n\omega_0\tau)]^2 + [\sin(n\omega_0\tau)]^2} \tag{14.5-35}$$

$$= \frac{\sqrt{2[1 - \cos(n\omega_0\tau)]}}{n2\pi} = \frac{\sin(n\omega_0\tau/2)}{n\pi}.$$

Some care must be taken with the signs of the terms here. In general, we must take the positive square root because the magnitude is, of necessity, nonnegative. The magnitude spectrum is plotted in Figure 14.51 for the dc value, the positive and negative of the fundamental, and the second through the fifth (positive and negative) harmonics. The light line is the "envelope," which is a plot of the magnitude of the Fourier transform of $x_0(t)$ divided by the period, $|X_0(\omega)|/T$, and the heavy lines terminated with the open circles represent the magnitude of the Fourier coefficients. Here is an important fact: if $\tau = T/2$, each of the even harmonics lies at a null (zero) of the envelope; hence, its value is zero. Stated another way, *all even harmonics are zero if $\tau = T/2$*. This can be seen by inserting $\tau = T/2$ into the formula for the Fourier coefficients, equation (14.5-35), because for this value of τ the sine in the numerator is zero for any even value of n.

Figure 14.51
Amplitude spectrum

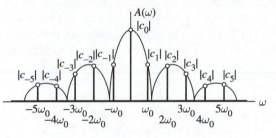

The phase spectrum is given by

$$\theta(\omega) = \angle\left[\frac{\sin(n\omega_0\tau) - j[1 - \cos(n\omega_0\tau)]}{n2\pi}\right] = \tan^{-1}\left[-\frac{1 - \cos(n\omega_0\tau)}{\sin(n\omega_0\tau)}\right] \tag{14.5-36}$$

$$= -\tan^{-1}\left[\frac{2\sin^2[n\omega_0\tau/2]}{2\sin(n\omega_0\tau/2)\cos(n\omega_0\tau/2)}\right] = -\tan^{-1}[\tan[n\omega_0\tau/2]] = -n\omega_0\tau/2.$$

This spectrum is plotted in Figure 14.52. The diagonal straight line is the envelope of the phase spectrum, namely $-\omega\tau/2$, and the solid lines capped with open circles the angles of the coefficients themselves. The Fourier series itself is given by

$$x(t) = \sum_{n=-\infty}^{\infty} c_n e^{jn\omega_0 t}, \tag{14.5-37}$$

where the c_n are given above, and the Fourier transform is

$$X(\omega) = \sum_{n=-\infty}^{\infty} 2\pi c_n \delta(\omega - n\omega_0). \tag{14.5-38}$$

---

[18] Note that $2[1 - \cos(x)] = 4[1 - \cos(x)]/2 = 4\sin^2(x/2)$.

Figure 14.52
Phase spectrum

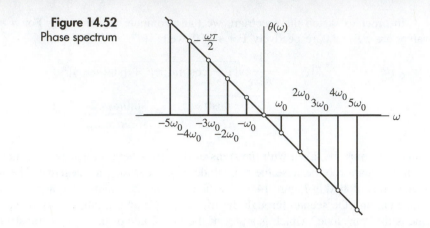

As you can see, the major problem is always computation of the c_n and plotting of the spectrum. The Fourier series and the Fourier transform are both simply written down.

The example we have just completed is an important one. It appears in many engineering applications; for example, it represents the waveform for a timing clock in a microprocessor. Additionally, it can also be used to find other transforms. For this reason, as well as to illustrate a number of points, we computed the coefficients directly. There are, however, ways of streamlining the process. One technique that is quite useful is to use the known relationships between the Fourier transform and its various derivatives. Recall that the single period waveform $x_0(t)$ associated with any periodic time function $x(t)$ has the property that $x_0(\pm\infty) = 0$. Thus, if $px_0(t) = y(t)$, then $X_0(\omega) = Y(\omega)/j\omega$; if $p^2 x_0(t) = y(t)$, then $X_0(\omega) = Y(\omega)/(j\omega)^2$; and so on. In this way, if we can find a simpler function that is some derivative of our given waveform, then we can find its transform easily—and then, of course, "sample it" to obtain the coefficients.

Figure 14.53   Derivative of example waveform

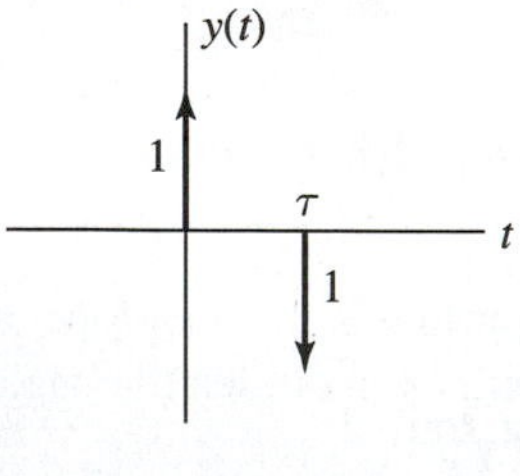

$y(t)$

As an illustration of the technique we have just outlined, notice that the first derivative of the single period form of the square wave in our last example is the waveform shown in Figure 14.53. It is quite easy to compute the Fourier transform of this one, because $y(t) = \delta(t) - \delta(t - \tau)$. Thus, the Fourier transform is $Y(\omega) = 1 - e^{-j\omega\tau}$. We quickly compute $X_0(\omega) = [1 - e^{-j\omega\tau}]/j\omega$. We then proceed as before to compute the c_n's. This we do without ever evaluating an integral! Notice, by the way, that the coefficient c_0 must, in general, be computed separately—as we have seen. However, we can take the limit as $\omega \to 0$ in our result above. This gives $X_0(0) = \tau$. We then divide by T to obtain c_0. This method works in general, though sometimes the limit computation is more difficult than computing the integral directly as an area.

Example 14.44   Find the Fourier coefficients of $x(t)$ in Figure 14.54.

Figure 14.54
An example waveform

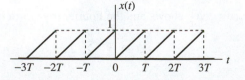

Solution   One period of our waveform is shown in Figure 14.55. We have chosen to use the value $a = 0$ to simplify our computation, though any other would suffice. If we compute the first derivative, we will get the waveform shown in Figure 14.56(a), for which we have al-

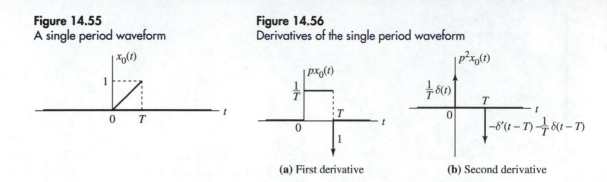

Figure 14.55
A single period waveform

Figure 14.56
Derivatives of the single period waveform

(a) First derivative   **(b)** Second derivative

ready computed the Fourier transform (of the nonsingular part, with $T = \tau$). However, if we compute the second derivative, we will have the waveform shown in Figure 14.56(b), a very simple one to work with. Notice that we have had to use the complete symbols in this sketch because there are two singularities occurring at the same point in time, $t = T$. The Fourier transform of $p^2x(t)$ is

$$\mathcal{F}\{p^2x_0(t)\} = (j\omega)^2X_0(\omega) \tag{14.5-39}$$

$$= \mathcal{F}\left\{\frac{1}{T}[\delta(t) - \delta(t - T) - T\delta'(t - T)]\right\} = \frac{1 - e^{-j\omega T} - j\omega Te^{-j\omega T}}{T},$$

which we can solve for $X(\omega_0)$:

$$X_0(\omega) = \frac{1 - e^{-j\omega T} - j\omega Te^{-j\omega T}}{(j\omega)^2T}. \tag{14.5-40}$$

Now, noting that $\omega_0 T = 2\pi$, we evaluate the c_n from our known formula, $c_n = X_0(n\omega_0)/T$. This gives

$$c_n = \frac{X_0(n\omega_0)}{T} = \frac{1 - e^{-jn2\pi} - (jn2\pi)e^{-jn2\pi}}{(jn2\pi)^2} = j\frac{1}{n2\pi} \tag{14.5-41}$$

We have used the fact that $e^{-jn2\pi} = 1$. For $n = 0$, we have (looking at Figure 14.55)

$$c_0 = \frac{1}{2}. \tag{14.5-42}$$

Because the Fourier coefficients are merely the Fourier transform of a one-period version of the original periodic waveform evaluated at $n\omega_0$,[19] we can use any of our Fourier transform properties to compute them. For instance, we know that an even time function has a real Fourier transform and an odd one has a purely imaginary transform. Furthermore, we know that a shift (delay) in the time waveform multiplies the transform by an exponential, and so on. The next example illustrates this.

Example 14.45   Find the Fourier coefficients of the waveform in Figure 14.57. (Note that the dashed lines mean that the function continues in a repetitive manner.)

---

[19] Or, in the jargon of electrical engineering, "sampled at the frequencies $n\omega_0$."

Figure 14.57
An example waveform

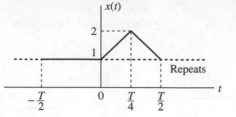

Solution   Let's first simplify the problem by subtracting one from $x(t)$. This yields the waveform in Figure 14.58. We then form a single period equivalent by taking the arbitrary constant $a = -T/2$, as shown in Figure 14.59(a). We see that if we shift this waveform by $T/4$ to the left, we will obtain the even time function shown in Figure 14.59(b).

Figure 14.58
A preliminary step

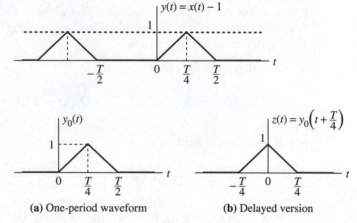

Figure 14.59
A one-period
waveform and a
delayed version

(a) One-period waveform   **(b)** Delayed version

Now let's get to work and differentiate! The first and second derivatives are shown in Figure 14.60(a) and (b). The second derivative consists merely of impulse functions, so its Fourier transform is easy to compute. It is

$$\mathcal{F}\{p^2 z(t)\} = \frac{4}{T}\left[e^{j\frac{\omega T}{4}} - 2 + e^{-j\frac{\omega T}{4}}\right] = \frac{4}{T}\left[2\cos\left(\frac{\omega T}{4}\right) - 2\right] \quad (14.5\text{-}43)$$

$$= -\frac{8}{T}\left[1 - \cos\left(\frac{\omega T}{4}\right)\right].$$

Thus, $$Z(\omega) = -\frac{8}{(j\omega)^2 T}\left[1 - \cos\left(\frac{\omega T}{4}\right)\right]. \quad (14.5\text{-}44)$$

Now, because $z(t) = y_0(t + T/4)$, we have $y_0(t) = z(t - T/4)$, so—applying the time

Figure 14.60
Derivatives of the
one-period waveform

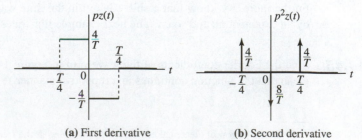

(a) First derivative   **(b)** Second derivative

shift property—we get

$$Y_0(\omega) = -\frac{8e^{-j\frac{\omega T}{4}}}{(j\omega)^2 T}\left[1 - \cos\left(\frac{\omega T}{4}\right)\right]. \tag{14.5-45}$$

We are now in a position to compute the Fourier coefficients of $y(t)$. They are

$$c'_n = \frac{Y_0(n\omega_0)}{T} = -\frac{8e^{-j\frac{n\pi}{2}}}{(jn2\pi)^2}\left[1 - \cos\left(\frac{n\pi}{2}\right)\right]. \tag{14.5-46}$$

This, of course, only holds if $n \neq 0$. If $n = 0$, we merely compute the area under one period of $y(t)$ and divide by the period. This gives

$$c'_0 = \frac{1}{4}. \tag{14.5-47}$$

We can simplify just a bit by noting that $j^2 = -1$. This transforms (14.5-46) into

$$c'_n = \frac{2e^{-j\frac{n\pi}{2}}}{n^2\pi^2}\left[1 - \cos\left(\frac{n\pi}{2}\right)\right]. \tag{14.5-48}$$

The Fourier series of $y(t)$ is

$$y(t) = \sum_{n=-\infty}^{\infty} c'_n e^{jn\omega_0 t}. \tag{14.5-49}$$

But what we want is the Fourier series of $x(t)$, not $y(t)$. However, this is easy; we just add the constant one to $y(t)$ to give $x(t)$. This modifies only the $n = 0$ Fourier coefficient. Thus, we finally have the Fourier coefficients of $x(t)$:

$$c_n = \frac{2e^{-j\frac{n\pi}{2}}}{n^2\pi^2}\left[1 - \cos\left(\frac{n\pi}{2}\right)\right] = 2e^{-j\frac{n\pi}{2}}\left[\frac{\sin(n\pi/4)}{n\pi}\right]^2; \quad n \neq 0 \tag{14.5-50}$$

$$c_0 = 1 + \frac{1}{4} = \frac{5}{4}.$$

In deriving the last form of c_n, we used the trigonometric relation

$$\sin^2(x) = \frac{1 - \cos(2x)}{2}. \tag{14.5-51}$$

Many texts on circuits and systems present elaborate lists of properties of the Fourier coefficients. They are, however, merely scaled multiples of the sampled version of the Fourier transform of the associated one-period waveform, so they enjoy all of the appropriate Fourier transform properties. We will, however, remind you that, providing $x(t)$ is a real time function,

$$c_{-n} = c_n^*. \tag{14.5-52}$$

This is in fact very obvious, given the relationship $c_n = X(n\omega_0)/T$, because this implies that $c_{-n} = X(-n\omega_0)/T = X^*(n\omega_0)/T = c_n^*$. We will use this relationship shortly.

Fourier Series:
Trigonometric Form

Historically, the complex form of the Fourier series was not the first to be studied. This position of honor is occupied by what is called the *trigonometric* Fourier series. It is, however, easy to derive from the complex form. For $n \neq 0$, we will write

$$c_n = \frac{a_n - jb_n}{2},$$ (14.5-53)

where

$$a_n = c_n + c_n^* = c_n + c_{-n}$$ (14.5-54)

and

$$b_n = j[c_n - c_n^*] = j[c_n - c_{-n}].$$ (14.5-55)

If $n = 0$, we will write

$$a_0 = c_0.$$ (14.5-56)

Now let's split the Fourier series representation into three terms:

$$x(t) = c_0 + \sum_{n=1}^{\infty} c_n e^{jn\omega_0 t} + \sum_{n=-\infty}^{-1} c_n e^{jn\omega_0 t}.$$ (14.5-57)

Changing the summation index in the second sum to $m = -n$, we have

$$x(t) = c_0 + \sum_{n=1}^{\infty} c_n e^{jn\omega_0 t} + \sum_{m=1}^{\infty} c_{-m} e^{-jm\omega_0 t}.$$ (14.5-58)

Thus, we have

$$x(t) = c_0 + \sum_{n=1}^{\infty} [c_n e^{jn\omega_0 t} + c_{-n} e^{-jn\omega_0 t}].$$ (14.5-59)

Let's look at the individual terms, express the c_n in terms of a_n and b_n (using the fact that $c_{-n} = c_n^*$ and assuming $x(t)$ is real and use Euler's formula to get

$$c_n e^{jn\omega_0 t} + c_{-n} e^{-jn\omega_0 t} = \left[\frac{a_n - jb_n}{2} \right] e^{jn\omega_0 t} + \left[\frac{a_n + jb_n}{2} \right] e^{-jn\omega_0 t}$$ (14.5-60)

$$= a_n \cos(n\omega_0 t) + b_n \sin(n\omega_0 t).$$

This, together with $c_0 = a_0$, gives us

$$x(t) = a_0 + \sum_{n=1}^{\infty} [a_n \cos(n\omega_0 t) + b_n \sin(n\omega_0 t)].$$ (14.5-61)

This is the version of the Fourier series referred to previously as the trigonometric form.

There is an explicit real version of the coefficients themselves. We simply use equations (14.5-54) and (14.5-55) to write

$$a_n = c_n + c_{-n} = \frac{1}{T} \int_a^{a+T} x(t)[e^{jn\omega_0 t} + e^{-jn\omega_0 t}] \, dt = \frac{2}{T} \int_a^{a+T} x(t) \cos(n\omega_0 t) \, dt$$ (14.5-62)

and

$$b_n = j[c_n - c_{-n}] = \frac{j}{T} \int_a^{a+T} x(t)[e^{-jn\omega_0 t} - e^{jn\omega_0 t}] \, dt = \frac{2}{T} \int_a^{a+T} x(t) \sin(n\omega_0 t) \, dt,$$ (14.5-63)

providing, of course, that $n \neq 0$. The zeroth-order coefficient is

$$a_0 = c_0 = \frac{1}{T} \int_a^{a+T} x(t) \, dt.$$ (14.5-64)

We will not work with the trigonometric form. Because we have derived the complex series from the transform with the attendant connection between the series coefficients and the transform itself, we feel that the complex Fourier series is more useful. If the need arises to use the trigonometric form, one merely splits the c_n coefficients into real and imaginary parts and regroups. The need for an entirely different mathematical framework is nonexistent. Furthermore, the complex exponential series has a more direct connection with the spectrum.

Fourier Series: Amplitude and Phase Form

There is another real form, on the other hand, that is quite useful in applications, for it expresses the series directly in terms of the amplitude and phase spectra. Too, it is much more quickly derived from the complex form. We simply write the complex coefficients in Euler form. This gives

$$c_n = A_n e^{j\theta_n}, \tag{14.5-65}$$

where

$$A_n = |c_n| \tag{14.5-66}$$

and

$$\theta_n = \angle c_n. \tag{14.5-67}$$

Thus (and this is the key point in our development) one still computes the complex coefficients and merely transforms them to Euler form—as one would anyway if the spectrum is to be sketched. Now we write

$$x(t) = a_0 + \sum_{n=1}^{\infty} [A_n e^{j[n\omega_0 t + \theta_n]} + A_{-n} e^{j[-n\omega_0 t + \theta_{-n}]}], \tag{14.5-68}$$

as before. We know, though, that $A_{-n} = A_n$ and $\theta_{-n} = -\theta_n$ because the magnitude of the spectrum is even and the phase is odd. Thus, we have

$$x(t) = a_0 + \sum_{n=1}^{\infty} [A_n[e^{j[n\omega_0 t + \theta_n]} + e^{-j[n\omega_0 t + \theta_n]}]] = a_0 + \sum_{n=1}^{\infty} 2A_n \cos(n\omega_0 t + \theta_n). \tag{14.5-69}$$

Finally, defining $A_0 = a_0$, we have

$$x(t) = A_0 + \sum_{n=1}^{\infty} 2A_n \cos(n\omega_0 t + \theta_n). \tag{14.5-70}$$

We will not present separate equations for the coefficients, for one merely computes the complex coefficients, then finds the magnitude and the angle and uses them in (14.5-66) and (14.5-67). More often than not, the A_n are redefined to incorporate the extra factor of 2 in (14.5-70).

Section 14.5 Quiz

Q14.5-1. Find the Fourier coefficients for the waveform in Figure Q14.5-1.

Q14.5-2. Find the Fourier coefficients for the waveform in Figure Q14.5-2.

Figure Q14.5-1

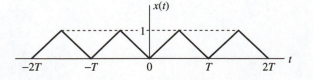

Figure Q14.5-2

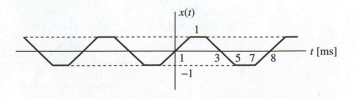

14.6 | Negative Frequency and the Hilbert Transform

The Concept of Frequency: A Preliminary Discussion

In this section we will discuss a few fundamental facts about Fourier transforms that find application in such diverse areas as communications and signal processing, automatic controls, and the advanced theory of circuits. We also discuss one answer to a question asked by many students: What is the meaning of negative frequency?

A beginner often asks this question upon his or her first exposure to the frequency domain concept. After all, the most basic definition of frequency is perhaps the number of, say, positive zero crossings of a periodic waveform per unit time. Such a number can only be positive. For the waveform $\cos(\omega_0 t)$, for instance, there is only one positive zero crossing in each period T. Thus, there are $1/T = \omega_0/2\pi = f_0$ such crossings per unit time. This concept applies, however, also to such a waveform as the square wave, which we discussed in Example 14.43 of Section 14.5. In this case, one still obtains $f_0 = \omega_0/2\pi$ as "the frequency." This is misleading, though, for we know that this waveform contains components at an infinite set of harmonically related frequencies. As it happens, this is the type of waveform used in many digital circuits, such as microprocessors, as a master "clock." Thus, one hears of (for instance) a new UltraChip computer using a 7X946 microprocessor chip "running at 500 MHz." In fact, this number refers only to the fundamental frequency. Other harmonics are present as well.

So how do we define frequency? In a very simplistic manner—which is often the best, anyway! We define it in terms of a *sinusoid*. We say that the constant ω_0 in the expression $\cos(\omega_0 t)$ is the "radian frequency" of the sinusoid and similarly that $f_0 = \omega_0/2\pi$ is the "hertz frequency." We do the same with the waveform $\sin(\omega_0 t)$. Because the complex exponential $e^{j\omega_0 t}$ is expressible as a linear combination of the preceding two, its frequency is also well defined. The Fourier transform expresses any waveform as a linear combination (sum or integral) of such waveforms, so we can discuss the "strength of the component" as the Fourier transform or coefficient evaluated at that frequency.

Negative Frequency

So much for the *definition* of frequency. What about negative frequencies? The simplest answer is to say that frequency, being defined as a mathematical constant, can have either sign. Thus, negative frequency makes sense, for $\cos(-\omega_0 t) = \cos(\omega_0 t)$ and $\sin(-\omega_0 t) = -\sin(\omega_0 t)$. This is not as simplistic as it perhaps sounds, for this fact is the basis for the conjugate symmetry property of the Fourier transform of a real waveform. If the transform were zero for $\omega < 0$, it would not possess this symmetry—which we have proven it must have.

But let's look at the problem in a more general way, one that will prove useful to you in other courses such as communications and signal processing. In Figure 14.61(a) we have sketched a stylized representation of the Fourier transform $X(\omega)$ of a time waveform $x(t)$, assuming that it is *band limited,* that is, that $X(\omega)$ is zero for $|\omega| > a$ for some positive real number a. In Figure 14.61(b) we have simply forced the negative frequency components to zero and have multiplied the positive ones by 2. The factor of 2 is actually a bit of hindsight, as so many things are, after having worked things through in advance.

Figure 14.61
The one-sided equivalent spectrum

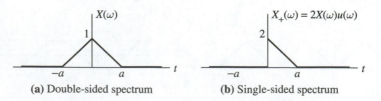

(a) Double-sided spectrum   (b) Single-sided spectrum

It will result in proper scaling of our final result. We now ask the following question: What is the nature of the signal $x_+(t)$ having $X_+(\omega)$ as its transform?

We answer this question by first writing the single-sided transform mathematically as

$$X_+(\omega) = 2X(\omega)u(\omega). \tag{14.6-1}$$

In Chapter 13, we defined the convolution of two waveforms $x(t)$ and $y(t)$ as the inverse Laplace transform of the products of their Laplace transforms $X(s)$ and $Y(s)$; that is,

$$\mathcal{L}\{x(t) \otimes y(t)\} = X(s)Y(s). \tag{14.6-2}$$

Because the Fourier transform is merely the Laplace transform evaluated for $s = j\omega$,

$$\mathcal{F}\{x(t) \otimes y(t)\} = X(\omega)Y(\omega), \tag{14.6-3}$$

or, equivalently,[20]

$$x(t) \otimes y(t) = \mathcal{F}^{-1}\{X(\omega)Y(\omega)\}. \tag{14.6-4}$$

Let's apply this last equation to $X_+(\omega) = X(\omega) \cdot 2u(\omega)$:

$$x_+(t) = \mathcal{F}^{-1}\{X_+(\omega)\} = x(t) \otimes \mathcal{F}^{-1}\{2u(\omega)\}. \tag{14.6-5}$$

Our only task, therefore, is to investigate the inverse Fourier transform of $2u(\omega)$. To do this, we will make use of the symmetry property:

$$x(t) \longleftrightarrow X(\omega) \implies X(t) \longleftrightarrow 2\pi x(-\omega). \tag{14.6-6}$$

We already know the transform of $2u(t)$. It is given by

$$2u(t) \longleftrightarrow 2\pi\delta(\omega) + \frac{2}{j\omega}. \tag{14.6-7}$$

The time reversal property says that

$$2u(-t) \longleftrightarrow 2\pi\delta(-\omega) + \frac{2}{j(-\omega)} = 2\pi\delta(\omega) + j\frac{2}{\omega} \tag{14.6-8}$$

because $\delta(\omega)$ is an even function. Thus, the symmetry theorem dictates that

$$2\pi\delta(t) + j\frac{2}{t} \longleftrightarrow 4\pi u(\omega). \tag{14.6-9}$$

Finally, we divide by 2π to obtain the transform we need:

$$\delta(t) + j\frac{1}{\pi t} \longleftrightarrow 2u(\omega). \tag{14.6-10}$$

Let's use this result, now, in (14.6-5) to find $x_+(t)$. It is

$$x_+(t) = x(t) \otimes \left[\delta(t) + j\frac{1}{\pi t}\right] = x(t) \otimes \delta(t) + x(t) \otimes j\frac{1}{\pi t}. \tag{14.6-11}$$

---

[20] Recall that the symbol $\otimes$ represents the operation of convolution (see Chapter 13, Section 13.8).

We have made use of the fact that $x(t) \otimes [y(t) + z(t)] = x(t) \otimes y(t) + x(t) \otimes z(t)$. The first term is merely

$$x(t) \otimes \delta(t) = \int_{-\infty}^{\infty} x(\alpha)\delta(t - \alpha)\, d\alpha = x(t) \tag{14.6-12}$$

at all points of continuity of $x(t)$ (which are the only ones we are concerned with in transform theory, as we have said). This is merely a restatement of the convolution property of the unit impulse. The second term is

$$x(t) \otimes j\frac{1}{\pi t} = j\int_{-\infty}^{\infty} \frac{x(\alpha)}{\pi(t - \alpha)}\, d\alpha = j\check{x}(t), \tag{14.6-13}$$

where

$$\check{x}(t) = \int_{-\infty}^{\infty} \frac{x(\alpha)}{\pi(t - \alpha)}\, d\alpha = \mathcal{H}\{x(t)\} \tag{14.6-14}$$

is called the *Hilbert transform* of $x(t)$. Thus, we have

$$x_+(t) = x(t) + j\check{x}(t). \tag{14.6-15}$$

Complex Time Signals Versus Negative Frequency: A Choice

Here is the importance of this result. *If we eliminate the negative frequency components from the Fourier transform of $x(t)$, we are forced to accept a complex time waveform as the result—with the real part being the original signal and the imaginary part being the Hilbert transform of the original signal.* Thus, one has an option: either grant the existence of negative frequencies or accept the fact that time signals are inherently all complex. As we have illustrated, one can treat physical signals (and hence real signals in the sense of "real world signals") as either real or complex, depending upon whether one wishes to deal with "one wire" or "two wire" waveforms. If we are dealing with only single wire signals, it seems more appropriate, however, to accept negative frequencies and keep the signals real.

Physical Realization of a Signal with Zero Negative Frequency Content

Let's explore the physical realization of our complex signal whose Fourier transform has only positive frequencies a bit more, however, for its interpretation leads to some interesting and practical results. Because $x_+(t) = x(t) + j\check{x}(t)$ we can compute the Fourier transform as

$$X_+(\omega) = X(\omega) + j\mathcal{F}\{\check{x}(t)\}. \tag{14.6-16}$$

More explicitly, we have

$$\mathcal{F}\{\check{x}(t)\} = \mathcal{F}\left\{x(t) \otimes \frac{1}{\pi t}\right\} = X(\omega)\,\mathcal{F}\left\{\frac{1}{\pi t}\right\}. \tag{14.6-17}$$

We have used the definition of the Hilbert transform and the fact that the transform of a convolution is the product of the transforms of the components. We know, however, that

$$\text{sgn}(t) \longleftrightarrow \frac{2}{j\omega} = -j\frac{2}{\omega}, \tag{14.6-18}$$

so the symmetry property gives

$$-j\frac{2}{t} \longleftrightarrow 2\pi\,\text{sgn}(-\omega). \tag{14.6-19}$$

Dividing by $-j2\pi$ gives

$$\frac{1}{\pi t} \longleftrightarrow j\,\text{sgn}(-\omega). \qquad (14.6\text{-}20)$$

The final result is

$$\mathcal{F}\{(\check{x}(t)\} = j\,\text{sgn}(-\omega)X(\omega). \qquad (14.6\text{-}21)$$

Physical interpretation in terms of a system diagram is shown in Figure 14.62. We should perhaps mention nomenclature: $x_+(t)$ is called the *analytic signal* associated with the real waveform $x(t)$. In passing, we mention that the approximation of the "Hilbert transformer" with the characteristics shown in Figure 14.62 is an important and difficult problem in circuit design—the mathematical theory of which is called "filter synthesis."

Figure 14.62
Generation of the
analytic signal

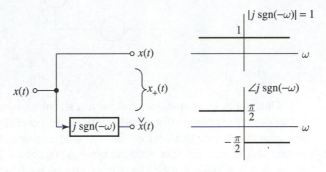

*Single Sideband
Amplitude Modulation*

Example 14.46 Sketch the spectrum of $y_+(t)$ in Figure 14.63 and derive the system diagram, using only real elements, for generating $y_+(t)$ from the real signal $x(t)$.

Figure 14.63
Generation of an SSB AM signal

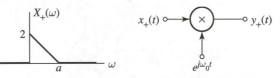

Solution This problem is actually straightforward, based on what we have already done. The first part simply uses the modulation property:

$$Y_+(\omega) = \mathcal{F}\{x_+(t)e^{j\omega_0 t}\} = X_+(\omega - \omega_0), \qquad (14.6\text{-}22)$$

which has the form shown in Figure 14.64. It is the same as $X_+(\omega)$ moved to the right by ω_0. The system diagram consists merely of a complex multiplier realized with real elements (as we discussed in Chapter 9), preceded by the simulation diagram of Figure 14.62, which generates the analytic signal $x_+(t)$. This is shown in Figure 14.65.

Figure 14.64
Spectrum of the output

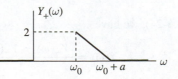

Figure 14.65
An SSB AM generator
(complex form)

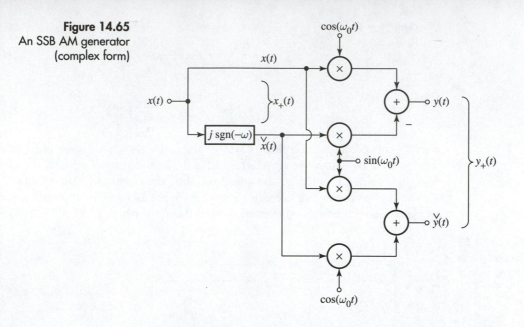

One could, of course, transmit such a signal on two wires (or over two channels, in general); however, in communications work one generally uses a single antenna, amplifying equipment, etc. Thus, let us discuss what happens if we decide to only use the real output $y(t)$. In fact, we have used the symbol $\check{y}(t)$ for the imaginary output, but how do we know that it is, indeed, the Hilbert transform of $y(t)$? Let's investigate these questions. If we merely drop the bottom two multipliers and the associated summer, as shown in Figure 14.66, we will retain the real part of the response, $y(t)$. This naturally simplifies the hardware realization.

Figure 14.66
An SSB AM generator
(real form)

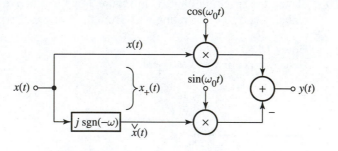

Let's investigate the spectrum of $y(t)$ by itself. We note that

$$y(t) = \text{Re}\{y(t) + j\check{y}(t)\} = \frac{y_+(t) + y_+^*(t)}{2}. \tag{14.6-23}$$

Thus,

$$Y(\omega) = \frac{\mathscr{F}\{y_+(t)\} + \mathscr{F}\{y_+^*(t)\}}{2}. \tag{14.6-24}$$

But

$$\mathscr{F}\{y_+^*(t)\} = \int_{-\infty}^{\infty} y_+^*(t)e^{-j\omega t}\,dt = \left[\int_{-\infty}^{\infty} y_+(t)e^{j\omega t}\,dt\right]^* = Y_+^*(-\omega). \tag{14.6-25}$$

Inserting this in (14.6-24), we have

$$Y(\omega) = \frac{Y_+(\omega) + Y_+^*(-\omega)}{2}. \tag{14.6-26}$$

We know the form of $Y_+(\omega)$ from Figure 14.64. Thus, we merely scale that frequency function down by a factor of 2 and add the result to the same frequency function conjugated and reflected about the vertical axis (that is, evaluated at negative values of the argument). This gives the result shown in Figure 14.67.

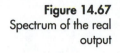

Figure 14.67
Spectrum of the real
output

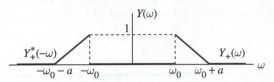

Notice the beauty of the relationships between real and imaginary parts and positive and negative frequencies! If we throw away the imaginary part, up pops the negative frequency component—one that is the conjugate symmetric part of the positive frequency one. Furthermore, we can now backtrack in our logic. If we reverse the process and throw away the negative frequency component, we automatically introduce the Hilbert transform of the time waveform as the imaginary part of our signal. But this imaginary part is simply the one we generated in Figure 14.65, so we were quite correct in labeling it with the symbol denoting the Hilbert transform of $y(t)$.

Section 14.6 Quiz

Q14.6-1. Find $\mathcal{H}\{\cos(\omega_0 t)\}$ and $\mathcal{H}\{\sin(\omega_0 t)\}$.

Q14.6-2. If $X_-(\omega) = 2X(\omega)u(-\omega)$, find the corresponding time function $x_-(t)$. Sketch a system for generating this waveform.

Chapter 14 Summary

In this chapter we have treated the *Fourier transform* as being equivalent to the Laplace transform evaluated on the imaginary axis, that is, as the Laplace transform with the substitution $s = j\omega$ made for the argument. Section 14.1 presents this definition and discusses the modifications that must be made if the waveform being transformed is not time-limited. A number of basic transforms are derived in this section.

Section 14.2 took up the problem of *inverting* the Fourier transform, that is, of finding the time function associated with a given function of ω. This led to the definition of an *inversion integral*, or *inverse Fourier transform*.

In Section 14.3, we discussed the concept of *instantaneous power spectrum*. This was defined to be the Fourier transform of the instantaneous power waveform. From this, the discussion led naturally into the idea of *energy density spectrum*.

Section 14.4 reviewed the *basic properties* of the Laplace transform, noting that they continue to hold for the Fourier transform with $s = j\omega$. It then *developed additional properties* that are peculiar to the Fourier transform alone in that they depend upon the argument being purely imaginary.

Section 14.5 then developed the idea of the *Fourier series* as a particular form of representation of a periodic time waveform. This followed from the fact that the Fourier transform of such a waveform is a train of impulse functions. A number of examples were worked, and several various forms of the series were presented.

Section 14.6 discussed an intriguing topic: that of *negative frequency*. We showed there that if one demands that a time waveform have zero negative frequency content, then one must accept the fact that such signals have complex time-varying components; furthermore, the imaginary part of such a signal is the *Hilbert transform* of the real part.

A signal having an imaginary part that is the Hilbert transform of the real part was defined to be the *analytic signal* associated with the real part waveform. These concepts were employed to discuss the concept of a *single sideband amplitude modulator*. This system was then, in turn, used to illustrate the mathematical process of forming the analytic signal from the real signal and, conversely, forming the real signal from the analytic signal by adding this frequency content back in.

Chapter 14 Review Quiz

RQ14.1. Find the Fourier transform of $x(t) = \sin(2\pi t)[u(t) - u(t - 1)]$ by using the defining integral.

RQ14.2. Find the Fourier transform of $x(t) = \sin(2\pi t)[u(t) - u(t - 1)]$ by differentiating twice and taking the transform of $p^2 x(t)$.

RQ14.3. Find the Fourier transform of $x(t) = 3\delta(t) - 2\delta'(t - 1) + t\delta''(t - 2)$.

RQ14.4. Find the Fourier transform of $x(t) = 4\sin^2(2\pi t)u(t)$.

RQ14.5. Find the Fourier transform of $x(t) = 4\sin^2(2\pi t)$. *Note:* $x(t)$ is not the same as the $x(t)$ in Question RQ14.4!

RQ14.6. Find the Fourier transform of the waveform $x(t) = 2\cos(2t)[u(t) - 4u(t - \pi)]$.

RQ14.7. Find the Fourier transform of the waveform $x(t) = 2\cos(2\pi t)\,\mathrm{sgn}(t)$.

RQ14.8. Find the Fourier transform of the waveform $x(t) = e^{-2|t|}\cos(2\pi t)$.

RQ14.9. Find the Fourier transform of the waveform sketched in Figure RQ14.1.

Figure RQ14.1

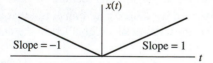

RQ14.10. Find the inverse transform of the frequency function $X(\omega) = 4\delta(\omega - 5)$.

RQ14.11. Find the inverse transform of $X(\omega) = u(\omega) - 2u(\omega - 2) + u(\omega - 3)$.

RQ14.12. Find the inverse transform of $X(\omega) = u(\omega) - 2u(\omega - 2) - 2u(\omega + 2)$. What is the *key* difference between $x(t)$ here and the one in Question RQ14.11?

RQ14.13. What value of the real parameter a makes $x(t) = \mathcal{F}^{-1}\{\delta(\omega - 1) + \delta(\omega - a)\}$ real?

RQ14.14. Find the instantaneous power spectrum of the waveform $x(t) = 2 + 4\sin(2t)$. What is the average power in $x(t)$?

RQ14.15. Find the energy density spectrum of the waveform $x(t) = u(t) - u(t - 1)$.

RQ14.16. If $x(t) = \sin(2\pi t)$ over the time interval $(-0.25, 0.25)$ and repeats periodically with period $T = 0.5$ outside that interval, find the Fourier coefficients of $x(t)$, determine the Fourier transform, and plot the amplitude and phase spectrum for the first five nonzero harmonics.

RQ14.17. The waveform $i_g(t) = 4 + 4\cos(2\pi t) + 3\sin(4\pi t)$ is applied to the input of the subcircuit shown in Figure RQ14.2. Find the steady-state response voltage $v(t)$. Assume that $v(t)$ is a stable response.

Figure RQ14.2

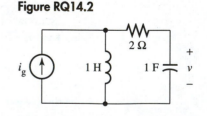

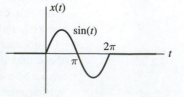

Chapter 14 Problems

Section 14.1 The Fourier Transform

14.1-1. Find the Fourier transform of the waveform shown in Figure P14.1-1.

Figure P14.1-1

14.1-2. Find the Fourier transform of the waveform shown in Figure P14.1-2.

Figure P14.1-2

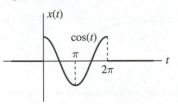

14.1-3. Find the Fourier transform of the waveform shown in Figure P14.1-3.

Figure P14.1-3

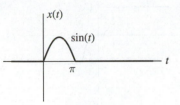

14.1-4. Find the Fourier transform of the waveform shown in Figure P14.1-4.

Figure P14.1-4

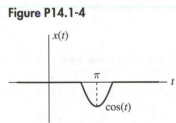

14.1-5. Find the Fourier transform of the waveform shown in Figure P14.1-5.

Figure P14.1-5

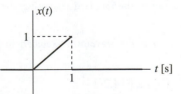

14.1-6. Find the Fourier transform of $x(t) = e^{-2t} \cos(3t)u(t)$.

14.1-7. Find the Fourier transform of $x(t) = e^{-t}u(t - 1)$.

14.1-8. Find the Fourier transform of $x(t) = te^{-2t}u(t)$.

14.1-9. Find the Fourier transform of $x(t) = e^{-2|t|} \cos(3t)$.

14.1-10. Find the Fourier transform of $x(t) = t^2 e^{-2|t|}$.

14.1-11. Find the Fourier transform of $x(t) = 3e^{-2t}u(t)$.

14.1-12. Find the Fourier transform of the waveform shown in Figure P14.1-6.

Figure P14.1-6

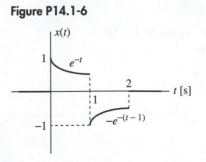

14.1-13. Find the Fourier transform of the waveform shown in Figure P14.1-7.

Figure P14.1-7

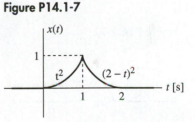

14.1-14. Find the Fourier transform of the waveform shown in Figure P14.1-8.

Figure P14.1-8

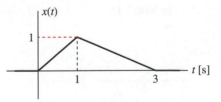

14.1-15. Find the Fourier transform of $x(t) = u(t - 2)$.

14.1-16. Find the Fourier transform of $x(t) = 3\delta(t) + 6\delta(t - 4) - 3\delta(t - 8)$.

14.1-17. For the circuit shown in Figure P14.1-9, find the response $v_0(t)$. Assuming that the natural response can be considered to be zero after five time constants, at what time (in, say, microseconds) does the response reach the ac steady state?

Figure P14.1-9

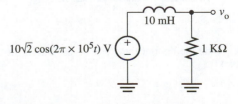

14.1-18. Find the Fourier transform of $x(t) = 4\delta(t + 2) - 4\delta(t - 2)$.

14.1-19. Find the Fourier transform of $x(t) = t^3 \delta(t + 2)$.

14.1-20. Find the Fourier transform of the waveform shown in Figure P14.1-10.

Figure P14.1-10

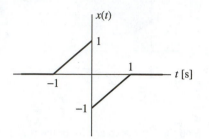

14.1-21. Find the Fourier transform of the waveform shown in Figure P14.1-11.

Figure P14.1-11

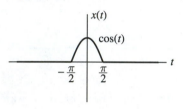

14.1-22. Find the Fourier transform of the waveform shown in Figure P14.1-12.

Figure P14.1-12

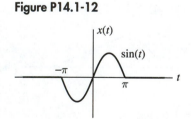

14.1-23. Find the Fourier transform of the "Batman" waveform shown in Figure P14.1-13.

Figure P14.1-13

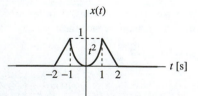

14.1-24. Find the Fourier transform of the waveform shown in Figure P14.1-14.

Figure P14.1-14

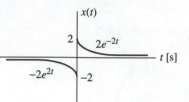

14.1-25. Find the Fourier transform of the waveform shown in Figure P14.1-15.

Figure P14.1-15

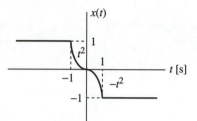

14.1-26. Find the Fourier transform of $x(t) = \sum_{n=-\infty}^{\infty} \delta(t - n)$.

14.1-27. Find the Fourier transform of $x(t) = \sum_{n=-\infty}^{\infty} r^n \delta(t - n)$, where $|r| < 1$.

Section 14.2 The Inverse Fourier Transform

14.2-1. If $X(\omega) = \cos(3\omega)$, find $x(t)$ using the inversion integral.

14.2-2. If $X(\omega) = \sin^3(2\omega)$ find $x(t)$ using the inversion integral and trigonometric identities.

14.2-3. If $X(\omega) = 1 + 0.3 \cos(3\omega) + 0.4 \sin(\omega)$, find $x(t)$ using the inversion integral.

14.2-4. If $X(\omega) = \sin(5\omega)$, find $x(t)$ using the inversion integral.

14.2-5. If $X(\omega)$ is the waveform sketched in Figure P14.2-1, find $x(t)$.

Figure P14.2-1

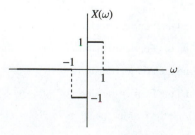

14.2-6. If $X(\omega)$ is the waveform sketched in Figure P14.2-2, find $x(t)$.

Figure P14.2-2

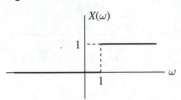

14.2-7. If $X(\omega)$ is the waveform sketched in Figure P14.2-3, find $x(t)$.

Figure P14.2-3

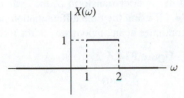

14.2-8. If $X(\omega)$ is the waveform sketched in Figure P14.2-4, find $x(t)$.

Figure P14.2-4

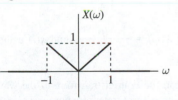

14.2-9. If $X(\omega)$ is the waveform sketched in Figure P14.2-5, find $x(t)$.

Figure P14.2-5

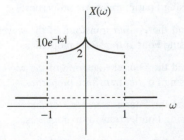

14.2-10. If $X(\omega)$ is the waveform sketched in Figure P14.2-6, find $x(t)$.

Figure P14.2-6

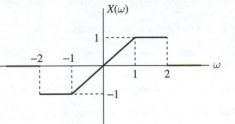

14.2-11. If $X(\omega)$ is the waveform sketched in Figure P14.2-7, find $x(t)$.

Figure P14.2-7

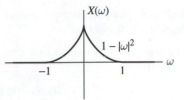

14.2-12. If $X(\omega) = \sum_{n=-\infty}^{\infty} \delta(\omega - n)$, find $x(t)$.

14.2-13. If $X(\omega) = \sum_{n=-\infty}^{\infty} r^n \delta(\omega - n)$, where $|r| < 1$, find $x(t)$.

Section 14.3 Spectral Analysis

14.3-1. Find and plot the spectrum of the waveform $x(t)$ in Problem 14.1-1.

14.3-2. Find and plot the spectrum of the waveform $x(t)$ in Problem 14.1-5.

14.3-3. Find and plot the spectrum of the waveform $x(t)$ in Problem 14.1-6.

14.3-4. Find and plot the spectrum of the waveform $x(t)$ in Problem 14.1-9.

14.3-5. Find and plot the spectrum of the waveform $x(t)$ in Problem 14.1-11.

14.3-6. Find and plot the spectrum of the waveform $x(t)$ in Problem 14.1-13.

14.3-7. Find and plot the spectrum of the waveform $x(t)$ in Problem 14.1-24.

14.3-8. Find and plot the spectrum of the waveform $x(t)$ in Problem 14.1-27. Then find the limit of this spectrum as $r \to 1-$ (that is, through values less than unity).

14.3-9. If $x(t) = 1 + \frac{1}{3} \cos(3t) + \frac{1}{5} \cos(5t) + \frac{1}{7} \cos(7t) + \frac{1}{9} \cos(9t)$:

    a. Plot $x(t)$ for two periods. (*Hint:* Use a calculator or a computer.)

    b. Find $X(\omega)$ and plot the spectrum.

14.3-10. If $x(t) = 1 + \frac{1}{3}\cos(3t)$:

    a. Plot $x(t)$ for two periods. (*Hint:* Use a calculator or a computer.)

    b. Find $X(\omega)$ and plot the spectrum.

14.3-11. If $x(t) = \cos(t)[u(t) - u(t - 2\pi)]$, find and plot the spectrum.

14.3-12. If $x(t) = \sin(t)\cos(10t)$, find and plot the spectrum.

14.3-13. Find and plot the spectrum of $y(t)$ in Figure P14.3-1.

Figure P14.3-1

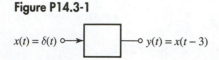

$x(t) = \delta(t) \circ\!\!\rightarrow \boxed{} \circ y(t) = x(t - 3)$

14.3-14. Find and plot the energy density spectrum of $x(t) = 0.5 + \cos(2\pi t) + 0.1\sin(3\pi t)$.

14.3-15. Find and plot the instantaneous power spectrum of $x(t)$ in Figure P14.1-10.

14.3-16. Let $x(t) = \cos(t) + \cos(\sqrt{2}t)$. Is $x(t)$ periodic? Find $X(\omega)$ and sketch the spectrum, the instantaneous power spectrum, and the energy density spectrum.

14.3-17. Using Parseval's theorem, find the total energy in the waveform $x(t) = 4e^{-2t}u(t)$.

14.3-18. Using Parseval's theorem, find the total energy in the waveform $x(t) = 4e^{-2|t|}$.

14.3-19. Find the Fourier transform of $x(t)y(t)$, if $x(t) = \cos(3t)$, and $y(t) = e^{-2t}u(t)$, using the frequency domain convolution property.

Section 14.4 Properties of the Fourier Transform

14.4-1. Use the time-frequency symmetry property to find $\mathcal{F}^{-1}\{\cos(a\omega)\}$.

14.4-2. Use the time-frequency symmetry property to find $\mathcal{F}^{-1}\{\sin(a\omega)\}$.

14.4-3. Use the time delay property and the known transform of $\cos(\omega_0 t)$ to determine the Fourier transform of $\sin(\omega_0 t)$.

14.4-4. Use the modulation property and the time-frequency symmetry property to find the Fourier transform of
$$x(t) = \frac{2a\cos(\omega_0 t)}{t^2 + a^2}.$$

14.4-5. Using the time differentiation property and the known transform of $x(t) = \sin(\omega_0 t)$, find the Fourier transform of $y(t) = \cos(\omega_0 t)$.

14.4-6. Find the Fourier transform of the waveform $x(t)$ in Figure P14.4-1 using the time differentiation property.

Figure P14.4-1

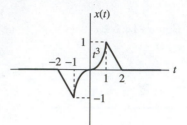

14.4-7. Find $\mathcal{F}\{u(-t)\}$ using the time reversal property.

14.4-8. Find $\mathcal{F}\{t^3 + 4t^2 + 2t + 3\}$ using any convenient properties you wish.

14.4-9. Find the Fourier transform of the *staircase* waveform $x(t)$ in Figure 14.4-2 using the time differentiation property. *Note:* the waveform continues in an obvious way to the right.

Figure P14.4-2

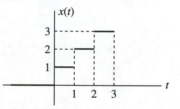

14.4-10. If $X(\omega) = \dfrac{1 - e^{-j2\omega}}{j\omega}$, find $x(t)$ using Fourier transform properties.

14.4-11. Find the Fourier transform of the waveform in Problem 14.1-2, using Fourier transform properties.

14.4-12. Find the Fourier transform of the waveform in Problem 14.1-5, using Fourier transform properties.

14.4-13. Find the Fourier transform of the waveform in Problem 14.1-13, using Fourier transform properties.

14.4-14. Find the Fourier transform of the waveform in Problem 14.1-23, using Fourier transform properties.

14.4-15. Find the Fourier transform of the waveform in Problem 14.1-24, using Fourier transform properties.

14.4-16. Find the Fourier transform of the waveform in Problem 14.1-25, using Fourier transform properties.

14.4-17. Find the Fourier transform of the waveform in Problem 14.2-7, using Fourier transform properties.

14.4-18. Find the Fourier transform of the waveform in Problem 14.2-8, using Fourier transform properties.

14.4-19. Find the Fourier transform of the waveform in Problem 14.2-10, using Fourier transform properties.

14.4-20. Find the Fourier transform of the waveform in Problem 14.2-11, using Fourier transform properties.

14.4-21. If $x(t) = t^2 e^{-3t} u(t)$, find $X(\omega)$.

14.4-22. If $X(\omega) = e^{-|\omega|} \cos(2\omega)$, find $x(t)$.

Section 14.5 Fourier Series

14.5-1. Find the Fourier coefficients for the periodic waveform shown in Figure P14.5-1.

Figure P14.5-1

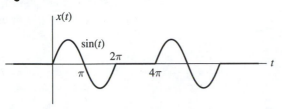

14.5-2. Find the Fourier coefficients for the periodic waveform shown in Figure P14.5-2.

Figure P14.5-2

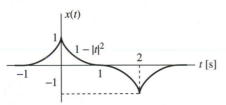

14.5-3. Find the Fourier coefficients for the periodic waveform shown in Figure P14.5-3.

Figure P14.5-3

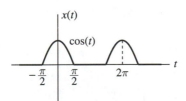

14.5-4. Find the Fourier coefficients for the periodic waveform shown in Figure P14.5-4.

Figure P14.5-4

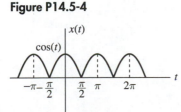

14.5-5. Find the Fourier coefficients for the periodic waveform shown in Figure P14.5-5.

Figure P14.5-5

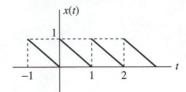

14.5-6. Find the Fourier coefficients for the periodic waveform shown in Figure P14.5-6.

Figure P14.5-6

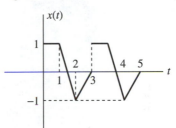

14.5-7. Find the Fourier coefficients of $y(t)$ in Figure P14.5-7 and sketch the spectrum. Find the instantaneous power spectrum and the energy density spectrum of $y(t)$.

Figure P14.5-7

$$x(t) = \cos(\omega_0 t) \longrightarrow \boxed{} \longrightarrow y(t) = |x(t)|$$

Section 14.6 Negative Frequency and the Hilbert Transform

14.6-1. If $x(t) = e^{-2t} u(t)$, find $\mathcal{H}\{x(t)\}$.

14.6-2. If $x(t) = e^{j\omega_0 t} u(t)$, find $\mathcal{H}\{x(t)\}$.

14.6-3. Suppose that $x(t) = 0$ for $t \leq 0$ (that is, $x(t)$ is one sided). Writing the Fourier transform in the form $X(\omega) = R(\omega) + jX(\omega)$, where $R(\omega)$ and $X(\omega)$ are both real functions, show that $X(\omega) = \mathcal{H}\{R(\omega)\}$. (This is called "real part sufficiency.")

14.6-4. Suppose that $X(\omega)$ is *band limited* to $|\omega| \leq \omega_0$, as shown in Figure P14.6-1. Define $X_-(\omega) = X(\omega)u(-\omega)$. Find the resulting time waveform $x_-(t)$.

Figure P14.6-1

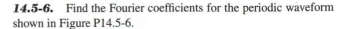

14.6-5. Find and sketch the spectrum of $x_+(t)e^{-j\omega_0 t}$, where $x_+(t)$ is the signal generated in Problem 14.6-4, and sketch a system diagram using real system elements for generating this signal. Then, simply remove the elements required to generate the imaginary part of this signal and thereby construct a *lower sideband generator.* Sketch the resulting spectrum.

Part IV

SELECTED TOPICS

15

Two-Port Subcircuits

This chapter presents in Section 15.1 the fundamental ideas regarding *equivalent circuits for two-ports:* subcircuits that have either three or four terminals and satisfy a certain constraint. The introduction discusses the description of a multiterminal subcircuit or element as an *affine* voltage-current relationship—or set of constraint relationships—among the various voltages and currents.

Sections 15.2 through 15.5 develop the specific *two-port parameter sets: impedance, admittance, hybrid,* and *chain (or transmission).* The *tee and pi equivalents* are presented, and a number of examples are worked out in some detail. In Section 15.6 the theory of two-ports is applied to solve a problem that is important in the application of circuit theory to electronics: the *analysis of feedback circuits.* In earlier sections it is shown that the various types of two-port parameter sets add for subcircuits connected in a corresponding manner: *series-series, series-shunt, shunt-series,* or *shunt-*

shunt. The exact analysis of such configurations is developed, and the usual "feedback approximation" is made in which one, the "amplifier," has high forward gain and very small reverse gain, and the other, the "feedback network," has small forward gain in comparison with that of the amplifier and very precise reverse gain. It is demonstrated that this results in a very precisely controlled set of characteristics for the overall network. Section 15.7 considers an aspect of two-ports that is not often treated: two-ports with internal independent sources and the resulting *two-port Thévenin, Norton,* and *mixed Thévenin/Norton equivalents.* Examples flagged with (O) or (T) pertain to op amps and transistors, respectively.

In this chapter, as in the last two, we use the *s-domain notation* for the most part. The reader should, according to his or her preference, interpret *s* as: the differential operator *p*, the Laplace variable $s = \sigma + j\omega$, the Fourier variable $s = j\omega$, or the phasor variable $s = j\omega$. When we wish to emphasize the time domain aspects of a problem, we will still use *p* in preference to *d/dt* and *1/p* rather than $\int_{-\infty}^{t}(\ )\,d\alpha$. We will also use the (perhaps somewhat unfortunate, but common) notation in which a resistor symbol is used for a general impedance function. Also, in the main, lowercase letters are used for both time and frequency variables for economy of presentation. This, too, is common practice in the literature.

15.1 | Introduction

In the preceding portions of this text, we have developed both time and frequency domain analysis of circuits and subcircuits constructed from two-terminal elements. We presented the idea of a subcircuit and its equivalent, primarily for the two-terminal type of subcircuit. In this chapter we generalize by discussing subcircuits that have three or four terminals, the latter with certain restrictions applied. We have already mentioned a special type of three-terminal subcircuit in Section 3.5 of Chapter 3, wherein we treated three-terminal subcircuits consisting only of sources; we considered others with active elements in Chapter 6. Here we extend our discussion to more general types of subcircuit.

Two-Terminal Elements and Subcircuits

Let's start by reviewing the idea of a two-terminal element or subcircuit in a slightly different light from that used in earlier chapters. We now know that one cannot distinguish a two-terminal circuit element, such as the one in Figure 15.1(a), from a two-terminal subcircuit, such as the one in Figure 15.1(b), as far as the rest of the circuit in which either is connected is concerned—*provided that the two have the same v-i characteristic.* When we draw a small box we are usually referring to an element, that is, a two-terminal subcircuit that cannot be decomposed into simpler two-terminal components; when we draw a larger box, we usually mean to indicate that the object is a subcircuit, that is, one that can

Figure 15.1
Two-terminal element and subcircuit

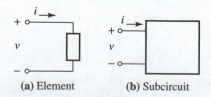

(a) Element   (b) Subcircuit

be decomposed into simpler objects. As far as the external circuit is concerned, though, they behave exactly the same if they have identical *v-i* characteristics.

Mathematical Relations

Let's explore the idea of the *v-i* characteristic a bit more by going back to basic mathematics. We recall that a *relation* in two variables is a set of ordered pairs of real numbers:

$$\mathcal{R} = \{(x, y): f(x, y) = 0\}. \tag{15.1-1}$$

Figure 15.2 Graph of example relation

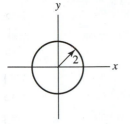

The two-variable expression $f(x, y) = 0$ is called the *constraint equation*. By the notation in (15.1-1) we mean that a given number pair (x, y) belongs to the relation $\mathcal{R}$ if it satisfies the constraint equation. As an example, consider the relation

$$C = \{(x, y): x^2 + y^2 - 4 = 0\}. \tag{15.1-2}$$

The *graph* of this relationship (that is, the relation depicted graphically on a pair of axes) is shown in Figure 15.2. It is a circle centered at the origin with radius 2.

Affine Relations

When we speak of the *v-i* characteristic of a two-terminal element or subcircuit, we simply mean a relation in the two variables *v* and *i*. Until now we have not been studying any and all relations; rather, we have been restricting ourselves to those of the form

$$av + bi + c = 0, \tag{15.1-3}$$

called an *affine* relationship. This is a basic assumption we have been making about two-terminal elements. For two-terminal *subcircuits* made up of *elements* having the relation (15.1-3) as their terminal characteristic, we can prove that (15.1-3) holds for the entire subcircuit using (for instance) nodal analysis. Though we will not pause to do a logical proof of this statement, the next two examples should make it plausible.

Example 15.1

Find the *v-i* characteristic of the subcircuit in Figure 15.3.

Figure 15.3 An example subcircuit

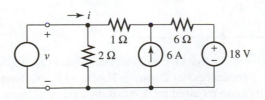

Solution

We apply an independent source to the two free terminals, as shown in Figure 15.4. We have not identified whether it is a v-source or an i-source on purpose—for the type does not matter. For some circuits a v-source will not work because it violates a KVL constraint; for others an i-source will not work because it violates a KCL constraint. We observe, however, that a source of one type or the other must be applicable to the terminals without violating a circuit constraint; otherwise, the subcircuit is simply not well defined. For our example here, either will work. If we choose a current source, we must write only one mesh equation (for the supermesh); however, two nodal equations will be required. If, on the other hand, we pick a voltage source, only one nodal equation will be required; in this case, we would have to write two mesh equations.

Figure 15.4 Testing the subcircuit

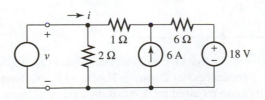

Arbitrarily choosing a voltage source, we prepare the circuit for nodal analysis, as shown in Figure 15.5. We must leave the voltage source value as a general literal variable, as we explained earlier in the text. The equation at the single essential node, whose voltage we have labeled v_x, is

$$\frac{v_x - v}{1} + \frac{v_x - 18}{6} = 6. \tag{15.1-4}$$

Solving for v_x, we obtain

$$v_x = \frac{6}{7}v + \frac{54}{7}. \tag{15.1-5}$$

Next, we use KCL at the top of the applied test source to obtain

$$i = \frac{v}{2} + \frac{v - v_x}{1} = \frac{9}{14}v - \frac{54}{7}. \tag{15.1-6}$$

Rearranging slightly, we have

$$9v - 14i - 108 = 0. \tag{15.1-7}$$

Thus, equation (15.1-7) is the v-i relationship for the subcircuit. This relation has the graph shown in Figure 15.6. The slope, of course, is the Thévenin equivalent resistance and the vertical intercept the Thévenin equivalent open circuit voltage (the horizontal intercept is the Norton equivalent short circuit current); however, that is not our objective here. We just wanted to show that the v-i relationship had the form of an affine relation holding between v and i—and we have done that.

Figure 15.5
The circuit prepared for nodal analysis

Figure 15.6
The v-i relation

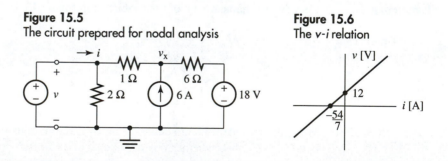

Example 15.2 Find the v-i relation for the circuit shown in Figure 15.7.

Figure 15.7
An example circuit

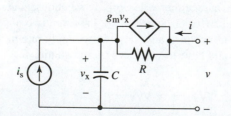

Solution Because there is a dynamic element (the capacitor), we draw the s-domain equivalent circuit with impedances, as shown in Figure 15.8. We have applied a test i-source because this results in *no* required mesh equations; we must, however, apply KVL around the bottom right mesh to obtain $V(s)$:

$$V(s) = \frac{1}{Cs}\{I(s) + I_s(s)\} + R\{I(s) + g_m V_x(s)\}. \qquad (15.1\text{-}8)$$

Because this circuit is so simple, we have not taped the dependent source; we would have done this, though, if the circuit were more complicated. Now writing the controlling voltage $V_x(s)$ in terms of the currents, we have

$$V_x(s) = \frac{1}{Cs}\{I(s) + I_s(s)\}. \qquad (15.1\text{-}9)$$

Solving (15.1-8) and (15.1-9) simultaneously (we will leave the algebra to you), we obtain

$$CsV(s) - [1 + g_m R + RCs]I(s) - (1 + g_m R)I_s(s) = 0. \qquad (15.1\text{-}10)$$

Again, this is an affine relationship—this time in the form of an s-domain equation. We remind you that if we interpret the variable s as the time differentiation operator p, we will have

$$Cpv(t) - [1 + g_m R + RCp]i(t) - [1 + g_m R]I_s(t) = 0. \qquad (15.1\text{-}11)$$

If either $v(t)$ or $i(t)$ is known, then we have a first-order differential equation with the dependent function of time as the other variable. Thus, affine relationships in s become linear constant-coefficient differential equations in t.

Figure 15.8
Testing the circuit:
s-domain equivalent

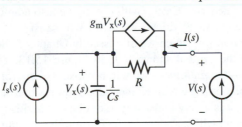

Affine Relations and Equivalent Subcircuits

The conclusion we can make from the last two examples is this: each and every two-terminal subcircuit consisting of R, L, C, independent sources, and dependent sources obeys the affine relationship:

$$av + bi + c = 0 \qquad (15.1\text{-}12)$$

in either the time domain or the s domain. If there are no energy storage elements—as in our first example—the coefficients a and b are real constants. If there *are* energy storage elements (L and/or C)—as in the second example—the coefficients are functions of s in the s domain or p in the time domain.[1] The parameter c is a constant if all the independent sources are dc and a time-varying function (or the Laplace transform of one) if these sources are time varying.

Now let's look at the v-i relationship (15.1-12) in more detail. We will investigate a number of different situations, one at a time.

---

[1] Or functions of $j\omega$ if we let $s = j\omega$ and consider the Fourier transform of $v(t)$ and $i(t)$.

Figure 15.9   The Thévenin equivalent subcircuit

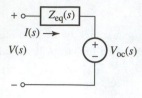

Figure 15.10   The Norton equivalent

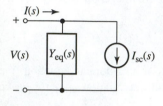

Figure 15.11   The equivalent passive element

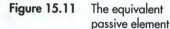

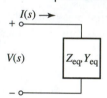

Figure 15.12   The norator

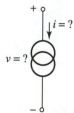

Case 1 ($a \neq 0$): If $a \neq 0$, we can solve for v. This gives

$$v = -\frac{b}{a}i - \frac{c}{a},$$ (15.1-13)

or in s-domain notation, $V(s) = Z_{eq}(s)I(s) + V_{oc}(s);$ (15.1-14)

that is, if $a = a(p)$, $b = b(p)$, and $c = c(p)$ are functions of the operator p, we write them as functions of the complex variable s. Do you recognize these equations? You should, for they describe our old friend the Thévenin equivalent subcircuit, shown in Figure 15.9.

Case 2 ($b \neq 0$): If $b \neq 0$, we can solve (15.1-12) for i, getting the time domain form

$$i = -\frac{a}{b}v - \frac{c}{b}$$ (15.1-15)

and the corresponding s-domain form

$$I(s) = Y_{eq}(s)V(s) + I_{sc}(s).$$ (15.1-16)

This is nothing more than the v-i characteristic of the Norton equivalent for our subcircuit, shown in Figure 15.10.

Case 3 ($c = 0$): In this case, assuming that either a or b is nonzero, the resulting equivalent circuit is the one shown in Figure 15.11—a single impedance or admittance. If $a = 0$ and $b \neq 0$, equation (15.1-12) gives $i = 0$. This describes an open circuit, with $Y_{eq}(s) = 0$. If, however, $a \neq 0$ and $b = 0$, equation (15.1-12) takes the form $v = 0$, the v-i relationship of a short circuit, and we have $Z_{eq}(s) = 0$.

Still considering Case 3 with $c = 0$, we also admit the possibility that $a = b = 0$. In this case, any pair of values of v and i will satisfy (15.1-12). The resulting equivalent is that of the *norator,* which we discussed in Chapter 5, Section 5.4, and depicted schematically with the symbol shown here in Figure 15.12. The other element we discussed in Chapter 5—the *nullator*—we will not discuss here for reasons of space. Suffice it to say that the nullator is a circuit element that requires two equations to describe its v-i relationship. Thus, it is a very unusual type of element and is found in practical applications only in conjunction with a norator in the form of the op amp model we discussed in Chapter 5: the (grounded) *nullor.*

There are three main ideas we have tried to illustrate here: (1) the behavior of an element or subcircuit is completely described by its terminal voltage-current relation, (2) that relation is an affine one, and (3) using the v-i relationship for a given subcircuit, we can find other equivalent subcircuits that have the same v-i relationship.

Port Constraints

Figure 15.13   Illustration of element voltage sufficiency assumption

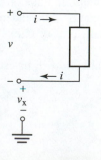

Before we consider more complicated subcircuits, let's pause to look at a basic fact about two-terminal elements and subcircuits. In Figure 15.13 we have sketched such an element, but this time we have assumed that the voltage reference is somewhere else in the circuit in which the device is found. Remember now that one of our basic assumptions about circuits is this: the voltage v *between* the element terminals and the terminal current i suffice to completely determine the behavior of the element when it is connected into a circuit. *The voltage v_x relative to the arbitrary reference does not affect this behavior.* Furthermore, we recall that the current out of the bottom terminal is the same as that entering the top one. We also remember that, according to the passive sign convention, if $P = vi$ is positive, then power is being delivered to the element through its terminals. Thus, we look upon the pair of terminals as a door or gateway, which we will call a *port*,[2] through which we can deliver power to the element or subcircuit.

---

[2] The French word for door is *porte;* the Spanish word is *puerto.*

Two-Port Subcircuits Suppose we now group two elements or subcircuits together in the manner indicated in Figure 15.14. This grouping is indicated by a conceptual box around the elements. We see that here there are two voltages v_x and v_y (relative to our arbitrary reference) that are completely irrelevant as far as the behavior of the two elements are concerned. The behavior of this group of two elements is completely described by the two voltages v_1 and v_2 *between* the two pairs of terminals; furthermore, we see that the current out of the bottom terminal of each element is the same as the current into the top one. We thus have two ports, each of which is capable of delivering power into the conceptual box. We will call this conceptual box a *two-port subcircuit,* or a *two-port.* In fact, *we will generalize this term to apply to any subcircuit having any number of elements internally and four terminals by which they interact with the external world,* as shown in Figure 15.15.

Figure 15.14
Two two-terminal elements considered as a two-port

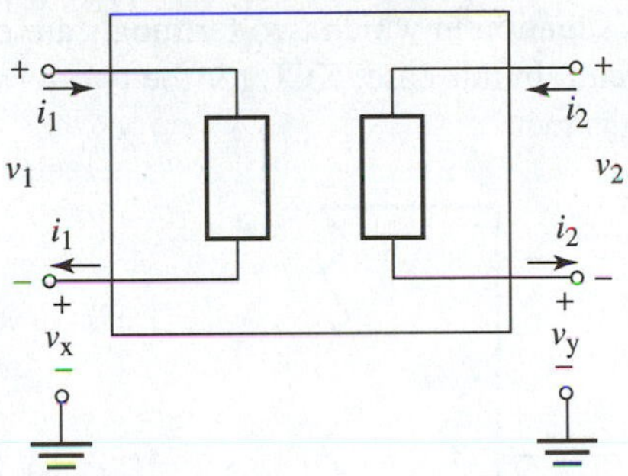

Figure 15.15
The general two-port subcircuit

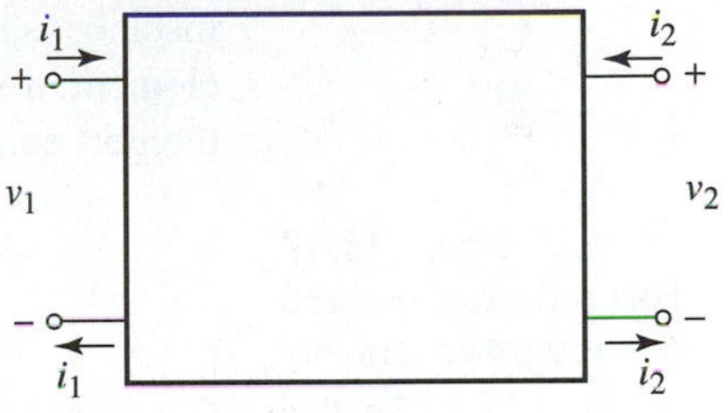

Any pair of terminals that are constrained by either the internal elements or the external circuit to carry equal and opposite currents will be called a port; furthermore, just as is the case for the particular subcircuit in Figure 15.14, we will assume that any two-port is completely described by two[3] *v-i* relationships holding among the port variables v_1, i_1, v_2, and i_2. We will also assume that they are affine relationships and write them as

$$a_1v_1 + a_2v_2 + a_3i_1 + a_4i_2 + c_1 = 0 \tag{15.1-17a}$$

$$b_1v_1 + b_2v_2 + b_3i_1 + b_4i_2 + c_2 = 0. \tag{15.1-17b}$$

For most of this chapter, we will assume that $c_1 = c_2 = 0$. Just as for the two-terminal subcircuit, this will mean that there are no independent sources in the equivalent subcircuit. Toward the end of the chapter we will investigate what happens when this is not the case. Assuming, therefore, that $c_1 = c_2 = 0$, the preceding equations become

$$a_1v_1 + a_2v_2 + a_3i_1 + a_4i_2 = 0 \tag{15.1-18a}$$

$$b_1v_1 + b_2v_2 + b_3i_1 + b_4i_2 = 0. \tag{15.1-18b}$$

We will investigate these equations in great detail in the remaining sections of this chapter, choosing various pairs of the variables as dependent variables. Each choice leads to a different model for the three-terminal subcircuit and to a different set of parameters.

---

[3] This is not really so obvious upon close examination. In fact, however, only specialized subcircuits fail to satisfy this assumption, and we will not be concerned with them. The same statement is true for the two-terminal subcircuit, which we have assumed is described by a single constraint equation. We will leave such investigations, however, to more advanced theoretical texts.

To get a better idea of the port constraint, look at Figure 15.16, which shows a terminal pair—the one on the left—that *does not* constitute a port. (Neither does the one on the right.) As you can readily see, the current into the top terminal is 4 A, and that coming from the bottom terminal is 0 A.

Figure 15.16
Violation of the port
constraint

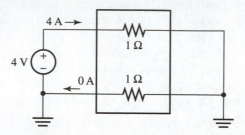

It is sometimes true that the external circuit imposes the port constraint on the terminal pair, rather than it being an intrinsic property of the subcircuit itself. Figure 15.17, for instance, shows a situation in which two terminals are connected to a single two-terminal element or subcircuit. In this case, KCL for the two-terminal subcircuit shows clearly that the port condition is met.

Figure 15.17
Port constraint ensured
by a two-terminal
element

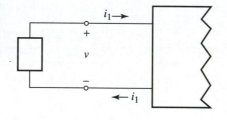

Example 15.3   Find the *v-i* constraint equations for the subcircuit in Figure 15.18.

Figure 15.18
An example subcircuit

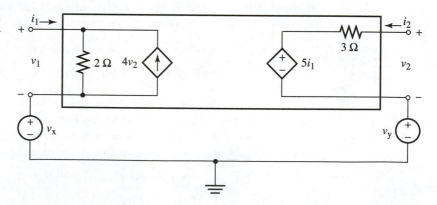

Solution   Note that the subcircuit of interest consists of the elements inside the solid line, that is, inside the box. Application of KCL, KVL, and Ohm's law gives

$$1v_1 - 2i_1 - 8v_2 + 0i_2 = 0 \tag{15.1-19a}$$

$$0v_1 - 5i_1 + 1v_2 - 3i_2 = 0. \tag{15.1-19b}$$

Now here is the important point: the constraint equations are independent of v_x and v_y. The only voltages appearing in the constraint relationships are v_1 and v_2—the terminal pair voltages. Thus, v_x and v_y may be varied at will without affecting the terminal relationships we have written in (15.1-19). This property often has practical import in isolating the ground reference of one circuit from that of another.

Three-Terminal
Subcircuits

Although the voltage relative to any arbitrary reference at one terminal of each port is irrelevant (that is, it does not enter into the constraint equations), in many cases the external circuit or the subcircuit itself constrains two of the terminal voltages to be identical. This converts a two-port into a three-terminal element or subcircuit, as we show in Figure 15.19. The constraint equations are exactly the same. We often draw the three-terminal subcircuit as in Figure 15.20. We cannot perform this operation, however, unless the bottom terminals are physically connected. For instance, if $v_x = 2$ V and $v_y = 10$ V in the circuit shown in Figure 15.18, we cannot connect the bottom terminals because the result would be a violation of KVL. Many electronic devices are of the three-terminal variety and many others are true two-ports. The theory we are going to develop holds equally well for both types.

Figure 15.19
Conversion of a two-port into a three-terminal subcircuit

Figure 15.20
A three-terminal subcircuit

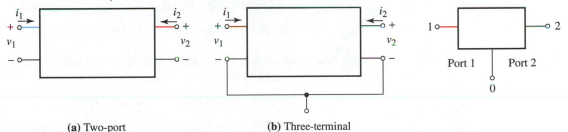

(a) Two-port

(b) Three-terminal

The Importance
of Two-Ports:
A Discussion

Before we close this section and turn to a consideration of specific representations of two-port and three-terminal subcircuits, let's pause to see why they are important. Until now, we have placed no restriction upon either the size or the complexity of a circuit. In practice, however, one does not conceive a circuit design as an entire whole. When the required complexity becomes larger than some ill-defined amount, the procedure is to break up the function of the circuit into smaller subfunctions. This is the procedure known as "top-down" design—an important technique in both electronic and software design. Thus, an automatic lamp control circuit might be conceptualized as in Figure 15.21: a photodiode light detector (a two-terminal element), followed by a voltage preamplifier, followed by a lowpass filter, followed by a power amplifier driving a mechanical actuator (a two-terminal device also). Each of the three large boxes is a two-port (or a three-terminal subcircuit if we choose to use a common ground reference for each terminal). By using this technique of top-down design, we break up the specification of a circuit into specifications of smaller one-ports and/or two-ports. These can then be designed and tested independently, then wired together to yield the overall circuit desired. Finally, we note that many electronic devices are *inherently* two-ports or three-terminal subcircuits.

Figure 15.21
Application of two-
ports

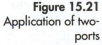
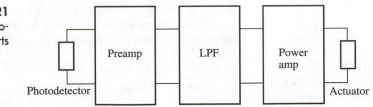

Photodetector

Preamp   LPF   Power
amp

Actuator

Two examples are the bipolar junction transistor and the magnetically coupled transformer (the latter to be studied in Chapter 16.) An electronic circuit consisting of bipolar transistors typically has the form shown in Figure 15.22. A knowledge of three-terminal networks is quite useful in analyzing such circuits.

Figure 15.22
A typical electronic circuit

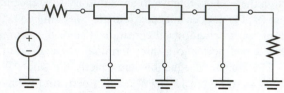

Some Comments on Two-Port Notation

We will close this section with a discussion of notation. We have been using lowercase letters for both the coefficients and variables of our two-port v-i constraint relations. If the two-ports are resistive, then the coefficients will be simple constants and the waveforms either constants or functions of time. If they contain inductors and/or capacitors, then the coefficients must be interpreted as differential operators. If we were to use our customary notation, we would write

$$A_1(p)v_1(t) + A_2(p)i_1(t) + A_3(p)v_2(t) + A_4(p)i_2(t) = 0 \qquad (15.1\text{-}20a)$$

$$B_1(p)v_1(t) + B_2(p)i_1(t) + B_3(p)v_2(t) + B_4(p)i_2(t) = 0. \qquad (15.1\text{-}20b)$$

On the other hand, one can (and we will, for the most part) perform either a Laplace or Fourier transformation on both sides, giving either

$$A_1(s)V_1(s) + A_2(s)I_1(s) + A_3(s)V_2(s) + A_4(s)I_2(s) = 0 \qquad (15.1\text{-}21a)$$

$$B_1(s)V_1(s) + B_2(s)I_1(s) + B_3(s)V_2(s) + B_4(s)I_2(s) = 0 \qquad (15.1\text{-}21b)$$

or

$$A_1(\omega)V_1(\omega) + A_2(\omega)I_1(\omega) + A_3(\omega)V_2(\omega) + A_4(\omega)I_2(\omega) = 0 \qquad (15.1\text{-}22a)$$

$$B_1(\omega)V_1(\omega) + B_2(\omega)I_1(\omega) + B_3(\omega)V_2(\omega) + B_4(\omega)I_2(\omega) = 0 \qquad (15.1\text{-}22b)$$

If we were considering the ac forced response of the two-port, we would use phasors:

$$A_1\overline{V}_1 + A_2\overline{I}_1 + A_3\overline{V}_2 + A_4\overline{I}_2 = 0 \qquad (15.1\text{-}23a)$$

$$B_1\overline{V}_1 + B_2\overline{I}_1 + B_3\overline{V}_2 + B_4\overline{I}_2 = 0. \qquad (15.1\text{-}23b)$$

You can see the problem. Until now we have been using lowercase letters for time-varying functions and uppercase ones for operators, transforms, and phasors. If we were to continue to do so, we would be forced to write four different notational forms for each set of constraint relationships. Therefore, to ease the notational load, we will simply present the equations in lowercase form. If we are concentrating on a Laplace transform solution, for example, we will then go to the more usual uppercase, and if it is to be an ac forced response calculation, we will return to the use of uppercase with a bar on top for phasors. Therefore, in general, each of our equations will be shorthand for any of the preceding types of variables.

For the same reason, we will often use a resistor symbol for any two-terminal element (though on other occasions we will continue to use a box). Thus, the symbol in Figure 15.23 could mean any of the four equations presented there.

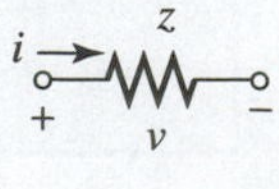

Figure 15.23
General passive element symbol

$$v(t) = z(p)i(t) \qquad V(\omega) = Z(j\omega)I(\omega)$$
$$V(s) = Z(s)I(s) \qquad \overline{V} = Z(j\omega)\overline{I}$$

Finally, until now we have defined dependent sources to always have a dependency relationship that is a real constant times a controlling voltage or current elsewhere in the circuit. For convenience, we now relax this and allow the "constant" multiplier to be an operator or an immittance (impedance or admittance) that depends upon p, s, or ω as appropriate. We show a couple of examples of this in Figure 15.24.

Figure 15.24
Generalized dependent sources

$i_c(t) = Y_m(p)v_x(t)$ $V_c(s) = Z_m(s)I_x(s)$

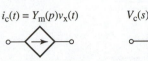   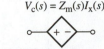

Section 15.1 Quiz

Q15.1-1. Sketch the relation $R - \{(x, y): y - 2(x - 1)^2 - 2 = 0\}$ in the x-y plane.

Q15.1-2. Determine whether or not the two two-terminal subcircuits in Figure Q15.1-1 are, by themselves, two-ports. That is,

Figure Q15.1-1

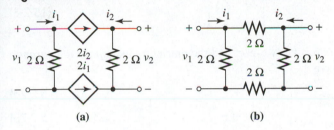

(a)   (b)

determine whether or not the port constraint is satisfied for arbitrary values of i_1 and i_2 and whether or not the constraint equations are functions of only the voltages v_1 and v_2 (as well, of course, of the two currents) for any circuit in which they are imbedded. Can either perform as a subcircuit when terminated by two-terminal elements?

Q15.1-3. A two-terminal subcircuit obeys the time domain constraint relationship $[(p + 2)/(p + 1)]v - 4i + 2\sin(3t) = 0$, where $p = d/dt$ is the differentiation operator. Find and sketch an equivalent subcircuit that also exhibits this characteristic. Indicate the element values. (*Hint:* First change to s-domain form, then derive $V(s)$ as a function of $I(s)$ and interpret the result as a Thévenin equivalent. Simplify to identify the elements in $Z_{eq}(s)$, and then transform back to the time domain.)

15.2 | The Impedance Parameters

As discussed in the last section, the general two-port subcircuit[4] in Figure 15.25 can be described by the following affine relationship between its terminal variables (providing it has no internal independent sources):

$$a_1v_1 + a_2v_2 + a_3i_1 + a_4i_2 = 0 \tag{15.2-1a}$$

$$b_1v_1 + b_2v_2 + b_3i_1 + b_4i_2 = 0. \tag{15.2-1b}$$

Let's sort the terminal variables into voltages and currents and transpose the currents to the right, giving

$$a_1v_1 + a_2v_2 = -a_3i_1 - a_4i_2 \tag{15.2-2a}$$

$$b_1v_1 + b_2v_2 = -b_3i_1 - b_4i_2. \tag{15.2-2b}$$

In matrix form, these equations can be written as

$$\begin{bmatrix} a_1 & a_2 \\ b_1 & b_2 \end{bmatrix} \begin{bmatrix} v_1 \\ v_2 \end{bmatrix} = \begin{bmatrix} -a_3 & -a_4 \\ -b_3 & -b_4 \end{bmatrix} \begin{bmatrix} i_1 \\ i_2 \end{bmatrix}. \tag{15.2-3}$$

---

[4] Or, as a special case, the equivalent three-terminal subcircuit obtained by connecting the bottom terminals.

In more compact notation, we write

$$A\bar{v} = B\bar{i},$$ (15.2-4)

with obvious definitions for the matrices A, B, $\bar{v}$, and $\bar{i}$.

Figure 15.25
The general two-port
subcircuit

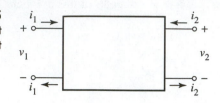

The z Parameters

Now *if the matrix A is nonsingular,* that is $|A| \neq 0$, we can multiply both sides by its inverse to obtain

$$\bar{v} = A^{-1}B\bar{i} = Z\bar{i}.$$ (15.2-5)

The matrix

$$Z = \begin{bmatrix} z_{11} & z_{12} \\ z_{21} & z_{22} \end{bmatrix}$$ (15.2-6)

is called the *impedance matrix* for the two-port and the entries z_{ij} its *impedance parameters*. In nonmatrix form, we have

$$v_1 = z_{11}i_1 + z_{12}i_2$$ (15.2-7a)

$$v_2 = z_{21}i_1 + z_{22}i_2.$$ (15.2-7b)

These parameters do not exist for all subcircuits, only for those for which the voltage coefficient matrix is nonsingular. If the subcircuit is resistive (that is, it only contains resistors plus, perhaps, dependent sources) the z_{ij} will be real numbers. If, on the other hand, there are capacitors and/or inductors, we interpret (15.2-7) as either operator equations, Laplace transform equations, or Fourier transform relationships. Of course, our notation suffers a bit because we normally write uppercase letters for operator, s-domain, or $j\omega$-domain quantities. It should be clear from the context which is meant.

The z-Parameter Equivalent Two-Port

If we notice that equations (15.2-7) are statements of KVL, we immediately see that the subcircuit shown in Figure 15.26(a) has the same impedance parameters. Thus, it is equivalent to the original one. If the complete circuit in which the two-port is located satisfies the additional condition that the bottom two terminals are at the same potential (because of constraints either external or internal to the subcircuit itself), then the equivalent circuit in Figure 15.26(b) results.

Figure 15.26
Impedance parameter
equivalents

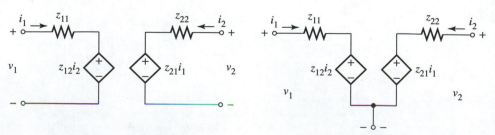

(a) Two-port equivalent   (b) Three-terminal equivalent

Determination of the z Parameters by Testing with Specified Port Conditions

The question now arises as to how one determines the impedance parameters for a given two-terminal subcircuit. The voltages are expressed in terms of the currents, so we can attach two current sources as in Figure 15.27 and compute the resulting voltages. Notice that the directions of the current sources must be consistent with our previous definitions: into the subcircuit on both sides. There are two independent sources, so we can use superposition. The partial circuit with the right-hand source deactivated is shown in Figure 15.28. Because $i_2 = 0$, our equations (15.2-7) take the form

$$v_1 = z_{11}i_1 \tag{15.2-8a}$$

$$v_2 = z_{21}i_1. \tag{15.2-8b}$$

We can now easily compute the associated impedance parameters:

$$z_{11} = \left[\frac{v_1}{i_1} \right]_{i_2=0} \tag{15.2-9}$$

and

$$z_{21} = \left[\frac{v_2}{i_1} \right]_{i_2=0}. \tag{15.2-10}$$

Notice that z_{11} is the *impedance of the two-terminal subcircuit that results when port 2 is open-circuited*. The port 2 terminals are irrelevant. This, however, is not true for z_{21}. We refer to it as the *transfer impedance from port 1 to port 2 with port 2 open-circuited*.

Figure 15.27
Testing the two-port

Figure 15.28
Port 2 open-circuited

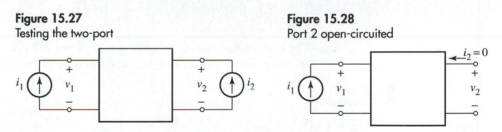

If we now deactivate the i_1 source and reactivate the i_2 source, we will have the partial circuit shown in Figure 15.29. The remaining two parameters can be quickly determined from the resulting equations:

$$v_1 = z_{12}i_2 \tag{15.2-11a}$$

$$v_2 = z_{22}i_2. \tag{15.2-11b}$$

They are

$$z_{12} = \left[\frac{v_1}{i_2} \right]_{i_1=0} \tag{15.2-12}$$

and

$$z_{22} = \left[\frac{v_2}{i_2} \right]_{i_1=0}. \tag{15.2-13}$$

z_{12} is the transfer impedance from port 2 to port 1 with port 1 open-circuited, and z_{22} is the impedance of the two-terminal subcircuit that results when port 1 is open-circuited.

Figure 15.29
Port 1 open-circuited

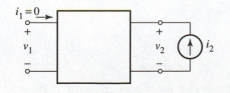

Notice that the voltages v_1 and v_2 are linear functions of i_1 and i_2 in the partial circuits; thus, for dc circuits, one can assume that i_1 and i_2 are alternately constants of 1 A in value. Similarly, for Laplace transform or ac steady-state frequency domain circuits, one can assume that these variables have unity value. In any case, this means that the impedance parameters are numerically equal to the voltages v_1 and v_2.

Example 15.4   Find the impedance parameters of the subcircuit in Figure 15.30.

Figure 15.30
An example subcircuit

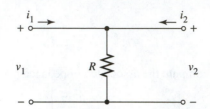

Solution   Let's just follow the procedure we have outlined. First, we open-circuit port 2 and apply a test current source to port 1. For convenience, let's adjust the value of this source to 1 A, as shown in Figure 15.31. We find that $v_1 = v_2 = R$; therefore, we see that $z_{11} = z_{21} = R$. Open-circuiting port 1 and driving port 2 with a 1-A current source gives the partial circuit in Figure 15.32. Here, we see that $v_1 = v_2 = R$. Thus, we have the values of z_{22} and z_{12}: $z_{22} = z_{12} = R$. Just for practice, let's write the impedance parameter equations in matrix form:

$$\begin{bmatrix} v_1 \\ v_2 \end{bmatrix} = \begin{bmatrix} R & R \\ R & R \end{bmatrix} \begin{bmatrix} i_1 \\ i_2 \end{bmatrix}. \tag{15.2-14}$$

The z-parameter equivalent subcircuit is shown in Figure 15.33. Notice that our original subcircuit is a three-terminal one, so the bottom two terminals are connected together.

Figure 15.31
Port 2 open-circuited

Figure 15.32
Port 1 open-circuited

Figure 15.33
The z-parameter equivalent subcircuit

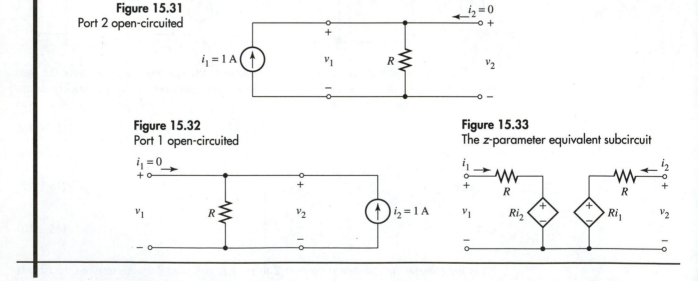

An Example:
A Two-Port Whose
z Parameters
Do Not Exist

Example 15.5   Find the z parameters for the subcircuit shown in Figure 15.34.

Figure 15.34
An example subcircuit

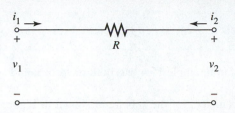

Solution Proceeding as in the last example, we first open-circuit port 2, as shown in Figure 15.35, and apply a 1-A test source to port 1. We now have a problem! The open-circuit condition at port 2 demands that the resistor current be zero, whereas the current source at port 1 insists that it be 1 A. We simply cannot find the z parameters associated with port 2 being open-circuited. (We would have the same problem with port 1 open-circuited.) To see what the problem is, let's simply refer back to the original subcircuit in Figure 15.34 and observe that the KCL and KVL constraints are

$$i_1 + i_2 = 0 \tag{15.2-15a}$$

and
$$v_1 = v_2 + Ri_1, \tag{15.2-15b}$$

respectively. We can rearrange these equations into the form

$$0v_1 + 0v_2 + 1i_1 - 1i_2 = 0 \tag{15.2-16a}$$

$$1v_1 - 1v_2 - Ri_1 + 0i_2 = 0. \tag{15.2-16b}$$

Now notice that the matrix of coefficients of the voltages is singular—it is

$$A = \begin{bmatrix} 0 & 0 \\ 1 & -1 \end{bmatrix}. \tag{15.2-17}$$

Recalling that the Z-parameter matrix is given by equation (15.2-5),

$$\bar{v} = A^{-1}B\bar{i} = Z\bar{i}, \tag{15.2-18}$$

we see that the Z matrix does not exist. This example was chosen to illustrate the fact that not all two-terminal subcircuits possess impedance parameters.

Figure 15.35
Port 2 open-circuited

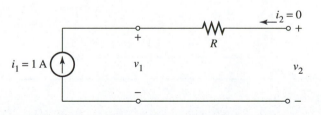

Reciprocity, Symmetry, and Antisymmetry Refer back to Example 15.4 now and notice that the end result was highly symmetric; in fact, all the z parameters were the same. This observation leads to the following definitions.

Definition 15.1 If $z_{12} = z_{21}$, the associated subcircuit is called *reciprocal*.

Definition 15.2 If a subcircuit is reciprocal *and* $z_{11} = z_{22}$, it is called *symmetric*.

Definition 15.3	If a subcircuit is reciprocal[5] *and* $z_{11} = 1/z_{22}$, it is *antimetric*.

The subcircuit in Example 15.4 was both reciprocal and symmetric.

Example 15.6   Find the z parameters for the subcircuit shown in Figure 15.36 and determine whether it is reciprocal, symmetric, and/or antimetric.

Figure 15.36
An example subcircuit

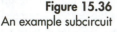
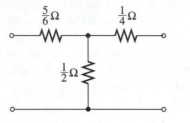

Solution   To find the values of z_{12} and z_{21}, we must still apply our current source test. Figure 15.37 shows a 1-A current source applied to port 1, with port 2 open-circuited. The value of v_2 is numerically equal to the value of z_{21}. Thus, z_{21} is 0.5 Ω. If we "flip the network around" and apply the test source to port 2 and observe the voltage at port 1, we will obtain exactly the same result. This should be apparent because the series resistors at the top of the subcircuit have no effect on the current through the vertical 0.5-Ω resistor—and, as the one connected to the open-circuited terminal has zero voltage drop, the voltage across the open-circuited port is identical to that across the 1/2-Ω resistor. Thus, $z_{12} = 0.5$ Ω also. The subcircuit is, therefore, reciprocal. We can check for symmetry and/or antimetry simply by computing the two-terminal impedances (resistances in this case) "looking into" port 1 with port 2 open-circuited, and vice versa. We easily find that $z_{11} = 5/6$ Ω $+ 1/2$ Ω $= 4/3$ Ω and $z_{22} = 1/4$ Ω $+ 1/2$ Ω $= 3/4$ Ω. Thus, we see at once that the subcircuit is antimetric, for $z_{11} = 1/z_{22}$. Clearly, it is not symmetric, for $z_{11} \neq z_{22}$.

Figure 15.37
Port 2 open-circuited

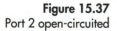
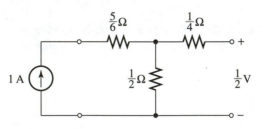

It is possible for a subcircuit to be both symmetric and antimetric. In fact, if we change the values of both the horizontal resistors at the top of the last subcircuit to 0.5 Ω, it *will* be both symmetric and antimetric because in that case $z_{11} = z_{22} = 1$ Ω. In fact, if $z_{11} = z_{22}$ and also $z_{11} = 1/z_{22}$, then $z_{11} = z_{22} = \pm 1$ Ω.

The Tee (or Wye) Equivalent for a Reciprocal Three-Terminal Subcircuit   Symmetry and antimetry are perhaps not exceedingly important concepts, but the idea of reciprocity is. In fact, we can obtain a great simplification of the equivalent for any *three-terminal subcircuit* that is reciprocal. To show this, we first note that the last example dealt with a subcircuit of a particular type. We show the general configuration in Figure 15.38. It is called a *tee* (as in the letter T) or a *wye* (as in the letter Y) because by

---

[5] Here, we mean that the numerical *values* are reciprocally related.

moving the horizontal impedances around a bit it can be made to look like either letter. Notice that it is a three-terminal subcircuit. We will leave it to you to go through the simple steps to show that

$$z_{11} = z_a + z_c, \tag{15.2-19a}$$

$$z_{22} = z_b + z_c, \tag{15.2-19b}$$

and

$$z_{12} = z_{21} = z_c. \tag{15.2-19c}$$

As $z_{12} = z_{21}$, the tee subcircuit is reciprocal.

Figure 15.38
A tee or wye subcircuit

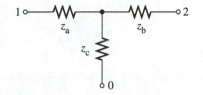

Now here is the item of interest about tee networks: *they can be made to represent any reciprocal three-terminal subcircuit.* Just consider the z-parameter equations for the general three-terminal subcircuit and rearrange them slightly, as follows:

$$v_1 = z_{11}i_1 + z_{12}i_2 = [z_{11} - z_{12}]i_1 + z_{12}[i_1 + i_2] \tag{15.2-20a}$$

and

$$v_2 = z_{12}i_1 + z_{22}i_2 = z_{12}[i_1 + i_2] + [z_{22} - z_{12}]i_2. \tag{15.2-20b}$$

Notice that we have used the reciprocity property in the second equation by setting $z_{21} = z_{12}$. Now inspect the tee subcircuit with the element values shown in Figure 15.39. As you can see, this subcircuit obeys exactly the same KVL equations as the original one.[6] In fact, one can quickly verify that it has the proper values for the various z parameters. We have thus shown that this subcircuit is equivalent to the original one.

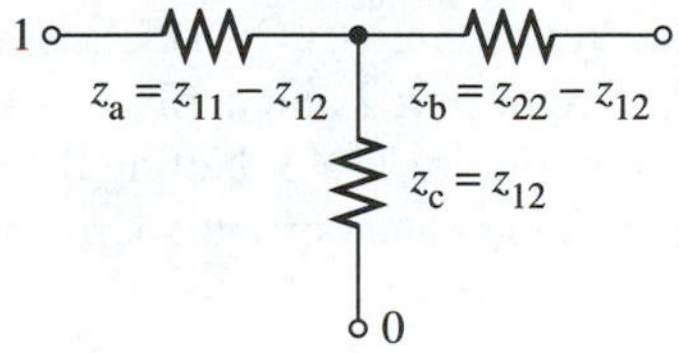

Figure 15.39
The equivalent tee or wye subcircuit

1 ○—⟋⟋⟋—•—⟋⟋⟋—○ 2
$z_a = z_{11} - z_{12}$ $z_b = z_{22} - z_{12}$
$z_c = z_{12}$
0

An Example: The Tee Equivalent for a Three-Terminal Pi (or Delta) Subcircuit

Example 15.7   Find the equivalent tee for the subcircuit in Figure 15.40.

Figure 15.40
A pi (or delta) subcircuit

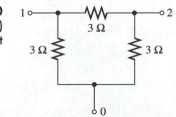

[6] We are, of course, assuming that i_1 enters terminal 1 (and so on) as usual.

Solution   This configuration for a subcircuit is termed either a pi (after the Greek letter π) or a delta (after the Greek capital letter Δ) because, with only a slight change in the shape, it can be made to look like either. We easily see that $z_{11} = z_{22} = 2\ \Omega$ and that $z_{12} = z_{21} = 1\ \Omega$. The tee subcircuit in Figure 15.41 is equivalent to the pi subcircuit in Figure 15.40 because $z_{11} - z_{12} = 1\ \Omega = z_{22} - z_{12}$. In the next section, we will show that any reciprocal three-terminal subcircuit has an equivalent pi subcircuit; thus, each pi subcircuit is equivalent to a tee—and vice versa.

Figure 15.41
The tee equivalent
subcircuit

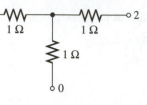

Example 15.8   Find the impedance parameters for the subcircuit in Figure 15.42.

Figure 15.42
An example subcircuit

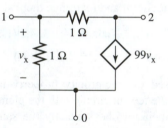

Solution   We must be a bit careful here. Even though this is a pi subcircuit, we cannot simply manipulate series and parallel impedances because one of the elements is not an impedance, but a voltage-controlled current source (VCCS). Thus, we must return to our basic testing procedure. First, we open-circuit port 2 and apply a 1-A current source to port 1, resulting in the partial circuit shown in Figure 15.43. Noting that $v_1 = v_x$ and that the top resistor is connected in series with the dependent current source, we can write one nodal equation at terminal 1:

$$\frac{v_1}{1} = 1 - 99v_1, \tag{15.2-21}$$

so that $v_1 = 0.01$ V. Because $i_1 = 1$ A, we have $z_{11} = 0.01\ \Omega$. Using KVL, we see that

$$v_2 = v_1 - 1 \times 99v_1 = -98 \times 0.01 = -0.98\ \text{V}, \tag{15.2-22}$$

so that $z_{21} = -0.98\ \Omega$. Now, we open-circuit port 1 and apply a 1-A test current source to port 2, as shown in Figure 15.44. A simple application of KCL suffices. We note that the current from right to left through the top 1-Ω resistor is $1 - 99v_x$ and that this is the same as the current through the left-hand 1-Ω resistor from top to bottom. Thus, we have

$$v_x = 1 \times [1 - 99v_x], \tag{15.2-23}$$

or $v_1 = v_x = 0.01$ V as before. Thus, $z_{12} = 0.01\ \Omega$. Also,

$$v_2 = v_x + 1 \times [1 - 99v_x] = 1 - 98v_x = 0.02\ \text{V}. \tag{15.2-24}$$

This implies that $z_{22} = 0.02\ \Omega$.

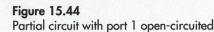

Figure 15.43
Partial circuit with port 2 open-circuited

Figure 15.44
Partial circuit with port 1 open-circuited

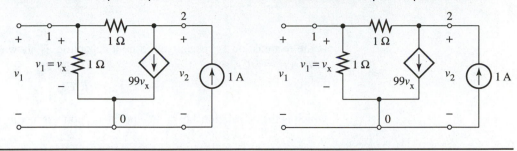

Notice that in the preceding example $z_{12} \neq z_{21}$; hence, the subcircuit was not reciprocal. *This is generally true of circuits that contain dependent sources of any type, whereas subcircuits containing only R, L, and C elements are reciprocal.*

The Gyrator The subcircuit shown in Figure 15.45(a) is called a *current-controlled gyrator.* It is useful in the design of electronic integrated circuits, as we will see shortly. Due to its usefulness, we use a special symbol for this subcircuit—the one in Figure 15.45(b). The constant R is called the *gyration resistance.* We will leave it to you to show (by the usual method of applying 1-A sources alternately to ports 1 and 2) that

$$Z = \begin{bmatrix} 0 & -R \\ R & 0 \end{bmatrix},$$ (15.2-25)

or

$$z_{11} = z_{22} = 0,$$ (15.2-26a)

$$z_{12} = -R,$$ (15.2-26b)

$$z_{21} = R.$$ (15.2-26c)

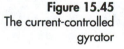

Figure 15.45
The current-controlled gyrator

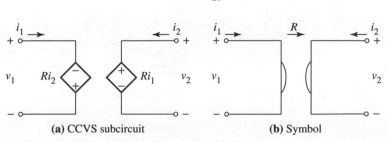

(a) CCVS subcircuit (b) Symbol

Inductance Simulation Using the Gyrator One reason for the importance of the gyrator in the design of electronic integrated circuits is its ability to simulate the behavior of an inductor. As we pointed out in Chapter 12, inductors are large and bulky and have low Q relative to that of a capacitor. Furthermore, to obtain large values of inductance, one must use a magnetic core upon which the coil is wound. Such magnetic materials are highly nonlinear (a nondesirable characteristic for most electronic circuits) and increase the loss, thereby reducing the Q. The resulting size of the inductor makes it impossible to use in a microelectronic integrated circuit. To get an idea of how the gyrator works, let's have a look at Figure 15.46(a), which shows a gyrator "terminated" by a general impedance z. We see that

$$v_1 = -Ri_2 = -R\left[-\frac{v_2}{z}\right] = -R\left[-\frac{Ri_1}{z}\right] = \frac{R^2}{z}i_1.$$ (15.2-27)

Therefore, the impedance presented to the "rest of the world" by the two terminals constituting port 1 is

$$z_1 = \frac{R^2}{z}. \tag{15.2-28}$$

If the termination is specialized to be a capacitor, as shown in Figure 15.46(b), we will have $z(s) = 1/Cs$, so

$$z_1 = R^2Cs, \tag{15.2-29}$$

which is the v-i relationship of an equivalent inductor having an inductance of

$$L_{eq} = R^2C. \tag{15.2-30}$$

The usefulness of this property is shown in the next example.

Figure 15.46
Inductance simulation

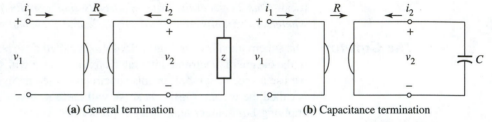

(a) General termination   (b) Capacitance termination

Example 15.9   If $R_p = 100$ kΩ, $C = 10$ nF, and $R = 1$ kΩ in the circuit shown in Figure 15.47, find the value of C_x that results in a resonance frequency of $\omega_0 = 100$ krad/s. Find the resulting value of Q_0 and the 3-dB bandwidth.

Figure 15.47
An example circuit

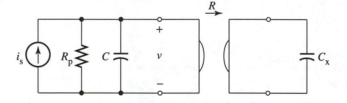

Solution   The gyrator with its capacitor termination C_x is equivalent to an inductor, as we show in Figure 15.48. Now, to make $\omega_0 = 100$ krad/s, we must make $L_{eq} = 1/(\omega_0^2C) = 1/(10^{10} \times 10^{-8}) = 10$ mH. Thus, $R^2C_x = 0.01$ — or $C_x = 0.01/R^2 = 0.01/10^6 = 10$ nF. For the parallel tuned circuit, we know that $Q_0 = \omega_0R_pC = 10^5 \times 10^5 \times 10^{-8} = 100$. Furthermore, we know that the bandwidth $= \omega_0/Q_0 = 10^5/100 = 1$ krad/s.

Figure 15.48
Equivalent circuit

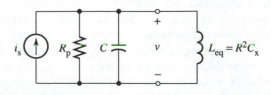

The Usefulness of   Many neophyte circuit analysts have a difficult time seeing how two-port parameters are
Two-Port Parameters   useful; this is understandable, because they are usually presented without a lot of motivational background. But this is also understandable because to do so would lead the dis-

cussion into fairly advanced topics in electronics, feedback control systems, and so on. We will try to do a bit better for you by discussing just a few of the applications as the opportunity arises. As it happens, the two-port concept, far from being idle mathematics, is actually one of the most important ideas circuit theory provides for practical applications.

Let's start by saying that if we have a circuit that is quite complicated, the two-port concept boils this complexity down into just four parameters: z_{11}, z_{22}, z_{12}, and z_{21}. If we illustrate this effect with a small circuit, the z parameters themselves might seem more complicated than the circuit itself; therefore, a practical illustration of complexity reduction will not be attempted here. We will, rather, concentrate on another topic of importance in practice.

Often one designs a subcircuit to "match" a given electronic item with a specified "load." An audio amplifier that matches a CD player (the source) into a set of speakers (the load) is a good example, so keep this metaphor in mind as we look at Figure 15.49. Think of the CD player as the v_g and R_g combination; that is, the v-source and resistor form the Thévenin equivalent of the CD player. The resistor labeled R_L is the equivalent resistance of the speaker (there is no signal coming from the speaker and being fed back into the amplifier). Also, we should mention that Figure 15.49 represents one channel of a stereo or quadraphonic system (there would be one such for each channel). The uppercase Z inside the two-port box means that we are assuming that the impedance parameters are known.

Figure 15.49
A two-port in action

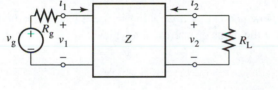

We would like to determine several things about the system, and we will mention them as we progress. For now, we will simply pose the problem as follows: given v_g, find all of the other voltages and currents in Figure 15.49. To do so, we will take a step that perhaps seems "counterintuitive." We will add a v-source in series with R_L, resulting in the setup shown in Figure 15.50. This simplifies our analysis because it makes the circuit symmetrical, as you can see by glancing at Figure 15.51. There, we have used the impedance parameter model for the two-port that we derived at the beginning of this section (see Figure 15.26). Suppose we ask for the impedance parameters of the two-port that results from enlarging our box in Figure 15.50 to include the source resistor R_g and load resistor R_L. Do you see the result? Because R_g is series connected with z_{11} and R_L with z_{22}, we merely add the two together to get the composite z parameters we are seeking. The values of z_{12} and z_{21} remain unchanged because addition of the two resistors does not change the relationship between the CCVS values and the two currents (remember z parameters are measured or computed with one port open circuited!). In matrix form, therefore, we can write

$$\begin{bmatrix} v_g \\ v_L \end{bmatrix} = \begin{bmatrix} z_{11} + R_g & z_{12} \\ z_{21} & z_{22} + R_L \end{bmatrix} \begin{bmatrix} i_1 \\ i_2 \end{bmatrix}. \tag{15.2-31}$$

Recalling that the inverse of a (2 × 2) matrix is given by

$$\begin{bmatrix} a & b \\ c & d \end{bmatrix}^{-1} = \frac{1}{\Delta} \begin{bmatrix} d & -b \\ -c & a \end{bmatrix}, \tag{15.2-32}$$

where $\Delta = ad - bc$ is the determinant of the matrix, we can solve (15.2-31) for i_1 and i_2:

$$\begin{bmatrix} i_1 \\ i_2 \end{bmatrix} = \frac{1}{\Delta} \begin{bmatrix} z_{22} + R_L & -z_{12} \\ -z_{21} & z_{11} + R_g \end{bmatrix} \begin{bmatrix} v_g \\ v_L \end{bmatrix}, \tag{15.2-33}$$

where $$\Delta = (z_{11} + R_g)(z_{22} + R_L) - z_{12}z_{21} \tag{15.2-34}$$

is the determinant of the (2×2) matrix in (15.2-31).

Figure 15.50
Adding a "load"
source

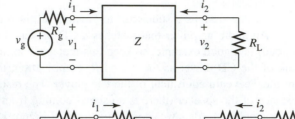

Figure 15.51
The z-parameter model

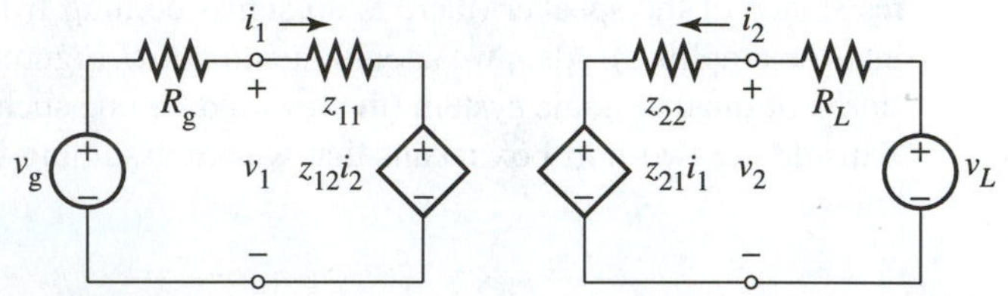

Now notice that if we let $v_L = 0$, we have solved our original circuit in Figure 15.49. One of the items of interest in a typical circuit will be the voltage gain, defined by[7]

$$A_v = \frac{v_2}{v_g} = -\frac{i_2 R_L}{v_g}. \tag{15.2-35}$$

Notice that this is the voltage gain from the source producing the signal to the load resistor. Letting $v_L = 0$ in (15.2-33) and solving, we obtain

$$A_v = \frac{z_{21}R_L}{\Delta} = \frac{z_{21}R_L}{(z_{11} + R_g)(z_{22} + R_L) - z_{12}z_{21}}. \tag{15.2-36}$$

This expression clearly shows how the various parameters affect the voltage gain—how one would modify them to achieve a desired goal.

There are a number of things of importance that we could derive, but we will defer these to the problems at the end of the chapter. All solutions are based on our general matrix solution in (15.2-33). What we will do, however, is make a couple of practical observations. An amplifier design usually consists of quite a large number of active devices such as BJTs, MOSFETs, and/or op amps. The problem is this: it is quite easy to design a circuit to produce very large intrinsic gains, but these gains are not very precise. In fact, the gain parameters of active devices can vary by a factor of 5 or 10 to 1 from one specific device to another of the same type. Furthermore, these gain parameters are quite sensitive to variations in temperature and they change as the component ages. It turns out that a very good design tactic is to design an amplifier with a very high gain—then "throw away" a lot of it in such a fashion that a very accurate gain is produced. (Look back at the op amp discussion in Chapter 5.) Let's see why.

Taking the voltage gain in equation (15.2-36) as an example, we let the parameter z_{21} (the *forward transfer impedance, or transimpedance*) become infinite. The voltage gain

---

[7] Be careful to notice the minus sign. It is there because the reference of i_2 points away from R_L.

becomes

$$A_v = \frac{R_L}{\dfrac{(z_{11} + R_g)(z_{22} + R_L)}{z_{21}} - z_{12}} \cong -\frac{R_L}{z_{12}}. \tag{15.2-37}$$

You may say, "But the voltage gain still depends on z_{12}, so we haven't solved our problem!" Perhaps it isn't clear yet, but we have! We will explain why soon. Here, we introduce a very, very important concept that we have only briefly touched upon in connection with op amp stability in Chapter 5: *feedback:*

The Effect of Feedback on the Two-Port Parameters: Series-Series Feedback

The impedance parameters are very convenient ones to use when we connect two two-ports in *series,* as shown in Figure 15.52. The reason this is called "feedback" is because the bottom two-port "samples" the output current i_2 from the top two-port and "feeds it back" as a voltage signal in series with the input of the top two-port. We could just as well consider the top two-port to be sampling the signal of the bottom one, but (as we will see) an amplifier designer has one two-port in mind as the "forward amplifier" and the other as the "feedback network," and the resulting design will not be symmetric. The properties of the two will be quite different.

Figure 15.52
Series-connected two-ports

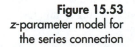

Our question now is this. Suppose we place a conceptual box around both two-ports with only the outermost terminals accessible. We then have a "larger" two-port. What are the impedance parameters of this "larger" two-port? To answer this question, let's use the two-port equivalent model of Figure 15.26 once again—this time for each constituent two-port. The result is shown in Figure 15.53. Do you see the advantage of the z parameters here? First, notice that $i_{1a} = i_{1b} = i_1$ and $i_{2a} = i_{2b} = i_2$; furthermore, we have

Figure 15.53
z-parameter model for the series connection

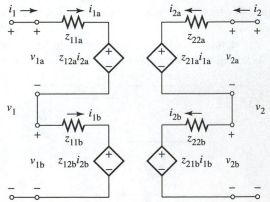

$v_1 = v_{1a} + v_{1b}$ and $v_2 = v_{2a} + v_{2b}$. Using these relationships and the z-parameter equations, we have

$$v_1 = (z_{11a} + z_{11b})i_1 + (z_{12a} + z_{12b})i_2 \tag{15.2-38a}$$

$$v_2 = (z_{21a} + z_{21b})i_1 + (z_{22a} + z_{22b})i_2, \tag{15.2-38b}$$

or in matrix form,

$$\begin{bmatrix} v_1 \\ v_2 \end{bmatrix} = \begin{bmatrix} (z_{11a} + z_{11b}) & (z_{12a} + z_{12b}) \\ (z_{21a} + z_{21b}) & (z_{22a} + z_{22b}) \end{bmatrix} \begin{bmatrix} i_1 \\ i_2 \end{bmatrix}. \tag{15.2-39}$$

Thus, the impedance parameters add for two-ports connected in series.

Here, now, is the payoff. Our amplifier designer designs the two two-ports such that z_{21a} is extremely large, and thus in particular $z_{21a} \gg z_{21b}$. Furthermore, he or she designs them so that $z_{12a} \ll z_{12b}$—and (we italicize this "and" because it is so important) so that z_{12b} is extremely precise. Then, using our modified z parameters in equation (15.2-37) for the voltage gain of the composite system, we get

$$A_v \cong -\frac{R_L}{z_{12b}}. \tag{15.2-40}$$

Thus, if R_L is precise (and resistors can be made so), the voltage gain can be adjusted as a precise ratio. The next example follows up on this idea a bit. Our series connection of two-ports, in connection with this design procedure, is called *series-series feedback*.

Example 15.10   Assume that the two-port parameters of the top two-port in Figure 15.54 are known, that $z_{12} \cong 0$, and z_{21} is very large ($z_{21} \to \infty$). Find the voltage gain $A_v = v_2/v_g$.

Figure 15.54
An example circuit

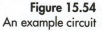
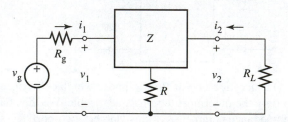

Solution   Notice that the box is actually a three-terminal network. We could analyze the circuit by merely using a single impedance parameter model for the three-terminal subcircuit and include the resistor on the bottom. We will, however, treat the bottom resistor as a two-port in its own right to show how our analysis fits within the framework we have already developed—which is more efficient for circuits with feedback subcircuits having greater complexity.

Before doing the analysis, we notice that the three-terminal subcircuit can be converted to its equivalent two-port form, as shown in Figure 15.55, with only a slight re-

Figure 15.55
The example circuit slightly rearranged

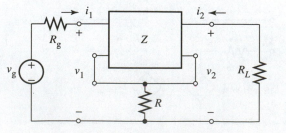

arrangement. At this point we recognize the shunt resistor subcircuit as the one whose z parameters we derived in Example 15.4 (Figure 15.30). The z-parameter model was shown in Figure 15.33. Using it in Figure 15.55, along with the z-parameter model for the top two-port, we obtain the equivalent shown in Figure 15.56. We have assumed that z_{12} for the top two-port is zero (as specified), thus making the equivalent voltage source a short circuit, as shown in the figure.

Figure 15.56
The impedance
parameter model

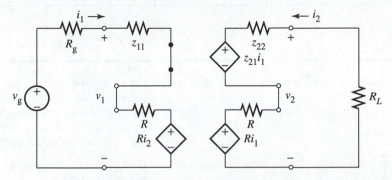

Let's try to be very clear about our approach to this problem. We are using the impedance parameter models because we know them to be particularly appropriate for series connections of two-ports—and we have recognized our topology as such a connection. We are not, however, simply going to "plug into" the earlier formula we derived, but will analyze the circuit "from scratch." We have kept z_{21} as a finite value, but we will let it become infinite after our analysis is complete.

We can write two mesh equations, now, in the form

$$(z_{11} + R + R_g)i_1 + Ri_2 = v_g \tag{15.2-41}$$

and

$$(z_{21} + R)i_1 + (z_{22} + R + R_L)i_2 = 0. \tag{15.2-42}$$

In matrix form, we have[8]

$$\begin{bmatrix} (z_{11} + R + R_g) & R \\ (z_{21} + R) & (z_{22} + R + R_L) \end{bmatrix}\begin{bmatrix} i_1 \\ i_2 \end{bmatrix} = \begin{bmatrix} v_g \\ 0 \end{bmatrix}. \tag{15.2-43}$$

Solving as usual, we get

$$\begin{bmatrix} i_1 \\ i_2 \end{bmatrix} = \frac{1}{(z_{11} + R + R_g)(z_{22} + R + R_L) - R(z_{21} + R)}\begin{bmatrix} (z_{22} + R + R_L) & -R \\ -(z_{21} + R) & (z_{11} + R + R_g) \end{bmatrix}\begin{bmatrix} v_g \\ 0 \end{bmatrix}, \tag{15.2-44}$$

so

$$i_2 = \frac{-(z_{21} + R)}{(z_{11} + R + R_g)(z_{22} + R + R_L) - R(z_{21} + R)}v_g. \tag{15.2-45}$$

The voltage gain, now, is

$$A_v = \frac{-i_2 R_L}{v_g} = \frac{(z_{21} + R)R_L}{(z_{11} + R + R_g)(z_{22} + R + R_L) - R(z_{21} + R)} \tag{15.2-46}$$

$$= \frac{\left(1 + \dfrac{R}{z_{21}}\right)R_L}{\dfrac{(z_{11} + R + R_g)(z_{22} + R + R_L)}{z_{21}} - R\left(1 + \dfrac{R}{z_{21}}\right)} \cong -\frac{R_L}{R}.$$

_____

[8] We could have just added the two impedance parameter matrices at the outset.

In the first line we used (15.2-45) and the definition of voltage gain; in the second, we divided by z_{21} (top and bottom). Finally, in getting the final approximation, we have let z_{21} become infinite—noticing that all terms divided by that term become zero. We stress again that the top two-port could, in reality, be a very complex amplifier subcircuit. Notice that the voltage gain is independent of the parameters of the top two-port; for this to be true, we only require that its z_{21} be very large and its z_{12} very small.

Section 15.2 Quiz

Q15.2-1. If a two-port subcircuit obeys the affine v-i relationship

$$2v_1 - 2v_2 + 20i_1 - 40i_2 = 0$$

$$1v_1 + 4v_2 - 20i_1 - 30i_2 = 0,$$

find and sketch the z-parameter equivalent subcircuit and label each value.

Q15.2-2. Find the z-parameters for the subcircuit in Figure Q15.2-1. Determine whether or not it is reciprocal, symmetric, and/or antimetric. Sketch the s-domain z-parameter equivalent subcircuit.

Figure Q15.2-1

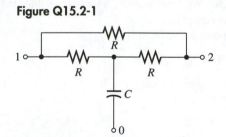

Q15.2-3. Assume that a two-port obeys the affine v-i relationship

$$2v_1 - 2v_2 + 2i_1 - 4i_2 = 0$$

$$4v_1 - 4v_2 + 4i_1 - 3i_2 = 0.$$

Let u_1 and u_2 denote any pair of variables from among v_1, v_2, i_1, and i_2 selected as the independent variables and w_1 and w_2 denote the other pair (the dependent variables). Determine which choices for each will result in a solvable set, that is, such that the preceding equations can be solved for w_1 and w_2 in terms of u_1 and u_2.

Q15.2-4. Assume that the two-port designated by the box in Figure Q15.2-2 has $z_{12} = 0$ and $z_{21} \to \infty$. Find the voltage gain $A_v = v_2/v_g$.

Figure Q15.2-2

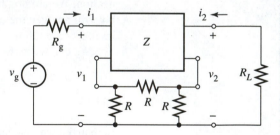

15.3 | The Admittance Parameters

We pointed out in the last section that the impedance parameters do not always exist for any arbitrary two-port, so we need alternate means of describing it. Therefore, we will develop another set of parameters in this section. Let's refresh our memory by looking once more at the general two-port subcircuit in Figure 15.57. This subcircuit or the three-terminal subcircuit that results from connecting the two bottom terminals, can be described by the following affine relationship between its terminal variables (providing it has no internal independent sources):

$$a_1v_1 + a_2v_2 + a_3i_1 + a_4i_2 = 0 \tag{15.3-1a}$$

$$b_1v_1 + b_2v_2 + b_3i_1 + b_4i_2 = 0. \tag{15.3-1b}$$

As in the last section, we will separate the terminal variables into voltages and currents and transpose the currents to the right, giving

$$a_1v_1 + a_2v_2 = -a_3i_1 - a_4i_2 \tag{15.3-2a}$$

$$b_1v_1 + b_2v_2 = -b_3i_1 - b_4i_2. \tag{15.3-2b}$$

In matrix form,

$$\begin{bmatrix} a_1 & a_2 \\ b_1 & b_2 \end{bmatrix}\begin{bmatrix} v_1 \\ v_2 \end{bmatrix} = \begin{bmatrix} -a_3 & -a_4 \\ -b_3 & -b_4 \end{bmatrix}\begin{bmatrix} i_1 \\ i_2 \end{bmatrix}. \tag{15.3-3}$$

In more compact notation, we write

$$A\bar{v} = B\bar{i}, \tag{15.3-4}$$

with obvious definitions for the matrices A, B, $\bar{v}$, and $\bar{i}$.

Figure 15.57
The general two-port
subcircuit

The y Parameters

So far, our development is identical to that for the z parameters; now, however, we note the following fact. If the determinant of B is nonzero, that is, the matrix B is nonsingular, we can write

$$\bar{i} = B^{-1}A\bar{v} = Y\bar{v}. \tag{15.3-5}$$

Notice that this is exactly the opposite of what we did in developing the z parameters. The matrix Y now has entries that are admittances:

$$Y = \begin{bmatrix} y_{11} & y_{12} \\ y_{21} & y_{22} \end{bmatrix}. \tag{15.3-6}$$

Thus, we can write

$$i_1 = y_{11}v_1 + y_{12}v_2 \tag{15.3-7a}$$

$$i_2 = y_{21}v_1 + y_{22}v_2. \tag{15.3-7b}$$

The y-Parameter Equivalent Two-Port

Equations (13.3-7) can be interpreted as KCL equations for the equivalent subcircuit shown in Figure 15.58. For a three-terminal subcircuit, of course, the bottom two terminals must be connected together in this equivalent subcircuit.

Figure 15.58
The y-parameter
equivalent subcircuit

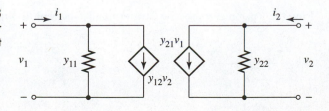

Determination of the y Parameters: Specified Port Conditions

To measure or compute these *admittance or y parameters,* it is convenient to use superposition and alternately drive the original subcircuit with a voltage source at each port, as shown in Figure 15.59(a) and (b). Note that a deactivated voltage source is a short circuit. When $v_2 = 0$, we use (15.3-7) to quickly compute y_{11} and y_{21}:

$$y_{11} = \left[\frac{i_1}{v_1}\right]_{v_2=0} \tag{15.3-8}$$

and

$$y_{21} = \left[\frac{i_2}{v_1}\right]_{v_2=0}. \tag{15.3-9}$$

Thus, y_{11} is the admittance "looking into" port 1 with port 2 shorted, and y_{21} is the transfer admittance from port 1 to port 2 with port 2 shorted. Notice that by taking the ratio, we can see at once that

$$A_{i21} = \left[\frac{i_2}{i_1}\right]_{v_2=0} = \frac{y_{21}}{y_{11}} \tag{15.3-10}$$

is the current gain from port 1 to port 2 with port 2 shorted. Similarly, if we deactivate v_1 and apply a voltage source to port 2, we see that

$$y_{12} = \left[\frac{i_1}{v_2}\right]_{v_1=0} \tag{15.3-11}$$

and

$$y_{22} = \left[\frac{i_2}{v_2}\right]_{v_1=0}. \tag{15.3-12}$$

Thus, y_{22} is the impedance "seen looking into" port 2 with port 1 shorted, and y_{12} is the transfer admittance from port 2 to port 1 with port 1 shorted. Taking their ratio produces

$$A_{i12} = \left[\frac{i_1}{i_2}\right]_{v_1=0} = \frac{y_{12}}{y_{22}} \tag{15.3-13}$$

as the current gain from port 2 to port 1 with port 1 shorted. Now let's apply these results by working some examples.

Figure 15.59
Application of superposition to compute the y parameters

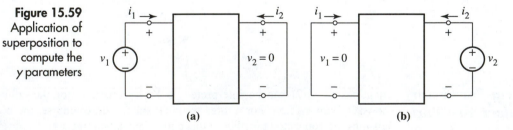

(a)   (b)

Example 15.11   Find the y parameters for the two-terminal subcircuit in Figure 15.60.

Figure 15.60
An example subcircuit

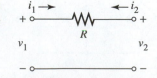

Solution Let's first short port 2 (the one on the right), as shown in Figure 15.61. Clearly,

$$y_{11} = \frac{i_1}{v_1} = \frac{1}{R}. \qquad (15.3\text{-}14)$$

Almost as clearly (but be careful about the sign!),

$$y_{21} = \frac{i_2}{v_1} = -\frac{1}{R}. \qquad (15.3\text{-}15)$$

If we short port 1 and apply a voltage source to port 2, the circuit shown in Figure 15.62 results. The resulting values of the y parameters are

$$y_{22} = \frac{i_2}{v_2} = \frac{1}{R} \qquad (15.3\text{-}16)$$

and

$$y_{12} = \frac{i_1}{v_2} = -\frac{1}{R}. \qquad (15.3\text{-}17)$$

Figure 15.61
Port 2 shorted

Figure 15.62
Port 1 shorted

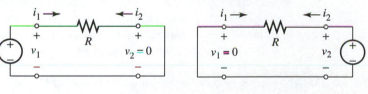

Notice that we could have adjusted the voltage sources each to 1 V in the last example had that reduced our computational effort.

We now provide an example for which the y parameters do not exist.

Example 15.12 Find the admittance parameters of the subcircuit in Figure 15.63.

Figure 15.63
An example subcircuit

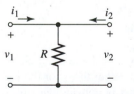

Solution Following our previous procedure, we apply a test voltage source to port 1 and short port 2, as in Figure 15.64. We see that this results in a circuit that violates KVL—unless the voltage source is adjusted to be identically zero, in which case the two currents are also

Figure 15.64
Port 2 shorted

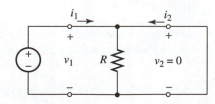

zero, and the y parameters become indeterminate ratios of the form 0/0. What is the problem?

To answer this question, let's simply write the KVL and KCL constraints for the original subcircuit in Figure 15.63:

$$1v_1 - 1v_2 + 0i_1 + 0i_2 = 0 \tag{15.3-18a}$$

$$1v_1 + 0v_2 - Ri_1 - Ri_2 = 0. \tag{15.3-18b}$$

We immediately see that the matrix B from equations (15.3-4) and (15.3-5) is

$$B = \begin{bmatrix} 0 & 0 \\ R & R \end{bmatrix}. \tag{15.3-19}$$

Its determinant is zero, so B cannot be inverted to obtain the Y matrix (and hence the y parameters, which are its elements).

The main purpose of the last example was simply to illustrate the fact that the admittance parameters do not always exist. In fact, a careful look at the last two examples shows that they are precisely the same as the first two that we worked for the z parameters in Section 15.2. The z parameters did not exist for the first, but did for the second—the opposite is true for the y parameters. Is this always the case? Not really. Some subcircuits have both types of parameter, still others have neither. The next example illustrates the last case: neither the z nor the y parameters exist.

Example 15.13   Show that neither z nor y parameters exist for the subcircuit in Figure 15.65.

Figure 15.65
An example subcircuit

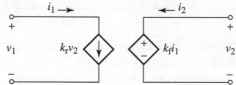

Solution   You can easily verify that the KVL-KCL constraints can be written as

$$0v_1 + k_r v_2 - 1i_1 + 0i_2 = 0 \tag{15.3-20a}$$

$$0v_1 + 1v_2 - k_f i_1 + 0i_2 = 0. \tag{15.3-20b}$$

We see at once that the matrix of voltage coefficients and the matrix of current coefficients are both singular; hence, neither the z nor the y parameters exist. It is true, however, that there *is* a square matrix that is nonsingular: the matrix of coefficients of v_2 and i_1. This is a *mixed* set of port variables that leads to the *hybrid parameters*. We will discuss them in a later section. It suffices to comment that *some* parameter set must always exist for any well-defined two-port. (Otherwise—by definition—it would not be well defined.)

Matrix Transformation
from z to y and
Vice Versa

For subcircuits that possess both z and y parameters, one can transform immediately from one parameter set to the other without retesting the subcircuit. To do this, recall that

$$\begin{bmatrix} v_1 \\ v_2 \end{bmatrix} = \begin{bmatrix} z_{11} & z_{12} \\ z_{21} & z_{22} \end{bmatrix} \begin{bmatrix} i_1 \\ i_2 \end{bmatrix}. \tag{15.3-21}$$

Letting

$$\Delta_z = \begin{vmatrix} z_{11} & z_{12} \\ z_{21} & z_{22} \end{vmatrix} = z_{11}z_{22} - z_{12}z_{21} \tag{15.3-22}$$

be the determinant of the z matrix and computing the inverse of that matrix, we have

$$\begin{bmatrix} i_1 \\ i_2 \end{bmatrix} = \frac{1}{\Delta_z} \begin{bmatrix} z_{22} & -z_{12} \\ -z_{21} & z_{11} \end{bmatrix} \begin{bmatrix} v_1 \\ v_2 \end{bmatrix}. \tag{15.3-23}$$

By the very definition of y parameters, however, we know that the currents are given by

$$\begin{bmatrix} i_1 \\ i_2 \end{bmatrix} = \begin{bmatrix} y_{11} & y_{12} \\ y_{21} & y_{22} \end{bmatrix} \begin{bmatrix} v_1 \\ v_2 \end{bmatrix}. \tag{15.3-24}$$

Thus, setting these two expressions equal results in

$$y_{11} = \frac{z_{22}}{\Delta_z}, \tag{15.3-25a}$$

$$y_{22} = \frac{z_{11}}{\Delta_z}, \tag{15.3-25b}$$

$$y_{12} = \frac{-z_{12}}{\Delta_z}, \tag{15.3-25c}$$

$$y_{21} = \frac{-z_{21}}{\Delta_z}. \tag{15.3-25d}$$

If the z parameters exist—and if $\Delta_z \neq 0$—then the y parameters also exist and are given by equations (15.3-25).

Here is one of the consequences of the transformation we have just derived: if $z_{12} = z_{21}$ for a given subcircuit, then $y_{12} = y_{21}$, and the converse is also true (provided, of course, that both parameter sets exist). Thus, one can test the admittance parameters to determine reciprocity. Too, if $z_{11} = z_{22}$, then $y_{11} = y_{22}$, and conversely. This shows that symmetry and antimetry can also be checked by referring to the y parameters.

The Pi (or Delta) Equivalent for Any Reciprocal Three-Terminal Subcircuit

Now let's have a look at our y parameter equations (15.3-7), repeated here for convenience *under the assumption that the subcircuit is a reciprocal three-terminal* one:

$$i_1 = y_{11}v_1 + y_{12}v_2 \tag{15.3-26a}$$

$$i_2 = y_{12}v_1 + y_{22}v_2. \tag{15.3-26b}$$

We rearrange the first by adding and subtracting $y_{12}v_1$ and the second by adding and subtracting $y_{12}v_2$. This gives

$$i_1 = [y_{11} + y_{12}]v_1 - y_{12}[v_1 - v_2] \tag{15.3-27a}$$

$$i_2 = -y_{12}[v_2 - v_1] + [y_{22} + y_{12}]v_2. \tag{15.3-27b}$$

(Notice that this only works if $y_{12} = y_{21}$.) We can interpret these two equations as KCL nodal equations for the equivalent subcircuit shown in Figure 15.66. Notice that this interpretation only works for three-terminal circuits because there must be a common voltage reference for both sides. This subcircuit is called a *pi subcircuit* because it resembles the Greek letter π. Sometimes it is also referred to as a *delta subcircuit* because it can easily be manipulated to look like the Greek letter Δ.

Figure 15.66
The equivalent pi
subcircuit

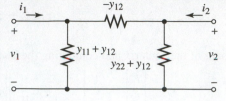

Example 15.14   Find the pi equivalent for the subcircuit shown in Figure 15.67.

Figure 15.67
A tee or wye subcircuit

Solution   This, of course, is the tee (or wye) subcircuit that we discussed in Section 15.2. We can easily compute the impedance parameters (or refer to equations (15.2-19a−c)) to see that

$$z_{11} = z_a + z_c, \tag{15.3-28a}$$

$$z_{22} = z_b + z_c, \tag{15.3-28b}$$

$$z_{12} = z_{21} = z_c. \tag{15.3-28c}$$

This subcircuit is, of course, both reciprocal and of the three-terminal variety. The determinant of the impedance parameter matrix is

$$\Delta_z = (z_a + z_c)(z_b + z_c) - z_c^2 = z_a z_b + z_a z_c + z_b z_c. \tag{15.3-29}$$

Therefore, the y parameters are

$$y_{11} = \frac{z_{22}}{\Delta_z} = \frac{z_b + z_c}{z_a z_b + z_a z_c + z_b z_c}, \tag{15.3-30a}$$

$$y_{22} = \frac{z_{11}}{\Delta_z} = \frac{z_a + z_c}{z_a z_b + z_a z_c + z_b z_c}, \tag{15.3-30b}$$

and

$$y_{12} = \frac{-z_{12}}{\Delta_z} = \frac{-z_c}{z_a z_b + z_a z_c + z_b z_c}. \tag{15.3-30c}$$

The element values of the equivalent pi subcircuit can be computed by applying the results shown in Figure 15.66; that is, the element values are

$$y_{11} + y_{12} = \frac{z_{22} - z_{12}}{\Delta_z} = \frac{z_b}{z_a z_b + z_a z_c + z_b z_c} = \frac{y_a y_c}{y_a + y_b + y_c}, \tag{15.3-31a}$$

$$y_{22} + y_{12} = \frac{z_{11} - z_{12}}{\Delta_z} = \frac{z_a}{z_a z_b + z_a z_c + z_b z_c} = \frac{y_b y_c}{y_a + y_b + y_c}, \tag{15.3-31b}$$

and

$$-y_{12} = \frac{z_{12}}{\Delta_z} = \frac{z_c}{z_a z_b + z_a z_c + z_b z_c} = \frac{y_a y_b}{y_a + y_b + y_c}. \tag{15.3-31c}$$

Perhaps the following statement is the best way to recall this result: *the admittance con-*

necting two terminals in the pi is equal to the product of the two admittances connecting these two terminals in the tee divided by the sum of all three tee admittances. The pi equivalent subcircuit is shown in Figure 15.68.

Figure 15.68
The equivalent pi subcircuit

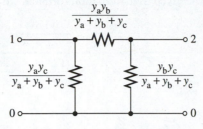

We will leave it to you to go through the same manipulations and show that the subcircuit in Figure 15.69(a) is equivalent to the one in Figure 15.69(b). In words, *the impedance connected to a given terminal in the tee is the product of the two impedances connected to the same terminal in the wye divided by the sum of all the impedances.*

At this point, we have shown that each reciprocal three-terminal subcircuit is equivalent to both a tee and a pi subcircuit and have presented the transformations for converting one to the other. Now let's work another example—one that includes dynamic elements.

Figure 15.69
The equivalent tee subcircuit

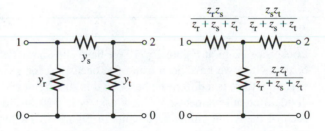

Example 15.15 Find the tee equivalent for the subcircuit shown in Figure 15.70.

Figure 15.70
An example subcircuit

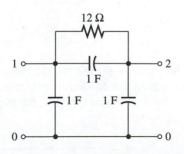

Solution We see that the subcircuit is a pi—providing one treats the 12-Ω resistor and its associated 1-F capacitor as an equivalent two-terminal subcircuit having a complex impedance of

$$Z_s(s) = \frac{12 \times \dfrac{1}{s}}{12 + \dfrac{1}{s}} = \frac{12}{12s + 1}. \tag{15.3-32}$$

Similarly,

$$Z_r(s) = \frac{1}{s} \tag{15.3-33}$$

and
$$Z_t(s) = \frac{1}{s}. \tag{15.3-34}$$

Thus, using our just-derived transformation formula, we have

$$Z_a = \frac{Z_r Z_s}{Z_r + Z_s + Z_t} = \frac{\dfrac{1}{s} \times \dfrac{12}{12s + 1}}{\dfrac{1}{s} + \dfrac{12}{12s + 1} + \dfrac{1}{s}} = \frac{6}{18s + 1} = Z_b \tag{15.3-35}$$

and
$$Z_c = \frac{Z_r Z_t}{Z_r + Z_s + Z_t} = \frac{\dfrac{1}{s} \times \dfrac{1}{s}}{\dfrac{1}{s} + \dfrac{12}{12s + 1} + \dfrac{1}{s}} = \frac{12s + 1}{2s(18s + 1)}. \tag{15.3-36}$$

The tee equivalent circuit, therefore, is as shown in Figure 15.71.

Figure 15.71
The tee equivalent

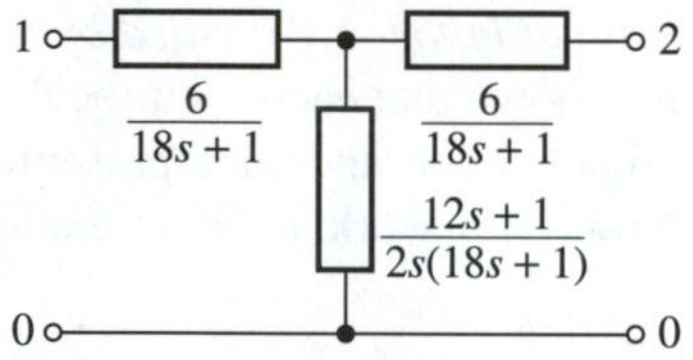

1 o—[$\frac{6}{18s + 1}$]—•—[$\frac{6}{18s + 1}$]—o 2

$$\frac{12s + 1}{2s(18s + 1)}$$

0 o———————————•———————————o 0

Two-Port Synthesis: A Brief Look

Look once again at Figure 15.71. We have shown the "elements" of the tee by our usual boxes, which we have been using (on occasion) for general impedances. We note, however, that there is a difference between this subcircuit and the original. The former had individual circuit elements. We now ask: Is it possible for us to obtain an equivalent for each of the two-terminal subcircuits in Figure 15.71 that consists of distinct circuit elements? Let's see. Suppose we rearrange equation (15.3-35):

$$Z_a = Z_b = \frac{1}{3s + \dfrac{1}{6}}. \tag{15.3-37}$$

Now, recognizing that the terms on the bottom are admittances, and that admittances add for parallel elements, we obtain the equivalent subcircuit shown in Figure 15.72. Determining Z_c is not so easy, for it is a more complicated function of s. Let's try doing a partial fraction expansion. This gives

$$Z_c = \frac{1}{2s} + \frac{1}{-6s - \dfrac{1}{3}}. \tag{15.3-38}$$

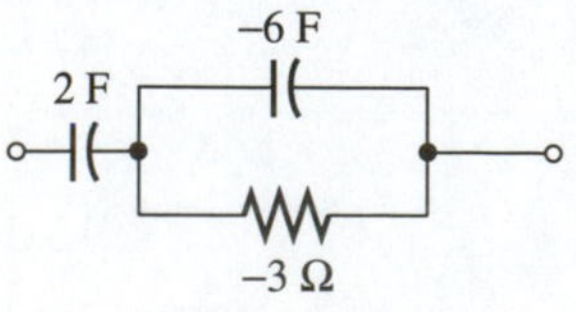

Again recalling that the inverse of an impedance is an admittance and that admittances add for parallel elements, whereas impedances of series elements add, we obtain the two-terminal subcircuit shown in Figure 15.73. The problem with this "realization" is that two of the elements are negative. Though such elements can be realized as the equivalent of subcircuits with dependent sources, this is not usually desirable because of complexity and accuracy considerations. Thus, we ask: Is it possible to realize $Z_c(s)$ in terms of positive R, L, C elements only? After all, we somewhat arbitrarily chose to perform a partial

Figure 15.72 Realization of Z_a and Z_b

3 F
—||—
—⋀⋀⋀—
6 Ω

Figure 15.73 Realization of Z_c

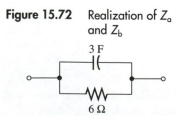

−6 F
2 F —||—
—||—
—⋀⋀⋀—
−3 Ω

fraction expansion, and other operations are possible. The general study of such questions and the techniques for realizing given impedance operators is called *network synthesis*. (As it happens, the results of network synthesis give a simple test that shows that our present $Z_c(s)$ is *not* realizable in terms of positive R, L, C elements.)

The Effect of Feedback on the Two-Port Parameters: Shunt-Shunt Feedback

In Section 15.2 we saw that the impedance parameters of two two-ports connected in series add to produce the impedance parameters of the composite two-port thus formed. We therefore suspect that the admittance parameters add for the parallel connection of two-ports shown in Figure 15.74—and we are right. Let's prove this statement. Our procedure will be to replace each of the constituent two-ports with the equivalent two-port model that we derived at the beginning of this section and drew in Figure 15.58. The result is shown in Figure 15.75. We see that the individual voltages are related to the composite ones by $v_{1a} = v_{1b} = v_1$ and $v_{2a} = v_{2b} = v_2$. The currents, however, sum: $i_1 = i_{1a} + i_{1b}$ and $i_2 = i_{2a} + i_{2b}$. Therefore, if we use the admittance parameters, we will obtain the following equations:

$$i_1 = (y_{11a} + y_{11b})v_1 + (y_{12a} + y_{12b})v_2 \qquad (15.3\text{-}39a)$$

$$i_2 = (y_{21a} + y_{21b})v_1 + (y_{22a} + y_{22b})v_2. \qquad (15.3\text{-}39b)$$

The equations, in matrix form, become

$$\begin{bmatrix} i_1 \\ i_2 \end{bmatrix} = \begin{bmatrix} (y_{11a} + y_{11b}) & (y_{12a} + y_{12b}) \\ (y_{21a} + y_{21b}) & (y_{22a} + y_{22b}) \end{bmatrix} \begin{bmatrix} v_1 \\ v_2 \end{bmatrix}. \qquad (15.3\text{-}40)$$

This proves our earlier statement: the admittance parameters of parallel-connected two-ports add to form the y parameters of the composite two-port. We note, once again, that in

Figure 15.74
Parallel-connected two-ports

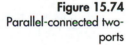

Figure 15.75
Admittance parameter model for the parallel-connected two-ports

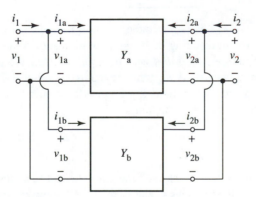

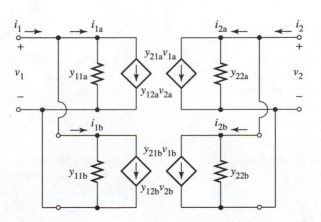

practice one often has $y_{12a} \cong 0$ and $y_{21a} \gg y_{21b}$. We will see shortly how this approximation yields a practical payoff.

As a practical example, let's consider the arrangement of Figure 15.76. There we show a two-port "terminated" at the input and output; that is, there is a nonideal source at the input and a load resistance at the output. The two-port is described by its admittance matrix Y. We will find the *current gain*

$$A_i = -\frac{i_2}{i_g}. \tag{15.3-41}$$

This is perhaps a good point to make a couple of observations. We have described the input source by its Norton equivalent. We could also have used the Thévenin equivalent; however, one often chooses the equivalent to make the mathematics simpler, and that is just what we have done here—because admittances of parallel elements add, and the admittance parameter model of Figure 15.58 is a parallel one. Our second observation is this: we have chosen to compute the current gain $-i_2/i_g$; we could just as well have chosen to compute a voltage gain, an impedance parameter, and so on. In other words, one can compute essentially any system function or response variable independently of the model chosen. We did not, however, discuss current gain in the last section, so we have chosen to deal with it here. Finally, we notice that the negative sign on i_2 in our definition of current gain is an arbitrary one, though one that is often used in practice.

Figure 15.76
An example circuit

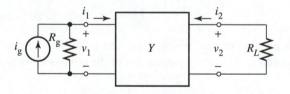

So to work! Let's replace the two-port box of Figure 15.76 with its admittance model of Figure 15.58, resulting in the configuration shown in Figure 15.77. We have also *added a current source in parallel with the load resistor*. We did this strictly for mathematical convenience, because if we combine the two parallel resistors at the input and do likewise for those at the output, and if we consider i_g and i_L to replace i_1 and i_2, we have the equivalent two-port formulation

$$\begin{bmatrix} i_g \\ i_L \end{bmatrix} = \begin{bmatrix} y_{11} + g_g & y_{12} \\ y_{21} & y_{22} + g_L \end{bmatrix} \begin{bmatrix} v_1 \\ v_2 \end{bmatrix}. \tag{15.3-42}$$

Notice that we have expressed the two external resistors in terms of their conductances. We solve for the voltages v_1 and v_L by inverting the y-parameter matrix:

$$\begin{bmatrix} v_1 \\ v_2 \end{bmatrix} = \frac{1}{\Delta} \begin{bmatrix} y_{22} + g_L & -y_{12} \\ -y_{21} & y_{11} + g_g \end{bmatrix} \begin{bmatrix} i_g \\ i_L \end{bmatrix}, \tag{15.3-43}$$

where Δ, as usual, is the determinant of the coefficient matrix in (15.3-43). Now we let $i_L = 0$, which describes the actual circuit we want to solve. We pick off the second equation in corresponding scalar form and write

$$v_2 = \frac{-y_{21}}{\Delta} i_g. \tag{15.3-44}$$

Recognizing that $v_2 = -i_2 R_L$, this equation becomes

$$A_i = -\frac{i_2}{i_g} = \frac{-y_{21}g_L}{\Delta} = \frac{-y_{21}g_L}{(y_{22} + g_L)(y_{11} + g_g) - y_{12}y_{21}}. \tag{15.3-45}$$

Let's now consider what happens in practice. The source perhaps represents a photo-diode, an electronic device that generates current in response to incident lumination. The parallel Norton equivalent resistance is very large, several hundred kilohms or several megohms. A designer then designs a current amplifier to boost up the current level. The output of the amplifier then might drive a mechanical actuator of some sort, which responds by activating when a beam of light shines on the photodiode. The actuator resistance is often quite small. The designer then takes the approach of designing "the box," the amplifier, such that y_{21} (the forward current gain) is very large and so that y_{12} is very precise. Using these assumptions in (15.3-45), we have

$$A_i = \frac{-g_L}{\dfrac{(y_{22} + g_L)(y_{11} + g_g)}{y_{21}} - y_{12}} \cong \frac{g_L}{y_{12}}. \qquad (15.3\text{-}46)$$

By setting y_{12} very precisely, then, our designer can adjust the overall current gain very accurately as well.

Figure 15.77
The admittance
parameter equivalent

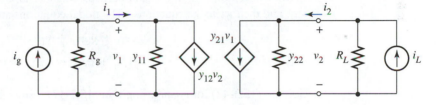

Now glance back at Figure 15.75 and the corresponding equations (15.3-39) and (15.3-40). If our designer uses a parallel connection of two two-ports, with the one on top (say) having a very large value for y_{21a} and a very small value of y_{12a}, then the corresponding composite admittance parameters will be $y_{21} \cong y_{21a}$ and $y_{12} \cong y_{12b}$. Then the top unit can be designed for $y_{21a} \to \infty$ and the bottom one for very precise y_{12b}, and (15.3-46) will describe the resulting parallel connection. In this context, by the way, the parallel configuration of two-ports is called *shunt-shunt feedback*. The following example illustrates the concept.

Example 15.16   Find the current gain $A_i = i_l/i_g$ for the circuit shown in Figure 15.78. Assume that $y_{21} \gg 1/R$ and $y_{12} \cong 0$.

Figure 15.78
An example circuit

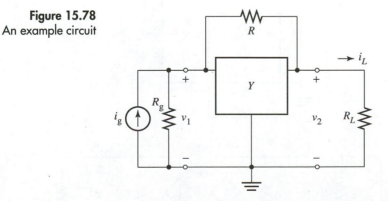

Solution   First, let's redraw the circuit as in Figure 15.79. Stop! Don't go on without spending some time on this figure! Trace out each connection, thereby making sure that the circuit is, indeed, equivalent to the one in Figure 15.78. The advantage of "bending the circuit around" as we have just done is this: it allows us to recognize that the two two-port sub-circuits are parallel connected and to identify the components that belong to each two-port.

Now we recall that we have already derived the admittance parameters for the bottom two-port in Figure 15.79. (Remember that the ground symbol shown just means that, prior to our rearrangement, we had agreed to measure all node voltages in the circuit relative to the node to which the symbol is attached. Therefore, it does not affect our two-port analysis). The two-port parameters of the feedback two-port, then, are those given in Example 15.11: $y_{11b} = y_{22b} = 1/R = g$ and $y_{12b} = y_{21b} = -1/R = -g$. The composite two-port model for our parallel connection then is the one shown in Figure 15.80, where the y_{ij} are now the *composite* parameters—the sum of the y parameters of both two-ports. Due to the approximations we were given in the problem statement of this example, we know that $y_{12} \cong y_{12b} = -g$, where $g = 1/R$, and that $y_{21} \cong y_{21a} \to \infty$ (the subscript "a" referring, as usual, to the top two-port in Figure 15.79). We can solve this circuit in a number of ways, but we will simply elect to use nodal analysis. (Note that we are not tap-ing the dependent sources simply to speed up our analysis. If you have problems follow-ing our development, you should perhaps consider taping them, calling them i_{c1} and i_{c2}, say, and then write the equations.) The nodal equations are

$$(y_{11} + g_g)v_1 + y_{12}v_2 = i_g \tag{15.3-47}$$

and
$$y_{21}v_1 + (y_{22} + g_L)v_2 = 0. \tag{15.3-48}$$

Solving (15.3-48) for v_1 and using it in (15.3-47) results in

$$v_2 = \cfrac{1}{y_{12} - \cfrac{(y_{11} + g_g)(y_{22} + g_L)}{y_{21}}}i_g \cong \frac{i_g}{y_{12}}, \tag{15.3-49}$$

because $y_{21} \to \infty$. (Alternatively, we could note that $y_{21} \to \infty$ in equation (15.3-48) im-

Figure 15.79
An equivalent circuit

Figure 15.80
The composite
admittance parameter
equivalent

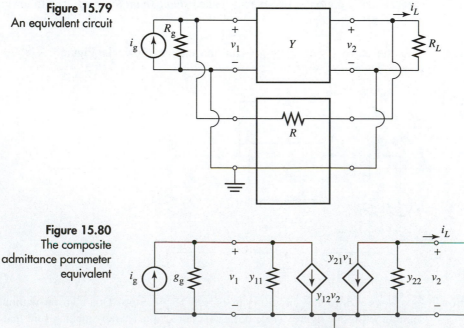

plies that $v_1 \to 0$, then use $v_1 = 0$ in equation (15.3-47)). Finally, we see that $i_L = g_L v_2$, so

$$A_i = \frac{i_L}{i_g} = \frac{g_L}{y_{12}} = \frac{R}{R_L}. \tag{15.3-50}$$

Section 15.3 Quiz

Q15.3-1. By performing direct tests, find the y parameters for the subcircuit in Figure Q15.3-1.

Figure Q15.3-1

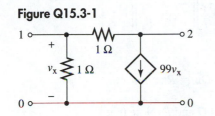

Q15.3-2. Find the z parameters for the two-port shown in Figure Q15.3-1 by inverting the Y matrix and thereby finding the Z matrix.

Q15.3-3. Find the equivalent π subcircuit (in the s domain) for the two-port shown in Figure Q15.3-2.

Figure Q15.3-2

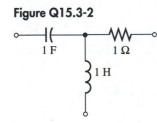

Q15.3-4. Find exact expressions for the current gain $A_i = i_0/i_g$ and the voltage gain $A_v = v_2/v_1$ for the circuit shown in Figure Q15.3-3. Then, assuming that $y_{12a} \cong 0$ and $y_{21a} \to \infty$, find the resulting values of current and voltage gain.

Figure Q15.3-3

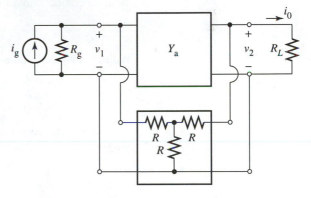

15.4 | The Transmission or Chain Parameters

In the last section we pointed out that for some two-ports neither the impedance parameters nor the admittance parameters exist. Therefore, we will develop still another set of parameters in this section. As usual, we will have the general two-port of Figure 15.81 (or the three-terminal network that results from connecting the two bottom terminals) in mind. We recall the affine relationship constraining its terminal variables:

$$a_1 v_1 + a_2 v_2 + a_3 i_1 + a_4 i_2 = 0 \tag{15.4-1a}$$
$$b_1 v_1 + b_2 v_2 + b_3 i_1 + b_4 i_2 = 0. \tag{15.4-1b}$$

Figure 15.81
The general two-port subcircuit

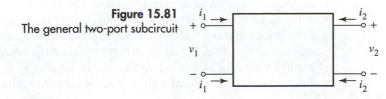

The Chain Parameters   As usual, we will separate the terminal variables into two groups of two each. For the impedance and admittance parameters, we selected the two voltages and the two currents as the two groups; the impedance parameters resulted when we solved for the voltages in terms of the currents, and the converse was true for the admittance parameters. Here, we will select the two groups in terms of the port with which they are associated. More specifically, we will select the *dependent variables* to be v_1 and i_1 and the *independent variables* to be v_2 and $-i_2$. (We change the sign on i_2 simply to be consistent with conventional usage, but we will comment on this a bit later.) Thus, we obtain

$$a_1 v_1 + a_3 i_1 = -a_2 v_2 + a_4(-i_2) \tag{15.4-2a}$$

$$b_1 v_1 + b_3 i_1 = -b_2 v_2 + b_4(-i_2). \tag{15.4-2b}$$

In matrix form,
$$\begin{bmatrix} a_1 & a_3 \\ b_1 & b_3 \end{bmatrix} \begin{bmatrix} v_1 \\ i_1 \end{bmatrix} = \begin{bmatrix} -a_2 & a_4 \\ -b_2 & b_4 \end{bmatrix} \begin{bmatrix} v_2 \\ -i_2 \end{bmatrix} \tag{15.4-3}$$

In more compact notation, we write

$$Q\bar{u} = R\bar{t}, \tag{15.4-4a}$$

with obvious definitions for the matrices Q and R. Note that

$$\bar{u} = \begin{bmatrix} v_1 \\ i_1 \end{bmatrix} \quad \text{and} \quad \bar{t} = \begin{bmatrix} v_2 \\ -i_2 \end{bmatrix}. \tag{15.4-4b}$$

Now if the determinant of Q is nonzero, that is, the matrix Q is nonsingular, we can write

$$\bar{u} = Q^{-1}R\bar{t} = T\bar{t}, \tag{15.4-5a}$$

or
$$\begin{bmatrix} v_1 \\ i_1 \end{bmatrix} = \begin{bmatrix} A & B \\ C & D \end{bmatrix} \begin{bmatrix} v_2 \\ -i_2 \end{bmatrix}. \tag{15.4-5b}$$

The matrix
$$T = \begin{bmatrix} A & B \\ C & D \end{bmatrix} \tag{15.4-6}$$

is called the *chain matrix,* or the *transmission matrix,* and the parameters A, B, C, and D are termed the *chain (or transmission, or A, B, C, D) parameters*. We are departing from the notation we established for the z and y parameters, for which we used lowercase subscripted versions of the letter we used for the matrix. The A, B, C, D notation, however, is conventional in the circuit theory literature. In nonmatrix form, we can write

$$\begin{aligned} v_1 &= Av_2 + B(-i_2) \\ i_1 &= Cv_2 + D(-i_2) \end{aligned}. \tag{15.4-7}$$

A Comment on the Port 2 Current Reference   This is perhaps the point to discuss the negative sign on i_2 a bit more fully. Referring back to Figure 15.81, note that we have defined i_2 to be going into the top terminal of port 2. Although this is conventional for any of the other parameter sets, the customary usage for the transmission parameters involves port 2 current in the opposite direction. This is due to the fact that the transmission parameters were developed to handle transmission problems in the telephone industry. There they were used to describe such things as a line filter in an audio cable. Typically, in that application, a positive input voltage v_1 resulted in positive values for i_1, and v_2, and a positive value of i_2 in an *outward* direction. Thus, in that context, i_2 was defined with its positive reference outward. In our case, we are developing all of the parameter sets together. Thus, it seems clearer to define all parameter sets relative to the same voltage and current references. Just remember that $-i_2$ is a current whose reference is outward, and you should not become confused.

Determination of the Chain Parameters: Specified Port Conditions

Let's have a look at the way we might either measure or compute the transmission parameters. Figure 15.82 shows the two-port with port 2 open-circuited, thus forcing the condition $-i_2 = 0$. In this case, we have

$$A = \left[\frac{v_1}{v_2} \right]_{-i_2=0} \tag{15.4-8}$$

and

$$C = \left[\frac{i_1}{v_2} \right]_{-i_2=0}. \tag{15.4-9}$$

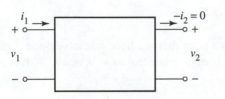

Figure 15.82
Port 2 open-circuited

Notice that we have not specified what is connected to port 1—we could drive it with a voltage source, a current source, or anything else, as long as port 2 is open-circuited. A is the reciprocal of the voltage gain from port 1 to port 2 with port 2 open-circuited; similarly, C is the reciprocal of the transfer impedance (transimpedance) from port 1 to port 2 with port 2 open-circuited. Notice, by the way, that if we take the ratio of these two parameters, we have the impedance "looking into" port 1 with port 2 open-circuited:

$$Z_{io} = \left. \frac{v_1}{i_1} \right]_{i_2=0} = \left. \frac{v_1/v_2}{i_1/v_2} \right]_{i_2=0} = \frac{A}{C}. \tag{15.4-10}$$

(Notice, by the way, that $Z_{io} = z_{11}$ and $C = 1/z_{21}$ in terms of the Z parameters.)

Figure 15.83 shows the two-port with port 2 short-circuited, thus forcing $v_2 = 0$. Here, we have

$$B = \left[\frac{v_1}{-i_2} \right]_{v_2=0} \tag{15.4-11}$$

and

$$D = \left[\frac{i_1}{-i_2} \right]_{v_2=0}. \tag{15.4-12}$$

Their ratio is the impedance "looking into" port 1 with port 2 shorted:

$$Z_{is} = \left. \frac{v_1}{i_1} \right]_{v_2=0} = \left. \frac{v_1/(-i_2)}{i_1/(-i_2)} \right]_{v_2=0} = \frac{B}{D}. \tag{15.4-13}$$

(Notice that $Z_{is} = 1/y_{11}$ and $B = -1/y_{21}$.)

Figure 15.83
Port 2 short-circuited

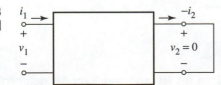

We will not present an equivalent circuit based on the chain parameters because all of them are of the "transfer" variety: they relate values of variables at one port to those at another. Thus, an equivalent subcircuit would not be too useful.

Examples: Finding the Chain Parameters

Example 15.17   Find the chain parameters for the subcircuit shown in Figure 15.84.

Figure 15.84
An example subcircuit

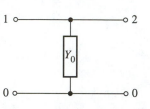

Solution   Let's run the two tests that we have just developed. Opening port 2 gives the circuit in Figure 15.85. It is easy to see that $v_2 = v_1$ and $i_1 = Y_0 v_1$, so that

$$A = 1 \tag{15.4-14}$$

and

$$C = Y_0. \tag{15.4-15}$$

The circuit shown in Figure 15.86 results when we short port 2. It is just as clear as before that now $i_1 = -i_2$ and that $v_1 = v_2 = 0$. Thus,

$$B = 0 \tag{15.4-16}$$

and

$$D = 1. \tag{15.4-17}$$

The transmission parameter matrix is

$$T = \begin{bmatrix} 1 & 0 \\ Y_0 & 1 \end{bmatrix}. \tag{15.4-18}$$

Figure 15.85
Port 2 open-circuited

Figure 15.86
Port 2 short-circuited

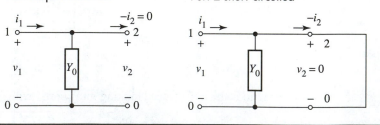

Example 15.18   Find the transmission parameter matrix for the subcircuit in Figure 15.87.

Figure 15.87
An example subcircuit

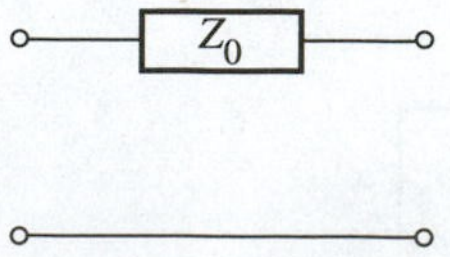

Solution   Identifying port 1 on the left and port 2 on the right, with the usual definitions of port voltages and currents, we first open-circuit port 2. This results in the circuit in Figure

15.88. Clearly $i_1 = -i_2 = 0$ and $v_1 = v_2$ (the voltage drop across Z_0 is zero), so

$$A = 1 \tag{15.4-19}$$

and

$$C = 0. \tag{15.4-20}$$

Figure 15.89 shows the same subcircuit with port 2 shorted. In this case, we still have $i_1 = -i_2$, but now $v_1 = Z_0(-i_2)$. Thus,

$$B = Z_0 \tag{15.4-21}$$

and

$$D = 1. \tag{15.4-22}$$

The chain parameter matrix is

$$T = \begin{bmatrix} 1 & Z_0 \\ 0 & 1 \end{bmatrix}. \tag{15.4-23}$$

Figure 15.88
Port 2 opened

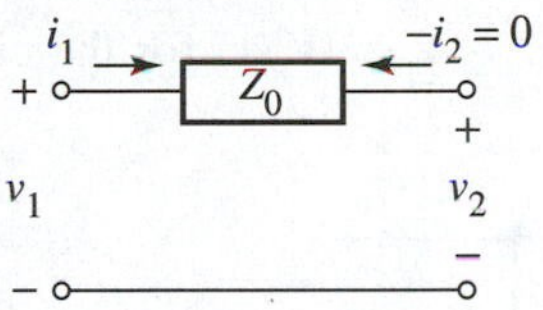

Figure 15.89
Port 2 shorted

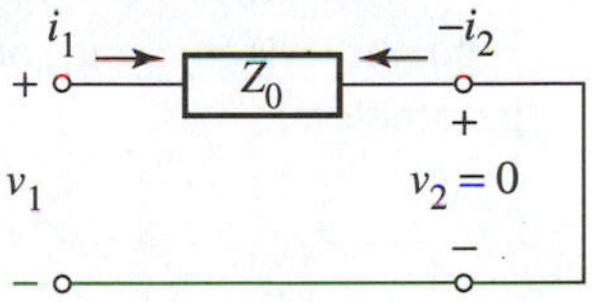

Notice that the chain parameters exist for both the subcircuits in the two preceding examples, whereas either the z or y parameters failed to exist for each (we did these examples in preceding sections—with resistors in the place of Y_0 and Z_0).

Direct Determination of the Chain Parameters

For simple circuits such as the ones we have just analyzed, it is often more efficient to forego the process of setting up the test circuits and simply do network analysis directly to express v_1 and i_1 in terms of v_2 and i_2. For instance, we show the subcircuit of Example 15.18 again in Figure 15.90 with the port variables identified. We can apply KVL and KCL at once to obtain the chain parameters by inspection:

$$v_1 = 1v_2 + Z_0(-i_2) \tag{15.4-24}$$

and

$$i_1 = 0v_2 + 1(-i_2). \tag{15.4-25}$$

In matrix form,

$$\begin{bmatrix} v_1 \\ i_1 \end{bmatrix} = \begin{bmatrix} 1 & Z_0 \\ 0 & 1 \end{bmatrix} \begin{bmatrix} v_2 \\ -i_2 \end{bmatrix}. \tag{15.4-26}$$

Figure 15.90
Example subcircuit again

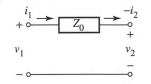

For more complicated subcircuits, such as those that contain dependent sources, for example, it is usually better to run the open and short circuit tests on port 2.

The Use of Chain Parameters to Analyze Cascade Connections of Two-Ports

We will now develop a property that makes the chain parameter matrices enormously useful in some applications. Figure 15.91 shows the *cascade (or tandem) connection* of two subcircuits, that is with port 2 of the first connected in parallel with port 1 of the second. This figure shows a very cogent reason for identifying the port 2 current to be coming out of port 2: it matches up very nicely with the sense of the port 1 current in the next subcircuit. We see at once that

$$\begin{bmatrix} v_1 \\ i_1 \end{bmatrix} = \begin{bmatrix} A & B \\ C & D \end{bmatrix} \begin{bmatrix} v_2 \\ -i_2 \end{bmatrix} = \begin{bmatrix} A & B \\ C & D \end{bmatrix} \begin{bmatrix} v_1' \\ i_1' \end{bmatrix} = \begin{bmatrix} A & B \\ C & D \end{bmatrix} \begin{bmatrix} A' & B' \\ C' & D' \end{bmatrix} \begin{bmatrix} v_2' \\ -i_2' \end{bmatrix}. \tag{15.4-27}$$

Figure 15.91
Cascade connection of two-ports

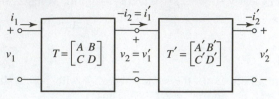

If we consider the entire cascade connection to be a *single* two-port, we see that its chain parameter matrix is simply the product of the two individual chain matrices (the subscript c stands for cascade):

$$T_c = \begin{bmatrix} A_c & B_c \\ C_c & D_c \end{bmatrix} = TT' = \begin{bmatrix} A & B \\ C & D \end{bmatrix} \begin{bmatrix} A' & B' \\ C' & D' \end{bmatrix}. \qquad (15.4\text{-}28)$$

*Example: Chain
Parameter Analysis of
a Ladder Circuit*

Example 15.19   Find the system function $H(s) = V_o(s)/V_g(s)$ for the circuit in Figure 15.92 using chain parameters.

Figure 15.92
An example circuit

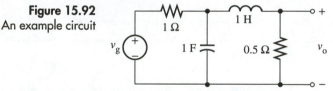

Solution   We represent the dynamic elements in terms of their impedance operators and break the circuit up into a cascade connection of four two-ports, as shown in Figure 15.93. We do not even have to indicate intermediate voltages and currents; we know that the two-ports are connected in a cascade arrangement. Further, we know that the matrices of the A, B, C, D parameters multiply to give the overall matrix for the set of four considered as a two-port in its entirety. We also can apply the results obtained in Examples 15.17 and 15.18. This gives

$$\begin{bmatrix} v_g \\ i_g \end{bmatrix} = \begin{bmatrix} 1 & 1 \\ 0 & 1 \end{bmatrix} \begin{bmatrix} 1 & 0 \\ s & 1 \end{bmatrix} \begin{bmatrix} 1 & s \\ 0 & 1 \end{bmatrix} \begin{bmatrix} 1 & 0 \\ 2 & 1 \end{bmatrix} \begin{bmatrix} v_o \\ 0 \end{bmatrix}. \qquad (15.4\text{-}29)$$

Figure 15.93
The circuit prepared
for analysis

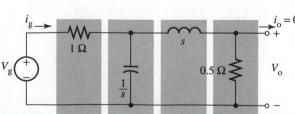

We have used the fact that $i_o = 0$ in the column matrix of port 2 variables. Now we need only multiply the matrices to obtain

$$\begin{bmatrix} v_g \\ i_g \end{bmatrix} = \begin{bmatrix} 2s^2 + 3s + 3 & s^2 + s + 1 \\ 2s^2 + s + 2 & s^2 + 1 \end{bmatrix} \begin{bmatrix} v_o \\ 0 \end{bmatrix}. \qquad (15.4\text{-}30)$$

The square (2×2) matrix is the transmission matrix for the cascade. A quick computation shows that $\Delta_T = |T| = 1$, so the two-port consisting of all elements except the source is reciprocal, as we will shortly prove. Finally, we obtain

$$v_g = (2s^2 + 3s + 3)v_o, \tag{15.4-31}$$

which we can solve to obtain

$$H(s) = \frac{v_o}{v_g} = \frac{1}{2s^2 + 3s + 3}. \tag{15.4-32}$$

Conversion of One Parameter Set to Another

In the last section, we discussed the conversion of y parameters to their z parameter counterparts and vice versa. We will do the same here. Note, however, that among three sets of parameters (z, y, and chain) there are six different transformations[9] that map one into another. Therefore, we will only develop one, the transformation from chain to z, and leave the others for you to derive.

We recall the chain parameter equations for convenience:

$$v_1 = Av_2 + B(-i_2) \tag{15.4-33a}$$

$$i_1 = Cv_2 + D(-i_2). \tag{15.4-33b}$$

Solving the second for v_2 gives

$$v_2 = \frac{1}{C}i_1 + \frac{D}{C}i_2. \tag{15.4-34}$$

Inserting this value in equation (15.4-33a) gives

$$v_1 = \frac{A}{C}i_1 + \left[\frac{AD}{C} - B\right]i_2. \tag{15.4-35}$$

Using these two expressions in the equations involving the z parameters gives

$$v_1 = z_{11}i_1 + z_{12}i_2 = \frac{A}{C}i_1 + \left[\frac{AD}{C} - B\right]i_2 \tag{15.4-36a}$$

$$v_2 = z_{21}i_1 + z_{22}i_2 = \frac{1}{C}i_1 + \frac{D}{C}i_2. \tag{15.4-36b}$$

Thus,

$$z_{11} = \frac{A}{C}, \tag{15.4-37}$$

$$z_{12} = \frac{AD}{C} - B = \frac{AD - BC}{C}, \tag{15.4-38}$$

---

[9] Each transformation maps a subject set into a different object set, the first of which can be selected in three ways, and the second of which can be selected in two ways. This gives six possible transformations.

$$z_{21} = \frac{1}{C}, \tag{15.4-39}$$

and
$$z_{22} = \frac{D}{C}. \tag{15.4-40}$$

Notice that $z_{12} = z_{21}$ if and only if $\Delta_T = AD - BC = 1$; that is, the two-port is reciprocal if and only if the determinant of the chain matrix is unity. Checking (15.4-30), only a simple polynomial multiplication was required to compute Δ_T and thus show that the two-port in our last example is reciprocal—as we noted there.

Example 15.20 Find the chain parameters for the subcircuit in Figure 15.94 and use them to check for reciprocity. Then find the z parameters by transforming the chain parameters.

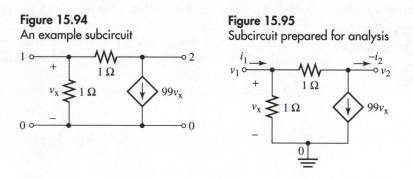

Figure 15.94
An example subcircuit

Figure 15.95
Subcircuit prepared for analysis

Solution This subcircuit is one for which we computed the z parameters in Section 15.2 and for which you were asked to compute the y parameters in the quiz at the end of Section 15.3; therefore, we will be able to check our work without too much difficulty. This circuit is relatively uncomplicated, so let's define our terminal variables and attempt to solve for the transmission parameters directly. We will refer to Figure 15.95 for our analysis. Our plan is to perform a nodal analysis for the controlling variable $v_1 = v_x$, so we have chosen our ground reference at the bottom in that figure. In terms of v_1, we have the following KCL equation at the 1-node:

$$\frac{v_1}{1} = v_1 = i_1 - (-i_2) - 99v_1. \tag{15.4-41}$$

Thus
$$v_1 = 0.01(i_1 + i_2). \tag{15.4-42}$$

However,
$$v_2 = v_1 - 1(-i_2 + 99v_1). \tag{15.4-43}$$

Solving the last two equations simultaneously for v_1 and i_1 gives

$$\begin{bmatrix} v_1 \\ i_1 \end{bmatrix} = \begin{bmatrix} -\dfrac{1}{98} & -\dfrac{1}{98} \\[2mm] -\dfrac{100}{98} & -\dfrac{2}{98} \end{bmatrix} \begin{bmatrix} v_2 \\ -i_2 \end{bmatrix}. \tag{15.4-44}$$

Thus, $A = B = -1/98$, $C = -100/98$, and $D = -2/98$. A quick computation shows that $A_T = AD - BC = -1/98 \neq 1$. Thus, the subcircuit is not reciprocal. Using the conversion equations (15.4-37) through (15.4-40) just derived, we have

$$z_{11} = \frac{A}{C} = 0.01\ \Omega, \tag{15.4-45}$$

$$z_{12} = \frac{AD - BC}{C} = \frac{-1/98}{-100/98} = 0.01 \ \Omega, \tag{15.4-46}$$

$$z_{21} = \frac{1}{C} = -0.98 \ \Omega, \tag{15.4-47}$$

and

$$z_{22} = \frac{D}{C} = 0.02 \ \Omega. \tag{15.4-48}$$

If you check the results of Example 15.8 in Section 15.2, you will see that these values are the same. Because $AD - BC = -1/98 \neq 1$, we know the subcircuit is not reciprocal, and this is validated by the fact that $z_{12} \neq z_{21}$ above.

The Reverse Chain Parameters

Before we finish this section, we will address one more topic. If we solve equation (15.4-4) for $\bar{t} = \begin{bmatrix} v_2 \\ -i_2 \end{bmatrix}$ instead of $\bar{u} = \begin{bmatrix} v_1 \\ i_i \end{bmatrix}$ as we did in developing the chain parameters, we will replace equation (15.4-5) by

$$\bar{t} = R^{-1}Q\bar{u} = [Q^{-1}R]^{-1}\bar{u} = T^{-1}\bar{u}. \tag{15.4-49}$$

Just as the admittance matrix is the inverse of the impedance matrix, this equation defines the *reverse chain matrix,* or the *reverse chain parameters.* In terms of port variables, we can write

$$\begin{bmatrix} v_2 \\ i_2 \end{bmatrix} = \begin{bmatrix} A_r & B_r \\ C_r & D_r \end{bmatrix} \begin{bmatrix} v_1 \\ -i_1 \end{bmatrix} = \begin{bmatrix} \dfrac{D}{\Delta_T} & \dfrac{B}{\Delta_T} \\[2mm] \dfrac{C}{\Delta_T} & \dfrac{A}{\Delta_T} \end{bmatrix} \begin{bmatrix} v_1 \\ -i_1 \end{bmatrix}. \tag{15.4-50}$$

Notice that we have reversed signs on both i_1 and i_2 relative to the "forward" chain parameter variables. The reverse chain parameters are not often used, so we will not work with them here.

Section 15.4 Quiz

Q15.4-1.   Find the chain parameters for the subcircuit in the box in Figure Q15.4-1. Use them to find the value of v_o if $i_s = 24$ A.

Figure Q15.4-1

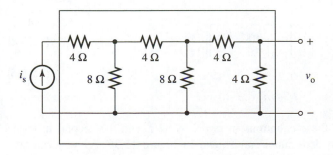

Q15.4-2.   Find $H(s)$ in the circuit shown in Figure Q15.4-2 relative to the response waveform $v_o(t)$ using chain parameters.

Figure Q15.4-2

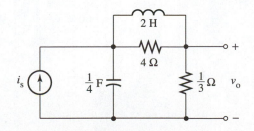

Q15.4-3.   Using the chain parameters derived in Question Q15.4-2, show that the circuit in Figure Q15.4-2 is reciprocal. Use the chain parameters to find the impedance parameters.

15.5 | Hybrid Parameters

By now we are very familiar with the affine relations for the general two-port network without internal sources:

$$a_1 v_1 + a_2 v_2 + a_3 i_1 + a_4 i_2 = 0 \tag{15.5-1a}$$

$$b_1 v_1 + b_2 v_2 + b_3 i_1 + b_4 i_2 = 0. \tag{15.5-1b}$$

For the z parameters, we selected the two voltages as the dependent variables and the two currents as the independent variables. For the y parameters, conversely, we chose the two currents to be dependent and the two voltages to be independent. In the case of the ordinary transmission or chain parameters, we selected the port 1 variables to be dependent and the port 2 variables to be independent, and conversely for the reverse chain parameters.

A Generalized Notational Scheme for Two-Port Parameters

Let's generalize our matrix equations now by writing

$$\begin{bmatrix} w_1 \\ w_2 \end{bmatrix} = \begin{bmatrix} m_{11} & m_{12} \\ m_{21} & m_{22} \end{bmatrix} \begin{bmatrix} u_1 \\ u_2 \end{bmatrix}, \tag{15.5-2}$$

or

$$\overline{w} = M\overline{u}, \tag{15.5-3}$$

where $\overline{w}$ and $\overline{u}$ are column matrices, each consisting of two port variables. In how many ways can we select these variables? We can select the top element of, say $\overline{u}$, in four ways because there are four different port variables (v_1, i_1, v_2, i_2) from which to choose. Once it is selected, there are three others from which to choose the bottom element. Thus, we can select $\overline{u}$ in 12 ways; however, there is no intrinsic difference between, for example, $\overline{u}_1 = \begin{bmatrix} v_1 \\ i_1 \end{bmatrix}$ and $\overline{u}_2 = \begin{bmatrix} i_1 \\ v_1 \end{bmatrix}$. The difference only amounts to renaming the entries in the M matrix (actually exchanging column 1 of that matrix with column 2). Thus, we must divide the number we arrived at above by 2, giving 6 different choices. As for $\overline{w}$, it is clear that after we have chosen the $\overline{u}$ variables the $\overline{w}$ variables are determined. Thus, our conclusion is this: there are exactly six different sets of two-port parameters (with the above-mentioned exceptions of trivial interchanges).

The Hybrid Parameters

We have, at this point, developed four of the six sets: impedance, admittance, chain, and reverse chain. We will now complete the sets of two-port parameters by selecting the variables in $\overline{u}$ such that there is one voltage variable and one current variable, each relative to a *different* port. For this reason, they will be called *hybrid parameters*.

The Hybrid h Parameters

The first set of hybrid parameters, called the *hybrid h parameters*, results when we choose $\overline{u} = \begin{bmatrix} i_1 \\ v_2 \end{bmatrix}$. We then write

$$\begin{bmatrix} v_1 \\ i_2 \end{bmatrix} = \begin{bmatrix} h_{11} & h_{12} \\ h_{21} & h_{22} \end{bmatrix} \begin{bmatrix} i_1 \\ v_2 \end{bmatrix}, \tag{15.5-4}$$

or

$$v_1 = h_{11} i_1 + h_{12} v_2 \tag{15.5-5}$$

$$i_2 = h_{21} i_1 + h_{22} v_2 .$$

We can interpret the first of these equations in terms of KVL and the second in terms of KCL to give us the equivalent subcircuit shown in Figure 15.96. As you can see,

Figure 15.96
The hybrid h-parameter
equivalent subcircuit

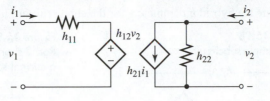

it is a *hybridized* version of the impedance and admittance equivalents, thus giving additional justification for our choice of names for these parameters. By the way, notice that, just as we did for the admittance equivalent, we are labeling *admittances,* wherever they occur, rather than following our previous custom of always labeling with impedances (for example, we have not inverted h_{22} in the diagram to make it an impedance).

In order to test a two-port under the assumption that the hybrid h parameter matrix exists, we note that it is the port 1 current and the port 2 voltage that are independent, so we drive the subcircuit with the test sources shown in Figure 15.97. We can then derive h_{11} and h_{21} by deactivating the voltage source, and, similarly, we can determine h_{12} and h_{22} by deactivating the current source. Using these conditions in equation (15.5-5), we find that

$$h_{11} = \left[\frac{v_1}{i_1}\right]_{v_2=0}, \tag{15.5-6}$$

$$h_{21} = \left[\frac{i_2}{i_1}\right]_{v_2=0}, \tag{15.5-7}$$

$$h_{12} = \left[\frac{v_1}{v_2}\right]_{i_1=0}, \tag{15.5-8}$$

and

$$h_{22} = \left[\frac{i_2}{v_2}\right]_{i_1=0}. \tag{15.5-9}$$

Figure 15.97
Testing the two-port

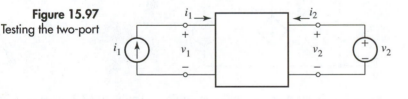

If you refer back to the appropriate sections, you will see that these results can be expressed in terms of various other parameters we have studied:

$$h_{11} = \frac{1}{y_{11}}, \tag{15.5-10}$$

$$h_{21} = \frac{1}{C}, \tag{15.5-11}$$

$$h_{12} = \frac{1}{A_r}, \tag{15.5-12}$$

and

$$h_{22} = \frac{1}{z_{22}}. \tag{15.5-13}$$

Example 15.21   Compute the hybrid h parameters for the subcircuit in Figure 15.98.

Figure 15.98
An example subcircuit

Figure 15.99
Port 2 shorted, port 1 driven by
a current source

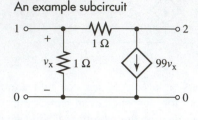

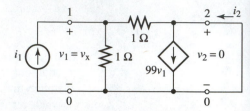

Solution   Let's test this subcircuit by first making $v_2 = 0$; that is, by shorting the right-hand port. This produces the circuit shown in Figure 15.99. We see that a simple application of parallel resistances gives

$$v_1 = 0.5i_1 \tag{15.5-14}$$

so

$$h_{11} = 0.5 \ \Omega. \tag{15.5-15}$$

Application of KCL at the terminal labeled 2 gives

$$i_2 = -0.5i_1 + 99 \times 0.5i_1 = 49i_1. \tag{15.5-16}$$

Thus,

$$h_{21} = 49. \tag{15.5-17}$$

If we test the subcircuit by deactivating i_1 (that is, open-circuiting port 1) and driving port 2 with a voltage source, we get the circuit shown in Figure 15.100. Then, by voltage division applied to the two resistors, we see that

$$v_1 = 0.5v_2, \tag{15.5-18}$$

so

$$h_{12} = 0.5. \tag{15.5-19}$$

Figure 15.100
Port 1 opened, port 2
driven by a voltage
source

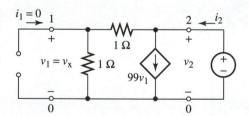

Notice that h_{12}, being a voltage ratio, is unitless. Applying KCL once again at terminal 2, we have

$$i_2 = 0.5v_2 + 99 \times 0.5v_2 = 50v_2. \tag{15.5-20}$$

Hence,

$$h_{22} = 50 \ \text{S (siemens)}. \tag{15.5-21}$$

The hybrid h-parameter equivalent subcircuit is shown in Figure 15.101.

Figure 15.101
The hybrid h-parameter
equivalent subcircuit

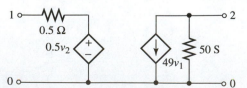

Transformation Between Hybrid *h* and Other Parameter Sets

Let's discuss in a bit more detail how to relate the hybrid *h* parameters to the others we have derived. Recall that we have already derived the transformation that maps the *z* parameters into the *y* parameters, and vice versa. We also related the chain parameters to the *z* parameters, leaving to you the task of relating them to the *y* parameters. In fact, one can transform from any of the six parameter sets to any other, provided that the resulting parameter set exists. Here we will derive the transformation for converting from the hybrid *h* parameters to the *z* parameters, as usual leaving the work of deriving the other transformations up to you.

For convenience, we repeat the equations for the hybrid *h* parameters:

$$v_1 = h_{11}i_1 + h_{12}v_2 \tag{15.5-22a}$$

$$i_2 = h_{21}i_1 + h_{22}v_2. \tag{15.5-22b}$$

Solving the second of these for v_2 results in

$$v_2 = -\frac{h_{21}}{h_{22}}i_1 + \frac{1}{h_{22}}i_2. \tag{15.5-23}$$

Using this result in equation (15.5-22a) gives

$$v_1 = \left[h_{11} - \frac{h_{12}h_{21}}{h_{22}}\right]i_1 + \frac{h_{12}}{h_{22}}i_2. \tag{15.5-24}$$

Thus, we have

$$z_{11} = h_{11} - \frac{h_{12}h_{21}}{h_{22}} = \frac{\Delta_h}{h_{22}}, \tag{15.5-25}$$

$$z_{12} = \frac{h_{12}}{h_{22}}, \tag{15.5-26}$$

$$z_{21} = -\frac{h_{21}}{h_{22}}, \tag{15.5-27}$$

and

$$z_{22} = \frac{1}{h_{22}}. \tag{15.5-28}$$

These equations, by the way, show that *a network is reciprocal if and only if* $h_{12} = -h_{21}$. The negative sign only has to do with our reference direction for the terminal currents. Otherwise, it is of the same form as $z_{12} = z_{21}$.

The Hybrid *g* Parameters

Our final parameter set consists of the *hybrid g parameters*. They are quite similar to the hybrid *h* parameters; for this reason, we will not develop them as fully. We select our equations according to

$$\begin{bmatrix} i_1 \\ v_2 \end{bmatrix} = \begin{bmatrix} g_{11} & g_{12} \\ g_{21} & g_{22} \end{bmatrix} \begin{bmatrix} v_1 \\ i_2 \end{bmatrix}, \tag{15.5-29}$$

or

$$i_1 = g_{11}v_1 + g_{12}i_2 \tag{15.5-30a}$$

$$v_2 = g_{21}v_1 + g_{22}i_2. \tag{15.5-30b}$$

Interpreting equation (15.5-30a) as a KCL equation and equation (15.5-30b) in terms of KVL gives the equivalent subcircuit shown in Figure 15.102. Notice that it, too, is a hy-

Figure 15.102
The hybrid g-parameter
equivalent subcircuit

Figure 15.103
Testing the two-port

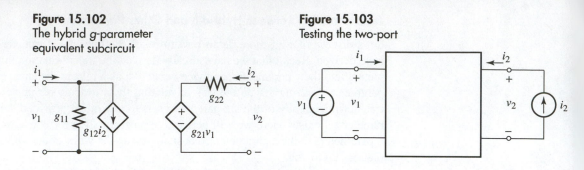

bridization of the z and y parameter equivalents. Again, we are labeling each parameter with either admittance or impedance, as appropriate.

When we write equations (15.5-29) and (15.5-30) we are assuming that v_1 and i_2 are independent variables; therefore, we can test the subcircuit as shown in Figure 15.103. Using zero values for v_1 and i_2 alternately in equations (15.5-30) gives

$$g_{11} = \left[\frac{i_1}{v_1}\right]_{i_2=0}, \qquad (15.5\text{-}31)$$

$$g_{21} = \left[\frac{v_2}{v_1}\right]_{i_2=0}, \qquad (15.5\text{-}32)$$

$$g_{12} = \left[\frac{i_1}{i_2}\right]_{v_1=0}, \qquad (15.5\text{-}33)$$

Table 15.1
The six two-port
parameter sets

Name of Parameters	Defining Equations	Reciprocity Condition	Equivalent Subcircuit
Impedance	$\begin{bmatrix} v_1 \\ v_2 \end{bmatrix} = \begin{bmatrix} z_{11} & z_{12} \\ z_{21} & z_{22} \end{bmatrix}\begin{bmatrix} i_1 \\ i_2 \end{bmatrix}$	$z_{12} = z_{21}$	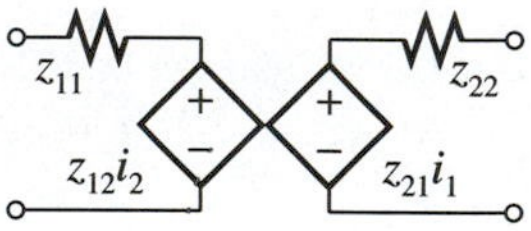
Admittance	$\begin{bmatrix} i_1 \\ i_2 \end{bmatrix} = \begin{bmatrix} y_{11} & y_{12} \\ y_{21} & y_{22} \end{bmatrix}\begin{bmatrix} v_1 \\ v_2 \end{bmatrix}$	$y_{12} = y_{21}$	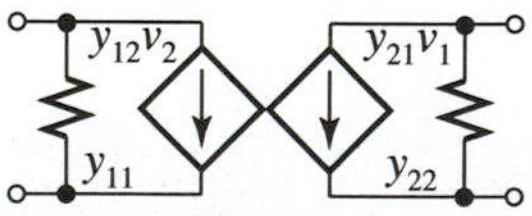
Hybrid h	$\begin{bmatrix} v_1 \\ i_2 \end{bmatrix} = \begin{bmatrix} h_{11} & h_{12} \\ h_{21} & h_{22} \end{bmatrix}\begin{bmatrix} i_1 \\ v_2 \end{bmatrix}$	$h_{12} = -h_{21}$	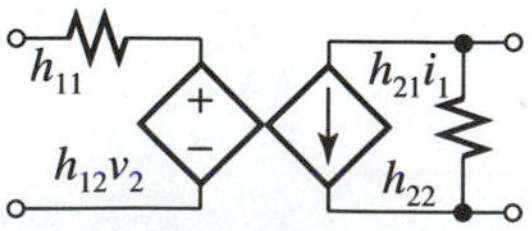
Hybrid g	$\begin{bmatrix} i_1 \\ v_2 \end{bmatrix} = \begin{bmatrix} g_{11} & g_{12} \\ g_{21} & g_{22} \end{bmatrix}\begin{bmatrix} v_1 \\ i_2 \end{bmatrix}$	$g_{12} = -g_{21}$	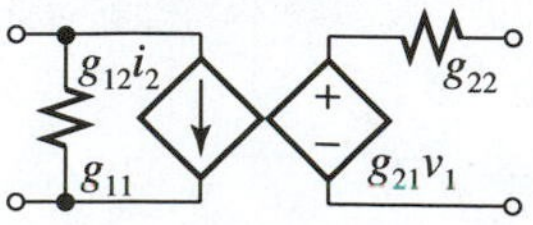
Chain	$\begin{bmatrix} v_1 \\ i_1 \end{bmatrix} = \begin{bmatrix} A & B \\ C & D \end{bmatrix}\begin{bmatrix} v_2 \\ -i_2 \end{bmatrix}$	$\Delta T = AD - BC = 1$	
Reverse Chain	$\begin{bmatrix} v_2 \\ i_2 \end{bmatrix} = \begin{bmatrix} A_r & B_r \\ C_r & D_r \end{bmatrix}\begin{bmatrix} v_1 \\ -i_1 \end{bmatrix}$	$\Delta T_r = A_r D_r - B_r C_r = 1$	

and
$$g_{22} = \left[\frac{v_2}{i_2}\right]_{v_1=0}.$$
(15.5-34)

Table 15.1 summarizes all of our parameter sets thus far: the defining matrix equation, the reciprocity condition, and the equivalent subcircuit. If the subcircuit has three terminals, of course, then the bottom two of each should be connected together.

Hybrid Parameter Port Connections

In Figure 15.104 we show two different connections of two-ports: one with a parallel connection at the output and a series connection at the input and the other with these connections reversed. If you reflect for just a moment, looking at the equivalent circuit summaries in Table 15.1, you will see that the hybrid h parameters add for the connection of Figure 15.104(a) and that the hybrid g parameters add for the connection in Figure 15.104(b). We will leave the verification of this statement as an exercise. We remind you that "shunt" is a synonym for "parallel."

Figure 15.104
Two two-port connections

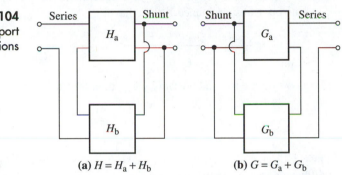

(a) $H = H_a + H_b$ (b) $G = G_a + G_b$

Circuit Analysis with the Hybrid h Parameters

As an illustration of the usefulness of our two last parameter sets, let's look at the circuit shown in Figure 15.105. As we have emphasized before, one can apply any of the parameter sets we have covered to analyze a circuit for any variable of interest, but it happens that some are more convenient for a given computation than others. In this case, we will use the hybrid h parameters to compute the *transadmittance* $Y_m = -i_2/v_g$. Again, the negative sign in our definition is a matter of choice. We now "unplug" our box symbol for the two-port and replace it with the hybrid h-parameter equivalent circuit, resulting in the circuit shown in Figure 15.106. We have added a current source connected in parallel with the load resistor to make the resulting circuit symmetric. This simplifies the analy-

Figure 15.105
An example circuit

Figure 15.106
The hybrid h-parameter equivalent

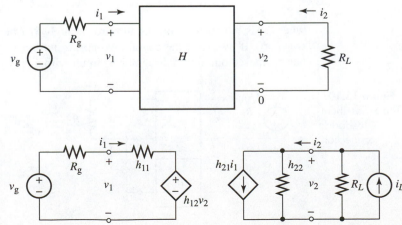

sis—as you will now see. We note that the source resistor and h_{11} add together (both are impedances), and h_{22} and $g_L = 1/R_L$ add (both are admittances). We will leave it to you to show that h_{12} and h_{21} are unaffected if we treat everything except the two independent sources as a composite two-port. We can then write the hybrid h-parameter equations for this resulting two-port in matrix form as

$$\begin{bmatrix} v_g \\ i_L \end{bmatrix} = \begin{bmatrix} h_{11} + R_g & h_{12} \\ h_{21} & h_{22} + g_L \end{bmatrix}\begin{bmatrix} i_1 \\ v_2 \end{bmatrix} \tag{15.5-35}$$

Solving, we get

$$\begin{bmatrix} i_1 \\ v_2 \end{bmatrix} = \frac{1}{\Delta}\begin{bmatrix} h_{22} + g_L & -h_{12} \\ -h_{21} & h_{11} + R_g \end{bmatrix}\begin{bmatrix} v_g \\ i_L \end{bmatrix}, \tag{15.5-36}$$

where $\Delta_h = (h_{11} + R_g)(h_{22} + g_L) - h_{12}h_{21}$ is the determinant of the square matrix in (15.5-35).

At this point, we simply let $i_L = 0$, thus replacing it with an open circuit in Figure 15.106. This results in a valid equivalent for our original circuit. We can now compute our transadmittance Y_m. We have

$$Y_m = \frac{-i_2}{v_g} = \frac{g_L v_2}{v_g} = \frac{-h_{21}g_L}{\Delta_h} = \frac{-h_{21}g_L}{(h_{11} + R_g)(h_{22} + g_L) - h_{12}h_{21}}. \tag{15.5-37}$$

If we now divide top and bottom by h_{21}, then let $h_{21} \to \infty$, we see that we will have

$$Y_m = -\frac{g_L}{h_{12}}. \tag{15.5-38}$$

Thus, in a practical setting, if we make h_{21} very large and h_{12} (and g_L) very precise, we will have a very accurate value for Y_m.

The Effect of Feedback on the Hybrid h Parameters

As you might suspect from the work we have already done with other parameter sets, we will achieve these goals with feedback. This time, it will be the connection that is so convenient to use with the hybrid h parameters: the series-shunt configuration of Figure 15.104(a). We simply replace the box in Figure 15.105 with the composite two-port of Figure 15.104(a), resulting in the circuit shown in Figure 15.107. We know that the hybrid h parameters add, so the analysis we have just done still works with each parameter replaced by the sum of the corresponding h parameters of the individual two-ports. If we therefore make $h_{12a} \cong 0$ and let $h_{21a} \to \infty$, we still have

$$Y_m = \frac{-i_2}{v_g} = -\frac{g_L}{h_{12b}}, \tag{15.5-39}$$

where i_2 is the current flowing upward through R_L (the port 2 current of the composite two-port). If we select the bottom two-port to have a very precise value for h_{12b}, we will have achieved our aim of designing a circuit with a very precise value of Y_m.

Figure 15.107
The series-shunt feedback connection

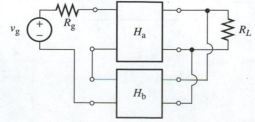

Example 15.22   Find the voltage gain $A_v = v_2/v_g$ for the circuit shown in Figure 15.108, then evaluate its value for $h_{12a} \cong 0$ and $h_{21a} \to \infty$.

Figure 15.108
An example circuit

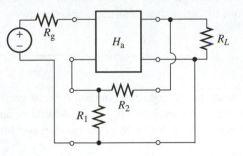

Solution   The hybrid h parameters for the box are assumed to be known. Therefore, we need only to find those for the resistor divider circuit on the bottom of the diagram. The best procedure in a case such as this is to write down the equations for the h parameters to refresh our memory. They are

$$v_1 = h_{11}i_1 + h_{12}v_2 \qquad (15.5\text{-}40a)$$

$$i_2 = h_{21}i_1 + h_{22}v_2. \qquad (15.5\text{-}40b)$$

We see that if we force v_2 to zero by shorting the right-hand terminals and apply an i-source to those on the left, we have the circuit shown in Figure 15.109, furthermore, equations (15.5-40) give us the values of h_{11b} and h_{21b}:

$$h_{11b} = \frac{v_1}{i_1}\bigg]_{v_2=0} = R_1 \parallel R_2 = \frac{R_1 R_2}{R_1 + R_2}, \qquad (15.5\text{-}41a)$$

and

$$h_{21b} = \frac{i_2}{i_1}\bigg]_{v_2=0} = -\frac{R_1}{R_1 + R_2}. \qquad (15.5\text{-}41b)$$

Figure 15.109
Testing the feedback
two-port

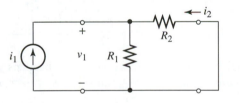

If we now apply a test voltage source to the right-hand terminals and open circuit those on the left, we obtain the test circuit shown in Figure 15.110. Equations (15.5-40) now give us the values of h_{12} and h_{22}:

$$h_{22b} = \frac{i_2}{v_2}\bigg]_{i_1=0} = \frac{1}{R_1 + R_2}, \qquad (15.5\text{-}42a)$$

and

$$h_{12b} = \frac{v_1}{v_2}\bigg]_{i_1=0} = \frac{R_1}{R_1 + R_2}. \qquad (15.5\text{-}42b)$$

Figure 15.110
The other test

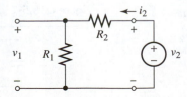

Now let's draw an equivalent circuit, recalling that the h parameters of the two two-ports add, to obtain the configuration in Figure 15.111. Be careful to keep in mind the fact that each of the h parameters is the sum of the h parameters of the two individual subcircuits. Now, by straightforward application of KVL, KCL, Ohm's law, current division, etc., we have

$$v_2 = -h_{21} \times \frac{1}{h_{22} + g_L} \times \frac{v_g - h_{12}v_2}{R_g + h_{11}}. \tag{15.5-43}$$

It is necessary in making such computations to keep in mind the dimensions of the various parameters; for example, h_{22} is an admittance and so it sums with $g_L = 1/R_L$ to produce the equivalent admittance of the parallel combination, and so on. Solving (15.5-43), we get

$$A_v = \frac{v_2}{v_g} = \frac{-h_{21}}{(h_{22} + g_L)(h_{11} + R_g) - h_{12}h_{21}}, \tag{15.5-44}$$

where the h parameters are the sum of those of the top two-port in Figure 15.107 (or 15.108) and those of the voltage divider given in equations (15.5-41) and (15.5-42). We will leave it to you to write out the (somewhat ghastly) complete expression after substitution. We will point out that we have already simplified our gain expression by noting that the h parameters add, and we will do even more. If you look carefully at Figure 15.111 you will see that R_g is connected in series with the input ports of both two-ports and that R_L is connected in parallel with both output ports. Thus, we can add R_g to h_{11} and g_L to h_{22} and form a composite overall two-port having h parameters that are simple sums of those of the two-ports and R_g and g_L. We can then see that the voltage gain expression in (15.5-44) is simply the open circuit voltage ratio of the composite two-port.

Figure 15.111
Composite h-parameter model

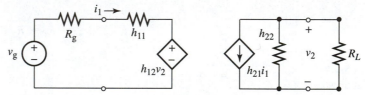

To complete our example, we ask what happens if we assume that $h_{12a} \cong 0$ and $h_{21a} \to \infty$. In this event, $h_{12} \cong h_{12b}$ and $h_{21} \cong h_{21a} \to \infty$, and equation (15.5-44) gives

$$A_v = \frac{1}{h_{12b}} = \frac{R_1 + R_2}{R_1} = 1 + \frac{R_2}{R_1}. \tag{15.5-45}$$

Does this expression look familiar? It should, for it is the voltage gain of the noninverting voltage amplifier topology using an operational amplifier that we investigated in Chapter 5.

Section 15.5 Quiz

Q15.5-1. Derive the z parameters in terms of the hybrid g parameters and find the condition on the latter for reciprocity.

Q15.5-2. Find the hybrid h parameters for the two-port tuned circuit shown in Figure Q15.5-1.

Figure Q15.5-1

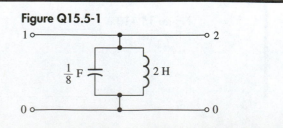

Q15.5-3. Find the hybrid *g* parameters for the two-port tuned circuit shown in Figure Q15.5-1.

Q15.5-4. Verify the matrix relations shown in Figure 15.104.

15.6 | Analysis of Feedback Circuits

Feedback Connections: Summary

We have considered feedback connections on several occasions in this chapter; now, however, we will tie things together by analyzing feedback circuits on a more unified basis. Before doing this, however, we draw your attention to Table 15.1 of the last section, which enumerated all of the parameter set equations and the corresponding equivalent subcircuits. We present it again here for ease of reference as Table 15.2, with the two chain parameter sets deleted because we will not be using them here.

Table 15.2
Four two-port parameter sets

Name of Parameters	Defining Equations	Reciprocity Condition	Equivalent Subcircuit
Impedance	$\begin{bmatrix} v_1 \\ v_2 \end{bmatrix} = \begin{bmatrix} z_{11} & z_{12} \\ z_{21} & z_{22} \end{bmatrix} \begin{bmatrix} i_1 \\ i_2 \end{bmatrix}$	$z_{12} = z_{21}$	
Admittance	$\begin{bmatrix} i_1 \\ i_2 \end{bmatrix} = \begin{bmatrix} y_{11} & y_{12} \\ y_{21} & y_{22} \end{bmatrix} \begin{bmatrix} v_1 \\ v_2 \end{bmatrix}$	$y_{12} = y_{21}$	
Hybrid *h*	$\begin{bmatrix} v_1 \\ i_2 \end{bmatrix} = \begin{bmatrix} h_{11} & h_{12} \\ h_{21} & h_{22} \end{bmatrix} \begin{bmatrix} i_1 \\ v_2 \end{bmatrix}$	$h_{12} = -h_{21}$	
Hybrid *g*	$\begin{bmatrix} i_1 \\ v_2 \end{bmatrix} = \begin{bmatrix} g_{11} & g_{12} \\ g_{21} & g_{22} \end{bmatrix} \begin{bmatrix} v_1 \\ i_2 \end{bmatrix}$	$g_{12} = -g_{21}$	

Generalized Analysis of Two-Ports

Now let's look once more at the general two-port shown in Figure 15.112. We have chosen to be noncommittal about the two-port parameters. We will allow the matrix *M* to be any of the matrices *Z*, *Y*, *H*, or *G*, with the m_{ij} being z_{ij}, y_{ij}, h_{ij}, or g_{ij}, respectively. (That is, we will not consider either of the chain parameter sets.) The two-port equations then

Figure 15.112
The general two-port subcircuit

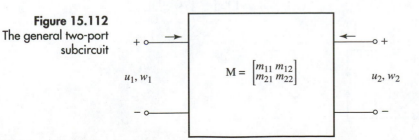

$$M = \begin{bmatrix} m_{11} & m_{12} \\ m_{21} & m_{22} \end{bmatrix}$$

u_1, w_1 u_2, w_2

read

$$\overline{w} = M\overline{u},$$ (15.6-1)

or

$$\begin{bmatrix} w_1 \\ w_2 \end{bmatrix} = \begin{bmatrix} m_{11} & m_{12} \\ m_{21} & m_{22} \end{bmatrix} \begin{bmatrix} u_1 \\ u_2 \end{bmatrix}.$$ (15.6-2)

We are considering u_1 and u_2 to be the independent variables and w_1 and w_2 the dependent variables. The subscripts refer to the port numbers, as usual: 1 means input port and 2 means output port. In scalar equation form, we write

$$w_1 = m_{11}u_1 + m_{12}u_2$$ (15.6-3a)

$$w_2 = m_{21}u_1 + m_{22}u_2.$$ (15.6-3b)

The Limiting Case: $m_{21} \to \infty$ Now let's see what happens when we allow m_{21} to become infinitely large, holding all others at their original finite values. If we divide the second equation by m_{21}, we have

$$\frac{1}{m_{21}} w_2 = u_1 + \frac{m_{22}}{m_{21}} u_2.$$ (15.6-4)

If we now let $m_{21} \to \infty$, we will have

$$u_1 = 0.$$ (15.6-5)

Using this value in equation (15.6-3a), we get

$$w_1 = m_{12}u_2,$$ (15.6-6)

which we will rearrange into

$$u_2 = \frac{1}{m_{12}} w_1.$$ (15.6-7)

Our two-port with $m_{21} \to \infty$ is now described by the equation

$$\begin{bmatrix} u_1 \\ u_2 \end{bmatrix} = \begin{bmatrix} 0 & 0 \\ 1/m_{12} & 0 \end{bmatrix} \begin{bmatrix} w_1 \\ w_2 \end{bmatrix}.$$ (15.6-8)

We will now interpret this result with equivalent two-port models, depending on the choices for the various variables.

Case 1 ($u_1 = i_1 = 0$): For this case we must clearly have $w_1 = v_1$. The input port must be an open circuit, and there are two different subcases, depending upon our choice for output variables.

   Subcase 1a ($u_2 = v_2$ and $w_2 = i_2$ ($M = H$): Then the two-port shown in Figure 15.113 is equivalent to the original one of Figure 15.112 with $\mu = 1/m_{12}$. Equation (15.6-8) now has the form

$$\begin{bmatrix} i_1 \\ v_2 \end{bmatrix} = \begin{bmatrix} 0 & 0 \\ \mu & 0 \end{bmatrix} \begin{bmatrix} v_1 \\ i_2 \end{bmatrix}.$$ (15.6-9)

Figure 15.113
A VCVS two-port

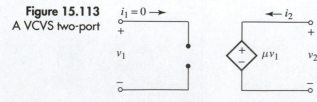

If you inspect Table 15.2, you will see that this is the hybrid g parameter representation under the condition that $h_{21} \rightarrow \infty$ and $h_{12} = 1/\mu$. (Notice that the subscripts do not match the original m parameters. we are *reinterpreting* our equations, now that we have chosen the voltage and current variables. Thus, $g_{21} = \mu = 1/m_{12}$ and $g_{12} = 0$.)

Subcase 1b ($u_2 = i_2$ and $w_2 = v_2$ ($M = Z$)): Then the two-port shown in Figure 15.114 is equivalent to the original one with $g_m = 1/m_{12}$. Our general two-port matrix relation of equation (15.6-8) now takes on the special form

$$\begin{bmatrix} i_1 \\ i_2 \end{bmatrix} = \begin{bmatrix} 0 & 0 \\ g_m & 0 \end{bmatrix} \begin{bmatrix} v_1 \\ v_2 \end{bmatrix}. \tag{15.6-10}$$

This you will recognize is merely the y parameter description of our two-port under the condition that $z_{21} \rightarrow \infty$.

Figure 15.114
A VCCS two-port

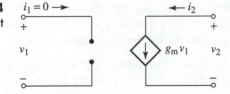

Now let's change our major cases (our choice of input controlling variable u_1).

Case 2 ($u_1 = v_1 = 0$): Again, no choice is left for w_1: it has to be i_1. The input must, this time, be a short circuit, and again there are two cases to consider corresponding to our choice of output variables.

Subcase 2a: Let's pick $u_2 = i_2$ and $w_2 = v_2$ ($M = G$) as in Subcase 1b. Then the two-port shown in Figure 15.115 is equivalent to the original one in Figure 15.112 with $\beta = 1/m_{12}$. The matrix relationship in equation (15.6-8) becomes

$$\begin{bmatrix} v_1 \\ i_2 \end{bmatrix} = \begin{bmatrix} 0 & 0 \\ \beta & 0 \end{bmatrix} \begin{bmatrix} i_1 \\ v_2 \end{bmatrix}. \tag{15.6-11}$$

But this is just the hybrid h parameter representation for the case $g_{21} \rightarrow \infty$.

Figure 15.115
A CCCS two-port

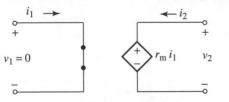

Subcase 2b: Let's pick $u_2 = v_2$ and $w_2 = i_2$ ($M = Y$). Then the two-port shown in Figure 15.116 is equivalent to the original one in Figure 15.112 with $r_m = 1/m_{12}$. The matrix relationship in equation (15.6-8) becomes

$$\begin{bmatrix} v_1 \\ v_2 \end{bmatrix} = \begin{bmatrix} 0 & 0 \\ r_m & 0 \end{bmatrix} \begin{bmatrix} i_1 \\ i_2 \end{bmatrix}. \tag{15.6-12}$$

But this is just the impedance parameter representation with $y_{21} \rightarrow \infty$.

Figure 15.116
A CCVS two-port

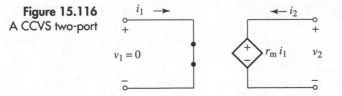

Summary:
Implications of
$m_{21} \rightarrow \infty$

To summarize our development thus far, we see that, depending on the parameter set chosen, allowing the "21 parameter" to become infinite has the effect of turning the given two-port into an ideal controlled source of the corresponding type. Of what value is this? A great deal! We discussed the practical applications of dependent sources in Chapter 5. For instance, an amplifier that boosts up the small signal from a sensor, such as a strain gauge, or from a source of information, such as a CD player, is ideally a dependent source. As it happens, active devices such as BJTs, MOSFETs, and op amps can be (and are) fabricated with *a very large intrinsic gain* (the m_{21} parameter); *unfortunately, however, the parameters of active devices are quite imprecise.* The dc gain of an op amp can vary from one unit to the next by a factor of 50% or more. The current gain β of a BJT can vary by a factor of 5 to 1 or so. All active device parameters change with temperature and with aging. The "12 parameter" of active devices, however, can be made so small that it can be considered to be zero (though it, like all other active device parameters, is highly imprecise). Resistor networks, on the other hand, have very precise parameter sets. Unfortunately, the forward gain parameter is not usually a suitable value (large enough) for resistive subcircuits in order for them to exhibit high gain.

Generalized Feedback
Analysis

Feedback amplifier design essentially involves designing an amplifier (that is, a dependent source) of one of the four types having a very precisely determined value of gain. The basic idea of the solution is shown in Figure 15.117. N_a (the subscript a stands for amplifier) is a two-port constructed with active devices and therefore has a very large value for the "21 parameter." N_f (the subscript f stands for feedback) is a resistive two-port having a very precise value for its "12 parameter." The subcircuits N_i and N_o are not two-ports: rather they have three terminal pairs and are called "connection subcircuits." They are chosen from the two topologies shown in Figure 15.118. If the series connection subcircuit is chosen for, say, N_i, then N_a and N_f will have their input ports connected in series. If the shunt connection subcircuit is chosen, their inputs will be connected in parallel.[10] Each choice for N_o, the output connection subcircuit, will result similarly in the given connection of the output ports of N_a and N_f. Notice, by the way, that we have enclosed everything in Figure 15.117 in a conceptual box labeled N, leaving only two terminals on the input side and two on the output side. This is to emphasize that the result is itself a two-port subcircuit.

Figure 15.117
A feedback subcircuit

Figure 15.118   Connection
subcircuits

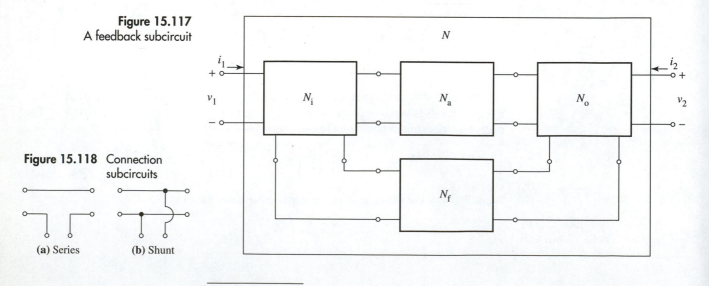

(a) Series   **(b)** Shunt

---

[10] "Shunt"—you will recall—is a synonym for "parallel."

Table 15.3
Feedback classification

Classification	N_i	N_o
1. series-series	series	series
2. series-shunt	series	shunt
3. shunt-series	shunt	series
4. shunt-shunt	shunt	shunt

We classify the composite two-port N in a manner determined by our choice of the two connection subcircuits N_i and N_o, as shown in Table 15.3. Thus, a shunt-series connection, for instance, is one for which N_i is a shunt connection subcircuit and N_o a series connection subcircuit.

What have we accomplished by connecting two such two-ports in this manner? Well, remember that the M matrices of the individual two-ports add, provided the M is chosen properly. For a series-shunt connection, for instance, the hybrid H parameter matrices add. We proved the corresponding fact for the impedance and admittance parameters in Sections 15.2 and 15.3 and asked you to do the same for the hybrid h and g parameters in Section 15.5. We show the results here in Figure 15.119. Here's the payoff. We can write the parameter matrix for any of the four connections in the generic form

$$\begin{bmatrix} w_1 \\ w_2 \end{bmatrix} = \begin{bmatrix} m_{11a} + m_{11f} & m_{12a} + m_{12f} \\ m_{21a} + m_{21f} & m_{22a} + m_{22f} \end{bmatrix} \begin{bmatrix} u_1 \\ u_2 \end{bmatrix}. \tag{15.6-13}$$

If we construct the amplifier two-port with active devices and make m_{21a} very large and m_{12a} very small—and if we make the feedback two-port out of resistors, thereby making m_{12f} extremely precise—we can make the approximations that $m_{21a} + m_{21f} \cong m_{21a}$ and $m_{12a} + m_{12f} \cong m_{12f}$. The matrix form of the relationship in (15.6-13) then becomes

$$\begin{bmatrix} w_1 \\ w_2 \end{bmatrix} = \begin{bmatrix} m_{11a} + m_{11f} & m_{12f} \\ m_{21a} & m_{22a} + m_{22f} \end{bmatrix} \begin{bmatrix} u_1 \\ u_2 \end{bmatrix}. \tag{15.6-14}$$

This is often called "the feedback approximation" or "the unilateral gain assumption," a two-port being unilateral if either m_{12} or m_{21} is zero. In this case, we divide the second scalar equation by m_{21a} and let that parameter become infinite. This transforms (15.6-14)

Figure 15.119
Catalog of feedback connections and their two-port relations

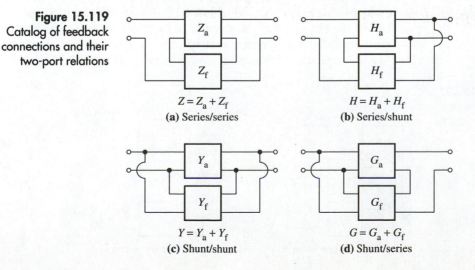

$Z = Z_a + Z_f$
(a) Series/series

$H = H_a + H_f$
(b) Series/shunt

$Y = Y_a + Y_f$
(c) Shunt/shunt

$G = G_a + G_f$
(d) Shunt/series

into

$$\begin{bmatrix} u_1 \\ u_2 \end{bmatrix} = \begin{bmatrix} 0 & 0 \\ 1/m_{12f} & 0 \end{bmatrix} \begin{bmatrix} w_1 \\ w_2 \end{bmatrix}. \tag{15.6-15}$$

Again, just as per our earlier derivation, this represents an ideal dependent source—which one depends upon our selection of port variables. Recalling that m_{12f} can be made a very precise quantity, we now have an *excellent* dependent source.

Most treatments of feedback in electronics texts make the two assumptions that $m_{21a} \gg m_{21f}$ and $m_{12a} \ll m_{12f}$ before beginning analysis. With our development, though, we can handle the general case just as easily without making these assumptions. Thus, when we analyze example circuits having numerical element values, we will do so exactly and then compare our results with those we obtain when we make the feedback approximation.

Source and Load (or "Termination") Resistors as Two-Ports

Two-ports are often "driven" by sources with finite equivalent resistances and are "loaded" with pure resistances. It turns out that it is convenient to consider these two resistors together as an "uncoupled resistance" two-port, as shown in Figure 15.120. Again, the box is only conceptual in nature. We can write the two-port equations as

$$v_1 = R_g i_1 \tag{15.6-16a}$$

$$v_2 = R_L i_2. \tag{15.6-16b}$$

As long as R_g and R_L are nonzero, each can be inverted. Therefore, each of the parameter descriptions exists for this two-port. For example, as we have written the two equations, we have the impedance parameters: $z_{11} = R_g$, $z_{22} = R_L$, and $z_{12} = z_{21} = 0$. We have called the two-port "uncoupled" because the 12 and 21 parameters are zero. We will leave it to you to show that the other parameter sets are as given in Figure 15.120.

Figure 15.120
An "uncoupled resistance" two-port

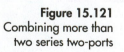

$$Z = \begin{bmatrix} R_g & 0 \\ 0 & R_L \end{bmatrix} \quad Y = \begin{bmatrix} g_g & 0 \\ 0 & g_L \end{bmatrix}$$

$$H = \begin{bmatrix} R_g & 0 \\ 0 & g_L \end{bmatrix} \quad G = \begin{bmatrix} g_g & 0 \\ 0 & R_L \end{bmatrix}$$

Two-Port Additivity: General Conditions

Before tackling our first example, we would like to point out that if one has *more than two* two-ports whose inputs are all connected in series or in parallel, and if the same is true of all of the outputs, then the appropriate parameter matrices add just as in the case of *two* two-ports that are so interconnected. Let's illustrate this with the z parameters and series-series connections. We show three two-ports in Figure 15.121(a) with their input

Figure 15.121
Combining more than two series two-ports

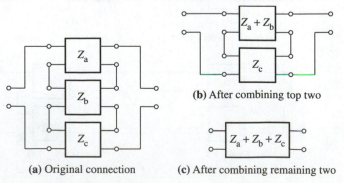

(a) Original connection

(b) After combining top two

(c) After combining remaining two

ports connected in series and their output ports also connected in this manner. Because the top two are series connected at both ports, we know by what we proved earlier that they can be replaced by a single two-port whose Z matrix is the sum of the two individual Z matrices. But this results in two two-ports that are connected as in Figure 15.121(b). These, in turn, can be combined with a Z matrix equal to the sum of the two individual Z matrices, as we show in Figure 15.121(c). Clearly, this argument works for any number of two-ports. We will leave it up to you to work through the other connections in the same fashion.

Series-Shunt
Feedback: A
Comprehensive
Example

Example 15.23   Analyze the series-shunt feedback network in Figure 15.122 to determine the values of voltage gain $A_v = v_2/v_g$, input impedance $Z_i = v_1/i_1$ and output impedance Z_o seen "looking away from R_L" with v_g deactivated.

Figure 15.122
The series-shunt
feedback
topology

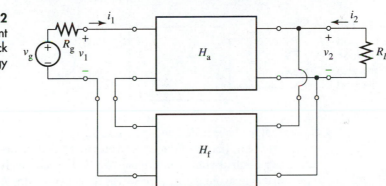

Solution   We have labeled the amplifier and feedback two-ports with the symbols for their h parameters, for these parameters add for the series-shunt connection.

General Analysis

Our first step will be to just rearrange the circuit a bit as shown in Figure 15.123. We have just moved R_g and R_L around so that they form an uncoupled resistor two-port that is series-shunt connected relative to the amplifier and feedback two-ports. Note that we have labeled the load resistor R_L with its conductance g_L, for this is the appropriate form for the hybrid h parameter representation. Also be careful to take note of the voltage variable v_1; we have had to take some care to maintain its original definition. Finally, carefully observe that i_2 is no longer the current into the composite output terminals; rather, it is the current upward through the load resistor in Figure 15.123. The input current i_1, however, has remained unaffected because the input is a series connection. Similarly, v_2 is still the output terminal voltage because the output terminals are all connected in parallel. Here's what we have accomplished with our rearrangement: we can represent the original circuit of Figure 15.122 with the single two-port of Figure 15.124, whose hybrid h parameters are the sums of those describing the three individual two-ports in Figure 15.123; that is, one has

$$\begin{bmatrix} v_g \\ i_L \end{bmatrix} = \begin{bmatrix} h_{11} & h_{12} \\ h_{21} & h_{22} \end{bmatrix}\begin{bmatrix} i_1 \\ v_2 \end{bmatrix} = \begin{bmatrix} h_{11a} + h_{11f} + R_g & h_{12a} + h_{12f} \\ h_{21a} + h_{21f} & h_{22a} + h_{22f} + g_L \end{bmatrix}\begin{bmatrix} i_1 \\ v_2 \end{bmatrix}. \quad (15.6\text{-}17)$$

Figure 15.123
The same circuit
rearranged

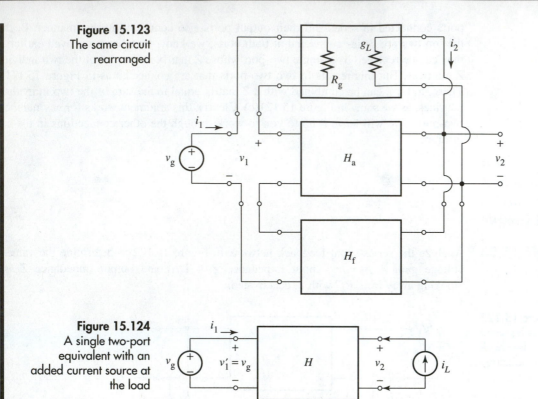

Figure 15.124
A single two-port
equivalent with an
added current source at
the load

Notice that we have added an i-source at the output purely to make the analysis easier. If we deactivate it, port 2 of our equivalent circuit will be an open circuit, and this will result in our original circuit once more. We will do our analysis in terms of the composite h parameters to avoid messy expressions, then express them in terms of the individual parameters only when the need arises.

The analysis is now quite easy. Using the inverse of the (2×2) H matrix in (15.6-17), we get

$$\begin{bmatrix} i_1 \\ v_2 \end{bmatrix} = \frac{1}{\Delta_h} \begin{bmatrix} h_{22} & -h_{12} \\ -h_{21} & h_{11} \end{bmatrix} \begin{bmatrix} v_g \\ i_L \end{bmatrix}. \tag{15.6-18}$$

Knowing i_1 and v_2, we can solve for all the parameters requested in the problem statement. For instance, the voltage gain from the input v-source to the load is

$$A_v = \frac{v_2}{v_g}\bigg]_{i_L=0} = -\frac{h_{21}}{\Delta_h}. \tag{15.6-19}$$

In deriving this expression, we simply set $i_L = 0$ in (15.6-18), then picked off the second scalar equation and solved for the desired voltage ratio.

Perhaps it is time to make a point. If we were working an example with numerical values, we would have followed our procedure of bending the resistors around, combining the h parameters, and adding a current source at the output. Then we would have written down the overall hybrid h parameter description and solved for i_1 and v_2. Now take care to note this: we would have had numerical values for h_{21} and Δ_h in (15.6-19), and the solution would have been a simple arithmetic computation.

Here, let us suppose that we want to know the effects of the various physical parameters (as opposed to equivalent combined values of these parameters) themselves. In this case, we use the definitions given in equation (15.6-17) in (15.6-19). Thus,

$$A_v = -\frac{h_{21a} + h_{21f}}{(h_{11a} + h_{11f} + R_g)(h_{22a} + h_{22f} + g_L) - (h_{21a} + h_{21f})(h_{12a} + h_{12f})}. \tag{15.6-20}$$

This is pretty complicated isn't it? Well, we will use this as an opportunity to stress our earlier point even harder: by picking our parameter set wisely, by rearranging the circuit into a standard form, and then by combining the individual parameters, we have turned a rather awesome piece of algebraic manipulation into a rather straightforward one.

Voltage Gain

Sometimes it is quite useful to use such expressions as (15.6-20), complex as they are. As an illustration, let us assume that our amplifier forward gain h_{21a} is very large. Dividing both numerator and denominator of (15.6-20) by this parameter, we obtain

$$A_v = -\frac{1 + h_{21f}/h_{21a}}{\dfrac{(h_{11a} + h_{11f} + R_g)(h_{22a} + h_{22f} + g_L)}{h_{21a}} - \left(1 + \dfrac{h_{21f}}{h_{21a}}\right)(h_{12a} + h_{12f})}. \tag{15.6-21}$$

Now, if we let $h_{21a} \rightarrow \infty$, the complexity disappears! The result is

$$A_v = \frac{1}{h_{12a} + h_{12f}}. \tag{15.6-22}$$

What we have just done is to impose *half* of the "feedback approximation." The other half stipulates that $h_{12a} \ll h_{12f}$. If we assume this as well, we will have

$$A_v = \frac{1}{h_{12f}}. \tag{15.6-23}$$

Now we could have obtained this result by using the more complex right-hand matrix in (15.6-17) and making these two approximations there. In following our procedure, however, we obtain the *exact* expression in (15.6-20) and can therefore use it to determine how close the approximation actually is.

Input Impedance

In order to compute the input impedance as defined in the problem statement, we will go back to our original circuit in Figure 15.122 and note that $v_1 = v_g - R_g i_1$. Thus,

$$Z_i = \left. \frac{v_1}{i_1} \right]_{i_L=0} = \left. \frac{v_g - R_g i_1}{i_1} \right]_{i_L=0} = \frac{\Delta_h}{h_{22}} - R_g. \tag{15.6-24}$$

Once again, if we have numerical values we simply compute it. Here, however, we will write out the complete expression to see how the various physical parameters affect Z_i. This results in

$$Z_i = \left. \frac{v_1}{i_1} \right]_{i_L=0} = \left. \frac{v_g - R_g i_1}{i_1} \right]_{i_L=0} \tag{15.6-25}$$

$$= \frac{(h_{11a} + h_{11f} + R_g)(h_{22a} + h_{22f} + g_L) - (h_{12a} + h_{12f})(h_{21a} + h_{21f})}{h_{22a} + h_{22f} + g_L} - R_g.$$

This is again a rather imposing expression. But once more we will eliminate the complexity by asking what happens if we allow $h_{21a} \to \infty$. Here, we simply note that the only appearance of h_{21a} is in the numerator of the fraction; hence, when it becomes infinitely large, so does Z_i. Thus, if $h_{21a} \to \infty$,

$$Z_i = \infty. \tag{15.6-26}$$

Before moving on to our computation of the one remaining parameter we were asked to find, let's talk about the definition of A_v and Z_i just a bit (for the finite-gain case). It might seem strange to "penalize the amplifier circuit" for the source resistance R_g when we compute the voltage gain. After all, R_g is large for a poor source (and A_v will consequently be small).

Our definition is fairly standard though, for it is often the most efficient quantity to work with—particularly in cascaded chains of many identical amplifiers, such as one finds in information transmission channels. Of course, we could have defined the voltage gain to be v_2/v_1 had we so desired, and we could have found it just as easily.

As far as the input resistance is concerned, our definition is more logical. Suppose we wanted, for instance, to know whether or not the source was delivering maximum power to the input of the amplifier. We would have then wanted exactly the quantity we computed.

Output Impedance

So far, we have used our result in (15.6-17) only with $i_L = 0$. Now, however, we must compute the output impedance of our composite feedback amplifier with the input voltage source deactivated; that is, with $v_g = 0$. We then need to reactivate the current source at the output and find the resulting ratio v_2/i_2.

When the current source is reactivated and the input voltage source deactivated, equation (15.6-18), our solution, becomes

$$\begin{bmatrix} i_1 \\ v_2 \end{bmatrix} = \frac{1}{\Delta_h} \begin{bmatrix} h_{22} & -h_{12} \\ -h_{21} & h_{11} \end{bmatrix} \begin{bmatrix} 0 \\ i_L \end{bmatrix} = \begin{bmatrix} -h_{12}/\Delta_h \\ h_{11}/\Delta_h \end{bmatrix} i_L. \tag{15.6-27}$$

Next, we flip back to Figure 15.122 and apply i_L at the output terminals (as we did in Figure 15.124, which is an equivalent circuit, to obtain the setup in Figure 15.125. From this diagram, we can obtain the desired output impedance:

$$Z_o = \frac{v_2}{i_2}\Bigg]_{v_g=0} = \frac{v_2}{i_L - g_L v_2}\Bigg]_{v_g=0} \tag{15.6-28}$$

$$= \frac{v_2/i_L}{1 - g_L v_2/i_L}\Bigg]_{v_g=0} = \frac{h_{11}/\Delta_h}{1 - g_L h_{11}/\Delta_h} = \frac{h_{11}}{\Delta_h - g_L h_{11}}.$$

Figure 15.125
The original circuit with an i-source added at the output port

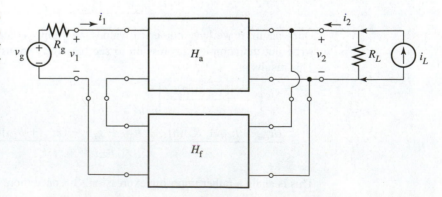

Finally, in terms of the individual parameters, we have

$$Z_o = \frac{h_{11a} + h_{11f} + R_g}{(h_{11a} + h_{11f} + R_g)(h_{22a} + h_{22f} + g_L) - (h_{12a} + h_{12f})(h_{21a} + h_{21f}) - g_L h_{11}}. \tag{15.6-29}$$

Now, if we let h_{21a} (which appears only in the denominator) go to infinity, we get

$$Z_o = 0. \tag{15.6-30}$$

Example Summary

What we have shown in this example is that series-shunt feedback—with high forward gain and low reverse gain for the amplifier and precise reverse gain for the feedback two-port—makes the composite two-port approach an ideal VCVS, that is, a dependent source with an infinite input impedance, a zero output impedance, and a precise voltage gain—the one illustrated in Figure 15.113.

Operational Amplifiers in Feedback Circuits

Our next example will be of a practical op amp circuit. In order to prepare the way for that example, we will now discuss a model of the op amp that is more realistic than our ideal model of Chapter 5. We show the model in Figure 15.126. Typical values are: $R_i = 1\ M\Omega$, $R_o = 100\ \Omega$, and $\mu = 10^5$ (though none are precise). We will use these values in the next example. In that example, we will need the hybrid h-parameter model for the op amp, which we present in Figure 15.127. The h parameters are easily obtained from the model in Figure 15.126 and are, in matrix form,

$$\begin{bmatrix} v_1 \\ i_2 \end{bmatrix} = \begin{bmatrix} R_i & 0 \\ -\mu R_i/R_o & g_o \end{bmatrix} \begin{bmatrix} i_1 \\ v_2 \end{bmatrix}. \tag{15.6-31}$$

We will leave the verification of this result up to you. It is important to keep in mind, by the way, that one of the output terminals is grounded. This is always true of the op amp.

Figure 15.126
A more realistic model
for the op amp

Figure 15.127
The hybrid h-parameter
model for the op amp

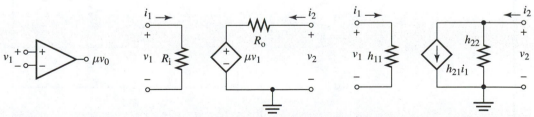

A Comprehensive Op Amp Example: Series-Shunt Feedback

Example 15.24(O)   Find the voltage gain $A_v = v_2/v_g$, the input impedance $Z_i = v_1/i_1$, and the output impedance "seen by" the load resistor R_L for the circuit in Figure 15.128. Assume typical values for the op amp parameters as given above.

Figure 15.128
An example circuit

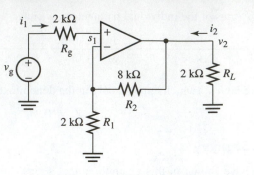

Solution   Before solving this circuit, let us (as before) add a current source at the output, simply to aid in our analysis. The resulting circuit is shown in Figure 15.129. The most difficult part in the analysis of essentially any feedback network is identification of the topology and of the constituent two-ports. We do this for our example circuit by rearranging the elements as shown in Figure 15.130. We see that there are three two-ports having series-connected inputs and shunt-connected outputs. (If you pause here to check this equivalent circuit in detail, your efforts will be amply repaid by a skill in handling feedback circuits.)

Figure 15.129
Adding an i-source at
the output

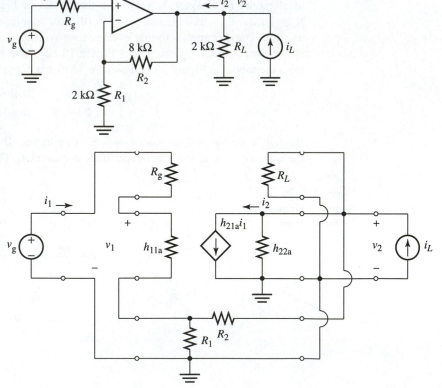

Figure 15.130
The circuit rearranged

Typical Values

Before proceeding, let's compute the values of the various hybrid parameters. We already know that $R_g = R_L = 2$ kΩ. We will need the reciprocal of R_L, which is $g_L = 0.5$ mS. For the op amp amplifier two-port, we have $h_{11a} = R_i = 1$ MΩ, $h_{22a} = 1/R_o = 0.01\ \Omega = 10$ mS, and $h_{21a} = -\mu R_i/R_o = -10^5 \times 10^6/100 = -10^9$ (unitless because it is a current ratio). The feedback two-port at the bottom was analyzed in Section 15.5. The results we derived there were $h_{11f} = R_1 \| R_2$ (the parallel combination of R_1 and R_2), $h_{22f} = 1/(R_1 + R_2)$, $h_{21f} = -R_1/(R_1 + R_2)$, and $h_{12f} = R_1/(R_1 + R_2)$. Using our given values of $R_1 = 2 \text{ k}\Omega$ and $R_2 = 8 \text{ k}\Omega$, we find that $h_{11f} = 1.6 \text{ k}\Omega$, $h_{22f} = 0.1 \text{ mS}$, $h_{21f} = -0.2$, and $h_{12f} = 0.2$. (Both h_{12f} and h_{21f} are unitless.)

The Composite Two-Port

This done, let's now recognize that the circuit in Figure 15.130 is equivalent to the one shown in Figure 15.131, where the h parameters of the composite two-port are the sums of the h parameters of the individual two-ports in Figure 15.130. The analysis goes exactly like that of the general case. We write the hybrid h-parameter relations as

$$\begin{bmatrix} v_g \\ i_L \end{bmatrix} = \begin{bmatrix} h_{11} & h_{12} \\ h_{21} & h_{22} \end{bmatrix} \begin{bmatrix} i_1 \\ v_2 \end{bmatrix}, \tag{15.6-32}$$

then invert to obtain

$$\begin{bmatrix} i_1 \\ v_2 \end{bmatrix} = \frac{1}{\Delta} \begin{bmatrix} h_{22} & -h_{12} \\ -h_{21} & h_{11} \end{bmatrix} \begin{bmatrix} v_g \\ i_L \end{bmatrix}. \tag{15.6-33}$$

Figure 15.131
Equivalent single two-port

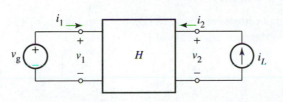

Now let's compute the numerical values of the parameters. They add, so

$$h_{11} = h_{11a} + h_{11f} + R_g = 1 \text{ M}\Omega + 1.6 \text{ k}\Omega + 2 \text{ k}\Omega \tag{15.6-34a}$$

$$= 1.0036 \text{ M}\Omega \cong 1 \text{ M}\Omega,$$

$$h_{22} = h_{22a} + h_{22f} + g_L = 10 \text{ mS} + 0.1 \text{ mS} + 0.5 \text{ mS} = 10.6 \text{ mS}, \tag{15.6-34b}$$

$$h_{21} = h_{21a} + h_{21f} = -10^9 - 0.2 \cong -10^9, \tag{15.6-34c}$$

and

$$h_{12} = h_{12a} + h_{12f} = 0 + 0.2 = 0.2. \tag{15.6-34d}$$

The determinant Δ thus has the value

$$\Delta = h_{11}h_{22} - h_{12}h_{21} = (1 \text{ M}\Omega)(10.6 \text{ mS}) - (0.2)(-10^9) \cong 0.2 \times 10^9. \tag{15.6-35}$$

Design Considerations

Before proceeding, let's look at these values. We notice that h_{11a} accounts for essentially all of the composite h_{11}, h_{22a} accounts for essentially all of the composite h_{22}, h_{21a} accounts for all of the composite h_{21}, and—in striking contrast—h_{12f} accounts for all of the composite h_{12}. This is typical. A designer designs the circuit so that the "loading" effects (input and output resistance and conductance) of the feedback two-port do not degrade the corresponding characteristics of the amplifier (one wants high input resistance and low output resistance (high output conductance) in order to approximate an ideal dependent voltage source). On the other hand, he or she wants the h_{12} to be set primarily by the added feedback network, as we have discussed previously in this section. Clearly the circuit we are analyzing has these characteristics.

Voltage Gain

We will now set i_L to zero and compute the voltage gain and the input impedance. Equation (15.6-33) thus assumes the form

$$\begin{bmatrix} i_1 \\ v_2 \end{bmatrix} = \frac{1}{\Delta} \begin{bmatrix} h_{22} & -h_{12} \\ -h_{21} & h_{11} \end{bmatrix} \begin{bmatrix} v_g \\ 0 \end{bmatrix} = \begin{bmatrix} h_{22}/\Delta \\ -h_{21}/\Delta \end{bmatrix} v_g. \tag{15.6-36}$$

Thus, the voltage gain is

$$A_v = \frac{v_2}{v_g}\bigg]_{i_L=0} = -\frac{h_{21}}{\Delta} = -\frac{-10^9}{0.2 \times 10^9} = +5. \tag{15.6-37}$$

If you flip back to Figure 15.128 and apply the ideal op amp analysis covered in Chapter 5, you will see that this is precisely the voltage gain one thus obtains. In this respect, at any rate, the op amp is quite close to being ideal when used in the circuit we are analyzing. If you look once more at equations (15.6-34) and (15.6-35), you will get an idea of the accuracy we are losing when we approximate the h parameters using the "approximately equal" signs.

Input Impedance

The input impedance is given by (refer to Figure 15.128 and note that v_1 is the node voltage on the right-hand side of R_g)

$$Z_i = \frac{v_1}{i_1}\bigg]_{i_L=0} = \frac{v_g - R_g i_1}{i_1}\bigg]_{i_L=0} = \frac{v_g}{i_1}\bigg]_{i_L=0} - R_g \tag{15.6-38}$$

$$= \frac{\Delta}{h_{22}} - R_g = \frac{0.2 \times 10^9}{10.7 \text{ mS}} - 2 \text{ k}\Omega = 1.9 \times 10^{10} \ \Omega.$$

This, of course, is an extremely high input impedance (ideal op amp analysis predicts an infinite input impedance). An alternative viewpoint to the computation of Z_i is to simply use (15.6-36) to find v_g/i_1, then subtract the series resistor of value $R_g = 2 \text{ k}\Omega$.

Output Impedance

Finally, to find the output impedance "seen by" the load resistor R_L, we must set $v_g = 0$ in equation (15.6-33) and reactivate i_L, resulting in

$$\begin{bmatrix} i_1 \\ v_2 \end{bmatrix} = \frac{1}{\Delta} \begin{bmatrix} h_{22} & -h_{12} \\ -h_{21} & h_{11} \end{bmatrix} \begin{bmatrix} 0 \\ i_L \end{bmatrix} = \begin{bmatrix} -h_{12}/\Delta \\ h_{11}/\Delta \end{bmatrix} i_L. \tag{15.6-39}$$

Referring back to Figure 15.129, we see that

$$Z_o = \frac{v_2}{i_2}\bigg]_{v_g=0} = \frac{v_2}{i_L - g_L v_2}\bigg]_{v_g=0} = \frac{1}{\dfrac{i_L}{v_2} - g_L}\bigg]_{v_g=0} \tag{15.6-40}$$

$$= \frac{1}{\dfrac{\Delta}{h_{11}} - g_L} = \frac{h_{11}}{\Delta - g_L h_{11}} = \frac{10^6}{0.2 \times 10^9 - (0.5 \times 10^{-3})(10^6)} = 5.0 \text{ m}\Omega.$$

Milliohms—a very low value! Thus, we see that our feedback hookup has forced the output impedance to become almost zero (the op amp output resistance, though fairly low at 100 Ω, was still much larger than this).

Example Summary

This example has served dual purposes. One was to outline the procedure for handling specific feedback amplifier circuits using known values; the other was to validate our earlier use of the ideal op amp model.

Series-Series Feedback: General Procedure

Figure 15.132 shows a series-series feedback connection. The input current for the amplifier two-port is the same as that of the feedback two-port and likewise for the output currents. The voltage at the input terminals of the composite two-port is the sum of the input terminal voltages of the two individual two-ports, and likewise for the output voltage. We know that the z parameters add for the two two-ports; furthermore, if we consider R_g and R_L as constituting an uncoupled resistive two-port (as explained above), we see that the resulting two-port is in series with the other two, giving the network shown in Figure 15.133. We have added an independent v-source at the output to make the analysis easier. When it is deactivated, the resulting circuit is the same as our original one in Figure 15.132.

Figure 15.132
Series-series feedback

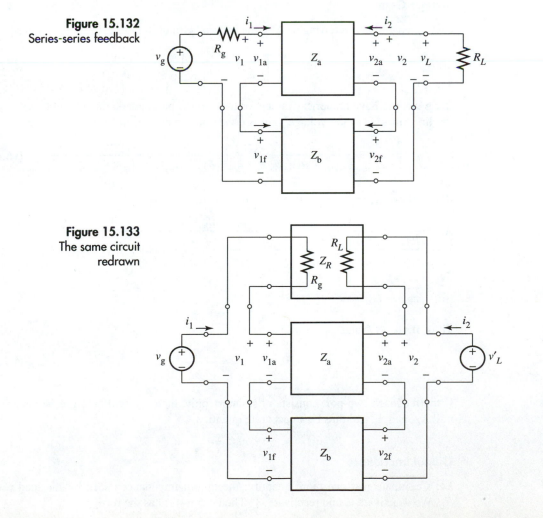

Figure 15.133
The same circuit redrawn

The Z matrices add for series-connected two-ports, so we can write

$$\begin{bmatrix} v_g \\ v'_L \end{bmatrix} = \begin{bmatrix} z_{11} & z_{12} \\ z_{21} & z_{22} \end{bmatrix} \begin{bmatrix} i_1 \\ i_2 \end{bmatrix},$$ (15.6-41)

where

$$z_{11} = z_{11a} + z_{11f} + R_g,$$ (15.6-42a)

$$z_{22} = z_{22a} + z_{22f} + R_L,$$ (15.6-42b)

$$z_{12} = z_{12a} + z_{12f},$$ (15.6-42c)

and

$$z_{21} = z_{21a} + z_{21f}.$$ (15.6-42d)

Just for the record, we note that the determinant of the Z matrix of the composite two-port is given by

$$\Delta = |Z| = (z_{11a} + z_{11f} + R_g)(z_{22a} + z_{22f} + R_L) - (z_{12a} + z_{12f})(z_{21a} + z_{21f}).$$ (15.6-42e)

Solving for the two unknown currents, we get

$$\begin{bmatrix} i_1 \\ i_2 \end{bmatrix} = \frac{1}{\Delta} \begin{bmatrix} z_{22} & -z_{12} \\ -z_{21} & z_{11} \end{bmatrix} \begin{bmatrix} v_g \\ v'_L \end{bmatrix}.$$ (15.6-43)

Voltage Gain

Letting $v_L = 0$ and referring back to the original circuit in Figure 15.132, we have

$$A_v = \frac{v_L}{v_g}\bigg]_{v'_L=0} = \frac{-i_2 R_L}{v_g}\bigg]_{v'_L=0} = \frac{z_{21} R_L}{\Delta}.$$ (15.6-44)

If we were to have numerical values for the various parameters, this would be the "end of the line" insofar as the voltage gain is concerned; however, in literal terms, we can write

$$A_v = \frac{(z_{21a} + z_{21f})R_L}{(z_{11a} + z_{11f} + R_g)(z_{22a} + z_{22f} + R_L) - (z_{12a} + z_{12f})(z_{21a} + z_{21f})}.$$ (15.6-45)

If we allow $z_{21a} \to \infty$, we will get

$$A_v = \frac{(1 + z_{21f}/z_{21a})R_L}{\dfrac{(z_{11a} + z_{11f} + R_g)(z_{22a} + z_{22f} + R_L)}{z_{21a}} - (z_{12a} + z_{12f})(1 + z_{21f}/z_{21a})} \longrightarrow$$ (15.6-46)

$$\frac{-R_L}{z_{12a} + z_{12f}}.$$

Thus, if $z_{12a} \cong 0$ and $z_{21a} \to \infty$, we have

$$A_v = \frac{-R_L}{z_{12f}}.$$ (15.6-47)

If the feedback two-port consists of resistors only, then z_{12f} will be a precise quantity, as will R_L, and A_v will thus be an accurate quantity.

Output Impedance

Let's compute just one more quantity: the output impedance "seen by" the load resistor R_L. We deactivate v_g and reactivate v'_L. Then (15.6-43) has the form

$$\begin{bmatrix} i_1 \\ i_2 \end{bmatrix} = \frac{1}{\Delta} \begin{bmatrix} z_{22} & -z_{12} \\ -z_{21} & z_{11} \end{bmatrix} \begin{bmatrix} 0 \\ v'_L \end{bmatrix} = \begin{bmatrix} -z_{12}/\Delta \\ z_{11}/\Delta \end{bmatrix} v'_L . \tag{15.6-48}$$

Referring to Figures 15.132 and 15.133, we have

$$Z_o = \frac{v_2}{i_2}\Bigg]_{v_g=0} - \frac{v_L = i_2 R_L}{i_2}\Bigg]_{v_g=0} = \frac{v_L}{i_2}\Bigg]_{v_g=0} - R_L \tag{15.6-49}$$

$$= \frac{\Delta}{z_{11}} - R_L = \frac{(z_{11a} + z_{11f} + R_g)(z_{22a} + z_{22f} + R_L) - (z_{12a} + z_{12f})(z_{21a} + z_{21f})}{z_{11a} + z_{11f} + r_g} - R_L.$$

If we allow z_{21a} to become infinite, we see that Z_o will approach infinity because z_{21a} is in the numerator of the fractional part of Z_o.

We will leave the computation of Z_i to you (it approaches infinity also as z_{21a} becomes infinite). We will, rather, work an example that is quite practical. In addition, it will demonstrate how to handle situations in which the natural selection of parameter sets is not easy to accomplish.

A BJT Example: Series-Series Feedback

Example 15.25(T) Find the voltage gain $A_v = v_2/v_g$ and the input impedance $Z_i = v_1/i_1$ for the BJT circuit shown in Figure 15.134. Assume that the BJT T has $\beta = 49$ and $r_\pi = 1\ k\Omega$. Also, let $R_g = 49\ k\Omega$, $R_E = 1\ k\Omega$, and $R_L = 10\ k\Omega$.

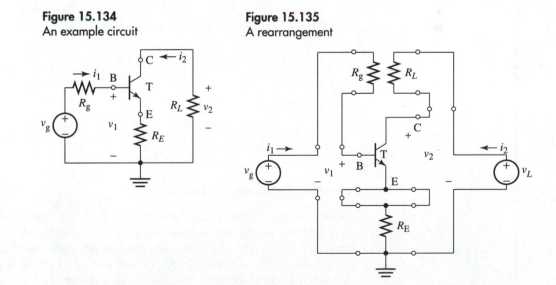

Figure 15.134
An example circuit

Figure 15.135
A rearrangement

Solution If you haven't covered the material in Chapter 5 on the BJT (bipolar junction transistor), don't panic! We will give you the equivalent circuit as soon as we discuss the topology a bit. We first rearrange the circuit as in Figure 15.135. We have already shown that the impedance parameters of the uncoupled resistive circuit are $z_{11f} = R_g$, $z_{22f} = R_L$, $z_{12f} = 0$, and $z_{21f} = 0$. We have also demonstrated that the impedance parameters of the feedback network (Example 15.4) are given by $z_{11f} = z_{22f} = z_{12f} = z_{21f} = R_E$—and that the admittance parameters do not exist.

The Hybrid π Model for the BJT: z Parameters

We developed the BJT model in Section 5.7 of Chapter 5. We repeat it here as Figure 15.136. We see a problem: the z parameters (which are those we need to analyze our se-ries-series feedback circuit in a natural manner) do not exist. More precisely, if you at-tempt to find z_{22} by connecting an independent i-source between the C and E terminals, you will find that KCL is, for general values of that source current, violated! So what can we do? The answer is to make the model for the BJT slightly more complicated by intro-ducing a resistor between C and E (in parallel with the CCCS), perform our circuit analy-sis, then let the value of that resistance become infinitely large.[11] We show the resulting model in Figure 15.137. We will leave it up to you to derive the z parameters for this more complex model: they are $z_{11a} = r_\pi$, $z_{22a} = r_c$, $z_{12a} = 0$, and $z_{21a} = -\beta r_c$.

Figure 15.136
The small-signal common emitter hybrid π model for the BJT

Figure 15.137
After adding a resistor

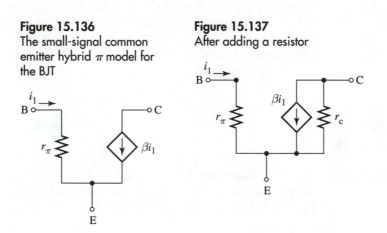

Referring to Figure 15.135 once more, we see that

$$\begin{bmatrix} v_g \\ v_L \end{bmatrix} = \begin{bmatrix} z_{11} & z_{12} \\ z_{21} & z_{22} \end{bmatrix} \begin{bmatrix} i_1 \\ i_2 \end{bmatrix}, \tag{15.6-50}$$

and the solution is

$$\begin{bmatrix} i_1 \\ i_2 \end{bmatrix} = \frac{1}{\Delta} \begin{bmatrix} z_{22} & -z_{12} \\ -z_{21} & z_{11} \end{bmatrix} \begin{bmatrix} v_g \\ v_L \end{bmatrix}. \tag{15.6-51}$$

Our Modus Operandi: A Discussion

Notice that we are advocating the repetition of the *individual steps* in the solution of such a problem—*not the commitment to memory of complicated formulas*. To perform this procedure efficiently, you must know the four parameter matrices listed at the beginning of this section and in Section 15.5 very well and have the inverse of (2×2) matrices well ingrained in your memory. You should also remember the small analytical device of adding an extra source at the output (sometimes a v-source, sometimes an i-source, de-pending upon the arrangement of the two-ports).

---

[11] In fact, a more extensive investigation of the BJT model shows that our added resistor should be present anyway.

Voltage Gain

Let's first set v_L to zero, giving

$$\begin{bmatrix} i_1 \\ i_2 \end{bmatrix} = \frac{1}{\Delta}\begin{bmatrix} z_{22} & -z_{12} \\ -z_{21} & z_{11} \end{bmatrix}\begin{bmatrix} v_g \\ 0 \end{bmatrix} = \begin{bmatrix} z_{22}/\Delta \\ -z_{21}/\Delta \end{bmatrix}v_g. \tag{15.6-52}$$

Thus, we have

$$A_v = \frac{v_2}{v_g}\bigg]_{v_L=0} = -\frac{i_2 R_L}{v_g}\bigg]_{v_L=0} = \frac{z_{21}R_L}{\Delta}. \tag{15.6-53}$$

If all of our elements had specified numerical values, we would simply compute the various parameters; as it is, we do not know the value of r_c, but must allow it to go to infinity. We note that the z parameters add for the three two-ports in Figure 15.135, so we can compute

$$z_{11} = r_\pi + R_g + R_E, \tag{15.6-54a}$$

$$z_{22} = r_c + R_L + R_E, \tag{15.6-54b}$$

$$z_{12} = R_E, \tag{15.6-54c}$$

and

$$z_{21} = -(\beta r_c + R_E). \tag{15.6-54d}$$

Thus

$$\Delta = (r_\pi + R_g + R_E)(r_c + R_L + R_E) + R_E(\beta r_c + R_E). \tag{15.6-54e}$$

Using these expressions in (15.6-53), we get

$$A_v = \frac{-(\beta r_c + R_E)R_L}{(r_\pi + R_g + R_E)(r_c + R_L + R_E) + R_E(\beta r_c + R_E)} \tag{15.6-55}$$

$$= \frac{-(\beta + R_E/r_c)R_L}{(r_\pi + R_g + R_E)[1 + (R_L + R_E)/r_c] + R_E(\beta + R_E/r_c)} \rightarrow \frac{-\beta R_L}{r_\pi + R_g + (\beta + 1)R_E}.$$

where we have let $\beta \rightarrow \infty$ in the last step. Inserting numerical values, one obtains $A_v = -(49 \times 10\text{ k}\Omega)/(1\text{ k}\Omega + 49\text{ k}\Omega + 50\text{ k}\Omega) = -4.9$.

The input impedance (again referring back to Figures 15.134 and 15.135) is

$$Z_1 = \frac{v_1}{i_1}\bigg]_{v_L=0} = \frac{v_g - i_1 R_g}{i_1}\bigg]_{v_L=0} = \frac{v_g}{i_1}\bigg]_{v_L=0} - R_g \tag{15.6-56}$$

$$= \frac{\Delta}{z_{22}} - R_g = \frac{(r_\pi + R_g + R_E)(r_c + R_L + R_E) + R_E(\beta r_c + R_E)}{r_c + R_L + R_E} - R_g$$

$$= r_\pi + R_E + \frac{R_E(\beta + R_E/r_c)}{1 + (R_L + R_E)/r_c} \rightarrow r_\pi + (\beta + 1)R_E.$$

Inserting our known values produces $Z_i = 1\text{ k}\Omega + 50 \times 1\text{ k}\Omega = 51\text{ k}\Omega$.

Section 15.6 Quiz

Q15.6-1. Consider the hybrid h parameter description of a two-port:

$$\begin{bmatrix} v_1 \\ i_2 \end{bmatrix} = \begin{bmatrix} h_{11} & h_{12} \\ h_{21} & h_{22} \end{bmatrix}\begin{bmatrix} i_1 \\ v_2 \end{bmatrix}.$$

Determine which of the z, y, h, or g parameter sets is appropriate when $h_{21} \rightarrow \infty$, draw the resulting dependent source model, and write down the corresponding parameter matrix equation.

Q15.6-2. Answer Question Q15.6-1, replacing the h parameters with the y parameters and letting y_{21} approach infinity.

Q15.6-3. Determine which parameter description for the individual two-ports (smaller boxes) in Figure Q15.6-1 is appropriate for the given connection, then find the corresponding parameter description for the composite two-port (larger box).

Q15.6-4. Add a current source of value i_g in parallel with a resistor of value R_g at the input of the composite two-port (big box) and a voltage source of value v_L in series with a load resistor of value R_L at the output of the composite two-port (big box) in Figure Q15.6-1 and compute:

Figure Q15.6-1

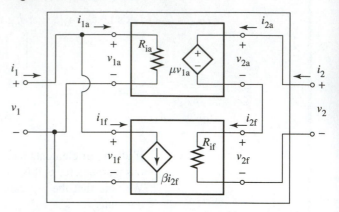

a. The overall current gain $A_i = \dfrac{-i_2}{i_g}\bigg]_{v_L=0}$

b. The input impedance "seen by" the input current source and its load resistor, $Z_i = \dfrac{v_1}{i_1}\bigg]_{v_L=0}$

c. The output impedance "seen by" the load current source and its resistor, $Z_o = \dfrac{v_2}{i_2}\bigg]_{i_g=0}$.

15.7 | Two-Ports with Internal Sources: Thévenin and Norton Equivalents

The Two-Port Thévenin Equivalent

Until now we have disallowed internal independent sources for our two-ports (and three-terminal subcircuits). If we do allow them, the fundamental constraint relationships are

$$a_1 v_1 + a_2 v_2 + a_3 i_1 + a_4 i_2 + c_1 = 0 \tag{15.7-1a}$$

$$b_1 v_1 + b_2 v_2 + b_3 i_1 + b_4 i_2 + c_2 = 0. \tag{15.7-1b}$$

If we select the voltages as the dependent variables and the currents as the independent variables—as in our development of the impedance parameters—and rearrange, we will obtain

$$\begin{bmatrix} a_1 & a_2 \\ b_1 & b_2 \end{bmatrix}\begin{bmatrix} v_1 \\ v_2 \end{bmatrix} = \begin{bmatrix} -c_1 \\ -c_2 \end{bmatrix} + \begin{bmatrix} -a_3 & -a_4 \\ -b_3 & -b_4 \end{bmatrix}\begin{bmatrix} i_1 \\ i_2 \end{bmatrix}, \tag{15.7-2}$$

which can be solved (assuming the determinant of the coefficient matrix of the voltages to be nonzero) to yield

$$\begin{bmatrix} v_1 \\ v_2 \end{bmatrix} = \begin{bmatrix} v_{oc1} \\ v_{oc2} \end{bmatrix} + \begin{bmatrix} z_{11} & z_{12} \\ z_{21} & z_{22} \end{bmatrix}\begin{bmatrix} i_1 \\ i_2 \end{bmatrix}. \tag{15.7-3}$$

If we set $i_1 = i_2 = 0$, then the voltages are equal to the voltages we have labeled with the subscripts "oc"; but setting the currents to zero is equivalent to open-circuiting the two ports. Thus, the matrix with the oc subscripts has as its entries the port voltages that result when both ports are open-circuited *simultaneously*. The matrix multiplying the currents is clearly the z parameter matrix. In nonmatrix form, we have

$$v_1 = v_{oc1} + z_{11} i_1 + z_{12} i_2 \tag{15.7-4a}$$

$$v_2 = v_{oc2} + z_{21} i_1 + z_{22} i_2. \tag{15.7-4b}$$

Interpreting these equations on a KVL basis results in the equivalent of Figure 15.138. We see that it consists of the Thévenin equivalent of each port (considered, individually, as a one-port) with the effect of the other port appearing as a dependent source. Thus, we can consider this equivalent to be the *two-port Thévenin equivalent* of the original two-port.

Figure 15.138
Two-port Thévenin
equivalent

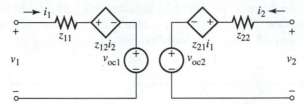

The Two-Port Norton Equivalent

If we rearrange (15.7-2) and solve for the currents, we will have (assuming that the determinant of the coefficient matrix of the currents is nonsingular)

$$\begin{bmatrix} i_1 \\ i_2 \end{bmatrix} = \begin{bmatrix} i_{sc1} \\ i_{sc2} \end{bmatrix} + \begin{bmatrix} y_{11} & y_{12} \\ y_{21} & y_{22} \end{bmatrix}\begin{bmatrix} v_1 \\ v_2 \end{bmatrix}. \tag{15.7-5}$$

If we short-circuit both ports simultaneously, both voltages will be forced to zero and the two port currents will be equal to their short-circuit values, which appear in the column matrix on the right-hand side of (15.7-5)! The matrix multiplying the voltages is clearly the admittance parameter matrix. In nonmatrix form, we have

$$i_1 = i_{sc1} + y_{11}v_1 + y_{12}v_2 \tag{15.7-6a}$$

$$i_2 = i_{sc2} + y_{21}v_1 + y_{22}v_2. \tag{15.7-6b}$$

Using a KCL interpretation, we get the *two-port Norton equivalent* in Figure 15.139.

Figure 15.139
Two-port Norton
equivalent

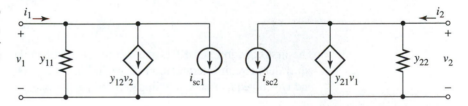

For both of the equivalents, one must deactivate all the internal independent sources in order to find (or test for) the dependent source and impedance/admittance values because deactivation of the internal sources reduces the values of c_1 and c_2 in equations (15.7-1) to zero, hence making the independent sources vanish in the Thévenin and Norton equivalents. Alternatively, one may simply attach two independent test current sources at the two ports, assign them *literal* values, and do the analysis.

Hybrid Thévenin/ Norton Two-Port Equivalents

If we pick the voltage v_1 and the current i_2 as our dependent variables (with i_1 and v_2 as the independent ones), we will be working with the hybrid h parameters. The rearranged version of (15.7-1) is

$$\begin{bmatrix} a_1 & a_4 \\ b_1 & b_4 \end{bmatrix}\begin{bmatrix} v_1 \\ i_2 \end{bmatrix} = \begin{bmatrix} -c_1 \\ -c_2 \end{bmatrix} + \begin{bmatrix} -a_3 & -a_2 \\ -b_3 & -b_2 \end{bmatrix}\begin{bmatrix} i_1 \\ v_2 \end{bmatrix}. \tag{15.7-7}$$

Solving, we obtain

$$\begin{bmatrix} v_1 \\ i_2 \end{bmatrix} = \begin{bmatrix} v_{oc1} \\ i_{sc2} \end{bmatrix} + \begin{bmatrix} h_{11} & h_{12} \\ h_{21} & h_{22} \end{bmatrix}\begin{bmatrix} i_1 \\ v_2 \end{bmatrix}, \tag{15.7-8}$$

or, in nonmatrix form,

$$v_1 = v_{oc1} + h_{11}i_1 + h_{12}v_2 \qquad (15.7\text{-}9a)$$

$$i_2 = i_{sc2} + h_{21}i_1 + h_{22}v_2. \qquad (15.7\text{-}9b)$$

This is a mixed set of equations; KVL applies for the first and KCL for the second. v_{oc1} is the voltage at port 1 and i_{sc2} the current at port 2 when port 1 is open circuited and port 2 shorted. The 2×2 matrix is clearly the hybrid h parameter matrix. The result, shown in Figure 15.140, *is a hybrid Thévenin-Norton equivalent.*

Figure 15.140
Two-port Thévenin-Norton hybrid h equivalent

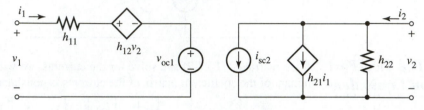

For the hybrid g parameters, we select i_1 and v_2 as our dependent variables (and, hence, v_1 and i_2 as the independent ones). This corresponds to solving

$$\begin{bmatrix} a_3 & a_2 \\ b_3 & b_2 \end{bmatrix}\begin{bmatrix} i_1 \\ v_2 \end{bmatrix} = \begin{bmatrix} -c_1 \\ -c_2 \end{bmatrix} + \begin{bmatrix} -a_1 & -a_4 \\ -b_1 & -b_4 \end{bmatrix}\begin{bmatrix} v_1 \\ i_2 \end{bmatrix}. \qquad (15.7\text{-}10)$$

We find that

$$\begin{bmatrix} i_1 \\ v_2 \end{bmatrix} = \begin{bmatrix} i_{sc1} \\ v_{oc2} \end{bmatrix} + \begin{bmatrix} g_{11} & g_{12} \\ g_{21} & g_{22} \end{bmatrix}\begin{bmatrix} v_1 \\ i_2 \end{bmatrix}. \qquad (15.7\text{-}11)$$

In nonmatrix form

$$i_1 = i_{sc1} + g_{11}v_1 + g_{12}i_2 \qquad (15.7\text{-}12a)$$

$$v_2 = v_{oc2} + g_{21}v_1 + g_{22}i_2. \qquad (15.7\text{-}12b)$$

The first equation is a KCL one and the second has the form of KVL. This gives us the equivalent circuit interpretation shown in Figure 15.141; it is another mixture of Thévenin and Norton equivalents, but flipped from the last one.

Figure 15.141
Two-port Thévenin-Norton hybrid g equivalent

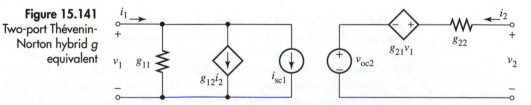

We point out that the bottom terminals would be connected together in each of the preceding equivalents if the original subcircuit were a three-terminal one.

Example 15.26   Find the two-port Thévenin equivalent for the two-port subcircuit shown in Figure 15.142. For what value(s) of μ does a Norton equivalent not exist?

Solution   We show the two-port with two attached test current sources at the two ports in Figure 15.143. We have taped the dependent source and prepared the resulting circuit for mesh analysis by defining mesh currents. There are three nonessential meshes—so *no* mesh equations are required. We need only untape the dependent source and express it in terms

Figure 15.142
An example two-port

Figure 15.143
Testing the two-port

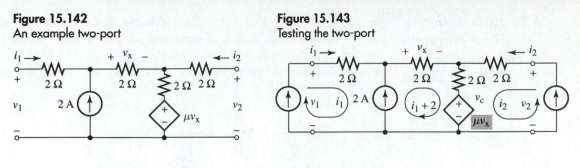

of i_1 and i_2, as follows:

$$v_c = \mu v_x = \mu \times 2 \times (i_1 + 2) = 2\mu i_1 + 4\mu. \tag{15.7-13}$$

We next merely write KVL equations to obtain two v-i relations:

$$v_1 = v_c + 2(i_1 + 2 + i_2) + 2(i_1 + 2) + 2i_1 \tag{15.7-14}$$

or, after untaping,

$$v_1 + 0v_2 - 2(\mu + 3)i_1 - 2i_2 - 4(\mu + 2) = 0, \tag{15.7-15}$$

and

$$v_2 = v_c + 2(i_1 + 2 + i_2) + 2i_2 \tag{15.7-16}$$

or, after untaping,

$$0v_1 + v_2 - 2(\mu + 1)i_1 - 4i_2 - 4(\mu + 1) = 0. \tag{15.7-17}$$

Note that these equations are in the standard form for our two-port v-i relations (15.7-1). The Thévenin equivalent two-port is shown in Figure 15.144, as can be verified by applying KVL to its two ports. This equivalent two-port is valid for any finite value of μ; however, we must compute the admittance parameters to find out which values (if any) cause the Norton equivalent to not exist.

Figure 15.144
The two-port Thévenin
equivalent

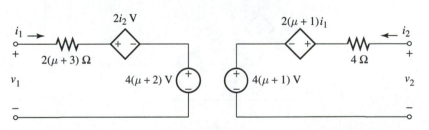

To do this, we just rearrange equations (15.7-15) and (15.7-17) in matrix form as

$$\begin{bmatrix} v_1 \\ v_2 \end{bmatrix} = \begin{bmatrix} 4(\mu + 2) \\ 4(\mu + 1) \end{bmatrix} + \begin{bmatrix} 2(\mu + 3) & 2 \\ 2(\mu + 1) & 4 \end{bmatrix} \begin{bmatrix} i_1 \\ i_2 \end{bmatrix}, \tag{15.7-18}$$

and note that the matrix multiplying the currents is the Z matrix. Its determinant is

$$|Z| = \begin{vmatrix} 2(\mu + 3) & 2 \\ 2(\mu + 1) & 4 \end{vmatrix} = 4\mu + 20 = 4(\mu + 5). \tag{15.7-19}$$

Because $Y = Z^{-1}$, the admittance parameters only exist if $|Z| \neq 0$—or, equivalently, if the dependent source voltage gain $\mu \neq -5$.

A Reexamination of Two-Terminal Thévenin/Norton Equivalents

We will now use our two-port Thévenin/Norton equivalents to discuss an issue that we brought up in Chapter 5: that of dependent sources in such equivalent transformations. We will study the issue with the circuit shown in Figure 15.145. There we see a circuit having two controlled sources and one independent source. We assume that the problem is to find the current i.

Figure 15.145
An example circuit

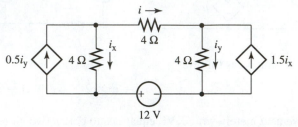

Now we can apply a number of techniques to obtain the solution, including the two-terminal Thévenin/Norton equivalent transformations. However, we choose to do things the following way to illustrate the subject of dependent sources. We break the circuit up into a two-port and two one-ports as shown in Figure 15.146. The question is: Are there actually two one-ports, or do they actually form a single two-port? In fact, we can consider N_1 and N_3 to constitute a single two-port because their terminal currents and voltages will be interdependent because of the presence of the controlled sources, each of which depends upon a variable in the other subcircuit. (Notice that we have defined two-terminal voltages and currents in Figure 15.146.) Thus, let's imagine ourselves to take wire cutters and snip out the two-port labeled N_2, then twist N_1 and N_3 around a bit. This gives the obvious two-port of Figure 15.147.

Figure 15.146
The example circuit partitioned into two-ports

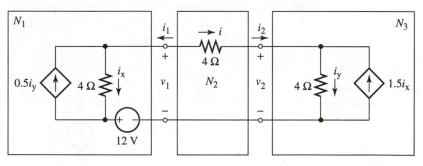

Figure 15.147
The composite two-port

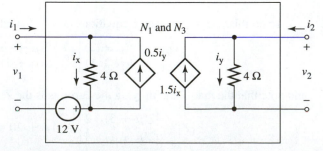

Let's once again attach two i-sources to the terminals and determine the two-port v-i relationship. The resulting circuit is shown in Figure 15.148. Applying KVL, KCL, and Ohm's law, we get

$$v_1 = 4i_x + 12 \tag{15.7-20}$$

and

$$v_2 = 4i_y \tag{15.7-21}$$

Figure 15.148
Testing the composite
two-port

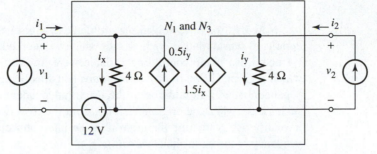

Next we express the controlling variables i_x and i_y in terms of the terminal currents i_1 and i_2. We write

$$i_x = i_1 + 0.5i_y \tag{15.7-22}$$

and

$$i_y = i_2 + 1.5i_x. \tag{15.7-23}$$

Solving these last two equations simultaneously, we get

$$i_x = 4i_1 + 2i_2 \tag{15.7-24}$$

and

$$i_y = 6i_1 + 4i_2. \tag{15.7-25}$$

Using these values in (15.7-20) and (15.7-21), we have

$$v_1 = 12 + 16i_1 + 8i_2 \tag{15.7-26}$$

and

$$v_2 = 24i_1 + 16i_2. \tag{15.7-27}$$

These last equations can be interpreted in terms of the dual Thévenin equivalent two-port of Figure 15.149. We have reinserted the two-port N_2 containing the resistor whose current we wanted to determine. A single KVL equation (noting that $i_1 = -i$ and $i_2 = i$)

Figure 15.149
The Thévenin
equivalent and the
resistor that was
removed

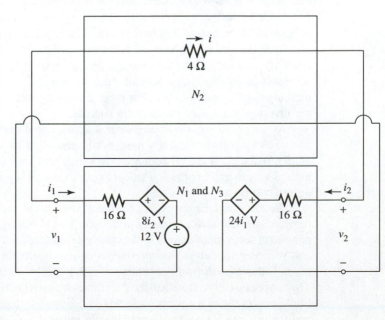

now gives the result we wanted:

$$12 + 8i + 16(-i) - 4i - 16i - 24(-i) = 0, \qquad (15.7\text{-}28)$$

or
$$i = 3 \text{ A}. \qquad (15.7\text{-}29)$$

Now that we have solved the problem, let's discuss it a bit. What we have actually done is to substantiate a method with which we worked in Chapter 5, Section 5.2. There, we noted that if the controlling variable were in a different subcircuit from the one for which a given Thévenin or Norton equivalent was being determined, one would obtain a dependent source in addition to the usual independent one. What we now see is that in such a case, when the original circuit is split into two two-terminal (one-port) subcircuits, in reality one is treating the network as an interconnected two-port—because of the dependency of the controlled source(s).

Section 15.7 Quiz

Q15.7-1. Find all four of the equivalents derived in this section for the subcircuit in Figure Q15.7-1. Assume that all resistors are 1 Ω in value and that the current source has a value of 10 A.

Figure Q15.7-1

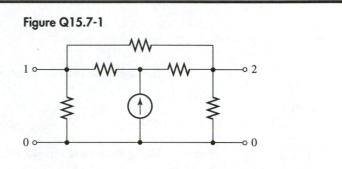

Chapter 15 Summary

We started this chapter in Section 15.1 with a discussion of the idea of a *port,* and the accompanying idea of a port *constraint.* We also defined the concepts of a *two-port subcircuit* and a *three-terminal subcircuit*—the latter being a special case of the former. We justified, on intuitive grounds, our assumption that a two-port is completely described by two constraint relations among the four port variables, v_1, v_2, i_1, and i_2. We also pointed out that two-ports whose constituent elements are those we have studied up until now (R, L, C, independent voltage and current sources, and dependent sources of all types) satisfy port variable constraints that have the form of an *affine relation.*

In Sections 15.2 through 15.5, we successively developed the z, y, *chain (or transmission) parameters* and the *hybrid h and g parameters*—in that order. Each parameter set results from selecting a particular pair of port variables to be independent, with the remaining two being dependent. For these parameter sets, we assumed that the two-port under investigation had no independent sources.

As we developed these parameter sets, we noted that a given interconnection of two or more two-ports is naturally more easily described by a specific parameter set; for example, if the ports are all connected in series with one another, the z-parameter set is a natural way to express the equivalent two-port consisting of all of the individual interconnected two-ports.

We discussed the concept of *feedback,* both in the sections devoted to the specific parameter sets, and also separately in Section 15.6—noting once again that a certain given parameter set is more natural to describe a given feedback connection than any of the others. We carefully defined an approximation that is often made in a somewhat vague manner in following electronics courses when they deal with the analysis of feedback circuits. This approximation is called the "feedback approximation" or the "unilateral assumption": one of the constituent two-ports is assumed to have a "12" parameter that is zero and the other to have a very large (ideally infinite) "21" parameter. We showed that two-ports connected in a given manner, however, can be *exactly* analyzed in a fashion that is not appreciably more complicated than those resulting from the approximation. Furthermore, if the feedback approximation is then made, one gains an appreciation for the approximation error involved.

Finally, we allowed our two-ports to include independent sources and developed two-port versions of the Thévenin and Norton equivalents developed earlier in Chapter 3. We used these more general two-port equivalents to justify our earlier results (in Chapter 5) for two-terminal subcircuits with dependent sources.

Chapter 15 Review Quiz

RQ15.1. Find the set of constraint relations in the form

$$a_1v_1 + a_1v_2 + a_3i_1 + a_4i_2 + c_1 = 0$$
$$b_1v_1 + b_2v_2 + b_3i_1 + b_4i_2 + c_2 = 0$$

that describe the two-port subcircuit shown in Figure RQ15.1. To make your answer unique, normalize your equations so that $a_1 = b_1 = 1$. Assume $\beta = 2$.

Figure RQ15.1

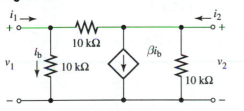

RQ15.2. Find the y parameters for the two-port in Figure RQ15.1 by manipulation of the equations you found in Question RQ15.1. Then confirm your result by running direct tests on the two-port. Assume $\beta = 2$.

RQ15.3. Find the z parameters for the two-port in Figure RQ15.1 by manipulation of the equations you found as the answer to Question RQ15.1. Then confirm your result by running direct tests on the two-port. Finally, invert the Y matrix you found as the answer to Question RQ15.2 and compare it with the first two methods. Assume $\beta = 2$.

RQ15.4. Find the h parameters for the two-port in Figure RQ15.1 by manipulation of the equations you found as the answer to Question RQ15.1. Then confirm your result by running direct tests on the two-port. Assume $\beta = 2$.

RQ15.5. Find the g parameters for the two-port in Figure RQ15.1 by manipulation of the equations you found as the answer to Question RQ15.1. Then confirm your result by running direct tests on the two-port. Finally, invert the H matrix you found as the answer to question RQ15.4 and compare it with the first two methods. Assume $\beta = 2$.

RQ15.6. Find the A, B, C, D parameters for the two-port in Figure RQ15.1 by manipulation of the equations you found as the answer to Question RQ15.1. Then confirm your result by running direct tests on the two-port. Assume $\beta = 2$.

RQ15.7. Find the A_r, B_r, C_r, D_r parameters for the two-port in Figure RQ15.1 by manipulation of the equations you found as the answer to Question RQ15.1. Then confirm your result by running direct tests on the two-port. Finally, invert the A, B, C, D matrix you found as the answer to Question RQ15.6 and compare it with the first two methods. Assume $\beta = 2$.

RQ15.8. Determine which parameter sets exist for the two-port in Figure RQ15.1 if $\beta \rightarrow \infty$. Write the matrix equation describing the resulting two-port and draw the equivalent model.

RQ15.9. Does a tee equivalent exist for the two-port in Figure RQ15.1? Why, or why not?

RQ15.10. Classify the feedback type for the circuit shown in Figure RQ15.2 (series-series, etc.).

Figure RQ15.2

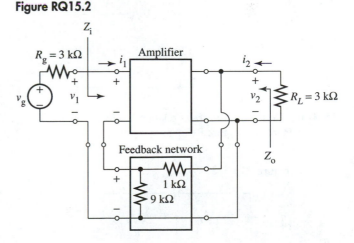

RQ15.11. The amplifier two-port shown in Figure RQ15.2 is the two-port shown in Figure RQ15.1 with a general value of β. Find general expressions for:

a. The voltage gain $A_v = v_2/v_g$
b. The voltage gain $A_{vo} = v_2/v_1$
c. The current gain $A_i = -i_2/i_1$
d. The input impedance $Z_i = v_1/i_1$
e. The output impedance Z_o "seen by" the load resistor R_L.

Evaluate these expressions for $\beta = 2$ and for $\beta \rightarrow \infty$.

Chapter 15 Problems

Section 15.1 Introduction

15.1-1.   Sketch the relationship

$$\mathcal{R} = \left\{ (x, y): \left[\frac{x-3}{2} \right]^2 + \left[\frac{y+2}{3} \right]^2 - 9 = 0 \right\}$$

in the x-y plane.

15.1-2.   The v-i characteristic of a two-terminal element is shown in Figure P15.1-1. Find two equivalent subcircuits for this element.

Figure P15.1-1

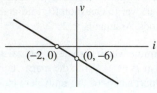

15.1-3.   Derive the constraint relationship for v_1, v_2, i_1, and i_2 for the subcircuit in Figure P15.1-2.

Figure P15.1-2

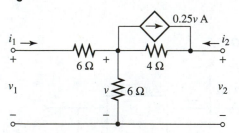

15.1-4.   Derive the constraint relationship for v_1, v_2, i_1, and i_2 for the subcircuit in Figure P15.1-3.

Figure P15.1-3

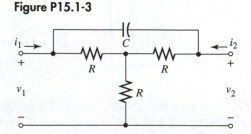

Section 15.2 The Impedance Parameters

15.2-1.   Find the z parameters for the subcircuit shown in Figure P15.2-1. Is it reciprocal? Is it symmetric?

Figure P15.2-1

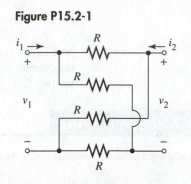

15.2-2.   Find the z parameters for the subcircuit shown in Figure P15.1-2. Is it reciprocal? Is it symmetric?

15.2-3.   Find the z parameters for the subcircuit shown in Figure P15.1-3. Is it reciprocal? Is it symmetric?

15.2-4.   Find the z parameters for the subcircuit shown in Figure P15.2-2. Is it reciprocal? Is it symmetric?

Figure P15.2-2

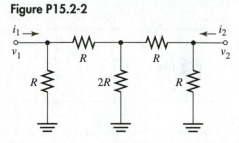

15.2-5.   Find the complex z parameters for the subcircuit shown in Figure P15.2-3. Is it reciprocal? Is it symmetric?

Figure P15.2-3

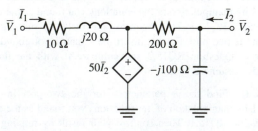

15.2-6.   Find the s-domain z parameters for the subcircuit shown in Figure P15.2-4. Is it reciprocal? Is it symmetric?

Figure P15.2-4

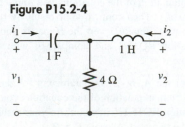

15.2-7. Find the *s*-domain *z* parameters for the subcircuit shown in Figure P15.2-5. Is it reciprocal? Is it symmetric?

Figure P15.2-5

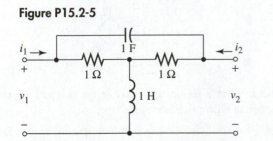

15.2-8. For the subcircuit shown in Figure P15.2-6, find the following quantities: $A_{v1} = v_2/v_1$, $A_{v2} = v_2/v_g$, $A_i = -i_2/i_1$, $Z_i = v_1/i_1$, and Z_o = impedance "seen by" R_L. Assume that $R_g = 1\ \Omega$ and $R_L = 2\ \Omega$.

Figure P15.2-6

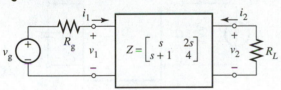

15.2-9. Find the tee equivalent for the subcircuit in Figure P15.2-2.

15.2-10. Why is there no tee equivalent for the subcircuit in Figure P15.2-1?

15.2-11. Find the *z* parameters for the subcircuit shown in Figure P15.2-7. Is it reciprocal? Is it symmetric?

Figure P15.2-7

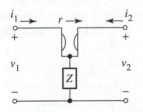

15.2-12. Find the *z* parameters for the subcircuit shown in Figure P15.2-8. Is it reciprocal? Is it symmetric?

Figure P15.2-8

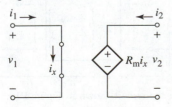

15.2-13. Find the *s*-domain *z* parameters for the boxed subcircuit in Figure P15.2-9. Then, using these *z* parameters, find the response $i_o(t)$ to the input voltage $v_s(t) = 30u(t)$ V. (You might find the following factorization useful, depending on how you choose to work the problem: $2s^3 + 7s^2 + 7s + 2 = (s + 1)(2s^2 + 5s + 2)$.)

Figure P15.2-9

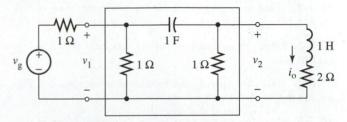

15.2-14. Find the overall *z* parameters for the *series connection* of the two two-ports shown in Figure P15.2-10 in terms of their individual *z* parameters.

Figure P15.2-10

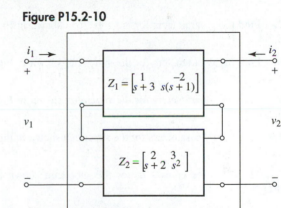

15.2-15. Find the indicated parameters in Figure P15.2-11. Also compute the voltage gain $A_v = v_2/v_1$.

Figure P15.2-11

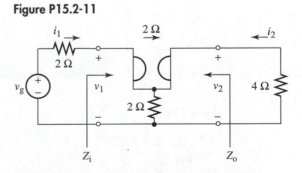

Section 15.3 The Admittance Parameters

15.3-1. Find R_a, R_b, and R_c to adjust the *y* parameters to $y_{12} = y_{21} = -1/9$ S, $y_{11} = y_{22} = 1/8$ S for the subcircuit shown in Figure P15.3-1.

Figure P15.3-1

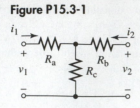

15.3-2. Find the y parameters for the subcircuit shown in Figure P15.1-2.

15.3-3. Find the y parameters for the subcircuit shown in Figure P15.1-3.

15.3-4. Find the y parameters for the subcircuit shown in Figure P15.2-1.

15.3-5. Find the y parameters for the subcircuit shown in Figure P15.2-2.

15.3-6. Find the y parameters for the subcircuit shown in Figure P15.2-3.

15.3-7. Find the y parameters for the subcircuit shown in Figure P15.2-4.

15.3-8. Find the y parameters for the subcircuit shown in Figure P15.2-5.

15.3-9. Find the y parameters for the subcircuit shown in Figure P15.2-7.

15.3-10. Find the y parameters for the subcircuit shown in Figure P15.2-8.

15.3-11. Find the tee equivalent for the subcircuit shown in Figure P15.3-2.

Figure P15.3-2

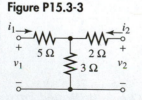

15.3-12. Find the π equivalent for the circuit shown in Figure P15.3-3.

Figure P15.3-3

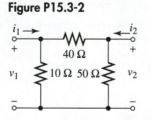

15.3-13. Find the y parameters for the subcircuit shown in Figure P15.3-4. For what value of a is $\Delta_y = 0$? (Δ_y is the determinant of the Y parameter matrix.) Notice that for this value of a the impedance parameters do not exist.

Figure P15.3-4

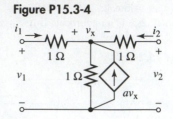

15.3-14. The Y parameter matrix of the two-port represented by the box in Figure P15.3-5 is given by

$$Y(s) = \begin{bmatrix} s & -1 \\ -1 & s+1 \end{bmatrix}.$$

Find the impulse response for the voltage v_o. (The current source waveform is the unit impulse function.)

Figure P15.3-5

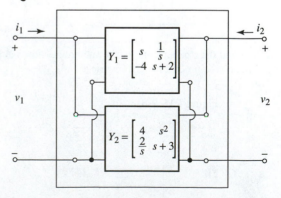

15.3-15. Find the s-domain y parameters (with conventional references for the port variables) for the subcircuit shown in Figure P15.3-6.

Figure P15.3-6

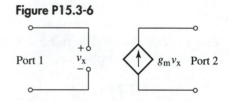

15.3-16. Find the overall y parameters for the *parallel connection* of the two two-ports shown in Figure P15.3-7 in terms of their individual y parameters.

Figure P15.3-7

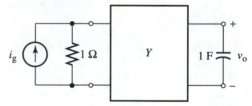

15.3-17. For the circuit shown in Figure P15.3-8, find the following quantities: $A_{i1} = -i_2/i_1$, $A_{i2} = -i_2/i_g$, $A_v = v_2/v_1$, $Z_i = v_1/i_1$, and $Z_o =$ impedance "seen by" R_L. Assume that $R_g = 1\ \Omega$ and $R_L = 2\ \Omega$.

Figure P15.3-8

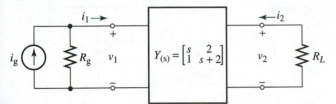

Section 15.4 The Transmission or Chain Parameters

15.4-1. Find the chain matrix for the subcircuit shown in Figure P15.1-2.

15.4-2. Find the chain matrix for the subcircuit shown in Figure P15.1-3.

15.4-3. Find the chain matrix for the subcircuit shown in Figure P15.2-1.

15.4-4. Find the chain matrix for the subcircuit shown in Figure P15.2-2.

15.4-5. Find the chain matrix for the subcircuit shown in Figure P15.2-3.

15.4-6. Find the chain matrix for the subcircuit shown in Figure P15.2-4.

15.4-7. Find the chain matrix for the subcircuit shown in Figure P15.2-5 and show that its determinant Δ_T is unity—and hence that the subcircuit is reciprocal.

15.4-8. Solve Problem 15.2-8 using the chain parameters.

15.4-9. Find the chain matrix for the subcircuit shown in Figure P15.2-7.

15.4-10. Find the chain matrix for the subcircuit shown in Figure P15.2-8.

15.4-11. Solve Problem 15.2-13 using the chain parameters.

15.4-12. Find the reverse chain matrix for the subcircuit shown in Figure P15.3-4.

15.4-13. Find the chain matrix for the two-port subcircuit shown in Figure P15.4-1. *Hint:* Find the chain parameters for each element and note that they are connected in cascade. Assume that each inductor has a value of 1 H and each capacitor, 1 F.

Figure P15.4-1

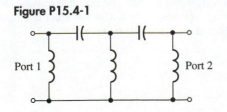

15.4-14. Find the reverse chain matrix for the subcircuit shown in Figure P15.4-1.

Section 15.5 Hybrid Parameters

15.5-1. Find the hybrid H matrix for the subcircuit shown in Figure P15.1-2.

15.5-2. Find the hybrid G matrix for the subcircuit shown in Figure P15.1-3.

15.5-3. Find the hybrid H matrix for the subcircuit shown in Figure P15.2-1.

15.5-4. Find the hybrid G matrix for the subcircuit shown in Figure P15.2-2.

15.5-5. Find the hybrid H matrix for the subcircuit shown in Figure P15.2-3.

15.5-6. Find the hybrid G matrix for the subcircuit shown in Figure P15.2-4.

15.5-7. Find the hybrid H matrix for the subcircuit shown in Figure P15.2-5.

15.5-8. Solve Problem 15.2-8 using the hybrid h parameters. (Convert the z parameters to h parameters, then solve.)

15.5-9. Find the hybrid H matrix for the subcircuit shown in Figure P15.2-7.

15.5-10. Solve Problem 15.2-13 using the hybrid h parameters.

15.5-11. Solve Problem 15.2-15 using the hybrid g parameters.

15.5-12. Find the hybrid h parameters for the composite two-port in Figure P15.5-1.

Figure P15.5-1

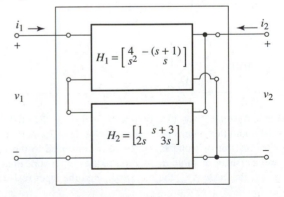

15.5-13. Add a voltage source of value v_g in series with a resistor of value $R_g = 2\ \Omega$ to the input terminals and a resistor of value $R_L = 1\ \Omega$ to the output terminals of the composite two-port in Figure P15.5-1. Compute the voltage gain v_2/v_g, the input resistance v_1/i_1, and the output resistance "seen by" the load resistor R_L.

15.5-14. Find the hybrid g parameters for the composite two-port in Figure P15.5-2 for $k = 1$. If k is allowed to become infinite, which sets of parameters are well defined for the composite two-port? Draw its two-port equivalent for $k = \infty$.

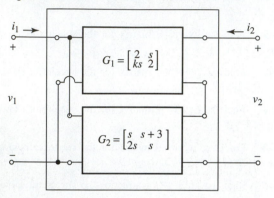

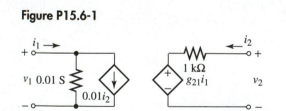

15.5-15. Add a current source of value i_g in parallel with a resistor of value $R_g = 2\,\Omega$ to the input terminals and a resistor of value $R_L = 3\,\Omega$ to the output terminals of the composite two-port in Figure P15.5-2, letting $k = 1$. Compute the current gain $-i_2/i_g$, the input resistance v_1/i_1, and the output resistance "seen by" the load resistor R_L.

Section 15.6 Analysis of Feedback Circuits

15.6-1. Figure P15.6-1 shows a two-port equivalent circuit. Assuming that g_{21} is allowed to become infinitely large, which of the four parameter sets studied in this section (z, y, h, g) exist? Draw the equivalent subcircuit(s) for it (them).

Figure P15.6-1

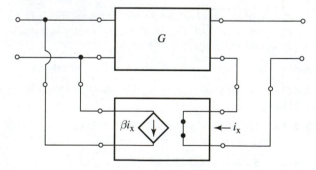

15.6-2. Assume that the two-port in Figure P15.6-1 has $g_{21} = 10$ S and that a current source of value $i_g = 2$ mA in parallel with a resistor of value $R_g = 100\,\Omega$ is connected to the input port, and a resistor of value $R_L = 1$ kΩ to the output port. Find the voltage v_2, the current i_2, the current i_1, and the input resistance $Z_i = v_1/i_1$.

15.6-3. Figure P15.6-2 shows the equivalent models of two interconnected two-ports. Classify the type of feedback, series-series, series-shunt, shunt-series, shunt-shunt (with the amplifier at the top and the feedback two-port at the bottom). Which parameter sets (z, y, h, g) exist for the composite two-port? Draw equivalent two-ports for those that do.

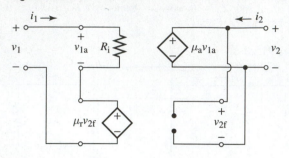

15.6-4. Let a voltage source v_g in series with a resistor R_g be connected to the input terminals of the composite two-port in Figure P15.6-2, while the output terminals remain open circuited. Compute the voltage gain $A_v = v_2/v_g$ and the input impedance $Z_i = v_1/i_1$.

15.6-5. Determine which of the parameter sets z, y, h, and g exist for the composite two-port in Figure P15.6-3 if g_{21} is allowed to approach infinity, assuming that $g_{12} = 0$. Determine the input impedance Z_{is} with the output terminals short circuited, the output impedance Z_{oo} with the input terminals open circuited, the voltage gain $A_v = v_2/v_1$ with the output terminals open circuited, and the current gain $A_i = -i_2/i_1$ with the output terminals shorted for $g_{21} = \infty$ and $g_{12} = 0$. Use the usual definitions for the port variables of the composite two-port.

Figure P15.6-3

15.6-6. Determine which of the parameter sets z, y, h, and g exist for the composite two-port in Figure P15.6-4 if z_{21} is allowed to approach infinity, assuming that $z_{12} = 0$. Determine the input impedance Z_{is} with the output terminals short circuited, the output impedance Z_{os} with the input terminals short circuited, the voltage gain $A_v = v_2/v_1$ with the output terminals open circuited, and the current gain $A_i = -i_2/i_1$ with the output terminals shorted. Assume that $z_{21} = \infty$ and $z_{12} = 0$. Use the usual definitions for the port variables of the composite two-port.

Figure P15.6-4

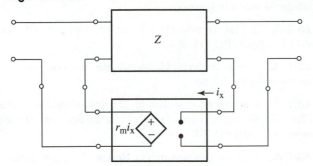

15.6-7. Determine which of the parameter sets z, y, h, and g exist for the composite two-port in Figure P15.6-5 if y_{21} is allowed to approach infinity, assuming that $y_{12} = 0$. Determine the input impedance Z_{io} with the output terminals open circuited, the output impedance Z_{oo} with the input terminals open circuited, the voltage gain $A_v = v_2/v_1$ with the output terminals open circuited, and the current gain $A_i = -i_2/i_1$ with the output terminals shorted. Assume that $y_{21} = \infty$ and $y_{12} = 0$. Use the usual definitions for the port variables of the composite two-port.

Figure P15.6-5

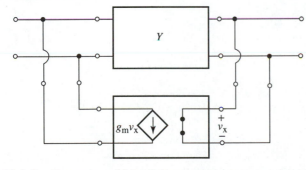

15.6-8. Determine which of the parameter sets z, y, h, and g exist for the composite two-port in Figure P15.6-6 if h_{21} is allowed to approach infinity, assuming that $h_{12} = 0$. Determine the input impedance Z_{io} with the output terminals open circuited, the output impedance Z_{os} with the input terminals short circuited, the voltage gain $A_v = v_2/v_1$ with the output terminals open circuited, and the current gain $A_i = -i_2/i_1$ with the output terminals shorted. Assume that $h_{21} = \infty$ and $h_{12} = 0$. Use the usual definitions for the port variables.

Figure P15.6-6

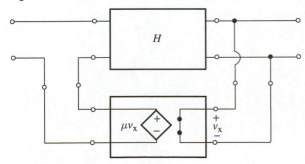

15.6-9. Derive the system function $H(s) = V_o(s)/V_g(s)$ for the circuit in Figure P15.6-7 using two-port techniques. Then find the value(s) of g_m such that the response $v_o(t)$ is stable, unstable, and marginally stable.

Figure P15.6-7

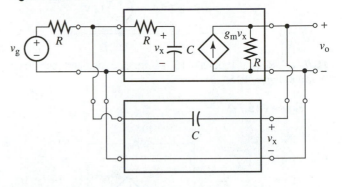

15.6-10. Find the current gain $A_i = -i_2/i_g$, the current gain $A_{io} = -i_2/i_1$, the voltage gain $A_v = v_2/v_1$, the input impedance Z_i shown and the output impedance Z_o shown—the last with i_g reduced to zero—for the circuit in Figure P15.6-8. (Compute all requested values for the limiting case: $g_{21} = \infty$ and $g_{12} = 0$.)

Figure P15.6-8

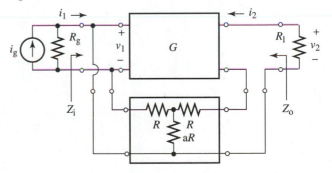

Section 15.7 Two-Ports with Internal Sources: Thévenin and Norton Equivalents

15.7-1. Find the two-port Thévenin equivalent for the two-port subcircuit shown in Figure P15.7-1.

Figure P15.7-1

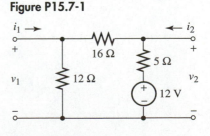

15.7-2. Repeat Problem 15.7-1 for the two-port subcircuit shown in Figure P15.7-2.

Figure P15.7-2

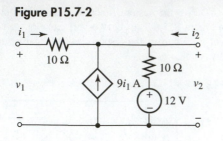

15.7-3. Repeat Problem 15.7-1 for the two-port subcircuit shown in Figure P15.7-3.

Figure P15.7-3

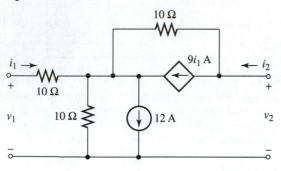

15.7-4. Find the two-port Norton equivalent for the two-port subcircuit shown in Figure P15.7-1.

15.7-5. Repeat Problem 15.7-4 for the two-port subcircuit shown in Figure P15.7-2.

15.7-6. Repeat Problem 15.7-4 for the two-port subcircuit shown in Figure P15.7-3.

15.7-7. Find the two-port hybrid Thévenin-Norton equivalent (Thévenin for port 1, Norton for port 2) for the two-port subcircuit shown in Figure P15.7-1.

15.7-8. Find the two-port hybrid Norton-Thévenin equivalent (Norton for port 1, Thévenin for port 2) for the two-port subcircuit shown in Figure P15.7-3.

16 The Transformer

Section 16.1 begins with a discussion of self-inductance from a physical point of view, including magnetic field intensity in the core, or coil form. It introduces the dot notation as an indication of direction of this field vector. Then a second coil is introduced, thus opening the subject of mutual inductance. The dot notation is extended and the equations are presented for the two-winding transformer. Some simple problems with mutual inductance are solved.

Section 16.2 briefly recalls the z parameters, discusses the z-parameter equivalent for the two-winding transformer, and derives the tee equivalent for transformers having two terminals connected. The inductance matrix is then inverted, assuming that it is nonsingular, to obtain the y parameters and the y-parameter equivalent. This model is used to derive the pi equivalent. Finally, initial condition models for the transformer are derived.

Section 16.3 consists of numerous examples illustrating the analysis of circuits with transformers using the basic equations, the z- and y-parameter equivalents, the tee and pi equivalents, and the initial condition models.

Section 16.4 discusses power and energy in the two-winding transformer. A rigorous proof of reciprocity ($L_{12} = L_{21}$) is given along the lines proposed by Dr. P.M. Lin of Purdue University, who has recently pointed out that the "proofs" found in extant texts are incorrect, and an upper bound for the mutual inductance is derived. This leads to a definition of the coefficient of coupling, followed by discussions of the perfectly coupled transformer and the ideal transformer. Finally, equivalent circuits for the nonideal transformer consisting of primary and secondary inductances (the latter referenced to the primary) and an ideal transformer are derived.

Section 16.5 introduces the SPICE syntax for mutual inductance. The special care that must be taken to avoid floating nodes is illustrated, and the simulation of circuits with ideal transformers is used to illustrate the process (already covered theoretically) of allowing inductances to become infinite to produce the ideal transformer model.

16.1 | Mutual Inductance

In this chapter we will introduce a new circuit element called the *transformer.* It finds application in the alternating current power distribution system, wherein voltages must be altered from their very high transmission levels to values consumers can use for lighting their homes, powering their computers, and running their refrigerators. It also has applications in audio amplifiers and control systems. We will not develop all of the more complicated models for the transformer that take into account various resistances in the transformer windings, power loss in the transformer core, and the like. Instead, we will concentrate on the basic principles of operation. With these basic principles, you will be equipped to understand the more involved models presented, for instance, in a course on electrical power.

Self-Inductance: A Physically Based Discussion

Earlier in the text, we defined inductance in a way that was purely mathematical. The same can be done for the transformer; however, in order to make the principle of transformer action more intuitive, we will review the basic idea of inductance just a bit from a physical point of view. If you haven't yet studied physics on the university level, you still should be able to follow the discussion based only upon your high school physics. It is only intended to be motivational in nature. All the definitions and analysis procedures will be purely mathematical. In Figure 16.1(a) we show a coil of wire wound on a "coil form" or core. In practice, the core is often made of a magnetic material (which can be nonlinear and exhibit memory effects characterized as "hysteresis"), but often it is constructed of a nonmagnetic material such as polystyrene or even air (that is, there is no coil form per se, but the coil shape is maintained by using wire that is stiff enough to support itself). We will assume that the core material is made of a "linear" material, which we will define in a moment.

Figure 16.1
Self-inductance

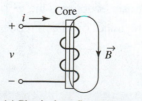

(a) Physical configuration   (b) Electrical symbol

Notice that our coil of wire has a direction associated with it—the direction of "wrap." The top conductor goes over the top of the coil form, back underneath it, and so on. It is important to keep this in mind as we proceed.

Suppose that we inject a current into the top wire; that is, we make i positive in the direction defined in Figure 16.1. It is found experimentally that a "field of force," the *magnetic field intensity* $\vec{B}$, is produced in the near vicinity of the coil. This force field is defined as follows. If we place a small compass (a pivoted magnetic needle) near the coil, it will experience a torque of a certain magnitude and will be deflected in a certain direction. The torque on the needle defines the magnitude of $\vec{B}$ and the direction of the needle the direction of $\vec{B}$, hence both together serve to define this vector field. If we map out the resulting force lines, we will get a series of concentric closed loops passing through the core. They come out of one end and trace a return path outside the core, entering at the other end. A direction is therefore associated with each loop. One such is shown in Figure 16.1(a). The direction of $\vec{B}$ inside the core is given by the "right-hand rule," which, in one form, says the following: if you curl your fingers in the direction in which the current flows through the conductor around the core then your thumb will point in the same direction as the $\vec{B}$ field inside the core.

Now pause to think about our next statement quite carefully: *if the direction of the current flow is reversed, the direction of $\vec{B}$ will reverse. If, however, we simultaneously reverse the wrap of the winding and inject a positive current into the bottom terminal, the direction of $\vec{B}$ will <u>not</u> reverse.* If we only reverse one—either the current or the wrap direction—the direction of $\vec{B}$ *will* reverse. The situation in which we have reversed both the current and the wrap direction is shown in Figure 16.2(a). Now look at the electrical inductor symbols shown in Figures 16.1(b) and 16.2(b). Because they are drawn in two dimensions rather than three and do not depict the core, we lose information about the direction of $\vec{B}$. One way we can solve this problem is by placing a black dot on the terminal of the inductor for which positive current produces a $\vec{B}$ field in a given direction in the core, say upward. This "dot notation" is shown in Figures 16.1(b) and 16.2(b).

Figure 16.2
Self-inductance:
opposite winding sense

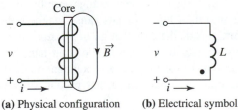

(a) Physical configuration   (b) Electrical symbol

Now, let's recall our earlier definition of *flux linkage.* It is related to voltage by

$$\lambda(t) = \frac{1}{p}\, v(t) = \int_{-\infty}^{t} v(\alpha)\, d\alpha. \tag{16.1-1}$$

There is a famous experimental principle associated with coils and magnetic fields known as *Faraday's law* after the great English electrical experimentalist Michael Faraday. It asserts that the magnitude of the flux linkage λ is directly proportional to the magnitude of the magnetic field intensity $\vec{B}$. Furthermore, another law[1] states that the magnitude of $\vec{B}$ is directly proportional to the current. (This is what we mean by a "linear core," one for which this latter statement is true.) In this case, it is clear that the magnitude of the flux

---

[1] Discovered by the American engineer Joseph Henry, Faraday (whose name appears in a very wide variety of electrical phenomena), and the French physicists Oersted, Biot, and Savart.

linkage is directly proportional to the current:

$$\lambda(t) = Li(t). \tag{16.1-2}$$

This is often called Henry's law after Joseph Henry. The constant L is called the *inductance,* or the *self-inductance* of the coil of wire.

Equations (16.1-1) and (16.1-2) give us the inductor v-i relationship, expressed as

$$i(t) = \frac{\lambda(t)}{L} = \frac{1}{L}\int_{-\infty}^{t} v(\alpha)\, d\alpha \tag{16.1-3}$$

(in integral form) or as

$$v(t) = p\lambda(t) = Lpi(t) \tag{16.1-4}$$

(in derivative form). The only question is, what is the sign of the voltage? (So far, we have been interpreting equations (16.1-1), (16.1-2), (16.1-3), and (16.1-4) as magnitudes.) As you might expect, there is a law covering the sign. It is called Lenz's law,[2] and it goes like this: if the inductor voltage (and flux linkage) reference has its positive sign on the terminal with the dot and the current reference points into the terminal with the dot, then equations (16.1-3) and (16.1-4) are valid—with the sign included. For instance, equation (16.1-4) says that if the current reference arrow points into the dotted terminal and the voltage reference has the plus sign on the dotted terminal, then $v(t)$ will be positive if di/dt is positive (that is, if $i(t)$ is increasing).

Adding a Second Winding: Coupled Inductors

Perhaps it seems that all of this explanation is unneeded because if we reverse the direction of both the current arrow and the plus voltage reference in Figures 16.1 and 16.2, we will simply cause both sides of equations (16.1-3) and (16.1-4) to be negative; hence, we can change them both back to positive and obtain the original forms again. The direction of $\vec{B}$ inside the core does not matter. True, but it will if we add another coil of wire in the near vicinity of the first. The situation is shown in Figures 16.3(a) and 16.3(b). At the top of Figure 16.3(a) we see that the windings are placed on a common core in such a fashion that currents into their top terminals cause magnetic field intensities (shown by the loop with the arrow) inside the core that go *in the same direction* around the core. The bottom sketch shows the situation in which the right-hand winding sense has been changed so that current into its top terminal causes a magnetic field intensity inside the core that is directed in the opposite direction to that produced by the winding on the left. We have,

Figure 16.3
The two-winding transformer

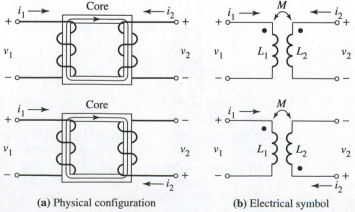

(a) Physical configuration   (b) Electrical symbol

[2] After Heinrich F.E. Lenz, who was born in Germany, but did his work in Russia.

however, sketched the situation that results when the current in the right-hand winding is reversed so that positive current goes into its bottom terminal rather than the top and the positive voltage reference is reversed so that it is on the bottom terminal as well.

In Figure 16.3(b) we show the corresponding electrical circuit diagrams for our *two-winding transformer*. The dots on the two windings indicate that—for either winding—if positive current is fed into the dotted terminal, then the direction of the magnetic field intensity around the core will be clockwise. The two situations depicted in Figure 16.3 exhaust all the possibilities. But what if both windings were reversed relative to Figure 16.3(a) (top)? As a matter of fact, this would result in exactly the same situation. To see why, just suppose you have meters connected so as to measure both currents and voltages in the arrangement at the top of Figure 16.3. Suppose that I ask you to close your eyes for a moment and, while your eyes are shut, I sneakily unhook the transformer terminals, flip the whole transformer upside down then reattach the terminals (to measure current into the new top terminals and positive voltage on them). You then open your eyes. Would you see a difference in the meter readings? If you think about it for just a moment, you realize that you will not—the transformer will look exactly the same as before the switch (even down to the direction of the field intensity inside the core)! We can go through the identical argument for the configuration at the bottom of Figure 16.3.

Mutual Inductance

How is a transformer different from an ordinary inductor? The answer is that if the two windings are a long way apart (imagine yourself to be holding one while I take the other and walk out of the room with it—perhaps we have to imagine an air core here), the two windings behave just like a pair of ordinary coils. If we move them closer and closer together, however, a new phenomenon occurs. The magnetic field intensity set up by the current in one winding adds to that produced by the other. Thus, each winding "sees" a modified field. This phenomenon is called *mutual inductance*. The current in a given winding sets up a magnetic field that consists of two parts: one goes through the core inside the winding, then out of the winding, but does not stay in the core; rather, it escapes out of the core and loops back through free space outside of the core to enter the other end. The other part goes around the core, through the other winding, and then back through the first again. We show this schematically in Figure 16.4. Following this line of reasoning, we see that the flux linkage of a given winding has two parts also: one that is proportional to the current in the same winding and another that is proportional to the current in the other winding. We call the effect of the first *self-inductance* and that of the second *mutual inductance*.

Figure 16.4
Mutual and
self-inductance

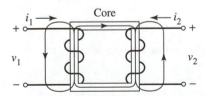

The Basic Transformer Equations

Assuming a linear core (as we have already discussed), we see that we can write

$$\lambda_1 = L_{11}i_1 + L_{12}i_2 \tag{16.1-5a}$$

and

$$\lambda_2 = L_{21}i_1 + L_{22}i_2. \tag{16.1-5b}$$

(The subscripts on the coefficients are chosen in the same way as the two-port parameters we developed in the last chapter.) In matrix form we have

$$\begin{bmatrix} \lambda_1 \\ \lambda_2 \end{bmatrix} = \begin{bmatrix} L_{11} & L_{12} \\ L_{21} & L_{22} \end{bmatrix} \begin{bmatrix} i_1 \\ i_2 \end{bmatrix} \tag{16.1-6}$$

or, more compactly, as

$$\bar{\lambda} = L\bar{i}.$$

(16.1-7)

In this equation $\bar{\lambda}$ is the (2×1) column matrix on the left side of (16.1-6), $\bar{i}$ the (2×1) column matrix on the right side of (16.1-6), and L the (2×2) matrix on the right side of (16.1-6). Equation (16.1-7) clearly shows that the terminal relationships for a transformer are a generalization of the (scalar) defining equation for self inductance: $\lambda = Li$.

We can simplify notation because it is a fact (to be proved later) that

$$L_{12} = L_{21} = M.$$

(16.1-8)

We have used a single symbol M to represent their common value: the *mutual inductance*. At the same time, we simplify our notation and let

$$L_1 = L_{11}$$

(16.1-9)

$$L_2 = L_{22}$$

(16.1-10)

We will call the left-hand winding of the transformer the *primary winding* and the right-hand one the *secondary winding*. Alternative terms are *input* and *output*, respectively. This is quite arbitrary because often we will turn the transformer around and refer to the windings in the opposite sense. In some applications a "power transformer" is plugged into the ac wall receptacle to transform[3] the voltages down to a lower voltage that can be applied to solid-state circuits. In this case, the "110-V side" would probably be called the input or the primary, and the lower voltage side the "load side," the secondary, or the output. If you were working for your local power company, though, and were concerned with noise or transients or radio frequency interference that a customer's solid-state devices were injecting into the power line, you might be tempted to call the 110-V side the output or secondary and the lower voltage side the input or primary. In any event, we will call L_1 the *primary inductance* and L_2 the *secondary inductance* and use the circuit symbols shown in Figure 16.3(b), repeated here for reference as Figure 16.5.

Figure 16.5
Circuit symbols for the two-winding transformer

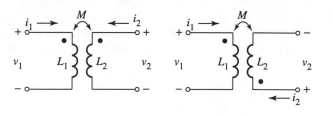

The equations that define the two-winding transformer, in our new notation, are

$$\lambda_1 = L_1 i_1 + M i_2$$

(16.1-11a)

and

$$\lambda_2 = M i_1 + L_2 i_2,$$

(16.1-11b)

or, in matrix form,

$$\begin{bmatrix} \lambda_1 \\ \lambda_2 \end{bmatrix} = \begin{bmatrix} L_1 & M \\ M & L_2 \end{bmatrix} \begin{bmatrix} i_1 \\ i_2 \end{bmatrix}.$$

(16.1-12)

Here is something that you should commit to memory right now, for it will be quite useful to you as you solve problems with transformers: *if you define both transformer current reference arrows <u>into</u> the dotted terminals and both voltage reference*

---

[3] Hence the term "transformer."

plus signs _on these same terminals_, then equations (16.1-11) and (16.1-12) hold with _positive signs for all variables and parameters_.

At this point we should perhaps point out that electrical circuits are solved in terms of voltage and current, so if we take note of the flux linkage-voltage relationship in equation (16.1-1), we can write equations (16.1-11) and (16.1-12) in the form

$$v_1 = L_1 p i_1 + M p i_2 \tag{16.1-13a}$$

and

$$v_2 = M p i_1 + L_2 p i_2, \tag{16.1-13b}$$

or, in matrix form,

$$\begin{bmatrix} v_1 \\ v_2 \end{bmatrix} = \begin{bmatrix} L_1 p & M p \\ M p & L_2 p \end{bmatrix} \begin{bmatrix} i_1 \\ i_2 \end{bmatrix}. \tag{16.1-13c}$$

If you are working with Laplace transforms, you need only change p to s; for Fourier transforms or phasors, replace p by $j\omega$. The resulting equations are still valid. In matrix form, they are

$$\begin{bmatrix} V_1(s) \\ V_2(s) \end{bmatrix} = \begin{bmatrix} L_1 s & M s \\ M s & L_2 s \end{bmatrix} \begin{bmatrix} I_1(s) \\ I_2(s) \end{bmatrix}, \tag{16.1-14a}$$

$$\begin{bmatrix} V_1(\omega) \\ V_2(\omega) \end{bmatrix} = \begin{bmatrix} j\omega L_1 & j\omega M \\ j\omega M & j\omega L_2 \end{bmatrix} \begin{bmatrix} I_1(\omega) \\ I_2(\omega) \end{bmatrix}, \tag{16.1-14b}$$

$$\begin{bmatrix} \overline{V}_1 \\ \overline{V}_2 \end{bmatrix} = \begin{bmatrix} j\omega L_1 & j\omega M \\ j\omega M & j\omega L_2 \end{bmatrix} \begin{bmatrix} \overline{I}_1 \\ \overline{I}_2 \end{bmatrix}. \tag{16.1-14c}$$

We also make the simple corresponding changes in the circuit symbols, as in Figure 16.6.

Figure 16.6
Laplace, Fourier, and phasor models

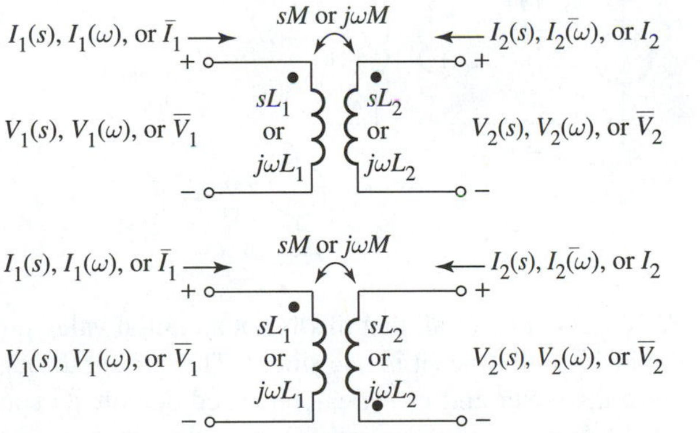

$I_1(s), I_1(\omega)$, or $\overline{I}_1 \longrightarrow$ sM or $j\omega M$ $\longleftarrow I_2(s), I_2(\omega)$, or I_2

$V_1(s), V_1(\omega)$, or $\overline{V}_1$ sL_1 or $j\omega L_1$ sL_2 or $j\omega L_2$ $V_2(s), V_2(\omega)$, or $\overline{V}_2$

$I_1(s), I_1(\omega)$, or $\overline{I}_1 \longrightarrow$ sM or $j\omega M$ $\longleftarrow I_2(s), I_2(\omega)$, or I_2

$V_1(s), V_1(\omega)$, or $\overline{V}_1$ sL_1 or $j\omega L_1$ sL_2 or $j\omega L_2$ $V_2(s), V_2(\omega)$, or $\overline{V}_2$

Examples

Example 16.1   Let $i_g(t) = 4e^{-2t}$ A in Figure 16.7. Find $v_1(t)$ and $v_2(t)$.

Figure 16.7
An example circuit

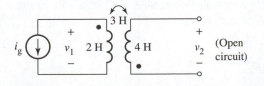

i_g v_1 2 H   3 H   4 H v_2 (Open circuit)

Solution   _Recall our system: define both transformer voltage references to be_ + _on the dotted terminals and both current reference arrows into the dots._ Doing so results in Figure 16.8. (Notice that the given definition of v_2 is contrary to this prescription, and our solution is to simply reverse it and change its sign.) After this "preprocessing" step, we can write

$$v_1 = 2p(-i_g) + 3p(0) = -2p\{4e^{-2t}\} = 16e^{-2t} \text{ V} \qquad (16.1\text{-}15a)$$

and $\qquad\qquad (-v_2) = 3p(-i_g) + 4p(0) = -3p\{4e^{-2t}\} = 24e^{-2t} \text{ V}. \qquad (16.1\text{-}15b)$

The second equation can then be solved for v_2:

$$v_2 = -24e^{-2t} \text{ V}. \qquad (16.1\text{-}16)$$

The secret here is to keep signs straight by forcing all currents and voltages to conform to our reference convention by reversing signs if necessary, then write both transformer equations (in the new variables) with positive signs.

Figure 16.8
The example circuit
after preprocessing
for analysis

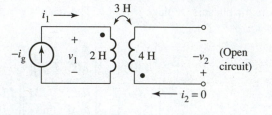

Example 16.2   Find $v(t)$ in Figure 16.9 if $v_g(t) = 8u(t)$ V.

Figure 16.9
An example circuit

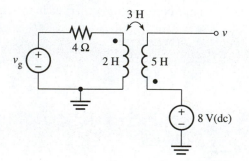

Solution   We observe first of all that this is not an initial value problem. The input voltage source is defined for all time (it is one sided). The current through the open-circuited secondary of the transformer and its series-connected dc voltage source is zero for all time (including $t \leq 0$). What we are illustrating here is that the transformer is a true two-port. The voltage reference for the secondary port is completely independent from the reference for the primary port and does not enter at all into the transformer equations.

Let's prepare the circuit for analysis by defining appropriate primary and secondary variables for the transformer and, instead of inductances, let's use complex impedances that are functions of the variable s. This results in the circuit of Figure 16.10. We have noted that the transform of $v_g(t)$ is

$$V_g(s) = \frac{8}{s}. \qquad (16.1\text{-}17)$$

We have also turned the dc source in the secondary into a one-sided one by multiplication by the unit step function—then we have taken the Laplace transform.[4] We can now write

---

[4] You can, as usual, interpret the variable s as the operator p. In this case, you should leave the sources in the time domain.

the *s*-domain equation:

$$\frac{8}{s} - 4I_1(s) = V_1(s) = 2sI_1(s) + 3s(0). \qquad (16.1\text{-}18)$$

Rearranging, we get

$$2(s + 2)I_1(s) = \frac{8}{s}. \qquad (16.1\text{-}19)$$

The solution is

$$I_1(s) = \frac{4}{s(s + 2)} = \frac{2}{s} - \frac{2}{s + 2}. \qquad (16.1\text{-}20)$$

(We performed a partial fraction expansion, or PFE, the details of which we leave for you to do.) Next, we write one KVL equation for the secondary, getting

$$V_2(s) = 3sI_1(s) + 5s(0) = \frac{12}{s + 2}. \qquad (16.1\text{-}21)$$

Inverting the two Laplace transforms, we have

$$i_1(t) = 2[1 - e^{-2t}]u(t) \text{ A} \qquad (16.1\text{-}22)$$

and

$$v(t) = 8 - v_2(t) = 8 - 12e^{-2t}u(t) \text{ V}. \qquad (16.1\text{-}23)$$

This result is only valid for $t > 0$, so our complete solution for $v(t)$ is

$$v(t) = \begin{cases} 8; & t \le 0 \\ 8 - 12e^{-2t}; & t > 0 \end{cases} \text{ V}. \qquad (16.1\text{-}24)$$

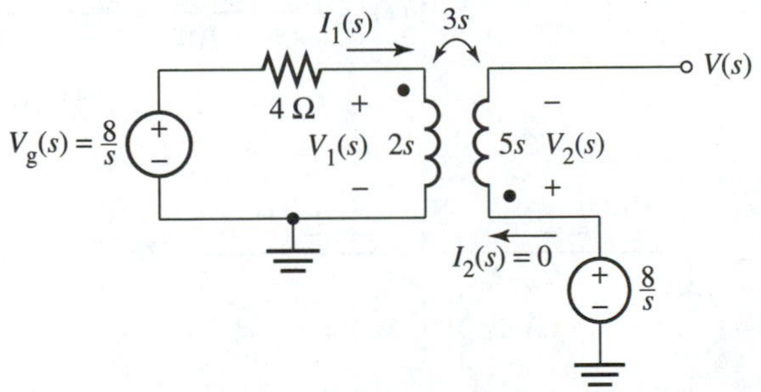

Figure 16.10
The *s*-domain
equivalent

The preceding example illustrates an important use for transformers: to isolate the ground reference of one circuit from that of another for safety reasons. There is no direct electrical connection between the two ports of the transformer—the electrical energy is coupled from one to the other magnetically.

Example 16.3   Find the response voltage $v(t)$ in Figure 16.11 if $v_g(t) = 20u(t)$ V.

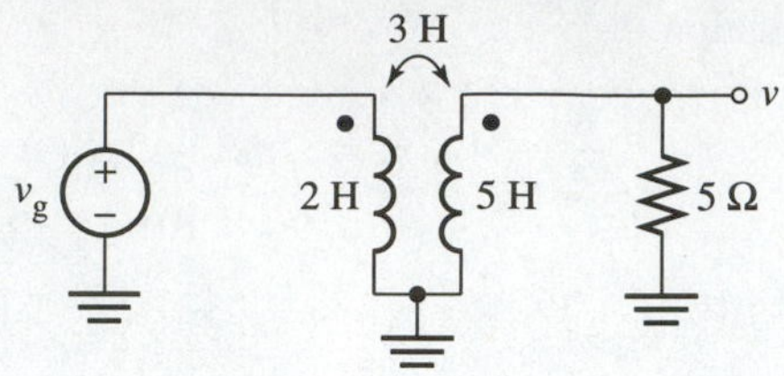

Figure 16.11
An example circuit

Solution   Notice that this is exactly the same transformer as in Example 16.2, with the secondary winding terminals reversed and used in a different circuit. We show the circuit prepared for analysis in Figure 16.12. We have again defined our terminal variables to be consistent with the dots. We write

$$V_1(s) = V_g(s) = 2sI_1(s) + 3sI_2(s) \qquad (16.1\text{-}25)$$

and

$$V(s) = V_2(s) = -5I_2(s) = 3sI_1(s) + 5sI_2(s). \qquad (16.1\text{-}26)$$

Thus

$$I_2(s) = \frac{-3s}{5(s+1)}I_1(s). \qquad (16.1\text{-}27)$$

Using this in (16.1-25) gives

$$V_g(s) = 2sI_1(s) + \frac{-9s^2}{5(s+1)}I_1(s) = \frac{s(s+10)}{5(s+1)}I_1(s), \qquad (16.1\text{-}28)$$

so

$$I_1(s) = \frac{5(s+1)}{s(s+10)}V_g(s). \qquad (16.1\text{-}29)$$

Thus,

$$V(s) = V_2(s) = -5I_2(s) = \frac{15s}{5(s+1)}I_1(s) = \frac{15}{s+10}V_g(s). \qquad (16.1\text{-}30)$$

Setting $V_g(s) = 20/s$, we get

$$V(s) = \frac{300}{s(s+10)} = \frac{30}{s} - \frac{30}{s+10}, \qquad (16.1\text{-}31)$$

or

$$v(t) = 30[1 - e^{-10t}]u(t) \text{ V}. \qquad (16.1\text{-}32)$$

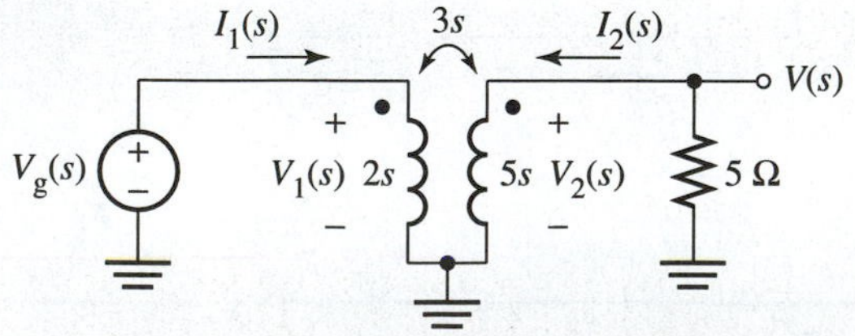

Figure 16.12
The example circuit
prepared for analysis

This solves the example problem, but we wish to remark upon one more thing before we finish. The solution, plotted in Figure 16.13 for positive values of time, is quite deceptive. We see that the voltage $v(t)$ approaches a constant 30 V as $t \to \infty$. But this implies that the current i_2 is constant, and so it has a zero derivative. As both independent variables for the transformer are currents—and, in fact, derivatives of currents—we see that the primary current must continue to change with time in order to produce a nonzero secondary voltage. We might, therefore, be interested in just how this happens, because we already know that the primary voltage is a constant for all positive time. Let's compute $i_1(t)$ and see. Its Laplace transform is

$$I_1(s) = \frac{5(s+1)}{s(s+10)}V_g(s) = \frac{100(s+1)}{s^2(s+10)} = \frac{9}{s} + \frac{10}{s^2} - \frac{9}{s+10}, \tag{16.1-33}$$

so

$$i_1(t) = [10t + 9 - 9e^{-10t}]u(t) \text{ A}. \tag{16.1-34}$$

This waveform is plotted for positive values of time in Figure 16.14. After initially rising nonlinearly, after a certain time interval it continues to increase linearly with time! This, of course, would finally lead to a destructive situation in practice. Linear models of our source and transformer would no longer be valid at high current levels.

If we look carefully at the system function relating the secondary voltage to the input voltage (the s-domain fraction multiplying $V_g(s)$ in (16.1-30)), we see that it is stable. That is reflected by the behavior of the response waveform in Figure 16.13. The system function relating $I_1(s)$ to $V_g(s)$ in (16.1-29), however, is only marginally stable because it has a pole at $s = 0$. This corresponds to the step response asymptotically going to infinity (after multiplication by $20/s$ there is a double pole at $s = 0$).

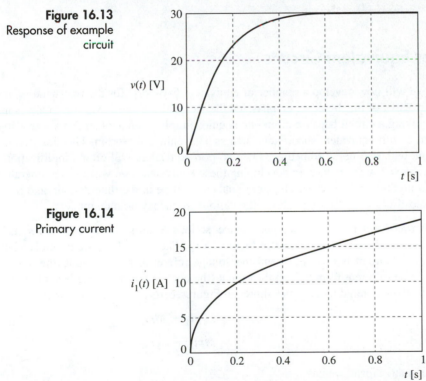

Figure 16.13
Response of example
circuit

Figure 16.14
Primary current

The main item of caution is simply that a voltage source placed in parallel with a winding of a transformer creates a special situation, exactly like the one in which a voltage source is placed in parallel with a simple inductor. The dc equivalent circuit for both elements is a short circuit (in the case of the transformer each port is equivalent to a short circuit), and a short circuit across an ideal voltage source is simply disallowed, for it violates KVL. In such a case, the dc steady state does not exist.

Section 16.1 Quiz

Q16.1-1. Find the secondary voltages for each of the circuits in Figure Q16.1-1. Assume, in each case, that $i(t) = -2tu(t)$ A.

Figure Q16.1-1

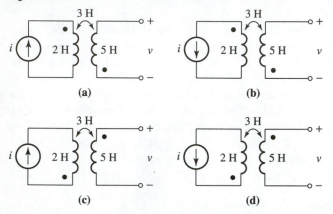

(a)   (b)

(c)   (d)

Q16.1-2. Solve for the current $i(t)$ and voltage $v(t)$ in Figure Q16.1-2 assuming that $v_g(t) = 4u(t)$ V. Do dc steady-state conditions exist in the circuit as $t \to \infty$?

Figure Q16.1-2

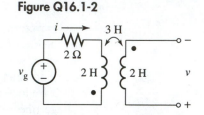

16.2 | Transformer Equivalent Circuits

We will now develop a number of equivalent two-ports for the two-winding transformer. In Part I of this book we emphasized the fact that one needs various analysis methods for solving a circuit because often one is quite simpler than another. This was also seen to be true with equivalent subcircuits such as Thévenin and Norton. The situation is no different with two-ports: sometimes one two-port equivalent will offer simplification that is not achieved with another. In developing these equivalents we will use the operator $p = d/dt$ in preference to s or $j\omega$. Thus, our analysis will be in the time domain and one need only substitute s or $j\omega$ for p to obtain the various frequency domain models.

The z-Parameter Transformer Equivalent

A transformer is a true two-port device, so let's recall the z parameters for such a subcircuit. We have reviewed them in Figure 16.15. The equivalent circuit holds whether or not the subcircuit is reciprocal, and the voltage reference for the input (the port on the left) can be different from that on the output (the port on the right). Let's apply these results to the two-terminal transformer shown in Figure 16.16(a). The equations are

$$v_1 = L_1 p i_1 + M p i_2 \qquad (16.2\text{-}1\text{a})$$

and

$$v_2 = M p i_1 + L_2 p i_2. \qquad (16.2\text{-}1\text{b})$$

In matrix form, we have

$$\begin{bmatrix} v_1 \\ v_2 \end{bmatrix} = \begin{bmatrix} L_1 p & M p \\ M p & L_2 p \end{bmatrix} \begin{bmatrix} i_1 \\ i_2 \end{bmatrix}. \qquad (16.2\text{-}2)$$

Thus, we have

$$z_{11} = L_1 p, \qquad (16.2\text{-}3\text{a})$$

$$z_{22} = L_2 p, \qquad (16.2\text{-}3\text{b})$$

and

$$z_{12} = z_{21} = M p. \qquad (16.2\text{-}3\text{c})$$

Interpreting equations (16.2-1a) and (16.2-1b) as the KVL equations of a two-loop subcircuit, we have the equivalent shown in Figure 16.16(b). Remember that we are now allowing gain parameters for dependent sources to contain the operator p (or s or $j\omega$).

Figure 16.15
Review of two-port z parameters

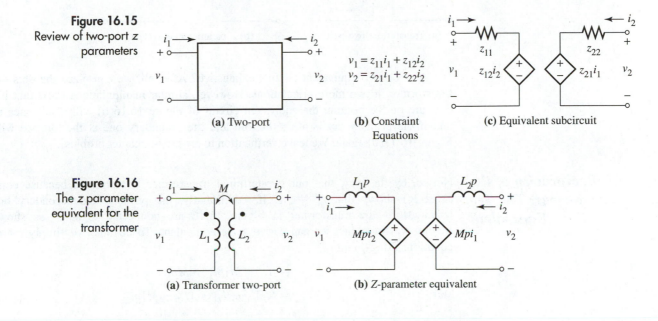

(a) Two-port

$$v_1 = z_{11}i_1 + z_{12}i_2$$
$$v_2 = z_{21}i_1 + z_{22}i_2$$

(b) Constraint Equations

(c) Equivalent subcircuit

Figure 16.16
The z parameter equivalent for the transformer

(a) Transformer two-port

(b) Z-parameter equivalent

Alternate Dot Orientations

You might object that we have assumed a particular dot orientation for our transformer—and rightly so. Suppose we change the dots to the pattern shown in Figure 16.17(a). What happens to the equivalent circuit? Recall that we have decided to always choose the terminal voltages on a transformer to be positive on the dots and the current reference arrows to be pointing into the dots. Thus, equations (16.2-1) and (16.2-2) continue to hold true; *however, our general two-port of Figure 16.15 assumes that both port voltages are positive at the top terminal and that both port currents are directed into those terminals.* This convention is consistent with our transformer variables for port 1 but not for port 2. Therefore, let's reverse the signs and references of the secondary winding variables by defining

$$v_2' = -v_2 \tag{16.2-4a}$$

and

$$i_2' = -i_2. \tag{16.2-4b}$$

Using these new variables in equations (16.2-1) gives

$$v_1 = L_1 p i_1 - M p i_2' \tag{16.2-5a}$$

and

$$v_2' = -M p i_1 + L_2 p i_2'. \tag{16.2-5b}$$

In matrix form, we have

$$\begin{bmatrix} v_1 \\ v_2' \end{bmatrix} = \begin{bmatrix} L_1 p & -Mp \\ -Mp & L_2 p \end{bmatrix} \begin{bmatrix} i_1 \\ i_2' \end{bmatrix}. \tag{16.2-6}$$

The effect has been to change the sign of the mutual term. This gives the equivalent subcircuit in Figure 16.17(b).

Figure 16.17
z-Parameter equivalent
for the transformer:
alternate dot
orientation

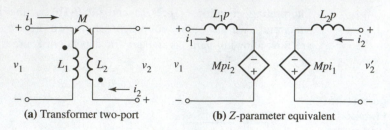

(a) Transformer two-port   (b) Z-parameter equivalent

Are the two equivalent circuits exhaustive? After all, we can place the dots on our transformer in two more orientations. However, similar manipulations show that if both dots are on the bottom, the equivalent circuit of Figure 16.16(b) will result once more, and if the primary dot is at the bottom and the secondary one at the top, we will get Figure 16.17(b) again. We leave verification to an end of chapter problem.

Reciprocity of the Transformer: The Tee Equivalent

Notice, by the way, that our two-terminal transformer is reciprocal because equation (16.2-3c) shows that $z_{12} = z_{21}$. If, additionally, the primary and secondary bottom terminals of the transformer in Figure 16.16 are connected together, as shown in Figure 16.18(a), then we can derive a tee equivalent. To do this, we simply rearrange equations (16.2-1) into

$$v_1 = [L_1 - M]pi_1 + Mp[i_1 + i_2] \tag{16.2-7a}$$

and

$$v_2 = Mp[i_1 + i_2] + [L_2 - M]pi_2. \tag{16.2-7b}$$

The references for the two port voltages are the same, so we can interpret these equations as the tee equivalent subcircuit shown in Figure 16.18(b). We have chosen to label the inductors with their inductance parameters, though we could have included the p operators (or s or $j\omega$) to form operator or complex impedances had we so desired.

Figure 16.18
The tee equivalent
subcircuit

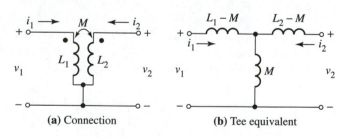

(a) Connection   (b) Tee equivalent

What about the transformer in Figure 16.17 with the alternate dot orientation? The easiest way to see what happens to our tee equivalent is to notice that the only difference between equations (16.2-1) corresponding to the transformer in Figure 16.16 and equations (16.2-5) describing the one in Figure 16.17 is a negative sign on the mutual inductance in the latter. Thus, we can draw the tee equivalent for the transformer connection in Figure 16.19(a) by simply reversing the signs on the M terms in Figure 16.18(b). We remind you that we have reversed the secondary terminal voltage and current to conform with our two-port impedance parameter convention. For this reason, they are labeled with primes. Because we are dealing with *equivalent* circuits in Figures 16.18 and 16.19, either of the two top inductors in Figure 16.18(b) could turn out to be negative—as is the vertical one in Figure 16.19(b). We will explore possible ranges of values for M in a later section of this chapter. We must, however, still admit negative inductances in our equivalent model.

Figure 16.19
The tee equivalent for
the alternate
connection

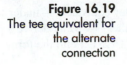
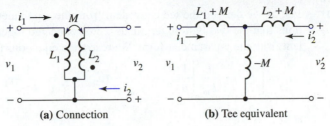

(a) Connection **(b) Tee equivalent**

The y-Parameter Equivalent for the Transformer

Just as we did for the general two-port subcircuit, we can derive a y-parameter equivalent subcircuit. We recall the matrix form of the transformer equations in (16.2-2):

$$\begin{bmatrix} v_1 \\ v_2 \end{bmatrix} = \begin{bmatrix} L_1 & M \\ M & L_2 \end{bmatrix} \begin{bmatrix} pi_1 \\ pi_2 \end{bmatrix}. \tag{16.2-8}$$

Notice that we have chosen to associate the differentiation operator $p = d/dt$ with the currents rather than keeping them in the matrix of coefficients. Assuming that the determinant of the inductance matrix is nonzero, that is

$$\Delta = L_1 L_2 - M^2 \neq 0, \tag{16.2-9}$$

we can invert to obtain

$$\begin{bmatrix} i_1 \\ i_2 \end{bmatrix} = \begin{bmatrix} L_2/\Delta & -M/\Delta \\ -M/\Delta & L_1/\Delta \end{bmatrix} \begin{bmatrix} \dfrac{1}{p} v_1 \\ \dfrac{1}{p} v_2 \end{bmatrix} = \begin{bmatrix} 1/L_a & -1/L_m \\ -1/L_m & 1/L_b \end{bmatrix} \begin{bmatrix} \dfrac{1}{p} v_1 \\ \dfrac{1}{p} v_2 \end{bmatrix}, \tag{16.2-10}$$

where

$$L_a = \Delta/L_2, \tag{16.2-11a}$$

$$L_b = \Delta/L_1, \tag{16.2-11b}$$

and

$$L_m = \Delta/M. \tag{16.2-11c}$$

In nonmatrix form, equation (16.2-10) becomes

$$i_1 = \frac{1}{L_a p} v_1 - \frac{1}{L_m p} v_2, \tag{16.2-12a}$$

and

$$i_2 = -\frac{1}{L_m p} v_1 + \frac{1}{L_b p} v_2. \tag{16.2-12b}$$

Interpreting these as KCL equations leads to the equivalent subcircuit shown in Figure 16.20. We have labeled the equivalent inductances with their inductance parameters; however, we have used the full operator notation for the transadmittances because they express current values in terms of voltages at another port. We would like to underscore the following caution: *the y-parameter equivalent only exists if* $\Delta = L_1 L_2 - M^2 \neq 0$.

Figure 16.20
The y-parameter
equivalent subcircuit

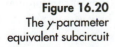
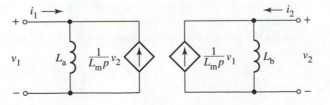

The Pi Equivalent for the Transformer

Just as we derived the tee equivalent from the z-parameter equivalent—under the conditions that the two bottom terminals were connected together—we can derive a pi equivalent from the y-parameter form. We rearrange equations (16.2-12) into the form

$$i_1 = \left[\frac{1}{L_a} - \frac{1}{L_m} \right] \frac{1}{p} v_1 + \frac{1}{L_m} \frac{1}{p}[v_1 - v_2] = \frac{1}{L_a' p} v_1 + \frac{1}{L_m p}[v_1 - v_2], \quad (16.2\text{-}13a)$$

and

$$i_2 = \frac{1}{L_m} \frac{1}{p}[v_2 - v_1] + \left[\frac{1}{L_b} - \frac{1}{L_m} \right] \frac{1}{p} v_2 = \frac{1}{L_m p}[v_2 - v_1] + \frac{1}{L_b' p} v_2. \quad (16.2\text{-}13b)$$

If we interpret these equations as KCL expressions, we have the equivalent subcircuit of Figure 16.21(b): the pi equivalent for the transformer. The element values are given by

$$L_a' = \frac{\Delta}{L_2 - M}, \quad (16.2\text{-}14a)$$

$$L_b' = \frac{\Delta}{L_1 - M}, \quad (16.2\text{-}14b)$$

and

$$L_m = \frac{\Delta}{M}. \quad (16.2\text{-}14c)$$

In deriving these values, we have simply matched coefficients in equations (16.2-13) and used the definitions in (16.2-11). Note that reversal of one of the dots in Figure 16.21(a) would merely reverse the sign of M in all of the preceding equations.

Figure 16.21
The pi equivalent subcircuit

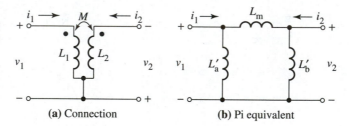

(a) Connection   (b) Pi equivalent

We could, of course, also derive the two hybrid equivalents at this point; however, we will not do this. Instead, we will look at the models we have just derived in the Laplace and frequency domains and discuss how initial conditions fit into our model structures.

Laplace and Fourier Models for the Transformer

The Laplace and Fourier domain models are very easy. We simply compute the Laplace transform of equations (16.2-1), obtaining

$$V_1(s) = L_1 s I_1(s) + M s I_2(s) \quad (16.2\text{-}15a)$$

$$V_2(s) = M s I_1(s) + L_2 s I_2(s). \quad (16.2\text{-}15b)$$

(We would do the same with the alternate dot connection of Figure 16.17; however, to conserve space, we will leave this as an exercise for you to do.) The only difference is that we now are dealing with nonoperator equations: functions of the complex variable s. The differentiation operation now becomes multiplication by s. If we evaluate these equations on the $j\omega$ axis (or take the limit as $\sigma \to 0+$, as we explained in Chapter 14), we will have

$$V_1(\omega) = j\omega L_1 I_1(\omega) + j\omega M I_2(\omega) \quad (16.2\text{-}16a)$$

$$V_2(\omega) = j\omega M I_1(\omega) + j\omega L_2 I_2(\omega). \quad (16.2\text{-}16b)$$

We simply use these algebraic equations in our circuit analysis procedures rather than the operator ones given in (16.2-1).

Initial Condition
Models for
the Transformer

Incorporation of initial conditions into our transformer model is somewhat more interesting. If you recall, our procedure in solving *RLC* circuits with initial conditions on the inductors and/or the capacitors was to force all waveforms to zero for negative time values. This introduced initial condition sources into the inductor and capacitor models. We will do the same with the transformer. Let's multiply both sides of (16.2-1) by the unit step function $u(t)$ to obtain

$$v_{1+} = v_1 u(t) = L_1 pi_1 u(t) + Mpi_2 u(t) \tag{16.2-17a}$$

and

$$v_{2+} = v_2 u(t) = Mpi_1 u(t) + L_2 pi_2 u(t). \tag{16.2-17b}$$

We recall that, for any waveform $x(t)$, we have

$$px_+(t) = p[x(t)u(t)] = [px(t)]u(t) + x(0)\delta(t). \tag{16.2-18}$$

Letting $x(t)$ be $v_1(t)$ and $v_2(t)$ sequentially and solving (16.2-18) for $[px(t)]u(t)$, then using the result in (16.2-17), we get the following:

$$v_{1+} = L_1 pi_{1+} + Mpi_{2+} - [L_1 i_1(0) + Mi_2(0)]\delta(t) \tag{16.2-19a}$$
$$= L_1 pi_{1+} + Mpi_{2+} - \lambda_1(0)\delta(t)$$

$$v_{2+} = Mpi_{1+} + L_2 pi_{2+} - [Mi_1(0) + L_2 i_2(0)]\delta(t) \tag{16.2-19b}$$
$$= Mpi_{1+} + L_2 pi_{2+} - \lambda_2(0)\delta(t).$$

We have recognized that the terms in brackets are the initial values of the primary and secondary flux linkages, respectively. The associated initial condition model is shown in Figure 16.22. If the dots are on opposite ends of the respective windings of the transformer, one approach is to simply recall that our positive transformer voltage reference is on the dot and the current into it. Thus, we merely flip either or both of the halves of our model upside down. We will leave it to you to go through the redefinition of port voltage and port current in order to derive the initial condition model of the transformer with the alternate dot locations.

Figure 16.22
The initial condition
transformer model

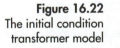
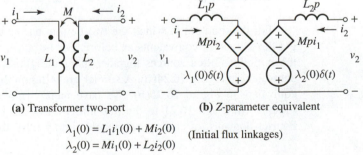

(a) Transformer two-port   **(b)** Z-parameter equivalent

$$\lambda_1(0) = L_1 i_1(0) + Mi_2(0)$$
$$\lambda_2(0) = Mi_1(0) + L_2 i_2(0)$$ (Initial flux linkages)

Now let's see how the initial conditions affect our *y*-parameter model. We first write equations (16.2-19) in matrix form:

$$\begin{bmatrix} v_{1+} \\ v_{2+} \end{bmatrix} = \begin{bmatrix} L_1 & M \\ M & L_2 \end{bmatrix}\begin{bmatrix} pi_{1+} \\ pi_{2+} \end{bmatrix} - \begin{bmatrix} L_1 & M \\ M & L_2 \end{bmatrix}\begin{bmatrix} i_1(0) \\ i_2(0) \end{bmatrix}\delta(t). \tag{16.2-20}$$

We now multiply by the inverse of the inductance matrix, assuming that its determinant is nonzero (that is, $\Delta = L_1 L_2 - M^2 \neq 0$), giving

$$\begin{bmatrix} i_{1+} \\ i_{2+} \end{bmatrix} = \begin{bmatrix} L_2/\Delta & -M/\Delta \\ -M/\Delta & L_1/\Delta \end{bmatrix} \begin{bmatrix} \dfrac{1}{p} v_{1+} \\ \dfrac{1}{p} v_{2+} \end{bmatrix} + \begin{bmatrix} i_1(0) \\ i_2(0) \end{bmatrix} u(t). \qquad (16.2\text{-}21)$$

As before, we let

$$L_a = \Delta/L_2, \qquad\qquad\qquad (16.2\text{-}22a)$$

$$L_b = \Delta/L_1, \qquad\qquad\qquad (16.2\text{-}22b)$$

and
$$L_m = \Delta/M. \qquad\qquad\qquad (16.2\text{-}22c)$$

Writing the equations out explicitly gives

$$i_{1+} = \frac{1}{L_a p} v_{1+} - \frac{1}{L_m p} v_{2+} + i_1(0)u(t) \qquad (16.2\text{-}23a)$$

and
$$i_{2+} = -\frac{1}{L_m p} v_{1+} + \frac{1}{L_b p} v_{2+} + i_2(0)u(t). \qquad (16.2\text{-}23b)$$

The matrix form of these equations is

$$\begin{bmatrix} i_{1+} \\ i_{2+} \end{bmatrix} = \begin{bmatrix} 1/(L_a p) & -1/(L_m p) \\ -1/(L_m p) & 1/(L_b p) \end{bmatrix} \begin{bmatrix} v_{1+} \\ v_{2+} \end{bmatrix} + \begin{bmatrix} i_1(0) \\ i_2(0) \end{bmatrix} u(t). \qquad (16.2\text{-}24)$$

Interpreting our equations in terms of KCL, we have the equivalent in Figure 16.23. We stress that this model only holds if $\Delta \neq 0$.

Figure 16.23
The y-parameter initial condition equivalent

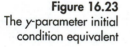

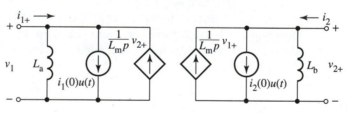

For the situation in which the two bottom terminals are connected together, we can get the tee and pi equivalents as before. For instance, we can group all elements except the initial condition sources together in Figure 16.22(b) to get the tee initial condition equivalent in Figure 16.24(b). A similar grouping on the one in Figure 16.23 gives the pi equivalent in Figure 16.25(b). We have merely applied the transformations shown in Figures 16.18 and 16.21 to the non-initial-condition elements alone. In the Laplace or Fourier domains, of course, one would simply label the inductance symbols with sL or

Figure 16.24
The tee initial condition equivalent

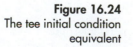

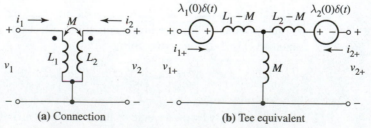

(a) Connection   (b) Tee equivalent

Figure 16.25
The pi initial condition
equivalent

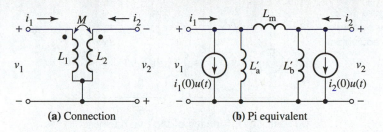

(a) Connection   **(b)** Pi equivalent

$j\omega L$, respectively. The initial condition sources would have values equal to the transform of the delta or step functions as appropriate. Notice that we have once again left the analysis of the alternate dot connection up to you.

Section 16.2 Quiz

All questions pertain to the two-terminal transformer shown in Figure Q16.2-1.

Figure Q16.2-1

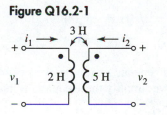

Q16.2-1. Draw the z-parameter equivalent and label all element values.

Q16.2-2. Draw the y-parameter equivalent and label all element values.

Q16.2-3. Assuming that the bottom terminals are connected together, draw the tee equivalent and label all element values.

Q16.2-4. Assuming that the bottom terminals are connected together, draw the pi equivalent and label all element values.

Q16.2-5. Now assume that initial currents of $i_1(0) = 2$ A and $i_2(0) = 3$ A are flowing in the transformer terminals. Answer Questions Q16.2-1 through Q16.2-4 using this information.

16.3 | Analysis of Circuits with Transformers

We will now illustrate how to incorporate transformers into the methods of circuit analysis that we have developed earlier in the text. The basic idea is to first write KVL or KCL equations (perhaps using mesh or nodal analysis) in terms of the transformer variables that have been defined relative to the dots on its windings, then to insert the transformer relations. Another alternative is to replace the transformer with one of its several equivalents containing two-terminal elements and to perform an analysis on the resulting circuit. We will now work some examples of both methods.

Example 16.4   Find the response voltage $v(t)$ for the circuit shown in Figure 16.26 if $v_g(t) = 4u(t)$ V.

Figure 16.26
An example circuit

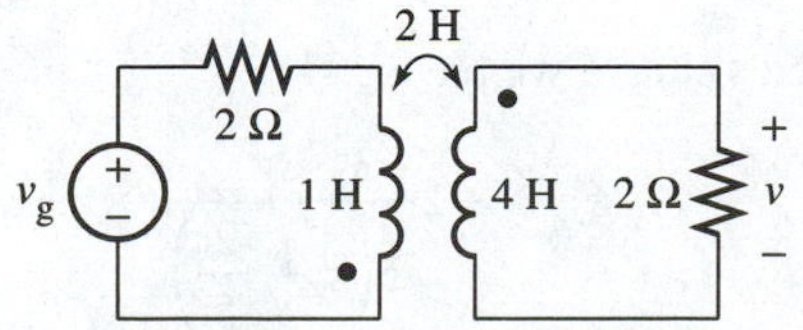

Solution   Our first step is to define the various terminal variables for the transformer. Anticipating doing the analysis work in the s domain, we will use impedance parameters to describe it. The resulting circuit is shown in Figure 16.27. We have defined the transformer currents *into* the dots and the transformer voltages plus *on* the dots. Choosing to do mesh analysis,

Figure 16.27
The s-domain
equivalent

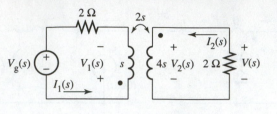

we write KVL for each of the two meshes:

$$V_1(s) + 2I_1(s) = -V_g(s) \qquad (16.3\text{-}1a)$$

$$V_2(s) + 2I_2(s) = 0. \qquad (16.3\text{-}1b)$$

Observe carefully that we have not yet written the transformer relationships; we have simply written KVL in terms of the transformer voltages and currents. Next, we express $V_1(s)$ and $V_2(s)$ in terms of the currents $I_1(s)$ and $I_2(s)$ using the transformer equations (which can be written with positive signs everywhere because of the way we defined the terminal variables). This results in

$$[sI_1(s) + 2sI_2(s)] + 2I_1(s) = -V_g(s) \qquad (16.3\text{-}2a)$$

and

$$[2sI_1(s) + 4sI_2(s)] + 2I_2(s) = 0. \qquad (16.3\text{-}2b)$$

Notice here how systematic we are being: in general, we use brackets or braces to set off the transformer voltages at this stage. Next, we simplify these equations into

$$(s + 2)I_1(s) + 2sI_2(s) = -V_g(s) \qquad (16.3\text{-}3a)$$

and

$$2sI_1(s) + 2(2s + 1)I_2(s) = 0. \qquad (16.3\text{-}3b)$$

In matrix form, these equations become

$$\begin{bmatrix} s + 2 & 2s \\ 2s & 2(2s + 1) \end{bmatrix} \begin{bmatrix} I_1(s) \\ I_2(s) \end{bmatrix} = \begin{bmatrix} -V_g(s) \\ 0 \end{bmatrix}. \qquad (16.3\text{-}4)$$

The solution for the mesh currents is

$$\begin{bmatrix} I_1(s) \\ I_2(s) \end{bmatrix} = \frac{1}{10\left(s + \dfrac{2}{5}\right)} \begin{bmatrix} 2(2s + 1) & -2s \\ -2s & s + 2 \end{bmatrix} \begin{bmatrix} -V_g(s) \\ 0 \end{bmatrix} = \begin{bmatrix} \dfrac{-(2s + 1)}{5\left(s + \dfrac{2}{5}\right)} \\ \dfrac{s}{5\left(s + \dfrac{2}{5}\right)} \end{bmatrix} V_g(s). \qquad (16.3\text{-}5)$$

Solving for the desired voltage, we have

$$V(s) = -2I_2(s) = -2\frac{s}{5\left(s + \dfrac{2}{5}\right)} \times \frac{4}{s} = \frac{-8}{5\left(s + \dfrac{2}{5}\right)}. \qquad (16.3\text{-}6)$$

Inverting, we get

$$v(t) = -1.6e^{-2t/5}\, u(t)\ \text{V}. \qquad (16.3\text{-}7)$$

Our next example shows the power of the method we have just presented when the signs are a bit trickier!

Example 16.5   Find the response voltage $v(t)$ for the circuit shown in Figure 16.28. Assume that $i_g(t) = 10u(t)$ A.

Figure 16.28
An example circuit

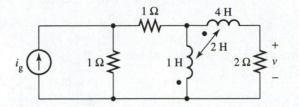

Solution   We note that there really is a transformer in this circuit! Even though they have been somewhat stretched and bent relative to the configuration we have been studying, the two inductors are coupled (thus forming a two-winding transformer), as shown by the arrow labeled with the mutual inductance of $M = 2$ H. We will again perform mesh analysis because nodal analysis is somewhat more difficult than mesh analysis when transformers are involved. We will discuss this in more detail a bit later in this section.

Here's our plan of attack. We see that there are three meshes, but the one with the current source is a nonessential one. Therefore, we must define an unknown mesh current in each of the other meshes. We will do so in such a manner that both enter the dots on the transformer terminals. We show the s-domain equivalent in Figure 16.29. Notice that we have defined our transformer voltages to be positive on the dots. In the following analysis, keep one thing firmly in mind: the total current into the primary dotted terminal *is not* $I_1(s)$, but the sum of $I_1(s)$ and $I_2(s)$—and we must deal with the total current. The KVL equation for the central mesh is, summing voltage drops,

$$V_1(s) + 1 \cdot I_1(s) + 1[I_1(s) + I_g(s)] = 0. \qquad (16.3\text{-}8a)$$

For the rightmost mesh we have, summing voltage drops,

$$2I_2(s) + V_1(s) + V_2(s) = 0. \qquad (16.3\text{-}8b)$$

Notice that both the transformer port voltages are drops in the direction defined by the mesh current $I_2(s)$.

Figure 16.29
The s-domain
equivalent
preprocessed for
mesh analysis

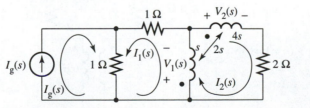

Our next step is to express the transformer port voltages in terms of the transformer terminal currents. (Do you notice the similarity of this process to the "taping and untaping" of dependent sources?) This gives

$$[s\{I_1(s) + I_2(s)\} + 2sI_2(s)] + 1 \cdot I_1 + 1[I_1(s) + I_g(s)] = 0 \qquad (16.3\text{-}9a)$$

and

$$2I_2(s) + [s\{I_1(s) + I_2(s)\} + 2sI_2(s)] + [2s\{I_1(s) + I_2(s)\} + 4sI_2(s)] = 0. \qquad (16.3\text{-}9b)$$

Please stop! Go back over these two equations one more time—then again and again until you are quite sure about where each term came from. The use of brackets or parentheses or braces is highly recommended. That way you can keep track of which terms are due to the transformer voltages and which are due to other elements in the circuit. After grouping terms, the preceding equations have the simpler forms

$$(s + 2)I_1(s) + 3sI_2(s) = -I_g(s) \tag{16.3-10a}$$

and

$$3sI_1(s) + (9s + 2)I_2(s) = 0. \tag{16.3-10b}$$

In matrix form these equations become

$$\begin{bmatrix} s + 2 & 3s \\ 3s & 9s + 2 \end{bmatrix} \begin{bmatrix} I_1(s) \\ I_2(s) \end{bmatrix} = \begin{bmatrix} -I_g(s) \\ 0 \end{bmatrix}. \tag{16.3-11}$$

Inversion gives

$$\begin{bmatrix} I_1(s) \\ I_2(s) \end{bmatrix} = \frac{1}{4(5s + 1)} \begin{bmatrix} 9s + 2 & -3s \\ -3s & s + 2 \end{bmatrix} \begin{bmatrix} -I_g(s) \\ 0 \end{bmatrix} = \begin{bmatrix} -\dfrac{9s + 2}{4(5s + 1)} I_g(s) \\ \dfrac{3s}{4(5s + 1)} I_g(s) \end{bmatrix}. \tag{16.3-12}$$

Thus, because $I_g(s) = 10/s$, we get

$$V(s) = 2I_2(s) = \frac{3s}{2(5s + 1)} \times \frac{10}{s} = \frac{3}{s + \dfrac{1}{5}}. \tag{16.3-13}$$

Finally, we have

$$v(t) = 3e^{-t/5} u(t) \text{ V}. \tag{16.3-14}$$

We will now discuss nodal analysis which, as we have said, is a bit more complicated than mesh analysis.

Example 16.6   Find the response current $i(t)$ for the circuit shown in Figure 16.30 using nodal analysis.

Figure 16.30
An example circuit

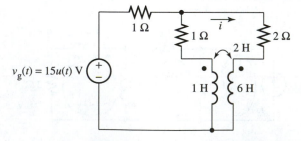

$v_g(t) = 15u(t)$ V

Solution   Our circuit is shown prepared for nodal analysis in the s domain in Figure 16.31. We have defined all the transformer variables consistently with our "dot convention," attached a ground reference, and labeled all the node voltages. We will treat the transformer as a single passive element. The KCL equation at the top node is

$$\frac{V_3(s) - V_g(s)}{1} + \frac{V_3(s) - V_1(s)}{1} + \frac{V_3(s) - V_2(s)}{2} = 0. \tag{16.3-15}$$

Multiplying both sides by 2, then collecting terms, gives

$$-2V_1(s) - V_2(s) + 5V_3(s) = 2V_g(s). \tag{16.3-16}$$

There is no voltage source connected to a transformer terminal, and both bottom terminals are attached to ground, so we will treat both remaining transformer terminals as essential nodes.

Figure 16.31
The example circuit prepared for nodal analysis in the s domain

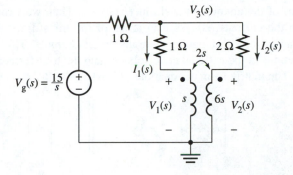

Here is a crucial point to remember: in nodal analysis, we must express the current in any passive element in terms of the node voltages. The transformer equations in s-domain impedance matrix form are

$$\begin{bmatrix} V_1(s) \\ V_2(s) \end{bmatrix} = \begin{bmatrix} s & 2s \\ 2s & 6s \end{bmatrix} \begin{bmatrix} I_1(s) \\ I_2(s) \end{bmatrix}, \tag{16.3-17}$$

so we invert them to obtain the admittance representation

$$\begin{bmatrix} I_1(s) \\ I_2(s) \end{bmatrix} = \begin{bmatrix} \dfrac{3}{s} & \dfrac{-1}{s} \\ \dfrac{-1}{s} & \dfrac{1}{2s} \end{bmatrix} \begin{bmatrix} V_1(s) \\ V_2(s) \end{bmatrix}. \tag{16.3-18}$$

(We could have "cheated" by leafing back through the last section, but it is much better to merely remember the basic transformer relationship, insert s or $j\omega$ as appropriate—recognizing the result as the z-parameter representation—and then convert to whatever form you need.) Writing (16.3-18) as two scalar equations, we have

$$I_1(s) = \frac{3}{s}V_1(s) - \frac{1}{s}V_2(s) \tag{16.3-19}$$

and

$$I_2(s) = -\frac{1}{s}V_1(s) + \frac{1}{2s}V_2(s). \tag{16.3-20}$$

Next, we write nodal equations at each of the nonreference transformer terminals:

$$\frac{V_1(s) - V_3(s)}{1} + I_1(s) = \frac{V_1(s) - V_3(s)}{1} + \frac{3}{s}V_1(s) - \frac{1}{s}V_2(s) = 0 \tag{16.3-21}$$

and

$$\frac{V_2(s) - V_3(s)}{2} + I_2(s) = \frac{V_2(s) - V_3(s)}{2} - \frac{1}{s}V_1(s) + \frac{1}{2s}V_2(s) = 0. \tag{16.3-22}$$

We can "massage" these two equations a bit to obtain

$$(s + 3)V_1(s) - V_2(s) - sV_3(s) = 0 \tag{16.3-23}$$

and

$$-2V_1(s) + (s + 1)V_2(s) - sV_3(s) = 0. \tag{16.3-24}$$

Let's pause for a moment to recap. What we have done is to treat the transformer as a single passive element connected to the top node. We have then written one KCL equation at the top terminal (note that we have not written any at the transformer terminals) in terms of the unknown transformer voltages. Then we wrote the transformer relationships with the currents expressed in terms of the transformer voltages and the external node voltages, thus eliminating the transformer currents. This gives an additional set of two equations in the external node voltages and the transformer port voltages.

The matrix form of equations (16.3-16), (16.3-23), and (16.3-24) is

$$\begin{bmatrix} -2 & -1 & 5 \\ (s + 3) & -1 & -s \\ -2 & (s + 1) & -s \end{bmatrix} \begin{bmatrix} V_1(s) \\ V_2(s) \\ V_3(s) \end{bmatrix} = \begin{bmatrix} 2V_g(s) \\ 0 \\ 0 \end{bmatrix} = \begin{bmatrix} \dfrac{30}{s} \\ 0 \\ 0 \end{bmatrix}. \tag{16.3-25}$$

We will leave it to you to apply your favorite algorithm to solve for the node voltages. They are

$$V_1(s) = \frac{30(s + 2)}{(2s + 1)(s + 5)}, \tag{16.3-26}$$

$$V_2(s) = \frac{30(s + 5)}{(2s + 1)(s + 5)} = \frac{30}{(2s + 1)}, \tag{16.3-27}$$

and

$$V_3(s) = \frac{30(s^2 + 4s + 1)}{s(2s + 1)(s + 5)}. \tag{16.3-28}$$

Notice that there is a pole-zero cancellation in the expression for $V_2(s)$, which often occurs. This means that $v_2(t)$ will be missing an exponential term that is present in $v_1(t)$ and $i_2(t)$. What we want, however, is the current $i(t) = i_2(t)$. In the s domain, therefore,

$$I(s) = \frac{V_3(s) - V_2(s)}{2} = \frac{-15(s - 1)}{s(2s + 1)(s + 5)}. \tag{16.3-29}$$

Therefore, the time domain current we are looking for is

$$i(t) = [3 - 5e^{-t/2} + 2e^{-5t}]u(t). \tag{16.3-30}$$

Now let's recap our nodal analysis procedure. We treat the nodes attached to transformer terminals as essential if no v-source or ground reference is attached to them, as nonessential if a grounded v-source is so attached, simply as a reference if the ground is attached, and as part of a supernode if a nongrounded v-source is attached. We next write appropriate nodal equations, using the y parameters of the transformer to express the transformer terminal currents in terms of the transformer port voltages. Note that this procedure only works if the y parameters exist (that is, if the determinant of the Z parameter matrix is nonzero).

Example 16.7   Find the response current $i(t)$ for the circuit in Example 16.6, repeated for reference here as Figure 16.32, using either a pi or a tee equivalent circuit.

Figure 16.32
An example circuit

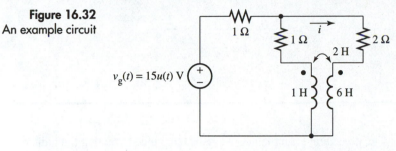

$v_g(t) = 15u(t)$ V

Solution   Let's first decide which equivalent subcircuit to use. The bottom terminals are connected together, so we can use either the pi or the tee equivalent. The pi equivalent would result in a circuit with three meshes and five nodes; the tee equivalent would give one with two meshes and five nodes. Not only is the latter simpler (one fewer meshes), but a number of elements appear in series. Therefore, we will work with the s-domain tee equivalent shown in Figure 16.33. (Glance back at the derivation in the last section to help justify the values we have used.) You should sketch the pi equivalent to verify our statements about relative simplicity. Notice, by the way, that one of the inductances in the tee equivalent is negative; this gives rise to the negative sign in one of the inductive impedances. One of the advantages of the tee (or pi) equivalent is simply that after converting to the equivalent we are not concerned with relative polarities or references for the two sets of transformer currents and voltages. In this case, we merely perform, say, mesh analysis on an RL circuit. The mesh equations are

$$(s + 2)I_x(s) - (1 - s)I(s) = V_g(s) \tag{16.3-31}$$

$$-(1 - s)I_x(s) + (3s + 3)I(s) = 0. \tag{16.3-32}$$

In matrix form we have

$$\begin{bmatrix} (s + 2) & -(1 - s) \\ -(1 - s) & (3s + 3) \end{bmatrix} \begin{bmatrix} I_x(s) \\ I(s) \end{bmatrix} = \begin{bmatrix} V_g(s) \\ 0 \end{bmatrix}. \tag{16.3-33}$$

Inversion gives

$$\begin{bmatrix} I_x(s) \\ I(s) \end{bmatrix} = \frac{1}{2s^2 + 11s + 5} \begin{bmatrix} (3s + 3) & (1 - s) \\ (1 - s) & (s + 2) \end{bmatrix} \begin{bmatrix} V_g(s) \\ 0 \end{bmatrix}. \tag{16.3-34}$$

Figure 16.33
The s-domain tee equivalent

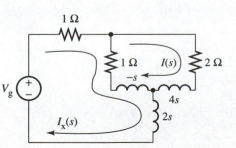

Thus,

$$I(s) = \frac{(1-s)}{2s^2 + 11s + 5}V_g(s) = \frac{(1-s)}{(2s+1)(s+5)} \times \frac{15}{s} \qquad (16.3\text{-}35)$$

$$= \frac{3}{s} - \frac{5}{s + \frac{1}{2}} + \frac{2}{s+5},$$

so

$$i(t) = [3 - 5e^{-t/2} + 2e^{-5t}]u(t) \text{ A.} \qquad (16.3\text{-}36)$$

The preceding example has, we hope, convinced you that the judicious use of equivalent models can be a computational time-saver.

Example 16.8   Find the system function $H(s)$ and the impulse response $h(t)$ for the voltage variable $v(t)$ in Figure 16.34.

Figure 16.34
An example circuit

Solution   This time, even though we could use either a tee or a pi, let's simply use the z-parameter equivalent circuit, as we show in Figure 16.35. Notice that we have taped the two dependent sources and labeled them V_{c1} and V_{c2}. Furthermore, notice that they have polarities depending upon the directions of the currents *into* the transformer terminals. Thus,

$$V_{c1} = s[I_3 - I_2] \qquad (16.3\text{-}37)$$

and

$$V_{c2} = s[I_1 - I_3]. \qquad (16.3\text{-}38)$$

(The mutual impedance is $Ms = s$.) The mesh equations are

$$2s(I_1 - I_3) = -V_{c1} + V_g = -s(I_3 - I_2) + V_g, \qquad (16.3\text{-}39)$$

$$1I_2 + s(I_2 - I_3) = V_{c2} = s(I_1 - I_3), \qquad (16.3\text{-}40)$$

and

$$\frac{1}{s}I_3 + s(I_3 - I_2) + 2s(I_3 - I_1) = V_{c1} - V_{c2} \qquad (16.3\text{-}41)$$

$$= s(I_3 - I_2) - s(I_1 - I_3) = s(2I_3 - I_1 - I_2).$$

Figure 16.35
The z-parameter
equivalent

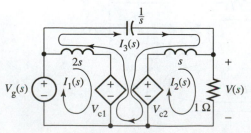

Combining terms, we get

$$2sI_1 - sI_2 - sI_3 = V_g, \tag{16.3-42}$$

$$-sI_1 + (s + 1)I_2 = 0, \tag{16.3-43}$$

and

$$-sI_1 + \left(\frac{1}{s} + s\right)I_3 = 0. \tag{16.3-44}$$

Solving for $V(s) = I_2(s)$, we obtain

$$V(s) = I_2(s) = \frac{s^2 + 1}{s^2 + s + 2} V_g(s), \tag{16.3-45}$$

so—because $V_g(s) = \mathcal{L}\{\delta(t)\} = 1$—the transfer function is

$$H(s) = \frac{s^2 + 1}{s^2 + s + 2}. \tag{16.3-46}$$

The impulse response is simply the inverse transform of $H(s)$, which we rewrite as

$$H(s) = \frac{s^2 + 1}{s^2 + s + 2} = 1 - \frac{s + \dfrac{1}{2} + \dfrac{1}{2}}{\left(s + \dfrac{1}{2}\right)^2 + \dfrac{7}{4}} \tag{16.3-47}$$

$$= 1 - \frac{s + \dfrac{1}{2}}{\left(s + \dfrac{1}{2}\right)^2 + \left(\dfrac{\sqrt{7}}{2}\right)^2} - \frac{1}{\sqrt{7}} \frac{\dfrac{\sqrt{7}}{2}}{\left(s + \dfrac{1}{2}\right)^2 + \left(\dfrac{\sqrt{7}}{2}\right)^2}.$$

Taking the inverse Laplace transform gives

$$h(t) = \delta(t) - e^{-t/2}\left[\cos\left(\frac{\sqrt{7}t}{2}\right) + \frac{1}{\sqrt{7}}\sin\left(\frac{\sqrt{7}t}{2}\right)\right]u(t)\ \text{V}. \tag{16.3-48}$$

Example 16.9   Rework Example 16.8 using nodal analysis and the y-parameter equivalent subcircuit for the transformer.

Solution   Again, rather than playing "cookbook engineer," we prefer to recall the Z matrix form for the transformer equations (glancing back at Figure 16.34 to refresh our memory),

$$\begin{bmatrix} V_1(s) \\ V_2(s) \end{bmatrix} = \begin{bmatrix} L_1 & M \\ M & L_2 \end{bmatrix}\begin{bmatrix} sI_1(s) \\ sI_2(s) \end{bmatrix} = \begin{bmatrix} 2 & 1 \\ 1 & 1 \end{bmatrix}\begin{bmatrix} sI_1(s) \\ sI_2(s) \end{bmatrix}, \tag{16.3-49}$$

where we have inserted our values of $L_1 = 2$ H, $L_2 = 1$ H, and $M = 1$ H. We have chosen to make things even easier than usual by factoring the common s from each impedance parameter and placing it in the column matrix of currents. The determinant of the inductance matrix is

$$\Delta = L_1L_2 - M^2 = 2 \times 1 - 1^2 = 1. \tag{16.3-50}$$

We invert the inductance matrix now and dividing by s, obtaining

$$
\begin{bmatrix} I_1(s) \\ I_2(s) \end{bmatrix} = \begin{bmatrix} \dfrac{1}{s} & \dfrac{-1}{s} \\ \dfrac{-1}{s} & \dfrac{2}{s} \end{bmatrix} \begin{bmatrix} V_1(s) \\ V_2(s) \end{bmatrix}.
\qquad (16.3\text{-}51)
$$

We can now interpret the corresponding scalar equations as KCL equations at the input and output nodes to draw the equivalent circuit shown in Figure 16.36. We have recognized that $V_1(s) = V_g(s)$ and $V_2(s) = V(s)$ in this circuit. The node driven by the input source is a nonessential node, so we need only one equation at the output node:

$$
s(V - V_g) + \frac{2}{s}V + \frac{V}{1} = \frac{1}{s}V_g,
\qquad (16.3\text{-}52)
$$

which we solve at once to get

$$
H(s) = \frac{V(s)}{V_g(s)} = \frac{s^2 + 1}{s^2 + s + 2}
\qquad (16.3\text{-}53)
$$

as before.

Figure 16.36
The *y*-parameter
equivalent

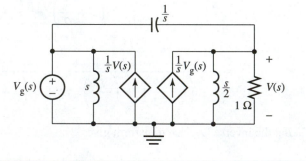

Is it clear that one equivalent subcircuit, when used in some given circuit, might be much more convenient than another? The nodal method in the last example was far less cumbersome than the mesh method in the preceding example. We will not, however, attempt to present any general rules for when one should use a given equivalent in preference to another. You must simply use your intuition, coupled with counting of nodes, branches, and so on, to determine the complexity of your analysis before starting. Let's continue with our examples by working one involving initial conditions.

Example 16.10   The circuit in Figure 16.37 is known to be operating in the dc steady state for $t \leq 0$. Find the response voltage $v(t)$ for all positive values of time.

Figure 16.37
An example circuit

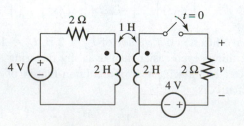

Solution   We start by noting that in the dc steady state all voltages and currents are constant. This implies that the transformer ports are equivalent to short circuits because, in that case, all time derivatives are zero. This implies that

$$v_1 = L_1 p i_1 + M p i_2 = 0 \tag{16.3-54a}$$

and

$$v_2 = M p i_1 + L_2 p i_2 = 0. \tag{16.3-54b}$$

Thus, for $t \leq 0$, the equivalent circuit is that shown in Figure 16.38. We see at once that $i_1 = 2$ A into the dot and $i_2 = 0$. Thus, the initial flux linkages are

$$\lambda_1(0) = L_1 i_1(0) + M i_2(0) = 2 \times 2 + 1 \times 0 = 4 \text{ Wb} \tag{16.3-55a}$$

and

$$\lambda_2(0) = M i_1(0) + L_2 i_2(0) = 1 \times 2 + 2 \times 0 = 2 \text{ Wb}. \tag{16.3-55b}$$

Figure 16.38
Dc steady-state
equivalent for $t \leq 0$

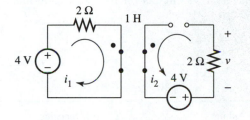

Removing the constraint of zero voltages in (16.3-54), then multiplying both sides by $u(t)$ and using our usual "one-sided truncation" technique to introduce initial conditions, we obtain

$$v_{1+} = L_1 p i_{1+} + M p i_{2+} - [L_1 i_1(0) + M i_2(0)]u(t) \tag{16.3-56a}$$

and

$$v_{2+} = M p i_{1+} + L_2 p i_{2+} - [M i_1(0) + L_2 i_2(0)]u(t). \tag{16.3-56b}$$

Using our known transformer values and the initial flux linkages from (16.3-55),

$$v_{1+} = 2 p i_{1+} + p i_{2+} - 4 \delta(t) \tag{16.3-57a}$$

and

$$v_{2+} = 1 p i_{1+} + 2 p i_{2+} - 2 \delta(t). \tag{16.3-57b}$$

Next, we transform these equations into the s domain, obtaining the $t > 0$ equivalent shown in Figure 16.39. We now write two mesh equations:

$$(2s + 2)I_1(s) + s I_2(s) = 4 + \frac{4}{s} \tag{16.3-58a}$$

and

$$s I_1(s) + (2s + 2)I_2 = 2 + \frac{4}{s}. \tag{16.3-58b}$$

Solving these two equations simultaneously for $I_2(s)$, we get

$$I_2(s) = \frac{8(s + 1)}{s(3s + 2)(s + 2)}. \tag{16.3-59}$$

Figure 16.39
s-Domain initial
condition model
for $t > 0$

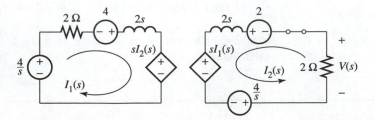

Writing $V(s) = -2I_2(s)$ and doing a partial fraction expansion, we get

$$V(s) = \frac{-16(s+1)}{s(3s+2)(s+2)} = \frac{-4}{s} + \frac{2}{s+\dfrac{2}{3}} + \frac{2}{s+2}. \qquad (16.3\text{-}60)$$

Inverting, we obtain

$$v(t) = [-4 + 2e^{-2t/3} + 2e^{-2t}]u(t) \text{ V}. \qquad (16.3\text{-}61)$$

Example 16.11   Find the equivalent subcircuits for the two two-terminal transformer subcircuits shown in Figure 16.40.

Figure 16.40
Two transformers
connected with their
windings in series

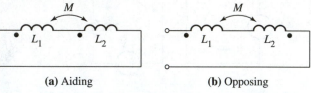

(a) Aiding   (b) Opposing

Solution   The operator equivalent for the connection shown in part (a) is shown in Figure 16.41. We have attached a test source that we have selected somewhat arbitrarily to be a current source. Notice that the current i goes into both dots in the "aiding" configuration. Thus, we have

$$v = v_1 + v_2 = L_1 pi + Mpi + Mpi + L_2 pi = (L_1 + L_2 + 2M)pi. \qquad (16.3\text{-}62)$$

Therefore, the subcircuit is equivalent to a single inductor having the inductance value

$$L_{\text{eq}} = L_1 + L_2 + 2M. \qquad (16.3\text{-}63)$$

Figure 16.41
Testing the subcircuits

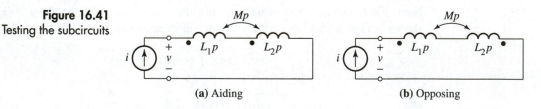

(a) Aiding   (b) Opposing

In the "opposing" configuration in Figures 16.40(a) and 16.41(b), on the other hand, we note that the current i goes into the left-hand winding dot and exits the dot on the right-hand winding. Thus, we must use the current $-i$ in computing the right-hand winding voltage, which will have its positive reference on the dot. Doing the computation, we have

$$v = v_1 - v_2 = L_1 pi + Mp(-i) - [Mpi + L_2 p(-i)] = (L_1 + L_2 - 2M)pi. \qquad (16.3\text{-}64)$$

Thus, the subcircuit is once again equivalent to a single inductance; this time, however, the equivalent inductance is

$$L_{\text{eq}} = L_1 + L_2 - 2M. \qquad (16.3\text{-}65)$$

Example 16.12   Investigate the stability properties of the transformer circuit shown in Figure 16.42.

Figure 16.42
An example circuit

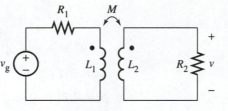

Solution   We recall that a circuit or system is stable if all of its responses have system functions whose poles are all in the left half-plane, unstable if one or more are in the right half-plane, and marginally stable if there are none in the right half-plane and one or more simple poles on the imaginary axis. Thus, we must compute all system functions. To do this, we use the s-domain model shown in Figure 16.43. Because the mesh currents determine all other variables (the voltages in the circuit, which are either real constants multiplied by these currents or derivatives of these currents), we need only derive the two transfer functions relating them to the independent source. The mesh equations are (we have practiced quite a bit, so we will write them directly)

$$R_1 I_1(s) + [L_1 s I_1(s) + Ms[-I_2(s)]] = V_g(s) \qquad (16.3\text{-}66\text{a})$$

and

$$R_2[-I_2(s)] + [Ms I_1(s) + L_2 s[-I_2(s)]] = 0. \qquad (16.3\text{-}66\text{b})$$

(Observe carefully that $-I_2(s)$ goes into the dot on the right-hand side.) Rewriting these equations in matrix form gives

$$\begin{bmatrix} L_1 s + R_1 & -Ms \\ -Ms & L_2 s + R_2 \end{bmatrix} \begin{bmatrix} I_1(s) \\ I_2(s) \end{bmatrix} = \begin{bmatrix} V_g(s) \\ 0 \end{bmatrix}. \qquad (16.3\text{-}67)$$

We have multiplied the second equation by -1 to show symmetry in the coefficient matrix. The solution is

$$\begin{bmatrix} I_1(s) \\ I_2(s) \end{bmatrix} = \frac{1}{(L_1 L_2 - M^2)s^2 + (R_1 L_2 + R_2 L_1)s + R_1 R_2} \begin{bmatrix} L_2 s + R_2 & Ms \\ Ms & L_1 s + R_1 \end{bmatrix} \begin{bmatrix} V_g(s) \\ 0 \end{bmatrix}. \qquad (16.3\text{-}68)$$

Thus, we have

$$\begin{bmatrix} I_1(s) \\ I_2(s) \end{bmatrix} = \begin{bmatrix} \dfrac{L_2 s + R_2}{(L_1 L_2 - M^2)s^2 + (R_1 L_2 + R_2 L_1)s + R_1 R_2} \\[3ex] \dfrac{Ms}{(L_1 L_2 - M^2)s^2 + (R_1 L_2 + R_2 L_1)s + R_1 R_2} \end{bmatrix} V_g(s). \qquad (16.3\text{-}69)$$

The desired transfer functions are the elements in the column matrix on the right-hand side of this equation. We see that the poles of the two are identical and are given by the

Figure 16.43
The s-domain
equivalent

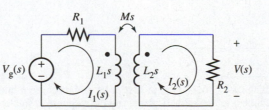

values of s for which

$$(L_1L_2 - M^2)s^2 + (R_1L_2 + R_2L_1)s + R_1R_2 = 0. \tag{16.3-70}$$

We write this in the form

$$as^2 + bs + c = 0, \tag{16.3-71}$$

where

$$a = L_1L_2 - M^2, \tag{16.3-72}$$

$$b = R_1L_2 + R_2L_1, \tag{16.3-73}$$

$$c = R_1R_2. \tag{16.3-74}$$

Next, we use the quadratic formula to find the roots of (16.3-70). They are at

$$s_{1,2} = \frac{-b \pm \sqrt{b^2 - 4ac}}{2a} = \frac{-b}{2a} \pm \sqrt{\left[\frac{b}{2a}\right]^2 - \frac{c}{a}}. \tag{16.3-75}$$

Now all of our elements—R_1, R_2, L_1, L_2, and M—are positive. Thus, b and c are both greater than zero; a, on the other hand, could conceivably be of either sign. There are only three cases to consider (unless $a = 0$):[5] the squared factor inside the radical is greater than, less than, or equal to c. In the first case, that of real roots, the radical term is positive and no greater than the first term $-b/2a$ in magnitude. This means that both roots have the same sign as a: if $a > 0$, both are in the left half-plane and the responses are stable, whereas if $a < 0$, both are in the right half-plane and both responses are unstable. In the second case, that of complex roots, the real parts of both have the same sign as a; once again, if $a > 0$, both are in the left half-plane and the responses are stable, whereas if $a < 0$, both are in the right half-plane and the responses are unstable. Finally, in the third case, the radical term is zero. The two roots are equal, but once again they both have the same sign as a and the same conclusions follow. Thus, stability depends upon the quantity

$$a = \Delta = L_1L_2 - M^2. \tag{16.3-76}$$

We will later see that $\Delta > 0$ is satisfied for all physical transformers and so all responses are stable—as far as the transformer is concerned.

Example 16.13   Find the ac steady-state voltage transfer function for the subcircuit shown in Figure 16.44 and sketch the gain function. Assume that $L_1 = 2$ H, $L_2 = M = 1$ H, $C = 1$ F, and $R = 1 \Omega$.

Figure 16.44
An example subcircuit

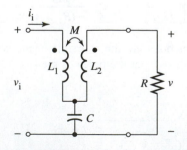

---

[5] In which case, (16.3-71) is of first order and, because b and c are both positive, there is a single pole, which is in the left half-plane, and the response is stable.

Solution   The transformer connections meet the conditions for using the tee equivalent shown in Figure 16.45, where we have used the phasor form. We have introduced a ground reference at the bottom node and an auxiliary node voltage at the top node in the middle. We now need only write one nodal equation (assuming the input voltage is known, say because we have attached a test voltage source):

$$\frac{\overline{V}}{1} + \frac{\overline{V}}{j\omega + \frac{1}{j\omega}} + \frac{\overline{V} - \overline{V}_i}{j\omega} = 0. \tag{16.3-77}$$

Solving for the transfer function $H(j\omega) = \overline{V}/\overline{V}_i$, we get

$$H(j\omega) = \frac{1 - \omega^2}{(1 - 2\omega^2) + j(1 - \omega^2)\omega}. \tag{16.3-78}$$

The gain is

$$A(\omega) = |H(j\omega)| = \frac{|1 - \omega^2|}{\sqrt{(1 - 2\omega^2)^2 + (1 - \omega^2)^2\omega^2}}, \tag{16.3-79}$$

and the phase shift is

$$\phi(\omega) = \angle H(j\omega) = \pi u(\omega^2 - 1) - \tan^{-1}\left[\frac{(1 - \omega^2)\omega}{1 - 2\omega^2}\right]. \tag{16.3-80}$$

The reason for the somewhat odd-looking first term is this: if $\omega > 1$, the numerator term in (16.3-78) is negative, and we have made it correspond to π radians of phase. The gain is plotted in Figure 16.46. Notice that there is a transmission zero at $\omega = 1$ rad/s corresponding to the series resonant frequency of the mutual inductance and the capacitor. At this frequency the series branch has zero impedance, so it shorts the signal to ground.

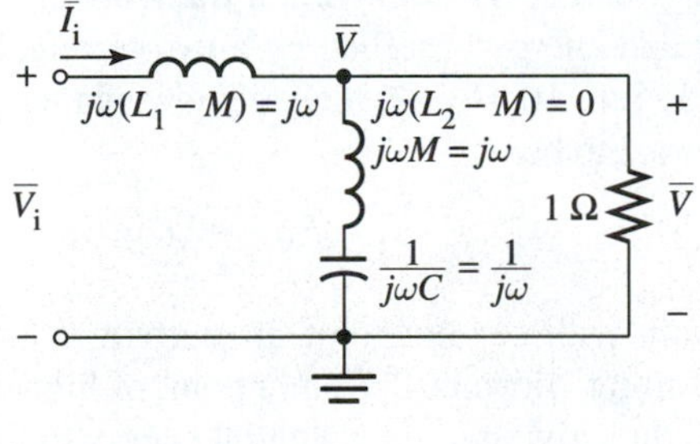

Figure 16.45
The phasor domain tee equivalent

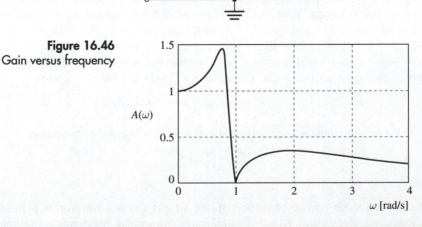

Figure 16.46
Gain versus frequency

Section 16.3 Quiz

Q16.3-1. Find the two port voltages v_1 and v_2 in terms of i_1 and i_2 in Figure Q16.3-1.

Figure Q16.3-1

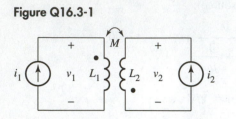

Q16.3-2. Find the two port voltages v_1 and v_2 in terms of i_1 and i_2 in Figure Q16.3-2.

Figure Q16.3-2

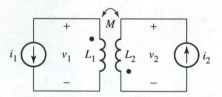

Q16.3-3. Find the impulse response for $v(t)$ in Figure Q16.3-3. Is it stable or unstable?

Figure Q16.3-3

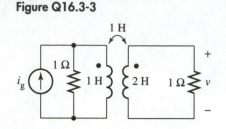

Q16.3-4. If $i_s(t) = 2\cos(t)$ in Figure Q16.3-3, find the ac forced response for $v(t)$.

16.4 | Energy Considerations and the Ideal Transformer

Power Absorption by Two-Ports

Before discussing the energy storage in a transformer, let's briefly review the concept of electrical power and energy absorbed by a two-terminal device or subcircuit, such as the one sketched in Figure 16.47(a), which we now can term a one-port. We have defined the instantaneous power to be

$$P(t) = v(t)i(t), \tag{16.4-1}$$

which is consistent with our image of the current $i(t)$ as consisting of small particles of positive charge falling "downhill" from a point of higher potential energy to one of lower potential energy, thus giving up its potential energy to the inside of the box (the element). We similarly think of two such streams of positive particles in the two-port of Figure 16.47(b). The port condition demands that the stream entering a top terminal leave the corresponding bottom terminal. Therefore, we simply add the two powers:

$$P(t) = v_1(t)i_1(t) + v_2(t)i_2(t). \tag{16.4-2}$$

We also recall that the energy delivered is the running integral of the power:

$$w(t) = \frac{1}{p}P(t). \tag{16.4-3}$$

We defined a one-port to be *passive* if, for all $v(t)$ and $i(t)$ waveforms permitted by the one-port v-i relationship, $w(t) \geq 0$ at all times. We extend this definition to hold for two-ports as well.

Figure 16.47
Power relations for
one-ports and
two-ports

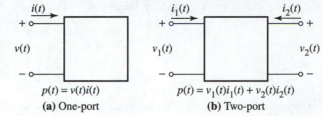

$$p(t) = v(t)i(t)$$
(a) One-port

$$p(t) = v_1(t)i_1(t) + v_2(t)i_2(t)$$
(b) Two-port

Reciprocity of the
Transformer: $L_{21} = L_{12}$

Now let's specialize by supposing that the two-port in Figure 16.47(b) is a transformer with $L_{12} \neq L_{21}$. Assuming that the dots are at the top terminals, we have

$$v_1(t) = L_1 p i_1(t) + L_{12} p i_2(t) \qquad (16.4\text{-}4a)$$

and

$$v_2(t) = L_{21} p i_1(t) + L_2 p i_2(t). \qquad (16.4\text{-}4b)$$

The instantaneous power absorbed by the transformer is

$$P(t) = v_1(t)i_1(t) + v_2(t)i_2(t) \qquad (16.4\text{-}5)$$

$$= [L_1 p i_1(t) + L_{12} p i_2(t)]i_1(t) + [L_{21} p i_1(t) + L_2 p i_2(t)]i_2(t)$$

$$= L_1 i_1(t) p i_1(t) + L_2 i_2(t) p i_2(t) + L_{12} i_1(t) p i_2(t) + L_{21} i_2(t) p i_1(t).$$

We know that

$$xpx = p\left[\frac{1}{2}x^2\right], \qquad (16.4\text{-}6)$$

so we can apply it to (16.4-5), first identifying x as i_1, then as i_2. Suppressing the (t) notation for compactness, we will then have

$$P(t) = p\left[\frac{1}{2}L_1 i_1^2 + \frac{1}{2}L_2 i_2^2\right] + L_{12} i_1 p i_2 + L_{21} i_2 p i_1. \qquad (16.4\text{-}7)$$

Under the assumption that $L_{12} \neq L_{21}$, we will express these two quantities as shown in Figure 16.48; that is, we will write

$$L_{12} = M + \frac{\delta M}{2} \qquad (16.4\text{-}8a)$$

and

$$L_{21} = M - \frac{\delta M}{2}. \qquad (16.4\text{-}8b)$$

Using these relations in equation (16.4-7) gives

$$P(t) = p\left[\frac{1}{2}L_1 i_1^2 + \frac{1}{2}L_2 i_2^2\right] + M[i_1 p i_2 + i_2 p i_1] + \frac{\delta M}{2}[i_1 p i_2 - i_2 p i_1]. \qquad (16.4\text{-}9)$$

Figure 16.48
Average and difference
decomposition

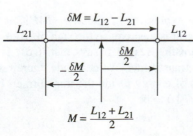

At this point, we recognize that, for any pair of variables x and y,

$$xpy + ypx = p(xy) \qquad (16.4\text{-}10)$$

and

$$xpy - ypx = x^2 p\left[\frac{y}{x}\right]. \qquad (16.4\text{-}11)$$

Using these relationships in equation (16.4-9) gives

$$P(t) = p\left[\frac{1}{2}L_1 i_1^2 + \frac{1}{2}L_1 i_1^2 + M i_1 i_2\right] + \frac{\delta M}{2}\left[i_1^2 p\left[\frac{i_2}{i_1}\right]\right]. \qquad (16.4\text{-}12)$$

Integrating (that is, multiplying by $1/p$), we get the energy delivered to the two-port:

$$w(t) = \frac{1}{p}P(t) = \frac{1}{2}L_1 i_1^2(t) + \frac{1}{2}L_2 i_2^2(t) + M i_1(t)i_2(t) \qquad (16.4\text{-}13)$$

$$+ \frac{\delta M}{2}\int_{-\infty}^{t} i_1^2(\alpha)p\left[\frac{i_2(\alpha)}{i_1(\alpha)}\right]d\alpha.$$

Let's pick the particular waveform for $i_1(t)$ shown in Figure 16.49. It goes from zero linearly up to 1 ampere at $t = 1$ s, then falls linearly to 0 ampere at $t = 2$ s and remains zero forever after. We define

$$i_2(t) = -ti_1(t). \qquad (16.4\text{-}14)$$

It has properties similar to those of $i_1(t)$—it is zero for $t \leq 0$, decreases to -1 ampere (though not linearly) at $t = 1$ s, then rises to zero again at $t = 2$ s and remains zero forever after. The waveform for $i_2(t)$ is shown in the right-hand graph of Figure 16.49.

Figure 16.49
Time variation of the transformer terminal currents

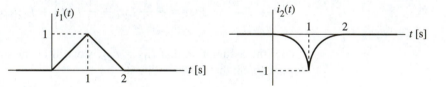

Now let's evaluate the energy absorbed by the transformer for all time when excited by current sources having these two waveforms. Noting that $i_1(\infty) = i_2(\infty) = 0$ and that $p[i_2(t)/i_1(t)] = p(-t) = -1$, we have

$$w(\infty) = \frac{1}{2}L_1 i_1^2(\infty) + \frac{1}{2}L_2 i_2^2(\infty) + M i_1(\infty)i_2(\infty) - \frac{\delta M}{2}\int_{-\infty}^{\infty} i_1^2(\alpha)\, d\alpha \qquad (16.4\text{-}15)$$

$$= -\frac{\delta M}{2}\int_{-\infty}^{\infty} i_1^2(\alpha)\, d\alpha.$$

The value of the last expression is actually $-\delta M/3$, as you can verify, but the important thing is that the integral is positive, so the sign of $w(\infty)$ is the opposite to that of δM. This is an extremely interesting result! It says that if δM is positive, then we have extracted net energy from the transformer. We conclude, therefore, that δM cannot be positive, for there are no sources of energy inside the transformer. However, if we return to equation (16.4-14) and define $i_2(t)$ to be

$$i_2(t) = ti_1(t) \qquad (16.4\text{-}16)$$

(the same as before except for a change in sign), the derivative of $i_2(t)/i_1(t)$ will now be $+1$ and the total energy delivered to the transformer over all time will be

$$w(\infty) = \frac{1}{2}L_1i_1^2(\infty) + \frac{1}{2}L_2i_2^2(\infty) + Mi_1(\infty)i_2(\infty) + \frac{\delta M}{2}\int_{-\infty}^{\infty}i_1^2(\alpha)\,d\alpha \qquad (16.4\text{-}17)$$

$$= +\frac{\delta M}{2}\int_{-\infty}^{\infty}i_1^2(\alpha)\,d\alpha.$$

But this says that if δM is *negative*, we have extracted net energy from the transformer. We are led inescapably to the conclusion that $\delta M = 0$; that is, we have proved that

$$L_{12} = L_{21} = M. \qquad (16.4\text{-}18)$$

Notice that we have *assumed* that the transformer is passive. This is an additional assumption that must be imposed.

The Energy Stored by a Transformer: An Upper Bound on M

Actually, we have shown a lot more. Equation (16.4-13), with $\delta M = 0$ shows that

$$w(t) = \frac{1}{2}L_1i_1^2(t) + \frac{1}{2}L_2i_2^2(t) + Mi_1(t)i_2(t). \qquad (16.4\text{-}19)$$

This gives us a simple, compact expression for the energy stored by the transformer at any instant of time as a function of the values of the two terminal currents. Equation (16.4-19) is the basic form to remember, but let's study it a bit more by rewriting it in the form

$$2w(t) = L_1i_1^2(t) + 2Mi_1(t)i_2(t) + L_2i_2^2(t) \qquad (16.4\text{-}20)$$

$$= L_1\left[i_1^2(t) + 2\frac{M}{L_1}i_1(t)i_2(t) + \frac{L_2}{L_1}i_2^2(t)\right]$$

$$= L_1\left[i_1(t) + \frac{M}{L_1}i_2(t)\right]^2 + \left[\frac{L_1L_2 - M^2}{L_1}\right]i_2^2(t).$$

Because $i_1(t)$ and $i_2(t)$ are independently specifiable for a transformer, let's make

$$i_1(t) = -\frac{M}{L_1}i_2(t). \qquad (16.4\text{-}21)$$

In this case we have

$$2w(t) = \left[\frac{L_1L_2 - M^2}{L_1}\right]i_2^2(t). \qquad (16.4\text{-}22)$$

Because $w(t)$ must be nonnegative, we have $L_1L_2 - M^2 \geq 0$. As $M \geq 0$, we have

$$0 \leq M \leq \sqrt{L_1L_2}. \qquad (16.4\text{-}23)$$

This gives an upper bound on the mutual inductance for physical transformers.

The Coefficient of Coupling

We want to strive for attainment of the upper bound on the mutual inductance because larger values for M couple more energy from the primary into the secondary. A measure of how close M is to the limit is the *coefficient of coupling*,

$$k = \frac{M}{\sqrt{L_1L_2}}. \qquad (16.4\text{-}24)$$

Thus, the upper bound on M is attained when $k = 1$. On the other hand, $k = 0$ indicates that $M = 0$ and therefore that the transformer consists merely of two separate inductors of values L_1 and L_2. In terms of the coupling coefficient, the basic transformer equations can be written

$$v_1(t) = L_1 p i_1(t) + M p i_2(t) = L_1 p i_1(t) + k\sqrt{L_1 L_2}\, p i_2(t) \tag{16.4-25a}$$

$$= L_1 p\left[i_1(t) + k\sqrt{\frac{L_2}{L_1}}\, i_2(t) \right]$$

and

$$v_2(t) = M p i_1(t) + L_2 p i_2(t) = k\sqrt{L_1 L_2}\, p i_1(t) + L_2 p i_2(t) \tag{16.4-25b}$$

$$= L_1 p\left[k\sqrt{\frac{L_2}{L_1}}\, i_1(t) + \frac{L_2}{L_1} i_2(t) \right].$$

The Perfectly Coupled Transformer

We define a *perfectly coupled* transformer to be one for which $k = 1$. In this case, the preceding equations assume the form

$$v_1(t) = L_1 p[i_1(t) + N i_2(t)] \tag{16.4-26a}$$

and

$$v_2(t) = N L_1 p[i_1(t) + N i_2(t)], \tag{16.4-26b}$$

where

$$N = \sqrt{\frac{L_2}{L_1}}. \tag{16.4-27}$$

The parameter N is called the *turns ratio*. The reason for this comes from the practical theory of inductor construction. It is a fact that inductance is proportional to the square of the number of turns of conductor used to fabricate the inductor. Thus, the L_2/L_1 ratio is proportional to the ratio of these squares, and its square root is proportional to the ratio of the turns. If we compute the ratio of v_2 to v_1 for the perfectly coupled transformer (that is, the ratio of (16.4-26b) to (16.4-26a)), we see that

$$\frac{v_2}{v_1} = N. \tag{16.4-28}$$

There is essentially only one dynamic equation, (16.4-26a), and one algebraic one, (16.4-28), which together specify the operation of the perfectly coupled transformer.

The Ideal Transformer

Now let's go one step further. We will start with our perfectly coupled transformer and assume that L_1 and L_2 approach infinity—*but in such a fashion that their ratio remains equal to the square of the turns ratio N, a fixed quantity*. In this case, we can rewrite equation (16.4-26a) by dividing both sides by $L_1 p$, then letting $L_1 \to \infty$, in the form

$$i_1(t) + N i_2(t) = \frac{v_1(t)}{L_1 p} \longrightarrow 0. \tag{16.4-29}$$

We will call the resulting two-port element an *ideal* transformer. As a consequence of the fact that it is perfectly coupled, equation (16.4-28) holds, and, as a consequence of the very large inductances, so does (16.4-29). Thus, the defining equations for the ideal transformer are

$$v_2 = N v_1, \tag{16.4-30a}$$

$$i_2 = -\frac{1}{N} i_1. \tag{16.4-30b}$$

The ideal transformer is specified by a single parameter, the turns ratio, therefore we will represent it by the symbol shown in Figure 16.50. It is the same symbol as that for the basic transformer, except that the turns ratio is indicated at the top by $1:N$. The two lines between the inductor symbols are inserted to remind us that these inductances are very large (∞, ideally). The symbol $1:N$ refers to the fact that the number of turns "used in making L_2" is N times that "used in making L_1." We have used quote marks around the phrase "used in making" because the ideal transformer is an idealization. It is a limiting equivalent circuit as these inductances become infinitely large; we allowed this to happen, though, in such a manner that the *ratio* of the two numbers of turns remained constant at N. Note that there is, therefore, a difference between the notation $1:N$ and $N:1$. Assuming that the right-hand port (the secondary) in Figure 16.50 is delivering a signal or energy to a load, the notation $1:N$ means a "step-up" transformer if $N > 1$ (that is v_2 is larger than v_1). The notation $N:1$, with $N > 1$, means that it is a "step-down" transformer. In this case, we reverse this specification on our circuit symbol in Figure 16.50. Observe that there are two vertical lines between the "coils" in Figure 16.50. You should always look for this on a circuit diagram for it is the tip-off that it is an ideal transformer. There is no arrow or symbol for mutual inductance—in fact the mutual inductance for an ideal transformer is infinite.

Figure 16.50
The ideal transformer symbol

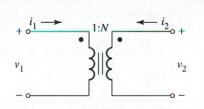

Transformers are used in circuit design to change voltage, current, and impedance levels. For instance, transistors and integrated circuits require low voltages. The most commonly accessible source of power, though, is the alternating current receptacle on the wall, which, as we have seen, delivers a sinusoidal voltage of about 160 volts peak. Thus, a step-down transformer is normally employed in the design of power supplies for an apparatus using such devices.

Impedance Transformation   Another use is in impedance transformation, as we indicate in Figure 16.51. In part (a) of the figure we show a transformer with its primary terminals attached to a resistor. The secondary voltage and current are related by

$$v_2 = Nv_1 = N[-i_1R] = N[Ni_2R] = N^2Ri_2. \qquad (16.4\text{-}31)$$

The voltage is proportional to the current; therefore, "looking into" the secondary terminals we see an equivalent resistor with value

$$R_{eq} = N^2R, \qquad (16.4\text{-}32)$$

as we have indicated in part (b) of the figure.

Figure 16.51
Impedance transformation

(a) Primary loading   (b) Secondary equivalent

Example 16.14   Find the Thévenin equivalent of the circuit to the left of terminals a-a', and compute the required value of turns ratio N for maximum power into the load resistor R_L in Figure 16.52.

Figure 16.52
An example circuit

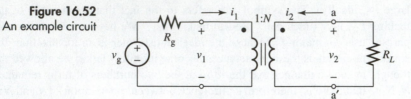

Solution   We will, as usual, first compute, let us say, the open circuit voltage. We use the circuit shown in Figure 16.53. Because $i_2 = 0$ due to the open circuit condition, we also have $i_1 = 0$ by the ideal transformer current constraint. But this implies that the voltage drop across R_g is zero, so $v_1 = v_g$. By the ideal transformer voltage constraint, we have

$$v_{oc} = Nv_g. \tag{16.4-33}$$

Now we must compute the Thévenin equivalent resistance. We attach a test source to the terminals a-a' (in this case, we have chosen a current source), as shown in Figure 16.54. We see that

$$v_2 = Nv_1 = N(-i_1 R_g) = N^2 R_g i_2. \tag{16.4-34}$$

Thus, $$R_{Th} = N^2 R_g. \tag{16.4-35}$$

The Thévenin equivalent, with the load resistance R_L replaced, is shown in Figure 16.55.

Figure 16.53
The open circuit voltage

Figure 16.54
The Thévenin equivalent resistance

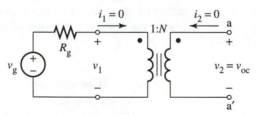

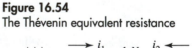

Figure 16.55
The Thévenin equivalent circuit

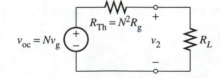

Now let's think about maximum power transfer just a bit. If the transformer turns ratio, v_g, and R_g were all fixed and R_L were variable, the solution would be

$$R_L = N^2 R_g; \tag{16.4-36}$$

however, in our problem statement we were asked to treat R_L as fixed and vary the turns ratio. We cannot simply make $N = 0$, thereby making $R_{Th} = 0$—as we might be tempted to do because we know that if R_{Th} is variable, its optimum value is zero. This is a different problem, however, because v_{oc} depends upon N! Let's therefore compute the power delivered to R_L. It is

$$P_{R_L} = v_2 i_2 = \left[\frac{R_L}{R_L + N^2 R_g} N v_g \right] \times \left[\frac{N v_g}{R_L + N^2 R_g} \right] = \frac{N^2 R_L v_g^2}{[R_L + N^2 R_g]^2}. \qquad (16.4\text{-}37)$$

We can rewrite this expression by dividing top and bottom by N^2, obtaining

$$P_{R_L} = \frac{R_L v_g^2}{\left[\dfrac{R_L}{N} + N R_g \right]^2}. \qquad (16.4\text{-}38)$$

Treating all other parameters as constants, we see that maximizing the load power is equivalent to minimizing the quantity inside the brackets in the denominator. We will leave it to you to show in the usual way (treating N as a continuous variable) that the optimum value is

$$N = \sqrt{\frac{R_L}{R_g}}. \qquad (16.4\text{-}39)$$

This is the same result we would obtain by simply demanding that the Thévenin equivalent resistance be equal to that of the load; we could not do this, however, because the circuit did not meet the assumptions we made in deriving this result in Chapter 5. There is, however, an easier solution. One can manipulate the circuit to cast it into the "variable load resistance" framework. Can you think of how to do it? We will ask you to do so—and provide a hint or two—in one of the end of chapter problems.

Losslessness of the Ideal Transformer

Before continuing, we would like to comment on an issue raised by the preceding example. In our solution we treated the voltages and currents as dc quantities in computing power. (If they were ac variables, of course, we would have been forced to include the usual factor of 1/2 in our power computations because of the effect of averaging the square of a sinusoid.) The issue is this: an ideal transformer is one with the *v-i* characteristic imposed by equations (16.4-30). This means that the currents and the voltages are related by a constant (the turns ratio or its inverse)—even for dc values! The problem is that the "real" transformer obeys equations (16.4-4), in which the two port voltages are expressed as linear combinations of derivatives of the port currents. In the dc forced response case, these two currents are constants, so their derivatives are zero. This means that the "real" transformer is equivalent to two short circuits at dc. So the thing to keep in mind is this: the ideal transformer model "works" at dc, but its relation to the "real" model is not very good at all there. This is a practical aspect of circuit theory, and you must develop a feel for when (under what circuit conditions) the ideal transformer is a "good enough" model for the real one with which you are working. One of the end of chapter problems explores this a bit.

Now let's compute the total power absorbed by an ideal transformer. Rather than going back to equation (16.4-19) for a nonideal one and then letting L_1 and L_2 become infinite, let's do our computation directly. The total instantaneous power absorbed is

$$P(t) = v_1(t) i_1(t) + v_2(t) i_2(t) \qquad (16.4\text{-}40)$$

$$= v_1(t) i_1(t) + [N v_1(t)] \left[-\frac{1}{N} i_1(t) \right] = 0.$$

Thus, the total instantaneous power absorbed by the ideal transformer is identically zero.

Many texts alter the ideal transformer model slightly. To discuss this modification, we will reproduce the ideal transformer model of Figure 16.50 as Figure 16.56(a) here for ease of reference. We notice that according to equation (16.4-30b),

$$i_2 = -\frac{1}{N}i_1. \tag{16.4-41}$$

The alternate model results when one "turns around" the positive reference for the current i_2, as shown in Figure 16.56(b). In this case, (16.4-41) is modified to

$$i_2 = \frac{1}{N}i_1. \tag{16.4-42}$$

The advantage is that one works with only positive signs. (In both cases, equation (16.4-30a) holds for the voltages.) We choose for the most part, however, to use the model in Figure 16.56(a) because the port variable conventions are the same as for the "realistic" transformer model—and the same as the z, y, h, and g parameter sets. (The alternate model of Figure 16.56(b) has the same convention as that of the A, B, C, D parameter set.)

Figure 16.56
The ideal transformer
model

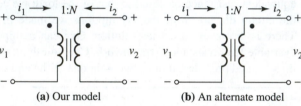

(a) Our model   (b) An alternate model

A Model for the Nonideal Transformer in Terms of the Ideal Transformer

We will end this section by developing a model for the nonideal transformer that is somewhat more practical for applications in that it incorporates the ideal transformer as an element. This allows the sources of nonideal behavior to become more readily apparent. We begin our development by repeating equations (16.4-25) for a nonideal transformer in terms of the coefficient of coupling here as equations (16.4-43):

$$v_1(t) = L_1 p[i_1(t) + kNi_2(t)] \tag{16.4-43a}$$

and

$$v_2(t) = NL_1 p[ki_1(t) + Ni_2(t)]. \tag{16.4-43b}$$

We rewrite them slightly in the following manner (adding and subtracting a term from each—$kL_1 pi_1$ for the first and $NkL_1 pi_2$ (inside the square brackets) for the second):

$$v_1(t) = (1 - k)L_1 pi_1(t) + kL_1 p[i_1(t) + Ni_2(t)] \tag{16.4-44a}$$

and

$$v_2(t) = N[kL_1 p[i_1(t) + Ni_2(t)] + (1 - k)NL_1 pi_2(t)]. \tag{16.4-44b}$$

If you use the ideal transformer constraints and KCL at the top of the vertical inductor, you will see that the subcircuit in Figure 16.57 obeys exactly these equations. Hence, it is equivalent to the original nonideal transformer. This is called the *equivalent referred to*

Figure 16.57
Equivalent referred to
the primary

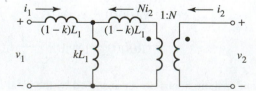

the primary because all the parasitic (nonideal) elements are on the primary side of the ideal transformer. An equivalent referred to the secondary is also possible.

In closing, we point out that we have not included many nonideal effects that are important in applications, such as winding resistance and eddy current and hysteresis loss in the transformer core. These topics, however, are quite complicated and are better left to later courses on ac power.

Section 16.4 Quiz

All questions pertain to the transformer in Figure Q16.4-1.

Figure Q16.4-1

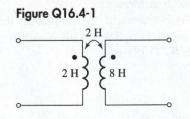

Q16.4-1. Suppose that the currents i_1 and i_2, respectively, are 2 A and 3 A (both dc values) relative to their conventional reference directions. Compute the stored energy.

Q16.4-2. Assume that the two terminal currents are $i_1(t) = 2 \sin(2t)$ A and $i_2(t) = 3 \cos(2t)$ A relative to their conventional reference directions. Find the terminal voltages (relative to their conventional references) and the instantaneous energy stored. Compute the average power dissipated in the transformer by averaging the instantaneous power.

Q16.4-3. Find the coefficient of coupling.

Q16.4-4. Find the equivalent referred to the primary.

16.5 | SPICE: Mutual Inductance

Mutual Inductance in SPICE

We will now consider how to generate SPICE models for circuits containing transformers. Perhaps the best way to start is with an example. Figure 16.58 shows a circuit that we analyzed in Section 16.3, Example 16.6. Although we want the response current $i(t)$ for the particular value of M shown ($M = 2$ H), let's think for a moment about the circuit as a function of M. Recall that the transformer v-i characteristic was defined by the equations

$$v_1(t) = L_1 pi_1(t) + Mpi_2(t) \tag{16.5-1}$$

and

$$v_2(t) = Mpi_1(t) + L_2 pi_2(t). \tag{16.5-2}$$

If $M = 0$, we see that these become

$$v_1(t) = L_1 pi_1(t) \tag{16.5-3}$$

and

$$v_2(t) = L_2 pi_2(t). \tag{16.5-4}$$

These are simply the equations of two uncoupled inductors.

Figure 16.58
An example circuit

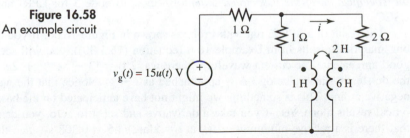

The way SPICE treats mutual inductance (and, thereby, transformers) is best explained by starting with equations (16.5-3) and (16.5-4). One writes a NETLIST as usual for the circuit having two inductors of values L_1 and L_2—*and then adds a statement* specifying the coefficient of coupling. Recall that the coefficient of coupling was defined in Section 16.4 by the formula

$$k = \frac{M}{\sqrt{L_1 L_2}}. \tag{16.5-5}$$

We showed there that the values of k are always in the range of 0 to 1, with $k = 1$ describing what we called a "perfectly coupled" transformer.

We will illustrate this additional SPICE statement by simulating the circuit of Figure 16.58 with $v_g(t) = 15u(t)$ as in Example 16.6. The circuit is shown prepared for SPICE analysis in Figure 16.59. The SPICE NETLIST is

```
MUTUAL INDUCTANCE EXAMPLE
VG     1     0     PWL (0, 0  0.001,15)
RA     1     2     1
RB     2     3     1
RC     2     4     2
LPRI   3     0     1
LSEC   4     0     6
K1     LPRI  LSEC 0.817
.PROBE
.TRAN 0.1  10   0      0.01
.END
```

Figure 16.59
The example circuit prepared for SPICE analysis

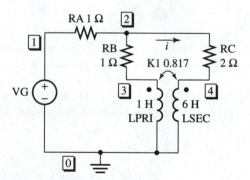

Most of it is self-explanatory. We have used a PWL source to simulate the step function and have assigned it a negligible rise time of 0.001 second. We have described the primary and secondary inductors as usual. The only new statement is the one for the coefficient of coupling. *Such statements always start with the letter k, mention the two inductors to be coupled by name, and then give the coefficient of coupling.* In our case, we have computed K1 = $M/\sqrt{\text{LPRI*LSEC}}$ = $2/\sqrt{6}$ = 0.817. Notice, by the way, that the inductors are *oriented. The dots are assumed to be attached to the node that comes first in the specification of the associated inductor*—thus, to node 3 for LPRI and node 4 for LSEC.

The result of a SPICE run with Probe is shown in Figure 16.60. If you compare with our analytical results from Example 16.6 (equation (16.3-30)), you will see that there is good agreement. The current waveform there was $i(t) = [3 - 5e^{-t/2} + 2e^{-5t}]u(t)$. Thus, the dc steady-state value of 3 A is approached as $t \to \infty$. Notice that the waveform goes negative at first. This is something we might not have anticipated on the basis of the analytical results alone. Yet, if you take a derivative and set it to zero, you can easily show that there is a relative minimum of $i(t)$ at $t = 2\ln(2)/4.5 = 0.308$ s—and this is verified by our graphical output. Notice that we have used Probe's math handling capability to

Figure 16.60
Response waveform

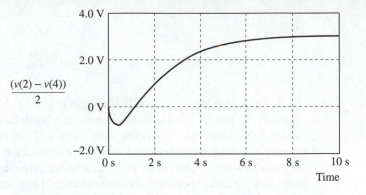

$$\frac{(v(2) - v(4))}{2}$$

compute a mathematical expression for our response current; we could just as easily have specified I(RC).

Now let's look at a simpler circuit, which—as you will see—is actually more difficult to simulate with SPICE. Figure 16.61 shows the circuit we analyzed in Example 16.4. We will, at first, somewhat mechanically (and blindly) attempt to do a SPICE simulation without recognizing any of the difficulties (but you can probably see one, can't you?). We show the circuit prepared for SPICE analysis in Figure 16.62. Notice that we have used the *coefficient of coupling* rather than the *mutual inductance* shown in the original circuit of Figure 16.61. The NETLIST is:

```
SECOND MUTUAL INDUCTANCE EXAMPLE
VG     1   0     PWL(0,0 0.001, 4)
RA     1   2     2
RB     3   4     2
LA     0   2     1
LB     3   4     4
K  LA  LB  1
.TRAN  0.1 20    0      0.01
.PROBE
.END
```

Figure 16.61
An example circuit

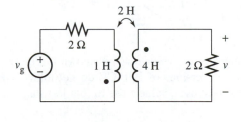

Figure 16.62
The example circuit prepared for SPICE analysis

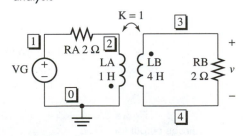

Here is the result of the SPICE compiler (there were errors in the NETLIST):

```
**** 11/14/95 14:53:22 ********** EVALUATION PSPICE (JULY 1993)
SECOND MUTUAL INDUCTANCE EXAMPLE
****    CIRCUIT DESCRIPTION
****************************************************************
                 VG    1   0    PWL(0, 0    0.001, 4)
                 RA    1   2    2
                 RB    3   4    2
                 LA    0   2    1
                 LB    3   4    4
                 K     LA  LB   1
```

```
.TRAN 0.1 20  0              0.01
.PROBE
.END

ERROR — NODE 3 IS FLOATING
ERROR — NODE 4 IS FLOATING
```

The problem here is that SPICE always expects a dc path from each and every node to ground. If you look back at our discussion of nodal analysis in Chapter 4, Section 4.2 (which SPICE uses, in a modified form), you will see that nodal analysis only works if there is a path between each and every pair of nodes in the deactivated circuit. Now if there are no sources, the circuit is, by definition, in the dc steady state, so the transformer looks like a pair of short circuits: one between the input terminals and one between the output terminals. But there is no path to ground from either of the two output terminals in our example circuit. One possible solution is to place a resistor with a very large value between one of them and ground, as we show in Figure 16.63. Here is the corresponding NETLIST:

```
SECOND MUTUAL INDUCTANCE EXAMPLE
*—FLOATING NODES REMOVED
VG    1   0    PWL(0,0 0.001, 4)
RA    1   2    2
RB    3   4    2
RHI   4   0    1E9
LA    0   2    1
LB    3   4    4
K    LA  LB    1
.TRAN    0.1  20      0       0.01
.PROBE
.END
```

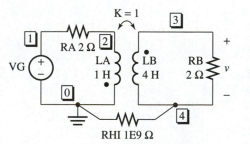

Figure 16.63
One solution of the floating node problem

The Probe output for the voltage $v(t)$ is shown in Figure 16.64. We should note here that some of the older versions of PSPICE do not allow a unit coefficient of coupling—in which case one must use a value close to (but less than) unity, such as 0.9999.

Figure 16.64
Probe output for SPICE run

$v(3, 4)$

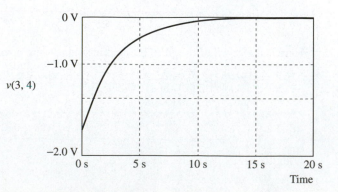

Ideal Transformers in SPICE

Our final SPICE topic will be the modeling of circuits containing ideal transformers. Here we will discover that our modeling procedure of Section 16.4 is illustrated quite well by the simulation procedure. Figure 16.65 shows an example circuit. We will assume that $N = 3$ and we will plot the frequency response for the response voltage $v_2(t)$. The circuit is shown prepared for SPICE analysis in Figure 16.66. Now we have emphasized the fact that an ideal transformer is not described by primary, secondary, and mutual inductances—because all are infinite.

Figure 16.65
An example circuit

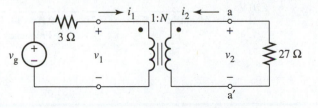

Figure 16.66
The example circuit prepared for SPICE analysis

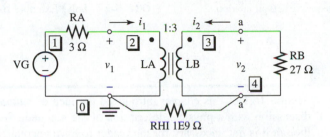

Why, then, have we labeled our ideal transformer with LA and LB? Simply because of the perverse nature of SPICE. *It does not allow ideal transformers.* Therefore, we recall how we derived the ideal transformer model in the first place. We simply let the primary and secondary inductances become infinitely large in such a way that the square root of their ratio—the turns ratio N—remained constant. In this case, we must make this ratio $N = 3$, and assign very large values to LA and LB. Therefore, we chose arbitrarily to let LA $= 100$ H and LB $= 900$ H. The SPICE NETLIST is:

```
MUTUAL INDUCTANCE EXAMPLE
VG   1    0     PWL(0,0 0.001,15)
RA   1    2     3
RB   3    4     27
RHI  4    0     1E9
LA   3    0     100
LB   4    0     900
K1   LA   LB    1
.PROBE
.TRAN     0.1 10  0    0.01
.END
```

The Probe output from a SPICE run is shown in Figure 16.67, where we have plotted the gain versus frequency. An ideal transformer would have exhibited a "flat" frequency response for all values of frequency. However, because we were forced to model our transformer as a "nonideal" one, we have selected our lower frequency limit for the simulation to be a very low frequency (10 mHz = 0.01 Hz) to show the "rolloff" effect at the lower frequencies due to the nonideal elements. In a more practical situation, you can do this with the actual values of LA and LB for a given transformer and determine the frequency range over which you can consider your transformer to be "ideal."

Figure 16.67
Probe results from
SPICE run

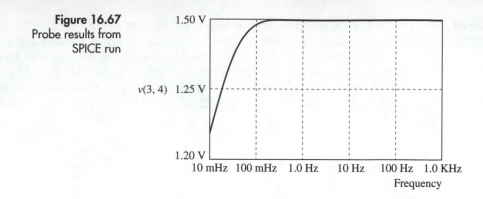

Section 16.5 Quiz

Q16.5-1. Solve Example 16.5 using SPICE simulation.   **Q16.5-2.** Solve Example 16.10 using SPICE simulation.

Chapter 16 Summary

Section 16.1 of this chapter introduced the idea of *mutual inductance.* We motivated the discussion with a physical description of the magnetic field surrounding a coil of wire—though it is not necessary for the reader to have encountered this concept before in order to appreciate the motivational aspects of our discussion. We introduced a *dot notation* to indicate the direction "around the core of the inductor" in which the magnetic field created by a given coil of wire is oriented. We showed that this is actually unnecessary in the case of a single inductor; however, when two coils are close together physically, there is an interaction between the two called *mutual inductance.* We discussed the behavior of the various ways in which two coils can be situated relative to each other exhaustively and showed that, with a very simple rule, the issue of winding orientation becomes trivial.

In Section 16.2, after reviewing the two-port z parameters, we developed a *z-parameter model* for the transformer, and showed how one can convert it into the *equivalent tee* configuration—provided that a terminal of each winding is connected to one terminal of the other winding. We also developed an *initial condition model* for the transformer based upon the z-parameter model. Following this, we repeated the same development, based this time, however, on the y parameters.

In Section 16.3, we worked numerous examples illustrating the analysis of circuits containing transformers—in the time domain, the Laplace transform (s) domain, and in the frequency domain (ω), as well as in phasor form.

In Section 16.4, we discussed the power and energy absorbed by a transformer and, based upon these concepts, gave a rigorous proof of reciprocity ($L_{12} = L_{21}$), based upon an argument proposed only recently by Dr. P. M. Lin of Purdue University. These ideas led to the concepts of *coefficient of coupling, unity coupled transformer,* and *ideal transformer.* Finally, we developed an equivalent for the nonideal transformer based upon the ideal transformer model.

In Section 16.5, we introduced the SPICE syntax for modeling mutual inductance. We discussed the care that must be taken to correct the modeling of circuits having disconnected parts caused by transformers. We took this occasion to further explore the process by which one derives the ideal transformer by letting the two self-inductances of a transformer become infinite.

Chapter 16 Review Quiz

RQ16.1. Place dots on the appropriate terminals of the three-winding transformer shown in Figure RQ16.1. Assume that positive magnetic field intensity is to the right in the core.

Figure RQ16.1

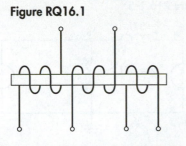

RQ16.2. If $i(t) = 2\cos(2t)$ A in the circuit shown in Figure RQ16.2, find $v_x(t)$.

Figure RQ16.2

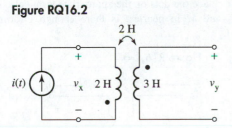

RQ16.3. Answer Question RQ16.2 for $v_y(t)$, rather than $v_x(t)$.

RQ16.4. Compute the flux linkage $\lambda_x(t)$ in Figure RQ16.2 if $i(t) = 2u(t)$ A.

RQ16.5. Compute the flux linkage $\lambda_y(t)$ in Figure RQ16.2 if $i(t) = 2u(t)$ A.

RQ16.6. Write the equations for the transformer shown in Figure RQ16.3 for the terminal variables shown.

Figure RQ16.3

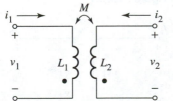

RQ16.7. Write the equations for the transformer shown in Figure RQ16.4 for the terminal variables shown.

Figure RQ16.4

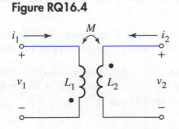

RQ16.8. Find the system function $H(s) = V(s)/I_g(s)$ for the circuit shown in Figure RQ16.5. Then use it to obtain the differential equation relating $v(t)$ to $i_g(t)$.

Figure RQ16.5

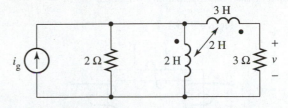

RQ16.9. Draw the equivalent circuit that results if the tee parameter equivalent is used for the transformer in Figure RQ16.5.

RQ16.10. Draw the equivalent circuit that results if the pi parameter equivalent is used for the transformer in Figure RQ16.5.

RQ16.11. Find $v(t)$ in the circuit shown in Figure RQ16.6 if $v_g(t) = 18u(t)$ V.

Figure RQ16.6

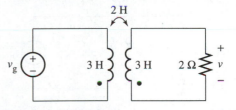

RQ16.12. If $i_a(0) = 2$ A and $i_b(0) = 3$ A in Figure RQ16.7, find $v_a(t)$ and $v_b(t)$ for $t > 0$.

Figure RQ16.7

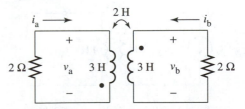

RQ16.13. Find the average power absorbed by the 32-Ω resistor in Figure RQ16.8 if the voltage source value is given by $v_g(t) = 10\sin(2t)$ V.

Figure RQ16.8

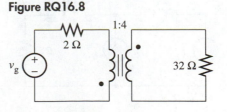

Chapter 16 Problems

Section 16.1 Mutual Inductance

16.1-1. A terminal of one winding of the four-winding transformer in Figure P16.1-1 is dotted. Place a dot on the appropriate terminal of each of the remaining windings.

Figure P16.1-1

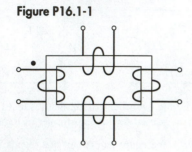

16.1-2. In Figure 16.1-2, $i_a(t) = 4\cos(2t)$ A and $i_b(t) = 2\sin(2t)$ A. Find $v_a(t)$ and $v_b(t)$.

Figure P16.1-2

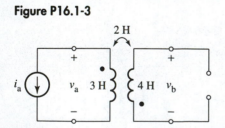

16.1-3. If $i_a(t) = 4tu(t)$ A in Figure P16.1-3, find and sketch $v_a(t)$ and $v_b(t)$.

Figure P16.1-3

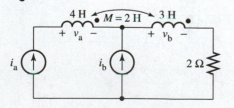

16.1-4. If $i_a(t) = 10e^{-t}u(t)$ A and $i_b(t) = 20e^{-t}u(t)$ A in Figure P16.1-4, find $v_a(t)$ and $v_b(t)$.

Figure P16.1-4

16.1-5. If $i_a(t) = 4tu(t)$ A and $i_b(t) = 3t^2u(t)$ A in Figure P16.1-5, find and sketch $v_a(t)$ and $v_b(t)$.

Figure P16.1-5

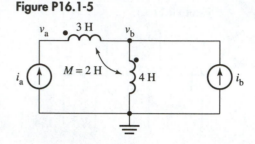

16.1-6. When $i_b(t) = 0$ and $i_a(t) = 2\cos(100t)$ A, the voltages are $v_a(t) = -800\sin(100t)$ V and $v_b(t) = 400\sin(100t)$ V in Figure P16.1-6. Place the dots on the appropriate terminals. Then determine L_1 and M in henries. Is there enough information to determine L_2?

Figure P16.1-6

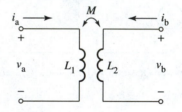

Section 16.2 Transformer Equivalent Circuits

16.2-1. Write the terminal equations in p operator form for the transformer of Figure P16.2-1 in terms of the defined terminal voltages and currents.

Figure P16.2-1

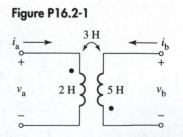

16.2-2. Find the s-plane impedance parameter matrix Z for the transformer of Figure P16.2-1 in terms of the defined port variables.

16.2-3. Find the s-plane Y parameter matrix for the transformer of Figure P16.2-1 in terms of the defined port variables.

16.2-4. Assuming that the bottom two terminals of the transformer in Figure P16.2-1 are connected, draw the tee equivalent and label its inductance values.

16.2-5. Assuming that the bottom two terminals of the transformer in Figure P16.2-1 are connected, draw the pi equivalent and label its inductance values.

16.2-6. Find an equivalent two-port for the subcircuit shown in Figure P16.2-2.

Figure P16.2-2

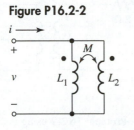

16.2-7. Find an equivalent two-port for the subcircuit shown in Figure P16.2-3.

Figure P16.2-3

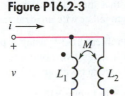

16.2-8. Replace the transformer in Figure P16.2-4 by its tee equivalent and find the input impedance Z at $\omega = 2$ rad/s.

Figure P16.2-4

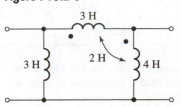

16.2-9. Find the equivalent tee for the pi network of inductors in Figure P16.2-5.

Figure P16.2-5

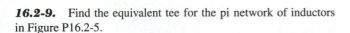

16.2-10. Find the equivalent pi for the tee network of inductors in Figure P16.2-6.

Figure P16.2-6

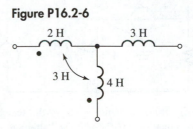

16.2-11. Find the impedance parameters (as functions of s) for the subcircuit shown in Figure P16.2-5 and draw the z-parameter equivalent circuit.

16.2-12. Find the admittance parameters (as functions of s) for the subcircuit shown in Figure P16.2-5 and draw the y-parameter equivalent circuit.

16.2-13. Find the impedance parameters (as functions of s) for the subcircuit shown in Figure P16.2-6 and draw the z-parameter equivalent circuit.

16.2-14. Find the admittance parameters (as functions of s) for the subcircuit shown in Figure P16.2-6 and draw the y-parameter equivalent circuit.

16.2-15. If $i_a(0) = 1$ A and $i_b(0) = 2$ A in Figure P16.2-7, draw the z-parameter initial condition model (expressing the z-parameters in terms of s).

Figure P16.2-7

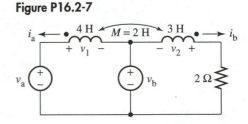

16.2-16. Repeat Problem 16.2-15, replacing the expression "z parameter" by the expression "y parameter."

16.2-17. Draw the tee initial condition model for the circuit in Figure P16.2-7, labeling all elements with their values (for the inductors, label the inductance values in henries).

16.2-18. Repeat Problem 16.2-17, replacing the word "tee" with the word "pi."

16.2-19. If $v_a(t) = 16e^{-t}$ V and $v_b = 32e^{-t}$ V (both for $t > 0$) in Figure P16.2-7, find $i_a(t)$ and $i_b(t)$ for all $t > 0$.

16.2-20. If $i_x(t) = 6e^{-2t}$ A, $i_y(t) = 16e^{-2t}$ A and $i_z(t) = 7e^{-2t}$ A in Figure P16.2-8, find $v_1(t)$ and $v_2(t)$ for all $t > 0$.

Figure P16.2-8

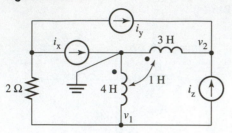

16.2-21. Verify the statements made in the text in the paragraph following equation (16.2-6) about "alternate dot orientations."

Section 16.3 Analysis of Circuits with Transformers

16.3-1. In Figure P16.3-1, let $v_g(t) = 12u(t)$ V and $M = 1/\sqrt{2}$ H. Find $v(t)$ for all t using the basic transformer equations.

Figure P16.3-1

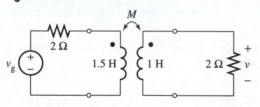

16.3-2. Let $M = \sqrt{3/2}$ H and $v_g(t) = 12u(t)$ V in the circuit shown in Figure P16.3-1. Find $v(t)$ for all t using the basic transformer equations.

16.3-3. Repeat Problem 16.3-1 using the tee equivalent (assuming that the bottom terminals of the transformer are connected together with an ideal conductor).

16.3-4. Repeat Problem 16.3-1 using the pi equivalent (assuming that the bottom terminals of the transformer are connected together with an ideal conductor).

16.3-5. If $i_g(t) = 10u(t)$ A in Figure P16.3-2, find $v(t)$ for all values of t using mesh analysis.

Figure P16.3-2

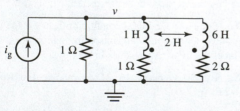

16.3-6. Repeat Problem 16.3-5 using nodal analysis.

16.3-7. Repeat Problem 16.3-5 using a tee equivalent.

16.3-8. Repeat Problem 16.3-5 using a pi equivalent.

16.3-9. The circuit in Figure P16.3-3 is operating in the dc steady state for $t \le 0$. Use mesh analysis to find $v(t)$ for all $t > 0$.

Figure P16.3-3

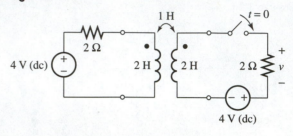

16.3-10. Repeat Problem 16.3-9 using nodal analysis. Assume that the bottom terminals of the transformer are connected together with an ideal conductor.

16.3-11. Repeat Problem 16.3-9 using a tee equivalent for the transformer. Assume that the two bottom terminals of the transformer are connected together with an ideal conductor.

16.3-12. Repeat Problem 16.3-9 using a pi equivalent for the transformer. Assume that the two bottom terminals of the transformer are connected together with an ideal conductor.

16.3-13. The circuit in Figure P16.3-4 is known to be in the dc steady state for all $t \le 0$. Find $v(t)$ for all $t > 0$ using mesh analysis.

Figure P16.3-4

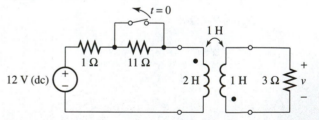

16.3-14. The circuit in Figure P16.3-5 is known to be in the dc steady state for all $t \le 0$. Find $v(t)$ for all $t > 0$ using nodal analysis.

Figure P16.3-5

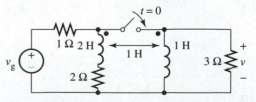

16.3-15. Assume that the circuit of Figure P16.3-6 is operating in the ac steady state. Use mesh analysis to find $v(t)$. Let $v_g(t) = 20 \sin(2t)$ V.

Figure P16.3-6

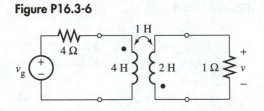

16.3-16. Assume that the circuit of Figure P16.3-7 is operating in the ac steady state. Use mesh analysis to find $v(t)$. Let $v_g(t) = 10 \cos(t)$ V.

Figure P16.3-7

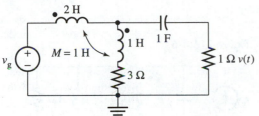

16.3-17. Find the voltage $v(t)$ in Figure P16.3-8 using mesh analysis. Assume that the two voltage sources are specified by $v_a(t) = 24 \cos(6t)u(t)$ V and $v_b(t) = 48 \cos(8t)u(t)$ V.

Figure P16.3-8

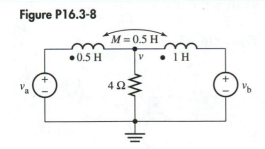

16.3-18. Find the ac steady-state voltage $v(t)$ in Figure P16.3-9 using mesh analysis. Assume that the voltage source waveform is $v_g(t) = 16 \cos(2t)$ V.

Figure P16.3-9

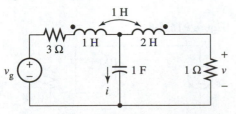

16.3-19. Find the system function $H(s) = I(s)/V_g(s)$ for the circuit in Figure P16.3-10. Plot the pole-zero diagram (PZD) and discuss the type of stability exhibited by the response waveform $i(t)$.

Figure P16.3-10

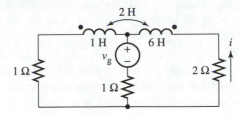

16.3-20. If $v_g(t) = 10u(t)$ V in Figure P16.3-19, find the current waveform $i(t)$ A.

16.3-21. If $i_g(t) = 10 \cos(2t)$ A in Figure P16.3-11, find the ac steady-state waveform $i(t)$.

Figure P16.3-11

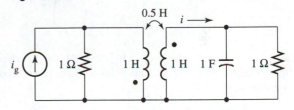

16.3-22. Determine the stability characteristics of the response $v(t)$ in Figure P16.3-12 as a function of the mutual inductance parameter M.

Figure P16.3-12

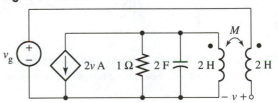

Section 16.4 Energy Considerations and the Ideal Transformer

16.4-1. The circuit in Figure P16.4-1 operates in the dc steady state for $t \leq 0$. Find the total amount of energy stored at $t = 0-$. Then find the total amount of energy delivered to R_L for all $t > 0$. Assume that $v_g(t) = 12$ V dc.

Figure P16.4-1

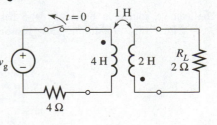

16.4-2. Suppose the mutual inductance in Figure P16.4-1 is changed to the general value M. For what value of M would a maximum amount of energy be delivered to R_L for all $t > 0$? What is the value of this maximum energy?

16.4-3. The transformer in Figure P16.4-2 is perfect; that is, the coefficient of coupling is $k = 1$. Find the transfer function $H(s) = I(s)/I_g(s)$ and use it to find $i(t)$ assuming that the current source has a waveform of $i_g(t) = 36u(t)$ A.

Figure P16.4-2

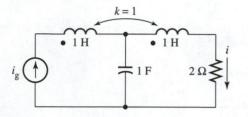

16.4-4. Find the rms phasors $\bar{V}_1$, $\bar{V}_2$, $\bar{I}_1$, and $\bar{I}_2$ for the circuit in Figure P16.4-3 assuming that the voltage source has a phasor rms value of 120 V at 0°.

Figure P16.4-3

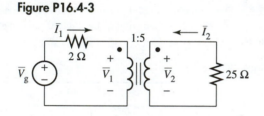

16.4-5. The voltage source in Figure P16.4-4 has a waveform of $v_g(t) = 120\sqrt{2} \cos(2t)$ V. Find the ac forced response waveform $v(t)$.

Figure P16.4-4

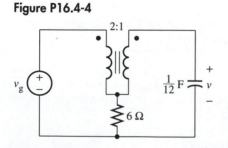

16.4-6. The voltage source in Figure P16.4-5 has a phasor value of $\bar{V}_g = 24\angle 0°$ rms. Find the rms phasor voltages $\bar{V}_1$ and $\bar{V}_2$.

Figure P16.4-5

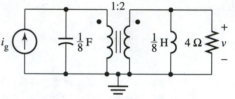

16.4-7. In Figure P16.4-6, $i_g(t) = \cos(\omega t)$ A. For what value(s) of ω is the voltage $v(t)$ the largest? What is the peak value of $v(t)$ at this frequency?

Figure P16.4-6

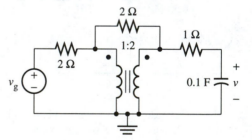

16.4-8. Find the ac steady-state waveform for $v(t)$ in Figure P16.4-7 if $v_g(t) = 12 \cos(2t)$ V.

Figure P16.4-7

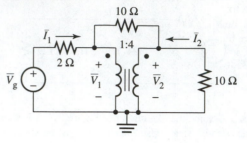

16.4-9. If the voltage source phasor in Figure P16.4-8 is $\bar{V}_g = 120\angle 0°$ V, find the rms value of $i(t)$.

Figure P16.4-8

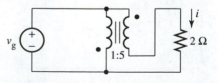

16.4-10. Assume that the turns ratio N and the Thévenin parameters are fixed in Figure P16.4-9 and find the values of R_L and X_L that result in maximum power absorbed by Z_L. Assume that R_L and R_{Th} are both nonnegative.

Figure P16.4-9

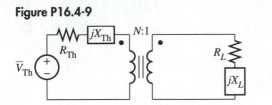

16.4-11. Repeat Problem 16.4-10 assuming that Z_L is fixed and the turns ratio N is variable. Assume that both resistance parame-

ters (R_L and R_{Th}) are nonnegative. (*Hint:* Draw the equivalent for the transformer and Z_L referred to the primary side.)

Section 16.5 SPICE: Mutual Inductance

16.5-1. Use SPICE to solve Problem 16.3-5.

16.5-2. Use SPICE to solve Problem 16.3-13.

16.5-3. Use SPICE to plot the frequency response for the voltage $v(t)$ in Problem 16.3-16. Use your results to find the steady state value of $v(t)$ when $v_g(t)$ has the waveform given in Problem 16.3-16.

Appendices

Appendix A offers a "mini-course" in linear algebra. The basic definitions and operations of matrix algebra are discussed, and the issue of solving a system of linear equations is developed thoroughly. The computation of the determinant of a matrix is presented, with attention to its use in solving linear equations using Cramer's rule and Gaussian reduction. The basic concepts of vector spaces are treated, among them such topics as linear dependence and independence and the concept of a basis for a vector space. This appendix develops all the concepts that one needs to read and understand this text—the principal use being in the development of nodal and mesh analysis in Chapter 4.

Appendix B contains a proof of Tellegen's theorem, one of the truly outstanding and fundamental cornerstones of circuit theory. It is used in the text to make the proof logically complete that nodal and mesh analysis always "work"—with only very minor restrictions to the circuit being analyzed. (A generalization of the nodal analysis method to "patch things up" is actually given in the text in Chapter 4.)

Appendix C presents an extremely effective table method for integration by parts, one of the main mathematical techniques used by engineers to compute integrals.

Finally, Appendix D contains an elementary, yet rigorous, development of generalized functions. This appendix is included for the more analytically oriented reader for whom the more intuitive treatment given in Chapter 6 is insufficient. We say that the development is "elementary," meaning that one can read and understand it with only a knowledge of first-year calculus; however, to be fair, we must also mention that it is more abstract than the material in the body of the text. This approach to generalized function theory is original with the author and suffices for the time domain analysis of circuits and other linear systems.

Appendix A | Linear Algebra, Determinants, and Matrices

Systems of Linear Equations and Matrices: Basic Definitions

The basic problem covered by this appendix is the solution of the system of linear equations such as

$$2x_1 + 2x_2 + 1x_3 = 10, \tag{A-1a}$$

$$1x_1 + 1x_2 + 1x_3 = 6, \tag{A-1b}$$

and

$$1x_1 + 0x_2 + 1x_3 = 5, \tag{A-1c}$$

that is, three linear equations in three unknowns, and its generalization to m equations in n unknowns. Of special interest will be the case in which $m = n$, that is, the case in which the number of equations is identical to the number of unknowns. We will use the system of equations we have just written to illustrate many of the concepts we will present.

The first thing we will do is introduce a more compact notational scheme. We will call

$$\bar{x} = \begin{bmatrix} x_1 \\ x_2 \\ x_3 \end{bmatrix} \tag{A-2a}$$

a (3×1) matrix, a column matrix with 3 rows, or a 3-vector. In general, we will say that

$$\bar{x} = \begin{bmatrix} x_1 \\ x_2 \\ \vdots \\ x_n \end{bmatrix} \tag{A-2b}$$

is an $(n \times 1)$ matrix, a column matrix with n rows, or an n-vector. Do you notice how much vertical space these symbols use? We will solve this notational problem by agreeing to write $\bar{x} = \text{col}(x_1, x_2, x_3)$ for a 3-vector and $\bar{x} = \text{col}(x_1, x_2, x_3, \cdots, x_n)$ for an n-vector. We define $\bar{x} = [x_1, x_2, x_3, \cdots, x_n]$ to be a *row vector*. Notice that it is intrinsically different from $\text{col}(x_1, x_2, x_3, \cdots, x_n)$, which is a column vector written from left to right rather than vertically. The "col" serves as a reminder, and we will often call an n-vector a column vector. The two types of vector are, however, linked. We define the *transpose* of a column vector to be a row vector and that of a row vector to be a column vector. That is, if $\bar{x} = \text{col}(x_1, x_2, x_3, \cdots, x_n)$, then the transpose of $\bar{x}$ is defined by

$$\bar{x}^t = [x_1, x_2, x_3, \cdots, x_n]. \tag{A-3}$$

On the other hand, if $\bar{x} = [x_1, x_2, x_3, \cdots, x_n]$, then

$$\bar{x}^t = \text{col}(x_1, x_2, x_3, \cdots, x_n). \tag{A-4}$$

As an example, suppose that $\bar{x} = \mathrm{col}(1, 2, 3)$. Then $\bar{x}^t = [1, 2, 3]$. If $\bar{x} = [1, 2, 3]$, then $\bar{x}^t = \mathrm{col}(1, 2, 3)$.

We still need a few definitions. To provide the symbols, we will let c be any arbitrary scalar (real or complex number), $\bar{x} = \mathrm{col}(x_1, x_2, x_3)$, and $\bar{y} = \mathrm{col}(y_1, y_2, y_3)$. We define $\bar{x} = \bar{y}$ if and only if $x_1 = y_1, x_2 = y_2$, and $x_3 = y_3$. We define the product of c and $\bar{x}$ by $c\bar{x} = \mathrm{col}(cx_1, cx_2, cx_3)$. Finally, we define the *scalar product,* the *inner product,* or the *dot product* (these terms all mean the same thing) by the equation

$$\bar{x}\cdot\bar{y} = \bar{x}^t\bar{y} = [x_1, x_2, x_3]\begin{bmatrix} y_1 \\ y_2 \\ y_3 \end{bmatrix} = x_1y_1 + x_2y_2 + x_3y_3. \tag{A-5}$$

Notice that we are also defining here the product of a row matrix and a column matrix. We have written the column matrix out vertically for emphasis. Note that

$$\bar{y}\cdot\bar{x} = \bar{y}^t\bar{x} = [y_1, y_2, y_3]\begin{bmatrix} x_1 \\ x_2 \\ x_3 \end{bmatrix} = y_1x_1 + y_2x_2 + y_3x_3 = \bar{x}\cdot\bar{y}. \tag{A-6}$$

Notice also that the scalar product of two vectors is a scalar—not a vector (or a row vector).

To see how our notation simplifies the writing of systems of linear equations, look at equations (A-1) once more. If we define $\bar{a}_1 = [2, 2, 1]$, $\bar{a}_2 = [1, 1, 1]$, $\bar{a}_3 = [1, 0, 1]$, $\bar{x} = \mathrm{col}(x_1, x_2, x_3)$ and $\bar{y} = \mathrm{col}(y_1, y_2, y_3) = \mathrm{col}(10, 6, 5)$, we can write equations (A-1) as

$$\bar{a}_1\cdot\bar{x} = y_1 = 10, \tag{A-7a}$$

$$\bar{a}_2\cdot\bar{x} = y_2 = 6, \tag{A-7b}$$

and $$\bar{a}_3\cdot\bar{x} = y_3 = 5. \tag{A-7c}$$

We can group these three equations into one in the following way:

$$\mathrm{col}(\bar{a}_1\cdot\bar{x}, \bar{a}_2\cdot\bar{x}, \bar{a}_3\cdot\bar{x}) = \bar{y}. \tag{A-8}$$

We can simplify our notation even further. We define the rectangular matrix A by

$$A = \mathrm{col}(\bar{a}_1, \bar{a}_2, \bar{a}_3) = \begin{bmatrix} \bar{a}_1 \\ \bar{a}_2 \\ \bar{a}_3 \end{bmatrix} = \begin{bmatrix} 2 & 2 & 1 \\ 1 & 1 & 1 \\ 1 & 0 & 1 \end{bmatrix}. \tag{A-9}$$

We call A a *rectangular* matrix with 3 rows and 3 columns or a "three by three" matrix (with the number of rows specified first, then the number of columns). We often write the latter expression as "A is a (3 × 3) matrix." Note that we do not use a bar over the symbol for a matrix that is not a column matrix (or column or row vector). Using this notation, our basic problem—to solve equations (A-1)—can be written in the form

$$\begin{bmatrix} 2 & 2 & 1 \\ 1 & 1 & 1 \\ 1 & 0 & 1 \end{bmatrix}\begin{bmatrix} x_1 \\ x_2 \\ x_3 \end{bmatrix} = \begin{bmatrix} 10 \\ 6 \\ 5 \end{bmatrix}. \tag{A-10}$$

If the dimensions are understood, one often writes the more symbolic expression,

$$A\bar{x} = \bar{y}. \tag{A-11}$$

We now generalize upon the preceding notational conventions in the following way. Let A be an $(m \times n)$ matrix (m rows and n columns). Then we often write

$$A = (a_{ij}), \tag{A-12}$$

where a_{ij} is the entry in the ith row and the jth column. Thus, if we say $a_{32} = 5$, we are saying that the element in the third row down from the top and in the second column from the left has the value 5. If we say that $a_{ij} = 2^i \sin(\sqrt{2}\pi j)$, we are giving the value to be found in the ith row and the jth column with $i = 1, 2, \cdots, m$ and $j = 1, 2, \cdots, n$.

Here is an example of a (4×3) rectangular matrix:

$$A = \begin{bmatrix} a_{11} & a_{12} & a_{13} \\ a_{21} & a_{22} & a_{23} \\ a_{31} & a_{32} & a_{33} \\ a_{41} & a_{42} & a_{43} \end{bmatrix}. \tag{A-13}$$

Now suppose that $\bar{x} = \text{col}(x_1, x_2, x_3)$. Then we can write

$$A\bar{x} = \begin{bmatrix} a_{11} & a_{12} & a_{13} \\ a_{21} & a_{22} & a_{23} \\ a_{31} & a_{32} & a_{33} \\ a_{41} & a_{42} & a_{43} \end{bmatrix} \begin{bmatrix} x_1 \\ x_2 \\ x_3 \end{bmatrix} = \begin{bmatrix} a_{11}x_1 + a_{12}x_2 + a_{13}x_3 \\ a_{21}x_1 + a_{22}x_2 + a_{23}x_3 \\ a_{31}x_1 + a_{32}x_2 + a_{33}x_3 \\ a_{41}x_1 + a_{42}x_2 + a_{43}x_3 \end{bmatrix}. \tag{A-14}$$

Thus, the product of a (4×3) matrix and a (3×1) matrix (column vector) is a (4×1) matrix. In general, the product of an $(m \times n)$ matrix and an $(n \times 1)$ matrix gives an $(m \times 1)$ matrix.

Thus far, we have defined the scalar product of two vectors (which only holds if the two vectors have the same number of elements), then we have used the scalar product to define the product of a rectangular matrix and a column matrix (which only works if the number of columns of the rectangular matrix is equal to the number of rows in the column matrix).

We now carry things to the final step, which is to define the product of two rectangular matrices. Let's carry out a specific example, and the general definition will then be clear by extension. Let

$$A = \text{col}(\bar{a}_1, \bar{a}_2) = \begin{bmatrix} \bar{a}_1 \\ \bar{a}_2 \end{bmatrix}, \tag{A-15}$$

where $\bar{a}_1 = [a_{11} \ a_{12} \ a_{13}]$ and $\bar{a}_2 = [a_{21} \ a_{22} \ a_{23}]$ are row vectors; also, let $B = [\bar{b}_1, \bar{b}_2]$, where $\bar{b}_1$ and $\bar{b}_2$ are column vectors:

$$\bar{b}_1 = \text{col}(b_{11}, b_{21}, b_{31}) = \begin{bmatrix} b_{11} \\ b_{21} \\ b_{31} \end{bmatrix} \quad \text{and} \quad \bar{b}_2 = \text{col}(b_{12}, b_{22}, b_{32}) = \begin{bmatrix} b_{12} \\ b_{22} \\ b_{32} \end{bmatrix}. \tag{A-16}$$

Then the product of the (2×3) matrix A and the (3×2) matrix B is defined by

$$AB = \begin{bmatrix} \bar{a}_1 \\ \bar{a}_2 \end{bmatrix} [\bar{b}_1 \ \bar{b}_2] = \begin{bmatrix} \bar{a}_1 \cdot \bar{b}_1 & \bar{a}_1 \cdot \bar{b}_2 \\ \bar{a}_2 \cdot \bar{b}_1 & \bar{a}_2 \cdot \bar{b}_2 \end{bmatrix} = \begin{bmatrix} a_{11} & a_{12} & a_{13} \\ a_{21} & a_{22} & a_{23} \end{bmatrix} \begin{bmatrix} b_{11} & b_{12} \\ b_{21} & b_{22} \\ b_{31} & b_{32} \end{bmatrix} \tag{A-17}$$

$$= \begin{bmatrix} a_{11}b_{11} + a_{12}b_{21} + a_{13}b_{31} & a_{11}b_{12} + a_{12}b_{22} + a_{13}b_{32} \\ a_{21}b_{11} + a_{22}b_{21} + a_{23}b_{31} & a_{21}b_{12} + a_{22}b_{22} + a_{23}b_{32} \end{bmatrix}.$$

We have defined the ijth element in the product to be the scalar product of the a row vector in row i of A times the column vector in column j of B. Notice that, for example, the 12 entry in AB is

$$\bar{a}_1 \cdot \bar{b}_2 = a_{11}b_{12} + a_{12}b_{22} + a_{13}b_{32}. \tag{A-18}$$

Look at this carefully and you will see the truth of the following general observation. If A is an $m \times k$ matrix and B a $k \times n$ matrix, and $C = AB$, then

$$C_{ij} = \bar{a}_i \cdot \bar{b}_j = \sum_{q=1}^{k} a_{iq}b_{qj}. \tag{A-19}$$

The element at the intersection of row i and column j in C is the scalar product of row i of the matrix A and column j of matrix B.

Adding matrices is very ordinary and simple, provided that the dimensions are the same. For example, if we let

$$A = \begin{bmatrix} a_{11} & a_{12} & a_{13} \\ a_{21} & a_{22} & a_{23} \end{bmatrix} \quad \text{and} \quad B = \begin{bmatrix} b_{11} & b_{12} & b_{13} \\ b_{21} & b_{22} & b_{23} \end{bmatrix}, \tag{A-20}$$

then $A + B$ is defined by

$$A + B = \begin{bmatrix} a_{11} + b_{11} & a_{12} + b_{12} & a_{13} + b_{13} \\ a_{21} + b_{21} & a_{22} + b_{22} & a_{23} + b_{23} \end{bmatrix}. \tag{A-21}$$

The necessity for the dimensions being equal should be clear: if they are not, there will be elements of one matrix "left over" with no mates when we try to add the two. Multiplication of a rectangular matrix by a scalar c is essentially no different than the multiplication of a vector by a scalar. If A is the matrix defined in equation (A-20), we have

$$cA = \begin{bmatrix} ca_{11} & ca_{12} & ca_{13} \\ ca_{21} & ca_{22} & ca_{23} \end{bmatrix}. \tag{A-22}$$

The generalization of these operations to other dimensions should be clear.

The transpose of a rectangular matrix is of some importance to us. If $A = (a_{ij})$, the transpose of A is defined to be $A^t = (a_{ji})$. The process of transposition interchanges the rows and the columns. Perhaps it is a bit clearer if we write $A = [\bar{a}_1, \bar{a}_2, \cdots, \bar{a}_n]$, where each of the $\bar{a}_i$'s is a column vector with m rows. Then we write the transpose of A as

$$A^t = \text{col}(\bar{A}_1^t, \bar{A}_2^t, \cdots, \bar{A}_n^t) = \begin{bmatrix} \bar{A}_1^t \\ \bar{A}_2^t \\ \cdots \\ \bar{A}_n^t \end{bmatrix}, \tag{A-23}$$

which is a rectangular matrix with n rows and m columns (each $\bar{A}_i^t$ is a row vector with m elements, or columns).

We will see that we need a matrix to serve the role of unity (the number 1) for matrices. The square matrix of dimension $(n \times n)$ having ones on the *principal diagonal* and zeros everywhere else is called the nth-order identity matrix:

$$I_n = \begin{bmatrix} 1 & 0 & \cdots & 0 \\ 0 & 1 & \cdots & 0 \\ \cdots & \cdots & \cdots & \cdots \\ 0 & 0 & \cdots & 1 \end{bmatrix}. \tag{A-24}$$

The *inverse* of a square matrix plays a large role in our development. The inverse of an $(n \times n)$ matrix A is defined to be the square matrix A^{-1} satisfying the equation

$$A^{-1}A = AA^{-1} = I_n. \tag{A-25}$$

Finding such a matrix will be one of our main goals, and we will tackle it a bit later.

Now, though, we will point out just a couple of facts. The multiplication of matrices is not commutative; that is, $AB \neq BA$ in general. To see this, just let $A = \begin{bmatrix} 1 & 2 \\ 3 & 4 \end{bmatrix}$ and $B = \begin{bmatrix} 0 & 1 \\ 0 & 0 \end{bmatrix}$. Then

$$AB = \begin{bmatrix} 1 & 2 \\ 3 & 4 \end{bmatrix}\begin{bmatrix} 0 & 1 \\ 0 & 0 \end{bmatrix} = \begin{bmatrix} 0 & 1 \\ 0 & 3 \end{bmatrix}, \quad \text{but} \quad BA = \begin{bmatrix} 0 & 1 \\ 0 & 0 \end{bmatrix}\begin{bmatrix} 1 & 2 \\ 3 & 4 \end{bmatrix} = \begin{bmatrix} 3 & 4 \\ 0 & 0 \end{bmatrix} \neq AB. \tag{A-26}$$

The inverse of a matrix is useful because it allows us to solve a system of n linear equations in n unknowns, providing that the inverse exists. For example, suppose that we are to solve the matrix equation

$$A\bar{x} = \bar{y}. \tag{A-27}$$

If we know the inverse of A, we simply multiply both sides by that matrix from the left:

$$A^{-1}(A\bar{x}) = A^{-1}\bar{y}. \tag{A-28}$$

Though matrix multiplication is not commutative, it *is* associative (though we will not prove it). Thus, we can write

$$A^{-1}(A\bar{x}) = (A^{-1}A)\bar{x} = I_n\bar{x} = A^{-1}\bar{y}. \tag{A-29}$$

Sorting out the important expressions, we have

$$I_n\bar{x} = A^{-1}\bar{y}. \tag{A-30}$$

Just what is $I_n\bar{x}$? To see, take the example of $I_3 \cdot \text{col}(x_1, x_2, x_3)$:

$$I_3\bar{x} = \begin{bmatrix} 1 & 0 & 0 \\ 0 & 1 & 0 \\ 0 & 0 & 1 \end{bmatrix}\begin{bmatrix} x_1 \\ x_2 \\ x_3 \end{bmatrix} = \begin{bmatrix} x_1 \\ x_2 \\ x_3 \end{bmatrix} = \bar{x}. \tag{A-31}$$

This, of course, is what we would expect from a quantity called the identity matrix. Thus, our solution is

$$\bar{x} = A^{-1}\bar{y}. \tag{A-32}$$

If we know the inverse of the coefficient matrix in our system of equations, we have the solution in the form of (A-32).

Let's now return to the basic system of equations with which we opened this appendix. In matrix form, these equations turned out to be equation (A-10), which we repeat here:

$$A\bar{x} = \begin{bmatrix} 2 & 2 & 1 \\ 1 & 1 & 1 \\ 1 & 0 & 1 \end{bmatrix}\begin{bmatrix} x_1 \\ x_2 \\ x_3 \end{bmatrix} = \bar{y} = \begin{bmatrix} 10 \\ 6 \\ 5 \end{bmatrix}. \tag{A-33}$$

We will derive the inverse matrix for A later, but at this point we will "fudge a little." It is

$$A^{-1} = \begin{bmatrix} 1 & -2 & 1 \\ 0 & 1 & -1 \\ -1 & 2 & 0 \end{bmatrix}. \tag{A-34}$$

(You should verify this here by computing AA^{-1} and $A^{-1}A$ and showing that both equal the (3×3) identity matrix.) Thus, we have

$$\bar{x} = A^{-1}\bar{y} = \begin{bmatrix} 1 & -2 & 1 \\ 0 & 1 & -1 \\ -1 & 2 & 0 \end{bmatrix}\begin{bmatrix} 10 \\ 6 \\ 5 \end{bmatrix} = \begin{bmatrix} 3 \\ 1 \\ 2 \end{bmatrix}. \tag{A-35}$$

If you insert these values into equation (A-33) or into the corresponding system in equations (A-1), you will see that they are, indeed, solutions.

   At this point, we see that finding an inverse matrix is equivalent to finding the solution of a system of linear equations, so we will devote some time to developing a procedure for computing it. First, however, we will need a little theory from linear algebra concerning vector spaces.

Vector Spaces   Suppose we consider the set of all possible vectors of the form $\bar{x} = \mathrm{col}(x_1, x_2, x_3)$ as the real numbers x_1, x_2, and x_3 take on all possible scalar values. We call this set Euclidean[1] 3-space, E^3. In general, the set of all possible n-vectors $\bar{x} = \mathrm{col}(x_1, x_2, \cdots, x_n)$ as $x_1, x_2, \cdots, x_n$ range through all possible scalar values is called Euclidean n-space E^n. The following concept about such spaces is so important that we will set it off formally as a definition.

Definition A.1   The set of vectors $\bar{x}_1, \bar{x}_2, \ldots, x_n$ is said to be *linearly independent* if the only scalars $a_1, a_2, \cdots a_n$ that satisfy the equation

$$a_1\bar{x}_1 + a_2\bar{x}_2 + \cdots + a_n\bar{x}_n = \bar{0}, \tag{A-36}$$

where $\bar{0} = \mathrm{col}(0, 0, \cdots, 0)$, are $a_1 = a_2 = \cdots = a_n = 0$. The combination of terms in (A-36) is called a *linear combination* of the vectors $\bar{x}_1$ through $\bar{x}_n$.

Consider the vectors $\bar{e}_1 = \mathrm{col}(1, 0, 0)$, $\bar{e}_2 = \mathrm{col}(0, 1, 0)$, and $\bar{e}_3 = \mathrm{col}(0, 0, 1)$—all of which are in E^3—as an example. We form a linear combination and set it equal to the zero vector:

$$a_1\bar{e}_1 + a_2\bar{e}_2 + a_3\bar{e}_3 = \bar{0}. \tag{A-37}$$

Then we ask: what are the possible values for the a's? To answer this question, we rewrite (A-37) in the more detailed form

$$a_1\begin{bmatrix} 1 \\ 0 \\ 0 \end{bmatrix} + a_2\begin{bmatrix} 0 \\ 1 \\ 0 \end{bmatrix} + a_3\begin{bmatrix} 0 \\ 0 \\ 1 \end{bmatrix} = \begin{bmatrix} 0 \\ 0 \\ 0 \end{bmatrix}. \tag{A-38}$$

Applying our definitions of sum and equality, we see that (A-38) is equivalent to the

---

[1] More exactly, we must also assume that the "size of" $\bar{x}$ is the "Euclidean norm": $|\bar{x}| = \sqrt{(x_1)^2 + (x_2)^2 + (x_3)^2}$.

equation $a_1 = a_2 = \cdots = 0$. This means that $\bar{e}_1, \bar{e}_2$, and $\bar{e}_3$ are linearly independent in E^3. In general, the set of vectors $\bar{e}_1, \bar{e}_2, \ldots, \bar{e}_n$, where

$$\bar{e}_i = \text{col}(0, 0, \ldots, 0, 1, 0, \ldots, 0),$$
$$\underset{i\text{th position}}{\uparrow} \tag{A-39}$$

is linearly independent in E^n, the space of all n-vectors.

The definition of linear independence must be coupled with the following one in order for us to see why the concept is so important.

Definition A.2   The set of all linear combinations

$$\bar{y} = a_1\bar{x}_1 + a_2\bar{x}_2 + \cdots a_n\bar{x}_n, \tag{A-40}$$

as $a_1, a_2, \ldots, a_n$ range over all scalar values, is called the space *spanned* by the vectors (whether or not the $\bar{x}_i$'s are linearly independent).

For an example, let's look at the space spanned by $\bar{e}_1, \bar{e}_2$, and $\bar{e}_3$ defined above. We form the linear combination

$$\bar{y} = a_1\bar{e}_1 + a_2\bar{e}_2 + a_3\bar{e}_3 = a_1\begin{bmatrix} 1 \\ 0 \\ 0 \end{bmatrix} + a_2\begin{bmatrix} 0 \\ 1 \\ 0 \end{bmatrix} + a_3\begin{bmatrix} 0 \\ 0 \\ 1 \end{bmatrix} = \begin{bmatrix} a_1 \\ a_2 \\ a_3 \end{bmatrix}. \tag{A-41}$$

As a_1, a_2, and a_3 range over all scalar values, we see that the vector $\bar{y}$ ranges over the entire space E^3. Thus, $\bar{e}_1, \bar{e}_2$, and $\bar{e}_3$ span E^3. Similarly, the set of vectors $\bar{e}_1, \bar{e}_2, \ldots, \bar{e}_n$ span E^n. Such sets of vectors are clearly important because any vector in the spanned space can be represented as a linear combination of those vectors—and we have reduced the problem of determining things about an infinite number of vectors to the consideration of the properties of a finite number. This clearly calls for another definition.

Definition A.3   A set of linearly independent vectors spanning a given space S is called a *basis* for S. If the set of linearly independent vectors has n members, then S is called an n-dimensional vector space.

Applying this last definition, we see that E^n is an n-dimensional vector space and that the set of vectors $\bar{e}_1, \bar{e}_2, \ldots, \bar{e}_n$ form a basis for that space. Thus, every vector in E^n can be written as a linear combination of $\bar{e}_1, \bar{e}_2, \ldots, \bar{e}_n$.

We should perhaps note that there are other basis sets for E^n. For instance, working with $n = 3$, the vectors $\bar{a}_1 = \text{col}(2, 1, 1), \bar{a}_2 = \text{col}(2, 1, 0)$, and $\bar{a}_3 = \text{col}(1, 1, 1)$ are a basis set for E^3. Do you recognize them? They are the columns of the coefficient matrix A in equations (A-1) and (A-10)—those of our ongoing example problem. They are linearly independent because if we set

$$\bar{y} = c_1\bar{a}_1 + c_2\bar{a}_2 + c_3\bar{a}_3 = c_1\begin{bmatrix} 2 \\ 1 \\ 1 \end{bmatrix} + c_2\begin{bmatrix} 2 \\ 1 \\ 0 \end{bmatrix} + c_3\begin{bmatrix} 1 \\ 1 \\ 1 \end{bmatrix} = \begin{bmatrix} 2c_1 + 2c_2 + c_3 \\ c_1 + c_2 + c_3 \\ c_1 + c_3 \end{bmatrix} = \begin{bmatrix} 0 \\ 0 \\ 0 \end{bmatrix}, \tag{A-42}$$

we get the equivalent system of equations

$$2c_1 + 2c_2 + c_3 = 0, \tag{A-43a}$$

$$c_1 + c_2 + c_3 = 0, \tag{A-43b}$$

and $\qquad$ $$c_1 + c_3 = 0. \tag{A-43c}$$

We have not yet developed organized methods for solving such systems; therefore, we will leave it to you to use the old method of substitution (say solving (A-43c) for c_3 in terms of c_1, using the result in (A-43b) to solve for c_2 in terms of c_1, and finally using both results in (A-43a) to solve for c_1), thereby showing that $c_1 = c_2 = c_3 = 0$. This means that the given column vectors of the matrix A are linearly independent. (Notice that we have used c's, rather than a's for the scalar coefficients to prevent confusion with the $\bar{a}$ vectors.)

Now let's check whether or not these same vectors span E^3. Our linear independence argument above showed that the zero vector could be so expressed—but now we must check whether or not this is true for an arbitrary 3-vector. Thus, we take an arbitrary 3-vector $\bar{y} = \text{col}(y_1, y_2, y_3)$ and ask whether it can be written as a linear combination of the three vectors $\bar{a}_1 = \text{col}(2, 1, 1), \bar{a}_2 = \text{col}(2, 1, 0)$, and $\bar{a}_3 = \text{col}(1, 1, 1)$. We can answer this if we can solve the matrix equation

$$c_1\bar{a}_1 + c_2\bar{a}_2 + c_3\bar{a}_3 = c_1\begin{bmatrix} 2 \\ 1 \\ 1 \end{bmatrix} + c_2\begin{bmatrix} 2 \\ 1 \\ 0 \end{bmatrix} + c_3\begin{bmatrix} 1 \\ 1 \\ 1 \end{bmatrix} = \begin{bmatrix} 2c_1 + 2c_2 + c_3 \\ c_1 + c_2 + c_3 \\ c_1 + c_3 \end{bmatrix} = \begin{bmatrix} y_1 \\ y_2 \\ y_3 \end{bmatrix} \tag{A-44}$$

for arbitrary values of y_1, y_2, and y_3. Again, (A-44) is equivalent to the system of equations

$$2c_1 + 2c_2 + c_3 = y_1, \tag{A-45a}$$

$$c_1 + c_2 + c_3 = y_2, \tag{A-45b}$$

and $\qquad$ $$c_1 + c_3 = y_3. \tag{A-45c}$$

We will again leave it to you to use the method (?) of substitution to show that this is possible and that these values are $c_1 = y_1 - 2y_2 + y_3$, $c_2 = y_2 - y_3$, and $c_3 = -y_1 + 2y_2$. This shows that the given set of column vectors spans E^3—and thus form a basis for it.

How does this little body of theory bear on our problem of solving the original set of linear equations in (A-1) and (A-10)? To find out, let's look at the matrix equation in (A-10), repeated here for convenience as (A-46):

$$\begin{bmatrix} 2 & 2 & 1 \\ 1 & 1 & 1 \\ 1 & 0 & 1 \end{bmatrix}\begin{bmatrix} x_1 \\ x_2 \\ x_3 \end{bmatrix} = \begin{bmatrix} 10 \\ 6 \\ 5 \end{bmatrix}. \tag{A-46}$$

The rectangular matrix of coefficients on the left-hand side of this equation can be written as

$$A = \begin{bmatrix} 2 & 2 & 1 \\ 1 & 1 & 1 \\ 1 & 0 & 1 \end{bmatrix} = [\bar{a}_1 \quad \bar{a}_2 \quad \bar{a}_3], \tag{A-47}$$

where $\bar{a}_1 = \text{col}(2, 1, 1), \bar{a}_2 = \text{col}(2, 1, 0)$, and $\bar{a}_3 = \text{col}(1, 1, 1)$. Then we can write (A-46) in the form

$$x_1\bar{a}_1 + x_2\bar{a}_2 + x_3\bar{a}_3 = \bar{y}, \tag{A-48}$$

where $\bar{y} = \text{col}(10, 6, 5)$. We want to find the values of the x's—and now we know whether or not a solution exists: *if $\bar{y}$ is in the space spanned by $\bar{a}_1$, $\bar{a}_2$, and $\bar{a}_3$, there is a solution.* If it is not, there will *not* be a solution. We already know that the column vectors $\bar{a}_1$, $\bar{a}_2$, and $\bar{a}_3$ span *all* of E^3, therefore, a solution certainly exists. More generally, let A be an $(m \times n)$ matrix and $\bar{x}$ and $\bar{y}$ be n-vectors (where $\bar{y}$ is specified and we are to find the value of $\bar{x}$). Then we want to solve the equation

$$A\bar{x} = \bar{y}. \tag{A-49}$$

This equation has a solution, provided that the vector $\bar{y}$ is contained in the space spanned by the column vectors of A; furthermore, it has a solution *for any arbitrary value* of $\bar{y}$ if the column vectors of A span all of E^n—that is, if they form a basis for E^n.

Determinants: Basic Definitions and Cramer's Rule

We now have a set of conditions allowing us to predict whether or not a given set of linear equations has a solution, but we still must develop a general method for finding that solution. We will do this by using determinants. First a small review of (2×2) determinants is in order.

We start by asking for the solution to the simultaneous linear equations

$$a_{11}x_1 + a_{12}x_2 = y_1 \tag{A-50a}$$

and

$$a_{21}x_1 + a_{22}x_2 = y_2. \tag{A-50b}$$

We will begin the "old-fashioned" way. We multiply (A-50a) by a_{22} and (A-50b) by a_{12}, then subtract the second equation from the first. This results in

$$(a_{11}a_{22} - a_{12}a_{21})x_1 = a_{22}y_1 - a_{12}y_2, \tag{A-51}$$

so

$$x_1 = \frac{a_{22}y_1 - a_{12}y_2}{a_{11}a_{22} - a_{12}a_{21}}. \tag{A-52}$$

This is true, of course, only as long as the denominator of (A-52) is nonzero. Similarly, we can multiply (A-50a) by a_{21} and (A-50b) by a_{11}—then subtract (A-50a) from (A-50b) to get

$$(a_{11}a_{22} - a_{12}a_{21})x_2 = -a_{21}y_1 + a_{11}y_2, \tag{A-53}$$

or

$$x_2 = \frac{-a_{21}y_1 + a_{11}y_2}{a_{11}a_{22} - a_{12}a_{21}}, \tag{A-54}$$

provided that the denominator is, again, nonzero.

Let's look at the entire problem and its solution from a matrix point of view. Equations (A-50) become the single matrix equation

$$\begin{bmatrix} a_{11} & a_{12} \\ a_{21} & a_{22} \end{bmatrix} \begin{bmatrix} x_1 \\ x_2 \end{bmatrix} = \begin{bmatrix} y_1 \\ y_2 \end{bmatrix}. \tag{A-55}$$

We have already found the solution. We now define the *determinant* of the (2×2) square matrix A by the formula

$$\det(A) = \begin{vmatrix} a_{11} & a_{12} \\ a_{21} & a_{22} \end{vmatrix} = a_{11}a_{22} - a_{12}a_{21}. \tag{A-56}$$

Now, assuming that $\det(A) \neq 0$, we can write our solution given in (A-52) and (A-54) in matrix form as

$$\begin{bmatrix} x_1 \\ x_2 \end{bmatrix} = \begin{bmatrix} \dfrac{a_{22}}{\det(A)} & \dfrac{-a_{12}}{\det(A)} \\ \dfrac{-a_{21}}{\det(A)} & \dfrac{a_{11}}{\det(A)} \end{bmatrix} \begin{bmatrix} y_1 \\ y_2 \end{bmatrix}. \tag{A-57}$$

This means that we have found an explicit form for the inverse matrix,

$$A^{-1} = \begin{bmatrix} a_{11} & a_{12} \\ a_{21} & a_{22} \end{bmatrix}^{-1} = \begin{bmatrix} \dfrac{a_{22}}{\det(A)} & \dfrac{-a_{12}}{\det(A)} \\ \dfrac{-a_{21}}{\det(A)} & \dfrac{a_{11}}{\det(A)} \end{bmatrix}. \tag{A-58}$$

A simple memory device is to swap the diagonal elements and change the signs of the off-diagonal elements, then divide by the determinant.

We will now derive a similar formula for the inverse of any $(n \times n)$ matrix having a nonzero determinant. To do so, however, we must give a general definition of the determinant and develop a few of its properties.

Let A be a square $(n \times n)$ matrix, that is, one having n rows and n columns. The determinant of A is a single scalar associated with A. If we modify A, we modify this number, so we can think of the determinant as a function of A. We will write this function in various ways; two such notational symbols are $\det(A)$ and $|A|$. If we are dealing with a particular matrix, for example, if

$$A = \begin{bmatrix} a_{11} & a_{12} & a_{13} \\ a_{21} & a_{22} & a_{23} \\ a_{31} & a_{32} & a_{33} \end{bmatrix}, \tag{A-59}$$

we will write

$$\det(A) = \begin{vmatrix} a_{11} & a_{12} & a_{13} \\ a_{21} & a_{22} & a_{23} \\ a_{31} & a_{32} & a_{33} \end{vmatrix}; \tag{A-60}$$

that is, we simply change the matrix brackets to vertical bars.

We already know that we can write any matrix as a row vector of column vectors, that is, in the form

$$A = [\bar{a}_1 \quad \bar{a}_2 \quad \cdots \quad \bar{a}_n]. \tag{A-61}$$

If A is a square matrix of dimensions $(n \times n)$, each of the entries in the row vector will be a column vector with n entries. For example, if

$$A = \begin{bmatrix} 1 & 2 & 3 \\ 4 & 5 & 6 \\ 7 & 8 & 9 \end{bmatrix}, \tag{A-62}$$

then we can write A as

$$A = [\bar{a}_1 \quad \bar{a}_2 \quad \bar{a}_3], \tag{A-63}$$

where $\bar{a}_1 = \mathrm{col}(1, 4, 7)$, $\bar{a}_2 = \mathrm{col}(2, 5, 8)$, and $\bar{a}_3 = \mathrm{col}(3, 6, 9)$. Thus, we can think of the determinant of A as a function of its column vectors. When we consider the determinant as such a function, it is helpful to have a more specific notation that brings this to light. Here is such a notation: if A is an $(n \times n)$ matrix written as in equation (A-61), we write

$$\det(A) = |A| = D(\bar{a}_1, \bar{a}_2, \ldots, \bar{a}_n). \tag{A-64}$$

In words, we are using the symbol D to represent the determinant function of n vector variables. We will use this symbolism to define the determinant function in general.

Definition A.4 Let $\bar{a}_1, \bar{a}_2, \ldots, \bar{a}_n$ be n column vectors in E^n and let us form the square $(n \times n)$ matrix $A = [\bar{a}_1, \bar{a}_2, \ldots, \bar{a}_n]$ from them. Then the determinant of A is a scalar-valued function $D(\bar{a}_1, \bar{a}_2, \ldots, \bar{a}_n)$ of the n column vectors having the following three properties:

1. $D(\bar{a}_1, \ldots, \bar{a}_i, \ldots, \bar{a}_j, \ldots, \bar{a}_n) = -D(\bar{a}_1, \ldots, \bar{a}_j, \ldots, \bar{a}_i, \ldots, \bar{a}_n).$

(This means that D is "skew-symmetric," or "antisymmetric." The idea here is that if we leave all of the vectors fixed except for the ith and jth—and reverse the positions of these two vectors—then the determinant will change sign.)

2. $D(\bar{a}_1, \ldots, c_1\bar{a}_i + c_2\bar{a}_i, \ldots, \bar{a}_n)$
$= c_1 D(\bar{a}_1, \ldots, \bar{a}_i, \ldots, \bar{a}_n) + c_2 D(\bar{a}_1, \ldots, \bar{a}_i, \ldots, \bar{a}_n).$

(In other words, the determinant function is linear in each argument.)[2]

3. $D(\bar{e}_1, \bar{e}_2, \bar{e}_3, \ldots, \bar{e}_n) = 1.$

(Looking at this another way, this says that $|I_n| = 1$ because the columns of I_n are $\bar{e}_1, \bar{e}_2, \bar{e}_3, \ldots, \bar{e}_n$—in that order from left to right.)

It turns out that these three properties completely define the determinant function. Don't let the notation scare you here; just read through each of the three properties and think of "flipping" the position of two vectors, of adding two and "sticking" them into one of the argument positions, and then putting the standard basis vectors into the function, one at a time.

Let's look at an example: the (2×2) case. We already know the value of this determinant, so we can check our answer. We compute

$$\begin{vmatrix} a_{11} & a_{12} \\ a_{21} & a_{22} \end{vmatrix} = D(\bar{a}_1, \bar{a}_2) = D\left(\begin{bmatrix} a_{11} \\ a_{21} \end{bmatrix}, \begin{bmatrix} a_{12} \\ a_{22} \end{bmatrix} \right) = D(a_{11}\bar{e}_1 + a_{21}\bar{e}_2, a_{12}\bar{e}_1 + a_{22}\bar{e}_2). \tag{A-65}$$

(Remember that $\bar{e}_1 = \text{col}(1, 0)$ and $\bar{e}_2 = \text{col}(0, 1)$.) Now hold on, for there are four terms! We use property 2 in Definition A.4 on the first argument to write

$$D(\bar{a}_1, \bar{a}_2) = a_{11}D(\bar{e}_1, a_{12}\bar{e}_1 + a_{22}\bar{e}_2) + a_{21}D(\bar{e}_2, a_{12}\bar{e}_1 + a_{22}\bar{e}_2). \tag{A-66}$$

Next, we use property 2 in Definition A.4 on the second argument to expand each component D function. This results in

$$D(\bar{a}_1, \bar{a}_2) = a_{11}a_{12}D(\bar{e}_1, \bar{e}_1) + a_{11}a_{22}D(\bar{e}_1, \bar{e}_2) + a_{21}a_{12}D(\bar{e}_2, \bar{e}_1) + a_{21}a_{22}D(\bar{e}_2, \bar{e}_2). \tag{A-67}$$

Now notice that if $\bar{a}$ is any (2×1) column vector, then $D(\bar{a}, \bar{a}) = -D(\bar{a}, \bar{a})$ by property 1 of Definition A.4 (switching the two arguments results in the same thing!). But this says that $2D(\bar{a}, \bar{a}) = 0$; in other words, $D(\bar{a}, \bar{a}) = 0$. Therefore, $D(\bar{e}_1, \bar{e}_1) = D(\bar{e}_2, \bar{e}_2) = 0$. Furthermore, property 1 says that $D(\bar{e}_2, \bar{e}_1) = -D(\bar{e}_1, \bar{e}_2)$. Finally, $D(\bar{e}_1, \bar{e}_2) = 1$ by property 3. Evaluating each term in (A-67) gives

$$D(\bar{a}_1, \bar{a}_2) = a_{11}a_{22} - a_{12}a_{21}, \tag{A-68}$$

---

[2] A function such as this is often called a *multilinear function*, or *tensor*.

which we already know to be true for the general (2×2) case by our earlier definition; thus, our more general definition above is consistent with the earlier one.

At this point, it should be clear that the three properties in Definition A.4 suffice to determine the determinant (pun intended) of any square matrix. We merely express each column vector as a linear combination of the standard E^n basis vectors $\bar{e}_1, \bar{e}_2, \ldots, \bar{e}_n$, which we know is possible because the latter are basis vectors, and apply linearity. This gives the determinant as a linear combination of values of the determinant function evaluated on the basic vectors. We then exchange the basis vectors using the antisymmetry property until they are in natural order—in which case we know the value is 1.

We can derive an explicit formula for the determinant. The notation is pretty messy in the general case, though, so let's look at the case in which A is a (4×4) matrix,

$$A = [\bar{a}_1 \quad \bar{a}_2 \quad \bar{a}_3 \quad \bar{a}_4] = \begin{bmatrix} a_{11} & a_{12} & a_{13} & a_{14} \\ a_{21} & a_{22} & a_{23} & a_{24} \\ a_{31} & a_{32} & a_{33} & a_{34} \\ a_{41} & a_{42} & a_{43} & a_{44} \end{bmatrix}. \tag{A-69}$$

We can express each column vector as a linear combination of the standard basis vectors $\bar{e}_1, \bar{e}_2, \bar{e}_3, \bar{e}_4$. For instance, the first column can be written in the form

$$\bar{a}_1 = a_{11}\bar{e}_1 + a_{21}\bar{e}_2 + a_{31}\bar{e}_3 + a_{41}\bar{e}_4. \tag{A-70}$$

The first index on the a's specifies the row number and the second the column (in this case, column 1). Using more compact notation, we write (for any column $j = 1, 2, 3, 4$)

$$\bar{a}_j = \sum_{i=1}^{4} a_{ij}\bar{e}_i. \tag{A-71}$$

We put this expression into our determinant function in the jth argument position to get

$$D\left(\sum_{i=1}^{4} a_{i1}\bar{e}_i, \sum_{k=1}^{4} a_{k2}\bar{e}_k, \sum_{m=1}^{4} a_{m3}\bar{e}_m, \sum_{n=1}^{4} a_{n4}\bar{e}_n \right). \tag{A-72}$$

(We have used different column indices in anticipation of what is to come.) Next, we apply linearity to each argument, one at a time. This results in

$$|A| = \sum_{i=1}^{4} \sum_{k=1}^{4} \sum_{m=1}^{4} \sum_{n=1}^{4} a_{i1}a_{k2}a_{m3}a_{n4} D(\bar{e}_i, \bar{e}_k, \bar{e}_m, \bar{e}_n). \tag{A-73}$$

Notice that each term is a product of four other terms—one taken from some entry in each column—times the determinant function evaluated on some arrangement of the basis vectors. Thus, if we can evaluate $D(\bar{e}_i, \bar{e}_k, \bar{e}_m, \bar{e}_n)$ for any values of i, k, m, and n, we will be able to evaluate the determinant.

To do this, we need the idea of a permutation. This concept is best illustrated as in Figure A.1. There we see the four basis vectors placed in four boxes, which we have numbered 1 through 4 from left to right. The problem is to swap the basis vectors a pair at a time until their subscripts match the box numbers. Figure A.2 shows one sequence of swaps that will do this. The arrow represents pairwise exchanges. We first exchange

Figure A.1
A permutation

1	2	3	4
$\bar{e}_3$	$\bar{e}_4$	$\bar{e}_1$	$\bar{e}_2$

Figure A.2
Performing pairwise swaps

1	2	3	4
$\bar{e}_3$	$\bar{e}_4$	$\bar{e}_1$	$\bar{e}_2$

$\rightarrow$

1	2	3	4
$\bar{e}_3$	$\bar{e}_2$	$\bar{e}_1$	$\bar{e}_4$

$\rightarrow$

1	2	3	4
$\bar{e}_1$	$\bar{e}_2$	$\bar{e}_3$	$\bar{e}_4$

$\bar{e}_2$ and $\bar{e}_4$, then $\bar{e}_1$ and $\bar{e}_3$. Thus, we have achieved the proper arrangement with two such operations. There are other possible sequences of swaps that will do the trick, but they all have one thing in common: *the number of swaps is always even or odd* for a given initial arrangement. We will not prove this. It is not difficult to do, but we will save space by asking you to repeat our example using different sequences to convince yourself of the truth of our assertion. We will denote this "evenness" or "oddness" by means of the *parity function:* $\sigma(i, k, m, n)$. If an odd number of swaps is required to rearrange i, k, m, n into the order 1, 2, 3, 4, we will define $\sigma(i, k, m, n) = -1$; if an even number of swaps is required, we will set $\sigma(i, k, m, n) = +1$.

Here is the advantage of our notation: it allows us to evaluate $D(\bar{e}_i, \bar{e}_k, \bar{e}_m, \bar{e}_n)$. Each pairwise swap of basis vectors changes the sign of the determinant, according to property 1 of Definition A.4. Thus the value of $D(\bar{e}_i, \bar{e}_k, \bar{e}_m, \bar{e}_n)$ is given by

$$D(\bar{e}_i, \bar{e}_k, \bar{e}_m, \bar{e}_n) = (-1)^{\sigma(i, k, m, n)}. \tag{A-74}$$

The only other thing we must note is this. If two column vectors are the same, the corresponding determinant is zero. To see this, just look at $D(\bar{e}_3, \bar{e}_4, \bar{e}_3, \bar{e}_1)$, where the basis vector $\bar{e}_3$ occurs twice. If we exchange the first and third arguments, we get the same thing (we are just exchanging $\bar{e}_3$ in position 1 with $\bar{e}_3$ in position 3). But this says that $D(\bar{e}_3, \bar{e}_4, \bar{e}_3, \bar{e}_1) = -D(\bar{e}_3, \bar{e}_4, \bar{e}_3, \bar{e}_1)$, or $2D(\bar{e}_3, \bar{e}_4, \bar{e}_3, \bar{e}_1) = 0$, which implies that

$$D(\bar{e}_3, \bar{e}_4, \bar{e}_3, \bar{e}_1) = 0. \tag{A-75}$$

Clearly, the same thing is true for any basis vector duplicated in any position. Using (A-74) and (A-75) in (A-73), we have an explicit evaluation of the determinant:

$$|A| = \sum_{i \neq k \neq m \neq n} a_{i1} a_{k2} a_{m3} a_{n4} (-1)^{\sigma(i, k, m, n)}. \tag{A-76}$$

We have used a shortened notation. The sum actually represents the sum over all indices in such a fashion that no two are equal.

One can extend (A-76) in an obvious way to write the determinant of any $(n \times n)$ matrix. We use n terms in our product, n indices, and the parity function for n integers. Though we will not do so, one can use the foregoing result to prove the following two properties of determinants:

$$|AB| = |A| \cdot |B| \tag{A-77}$$

and

$$|A^t| = |A|; \tag{A-78}$$

that is, the determinant of a product is the product of the two individual determinants, and the determinant of the transpose (all rows and columns exchanged) is the same as the determinant of the original matrix itself.

So our definition always works but it entails a lot of effort. We will leave it to you to use equation (A-76) to evaluate the general (3×3) determinant. It has nine terms, so allow yourself a bit of time, a nice comfortable desk, and some refreshment to keep you going—but do it! You will get an innate feel for the definition, and you will need this before going on.

We will develop a way of avoiding some of the labor. As it happens, it is the linearity property that we capitalize upon. Here's how it goes for the (4×4) case that we have just been studying.

$$|A| = D(\bar{a}_1, \bar{a}_2, \bar{a}_3, \bar{a}_4) = D(\bar{a}_1, \bar{a}_2, \sum_{i=1}^{4} a_{ij}\bar{e}_i, \bar{a}_4) = \sum_{i=1}^{4} a_{ij} D(\bar{a}_1, \bar{a}_2, \bar{e}_i, \bar{a}_4). \tag{A-79}$$

Let's think about the determinant on the right of (A-79)—the general one containing three of the original columns, but with the third column replaced by the basis vector $\bar{e}_i$. Written out in "vertical bar" notation for the particular case in which $\bar{e}_i = \bar{e}_2$, it is

$$D(\bar{a}_1, \bar{a}_2, \bar{e}_2, \bar{a}_4) = \begin{vmatrix} a_{11} & a_{12} & 0 & a_{14} \\ a_{21} & a_{22} & 1 & a_{24} \\ a_{31} & a_{32} & 0 & a_{34} \\ a_{41} & a_{42} & 0 & a_{44} \end{vmatrix}. \tag{A-80}$$

Equation (A-76) tells us to take an element from the first row (any column), then one from the second (any column except the one selected for the first choice because the resulting term would then be multiplied by zero), then one from the third row (any column except for the ones selected for the first two), and finally one from the fourth row (any column not already selected—and there will be only one choice here!). However, all terms will be zero unless the second row choice is column 3. This eliminates the other elements in row 2 from ever appearing in the determinant. If you think about this just a bit, you will see that this means that

$$D(\bar{a}_1, \bar{a}_2, \bar{e}_2, \bar{a}_4) = \begin{vmatrix} a_{11} & a_{12} & 0 & a_{14} \\ a_{21} & a_{22} & 1 & a_{24} \\ a_{31} & a_{32} & 0 & a_{34} \\ a_{41} & a_{42} & 0 & a_{44} \end{vmatrix} = (-1)^{3-2} \begin{vmatrix} a_{11} & a_{12} & a_{14} \\ a_{31} & a_{32} & a_{34} \\ a_{41} & a_{42} & a_{44} \end{vmatrix} = - \begin{vmatrix} a_{11} & a_{12} & a_{14} \\ a_{31} & a_{32} & a_{34} \\ a_{41} & a_{42} & a_{44} \end{vmatrix}. \tag{A-81}$$

The minus sign occurs because we have the basis vector $\bar{e}_2$ in the third column, so it will take $3 - 2 = 1$ swap to put it in its proper place. (In general, $\bar{e}_i$ in the jth column will require $|j - i|$ swaps.)

Example A.1   Find the determinant of the matrix

$$A = \begin{bmatrix} 1 & 2 & 3 \\ 4 & 5 & 6 \\ 7 & 8 & 9 \end{bmatrix}. \tag{A-82}$$

Solution   We select any column, let's say the second, and apply (A-79). This gives

$$|A| = \begin{vmatrix} 1 & 2 & 3 \\ 4 & 5 & 6 \\ 7 & 8 & 9 \end{vmatrix} = 2 \times (-1)^{2-1} \begin{vmatrix} 4 & 6 \\ 7 & 9 \end{vmatrix} + 5 \times (-1)^{2-2} \begin{vmatrix} 1 & 3 \\ 7 & 9 \end{vmatrix} + 8 \times (-1)^{2-3} \begin{vmatrix} 1 & 3 \\ 4 & 6 \end{vmatrix} \tag{A-83}$$

$$= -2(36 - 42) + 5(9 - 21) - 8(6 - 12) = 0.$$

Even though we did not stop to prove them, the properties in equations (A-77) and (A-78) are important ones. The next example illustrates this.

Example A.2   Find the determinant of the matrix

$$A = \begin{bmatrix} 1 & 2 & 3 \\ 4 & 5 & 6 \\ 0 & 8 & 0 \end{bmatrix}. \tag{A-84}$$

Solution   Before we tackle the numerics, let's pause for an inspection of our matrix. It is somewhat special, for it has two zero elements in the third row. If the row were a column, we could expand along that column and only have one term in that expansion. How do we convert rows to columns? By taking the transpose of the matrix—and, to make things nice, the property in equation (A-78) says that the resulting determinant is the same as that of the one we want. So let's first take the transpose, then compute the determinant. The transpose of A is obtained by making the first row the first column, the second row the second column, and the third row the third column:

$$A^t = \begin{bmatrix} 1 & 4 & 0 \\ 2 & 5 & 8 \\ 3 & 6 & 0 \end{bmatrix}. \tag{A-85}$$

Now we can expand along the third column, to obtain

$$|A^t| = \begin{vmatrix} 1 & 4 & 0 \\ 2 & 5 & 8 \\ 3 & 6 & 0 \end{vmatrix} = 0 \times (-1)^{(3-1)} \begin{vmatrix} 2 & 5 \\ 3 & 6 \end{vmatrix} + 8 \times (-1)^{(3-2)} \begin{vmatrix} 1 & 4 \\ 3 & 6 \end{vmatrix} + 0 \times (-1)^{(3-3)} \begin{vmatrix} 1 & 4 \\ 2 & 5 \end{vmatrix} \tag{A-86}$$

$$= -8(6 - 12) = 48.$$

Actually, because the determinant of any transposed matrix is the same as the matrix itself, we can forego the operation of transposition and simply expand along a row, rather than along a column. You might rework the last example using this method for the practice. Both procedures are called Laplace expansion—named for the same Laplace as that of the Laplace transformation covered in Chapter 13, but there is no connection between the two topics.

We will now develop a formula for the inverse of a matrix that generalizes the one we derived in equation (A-58). We recall that if the matrix equation $A\bar{x} = \bar{y}$ has a solution, then the right-hand vector $\bar{y}$ can be written as a linear combination of the column vectors of A:

$$\bar{y} = x_1\bar{a}_1 + x_2\bar{a}_2 + \cdots + x_n\bar{a}_n, \tag{A-87}$$

where $x_1, x_2, \ldots, x_n$ are the components of the $\bar{x}$ vector. Let's simply take this expression and substitute it for the column vector $\bar{a}_i$ in the determinant function. This gives

$$D(\bar{a}_1, \bar{a}_2, \ldots, \bar{a}_{i-1}, \bar{y}, \bar{a}_{i+1}, \ldots, \bar{a}_n) = x_1 D(\bar{a}_1, \bar{a}_2, \ldots, \bar{a}_{i-1}, \bar{a}_1, \bar{a}_{i+1}, \ldots, \bar{a}_n) \tag{A-88}$$
$$+ x_2 D(\bar{a}_1, \bar{a}_2, \ldots, \bar{a}_{i-1}, \bar{a}_2, \bar{a}_{i+1}, \ldots, \bar{a}_n) + \cdots + x_i D(\bar{a}_1, \bar{a}_2, \ldots, \bar{a}_{i-1}, \bar{a}_i, \bar{a}_{i+1}, \ldots, \bar{a}_n)$$
$$+ x_{i+1} D(\bar{a}_1, \bar{a}_2, \ldots, \bar{a}_{i-1}, \bar{a}_{i+1}, \bar{a}_{i+1}, \ldots, \bar{a}_n) + \cdots + x_n D(\bar{a}_1, \bar{a}_2, \ldots, \bar{a}_{i-1}, \bar{a}_n, \bar{a}_{i+1}, \ldots, \bar{a}_n).$$

It is unfortunate that the notation is so involved because the idea is simple. We insert (A-87) in the ith position in the determinant function and use linearity to break it up into the sum of n terms, each of which has an $\bar{a}$ vector from (A-87) in the ith position. Now in all cases but one, this means that the D function has two arguments that are identical—and just as we argued in our derivation of equation (A-76), this means that the resulting value is zero because reversing the two identical arguments changes the sign and, at the same time, results in the *same* determinant value. The only number that is equal to its negative is zero. Hence, all the terms in (A-88) are zero except for the one involving $\bar{a}_i$. Because it occurs in its proper place, the corresponding determinant value is that of the original determinant $D(\bar{a}_1, \bar{a}_2, \ldots, \bar{a}_{i-1}, \bar{a}_i, \bar{a}_{i+1}, \ldots, \bar{a}_n)$. Thus, we can write

$$D(\bar{a}_1, \bar{a}_2, \ldots, \bar{a}_{i-1}, \bar{y}, \bar{a}_{i+1}, \ldots, \bar{a}_n) = x_i D(\bar{a}_1, \bar{a}_2, \ldots, \bar{a}_{i-1}, \bar{a}_i, \bar{a}_{i+1}, \ldots, \bar{a}_n). \tag{A-89}$$

In simpler notation, this becomes

$$|\bar{a}_1\,\bar{a}_2 \cdots \bar{a}_{i-1}\,\bar{y}\,\bar{a}_{i+1} \cdots \bar{a}_n| = x_i\,|\bar{a}_1\,\bar{a}_2 \cdots \bar{a}_{i-1}\,\bar{a}_i\,\bar{a}_{i+1} \cdots \bar{a}_n|. \tag{A-90}$$

The left side is the determinant formed by replacing the *i*th column of the matrix *A* by the right-hand side vector in $A\bar{x} = \bar{y}$. The one on the right side is, of course, merely the determinant of *A*. Now—*if the determinant of A is not zero*—we can write

$$x_i = \frac{|\bar{a}_1\bar{a}_2 \ldots \bar{a}_{i-1}\bar{y}\,\bar{a}_{i+1} \ldots \bar{a}_n|}{|A|}. \tag{A-91}$$

This is the formula often referred to as Cramer's rule. (Cramer was a French mathematician, and his name is pronounced "Cra-mehr," with a short a as in "ah"; Floyd Cramer, pronounced with a long a, is the name of a U.S. pop music pianist!)

If the determinant of *A* is zero, formula (A-89) still holds, and we can make the following observations. If the determinant on the left-hand side with $\bar{y}$ replacing $\bar{a}_i$ is nonzero, there is no solution because we would have the equation "zero equals a nonzero number." In this case, we say that the system is inconsistent. If we were to trace it back a bit farther, we would see that this occurs when $\bar{y}$ is not in the space spanned by the column vectors—and this can only occur if the column vectors of *A* are linearly dependent. (In this case, one can easily show that $|A| = 0$ because one column can be expressed as a linear combination of the others; expansion along this column then results in determinants with repeated columns.) If, on the other hand, the left-hand determinant in (A-89) is zero, then any value is a possible solution for x_i. Thus, in summary, *there is a unique solution to our original set of equations if and only if the determinant of the coefficient matrix is nonzero.* We used this fact in our discussion of solvability of nodal and mesh equations in Chapter 4.

Let's use Cramer's rule to solve our original example system of equations in (A-1) and (A-10), the latter of which we repeat once more here for reference:

$$\begin{bmatrix} 2 & 2 & 1 \\ 1 & 1 & 1 \\ 1 & 0 & 1 \end{bmatrix} \begin{bmatrix} x_1 \\ x_2 \\ x_3 \end{bmatrix} = \begin{bmatrix} 10 \\ 6 \\ 5 \end{bmatrix}. \tag{A-92}$$

Thus,

$$A = [\bar{a}_1 \quad \bar{a}_2 \quad \bar{a}_3] = \begin{bmatrix} 2 & 2 & 1 \\ 1 & 1 & 1 \\ 1 & 0 & 1 \end{bmatrix}. \tag{A-93}$$

Applying Cramer's rule and expanding along the first column, we have

$$x_1 = \frac{\begin{vmatrix} 10 & 2 & 1 \\ 6 & 1 & 1 \\ 5 & 0 & 1 \end{vmatrix}}{\begin{vmatrix} 2 & 2 & 1 \\ 1 & 1 & 1 \\ 1 & 0 & 1 \end{vmatrix}} = \frac{10(1 \times 1 - 1 \times 0) - 6(2 \times 1 - 1 \times 0) + 5(2 \times 1 - 1 \times 1)}{2(1 \times 1 - 1 \times 0) - 1(2 \times 1 - 1 \times 0) + 1(2 \times 1 - 1 \times 1)} = \frac{3}{1} = 3, \tag{A-94}$$

$$x_2 = \frac{\begin{vmatrix} 2 & 10 & 1 \\ 1 & 6 & 1 \\ 1 & 5 & 1 \end{vmatrix}}{\begin{vmatrix} 2 & 2 & 1 \\ 1 & 1 & 1 \\ 1 & 0 & 1 \end{vmatrix}} = \frac{2(6 \times 1 - 5 \times 1) - 1(10 \times 1 - 5 \times 1) + 1(10 \times 1 - 6 \times 1)}{2(1 \times 1 - 1 \times 0) - 1(2 \times 1 - 1 \times 0) + 1(2 \times 1 - 1 \times 1)} = \frac{1}{1} = 1, \tag{A-95}$$

and

$$x_3 = \frac{\begin{vmatrix} 2 & 2 & 10 \\ 1 & 1 & 6 \\ 1 & 0 & 5 \end{vmatrix}}{\begin{vmatrix} 2 & 2 & 1 \\ 1 & 1 & 1 \\ 1 & 0 & 1 \end{vmatrix}} = \frac{2(1 \times 5 - 6 \times 0) - 1(2 \times 5 - 10 \times 0) + 1(2 \times 6 - 10 \times 1)}{2(1 \times 1 - 1 \times 0) - 1(2 \times 1 - 1 \times 0) + 1(2 \times 1 - 1 \times 1)} = \frac{2}{1} = 2. \quad \text{(A-96)}$$

Notice that we only had to compute $|A|$ once (for x_1) because we could then reuse it in the last two expressions.

The (Closed Form) Inverse of a Square Matrix: The Adjoint Matrix

We will now use Cramer's rule to develop a formula for the inverse of any square matrix A. The inverse of A, A^{-1}, is the solution for X in the matrix equation

$$AX = I_n, \quad \text{(A-97)}$$

where A and X are both $(n \times n)$ matrices and I_n is, as usual, the $(n \times n)$ identity matrix. We will write X and I_n in terms of their columns, obtaining the following equation:

$$A[\bar{x}_1 \, \bar{x}_2 \ldots \bar{x}_n] = [\bar{e}_1 \, \bar{e}_2 \ldots \bar{e}_n]. \quad \text{(A-98)}$$

Now observe that the dimensions of A and each $\bar{x}_i$ are compatible for multiplication (A is $(n \times n)$ and $\bar{x}_i$ is $(n \times 1)$, so the result is $(n \times 1)$, thus we can write

$$[A\bar{x}_1 \, A\bar{x}_2 \ldots A\bar{x}_n] = [\bar{e}_1 \, \bar{e}_2 \ldots \bar{e}_n]. \quad \text{(A-99)}$$

We can then set each column vector in the matrix on the left equal to its corresponding column vector on the right. Thus, we have

$$A\bar{x}_j = \bar{e}_j, \quad \text{(A-100)}$$

for each $j = 1, 2, \ldots, n$.

Assuming that the determinant of A is nonzero, we can solve (A-100) for each element $x_{ij}(i = 1, 2, \ldots, n)$ using Cramer's rule. This gives

$$x_{ij} = \frac{D(\bar{a}_1, \bar{a}_2, \ldots, \bar{a}_{i-1}, \bar{e}_j, \bar{a}_{i+1}, \ldots, \bar{a}_n)}{|A|} = \frac{D_{ij}}{|A|}. \quad \text{(A-101)}$$

To find the ith row of the column matrix $\bar{x}_i$, we insert the basis vector $\bar{e}_j$ in column i of the matrix A—and then compute the ratio of the resulting determinant to the determinant of A. Notice that we have called the numerator quantity D_{ij}, the first subscript referring to the *column* of A into which the basis vector is inserted and the second referring to *the basis vector being inserted*. The solution x_{ij}, then, is the ijth element in the inverse matrix A^{-1}:

$$A^{-1} = [x_{ij}] = \frac{1}{|A|}[D_{ij}] = [D_{ij}/|A|]. \quad \text{(A-102)}$$

The matrix $\tilde{A} = [D_{ij}]$, by the way, is called the *adjoint* matrix of A.

Let's do a (2×2) example. We write

$$A = \begin{bmatrix} a_{11} & a_{12} \\ a_{21} & a_{22} \end{bmatrix}. \quad \text{(A-103)}$$

Then we successively compute

$$D_{11} = \begin{vmatrix} 1 & a_{12} \\ 0 & a_{22} \end{vmatrix} = a_{22}, D_{12} = \begin{vmatrix} 0 & a_{12} \\ 1 & a_{22} \end{vmatrix} = -a_{12}, D_{21} = \begin{vmatrix} a_{11} & 1 \\ a_{21} & 0 \end{vmatrix} = -a_{21}, \text{ and } D_{22} = \begin{vmatrix} a_{11} & 0 \\ a_{21} & 1 \end{vmatrix} = a_{11}. \quad \text{(A-104)}$$

Thus, the inverse matrix is given by

$$A^{-1} = \frac{1}{|A|} \begin{bmatrix} D_{11} & D_{12} \\ D_{21} & D_{22} \end{bmatrix} = \begin{bmatrix} \dfrac{a_{22}}{|A|} & \dfrac{-a_{12}}{|A|} \\ \dfrac{-a_{21}}{|A|} & \dfrac{a_{11}}{|A|} \end{bmatrix}. \quad \text{(A-105)}$$

Example A.3   Solve the matrix equation

$$\begin{bmatrix} 2 & 2 & 1 \\ 1 & 1 & 1 \\ 1 & 0 & 1 \end{bmatrix} \begin{bmatrix} x_1 \\ x_2 \\ x_3 \end{bmatrix} = \begin{bmatrix} 10 \\ 6 \\ 5 \end{bmatrix} \quad \text{(A-106)}$$

using the adjoint matrix.

Solution   Because

$$A = \begin{bmatrix} 2 & 2 & 1 \\ 1 & 1 & 1 \\ 1 & 0 & 1 \end{bmatrix}, \quad \text{(A-107)}$$

we must compute nine D_{ij} determinants. They are

$$D_{11} = \begin{vmatrix} 1 & 2 & 1 \\ 0 & 1 & 1 \\ 0 & 0 & 1 \end{vmatrix} = 1, D_{12} = \begin{vmatrix} 0 & 2 & 1 \\ 1 & 1 & 1 \\ 0 & 0 & 1 \end{vmatrix} = -2, \quad \text{(A-108)}$$

Continuing, $\quad D_{13} = \begin{vmatrix} 0 & 2 & 1 \\ 0 & 1 & 1 \\ 1 & 0 & 1 \end{vmatrix} = 1, \quad D_{21} = \begin{vmatrix} 2 & 1 & 1 \\ 1 & 0 & 1 \\ 1 & 0 & 1 \end{vmatrix} = 0, \quad D_{22} = \begin{vmatrix} 2 & 0 & 1 \\ 1 & 1 & 1 \\ 1 & 0 & 1 \end{vmatrix} = 1,$

$$D_{23} = \begin{vmatrix} 2 & 0 & 1 \\ 1 & 0 & 1 \\ 1 & 1 & 1 \end{vmatrix} = -1, \quad D_{31} = \begin{vmatrix} 2 & 2 & 1 \\ 1 & 1 & 0 \\ 1 & 0 & 0 \end{vmatrix} = -1, \quad D_{32} = \begin{vmatrix} 2 & 2 & 0 \\ 1 & 1 & 1 \\ 1 & 0 & 0 \end{vmatrix} = 2, \quad \text{and}$$

$$D_{33} = \begin{vmatrix} 2 & 2 & 0 \\ 1 & 1 & 0 \\ 1 & 0 & 1 \end{vmatrix} = 0.$$

(Remember that D_{ij} denotes $\bar{e}_j$ inserted in column i. This is a bit hard to keep straight!) Thus, using the determinant that we computed earlier to be $|A| = 1$, we can write the inverse matrix as

$$A^{-1} = \left(\frac{D_{ij}}{|A|} \right) = \begin{bmatrix} 1 & -2 & 1 \\ 0 & 1 & -1 \\ -1 & 2 & 0 \end{bmatrix}. \quad \text{(A-109)}$$

You can check this by forming the product $A^{-1}A$ and AA^{-1} and showing that both are equal to zero.

*The (Algorithmic)
Inverse of a Square
Matrix: Gaussian
Reduction*

Cramer's rule is very handy to have around because it allows us to write the solution to matrix equations in a very compact way. In closed form, however, it still leaves much to be desired as a computational tool because the computation of the inverse of an $(n \times n)$ matrix requires the computation of n^2 submatrices of order $n - 1$ as well as the determinant of the matrix itself. We need a procedure that is numerically less complicated, and that is what we will now develop. The basic process is called Gaussian reduction, or (sometimes) row reduction, though there are many variations having different names than these.

We start with our ongoing basic example again, that of solving equations (A-1) for the x variables. We present it again here as equations (A-110):

$$2x_1 + 2x_2 + 1x_3 = 10, \tag{A-110a}$$

$$1x_1 + 1x_2 + 1x_3 = 6, \tag{A-110b}$$

and

$$1x_1 + 0x_2 + 1x_3 = 5. \tag{A-110c}$$

We next define *elementary row operations:*

1. Replace equation i by the product of any scalar c times equation i. We will agree to write this symbolically as

$$q_i \longleftarrow cq_i. \tag{A-111}$$

2. Replace equation i by the sum of equation i and c times equation j. Symbolically, we write

$$q_i \longleftarrow q_i + cq_j. \tag{A-112}$$

3. Interchange equation i and equation j. Symbolically,

$$q_i \longleftrightarrow q_j. \tag{A-113}$$

By adding two equations or multiplying an equation by a scalar we mean adding corresponding terms of two equations or multiplying each term by a constant—on both sides of the equation. The matrix form of equations (A-110) (equation (A-10) repeated) is

$$\begin{bmatrix} 2 & 2 & 1 \\ 1 & 1 & 1 \\ 1 & 0 & 1 \end{bmatrix} \begin{bmatrix} x_1 \\ x_2 \\ x_3 \end{bmatrix} = \begin{bmatrix} 10 \\ 6 \\ 5 \end{bmatrix}. \tag{A-114}$$

The elementary row operations on equations clearly correspond to corresponding operations on the rows of the coefficient matrix on the left side and of the constant vector on the right side.

We will apply elementary row operations to both sides of (A-114), but in order to keep things a bit neater, we define the *augmented matrix:*

$$A_a = \begin{bmatrix} 2 & 2 & 1 & 10 \\ 1 & 1 & 1 & 6 \\ 1 & 0 & 1 & 5 \end{bmatrix}. \tag{A-115}$$

We are *not* subtracting the right-side vector from the left or anything of this sort, we are just providing a convenient set of pockets in which to carry the right side along so that we can perform the same operations on it that we perform on the coefficient matrix. We will apply row operations on the augmented matrix to reduce the coefficient matrix to a simpler form. Here's how it goes. We apply the following sequence of row operations:

$$
A_a = \begin{bmatrix} 2 & 2 & 1 & | & 10 \\ 1 & 1 & 1 & | & 6 \\ 1 & 0 & 1 & | & 5 \end{bmatrix} \xrightarrow[q_3 \leftarrow q_3 - 0.5q_1]{q_2 \leftarrow q_2 - 0.5q_1} \begin{bmatrix} 2 & 2 & 1 & | & 10 \\ 0 & 0 & 0.5 & | & 1 \\ 0 & -1 & 0.5 & | & 0 \end{bmatrix} \xrightarrow{q_3 \leftrightarrow q_2} \begin{bmatrix} 2 & 2 & 1 & | & 10 \\ 0 & -1 & 0.5 & | & 0 \\ 0 & 0 & 0.5 & | & 1 \end{bmatrix}. \quad \text{(A-116)}
$$

You should read this "equation" the following way: after applying the elementary row operations $q_2 \leftarrow q_2 - 0.5q_1$ and $q_3 \leftarrow q_3 - 0.5q_1$, we get the matrix at the point of the first long arrow pointing right, and after the operation $q_3 \leftrightarrow q_2$, we get the last matrix.

Now notice the structure of this last matrix. If we put the rightmost column back on the right where it belongs, we get the matrix equation

$$
\begin{bmatrix} 2 & 2 & 1 \\ 0 & -1 & 0.5 \\ 0 & 0 & 0.5 \end{bmatrix} \begin{bmatrix} x_1 \\ x_2 \\ x_3 \end{bmatrix} = \begin{bmatrix} 10 \\ 0 \\ 1 \end{bmatrix}. \quad \text{(A-117)}
$$

The coefficient is said to be in *triangular form* because its nonzero elements form a triangle; more specifically, it is sometimes called *upper triangular* because the nonzero triangle is in the upper half of the matrix. Let's look at the equations in scalar form, taking the last row first. We get $0.5x_3 = 1$, which we can solve for $x_3 = 2$. Next, we read off the second row: $-x_2 + 0.5x_3 = 0$, and if we use our result for x_3 just obtained, we get $x_2 = 1$. Last, we read off the first equation from the first row, namely $2x_1 + 2x_2 + x_3 = 10$, which, together with our known values for x_3 and x_2, gives $x_1 = 3$. We call this process *back substitution* because we are reading the equations off backward. The last equation will always (if $|A| \neq 0$) have only one unknown, the next to last only two, etc., so we solve the system backward.

Here is another piece of information we can get from (A-117): the determinant of the coefficient matrix A. To do this, let's look at the operation $q_3 \leftarrow q_3 - 0.5q_1$ by recalling that the determinant of a matrix is the same as that of its transpose. Thus, just for convenience (because we have worked with column manipulations and have not illustrated row manipulations for determinants), we will look at the effect on the determinant of doing the same operation on the columns of the matrix. Thus, we look at the determinant resulting from adding c times the first column to the third column. We have

$$
D(\bar{a}_1, \bar{a}_2, \bar{a}_3 + c\bar{a}_1) = D(\bar{a}_1, \bar{a}_2, \bar{a}_3) + cD(\bar{a}_1, \bar{a}_2, \bar{a}_1) = D(\bar{a}_1, \bar{a}_2, \bar{a}_3). \quad \text{(A-118)}
$$

We have used the fact that the second determinant after the first equal sign has two identical columns, and hence is zero. This shows that adding a constant multiple of a given column to another has no effect on the determinant, and—by our transpose property—the same thing holds for the corresponding row operation. We can likewise show that exchanging two rows changes the sign of the determinant because this is true for column exchanges. Finally, if we multiply a row by c, this is equivalent to multiplying a column by c; therefore, by the linearity property, it multiplies the determinant by c.

What we have shown is this: if we keep track of the type of operations we perform, we can compute the determinant in terms of the determinant of the row-reduced triangular matrix. For our preceding example, we have applied two operations that did not change the determinant, then one that altered its sign. Thus, the determinant of the original coefficient matrix is the negative of the upper triangular one we have produced. What is its determinant? Well, glance back at equation (A-117) once more and think of expanding the determinant of the coefficient matrix along the first column. There will only be a single term, which is twice the determinant formed by erasing the first row and the first column. The corresponding matrix is still upper triangular; specifically, it is $\begin{bmatrix} -1 & 0.5 \\ 0 & 0.5 \end{bmatrix}$.

We can, of course, just compute its determinant, but for the sake of generalization, we expand again along the first column. We then get -1 times the determinant of a (1×1) matrix (scalar), which is 0.5. This argument shows the following to be true: *the determinant of a triangular matrix is the product of its diagonal element values.* In our case, this determinant is $2 \times (-1) \times (0.5) = -1$. But we have already noted that we must take the negative to determine the determinant of the original matrix so $|A| = 1$.

Here is a summary. Perform elementary row operations to reduce the augmented matrix to triangular form, keeping track of the row operations applied. Then, to obtain the solution to the original matrix equation, move the last column back to the right-hand side and solve by back substitution. To obtain the determinant of the original coefficient matrix, multiply the diagonal elements of the triangular matrix (after removing the last column, of course) to obtain *its* determinant. Then modify its value depending on the row operations applied to compute the original determinant.

Computing the Inverse of a Square Matrix Using Gaussian Reduction

Now let's look at the use of Gaussian reduction to obtain the inverse of a matrix. Suppose that the matrix to be inverted is A, an arbitrary $(n \times n)$ square matrix. We have already seen (in developing the inverse by using Cramer's rule) that A^{-1} is that $(n \times n)$ matrix X that is the solution to

$$AX = I_n; \tag{A-119}$$

in other words, the columns of A^{-1} are the n-vector $\bar{x}_j$ (columns of X) that are solutions to the equation

$$A\bar{x}_j = \bar{e}_j, \tag{A-120}$$

for $j = 1, 2, \ldots, n$—and we know how to solve this problem using row reduction. But we can do even better by doing them all at one time. Here's how. Instead of augmenting the A matrix with only one column, we do it with n—all n of the $\bar{e}_j$ at once (which of course is the $(n \times n)$ identity matrix I_n). That is, we form the augmented matrix $[A : I_n]$ and apply row reduction operations until A is transformed into the unit matrix I_n.

Let's look at an example. We will use our old friend

$$A = \begin{bmatrix} 2 & 2 & 1 \\ 1 & 1 & 1 \\ 1 & 0 & 1 \end{bmatrix} \tag{A-121}$$

with which we have already worked on several occasions. The augmented matrix is

$$A_a = \begin{bmatrix} 2 & 2 & 1 & 1 & 0 & 0 \\ 1 & 1 & 1 & 0 & 1 & 0 \\ 1 & 0 & 1 & 0 & 0 & 1 \end{bmatrix}. \tag{A-122}$$

So let's begin doing our row reduction. We do the same operations as before, obtaining

$$A_a = \begin{bmatrix} 2 & 2 & 1 & 1 & 0 & 0 \\ 1 & 1 & 1 & 0 & 1 & 0 \\ 1 & 0 & 1 & 0 & 0 & 1 \end{bmatrix} \xrightarrow[q_3 \leftarrow q_3 - 0.5q_1]{q_2 \leftarrow q_2 - 0.5q_1} \begin{bmatrix} 2 & 2 & 1 & 1 & 0 & 0 \\ 0 & 0 & 0.5 & -0.5 & 1 & 0 \\ 0 & -1 & 0.5 & -0.5 & 0 & 1 \end{bmatrix} \tag{A-123}$$

$$\xrightarrow{q_3 \leftrightarrow q_2} \begin{bmatrix} 2 & 2 & 1 & 1 & 0 & 0 \\ 0 & -1 & 0.5 & -0.5 & 0 & 1 \\ 0 & 0 & 0.5 & -0.5 & 1 & 0 \end{bmatrix}.$$

This time, however, we will continue with our row operations to reduce the matrix to the left of the vertical bar to the $(n \times n)$ identity matrix. We can do this in a lot of ways. One sequence of row operations to accomplish this is

$$\begin{bmatrix} 2 & 2 & 1 & 1 & 0 & 0 \\ 0 & -1 & 0.5 & -0.5 & 0 & 1 \\ 0 & 0 & 0.5 & -0.5 & 1 & 0 \end{bmatrix} \xrightarrow{q_3 \leftarrow 2q_3} \begin{bmatrix} 2 & 2 & 1 & 1 & 0 & 0 \\ 0 & -1 & 0.5 & -0.5 & 0 & 1 \\ 0 & 0 & 1 & -1 & 2 & 0 \end{bmatrix} \qquad \text{(A-124)}$$

$$\xrightarrow{q_1 \leftarrow q_1 + 2q_2} \begin{bmatrix} 2 & 0 & 2 & 0 & 0 & 2 \\ 0 & -1 & 0.5 & -0.5 & 0 & 1 \\ 0 & 0 & 1 & -1 & 2 & 0 \end{bmatrix} \xrightarrow[q_2 \leftarrow q_2 - 0.5q_3]{q_1 \leftarrow q_1 - 2q_3} \begin{bmatrix} 2 & 0 & 0 & 2 & -4 & 2 \\ 0 & -1 & 0 & 0 & -1 & 1 \\ 0 & 0 & 1 & -1 & 2 & 0 \end{bmatrix}$$

$$\xrightarrow[q_2 \leftarrow -1q_2]{q_1 \leftarrow 0.5q_1} \begin{bmatrix} 1 & 0 & 0 & 1 & -2 & 1 \\ 0 & 1 & 0 & 0 & 1 & -1 \\ 0 & 0 & 1 & -1 & 2 & 0 \end{bmatrix}$$

Remember that our elementary row operations have not changed the solution set, so our original equation in (A-119) has been transformed into the equivalent equation

$$I_n X = X = B, \qquad \text{(A-125)}$$

where B is the rightmost square matrix in the last result of equation (A-124). We already know that this solution X is the inverse we are looking for, so

$$A^{-1} = \begin{bmatrix} 1 & -2 & 1 \\ 0 & 1 & -1 \\ -1 & 2 & 0 \end{bmatrix}. \qquad \text{(A-126)}$$

If you check with our Cramer's rule computation in Example A.3, you will see that they are in agreement.

It should be quite clear by now that row reduction is a method that is far superior to Cramer's rule, computationally speaking. Not only is it numerically efficient, but we can use it to simply compute a solution if the right-hand side vector is known, we can compute the inverse if we need the solution to a number of different right-hand side vectors, and (as we have shown) we can compute the determinant of A.

But wait—there is even more. Row reduction will also give us some additional information, as we will now show.

Suppose we are to solve the system of equations

$$2x_1 + 2x_2 + 4x_3 = 10, \qquad \text{(A-127a)}$$

$$1x_1 + 1x_2 + 2x_3 = 5, \qquad \text{(A-127b)}$$

and

$$1x_1 + 0x_2 + 1x_3 = 2. \qquad \text{(A-127c)}$$

We express these equations in matrix form:

$$\begin{bmatrix} 2 & 2 & 4 \\ 1 & 1 & 2 \\ 1 & 0 & 1 \end{bmatrix} \begin{bmatrix} x_1 \\ x_2 \\ x_3 \end{bmatrix} = \begin{bmatrix} 10 \\ 5 \\ 2 \end{bmatrix}. \qquad \text{(A-128)}$$

(This is not absolutely necessary, but it is recommended.) We next form the augmented matrix (we will not seek the inverse, only the solution) and set to work with row reduction:

$$\begin{bmatrix} 2 & 2 & 4 & | & 10 \\ 1 & 1 & 2 & | & 5 \\ 1 & 0 & 1 & | & 2 \end{bmatrix} \xrightarrow[\substack{q_2 \leftarrow q_2 - 0.5q_1 \\ q_3 \leftarrow q_3 - 0.5q_1}]{} \begin{bmatrix} 2 & 2 & 4 & | & 10 \\ 0 & 0 & 0 & | & 0 \\ 0 & -1 & -1 & | & -3 \end{bmatrix} \xrightarrow[\substack{q_3 \leftarrow -1q_3 \\ q_3 \leftrightarrow q_2}]{} \begin{bmatrix} 2 & 2 & 4 & | & 10 \\ 0 & 1 & 1 & | & 3 \\ 0 & 0 & 0 & | & 0 \end{bmatrix}. \tag{A-129}$$

The interesting thing here is that there is a complete row of zeros—the second row—that we swap with the third row to make it the last.

In this case there is not a unique solution. Here is why. We move the rightmost column back to the right-hand side and reintroduce the x variables to get

$$\begin{bmatrix} 2 & 2 & 4 \\ 0 & 1 & 1 \\ 0 & 0 & 0 \end{bmatrix} \begin{bmatrix} x_1 \\ x_2 \\ x_3 \end{bmatrix} = \begin{bmatrix} 10 \\ 3 \\ 0 \end{bmatrix}. \tag{A-130}$$

We note first that the determinant of the upper triangular matrix is 0; therefore, so is the determinant of the original coefficient matrix. Second, we see that we cannot solve the last equation for x_3 as we could if the determinant were nonzero; so we go back to the second equation, obtaining $x_2 + x_3 = 3$, or $x_2 = 3 - x_3$. Using this in the first equation, we obtain $2x_1 + 2x_2 + 4x_3 = 10$, or $x_1 = 5 - 2x_3 - (3 - x_3) = 2 - x_3$. Thus, we have an infinite number of solutions for x_1, x_2, and x_3: one for each arbitrary choice of value for x_3. There is a solution, but not a *unique* solution.

In general, if we use row reduction we can obtain a set of r, $0 \le r \le n$, rows that are not identically zero, that have a one in their first nonzero column, and have the property that all of the entries in the matrix below the one in the same column are zeros. The remaining $n - r$ rows are identically zero as far as the coefficient matrix itself is concerned; however, the "augmented" column (the one added) might or might not have a zero in the "all-zero columns" of the row reduced coefficient matrix. If it does have zeros there—making the last $n - r$ rows of the augmented matrix identically zero—there are a multiplicity of solutions. If $r = n$, there is a unique solution. If $r < n$ and the "augmented column" does not have zero entries in the last $n - r$ columns, there is no solution, and the equations are said to be inconsistent.

Let's look at an inconsistent set of equations. To illustrate this point, we will use the example we just worked, but change the right-hand side constant vector. Therefore, assume that the matrix form of the equations to be solved is

$$\begin{bmatrix} 2 & 2 & 4 \\ 1 & 1 & 2 \\ 1 & 0 & 1 \end{bmatrix} \begin{bmatrix} x_1 \\ x_2 \\ x_3 \end{bmatrix} = \begin{bmatrix} 10 \\ 7 \\ 2 \end{bmatrix}. \tag{A-131}$$

We use exactly the same sequence of row reduction steps to obtain

$$\begin{bmatrix} 2 & 2 & 4 & | & 10 \\ 1 & 1 & 2 & | & 7 \\ 1 & 0 & 1 & | & 2 \end{bmatrix} \xrightarrow[\substack{q_2 \leftarrow q_2 - 0.5q_1 \\ q_3 \leftarrow q_3 - 0.5q_1}]{} \begin{bmatrix} 2 & 2 & 4 & | & 10 \\ 0 & 0 & 0 & | & 2 \\ 0 & -1 & -1 & | & -3 \end{bmatrix} \xrightarrow[\substack{q_3 \leftarrow -1q_3 \\ q_3 \leftrightarrow q_2}]{} \begin{bmatrix} 2 & 2 & 4 & | & 10 \\ 0 & 1 & 1 & | & 3 \\ 0 & 0 & 0 & | & 2 \end{bmatrix}. \tag{A-132}$$

Do you see what has happened? We pass from having a multiplicity of solutions to a situation in which the last equation reads $0 = 2$! Clearly there is no solution.

To summarize, after row reduction we have one of three different situations:

1. $r < n$ and the reduced right-hand side column has zeros in the last $n - r$ rows. In this case, there is a solution, but it is not unique; rather, there are many solutions and the determinant is zero.

2. $r < n$ and the reduced right-hand side column has at least one nonzero entry in the last $n - r$ rows. In this case, there is no solution (the original set of equations is inconsistent), and the determinant is zero.

3. $r = n$. In this case, the determinant is nonzero, and there is a unique solution.

Notice that if the right-hand side vector consists of all zeros, then only situations (1) or (3) can hold; that is, there is always a solution ($\bar{x} = \bar{0}$) to the matrix equation $A\bar{x} = \bar{0}$. Furthermore, we see that $\bar{x} = \bar{0}$ is the *unique* solution to the equation $A\bar{x} = \bar{0}$ if and only if $\det(A) \neq 0$—because row reduction always results in situation 3. We use this fact in our proof, in Chapter 4, that nodal and mesh analysis always work for connected and coupled networks, respectively.

If you have not yet covered differential operators in Part II of the text, you should skip the next example and discussion—unless you have covered the material on the Laplace transform in Part III, in which case you should change all the p's to s's, the delta functions to the number 1 (which is the transform of $\delta(t)$), and the time functions to transforms (functions of s).

Row reduction can be applied to systems of differential equations because—assuming one-sided signals and causality of all operations—each and every differential operator has a unique inverse. Let's take the example system

$$(p + 1)x(t) - y(t) - z(t) = \delta(t) \tag{A-133a}$$

$$-x(t) + (p + 1)y(t) - z(t) = \delta(t) \tag{A-133b}$$

$$-x(t) - y(t) + (p + 1)z(t) = \delta(t). \tag{A-133c}$$

In matrix form, these equations become

$$\begin{bmatrix} p+1 & -1 & -1 \\ -1 & p+1 & -1 \\ -1 & -1 & p+1 \end{bmatrix} \begin{bmatrix} x(t) \\ y(t) \\ z(t) \end{bmatrix} = \begin{bmatrix} 1 \\ 1 \\ 1 \end{bmatrix} \delta(t). \tag{A-134}$$

Next we form the augmented matrix (omitting the $\delta(t)$) and row reduce, getting

$$\left[\begin{array}{ccc|c} p+1 & -1 & -1 & 1 \\ -1 & p+1 & -1 & 1 \\ -1 & -1 & p+1 & 1 \end{array}\right] \xrightarrow{q_1 \leftrightarrow q_3} \left[\begin{array}{ccc|c} -1 & -1 & p+1 & 1 \\ -1 & p+1 & -1 & 1 \\ p+1 & -1 & -1 & 1 \end{array}\right] \tag{A-135}$$

$$\xrightarrow[q_3 \leftrightarrow q_3+(p+1)q_1]{q_2 \leftrightarrow q_2-q_1} \left[\begin{array}{ccc|c} -1 & -1 & p+1 & 1 \\ 0 & p+2 & -(p+2) & 0 \\ 0 & -(p+2) & p(p+2) & p+2 \end{array}\right]$$

$$\xrightarrow{q_1 \leftrightarrow q_3+q_2} \left[\begin{array}{ccc|c} -1 & -1 & p+1 & 1 \\ 0 & p+2 & -(p+2) & 0 \\ 0 & 0 & (p-1)(p+2) & p+2 \end{array}\right]$$

Shifting the last column back to the right-hand side and solving, we have for the last equation,

$$(p - 1)(p + 2)z(t) = (p + 2)\delta(t). \tag{A-136}$$

We cancel the common factor and easily solve the reduced equation $(p - 1)z(t) = \delta(t)$ to obtain

$$z(t) = \frac{1}{p - 1}\delta(t) = e^t u(t). \tag{A-137}$$

The second equation is $(p + 2)y(t) - (p + 2)z(t) = 0$, or

$$y(t) = z(t) = e^t u(t). \tag{A-138}$$

The first equation is $-x(t) - y(t) + (p + 1)z(t) = \delta(t)$, or $x(t) = -y(t) + (p + 1)z(t) - \delta(t)$. Thus, as we already know that $y(t) = z(t)$, we can write

$$x(t) = pz(t) - \delta(t) = e^t u(t) + \delta(t) - \delta(t) = e^t u(t). \tag{A-139}$$

Thus, all of the response waveforms are identical.

There are many more aspects to the subject we have treated in this appendix. It is usually referred to as linear algebra, and you will undoubtedly find that you have need for much more in-depth knowledge of its subject matter as you deal with more theoretically involved aspects of electrical engineering. The author urges you to consider taking a course in the subject if the opportunity presents itself.

Appendix B | Tellegen's Theorem

We will devote this short appendix to an extremely important result in circuit theory. Not only does it provide the missing link we relied upon in Chapter 4 on nodal and mesh analysis, but it is also essential in working with complex power in Chapter 11.

A Motivational Example

Figure B.1 shows two circuits that have the same topology; that is, their elements might be different (consequently, so will their currents and voltages), but they have exactly the same number of elements that are connected together in exactly the same way. We show the element voltages in Figure B.1(a) as v_a, v_b, and v_c. The element currents are i_a, i_b, and i_c, and the node voltages are v_1 and v_2. For the circuit in Figure B.1(b) we use the same symbols except with primes; thus, we will refer to this circuit as the "primed network" and the one in Figure B.1(a) as the "unprimed network." We are now going to investigate the value of the sum of the products of the element voltages in the unprimed network and the element currents in the primed network. We can use KVL to express the element voltages in the unprimed network in terms of its node voltages, thus obtaining

$$v_a i_a' + v_b i_b' + v_c i_c' = v_1 i_a' + (v_1 - v_2)i_b' + v_2 i_c' \tag{B-1}$$

$$= v_1(i_a' + i_b') + v_2(-i_b' + i_c') = v_1 \times 0 + v_2 \times 0 = 0.$$

Figure B.1
Two networks with the same topology

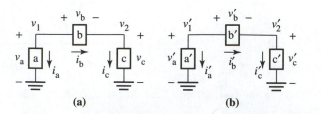

(a)   (b)

This is a remarkable result! The product of the element voltages in one circuit times the corresponding element currents in a *different* network sum to zero identically! The nature of the specific elements does not matter—the elements could even be nonlinear. The only thing that matters is the circuit topology; namely, the way in which the elements are connected. The question is: Does this relationship hold in general, or is it only a fluke occurrence for our specific example? It is, in fact, a general result that we will now prove.

A General Proof of Tellegen's Theorem

Figure B.2(a) shows a general element in a given network (the unprimed network), and Figure B.2(b) shows the corresponding element in another network having exactly the same topology (the primed network). Because the topologies are the same, for each element connected between nodes i and j, say, in the unprimed network there is a corresponding element connected between the same two nodes in the primed network. We investigate the sum

$$\sum_{k=1}^{B} v_{bk} i'_{bk} \tag{B-2}$$

We are assuming that there are B elements in each network. The voltage v_{bk} is the voltage across element e_k in the unprimed network, and i'_{bk} is the current through the corresponding element in the primed network, and we sum over all possible elements. We first use KVL to express the element voltage in terms of the node voltages:

$$v_{bk} = v_i - v_j. \tag{B-3}$$

Figure B.2
Two general network elements for the proof of Tellegen's theorem

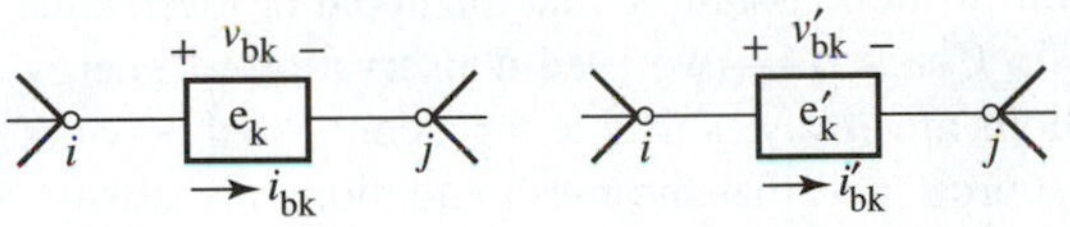

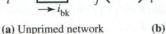

(a) Unprimed network   **(b)** Primed network

Next, we notice that for each element there is a pair of nodes between which the element is connected. Thus, we can rewrite the sum in (B-2) as

$$\sum_{k=1}^{B} v_{bk} i'_{bk} = \frac{1}{2} \sum_{i=1}^{N} \sum_{j=1}^{N} (v_1 - v_j) i'_{ij}. \tag{B-4}$$

In this equation, we have used the symbol i'_{ij} for the current from node i to node j through the given element in the primed network. The reason we used the 1/2 factor is that each term gets counted twice: once for node i and again for node j. Notice here that

$$(v_i - v_j) i'_{ij} = (v_j - v_i) i'_{ji}. \tag{B-5}$$

If there happens to be no element between a given pair of nodes, then the corresponding current is zero. We can now rearrange (B-4) into the form

$$\sum_{k=1}^{B} v_{bk} i'_{bk} = \frac{1}{2} \sum_{i=1}^{N} v_i \sum_{j=1}^{N} i'_{ij} - \frac{1}{2} \sum_{j=1}^{N} v_j \sum_{i=1}^{N} i'_{ij} \tag{B-6}$$

$$= \frac{1}{2} \sum_{i=1}^{N} v_i \sum_{j=1}^{N} i'_{ij} + \frac{1}{2} \sum_{j=1}^{N} v_j \sum_{i=1}^{N} i'_{ji}.$$

In the second equality we have used the fact that $i'_{ji} = -i'_{ij}$. Here is the crucial observation: each of the "inside sums" represents a KCL equation at a node—the first at node i and the second at node j. Thus, *both these sums are zero.* We have thus proved

$$\sum_{k=1}^{B} v_{bk} i'_{bk} = 0, \tag{B-7}$$

and it is this result that we call *Tellegen's theorem.*

Perhaps the most useful form of Tellegen's theorem occurs when the primed and unprimed networks are identical; that is, when the element voltages and currents are from the same network. In this case, we see that the sum in question is merely the total power absorbed by the network. Thus, we have

$$P_{\text{total absorbed}} = \sum_{k=1}^{B} v_{bk}i'_{bk} = 0;$$ (B-8)

that is, the sum of the powers absorbed by *all elements* (including sources as well as resistors) is identically zero. The integrated form of (B-8) is merely *energy conservation*.

A Summary: The Logical Independence of Circuit Theory Relative to Physics

Let's recap what we accomplished in the first four chapters of this text. We started with two basic concepts: element and conductor. We then defined three fundamental physical variables that can be measured: time, current, and voltage. All other ideas were defined in terms of these variables. Thus, from element and conductor, we defined such concepts as node, loop, and mesh. From time, voltage, and current, we defined a number of derived variables: charge, flux linkage, power, and energy. Tellegen's theorem now tells us that conservation of energy holds for all circuits—without our having to rely on any concepts from physics that we have not defined. As a matter of fact, we have derived Tellegen's theorem without assuming that the nodal or mesh equations can be solved. This is crucial, for—in Chapter 4—we used conservation of energy in deriving the fact that the nodal equations are always solvable for a connected network having only resistors and independent sources and that the mesh equations are always solvable for a coupled planar network having only resistors and independent sources. Tellegen's theorem assures us that our proofs were valid. In other chapters of this text, we rely upon Tellegen's theorem in a number of derivations.

Example B.1   Show directly that Tellegen's theorem holds for the two circuits in Figure B.3. Then show, in particular, that conservation of energy holds for each.

Solution   Using superposition, we can compute the values in Table B.1. Thus, the Tellegen sum is

$$v_a i'_a + v_b i'_b + v_c i'_c + v_d i'_d = 18 \times 4 + 18 \times 2 + (-18) \times (-6) + 36 \times -6 = 0.$$ (B-9)

If we compute the total power absorbed by all the elements of the unprimed circuit, we

Figure B.3
Two example circuits having the same topology

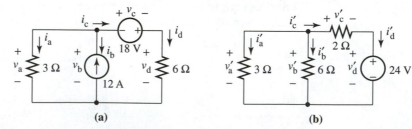

(a)   (b)

Table B.1
Element voltages and currents

Unprimed Circuit	Primed Circuit
$v_a = 18$ V, $i_a = 6$ A	$v_a' = 12$ V, $i_a' = 4$ A
$v_b = 18$ V, $i_b = -12$ A	$v_b' = 12$ V, $i_b' = 2$ A
$v_c = -18$ V, $i_c = 6$ A	$v_c' = -12$ V, $i_c' = -6$ A
$v_d = 36$ V, $i_d = 6$ A	$v_d' = 24$ V, $i_d' = -6$ A

will find that

$$P_{\text{tot}} = v_a i_a + v_b i_b + v_c i_c + v_d i_d = 18 \times 6 + 18 \times (-12) + (-18) \times 6 + 36 \times 6 = 0. \quad \text{(B-10)}$$

Finally, the total absorbed by all the elements in the primed circuit is

$$P_{\text{tot}} = v_a' i_a' + v_b' i_b' + v_c' i_c' + v_d' i_d' = 12 \times 4 + 12 \times 2 + (-12) \times (-6) + 24 \times (-6) = 0. \quad \text{(B-11)}$$

We will leave it to you to show that, additionally, the product of the primed network element voltages and the unprimed network currents also sums to zero.

Review Problems

Problem B.1.   Verify Tellegen's theorem for the two networks in Figure PB.1 by directly forming the appropriate sum of non-primed network element currents and primed network voltages. Assume that all vertical element voltages are positive at the top and all vertical element currents have their references downward. Define the horizontal element voltage to be positive on the left and its element current to flow from left to right.

Figure PB.1

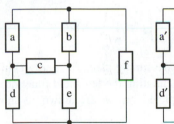

Problem B.2.   Solve the circuits in Figure PB.2 for the element voltages and currents and show that conservation of energy holds for each.

Figure PB.2

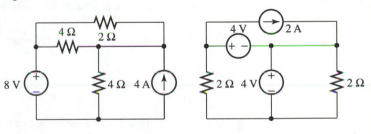

Appendix C | Integration by Parts (Tabular Method)

It is often important to be able to integrate simple functions rapidly. For this reason, we present here a fast method for the integration of simple functions.

We first recall the well-known integration by parts formula:

$$\int u \, dv = uv - \int v \, du. \quad \text{(C-1)}$$

The integrals here are taken to be of the indefinite variety; that is, functions whose derivatives are the functions inside the integral signs. We apply this formula to the computation of

$$I = \int f(t) g(t) \, dt. \quad \text{(C-2)}$$

For convenience in our development we will use the following notation:

$$p^k f(t) = \frac{d^k f(t)}{dt^k} \quad \text{(C-3)}$$

and
$$\frac{1}{p^k}g(t) = \underbrace{\int\int\cdots\int}_{k \text{ times}} g(t)\,dt, \tag{C-4}$$

with
$$p^0 f(t) = f(t) \tag{C-5}$$

and
$$\frac{1}{p^0}g(t) = g(t). \tag{C-6}$$

Therefore, the integration by parts formula gives

$$I = \int f(t)g(t)\,dt = f(t)\frac{1}{p}g(t) - \int pf(t)\frac{1}{p}g(t)\,dt. \tag{C-7}$$

Using this same formula for the integral on the right-hand side yields

$$I = f(t)\frac{1}{p}g(t) - pf(t)\frac{1}{p^2}g(t) + \int p^2f(t)\frac{1}{p^2}g(t)\,dt. \tag{C-8}$$

If we continue in this fashion, we obtain after n integrations,

$$\begin{aligned} I = {} & f(t)\frac{1}{p}g(t) - pf(t)\frac{1}{p^2}g(t) + p^2f(t)\frac{1}{p^3}g(t) - \cdots \\ & + (-1)^{n-1}(-p^{n-1}f(t)\frac{1}{p^n}g(t) + (-1)^n\int p^nf(t)\frac{1}{p^n}g(t)\,dt. \end{aligned} \tag{C-9}$$

This result is easier to see with the help of Table C.1. We simply place $f(t)$ and $g(t)$ in the first row; after that, each new row is derived by differentiating the left-hand function and integrating the right-hand one. Once the table is completed, arrows are drawn, as shown.

Table C.1 Integration by parts table

k	$p^kf(t)$		$\frac{1}{p^k}g(t)$
0	$f(t)$	$+$	$g(t)$
1	$pf(t)$	$-$	$\frac{1}{p}g(t)$
2	$p^2f(t)$	$+$	$\frac{1}{p^2}g(t)$
3	$p^3f(t)$	$-$	$\frac{1}{p^3}g(t)$
⋮	⋮		⋮
$n-1$	$p^{n-1}f(t)$	$(-1)^{n-1}$	$\frac{1}{p^{n-1}}g(t)$
n	$p^nf(t)$	$(-1)^n$ $\frac{1}{p}$ →	$\frac{1}{p^n}g(t)$

The first n (zero through $n - 1$) are drawn diagonally, the last (n) horizontally. Signs are then associated with each arrow on an alternating basis, starting with a plus. Finally, one expresses the integral being computed as the sum of the *products* of the functions at the ends of the same diagonal arrow, with the sign noted on the arrow; to this, one adds the *integral* of the product of the last two elements in the table with the sign noted above the horizontal arrow.

Example C.1

Find $I = 1/p[e^t \sin(t)]$.

Solution

Let $f(t) = e^t$ and $g(t) = \sin(t)$. The integration by parts table is given in Table C.2. Letting I be the integral of interest, we see that

$$I = -e^t \cos(t) + e^t \sin(t) - \int e^t \sin(t\ dt), \tag{C-10}$$

but the final term is simply the original integral, I, itself. Thus,

$$I = \frac{1}{2} e^t[\sin(t) - \cos(t)] + K. \tag{C-11}$$

We have added an arbitrary constant because the integrand is not necessarily one sided.

Table C.2   Table for Example C.1

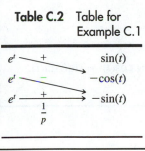

The preceding example illustrates an important case: that for which, after a given number of steps, the integrand reappears. This happens often for integrands involving trigonometric functions. Another important case is one for which a function becomes zero after a finite number of differentiations (that is, one of the functions is a polynomial).

Example C.2

Find $1/p[t^3 e^{2t}]$.

Solution

The integration table is given in Table C.3. This time we just write the integral:

$$I = \frac{1}{2}t^3 e^{2t} - \frac{3}{4}t^2 e^{2t} + \frac{6}{8}te^{2t} - \frac{6}{16}e^{2t} + K. \tag{C-12}$$

Table C.3   Table for Example C.2

t^3	$+$	e^{2t}
$3t^2$	$-$	$\frac{1}{2}e^{2t}$
$6t$	$+$	$\frac{1}{4}e^{2t}$
6	$-$	$\frac{1}{8}e^{2t}$
0	$+$ $\frac{1}{p}$	$\frac{1}{16}e^{2t}$

In many cases we would like to find the integral of a function $f(t)$ that is zero for nonpositive values of t. We note that if $f(t) = 0$ for $t \leq 0$, then

$$\int_{-\infty}^t f(\alpha)\ d\alpha = \left[\int_0^t f(\alpha)\ d\alpha\right] u(t). \tag{C-13}$$

In this case, we must evaluate a definite integral. To do this, recall that if $F(t)$ is an indefinite integral of $f(t)$, then by the Fundamental Theorem of Calculus,

$$\int_0^t f(\alpha)\, d\alpha = F(t) - F(0). \tag{C-14}$$

Example C.3

Find $1/p[t \sin(t)u(t)]$.

Solution

Let t be the function to be differentiated and $\sin(t)$ the one to be integrated. Then the integration table is the one shown in Table C.4. We have left the arrows for you to draw this time. The indefinite integral is given by $F(t) = -t\cos(t) + \sin(t)$. Hence, because $F(0) = 0$, we have

Table C.4
Table for Example C.3

t	$\sin(t)$
1	$-\cos(t)$
0	$-\sin(t)$

$$\frac{1}{p}[t \sin(t)u(t)] = [F(t) - F(0)]u(t)$$

$$= [-t\cos(t) + \sin(t) - 0]u(t) = [-t\cos(t) + \sin(t)]u(t). \tag{C-15}$$

Review Problems

Problem C.1. Find $1/p[e^{-2t}\sin(4t)]$.

Problem C.2. Find $1/p[3e^{-4t}u(t)]$.

Problem C.3. Find $1/p[e^t \sin(t)u(t)]$.

Appendix D | Generalized Functions

This appendix develops a rigorous elementary theory of generalized functions. "Elementary" means that only a first-year calculus background is required to understand it. It is very intuitive and conceptual in nature, much simpler than former developments, which required a high level of abstract mathematics to assimilate. It has not heretofore been presented because it is the author's original work. For this reason it is given here as a reference—as well as for the use of those wishing to see a more rigorous development than that of the text proper. There is some overlap of topics here with the material of Chapter 13.

Jump Discontinuities

We first consider the physical measurement of waveforms. Figure D.1(a) shows a mathematically ideal waveform: one with a discontinuity at t_0. If we attempt to investigate such a waveform in the laboratory with an oscilloscope, the picture we actually see is something like the one in Figure D.1(b); it is continuous with a finite transition time ϵ between the two points at which the ideal waveform has a jump. If we look at the waveform with a better oscilloscope, however, the transition time will be shorter (smaller ϵ). In the limit as

Figure D.1
"Physical" waveforms
(idealization and observable
waveform)

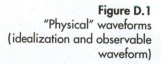

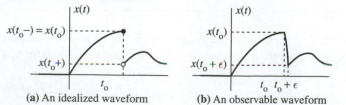

(a) An idealized waveform   (b) An observable waveform

the quality of our observation instrument approaches ideal, the observable waveform will approach the one in Figure D.1(a). Notice, however, that it will always be continuous from the left; that is,

$$\lim_{\epsilon \to 0+} x(t_o - \epsilon) = x(t_o-) = x(t_o). \tag{D-1}$$

For this reason, we have shown the value of the idealized waveform at t_o with a solid dot at the top and an open one at the bottom; this merely serves as a reminder that a finite transition time is required in practice and that the idealized waveform is continuous from the left. We define the "jump" in $x(t)$ at t_o as

$$x(t_o+) - x(t_o-) = x(t_o+) - x(t_o), \tag{D-2}$$

where $x(t_o+)$ is the right-hand limit

$$x(t_o+) = \lim_{\epsilon \to 0+} x(t_o + \epsilon). \tag{D-3}$$

We also recognize that we can build a laboratory approximation to a differentiator and hook several up as shown in Figure D.2 to look at derivative waveforms of various orders. If we were to apply the waveform in Figure D.1 to the input, we would expect to see waveforms at each output point having the same general characteristics as those of Figure D.1; that is, the "physical" waveforms would all be continuous at each time point and would approach left-continuous functions as our differentiators and observing instrument came closer and closer to ideal.

Figure D.2
A chain of differentiators

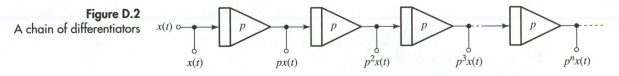

Finally, notice that we have depicted the waveforms in Figure D.1 as being identically zero for all $t \le 0$; this corresponds to the physical idea that a signal generator is "turned on" at some time point, which we arbitrarily take to be $t = 0$. We call such waveforms *one sided*. We now provide the following definition of a basic set of waveforms.

Definition D.1   A time function is called a K_∞ waveform if it and each of its derivatives (of any order) have the following properties:

1. It is one sided; that is, it is identically zero for $t \le 0$.

2. It is continuous everywhere except at a finite set of "singular points," of which there are only a finite number in each interval of finite length, at which it fails to be continuous. At each of these points it is left continuous and has at most a finite jump discontinuity. (This means that $x(t_o+) - x(t_o-) = x(t_o+) - x(t_o)$ is finite.)

We will assume henceforward (unless we specifically state otherwise) that all waveforms are K_∞ functions.

The Fundamental Theorem of Calculus and a Generalization   Discontinuous functions are not often treated at any depth in calculus courses, and there is a very important consequence: the "failure" of the Fundamental Theorem of Calculus (FTC). The FTC says that if $x(t)$ is differentiable on the open interval (a, b) and its derivative is continuous on the closed interval $[a, b]$,[1] then

$$\int_a^b px(t)\, dt = x(b) - x(a), \tag{D-4}$$

where $px(t) = dx(t)/dt$. The limitation that the derivative $px(t)$ be continuous means that the function $x(t)$ itself must be continuous. *Neither of these limitations is necessary.*

We will now state and prove a Modified Fundamental Theorem of Calculus (MFTC) that is more general than the one just stated and that will apply to K_∞ waveforms in an important way.

Theorem D.1 (MFTC)   Let $x(t)$ be a waveform that is defined on some open interval containing the closed interval $[a, b]$.[2] Assume that there is at most a finite set of time points $t_o < t_1 < \cdots < t_N$ in $[a, b]$ at which either $x(t)$ or its derivative $px(t) = x'(t)$ (or both) fail to exist and/or be continuous. If $x(t)$ or $px(t)$ do not exist at one of these points, we "fill in" its value by left-continuity, but we will assume that both $x(t)$ and $px(t)$ are continuous everywhere else in $[a, b]$. Then, if Δ_i is the jump in $x(t)$ at $t = t_i(i = 0, 1, \ldots, N)$,

$$\int_a^b px(t)\, dt + \sum_{i=0}^N \Delta_i = x(b) - x(a). \tag{D-5}$$

Proof   We will first assume that there is only one of the "singular points" at $t = t_i$, with $a < t_i < b$. Because $px(t)$ is discontinuous at t_i (if it does not exist there, we simply "fill in" its value using left-continuity as we noted in the statement of the theorem), the integral in (D-5) can be interpreted as an improper integral at t_i, that is, as

$$\int_a^b px(t)\, dt = \lim_{\epsilon \to 0+} \int_a^{t_o - \epsilon} px(t)\, dt + \lim_{\epsilon \to 0+} \int_{t_o + \epsilon}^b px(t)\, dt. \tag{D-6}$$

Next, we notice that $px(t)$ is continuous over each of the integration intervals. This allows us to use the usual FTC (as stated above). This results in

$$\int_a^b px(t)\, dt = \lim_{\epsilon \to 0+} [x(t_i - \epsilon) - x(a)] + \lim_{\epsilon \to 0+} [x(b) - x(t_i + \epsilon)] = x(b) - x(a) - \Delta_i. \tag{D-7}$$

(We have evaluated the ϵ limit as $x(t_i+) - x(t_i-)$ and recognized that this is just Δ_i.) Moving the jump Δ_i to the left side gives (D-5). If $t_i = a$ or $t_i = b$, a similar argument yields the same result, applying the FTC over the intervals $[a + \epsilon, b]$ and $[a, b - \epsilon]$, respectively.

If we now allow our "singularity set" to have more than one point, we simply repeat the preceding argument, removing a small interval around each of the time points t_i, then computing the limit, to arrive at the general expression in (D-5).   ∎

---

[1] Math courses generally allow the function and its derivative to not even be defined outside the interval being considered, and a derivative requires that a two-sided limit of the difference quotient exist. This means that the derivative can only be considered (in this context) to be well defined on the *open* interval.

[2] This just means that the derivative can be considered at the end points of $[a, b]$.

The MFTC we have just proved holds for more general functions than those of the K_∞ variety, but we will only use it for the latter. The theory of generalized functions (as well as several theories of integration) was developed to make the original FTC "work." We will do this for K_∞ waveforms by developing the mathematics to remove the sum of the jumps on the left side of (D-5). We do this by generalizing our concept of function a bit to carry this information along as we differentiate. We will sometimes refer to K_∞ waveforms as *ordinary functions* (as opposed to generalized functions).

Definition D.2   The number triple (t_i, m_i, Δ_i) will be called a *singularity*. When associated with an ordinary waveform $x(t)$, it means that after $m_i + 1$ integrations of $x(t)$ one adds a jump (a step function) at $t = t_i$ of height Δ_i. We say that t_i is the *occurrence time*, m_i is the *order*, and Δ_i the *strength*, or *amplitude*, or *jump* at $t = t_i$.

Figure D.3 Singularity symbol

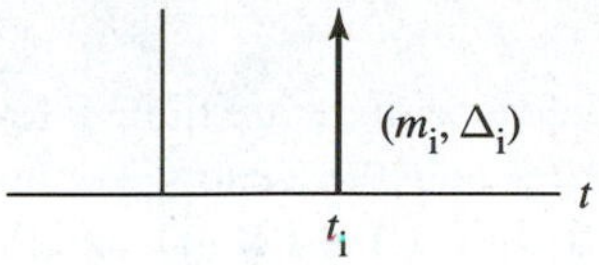

(m_i, Δ_i)

t_i

The symbol we will use for a singularity is an arrow, as shown in Figure D.3. The time of occurrence is shown on the t axis, therefore, we merely mark the order and the strength in parentheses next to the vertical arrow. If the jump Δ_i is negative, we will often use the absolute value of Δ_i in the parentheses and point the arrow downward, as shown in Figure D.4. There, we show the symbol for the singularity $(t_i, 2, -5)$. We will quite often have the need to consider sets of singularities, so just to fix notation, look at the singularity set

$$S = \{(0, 1, 1), (5, 3, -2)\}. \tag{D-8}$$

There are two singularities: one with an occurrence time $t_o = 0$, order $m_o = 1$, and strength $\Delta_o = 1$ and another with an occurrence time $t_1 = 5$, order $m_1 = 3$, and strength $\Delta_1 = -2$. These singularities are plotted in Figure D.5. If these singularities were associated with an ordinary function, that function would be drawn on the same graph. Recalling the definition of *set union*, we can write this example singularity set as

$$S = \{(0, 1, 1)\} \cup \{(5, 3, -2)\}, \tag{D-9}$$

that is, as the union of two *singleton* sets. (A singleton set is a set with only one element.) In general, we will need to write a general singularity set as the union of a finite or countable number of singleton sets:

$$S = \bigcup_{i=0}^{\infty} \{(t_i, m_i, \Delta_i)\}. \tag{D-10}$$

The braces are necessary to denote that we are treating each singularity as a set.

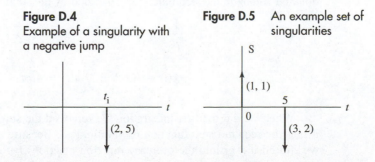

Figure D.4
Example of a singularity with
a negative jump

t_i

$(2, 5)$

Figure D.5   An example set of
singularities

S

$(1, 1)$

5

0

$(3, 2)$

Generalized Functions: Definition and Example

> **Definition D.3**   A *generalized function* (abbreviated gf) is the ordered pair
>
> $$\hat{x}(t) = (x(t), S_x), \tag{D-11}$$
>
> where the first element $x(t)$ is an ordinary (i.e., K_∞) function and S_x is a countable (discrete) set of singularities, none of whose orders are greater than some fixed nonnegative integer N (called the *order* of the generalized function).

Example D.1   Let $\hat{x}(t) = (12tu(t), \{(0, 0, 3), (2, 1, -4)\})$. Plot this generalized function. (The hat over the x denotes the fact that it is a *generalized function:* there is no hat over the ordinary function inside the parentheses.)

Solution   This example is mainly for the purpose of stressing notation because there are quite a few parentheses and braces. Notice that the outermost parentheses and the central comma mark off the entries in the gf: the ordinary function associated with $\hat{x}(t)$ is $12tu(t)$, and the singularity set has two singularities, $(0, 0, 3)$ and $(2, 1, -4)$. The plot of $\hat{x}(t)$ is shown in Figure D.6. Notice that we have simply sketched the ordinary function "on top of" the singularities on the same graph.

Figure D.6
An example set of singularities

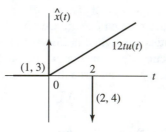

We will discuss derivatives and integrals of generalized functions in a moment. For now, however, let's recall the significance of a singularity. The one at $t = 0$ means that after we integrate $x(t) = 12tu(t)$ once (the order of the singularity at $t = 0$ is zero), we "add in" a step function at the occurrence time ($t = 0$) with an amplitude of 3 units. The one at $t = 2$ says that we must subtract (because the arrow is pointing downward) a step of jump 4 units after *two* integrations. The first integral of $12tu(t)$ is $6t^2 u(t)$, and the second integral is $2t^3 u(t)$. Piecing the parts together (the results of integrating the ordinary function and adding in the step functions generated by the singularities) gives

$$\frac{1}{p}\hat{x}(t) = ((6t^2 + 3)u(t), \{(2, 0, -4)\}) \tag{D-12}$$

and

$$\frac{1}{p^2}\hat{x}(t) = ((2t^3 + 3t)u(t) - 4u(t - 2), \varnothing). \tag{D-13}$$

When we did the first integration we removed the singularity at $t = 0$, transforming it into the required step function of amplitude 3. Because a step is an ordinary function, we shifted it over into the ordinary function position (before the separating comma). At the same time, we decreased the order of the singularity at $t = 2$ by one because now (after one integration) only one more integration is required to produce its associated step function.

When we performed this second integration, we had to integrate the step function at $t = 0$ that was generated by the singularity at $t = 0$ on the first integration, giving the $3tu(t)$ term in the ordinary function position in (D-13). At the same time, we added a step function of -4 units at $t = 2$ as required by the singularity at $t = 2$ and removed the latter from the singularity set. Because there are no longer any singularities, the singularity set becomes the empty set $\varnothing$ as shown in equation (D-13). The two integrated waveforms are shown in Figure D.7.

Figure D.7
The first two running
integrals

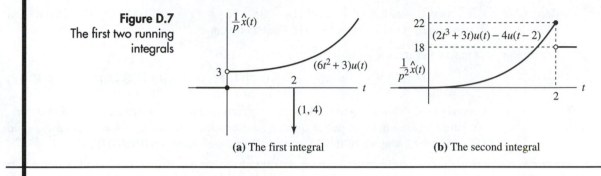

(a) The first integral   **(b)** The second integral

The Fundamental Idea of a Generalized Function: An Intuitive Summary

We would like to call your attention here to a couple of important facts. First, the singularities produce discontinuities in the integrated waveform (turning things around, when we differentiate a discontinuous waveform, we can "store the jumps" in the singularity set S_x). Second, after a finite number of integrations, the result is an ordinary function—with an empty singularity set $\varnothing$—because the order of each singularity is no greater than some finite fixed integer N. Thus, a generalized function will be the derivative (of some finite order) of an ordinary waveform.

(As we are dealing with waveforms having *jumps* and providing a *pouch* to store them in, we might call our generalized functions *kangaroo* functions. We will, however, simply call them K_∞ *generalized functions* and grin a bit when we write the K!)

Development of the Machinery of Generalized Functions

If you have followed each step of our solution in the last example, you already have grasped the basic idea of generalized functions very well. The only thing left to do is to develop the formal machinery for manipulation; however, "only" might be misleading, for the machinery is important. We would like to be able to perform on generalized functions all of the same operations that we perform on ordinary functions. To do so, we must provide the appropriate definitions for the basic operations.

Definition D.4   Let $\hat{x}(t) = (x(t), S_x)$ be an arbitrary generalized function and let c be an arbitrary real or complex number. Then we define

$$c\hat{x}(t) = (cx(t), cS_x), \tag{D-14}$$

where cS_x is defined as follows: each singularity (t_i, m_i, Δ_i) in S_x is replaced by the singularity $(t_i, m_i, c\Delta_i)$ in cS_x.

Definition D.5   Let $\hat{x}(t) = (x(t), S_x)$ and $\hat{y}(t) = (y(t), S_y)$ be arbitrary generalized functions. Then the *sum* of $\hat{x}(t)$ and $\hat{y}(t)$ is defined by

$$\hat{x}(t) + \hat{y}(t) = (x(t) + y(t), S_x + S_y), \tag{D-15}$$

where $x(t) + y(t)$ is the usual sum of the ordinary functions $x(t)$ and $y(t)$ and $S_x + S_y$ is the set union $S_x \cup S_y$—unless S_x and S_y contain one or more singularities with the same occurrence time and the same order. Then, if (t_i, m_i, Δ_{xi}) is a singularity in S_x and (t_i, m_i, Δ_{yi}) is a singularity in S_y, then the two are removed and replaced by the single singularity $(t_i, m_i, \Delta_{xi} + \Delta_{yi})$ in $S_x + S_y$.

This definition says that the singularities "add" almost like a set union.

Example D.2 Let two generalized functions be $\hat{x}(t) = (e^{-2t}u(t), \{(1, 1, 3), (2, 2, 1)\})$ and $\hat{y}(t) = (tu(t), \{(1, 1, -2), (2, 3, 2)\})$. Find their sum.

Solution We just carry out the prescription in Definition D.5:

$$\hat{x}(t) + \hat{y}(t) = ((t + e^{-2t})u(t), \{(1, 1, 1), (2, 2, 1), (2, 3, 2)\}). \tag{D-16}$$

Because $(1, 1, 3)$ in $\hat{x}(t)$ and $(1, 1, -2)$ in $\hat{y}(t)$ have common occurrence times and orders, we simply added the two jumps; then we formed the remainder of the singularity set in $\hat{x}(t) + \hat{y}(t)$ by listing all of the remaining singularities in both $\hat{x}(t)$ and $\hat{y}(t)$.

Sometimes it is useful to consider infinite series of ordinary functions, and the situation is no different with generalized functions. The following is a first definition, one that we will later generalize.

Definition D.6 Let $\hat{x}_n(t) = (x_n(t), S_{x_n})$ be an arbitrary generalized function for each value of n having the property that the series of ordinary functions

$$\sum_{n=0}^{\infty} x_n(t) = x(t) \tag{D-17}$$

converges, as does the series of singularity sets

$$\sum_{n=0}^{\infty} S_{x_n} = S_x. \tag{D-18}$$

(We define the convergence of singularity sets as follows. If the S_{x_n} have no singularities having common occurrence times and orders, then S_x is merely a set union; hence, there is no question of convergence in the usual sense because the union in (D-18) always exists. If there are only a finite number of the S_{x_n} containing singularities of this sort, the infinite sum is again well defined. If there are an infinite number of the S_{x_n} having singularities with the same occurrence time t_i and order m_i, then we replace the entire set of such singularities by the single singularity

$$\left(t_i, m_i, \sum_{n=0}^{\infty} \Delta_{x_{ni}} \right), \tag{D-19}$$

where $\Delta_{x_{ni}}$ is the jump of the given singularity in S_{x_n}. One says that the series of singularity sets S_{x_n} converges in either of the first two cases, and that it converges in the third if the series of jumps $\sum_{n=0}^{\infty} \Delta_{x_{ni}} = \Delta$ converges.)

Example D.3 Let $\hat{x}_n(t) = (2^{-n}u(t), \{(nT, 1, 1): n = 0, 1, \cdots\})$ Plot the generic generalized function $\hat{x}_n(t)$. Then determine whether the series $\sum\limits_{n=0}^{\infty} \hat{x}_n(t) = \hat{x}(t))$ converges and, if it does, find the sum $\hat{x}(t)$.

Solution The indicated series, though easy to sum, is somewhat more complex notationally, so let's look at the individual term $\hat{x}_n(t)$ a bit more closely. That is, we fix n to be some arbitrary integer and try to picture the resulting gf. We see that the singularity set for $\hat{x}_n(t)$ is the singleton set $S_{x_n} = \{(nT, 1, 1)\}$, a single singularity at the location $t = nT$. The ordinary function $x_n(t) = 2^{-n}u(t)$ is a step function. Thus, we can plot $\hat{x}_n(t)$ as in Figure D.8. The sum of the ordinary functions is basically that of the *geometric* series,

$$\sum_{n=0}^{\infty} a^n = \frac{1}{1-a}, \tag{D-20}$$

with $a = 1/2$. This sum, therefore, is

$$\sum_{n=0}^{\infty} \left(\frac{1}{2}\right)^n u(t) = \frac{1}{1-\dfrac{1}{2}} u(t) = 2u(t). \tag{D-21}$$

We merely "unionize" the preceding singleton sets to get the singularity set for the sum,

$$S_x = \bigcup_{n=0}^{\infty} \{(nT, 1, 1)\}. \tag{D-22}$$

It consists of singularities of the first order with jumps of one unit, equally spaced at intervals of length T from $t = 0$ to $t = \infty$ (see Figure D.9).

Figure D.8
The typical term

Figure D.9
The series sum

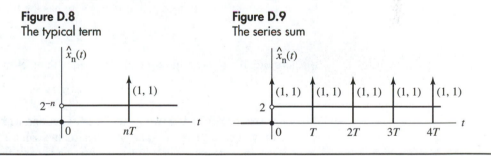

The preceding example illustrates very well the idea that the sum of generalized functions having singularity sets with quite different singularities only involves set unions for the sum of the individual singularity set sums. The following definition gives us another simple, though needed, concept: that of the value of a gf.

Several Important Special Types of Generalized Functions

Definition D.7 Let $\hat{x}(t) = (x(t), S_x)$ be an arbitrary generalized function and let a be an arbitrary value of time. Then we define the *value* of $\hat{x}(t)$ at $t = a$ to be the value of its corresponding ordinary function there:

$$\hat{x}(a) = x(a). \tag{D-23}$$

Definition D.8   The *zero* generalized function is defined by

$$\hat{0} = (0, \varnothing). \qquad \text{(D-24)}$$

The zero generalized function certainly has the value zero at each value of t; however, it is not the only one having this property.

Definition D.9   The *unit impulse function* (or *unit delta function*) at $t = 0$ is defined by

$$\delta(t) = (0, \{(0, 0, 1)\}). \qquad \text{(D-25)}$$

Notice that $\delta(t)$ has the value zero for each value of t, but it is not the same as the zero generalized function. Notice, too, that we have deviated from our established notation of using a caret (hat) over all generalized functions because the delta function is already so commonly used in the circuits and systems literature without it.

Definition D.10   We will call any generalized function with an ordinary function component that is identically zero,

$$\hat{x}_s(t) = (0, S), \qquad \text{(D-26)}$$

a *singularity function.*

The unit impulse function, for instance, is a singularity function of order zero.
     Now suppose that $\hat{x}(t) = (x(t), \varnothing)$, $\hat{y}(t) = (y(t), \varnothing)$ are two generalized functions having an empty singularity set. Then their sum has the form

$$\hat{x}(t) + \hat{y}(t) = (x(t) + y(t), \varnothing); \qquad \text{(D-27)}$$

furthermore, if c is any arbitrary real or complex number, we have

$$c\hat{x}(t) = (cx(t), \varnothing). \qquad \text{(D-28)}$$

Thus, we see that if we add two generalized functions with an empty singularity set or if we multiply such a gf by a scalar, another of the same sort always results. This is the same as saying that the general function notation is merely an alternate name for the ordinary function involved. Thus, we identify the two by writing

$$x(t) = (x(t), \varnothing). \qquad \text{(D-29)}$$

This use of the equality sign does not mean the equality of two generalized functions, but merely our agreement to use $x(t)$ as the symbol for the gf $(x(t), \varnothing)$. Equality of two generalized functions, in fact, is something we will now define.

Definition D.11   Let $\hat{x}(t) = (x(t), S_x)$ and $\hat{y}(t) = (y(t), S_y)$ be arbitrary generalized functions. Then we write

$$\hat{x}(t) = \hat{y}(t) \qquad \text{(D-30)}$$

if and only if $x(t) = y(t)$ for each value of t (which is the same thing as saying that the *values* of $\hat{x}(t)$ and $\hat{y}(t)$ are equal for each value of t and $S_x = S_y$; that is each singularity in S_x is also a singularity to be found in S_y, and vice versa.

Two Special Decompositions of an Arbitrary Generalized Function

We will now begin to develop some ideas that make generalized functions look notationally very much like ordinary functions with which we are already acquainted. To do so, we will take an arbitrary generalized function and write it as follows:

$$\hat{x}(t) = (x(t), S_x) = (x(t), \varnothing) + (0, S_x) = x(t) + x_s(t). \tag{D-31}$$

We have written $x(t)$ in the place of the gf $(x(t), \varnothing)$ as per our earlier agreement and have used the subscript s on the singular function symbol $x_s(t) = (0, S_x)$ to denote the fact that it is a singularity function. We will not write the caret on top of the function when we use the s subscript. We refer to $x(t)$ as the *ordinary part* of $\hat{x}(t)$ and to $x_s(t)$ as its *singular part*. If we wished, we could also recognize that $x(t)$, being a K_∞ waveform, can be written as the sum of a continuous part $x_c(t)$ and a jump (or discontinuous) part $x_j(t)$ and write (D-31) in the form

$$\hat{x}(t) = x_c(t) + x_j(t) + x_s(t). \tag{D-32}$$

We will develop another standard way of writing a gf after we have developed a bit more notation.

The Derivative of an Arbitrary Generalized Function

Definition D.12

Let $\hat{x}(t) = (x(t), S_x)$ be an arbitrary generalized function. Then the *derivative* of $\hat{x}(t)$ is defined by

$$p\hat{x}(t) = (px(t), pS_x), \tag{D-33}$$

where $px(t)$ is the ordinary (K_∞) derivative, with the value at discontinuity points "filled in" by left-continuity, and where pS_x is a singularity set defined as follows. If (t_i, m_i, Δ_i) is a singularity in S_x, then it is replaced by $(t_i, m_i + 1, \Delta_i)$ in pS_x; furthermore, if t_i is a point at which $x(t)$ has a jump discontinuity (the only type possible for a K_∞ function) of Δ_i, then the zeroth-order singularity $(t_i, 0, \Delta_i)$ is incorporated as an element of pS_x.

Example D.4

Find the derivative of the unit step function $u(t)$.

Solution

We identify $u(t)$ as a gf by writing $u(t) = (u(t), \varnothing)$.[3] Because the ordinary function $u(t)$ has a single discontinuity of amplitude one at $t = 0$, and because the ordinary derivative is zero for each $t \neq 0$, we have (filling in the value at $t = 0$ by left-continuity,

$$pu(t) = (0, \{(0, 0, 1)\}) = \delta(t). \tag{D-34}$$

---

[3] Look back at our comment right after equation (D-29) regarding this use of the equal sign.

*The Running Integral
of an Arbitrary
Generalized Function*

Definition D.13   Let $\hat{x}(t) = (x(t), S_x)$ be an arbitrary generalized function. Then we define the running integral of $\hat{x}(t)$ by

$$\frac{1}{p}\hat{x}(t) = \left(\frac{1}{p}x(t) + x_j(t), \frac{1}{p}S_x\right), \tag{D-35}$$

where the elements on the right side are defined as follows. The function $\frac{1}{p}x(t)$ is the usual running integral of $x(t)$, which always exists and is continuous because $x(t)$ is a K_∞ function. The singularity set $\frac{1}{p}S_x$ is defined as follows: if (t_i, m_i, Δ_i) is a singularity in S_x with $m_i \geq 1$, then it is replaced by $(t_i, m_i - 1, \Delta_i)$ in $\frac{1}{p}S_x$, and each of the singularities in S_x of zeroth order ($m_i = 0$) is simply deleted in forming $\frac{1}{p}S_x$. The zeroth-order singularities are used to form the jump function $x_j(t)$ as follows: if $(t_i, 1, \Delta_i)$ is a zeroth-order singularity in S_x for each i, then one writes $x_j(t) = \sum_{i=0}^{\infty} \Delta_i u(t - t_i)$.

Now look carefully back over the definitions of derivative and integral. You will see that these two operations are exactly the reverses of one another: the discontinuities of $x(t)$ are used to form the first-order singularities in $p\hat{x}(t)$, and, conversely, the first-order singularities of $\hat{x}(t)$ are deleted and used to form the discontinuous part of the ordinary function when integrating. Otherwise, one increases the order of all singularities by one in differentiation and decreases them by one in integration. Thus, we see that integration and differentiation are true inverses of one another and that each generalized function has derivatives of all orders (for the process can be repeated any number of times). We state this formally as a theorem. No proof is required, for the theorem follows immediately from the definition.

Theorem D.2   Let $\hat{x}(t)$ be an arbitrary K_∞ generalized function. Then the derivative $p\hat{x}(t)$ and the integral $(1/p)\hat{x}(t)$ both exist as generalized functions; furthermore, the two operations are inverses of one another:

$$\frac{1}{p}\{p\hat{x}(t)\} = p\left\{\frac{1}{p}\hat{x}(t)\right\} = \hat{x}(t). \tag{D-36}$$

Our remark that the derivative of any order of $\hat{x}(t)$ exists follows from the fact that the derivative of any gf is another gf, which, in turn, is also differentiable.

Example D.5   Find the derivative of the gf $\hat{x}(t) = (2u(t), \{(nT, 0, 1): n = 0, 1, \cdots\})$. Then compute the integral of the derivative and show that it is the same as $\hat{x}(t)$.

Solution   The ordinary part of $\hat{x}(t)$, namely $2u(t)$, is discontinuous at $t = 0$. Its derivative is zero everywhere except at $t = 0$, at which point we fill in the value by left-continuity. This means that the singularity set pS_x is derived by simply increasing the order of each singularity already in S_x by one and converting the jump at $t = 0$ to a zeroth-order singularity and adding it to those already in pS_x. Therefore, we can write

$$p\hat{x}(t) = (0, \{(0, 0, 2)\} \bigcup_{n=0}^{\infty} \{(nT, 1, 1)\}). \tag{D-37}$$

Thus, $p\hat{x}(t)$ is a singularity function with equally spaced singularities spaced T seconds apart.

To form the integral of the derivative, we simply reduce the order of all of the singularities by one—except for the zeroth-order one at $t = 0$, which we convert back to the step function $2u(t)$ and add to the identically zero ordinary function in (D-37) to get $\hat{x}(t)$ back.

The Time-Shifted Form of a Generalized Function

The operation of time shifting an ordinary function is a valuable one, so we will define such a procedure for gf's as well.

Definition D.14   Let $\hat{x}(t) = (x(t), S_x)$ be an arbitrary generalized function and let a be an arbitrary value of time. Then the (right) *time shift* of $\hat{x}(t)$ is defined to be the gf

$$\hat{x}(t - a) = (x(t - a), S_x^a), \tag{D-38}$$

where S_x^a is the set of singularities S_x shifted to the right by a units; that is, if (t_i, m_i, Δ_i) is in S_x, then it becomes $(t_i + a, m_i, \Delta_i)$ in S_x^a.

As an example, suppose that the gf under consideration is $\delta(t) = (0, \{(0, 0, 1)\})$. Then we have

$$\delta(t - a) = (0, \{(a, 0, 1)\}). \tag{D-39}$$

A Canonical Form for an Arbitrary Generalized Function

Definition D.14 gives rise to a very suggestive and important way of writing any generalized function, which we referred to earlier. Suppose that $\hat{x}(t) = (x(t), S_x)$ is an arbitrary generalized function. The definition of a generalized function required S_x to be countable (that is, it either has a finite number of elements or it has an infinite number that can be enumerated as a sequence). Thus, in general, we can write S_x in the form of a union of singleton sets of singularities, as follows:

$$S_x = \bigcup_{i=0}^{\infty} \{(t_i, m_i, \Delta_i)\}, \tag{D-40}$$

where $m_i \geq 0$ for each i. Note that if there are only a finite number of singularities, we would merely replace the infinite "upper limit" on the union symbol with a finite integer. In any case, we can write the generalized function $\hat{x}(t) = (x(t), S_x)$ as

$$\hat{x}(t) = (x(t), \varnothing) + (0, S_x) = x(t) + (0, \bigcup_{i=0}^{\infty} \{(t_i, m_i, \Delta_i)\}). \tag{D-41}$$

Thus, we have written the gf as the sum of an ordinary function and a series of singularity functions. Exactly what are these singularity functions? To see this, just remember that differentiation increases the order by one and use equation (D-39) to see that

$$p^{m_i}\delta(t - t_i) = p^{m_i}(0, \{(t_i, 0, 1)\}) = (0, \{(t_i, m_i, 1)\}). \tag{D-42}$$

Thus, the singleton sets in (D-41) are the same as the one in (D-42) except for the constant Δi, which we can factor out to obtain

$$\hat{x}(t) = x(t) + \sum_{i=0}^{\infty} \Delta_i \delta^{(m_i)}(t - t_i). \tag{D-43}$$

As a bow to conventional notation, we have written

$$\delta^{(m_i)}(t - t_i) = p^{(m_i)}\delta(t - t_i). \tag{D-44}$$

We stress once again that there is in actuality no question of convergence of the singularity functions because the series is actually a somewhat disguised form of set union. We will call (D-43) the *canonical representation* of the gf $\hat{x}(t)$.

Example D.6 Find the canonical form derivative of the waveform in Figure D.10.

Figure D.10
An example waveform

Solution We consider the given waveform as a generalized function $\hat{x}(t) = (x(t), \varnothing)$. It has jumps of $+1$ at $t = 0, 2, 4, \ldots$, and jumps of -1 at $t = 1, 3, 5, \ldots$, and so on. These discontinuities turn into singularities of strengths ± 1 at $t = 0, 1, 2, \ldots$; also, the ordinary K_∞ derivative of $x(t)$ is zero everywhere except at these points. Though the ordinary derivative of $x(t)$ does not exist at these points, we "fill them in" using left-continuity. This gives an ordinary function component of the derivative that is zero everywhere. The generalized derivative, then, is the generalized function given by

$$p\hat{x}(t) = (0, \{(2i, 0, 1): i = 0, 1, \ldots\} \cup \{((2i + 1), 0, -1,): i = 0, 1, \ldots\}) \tag{D-45}$$

$$= \sum_{i=0}^{\infty} (-1)^i \delta(t - i).$$

Thus, $p\hat{x}(t)$ is a singularity function with equally spaced singularities spaced 1 second apart and having unit magnitude strengths with alternating sign. It is plotted in Figure D.11. Notice the solid baseline, which is the graph of the ordinary part of $p\hat{x}(t)$.

Figure D.11
The derivative waveform

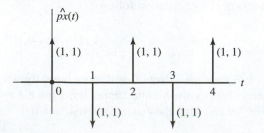

The Definite Integral of an Arbitrary Generalized Function

Definition D.15 Let $\hat{x}(t) = (x(t), S_x)$ be an arbitrary generalized function and let its running integral be denoted by $\hat{X}(t) = \dfrac{1}{p}\hat{x}(t)$. Then the definite integral of $\hat{x}(t)$ over the interval $[a, b]$ is defined by

$$\int_a^b \hat{x}(t)\, dt = \hat{X}(b) - \hat{X}(a), \tag{D-46}$$

where $\hat{X}(b)$ and $\hat{X}(a)$ are the *values* of the gf $\hat{X}(t)$ at $t = b$ and at $t = a$.

Example D.7 Compute the (definite) integral of $\delta(t - t_o)$ between the limits a and b, where t_o is an arbitrary value of time. Assume that $a < b$.

Solution We know that $\delta(t - t_o)$ is the derivative of $u(t - t_o)$; conversely, we know that $u(t - t_o)$ is the running integral of $\delta(t - t_o)$. Therefore, we have

$$\int_a^b \delta(t - t_o)\, dt = u(b - t_o) - u(a - t_o) = \begin{cases} 0; & t_o < a \\ 1; & a < t_o \le b. \\ 0; & t_o > b \end{cases} \tag{D-47}$$

Example D.8 Compute the (definite) integral of $p\delta(t - t_o) = \delta'(t - t_o)$ between the limits a and b, where t_o is an arbitrary value of time. Assume that $a < b$.

Solution We know that $\delta'(t - t_o)$ is the derivative of $\delta(t - t_o)$; conversely, we know that $\delta(t - t_o)$ is the running integral of $\delta'(t - t_o)$. Therefore, we have

$$\int_a^b \delta'(t - t_o)\, dt = \delta(b - t_o) - \delta(a - t_o) = 0. \tag{D-48}$$

The unit impulse function, being a singularity function, has the *value* zero for each value of t.

The Product of Two Generalized Functions

We will now develop a definition of the product of two generalized functions. Surprisingly, it is rather complicated to do (though it still does not require mathematics more advanced than first-year calculus), so we will break it up into several steps.

We will make the first step graphically by looking at Figure D.12. There we show a generalized function having a couple of singularities. Following the dark arrow to the right, we first differentiate, then multiply by a step function $u(t - 2)$. The result is to remove the singularity function to the left of $t = 2$—and to introduce a jump at $t = 2$. (This is in addition, of course, to the usual process of differentiation, which is to increase the order of the singularities by one, differentiate the ordinary component, etc.). If we do things in the reverse order, we follow the arrows downward in the figure. What is the result? The two resulting functions are identical, except for one thing: *"truncation, then*

Figure D.12
"Truncation, then differentiation" versus "differentiation, then truncation"

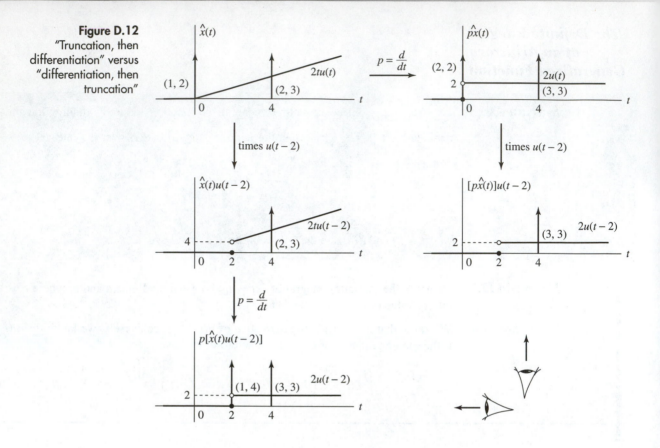

differentiation" inserts an extra impulse function due to the discontinuity introduced into the ordinary function before it is differentiated. We could make this idea completely rigorous symbolically; unfortunately, the notation becomes quite complicated to write. We will, therefore, content ourselves with the "graphical proof" we have just given.

Theorem D.3   If $\hat{x}(t)$ is a generalized function whose ordinary component $x(t)$ is continuous at $t = a$ and if it has no singularities occurring at $t = a$, then

$$p[\hat{x}(t)u(t - a)] = [p\hat{x}(t)]u(t - a) + \hat{x}(a)\delta(t - a). \qquad (D-49)$$

As we have just said, we will not write out an algebraic proof; instead, we merely note that the left-hand side of (D-49) represents the "downward" path in Figure D-12, whereas the first term on the right-hand side represents the "rightward, then downward" path. We must add the extra impulse function, whose strength is the value of $\hat{x}(t)$ at $t = a$. Notice that we could replace the symbol $\hat{x}(a)$ with $x(a)$, if we like, because the latter is by definition the value of the gf at $t = a$.

It would perhaps have been a bit more rigorous to have given Theorem D.3 as a definition, rather than as a theorem, because as yet we have not defined the product of two generalized functions—nor of a generalized function and an ordinary one. We have chosen the latter course because it is clear that multiplication of a singularity function by the unit step (when the jump occurs at some time other than the occurrence time of the singularity) is equivalent to multiplication of the former by a constant, and we have already defined this operation.

Now let's recall the Leibniz rule for differentiation. In the usual terminology, it says that if $f(t)$ and $g(t)$ are two ordinary functions having derivatives, then

$$p[f(t)g(t)] = f(t)pg(t) + g(t)pf(t). \tag{D-50}$$

We will now *define* the product of a unit impulse function and a generalized function so that the Leibniz rule holds for this operation (assuming that the ordinary part of the gf is continuous at the occurrence time of the unit impulse function). Thus, assuming that (D-50) is valid, we use (D-49) to write

$$p[\hat{x}(t)u(t-a)] = [p\hat{x}(t)]u(t-a) + \hat{x}(a)\delta(t-a) \tag{D-51}$$

$$= [p\hat{x}(t)]u(t-a) + \hat{x}(t)\delta(t-a).$$

The first equality is just (D-49) and the second is (D-50) with $f(t) = \hat{x}(t)$ and $g(t) = u(t-a)$. We have also recognized that $pu(t-a) = \delta(t-a)$. This results in the *sampling property* for the unit impulse function[4] (and defines the product of $\hat{x}(t)$ and $\delta(t-a)$):

$$\hat{x}(t)\delta(t-a) = \hat{x}(a)\delta(t-a). \tag{D-52}$$

We caution you that this is only valid if the ordinary part of $\hat{x}(t)$ is continuous at $t = a$.

Example D.9 Find the derivative of $x(t) = \cos(2t)u(t)$.

Solution We are assuming that the Leibniz rule is valid, so we compute

$$p[\cos(2t)u(t)] = -2\sin(2t)u(t) + \cos(2t)pu(t) = -2\sin(2t)u(t) + \cos(2t)\delta(t) \tag{D-53}$$

$$= -2\sin(2t)u(t) + \cos(0)\delta(t) = -2\sin(2t)u(t) + \delta(t).$$

Here we have applied the Leibniz rule, noticed that the derivative of $u(t)$ is $\delta(t)$, and then used the sampling property in (D-52) to evaluate the cosine at $t = 0$.

Sometimes it is necessary in circuits and systems work to use the product of a generalized function and a singularity function. For this we have the following definition.

Definition D.16 Let $\hat{x}(t)$ be a generalized function whose ordinary part $x(t)$ has a continuous nth derivative at $t = a$. Then we define the product of $\hat{x}(t)$ and $\delta^{(n)}(t-a)$ (the latter being the nth derivative of the unit impulse function) to be the expression[5]

$$\hat{x}(t)\delta^{(n)}(t-a) = \sum_{k=0}^{n}\binom{n}{k}(-1)^k\hat{x}^{(k)}(a)\delta^{(n-k)}(t-a), \tag{D-54}$$

where, for compactness, we have used the notation $[p^k\hat{x}(t)]_{t=a} = \hat{x}^{(k)}(a)$, which is, of course, the same as $[p^k\hat{x}(t)]_{t=a} = x^{(k)}(a)$.

---

[4] Often given in other texts as the formula $\int_{-\infty}^{\infty}\hat{x}(t)\delta(t-a) = \hat{x}(a)$.

[5] Recall that $\binom{n}{k} = \dfrac{n!}{k!(n-k)!}$, by definition.

Unlike those we have dealt with up until now, this definition might look singularly unmotivated; however, the next theorem shows why we have to do things in such a complicated way. It has a tedious, but straightforward, proof.

Theorem D.4   Definition D.16 is the only one possible, assuming that the sampling property in equation (D-52) and the Leibniz rule both hold.

Proof   We will prove this by assuming the truth of equation (D-52) and the Leibniz rule, and then showing that the truth of (D-54) is a logical consequence.

Our proof is by induction. We first observe that equation (D-54) becomes (D-52) when we set $n = 0$. Next, we assume that (D-54) is true for a general value of n and take the derivative of both sides (we can do this because, as we have already shown, the derivative of any gf exists as another gf), obtaining the equation

$$p[\hat{x}(t)\delta^{(n)}(t - a)] = \sum_{k=0}^{n} \binom{n}{k}(-1)^k \hat{x}^{(k)}(a)\delta^{(n-k+1)}(t - a). \tag{D-55}$$

To see that this is true, just notice that the delta functions are the only things to differentiate on the right-hand side (the only other items being multiplicative constants), and doing so increases the order of each by one.

Next, we apply the Leibniz rule to the left-hand side. This gives

$$p[\hat{x}(t)\delta^{(n)}(t - a)] = [p\hat{x}(t)]\delta^{(n)}(t - a) + \hat{x}(t)\delta^{(n+1)}(t - a). \tag{D-56}$$

Both terms on the right-hand side are well-defined gf's; furthermore, we know that the first term on the right-hand side is just (D-54) with $\hat{x}(t)$ replaced by $p\hat{x}(t)$. Setting the right-hand side of (D-56) equal to the right-hand side of (D-55), now, and using the comment we have just made gives

$$\sum_{k=0}^{n} \binom{n}{k}(-1)^k \hat{x}^{(k+1)}(a)\delta^{(n-k)}(t - a) + \hat{x}(t)\delta^{(n+1)}(t - a) \tag{D-57}$$

$$= \sum_{k=0}^{n} \binom{n}{k}(-1)^k \hat{x}^{(k)}(a)\delta^{(n-k+1)}(t - a).$$

Our next step is to solve for $\hat{x}(t)\delta^{(n+1)}(t - a)$:

$$\hat{x}(t)\delta^{(n+1)}(t - a) = \sum_{k=0}^{n} \binom{n}{k}(-1)^k \hat{x}^{(k)}(a)\delta^{(n-k+1)}(t - a) \tag{D-58}$$

$$- \sum_{k=0}^{n} \binom{n}{k}(-1)^k \hat{x}^{(k+1)}(a)\delta^{(n-k)}(t - a).$$

We next split off the $k = 0$ term in the first sum and change the summation index in the second from k to $q = k + 1$, to obtain

$$\hat{x}(t)\delta^{(n+1)}(t - a) = \hat{x}(a)\delta^{(n+1)}(t - a) + \sum_{k=1}^{n} \binom{n}{k}(-1)^k \hat{x}^{(k)}(a)\delta^{(n-k+1)}(t - a) \tag{D-59}$$

$$+ \sum_{q=1}^{n+1} \binom{n}{q-1}(-1)^q \hat{x}^{(q)}(a)\delta^{(n-q+1)}(t - a).$$

Because q is merely a dummy variable in the second sum, we can change it back to the symbol k, then split off the $(n + 1)$st term in it to give

$$\hat{x}(t)\delta^{(n+1)}(t - a) = \hat{x}(a)\delta^{(n+1)}(t - a) + \sum_{k=1}^{n}\left\{\binom{n}{k} + \binom{n}{k - 1}\right\}(-1)^{k}\hat{x}^{(k)}(a)\delta^{(n-k+1)}(t - a) \quad \text{(D-60)}$$

$$+ (-1)^{n+1}\hat{x}^{(n+1)}(a)\delta(t - a).$$

We are getting close. We next look at the sum of the two combinatorial functions in (D-60), writing out the definition of each and recombining to get

$$\binom{n}{k} + \binom{n}{k - 1} = \frac{n!}{k!(n - k)!} + \frac{n!}{(k - 1)!(n - k + 1)!} = \frac{n!}{(k - 1)!(n - k)!}\left[\frac{1}{k} + \frac{1}{n - k + 1}\right] \quad \text{(D-61)}$$

$$= \frac{n!}{(k - 1)!(n - k)!} \times \frac{n + 1}{k(n - k + 1)} = \frac{(n + 1)!}{k!(n + 1 - k)!} = \binom{n + 1}{k}.$$

Using this result in (D-60) and recombining the two "extra" terms into the sum, we get

$$\hat{x}(t)\delta^{(n+1)}(t - a) = \sum_{k=0}^{n+1}\binom{n + 1}{k}(-1)^{k}\hat{x}^{(k)}(a)\delta^{(n-k+1)}(t - a). \quad \text{(D-62)}$$

But this is just equation (D-54), which we were trying to prove, with n replaced by $n + 1$.

To summarize, we have shown that if we assume that the sampling property in equation (D-52) and the Leibniz rule are both valid, then (D-54) results as a logical consequence. Thus, our definition in Definition D.16 is the only one consistent with these assumptions. ∎

We should stress one point about Definition D.16 and Theorem D.4. We have actually used the fact that if the ordinary part of $p^n\hat{x}(t)$, namely $x^{(n)}(t)$, is continuous at $t = a$, then this property is also true for all lower-order derivatives as well.

Example D.10   Let $\hat{x}(t) = e^{-at}u(t)$ and $\hat{y}(t) = \delta^{(2)}(t - 1)$. Find their product.

Solution   We must compute the ordinary part of the first two derivatives of $\hat{x}(t)$. We find that

$$p\hat{x}(t) = -ae^{-at}u(t) + \delta(t) \quad \text{(D-63)}$$

and $\qquad\qquad p^2\hat{x}(t) = a^2e^{-at}u(t) - a\delta(t) + \delta'(t). \quad \text{(D-64)}$

In computing these generalized derivatives, we have simply followed the procedure we outlined in Example D.9. We see that the values of the ordinary parts of $\hat{x}(t)$ and its first two derivatives evaluated at $t = 1$ are

$$x(1) = e^{-a}, \quad \text{(D-65)}$$

$$x'(1) = -ae^{-a}, \quad \text{(D-66)}$$

and $\qquad\qquad x''(1) = a^2e^{-a}. \quad \text{(D-67)}$

(We have also used the fact that the values of the singularity functions $\delta(t)$ and $\delta'(t)$ are both zero.) Next, we write out the equation for $\hat{x}(t)\delta^{(2)}(t)$ using (D-54) with $n = 2$:

$$\hat{x}(t)\delta^{(2)}(t - 1) = \sum_{k=0}^{2} \binom{2}{k}(-1)^k x^{(k)}(1)\delta^{(2-k)}(t - 1) \tag{D-68}$$

$$= x(1)\delta^{(2)}(t - 1) - x'(1)\delta'(t - 1) + x''(1)\delta(t - 1)$$

$$= e^{-a}\delta^{(2)}(t - 1) + ae^{-a}\delta'(t - 1) + a^2 e^{-a}\delta(t - 1).$$

Now that we have defined the product of a generalized function and the general derivative of a singularity function, we can define the product of two generalized functions in the following way.

Definition D.17   Let $\hat{x}(t)$ and $\hat{y}(t)$ be two generalized functions with the canonical decompositions

$$\hat{x}(t) = x(t) + \sum_{i=0}^{\infty} \Delta_{xi}\delta^{(m_{xi})}(t - t_{xi}) \tag{D-69}$$

and

$$\hat{y}(t) = y(t) + \sum_{i=0}^{\infty} \Delta_{yi}\, d^{(m_{yi})}(t - t_{yi}), \tag{D-70}$$

respectively. Assume that these gf's have no singularities with the same occurrence times, that any discontinuities in $x(t)$ and $y(t)$ occur at different times, and that these discontinuities of both $x(t)$ and $y(t)$ occur at different times than the occurrence times of the singularities in the other gf. Then we define the product of $\hat{x}(t)$ and $\hat{y}(t)$ by

$$\hat{x}(t)\hat{y}(t) = x(t)y(t) + \sum_{i=0}^{\infty} \Delta_{xi}y(t)\delta^{(m_{xi})}(t - t_{xi}) + \sum_{i=0}^{\infty} \Delta_{yi}x(t)\delta^{(m_{yi})}(t - t_{yi}). \tag{D-71}$$

Each of the terms on the right-hand side of (D-71) is well defined; therefore, our definition is consistent both rigorously and intuitively with Definition D.16. In fact, if we write our gf's as $\hat{x}(t) = x(t) + x_s(t)$ and $\hat{y}(t) = y(t) + y_s(t)$, where $x_s(t)$ and $y_s(t)$ are the singular components of the two gf's $\hat{x}(t)$ and $\hat{y}(t)$, respectively, we see that Definition D.17 is equivalent to defining $\hat{x}(t)\hat{y}(t) = x(t)y(t) + x_s(t)y(t) + x(t)y_s(t) + x_s(t)y_s(t)$, where the last term $x_s(t)$ and $y_s(t)$ is defined to be the zero gf, $\hat{0} = (0, \varnothing)$. The complexity in carrying out the definition for any pair of gf's is the possible complexity in the cross product terms (for higher orders of singularities).

Integration by Parts for Generalized Functions

Theorem D.5 (Integration by Parts)   Let $\hat{x}(t)$ and $\hat{y}(t)$ be two generalized functions having the property that $\hat{x}(t)p\hat{y}(t)$ and $\hat{y}(t)p\hat{x}(t)$ are both well defined. Then

$$\frac{1}{p}[\hat{x}(t)p\hat{y}(t)] = \hat{x}(t)\hat{y}(t) - \frac{1}{p}[\hat{y}(t)p\hat{x}(t)]. \tag{D-72}$$

In terms of definite integrals, evaluating (D-72) at the end points of an interval $[a, b]$ gives

$$\int_a^b \hat{x}(t)p\hat{y}(t)\, dt = \hat{x}(t)\hat{y}(t)]_a^b - \int_a^b \hat{y}(t)p\hat{x}(t)\, dt. \tag{D-73}$$

Proof   The proof is quite easy. We just make the observation that if the two products $\hat{x}(t)p\hat{y}(t)$ and $\hat{y}(t)p\hat{x}(t)$ are both well defined, then so is the product $\hat{x}(t)\hat{y}(t)$ because the derivative of any gf has a singularity set that contains at least all of the singularities of the original function; furthermore, any discontinuities that are introduced in this process must have occurrence times that are different from those of either the discontinuities or the singularity functions of the other gf in the product. This means that the product $\hat{x}(t)\hat{y}(t)$ is well defined. Because it then certainly has a derivative (all gf's do)—and we are assured that the Leibniz rule holds by our definition of the product of two gf's—we can differentiate the product $\hat{x}(t)\hat{y}(t)$ to obtain

$$p[\hat{x}(t)\hat{y}(t)] = \hat{x}(t)p\hat{y}(t) + \hat{y}(t)p\hat{x}(t). \tag{D-74}$$

Taking the running integral of both sides, noting that the integral of the derivative on the left-hand side of (D-74) gives the product back, then moving the second term on the right side of the resulting equation to the left side gives (D-72). Finally, if we simply evaluate the result at $t = b$ and $t = a$ and subtract the latter from the former, we will obtain (D-73).   ∎

Example D.11   Compute the definite integral

$$\int_a^b \hat{x}(t)\delta^{(n)}(t - t_o)\, dt, \tag{D-75}$$

where the ordinary part of the nth derivative of the gf $\hat{x}(t)$, $p^n\hat{x}(t)$, has no singularities at $t = t_o$ and whose ordinary part $x^{(n)}(t)$ exists and is continuous at $t = t_o$. Assume $a < t_o < b$.

Solution   We simply use integration by parts, obtaining

$$\int_a^b \hat{x}(t)\delta^{(n)}(t - t_o)\, dt = \hat{x}(t)\delta^{(n-1)}(t - t_o)\Big]_a^b \tag{D-76}$$

$$-\int_a^b [p\hat{x}(t)]\delta^{(n-1)}(t - t_o)\, dt = -\int_a^b [p\hat{x}(t)]\delta^{(n-1)}(t - t_o)\, dt.$$

We have used the facts that all of the products involved are well defined by the assumptions in our theorem and that the integrated term has a value of zero at the two limits. Repeating this n times gives

$$\int_a^b \hat{x}(t)\delta^{(n)}(t - t_o)\, dt = (-1)^n\int_a^b [p^n\hat{x}(t)]\delta(t - t_o)\, dt = (-1)^n[p^n\hat{x}(t_o)u(t - t_o)]_a^b. \tag{D-77}$$

In the last step, we used the sampling property to express $p^n\hat{x}(t)\delta(t - t_o)$ as $p^n\hat{x}(t_o)\delta(t - t_o)$, then factored the constant term $p^n\hat{x}(t_o)$ from the integral. Then we observed that the running integral of the delta function is the unit step function. Recognizing that the constant t_o is between a and b gives $u(b - t_o) = 1$ and $u(a - t_o) = 0$; furthermore, we know that the value of $\hat{x}(t_o) = x^{(n)}(t_o)$. Thus, we have[6]

$$\int_a^b \hat{x}(t)\delta^{(n)}(t - t_o)\, dt = (-1)^n p^n\hat{x}(t_o) = (-1)^n x^{(n)}(t_o). \tag{D-78}$$

---

[6] This is a generalization of the formula given in footnote 4 in this appendix.

The Concept of a
Generalized Limit

The conventional approach used in most circuits and systems texts to define the unit impulse function is to treat it as a limit of a sequence of ordinary approximating functions. This limit does not exist in the usual sense, so that approach is not rigorous at all; however, it does provide an intuitive feeling for how one might approximate an impulse function in the laboratory. Thus, we will develop a rigorous justification of this idea as our final topic in this appendix.

Our first step in defining the limit of a gf is to note that every generalized function is the derivative (of some order) of an ordinary (K_∞) function. To see this, consider

$$\hat{x}(t) = u(t) + \delta'(t). \tag{D-79}$$

This gf is of order $N = 2$ because two integrations are necessary to convert it to an ordinary K_∞ waveform. We have not made much explicit use of the order before, so let's compute the first two running integrals and see what happens. We obtain

$$\frac{1}{p}\hat{x}(t) = tu(t) + \delta(t) \tag{D-80}$$

and

$$\frac{1}{p^2}\hat{x}(t) = 0.5t^2u(t) + u(t). \tag{D-81}$$

Thus, after $N = 2$ integrations the resulting waveform is an ordinary K_∞ function. Then two derivatives of this function gives us the original generalized function again. Though this result is quite obvious, it is important, so it deserves to be called a theorem.

Theorem D.6   Each generalized function of order N is the Nth derivative of an ordinary K_∞ waveform $f(t)$; that is, there exists a K_∞ waveform $f(t)$ such that

$$\hat{x}(t) = p^N f(t). \tag{D-82}$$

Before we discuss how to apply this result in the computation of limits, we first ask ourselves why we cannot define limits directly in terms of the gf itself. Here is the reason. Consider the generalized function

$$\hat{x}_\epsilon(t) = \delta(t) + \delta(t - \epsilon). \tag{D-83}$$

It is plotted in Figure D.13. What would be the result of attempting to take the limit as ϵ approaches zero? Nothing much, actually, because two impulse functions—even if they are close together—are still two impulse functions, and nothing else. Suppose, however, that we compute the first running integral (noting that $\hat{x}_\epsilon(t)$ is of order zero). We get

$$\frac{1}{p}\hat{x}_\epsilon(t) = \frac{1}{p}\delta(t) + \frac{1}{p}\delta(t - \epsilon) = u(t) + u(t - \epsilon). \tag{D-84}$$

Figure D.13
A parameterized gf

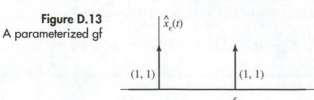

Figure D.14
After one integration

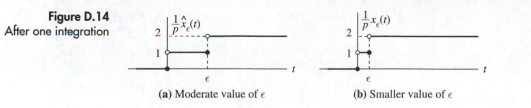

(a) Moderate value of ϵ **(b)** Smaller value of ϵ

This waveform is sketched in Figure D.14. Can you see the improvement in what we would normally consider to be a limiting operation? Now, as $\epsilon \to 0$, we see that the subject waveform does, indeed, approach a conventional waveform more and more closely: $2u(t)$. We would like to say that the limit of $\hat{x}_\epsilon(t)$ is $2\delta(t)$, but we must work with the integrated waveform to do so. In the process, we are generalizing our concept of limit, just as we had to generalize our concept of function in order to make the FTC work. We will now do that, in the form of the following definition.

Definition D.18   Let $\hat{x}_\epsilon(t)$ be a generalized function for each value of the parameter ϵ. Suppose that there is a smallest integer m (independent of ϵ) for which the following two properties are true:

1. $f_\epsilon(t) = \dfrac{1}{p^m}\hat{x}_\epsilon(t)$ is an ordinary K_∞ waveform for each value of ϵ.

2. $f(t) = \lim\limits_{\epsilon \to \epsilon_0} f_\epsilon(t)$ exists, where $f(t)$ is an ordinary K_∞ waveform.

Then we say that $\hat{x}(t) = \lim\limits_{\epsilon \to \epsilon_0} \hat{x}_\epsilon(t)$ exists, where $\hat{x}(t) = p^m f(t)$.

We would like to point out that m might not be the order of the generalized function $\hat{x}_\epsilon(t)$. More integrations might be necessary in order to have convergence occur.

Example D.12   Determine whether or not the limit of the parameterized waveform

$$x_\epsilon(t) = \frac{u(t) - u(t - \epsilon)}{\epsilon} \tag{D-85}$$

exists. If it does, compute the limit.

Solution   Notice that this is the difference quotient of the unit step function. The difference quotient exists for any waveform—and the limit, if one exists in the usual sense, is the ordinary derivative of the waveform in question. The waveform for $x_\epsilon(t)$ is shown in Figure D.15. If you pick any $t \neq 0$, you will see that the ordinary limit as $\epsilon \to 0$ is zero. (We can assume that this is true at $t = 0$ also because of our left-continuity assumption.) Thus, the ordinary derivative of $u(t)$ is the identically zero function. What we would like to show is that the *generalized* limit and derivative do, indeed, exist—everywhere.

Figure D.15
The difference quotient of the unit step function and its integral

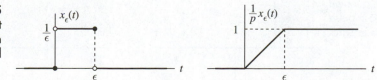

Let's see what happens after we integrate one time. We get

$$\frac{1}{p}\hat{x}_\epsilon(t) = \frac{1}{p}\left\{\frac{1}{\epsilon}u(t) - \frac{1}{\epsilon}u(t - \epsilon)\right\} \tag{D-86}$$

$$= \frac{1}{\epsilon}tu(t) - \frac{1}{\epsilon}(t - \epsilon)u(t - \epsilon)$$

$$= \frac{1}{\epsilon}t\{u(t) - u(t - \epsilon)\} + u(t - \epsilon).$$

This waveform is plotted in Figure D.15. It clearly has the limit

$$\lim_{\epsilon \to 0}\frac{1}{p}x_\epsilon(t) = u(t). \tag{D-87}$$

(Notice that the value is zero at $t = 0$, consistent with our requirement that all waveforms be continuous from the left.) Therefore, the limit in question does, indeed, exist, and its value is

$$\hat{x}(t) = \lim_{\epsilon \to 0}\hat{x}_\epsilon(t) = pu(t) = \delta(t). \tag{D-88}$$

Did you notice that the parameterized waveform in the last example had order $N = 0$ for each value of ϵ? It was an ordinary function, so no integrations were required to turn it into one. However, one integration was required to make the ordinary limit exist everywhere, so the limit was a generalized function (in this case a pure singularity function) of order one. We can see now that if ϵ is small, $x_\epsilon(t)$ in our example makes a good approximation to the unit impulse function. We note here that our definition of the limit of a generalized function provides the generalization of infinite series promised earlier because an infinite series is the limit of a sequence of partial sums.

Example D.13   Determine whether or not the limit of the parameterized generalized function

$$\hat{x}_\epsilon(t) = \frac{\delta(t) - \delta(t - \epsilon)}{\epsilon} \tag{D-89}$$

exists. If it does, compute the limit.

Solution   Notice that this is the difference quotient of the unit impulse function. Its waveform is shown in Figure D.16. You undoubtedly already know what the limit will be, but let's dot the i's and cross the t's. We have already observed that waveforms such as those of $\hat{x}_\epsilon(t)$ do not have an ordinary limit. Because both impulses are singularity functions of first order, we try integrating once. The result is shown in Figure D.16 on the right. We have al-

Figure D.16
The difference quotient
of the unit impulse
function and its integral

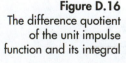
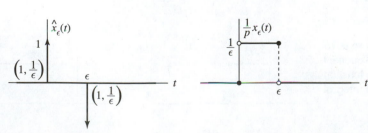

ready seen in our last example that the ordinary limit of this waveform does not exist at $t = 0$; however, we did the *integral* of that waveform in that example and showed that its limit was $u(t)$, the unit step function. Thus, we now see that the *second* integral of our present parameterized generalized function $\hat{x}_\epsilon(t)$ is the first integral of the difference quotient of the unit step function,

$$\frac{1}{p^2}\hat{x}_\epsilon(t) = \frac{1}{p^2}\left[\frac{1}{\epsilon}\delta(t) - \frac{1}{\epsilon}\delta(t - \epsilon)\right] = \frac{1}{p}\frac{u(t) - u(t - \epsilon)}{\epsilon}; \qquad \text{(D-90)}$$

furthermore, we have already shown that the limit of this K_∞ function is the unit step function. Thus, for our present waveform, we have

$$\lim_{\epsilon \to 0}\frac{1}{p^2}x_\epsilon(t) = u(t). \qquad \text{(D-91)}$$

Therefore, the generalized limit exists and

$$\hat{x}(t) = \lim_{\epsilon \to 0}\hat{x}_\epsilon(t) = p^2 u(t) = \delta'(t). \qquad \text{(D-92)}$$

That is, our limit is the first derivative of the unit impulse function.

We could readily use what we have developed thus far to show that the limit of the difference quotient of *any* generalized function exists in the generalized sense and that this limit is the generalized derivative of the given gf. We will, however, leave this for you to do. (If you have read and understood what we have done in this appendix, you have the interest and the ability to prove this, and the author offers his congratulations!)

CHAPTER 1

Section 1.1

1.1-1.

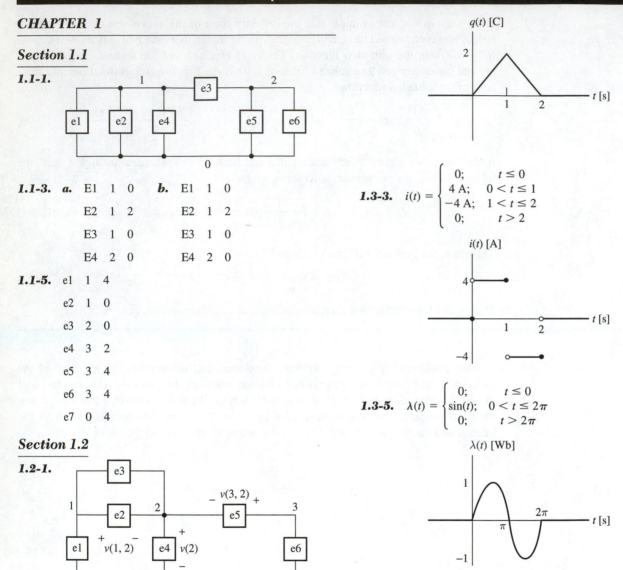

1.1-3. **a.**

E1	1	0
E2	1	2
E3	1	0
E4	2	0

b.

E1	1	0
E2	1	2
E3	1	0
E4	2	0

1.1-5.

e1	1	4
e2	1	0
e3	2	0
e4	3	2
e5	3	4
e6	3	4
e7	0	4

Section 1.2

1.2-1.

1.2-3. 3×10^{-15} MV, 3×10^{-12} kV, 3×10^{-6} mV, 3×10^{-3} μV, 3×10^{3} pV

1.2-5. 10^{13} pA

1.2-7. $t = 0.843 \times 10^{-10}$ S, 300 m

1.2-9.

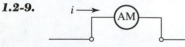

Section 1.3

1.3-1. $q(t) = \begin{cases} 0; & t \le 0 \\ 2t; & 0 < t \le 1 \\ -2t; & 1 < t \le 2 \\ 0; & t > 2 \end{cases}$

1.3-3. $i(t) = \begin{cases} 0; & t \le 0 \\ 4\ \text{A}; & 0 < t \le 1 \\ -4\ \text{A}; & 1 < t \le 2 \\ 0; & t > 2 \end{cases}$

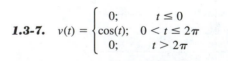

1.3-5. $\lambda(t) = \begin{cases} 0; & t \le 0 \\ \sin(t); & 0 < t \le 2\pi \\ 0; & t > 2\pi \end{cases}$

1.3-7. $v(t) = \begin{cases} 0; & t \le 0 \\ \cos(t); & 0 < t \le 2\pi \\ 0; & t > 2\pi \end{cases}$

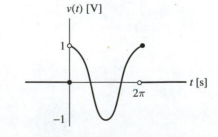

1.3-9. 10 nC

1.3-11. $q(t) = \dfrac{3t^2 - 10t + 15}{2}$

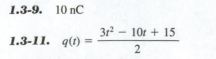

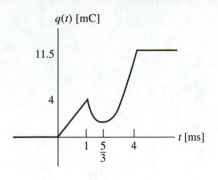

1.3-13. 0.35 s

1.3-15. $v(t) = \begin{cases} 0; & t \le 0 \\ 8\,v; & 0 < t \le 1 \\ -8\,v; & 1 < t \le 2 \\ 0; & t > 2 \end{cases}$

$v(t)$ [kV]

8

t [μs]

1   2

-8

Section 1.4

1.4-1. $P_{abs}(t) = 20\,u(t)$ W

1.4-3. $P(0.5\ \text{s}) = 10e^{-1}$ mW;
$w(t) = 5(1 - e^{-2t})\,u(t)$ mJ $- 5$ mJ

$w(t)$ [mJ]

5

t [s]

1.4-5. $\lambda(t) = tu(t)$ Wb; $q(t) = 5(1 - e^{-2t})\,u(t)$ mC

1.4-7. $P_{abs}(t) = -20\,u(t)$ mW

1.4-9. $P_{abs}(t) = 2i^2(t)$ W; $w(t) = (2/3)\,i^3(t)$ J

1.4-11. $P(t) = 10\,i(t)\,u(t)$

$i(t) = \begin{cases} 0; & t \le 0 \\ (1/10)\,P(t); & 0 < t \le 2 \\ \text{undetermined}; & t > 2 \end{cases}$

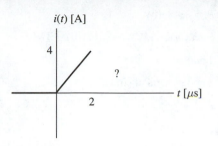

1.4-13. $w(t) = -2t\,u(t)$ J < 0, $t > 0$

Section 1.5

1.5-1. **a.** Allowed   **b.** Disallowed   **c.** Allowed
d. Disallowed

1.5-3. $i(e3) = -8$ A, $i(e2) = 8$ A, $i(e1) = -18$ A

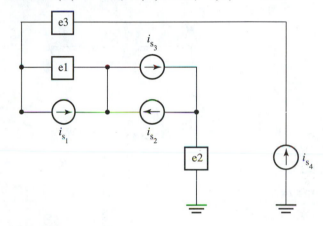

1.5-5. i_{s_1} 2   3   2

i_{s_2} 0   4   4

i_{s_3} 2   1   6

e_1 2   0

e_2 3   4

e_3 3   4

CHAPTER 2

Section 2.1

2.1-1. $5 \times 10^5\ \Omega$,   500 kΩ,   0.5 MΩ

2.1-3. $0.1\,\dfrac{\Omega}{m}$, $0.03\,\dfrac{\Omega}{ft}$

2.1-5. **a.** 1000 V   **b.** 10^6 V   **c.** 10^{15} V
d. $v = \infty$ as $R \to \infty$ **e.** Only if driven by i-source
f. $V = 0$ as $R \to 0$

2.1-7. Req $= -3$ kΩ

2.1-9. $V_s = -3$ V

2.1-11. $P = -18$ mW

2.1-13. $\cos^2(10t)u(t)$ W, $-\cos^2(10t)u(t)$ W

2.1-15. 4 nodes

2.1-17. 2 nodes, parallel circuit

Section 2.2

2.2-1. $i_x = -4$ A

2.2-3. $i_x = 2$ A, $v_x = 8$ V

2.2-5. $i_x = 0$ A

2.2-7. $v_y = 38$ V

2.2-9. $i_y = 0$ A

2.2-11. **a.** 12 A, 1200 A, 12 kA
b. 12 A regardless of R

2.2-13. **a.**   **b.**   **c.**

2.2-15. $P_{SR} = 20$ W; $P_{SL} = 80$ W; $P_4 = 400$ W;
$P_{6.25} = 400$ W; $P_{60\,V} = -720$ W; $P_{6\,A} = -180$ W;
$P_{tot} = 0$ W

2.2-17. -4 A, 8 V

Section 2.3

2.3-1. (b)

2.3-3. 6 V

2.3-5. 2 V

2.3-7. $v_x = 8$ V; $v_y = -8$ V; $v_{xy} = 16$ V

2.3-9. $v_7 = 14$ V; $v_4 = 8$ V; $v_2 = 4$ V;

Section 2.4

2.4-1. Open circuit

2.4-3. Req $= 2\,\Omega$; $i_x = -3$ A

2.4-5. $v_x = 9$ V, $v_4 = 12$ V

2.4-7. $2\,\Omega$

2.4-9. $i_x = i_s$; $i_y = 0$

2.4-11. $i_{4L} = 3$ A; $i_{4R} = 3$ A

CHAPTER 3

Section 3.1

3.1-1. $i_x = 3$ A

3.1-3. $i_x = 1$ A

3.1-5. $R_{eq} = 8\,\Omega$; $v_s = -6$ V

3.1-7. 6 V

Section 3.2

3.2-1. $i_x = 3$ A

3.2-3. $i_x = -4$ A

3.2-5. $i = -6$ A

Section 3.3

3.3-1. **a.** $v_{oc} = 6$ V, $R_{eq} = 4\,\Omega$

b. $i_{sc} = 1.5$ A, $R_{eq} = 4\,\Omega$

3.3-3. $i_x = -4$ A

3.3-5. $i = -6$ A

3.3-7. $v_{oc} = -16$ V

3.3-9. $R_{eq} = 2\,\Omega$

Section 3.4

3.4-1. **a.** 0 W   **b.** 0.22 kW   **c.** 0.25 kW
d. 0.24 kW   **e.** 0.22 kW

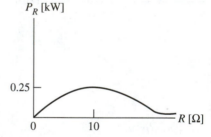

3.4-3. $R = 20\,\Omega$

Section 3.5

3.5-1. $i_x = 3$ A

3.5-3. $v = 0$

3.5-5. 8 V

Section 3.6

3.6-1. $v_s = -36$ V

CHAPTER 4

Section 4.1

4.1-1. **a.** $\dfrac{v_1}{4} = 2$ **b.** $\dfrac{v_0}{4} = -2$

4.1-3. **a.** $\dfrac{v_1}{8} + \dfrac{v_1 - v_2}{3} = 3$; $\dfrac{v_2}{2} + \dfrac{v_2 - v_1}{3} = -1$

b. $\dfrac{v_0}{8} + \dfrac{v_0 - v_2}{2} = 1 - 3$; $\dfrac{v_2}{3} + \dfrac{v_2 - v_0}{2} = -1$

c. $\dfrac{v_1}{3} + \dfrac{v_1 - v_0}{8} = 3;\quad \dfrac{v_0}{2} + \dfrac{v_0 - v_1}{8} = 1 - 3$

4.1-5. a. $\dfrac{v_1}{2} + \dfrac{v_1 - v_2}{3} = 3;\quad \dfrac{v_2 - v_1}{3} + \dfrac{v_2 - v_3}{3} = -4;$

$\dfrac{v_3}{4} + \dfrac{v_3 - v_2}{3} = 4$

b. $\dfrac{v_0}{2} + \dfrac{v_0 - v_3}{4} = -3;\quad \dfrac{v_3 - v_0}{4} + \dfrac{v_3 - v_2}{3} = 4;$

$\dfrac{v_2}{3} + \dfrac{v_2 - v_3}{3} = -4$

c. $\dfrac{v_0 - v_1}{2} + \dfrac{v_0 - v_3}{4} = -3;\quad \dfrac{v_1 - v_0}{2} + \dfrac{v_1}{3} = 3;$

$\dfrac{v_3}{3} + \dfrac{v_3 - v_0}{4} = 4$

d. $\dfrac{v_0}{4} + \dfrac{v_0 - v_1}{2} = -3;\quad \dfrac{v_1 - v_0}{2} + \dfrac{v_1 - v_2}{3} = 3;$

$\dfrac{v_2}{3} + \dfrac{v_2 - v_1}{3} = -4$

4.1-7. No unique solution because deactivated network not connected

Section 4.2

4.2-1. a. 1 ordinary, 1 generalized

b. 1 essential, 1 nonessential

c. 1 supernode

4.2-3. $v_0 = -26$ V

4.2-5. $v_0 = 32$ V, $\quad v_2 = 24$ V

4.2-7. $v_1 = 4$ V, $\quad v_0 = -4$ V

4.2-9. $\dfrac{108}{19}$ V, $\quad$ 8 kΩ

Section 4.3

4.3.1. a. $3i = 9,\quad i = 3$ A
b. $3i = -9,\quad i = -3$ A
4.3-3. a. 2 nodal equations; $\quad$ 3 mesh equations
b. $i_1 = 4$ A; $\quad i_2 = 2$ A; $\quad i_3 = 1$ A
c. 4 V

4.3-5. -4 A, $\quad$ 12 V

4.3-7. $\dfrac{108}{19}$ V, $\quad$ 8 kΩ

Section 4.4

4.4-1. a. $N_{NE} = 3$
b. $N_{ME} = 1$
c. 1 essential, 3 nonessential
d. $2(i - 3) + 3(i - 2) + 3(i + 4) + 2i = 0$
e. $i = 0,\quad v = -6$ V

4.4-3. 48 V

4.4-5. 16 V

4.4-7. -26 V

Section 4.5

4.5-1. 8 V, $\quad$ 2 V

4.5-3. 26 V

4.5-5. 4 V

4.5-7. -16 V

4.5-9. 4 A, $\quad$ 2 A, $\quad$ 1 A; $\quad$ 4 V

4.5-11. 12 V, $\quad -4$ A

4.5-13. 5 A, $\quad$ 2 A

4.5-15. -4 A

4.5-17. $i_1 = -1$ A, $i_2 = -\dfrac{5}{2}$ A, $v = -3$ V

4.5-19. $v = -3$ V

CHAPTER 5

Section 5.1

5.1-1. $i = -1$ A

5.1-3. $v_0(t) = 4 \sin(\omega t)$ V

5.1-5. a. $i = -1$ A $\qquad$ **b.** $\mu = -3$

5.1-7. a. $i = 1$ A $\qquad$ **b.** Solvable for all μ

Section 5.2

5.2-1. 6 V

5.2-3. 3 A

5.2-5. 6 V

5.2-7. 6 Ω

5.2-9. $R_{eq} = -1\ \Omega$, $\quad V_{oc} = -6$ V

5.2-11. 78 V

Section 5.3

5.3-1. $v = 6$ V

5.3-3. 78 V

5.3-5. $R_{eq} = -1\ \Omega$, $\quad V_{oc} = -6$ V

5.3-7. 6 V

5.3-9. 78 V

Section 5.4

5.4-1. 0 V

5.4-3. 5 kΩ

5.4-5. $R_{eq} = \dfrac{R(-R_x)}{R - R_x}$

Section 5.5

5.5-1. Stable

5.5-3. Stable for both cases

Section 5.6

5.6-1. 28 V

5.6-3. 8 V

5.6-5. $k < 5$

5.6-7. 24 V

5.6-9. $v_0 = \dfrac{k}{5 - k}\, v_i$

Section 5.7

5.7-1. 0.334 mA

5.7-3. 0.645 V

5.7-5. $A_v = 9.43$

5.7-7. $R_m = -100 \text{ k}\Omega \qquad R_{in} = 225\ \Omega$

CHAPTER 6

Section 6.1

6.1-1. *a.* $i_x = \dfrac{1}{90} v_s + \dfrac{1}{3} i_s$ *b.* $i_x(t) = 3\, u(t)$ A

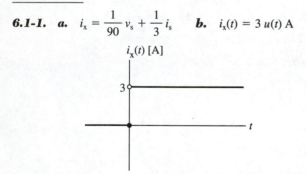

6.1-3. $v = \dfrac{1}{2} v_c + i_s; \quad v = \dfrac{3}{2} i_s; \quad v(t) = 6 \sin(3t)$ V

Section 6.2

6.2-1. $q = 20\ \mu C; \quad w = 100\ \mu J$

6.2-3. $i(t) = 1000\, u(t)$ mA $w(t) = 80$ nJ

6.2-5. $p\, x(t) = 10\, u(t)$

$$\frac{1}{p}\,[p\, x(t)] = 10t\, u(t)$$

$$\frac{1}{p}\, x(t) = \begin{cases} 0, & t \le 0 \\ 5t^2, & t > 0 \end{cases}$$

$$p\left[\frac{1}{p}\, x(t)\right] = 10t\, u(t)$$

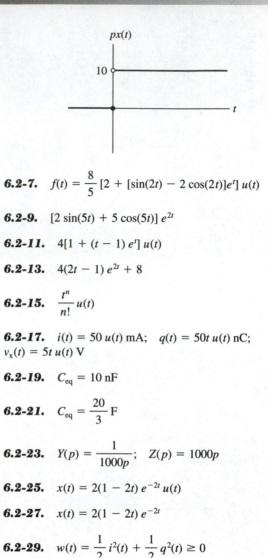

6.2-7. $f(t) = \dfrac{8}{5}\,[2 + [\sin(2t) - 2\cos(2t)]e^t]\, u(t)$

6.2-9. $[2\sin(5t) + 5\cos(5t)]\, e^{2t}$

6.2-11. $4[1 + (t - 1)\, e^t]\, u(t)$

6.2-13. $4(2t - 1)\, e^{2t} + 8$

6.2-15. $\dfrac{t^n}{n!}\, u(t)$

6.2-17. $i(t) = 50\, u(t)$ mA; $q(t) = 50t\, u(t)$ nC; $v_x(t) = 5t\, u(t)$ V

6.2-19. $C_{eq} = 10$ nF

6.2-21. $C_{eq} = \dfrac{20}{3}$ F

6.2-23. $Y(p) = \dfrac{1}{1000p}; \quad Z(p) = 1000p$

6.2-25. $x(t) = 2(1 - 2t)\, e^{-2t}\, u(t)$

6.2-27. $x(t) = 2(1 - 2t)\, e^{-2t}$

6.2-29. $w(t) = \dfrac{1}{2}\, i^2(t) + \dfrac{1}{2}\, q^2(t) \ge 0$

Section 6.3

6.3-1. $\lambda = 4$ Wb

6.3-3.

6.3-5. $L_{eq} = 6$ H

6.3-7. $i(t) = 2 \sin(2t) \, u(t)$ A

$P(t) = 2\delta(t) + 8\sin(4t) \, u(t)$ W

$w(t) = 2[1 - \cos(4t)] \, u(t)$ J

Section 6.4

6.4-1. $v = 16$ V

$i = 4$ A

6.4-3. $i = 4$ A

6.4-5. 12 V

6.4-7. No dc steady state

Section 6.5

6.5-1. $v(t) = 5t \, u(t)$ V

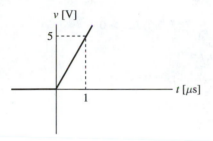

6.5-3. $v(t) = \begin{cases} 6 \text{ V}; & t \le 0 \\ \dfrac{10}{3} u(t) \text{ V}; & t > 0 \end{cases}$

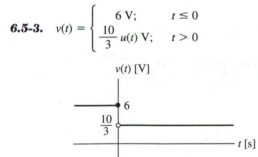

6.5-5. $v(t) = \begin{cases} 36 \text{ V}; & t \le 0 \\ 24 \, u(t) \text{ V}; & t > 0 \end{cases}$

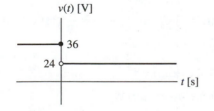

6.5-7. $w(t) = \begin{cases} \dfrac{1}{2} C V_0^2; & t = 0- \\ \dfrac{1}{4} C V_0^2; & t = 0+ \end{cases}$

Section 6.6

6.6-1. $\dfrac{1}{p} x(t) = 1$

6.6-3. $v(t) = 24 \, \delta(t)$

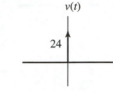

6.6-5. $2 \, \delta(t - 3)$

6.6-7. $2 \, u(t - 1) - 4 \, u(t - 2) + 2 \, u(t - 3)$

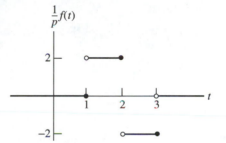

6.6-9. $-4e^{-(t-2)} u(t - 2) + 4 \, \delta(t - 2)$

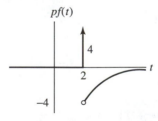

6.6-11.

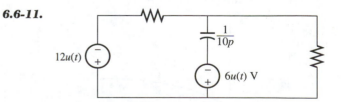

6.6-13. **a.** $v_{oc} = 12 \, \delta(t)$ V, $R_{eq} = \dfrac{14}{9} p$

b. $i_{sc} = \dfrac{54}{7} u(t)$ A, $R_{eq} = \dfrac{14}{9} p$

6.6-15. $u(t) + 2u(t-1) + u(t-2) - u(t-3) - 2u(t-4) - u(t-5)$

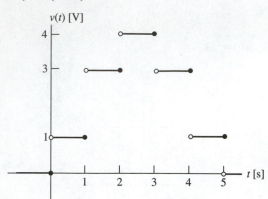

Section 6.7

6.7-1. $L_{eq} = (\beta + 1)\,\mu H$

$L = 1\,H, \quad L = 0\,H, \quad L = -1\,H$

6.7-3.

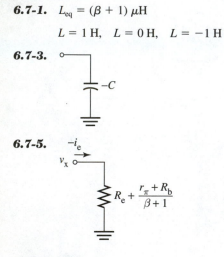

$-C$

6.7-5.

$-i_e$

v_x

$R_e + \dfrac{r_\pi + R_b}{\beta + 1}$

CHAPTER 7

Section 7.1

7.1-1. $s(t) = (1 - e^{-4t})\,u(t)\,A; \quad \tau = \dfrac{1}{4}\,sec$

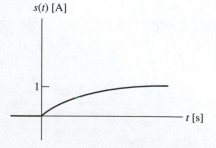

$s(t)$ [A]

7.1-3. $i(t) = 16t\,e^{-4t}\,u(t)\,A$

7.1-5. $y(t) = \dfrac{1}{4}[(2t^2 - 2t + 1) - e^{-2t}]\,u(t); \quad \tau = \dfrac{1}{2}\,sec$

7.1-7. $s(t) = \dfrac{1}{3}(1 - e^{-2t})\,u(t)\,V; \quad \tau = \dfrac{1}{2}\,sec$

7.1-9. $v(t) = 2\,[\sin(2t) - \cos(2t) + e^{-2t}]\,u(t)\,V$

7.1-11. $10\,V; \quad 18.42\,sec$

7.1-13. $v(t) = 10\,[u(t) - u(t-1)] - 10\,[u(t) - e^{1/4}\,u(t-1)]\,e^{-t/4}\,V$

$\qquad i(t) = 5\,[u(t) - e^{1/4}\,u(t-1)]\,e^{-t/4}\,A$

7.1-15. $v(t) = 2(1 + e^{-2t})\,u(t)\,V$

7.1-17. **a.** $(1 + e^{-4t})\,u(t)\,V \qquad$ **b.** $(4 - 2e^{2t})\,u(t)\,V$
c. $2(1 - 2t)\,u(t)\,V \qquad$ **d.** $R = 2\,\Omega$

7.1-19. **a.** $E_{Left\,C} = \dfrac{1}{2}\,CV^2,\ E_{Right\,C} = 0$

b. $i(t) = 0\,A; t \le 0$

$\qquad i(t) = \dfrac{V}{R}\,e^{(-2/RC)t}u(t)\,A; t > 0$

c. $E_R = 0; t \le 0$

$\qquad E_R = -\dfrac{CV^2}{4}\,[e^{(-4/RC)t} - 1]\,u(t); t > 0$

Section 7.2

7.2-1. **a.** $y(t) = 6\,e^{-3t}\,u(t)$
b. $y(t) = 12(1 - 3t)\,e^{-3t}\,u(t)$

7.2-3. $y_N(t) = 18\,e^{-3t}\,u(t); \quad y_F(t) = -6\,e^{-t}\,u(t)$

7.2-5. $v(t) = 6\,I_M\left[\dfrac{\omega^2\cos(\omega t) - 3\,\omega\sin(\omega t)}{\omega^2 + 9}\right]u(t) +$

$\qquad \dfrac{54\,I_m}{\omega^2 + 9}\,e^{-3t}\,u(t)$

7.2-7. $v(t) = e^{-t/20}\,u(t)\,V$

Section 7.3

7.3-1. $v(t) = \dfrac{1}{2}\,\dfrac{p+4}{p+2}\,v_s$

7.3-3. $y_N(t) = \dfrac{2}{9}\,e^{-3t}\,u(t), \quad y_F(t) = \dfrac{1}{9}(15t - 2)\,u(t); \quad$ Yes, No

7.3-5. $v(t) = e^{-2t}\,u(t)\,V$

7.3-7. $4\,\delta(t) - 16\,e^{-4t}\,u(t)\,V$

7.3-9. $s(t) = [1 + e^{-2t}]\,u(t); \quad h(t) = 2\,\delta(t) - 2\,e^{-2t}\,u(t)$

7.3-11. $i(t) = \begin{cases} \dfrac{1}{2}; & t \le 0 \\[2mm] 2 - \dfrac{3}{2}\,e^{-4t}; & t > 0 \end{cases}$

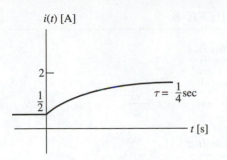

Section 7.4

7.4-1.

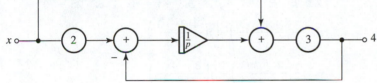

7.4-3.

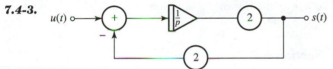

7.4-5. Stable; $H(p) = \dfrac{1}{(p+4)^3}$

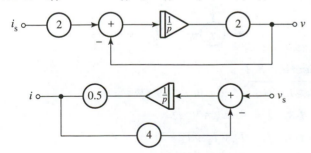

7.4-7. $i(t) = 4t\, e^{-2t}\, u(t)$ A, $\quad v(t) = 2(1 - e^{-2t})\, u(t)$ V

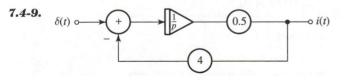

7.4-9.

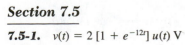

Section 7.5

7.5-1. $v(t) = 2\,[1 + e^{-12t}]\, u(t)$ V

7.5-3. $H(p) = -2\left(\dfrac{p + \dfrac{1}{80}}{p + \dfrac{1}{400}} \right)$

$s(t) = -(10 - 8e^{-t/400})\, u(t)$ V

$h(t) = -2\,\delta(t) + 0.02\, e^{-t/400}\, u(t)$ V

$v(t) = -4e^{-t/400}\, u(t)$ V

7.5-5. $H(p) = \dfrac{p - 1}{p + 1}$

$h(t) = \delta(t) - 2e^{-t}\, u(t)$ V

$s(t) = (2e^{-t} - 1)\, u(t)$ V

$v(t) = (1 - 2t)\, e^{-t}\, u(t)$ V

7.5-7. **a.** $v_x(t) = -245(2 - e^{-t})\, u(t)$ mV,
$v_y(t) = -245\, e^{-t}\, u(t)$ mV
b. $v_x(t) = -245(1 + t)\, e^{-t}\, u(t)$ mV,
$v_y(t) = -245(1 - t)\, e^{-t}\, u(t)$ mV

7.5-9. $h(t) = \dfrac{-C_o}{C_\pi + C_o}\,\delta(t) +$

$\left(g_m + \dfrac{C_o}{C_\pi + C_o}\, g_\pi \right) e^{-(g_\pi/c_\pi + c_o)t} u(t)$

Section 7.6

7.6-1. $500t\, u(t)$ A

$100(1 - e^{-5t})\, u(t)$ A

$0 < t \ll \dfrac{1}{5}$ sec

Section 7.7

7.7-1. $v(t) = 4(1 - e^{-2t})$ V

$i(t) = 2(3 + e^{-t})$ A

7.7-3. **a.** Stable

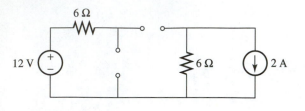

b. 12 V

c. $i(t) = \begin{cases} -2 \text{ A}, & t \leq 0 \\ 2e^{-2t} \text{ A}, & t > 0 \end{cases}$

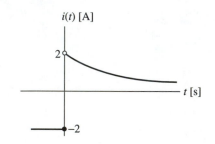

7.7-5. $v(t) = \begin{cases} 72 \text{ V}, & t \leq 0 \\ \dfrac{8}{3}(35 + 4e^{-2t}) \text{ V}, & t > 0 \end{cases}$

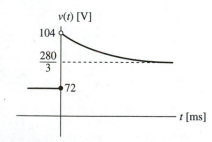

7.7-7. $v(t) = \begin{cases} 20 \text{ mV}, & t \leq 0 \\ 10(3 - 2e^{-t/20})\, u(t) \text{ mV}, & t > 0 \end{cases}$

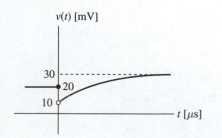

CHAPTER 8

Section 8.1

8.1-1. $z = \pm j$

8.1-3. $z = -\dfrac{1}{2} \pm j\dfrac{\sqrt{3}}{2}$

Section 8.2

8.2-1. $z_1 = -1 - j$, $z_2 = 1 + j$, $z_3 = 3 - j4$, $z_4 = -\sqrt{3} + j$

a. $0 + j0$ **b.** $0 - j2$ **c.** $6 + j0$

d. $(7\sqrt{3} - 1) + j(7 + \sqrt{3})$

8.2-3. **a.** $-7 + j$ **b.** $-1 + j0$ **c.** $25 + j0$

d. $-\left(\dfrac{4 + 3\sqrt{3}}{4}\right) + j\left(\dfrac{4\sqrt{3} - 3}{4}\right)$

e. $-\left(\dfrac{4 - 3\sqrt{3}}{2}\right) - j\left(\dfrac{3 + 4\sqrt{3}}{2}\right)$

8.2-5. $z_1 = 2e^{j(\pi/5)}$, $z_2 = 2e^{j(3\pi/5)}$, $z_3 = 2e^{j\pi}$, $z_4 = 2e^{j(7\pi/5)}$, $z_5 = 2e^{j(9\pi/5)}$

Section 8.3

8.3-1. $w(t) = e^{-4t}\sin(4t)\,u(t) + je^{-4t}[1 - \cos(4t)]\,u(t)$

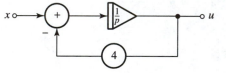

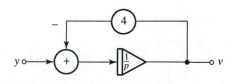

Section 8.4

8.4-1. $v(t) = 5\cos(10t - 36.9°)$ V

8.4-3. $\bar{I} = -10\sqrt{2} + 10\sqrt{2}$ A

8.4-5. $\bar{I}_1 + \bar{I}_2 + \bar{I}_3 + \bar{I}_4 = \bar{O}$

CHAPTER 9

Section 9.1

9.1-1. **a.** $y(t) = \dfrac{1}{3}(e^{-t} - e^{-4t})\,u(t)$

b. $y(t) = te^{-2t}\,u(t)$ **c.** $y(t) = \dfrac{1}{\sqrt{3}}e^{-t}\sin(\sqrt{3}t)\,u(t)$

9.1-3. $\dfrac{d^2v}{dt^2} + 5\dfrac{dv}{dt} + 4v = 40\,i_s$

$s(t) = \dfrac{10}{3}(3 - 4e^{-t} + e^{-4t})\,u(t)\ \text{V}$

9.1-5. $z_{eq}(p) = \dfrac{2(p^2 + 4)}{p}$ $\qquad v_{oc}(t) = \dfrac{8}{p}\,i_s$

$\dfrac{d^2v}{dt^2} + 5\dfrac{dv}{dt} + 4v = 40i_s$

9.1-7. $z_{eq}(p) = \dfrac{z(p^2 + 4)}{p}$

$i_{sc}(t) = \dfrac{4}{p^2 + 4}\,i_s$

$\dfrac{d^2v}{dt^2} + 5\dfrac{dv}{dt} + 4v = 40i_s$

9.1-9. $\dfrac{dv}{dt} + v = i_{s_1} + \dfrac{di_{s_2}}{dt} + i_{s_2}$

Section 9.2

9.2-1. $R = \dfrac{1}{2}\sqrt{\dfrac{L}{C}}$

9.2-3. $5\ \Omega$

Section 9.3

9.3-1. $y(t) = \left[\dfrac{1}{6} - \dfrac{1}{2}e^{-2t} + \dfrac{1}{3}e^{-3t}\right]u(t)$

9.3-3. $y(t) = (e^{-2t} + e^{-4t})\,u(t)$

9.3-5. $y(t) = [e^{-t} - e^{-2t}]\,u(t) - [e^{-(t-T)} - e^{-2(t-T)}]\,u(t - T)$

9.3-7. $v(t) = \dfrac{4}{7}(27 - 21e^{-t} - 6e^{-7t})\,u(t)\ \text{V}$

9.3-9. $i(t) = 2(e^{-2t} - e^{-7t})\,u(t)\ \text{A}$

Section 9.4

9.4-1. $y(t) = [1 - e^{-t} - te^{-t}]\,u(t)$

9.4-3. $y(t) = \dfrac{t^4}{24}e^{-3t}\,u(t)$

9.4-5. $x(t) \circ\!\!-\!\!\boxed{\dfrac{1}{p+1}}\!\!-\!\!\boxed{\dfrac{1}{p+1}}\!\!-\!\!\circ\, y(t)$

9.4-7. $x(t) \circ\!\!-\!\!\boxed{\dfrac{1}{p+4}}\!\!-\!\!\circ\, y(t)$

9.4-9. $v(t) = -4[13 + (6t - 13)e^{-2t}]\,u(t)\ \text{V}$

9.4-11.

$u(t) \circ\!\!-\!\!\boxed{\dfrac{1}{p+10}}\!\!-\!\!\boxed{\dfrac{1}{p+10}}\!\!-\!\!\bigcirc\!\!600 \to \overset{-}{\boxed{+}} \to\circ\, i(t)$

$\bigcirc 10$

Section 9.5

9.5-1. $h(t) = \sin(\omega_0 t)\,u(t)$

9.5-3. $y(t) = 8u(t) + 8e^{-4t}[-\cos(2t) + 3\sin(2t)]\,u(t)$

9.5-5. $s(t) = \dfrac{1}{13}\left[1 + \dfrac{1}{3}e^{-2t}\{-3\cos(3t) - 2\sin(3t)\}\right]u(t)$

9.5-7.

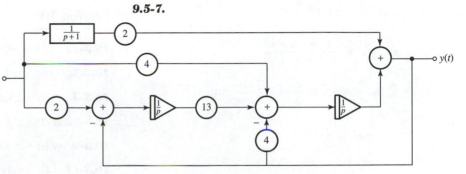

9.5-9. $h(t) = \dfrac{3}{2}e^{-3t}[4\cos(4t) - 3\sin(4t)]\,u(t)$

Section 9.6

9.6-1. $y(t) = [(2t - 1)e^{-2t} + e^{-4t}]\,u(t)$

$\delta(t) \circ\!\!-\!\!\boxed{\dfrac{1}{p+2}}\!\!-\!\!\boxed{\dfrac{1}{p+2}}\!\!-\!\!\boxed{\dfrac{1}{p+4}}\!\!-\!\!\bigcirc\!\!4\!\!-\!\!\circ\, y(t)$

9.6-3. $[(20 - 15t)e^{-t} - 20e^{-4t}]\,u(t)$

9.6-5. $i_0(t) = [\dfrac{5}{2} - \dfrac{8}{3}e^{-t} + \dfrac{1}{6}e^{-4t}]\,u(t)\ \text{A}$

9.6-7. $v(t) = e^{-3t}[-\cos(4t) + \dfrac{3}{4}\sin(4t)]\,u(t) + \cos(5t)\,u(t)\ \text{V}$

9.6-9. $y(t) = \dfrac{t^2}{2}e^{-2t}\,u(t)$

Section 9.7

9.7-1. $i(t) = \dfrac{i_{s_1} - i_{s_2}}{p + 2}$

9.7-3. $i(t) = 3\sin(2t)\,u(t)\ \text{A}$

CHAPTER 10

Section 10.1

10.1-1. $H(s) = \dfrac{1}{s^2 + 3s + 2}$ $\qquad y_f(t) = 4e^{-t/2}\,u(t)$

10.1-3. $H(s) = \dfrac{s + 1}{(s + 2)^2}$, $y_f(t) = -8e^{-3t}\,u(t)$

10.1-5.

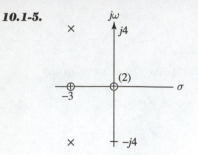

10.1-7. **a.** $Y(s) = \dfrac{s^2 + 2s + 4}{4(s^2 + 4s + 8)}$

b. $4\dfrac{d^2i}{dt^2} + 16\dfrac{di}{dt} + 32i = \dfrac{d^2v}{dt^2} + 2\dfrac{dv}{dt} + 4v$

c.

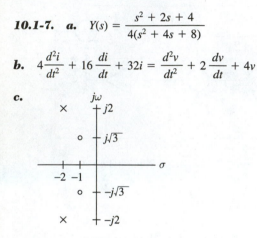

d. $i_f(t) = 0.243e^{-t}\cos(\sqrt{2}t - 43.3°)$ A

10.1-9. $Cp\,\delta(t) + \dfrac{1}{L}u(t)$

Section 10.2

10.2-1. $\mu > 0 \longrightarrow$ Stable $\mu < 0 \longrightarrow$ Unstable
$\mu = 0 \longrightarrow$ Marginally stable

10.2-3.

Response is stable.

10.2-5.

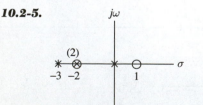

Response is marginally stable.

10.2-7. $\mu < 3 \longrightarrow$ Stable $\mu = 3 \longrightarrow$ Marginally stable
$\mu > 3 \longrightarrow$ Unstable

10.2-9. 2 RHP roots, 2 LHP roots

10.2-11. 2 LHP roots, 2 imaginary axis roots

10.2-13. Unstable for all values of μ

Section 10.3

10.3-1. $y_f(t) = -2e^{2t}u(t)$, unstable response

10.3-3. $y_f(t) = \sqrt{2}\cos(2t - 45°)u(t)$, stable response

10.3-5. $v_f(t) = 2\sqrt{2}\cos(2t - 135°)$ V

10.3-7. $v_f(t) = 3.6\cos(2t - 63.4°)$ V

10.3-9. $v_f(t) = 8\cos(2t - 90°)$ V

10.3-11. Dc steady state exists.

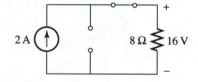

Section 10.4

10.4-1. $V_s = 20$ V $I = 2$ A $V = 12$ V

10.4-3. $v(t) = (1 - e^{-200t})u(t)$

10.4-5. **a.** 10 μF, 10^5 H   **b.** 10 pF, 0.1 H

Section 10.5

10.5-1. **a.** $v(t) = \left(t - \dfrac{3}{2}\right)u(t) + \left(2e^{-t} - \dfrac{1}{2}e^{-2t}\right)u(t)$

b. $v(t) = u(t) - 2e^{-t}u(t) + e^{-2t}u(t)$

10.5-3. $v(t) = \dfrac{1}{2}[3 - 3e^{-2t} - e^{-t}\sin(2t)]u(t)$ V

CHAPTER 11

Section 11.0

11.0-1. **a.** $5\sqrt{3} + j5$ **b.** $-3\sqrt{3} + j3$
c. $-4 - j4\sqrt{3}$ **d.** $-10\sqrt{2} + j10\sqrt{2}$ **e.** $13\angle112.6°$
f. $20\angle-36.9°$ **g.** $15.03\angle-93.8°$ **h.** $10.01\angle177.1°$
i. $27.78\angle125.35°$ **j.** $\sqrt{2}\angle135°$ **k.** $0.23\angle-85.1°$
l. $81.6\angle157.7°$ **m.** $0.316\angle66.2°$

11.0-3.

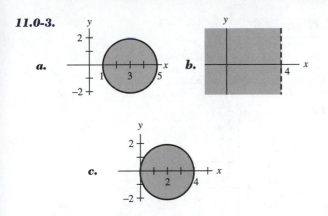

a.

b.

c.

Section 11.1

11.1-1. $i(t) = 5 \cos (100t + 36.9°)$ A for both cases.

11.1-3. *a.* $v_f(t) = 4 \sin (4t)$ V   *b.* $i_f(t) = -4 \sin (4t)$ A

c. $v_f(t) - 108 \sin (6t)$ V   *d.* $i_f(t) = \frac{1}{3} \sin (6t)$ A

11.1-5. $i_{xf}(t) = 1.2 \cos (8t - 53.1°)$ A

$H(j\omega) = 0.6 \angle -53.1°$

11.1-7. $i_f(t) = -\sin (2t)$ A

Section 11.2

11.2-1. $v_f(t) = 2 \cos (2t - 135°)$ V

11.2-3. $v_f(t) = 2 \cos (2t - 135°)$ V

11.2-5. $v_f(t) = 22.4 \cos (2t - 63.4°)$ V

11.2-7. $v_f(t) = 3\sqrt{2} \cos (4t - 135°)$ V

11.2-9. $i_f(t) = 2.06 \cos (5000t + 28.8°)$ mA

11.2-11. $v_f(t) = 3 \cos (2t) + 8 \cos (4t - 6.9°)$ V

11.2-13. $v_f(t) = 4 \cos (2t)$ V

Section 11.3

11.3-1. $<x(t)> = 1$

11.3-3. PF = 0.707 leading $P = 18$ W

11.3-5. $x_{rms} = 0.730$

11.3-7. *a.* $P_{v_s} = -4$ W $P_{L1\Omega} = 4$ W

$P_{R1\Omega} = 0$ W $P_c = 0$ W $P_L = 0$ W

b. $P_{L1\Omega} = 4$ W $P_X = 0$ W

Section 11.4

11.4-1. *a.* $\omega = 9000$ rad/sec

b. $S_R = 24 \angle 0°$, $S_c = 4 \angle 90°$, $S_L = 36 \angle -90°$

Max $E_c = 0.004$ mJ, Max $E_L = 0.036$ mJ $P_R = 24$ W

11.4-3. PF = 0.6 lagging. Capacitor.

11.4-5. $P = 10$ W

11.4-7. $P = 7$ W $Q = -1$ VAR $A = \sqrt{50}$ VA

$S = 7 - j$

CHAPTER 12

Section 12.1

12.1-1. $x(t) = \frac{2}{\pi} \sin(2\pi t) - \frac{2}{\pi} \sin(4\pi t) + \frac{2}{3\pi} \sin(6\pi t) -$

$\frac{1}{\pi} \sin(8\pi t) + \frac{2}{5\pi} \sin(10\pi t)$

Section 12.2

12.2-1. $H(j\omega) = \dfrac{1}{\sqrt{1 + \left(\dfrac{\omega L}{R}\right)^2}} \angle -\tan^{-1}\left(\dfrac{\omega L}{R}\right)$

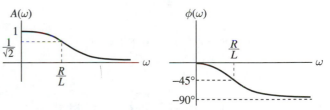

12.2-3. $H(j\omega) = \dfrac{1}{\sqrt{1 + (\omega RC)^2}} \angle -\tan^{-1}(\omega RC)$

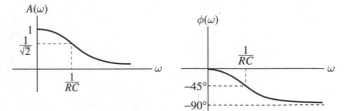

12.2-5. $Z(j\omega) = \dfrac{R}{\sqrt{1 + (\omega RC)^2}} \angle -\tan^{-1}(\omega RC)$

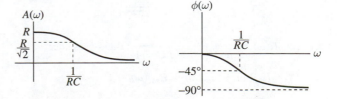

12.2-7. $y(t) = 9.95 \cos(2\pi \times 10^4 t - 5.7°) +$ 0.05 $\sin(20\pi t - 89°)$

Section 12.3

12.3-1. $x(\omega) = \dfrac{10^{-4}\omega}{1 - \left(\dfrac{\omega}{10^7}\right)^2}$

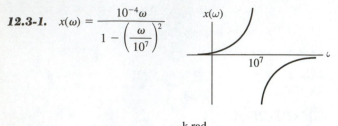

12.3-3. *a.* 200 kΩ   *b.* 50 $\dfrac{k\,rad}{s}$

Section 12.4

12.4-1.

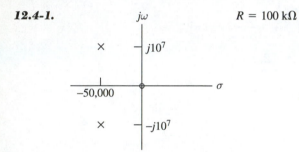

$R = 100$ kΩ

12.4-3. $v(t') = (1 - e^{-t'/4}) \cos(100t') u(t)$ V

Section 12.5

12.5-1. 0.253 mΩ   *a.* 2%   *b.* −17.4%

12.5-3. $B(\omega) = \dfrac{j10^{-4}\,\omega}{1 - \left(\dfrac{\omega}{\omega_0}\right)^2}$

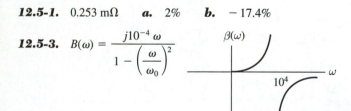

12.5-5. $R = 6.67\Omega$

12.5-7.

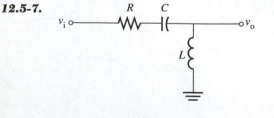

12.5-9. $H(\omega) = \dfrac{j\omega RC}{j\omega RC - \omega^2 LC + 1}$. Band-pass filter

12.5-11. $H_{LP}(\omega) + H_{BP}(\omega) + H_{MP}(\omega) = 1$

Section 12.6

12.6-1. 31.1 dB, 43.8 dB, 39.8 dB, 80 dB

12.6-3. 2 decades, 2 decades

12.6-5.

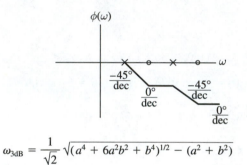

$$\omega_{3dB} = \frac{1}{\sqrt{2}} \sqrt{(a^4 + 6a^2b^2 + b^4)^{1/2} - (a^2 + b^2)}$$

12.6-7.

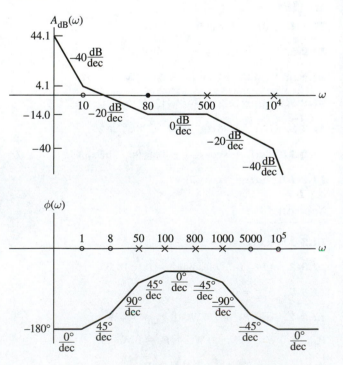

Section 12.7

12.7-1. $H(j\omega) = -0.1 \dfrac{j\omega + 50000}{j\omega + 500}$

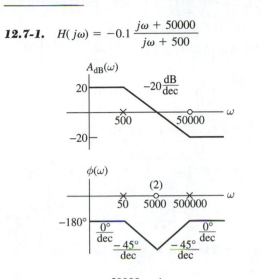

12.7-3. $H(j\omega) = \dfrac{50000 - j\omega}{50000 + j\omega}$

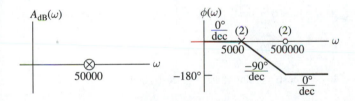

CHAPTER 13

Section 13.1

13.1-1. $y(t) = \dfrac{1}{3 + j4}[e^{(3+j4)t} - 1]\, u(t)$ $Y(s, t) = \dfrac{1}{3 + j4}$

13.1-3. $X(j\omega, t) = \dfrac{1}{2 + j\omega}$

13.1-5. $X(s) = \dfrac{(1 - e^{-s})^2}{s}$; ROC = Entire s-plane

13.1-7. $X(s) = \dfrac{1}{(s + 2)^2}$; ROC: $\sigma > -2$

13.1-9. $X(s) = \dfrac{8}{s^2 + 4s + 8}$

13.1-11. $X(s) = \dfrac{24}{(s - 2)^5}$

13.1-13. $x(t) = e^{-2t}\cos(3t)\, u(t)$

Section 13.2

13.2-1. $2[1 - e^{-2t}]\, u(t)$, $[2t - 1 + e^{-2t}]\, u(t)$

13.2-3. $\delta(t) - 2\delta(t - 1) + \delta(t - 2)$,
$u(t) - 2u(t - 1) + u(t - 2)$

13.2-5. $(1 - 2t)\, e^{-2t}\, u(t)$, te^{-2t}

13.2-7. $x_c(t) = 2t\, u(t) - (2t - 2)\, u(t - 2)$;
$x_j(t) = -4u(t - 1)$ $px(t) = 2u(t) - 2u(t - 2) - 4\delta(t - 1)$

13.2-9. $x_c(t) = 2tu(t)$; $x_j(t) = -2\sum_{n=1}^{\infty} u(t - n)$

$px(t) = 2u(t) - 2\sum_{n=1}^{\infty} \delta(t - n)$

13.2-11.

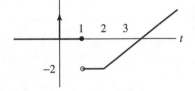

$px(t) = 4\delta'(t) - 2\delta(t - 1) + 2u(t - 2)$

$\dfrac{1}{p}x(t) = 4u(t) - 2(t - 1)\, u(t - 1) + (t - 2)^2\, u(t - 2)$

13.2-13. $-12e^{-3}$

Section 13.3

13.3-1. $X(s) = \dfrac{2(1 - 2se^{-s} - e^{-2s})}{s^2}$

13.3-3. $x(t) = (2t - 1 + e^{-2t})\, u(t)$

13.3-5. $X(s) = 4 - 3e^{-2s}$

13.3-7. $X(s) = \dfrac{2}{s^2(1 + e^{-s})}$

13.3-9. $y(t) = \dfrac{t^2}{2}e^{-2t}\, u(t)$

13.3-11. $x(t) = (t - 2)\, e^{-2(t-2)}\, u(t - 2)$

13.3-13. **a.** $\dfrac{e^{-s}}{s^2} + \dfrac{e^{-s}}{s + 1}$ **b.** $\dfrac{e^{-s}}{s^2} - \dfrac{e^{-s}}{s} + \dfrac{e^{-s+1}}{s + 1}$

c. $\dfrac{2e^{-(s+3)}}{(s + 3)^2} + \dfrac{2e^{-(s+3)}}{s + 3}$

d. $\dfrac{2e^{-2s-6}}{[(s + 3)^2 + 16]^2}[(s^2 + 3s - 16)\cos(8) - 8(s + 3)\sin(8)]$

$+ \dfrac{4e^{-2s-6}}{(s + 3)^2 + 16}[(s + 3)\cos(8) - 4\sin(8)]$

e. $2s^2 + 4se^{-s} - 3e^{-2s} + \dfrac{4}{s}e^{-3s}$

13.3-15. $X(s) = \dfrac{1}{s(1 - e^{-s})}$

Section 13.4

13.4-1. $[10 - 6e^{-8t} + 4e^{-6t}]\, u(t)$

13.4-3. $[3e^{-2t} + 3(6t - 1)]\, u(t)$

13.4-5. $2[(6t + 1)\, e^{-t} - e^{-4t}]\, u(t)$

13.4-7. $6[1 - 5e^{-2t} + e^{-t}\{4\cos(t) - 2\sin(t)\}]\, u(t)$

Section 13.5

13.5-1. **a.** $e^{-t}\sin(t)\, u(t);\ [\dfrac{1}{2} - \dfrac{1}{2}e^{-t}\{\cos(t) + \sin(t)\}]\, u(t)$

b. $\dfrac{1}{5}[1 - e^{-5/2t}]\, u(t);\ \dfrac{1}{25}[(5t - 2) + 2e^{-5/2t}]\, u(t)$

c. $\dfrac{1}{8}[e^{4t} - e^{-4t}]\, u(t);\ \dfrac{1}{32}[-2 + e^{4t} + e^{-4t}]\, u(t)$

d. $\dfrac{1}{4\sqrt{2}}[e^{2\sqrt{2}t} - e^{-2\sqrt{2}t}]\, u(t);\ \dfrac{1}{16}[-2 + e^{2\sqrt{2}t} + e^{-2\sqrt{2}t}]\, u(t)$

e. $\dfrac{1}{3}\sin(3t)\, u(t);\ \dfrac{1}{9}[1 - \cos(3t)]\, u(t)$

f. $\dfrac{1}{2}\sin\left(\dfrac{t}{2}\right)u(t);\ \left[1 - \cos\left(\dfrac{t}{2}\right)\right]u(t)$

g. $[e^{2t} - e^{t}]\, u(t);\ \dfrac{1}{2}[1 - 2e^{t} + e^{2t}]\, u(t)$

h. $\dfrac{1}{5}[-e^{-2t} + e^{3t}]\, u(t);\ \dfrac{1}{30}[-5 + 3e^{-2t} + 2e^{3t}]\, u(t)$

i. $te^{-4t}\, u(t);\ \dfrac{1}{16}(1 - (4t + 1)\, e^{-4t}]\, u(t)$

j. $te^{5t}\, u(t);\ \dfrac{1}{25}[1 + (5t - 1)e^{5t}]\, u(t)$

k. $\dfrac{1}{5}[e^{t} - e^{-4t}]\, u(t);\ \dfrac{1}{20}[-5 + 4e^{t} + e^{-4t}]\, u(t)$

l. $\dfrac{1}{6}[-e^{-t/2} + e^{t/4}]\, u(t);\ \dfrac{1}{3}[-3 + e^{-t/2} + 2e^{t/4}]\, u(t)$

m. $\dfrac{1}{30}[-6 + 5e^{-t} + e^{5t}]\, u(t);\ \left[-\dfrac{1}{5}t - \dfrac{1}{6}e^{-t} + \dfrac{1}{150}e^{5t}\right]u(t)$

n. $\left[e^{-t} + e^{-t/2}\left\{\cos\left(\dfrac{\sqrt{3}}{2}t\right) + \dfrac{1}{\sqrt{3}}\sin\left(\dfrac{\sqrt{3}}{2}t\right)\right\}\right]u(t);$

$\left[1 - e^{-t} - \dfrac{2}{\sqrt{3}}e^{-t/2}\sin\left(\dfrac{\sqrt{3}}{2}t\right)\right]u(t)$

o. $\dfrac{1}{16}[(4t - 1)\, e^{3t} + e^{-t}]\, u(t);$

$\dfrac{1}{144}[16 - 9e^{-t} + (12t - 7)\, e^{3t}]\, u(t)$

p. $\dfrac{1}{8}\left[e^{2t} - e^{-t/2}\left\{\cos\left(\dfrac{\sqrt{7}}{2}t\right) + \dfrac{5}{\sqrt{7}}\sin\left(\dfrac{\sqrt{7}}{2}t\right)\right\}\right]u(t);$

$\dfrac{1}{16}\left[-4 + e^{2t} + e^{-t/2}\left\{3\cos\left(\dfrac{\sqrt{7}}{2}t\right) - \dfrac{1}{\sqrt{7}}\sin\left(\dfrac{\sqrt{7}}{2}t\right)\right\}\right]u(t)$

13.5-3. **a.** $\dfrac{1}{2}[(t + 2) - (5t + 2)e^{-2t}]\, u(t)$

b. $\left[\dfrac{2}{3}t^4 - \dfrac{8}{3}t^3 + 8t^2 - 16t + 16 - 16e^{-t}\right]u(t)$

c. $\dfrac{1}{49}[(7t - 1)e^{4t} + e^{-3t}]\, u(t)$

d. $\dfrac{1}{4}[12 - 4e^{t} - 7e^{-t} - e^{3t}]\, u(t)$

e. $\dfrac{1}{16}[(4t - 3) + (24t - 8)e^{-2t} + 11e^{-4t}]\, u(t)$

f. $\dfrac{1}{4}\left[\sin(t) - \dfrac{1}{5}\sin(5t)\right]u(t)$

g. $\dfrac{4}{3}[\cos(t) - \cos(2t)]\, u(t) + \left[\sin(t) - \dfrac{1}{2}\sin(2t)\right]u(t)$

h. $\dfrac{1}{6}e^{t}[2\sin(t) - \sin(2t)]\, u(t)$

i. $-[t\cos(t) - \sin(t)]\, u(t) + 2\cos(t)\, u(t) -$

$2e^{-t/2}\left[\cos\left(\dfrac{\sqrt{3}}{2}t\right) + \dfrac{1}{\sqrt{3}}\sin\left(\dfrac{\sqrt{3}}{2}t\right)\right]u(t)$

j. $\dfrac{11}{4096}[28t^3 - (21/2)t^2 - 8t + 1 - e^{-8t}]\, u(t)$

13.5-5. **a.** $2\cos(4t) - \dfrac{1}{2}\sin(4t)$ **b.** $5(t + 1)e^{t}$

c. $\dfrac{3}{4}(e^{-t} - e^{-5t})$ **d.** $\dfrac{1}{8}[-2 + 5e^{8t} + 5e^{-8t}],\ t > 0$

e. $\dfrac{1}{125}[-41 + 45t - 12.5t^2 + 41e^{5t}],\ t > 0$

f. $\dfrac{1}{20}[4 + 5e^{-2t} - 29e^{6t}],\ t > 0$

g. $-5 + 3.6e^{-t} + 0.4\cos(2t) + 0.8\sin(2t) + 2\sin(t)$

13.5-7. **a.** $y(0)e^{-2t};\ 4(1 - e^{-2t})$

b. $[y'(0) + 2y(0)]e^{-t} - [y'(0) + y(0)]e^{-2t};\ 4[1 - e^{-t}]$

c. $y(0)e^{-2t};\ 6[-e^{-2t} + \cos(2t) + \sin(2t)]$

d. $\dfrac{1}{5}[y'(0) + 3y(0)]e^{2t} + [y'(0) + 4y(0)]e^{-3t};$

$4\left[-1 + \dfrac{3}{5}e^{2t} + \dfrac{2}{5}e^{-3t}\right]$

e. $[[y'(0) - 2y(0)]t + y(0)]e^{2t}; \dfrac{1}{4}te^{2t} - \dfrac{1}{8}\sin(2t)$

Section 13.6

13.6-1. $s(t) = \sqrt{\dfrac{L}{C}}\sin\left(\dfrac{1}{\sqrt{LC}}t\right)u(t)$ V

$h(t) = \dfrac{1}{C}\cos\left(\dfrac{1}{\sqrt{LC}}t\right)u(t)$ V

13.6-3. **a.** Critically damped: $R = \dfrac{1}{2}\sqrt{\dfrac{L}{C}}$

Underdamped: $R < \dfrac{1}{2}\sqrt{\dfrac{L}{C}}$

Overdamped: $R > \dfrac{1}{2}\sqrt{\dfrac{L}{C}}$

$H(s) = \dfrac{\dfrac{1}{C}s}{s^2 + \dfrac{1}{RC}s + \dfrac{1}{LC}}$

Critically damped   Underdamped   Overdamped

b. $s(t) = \dfrac{1}{C}te^{-t/\sqrt{LC}}u(t)$ V

$h(t) = \dfrac{1}{C}\left[1 - \dfrac{1}{\sqrt{LC}}t\right]e^{-t/\sqrt{LC}}u(t)$ V

13.6-5. **a.** Critically damped: $R = 2\sqrt{\dfrac{L}{C}}$

Underdamped: $R < 2\sqrt{\dfrac{L}{C}}$

Overdamped: $R > 2\sqrt{\dfrac{L}{C}}$

$H(s) = \dfrac{s^2}{s^2 + \dfrac{R}{L}s + \dfrac{1}{LC}}$

b. $s(t) = \left[1 - \dfrac{1}{\sqrt{LC}}t\right]e^{-t/\sqrt{LC}}u(t)$ V

$h(t) = \delta(t) - \dfrac{1}{\sqrt{LC}}\left[2 - \dfrac{1}{\sqrt{LC}}t\right]e^{-t/\sqrt{LC}}u(t)$ V

13.6-7. $v(t) = [2 - (3t + 2)e^{-2t}]u(t)$ V

13.6-9. $i(t) = [7 - (6t + 7)e^{-2t}]u(t)$ A

13.6-11. **a.** $v_{oc}(t) = 4(1 - 2t)e^{-2t}$ V   **b.** $i_{sc}(t) = 1$ A

13.6-13. **a.** $v_{oc}(t) = (5.61e^{-3.73t} - 1.61e^{-1.07t})$ V
b. $i_{sc}(t) = 1.2$ A

13.6-15. **a.** $R = 1\,\Omega, C = 1$ mF   **b.** $R = 1\,\Omega, C = 1$ F

c. $H(s) = \dfrac{1}{s^2 + (3 - \mu)s + 1}$

d. $\mu = 5$: Unstable, $\mu = 3$: Marginally stable, $\mu = 2$: Stable, $\mu = 1$: Stable

e. $v(t) = 0.01[1 - (1000t + 1)e^{-1000t}]\,u(t)$ A

f. $v(t) = 0.001\left[-\dfrac{2}{\sqrt{3}}e^{-500t}\sin(500\sqrt{3}\,t) + \sin(1000t)\right]u(t)$ A

13.6-17. $s(t) = \dfrac{1}{9}e^{-50/9t}\,u(t)$

Section 13.7

13.7-1. $v(t) = 6\,u(t)$

13.7-3. $i(t) = \dfrac{24}{\sqrt{8}}\sin(\sqrt{8}t)\,u(t)$ A

13.7-5. $i(t) = 4$ A

13.7-7. $v(t) = 6\delta(t) + 5[6 + e^{-5t}]\,u(t)$ V

13.7-9. $\mu = 1$: Stable, $\mu = 2$: Stable, $\mu = 3$: Stable

$\mu = 1: v(t) = \begin{cases} \dfrac{4}{3}[4e^{-t} - e^{-4t}]; t > 0 \\ 4; t \le 0 \end{cases}$ V

$\mu = 2: v(t) = \begin{cases} 4(2t + 1)e^{-2t}; t > 0 \\ 4; t \le 0 \end{cases}$ V

$\mu = 3: v(t) = \begin{cases} e^{-3/2t}\left[4\cos\left(\dfrac{\sqrt{7}}{2}t\right) + \dfrac{12}{\sqrt{7}}\sin\left(\dfrac{\sqrt{7}}{2}t\right)\right]; t > 0 \\ 4; t \le 0 \end{cases}$ V

13.7-11. $v(t) = \begin{cases} -2[1 + 8te^{-4t} - 2e^{-4t}]; t > 0 \\ -2; t \le 0 \end{cases}$ V

Section 13.8

13.8-1.

$x(t) \otimes y(t) = \begin{cases} 0; & t \le 0 \\ \dfrac{1}{2}t^2; & 0 < t \le 1 \\ \dfrac{1}{2}; & t > 1 \end{cases}$

13.8-3. $x(t) \otimes y(t) = \dfrac{t^2}{2}u(t) - \dfrac{(t - 1)^2}{2}u(t - 1) -$
$(t - 1)\,u(t - 1)$

13.8-5. $\quad 0; \quad t \le 0$

$\qquad te^{-t}; \quad t > 0$

13.8-7. $\quad te^{-t} u(t)$

CHAPTER 14

Section 14.1

14.1-1. $\quad X(\omega) = \dfrac{1 - e^{-j2\pi\omega}}{1 - \omega^2}$

14.1-3. $\quad X(\omega) = \dfrac{1 + e^{-3\pi\omega}}{1 - \omega^2}$

14.1-5. $\quad X(\omega) = \dfrac{1}{\omega^2}[(1 + j\omega)e^{-j\omega} - 1]$

14.1-7. $\quad X(\omega) = \dfrac{e^{-(1+j\omega)}}{1 + j\omega}$

14.1-9. $\quad X(\omega) = \dfrac{4(\omega^2 + 13)}{\omega^4 - 10\omega^2 + 169}$

14.1-11. $\quad X(\omega) = \dfrac{3}{j\omega + 2}$

14.1-13. $\quad X(\omega) = \dfrac{4}{\omega^2}e^{-j\omega} + j\dfrac{2}{\omega^3}(1 - e^{-j2\omega})$

14.1-15. $\quad X(\omega) = \dfrac{e^{-j^2\omega}}{j\omega}$

14.1-17.

$v_0(t) = \dfrac{10\sqrt{2}}{1 + 4\pi^2}[\cos(2\pi \times 10^5 t) + 2\pi \sin(2\pi \times 10^5 t)]\, u(t)$

$\qquad - \dfrac{10\sqrt{2}}{1 + 4\pi^2} e^{-10^5 t}\, u(t) \qquad t = 50\ \mu s$

14.1-19. $\quad X(\omega) = -8e^{j^2\omega}$

14.1-21. $\quad X(\omega) = \dfrac{2\cos\left(\dfrac{\pi}{2}\omega\right)}{1 - \omega^2}$

14.1-23. $\quad X(\omega) = -\dfrac{2}{\omega^2}\cos(2\omega) - \dfrac{4}{\omega^3}\sin(\omega)$

$\qquad + \dfrac{1}{\omega^2}(e^{j\omega} + 3e^{-j\omega})$

14.1-25. $\quad X(\omega) = j\dfrac{4}{\omega^3}[\cos(\omega) + \omega \sin(\omega) - 1]$

14.1-27. $\quad X(\omega) = \dfrac{1}{1 - re^{-j\omega}}$

Section 14.2

14.2-1. $\quad \dfrac{1}{2}[\delta(t + 3) + \delta(t - 3)]$

14.2-3. $\quad \delta(t) + 0.15\,\delta(t + 3) + 0.15\,\delta(t - 3) - j0.2\,\delta(t + 1)$

$\qquad + j0.2\,\delta(t - 1)$

14.2-5. $\quad j4\dfrac{\sin^2\left(\dfrac{t}{2}\right)}{t}$

14.2-7. $\quad \dfrac{\sin\left(\dfrac{t}{2}\right)}{\pi t}e^{j3/2t}$

14.2-9. $\quad \dfrac{10}{\pi(1 + t^2)}[1 + te^{-1}\sin(t) - e^{-1}\cos(t)]$

$\qquad - 2e^{-1}\dfrac{\sin(t)}{t}$

14.2-11. $\quad 2\dfrac{\sin(t) - t\cos(t)}{\pi t^3}$

14.2-13. $\quad \dfrac{1}{2\pi(1 - re^{jt})}$

Section 14.3

14.3-1. $\quad A(\omega) = 2\left|\dfrac{\sin(\pi\omega)}{1 - \omega^2}\right| \qquad \phi(\omega) = -\dfrac{\pi\omega}{2} \pm \dfrac{\pi}{2}$

14.3-3. $\quad A(\omega) = \sqrt{\dfrac{4 + \omega^2}{169 - 10\omega^2 + \omega^4}}$

$\phi(\omega) = \tan^{-1}\left(\dfrac{\omega}{2}\right) - \tan^{-1}\left(\dfrac{4\omega}{13 - \omega^2}\right)$

14.3-5. $\quad A(\omega) = \dfrac{3}{\sqrt{\omega^2 + 4}}; \quad \phi(\omega) = -\tan^{-1}\left(\dfrac{\omega}{2}\right)$

14.3-7. $\quad A(\omega) = \dfrac{4\omega}{4 + \omega^2} \qquad \phi(\omega) = -90°$

14.3-9. $\quad x(\omega) = 2\pi\,\delta(\omega) + \dfrac{1}{3}[\pi\delta(\omega - 3) + \pi\delta(\omega + 3)]$

$\qquad + \dfrac{1}{5}[\pi\delta(\omega - 5) + \pi\delta(\omega + 5)] + \dfrac{1}{7}[\pi\delta(\omega - 7)$

$\qquad + \pi\delta(\omega + 7)] + \dfrac{1}{9}[\pi\delta(\omega - 9) + \pi\delta(\omega + 9)]$

14.3-11. $\quad A(\omega) = \left|\dfrac{2\omega\sin(\pi\omega)}{1 - \omega^2}\right| \qquad \phi(\omega) = -\pi\omega + \pi$

14.3-13. $\quad Y(\omega) = e^{-j3\omega}$

14.3-15. $\quad s(\omega) = \dfrac{4[\omega - \sin(\omega)]}{\omega^3}$

14.3-17. $s(0) = 4$

14.3-19. $\dfrac{2 + j\omega}{13 + j4\omega - \omega^2}$

Section 14.4

14.4-1. $\dfrac{1}{2}\delta(t + a) + \dfrac{1}{2}\delta(t - a)$

14.4-3. $j\pi\delta(\omega + \omega_0) - j\pi\delta(\omega - \omega_0)$

14.4-5. $\pi\delta(\omega + \omega_0) + \pi\delta(\omega - \omega_0)$

14.4-7. $\pi\delta(\omega) - \dfrac{1}{j\omega}$

14.4-9. $\dfrac{1}{j\omega}\displaystyle\sum_{n=0}^{\infty} e^{-j\omega n}$

14.4-11. $\dfrac{j\omega(1 - e^{-j\omega})}{1 - \omega^2}$

14.4-13. $\dfrac{2 - 2(1 + j2\omega)e^{-j\omega} + 2e^{-j2\omega}}{(j\omega)^3}$

14.4-15. $\dfrac{-j4\omega}{\omega^2 + 4}$

14.4-17. $\dfrac{\delta(\omega - 1) - \delta(\omega - 2)}{q}$

14.4-19.
$\dfrac{-\delta'(\omega + 2) + \delta(\omega + 1) - \delta(\omega - 1) - \delta'(\omega - 2)}{q^2}$

14.4-21. $\dfrac{2}{(j\omega + 3)^3}$

Section 14.5

14.5-1. $c_n = \dfrac{4}{n^2 - 4}[(-1)^n - 1]$

14.5-3. $c_n = \dfrac{\cos\left(\dfrac{n\pi}{2}\right)}{\pi(1 - n^2)}$

14.5-5. $c_n = \dfrac{1}{j2\pi n}$

14.5-7. $c_n = \dfrac{(-1)^n}{\pi\omega_0(1 - 4n^2)}$

Section 14.6

14.6-1. $x(t) = \dfrac{j}{2\pi}\displaystyle\int_{-\infty}^{\infty} \dfrac{\operatorname{sgn}(-\omega)}{2 + j\omega}e^{j\omega t}d\omega$

14.6-3. $X(\omega) = \mathcal{H}\{R(\omega)\}$

CHAPTER 15

Section 15.1

15.1-1. Ellipse centered at $(3, -2)$

15.1-3. $8v_1 - 3v_2 - 60i_1 + 0i_2 = 0$
$v_1 - v_2 + 0i_1 + 10i_2 = 0$

Section 15.2

15.2-1. $z_{11} = z_{22} = R$, $z_{12} = z_{21} = 0$
Reciprocal and symmetric.

15.2-3. $z_{11} = z_{22} = \dfrac{R(3RCs + 2)}{2RCs + 1}$,

$z_{12} = z_{21} = \dfrac{R(3RCs + 1)}{2RCs + 1}$ Reciprocal and symmetric

15.2-5. $z_{11} = 10 + j20$, $z_{12} = 50$, $z_{21} = 0$,

$z_{22} = \dfrac{250}{1 + j2}$ Not reciprocal or symmetric

15.2-7. $z_{11} = z_{22} = \dfrac{2s^2 + 2s + 1}{2s + 1}$, $z_{12} = z_{21} = \dfrac{2s(s + 1)}{2s + 1}$

Reciprocal and symmetric

15.2-9.

15.2-11. $z_{11} = z - r$, $z_{12} = -r$, $z_{21} = r$, $z_{22} = z + r$

Not reciprocal or symmetric

15.2-13. $z_{11} = z_{22} = \dfrac{s + 1}{2s + 1}$, $z_{12} = z_{21} = \dfrac{s}{2s + 1}$

$i_o(t) = 10(e^{-t} - e^{-2t})u(t)$ A

15.2-15. $z_i = 2\ \Omega$, $z_o = 2\ \Omega$ $A_v = \dfrac{4}{3}$

Section 15.3

15.3-1. $R_a = R_b = \dfrac{72}{17}\ \Omega$, $R_c = \dfrac{576}{17}\ \Omega$

15.3-3. $y_{11} = y_{22} = \dfrac{3RCs + 2}{3R}$

$y_{12} = y_{21} = \dfrac{-(3RCs + 1)}{3R}$

15.3-5. $y_{11} = y_{22} = \dfrac{8}{5R}$, $y_{12} = y_{21} = -\dfrac{2}{5R}$

15.3-7. $y_{11} = \dfrac{s(s + 4)}{4s^2 + s + 4}$, $y_{12} = y_{21} = \dfrac{-4s}{4s^2 + s + 4}$,

$y_{22} = \dfrac{4s + 1}{4s^2 + s + 4}$

15.3-9. $y_{11} = \dfrac{z + r}{z^2}$, $y_{12} = \dfrac{r}{z^2}$

$y_{21} = -\dfrac{r}{z^2}$, $y_{22} = \dfrac{z - r}{z^2}$

15.3-11.

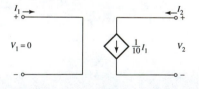

4 Ω   20 Ω
5 Ω

15.3-13. $y_{11} = \dfrac{2}{3+a}, \quad y_{12} = -\dfrac{1}{3+a}$

$y_{21} = -\dfrac{(1+a)}{3+a}, \quad y_{22} = \dfrac{2+a}{3+a} \quad \Delta y \neq 0 \text{ for any } a$

15.3-15. $y_{11} = 0, \quad y_{12} = 0$

$y_{21} = -g_m, \quad y_{22} = 0$

15.3-17. $Ai_1 = \dfrac{-1}{2s^2 + 5s - 4}, \quad Ai_2 = \dfrac{-1}{2s^2 + 7s + 1}$

$A_v = \dfrac{-2}{2s+5} \quad Z_i = \dfrac{2s+5}{2s^2+5s-4} \quad Z_o = \dfrac{s+1}{s^2+3s}$

Section 15.4

15.4-1.
$$\begin{bmatrix} \dfrac{1}{6} & 0 \\ -\dfrac{1}{60} & \dfrac{1}{20} \end{bmatrix}$$

15.4-3. Chain parameters do not exist.

15.4-5. Chain parameters do not exist.

15.4-7.
$$\begin{bmatrix} \dfrac{2s^2+2s+1}{2s(s+1)} & \dfrac{2s+1}{2s(s+1)} \\ \dfrac{2s+1}{2s(s+1)} & \dfrac{2s^2+2s+1}{2s(s+1)} \end{bmatrix}$$

15.4-9.
$$\begin{bmatrix} \dfrac{z}{r}-1 & \dfrac{z^2}{r} \\ \dfrac{1}{r} & \dfrac{z}{r}+1 \end{bmatrix}$$

15.4-11.
$$\begin{bmatrix} \dfrac{3(s+1)}{s} & \dfrac{s+2}{s} \\ \dfrac{2s^2+6s+3}{s(s+2)} & \dfrac{s+1}{s} \end{bmatrix}$$

$i_o(t) = 10(e^{-t} - e^{-2t})\,u(t) \text{ A}$

15.4-13.
$$\begin{bmatrix} 1+\dfrac{3}{s^2}+\dfrac{1}{s^4} & \dfrac{2}{s}+\dfrac{1}{s^3} \\ \dfrac{3}{s}+\dfrac{4}{s^3}+\dfrac{1}{s^5} & 1+\dfrac{3}{s^2}+\dfrac{1}{s^4} \end{bmatrix}$$

Section 15.5

15.5-1.
$$\begin{bmatrix} \dfrac{15}{2} & \dfrac{3}{8} \\ -\dfrac{3}{4} & \dfrac{1}{16} \end{bmatrix}$$

15.5-3.
$$\begin{bmatrix} R & 0 \\ 0 & \dfrac{1}{R} \end{bmatrix}$$

15.5-5.
$$\begin{bmatrix} 10+j20 & \dfrac{1+j2}{5} \\ 0 & \dfrac{1+j2}{250} \end{bmatrix}$$

15.5-7.
$$\begin{bmatrix} \dfrac{2s+1}{2s^2+2s+1} & \dfrac{2s(s+1)}{2s^2+2s+1} \\ \dfrac{-2s(s+1)}{2s^2+2s+1} & \dfrac{2s+1}{2s^2+2s+1} \end{bmatrix}$$

15.5-9.
$$\begin{bmatrix} \dfrac{z^2}{r+z} & \dfrac{-r}{r+z} \\ \dfrac{-r}{r+z} & \dfrac{1}{r+z} \end{bmatrix}$$

15.5-11. $Z_i = 2\,\Omega, \quad Z_o = 2\,\Omega, \quad A_v = \dfrac{4}{3}$

15.5-13. $A_v = \dfrac{s(s+2)}{2s^2-24s-7}, \quad Z_i = \dfrac{-2s^2+16s+5}{4s+1},$

$Z_o = \dfrac{-7}{2s(s+2)}$

15.5-15. $A_i = \dfrac{-6s}{10s^2+3s-25}, \quad Z_i = \dfrac{-(s+5)}{5s^2+2s-10},$

$Z_o = \dfrac{-10s^2-9s+10}{2s+5}$

Section 15.6

15.6-1. h parameters exist.

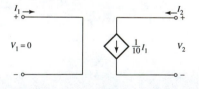

$I_1 \rightarrow$ $\leftarrow I_2$

$V_1 = 0$ $\dfrac{1}{10}I_1$ V_2

15.6-3. Series/shunt
z parameters exist.

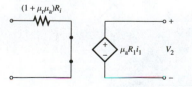

$(1+\mu_r\mu_a)R_i$

$\mu_a R_1 i_1 \quad V_2$

g parameters exist.

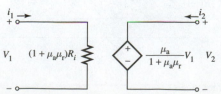

$i_1 \rightarrow$ $\leftarrow i_2$

$V_1 \quad (1+\mu_a\mu_r)R_i \quad \dfrac{\mu_a}{1+\mu_a\mu_r}V_1 \quad V_2$

15.6-5. h parameters exist.

$$z_{is} = 0, \quad z_{oo} = \infty, \quad A_v = \infty, \quad A_i = -\frac{1}{\beta}$$

15.6-7. z parameters exist.

$$z_i = \frac{1}{y_{11}}, \quad Z_o = \frac{1}{y_{22}}, \quad A_v = \infty, \quad A_i = \infty$$

15.6-9. $H(s) = \dfrac{(RCs)^2 + (RCs) + R_{g_m}}{3(RCs)^2 + (4 - R_{g_m})RCs + 1}$

$$g_m < \frac{4}{R} \longrightarrow \text{Stable} \quad g_m > \frac{4}{R} \longrightarrow \text{Unstable}$$

$$g_m = \frac{4}{R} \longrightarrow \text{Marginally stable}$$

Section 15.7

15.7-1.

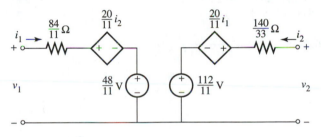

15.7-3.

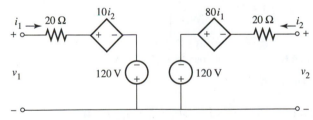

15.7-5.

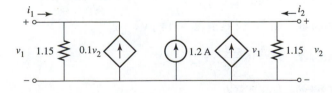

15.7-7.

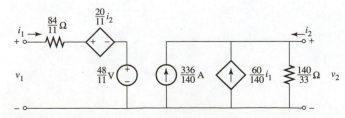

Section 16.1

16.1-1.

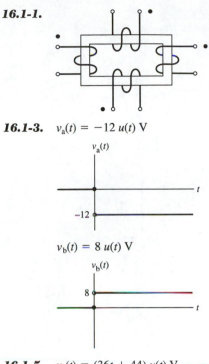

16.1-3. $v_a(t) = -12\, u(t)$ V

$v_b(t) = 8\, u(t)$ V

16.1-5. $v_a(t) = (36t + 44)\, u(t)$ V

$v_b(t) = (24t + 24)\, u(t)$ V

Section 16.2

16.2-1. $v_a = 2pi_a - 3pi_b \quad v_b = -3pi_a + 5pi_b$

16.2-3. $Y(s) = \begin{bmatrix} \dfrac{5}{s} & \dfrac{3}{s} \\ \dfrac{3}{s} & \dfrac{2}{s} \end{bmatrix}$

16.2-5.

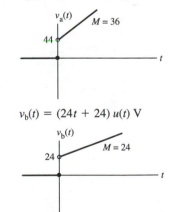

16.2-7.

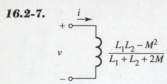

$$\frac{L_1 L_2 - M^2}{L_1 + L_2 + 2M}$$

16.2-9.

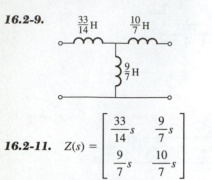

16.2-11. $Z(s) = \begin{bmatrix} \dfrac{33}{14}s & \dfrac{9}{7}s \\[2mm] \dfrac{9}{7}s & \dfrac{10}{7}s \end{bmatrix}$

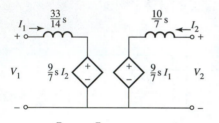

16.2-13. $Z(s) = \begin{bmatrix} s & s \\ 2s & 7s \end{bmatrix}$

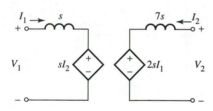

16.2-15.

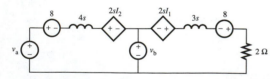

16.2-17.

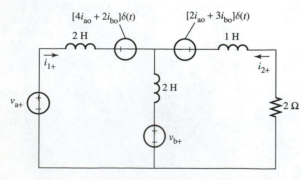

16.2-19. $i_a(t) = (4 - 6te^{-t} - 4e^{-t})\, u(t)$ $i_b(t) = 12te^{-t}\, u(t)$

Section 16.3

16.3-1. $v(t) = 4\sqrt{2}\,[e^{-t} - e^{-4t}]\, u(t)$ V

16.3-3. $v(t) = 4\sqrt{2}\,[e^{-t} - e^{-4t}]\, u(t)$ V

16.3-5. $v(t) = \dfrac{1}{3}(12 + 10e^{-t/2} + 8e^{-5t})\, u(t)$ V

16.3-7. $v(t) = \dfrac{1}{3}(12 + 10e^{-t/2} + 8e^{-5t})\, u(t)$ V

16.3-9. $v(t) = (-4 + 2e^{-2t/3} + 2e^{-2t})\, u(t)$ V

16.3-11. $v(t) = (-4 + 2e^{-2t/3} + 2e^{-2t})\, u(t)$ V

16.3-13. $v(t) = \dfrac{66}{\sqrt{5}}\,[-e^{-(9-3\sqrt{5})t} + e^{-(9+3\sqrt{5})t}]\, u(t)$ V

16.3-15. $v(t) = \cos(2t + 36.9°)$ V

16.3-17. $v(t) = 14.2\cos(6t - 8.5°) +$
18.8 $\cos(8t - 11.3°)$ V

16.3-19. $H(s) = \dfrac{s - 1}{2s^2 + 11s + 5}$

16.3-21. $i(t) = 3.54\cos(2t + 135.0°)$ A

Section 16.4

16.4-1. $w(0^-) = 18$ J $w(t) = 2.25$ J, $t > 0$

16.4-3. $H(s) = \dfrac{-s^2 + 1}{(s + 1)^2}$ $i(t) = 36(1 - 2e^{-t})\, u(t)$ A

16.4-5. $v(t) = 82.3\cos(2t - 14°)$ V

16.4-7. $w = 16\dfrac{\text{rad}}{\text{sec}}$ $v_{\text{peak}} = 2$ V

16.4-9. $I_{\text{RMS}} = 169$

16.4-11. $R_L = N^2 R_{\text{Th}}$ $X_L = -N^2 X_{\text{Th}}$

Important definitions appear on pages that are printed in **boldface** type.